TRANSACTIONS

OF THE

INTERNATIONAL

ASTRONOMICAL UNION

VOLUME XXA

REPORTS

INTERNATIONAL COUNCIL OF SCIENTIFIC UNIONS
INTERNATIONAL ASTRONOMICAL UNION
UNION ASTRONOMIQUE INTERNATIONALE

TRANSACTIONS

OF THE

INTERNATIONAL ASTRONOMICAL UNION

VOLUME XXA

REPORTS

ON

ASTRONOMY

Edited by

JEAN-PIERRE SWINGS

General Secretary of the Union

KLUWER ACADEMIC PUBLISHERS

DORDRECHT / BOSTON / LONDON

Library of Congress Cataloging in Publication Data

Reports on astronomy.

   (Transactions of the International Astronomical
Union ; v. 20A)
   "Published on behalf of the International Astronomical
Union"--T.p. verso.
   1. Astronomy--Congresses. 2. Astrophysics--Con-
gresses. I. Swings, J.-P. II. International
Astronomical Union. III. Series.
QB1.I6 vol. 20A   520 s [520]   88-6796

ISBN-13: 978-94-010-7839-9   e-ISBN-13: 978-94-009-2981-4
DOI: 10.1007/ 978-94-009-2981-4

Published on behalf of
the International Astronomical Union
by
Kluwer Academic Publishers, P.O. Box 17, 3300 AA Dordrecht, The Netherlands.

Kluwer Academic Publishers incorporates
the publishing programmes of
D. Reidel, Martinus Nijhoff, Dr W. Junk and MTP Press.

Sold and distributed in the U.S.A. and Canada
by Kluwer Academic Publishers,
101 Philip Drive, Norwell, MA 02061, U.S.A.

In all other countries, sold and distributed
by Kluwer Academic Publishers Group,
P.O. Box 322, 3300 AH Dordrecht, The Netherlands.

Foreword

A few months ago, I wrote to the 41 Presidents of the International Astronomical Union Commissions, requesting them to start preparing their contribution to these "Reports on Astronomy" which represent volume XXA of the Transactions. I specifically asked them to try to highlight the astronomical research relevant to their Commission, that took place between 1984 and 1987. During this exciting period we witnessed a few fascinating astronomical objects that appeared in the sky, such as Halley's comet, the supernova in the LMC SN 1987 A, new gravitational lenses, etc.., so that these reports also contain preliminary results.

The topics of IAU Commissions deal with all of contemporary astronomy and astrophysics, so that the reports presented here should be most useful for those wishing to gain an overview of a certain field, not necessarily near their own research area. In some cases, extensive bibliographical data are provided as well. It is intended to make some publicity about the "Reports on Astronomy" so as to attempt to distribute them more widely.

Each Commission President was requested, by late 1986, to begin preparations for the compilation of his/her Commission report. Commission members were asked to supply details about their individual research programmes. The ensuing very substantial task of editing large amounts of valuable material was undertaken by the Presidents, often supported by a team of authors, all recognized authorities in their fields. In order to preserve some measure of uniformity -which is in any case very difficult with so many authors- editorial guidelines were sent to the Presidents, including the number of pages allotted to each Commission.

I am most thankful to all involved, Presidents and Members of Commissions, for having put so much work into the preparation of these reports and for having complied so well with the guidelines. It is due to their great efforts that it is possible to make this volume available well before the XXth General Assembly : it will serve as a valuable basis for the Commission meetings there as well.

I look forward to receiving your comments on these Transactions so as to improve them in the future, and I hope to meet you on the occasion of the XXth General Assembly in Baltimore, next August 2-11.

Jean-Pierre Swings
General Secretary, IAU
November 1987.

CONTENTS

REPORTS OF COMMISSIONS

4. EPHEMERIDES (EPHEMERIDES)

PRESIDENT : B. Morando

VICE-PRESIDENT : P. K. Seidelmann

ORGANIZING COMMITTEE : V. K. Abalakin, S. Aoki, J. Chapront, R. L. Duncombe, T. Lederle, J. H. Lieske, B. D. Yallop, A. Yamazaki.

I. Introduction

This report covers the period from 1 July 1984 to 30 June 1987. The ephemerides that have been published during this period have made use of the new system of astronomical constants adopted at the XVIth General Assembly of the IAU in Grenoble. Yet some difficulties have arisen because of the lack of catalogues or maps of the heavens established for the epoch J2000.0. This is awkward for observers of comets and minor planets and, for that reason, Commission 20 decided, at the New Delhi meeting of IAU that there would be a gradual introduction of the J2000.0 system as far as those bodies are concerned.

The first issue of an IAU Commission 4 Circular has been sent to all the members of the Commission in May 1987. The aim of it is to exchange information in between general assemblies and generally to establish a link between the members of the commission. It is hoped that these members will contribute to the next issues of this circular.

II. International and National Ephemerides

1. THE FUNDAMENTAL SYSTEM

Starting with the volume for the year 1988 the Apparent Places of Fundamental Stars (APFS) are based on the mean positions and proper motions which will be published in the Fifth Fundamental Catalogue (FK5). The basic part of this catalogue, containing the classical 1535 fundamental stars, will become available for general use with its publication which is scheduled for the end of 1987. Further details concerning the status of the work on the the FK5 are given in the report of commission 8 in this volume.

In order to provide the users of the already published Apparent Places of Fundamental Stars volumes for 1984 through 1987 with positions on the basis of the FK5, corrections FK5-FK4 to the FK4 stars are given in an appendix to the volume of the Apparent Places of Fundamental Stars for the year 1988. The corrections are the result of the systematic and individual improvement of the FK4, computed for the epoch and the equinox of the beginning of the years 1984 to 1988.

2. PRINTED EPHEMERIDES

H.M. Nautical Almanac Office, Royal Greenwich Observatory, Herstmonceux Castle, United Kingdom and the Nautical Almanac Office, U. S. Naval Observatory, Washington, D.C., USA have continued to cooperate in the production and publication of the unified almanacs, namely, the Astronomical Almanac, the Nautical Almanac, the Air Almanac and the Astronomical Phenomena. Beginning with 1987, the Air Almanac was published as a single issue for a full year. Sight Reduction Tables for Air Navigation, vol. 1, epoch 1990 have also been published. The data, either composed or in machine readable form, have been supplied upon request for other countries. The Nautical Almanac Office of the U.S. Naval Observatory has continued to publish annually the Almanac for Computers containing the polynomial coefficients for computing the positions of the Sun, Moon and planets to the accuracy desired for anytime during the year. H. M. Stationery Office have published The Star Almanac for Land Surveyors for 1986, 1987 and 1988.

Bureau des Longitudes, Paris France, have published yearly the Connaissance des Temps, the Ephémérides astronomiques-Annuaire du Bureau des Longitudes and the Ephémérides Nautiques. The Ephémérides astronomiques-Annuaire du Bureau des Longitudes contain, starting from 1986, ephemerides of the minor planets whose magnitude is smaller than 10, at opposition during the year and, also, starting from 1987, ephemerides of periodic comets going through perihelion during the year. Bureau des Longitudes published also three Supplement to Connaissance des Temps : Ephémérides des satellites de Jupiter, Saturne et Uranus (Editions de Physique) that give the differential coordinates of the satellites and tables for the computation of the mutual phenomena of the Galilean satellites of Jupiter - Phénomènes et Configuration des Satellites Galiléens de Jupiter, since 1980 - Configuration des Huit Premiers Satellites de Saturne, since 1985. Bureau des Longitudes have also produced the following ephemerides, published in the Notes Scientifiques et Techniques du Bureau des Longitudes (designated by S followed by a number) : Ephéméride de la comète de Halley pour 1985-1986-Courbes de visibilité (S006), Observation de la comète de Halley en 1985-1986 (S007), Excentricités et inclinaisons moyennes des orbites des satellites galiléens de Jupiter (S009), Ephémérides des petites planètes de 1986 à 1988 (S010, S011, S016), Détermination d'orbites de comètes pour 1986 et 1988 (S014, S017, S019).

The Institute for Theoretical Astronomy, Leningrad, USSR, have published The Astronomical Yearbook and the Ephemerides of Minor Planets.

The Japanese Ephemeris, the Nautical Almanac and the Abridged Nautical Almanac have continued to be published for the years 1986, 1987 and 1988 by the Hydrographic Department of Japan, Tokyo. All the volumes are compiled in accordance with the recommendations of IAU. No serious changes in the contents have been made after the volume for 1985 in which was introduced a new scheme of computation based on the fundamental reference frame of FK5, numerical integration for the coordinates of the bodies in the solar system, day numbers in rectangular coordinates etc. The Japanese Ephemeris from 1985 onwards contains the explanation of the method on which the new ephemerides are based. The Japanese Ephemeris for 1988 contains a new table " The orbital longitude and latitude of the Moon reduced from the lunar occultation observations " which replaces the former " Redduction from Ephemeris Time to Universal Time". The Polaris Almanac for Azimuth Determination, the Altitude and Azimuth Observation Almanac for Antartic Observation and the Abstract from the Japanese Ephemeris have also been published yearly by the Hydrographic Department of Japan.

The Indian Astronomical Ephemeris, Tables of Sunrise, Sunset, Moonrise, Moonset and Rashtriya Panchang, in thirteen languages giving details on the indian calendar and festival dates have been published by the Positional Astronomy Center, Calcutta, India.

3. EPHEMERIDES ON FLOPPY DISKS

For 1986 the Floppy Almanac was introduced by the Nautical Almanac Office, U. S. Naval Observatory. This is a disk that can be run on IBM compatible PCs and provides in an user friendly mode capability for computing in various coordinate systems the positions of the Sun, Moon, planets and stars. Topocentric phenomena and positions can be computed. The program is also available for Microvax Computers on RX 50 diskettes, or for VM CMS systems for the IBM 370, 4300, 3000 computers. A user's guide is available to accompany the software. The Floppy Almanac can be used to compute the data for other almanacs. Using projected values of delta T, issues of the Floppy Almanac are available through the year 2000.

Bureau des Longitudes have issued the following softwares on floppy disks : Programme de Concordance des Calendriers et Fêtes Religieuses, Ephémérides de redécouverte des Comètes, Ephémérides des Satellites de Jupiter, Saturne et Uranus (for IBM compatible PCs).

4. BASIS OF THE EPHEMERIDES

The Jet Propulsion Laboratory (JPL) Ephemerides continue to be improved as newer and more accurate observational data become available. The ephemerides are now fitted to over 80 000 observations- meridian transits, astrolabes, photometry, radar, S/C tracking and ranging, ring and disk occultation timings and radio measurements of thermal emissions. Comparisons of the various datatypes have led to increased understanding of the systematic errors present in some of the earlier optical data types. As such, present ephemerides, for the outer planets in particular, are significantly improved over those of DE200.

Reference frame studies have included the establishment of the JPL Radio Frame and the Dynamical Reference Frame of the Lunar/Planetary Ephemerides, determination of ties between the various reference systems and development of the concept of the dynamical equinox as a reference point for the modern ephemerides and the connection of other coordinate systems to this reference point. Ties between VLBI and optical frames have been established via the observations of radio stars by JPL in collaboration with French astronomers. A link has been determined between the VLBI and the Ephemeris frame through the use of differential VLBI. Very Large Array measurements of Jupiter, Saturn, Uranus and Neptune provide both a tie between the outer planet ephemerides and the radio frame and a means of improving the ephemerides themselves.

5. LONG-TERM EPHEMERIDES

Bureau des Longitudes have published, on floppy disks for Macintosh microcomputers Solution Approchée (ELP 2000-85) du mouvement de la Lune valable sur plusieurs milliers d'années . P. Bretagnon and J.-L. Simon, from Bureau des Longitudes have published Planetary Programs and Tables from -4000 to +2800 (Willmann-Bell, Inc) that provide time-dependent expansions of the longitude and radius vector of the Sun and the heliocentric coordinates of the planets. Bureau des Longitudes have also published Tables des Positions du Soleil, des Planètes et de la Lune entre 1950 et 2020 (Note scientifique et technique du Bureau des Longitudes n°S012).

The following U. S. Naval Observatory Circulars have been prepared : N° 169, Phases of the Moon 2000-2049, n°170, Solar Eclipses 1991-2000.

B. D. Yallop and C. Y. Hohenkerk, of the Royal Greenwich Observatory have published, in 1985, Compact Data for Navigation and Astronomy for the years 1986 to 1990.

J. Laskar and R. A. Jacobson have published An Analytical Ephemeris of the Uranian Satellites, fitted on earth-based and Voyager data (*Astron. Astrophys.* in press).

III. Theoretical work related to the ephemerides

B. Guinot, of the Bureau International des Poids et Mesures and P. K. Seidelmann of the U.S. Naval Observatory have circulated a reprint of a paper titled " Timescales, their history, definition and interpretation". This paper was prepared to help achieve the agreement required of the working Group on reference frames established by the IAU in 1985.

J. Laskar (Bureau des Longitudes) has obtained a new solution for the motion of the pole of the ecliptic using his own general planetary theory, the VSOP82 theory by P. Bretagnon and the theory of the rotation of a rigid earth established in 1977 by Kinoshita. He obtains new formulae for the precession valid for 10 000 years

(Secular terms of classical planetary theories using the results of general theory- *Astron. Astrophys.* , 1986, 157, 59).

M. Chapront-Touzé and J. Chapront have computed the secular variations of the fundamental arguments of the lunar theory to the fourth power of the time (ELP2000-85 : A semi-analytical lunar ephemeris adequate for historical times. *Astron. Astrophys.* in press).

A test for the continuity requirement at the 1984 changeover has been performed. Simultaneously the connection between the radio system and optical system has been discussed by Aoki et al. (*astrometric Techniques*, 123-131). Th. Hirayama, H. Kinoshita, M.-K. Fujimoto and T. Fukushima have found an analytical expression of TDB-TDT with an accuracy at the 5ns level (in press in the proceedings of the IAG symp. 1987).

Gutzwiller and Schmidt have published in the *Astronomical Papers prepared for the use of the American Ephemeris and Nautical Almanac* vol. XXIII, Part I, a paper on "The motion of the Moon as computed by the method of Hill, Brown and Eckert".

The following papers have been issued as Technical Notes of H. M. Nautical Office : n° 57 B. D. Yallop, 1986. Ground Illumination ; n°62 B. D. Yallop and C. Y. Hohenkerk ; 1985. Coefficients for calculating the Greenwich Hour Angle and Declination of stars ; n°63 C. Y. Hohenkerk and A. T. Sinclair, 1985. The computation of angullar atmospheric refraction at large zenith angles ; n°64 B. D. Yallop, 1986. Algorithms for calculating the dates of Easter ; n°65 C. Y. Hohenkerk, 1986. Determination of polynomial coefficients from B-spline coefficients. Astronomical and calendarial data up to to 1992 were published in the Royal Greenwich Observatory Astronomical Information Sheets, as were also the following notes : n°48 B. D. Yallop and C. Y. Hohenkerk, 1985. Closest approach of Polaris to the North Celestial Pole in AD 2100 ; n°50 B. D. Yallop, 1987. Earliesr sighting of the New Moon in 1987.

A. S. Sochilina, of the Institute for Theoretical Astronomy in Leningrad, USSR, has published a paper On the Choice of Reference Frame in Investigations of High Satellites Motions in *Bull. Inst. Astr. Leningr.* 15, n°9 (172), 481-485, 1986.

M. Ilyas (Malaysia) has studied how to unify the various lunar calendars and how to predict the earliest visibility of the lunar crescent in the context of Islamic calendar (*J. Roy. Astron. Soc. Can.* 80, 1986, 134-141, 328-335).

IV. Observations in view of improving the ephemerides

D. Pascu has made photographic observations of the Martian moons, of Jupiter and the Saturnian satellites I-VIII with the 26-inch refractor of the U.S. Naval Observatory in Washington . D. Pascu and P. K. Seidelmann have continued to make observations of Jupiter XIV, Saturn XII, XIII and XIV, Uranus I-V and Neptune I and II with the Mark 4 CCD Camera of the Space Telescope Widefield Planetary Camera Team on the 61-inch telescope at Flagstaff, Arizona. K. J. Johnson, of the Naval Research Laboratory, C. M. Wade of National Radio Astronomy Observatory and G. H. Kaplan, T. S. Carrol and P. K. Seidelmann of the U. S. Naval Observatory have made observations of minor planets 1, 2, 4 and 10 with the Very Large Array in Soccoro, New Mexico.

The last three years have seen improvement in Lunar Laser Ranging data quality and the development of Lunar Laser Ranging Network. Recent equipment and software improvements at the stations have resulted in approximately 5cm ranges (the data prior to 1984 had ranges over 10cm accuracy) ; data are currently being acquired from three stations : CERGA site (France), Manui (Hawaii-USA) and McDonald (Texas-USA). An analysis of the seventeen-year Lunar Laser Ranging data set yields a value for the GM of the Earth of 398 600.437 +/- 0.006 km^3/s^2 in the solar system barycentric frame and 398 600.443 +/- 0.006 km^3/s^2 in the geocentric system, comparable to Lageos (the Lunar Laser Ranging result agrees with the Lageos result within one standard deviation of the error estimate). The IAU adopted values of the 18.6-year nutation and the precession have been checked against the Lunar Laser Ranging data. The increased accuracy of Lunar Laser Ranging should result in an improved lunar ephemeris. Many papers related to Lunar Laser Ranging or on reference frames, by G. L. Berge, J. O. Dickey, P. B. Esposito, J. L. Fanselow, J. F. Lestrade, R. P. Linfield, W. G. Melbourne, D. O. Muhleman, X. X. Newhall, A. E. Niell R. A. Preston, M. Rapaport, Y. Requième, D. J. Rudy, E. M. Standish, J. G. Williams will be found in the following publications : Proceedings of the IAU Symposium n°128, *The Earth's Rotation and Reference Frames for Geodesy and Geodynamics*, editors G. Wilkins and A. Babcock, D. Reidel, Boston, 1987, in press. The book *Reference Frames*, editors, B. Kolaczek, J. Kovalevsky, I. Mueller, D. Reidel, 1987, in press. The Proceedings of the IUGG Symposium, *Relativistic Effects in Geodesy XIX General Assembly*, Vancouver, 1987.Report on the *MERIT-COTES Campaign on Earth's Rotation and Reference System, Part I : Proceedings of the third MERIT workshop and the Joint MERIT-COTES Working Group Meetings*, editor G. Wilkins, Royal Greenwich Observatory, 1987, in press. Report on the *MERIT-COTES Campaign on Earth Rotation and Reference Systems, Part II : Proceedings of the International Conference on Earth Rotation and the Terrestrial Reference Frame*, editor I. Mueller, Ohio State University, vol. 2, 1985. Proceedings of the IAU Symposium n°109 : *Astrometric Techniques* (Gainesville, Florida, 1984), editors H. K. Eichhorn and R. J. Leacock, Reidel, Dordrecht-Holland, 1986. *Bulletin of the American Astronomical Society*, 1986. Proceedings of the Joint discussion on Reference Frames at th XIX General Assembly of the IAU, *Highlights of Astronomy*, vol. 7 editor J. P. Swings, Reidel, Dordrecht-Holland, 1986. Proceedings of the IAU colloquium n°114 : *Relativity in Celestial Mechanics and Astrometry*, Reidel, Boston, 1985. Special LAGEOS issue of the *Journal of Geophysical Research*, vol. 90, 1985. Proceedings of the international symposium : Figure and Dynamics of the Earth, Moon and PLanets, special issue of the *Monograph Series of the Research Institute of Geodesy, Topography and Cartography*, editor P. Holota, in press, 1987. *Celestial Mechanics*, 37, 329-337, 1985. *Transactions of the American Geophysical Union*, EOS, 67, 16, 259, 1986. Proceedings of the *Fifth International Workshop on Laser Ranging Instrumentation*, editor J. Gaignebet, vol. 1, 19-28, 1985.

The services of the International Lunar Occultation Centre, in Japan, have been continued since 1981. The number of the timing data collected at the centre was 31 370 from 35 countries during the years 1984 to 1986. Reports containing all the reduced data as well as the station coordinates related are published annually.

A campaign for the observations of the mutual phenomena of the Galilean satellites of Jupiter in 1985 has been organized under the auspices of Bureau des Longitudes. Many good observations were made in France, Italy, Spain, Brazil and at the ESO observatory in Chile. Preliminary results will be found in the proceedings of a colloquium held at Bagnères-de Bigorre (*Annales de Physique*, vol.12, 1987).

V. Working Groups having to do with ephemerides

The IAU/IAG/COSPAR Working Group on Cartographic Coordinates and Rotational Elements of the Planets and Satellites has continued to gather informations and will present an updated set of data in the report which will be presented at the Baltimore General Assembly of the IAU. The main changes will affect the radii of Jupiter, Saturn, Uranus, Neptune, Io, Mimas and the Uranian satellites.

A Working Group on astronomical constants has been formed following Resolution C1 of the IAU by the presidents of IAU Commissions 4, 7, 8, 19 and 31. The purpose is to review the current determinations of the astronomical and geodetic constants, provide best estimates, accuracies and sources. The Working Group is composed as follows : B. Morando, president, V. K. Abalakin, W. E. Carter, J. Chapront, B. Chovitz, H. Kinoshita, J. Lieske, J. Schubart, P. K. Seidelmann, E. M. Standish, J. M. Wahr, G. Wilkins, Ya. S. Yatskiv. The working group will send a preliminary report to the General Secretary of the IAU before the end of 1987.

Following Resolution C2 of the IAU a Working Group on Reference Systems has been formed by the presidents of IAU Commissions 4,7,8 19, 20, 24, 31, 33 and 40. The president of the working group is James A. Hughes. This working group and the working group on astronomical constants met together in Paris in June 1987.

B. MORANDO
President of the Commission.

COMMISSION 5: DOCUMENTATION AND ASTRONOMICAL DATA
DOCUMENTATION ET DONNEES ASTRONOMIQUES

PRESIDENT: G. A. Wilkins
VICE-PRESIDENT: B. Hauck
ORGANISING COMMITTEE: O. B. Dluzhnevskaya, W. D. Heintz, C. O. Jaschek,
                          P. Lantos, S. Mitton, F. Spite, P. A. Wayman,
                          G. Westerhout, C. E. Worley

REPORT BY THE PRESIDENT FOR THE PERIOD 1985 DECEMBER 1 TO 1987 OCTOBER 30

The aims of this report are, firstly, to review the activities of Commission 5 during the period since the IAU General Assembly in Delhi in November 1985 and, secondly, to draw attention to other relevant activities. It is based mainly on contributions from the Chairmen of the working groups and other members of the Commission, but it also includes some items of general interest that have been taken from the Commission's Newsletter. The Working Groups and their Chairmen are as follows:

Astronomical Data	G. Westerhout
Designations	C. O. Jaschek
Classification	P. Lantos
Abstracting Guidelines	L. D. Schmadel

The Newsletter, which was issued in March 1986 and July 1987, was primarily intended to provide a means of communication between the members and consultants of the Commission. It was hoped, however, that other astronomers would find items of interest in it and so it was reproduced in the information bulletin of the astronomical data centre at Strasbourg. Requests to be included in the distribution were received from librarians and others not previously associated with Commission 5. The frequency of issue will be increased if more items of information or comment are sent to the Editor.

The announcement of a Workshop on "Library and Information Services in Astronomy" to be held in Washington, D.C., just prior to the 1988 General Assembly attracted much interest and the IAU Executive Committee agreed that it shall be IAU Colloquium No. 110. A complementary Joint Discussion on "Documentation, Data Services and Astronomy" will be held early in the Assembly to bring together the users and providers of information services in considering the problems and opportunities presented by the availability of new techniques.

ASTRONOMICAL DATA

Much information about activities in data compilation and data handling is published bi-annually in the Bulletin d'Information du Centre de Données Stellaires (CDS) in Strasbourg, France. Issue No. 31 contains the papers presented at meetings organized by the CDS on "Archiving Astronomical Observations" and on "Astronomical Data Networks". The proceedings of the Tblisi Colloquium on "Stellar Catalogs: Data Compilation, Analysis, Scientific Results" were published in Bulletin 59, Abastumani Astrophysical Observatory, Tblisi, 1985. A large symposium (over 110 participants) organized by the ESA/ESO Space Telescope European Coordinating Facility, was held in Garching, FRG, on 12-14 October 1987 on "Astronomy from Large Data Bases: Scientific Objectives and Methodological Approaches." The Working Group on Astronomical Data held an ad hoc meeting during this symposium.

The IAU Executive Committee has requested that the Commission give particular attention to the problems of data archiving and refreshing, and so this topic will be discussed during the ordinary meetings of the Commission at Baltimore as well as at Colloquium No. 110 and at the Joint Discussion.

CODATA activities

The 10th International CODATA Conference which was held in July 1986 in Ottawa, Canada, on the topic of "Computer Handling and Dissemination of Data", was attended by only four astronomers even though it is an excellent forum for interdisciplinary scientific exchange. It was attended by 270 people and among other things contained excellent discussions on data dissemination (CD-ROMS, etc) and expert systems. CODATA Bulletin No. 64 contains selected papers. Bulletin No. 63 contains the "Adjustment of the Fundamental Physical Constants". The 11th CODATA Conference will be held in Karlsruhe, FRG, 26-29 September 1988, with the theme "Scientific and Technical Data in a New Era". Among the topics to be covered is "Geo- and space sciences", with three invited papers on astronomical data. The IAU representative on CODATA is G. Westerhout.

Centre de Données Stellaires: Strasbourg

The most important development concerning the SIMBAD database of the Stellar Data Centre (CDS) at the University of Strasbourg is its enhanced accessibility through networks. In October 1987 it can be accessed from 122 astronomical institutes in 21 countries, and the number of users keeps growing rapidly. The transfer of data to users has been improved, so that samples may be easily copied for further treatment. Projects are underway to incorporate the whole Durchmusterungen into SIMBAD, as well as the IRAS Point Source Catalogue and the brightest stars of the Guide Star Catalogue. Finally, the number of catalogues available has also increased through exchange with other data centres, and is now of the order of 500.

Directories of astronomical associations, societies and professional institutions have been issued as special publications by CDS (Heck and Manfroid, 1985 and 1986); new editions are in preparation. A new survey of astronomical data sources has been made by Jaschek (1987).

Astronomical Data Center: NASA/GSFC

During this period the Astronomical Data Center (ADC) at the Goddard Space Flight Center completed 1460 requests for data and/or information, with 1465 machine-readable catalogues disseminated. The archive of astronomical data now contains more than 500 catalogues. A machine-readable version of the Bonner Durchmusterung (zones +89 to -23) was completed through an international collaborative effort, while a new catalogue of WDS-DM-HD-ADS cross identifications was prepared. The Infrared Source Cross Index (Schmitz et. al., 1987) and a second edition of the Catalog of Infrared Observations (CIO; Gezari et. al., 1987) were prepared and published; the latter now contains all IRAS data for sources in the CIO database. An updated version of the Bibliographical Index of Objects Observed by IUE (Mead et. al., 1987) was completed and the first and second versions of the machine-readable Data Inventory of Space-based Celestial Observations (DISCO; Brotzman et. al., 1987) were prepared. An electronic network distribution service for data was begun using BITNET and the Space Physics Analysis Network (SPAN). An ADC Online Information System was also developed and implemented to provide a search capability by keywords and an interactive ordering service.

Central Institute of Astrophysics: Potsdam

The cooperation between the CDS at Strasbourg and the Central Institute for Astrophysics (CIAP) at Potsdam was continued, and CIAP now holds 364 catalogues that have been obtained from the CDS in order to make them available to scientists of the GDR and other socialist countries. The Bibliographical Catalogue of Variable Stars was extended so that it now contains nearly 323000 records for about 30000 stars. A bibliographical catalogue of suspected variable stars is in preparation. New data on variable stars and equivalent widths (Friedemann, 1987) were sent to the CDS. Three thematical databases (STAR, containing basic information on stars; CAL, on non-stellar objects; and TPK, the Tautenburg plate catalogue) were developed on the basis of the Swedish database system MIMER and are now available on-line.

FITS Task Force

The FITS Task Force was created by the Working Group on Astronomical Data during the IAU General Assembly in 1982 when the FITS tape format was recommended for the interchange of image data between observatories. The two main functions of the Task Force are: (1) to channel comments and suggestions on the usage of FITS for the interchange of data, and (2) to investigate the extension of FITS for use in the exchange of catalogues. The FITS Task Force consists at present of: P. Grosbol (ESO, Chairman), F, Ochsenbein (ESO), W. H. Warren, (NASA), and D. Wells (NRAO).

A European FITS Committee was created with members from major institutes in most European countries to act as a local forum for discussions of FITS matters. After extensive tests, a proposal for a Generalized Extension of FITS including a Table Extension was finalized and presented to the AAS-WGAS FITS Group and to the European FITS Committee; both groups accepted the proposal. The final text of the proposal was accepted for publication in Astronomy and Astrophysics Supplement Series. The Task Force also made a proposal for the physical blocking of FITS data files to improve efficiency, while maintaining the logical blocksize of 2880 bytes; it allows a blocking factor of 1 to 10 on nine-track magnetic tapes. This proposal was also accepted, and both came into use from 1 January 1987.

An electronic mailbox for FITS was set up at NRAO by D Wells. It is accessible from all major computer networks and will distribute mail messages to all major institutes using FITS. It enables a faster and more general discussion of the FITS standard in the community.

DESIGNATIONS

The general resolution (C3) on astronomical designations that was adopted by the Commission at Delhi was published in IAU Information Bulletin No. 55 (February 1986, pp. 19-21), as were the resolutions of Commission 28 on the designation of supernovae (C10, pp. 22-23) and Commission 40 on radio-source nomenclature (C12, pp. 23-24); they were also published in the Transactions of the IAU (19B, 40-44, 48, 49-51, 1986). It is too early to assess their influence on authors and editors, but the "clearing house" of Dickel, Jaschek, Lortet, Mead and Warren has operated successfully in several cases. The Working Group on Nomenclature of Commission 34 (Interstellar Matter) has subsequently published its recommendations (Dickel et. al., 1987). Further valuable dictionaries of designations have been published by Lortet (1986a, b) and Lortet and Spite (1986).

The astronomical community at large is only gradually becoming aware of the importance of the use of proper practices for the designation of astronomical objects, but the rapidly growing use of data centres may help to enforce them. The attention of authors, editors and referees will also be drawn to them in the new IAU Style Book.

CLASSIFICATION AND KEYWORDS

There has been no progress of note in the revision of the Universal Decimal Classification for Astronomy (UDC 52). A revised draft list of keywords has been prepared by P. Lantos. The preparation of a thesaurus is now in hand by a group led by R. M. Shobbrook (Librarian of the Anglo-Australian Observatory). Questionnaires were distributed to 96 librarians to seek assistance and information about current lists of subject headings. It is hoped that a draft listing will be available for comment during IAU Colloquium No. 110.

The American Institute of Physics has published (AIP, 1987) a physics and astronomy classification scheme for use in its publications; it is based on the 1977 ICSU/AB system, which is now under revision by the Physics Working Group of ICSTI (see below).

ABSTRACTING

The Astronomisches Rechen-Institut (ARI) at Heidelberg, which produces Astronomy and Astrophysics Abstracts (AAA), will cooperate with the Fachinformationszentrum (FIZ) at Karlsruhe in the fields of astronomy and astrophysics; FIZ is responsible for abstracting and other information services for physics and many other areas. FIZ has cancelled its "Monthly Service" of astronomical abstracts and ARI will not distribute the AAA abstracts on magnetic tape, but instead these abstracts will be included in the FIZ database "Physics Briefs" and so will be accessible on-line prior to publication in AAA.

The second report on "Guidelines for Abstracts" (Schmadel, 1985) is being used in the preparation of the new IAU Style Book (see below), while the appended list of keywords serves as a basis for the production of the IAU thesaurus (see above). The increasing use of on-line searching of abstracts makes it all the more important that all authors use unambiguous designations for the astronomical objects referred to in astronomical papers and catalogues. Studies are being made to ensure that the systems used by AAA, FIZ and CDS (Strasbourg) are compatible with each other and consistent with the IAU recommendations. Already a high degree of correspondence has been demonstrated.

IAU STYLE BOOK

A first draft of the new IAU Style Book was distributed for comment on April 1986 to about 30 persons who were believed to have an active interest in its recommendations; some of the copies were circulated to other astronomers and editorial staff. The large number of comments that were returned represented the views of a good cross-section of those concerned in the preparation, publication and use of the astronomical literature. The draft, which had the subtitle "A Manual for use on the Preparation of Astronomical Reports and Papers" was intended primarily for use in the preparation of typewritten camera-ready copy for IAU publications, especially for the Transactions and the proceedings of symposia and colloquia. The General Secretary of the IAU suggested that an attempt should be made to seek the agreement of the editors of the principal astronomical journals in order to reduce as far as practicable the differences between the requirements of these publications. This increased the difficulty of the task by an order of magnitude and made it impossible to issue the manual in 1986 as had been intended. It has also become clear that the manual must contain recommendations for use with desk-top systems capable of producing copy of type-set quality. It is hoped that a second draft will be discussed at a meeting of editors early in 1988 and that an agreed version will be available at the General Assembly.

ICSTI

The Officers of the IAU have decided that the IAU should, for the time being, no longer participate in the activities of the International Council for Scientific and Technical Information (ICSTI) with effect from 1 January 1988. ICSTI was established in June 1984 as the successor to the Abstracting Board of the International Council of Scientific Unions (ICSU AB). Its activities are of general interest to astronomy and include the revision of the international classification system for physics and the preparation of a directory of numerical databases, but at present the value to the IAU of membership does not match the costs for dues and attendance at the annual meeting. Astronomy and Astrophysics Abstracts is a member of ICSTI.

UNION LISTS AND OTHER PUBLICATIONS

A compilation of the serial holdings of 14 astronomy collections in the USA was prepared by J. A. Lola (Yerkes Observatory) and a list for the United Kingdom was prepared by A. R. Macdonald (Royal Observatory, Edinburgh). The guide to information services in astronomy by Rey (1983) is under revision. Various lists of observatories were published by Howse (1986) and Vercoutter (1986). An English-Chinese Dictionary of Astronomy containing 16000 terms and seven lists of specialised terms was published in China in 1986, and the multilingual dictionary that is being prepared by J. Kleczek is in press.

OTHER MATTERS

The problems caused by the increases in the costs of journals and books have been exacerbated for many libraries by reductions in budgets and by the extra costs of providing new on-line services. As a consequence, subscriptions to many journals, particularly the less heavily used smaller journals, including translations from other languages into English, are liable to be cancelled, and hence the viability of such journals is in doubt. The Commission may wish to consider whether it could take any action that might alleviate these problems. The Commission also needs to follow up the meeting that was held in Delhi (Trans. IAU 19B, 100) to consider the problems of "developing astronomical institutions". These need access to past and current literature and to the new information-retrieval services if they are to carry out their programmes effectively. It is hoped that it will be possible for such institutions to be represented at IAU Colloquium No. 110 and that better cooperative arrangements may be established as a consequence of the discussions at the meeting.

The programme of the meetings of the Commission at Baltimore will need to be very extensive in order to cover adequately all of the many topics discussed in this report as well as others that for various reasons have not been mentioned. It is clear that the increased activity in astronomical-data topics that has taken place in recent years is about to be matched by a corresponding increase in activity in the other fields of concern to the Commission. It is of interest to note that some of these topics, such as designation and information retrieval techniques, are not only of concern to all of the members of the Commission but that their importance is being recognized by a much higher proportion of the astronomical community. I would like to conclude my report by thanking all those who have contributed to this report and to the activities of the Commission during this period.

REFERENCES

AIP, 1987.   Physics and astronomy classification scheme - 1987.   American
    Institute of Physics, New York: AIP Publ. R-261.7.

H.R. Dickel, M. C. Lortet & K. S. de Boer, 1987. Designation and nomenclature
    for astronomical sources of radiation. Astron. Astrophys. Supp. Ser 68,
    75-80.

C. Friedemann, 1987.   A second catalogue of equivalent widths of the
    interstellar 2170 A band. Obs. Strasbourg: Bull. Inform. CDS 32, 77-80.

A. Heck & J. Manfroid, 1985.   International directory of astronomical
    associations and societies (IDAAS 1986). Obs. Strasbourg: Publ. Spec.
    CDS 8.

A. Heck & J. Manfroid, 1986.   International directory of professional
    astronomical institutions (IDPAI 1987). Obs. Strasbourg: Publ. Spec.
    CDS 9.

D. Howse, 1986.   The Greenwich list of observatories: a world list of
    astronomical observatories, instruments and clocks, 1670-1850. J. Hist.
    Astron. 17 Part 4, i-iv, 1-100.

C. O. Jaschek, 1987. (List of astronomical data sources). Obs. Strasbourg:
    Bull. Inform. CDS 33.

M. C. Lortet, 1986a.   Nomenclature for objects in the Magellanic Clouds.
    Astron. Astrophys. Supp. Ser. 64, 303-324.

M. C. Lortet, 1986b.   Nomenclature for objects in the Galaxy M33.   Astron.
    Astrophys. Supp. Ser. 64, 325-328.

M. C. Lortet & F. Spite, 1986. First supplement to the first dictionary of the
    nomenclature of celestial objects. Astron. Astrophys. Supp. Ser. 64, 329-
    389.

J. M. Mead, L. E. Brotzman & Y. Kando, 1987. Bibliographical index of objects
    observed by IUE. IUE Newsletter No. 30.

J. M. Rey, 1983.   Information sources and services in astronomy, astrophysics,
    and related space sciences.   Smithsonian Astrophysical Observatory,
    Cambridge, Mass: Smithsonian Inst. Libr. Res. Guide 2.

L. D. Schmadel, 1985. Guidelines for abstracts and other bibliographic items in
    the astronomical literature. Obs. Strasbourg: Bull. Inf. CDS 28, 95-109.

M. Schmitz, J. M. Mead & D. Y. Gezari, 1987. The infrared source cross-index.
    NASA Ref. Publ. 1182.

P. A. Vercoutter, 1986.   Directory of European Observatories 1986. Published
    privately by author at Jan Van Eyckdreef 7, B-8900 IEPER, Belgium.

6. ASTRONOMICAL TELEGRAMS (TÉLÉGRAMMES ASTRONOMIQUES)
(Committee of the Executive Committee)

PRESIDENT: A. Mrkos
VICE-PRESIDENT: E. Roemer
ORGANIZING COMMITTEE: B. G. Marsden
DIRECTOR OF THE BUREAU: B. G. Marsden, Smithsonian Astrophysical Observatory, 60 Garden Street, Cambridge, MA 02138, USA (TWX 710-320-6842 ASTROGRAM CAM)
ASSOCIATE DIRECTOR OF THE BUREAU: vacant
ASSISTANT DIRECTORS OF THE BUREAU: C. M. Bardwell, D. W. E. Green

I. INTRODUCTION

The commission has been able to continue its work during the last triennium with a minimum of effort. Thanks for this and for the untiring dedication are due to the Director of the Bureau, B. G. Marsden. We are also grateful to the Smithsonian Astrophysical Observatory for its general support.

According to the annoucement published in the IAU Information Bulletin no 58, K. A. Thernöe died on 1987 March 7. He was the director of the Central Telegram Bureau during the years 1960 - 1964 in Copenhagen and for many years he served as principal substitute for Julie Vinter Hansen when she ran the Bureau.

Commission 6 was the principal sponsor in IAU Colloquium "The Contribution of Amateurs Astronomers to the Astronomy". The Colloquium took place in Paris from June 20 to June 24, 1987. Nine commissions of IAU supported the meeting which stressed the importance of amateurs for astronomy.

A. Mrkos
President of the Commission

II. REPORT OF THE CENTRAL BUREAU FOR ASTRONOMICAL TELEGRAMS

During the triennium the Central Bureau for Astronomical Telegrams issued telegram books and Circulars on the following occasions:

Telegrams		Circulars	
1984 (from 1 Nov.)	14	16	(Nos. 4003-4024)
1985	35	59	(Nos. 4025-4156)
1986	41	61	(Nos. 4157-4191)
1987 (to 30 June)	26		(Nos. 4292-4412)

Much of the Bureau's rather high activity was due to Halley's comet, news of which was published in the Circulars at a steady rate during the first two years of the triennium. The appearance of supernova 1987A in the Large Magellanic Cloud in February 1987 resulted in a record-breaking surge in the rate of publication of the Circulars, with as many as 31 being issued during the 20 days following the discovery, and the above tabulation shows almost as many Circulars during the first half of 1987 as during a normal complete year.

It is clear that SN 1987A has had a permanent effect on the manner in which the Central Bureau operates. As mentioned in the last triennial report, early in 1984 the Bureau initiated its computer service, one of the features of which allows subscribers to see the Circulars as soon as they are issued.

However, the incompatibility of international modem protocols and the expense of commercial telephone lines meant that by February 1987 there were only five non-North American subscribers out of a total of about 50. Although it had been hoped shortly to make the computer service accessible over the SPAN network anyway, the extreme interest in the supernova accelerated the process, and with the support of NASA arrangements could be quickly made for international tie-ins to SPAN. By mid-1987 the total number of subscribers to the computer service had doubled, and the number of subscribers outside North America had quadrupled, as had the total actual usage of the service. As a concession to the computer-service users Circulars would no longer be issued only when they could be printed; they are now prepared as appropriate on any day of the week and at any time, und unless they are particularly urgent they can be delayed several days before printing.

Although the electronic dissemination of the Circulars tends to obviate the often need for the Bureau's telegram service, the latter is still essential for astronomers outside North America, western Europe, Japan, Australia, New Zealand and South Africa, as well as for astronomers who choose not to check the computer service every day. In March 1987 the per-message charges for receipt of the telegrams by domestic public message were increased from $7.00 to $9.00, by domestic mailgram from $3.00 to $4.00 and by overseas full-rate cablegram from $11.00 to $14.00 (with a possible surcharge for very long messages). For subscribers in Australia and New Zealand the charge remains at $2.00, thanks to the generosity of the Ionospheric Prediction Service in Sydney, whence P. Davies redisseminates the telegrams.

Following an increase in the per-issue subscription rates for the Circulars in November 1984 to 65c for 'regular' accounts and 39c for 'special' (not-invoiced) accounts, the number of subscribers dropped from 780 to 760, where it remained until April 1986, when the billing system was instead put on a monthly basis. The monthly subscription rate is $7.50 for regular accounts, $4.50 per special accounts, and there are additional charges of the same amounts for the computer service. Although the billing change was in response to many requests, the number of subscribers then dropped to 730 and below. Following the appearance of SN 1987A the number increased to 750.

B. G. MARSDEN
Director of the Bureau

7 - MECANIQUE CELESTE (CELESTIAL MECHANICS)

PRESIDENT : V.A. Brumberg
VICE-PRESIDENT : J. Henrard
ORGANIZING COMMITTEE : Ju.V. Batrakov, K.B. Bhatnagar, J. Chapront, A. Deprit, S. Ferraz-Mello, J.D. Hadjidemetriou, H. Kinoshita, J. Kovalevsky, H. Scholl, P.K. Seidelmann, A. Sinclair, Z.H. Yi.

A. Introduction

During 1985-1987 Celestial Mechanics has been intensively developed in all its branches embracing physical bases, mathematical aspects, computational techniques and astronomical objectives. Commission 7 has organized three IAU conferences : Symposium No. 114 "Relativity in Celestial Mechanics and Astrometry" (Leningrad, May 1985), Colloquium No. 96 "The Few Body Problem" (Turku, June 1987) and Topical Session "Resonances in the Solar System" of the X-th European Regional Astronomy Meeting (Prague, August 1987). Members of the commission have broadly participated in the NATO Advanced Study Institute "Long-Term Dynamical Behaviour of Natural and Artificial N-Body Systems" (Cortina d'Ampezzo, August 1987) and some other international and regional conferences. Prospects of the actual celestial mechanics investigations have been discussed at a session of Commission 7 at the XIX-th IAU General Assembly (New Delhi, November 1985). Three papers dealing with the unsolved problems of celestial mechanics were primarily addressed to the rising generation of celestial mechanicians (V.A. Brumberg and J. Kovalevsky, CM. 39, 133, 1986; P.K. Seidelmann, CM. 39, 141, 1986). Several books and specific topic proceedings in celestial mechanics were published during these years :

E.P. Aksenov, Special Functions of Celestial Mechanics, Nauka, Moscow, 1986 (in Russian).

V.I. Arnold, V.V. Kozlov and A.I. Nejshtadt, Mathematical Aspects of Classical and Celestial Mechanics, VINITI, Moscow, 1985 (in Russian).

V.G. Golubev and E.A. Grebenikov, The Problem of Three Bodies in Celestial Mechanics, Moscow State University, Moscow, 1985 (in Russian).

K.V. Kholshevnikov, Asymptotic Methods of Celestial Mechanics, Leningrad State University, Leningrad, 1985 (in Russian).

BHA : Space Dynamics and Celestial Mechanics (ed. K.B. Bhatnagar), Reidel, 1986.

BUM : Natural Satellites (eds J.A. Burns and M.S. Matthews), Univ. Arizona Press, 1986.

CAH : Gravitation in Astrophysics (eds. B. Carter and J.B. Hartle), Plenum Press, 1987.

DDM : The Stability of Planetary Systems (eds. R.L. Duncombe, R. Dvorak and P.J. Message), CM. 34, Nos. 1-4, 1984.

DLS : Fundamental Astronomy and Solar System Dynamics (eds. R.L. Duncombe, J.H. Lieske and P.K. Seidelmann), CM. 37, No. 3, 1985.

EIL : Astrometric Techniques, IAU Symp. No. 109 (eds. H.K. Eichhorn and R.J. Leacock), Reidel, 1986.

FMS : Resonances in the Motion of Planets, Satellites and Asteroids (eds. S. Ferraz-Mello and W. Sessin), Univ. of Sao Paulo, 1985.

HAI : 300 Years of Gravitation (eds. S.W. Hawking and W. Israel), Cambridge Univ. Press, 1987.

KKM : Reference Systems (eds. B. Kolaczek, J. Kovalevsky and I.I. Mueller), Reidel, 1987.

KNY : Proc. of the Twentieth Symposium on Celestial Mechanics (eds. H. Kinoshita, H. Nakai and M. Yoshikawa), Tokyo, 1987.

KOB : Relativity in Celestial Mechanics and Astrometry, IAU Symp. No. 114 (eds. J. Kovalevsky and V.A. Brumberg), Reidel, 1986.

SZB : Stability of the Solar System and its Minor Natural and Artificial Bodies (ed. V. Szebehely), Reidel, 1985.

WIB : The Earth's Rotation and Reference Frames for Geodesy and Geodynamics (eds. G. Wilkins and A. Babcock), IAU Symp. No. 128, Reidel, 1986.

The amount of periodical publications in celestial mechanics is abundant. It is sufficient to look through "Celestial Mechanics" Journal (3 volumes or 12 issues per annum). So, following the example of the 1985 report by J. Kovalevsky, this report will not claim to cover all fields of interest of celestial mechanics and to give a detailed bibliography. In particular, taking into account the incoming symposium on "Applications of Computer Technology to Dynamical Astronomy" (Gaithersburg, July 1988) no mention is made here on the computational techniques. The main attention is given to domains not included in the previous report, such as relativistic celestial mechanics, resonance theory, asymptotic methods, artificial satellite motion, etc.

In preparing this report the reviews by N. Borderies, J. Chapront, T. Damour, A. Deprit, N.V. Emel'yanov, S. Ferraz-Mello, J.D. Hadjidemetriou, J. Henrard, K.V. Kholshevnikov, P.K. Seidelmann, M. Soffel, Z.H. Yi and individual reports of many other commission members are highly appreciated.

In the course of the report, Astronomy and Astrophysics Abstracts numbers are used except for the papers published in the most common journals or the proceedings indicated above.

B.   Relativistic Celestial Mechanics

(Main contributor : V.A. Brumberg)

In recent years the general theory of relativity (GRT) in its most simple applications is no longer seen as a theory under verification, but must be considered as a necessary framework in the discussion of high-precision observations and for the construction of accurate dynamical ephemerides. Disregarding the importance of general relativity as the physical foundation of celestial mechanics and from a purely operational point of view the distinction between relativistic and Newtonian description of the motion of celestial bodies is displayed 1) mathematically by the structure of the field equations and the equations of motion and 2) physically by the way we compare the theoretical and observational data. The first question is the subject of relativistic celestial mechanics, the second one belongs to relativistic astrometry. Only the consistent simultaneous solution of the two questions leads to physically meaningful results. The essence of the second question is the fact that in contrast to the inertial coordinates of Newtonian mechanics the coordinates of general relativity have no physical meaning and cannot be considered as the measurable quantities. Therefore, expressed in terms of the coordinates the conclusions of the relativistic dynamical theories are not unique, depend on the type of the coordinates employed and cannot be directly confirmed or refuted by observations. Only in terms of the measurable quantities these conclusions become unique and may be compared with observations. The IAU Symposium 114 mentioned above gave an insight into these both questions and was stimulatory to further investigations in this domain. The crucial problems on this way are as follows :

I.   Experimental Verifications

The vast program of the parametrized post-Newtonian formalism performed during the past two decades by C.M. Will with the aim to test general relativity and other theories of gravitation had led to the experimental verification of general relativity as the theory best fitted to observations (KOB, 355). As reported by P.K. Seidelmann, E.J. Santoro and K.F. Pulkinnen (KOB, 99) there are some discrepancies between the planetary ephemerides and the observational data but they may be due either to systematic discrepancies in the observational data or inadequacies in our knowledge of a model of the Solar system.

II.   Gravitational Radiation

For the first time after Einstein, Infeld and Fock a significant progress was achieved in formulating the equations of motion of celestial bodies taking into account the gravitational radiation. The culminating contributions were made by T. Damour (33.066.077; CAH, in press) and S.M. Kopejkin (40.066.087). This made it possible to investigate the orbital motion of the binary pulsar PSR 1913+16 by the rigorous celestial mechanics methods (solution of the relativistic two body problem with account of the gravitational radiation terms) and to confirm the previous results obtained on the basis of the approximate linearized theory of gravitational radiation. Besides, these investigations give an insight into the general structure of the equations of motion in general relativity.

III. Matching

The modern approaches to derive the equations of motion do not attempt to present solution in the global form valid for the whole space. Instead, the space-time is splitted up into several sub-domains with specific physical characteristics, for example : the internal region with dominant influence of the gravitational field of the body under consideration, the buffer region where the internal and external fields have comparable effects and the external region where the gravitational field of other bodies dominates. The solutions valid for each region separately are combined by the matching technique. This driving idea has been exposed in details by K.S. Thorne and J.B. Hartle (40.067.014). T. Damour has presented a general review of the problem of motion in Newtonian and Einsteinian gravity (HAI, 128). This review may be consulted as an excellent introduction into the modern state of the problem of motion in general relativity. The matching procedure allowed to derive the equations of motion not only for the Solar system bodies but for the condensed objects, like the black holes, as well. In the modified form the matching procedure is applied by S.M. Kopejkin (Sternberg Astron. Inst. Trans., 59, in press) to obtain the equations of motion of celestial bodies taking into account their shape, axial rotation and internal structure. The physical characteristics of the bodies (sphericity, rigid body distribution of velocities inside the body, etc.) are described in the body's proper reference frame and this local solution is matched to the global one. This procedure removes the terms of non-physical origin and makes the whole technique more rigorous mathematically and more meaningful physically.

IV. Reference Frames

Relativistic treatment of the reference frames is at present of particular interest. The concept of reference frame is often differently used in physics and astronomy leading sometimes to misunderstanding. For astronomical purposes it is suitable to follow the operational definition by J. Kovalevsky (Bull. Astr., 10, 87, 1985) considering that a reference frame results from materialization of a coordinate system by means of some reference astronomical objects. The most widespread approach in the relativistic theory of astronomical reference frames is to construct the proper reference frame of a fictitious or actual observer with the aid of the Fermi normal coordinates (the time axis is the proper time of the observer, the three space axes are the spacelike geodesics orthogonal to the world-line of the observer). In application to the geocentric frame one has at first to construct the proper reference frame for the massless Earth and then to substitute the corresponding coordinate transformation into the full metric incorporating the gravitational influence of the Earth. This approach has been developed in papers presented at the IAU Symposium No 114 by B. Bertotti, C. Boucher, T. Fukushima et al., M. Fujimoto and E. Grafarend (KOB, 233, 241, 145, 269) and reviewed by N. Ashby and B. Bertotti (Phys. Rev., D34, 2246, 1986). These reference frames were examined and applied to specific astronomical problems by M. Soffel et al. (Veröff. Bayerisch. Komm. intern. Erdmessung, 48, 237, 1986). Another, but somewhat similar way based on a finite linear transformation from barycentric to geocentric coordinates is exposed by N.V. Pavlov (38.066.211.212). In discussing the astronomical observations the infinitesimal transformations are sometimes used based on the possibility of local splitting of the space-time at the point of observation into the time axis and the three dimensional space. In this manner R.W. Hellings (41.066.196) and V.A. Brumberg (EIL, 19) introduce the local inertial coordinate system in the infinitesimal vicinity of the point of observation and derive the formulas for the relativistic reduction of observations.

Evidently, one may use any coordinates in constructing the reference frames. But if a coordinate system is not dynamically adequate to the problem under consideration both the solution of the dynamical problem and the transformation to the observational data will contain a number of extra terms caused only by the inadequate choice of the reference frame. These terms cancel out in the expressions of measurable quantities and the resulting relativistic effects turn out to be much smaller than the relativistic perturbations in the coordinate description of the dynamical problem. On the contrary, if the coordinate system is dynamically adequate the coordinate solution of the dynamical problem will not contain any large term of non-dynamical origin and will insignificantly change in converting to the measurable quantities. Reasoning from these considerations V.A. Brumberg and S.M. Kopejkin (KKM, in press) have developed a relativistic theory of the reference frames satisfying two conditions : 1) all basic non-rotating reference systems are built in the harmonic coordinates, 2) their metric tensors represent the dynamically adequate solutions of the Einstein field equations for the corresponding problems. In such a way, the barycentric reference systems (BRS), the geocentric reference system (GRS), the topocentric reference system (TRS) and the satellite reference system (SRS) have been constructed.

V.   Time Scales

The theory of reference frames is in close relation with the investigations on the time scales and the system of astronomical units. In many respects this is a question of definitions and agreements. The operational definition of TAI is exposed by B. Guinot (CM, 38, 155, 1986). The relation TDB-TDT based on the BDL analytical planetary and lunar theories of motion has been obtained by L. Fairhead, P. Bretagnon and J.-F. Lestrade (WIB, in press) and by T.H. Hirayama et al. (KNY, 75). Using the JPL numerical ephemerides this relation was studied by D.C. Backer and R.W. Hellings (Ann. Rev. Astron. Astroph., 24, 537, 1986). The dependence of the astronomical units of measurement upon the reference frame has been investigated by T. Fukushima et al. (CM,38, 215, 1986) and R.W. Hellings (41.006.196). The consistent theory of transformations BRS→GRS→TRS (SRS) enables to treat these questions in straightforward manner.

VI.   Modern Theories of Motion in the Solar System

Relativistic terms are taken into account in the planetary theories in the barycentric harmonic coordinates in numerical (JPL, ITA) or analytical (BDL) forms (see KOB). Relativistic terms in the lunar theory by J.F. Lestrade and M. Chapront-Touzé (32.094.028) and by V.A. Brumberg and T.V. Ivanova (39.094.017) as well as relativistic terms in the artificial satellite theory by C.F. Martin, M.H. Torrence and C.W. Misner (40.052.033) are found in the BRS and their amplitude may attain several meters. In converting to the GRS their amplitude reduces to the centimeter level. This was shown explicitly by M. Soffel, H. Ruder and M. Schneider for the Moon (41.094.009) and by S.Y. Zhu et al. for Earth satellites (IAU Colloq. No. 96). Considering that the equations of lunar and satellite motion in BRS are known it is easy to obtain the equations in GRS with the help of the transformation BRS → GRS. Formulas for the relativistic reduction of observations are given within sufficient practical accuracy in Japanese Ephemeris for 1985. More rigorous theory may be developed using the transformations BRS →

GRS →TRS (SRS) admitting the rigid body rotations of the systems. Particular attention is drawn now to the pulsar timing formulas. The most detailed results are obtained by T. Damour and N. Deruelle (Ann. Inst. H. Poincaré, 44, 263, 1986) and D.C. Backer and R.W. Hellings (Ann. Rev. Astron. Astroph. 24, 537, 1986).

C. Analytical Methods and Orbital Resonance

(Main contributors : S. Ferraz-Mello and J. Henrard)

Perturbation theory as a tool of investigation and resonance phenomena in the Solar System as a reservoir of interesting problems are continuing to attract many investigators.

Perturbation theories used in Celestial Mechanics are now almost exclusively based upon Lie Transforms methods as a technique and K.A.M. theory as a justification. In that frame two aspects have received continued attention : the formal or algebraic aspect (how to set up a perturbation theory) which we review under the title "Intrinsic Perturbation Theories" and the convergence aspect (what does a perturbation theory tell us) which we review under the title "General Estimates". The theoretical results relative to this second aspect have to be supplemented with the information given by numerical experiments (see "Planetary Theories") and the study of the chaotic behaviour of Hamiltonian systems (see "Chaotic Motions").

Applications of Perturbation Theories to the Solar System are broken down in this report in the three traditional chapters : Planetary Theories, Satellite Theories and Asteroid motions, each with its own physical but also mathematical peculiarities, although they have many similarities. A new chapter quite different from them : the Planetary Rings has been added recently and has already blossomed into a very rich field of research.

We have briefly covered under special headings two particular aspects as they do not enter readily in the adopted classification. They are "Resonance Sweeping" by which are analysed the effects of small non-conservative forces in shaping the Solar System as we see it now and "Chaotic Motions", still in his infancy as a topic of research in Celestial Mechanics, but which is likely to grow in importance both for theoretical and practical investigations.

I.   Intrinsic Perturbation Theories

Several contributions have emphasized in the last few years the intrinsic characteristics of general perturbation theories. Perturbation theories have usually been specialized to selected sets of phase variables or to selected types of perturbations and it is not always easy to decide if the difficulties encountered in some procedures are physically significant or the reflect of an unappropriate choice of coordinates or of unperturbed first approximation. Ferraz-Mello (CM. 35.209, CM. 35.221) developed the equations of the method of Hori for first order resonances and emphasizes the fact that the unperturbed first approximation should already reflect the topology of the full system (Ferraz-Mello, Sessin CM.34.453). Deprit in a series of papers (CM.

24.111, CM. 26.9, CM. 29.229, 39.042.059) and Kummer (41.042.087) use a group theoretical approach to describe the intrinsic properties of general perturbation techniques. This led Coffey et al (CM.39.365) to propose spherical charts rather than the usual cylindrical ones related to Delaunay's elements to describe the averaged main problem of an artificial satellite. The structure of the phase space in the vicinity of the critical inclination can then be fully explained. Action-Angle variables have been used successfully by Sessin and Ferraz-Mello (CM.32.307), Wisdom (IC. 63.272) and Henrard-Lemaitre (IC. 69.266, CM. 39.213) to treat a double resonance problem (see below under the heading : Asteroid motion). Pauwels (CM. 30.229) has shown how the phase space of the secular orbit-orbit coupling should be represented on a sphere rather than a plane. Sessin (CM.39.173) has proposed a new theory of integration for the so-called singularity of Poincaré.

The Ideal Resonance Problem (Garfinkel A.J. 76.157) has been revisited in an effort to bypass the difficulties of the Bohlin procedure. Jupp and Abdulla (CM. 34.411, CM. 37.183) use Lie series in the libration domain and Sessin (CM. in press) solves it using the method of Delaunay. Deprit (Adv. Astro. Sci. 46.521) shows how the "disencumbering" of the problem makes its physical mechanisms more transparent. Howland (CM. subm.) and Henrard-Wauthier (CM. subm.) use non-canonical transformations to reduce the problem to either a classical problem in perturbation theory or to the pendulum itself.

In the case of the elliptic restricted problem, the technique of Sessin allowing the reduction of a truncated averaged Hamiltonian to the circular case, through a suitable change of variables, was reconsidered by Tsuchida (FMS.149), Wisdom (CM.38.175), Henrard et al (CM.38.335) and Ferraz-Mello (FMS.37 and AJ. 94.808).

II.   General Estimates

The well-known K.A.M. theory states that the series expansions of general perturbation methods gives information for all time but only on a peculiar closed subset of the phase space (the "good" tori). Nekhoroshev (Russ. Math. Surveys 32.1) using the same techniques has obtained important results valid on open subsets of the phase space by renouncing to have estimates for all times. These results have recently been refined and described in the context of Celestial Mechanics by Benettin, Galgani and Giorgilli (CM. 37.1, CM. 37.95).

III.   Planetary Theories

The interest in Planetary Theories has continued to be strong especially in general planetary theories with a validity covering a time interval of the order of the million years or more. Classical theories have been revisited by Sukhotin (39.042.086) and by Knezevic (CM. 38.123). Laskar has continued his work on the numerical integration of the analytically averaged "autonomous" system (AA. 144.133, AA. 157.59, 41.042.002). An interesting new method called "synthetic theory" by Milani and Nobili has been developed. It consists in extracting by numerical averaging from the output of long term numerical integrations, information about frequencies, possible locking in resonance, exchanges in energy or momentum between planets, etc. The numerical integrations of Kinoshita and Nakai (5 outer planets for 5 million years - CM. 34.203) have been used for this purpose (Milani, Nobili 38.091.004, CM.35.269). Further progresses along these lines are coming from the LONGSTOP project (5 outer planets for 100 million years -Milani et al. 41.091.003, AA. 172.265, Carpino et al. AA. 181.182). Applegate et al (AJ.92.176) have also performed a numerical integration of the 5 outer planets for 210 million years.

Some perturbative terms like the 1.1 million years oscillation in the semi-major axis of Uranus and Neptune have been computed analytically only after being detected in the numerical integration (Milani et al. AA.172.265). While at some level of accuracy these results show a regular structure, the resolution is now such that one can wonder about the intrinsic numerical convergence of any planetary theory. The solar system could be chaotic with a very slow time constant (Milani et al. subm.). Mixing numerical and analytical methods has also been used successfully by Bretagnon (CM. 38.191) in order to improve the short period ephemeris of the planets.

IV.   Satellite Theory

Satellite theories (except the theory of the Moon which was reviewed three years ago) are not yet encountering these difficulties. The field has been reviewed by Ferraz-Mello (CM. 34.223 and FMS.37). Work has been continuing on the theory of the Galilean satellites of Jupiter (Thuillot CM. 34.245, Henrard CM. 34.255, Vu Thesis-Paris 1986). The Uranian satellites were studied with emphasis in the great inequalities of long period able to affect ephemerides and mass determinations. They were studied by Lazzaro et al. (38.101.018, 40.101.062 and AA. 182.150) with the classical Laplace Souillart theory and by Laskar (AA. 166.349) using his version of the method of general planetary theories (39.042.006). The theory of Laskar was compared by himself and Jacobson (AA. in press) to the existing earth based observations and allowed them a new determination of the masses within 15 % of the values found from Voyager II data.

Orbital resonances among Saturn's satellites were reviewed by Greenberg (38.100.143). Sato and Ferraz-Mello (FMS.105) and Salgado-Sessin (FMS.099) studied the resonance Enceladus-Dione. The resonance involving Hyperion is made more difficult by the great forced eccentricity (0.104) and present theories are simply not accurate enough to represent the motion of Hyperion as was shown by Taylor (38.100.078), Sinclair and Taylor (AA. 147.241), Taylor et al. (AA. in press) and Dourneau (Thesis-Bordeaux 1987). The evolution and the stability of the coorbital satellites of Saturn present new challenges. They have been studied by Lissauer et al. (IC. 64.425) and Sinclair (38.100.002). Hill's problem revisited by Henon and Petit (CM. 38.67, IC. 66.536) and Spirig and Waldvogel (40.042.029) could be a good model to analyze some of their properties. Greenberg (IC.70.334) studied the Laplacian libration and found a branch of solutions where the critical angle is different from 0 or π.

V.   Asteroid Motion

The distribution of asteroids in the asteroid belt has continued to attract much attention. The field has been reviewed by Scholl (39.098.044, 40.042.103). The depletion of the outer asteroid belt seems to be well explained by the perturbation of Jupiter leading except in a few "protected" cases to close approaches (Milani, Nobili AA. 144.261, 39.098.029). The identification of families (Farinella et al. 41.098.038) and their possible explanation through collisions (Zappala et al. IC. 59.261) necessitate better theories (Knezevic 41.098.041). The role of secular resonances in shaping the belt has been studied by Kozai (CM. 36.47, 39.098.033), Nakai and Kinoshita (CM. 36.391) and Scholl and Froeschlé (41.098.042). Zellner et al. (39.098.094) have drawn the attention on the rather sharp upper limits in inclinations and eccentricities in the distribution of the Asteroids. Froeschlé and Scholl (AA.158.259) have shown that resonances can break up meteor stream into arcs of different sizes with distinctly different dynamical evolutions.

Numerical studies of Wisdom (AJ. 87.577, IC. 56.51), Murray and Fox (IC. 59.221) and Sidlichovsky and Melendo (41.098.009) have shown how the 3/1 and 5/2 Kirkwood gaps could be produced by close encounters with Mars due to large eccentricities induced by the Jupiter perturbations. In part of the phase space these large forced eccentricities are intermittent and a result of the chaotic character of the motion. Wisdom (IC. 63.272) has identified the mechanism producing this chaotic behavior as the separatrix crossings in the two critical arguments problem describing the averaged elliptic three body problem and not as previously thought as the effect of the short periodic terms. The effect of the eccentricity of the perturbing body has also been investigated numerically by Ferraz-Mello and Dvorak (AA.179.304) in another context (Resonance Enceladus-Dione). The 2/1 Kirkwood gap has been investigated numerically by Murray (IC. 65.70) and analytically by Henrard and Lemaitre (IC. 69.266, CM. 39.213). In the planar problem at least, a significant part of the volume of the phase space seems to be protected against increases in eccentricity large enough to induce close encounters with Mars. These resonances have also been studied by Gerasimov (A.Zh. 63.567) when second order harmonics are included (37.042.070) or not (A.Zh.63.763).

VI.   Planetary Rings

(main contributor : N. Borderies)

1. <u>Ring-satellite interactions</u> : Encounters between two small bodies orbiting around a planet can be dynamically described by the Hill's problem. Petit and Henon (IC. 66.536) have shown that the number of parameters of the problem can be reduced to one, the reduced impact parameter, and have studied in details the family of solutions. Analytical developments relevant to this study are presented by these authors (CM. 38.67). In a subsequent paper (AA. 173.389), Petit and Henon generalized the interaction model by considering particles with finite sizes. This study leads to a numerical simulation of planetary rings (Petit, Ph.D. Thesis). Roques, in her thesis, has developed a different numerical model of the interactions between a satellite and a ring, in which she studied the response of the particles to the instantaneous perturbation exerted by a satellite. Theoretical work on disk-satellite interaction has been pursued by Sicardy (FMS. 167) and Meyer-Vernet and Sicardy (IC. 69.157). These papers deal with the torque exerted by a satellite on a disk at a Lindblad resonance. A powerful approach in the study of planetary rings is based on the concept of streamlines of the flow of particles. Borderies and Longaretti clarified certain problems of Celestial Mechanics posed by this approach (IC., in print).

2. <u>Modelizations of the collisions effects</u> : Theoretical work on the dynamics of a disk of colliding particles has been developed in several directions. Brophy and Esposito (IC., in print) have solved the Boltzmann equations for the equilibrium velocity distribution and have compared their solution with the approximate analytical solution of Goldreich and Tremaine. Shu and Stewart (IC. 62.360) used the Krook equation, which is simpler than the Boltzmann equation. This simplicity allowed them to solve analytically the problem for an unperturbed ring. Bor-

deries, Goldreich and Tremaine (IC. 63.406) have studied the case where the ring particles are closely packed together, such that their collective behaviour resembles that of a liquid. They found that density waves are unstable in such a ring and that eccentric modes can spontaneously develop in narrow rings. Araki and Tremaine (IC. 65.83) have studied the dynamics of dense particle disks by adapting the Enskog theory of dense hard sphere gases to inelastic spheres. They suggested that the complex ringlet structure in the B ring of Saturn represents adjacent solid and liquid phases. Collision and accretion process determine the distribution of particle sizes in a ring. A theoretical model has been recently developed in his thesis by Longaretti. He found that Saturn's rings could not consist of hard spheres because accretion would be too efficient and was lead to the conclusion that individual particles in these rings have a brief survival time.

3. <u>Waves</u> : Many density waves have been found in Saturn's rings. They are associated with the resonant forcing of the particle eccentricities. A smaller number of bending waves, associated with the resonant forcing of the inclinations, are also present. The main problem concerning density waves is that most of those which are observed in Saturn's rings are nonlinear : the perturbation in surface density is not small compared to the unperturbed surface density. Hence a lot of effort has been put in constructing nonlinear theories of density waves (Shu et al., APJ. 291.356, APJ. 299.542 and Borderies et al. IC., 68.522). The analysis of a density wave profile (Longaretti and Borderies, IC. 67.211) has allowed to check certain theoretical results such as the nonlinear dispersion relation which had been derived independently by the two groups mentioned above. Work on bending waves has been accomplished by Gresh et al. (IC. 68.481). Bending waves conform more closely than density waves to the linear model.

4. <u>Ring confinement</u> : Voyager images have revealed radial undulations at the edges of the Encke gap in Saturn's rings. These waves indicate that a satellite is present in the gap. Cuzzi and Scargle (AJ. 292.276) and Showalter et al. (IC. 66.297) have determined the location and size of the satellite responsible for the undulations and have discussed the implications of their analysis for the shepherding theory. The new data acquired by Voyager during its encounter with Uranus have permitted new tests of the theories. French et al. (SC. 231.480 and IC., in print) discovered that a normal mode $m = 2$ was spontaneously excited in the δ ring. Goldreich and Porco (AJ., in print) studied the sheperding of the Uranian rings. They found that the shepherding of the ϵ ring was consistent with the theory of confinement while, on the other hand, the α and β rings pose problems which lead the authors to question the self-gravity model for the rigid precession of these rings.

5. <u>Neptune's arcs rings</u> : With the discovery by Hubbard and Brahic (NA. 319.636) of incomplete rings around Neptune, the problem for the theoreticians has been to explain this observation. Lissauer (NA. 318.544) has proposed that the arcs could lie at the Trojan point of an undiscovered satellite of Neptune. Another satellite would be responsible for the radial confinement of the arc. An alternate theory has been proposed by Goldreich et al. (AA. 92.490). They invoke corotation resonances with a satellite moving on an inclined orbit. The same satellite would be responsible for the confinement of the arcs.

VII. Resonance Sweeping

The effect of small nonconservative forces in shaping the solar system as we see it now has first been brought to the attention of Celestial Mechanicians by Goldreich (MNRAS 130.159). This mechanism is particularly important when a resonance is encountered. It has continued to receive much attention in the last three years. Analytical tools have been developed by Lemaitre (CM. 32.109), Borderies and Goldreich (CM. 32.127) and Hadjidemetriou (ZAMP 37.776). Applications to the evolution of satellites due to tidal dissipation has been reviewed by Peale (BUM), Kovalevsky (39.091.038) and Henrard (40.091.031). Further contributions to the tidal evolution of the Moon are presented by Kovalevsky (40.042.098). Spin orbit resonances and their tidal evolution have been investigated by Sidlichovsky (39.042.044), Henrard (FMS 19) and Beletsky (40.042.091). The effects of this mechanism upon the accretion of planets has been investigated by Weidenshilling and Davis (IC. 62.12). Applications to the asteroid belt structure (D'yakov and Reznikov 37.098.011, 41.042.062, Gonczi et al. IC.51.633, Lemaitre CM. 34.329) shows that this mechanism provides an alternative explanation for the Kirkwood gaps.

VIII. Chaotic Motions

Chaotic behavior in Hamiltonian systems has been the subject of intensive research in Theoretical Physics for some times now. Henon (AJ. 69.73) and Froeschlé (AA. 9.15) were apparently the first to study it in the context of Celestial Mechanics leading to applications to galactic dynamics and to the asteroid belt (Froeschlé-Scholl AA. 48.389). It is only quite recently however that it was realized that this kind of behavior can actually be seen in the Solar System in the rotation of Hyperion (Wisdom et al. IC. 58.137, Peale 38.100.071) or could be an important mechanism in the explanation of the Kirkwood gaps (Wisdom IC. 56.51, IC. 63.272) or the depletion of the outer asteroid belt (Milani, Nobili AA. 144.261). The theory and computation of the Lyapunov characteristic exponents has been reviewed by Froeschlé (CM. 34.95) and the use of the Kolmogorov entropy as an estimate of the disorder of a system tested by Gonczi et al. (CM. 34.117). The threshold of chaotic and regular behavior has been analyzed by Innanen (AJ. 90.2377), Yokoyama (37.042.101), Galgani (40.042.130) and Simo (40.042.124).

D. Artificial Satellite Theory

(Main Contributor : André Deprit)

The scientific adventure speaks here in profusion : the refereed literature in periodic journals is but a fraction of what comes out in the form of technical memoranda, internal notes, prepublication of conference proceedings, and contractor's reports. One way to deal with such a wealth of information is to survey the major themes. Having gotten the lay of the land, the reader will find it easy to track the precise trails through the professional compilations of abstracts.

For the past thirty years, in fact since the seminal studies of Orlov, Robertson, Blitzer and Breakwell in the mid-50s, artificial satellite theory has lived under a dark cloud. Was the critical inclination an essential feature of the phase space or simply a spurious singularity caused by the coordinates or the perturbation algorithm ? The enigma has now been solved. In the main problem (J_2 only), along the family of mean circles of a given radius, there occurs a small segment where the orbits are unstable. The endpoints of that interval correspond to circles in orbital planes inclined over the equator at angles slightly above or below the critical value $\arctan 2 = 63 \deg .43$. These singular orbits owing to their resulting from a Hopf bifurcation testify to the intrinsic nature of the critical inclination (S.L. Coffey, R. Cushman, A. Deprit and B.R. Miller). The main problem of artificial satellite theory and, for the matter, any perturbed Keplerian system having a rotational symmetry, has now been shown to admit a two-dimensional bundle of spheres as its reduced phase space. Numerous systems have been analyzed in that framework : the Stark effect (A. Deprit), the Zeeman effect (S.L. Coffey, A. Deprit, B.R. Miller and C.A. Williams), the effect of radiation pressure on dust orbiting a planet (R. Cushman, A. Deprit and J.C. Van der Meer), the Gylden systems or Keplerian systems whose Gaussian parameter varies with time (C.A. Williams), the relativistic Two Body problem (D. Richardson).

In the global picture of the average phase space, circular solutions emerge as equilibrium positions; mission engineers know them as "frozen orbits". Since the time when these singular solutions were shown to present great advantages for controlling the altitude of low satellites (E. Cutting, G. Born and J. Frautnick, 1978), they have been set as nominal orbits for SEASAT-A, GEOSAT, NROSS, TOPEX/POSEIDON, ERS-1 and other spacecrafts carrying altimeters around the earth. Uphoff (1985) has proposed to keep orbiters on "frozen orbits" around Venus. At inclinations where other equilibria arise in the neighborhood of a frozen orbit, Brouwer's third reduction by eliminating the argument of perigee will fail (C.C. Tang, 1987). Furthermore, numerical contouring reveals that the position of a frozen orbit in the reduced phase space and its stability are affected significantly by high degree zonal harmonics in the gravity field of the planet (W.D. McClain, 1986) and by the luni-solar perturbations (M.E. Hough). Catastrophe Theory is likely to offer the means of exploring these intricate problems without getting lost in an avalanche of parameters and perturbations (K.R. Meyer).

The appeal to simplification in analytical theories never ages. Undoubtedly Brouwer's solution for the main problem in closed form as far eccentricities and inclinations are concerned marked a decisive break with the tradition. It prepared celestial mechanics to the impact of Geometric Dynamics. Intermediaries obtained by stamping out separable Hamiltonians are replaced by "natural" intermediaries resulting from intrinsic reductions in the sense of Meyer, Marsden and Weinstein. The question now is how much symmetry we need or want in the Hamiltonian of artificial satellite theory. The elimination of the parallax (A. Deprit) and the elimination of the perigee (K.T. Alfriend and S.L. Coffey) are now interpreted as two steps in a Lie transformation meant to confer spherical symmetry to the zonal problem of artificial satellite theory. Seeing Cid's inchoate attempt at producing a radial intermediary blossom into a second order problem integrable simply in terms of elliptic integrals (Cid, Ferrer, Sein-Echaluce) arouses expectations that major theories like those of the moon or the solar system will undergo the kind of disen-cumbering that Delaunay and Hill called for hundred years ago. In the mean time, theorists go on improving the techniques inherited from the XIXth century, witness a revision of Hansen's recurrences for obtaining the coefficients of high degree in the formulas of the Two-Body problem developed according to the powers of the eccentricity (R.Proulx).

The International Sun-Earth Explorers (ISEE), a collaboration between NASA and ESA, have reinforced the special emphasis that Molniya type satellites for communication over long distances have put on orbits with very large eccentricities. One would have believed, at first glance, the dominant perturbation is alternatively the drag on the lower part of the loops and the lunar and solar attractions on the upper part. A careful analysis, both numerical and analytical, has revealed, though, that the lunar attraction is mainly responsible for the decay of ISEE-1 and ISEE-2 (J.P. Amenabar et al. at the Computer Sciences Corporation, Silver Spring, MD, U.S.A.). This is but one of many instances where classical celestial mechanics is called upon nowadays to contribute to control theory in maneuvering satellites to extend their lifetime or to keep them close to their specified paths. On the other hand, tracking engineers are increasingly involved with the problem of extending extrapolations over very long arcs. Clearing dead satellites from the equatorial geosynchronous belt is one of the practical issues to which astrodynamicists contribute imaginative solutions. In those areas, competition for attention, merit, speed, versatility and portability is as vigorous as ever between general purpose numerical integrators, semi-analytical theories - numerical integration of numerically averaged equations advocated by P.J. Cefola and his group at the Charles Stark Draper Laboratory -, the stroboscopic method - devised by C. Lubowe in the late 50s at the Bell Telephone Laboratories, but re-invented by E.A. Roth at the European Space Operations Center (ESOC) in Darmstadt.

Yet, much of the computational story of these past three years has been the tremendous performance of UTOPIA and GEODYN in the midst of giant electronic developments. Satellite Laser Ranging (SLR) has emerged as a technology capable of providing geodesists and oceanographers with a vast array of new measurements and global coverage. The most fastidious orbit correctors operating in batch mode, UTOPIA and GEODYN, have now taken residence in the fastest vector preocessors available, respectively a CRAY at the University of Texas in Austin (UT) and a CYBER 205 at NASA Goddard Space Flight Center (GSFC). The transfer in both cases took two years; it was completed for the most part in the early months of 1986. In the new computer environment. David E. Smith and the Laboratory for Geophysics at GSFC set themselves to the task of analyzing in one batch all laser rangings to 17 satellites, deriving therefrom the first "universal" gravity model of the earth, GEM-T1. For his part, Byron Tapley with his team at the Center for Space Research at UT processes observations in single batches extending over 4000 days ($\approx$ 11 years) as a matter of routine. Copies of GEODYN and UTOPIA have been distributed to major institutions in the U.S.A. and outside. Nonetheless, independent work on the determination and analysis of precise orbits is being carried out at the University of Nottingam (V. Ashkenazi), at the Technische Hogeschool Delft (K.F. Wakker), at the Centre National d'Etudes Spaciales (CNES) of Toulouse (F. Nouel), to list but a few.

Vector processing is just one of the steps SLR is taking to improve its analyses. Refining the modelling of physical effects affecting the orbit of a close satellite is another one. In this regard, one should mention a flurry of studies caused by two unexpected phenomena : 1) a secular deceleration in LAGEOS's semimajor axis (D.P. Rubincam, F. Barlier) and 2) a secular acceleration of its node (C. F. Yoder et al.).

When the European Space Agency puts ERS-1 in orbit and, somewhat later, after NASA and CNES launch TOPEX/POSEIDON, a new era in orbit determination will be entered. The altimetry equipment on these platforms will produce information about the surface of the oceans with unprecedent accuracy at the level of 10 cm. Preparation of the missions is eliciting numerous studies ranging from a refitting of Kaula's solution of the variational equations in artificial satellite theory (Rosborough) to the possibility of separating permanent sea surface topography from time-varying features by comparing data at "crossover points" (G.H. Born, R.E. Cheney, J.G. Marsh, C.K. Shum, B.D. Schutz, B.D. Tapley, etc.). Unrelentingly, the orbit of SEASAT, the first satellite dedicated to studying the surface of the oceans, and that of GEOSAT-ERM are recomputed to evaluate the most recent models of the earth's gravity field in conjunction with Wahr's model of tides for the solid earth and that of Schwiderski for ocean tides.

The Global Positioning System (GPS) adds a purely geometric element to the problem of orbit determination from observations of close satellites. Studies of triangulation by clock carrying spacecrafts are pursued vigorously with a view of identifying the factors most likely to affect the accuracy at the level of 1m and below. Most critical among them would be the solar radiation pressure and the stochastic tropospheric delay at the zenith of the receiving station (S.M. Lichten at the Earth Orbiter and Navigation Systems Group, Jet Propulsion Laboratory). A GPS receiver on TOPEX/POSEIDON would bring the errors below the 10cm level (B.G. Williams, loc. cit.). Optimists talk with tongue in cheek of coming to 1mm by the mid-90s. Orbit determination by numerical integration of satellite orbits and differential corrections is now firmly established on the highest grounds of perfection in numerical engineering.

V.A. Brumberg,
President of the Commission.

COMMISSION 8 : <u>POSITIONAL ASTRONOMY (ASTRONOMIE DE POSITION)</u>

PRESIDENT : Y. Requième
VICE-PRESIDENT : M. Miyamoto
ORGANIZING COMMITTEE : P. Benevides-Soares, D. Duma, L. Helmer, J. Hugues,
         L. Lindegren, Luo Ding-Jiang, F. Noel, G. Pinigin,
         L. Quijano, S. Sadzakov, H. Schwan, C. Smith, M. Yoshizawa

  The Commission deeply regrets the death of our colleague G. Teleki who passed away on February 1987.

  After the retirement of T. Dambara, S. Lautsen and W. Gliese and the arrival of C. Turon-Lacarrieu as a new member, the membership of our commission is 142.

I. <u>INTRODUCTORY REMARKS</u>

  Due to the size of the reports received from the observatories and according to the recommendations of the I.A.U. Executive Committee this triennal report will be (instead of a list of topics worked on by Commission members) a comprehensive review around the main objectives of our commission : Galactic reference frame, link to the Extragalactic and Solar reference frames, related instrumental research and reduction methods.

  The overlap of the areas of interest and techniques of Commission 8 and Commission 24 is more and more important and it would be the time to clarify the proper responsibilities of our individual commissions. Renaming of Commission 8 into "Stellar (or Galactic) Reference Frame" could be the best way to state clearly our main practical objective in front of the astronomical community.

  One of the milestones of this triennum was during the last General Assembly at Delhi the creation of a pluridisciplinary Working Group to give recommendations for the definition of the Conventional Celestial Reference System and ways of specifying practical realizations of this system. The establishment and the maintenance of the future primary extragalactic reference frame based on VLBI groups involved in determination of Earth rotation parameters should be obviously under the responsibility of Commission 19 whereas the control of the optical counterpart of this primary frame could be the duty of Commission 24 within its Working Group "Optical-radio reference frame". But improvement and extension of the stellar system to fainter stars with link to the planetary and extragalactic systems will remain the huge task of our Commission 8.

II. <u>INSTRUMENTATION</u>

  During the past three years an intensive effort of modernization and automation of the classical instruments was made by many observatories, opening out new horizons for the ground-based astrometry : improved accuracy, increase in observing speed and limiting magnitude. New instruments are under development or nearly completed. Waiting for the success of space astrometry we enter already in an exciting period of renewal when the asymptotic limit of 20-30 milliarcseconds seems to be within reach for most of our photoelectric instruments even up to the magnitude 12-13 for several ones.

  It is important to mention that progress in modernization is also beginning in the Southern hemisphere and these efforts should be strongly encouraged.

1. Automatic Meridian Circles

The *Carlsberg automatic meridian circle* was run jointly by Copenhagen University Observatory, the Royal Greenwich Observatory and the Instituto y Observatorio of Marina at the Spanish Observatorio del Roque de los Muchachos situated on the island of La Palma (lat. 28.8° Nord, alt. 2 400 m). The telescope was operated completely automatically under the control of two H.P. mini-computers. Instrumental calibrations were carried out hourly using collimators, nadir pool and since 1986 azimuth marks (40.032.021). Also for azimuth determination, 29 polar pairs and 3 faint polarissimae were introduced to the regular observing programme in 1986. Internal refraction in the telescope tube was greatly diminished by tangetial ventilation following an investigation of Høg (41.032.069). The method of observation was essentially differential, but only in the sense that one rigid rotation in R.A. and DEC was carried out nightly to fit to the FK4 system, the drift of the instrument having been removed using the hourly calibrations. From May 1984 to December 1986, more than 200 000 star positions, 840 planet positions and 4 800 asteroid positions were obtained. The internal standard error for one observation was found to be : R.A. 0.19" (sec z)$^{0.6}$, DEC 0.18" (sec z)$^{0.9}$, MAG 0.06 sec z (Morrison et al.). Futhermore a new photoelectric micrometer is under development.

The *Tokyo photoelectric meridian circle* has been in regular operation since the end of 1984. The operation and data acquisition, together with meteorological data acquisition, are fully automatized. Stars brighter than V = 8 are observed with the oscillating slit micrometer during 30 seconds whereas fainter stars are scanned during a longer time. About 40-45 objects up to V = 12.2 are observed every hour with regular determinations of the instrumental constants, including horizontal and vertical flexure, and monitoring of the pivot irregularities by means of axial collimators. From the first catalogue the internal standard error of a single observation depends on the apparent magnitude : R.A. 0.18" to 0.22", DEC 0.20" to 0.27", MAG 0.07 (Yoshizawa and Miyamoto). Running in parallel is the development of a two-dimensional CCD micrometer which is expected to increase the productivity of future observations.

The *Bordeaux automatic meridian circle* is also in regular operation. Observations are strictly differential (without determination of physical constants) for objects up to V = 12.5 with internal standard errors of 0.11" in R.A. and 0.16" in DEC. The apparent magnitude is also obtained with a standard error of ± 0.15, providing a sure identification of faint stars.

The *Washington seven-inch automatic transit circle* at the New Zealand Black Birch Station became fully operational with solar system objects being regularly observed since late 1986.

The automation of the *Pulkovo horizontal meridian circle* (Pinigin et al., 40.032.001, 40.034.205, 40.034.207) and of the *Pulkovo and Nikolaev classical meridian circle* (Ershov, Streletsky et al., 40.034.209, 40.034.210) were completed and regular observations have begun.

2. Automatic Astrolabes

In *China* the photoelectric astrolabe Mark III with its 260/5000 mm optical system and photon counting micrometer working up to V = 11-12 is in the debugging phase. Meanwhile the classical instruments still in operation are being improved for automatic observation indluding planets. After 1988 limiting magnitude of photoelectric astrolabes will be increased to V = 9 (Luo et al.).

In *France* the CERGA photoelectric astrolabe is now completely automatic (Billaud).

3. New Instruments

Li, Hu and Høg report that an agreement between three astronomical institutes in China (Shaanxi and Nanjing) and Denmark has been signed about the development and operation of the *automatic glass meridian circle.* The instrument with 24 cm aperture is being manufactured in China after some of its components have been developed and tested in Denmark. After first tests of the instrument in Nanjing it is planned to erect it in Brorfelde for an optimization period of a year. It shall be finally operated in Shaanxi where a new observatory site is presently beeing selected in the Li Shan mountains near Xian.

At Goloseevo and Nikolaev new *axial meridian circles* were designed, built, investigated and tested (Kharin, Minyailo, Lazorenko, Shorik, Pinigin, Shornikov, Konin, 40.032.070).

At Pulkovo new designs for a pentag meridian circle (Nemiro, 39.032.52) and a reflecting meridian circle (Nemiro, Streletsky) were proposed.

The U.S. Naval Observatory is undertaking the construction of a phase coherent, *astrometric optical interferometer.* Building upon the experience gained from the Mark III interferometer at Mount Wilson (Shao et al., 42.034.028), the U.S.N.O. instrument is currently in the early stages of conceptualization and initial design. Accuracies of 0.02" up to V = 12 are expected but the final goal could be the milli-arcsecond level.

4. Modernization of Meridian Instruments

Automation of the San Fernando meridian circle, in a similar way to the Carlsberg meridian circle, started in September 1986. The photoelectric micrometer and other major parts for the automatic telescope setting are being manufactured in the Brorfelde workshop (Quijano, Sanchez).

The Pulkovo photographic vertical circle (Bagildinsky, Shkutov et al.) and the Odessa Repsold meridian circle (Volyanskaya et al., 40.032.073) have been modernized.

CCD matrices were incorporated in circle reading devices at Moscow (Golovko) and Pulkovo (Ershov et al., 42.032.004).

A photodiode array device (Reticon 512) to measure automatically the circle readings will be installed on the Santiago Repsold meridian circle (Carrasco) and some similar device is planned for the San Juan meridian circle (Carestia).

Considerable progress has been realized on the Washington six-inch transit circle. An average of 35 000 observations per year were made during the first two years.

5. Solar Astrolabes

In France the CERGA solar astrolabe observes the Sun at 10 different zenith distances in order to follow the whole apparent orbit. Replacement of the visual micrometer by a CCD camera with real-time image processor gave encouraging results (Laclare, Journet, 40.080.004, 42.041.004).

Modifications of classical astrolabes for solar and planetary observations are being introduced at many observatories. The classical prism is replaced by several "cervit angles" which allow observations at different zenith distances. Sun observations are planned with a convenient solar filter at Paris (Chollet), San Fernando (Sanchez), Cagliari (Proverbio), Santiago (Noel), Sao-Paulo (Benevides).

6. The Hipparcos-Tycho Space Mission

The Hipparcos mission preparation has progressed in a very satisfactory manner since the last IAU General Assembly. In 1986, the project passed successfully the critical Design Review, and the construction of the engineering and of the flight models has followed the fixed schedule, so that it is still expected that the satellite will be finished in March 1988. But, due to the delays of the Ariane launcher, the launch is now scheduled only in 1989, for a 2.5 year mission. This means pratically one year delay in delivering the final Hipparcos and Tycho catalogues, so that their delivery to the general scientific community is to be expected only in 1995.

The supplementary accuracy analyses and the results of tests on the various subsystems already constructed confirmed the expected accuracies that should not exceed 0.002" in positions and parallaxes and 0.002" per year in proper motions for stars brighter than visual magnitude 9 with a degradation of a factor 2 for magnitude 12, provided that the next solar maximum that will occur during the mission is not exceptionally active.

The structure and the tasks of the four consortia responsible for producing the input catalogue and for reducing the data have been presented to the New-Delhi General Assembly (Transaction XIXth IAU General Assembly, Highlights, vol. XIX) and to the IAU Symposium 109 (H. Eichhorn ed., Gainesville) and this will not be repeated in this report.

Since 1985, the two main mission data reduction consortia, FAST and NDAC, have finalized their reduction algorithms and are preparing the final engineered softwares that will be used during the actual data reduction. An important exchange of simulated data between the two consortia and the comparison of the results obtained in using them have permitted to verify that the algorithms, although different, give comparable results and no systematic differences seem to exist. These comparisons are very important, since the final result of the reduction should be a combined unique catalogue.

The TDAC Consortium is now finalizing the preliminary softwares and checking the adopted algorithms for the reduction of the continuous flow of data produced by the star mapper. The TDAC input catalogue consisting of some $1 \frac{1}{2}$ or 2 million stars up to V = 11 from the Guide Star Catalogue and the CDS (Centre de Données de Strasbourg) data bank should be ready in 1988 and will be used to identify the stars giving detectable signals.

The INCA Consortium, responsible for the production of the Hipparcos input catalogue, i.e. the selection of about 115 000 stars that will be observed by the Hipparcos main mission among the 210 000 proposed stars, has produced a first tentative version of the input catalogue, result of numerical mission simulations taking into account the scientific requirements from the about 200 proposals but also the observational constraints inherent in the Hipparcos satellite operation. The preliminary ground-based observations are nearly over : about 9 000 stars observed with the automatic meridian circles of Bordeaux and La Palma, nearly 100 000 star positions measured on ESO and SRC Sky Survey plates, about 8 000 stars observed in multicolour photoelectric photometry. It is intended to publish by 1990 the 210 000 stars proposed for Hipparcos, including old and new acquired astrometric and astrophysical data for the stars. This will hopefully encourage further ground-based observations of these stars (Høg, Kovalevsky, Turon).

III. STELLAR REFERENCE FRAME AND OBSERVING PROGRAMMES

The main event of this triennal period was the completion of the basic FK5, resulting in a spectacular improvement (as already confirmed by the La Palma and

Bordeaux meridian circles from a preliminary version). The extension of the FK5 to fainter stars, the final compilation of the SRS and the observing programmes related to the whole IRS are also important steps towards the improvement of the stellar reference frame.

1. The Basic FK5

At the Astronomisches Rechen-Institut (Heidelberg) the work on the FK5 has been continued and partially finished. Systematic corrections to the FK4 positions and proper motions were determined on the basis of suitable absolute and quasi-absolute observations with mean epochs later than 1900, altogether about 70 (50) catalogues in right ascension (declination). These corrections were determined using the analytical method developed by Bien et al. (25.041.067) for the determination of the systematic relations Cat-FK4, and by Schwan (34.041.052) for the derivation of the resulting corrections FK5-FK4. The systematic errors in the FK4 as indicated by the deviations Cat-FK4 for the modern observations which are not yet included in the FK4, were demonstrated by Fricke (40.041.028). Individual corrections to the FK4 positions and proper motions were derived by incorporating the new observations which have become available after the completion of the FK4 into the FK4 mean positions and proper motions. Altogether about 90 catalogues could be used for the individual improvement of the FK4. Weights were assigned to the catalogue positions according to a method developed by Bien (Astron. Astrophys., in press). The final mean positions and proper motions in the FK5 were computed for the mean equinox and equator of J2000.0 in accordance with the resolutions adopted by the IAU. The publication of the basic part of the FK5 consisting of the classical 1535 fundamental stars is in preparation (Fricke, Schwan, Lederle et al.), and the printed version of this catalogue will become available in the near future.

2. The FK5 Extension

The selection of the new fundamental stars and the determination of mean positions and proper motions for these stars is being made jointly by the U.S. Naval Observatory, Washington, and Heidelberg.

Mean positions and proper motions for about 1000 FK4 Sup stars with apparent magnitudes 5.5 to 7.0 will be derived at Heidelberg. The major part of this work, the identification of the observations in about 300 catalogues which have been considered, and the determination of the systematic relations Cat-FK4 for the relevant catalogues, is near to completion.

A list of more than 2 000 fainter stars was selected at the U.S.N.O. (Corbin, 41.041.022) in the IRS catalogue according to distribution in magnitude and spectral type, quality of the observational history, mean error of the proper motion and distribution over the sky, absence of component. These criteria were difficult to satisfy at the faint end of the magnitude range and south of -50°. The list of faint fundamental stars was completed and forwarded to Heidelberg.

It is expected that the average accuracy of the new fundamental stars will be inferior to that of the classical fundamental stars in the FK5 at least by a factor of two. For this reason the new fundamental stars will be published in a separate volume, presumably in the course of the year 1988.

Absolute and quasi-absolute observations of the FK5 and of its extension have obviously priority in the programmes for many meridian circles, but also for photo-electric transit instruments and various astrolabes in U.S.S.R., China and South America, resulting in many individual accurate catalogues. A general catalogue of astrolabes is in construction for the Southern hemisphere.

3. The IRS Catalogue

Compilation of the SRS. The joint work by the U.S.N.O. and the Pulkovo Observatory
on the SRS catalogue of positions has progressed through several stages : compilation of preliminary catalogues independently at both observatories, exchange of
the preliminary catalogues and of the data bases of FK4 and SRS observations made
at the individual observatories during the SRS observing campaign, comparison and
rectification of differences between the compiled catalogues and the data bases
and discussion of discordant observations (Zverev, Polojentsev et al., 40.002.054,
Hugues, Smith et al., 41.041.021).

Definitive compilations were exchanged recently. It has been agreed that the
Pulkovo SRS right ascension system should be brought closed to the FK4 by first
making a detailed reduction to the Washington SRS right ascension system for right
ascension and declination dependent systematic differences. Then the Washington
and revised Pulkovo right ascensions can be combined with equal weight without any
revisions.

It is expected that the joint Washington-Pulkovo work on the compilation of
the SRS catalogue will be completed by the end of 1987. Proper motions for this
catalogue are always under study at the U.S.N.O.

Re-observation of the IRS. In U.S.S.R. the following observatories have given their
consent to participate in new observation of the IRS programme : Pulkovo, Moscow,
Kiev, Engelhardt, Nikolaev. Meridian observations have been already started at
Pulkovo and Kiev.

At the U.S.N.O., the Six-Inch transit circle is engaged in the observation of
the AGK3R while the Seven-Inch installed in New-Zealand begins to observe the whole
SRS.

On the other hand, 35 900 IRS stars are in the observing programme of the La
Palma meridian circle and 20 000 AGK3R stars will be observed by the Tokyo meridian
circle.

At last, the San Juan Repsold meridian circle is used for a re-observation of
the SRS San Juan 72 catalogue (Carestia).

4. Bright Stars

In China, as a suggestion from Purple Mountain Observatory, a general programme will be put into practice with the classical instruments at Beijing, Shaanxi,
Shangai, Wuchang and Yunnan, including bright and faint stars in a fundamental catalogue. The observations of the brightest part, i.e. about 2 000 stars of geodesic
and other catalogues, will be completed before the end of 1989.

In U.S.S.R., a catalogue of 4 949 bright stars (geodetic) has been completed.
The standard errors are ± 0.004 s and $\pm 0.11''$ on positions, ± 0.012 s/cy and $\pm 0.17''$/cy
on proper motions. This catalogue has been reduced to the FK5 (Khrutskaya, 40.041.
038).

At Pulkovo and Leningrad new reduction of original U.S.S.R. Time Service catalogues concerning 300 000 positions was performed (Afanasieva et al., 39.002.032).
The result is an accurate catalogue of right ascensions KSV2.

5. Faint Stars

Photoelectric instruments are now or will be shortly able to observe stars up
to $V = 12$-13. On the other hand, for photographic astrometry it would be important

to have a reference system in that range of magnitude. Compilation of a list of such faint stars with a density of 1 star per square degree is currently carried out, jointly by Brorfelde and the Royal Greenwich Observatory. First epoch positions are to be found in the Astrographic catalogue and other photographic catalogues.

6. Other Programmes and Catalogues

Double stars. A programme of 2 500 double stars unsuitable for photoelectric instruments is currently observed at Kiev, Moscow, Karkov, Odessa, Engelhardt, Belgrade and Cerro-Calan. Two DS catalogues are already published (Chernega et al., Golovko).

High luminosity stars (39.002.107). About 3 000 stars with magnitudes between 6.5 and 9.5 are observed at Nikolaev, Moscow, Kharkov, Odessa, Kiev and Belgrade. At Moscow a first HLS catalogue was published (42.002.046).

Polar stars. In U.S.S.R., corrections to the declinations and proper motions of 524 polar stars were obtained at Karkov (Dergach). A catalogue of declinations for 1 407 polar stars has been completed (Pavlenko et al., 41.041.039). Observations of 308 bright polar stars were completed with the Large transit instrument of Belgrade (Pakvor).

Equatorial stars. Observations of 1 300 equatorial stars within ± 1.5° have been made by the photoelectric meridian circle at Nikolaev.

Latitude stars. At Engelhardt a catalogue of declinations of 2 910 latitude programme stars using 75 original catalogues obtained between 1930 and 1980 has been compiled (standard errors ± 0.07" on positions, ± 0.28"/cy on proper motions). Its application to the latitude observations reduces the annual wave in the mean latitude (Urasina, 41.002.037-038).

The Pulkovo catalogue of latitude stars was reduced and published (Bagildinsky et al.).

O-B stars and nearby stars. A catalogue of 1 059 O-B stars observed by the visual meridian circle at Tokyo has been published (Yasuda et al., 41.002.057). Furthermore 5 000 O-B stars are in the observing programme of the Tokyo photoelectric meridian circle.

The La Palma programme includes also 5 700 O-B stars and 10 500 F stars within 100 pc. Positions are to be found in the CAMC catalogues published yearly by San Fernando (catalogues 1984, 1985 and 1986 are available, 40.002.094, 42.002.073).

Miscellaneous programmes. A preliminary catalogue of 931 FKSZ stars for the whole sky has been constructed (Yatskiv, Polojentsev). At Goloseevo and Leningrad new proper motions of the PFKSZ-2 catalogue were obtained giving an improvement of the accuracy by a factor 1.5. A right ascension catalogue of 549 FKSZ was compiled at Tashkent (Boroditsky). Observations of 651 FKSZ were completed at Cerro-Calan with the Repsold meridian circle (Carrasco).

Absolute observations of fundamental bright and faint stars declinations were completed with the Wanschaff vertical circle at Goloseevo (Kharin et al.).

At Pulkovo a catalogue of absolute declinations of 1 425 stars was compiled on the basis of observations with the Zverev photographic vertical circle made in Chile (1965-1966).

IV. LINK TO THE EXTRAGALACTIC AND PLANETARY REFERENCE FRAMES

The connection between the different reference frames appears as one of the main problems of this decade (see Joint Discussion on Reference frames, New-Delhi General Assembly). Many observatories have already begun observing programmes related to this question : radiostars and reference stars around extragalactic sources with optical counterpart, Sun, major and minor planets. Construction of the future primary extragalactic reference frame and compilation of a suitable list of radiostars are reported by the Working Group on Optical-Radio Reference frame (to be found in the Commission 24 report).

1. Radiostars

Various lists are in the programmes of the observatories. An extensive masterfile including about 800 confirmed, suspected or potential radiostars has been compiled at Heidelberg (Walter) from different Hipparcos proposals and preliminary list from De Vegt, Florkowski and Johnston. But the data continuously updated have to be submitted to a severe critical examination (40.041.018). A subset of this list including 152 radiostars has been integrated in the Hipparcos Input Catalogue and classified as "superpriority stars".

Bright radiostars are being observed by visual astrolabes and meridian circles (Paris, San Fernando, San Juan, Sao Paulo, Santiago, Cerro-Calan, Torino). At Beijing, Shaanxi and Yunnan observatories, these objects are also observed with photoelectric astrolabes. The initial result at Beijing shows a precision of about ± 0.05".

A list of 275 radiostars is regularly observed with the La Palma automatic meridian circle whereas 300 Mira variable stars (H2O maser sources) are in the programme of the Tokyo photoelectric meridian circle. The Bordeaux Observatory is completing observations of 180 radiostars with its automatic meridian circle : with 20-30 observations per star the expected accuracy should be at least ± 0.05". Comparisons with JPL VLBI positions shows a satisfactory agreement (40.041.020).

Radiostars have been also incorporated into the U.S.N.O. Washington and New-Zealand absolute observing programmes.

At last, study on connection between optical and radio reference frames is being actively carried out at Shangai (40.043.005, 42.041.028), Purple Mountain and Beijing.

2. Reference Stars around Extragalactic Sources

At Odessa, Kiev and Belgrade a list of about 1 500 reference stars was observed and first results were published (Chernega et al.).

The La Palma meridian circle programme includes 11 400 faint stars (11 < B < 12) around 360 extragalactic sources (20 to 30 stars per field).

Observations of radiosource reference stars continue on the 8-inch transit circle in Flagstaff. Comparisons with photographic positions show results quite successful.

A list of 164 "super high priority" stars has been incorporated in the Hipparcos Input Catalogue in order to link the Hipparcos sphere to the extragalactic reference frame. The fine guidance sensors of the Hubble Space Telescope are expected to measure the angular separation between one of these selected stars and the nearby extragalactic object (within 18') with an accuracy of 2 milliarcseconds.

3. Sun and Major Planets

In U.S.S.R. day time meridian observations of the Sun, Mercury, Venus and Mars are made traditionally at Nikolaev (41.041.016), Tachkent (40.041.061), Goloseevo (39.041.014) and Kharkov. Since 1984 regular meridian observations of these bodies have been carried out at the Kislovodsk high altitude station of the Pulkovo Observatory (h = 2 210 m). The accuracy of these observations is 1.5 times better than those made at Pulkovo.

Uranus and Neptune have been observed regularly by the automatic meridian circles of La Palma and Bordeaux. The results show clearly a systematic deviation on right ascensions of these planets of about -0.3" with respect to the standard ephemerides DE200 (Rapaport et al.).

Sun and major planets are always observed in the absolute programmes of the U.S.N.O. transit circles and now with the Tokyo photoelectric meridian instrument.

As already mentioned many modified astrolabes are now participating in these observations and should make an important contribution to the link between the FK5 system and the planetary reference frame.

The analysis of the Sun observations made during the past ten years by the C.E.R.C.A. solar astrolabe (Laclare and Journet) shows a satisfactory definition of the equinox and of some elements of the orbit. Furthermore this kind of observation gives a good determination of the solar diameter : variations appear during the measurement period with possible correlations with some activity parameters.

4. Minor Planets

The brightest minor planets were observed many times by the classical instruments. With limiting magnitude of 12-13 photoelectric meridian circles can observe fainter minor planets. A list of 63 minor planets introduced in the Hipparcos Input Catalogue for link to the planetary frame was incorporated in the programmes of La Palma (4 800 observations) and Bordeaux (2 200 observations).

At Leningrad (ITA) work for organization and reduction of 30 000 observations of 20 minor planets according to the ITA programme has been continued for the improvement of zero-points and determination of systematic errors of the FK4 and other catalogues. Forty observatories in the world participated in this programme (Batrakov et al.).

The systematic errors of Yale, SAO and AGK3 catalogues were determined using 424 minor planet observations at Goloseevo (Maior, 39.041.010).

V. MOVING BACK THE LIMITS OF GROUND-BASED ASTROMETRY

The ground-based astrometry is mainly limited by refraction. During these last three years, research in that domain was taking steps forward. Meanwhile meridian workers realized the necessity to monitor regularly the division errors of their declination circles which may introduce random and systematic errors in their catalogues. At last theoretical studies were beginning to show the interest of global reduction and statistical methods for best processing of the astrolabe and meridian circle measurements.

1. Refraction

New refraction tables of Pulkovo Observatory were published in 1985 (39.082. 045). These tables were compiled using modern results on the physical and optical

properties of the Earth atmosphere. It is the result of intensive research made at
Pulkovo and Engelhardt (Guseva, Nefedeva, Kolchinsky et al., 40.082.106-108, 41.082.
029, 41.082.059-060). Inclinations of atmospheric layers of equal density were calculated. New computing method of the Hartmann refraction was proposed. At Pulkovo
chromatic refraction was carefully studied (Bagildinsky, Zhilinsky, Shkutov, 40.
082.107).

In China study of the effects of refraction on astrometric observations was
going further. Experiments have been done on astronomical instruments in Misuzawa,
Tokyo, Wuchang and Tianjin (Hu et al., 39.082.050, 41.082.072).

A new efficient method to evaluate by numerical integrations the astronomical
refraction as a function of the wavelength was applied to the Tokyo PMC catalogues
(Fukaya and Yoshizawa, 40.082.086).

Test observations were made at the high altitude Moscow Observatory station
in central Asia with a Repsold meridian circle and seeing was investigated (Golovko,
Guliaev et al., 38.041.037, 40.032.071, 41.041.027). Furthermore the high-altitude
Kislovodsk station of Pulkovo Observatory was equipped by the Struve-Ertel vertical
circle and large transit instrument.

A practical method to detect and correct the room refraction in the pavilion
was investigated at Tokyo, the benefit of this method to improve the observational
accuracy is now under examination.

Influence of anomalous refraction was the object of interesting studies
(Gubanov, 42.082.056), particularly the effect on day time observations (Fedorov,
39.082.102) and on time determinations (Yang, 41.044.066). Furthermore the influence
of detection noise and atmospheric turbulence on the accuracy of stellar position
measurements has been studied at CERGA (Gay and Pochet, 41.036.012-013).

Conclusive and quite surprising results about internal refraction (INR) in the
telescope tube of conventional meridian circles have been reached. Evidence of a
variable INR of 1" to 2" has been found in three meridian circles at U.S.N.O., La
Palma and Tokyo and in the vertical circle at Kiev, Golossejevo. It is shown that
all determinations of the flexure term (F sinz) have been completely dominated by
INR - although everybody has believed to be measuring mechanical flexure. The INR
has been removed in the MC at La Palma by a simple ventilation inside the tube
installed in 1985. As a further result determination of the atmospheric refraction
constant of each night by means of the FK5 stars has become more accurate and now,
in fact, agrees with the one calculated from meteorological data within ± 0.09" rms.
The variable INR has corrupted all previous determinations of the refraction constant (Høg and Miller, 42.032.051, Høg and Fabricius, submitted to Astron. Astrophys.).

2. Meridian Declination Circle Errors

Application of an efficient method for measuring graduation errors to the
Tokyo automatic meridian circle showed that the annual change of the glass circle
error amounts up to about 0.05" for some circle readings (Miyamoto et al., 39.032.
024, 42.036.078). The yearly corrections of the annual change are stored in computer and applied to the observations. A quick scheme to monitor a possible diurnal
change was developed but no meaningful change of the graduation error was found
within a clear day.

An efficient method was also elaborated and applied to a full determination of
the division errors of the Moscow meridian circle (Tauber, 42.034.007). Complete
determinations of the division errors were also carried out and repeated at different ambient temperatures on the Bordeaux automatic meridian circle (Requième et
al.) and on the Pulkovo horizontal meridian circle (Gumerov, Pinigin et al., 42.
032.043).

Results of determinations of graduation errors by different circle scanning systems of the U.S.N.O. are reported in 42.032.001 (Rafferty et al.). Analysis of the effect of errors on the reading of a graduated circle is also reported in 41. 036.143 (Mao). Comparison of classical methods of determination were carried out in Bordeaux (Rapaport, 39.036.079).

3. Reduction Methods

A method of global reduction of fundamental astrometric data was at first proposed at Sao-Paulo (41.036.212) and applied to astrolabe observations (Benevides and Clauzet). Application to global adjustment of star position in the compilation of meridian circle catalogues is under investigation at Sao-Paulo, Bordeaux and Tokyo.

Techniques for dealing with discordant observations were proposed (Branham, 41.041.006). Error estimates with solution in the L_1 norm of overdetermined systems were given (Branham, Celes. Mech., in press).

At Pulkovo a method was elaborated for a parametric adjustment of the absolute astrometric observations taking into consideration the errors of the initial data (Gubanov). Optimum stable estimations of the errors of astrometric measurements were proposed at Engelhardt (Tokhtaseva).

Statistical problems about astrometric reductions were investigated at Paris with application to the detection of poor observations and catalogue errors in the Paris astrolabe data (Bougeard).

VI. MISCELLANEOUS

The rediscussion of older meridian circle catalogues has been started at Heidelberg. The aim is to establish a rather complete astrometric data bank and to derive from this material improved proper motions for a large number of stars. Even the Hipparcos proper motions could be improved by combining the Hipparcos data with suitable ground-based measurements (Wielen et al.).

Data collected by the U.S.N.O. from meridian circle and astrographic catalogues should have about 360 000 stars with an average of about 4 catalogue positions per star. Work has been started to adapt Schwan analytical method to the systematic reductions (Smith et al.).

Other extensive researches around the reference frames are scattered in the Commission 4, 19, 24 and 40 reports. A Joint Discussion will be held at the next Baltimore General Assembly (chairman : P. Seidelmann) on observations, theory and computation at the milliarsecond level, use of millisecond pulsars, report of the Working groups on reference frames, nutation and astronomical constants. Members of Commission 8 are obviously invited to participate actively in that important meeting.

Acknowledgements. The author wishes to thank those of his colleagues who have assisted with the compilation of this report through submitting information on their activities.

Yves Requième
President of the Commission

COMMISSION 9:   INSTRUMENTS AND TECHNIQUES (INSTRUMENTS ET TECHNIQUES)

PRESIDENT: C.M. Humphries   VICE-PRESIDENT: J. Davis

ORGANISING COMMITTEE: J.C. Bhattacharyya, O. Engvold, B.P. Fort, Hu Ning-Sheng,
W.C. Livingston, D.F. Malin, N.V. Steshenko, R.G. Tull, G.A.H. Walker,
W.L. Wilcock, G. Wlérick.

I. INTRODUCTION

The technology leading to very large aperture telescopes and their optics
has progressed well in the period since 1984 and plans for many new large aperture telescopes have been made. Focal plane instrumentation continues to become
more sophisticated or more efficient: multi-object capabilities, automatic
instrument control and operation, and increasing use of CCDs are examples of
areas to which this applies. The proportion of time devoted to observations
using two-dimensional photoelectronic detectors has grown substantially at many
observatories, particularly with telescopes of moderate aperture; and the use
of high quantum efficiency array detectors is now being extended into the infrared spectral region. Important advances have also been made in instrumentation
and techniques for ground-based high angular resolution interferometry.

Meetings and workshops on telescopes and their instrumentation that have
been held since 1984 include:

- An IAU symposium on Instrumentation and Research Programmes for Small Telescopes took place in Christchurch, New Zealand, in December 1985 (IAU Symposium No. 118, ed. J.B. Hearnshaw and P.L. Cottrell).

- Meetings organised by the International Society for Optical Engineering
  (SPIE) were held in Tucson, Arizona, in March 1986 (Proc. SPIE, Vol.627,
  Instrumentation in Astronomy VI, 1986; Proc. SPIE, Vol.628, Advanced Tech-
  nology Optical Telescopes III, 1986).

- In January 1987, a joint ESO/NOAO workshop on High-Resolution Imaging from
  the Ground using Interferometric Techniques took place at Oracle, Arizona
  (Proceedings, ed. J. Goad, NOAO, Tucson, 1987).

- In July 1987, a workshop on Instrumentation for Ground-Based Optical Astronomy was held at the University of California, Santa Cruz.

Other meetings that have taken place in the period under review are cited
in the text below.

Commission 9 records its tribute here to Dr. D.S. Brown of Durham University, England, whose untimely death occurred on 17 July 1987. Although not a
member of IAU, David Brown was known to many astronomers throughout the world
and was one of the most respected and influential figures in the design and
manufacture of astronomical optics. As Optical Manager of Sir Howard Grubb
Parsons Co. Ltd. of Newcastle-upon-Tyne, and later as Technical Director,
David Brown supervised the optical manufacture of many 2-metre class telescopes
in the 1950s and 1960s (including those at Sutherland, South Africa; David Dun-
lap Observatory, Canada; Mount Stromlo, Australia; Helwan, Egypt; and Okayama,
Japan). In subsequent years a further succession of high quality large optics
was designed, manufactured and tested under his guidance and control, including

the 1.0m and 1.2m achromatic correctors for the ESO and Palomar Schmidt telescopes, respectively, and all of the recent large British optical telescopes
(the 2.5m Isaac Newton Telescope, the 3.9m Anglo-Australian Telescope, the 1.2m
U.K. Schmidt Telescope, the 3.8m U.K. Infrared Telescope, and the 4.2m William
Herschel Telescope).

II. NEW TELESCOPES - RECENTLY COMPLETED OR BEING DEVELOPED

With the building and dome of the Keck 10m telescope nearing completion on
Mauna Kea, commercial production of the 36 segments of the primary mirror has
started. Once the stress-polishing technology has become commercially routine,
the aim is to produce one finished segment every three weeks; when these are
assembled and aligned to form the complete mirror at the end of 1989, the specification for the combined image is 0.32 arcsec (80% encircled energy diameter;
500nm wavelength) at the f/1.75 prime focus, and 0.4 arcsec at the Nasmyth/Cass-
egrain foci, excluding seeing contributions. TIW of Sunnyvale, California, is
the main contractor for the telescope structure; the dome was manufactured by
Coast Steel Ltd (Vancouver); the mirror support structure is managed by the
University of California, Berkeley; and the mirror segments are being produced
from Schott Zerodur by Itek Corporation. First light with the new telescope is
scheduled for 1990.

Development of the Japanese 7.5m telescope will have the highest priority
for the new National Astronomical Observatory when it is formed in 1988 (incorporating the present Tokyo Astronomical Observatory). The performance of a 7.5m
diameter, 0.2m thick, meniscus primary has been studied using finite element
analysis, and wind tunnel tests of the dome have been performed at the University of Kyoto to investigate the effect of wind gusts on the primary mirror
support system. Engineering tests on the servo system of the mirror support,
using a Shack-Hartmann monitor, are being performed with a 0.6m mirror of thickness 2cm. The site proposed for the telescope is Mauna Kea.

A workshop on the 4 x 8m VLT array of the European Southern Observatory
was held in Venice in September 1986 and the Proceedings of that meeting
(ed. S. D'Odorico and J.-P. Swings) are obtainable from ESO and give the status
of the VLT project at that time. Site testing measurements in Chile are
currently being obtained at Cerro Paranal (in the Atacama desert region to the
north of La Silla) and these are being compared with measurements from La Silla.
Meanwhile, the thin 3.58m Zerodur primary mirror for the ESO New Technology
Telescope is being optically figured at Zeiss Oberkochen, and pre-assembly of
the mechanical structure of the NTT has been completed at INNSE in Brescia,
Italy, prior to shipment to La Silla in 1988.

The 4.2m William Herschel Telescope at La Palma saw first light in mid-1987
and has started its operational phase. Final engineering tests confirmed the
high optical quality of the Cervit primary mirror, which is capable of producing
0.3 arcsec images (85% encircled energy diameter). A three-element correcting
lens at prime focus gives a 40 arcmin unvignetted field at an effective focal
ratio of f/2.8; Cassegrain and Nasmyth foci are also provided.

Also sited at La Palma, the 2.56m Nordic Optical Telescope has been completed mechanically and electrically, and the optics (Schott Zerodur; optical
work by Tuorla Optics Laboratory, Turku, Finland) are due to be installed in
1988. The telescope is the result of co-operation between astronomical institutes in Denmark, Finland, Norway and Sweden, together with Iceland.

In India, the new 2.34m telescope became operational in November 1985 and
was inaugurated as the Vainu Bappu Telescope in January 1986. A 1.22m infrared

telescope has been manufactured and is awaiting installation at a site near Mt. Abu (24 degrees 40 min N; 72 degrees 45 min E), and should be operational in 1988. A project for the construction of a 10m millimetre-wave dish and receiving equipment is making good progress, and the design of a new aperture-synthesis radio telescope has started. This will consist of 30m steerable dishes spaced over baselines extending to 27km in the region near Pune (19 degrees 05 min N; 74 degrees 03 min E) and the anticipated completion date is in the mid-1990s.

Three new telescopes have been built in China in the period under review, each with Ritchey-Chretien optics. A 1.26m infrared telescope with f/2.2 primary was made for Beijing Observatory by Nanjing Astronomical Instrument Factory (NAIF) and has been in operation since 1986; a 1.56m astrometric telescope has been produced in Shanghai for Shanghai Observatory and started operation in 1987; and a 2.16m stellar telescope is being tested at NAIF for Beijing Observatory. Future plans include a 1.5m Schmidt f/2.5 telescope being designed at NAIF and funded by Academia Sinica; and a 60cm solar telescope, with five telescope tubes, is being constructed at NAIF for Beijing Observatory.

In the development programme leading to the manufacture of 8m borosilicate honeycomb mirrors, several blanks up to 1.8m diameter were produced by Angel's group with a 2-metre furnace at Tucson. One of these, which was spun-cast to a focal ratio f/1, will be stressed-lap polished as a demonstration of techniques to be applied to subsequent larger mirrors, and will be used in an advanced technology telescope to be developed by the Vatican and Steward observatories.

The new mirror spin-casting laboratory at Tucson is now complete and has a 12-metre turntable currently holding a 6-metre furnace. This will be used initially to produce two 3.5m mirrors, one of which will have a focal ratio f/1.75 for the Apache Pt. telescope, New Mexico, built by ARC (a consortium of four universities in the U.S.A.); the other will be used by a consortium comprising the Universities of Arizona, Indiana and Wisconsin.

Present plans include the production of five still larger mirrors. A 6.5m will be used with the Multiple Mirror Telescope. Two 8m mirrors will supplied for the Columbus project (a 2 x 8m multiple mirror telescope proposed for Mt. Graham, being developed by the Universities of Arizona, Chicago and Ohio State in partnership with Arcetri Astrophysical Observatory, Italy). Another 8m mirror will be used for the Magellan project (University of Arizona in conjunction with the Carnegie Institute and John Hopkins University), a single-mirror telescope to be sited at Las Campanas, Chile. And a further 8m mirror will be used by the Arizona groups in partnership with the Smithsonian Astrophysical Observatory for a re-defined National New Technology Telescope, with the northern hemisphere one probably to be sited at Mauna Kea.

Development of a telescope using a liquid mirror has been been the goal at Universite Laval (Borra E.F. et al. 1985, Publ.Astr.Soc.Pac. 97, p.454). The intention is to construct a zenith-pointing transit telescope with the primary mirror formed from a rotating container of mercury. The turntable for such a rotating mirror can be provided by a circulating water flotation system operated by a pump; moving, deformable, auxiliary optics would permit the tracking of a star over a wide field. So far, a 1.65m mirror has been made in this way and apertures of 15m or larger are considered to be feasible eventually.

Dominion Astrophysical Observatory, in conjunction with the University of Montreal, have made a proposal for a 2.5m, high angular resolution telescope, the tube of which would be filled with helium to optimise heat transfer and minimise internal seeing effects (as in the solar telescope LEST). Another design being pursued at DAO is the Floating Boule Telescope, a low-cost structure considered capable of supporting mirrors up to 10m diameter. Also at DAO, the Canadian Space Astronomy Data Centre (CSADC) is being established there primarily for reduction and archiving of data from the Space Telescope, but also for handling data from the Canada-France-Hawaii Telescope.

Two large submillimetre telescopes have been constructed in the past two years: the Swedish/ESO 15m telescope (SEST) at La Silla, and the joint British and Dutch 15m James Clerk Maxwell telescope (JCMT) at Mauna Kea. The SEST reflector has a surface profile accuracy of about 50 microns r.m.s. (each of the 176 individual panels has an accuracy of 16 microns r.m.s) and will operate coherently at wavelengths down to 0.8mm (375GHz). It is identical to three other movable telescopes being built in France by the Institut de Radioastron- omie Millimetrique, Grenoble, and is designed to operate without a radome or other enclosure in winds up to 50 km per hour.

The JCMT reflector has an overall surface accuracy of about 30 microns and should allow efficient operation down to 0.5mm wavelength. To protect the telescope and allow the surface accuracy to be maintained, it is contained in a rotating enclosure; the viewing aperture is covered with a membrane of woven polytetrafluoroethylene and will allow operation in winds up to 70 km per hour.

III. WIDE-FIELD TELESCOPES

The main facility at Kiso Observatory in Japan is the 1.05m Schmidt tele- scope there. The control system has been renewed, and a PDS 2020 microdens- itometer and upgraded computing and graphic display facilities were installed in 1985. In 1984 (September 14) an earthquake of magnitude 6.8 caused damage at the observatory; the epicentre of the quake was only 8km north of the observatory and 2km below ground, causing damage to the concrete pier of the Schmidt and to the dome shutter. Fortunately the telescope optics were found to be unharmed and no significant change was found in the pointing direction of the telescope polar axis. Nevertheless, it required some six months for the repairs to be completed.

A 45cm Schmidt telescope has been fabricated at the workshops of the Indian Institute of Astrophysics at Kavalur, and became operational in March 1986.

The commissioning by Caltech of the new achromatic corrector plate for the Palomar 1.2m Schmidt telescope prepares the way for a new-epoch multicolour survey of the northern hemisphere skies.

The Nanjing Astronomical Instrument Factory is currently designing a 1.5m f/2.5 Schmidt telescope with achromatic corrector plate. When this is built it will become the largest Schmidt telescope in the world (the current largest being the 1.34m Tautenburg Schmidt).

Interest from other countries has also been shown in building wide-field telescopes of aperture larger than those that exist at present. At a workshop held in Tucson, February 1987, on Instrumentation for Cosmology, the recommend- ation of the workshop was that a wide-field 4m class telescope should be con- structed to carry out a redshift survey of galaxies down to B=19 magnitude over a steradian of sky near the north galactic pole. A field of at least 2 degrees would be required for this and approx. 400 fibres would be used to feed multiple spectrographs.

At another workshop on wide-field telescopes held in April 1986 in Cambridge, England, the three-mirror telescope proposed by Willstrop (1984, Mon. Not. R. astr. Soc. 210, 597; and 1985, 216, 411) was discussed, and some practical work with a scaled prototype is in progress. A mounting system for very large tele-

scopes, with particular application to the classical Schmidt, has been described
by Reddish and Simmonds (1987, Mon. Not. R. astr. Soc. 228, 537); pointing in
elevation is provided by a flat mirror inclined at 45 degrees to the horizontal;
the design has the appearance of a siderostat and the behaviour of an altazimuth
telescope.

IV. SOLAR TELESCOPES
(W.C. Livingston)

The goal of high spatial resolution dominates new solar facilities. At
La Palma, the Swedish 50cm telescope installed in 1985 has established new
performance standards; interferometric tests and speckle observations of stars
show that total wavefront errors are 1/15 wave peak-to-peak. White-light
granulation sequences confirm this image quality. At Tenerife, the German
70cm vacuum tower is going into operation with the aim of feeding a large and
efficient echelle spectrograph. Engineers at LEST (Large European Solar Telescope) have demonstrated that a helium filled telescope is seeing-free and that
thin entrance windows are practical.

Other efforts are towards low polarization telescopes. THEMIS (Telescope
Heliographique pour l'Etude du Magnetisme at des Instabilites Solares) is now
under construction, and LEST is in the planning and site selection phase with
a difficult choice between La Palma and Mauna Kea. For solar oscillation
studies, the prototype Global Oscillations Network Group (GONG) instrument is
under test and a series of experiments are being conducted at the South Pole
where, with luck, continuous runs of several days will be realised.

V. TECHNIQUES

Limitations of space permit mention here only of a few, highly selected,
developments in techniques and focal plane instrumentation.

Many observatories now have fibre optic systems that are in routine use
on telescopes for multiple-object spectroscopy or photometry. Examples of
existing systems are: the MX fibre-coupled spectrometer (with automated and
remote controlled fibre positioning) developed at Steward Observatory, Arizona;
the Anglo-Australian Telescope fibre system with AUTOFIB (also automated and
remote controlled); OPTOPUS at the ESO 3.6m telescope; and FLAIR, a wide-field
(40 square degree) multi-object system in use at the U.K. Schmidt Telescope.

There are many other applications of fibres that are now in routine use at
some observatories, such as image scrambling, image slicing, extended-object
image sampling, and image transfer from the telescope to a coude station or to
floor-mounted instruments.

A new development by the Advanced Concept Exploration group of NOAO is a
two-beam interferometer using single-mode fibres with which white-light fringes
have been obtained. Thus, long baseline interferometry may become possible
using single-mode fibres as waveguides.

Adaptive optics compensation of atmospheric seeing effects has resurfaced
as a real possibility for observational optical astronomy; this follows a
period when practical limitations (guide star brightness requirements and iso-
planatic field restrictions) were seemingly intractable.

Good progress has been made with wavefront-tilt compensation, particularly
at the University of Hawaii, and several other institutes (e.g. NOAO and ESO)

are devoting effort to this. At the DAO, a high angular resolution CCD camera
is being developed for the CFHT prime focus with an optical package that
includes a piezo-electric wavefront-tilt compensator and a fast shutter system
for discriminating automatically between moments of good and poor atmospheric
seeing, and using only the best.

In 1985, Foy and Labeyrie (Astron. Astrophys. 152, L29) suggested that
backscattered laser light from atmospheric layers at altitudes up to 100km would
give enough photons to activate a fast adaptive optical system, and that in this
way an artificial guide star could be created. The first experiments using this
method have now been reported by L.A. Thompson and C.S. Gardner (1987, Nature
328, p. 229) using University of Hawaii 2.2m and 0.6m telescopes at Mauna Kea
and lidar equipment developed by the University of Illinois.

Another method of obtaining diffraction-limited images with large ground-
based telescopes is to apply aperture synthesis and phase closure techniques to
short exposure images taken with an array of apertures placed over the primary
mirror, each sub-aperture being no larger than the scale size r(0) of the atmospheric fluctuations. The first reconstructed images obtained by this method at
optical wavelengths using the 2.5m Isaac Newton Telescope at La Palma have been
reported by Haniff et al.(1987, Nature 328, p. 694).

Two dimensional infrared detector arrays are now available and have applications similar to those of CCDs at visible wavelengths, i.e. high quantum
efficiency direct imaging and spectroscopy. IRCAM is an IR imager developed
at the Royal Observatory, Edinburgh, and now in operation at UKIRT, Hawaii. It
employs a thinned, back-illuminated, silicon-hybridised 62 x 58 array of photo-
voltaic indium antimonide diodes and gives quantum efficiencies greater than 60
per cent at wavelengths from 1 to 5 microns. Similar technology is being used
by NOAO to produce IR imagers for the telescopes at Kitt Peak and CTIO, Cerro
Tololo. The ESO system (IRSPEC) that has been in use since 1985 is a cooled
grating spectrometer equipped with an array of 32 indium antimonide diodes;
ESO are currently experimenting with 64 x 64 mercury-cadmium-telluride arrays
for the 1 to 4.2 micron region with IRSPEC. 128 x 128 arrays are also becoming
available now. An account of developments and applications of IR arrays is
given in the Proceedings of a workshop on Ground-Based Astronomical Observations
with IR Array Detectors held in Hilo, Hawaii, in 1987 (Infrared Astronomy with
Arrays, ed. C.G. Wynn-Williams and E.E. Becklin, published by University of
Hawaii, Hilo, 1987).

Supernova 1987A triggered the occasional ad hoc construction of some
special instruments as well as harnessing large amounts of observing time with
existing ones. At the Anglo-Australian Telescope at Siding Spring Observatory,
advantage was taken of the supernova's brightness to build an ultra-high resolution spectrograph. This used a Littrow-mounted prism-dispersed 204 x 408mm
echelle grating (79 g/mm) to reach an attained resolving power of 570,000, or
10 milliangstroms at sodium D.

VI. CCD DETECTORS
(Report by G.A.H. Walker, Chairman of the Working Group
on Photoelectronic Detectors)

Very valuable and complete compilations on the state of CCD technology and
their critical application to astronomy are contained in the Proceedings of a
conference held at Observatoire d'Haute Provence in June 1986, published by and
available from ESO; and in three special issues of Optical Engineering Vol.26,
Nos.8,9,10 (available from SPIE, PO Box 10, Bellingham, Washington 98227-0010,
U.S.A.). These are highly recommended to anyone involved in the use of CCDs.

Progress in the development of new, low-noise, large-format CCDs suitable for astronomy has not lived up to the promise held out at the 1985 IAU General Assembly meeting in Delhi. Any group outside the U.S.A. that plans to build a CCD detection system for either in-focus photometry or for spectroscopy has little choice. The withdrawal of RCA from CCD production was known at the time of the Delhi meeting. The few available double density RCA 640x1024 devices are being used in both thinned and unthinned versions. The former tends to crinkle which makes it unsuitable for fast optics.

Tektronix has been unable to deliver grade one 512x512 or 2048x2048 devices in either front or rear illuminated versions. Those products up to mid-1987 contained an unacceptably large number of potential pockets that inhibit the complete transfer of charge during readout. Such pockets are also found in even the best CCDs from other manufacturers. At the time of writing, Tektronix has moved production to a new facility where they hope that, with improved quality control, grade one 512M devices may become available in late 1987.

Thomson CSF and English Electric Valve Co. (ex-GEC) remain reliable sources for the TH31133 (384x576, 23 micron pixel) and P8603A (385x576, 22micron pixel) CCDs, respectively. Both devices are operated with a readout noise of between 10 and 20 electrons rms. They are unthinned with consequently lower peak quantum efficiency (30 per cent) than thinned devices. Both companies are experimenting with large format devices. EEV is developing a single, large device while Thomson is considering a closely butted mosaic of special versions of their 384x576 device.

A fluorescent coating developed at ESO (Cullum, M. et al. 1985, Astron. Astrophys. 153, L1) has been used successfully to extend the sensitivity of these devices into the uv at a level of about 20 per cent. A uv flooding technique has also been used at Steward Observatory (see: Leach, R.W. and Lesser, M.P. 1987, Publ.Astr.Soc.Pac. 99, p.668), from original work at JPL, to improve blue and uv responses.

Kodak is producing a 1035x1320 CCD with 7 micron pixels that has interesting possibilities for astronomy. The quantum efficiency curve has several strong peaks and valleys but is otherwise typical for a thick device. Some reports suggest that the readout noise could be less than 30 electrons rms. E.G.& G. Reticon is also developing CCDs suitable for astronomy with a view to marketing them eventually.

Linear arrays of silicon diodes (Reticons) are still used routinely for spectroscopy at many observatories and a continued supply of these devices seems assured. Several new systems have been commissioned since the Delhi meeting. Tull at the University of Texas has commissioned the Octicon in the coude spectrograph of the 2.5m telescope. It consists of eight 1872F/30 E.G.& G. Reticons in line. Several groups are building detection systems based on the 4096 Reticon array. Thomson CSF has developed a 2x2048 diode array in which the linear arrays are separated by 500 micron and the diode dimensions are 13x750 microns. It uses CCD output multiplexers and noise levels of 350 electrons rms are reported. Several new intensified Reticons have been commissioned for photon counting spectroscopy in the spectrograph mode since the Delhi meeting.

While Reticons can, with care, be routinely calibrated photometrically to better than 0.1 per cent, it has not been clearly demonstrated that CCDs can be calibrated to better than 1 per cent. This matter will be an important one for discussion at the Baltimore General Assembly meeting in 1988.

VII. ASTRONOMICAL PHOTOGRAPHY
(Report by the Chairman of the Working Group, D.F. Malin)

The principal activity since the General Assembly in New Delhi was a work-shop meeting of the Working Group on Astronomical Photography held in Jena in the German Democratic Republic between April 21 and 24, 1987. The scientific sessions were conducted in the Hall of the University of Jena and were attended by over 60 delegates from 13 countries. More than half of the attendees were from Eastern Europe with 26 from the host country and 14 from West Germany.

During the scientific sessions 36 papers were presented on the usual wide range of topics covered at such meetings. While many of these concerned the practicalities of hypersensitising and processing photographic materials, con-siderable attention was given to calibration problems and to the machine measurement of plates. These topics were also reflected in a series of poster papers that were available for inspection in the entrance to the Hall.

The town of Jena was a very appropriate setting for our meeting. Apart from the formal scientific presentations, delegates had the opportunity to make an evening visit to the Karl Schwarzschild Observatory, and one full day was devoted to a visit to the nearby historic town of Weimar and to the optical factory of Carl Zeiss Jena where we saw the assembly of a very wide range of equipment with astronomical applications, from planetarium projectors to 2m telescopes.

Apart from the scientific activities we also held a short meeting of the Working Group Organising Committee on April 23. Present were:

David Malin (Australia) - Chairman; Siegfried Marx (GDR) - Secretary;
Olga Dokuchaeva (USSR); Keiichi Ishida (Japan); Jean-Louis Heudier (France);
Milcho Tsvetkov (Bulgaria); Richard West (FRG); Olga Zichova (Czechoslovakia).

1. Report of the Organising Committee Meeting:
A) We have received information from Schott Glaswerke in Mainz that, contrary to rumour, they did intend to continue the manufacture of filter glasses in the large sizes suitable for large telescopes. However, it was clear that not all coloured glass filters would be available at all times due to a rationalisation of their manufacturing programme. A letter to Carl Zeiss Jena enquiring about their range of filters produced no response. A similar letter to the Hoya Corporation in Tokyo revealed that this company do make some glass filters of interest to astronomers but their maximum size is 165 x 165mm.

B) Though at least one observatory (U.K. Schmidt) has experimented with large sheet film in place of glass plates, there are as yet no spectroscopic emulsions coated on sheet films. The U.K. Schmidt experiments were focus tests using duplicating film cut down to 355mm squares. There is no evidence that new or existing spectroscopic emulsions (from Kodak) will be coated on anything other than glass, though it has been noted that Kodak's Technical Pan film Type 2415 (not strictly a spectroscopic emulsion) is now available to special order in 14 x 14 inch size sheets of film.

C) A survey to identify those batches of Kodak plates that respond well to hydrogen hypersensitising has not been very successful. Only one reply was received from the 15 circulars sent to people and organisations that were thought to treat hypering seriously. At least two more correspondents have promised results however, so these are awaited with interest.

2. The Future of the Working Group:
This topic was discussed at some length especially in the light of the AAS photographic group's decision to broaden their remit to include all types of

two-dimensional detectors. No firm decision was made on this but my feeling is
that the Photographic Working Group should remain as it is, particularly since
groups considering other 2-D detectors are well catered for in Commission 9.

3. Plate Costs:

This subject is raised with increasing vigour at every meeting. It is
clear that overall, the cost of plates is now a serious problem and it has
reached the point where it is inhibiting research. There was also evidence
presented that plate costs varied greatly from country to country. The most
extreme example of this came from Prof. Ishida who said that one plate for the
Kiso Schmidt now cost 30,000 yen, or about 200 US dollars.

It became evident at the meeting that the East German manufacturer Orwo
had at least one emulsion on glass that may be of interest to astronomers.
This material is identified as ZU21 and is a blue-sensitive emulsion with
properties somewhere between 103aO and IIaO. However its most attractive
feature is its cost, which is apparently much less than the Kodak equivalent.

4. New Photographic Emulsions:

In the early part of this year the photographic grapevine buzzed with
rumours of a new astronomical emulsion from Eastman Kodak based on the T-grain
technology which has been so successful, firstly in colour film, more recently
on glass. As yet I am not aware of any name or code number for the new material. The following report is derived in part from my own experience with the
samples supplied and from private communications from colleagues at the U.K.
and Palomar Schmidt telescopes.

The material released for test appears to be quite similar to the emulsion
that is marketed as T-max 100. It has a spectral response that terminates in
the red at about 630nm, has poor blue sensitivity (compared with IIIaF) and has
a number of prominent sharp peaks in the green and yellow parts of its sensitivity curve. In this regard it is quite unlike any of the normal astronomical
emulsions.

When processed for 5 minutes in D19 the characteristic curve is long,
straight and shallow, i.e. the emulsion has low contrast (gamma = 1.2 to 1.5),
wide dynamic range and small toe region. It is likely therefore that its peak
quantum efficiency for astronomical use will occur at a low density and the DQE
curve will be broad, unlike the IIIa series. Of course, the low contrast
(which might be increased by different developers or post-process copying) is
of no advantage for the detection of faint objects. Again the gross properties
are those of a general purpose camera film.

The T-grain emulsion on glass is extremely slow for long exposures. With
a 15 minute exposure to 3200 degree K tungsten light through a GG 385 filter,
the unhypered material is a factor of 3 or 4 times slower than an unhypered
IIIaF plate (batch 1L5) and a factor of 2 slower than an unhypered IIIaJ (batch
1C5). However, the T-grain emulsion is about 5 times faster than 2415 on glass
(153-01, manufactured 03/85) exposed as above. Like 2415, T-grain on glass also
responds very slowly to hydrogen hypering, but worthwhile speed increases are
obtained with drastic treatment (e.g. a 2 hour bake in pure hydrogen, at 65 degrees C). The best white-light speed we obtained was about that of an unhypered
IIaO at a desity of 0.6 above a fog of 0.2 (up from an as-received fog of 0.08),
a 5-fold speed increase. Interestingly, the same extreme hypering recipe produced a similar speed for 2415 on glass, a gain of about 12 times. Subjectively,
the grain structure of the new emulsion is similar to that of the IIIa products
but some measurements indicate much better signal-to-noise than IIIa materials,
though more careful measurements will be needed to confirm this.

Overall, however, the low speed, low contrast, and peculiar spectral sensitisation of T-grain on glass will not commend it to professional astronomers
though much more needs to be done to explore the material fully, especially if
the spectral sensitivity is either improved in the blue or extended in the red.

5. The Next Meeting:
It is planned to hold the next Working Group meeting in Baltimore as part
of the activities of Commission 9 during the 1988 IAU General Assembly. Preliminary plans are that we hold two half-day sessions, one of these for the
presentation of mainly photographic results to, and by, those of us who are
interested in photographic astronomy; and the other with the much wider purpose
of presenting astronomical results derived by photographic means. It was agreed
that it will be important to show that photography is still an important source
of astronomical data and discovery, and is entirely competitive with other
techniques and in some cases superior. To this end, we will encourage our
colleagues to bring to Baltimore large numbers of posters illustrating new and
exciting applications of photography to capture the interest of what will
undoubtedly be a very large and important meeting.

VIII. HIGH ANGULAR RESOLUTION INTERFEROMETRY
(Report by the Chairman of the Working Group, J. Davis)

a) Introduction:
Interest in high angular resolution interferometry has increased
dramatically in the period 1984-7. Significant advances in instrumentation
have occurred and several important observational demonstrations of instruments
and techniques have been made. These include angular diameter measurements to
the order of 1% accuracy by amplitude interferometry in the infrared [1] and
the visual [2], fringe detection between two large aperture telescopes [3],
fringe tracking and large angle astrometry with accuracy of the order of 70 mas
in a highly automated prototype astrometric interferometer [4], the development
of double Fourier spatio-spectral interferometry [5], and the measurement of
closure phase at optical wavelengths [6].

The field is expanding rapidly and to assist members of the Working Group
and those entering the field, lists of publications in high angular resolution
interferometry have been prepared, based on responses from members of the Group,
covering the years 1983-5 and 1986. Although the lists are not exhaustive, due
to an incomplete response from members of the Working Group, the lists have
proved useful and are available from the Chairman of the Group. It is intended
that these lists be produced early in each year to cover publications in the
previous year.

b) Single-Aperture Interferometry:
Speckle interferometry is now regarded as a well established technique
enabling the full diffraction limited resolution of a telescope to be realised
in measurements of angular dimensions of single and multiple objects. The field
was reviewed by Dainty in 1984 [7] who noted that the technique was by then over
a decade old and that some 80 papers primarily concerned with astronomical
results had been published. Examples of applications include the study of binary
stars where the technique has been particularly productive, the determination of
diameters for satellites of Pluto, Uranus and Neptune, Pluto itself and of
asteroids, and studies of solar granulation. Extragalactic applications include
the resolution of components of a quasar system, the determination of the diameters of planetary nebulae in the SMC and LMC, and more recently the study of
SN1987a in the LMC.

The reconstruction of images from interferometric data has been the subject

of considerable activity, both theoretical and practical, and has been included
in the programs of several meetings [8,9,10]. The difficulties encountered in
efforts to reconstruct images from the modulus of the fringe visibility alone
have led to an appreciation of the need to preserve phase information. As a
result, considerable attention has been given to the development of methods that
yield object phase information including speckle holography, shift-and-add,
Knox-Thompson and the triple correlation technique. Roddier has given a concise
review of the various methods [11] and has also prepared for publication a major
review on "Interferometric Imaging in Optical Astronomy", which includes a near
exhaustive reference list. Images of astronomical objects yielding important
astrophysical results are now being obtained and these include images of α Ori
and its envelope, images revealing multiple stars in objects such as η Car,
images of asteroids, and of circumstellar dust shells and emission regions, etc.
These high resolution imaging techniques will be extremely important for the new
very large optical telescopes such as the Keck Telescope.

The Multiple-Mirror Telescope (MMT) has been operated with all six mirrors
optically co-phased to give the complete u-v plane cover of an appropriately
masked 6.86m aperture telescope and diffraction limited images have been
obtained [12].

Observations using masks with a number of r(0) diameter apertures covering
large telescope apertures have succeeded in measuring closure phases at optical
wavelengths [6]. Image reconstruction techniques developed for aperture synthesis in radio astronomy have been applied successfully to optical closure phase
data [13]. This work has excited considerable interest in the prospect of using
closure phases with apertures separated by long baselines. This will be
discussed in the next section.

An optical aperture synthesis technique (TOAST) applicable to ground-based
telescopes, and which would provide model independent image reconstruction, has
been proposed [14].

The use of adaptive optics to provide diffraction limited images for large
aperture telescopes is under active investigation [15] and it has been suggested
that an artificial reference source might be provided by back-scattering of a
lasar beam from high altitude [16].

c) Two-Aperture Interferometry:
    Two prototype amplitude interferometers with effective apertures r(0) and
incorporating piezo-electrically driven star tracking systems became operational
in 1985 and 1986.

The 11.4m N-S baseline prototype stellar interferometer developed by the
University of Sydney commenced observations in October 1985. It is being used
in a test and development program for a very high angular resolution instrument
but as a demonstration of its performance the angular diameter of Sirius has
been measured with an accuracy of +/- 1.5% [2]. In parallel with this program,
and based on the experience gained from it, construction of a new Sydney University Stellar Interferometer (SUSI) [17] has commenced at Culgoora, northern
NSW, Australia. The new instrument will have siderostats with effective apertures of 14cm and N-S baselines covering the range 5-640 metres.

The Mark III astrometric interferometer on Mt. Wilson [4,18] first
observed fringes in September 1986 with an initial 12m N-S baseline. The
instrument is the result of collaboration between the Smithsonian Astrophysical
Observatory, MIT, the U.S. Naval Research Laboratories and the U.S. Naval
Observatory. It is highly automated with computer controlled star acquisition
and it features a fringe-tracking path equalisation system. Early in its test

program 165 measurments of 23 stars were obtained in one night and large angle
astrometric measurements in one color repeat to about 70mas from night to night.
Preliminary two-color measurements indicate that they can reduce the atmospheric
error by approximately a factor of five. Additional metrology systems, the
extension of the baseline to 20m N-S and the addition of 9m NE-SW and NW-SE
baselines are being implemented.

The original two telescope (0.26m aperture) interferometer at CERGA known
as I2T has evolved and been extended since it was started in 1974. The current
N-S baseline range of 6-67m is being extended to give baselines up to 144m.
The beam-combining optics have been replaced and 0.25nm spectral resolution is
now available [19]. The importance of the high spectral resolution has been
demonstrated in measurements of the Be star γ Cas in the continuum and in the
Hα emission line [20] from which the emission envelope can be distinguished from
the star. Observations made with the PAPA detector have allowed the effect of
sampling time on the measured visibility to be investigated [19]. The instrument
has also been used in the infrared to provide angular diameter measurements of
extremely high precision (1%) [1].

The larger CERGA interferometer, GI2T, has observed fringes with the full
1.5m apertures of its two telescopes [21] following initial fringe observations
with reduced apertures (0.42m) and a baseline of 13.8m [3]. Work is under way
to reduce problems, including vibration, encountered in the first observations
and to develop an improved fringe detection system.

In the infrared a two telescope 10 micron heterodyne interferometer is
under construction at Berkeley [22]. Townes reports that the first 1.65m aperture telescope is performing well in tracking tests and the second telescope is
complete. A site on Mt. Wilson has been prepared for the installation of the
telescopes as an interferometer. The insrument will be used for both high
angular resolution measurements and precision astrometry.

d) Multi-Aperture Interferometry:
    The demonstration of closure phase measurements at optical wavelengths [6]
has brought the prospect of imaging from multi-aperture interferometry closer to
reality. Several groups are now constructing or carrying out feasibility/design
studies of multi-aperture instruments with high resolution imaging in mind.
Included are:

1. Development of the CERGA I2T into I4T [19]. Plans are in hand to add a 60m
range of E-W baselines to form a cross configuration with the N-S baselines of
the I2T and to add two telescopes to give a total of four 0.26m telescopes. The
completed instrument will be used to measure closure phases and to reconstruct
images.

2. The Sydney University Stellar Interferometer (SUSI) [17] now under construction. The first station of an E-W array of siderostats is included in the initial construction of SUSI and the design will permit the future combination of
up to 4 beams for the measurement of closure phase and to reconstruct images.

3. The Cambridge Optical Aperture Synthesis Telescope (COAST) project [13].
The design is based on an array of four siderostats with effective apertures of
40cm to give six baselines up to 100m. The instrument will initially operate
in the spectral range 800-1600nm and is expected to reach magnitudes of at least
10 and possibly 14.

4. The ESO Very Large Telescope project. The VLT includes four 8m telescopes
distributed over a range of about 100m. It is planned to phase the four telescopes for interferometric operation. Two additional smaller, movable telescopes

are proposed to provide improved u-v plane coverage for very high resolution imaging.

5. Design study by the Centre for High Angular Resolution Astronomy (CHARA) at Georgia State University. A detailed design study is being conducted and is considering an array of 1m telescopes to give a maximum baseline in the 300-400m range.

In addition there are several projects starting in 1987. These include a preliminary design study for a United States Naval Observatory astrometric array based on experience with the Mk. III prototype astrometric interferometer [18], a collaborative project between the Smithsonian Astrophysical Observatory and the Universities of Massachusetts and Wyoming to develop an Interferometric Optical Telescope Array (IOTA) for optical and infrared wavelengths on Mount Hopkins in Arizona, and a National Optical Astronomy Observatories project to develop an interferometric array for the infrared/optical. Proposals in the discussion stage include the ambitious Optical Very Large Array (OVLA) [21].

e) General Developments:
    There have been a number of developments with implications for high angular resolution interferomtry that are relevant to many or all of the preceding sections. These include experimental work on the use of optical fibers to link telescopes for coherent operation [23] and an analysis of group delay fringe tracking for faint operation of long baseline interferometers [24]. The importance of two dimensional photon-counting cameras for all forms of high angular resolution interferometry has been clearly demonstrated by the application of the PAPA camera to speckle interferometry [25], fringe visibility measurements [19], and its use in a star tracker for a long baseline stellar interferometer [26].

    The development and demonstration of double Fourier interferometry [5] to provide spatial information for all spectral elements within the observed bandpass simultaneously is another important development. It is particularly well suited for differential measurements.

f) Interferometry From Space:
    A number of approaches to achieving high angular resolution in space have been proposed including those with the acronyms COSMIC, OASIS, SAMSI, SPI, and TRIO. These have been developed to varying degrees and have been discussed at a number of colloquia and workshops devoted to interferometric imaging from space [8, 9, 27, 28]. Attention is currently directed towards connected telescope arrays using either the COSMIC [29] or OASIS [30] approaches, and critical technical issues such as structure stiffness etc., are the subject of a NASA study at JPL. No decision has yet been made regarding which approach should be chosen for intensive study but efforts to reach such a decision are being made through regular ESA-NASA workshops, the most recent being the ESA Workshop on "Optical Interferometry in Space" held at Granada in June 1987 and to be published by ESA [31].

    A micro-arcsecond astrometric interferometer for space, known as POINTS (Precision Optical INTerferometer in Space), has been proposed [32] and collaboration between the Harvard-Smithsonian Center for Astrophysics and the Perkin-Elmer Corporation has been established for the development, manufacture and laboratory testing of critical optical components.

References

1. Di Benedetto, G.P. and Foy, R. 1986, Astron. Astrophys. 166, 204.
2. Davis, J. and Tango, W.J. 1986, Nature 323, 234.
3. Labeyrie, A. et al. 1986, Astron. Astrophys. 162, 359.
4. Shao, M. and Colavita, M.M. 1987, in ref. 10, p.115.
5. Ridgway, S. and Mariotti, J.M. 1987, in ref. 10, p.93.
6. Baldwin, J.E. et al. 1986, Nature 320, 595.
7. Dainty, J.C. 1984, "Stellar Speckle Interferometry", Topics in Applied Physics 9, pp.254-320.
8. Proceedings of Workshop on "High Angular Resolution Optical Interferometry from Space" (Baltimore, U.S.A.), Bull. Am. astr. Soc. 16, No. 3 (Part II), 1984.
9. Proceedings of Colloquium on "Kilometric Optical Arrays in Space" (Cargese, France), ESA Pub. SP-226, 1985.
10. "Interferometric Imaging in Astronomy", Proceedings of Joint ESO-NOAO Workshop on "High-Resolution Imaging from the Ground Using Interferometric Techniques (Oracle, Arizona, U.S.A.), ed. J. Goad, NOAO, 1987.
11. Roddier, F. 1987, in ref. 10, p.1.
12. Hebden, J.C., Hege, E.K. and Beckers, J.M. 1986, Proc. SPIE 628, 42.
13. Baldwin, J.E. 1987, in ref. 10, p.139.
14. Greenway, A.H. 1986, Optics Comm. 58, 149.
15. Beckers, J. et al. 1986, Proc. SPIE 628, 290.
16. Foy, R. and Labeyrie, A. 1985, Astron. Astrophys. 152, L29.
17. Davis, J. and Tango, W.J. 1985, Proc. astr. Soc. Australia 6, 38.
18. Shao, M., Colavita, M.M. and Staelin, D.H. 1986, Proc. SPIE 628, 250.
19. Koechlin, L. et al. 1987, in ref. 10, p.109.
20. Thom, C., Granes, P. and Vakili, F. 1986, Astron. Astrophys. 165, L13.
21. Labeyrie, A., Bosc, I. and Mourard, D. 1987, in ref. 10, p.97.
22. Townes, C.H. et al. 1986, Proc. SPIE 628, 281.
23. Shaklan, S.B. and Roddier, F. 1987, Appl. Opt. 26, 2159.
24. Nisenson, P. and Traub, W. 1987, in ref. 10, p.129.
25. Papaliolios, C., Nisenson, P. and Ebstein, S. 1985, Appl. Opt. 24, 287.
26. Clark, L.D., Shao, M. and Colavita, M.M. 1986, Proc. SPIE 627, 838.
27. Report of Meetings of Working Group on High Angular Resolution Interferometry (1985), Trans. I.A.U., XIXB, 128, 1986.
28. Cambridge Workshop on "Imaging Interferometry" (October 1985), Final Report, NASA, March 1987.
29. Traub, W.A. and Carleton, N.P. 1985, in ref. 9, p.43.
30. Noordam, J.E., Atherton, P.D. and Greenway, A.H. 1985, in ref. 9, p.63.
31. Proceedings of ESA Workshop on "Optical Interferometry in Space", ESA Pub. SP-273, 1987.
32. Reasenberg, R.D. 1984, in ref. 8., p.758.

IX. Acknowledgements

The unattributed sections in this report have been prepared with the help of information supplied by many contributors, including the following:
A. Ardeberg, H. Ando, J.R.P. Angel, J.M. Beckers, J.C. Bhattacharyya, B.P. Fort, P.R.Gillingham, W. Grundmann, Hu Ning-sheng, K. Ishida, T.J. Lee, I.S. McLean, J. Nelson, E.H. Richardson, M.G. Smith, F.G. Watson, R.M. West, Y. Yamashita.

                                              C.M. Humphries
                                              President of Commission

COMMISSION 10. SOLAR ACTIVITY (ACTIVITE SOLAIRE)

President : M. Pick   Secretary : S.R. Kane
Vice President : E. Priest
Organizing Committee : C. Alissandrakis, R. Falciani, T. Hirayama,
                       P. Kaufman, V. Krishan, G.V. Kuklin,
                       E. Tandberg-Hanssen, E. Valnicek, Z.D. Zhang.

In preparing the present report, which covers the period July 1, 1984, to June 30, 1987, close collaboration has taken place between Commission 10 and 12, the two solar commissions, in order to avoid duplications and to insure that pertinent subjects are treated. The reader is referred to the report of Commission 12 for further solar topics. The proceedings are found at the beginning of the references for each section, followed by the usual alphabetical listing. In some sections this listing refers to the previous proceedings by their numbers ; in others we retain the conventional reference. It is a pleasure to acknowledge the excellent work of the reviewers who wrote the different sections of this report, and all the members of the commission who provided information on research to be included.

I. Ground - Based Optical Solar Instrumentation
(W. Mattig)

This report is necessarily incomplete. However, it is to be noted that the IAU has also a special Commission for Instruments and Techniques (Commission 9), and that a section on solar instrumentation is also contained in the report of commission 12.

A great part of the instrumental development in recent years has been reported at special meetings. The proceedings of these meetings are a rich source of information, e.g. the meetings in Kunming 1983 (I), in Huntsville 1984 (II), in Toulouse 1984 (III), and in Tenerife 1986 (IV). On the occasion of the ceremonial inauguration of the observartories on the Canary Islands a special issue of the "Vistas in Astronomy" has been published (V).

A. TELESCOPES

The advantages and disadvantages of possible telescope configurations for high resolution solar observations have been discussed in a review by Dunn (1985). At the Hida Observatory (Japan) a new domeless solar tower telescope was installed. This system includes a Gregory telescope with vertical and horizontal spectrographs. The best resolution achieved is 0.3 arcsec (Nakai and Hattori, 1985). At the Kunming Observatory a SPAR with double tubes of 26 cm aperture has been installed for white light photographs and for chromospheric studies with a Hα filter of 0,25 λ passband.

For the Beijing Observatory a refractive imaging system with an effective aperture of 35 cm was installed to measure the solar magnetic field (Li, 1985). A new 60 cm Gregory telescope for the same observatory is under construction. The solar tower telescope of the Nanjing University and its multichannel spectrograph for studying solar flares and prominences are described by Fang and Huang (1985). Beckers (1985) has discussed future techniques and instrumentation in solar-stellar physics ; the National Solar Observatory is taking initiative to make available the McMath Solar Telescope at Kitt Peak for fully scheduled nighttime observations which pertain to "solar-stellar-connection" astronomy (Smith, Jaksha, 1984).

In Cuba a Horizontal Solar Telescope is being installed by the Pulkovo Observatory staff (Nikonov et al., 1984).

Great progress has been made to put into operation the solar instruments on the Canary Islands. On La Palma, the new Swedish tower telescope with a 50 cm diameter lens, described by Wyller and Scharmer (1985) and Scharmer and Brown (1985) is operating now and has demonstrated the high quality of the instrument and the site. From the German solar telescopes on Tenerife described by Schroter, Soltau and Wiehr (1985) the Gregory Coudé telescope is now operating. The building for the 60 cm Vacuum

Tower Telescope is ready. The beginning of the scientific work, in particular the high resolution spectroscopy with an echelle spectrograph, is scheduled for 1988. At the same "Observatorio del Teide" the Spaniards have built a new solar laboratory mostly for global oscillations of the sun (Roca Cortes, 1984). France is planning to build the THEMIS solar telescope. The 90 cm polarization-free instrument is expected to be finished by 1990 (Mein and Rayrole, 1985).

Extended studies for the LEST (formerly "Large European Solar Telescope", now renamed to "Large Earth-based Solar Telescope") have been carried out in the last years. They are published in the series of LEST Foundation Technical Reports 1-24. Many fundamental problems and technical details are discussed in this series. General reviews have been given by the LEST Foundation and, e.g., by Stenflo (1985b) and Engvold (1985). The design of the 2,4 m aperture telescope is "polarization-free". Helium-filling and adaptive optics are used with the aim of achieving a 0.1 arcsec spatial resolution. Beside all these new installations many improvements of existing instruments have been developed. Pierce (1987) has built a windscreen around the heliostat of the McMath telescope, resulting in a reduction of the declination windshake amplitude by at least a factor of 10.

The use of dynamical dampers for reducing the vibration level of a solar telescope is reported by Korenev and Kitov (1984).

A scanning heliometer measuring the fluctuations of the maximun brightness gradient all around the solar limb is being developed by Rosch and Yerle (1984). Sofia et al. (1984) have drafted the concept of an instrument called the solar disk sextant to be used in space to measure the shape and the size of the sun and their variations.

B. MAGNETOGRAPHS

The vector magnetograph of the Okayama Astrophysical Observatory (Makita et al, 1985 a,b) is in operation since the autumn of 1982. It is fed by the 65 cm solar Coudé telescope with a 10 m Littrow Spectrograph.

A detailed description of Sayan Solar Observatory (Irkutsk) is given by Grigoryev (1985).

At the Beijing Observatory a video-magnetograph is in operation ; a specially designed birefringent filter with a FWHM of 0.15 λ is used. (Li 1985, Ai and Hu, 1985, Hu and Ai, 1985a, 1986).

A new video Stokes polarimeter is designed by Richter et al. (1985) for the 61 cm Vacuum telescope of the San Fernando Observatory. By using a TV camera system simultaneous scans of six polarization components of a given line profile are possible.

Recent improvements of the Haleakala polarimeter include a high resolution echelle spectrometer coupled to telescope by optical fibers and 128-element diode array detectors (Mickey, 1985 a,b). A new apparatus is described by Scholiers and Wiehr (1985 a,b) which measured the Stokes profiles by means of a two dimensional 100*000 detector array. The spatial range covers 75 arcsec; the spectral window is about 1.8 λ. The Marshall Space Flight Center magnetograph has undergone an extensive upgrading in both electronic control of the magnetograph hardware and in polarisation optics (West,1985, Hagyard et al., 1985).

The new solar magnetograph for the Canary Islands Observatory, THEMIS, has been designed for measurements of the magnetic field without interference with the local variations of the thermodynamical parameters (Rayrole, 1985, Rayrole and Ribes, 1985). A number of new instrument concepts for magnetic and velocity field measurements is proposed by Stenflo (1985a): (a) schemes for simultaneous recording of all four Stokes parameters requiring only one piezoelastic modulator; (b) a system using a solid polarizing Michelson interferometer in combination with broad-band prefilters and a piezoelastic modulator.

C. BIREFRIGENT FILTERS

A new birefringent filter of the Lyot type (FPSS) with a passband of 0.13 λ or a spectral resolution of 45000 is used with a 28 cm solar refractor at the Observatoire de Meudon for radial velocity imaging (Dollfus et al., 1985, Dollfus, 1985). A Stokes parameter modulator is placed at the entrance of the filter to produce vector polarisation mappings over the solar surface.

The Beijing solar telescope is equipped with a birefringent filter with two

passbands (5234 Å and Hβ) with a half-width of 0.15 Å ; also Solc elements are used
(Ai et al., 1984, 1985) for measurements of both the solar magnetic vector field and
the sight-line velocity field. The law of conservation of the integrated transmission
of a tunable Lyot birefringent filter is described by Hu and Ai (1985).

By using polarizing beam splitters a 5-resp. 9-channel birefringent filter is
under construction for observations with the newly planned 60 cm Vacuum Gregory
Telescope at the Huairou reservoir station of Beijing observatory (Ai, and Hu, 1985,
Hu and Ai, 1985b, 1986).

On the Pamirs solar telescope of the Pulkovo Observatory an additional optical
channel with a birefringent filter for Hα with 0.5 Å Bandwidth was built (Krundal,
1986).

The Arcetri group is planning solar two-dimensional spectroscopy with a Universal
Birefringent Filter (UBF) and a Fabry-Perot-Interferometer (Bonaccini et al., 1985).
A progress report on the updating of the Zeiss UBF for the Vacuum Tower Telescope in
Izana, Tenerife, is given by Cavallini et al.,(1985).

A Spectrophotometer was newly installed at the Arcetri Observatory Solar Tower
(Cavallini et al., 1987). This instrument basically consists of a Fabry-Perot
Interferometer mounted in tandem with a medium-sized grating spectrograph.

In connection with narrow band filter the temperature variation of ADP
birefringence is measured and discussed by Harvey (1987).

Rust et al. (1986) have designed and fabricated a tunable Fabry-Perot filter from
a 50 mm wafer of optical-quality lithium Niobate crystal. The passband (FWHM) is 0.017
nm, the effective finesse 18,6. A design consideration for a satellite-borne
magnetograph with the Fabry-Perot filter has been done also by Rust (1985).

An imaging triple Fabry-Perot interferometer has been developed by Winter (1984,
1985). It is tunable over the range 400 - 680 nm with a bandwidth of 20 - 50 mÅ. A
multiloop servocontrol system maintains the parallelism and spacing of the plates to
an accuracy of λ/500.

D. MISCELLANEOUS

Semel (1987) has derived the laws of refraction for the general case of plane waves
incident on uniaxial crystals. Optical aberrations and their effects on observations
with high spatial resolution are discussed.

To improve the image quality in large instruments, sunspot trackers have been
built by von der Luhe and Title. The instruments have been tested and are used at the
Vacuum Newton Telescope in Izana and at the Sacramento Peak Observatory.

Correlation trackers are under construction by Dunn, von der Luhe and Title.

A 57-actuator active mirror has been developed by Smithson et al. (1984).
Laboratory tests and seeing measurements show that the mirror is potentially capable
of reaching the diffraction limit of a 20 to 40 inch aperture telescope in uncorrected
seeing of about 2 arsec or better.

Adaptive optical systems for LEST have been proposed by von der Luhe (1983).

References

I : Proceedings of Kunming Workshop on Solar Physics and Interplanetary Travelling
    Phenomena, Science Press, Beijing, China, de Jager, C., and Chen Biao (eds), 1985.
II : Measurements of Solar Vector Magnetic Fields. NASA Conference Publication 2374,
    Hagyard, M.J. (ed), 1985.
III : High Resolution in Solar Physics. Lecture Notes in Physics 233, Springer,
    Heidelberg, Muller, R. (ed), 1985.
IV : The Role of Fine Scale Magnetic Fields on the Structure of the Solar Atmosphere.
    Proceedings of the Inaugural Workshop for the D-E-S Telescope Installations on the
    Canary Islands, Cambridge, Schroter, E.H., Wyller, A.A., and Vazquez, M. (ed),
    1987.
V : Vistas in Astronomy 28, 411-576, Murdin, P. and Beer, P. (ed), 1985.
Ai, G., Hu, Y., Li, T., He, F., Hou, H., Gu, Z., Ni, H.: 1984, Scienta Sinica (Series A)
    27, 173, 1985, 1, 1236.
Ai, G., Hu Y.: 1985, II, 257.
Beckers, J.M.: 1985, Australian J. Phys., 38, 791.
Bonaccini, D., Cavallini, F., Ceppatelli, G., Righini, A.: 1985, III, 118.

Cavallini, F., Ceppatelli, G., Paloski, S., Tantulli, F: 1985, JOSO Ann. Rep., p. 58.

Cavallini, F., Ceppatelli, G., Righini, A., Meco, M., Paloschi, S., Tantulli, F.: 1987, Astron. Astrophys., in press.

Dollfus A., Colson, F., Crussaire, D., Lannay, F.: 1985, Astron. Astrophys. 151, 235.

Dollfus, A.: 1985, II, 192.

Dunn, R.B.: 1985, Solar Physics 100,1.

Engvold, O.: 1985, III, 15.

Fang, C., Huang, Y.: 1985, I, 1177.

Grigoryev, V.M., Kobanov, N.I., Osak, B.F., Selivanov, V.L., Stepanov, V.E.: 1985, II, 231.

Hagyard, M.J., Cumings, N.P., West, E.A.: 1985, I, 1216.

Harvey, J.: 1987, Appl. Opt., 26, 2057.

Hu, Y., Ai G.: 1985a, I, 1236.

Hu, Y., Ai G.: 1985b, I, 1252.

Hu, Y., Ai G.: 1986, Scienta Sinica, 8, 889, (in Chinese).

Korenov, B.G., Kitov, A.K.: 1984, Issled. Geomagn. Aehron. Fiz. Solntsa, Moskva, 69, 197.

Krundal, A.V.: 1986, Soln. Dannye 1986, N° 3, 85.

Li, T.: 1985, I, 1204.

Luhe, O.V.D.: 1983, LEST Technical Report N° 2.

Makita, M., Hamana, S., Nishi K.: 1985a, II, 173.

Makita, M., Hamana, S., Nishi, K., Shimizu, M., Koyano, H., Sakurai, T., Komatsu, H.: 1985b, Publ. Astr. Soc. Jpn. 37, 561.

Mein, P., Rayrole, J.: 1985, Vistas in Astronomy, 567.

Mickey, D.L.: 1985a, II, 183.

Mickey, D.L.: 1985b, Solar Physics 97, 223.

Nakai, Y., Hattori, A.: Contrib. Kwasan, Hida Obs., Univ. Kyoto, N° 260.

Nikonov, O.V., Kulish, A.P., Lebedeva, L.A., Muzalevskij, Yu.S.: 1984, Soln. Dannye 1983, N° 12, 70.

Pierce, A.K.: 1987, Solar Physics 107, 397.

Rayrole, J.: 1985, II, 219.

Rayrole, J., Ribes, E.: 1985, I, 1227.

Richter, P.H., Zeldin, L.K., Loftin, T.A.: 1985, II, 202.

Roca Cortes, T.: 1984, JOSO Ann. Rep., p. 42.

Rosch, J., Yerle, R.: 1984, ESA Spec. Publ. 220, 217.

Rust, D.M.: 1985, II, 141.

Rust, D.M., Burton, C.H., Leistner, A.J.: 1986, SPIE 627, 39.

Scharmer, G.B., Brown, D.S.: 1985, Appl. Opt. 24, 2558.

Schroter, E.H., Soltau, D., Wiehr, E.J.: 1985, V, 519.

Scholiers, W., Wiehr, E.: 1985a, Solar Physics 99, 349.

Scholiers, W., Wiehr, R.: 1985b, II, 153.

Semel, M.: 1987, Astron. Astrophys. 178, 257.

Smith, M.A., Jaksha, D.B.: 1984, Cool stars, stellar systems and the Sun, p. 182.

Smithson, R.C., Marshall, N.K., Sharbaugh, R.J., Pope, T.P.: 1984, Small-scale dynamical processes in quiet stellar atmospheres, p. 66.

Sofia, S., Chiau, H.Y., Maier, E., Schatten, K.H., Minott, P., Endal, A.S.: 1984, Appl. Opt., 23, 1235 ; see also Appl. Opt. 23, 1226 and 1230.

Stenflo, J.O.: 1985a, I, 1139, 1272, 1275.

Stenflo, J.O.: 1985b, V, 571.

West, E.A.: 1985, II, 160.

Winter, J.G.: 1984, Optica Acta, 31, 823.

Winter, J.G.: 1985, J. Phys. E.: Sci. Instr., 18, 505.

Wyller, A.A., Scharmer G.: 1985, V, 467.

II. Spots and Intense Flux Tubes
(J.C. Henoux)

The development of research on starspots, stellar activity, and the suspected relationship between coronal heating and magnetic field have reenforced the interest of the study of the solar magnetic field and the study of the associated thermodynamic

structures. Several proceedings of scientific meetings appeared from 1984 to 1987
(Measurements of Solar Vector Magnetic Fields, 1985 (I) ; The Hydrodynamics of the
Sun, 1984 (II) ; High Resolution in Solar Physics, 1985 (III) ; Theoritical Problems in
High Resolution Solar Physics, 1985 (IV) ; Small Scale Magnetic Flux Concentration in
the Solar Atmosphere, 1986 (V)). The finding that the solar irradiance in affected by
solar activity has renewed interest in photometry of sunspots and faculae. Sunspots
have been used for investigating solar differential and meridional motions. Some
results are also found in Section III.

A. MAGNETIC FIELDS IN SUNSPOTS AND CONCENTRATED FLUX TUBES

Vector magnetic field measurements are required to determine the strains imposed
on the magnetic field of active regions (Parker 1985). Techniques for measurements of
vector magnetic fields are reviewed in Measurements of Solar Magnetic Fields (Hagyard,
ed., 1985). The Stokes profiles are affected by magneto optic, damping, parasitic
light and solar velocity field (Landolfi et al., 1984 ; Landi Degl'Innocenti, 1985 ;
Stenflo, 1985a). Numerical methods for solving the radiative transfer equations in the
presence of velocity gradients have been published by Skumanich et al. (1985) and
Rachkowsky (1986). Makita (1986a), Ye and Jin (1986) investigated Faraday rotation in
sunspots. New observations of broad band circular polarization in sunspots were made
by Henson and Kemp (1984) and Makita and Ohki (1986). To explain these observations
Makita (1986b) assumed a right handed differentially twisted magnetic field. From
transverse magnetic field measurements, electric current densities were inferred by
Ding et al. (1985), Hagyard et al. (1985), Gopasyuk et al. (1985) and Gary et al.
(1987). Isotropic, Pederson, Hall and Cowling photospheric conductivities were
computed by Kubat and Karlicky (1986).

The diagnostic foundation for the determination of the magnetic field of
unresolved structures was reviewed by Stenflo (1986a,b) and Semel (1986). From the
recording of polarized line profiles with high spectral resolution and signal to noise
ratio over large portions of the solar spectrum, the KG field strengths, at
photospheric level, in plages and network were confirmed (Stenflo, 1984, 1985b ;
Stenflo et al. 1984 ; Solanki and Stenflo, 1984).Consistent with the observed center
to limb variation of the field strength, the magnetic field strength inferred from the
infra-red 1.56 Å line, is larger than that found from lines formed higher in the
photosphere (Stenflo et al. 1987a,b). The thermodynamic properties of flux tubes
presented in papers listed in B, were derived assuming a tubular geometry. Pneuman et
al. (1985) used an expansion technique to take into account the effect of field line
curvature. Pizzo (1986) solved numerically the magnetostatic equations for thick flux
tubes, and showed that vertical flux tubes of small horizontal scale might have highly
divergent magnetic topologies. A review on the theoritical modeling of solar magnetic
structures was given by Low (1985), and the causes of magnetic structuring were
discussed by Roberts (1984). At the top of a flux tube, the magnetic field lines are
concave towards the external gas and therefore liable to the interchange instability.
Schussler (1984) showed that a whirl can stabilize the tube.

B. THERMODYNAMIC MODELS OF SUNSPOTS AND CONCENTRATED FLUX TUBES, AND MOTIONS

Moore and Rabin (1985) reviewed sunspots properties. Assuming that sunspots
temperatures and densities are independant of sunspots ages, sizes, and types, new
reference sunspot models have been published (Staude et al., 1984 ; Yun and Beed,
1986). Three semi-empirical photospheric sunspot models corresponding to the early,
middle and late phases of the solar cycle were derived by Maltby et al. (1986), from
continuum intensity observations, from 1968 to 1984. On the other hand, on the basis of
line observations, Sobotka (1985a,b) proposed 12 semi-empirical models of sunspots of
different sizes and ages. The contribution of molecules to the opacity, in the infrared, and the centre to limb variation of the rotational temperature of SIO fundamental
band, for various models, were computed by Punetha and Joshi (1984a,b). The brightness
in the continuum is not uniform in umbrae. However Wiehr (1985a) could not explain the
variations of brightness with solar cycle by the influence of umbral dots. The size,
temperature and half-life time of umbral dats were measured respectively by Grossmann-
Doerth et al. (1986) and by Kusoffsky and Lundstedt (1986). Knobloch and Weiss (1984)
estimated the periods of non linear convective oscillations that could look like

umbral dots. Similar interpretation was made by Choudhuri (1986) in his numerical study of the dynamic of magnetically trapped fluids. On the other hand Kitai (1986) found that umbral dots extend to chromosphere and suggested that they are excited by infalling matter from corona. The average temperature increase in plage and network flux tubes was derived, by Solanki and Stenflo (1985), from a statistical analysis of Stokes I and V line profiles.

Motions are present in magnetic flux tubes. The strong asymmetry of the stokes V profiles implies the existence of line of sight gradients of the velocity in flux tubes. Systematic downflows could explain stockes V profiles observations with moderate field strength (Ribes et al. 1985a,b). Downflows could also provide an efficient mechanism for heating flux tubes by entropy transport (Hasan and Schussler, 1985). On one hand large stokes V zero crossing wavelength shifts were reported by Wiehr (1985b) and Scholiers and Wiehr (1985). On the other hand no Doppler shift of the stokes V zero crossing position greater than 200 m s^{-1} was observed on high resolution polarized spectra (Solanki, 1986). Cavallini et al. (1985, 1987) interpreted observed asymmetries and shifts as due to convection weakening in flux tubes. By convoluting spectrally resolved stokes V profiles with various model instrumental profiles, Solanki and Stenflo (1986) concluded that the published V zero crossing wavelength shifts are consistent with the absence of steady downflows during their 30 minutes integration time. Then they explained the large asymmetries, large non-thermal broadenings, and small Doppler shifts by oscillations. Oscillations in flux tubes might be produced by convective instability (Hasan 1984, 1985) or might be resonant oscillations induced by external pressure fluctuations (Venkatakrishnan 1986).

Recent observational and theoritical work on oscillations in sunspots has been reviewed by Thomas (1985). Photoelectric observations of chromospheric sunspot oscillations were reported by Lites (1985). Lites and Thomas (1985) interpreted the observed 3 minutes oscillations as a resonant mode of fast magnetoatmospheric waves in the sunspot photosphere. At the same time Staude et al. (1985) explained the oscillations observed, on the SMM spacecraft, in transition zone lines, as a resonant mode of slow magnetoatmospheric waves above the temperature minimum region. Similar explanation was used by Lites (1986a,b) who suggested non linear interaction between the two kinds of waves. Further evidence for the resonant chromosphere model were reported by Gurman (1987) and Thomas et al. (1987). New observations of the interaction of solar modes with a sunspot were reported by Abdelatif and al. (1986). The steady-state characteristics of the Evershed flow as been investigated, in the photosphere by Makita an Kawakami (1986), Kuveler and Wiehr (1985), Wiehr et al. (1986) and Wiehr and Stellmacher (1985), and in both photosphere and chromosphere by Dialetis et al. (1985). Long period (45 minutes) photospheric velocity oscillations were observed by Gopasyuk (1985) and by Berton and Rayrole (1985) who also observed associated magnetic field torsional oscillations. The correlation between flare activity and sunspot rotation was confirmed by Gopasyuk and Lazareva (1986).

C. ORIGIN AND DECAY OF SUNSPOTS AND CONCENTRATED FLUX TUBES

A widely used working hypothesis in the investigation of sunspot formation is the emergence of intense magnetic flux tubes from beneath the photosphere. Zwaan (1985a,b) and Garcia de la Rosa (1984) explained sunspots appearance by the emergence of a loop shaped bundle of many flux tubes. Interaction of these flux tubes with giant cells and supergranules is suggested by Mc Intosh and Wilson (1985). Monero-Insertis (1984) confirmed the imposibility of keeping toroidal flux tubes in the deep convection zone for any time longer than a small fraction of the solar cycle period. Following Schussler (1984) the bottom of the convective zone is a favorable region for magnetic field concentration. Giovanelli (1985) revisited the interaction of flux tubes with the convection zone pointing out the importance of the potential energy and of the gas entry in the tube. The cooling time scales of growing sunspots were investigated by Chou (1987). New observations of submergence of magnetic field without spreading or diffusion were reported (Rabin et al., 1984 ; Zirin, 1985), suggesting that the magnetic field is help by subsurface forces (Parker, 1984). On the other hand De Vore et al. (1985) and Sheeley et al. (1986) simulated the evolution of a sample of magnetic regions and of the gross solar magnetic field, by solving a transport equation at the photosphere and Grossmann-Doerth et al. (1987) suggested a spurious variation of photospheric magnetic flux. 3 D numerical simulations of the effect of convection on buoyant magnetic flux tubes (Nordlund, 1985 ; Schmidt et al., 1985 ; Weiss, 1985) lead to qualitative agreement with observations that small flux tubes are swept to the cell boundary, while larger are dragged to the axis of the cell. Mechanism of generation of magnetic fields in photospheric layers have been reviewed (Wilson, 1986). The idea that cyclonic motions, in large scale magnetic field patterns, are at the origin of sunspots was reexamined : Akasofu (1984a,b, 1985) qualitative model of sunspot cyclonic origin is based on the photospheric ambipolar dynamo effect that is also invoked by Simon and Wilson (1985) to explain 20-40 minutes flux changes in small magnetic regions ; Schatten and Mayr (1985) reconsidered sunspot cooling by superadiabatic effect in the presence of cyclonic motions. A statical study of the spiral spots was made by Ding et al. (1987). The apparent E.W. asymmetry of appearance and disappearance of short lived sunspot groups was explained by Kopecky (1985). The distribution Density Function of sunspot groups as a function of lifetime was determined by Kuklin (1985). A quasi-biennal oscillation in sunspot activity, over the period 1749-1981, was found by Apostolov (1985), and Wilson (1986) reported an even-odd cycle distribution in Mt. Wilson number of spots data. Castenmiller et al. (1986) investigated the properties of the Gaizauskas type of complex of activity.

D.   SUNSPOT MOTIONS

No clear picture comes out yet of the published results of investigations on sunspot velocity. Sunspot velocity depends of age (Gokhale and Hiremath, 1984) and type. Differential rotation occurs between sunspots within a group (Gilman and Howard, 1986). From the observed correlation between longitude and latitude motions, Gilman and Howard (1984, 1985) concluded that angular momentum might be transported towards the equator by sunspots. Results on meridional circulation are still controversial. A general drift towards south for spots in the equator near regions was found by Hanslmeier and Lustig (1986) and Lustig and Hanslmeier (1987). Other authors observed an average midlatitude northward flow (Howard and Gilman, 1986) or did not find any meridional motion significantly different from zero (Balthasar et al. 1986). Restraining their investigation to young sunspots, Ribes et al. (1985c) found four latitude bands with alternate direction of circulation that reverse during the solar cycle. Tuominen and Vitanen (1984) related the meridional flow to the cycle and to torsional oscillations.

References

I : **Measurements of Solar Vector Magnetic Fields**. NASA Conference Publication 2374, Hagyard, M.J. (ed), 1985.
II : **The Hydrodynamics of the Sun**, Noordwijkerout, ESA SP 220, 1984.
III : **High Resolution in Solar Physics**, Lecture Notes in Physics 233, Springer, Heidelberg, Muller (ed), 1985.
IV : **Theoritical Problems in High Resolution solar physics**, Schmidt (ed), 1985.
V : **Small scale magnetic flux concentration in the solar atmosphere**, Deinzer, Knolker and Voigt (eds), 1986.
Abdelatif, T.E., Lites, B.W., and Thomas, J.H.: 1986, Astrophys. J. 311, 1015.
Akasofu, S.I.: 1984a, Planet. Space Sci., 32, 1257.
Akasofu, S.I.: 1984b, Planet. Space Sci., 32, 1469.
Akasofu, S.I.: 1985, Planet. Space Sci. 33, 275.
Apostolov, E.M.: 1985, Bull. Astron. Inst. Czechosl. 36, 97.
Balthazar, H., Vazquez, M., Wohl, H.: 1986, Astron. Astrophys. 155, 87.
Balthazar, H., Lustig, G., Stark, D., Wohl, H.: 1986, Astron. Astrophys. 160, 277.
Berton, R., and Rayrole, J.: 1985, Astron. Astrophys. 152, 219.
Castenmiller, M.J.M., Zwaan, C., and Van der Zalm, E.B.J.: 1986, Solar Phys. 105, 237.
Cavallini, F., Ceppatelli, G., and Righini, A.: 1985, Astron. Astrophys. 143, 116.
Cavallini, F., Ceppatelli, G., and Righini, A.: 1987, Astron. Astrophys. 173, 155.
Chou, D.Y.: 1987, Astrophys. J. 312, 955.
Choudhuri, A.R.: 1986, Astrophys. J. 302, 809.
De Vore, C.R., Sheeley, Jr, N.R., Boris, J.P., Young, Jr, T.R., Harvey, K.L.: 1985, Solar Phys. 102, 41.

Dialetis, D., Mein, P., Alissandrakis, C.E.: 1985, Astron. Astrophys. 147, 93.

Ding, Y.T., Hong, Q.F., Hagyard, M.J. and De Loach, A.C.: 1985, Measurements of Solar Vector Magnetic Fields, Hagyard ed., 379-398.

Ding, Y.T, Hong, Q.F., and Wang, H.Z.: 1987, Solar Phys. 107, 221.

Gakhale, M.H., and Hiremath, K.M.: 1984, Bull. Astr. Soc. India, 12, 398.

Garcia de la Rosa, J.I.: 1984, The Hydromagnetics of the Sun, ESA SP 220, 229-230.

Gary, G.A., Moore, R.L., Hagyard, M.J., and Haisch, B.M.: 1987, Astrophys. J. 314, 782.

Gilman, P.A., and Howard, R.: 1984, Solar Phys. 93, 171.

Gilman, P.A., and Howard, R.: 1985, Astrophys. J. 295, 233.

Gilman, P.A., and Howard, R.: 1986, Astrophys. J. 303, 480.

Giovanelli, R.G. : 1985, Australian J. Phys. 38, 1067.

Gopasyuk, S.I.: 1985, Yzv.Krymskoj Astrofiz. Obs. 73, 9.

Gopasyuk, S.I., and Lazareva, L.F.: 1986, Yzv. Krymskoj Astrofiz. Obs. 74, 84.

Gopasyuk, S.I., Kalman, B., Romanov, V.A.: 1985, Yzv. Krymskoj Astrofiz. Obs. 72, 171.

Grossmann-Doerth, U., Schmidt, W and Schroter, E.H.: 1986, Astron. Astrophys. 156, 347.

Grossmann-Doerth, U., Pahlke, K.D. and Schussler, M.: 1987, Astron. Astrophys. 176, 139.

Gurman, J.B.: 1987, Solar Phys. 108, 61.

Hagyard, M.J., West, E.A., Smith, Jr, J.B.: 1985, Proceedings of Kunming Workshop an Solar Physics and Interplanetary Travelling Phenomena, 204-211.

Hanslmeier, A., and Lustig, G.: 1986, Astron. Astrophys. 154, 227.

Hasan, S.S.: 1984, Astrophys. J. 285, 851.

Hasan, S.S.: 1985, Astron. Astrophys. 143, 39.

Hasan, S.S., and Schussler, M.: 1985, Astron. Astrophys. 151, 69.

Henson, G.H., and Kemp, J.C.: 1984, Solar Phys. 93, 289.

Herbold, G., Ulmschneider, P., Sprint, H.C., and Rosner, R.: 1985, Astron. Astrophys. 145, 157.

Howard, R., and Gilman, P.A.: 1986, Astrophys. J. 307, 389.

Kitai, R. : 1986, Solar Phys. 104, 287.

Knobloch, E., and Weiss, N.O.: 1984, MNRAS 207, 203.

Kopecky, M.: 1985, Bull. Astron. Inst. Czech. 36, 359.

Kubat, J., and Karlichy, M.: 1986, Bull. Astron. Inst. Czech. 37, 155.

Kuklin, G.V.: 1985, Bull. Astron. Inst. Czech. 36, 284.

Kusoffsky, U., and Lundstedt, H.: 1986, Astron. Astrophys. 160, 51.

Kuveler, G. and Wiehr, E.: 1985, Astron. Astrophys. 142, 205.

Landi Degl'Innocenti, E.: 1985, Measurements of Solar Vector Magnetic Fields, Hagyard, ed., 279-299.

Landolfi, M., Landi Degl'Innocenti, E. Arena, P.: 1984, Solar Phys. 93, 269.

Lites, B.: 1984, Astrophys. J. 277, 874.

Lites, B.: 1986a, Astrophys. J. 301, 992.

Lites, B.: 1986b, Astrophys. J. 301, 1005.

Lites, B., and Thomas, T.H.: 1985, Astrophys. J. 294, 682.

Low, B.C.: 1985, Measurements of Solar Vector Magnetic Fields, Hagyard, ed., 49-65.

Lustig, G., and Hanslmeier, A.: 1987, Astron. Astrophys. 172, 332.

Mc Intosh, P.S., and Wilson P.R.: 1985, Solar Phys. 97, 59.

Makita, M.: 1986a, Solar Phys. 103, 1.

Makita, M.: 1986b, Solar Phys. 106, 269.

Makita, M., and Kawakami, H.: 1986, Publ. Astron. Soc. Japan 38, 257.

Makita, M., and Ohki, Y.: 1986, Ann. Tokyo Astron. Obs. 21, 1.

Maltby, P., Avrett. E.H., Carlsson. M., Kjeldseth-Moe, O., Kurucz, R.L., and Loeser, R.: 1986, Astrophys. J. 306, 284.

Moore, R., Rabin, D.: 1985, Ann. Rev. Astron. Astrophys. 23, 239-266.

Moreno-Insertis, F.: 1984, The Hydrodynamics of the Sun ESA SP 220, 81-84.

Nordlund, A.: 1985, Theoritical problems in high resolution Solar Physics, MPA 212, 101-423.

Parker, E.N.: 1984, Astrophys. J. 280, 423.

Parker, E.N.: 1985, Measurements of Solar Vector Magnetic Fields, Hagyard, ed., 7-16.

Pizza, V.J.: 1985, Astrophys. J. 302, 785.

Pneuman, G.W., Solanki, S.K., and Stenflo, J.O.: 1986, Astron. Astrophys. 154, 231.
Punetha, L.M., Joshi, G.C.: 1984a, Bull. Astr. Soc. India, 12, 233.
Punetha, L.M., Joshi, G.C.: 1984b, Bull. Astr. Soc. India, 12, 229.
Rabin, D., Moore, R., and Hagyard, M.J.: 1984, Astrophys. J. 287, 404.
Rachkovsky, D.N.: 1986, Izr. Krymskoj. Astrofiz. Obs. 74, 158.
Ribes, E., Rees, D.E., and Cheng Fang.: 1985, Astrophys. J. 296, 268.
Ribes, E., Rees, D.E., and Fang Cheng.: 1985, Measurements of Solar Vector Magnetic
    Fields, Hagyard ed., 300-305.
Ribes, E., Mein, P., and Mangeney, A.: 1985, Nature 318, 170.
Roberts, B.: 1984, Adv. Space Res., Vol. 4, No 8, 17-27.
Schatten, K.H., and Mayr, H.G.: 1985, Astrophys. J. 299, 1051.
Schmidt, H.U., Simon, G.W., Weiss, N.O.: 1985, Astron. Astrophys. 148, 191.
Scholiers, W., Wiehr, E.: 1985, Solar Phys. 99, 349.
Schussler, M.: 1984, The Hydromagnetics of the Sun, ESA SP 220, 67-76.
Schussler, M.: 1984, Astron. Astrophys. 140, 453.
Semel, M.: 1986, Small Scale Magnetic Flux concentrations in the Solar Photosphere, W.
    Deinzer, M. Knolher and H.H. Voigt Editors, 39-58.
Simon, G.W., and Wilson, P.R.: 1985, Astrophys. J. 295, 241.
Skumanich, A., Rees, D.E., and Lites, B.W.: 1985, Measurements of Solar Vector
    Magnetic Fields, Hagyard ed., 306-331.
Sobotka, M.: 1985a, Bull. Astron. Inst. Czech. 36, 230.
Sobotka, M.: 1985b, Sov. Astr. 29, 995.
Solanki S.K.: 1986, Astron. Astrophys. 168, 311.
Solanki, S.K., and Stenflo, J.O.: 1984, Astron. Astrophys. 140, 185.
Solanki, S.K., and Stenflo, J.O.: 1986, Astron. Astrophys. 170, 120.
Staude, J., Furstenberg, F., Hildebrandt, J., Kruger, A., Jakimiec, J., Obridko, V.N,
    Siarkowski, M., Sylwester, B., Sylwester, J.: 1984, Sov. Astron. 28, 564.
Staude, J., Zugzada, Y.D., and Loncans, V.: 1985, Solar Phys. 95, 37.
Stenflo, J.O.: 1984, Adv. Space Res. 4, 5.
Stenflo, J.O.: 1985a, Measurements of Solar Vector Magnetic Fields, Hagyard, ed., 263-
    278.
Stenflo, J.O.: 1985b, Solar Phys. 100, 189.
Stenflo, J.O.: 1986a, Mitt. Astron. Ges., 65, 25.
Stenflo, J.O.: 1986b, Small Scale Magnetic Flux concentrations in the Solar
    Photosphere, W. Deinzer, M. Knolher and H.M. Voigt Editors, 59, 75.
Stenflo, J.O., Harvey, J.W., Brault, J.W., and Solanki, S.K.: 1984, Astron. Astrophys.
    131, 333.
Stenflo, J.O., Solanki, S.K., and Harvey, J.W.: 1987a, Astron. Astrophys. 173, 167.
Stenflo, J.O., Solanki, S.K., and Harvey, J.W.: 1987b, Astron. Astrophys. 173, 305.
Thomas, J.H.: 1985, Australian J. Phys. 38, 811.
Thomas, J.H., Lites, B.W., Gurman, J.B., and Ladd, E.: 1987, Astrophys. J. 312, 457.
Tuominen, I., and Virtanen, H.: 1984, Astron. Nachr. 305, 225.
Venkatakrishnam, P.: 1986, Solar Phys. 104, 347.
Weiss, N.O.: 1985, High Resolution in Solar Physics, Muller, R., ed., 217, 260.
Wiehr, E.: 1985a, High Resolution in Solar Physics, Muller, R.; ed., 254, 260.
Wiehr, E.: 1985b, Astron. Astrophys. 149, 217.
Wiehr, E., and Stellmacher, G.: 1985, High Resolution in Solar Physics, Muller, R.,
    ed., 198-202.
Wiehr, E., Stellmacher, G., Knolher, M., and Grosser, H.: 1986, Astron. Astrophys.
    155, 402.
Wilson, P.R.: 1987, Solar Phys. 106, 1.
Wilson, R.M.: 1986, Solar Phys. 106, 29.
Ye Shi-Hui and Jin Jie-Hai.: 1986, Solar Phys. 104, 273.
Yun, H.S., Beeb, H.A.: 1986, Astrophys. Space Sci. 118, 173.
Zurin, H.: 1985, Astrophys. J. 291, 858.
Zwaan, C.: 1985, Solar Phys. 100, 397.
Zwaan, C.: 1985, High Resolution in Solar Physics, Muller, R., ed., 263-276.

III. <u>Active Regions</u>
(R. Falciani)

During the past three years arguments related to the AR physics can be found in the proceedings of the following meetings : Chromospheric Diagnostics and Modelling, NSO Sacramento Peak, 1984 (I) ; The Hydromagnetics of the Sun, Noordwijkerout, ESA, 1984 (II) ; Solar and Stellar Physics, Titisee, 1987 (III). Many aspects concerning the high activity levels of AR's can also be found in the proceedings of large number of meetings organized on solar flares. Microwave mapping is reported in Section VII. Some results are also reported in Section II.

A. STRUCTURE EVOLUTION AND ENERGY BALANCE OF SOLAR ACTIVE REGIONS.

Schroeter (1984) critized the interpretation of the 13.1—day periodicity in the solar global oscillation measurements as to be a direct evidence for fast rotating solar core. He suggests that the rotation of AR's, occurring predominantly in long lived active longitudes 180 degrees apart, can simulate an AR—weighted Doppler signal in good agreement with the global oscillation measurements.

New measurements of the contrast of solar faculae have been reported by Libbrecht and Kuhn (1984, 1985). A decrease in temperature gradient above optical depth =1 at 0.5μ, from quiet sun to faculae, was observed by Foukal and Duvall (1985). These observations are respectively in qualitative agreement with contrast computations, made by Chapman and Gingell (1984) for a "hot wall" type model, and with the hot wall model of Deinzer et al. (1984a,b). Lawrence et al. (1985) determined that, for visible wavelengths, the facular luminosity excess is about 0.5 of the sunspot luminosity deficit for one AR transit on the solar disk (Aug. 1982). On the other hand Chapman and Meyer (1986) found that, for five AR's transits, the sunspot energy deficit can be roughly balanced by the facular energy excess. Near the disk center a very low net average excess intensity, for active regions, was measured by Hirayama et al. (1985). Pap (1985) investigated the differences of the characteristics of the correlation between the variations of the solar constant and active and old sunspot area. The very important problem of the energy balance between sunspots and faculae in AR's seems not to be solved.

Schrijver et al. (1985) studied UV and X—ray emission of AR's of various sizes and in various stages of evolution in order to clarify many aspects of the stellar outer atmospheres observations. Areas at chromospheric levels are larger than areas at TR levels ; near the edge the TR seems to be much thinner than well inside the AR. The average line intensities show a small spread (factor of 2 in chromospheric lines, factor of 4 in coronal ones). Brightenings greater than a factor of 2 in the Si— IV line intensity have been observed by Hayes and Shine (1987) in a number of AR's. Their emission time profiles, density increase profiles and slight redshifts (in some cases) suggest them that these UV bursts might be due to energy release mechanisms similar to flares but on a smaller scale. In some cases a secondary gradual brightening is observed after the main impulsive burst and the authors suppose that it could be the analog of the flare thermal phase. Impulsive EUV bursts from a small growing AR are observed by Withbroe et al. (1985), morphologically similar to those observed in the coronal bright points. This fact bring them to the conclusion that the impulsive heating (probably due to rapid release of magnetic energy) is also an important heating source for the upper chromosphere and lower coronal layers in small bipolar AR's. The emergence of magnetic flux from lower levels might be the triggering mechanism.

Simultaneous and cospatial FTS measurements of the CO bands and Ca—II/K central line intensities in AR's by Ayres et al. (1985) show an "anomalous" behaviour because the line core brightness temperature of the strongest CO features are lower in the plage areas than the minimum temperature of the solar atmosphere empirical models. The authors then confirm their previous hypothesis, that the solar outer atmosphere might be thermally bifurcated. However this conclusion led to some controversies, since this bifurcated model seems inconsistent with the Ca—II, far IR and UV observations (some more details could be found in the Commission 12 report).

B. EMERGING FLUX REGIONS (EFM)

Excellent observing material enables Brants (1985a,b), and Brants and Steenbeek (1985) to outline more precise ideas on the physical behaviour of an EFR. Roundish darkened patches within the intergranular lanes (protopores) may precede the birth of a pore. A fast growing pore coincides with the area of a downflow of 1.5 Km/sec. The majority of the pores grow in area and darken ; they tend to coalesce to form a spot, particularly near the leading and following edges of the EFR, where strong velocity shears are also observed. Magnetic flux is found over most of the EFR area, with values ranging between 100 and 1700 G outside pores. The complicated behaviour and interacting signatures, regarding the birth of AR's, are nicely reviewed by Zwaan (1985). From the analysis of all the EFR's found in the BBSO material during 1978 Liggett and Zirin (1985) derived that the rate of flux emergence is 10 times higher in AR's than in quiet regions.

From the comparison of C IV Dopplergrams and Hα images Athay et al. (1986a, b) infer that magnetic shears are a common property of AR's. Even if magnetic shears are often associated with flaring activity, they seem not to be a totally sufficient condition to produce flares.

Topka et al. (1986) considered the small scale changes occurring in magnetic flux elements and they derived no evidence for migration or diffusion when magnetic cancellation occurs. The comparison of the SMM-UVSP spectroheliograms and the distributions of vertical electric current density , calculated from vector magnetographic measurements, enable de Loach et al. (1984) and Haisch et al. (1986) to claim that, in spite of some determined correlations, the expected scaling relations between simple ohmic heating and radiative losses cannot be established.

C. VELOCITY FIELDS AND FLOWS

The analysis of AR's loop flow as derived from SMM observations has been reviewed by Poletto and Kopp (1984). Lites et al. (1985) showed that He-I 10830 A line Doppler shifts maps indicate persistent flow pattern in AR's upper chromosphere near the limb. These features are very similar to those already observed in the TR C-IV Dopplergrams. Time series of two-dimensional maps of V and B have been statistically analyzed by Berton (1986). The improvement of the existing methods for the determination of the V and B vector maps from time sequence of AR observations and the ambiguity of the various solutions are also discussed by Berton (1987). C-IV spectroheliograms and Dopplergrams enable Kopp et al. (1985) to derive empirical velocity and EM profiles all along AR limb loops. They compare these parameters with the theoretical values obtained from three different simple models (steady-state loop with siphon flow, steady downflow in both loop legs, not steady downflow) and conclude that none of these is capable to reproduce the observed parameters behaviour. Improved quality observations seem to be needed to provide reliable constraints for developing more realistic models of cool loop dynamics.

D. LOOP STRUCTURES

As loops are believed to trace out the magnetic lines of force connected with the AR, notable efforts have been given to study this type of structures.

In their long series of papers on high resolution observations of the solar chromosphere Bray and Loughhead (1985, 1986), Loughead and Bray (1984, 1985) carefully determined many parameters of AR Hα coronal loops. The loop material rises along one leg with an increasing upward velocity (despite the retarding effect of the gravity). In the other leg the material descends under the effect of a retarding force since it does not attain the free fall velocity. All the derived velocities are supersonic (Mach numbers in the range 1.5 - 5.). True widths are of the order of 0.7 - 0.8 arcsec, with a central axis contrast of the order of 0.8 and longitudinal Hα optical depth ranging from 0.06 to 1.3. Detailed analysis of the variations of the thermodynamic parameters (Ne, Te, P etc.) along the loop axis exhibit a wavelike nature, with an apparent wavelength of the order of 50000 - 60000 Km. Cram (1986) examined the consistency between the values of the Hα source function derived from the cloud model interpretation of the measured Hα contrast of coronal loops and the values calculated with NLTE. A generally satisfactory agreement is found.

EUV spectra of AR loop structures have been studied by Doyle et al. (1986). Physical parameters derived from low temperature lines ($< 10^5$ K) seem to be inconsistent with the values inferred from scaling laws based on static loop models. Only the assumption that the Te and EM values, obtained from Mg-X data, remain constant up to 5.10^6K allow a better agreement between the derived parameters and the model ones. Narain and Kumar (1985) contributed to the study of the heating and cooling mechanisms in AR loops.

References

I : Chromospheric Diagnostic and Modelling, NSO, Sacramento-Peak, Lites (ed), 1984.

II : The Hydrodynamics of the Sun, Noordwijkerout, ESA, SP 220, 1984.

III : Solar and Stellar Physics, Titisee, 1987.

Athay, R.G., Jones, H.P. and Zirin, H.: 1986a, Astrophys. J. 303, 877.

Athay, R.G., Klimchuk, J.A., Jones, H.P. and Zirin, H.: 1986b, Astrophys. J. 303, 884.

Ayres, T.R., Testerman, L. and Brault, J.W.: 1985, Astrophys. J. 304, 542.

Berton, R.: 1986, Astron. Astrophys. 161, 381.

Berton, R.: 1987, Astron. Astrophys. 175, 238.

Brants, J.J.: 1985a, Solar Phys. 95, 15.

Brants, J.J.: 1985b, Solar Phys. 98, 197.

Brants, J.J. and Steenbeek, J.C.M.: 1985, Solar Phys. 96, 229.

Bray, R.J. and Loughhead, R.E.: 1985, Astron. Astrophys. 142, 199.

Bray, R.J. and Loughhead, R.E.: 1986, Astrophys. J. 301, 989.

Chapman, G.A., and Gingell, T.W.: 1984, Solar Phys. 91, 243.

Chapman, G.A. and Meyer, A.D.: 1986, Solar Phys. 103, 21.

Cram, L.E.: 1986, Astrophys. J. 300, 830.

Deinzer, W., Hensler, G., Schussler, M. and Weisshaar, E.: 1984a, Astron. Astrophys. 139, 426.

Deinzer, W., Hensler, G., Schussler, M. and Weisshaar, E.: 1984b, Astron. Astrophys. 139, 435.

de Loach, A.C., Hagyard, M.J., Rabin, D., Moore, R.L., Smith, Jr., J.B., West, E.A. and Tandberg-Hanssen, E.: 1984, Solar Phys. 91, 235.

Doyle, J.G., Mason, H.E. and Vernazza, J.E.: 1986, Astron. Astrophys. 150, 69.

Foukal, P., and Duvall, Jr, T.: 1985, Astrophys.. J. 196, 739.

Haisch, B.M., Bruner, M.E., Hagyard, M.J. and Bonnet, R.M.: 1986, Astrophys. J. 300, 428.

Hayes, M. and Shine, R.A.: 1987, Astrophys. J. 312, 943.

Hirayama, T., Hamana, S., and Mizugaki, K.: 1985, Solar Phys. 99, 43.

Kopp, R.A., Poletto, G., Noci, G. and Bruner, M.: 1985, Solar Phys. 98, 91.

Lawrence, J.K., Chapman, G.A., Herzog, A.D. and Shelton, J.C.: 1985, Astrophys. J. 292, 297.

Libbrecht, K.G., and Kuhn, J.R.: 1984, Astrophys. J. 292, 297.

Libbrecht, K.G., and Kuhn, J.R.: 1985, Astrophys. J. 299, 1047.

Liggett, M. and Zirin, H.: 1985, Solar Phys. 97, 51.

Lites, B.W., Keil, S.L., Scharmer, G.B. and Wyller, A.A.: 1985, Solar Phys. 97, 35.

Loughhead, R.E. and Bray, R.J.: 1984, Astrophys. J. 283, 392.

Loughhead, R.E., Bray, R.J. and Wang, J.L.: 1985, Astrophys. J. 294, 697.

Narain, U. and Kumar, M.: 1985, Solar Phys. 99, 111.

Pap, J.: 1985, Solar Phys. 97, 21.

Poletto, G. and Koop, R.A.: 1984, Mem. It. Ast. Soc. 55, 773.

Schrijver, C.J., Zwaan, C., Maxon, C.W. and Noyes, R.W.: 1985, Astron. Astrophys. 149, 123.

Schroeter, E.H.: 1984, Astron. Astrophys. 139, 538.

Topka, K.P.,Tarbell, T.D. and Title, A.M.: 1986, Astrophys. J. 306, 304.

Withbroe, G.L., Habbal, S.R. and Ronan, R.: 1985, Solar Phys. 95, 297.

Zwaan, C.: 1985, Solar Phys. 100, 397.

IV. Prominences and Filaments
(J.M. Malherbe)

This report is a summary of the work done on the physics of prominences and filaments . An overview of modern observations has been written by Hirayama (1985). The behaviour of solar Coronal and Prominence Plasmas (CPP) has been the topic of two SMM meetings, organized by A. Poland in 1985 and 1986, in Goddard Space Flight Center. An extensive review of the subject (formation, spectroscopic diagnostics and instabilities, with a large amount of contributed papers) can be found in the proceedings (Poland, 1986 (I)). Also, the physics and the structure of prominences will be discussed, from a more theoretical point of view, during a next workshop, in Palma de Mallorca, organized by E. Priest and J.L. Ballester in November 1987.

Recent progresses in the understanding of the formation, the structure, the equilibrium and support, and the instabilities ("Disparitions Brusques") of solar prominences will be summarized in this paper, and results of spectroscopic diagnostics (measurements of magnetic fields, velocities, temperatures and densities) will be stressed, from both a theoretical and observational point of view.

Prominences are structures of the solar corona ; but it is commonly thought that they are 100 times denser and cooler (chromospheric like conditions). They are anchored in the photosphere by feet ; the MHD mechanisms which operate in the formation, equilibrium and instabilities of these objects are complex and involve non linear effects due to interactions between heating, conduction, radiation, magnetic field reconnection and gravity, which make fascinating these beautiful features of the Sun.

A. FORMATION

It is well known that the mass of a quiescent prominence is an appreciable part of the mass of the entire corona (roughly one tenth or more), which makes difficult to form these structures only by coronal condensation. Hence, possible mechanisms of formation are now divided in two categories, namely condensation (of the coronal plasma) and injection (of the chromospheric material into the corona). Martin (1986), from recent observations, shows that filament formation is characterized by a continuous accumulation of dense threads along the channel between large scale opposite magnetic polarities, and by small scale magnetic field cancellation.

1. Condensation.

Sparks and Van Hoven (1985) considered thermal instabilities of the coronal plasma in sheared magnetic fields, without conduction; Van Hoven et al (1984) studied the interaction between radiation and tearing, while Tachi et al (1985) investigated the effects of viscosity in radiative and reconnection instabilities of sheared fields; Van Hoven and Mok (1984), and Van Hoven et al (1986), incorporated the effect of anisotropic thermal conduction and found unstable modes due to the perpendicular component. Malherbe (1987) and Forbes and Malherbe (1986) have numerically solved the 2D resistive-radiative MHD equations in line-tied current sheets and discovered a shock condensation mechanism. An (1984, 1985) looked at the condensation modes in cylindrical plasmas (loops) and examined the effect of the shear (An, 1986) (see also section V A).

2. Injection.

Injection processes can be subdivided into surge-like and evaporation-like models. An et al (1986) suggested that material is ballistically launched from the chromosphere to the corona at the speed of spicules, while Poland and Mariska (1986) showed that a sustained heat release in a loop may give rise to a solar evaporation and thermal instability at the top of the loop.

B. SPECTROSCOPIC DIAGNOSTICS

1. Velocity field.

It is now clear that the equilibrium of filaments is not static, but dynamic. Upward motions were found by Schmieder et al (1984a) in chromospheric lines, by Athay et al (1985) and Engvold et al (1985) in UV lines. A correlation between these flows and large convective motions in the photosphere (giant cells) is suspected by Schmieder et al (1984b). Ioshpa et al (1986) have compared Hα and photospheric velocities around a filament. Small scale motions were studied by Landman (1985a). Oscillations in prominences were searched by Bashkirtsev and Mashnich (1984) (long periods), while short periods were found by Tsubaki and Takeuchi (1986) and Tsubaki et al (1987). On the contrary, no chromospheric oscillations were observed by Malherbe et al (1987) in filaments.

Large scale motions in loop prominences have also been reported by Loughead and Bray (1984), Cui et al (1985), while oscillations in loops have been studied by Solovev (1985).

2. Magnetic fields.

Prominences are composed of a set of dense and cold threads of electric currents supported against gravity (and partly isolated from heating by conduction and wave dissipation) by magnetic fields. Longitudinal fields were recorded in filaments by Maksimov and Ermakova (1985), and in prominences by Nikolsky et al (1984, 1985) using Zeeman effect. Kim et al (1986) have studied correlations between line of sight magnetic and velocity fields.

Landolfi and Landi (1985) investigated line polarization as a function of the vector magnetic field. Also Hanle effect was theoretically studied by Bommier (1987) and used for magnetic vector measurements in quiescent prominences by Gornyj et al (1984), Leroy et al (1984) and Querfeld et al (1985). It is now well established that the angle between the field and the filament axis is small (25 degrees), and that the topology is more often of the Raadu–Kuperus type (1973) than the Kippenhahn–Schluter one (1957). At last, Ballester and Kleckzek (1984) and Ballester (1984) give a model derived from observations of magnetic and electric fields in prominences.

3. Pressure, temperature, and densities.

Radio emission was investigated above filaments by Stewart (1984), Apushkinskij (1985) and Hiei et al (1986). Zirker (1985) observed infra-red prominence H lines. Smartt and Zhang (1984) analysed soft X-ray data of a quiescent prominence. Loughead et al (1985) give a complete description of MHD physical quantities in an active loop.

Radiative transfer in prominences has also been extensively studied by several authors : Fontenla and Rovira (1985), Landman (1984, 1985b), Cram (1986), Zarkhova (1984) present new calculations. Semi-empirical models are proposed by Zhang and Fang (1987), including transfer and hydrostatic equilibrium. The gas pressure in prominences has been determined by Hellwig et al (1984). Stellmacher et al (1986) have measured electron densities in faint prominences. Some discrepancies exist in the density measurements : Bommier et al (1984) found densities of 10^{10} cm^{-3} and less, while Landman (1984) obtains values bigger than 10^{11} cm^{-3} ; this last result makes questionable condensation mechanisms of the coronal plasma to account for the huge mass contained in prominences.

The thermal structure of filaments is still unclear: Kundu (1986) and Gary (1986) suggested hot threads around a cold core, while Schmahl and Orrall (1986) suggested non isothermal loops at various temperatures.

C. OVERALL STRUCTURE, EQUILIBRIUM AND SUPPORT

1. Large scale structure.

The overall structure of solar filaments was studied by Soru–Escaut et al (1985) and singularities in the solar rotation were found. The behaviour of filaments as a function of the solar cycle was investigated by Gnevyshev and Makarov (1985), Makarov (1985) and Fujimori (1984). Liggett and Zirin (1984) have studied proper motions of prominences and gave evidence of rotational motions. It becomes now urgent for a better understanding of prominence structure and formation to collect data on feet, which connect filaments to the solar convective zone.

2. Equilibrium and support.

New equilibrium (magnetostatic) models for the support of prominences (which are often considered as thin current sheets) against gravity have been proposed: Anzer and Priest (1985) considered the equilibrium of Kuperus–Raadu type prominences imbedded in potential coronal magnetic fields; Amari and Aly (1986) investigated the equilibrium of massive lines of current in sheared force–free coronal fields. Also, the externally driven quasi–static evolution of magnetostatic equilibria (especially current sheets) has been reviewed by Aly (1986), while Wu et al (1986) suggested that induced mass and wave motions could be driven by converging or diverging photospheric flows.

Low (1984) put forward three–dimensional magnetostatic models relevant to prominence support ; and Osherovich (1985) obtained different magnetic supports based on eigenvalue solutions and examined the behaviour of filaments in horizontal external magnetic fields. Galindo–Trejo and Shindler (1984) investigated the MHD stability of sheet equilibria, relevant to the Kippenhahn– Schluter model. Jensen (1986) suggested that Alfvén wave dissipation could provide a support mechanism for prominences. At last, a thermal model for the fine structure of prominences has been proposed by Démoulin et al (1987).

D. INSTABILITIES AND DISAPPEARANCES

It is now well established that disappearances of filaments can be subdivided in two classes: thermal disappearances (corresponding to plasma heating) and dynamic ones (characterized by ejecta). Eruptive filaments may reform later after instabilities, and a lot of events are assiociated to flares (and two ribbon flares). What causes filament instability is a difficult problem : the emergence of new flux (followed by magnetic reconnection), the variation of the shear, thermal non–equilibrium, bifurcations between multiple equilibria under changes of boundary conditions are possible candidates. The association between flares and filament eruptions was investigated by Tang (1986). Interaction between filaments and flares was also analysed by Rust (1984). Apushkinskij and Topchilo (1984) have suggested that deformations due to differential rotation may trigger prominence instabilities.

Particle acceleration during filament disappearances was studied by Kahler et al (1986). Fast material ejecta during instabilities have been modelled by Raadu et al (1987). Schmieder et al (1985), Wang (1985) and Kurokawa et al (1987) reported twisting and untwisting motions in activated filaments and prominences. Pneuman (1984) has presented a model for rising helical prominences during activation. The dynamics of active prominences was also analysed by Klepikov and Platov (1985).

Coronal mass ejection, in relation with eruptive prominences, is also studied by Illing and Athay (1986), Athay and Illing (1986) and Illing and Hundhausen (1986). Heating processes have been examined by Mouradian and Martres (1986) and by Simon et al (1984) ; in this last case, the observed event is suggested to be the consequence of a new emerging flux and magnetic reconnection which ensues.

The fine structure of an active prominence was studied by Nikiforova (1985). At last, the density of post flare loops, dense and cold features which are observed to form during the gradual phase of two ribbon flares, is studied by Hanakoa and Kurokawa (1986); the results are consistent with reconnection models proposed by Kopp and Pneuman in 1976.

References

I : Proceedings of Coronal and Prominence Plasmas (CCP) Workshops, NASA Conference Publication 2442, Poland, K.V. (ed), 1986.

Aly, J.J.: 1986, Proceedings of the Workshop on "Interstellar Magnetic Fields", Schloss Ringberg, Germany, in press.

Amari, T., Aly, J.J.: 1986, proceedings of the workshop on "Interstellar Magnetic Fields", Schloss Ringberg, Germany, in press.

An, C.H.: 1984, Astrophys. J., 276, 755.

An, C.H.: 1985, Astrophys. J., 298, 409.

An, C.H.: 1986, Astrophys. J., 304, 532.

An, C.H., Bao J.J., Wu S.T.: 1986, CPP proceedings, A. Poland Ed., 51.

Anzer, U., Priest, E.R.: 1985, Solar Phys., 95, 263.

Apushkinskij, G.P.: 1985, Astron. Z., 62, 992.

Apushkinskij, G.P., Topchilo, N.A.: 1984, Astron. Z., 61, 1150.
Athay, G., Jones, H., Zirin, H.: 1985, Astrophys. J., 288, 363.
Athay, G., Illing, R.M.E.: 1986, J. G. R., 91, 10961.
Ballester, J.L.: 1984, Solar phys., 94, 151.
Ballester, J.L., Kleckzek, J.: 1984, Solar Phys., 90, 37.
Bashkirtsev, V.S., Mashnich, G.P.: 1984, Solar Phys., 91, 93.
Bommier, V.: 1987, thèse de Doctorat d'Etat, Université de Paris VII.
Cui, L.S., Hu, J., Ji, G.P., Ni, X.B., Huang, Y.R., Fang, C.: 1985, Chin. Astron.
     Astrophys., 9, 49.
Cram, L.E.: 1986, Astrophys. J., 300, 830.
Démoulin, P., Raadu, M.A., Malherbe, J.M., Schmieder, B.: 1987, Astron. Astrophys.,
     in press.
Engvold, O., Tandberg-Hanssen, E., Reichmann, E.: 1985, Solar Phys., 96, 35.
Fontenla, J.M., Rovira, M.: 1985, Solar Phys., 96, 53.
Forbes, T.G., Malherbe, J.M.: 1986, Astrophys. J. letters, 302, L67.
Fujimori, K.I.: 1984, Publ. Astron. Soc. Japan, 36, 189.
Galindo-Trejo, J., Schindler, K.: 1984, Astrophys. J., 277, 422.
Gary, D.: 1986, in CPP proceedings, A. Poland Ed., 121.
Gornyj, M.B., Kupriyanov, D.V., Matisov, B.G.: 1984, Astron. Z., 61, 1158.
Gnevyshev, M.N., Makarov, V.I.: 1985, Solar Phys., 95, 189.
Hanaoka, Y. and Kurokawa, H.: 1986, Solar Phys., 105, 133.
Hellwig, J., Stellmacher, G., Wiehr, E.: 1984, Astron. Astrophys., 140, 449.
Hiei, E., Ishiguro, M., Kosugi, T., Shibasaki, K.: 1986, in proceedings of CPP
     meeting, A. Poland Ed., 109.
Hirayama, T.: 1885, Solar Phys., 100, 415.
Illing, R.M.E., Athay, G.: 1986, Solar Phys., 105, 173.
Illing, R.M.E., Hundhausen, A.J.: 1986, J. G. R., 91, 10951.
Ioshpa, B.A., Kozhevatov, I.E., Kulikova, E.K., Mogilevskij, E.I.: 1986, Soln.
     Dannye, 5, 68.
Jensen, E.: 1986, CPP proceedings, A. Poland Ed., 63.
Kahler, S.W., Cliver, E.W., Cane, H.V., Mc Guire, R.E., Stone, R.G., Sheeley,
     N.R.:1986, Astrophys. J., 302, 504.
Kim, I., Koutchmy, S., Stellmacher, G., Stepanov, A.I.: 1986, in proceedings of "the
     role of fine-scale magnetic fields on the structure of the Solar atmosphere"
     meeting, Tenerife, Canary Islands, in press.
Kippenhahn, R., Schluter, A.: 1957, Astrophys. J., 43, 36.
Klepikov, V., Platov, V.: 1985, Astron. Z., 62, 983.
Kundu, M.R.: 1986, in CPP proceedings, A. Poland Ed., 117.
Kurokawa, H., Hanaoka, Y., Shibata, K., Uchida Y.: 1987, Solar Phys., in press.
Landolfi, M., Landi, E.: 1985, Solar Phys., 98, 53.
Landman, D.A.: 1984, Astrophys. J., 279, 438.
Landman, D.A.: 1985a, Astrophys. J., 295, 220.
Landman, D.A.: 1985b, Astrophys. J., 290, 369.
Leroy, J.L., Bommier, V., Sahal-Brechot, S.: 1984, Astron. Astrophys., 131, 33.
Liggett, M., Zirin, H.: 1984, Solar Phys., 91, 259.
Loughead, R.E., Bray, R.J., Jia-Long Wang: 1985, Astrophys. J., 294, 697.
Loughead, R.E., Bray, R.J.: 1984, Astrophys. J., 283, 392.
Low, B.C.: 1984, Astrophys. J., 277, 415.
Malherbe, J.M.: 1987, thèse de Doctorat d'Etat, Université de Paris VII.
Malherbe, J.M., Schmieder, R., Mein, P., Tandberg-Hanssen, E.: 1987, Astron.
     Astrophys., 172, 316.
Makarov, V.I.: 1984, Solar Phys., 93, 393.
Maksimov, V.P., Ermakova, L.V.: 1985, Astron. Z., 62, 558.
Martin, S.F.: 1986, in CPP proceedings, A. Poland Ed., 73.
Mouradian, Z., Martres, M.J.: 1986, in proceedings of the CPP meeting, A. Poland Ed.,
     221.
Nikiforova, T.P.: 1985, Soln. Dannye, 8, 79.
Nikolsky, G.M., Kim, I.S., Koutchmy, S., Stellmacher, G.: 1984, Astron. Astrophys.,
     140, 112.

Nikolsky, G.M., Kim, I.S., Koutchmy, S., Stepanov, A.I., Stellmacher, G.: 1985,Astron. Z., 62, 1147.

Osherovich, V.A.: 1985, Astrophys. J., 297, 314.

Pneuman, G.W.: 1984, Solar Phys., 94, 299.

Poland, A.I., Mariska, J.T.: 1986, Solar Phys., 104, 303.

Querfeld, C.W., Smartt, R.N., Bommier, V., Landi, E., House, L.L.: 1985, Solar Phys., 96, 277.

Raadu, M.A., Kuperus, M.: 1973, Solar Phys., 28, 77.

Raadu, M., Malherbe, J.M., Schmieder, B., Mein, P.: 1987, Solar Phys., in press.

Rust, D.M.: 1984, Solar Phys., 93, 73.

Schmahl, E., Orrall, F.: 1986, in CPP proceedings, A. Poland Ed., 127.

Schmieder, B., Malherbe, J.M., Mein, P., Tandberg-Hanssen, E.: 1984a, Astron. Astrophys., 136, 81.

Schmieder, B., Ribes, E., Mein, P., Malherbe, J.M.: 1984b, Mem. S.A. It., 55, 319.

Schmieder, B., Raadu, M.A., Malherbe, J.M.: 1985, Astron. Astrophys., 142, 249.

Simon, G., Mein, N., Mein, P., Gesztelyi, L.: 1984, Solar Phys., 93, 325.

Smartt, R.N., Zhang, Z.: 1985, Solar phys., 90, 315.

Solovev, A.A.: 1985, Soln. Dannye, 9, 65.

Soru-Escaut, I., Martres, M.J., Mouradian, Z.: 1985, Astron. Astrophys., 145, 19.

Sparks, L., Van Hoven, G.: 1985, Solar Phys., 97, 283.

Stellmacher, G., Koutchmy, S., Lebecq, C.: 1986, Astron. Astrophys., 162, 307.

Stewart, R.T.: 1984, Solar Phys., 94, 379.

Tachi, T., Steinolfson R.S., Van Hoven G.: 1985, Solar Phys., 95, 119.

Tang, F.: 1986, Solar Phys., 105, 2.

Tsubaki, T., Ohnishi, Y., Suematsu, Y.: 1987, Publ. Astron. Soc. Japan, 39, 179.

Tsubaki, T., Takeuchi, A.: 1986, Solar Phys., 104, 2.

Van Hoven, G., Mok, Y.: 1984, Astrophys. J., 282, 267.

Van Hoven, G., Sparks ,L., Tachi, T.: 1986, Astrophys. J., 300, 249.

Van Hoven, G., Tachi, T., Steinolfson, R.S.: 1984, Astrophys. J., 280, 391.

Wang, J.L.: 1985, Scientia Sinica, 28, 1308.

Wu, S.T., Hu, Y.Q., Nakagawa, Y., Tandberg-Hanssen, E.: 1986, Astrophys. J., in press.

Zarkhova, V.V.: 1984, Astron. Astrophys., 51, 45.

Zhang, Q.Z., Fang, C.: 1987, Astron. Astrophys., 175, 277.

Zirker, J.B.: 1985, Solar Phys., 102, 33.

V. Solar Maximun Analysis
(Z. Svestka)

The basic results of the SMA were presented and summarized in two SMA Symposia Proceedings, edited by Simon (1984 (I) we will refer to this publication using abbreviation SIM) and by De Jager and Svestka (1986, DJS (II)). Other papers discussing SMA results can be found in Proceedings of SMM Workshops (Kundu and Woodgate, 1986 – KAW (III) ; Poland, 1986 – POL (IV) ; Dennis et al., 1987 (DOK (V)) and three other meetings held at Kunming (De Jager and Chen Biao, 1985, JCB (VI)) Irkutsk (Stepanov and Obridko, 1986 (SAO (VII)) and Sacramento Peak (Neidig, 1986, NEI (VIII)). A summary of Soviet contributions to the SMA (until 1985) was prepared by Stepanov (in SAO, p. 5). Kindly note that this Section does not mention most of the SMA results that belong to the topics of the subsequent sections of this report : thus the SMA results of space observations should be found in Section VI, radio observations in Section VII, etc. The present report will be mostly limited to those results of the SMA which use or interpret optical ground-based data.

A. FLARE BUILD -UP

A useful discussion of preflare activity can be found in Priest et al. (in KAW, Chapter 1). Kuin and Martens (Martens, 1986 and ref. there) extended the scenario for preflare energy build-up in a filament circuit to a 3D-model. Energy storage in sheared magnetic field structures has been discussed by Zwingmann et al. (1985) and Hofman et al. (1987). Peres-Enriquez (1985) revived and reexamined the old Elliot's idea of a build-up in the form of energetic particles stored in magnetic loops. Lin Xinping and Chang Guohua (in JCB, p. 436) suggest a transformation of kinetic energy from convection as the source of energy storage. Wu et al. (in SAO, p. 393 and DJS p. 53) have developed a numerical MHD model to study energy buil-up in the form of electric current.

Several Chinese authors in JCB (pp. 219 - 317) modeled force-free fields with constant α in flaring active regions. Seehafer (1985) considers a spatially constant, but time-variable α and believes to find the topology responsible for a series of homologous flares. Also several papers in SIM discuss various aspects of the flare build-up and problems related to homologous flares. Machado (1985) has found that changes in flare homology correlate with the development of magnetic shear.

The earlier idea (Hagyard et al. in SIM, P. 71) that a flare is triggered when the shear exceeds a critical limit has not been confirmed: The shear is a necessary but not sufficient preflare condition (Hagyard and Rabin in DJS p. 7). Strong shear can be present without concurring flaring (Athay et al., 1986a). The most energetic flares do not always occur at the places of the highest shear, though the flare onset is commonly observed there (Machado et al. in DJS, p. 33). Lin Yuanzhang and Gaizauskas (1987) find that Hα kernels in a flare agree with peak values of the longitudinal electric current. The best observed mechanism for the formation of shear is sunspot motion (Gezstelyi and Kalman in DJS, p. 21 ; Neidig et al., DJS, p. 25 ; Kovocz and Dezso, DJS, p. 29 ; Gezstelvi, in NEI, p. 163). Sometimes shear can develop from head-on collision of spots (Gaizauskas and Harvey in DJS, p. 17), and also large-scale plasma flows can produce shear (Athay et al., 1986b). Other sources of shear can be flux emergence, submergence, and even flux cancellation (Martin in POL, p.73).

Martin et al. (in SIM, p. 61) have identified and analysed the evolution of flare sites where new magnetic flux emerged. Flux emergence prior to the occurrence of three major flares was reported by Guo Quanshi et al. (in JCB, p. 642). Kundu (in DJS, p. 63) and Lang and Wilson (in DJS, p. 97) present high-resolution microwave observations of tentatively emerging new flux. Rising and twisting of magnetic loops around the site of the subsequent flare was reported by Kundu et al. (1985). Forbes (in SIM, p. 53) and Forbes and Priest (1984) have made numerical simulations of reconnection in an emerging magnetic field Priest (in DJS, p. 73) also gives a review of the role of magnetic reconnection in flares : it can give rise to small flares, trigger larger events, and release magnetic energy in major flares, Hong et al. (1987) consider the development of microinstabilities during emergence of new flux and propose four different types of reconnection that lead to different kinds of activity. Magnetic reconnection in a high-temperature turbulent sheat was considered by Somov and Titov (1985), and Tachi et al. (1985) expanded their earlier treatment of reconnection by including effects of compressibility and viscosity (see also section IV A).

Van Hoven and Hurford (in SIM, p. 95 and DJS, p. 83) have summarized the most recent progress in the study of flare precursors and provided new theoretical results on filament formation and eruption. They find ample evidence that energetic processes are already at work minutes before the onset of the flare impulsive phase. Many other papers also discussed the preflare filament activation and eruption (see section IV D) and some describe or model the untwisting and relaxation of the rising filament (Gaizauskas in JCB, p. 710 ; Kurokawa et al., 1987 ; see section IV D). As the flare onset is concerned, the reader is referred to Section VI.

B. ENERGY RELEASE

Henoux (in SIM, p. 227) and later on several other authors have reviewed the recent advances in the energy release in flares : Somov (in DJS, p. 177) emphasized the increasing evidence for reconnection in flares ; Kundu (in DJS, p. 207) stressed the evidence for simultaneous production of MeV electrons and protons; Machado and Moore (in CDS, P. 217) have discussed the association of the energy release with the magnetic environment ; Sturrock et al. (1984) examined observational evidence concerning energy release in flares and proposed different processes that may be operative in flares on different time scales, Simnett (1986) considers a model in which the bulk of energy released during the impulsive phase resides in non-thermal protons. The impulsive phase energy transport was discussed extensively by Canfield et al. (in KAW, chapter 3). The SMA also produced many papers on particle acceleration, but we leave all the discussion of this topic to Sections VI and VII.

Great attention has been paid to the problem of chromospheric evaporation manifested by asymmetrical profiles in flare spectra. Blue asymmetry in coronal lines found by Antonucci et al. (1985) and red asymmetry commonly seen in chromospheric lines (e.g., Tchimito and Kurokawa, 1984) was for the first time observed simultaneously in the same flare (Zarro et al., in DJS,p.155). While Canfield and Gunkler (1985) have found that evaporation is predominantly caused through conduction, Fisher et al. (1985) demonstrate that explosive evaporation is produced by electron beams which are turned on instantaneously. When the conductive flux out of the explosively evaporated plasma becomes comparable to the deposited electron heating flux, evaporation ceases to be explosive and is driven there after by conduction (Fisher, 1987). Theoretical red-shifted Hα profiles during explosive evaporation were computed by Canfield and Gayley (in DOK, p. 249). The impulsive phase gas dynamic has been thoroughly reviewed by Canfield (in NEI, p. 10 and DJS, p.167). A numerical simulation for dynamics of non-thermal electrons injected into coronal loop has been made by Takakura (1986) who concludes that the column density in the loop decides whether the resulting energy release is impulsive or gradual.

Close association of Hα kernels with individual peaks of energy release in hard X-rays was demonstrated by Loughhead et al. (1985), Kurokawa (in NEI, p. 51), and Wulser an Kampfer (in DOK, p. 301). The Hα-line asymmetry in flaring kernels was studied by Gaizauskas (in NEI, p. 37), First images of flare kernels in λ4686 line of He II were obtained by Zirin and Hirayama (1985). A kernel model in which chromosphere largely evaporates and Hα emission is radiated by a thin layer at the kernel basis has been proposed by de Jager (1985). Very small geometrical thickness of the emitting flare layer has been confirmed by Hirayama and Nakita (in NEI, p. 298). Gan and Fang (1987) analyzed chromospheric spectra of a major flare and time variations in the flaring chromosphere. Sylwester et al (1986) discussed the intensity ratio of the components of the Mg XII 8.42λ doublet in flare spectra obtained in Intercosmos 7.

Great progress has been made in observation and interpretation of the γ-ray flares, but this whole topic, though being one of the most fruitful fields of the SMA, clearly belongs to Section VI. For a recent review, the reader is referred to Hudson (1985) and Vlahos et al. (in KAW, Chapter 2). Another impulsive phase phenomenon intensely studied during SMA was the white-light (WL) emission from flares. Several papers on this topic were published in NEI (pp. 128 – 282) : deep atmospheric heating during flares, and WL flares in particular, was discussed by Emslie (p. 182), and Rust (p. 282) ; Gesztelyi et al. (p. 163) proposed a WL flare interpretation by means of current dissipation deep in the atmosphere. Lites et al. (p. 101) established chromospheric origin of a WL flare from measurements of He I and He II lines in its spectra. Neidig (pp. 142 and 152) demonstrated the possibility of a purely chromospheric origin for the WL emission and Avrett et al. (p. 216) computed chromospheric flare models with particular emphasis on the WL emission. Impulsive and gradual components in WL emission correlating with similar components in hard X-rays and microwaves have been demonstrated by Kane et al. (1985). In several papers, Aboudarham and Henoux (1986) have considered the energy deposit by electron bombardment during WL flares, the subsequent radiative heating of the photosphere and the energetic equilibrium of a flaring chromosphere. Hiei (in DJS, p. 227) presents continuous spectra of several WL flares and suggests that there exist three different types of the WL emission. A list of 45 WL flares and their characaeristics was published by Chen and Wang (in JCB, p. 735).

C. GRADUAL PHASE AND FLARES IN GENERAL

Several papers on Hα observations of major two-ribbon flares and spectra of post-flare loops were published in JCB. Motions in post-flare loops were studied by Gu et al. (in JCB, p. 669), Din et al. (p. 673), Li and Zhang (p. 685) and by Xu (in SAO p. 147). Tang (1985) demonstrated various flare topologies from Hα observations including multiple- ribbon flares. A study of growing loops at various temperatures (Hα to X-rays) has revealed that the (post-) flare loops must be shrinking and their density must be increasing during their cooling (Svestka et al., 1987). Indeed, Hanaoka et al. (1986) found electron densities of 8×10^9 and 10^{11} cm^{-3} in the hot and cold loops, respectively. According to Sakurai (1985), potential-field modeling gives a reasonable fitting in the case of two-ribbon flares ; but other kinds of flares, and

impulsive phase of flares, do not fit. Forbes and Malherbe (in NEI, p. 443) generalized
the Kopp and Pneuman model for coronal condensations in the two-ribbon flares. Kopp
and Poletto (in SAO, p. 103 and POL, p. 235) present a numerical technique which
enables one to reconstruct the reconnected magnetic field after two-ribbon flares and
computed the coronal electric field in the Kopp and Pneuman model (in NEI, p. 453). Aly
(1985) has studied the evolution of a force-free field toward an open field, driven by
shearing motions, and applies it to the theory of two-ribbon flares and giant post-
flare arches.

The energetics of flares was discussed by Somov (in SAO, p. 181) and by Wu et al.
(in KAW, Chapter 5). Various factors causing varieties in flares were discussed by
Svestka (in NEI, p. 332), flare observations in helium lines were summarized by Zirin
(in NEI, p. 78), and an atlas of Hα images flares of different types was published by
Ambastha and Bhatnagar (1985). Several extensive papers have presented complex
analyses of major flares, based on data obtained at various wavelenghts Tanaka and
Zirin, 1985 ; De Jager and Svestka, 1985 ; McCabe et al., 1986 ; Wang Jialong et al.,
1986). At the other end of the flare-size spectrum, Schadee and Gaizauskas (in SIM, p.
117) identified two extremely weak X-ray bursts with Hα miniflares, while the long-
lived weak X-ray enhancements have been tentatively ascribed to the collective result
of a manifold of "nanoflares" (Schadee, in DJS, p. 41).

A real progress has been made in our understanding of mass ejections from the Sun,
but once again, the reader is referred of the Section VI for reports on this topic. The
scientific highlights of STIP intervals during the SMY/SMA period have been summarized
by Dryer and Shea (in DJS, p. 343). Advances made in CME research during the SMA period
were reviewed by Hildner (in DJS, p. 297). The second volume of JCB contains many
papers concerning CMEs. The unique method of CME observations with zodiacal light
photometers of Helios spacraft has been described by Jackson (1985).

References

I :   Solar Maximun Analysis, Adv. Space Research 4, N°. 7 (SIM), Simon, P.A. (ed),
      1984.
II :   The Physics of Solar Flares, Adv. Space Research 6, No. 6 (DJS), de Jager and
      Svestka, Z. (eds.), 1986.
III : Energetic Phenomena on the Sun, NASA CP 2439 (KAW), Kundu, M., and Woodgate, B.
      (eds,), 1986.
IV : Coronal and Prominence Plasmas, NASA CP 2442 (POL), Poland, A.I. (ed.), 1986.
V : Rapid Fluctuations in Solar Flares, NASA CP 2449 (DOK), Dennis, B.R., Orwig,
      L.E., and Kiplinger, A.L (eds.), 1987.
VI : Proc. Kunming Workshop on Solar Physics and Interplanetary Travelling Phenomena,
      Beijing, (JCB), De Jager, C. and Chen Biao (eds), 1985.
VII :Solar Maximun Analysis, Proceedings SMA Worksop Irkutsk, VNU Science Press (SAO),
      Stepanov, V.E. and Obridko, V.N. (eds.), 1986.
VIII : Proc. NSO/SMM Symposium, AURA (NEI), Neidig, D.L. (ed.), 1986.
Aboudarham, J. and Henoux, J.C: 1986, Astron, Astrophys, 156, 73 and 168, 301, 1987,
      Astron, Astrophys. 174, 270.
Aly, J.J: 1985, Astron, Astrophys. 143, 19.
Ambastha, A and Bhatnagar, A.: 1985, Photographic Atlas of the Solar Chromosphere,
      1976 - 1984, Udaipur Solar Observatory.
Antonucci, E., Dennis, B.R., Gabriel, A.H., and Simnett, G.M.: 1985, Solar Phys, 96,
      129.
Athay, R.G., Jones, H.P., and Zirin, H.: 1986a, Astrophys. J. 303, 877.
Athay, R.G., Klimchuk, J.A., Jones, A.P., and Zirin, H.: 1986b, Astrophys. J. 303,
      884.
Canfield, R.C, and Gunkler, T.A.: 1985, Astrophys. J. 228, 353.
de Jager, C.: 1985, Solar Phys. 98, 267.
de Jager and Svestka, Z.: 1985, Solar Phys. 100, 435.
Fisher, G.H.: 1987, Astrophys. J. 317, 502.
Fisher, G.H., Canfield, R.C., and McClymont, A.N.: 1985, Astrophys. J. 289, 425 and
      434.
Forbes, T.G. and Priest, E.R.: 1984, Solar Phys. 94, 315.

Gan, W.Q. and Fang, C.: 1987, Solar Phys. 107, 311.

Hanaoka, Y., Kurokawa, H., ans Saito, S.: 1986, Solar Phys. 108, 151.

Hofmann, A., Rendtel, J., Aurass, H., and Kalman, B.: 1987, Solar Phys. 108, 151.

Hong, W.L., Pallavicini, R., and Cheng, C.C.: 1987, Solar Phys. 107, 271.

Hudson, H.S.: 1985, Solar Phys. 93, 105.

Ichimoto, K. and Kurokawa, H.: 1984, Solar Phys. 93, 105.

Jackson, B.V: 1985, Solar Phys. 100, 563.

Kane, S.R., Love, J.J., Neidig, D.P., and Cliver. E.V.: 1985, Astrophys. J. 290, L45.

Kundu, M.R. and 5 co-authors: 1985, Astrophys. J. Suppl. 57, 621.

Kurokawa, H., Hanaoka, Y., Shibata, K., and Uchida, Y.: 1987, Solar Phys. 108, 251.

Lin, Y. and Gaizauskas, V.: 1987, Solar Phys. 109, 81.

Loughhead, R., Blows, G., and Wang, J.: 1985, Publ. Astron. Soc. Japan 37, 619.

Machado, M.E.: 1985, Solar Phys. 99, 159.

Martens, P.C.H.: 1986, Solar Phys. 107, 95.

McCabe, M.K. and 4 co-authors: 1986, Solar Phys. 103, 399.

Perez-Enriquez, R.: 1985, Solar Phys. 97, 131.

Sakurai, T.: 1985, Solar Phys. 95, 311.

Seehafer, N.: 1985, Solar Phys. 96, 307.

Simnett, G.M.: 1986, Solar Phys. 106, 165.

Somov, B.V. and Titov, V.S.: 1985, Solar Phys. 95, 141, and 102, 79.

Sturrock, P.A., Kaufmann, P., Moore, R.L., and Smith, D.F.: 1984, Solar Phys. 94, 341.

Svestka, Z. and 5 co-authors : 1987, Solar Phys. 108, 237.

Sylwester, B. and 8 co-authors: 1986, Solar Phys. 103, 67.

Tachi, T., Steinolfson, R.S., and Van Hoven, G.: 1985, Solar Phys. 95, 119.

Takakura, T.: 1986, Solar Phys. 104, 363.

Tanaka, K. and Zirin, H.: 1985, Astrophys. J. 299, 1036.

Tang, F.: 1985, Solar Phys. 102, 131.

Wang, J. and 4 co-authors: 1986, Acta Astrophys. Sinica 6, 26.

Zirin, H. Hirayama, T.: 1985, Astrophys. J. 299, 536.

Zwingmann, W., Schindler, K., and Birn, J.: 1985, Solar Phys. 99, 133.

VI. Highlights of the Solar Activity Studies Made with Instruments Aboard Spacecraft

(S. R.Kane)

A. INTRODUCTION

Although the solar activity began to decrease rapidly after 1983, the analysis and interpretation of the observations made with instruments aboard the SMM, Hinotori, P78-1, ISEE-3 (ICE), PVO, Venera, and PROGNOZ spacecraft continued to produce important scientific results. The observational results inspired new theoretical studies or extensions of the earlier studies. Several symposia and workshops were organized for presentation and discussion of coordinated studies or studies in progress.

Symposia on the results from the Solar Maximum Data Analysis (SMA) were held in Gratz, Austria (Simon 1984 (I)) and Toulouse, France (de Jager and Svestka 1986 (II)). The SMM workshops helped to bring together many solar physicists from many countries to study specific aspects of solar activity. The participants included both ground-based observers and those associated with instruments aboard spacecraft. The proceedings of the following three SMM workshops have now been published : Energetic Phenomena on the Sun (Kundu and Woodgate 1986 (III)), Coronal and Prominence Plasmas (Poland 1986 (IV)) and Rapid Fluctuations in Solar Flares (Dennis et al. 1986 (V)). The proceedings of the National Solar Observatory/SMM symposium on the Lower Atmosphere of Solar Flares have also been published (Neidig 1986 (VI)). Results related to the Sun and the Heliosphere in Three Dimensions are available in the proceedings of the 19th ESLAB Symposium (Marsden 1985 (VII)). Studies related to solar flares and particle acceleration have been reviewed by de Jager (1986).

The largest number of spacecraft observational studies were related to solar flares. This is to be expected since the instruments aboard spacecraft such as SMM and

Hinotori were designed primarily to observe solar flares. The emphasis was on the observations of high energy phenomena (involving non-thermal particles and/or high temperature plasma) with high spatial, spectral and temporal resolution. Many new results were obtained through comparative studies involving spacecraft observations of the X-ray, gamma-ray and UV emission and the ground-based observations at optical and radio wavelengths. The highlights of some of the significant results are presented below.

B.   HINOTORI SATELLITE

A large number of results have been obtained with X-ray instruments on the Japanese satellite Hinotori. The impact of these observations on solar flare research has been reviewed by Tanaka (1987). Three types of flares (A, B, C) and five emission components have now been identified (Tsuneta 1984 ; Takakura et al. 1984). Evidence for chromospheric evaporation has also been found. The flare classification is based on the X-ray time profile, spectrum and morphology. The five components in the 5-40 keV X-ray emission are impulsive, gradual-hard, soft X-ray thermal, hot thermal, and quasi-thermal.

Thermal flare with mostly gradual rise and fall in the soft X-rays and low energy (<40 keV) hard X-rays, and intense emission in Fe XXV lines is called type A flare. The microwave emission is relatively weak in this flare. The hard X-ray source is compact (<5000 km) and the spectrum is very steep above 40 KeV. Very effective heating but insufficient particle acceleration in relatively large (300 G) magnetic fields seems to result in type A flares.

The impulsive flare giving rise to rapidly varying impulsive spikes and gradual hard X-ray emission is classified as type B flare. The impulsive spectrum is hard and the source extends from the low corona to the footpoints. The gradual hard X-ray emission, if present, has a comparatively softer spectrum and the source is relatively compact and is located at higher altitudes. The associated microwave emission is intense. The flare appears to be associated with an eruptive filament. It seems to occur in complex sheared low lying magnetic fields after separation from the eruptive filament. The acceleration of particles is very rapid. The gradual phase presumably results from progressive energy release in the higher magnetic loops of an arcade.

The gradual hard X-ray flare, called type C, is a longenduring (>30 min.) hard X-ray burst with broad peaks but no impulsive emission. The X-ray spectrum is very hard and hardens systematically with time. The source is located at high altitudes (>40000 km) in the corona. The microwave emission is usually very intense. Type C flare represents a very efficient, but slow particle acceleration in relatively weak (50 G) magnetic fields.

Intense emission of iron Kα lines at 1.936 angstrom and 1.940 angstrom associated with a hard X-ray burst was observed with the Bragg crystal spectrometer. The continuum X-ray spectrum in 1.5-12.5 keV range was also measured with high spectral resolution. The observations were compared with the Kα emission expected from the collisional impact of a beam of energetic electrons and the X-ray fluorescence. It was found that the observed Kα emission can be explained by the irradiation of the photosphere by the X-ray continuum with the observed power law spectrum extending down to the threshold energy (7 keV) for K-shell ionization (Tanaka et al. 1984).

High time resolution (62.5 ms) observations of the hard X-ray emission from a flare were compared with the simultaneous observations 22 GHz radio emissions made with 1 ms time resolution at the Itapetinga Radio Observatory in Brazil. It was found that throughout the duration of the burst all the radio burst structures were delayed by 0.2-0.9 s with respect to the hard X-ray flux structures. Different burst structures showed different delays indicating independent emission sources. Also the time structure of the degree of radio polarization preceded the structure in the total microwave flux by 0.1-0.5 s (Costa et al. 1984). Rapid variation of the turn-over frequency of the microwave spectrum has also been observed in the course of some solar bursts (Zodi et al. 1986).

C. SOLAR MAXIMUM MISSION

A large variety of results regarding the energy release and energetic particle transportation processes in solar flares have been obtained with instruments aboard the SMM satellite. Many of these results have been recently reviewed by Dennis (1985).

Some of the X-ray imaging observations made with the SMM HXIS instrument have been further examined and/or reinterpreted. SMM seems to have observed mostly type B and type C flares (Hinotori classification). There are, however, several differences. In some events, type B and type C characteristics occur on different X-ray peaks of the same flare. Impulsive (type B) and gradual (type C) phases probably occur in all energetic flares, the relative importance of the two phases varying from one flare to another (Dennis 1985).

Two classes of gamma ray/proton flares have been identified by Bai (1986): impulsive and gradual. They presumably differ in the location and mechanism for acceleration of protons. Whereas the impulsive flares have the "first phase and second-step" acceleration in low lying closed magnetic loops, the acceleration in the gradual flares is by shock waves high in the corona.

A different classification of flares has been suggested by Svestka (1986). It is based on two assumptions : (1) A flare is a short-lived release of energy resulting from a rearrangement of the magnetic structure, and (2) the mode of energy release is a reconnection of magnetic field lines. They lead to two classes of flares: dynamic flares and confined flares. The large variety of observed flares are presumably caused by the differences in the boundary conditions of the flare process.

Characteristics of pre-impulsive and impulsive phases of large hard X-ray flares have been studied by Klein et al. (1986). They have found that during the pre-impulsive phase the hard X-ray emission is restricted to energies <200 keV. On the other hand, during the impulsive phase the hard X-rays over a wide energy range are emitted simultaneously. Impulsive photon emission from 40 keV to 40 MeV has been found to be essentially simultaneous within 1 s. Thus there is evidence that both non-relativistic and relativistic particles are accelerated within 1 s during the impulsive phase (Kane et al. 1986).

An analysis of the long-term variations in the occurrence frequency of energetic flares during solar cycle 21 indicates a 154 day periodicity for gamma-ray flares (Rieger et al. 1984) and a comparable period for hard X-ray flares (Kiplinger et al. 1984). Similar periodicities have been found in the soft X-ray flares observed by GOES (Rieger et al. 1984) and solar microwave bursts (Bogart and Bai 1985). At present, there is no clear understanding of this 152-158 day periodicity. Bai and Sturrock (1987) have argued that this is not a local phenomena such as interaction and alignment of local hot spots rotating at different rates.

Variation of gamma-ray flare occurrence frequency with the heliocentric angle of the flare has been used to measure the directivity of the radiation source and hence the anisotropy of the emitting particles. The earlier result for >10 MeV photons has now been extended to lower energies. Vestrand et al. (1987) have found significant center-to-limb variation in 300 KeV-1 MeV flares, both in terms of flare occurrence and spectral index, indicating an anisotropic electron distribution such as downwardly directed Gaussian beam or a "pancake" distribution that peaks in directions parallel to the photosphere. These findings are consistent with the results of the earlier computational studies by Petrosian (1985) and Dermer and Ramaty (1986).

Simultaneous observations of UV and hard X-ray emission with high (<0.1 s) time resolution have provided a better understanding of the propagation of energetic electrons from the corona to the chromosphere. The peaks in the hard X-ray, O V line, and UV continuum emissions have been found to occur simultaneously within < 0.1 s (Orwig and Woodgate 1986). Since the UV continuum is expected to be produced at depths where density ne > 10^{14} cm^{-3} and hence 20-100 keV electrons are unable to penetrate to the UV continuum source in the normal solar atmosphere, a "hole boring" scenario has been suggested. In this process, a high intensity narrow beam of electrons moving along magnetic field lines causes rapid local heating which produces laterally moving shock waves. This, in turn, creates a low density region (hole) through which the remainder of the electrons can penetrate much deeper than normally possible. An alternative photo-ionization process has been proposed by Machado and Mauras (1986) to explain the UV continuum. In this process the EUV line emission, produced by the energy deposition by electrons in the transition region, "shines" on the deeper chromospheric layers and alters the Si ionization balance. This, in turn, is expected to cause enhanced UV continuum emission.

Simultaneous high time resolution ($\leqslant$ 0.01 s) observations of the hard X-ray and microwave emission from an intense and fast spike-like solar burst indicate that the emission consisted of short time scale structure superimposed on an underlying gradual emission (Kaufmann et al. 1984). The radio measurements were made at the Itapetinga Radio Observatory in Brazil and the Owens Valley Radio Observatory and Sagamore Hill Radio Observatory in the U.S.A. The repetition rate for the radio line structure is 30-60 ms at the peak or 16-20 pulses per sec. The observations have been explained in terms of elementary flare bursts from multiple sources associated with interacting magnetic loops. Quasi-quantization of injected energy is also possible. In some fast radio pulses coincident with hard X-rays the radio spectrum peaks at a very high frequency ($\geqslant$ 100 GHz) (Kaufmann et al. 1985, 1986). It has been suggested that in these pulses ultra-relativistic electrons produce microwave emission through synchrotron mechanism and the hard X-rays through inverse Compton scattering of the electrons on the synchrotron photons (Correia et al. 1986).

D.   ISEE-3 (ICE) AND PVO SPACECRAFT

Stereoscopic observations of hard X-ray and low energy gamma-ray emission from solar flares, made with instruments aboard the Third International Sun Earth Explorer (ISEE-3/ICE) and Pioneer Venus Orbiter (PVO) spacecraft, have been used to study the spatial structure of the hard X-ray source and low energy cut-off of the non-thermal electron spectrum. Significant impulsive hard X-ray and soft X-ray emission has been observed at coronal altitudes $\geqslant$ 1.5 x 10 5 km (Kane and Urbarz 1986). The impulsive X-ray spectrum has been found to extend down to 1 keV indicating a low energy cut-off at comparable energy for the non-thermal electron spectrum.

Properties of electron acceleration especially the density structure in the electron acceleration region, have been deduced through correlative studies of the hard X-ray bursts and decimetric type III and spike bursts. It appears that the acceleration occurs in a highly inhomogeneous region with large density variations (Benz and Kane 1986). In some flares, the electron acceleration is delayed and appears to occur in post flare loops (Kai et al. 1986 ; Cliver et al. 1986) (see also section VII).

Comparison of the hard X-ray and white light observations suggest that the hard X-ray source consists of several kernels, few arc sec in size. Energy disposition by >25 keV electrons can explain the white light observations provided the density structure in the flare is not homogeneous (Kane et al. 1985).

Several results have been obtained regarding the composition of the interplanetary solar flare particles. The heavy-ion enrichment is found to be the same (within a factor of 2) for all flares. However, the degree of heavy-ion enrichment is uncorrelated with the ^{3}He enrichment indicating an acceleration mechanism in which ^{3}He enrichment process is not responsible for the heavy-ion enrichment (Mason et al. 1986).

E.   P78-1 SATELLITE

High resolution observations of 5.5-12 angstrom X-ray spectrum under a variety of solar conditions, such as flare onset, flares and non-flaring active regions, have been made with instruments aboard the US Air Force satellite P78-1. The resulting compilation of spectral lines includes many lines useful for plasma diagnostics, such as the high temperature lines of Fe XXII-XXIV and density sensitive lines Mg XI and Si XIII (McKenzie et al. 1985 ; Keenan et al. 1985). Comparison with theoretical calculations shows partial agreement.

F.   VENERA AND PROGNOZ SPACECRAFT

Solar bursts of hard X-rays >100 keV have been observed simultaneously with high time resolution (2 ms) instruments aboard the Soviet spacecraft Venera 11,12,13,14, and Prognoz. A comparison of the observed X-ray flux modulation period of <1.6 s found in some solar flares with that expected from a MHD model of compact magnetic loops shows good agreement (Desai et al. 1986).

Soft and hard X-ray emission from the sun has been recorded with instruments aboard Prognoz 5-10. The data have been published in a series of publications, the latest publication being related to the data from Prognoz 10 satellite (Valnicek et al. 1987). Problems associated with the calibration of different X-ray detectors have been discussed by Valnicek et al. (1984).

References

I : Solar Maximum Analysis, Adv. in Space Res., Vol 4, N° 7, Simon, P.A. (ed), 1984.

II : The Physics of Solar Flares, Adv. in Space Res., Vol 6, N° 6, de Jager, C. and Svestka, Z. (eds) 1987.

III : Energetic Phenomena on the Sun, NASA Conf. Publ. 2439, Kundu, M., and Woodgate, B. (eds), 1986.

IV : Coronal and Prominence Plasmas, NASA Conf. Publ. 2442, Poland, A.L. (ed), 1986.

V : Rapid Fluctuations in Solar Flares, NASA Conf. Publ. 2449, Dennis, B.R., Orwig, L.E., and Kiplinger, A.L. (eds), 1986.

VI : The Lower Atmosphere of Solar Flares, Publ. of National Solar Observatory/Sacramento Peak, Sunspot, New Mexico, Neidig, D.F. (ed), 1986.

VII : The Sun and Heliosphere in Three Dimensions, D. Reidel Publ. Co., Marsden, R.G. (ed), 1985.

Bai, T.: 1986, Astrophys. J., 308, 912.

Bai, T., and Sturrock, P.A.: 1987, Preprint, Center for Space Sciences and Astrophys., Stanford University, Nature (in press).

Benz, A.O., and Kane, S.R.: 1986, Solar Phys., 104, 179.

Bogart, R.S., and Bai, T.: 1985, Astrophys. J. Lett., 299, L51.

Cliver, E.W., Dennis, B.R., Kiplinger, A.L., Kane, S.R., Neidig, D.F., Sheeley, Jr. N.R., and Koomen, M.J.: 1986, Astrophys. J., 305, 920.

Correia, E., Kaufmann, P. Costa J.E.R., Zodi Vaz, A.M., and Dennis, B.R.: 1986, Rapid Fluctuations in Solar Flares, NASA Conference 2449, Dennis B.R., Orwig, E. and Kiplinger (eds).

Costa, J.E.R., Kaufmann, P., and Takakura, T.: 1984, Solar Phys., 94, 369.

Dennis, B.R.: 1985, Solar Phys., 100, 435.

Dermer, C.D., and Ramaty, R.: 1986, Astrophys. J., 301, 962.

de Jager, C.: 1986, Space Sci. Rev., 44, 43.

Desai, U.D., Kouveliotou, C., Barat, C., Hurley, K., Niel., M., Talon, R., and Vedrenne, G.: 1986, Rapid Fluctuations in Solar Flares, eds., B.R. Dennis, L E. Orwig, and A.L. Kiplinger, NASA Conf. Publ. 2449, p. 43.

Kai, K., Nakajima, H., Kosugi, T., Stewart, R.T., Nelson, G.J., and Kane, S.R.: 1986, Solar Phys., 105, 383.

Kane, S R., Evenson, P., and Meyer, P.: 1985, Astrophys. J. Lett., 299, L07.

Kane, S.R., Chupp, E.L., Forrest, D.J., Share, G.H., and Rieger, E.: 1986, Astrophys. J. Lett., 300, L95.

Kane, S.R., and Urbarz, H.W.: 1986, Proc. STIP Symposium on Retrospective Analyses, (eds) M.A. Shea and D.F. Smart, Book Crafters, Inc.

Kaufmann, P., Correia, E., Costa, J.E.R., Dennis, B.R., Hurford, G.J., and Brown, J.C.: 1984, Solar Phys., 91, 356.

Kaufmann, P., Correia, E., Costa, J.E.R., Zodi Vaz, A.M.: 1986, Solar Flares and Coronal Physics using P/OF as a Research Tool, E. Tandberg-Hanssen and H.S., Hudson (eds), NASA Conf. Publ. 2421, p. 208.

Kaufmann, P., Correia, E., Costa, J.E.R., Zodi Vaz, A.M., and Dennis, B.R.: 1985, Nature, 313, 380.

Keenan, F.P., Kingston, A.E., and McKenzie, D.L.: 1985, Astrophys. J., 291, 855.

Kiplinger, A.L., Dennis, B.R., Orwig, L.E.: 1984, Bull. Am. Astron. Soc. 16, 4.

Klein, K.-L., Pick, M., and Magun, A.: 1986, V, p. 79.

Machado, M.E., and Mauras, P.: 1986, V, p. 271.

Mason, G.M., Reames, D.V., Klecker, B., Hovestadt, D., and Von Rosenvinge, T. T.: 1986, Astrophys. J., 303, 849.

McKenzie, D.L., Landecker, P.B., Feldman, U., and Doschek, G.A.: 1985, Astrophys. J., 289, 849.

Orwig, L., Woodgate, B., and Nakada, M.P. : 1985, Bull. Am. Astron. Soc.

Petrosian, V.: 1985, Astrophys. J., 299, 987.

Rieger, E., Share G.H., Forrest, D.J., Kanbach, G., Keppin, C., Chupp, E.L.: 1984, Nature, 312, 5995.

Svestka, Z.: 1986, The lower Atmosphere of Solar Flares, D. F. Neidig (ed), National Solar Observatory/Sacramento Peak, Sunspot, New Mexico , P. 332.

Takukura, T., Tanaka, K., and Hiei, E.: 1984, Adv. Space Res., 4, 143.

Tanaka, K.: 1987, Publ. Astron. Soc Japan, 39, 1.

Tanaka, K., Watanabe, T., and Nitta, N.: 1984, Astrophys. J., 282, 703.

Tsuneta, S.: 1984, Proceedings of Japan-France Seminar on Active Phenomena in the Outer Atmosphere of the Sun and Stars, Pecker, J.C. and Uchida Y. (eds).

Valnicek, B.., Farnik, F., Sylwester, B., and Sylwester, J.: 1984, Adv. Space Res., 4, 121

Valnicek, B., Farnik, F., Sudova, J., Kormarek, R., Likin, O., and Pisarenko, N.: 1987, Pognoz Data Part IV, Astron. Inst. Ozech. Acad. Sc, Publ. No. 63.

Vestrand, W.T., Forrest, D.J., chupp, E.L., Rieger, E., and Share, G.H.: 1987, Univ. of New Hampshire preprint, Astrophys. J. (in press).

Zodi, Vaz, A.M., Kaufmann, P., Correia E., Costa, J.E.R., Clived, E.W., Takakura, T.: 1986, V, p. 171.

VII. RESULTS FROM RADIO OBSERVATIONS
(C. Alissandrakis)

This section does not mention some results reported in sections V and VI. Other papers can be found in proceedings of CESRA Workshops (Benz 1985 (I), Pick and Trottet 1986 (II)).

A. INSTRUMENTATION

During the last three years important developments in instrumentation have taken place. A new antenna has been added to the Nobeyama solar radio interferometer increasing its one-dimensional resolution to 25" at 17 GHz (Nakajima et al, 1984). A frequency-agile receiver system has been added to the Owens Valley 2-element interferometer, operating at up to 86 frequencies in the 1-18 GHz range (Hurford et al, 1984), while a new digital, high time resolution spectrometer is in operation at Bern. The VLA and the RATAN-600 have continued to give high spatial resolution data at cm wavelengths, while the E-W branch of the Siberian Solar Radio Telescope is fully operational, giving one-dimensional scans at 5.2 cm with a resolution up to 17" (Smolkov et al, 1986). In the long wavelength range the Nançay Radioheliograph has been used as an earth rotation aperture synthesis instrument (Allissandrakis et al, 1985), while the Clark Lake Radioheliograph is operating regularly at decametric wavelengths providing instantaneous maps. A multi-frequency system has been installed in the N-S branch of the Nançay Radioheliograph. In spite of these developments the community has suffered greatly from the loss of the Culgoora Radioheliograph.

B. QUIET SUN

Just after the solar maximum, few publications refer to the quiet sun, Kosugi et al (1986) detected polar cap emission of 3-7% at 36 GHz apparently associated with coronal holes. Alissandrakis et al (1985) presented maps of the corona at 169 MHz. The maps show coronal holes with $T_b \sim 10^5$ K and apparent height of .03 R_O ; the distribution of brightness at 1.15 R_O shows a gross similarity with the coronal green line. Bright features are sources of noise storm continua, while weaker emission regions are associated with neutral lines of the magnetic field. Limb synoptic charts (Lantos and Alissandrakis, 1986) are very similar to K-coronameter charts and show very well the base of the heliosheet. Kundu et al (1987a) presented coronal maps at 30.9, 50 and 73.8 MHz for one rotation. A coronal hole shows an eastward displacement at the lower frequencies. Elongated features at the limb correspond well to white light streamers. Radio synoptic charts show overall similarities with K-coronameter charts. The solar diameter at decameter wavelengths was measured by Gergely et al (1985). Observations of bright points with a lifetime of a few minutes have been reported by Habbal et al (1986) at 20 cm and Fu et al (1987) at 6 cm. Benz and Furst (1987) found little correlation in simultaneous observations of solar fluctuations at 4.75 GHz from Arecibo and Effelsberg. New observations of filaments at 6 and 20 cm with the VLA have been reported by Kundu et al (1986a).

C. ACTIVE REGIONS

The emission of $\mu-\lambda$ sunspot associated sources in terms of the gyroresonance process is well understood. Alissandrakis and Kundu (1984) observed a simple, sunspot associated source over 6 days ; they identified the region where the magnetic field is parallel to the line of sight and they mapped the B vector. Kundu and Alissandrakis (1984) detected islands of o-mode polarisation inside sunspot associated sources and interpreted them in terms of hotter than average regions at the height of the second harmonic. A detailed study of the inversion of circular polarisation gave a height of 0.16–0.19 R_O and a magnetic field of 10–20 G for the depolarisation region. The polarisation inversion was also studied by Gelfreikh et al (1987) who found similar values for the depolarisation region. Kruger et al (1986) computed model parameters from spectral observations of 36 sources ; they found values of $(7.4 \pm 2.3) \times 10^8$ cm for the scale height of the magnetic field and a value of B ~ 1750 G at a height of 2000 km. Model computations for sunspot-associated sources were given by Kruger at al (1985), while Siarkowsky (1984) pointed out that the steep temperature gradient in the transition region results in a non-Maxwellian distribution which may increase the observed brightness of a sunspot-associated source. Computations of gyroresonance emission from a hot coronal loop were made by Holman and Kundu (1985). Strong sources associated with neutral lines were studied by Kundu and Alissandrakis (1984) and Akhmedov et al (1986a, 1986b) ; Kundu and Alissandrakis pointed out their association with soft X-ray loops and arch filament systems and their persistence for at least 6 days ; Akhmedov et al pointed out their very steep spectra which cannot be interpreted in terms of traditional models of thermal emission. The emission mechanism of this type of sources is still uncertain.

Multi-λ observations of active regions were reported by Schevgaonkar and Kundu (1984, 1985a, 1985b) and Gary and Hurford (1987). Using the VLA and the Owens Valley frequency agile interferometer in the range of 1.4 to 8 GHz during an eclipse, the latter found a transition from sunspot associated sources at short wavelengths to a source associated with a large loop at long wavelengths. Maps of active regions at closely spaced frequencies were obtained by Schmahl et al (1984) at 3 frequencies near 6 cm and by Willson (1985) at 10 frequencies near 20 cm ; Schmahl et al found a source with a spectral slope of 6 and they proposed interpretations in terms of neutral current sheets and cyclotron lines ; Willson found a spectral peak with a bandwidth of ~100 MHz which he interpreted as a cyclotron line due to a hot region located at the 4th harmonic of the gyrofrequency. Willson and Lang (1986) observed compact, variable sources inside active regions at 2 cm.

D. MICROWAVE BURSTS

Observations with the VLA have provided additional evidence for pre-burst changes in the active region (Willson, 1984 ; Melozzi et al, 1985 ; Kundu and Shevgaonkar, 1985) ; these include heating of the burst region, sudden change of polarisation, change of the zero polarisation line and appearance of new sources (Kundu, 1986). Nakajima et al (1985) found sympathetic $\mu-\lambda$ and X-ray bursts separated by up to 10^6 km, implying exciter speeds of 10^4–10^5 km/sec. Kundu et al (1987b) presented the time evolution of a complex unipolar burst at 6 cm. Multi-frequency observations as well as simultaneous imaging observations in the microwave and X-ray range have given the opportunity to compare the structure of bursts at different frequencies. Melozzi et al (1985) found the 6 cm emission near the legs of flaring loops, while the 20 cm emission was complex and displaced with respect to the 6 cm sources. Shevgoankar and Kundu (1985b) found that in some impulsive peaks the 2 cm sources occured at the same location as the maximum of the 6 cm source, while in others they were located at the footpoints of the 6 cm loop. Dulk et al (1986a) observed an event near the limb, in which the 6 cm source spanned the Hα emission patches, while the 2 cm source was located at the edge of the 6 cm source. Takakura et al, (1985) observed a burst, apparently associated with an active region behind the limb, at 6 cm and 20–30 keV ; the centroid of both sources was above the limb but the 6 cm source located 3×10^4 km

further out than the X-ray source, Schmahl et al, (1985) observed a limb event with the
6 cm emission in the form of a loop with peaks at the footpoints and the 20-30 keV
emission close to one of the footpoints. The ensemble of these observations does not
give a very clear picture, however it appears that quite often the long wavelength $\mu-\lambda$
emission comes from the entire flaring loop, while the short wavelength $\mu-\lambda$ emission
and the hard X-ray (HXR) emission comes from the footpoints. This is in agreement with
the models of Alissandrakis ans Preka-Papadema (1984) and Klein and Trottet (1984) ;
the former also pointed out the importance of propagation effects in the observed
sense of circular polarisation.

By comparing the rate of increase of 17 GHz and HXR bursts, Nitta and Kosugi (1986)
concluded that the former are due to electrons with energy less than a few hundred keV.
Mac Kinnon et al, (1986) reported a $\mu-\lambda$ burst with two peaks, of which only one showed
HXR emission (see also section VI, C). In a study of gradual $\mu-\lambda$ and soft X-ray (SXR)-
events Schmahl et al (1986) found that the thermal plasma parameters deduced from SXR
cannot account for the $\mu-\lambda$ emission and proposed a non-isothermal model. The question
of the number of electrons required for the $\mu-\lambda$ and HXR emission was studied by Gary
and Tang (1985), Gary (1985), Klein et al, (1986) and Kai (1986) who concluded that the
numbers can be reconciled, while Schmahl et al (1985) found differences of a factor of
10^3. The interpretation of the impulsive phase in terms of thermal emission was
studied by Batchelor et al, (1985) and Wiehl et al, (1985a).

E.   SHORT SCALE TIME VARIATIONS IN BURST SOURCES

There is a growing interest in fast structures from the mm to the m range. Loran et
al, (1985) presented an interpretation of the ripple structure in $\mu-\lambda$ events in terms
of the elementary burst concept. Kaufmann et al (1985) analysed observations of 17
simple microbursts at 22 GHz ; their peak fluxes were 0.9 to 8.8 sfu, their time scales
$0.05 < t \ll 1s$ and $0.5 < t < 2s$ and they were 7-67% polarised.

Wiehl et al, (1985b) studied 664 decimetric pulsation events, classified them in
short ($<1s$) and long (5-300s) and interpreted them in terms of transient and trapped
electrons respectively. Benz (1985) studied events rich in spikes with relative
bandwidth down to 1.5 % ; he estimated an upper limit to their size of 200 km.
Aschwanden and Benz (1986) found that pulsations exhibit three times higher drift
rates at 650 MHz than type III's. Stahli and Magun (1986) found that msec spikes
observed in the long $\mu-\lambda$ range extend up to 5.2 MHz. They occur during $\mu-\lambda$ continuum
emission, mostly in the rise and maximum phase ; they are polarised from −100% to 100%,
with no correlation with the continuum polarisation. In the 3570-3370 MHz range their
bandwidth can be < 0.5 MHz. Jin et al (1986) analysed 250 spike events at 2.84 GHz ;
they have shorter duration and higher flux than previously known and many were not
resolved at 1 ms. Enome and Orwig (1986) found that events with a very spiky structure
had relatively high associated HXR emission (see also section VI, D). Stahli and Benz
(1987) observed drifting structures in the 3100-5205 MHz range with a duration of 25-
200 ms, bandwidth > 150 MHz, slightly circularly polarised ; they had reverse
frequency drifts with an average of 8100±4300 MHz/s and were interpreted as signatures
of electron beams.

Electron cyclotron masering (ECM) is the favorite candidate for the
interpretation of high brightness temperature fast structures. The growth rates have
been studied by Sharma and Vlahos (1984), Winglee (1985a), Li (1986) and Zhao and Shi
(1986). The effects of finite plasma temperature and superthermal tails have been
studied by Sharma and Vlahos (1984), Winglee (1985b) and Vlahos and Sharma (1985). The
non-linear evolution of the instability in a ring distribution was studied by Sprangle
and Vlahos (1986), while White et al, (1986a) studied the formation of a loss cone in a
free steaming approximation, Although ECM can produce the required T_b, its applicability is still uncertain due to the sensitivity of the growth rates to ω_p/Ω_e and the
problems of possible gyroresonance absorption.

At longer wavelengths Bakunin and Chernov (1985) have studied broad-band spike
bursts in the 175-235 MHz range ; they are rare in noise storms, while in type IV's
series of non-periodic spikes predominate. Elgaroy (1986) studied 1552 pulses in a
type IV event and found an average period of 0.09s at 300 and 500 MHz. The polarisation
of type IV associated spikes varies widely, but it is almost constant for the same
group (Nonino et al, 1986), while their decay phase is similar to that of type III's

(Abrami et al, 1986). Zaitsev et al, (1984) and Rozenraukh and Stepanov (1986) proposed an interpretation in terms of MHD pulsations for type IV pulse trains, while Zaitsev and Zlotnik (1986) suggested that ms time structure at m–λ is similar to Jovian s–bursts.

F. METRIC BURST EMISSION

Stewart (1985) found that storm sources occur in large unipolar cells, which after the solar maximum form a longitudinal sector pattern. Stewart etal (1986) found a close association between changes of the large scale photospheric magnetic field and the onset and cessation of noise storms ; the new flux diffuses slowly in the corona, producing storms 1–2 days later. Concerning the theoretical interpretation of the emission, Malara et al (1984) proposed that the non–linear coupling of ion acoustic waves and high frequency waves might explain the o–mode polarisation. Levin and Rapoport (1985) considered the streaming of diffusely distributed super–thermal electrons along inhomogeneous magnetic field. Wentzel (1985) interpreted the type I continuum in terms of a non relativistic electron gap distribution. Wentzel et al (1986) suggested that all type I bursts are emitted fully polarised and depolarisation occurs due to large angle scattering. Wentzel (1986) suggested that type I continuum may arise from the combination of 2 electrostatic waves, nearly normal to the magnetic field.

Robinson (1985) measured type II velocities from dynamic spectra and found a distribution which peaks at 500–700 km/s. Robinson and Stewart (1985) found that type II's can occur at the leading edge of a coronal mass ejection (CME), well behind the leading edge, or in the absence of a CME ; the first kind is probably generated by the CME, the last by a blast wave. Gergely (1986) made a statistical analysis of CME's and moving type IV's and found that 1/3–1/2 of the latter are associated with CME's but only 5% of CME's show the inverse association; the mean speed of the type IV is smaller than that of the CME. Lengyel–Frey et al (1985) gave observations of interplanetary (IP) type II's ; the fundamental was more intense and variable, had a larger size and a smaller bandwidth. In a model of type II emission, Krasnosel'skikh et al (1985) suggest that nonmaxwellian electron tails can form from low frequency waves driven by unstable ion beams produced in a collisionless shock wave. In a different approach, Wu et al (1986) considered the formation of a hollow beam electron distribution in the shock wave, which attains larger pitch angles as it propagates to low altitudes ; the result is a maser instability which amplifies unpolarised or weakly polarised radiation. Lakhina and Buti (1985) suggested that type IV radiation is emitted from the interaction of slow whistler solitons with upper hybrid waves in a region with a transient density increase. Cane and Stone (1984) investigated the relationship between interplanetary and metric type II bursts, interplanetary shocks and energetic particles.

Two–dimensional maps of microbursts (weak type III) at decametric wavelengths have been given by Kundu et al (1986b), White et al (1986b) and Gopalswamy et al (1987) ; they are short duration events (2–10s) with a peak T_b sometimes less than 10^6K and a size of ~ 15' at 50 MHz. Their brightness is too low to be interpreted in terms of current type III theories ; their size increases at lower frequencies.

Poquerusse et al (1984) studied the decay of U bursts and concluded that the process is beam dependent. Leblanc and Hoyos (1985) found a close relation between type U groups and type II's. Eremin (1985) presented examples of radio echos observed during a type III storm. Pick and Ji (1987) studied the relative position of spatial components of type III's at 169 MHz and concluded that the angular extent of the electron injection region was ~35°. Mein and Avignon (1985) and Chiuderi–Drago et al (1986) found an association of type III's with ascending Hα features. Raoult et al (1985) studied 55 type III groups associated with HXR bursts ; they found that those accompanied with type V continuum longer than 30s were associated with much more intense HXR bursts, with spectra extending to ≥100 keV. They also found that comparative hard x–ray and metric radio observations indicate interaction of magnetic structures and subsequent magnetic reconnection in the vicinity of the electron acceleration/ injection region giving rise to the impulsive phase of a flare. In a study of IP type III's Dulk et al (1984) found that the emission begins at the same time as the Langmuir waves and it is at the fundamental. Steinberg et al (1984) concluded that propagation effects, group delays, ducting and scattering are important in IP type III's. Steinberg et al (1985) attributed the large size of IP type III's to scattering ; scattering makes them visible even if they originate behind the sun (Dulk et al, 1985). Dulk et al, (1986b) traced the lines of force of the IP magnetic field using type III trajectories. Reiner and Stone (1986) reconstructed type III trajectories using the frequency drift, in addition to the position of the source. Fokker (1984) examined the deviation of electron stream trajectories from a smooth path due to irregularities in the magnetic field lines and pitch angle variations. Lin et al (1986) gave evidence of non-linear wave-wave interactions in IP type III's ; they observed spiky Langmuir waves to be driven by electron beams associated to type III's ; low frequency, apparently ion-acoustic wave were often observed, coincident in time with the more intense Langmuir spikes. Grognard (1984) proposed a simple test of the quasilinear theory and estimated the initial temperature of the associated distributions. Levin (1985) suggested that plasma turbulence behind the leading edge of an electron stream leads to quasilinear relaxation ; Melrose et al (1986) presented a model of IP type III emission, in which Langmuir wave growth is suppressed by refraction in density irregularities except near density minima where clumps of waves form. A two component model for the generation of type III fundamental was proposed by Eremin and Zaitsev (1985).

References

I : Radio continua during Solar Flares, Solar Phys. 104, 1, Benz A. (ed), 1986.
II : Particle Acceleration and Trapping in Solar Flares, Solar Phys. (under press), Pick M. and Trottet, G. (eds), 1987.
Abrami, A., Messerotti, M., Zlobec, P.: 1986, Solar Phys., 104, 51.
Akhmedov, Sh.B., Borovik, V.N., Gelgreikh, G.B., Bogod, V.M., Korzhavin, A.N., Petrov, Z.E., Dikij, V.N., Lang, K.R. and Willson, R.F.: 1986a, Astrophys. J., 301, 460.
Akhmedov, Sh.B., Bogod, V.M., Gelfreikh, G.B., Hildebrandt, J., Kruger, A.: 1986b, Contr. Skaln. Pleso, 15, 339.
Alissandrakis, C.E., Kundu, M.R.: 1984, Astron, Astrophys., 179, 271.
Alissandrakis, C.E., Preka-Papadema, P.: 1984, Astron. Astrophys., 139, 507.
Alissandrakis, C.E., Lantos, P., Nicolaidis, E.:1985, Solar Phys., 97, 267.
Aschwanden, M.J., Benz, A.O.: 1986, Astron. Astrophys., 158, 102.
Bakunin, L.M., Chernov, G.P.: 1985, Astron. Zh., 62, 973.
Batchelor, D.A., Crannell, C.J., Wiehl, H.J., Magun, A.: 1985, Astrophys. J., 298, 258.
Benz, A.O.: 1985, Solar Phys., 96, 357.
Benz, A.O., Furst, E.: 1987, Astron. Astrophys., 175, 282.
Cane, H.V., Stone, R.G.: 1984, Astrophys. J., 282, 339.
Chiuderi-Drago, F., Mein, N., Pick, M.: 1986, Solar Phys., 103, 235.
Dulk, G.A., Steinberg, J.L., Hoang, S.: 1984, Astron, Astrophys., 141, 30.
Dulk, G.A., Steinberg, J.L., Lecacheux, A., Hoang, S., MacDowall, R.J.: 1985, Astron, Astrophys., 150, L28.
Dulk, G.A., Bastian, T.S., Kane, S.R.: 1986a, Astrophys. J., 300, 438.
Dulk, G.A., Steinberg, J.L. Hoang, S., Lecacheux, A.: 1986b in R. G. Marsden (ed) " The Sun and the Heliosphere in Three Dimensions". D. Reidel, P. 229.
Elgaroy,O.: 1986, Solar Phys., 104, 43.
Enome, S., Orwig, L.E.: 1986, Nature, 321, 421.
Eremin, A.B: 1985, Pisma Astr. Zh., 12, 338.
Eremin, A.B., Zaitsev, V.V.: 1985, Solar Phys., 99, 110.
Fokker, A.D: 1984, Solar Phys., 93, 379.
Fu, Q., Kundu, M.R., Schmahl, E.J.: 1987, Solar Phys., 108, 99.
Gary, D.E.: 1985, Astrophys. J., 297, 799.
Gary, D.E., Tang, F.: 1985, Astrophys. J., 288, 385.
Gary, D.E., Hurford, G.J.: 1987, Astrophys. J., 317, 522.
Gelfreikh, G.B., Peterova, N.G., Ryabov, B.L.: 1987, Solar Phys., 108, 89.
Gergely, T.E., Gross, B. P., Kundu, M. R.: 1985, Solar Phys.. 99, 323.

Gergely, T.E.: 1986, Solar Phys., 104, 175.
Gopalswamy, N., Kundu, M.R., Szabo, A.: 1987, Solar Phys., 108, 333.
Grognard, R.J.-M.: 1984, Solar Phys., 94, 165.
Habbal, S.R., Ronan, R.S., Withbroe, G.L., Shevgaonkar, R.K., Kundu, M.R.: 1986, Astrophys, J., 307, 740.
Holman, G.D., Kundu, M.R.: 1985 Astrophys. J., 292, 291.
Hurford, G.J., Read, R.B., Zirin, H.: 1984, Solar Phys., 94, 413.
Jin, S.Z., Zhao, R.Y., Fu, O.J.: 1986, Solar Phys., 104, 391.
Kai, K.: 1986, Solar Phys., 104, 235.
Klein, K.-L., Trottet, G.: 1984, Astron. Astrophys., 141, 67.
Klein, K.-L., Trottet, G., Magun, A.: 1986, Solar Phys., 104, 243.
    1985a, Nature, 313, 380.
Kaufmann, P., Correia, E., Costa, J.E.R., Sawant, H.S., Zodi Vaz, A.M.: 1985, Solar Phys., 95, 185.
Kosugi, T., Ishiguro, M., Shibasaki, K.: 1986, Publ. Astron. Soc. Japan. 38, 1.
Krasnosel'skikh, V.V., Kruchina, E.N., Thejappa, G., Volokitin, A.S.: 1985, Astron. Astrophys.. 149, 323.
Kruger, A., Hildebrandt, J., Furstenberg, F.: 1985, Astron. Astrophys.. 143, 72.
Kruger, A., Hildebrandt, J., Bogod, V.M., Korzhavin, A.N., Akhmedov, Sh.B., Gelfreikh, G. B.: 1986, Solar Phys., 105, 111.
Kundu, M.R.: 1986, Adv. Space Res., 6, 93.
Kundu, M.R., Alissandrakis, C.E.: 1984, Solar Phys.. 94, 249.
Kundu, M.R., Shevgaonkar, R.K.: 1985, Astrophys. J., 291, 860.
Kundu, M.R., Melozi, M., Shevgaonkar, R.K.: 1986a, Astron. Astrophys., 167, 166.
Kundu, M.R., Gergely, T.E., Szabo, A., Loicono, R., White, S.M.: 1986b, Astrophys. J., 308, 436
Kundu, M.R., Gergely, T.E., Schmahl, E.J., Szabo, A., Loiacono, R., Wang, Z.: 1987a, Solar Phys., 108, 113.
Kundu, M.R., McConnel, D., White, S.M., Shevgaonkar, R.K.: 1987b, Astron. Astrophys., 176, 131.
Lakhina, G.S., Buti, B.: 1985, Solar Phys., 99, 277.
Lantos, P., Alissandrakis, C.E.: 1986, NASA CP-2442, 257.
Leblanc, Y., Hoyos, M.: 1985, Astron. Astrophys., 143, 365.
Lengyel-Frey, D., Stone, R.G., Bougeret, J.L.: 1985, Astron. Astrophys., 151, 215.
Levin, B.N.: 1985, Astron. Astrophys., 143, 54.
Levin, B.N., Rapoport, V. O.: 1985, Solar Phys., 96, 371.
Li, H.W.: 1986, Solar Phys., 104, 131.
Lin, R.P., Levedahl, W.K., Lotko, W., Gurnett, D.A., Scarf, F.L.: 1986, Astrophys. J., 308, 954.
Loran, J.M., Brown, J.C., Correia, E., Kaufmann, P.: 1985, Solar Phys., 97, 363.
MacKinnon, A.L., Costa, J.E.R., Kaufmann, P., Dennis, B.R.: 1986, Solar Phys., 104, 191.
Malara, F., Veltri, P., Zimbardo, G.: 1986, Astron. Astrophys., 139, 71.
Mein, M., Avignon, Y.: Solar Phys., 95, 331.
Melozzi, M., Kundu, M.R., Shevgaonkar, R.K.:1985, Solar Phys., 97, 345.
Melrose. D.B., Dulk, G.A., Chairns, I. H.: 1986, Astron. Astrophys., 163, 229.
Nakajima, H., Dennis, B.R., Hoyng, P., Nelson, G., Kosugi, T., Kai, K.: 1985, Astrophys. J., 288, 806.
Nakajima, H., Sekiguchi, H., Kosugi, T., Shiomi, Y., Sawa, M., Kawashima, S., Kai, K.: 1984, Publ. Astron. Soc. Japan. 36, 383.
Nitta, N., Kosugi, T.: 1986, Solar Phys., 105, 73.
Nonino, N., Abrami., A., Comari, M., Messerotti, M., Zlobec, P.: 1986, Solar Phys., 104, 111.
Poquerusse, M., Bougeret, J.L., Caroubalos, C.: 1984, Astron. Astrophys., 136, 10
Pick, M., Ji, S.C.: 1987, Solar Phys., 107, 159.
Raoult, A., Pick, M., Dennis, B.R., Kane S.R.: 1985, Astrophys. J. 209, 1027.
Reiner, M.J., Stone, R.G.: 1986, Solar Phys., 106, 397.
Robinson, R.D.: 1985, Solar Phys., 95, 343.
Robinson, R.D., Stewart, R.T.: 1985, Solar Phys., 97, 145.
Rozenraukh, Yu, M., Stepanov, A.V.: 1986, Contr. Skaln. Pleso., 15, 409.

Schmahl, E.J., Shevgaonkar, R.K., Kundu, M.R., Mc Connell, D.: 1984, Solar   Phys.,
    93, 305.
Schmahl, E.J., Kundu, M.R., Dennis, B.R.: 1985, Astrophys. J., 299, 1017.
Schmahl, E.J., Kundu, M.R., Erskine, F.T.: 1986, Solar Phys., 105, 87.
Sharma, R.R., Vlahos L.: 1984, Astrophys. J., 283, 413.
Shevgaonkar, R.K., Kundu, M.R.: 1984, Astrophys. J., 283, 413.
Shevgaonkar, R.K., Kundu, M.R.: 1985a, Solar Phys., 98, 19.
Shevgaonkar, R.K., Kundu, M.R.: 1985b, Astrophys. J., 292, 733.
Siarkowski, M.: 1984, Solar Phys., 94, 105.
Smolkov, G.Ya., Pistolkors, A.A., Treskov, T.A., Krissinel, B.B., Putilov, V.A.,
    Potapov, N.N.: 1986, Astrophys. Space Sci., 119, 1.
Sprangle, P., Vlahos, L.: 1986, Phys. Rev. A., 33, 1261.
Stahli, M., Magun, A.: 1986, Solar Phys. 104, 117.
Stahli, M., Benz, A.O.: 1987, Astron. Astrophys., 175, 271.
Steinberg, J.L., Dulk, G.A., Hoang, S.,, Lecacheux, A., Aubier, M.: 1984, Astron.
    Astrophys., 140, 39.
Steinberg, J.L., Hoang, S., Dulk, G.A.: 1985, Astron. Astrophys., 150, 205.
Stewart, R.T.: 1985, Solar Phys., 96, 381.
Stewart, R.T., Brueckner, G.E., Dere, G.E.: 1986, Solar Phys., 106, 107.
Takakura, T., Kundu, M.R., McConnell, D., Okhi, K.: 1985, Astrophys. J., 298, 431.
Vlahos, L., Sharma, R.R.: 1985, Astrophys. J., 290, 347.
Wentzel, D.G.: 1985, Astron. Astrophys., 296, 278.
Wentzel, D.G.: 1986, Solar Phys., 103, 141.
Wentzel, D.G., Zlobec, P., Messerotti, M.: 1986, Astron. Astrophys., 159, 40.
White, S.M., Melrose, D.B., Dulk, G.A.: 1986a, Astrophys. J., 308, 424.
White, S.M., Kundu, M.R., Szabo, A.: 1986b, Solar Phys., 107, 135.
Wiehl, H.J., Batchelor, D.A., Crannell, C.J., Dennis, B.R., Price, P.N., Magun, A.:
    1985a, Solar Phys., 96, 339.
Wiehl, H.J., Benz, A.O. and Aschwanden, M.J.: 1985b, Solar Phys. 95, 167.
Willson, R.F.: 1984, Solar Phys., 92, 189.
Willson, R.F.: 1985, Astrophys. J., 298, 911.
Willson, R.F., Lang, K.R.: 1986, Astrophys. J., 3085, 443.
Winglee, R.M.: 1985a, JGR, 90, 9663.
Winglee, R.M.: 1985b, Astrophys. J., 291, 160.
Wu, C.S., Steinelfson, R.S., Zhou, G.C.: 1986, Astrophys. J., 309, 392.
Zaitsev, V.V., Stepanov, A.V., Chernov, G.P.: 1984, Solar Phys., 93, 363.
Zaitsev, V.V., Zlotnik, E.Ya.: 1986, Pisma Astron. Zh., 12, 311.
Zhao, R.Y., Shi, J.K.: 1986, Solar Phys., 104, 137.

VIII. Theory of Solar Flares
(E.R. Priest)

A.   INTRODUCTION

By far the most significant event for Solar Flares as a whole over the past 3 years
has been the operation of the Solar Maximum Mission Satellite, together with the
accompanying data analysis, ground-based support and theoretical modelling. This has
culminated in the series of SMM Flare Workshops, whose proceedings have now appeared
(Kundu and Woodgate 1986 (I)), with chapters on a wide variety of topics which indicate
the enormity and complexity of the flare problem.

Here we shall review only one aspect, namely the MHD theory of a flare (Priest
1986a), focussing on two topics. These are the instability or nonequilibrium process
which initiates a large flare and the magnetic reconnection process whereby the stored
magnetic energy is released. However, one should bear in mind the subtle interaction
between the MHD and the microscopic plasma physics of the flare : the MHD provides the
environment (the current sheets, shock waves and turbulent medium) where particles can
be accelerated, whereas microscopic processes will determine the turbulent transport
coefficients. Furthermore, the MHD coupling between a plasma and a magnetic field is
much more complex and represents quite different physics from simply the
electromagnetism of circuits, so it can often be misleading and dangerous to use
circuit theory analogues.

A large solar flare has three phases : a preflare phase, during which a flux-tube (a prominence) starts to rise slowly ; a rise phase, when the prominence suddenly erupts rapidly and reconnection is initiated ; a main phase, when the reconnection continues and produces separating Hα ribbons and rising "post"-flare loops.

The role of reconnection is : to create small flares by emerging flux (Park et al. 1984) or by lateral motion or by reconnection submergence in cancelling magnetic features (Martin et al. 1985, Priest 1986a,b) ; to initiate the energy release at the rise phase of a large flare (below the rising prominence) and to continue the energy release through the main phase ; it may even trigger the eruption.

B. BASIC RECONNECTION THEORY

Reconnection is important for heating the plasma as well as producing the intense and localised electric fields to accelerate fast particles. The basic theory is split into two parts (Priest 1984, Dubois et al. 1985, Hones 1984, Pudovkin and Semenov 1985) : the tearing mode instability of a current sheet or sheared magnetic field, whereby an equilibrium goes linearly unstable to the breaking and reconnecting of field lines; the fast nonlinear state of steady reconnection. Tearing theory has been extended in many ways : e.g. by including viscosity (Park et al. 1984) ; by studying its non linear evolution at very large magnetic Reynolds number (Steinolfson and Van Hoven 1984a), which shows that the islands can become much wider than previously thought and that secondary islands can be generated ; and by coupling with optically thin radiation (Steinolfson and Van Hoven 1984b, see also references in Section IV A-1). The latter shows that radiative tearing can operate much faster than ordinary tearing and that perpendicular conduction produces thermal ripples (Steinolfson 1984).

Detailed numerical experiments have shown that, in its nonlinear development, tearing can evolve into a fast steady regime, but often such regimes are rather different from the classical modes. Analytically, a new unified theory for fast steady reconnection has been discovered (Priest and Forbes 1986) which includes the distinct classical models due to Sweet-Parker, Petschek and Sonnerup as special cases; it also possesses many new regimes, such as flux pile-up with a reconnection rate much faster than Petschek, up to the Alfven speed. The new regimes explain many previously puzzling features of numerical experiments (Forbes and Priest 1987).

The most detailed numerical experiments (Biskamp 1986) show three new features:the new inflow regimes of Priest and Forbes (1986) ; jets of plasma expelled along the separatrices, for which a theory exists (Soward and Priest 1986); and reversed current spikes or fast shocks at or near the ends of the central diffusion region (see also Forbes and Priest 1984, Forbes 1986). In addition, reconnection may sometimes have a filamentary and turbulent structure (Mattheus and Lamkin 1986), and a series of simulations of reconnection in the geomagnetic tail are also of importance for flares (Birn 1984, Birn and Schindler 1986, Birn et al.1987, Lee et al. 1985, Lee and Fu 1986, Sholer and Roth 1987) .

C. CAUSE OF ERUPTION

Theoretical proposals for the onset of a prominence eruption include : an eruptive MHD instability (of kink type) of the preflare magnetic configuration, modelled either as a flux tube or as a coronal arcade (Priest 1986) ; a lack of magnetic equilibrium reached after a slow evolution through a series of force free or magnetostatic equilibria (Priest 1986). Substantial progress has been made recently on the analysis of such instability and nonequilibrium thresholds, as follows. Nonequilibrium of a flux tube acted on by magnetic tension and magnetic buoyancy may occur .if the tube becomes twisted too much or its footpoints become too widely separated (typically a few times the coronal scale height) (Browning and Priest 1984, Browning and Priest 1986). A most important analysis of arcade equilibria has also been completed with the footpoint positions and base pressure imposed (Moreno-Insertis 1986) ; it shows how magnetic catastrophes can occur with certain combinations of base conditions. Earlier work on the ideal stability of magnetic arcades has been greatly extended. A crucial stabilising effect that is now included is that of photospheric line-tying, and it has been shown by that it is best simulated by assuming "rigid wall" conditions so that all components of a coronal disturbance vanish at the photospheric boundary (Hood

1986b). New separable magnetohydrostatic solutions have been discovered (Melville et al. 1984), and sufficient conditions produced for their stability, which also become necessary when the axial field vanishes (Hood 1984a, Hood 1984b, Melville et al. 1986). Ideal ballooning instabilities which are driven by pressure gradients have been analysed (Hood 1986a) and resistive ballooning modes are found always to be unstable when the plasma pressure has a maximum on the arcade axis - this demonstrates that line-tying is not sufficient to stabilise resistive modes (Velli and Hood 1986). Both ballooning and thermal condensation modes are likely to be important in creating small-scale filamentation in the corona and in enhancing the global transport coefficients (Van Hoven et al 1987, Bodo et al 1985, Bodo et al 1987, see also references in section IV A-1). A study of a simple sheared arcade demonstrates how a pressure gradient can destabilise the arcade, but three force-free arcades were found to be stable to all the perturbations that were tried (Cargill et al 1986). The conclusion therefore is that eruptions may result when either the pressure build-up or the size of the magnetic island associated with the presence of a prominence are too great.

A new method for calculating force-free fields numerically has been set up (Yang et al 1986) which suggests that a closed field always has less energy than a completely open field, in agreement with Aly's conjecture(Aly 1984), but an example of a catastrophic opening to partly open field has been discovered (Low 1986). Global magnetostatic fields may also be buoyantly unstable (Low 1984).

D.   MAIN PHASE OF ENERGY RELEASE
Much more detailed models have been developed of the main phase reconnection process as the field lines close down and create the hot "post"-flare loops and Hα ribbons. The kinematic model for deducing the electric field in the reconnection region from observed ribbon or loop motions has been greatly extended and applied to particular flares (Kopp and Poletto 1984, Kopp and Poletto 1985, Poletto and Kopp 1986). Furthermore, the fully MHD numerical model has several new interesting features (Forbes 1986, Forbes and Malherbe 1986, see also references in section IV A-1) : reconnection develops into an impulsive bursty regime, which explains the sudden observed jumps in loop height ; the presence of a fast-mode shock standing in the downflow from the reconnection site reduces the flow speed by a factor of four - it may be a source of fast particles and triggers a radiative condensation for the cool loops ; a reversed deflection current deflects the flow around the stagnation region ; most of the energy released at the slow shocks is conducted down to the chromosphere where it drives plasma upwards by evaporation.

References

I : Proc. SMM Flare Workshop "Energetic Phenomena on the Sun", NASA CP 2439, Kundu, M.R. and Woodgate, B. (eds), 1986.
Aly, J.: 1984, Astrophys. J. 283, 349.
Birn, J.: 1984, Adv. Space Res. 4, 449.
Birn, J. and Schindler, K.: 1986, J. Geophys. Res. 91, 8817.
Birn, J., Hones, E., Schindler, K.: 1987, J. Geophys. Res. 91, 1116.
Biskamp, D.: 1986, Phys. Fluids 29, 1520.
Bodo, G., Ferrari, A., Massaglia, S., Rosner, R., Vaiana, G.: 1985, Astrophys. J. 291, 798.
Bodo, G., Ferrari, A., Massaglia, S., Rosner, R.: 1987, Astrophys. J. 313, 432.
Browning, P. and Priest, E.: 1984, Solar Phys. 92, 173.
Browning, P. and Priest, E.: 1986, Solar Phys. 106, 335.
Cargill, P., Hood, A., Migliuolo, S.: 1986, Astrophys. J. 309, 402.
Dubois, M., Gresillon, D., Bussac, M.: 1985, Magnetic Reconnection and Turbulence, Ed. de Phys. Paris.
Forbes, T. and Priest, E.: 1984, Solar Phys. 94, 315.
Forbes, T.: 1986, Astrophys. J. 305, 553.
Forbes, T. and Priest, E.: 1987, Rev. Geophys. 25, 7.
Forbes, T. and Malherbe, J.: 1986, The lower atmosphere of solar flares, D. Neidig (ed), p. 443.

Hones, E.: 1984, "Magnetic Reconnection in Space and Lab. Plasmas" AGU, Washington.

Hood, A.: 1984a, Geophys. Astrophys. Fluid Dynamics 28, 223.

Hood, A.: 1984b, Adv. Space Res. 4, 49.

Hood, A.: 1986a, Solar Phys. 103, 329.

Hood, A.: 1986b, Solar Phys. 105, 307.

Kopp, R. and Poletto, G.: 1984, Solar Phys. 50, 85.

Kopp, R. and Poletto, G.: 1985, Proc. IAU Colloq. Nº 86, G. Doschek (ed) p. 17.

Lee, L., Fu, Z., Akasofu, S.: 1985, J. Geophys. Res. 90, 10896.

Lee, L. and Fu, Z.: 1986, J. Geophys. Res. 91, 4551 ; 91, 6807.

Low, B.: 1984, Astrophys. J. 286, 772.

Low, B.: 1986, Astrophys. J. 307, 205.

Martin, S., Livi, S., Wang, J.: 1985, Australian J. Phys. 38, 929.

Mattheus, W. and Lamkin, S.: 1986, Phys. Fluids 29, 2513.

Melville, J., Hood, A., Priest, E.: 1984, Solar Phys. 92, 15.

Melville, J., Hood, A., Priest, E.: 1986, Solar Phys. 105, 291.

Moreno-Insertis, F.: 1986, Astron. Astrophys. 166, 335.

Park, W., Monticello, D., White, R.: 1984, Phys. Fluids 27, 137.

Poletto, G. and Kopp, R.: 1986, "The Lower Atmosphere of Solar Flares", D. Neidig (ed)
    p. 453.

Priest, E.: 1984, Rep. Prog. Phys. 48, 955.

Priest, E.: 1986a, Solar Phys. 104, 1.

Priest, E.: 1986b, Proc. Teneriffe Workshop on Small-scale Magnetic Fields.

Priest, E. and Forbes. T.: 1986, J. Geophys. Res. 91, 5579.

Pudovkin, M. and Semenov, V.: 1985, Space Sci. Rev. 41, 1.

Sholer, M. and Roth. D.: 1987, J. Geophys. Res. 92, 3223.

Soward, A. and Priest, E.: 1986, J. Plasma Phys. 35, 333.

Steinolfson, R.: 1984, Astrophys. J. 281, 854.

Steinolfson, R. and Van Hoven, G.: 1984a, Phys. Fluids 27, 1207.

Steinolfson, R. and Van Hoven, G.: 1984b, Astrophys. J. 276, 391.

Van Hoven, G., Sparks, L., Schnack, D.: 1987, Astrophys. J. 317, 91.

Velli, M. and Hood, A.: 1986, Solar Phys. 106, 353.

Wu, S. et al.: 1986, "Flare Energetics" Ch. 5 of ref. I.

Yang, W., Sturrock, P., Antiochos, S.: 1986, Astrophys. J. 309, 383.

M. Pick,
President of Commission.

COMMISSION 12: RADIATION AND STRUCTURE OF THE SOLAR ATMOSPHERE
(RADIATION ET STRUCTURE DE L'ATMOSPHERE SOLAIRE)

President: M. Kuperus   Vice president: J. Harvey

I. INTRODUCTION

(M. Kuperus)

Solar Physics has been traditionally divided into Structure and Radiation of the Solar Atmosphere (commission 12) and Solar Activity (commission 10). There has been increasing evidence that solar activity, which is basically of magnetic origin, occurs on a great variety of scales and thus immediately touches upon the structure of the solar atmosphere as well as the structure and dynamics of the convection zone. As a consequence progress in the field of origin and evolution of solar magnetic fields from a large scale, 'the dynamo', to small scale is included in this report. In the past few years particular attention has been paid to the fact that the fluctuations in the magnetic field are much larger than the mean field and that the dynamo modes may be stochastically excited. The question whether there is a magnetic reservoir at the bottom of the convection zone still remains to be resolved. The interaction of the convection and the magnetic field resulting in an enhancement of the magnetic field in the intergranular lanes is studied by numerical modelling.

A real understanding of the magnetohydrodynamics of the subphotospheric layers requires a detailed study of the solar oscillations. The excitation of the so-called p-modes is likely to take place in the turbulent convection zone. Helioseismology will make it possible to study the thermodynamic and magnetohydrodynamic structure of the solar interior, thus preparing a foundation for stellar seismology, a field of growing interest, which will lead to a new understanding of stellar interior structure.

The outer solar atmosphere seems to consist primarily of structures that are shaken or sheared by photospheric motions. There is some evidence that the powering of the outer layers is magnetic of origin, though the actual mechanism is still a matter of debate.

It is of great importance to Astronomy and Astrophysics that the above mentioned outstanding problems of solar interior structure, solar magnetohydrodynamics and solar outer atmosphere are understood so that further progress can be made in stellar physics. For this to occur new resolution solar instrumentation is needed.

The president of commission 12 thanks the authors of this report for their contributions. He also wishes to thank the organizing committee for their support during the last three years.

II. SOLAR INTERIOR STRUCTURE

(Ken C. Libbrecht)

A great deal of progress has been made in recent years in the field of helioseismology (for a review of the field see Deubner and Gough 1984, Christensen-Dalsgaard et al. 1985, and Leibacher et al. 1985). Although it's been less than a decade since individual p-modes were first identified in solar velocity data (Grec et al. 1980), measurements have recently been made (unpublished) which isolated thousands of p-modes and determined each of their frequencies with accuracies as high as a few tenths of a microhertz, or to a part in 10^4. While in the past

makers of the standard solar model were constrained to duplicate little other than
the measured mass, radius, and luminosity of the sun, measured p-mode frequencies
now provide thousands of additional model constraints. Solar p-mode frequencies
have quickly risen to be among the most accurately determined physically
significant quantities in all of astrophysics. (For contrast, pulsar period
measurements have been made to accuracies of 10^{12} or greater, but this reflects
simply a rotation rate of a neutron star, which returns no physical understanding
of the phenomenon. Pulsar period derivatives are physically interesting, but their
measurements are not nearly so precise, to a part in 10^4 with the binary pulsar
1913+16, for example.

Using the p-mode frequency measurements to determine the interior structure
of the sun promises to be a very interesting but difficult task. Current standard
solar models are able to reproduce most of the mode frequencies to of order one
percent (unpublished) which confirms our basic understanding of the oscillations
but is far from reproducing the frequencies at the level of their measurement
uncertainties. Theoretical efforts are currently under way to better understand
the basic properties of the solar interior, such as the equation of state and the
opacity of the solar plasma, in order to produce an improved solar model which
will better fit the observations.

Another approach to using the p-mode data is to invert the measured frequencies into a measurement of the speed of sound as a function of radius in the solar
interior. A first result in this direction has been given by Christensen-Dalsgaard
et al. (1985), where the sound speed determined from p-modes agreed quite well
with that from the standard model calculation. Although the p-mode frequencies do
not provide a very sensitive probe of the deep solar interior below one-half to
one-third of a solar radius, they do provide a very good probe of the convection
zone and the transition region between the radiative zone and the convection zone.

Measurements of the rotatonal splitting of p-mode frequencies have also been
proceeding apace, with recent contributions by Duvall et al. (1986) and Brown and
Morrow (1987). Efforts to invert the measurements to infer the solar rotation rate
as a function of depth and latitude are still in progress, but it is likely that
helioseismology will provide a quite accurate determination of the rotation rate
throughout the convection zone. Such a measurement will be invaluable for
comparison with computer models of the solar convection zone (Glatzmaier 1985),
and is important input for understanding the solar dynamo. Recent measurements of
the surface rotation rate determined by the 60-year Mt. Wilson white-lightplate
collection (Gilman and Howard 1984) showed a one percent increase in the surface
rotation rate during solar minimum. If true it should be possible to measure this
increase using p-modes as well.

While the p-mode frequencies provide input for understanding the solar
interior structure, their measured amplitudes reflect the dynamics of the mode
excitation and damping mechanisms. Calculations of overstability mechanisms, such
as the κ-mechanism, are still unable to confidently determine if the p-modes are
stable or overstable, owing to our poor understanding of turbulent viscosity in
the convection zone. However it is beginning to appear that the modes must be
stable, since if they were unstable it is likely that a few modes would grow to
very large amplitudes, as is seen in other oscillating stars such as the Cepheids,
in contrast to the millions of low-amplitude p-modes observed in the sun.

A more likely mechanism for exciting the p-modes is via turbulent convection,
originally proposed by Goldreich and Keeley (1977). The acoustic noise generated
in the convection zone is trapped inside the solar acoustic cavity, and results in
the excitation of the sun's normal modes. Recent work by Goldreich and Kumar
(1986) predicts the energy E of a solar p-mode is given by the familiar-looking
formula $E = mc^2$, where m is the mass of a resonant turbulent eddy, and c is the

sound speed inside the eddy. For 5-minute oscillations, a resonant eddy is simply a solar granule, and the predicted energy is of order 10^{28} ergs, which is in fairly good agreement with observation (Libbrecht et al. 1986).

The theoretical work by Goldreich and Kumar represents a fundamental improvement in our understanding of the interaction of turbulence with sound waves, and is good example of how "basic" research in astrophysics can have broad implications. This work, which describes the emission and absorption of sound waves by turbulence, is an extension of the seminal work by Lighthill (1952) which described the emission of sound by turbulence, with the most common application being none other than airport noise.

Other helioseismology results include the report by Woodard and Noyes (1985) of the detection of solar cycle shifts of p-mode frequencies of 0.1 µHz per year between 1980 and 1984. This result has yet to be confirmed, however, and other workers (unpublished) have placed upper limits of approximately 0.1 µHz/yr on p-mode frequency shifts. While it appeared in the past that g-modes had also been detected on the sun (see Solar Phys. vol. 82 1983, Delache and Scherrer 1983), more recent measurements have turned up negative, and the general agreement of the solar community is that g-mode detections have not been confirmed.

The solar neutrino problem remains a problem, but recently the MSW theory of neutrino oscillations (Mikheyev and Smirnov 1986, Wolfenstein 1979) has been accepted as an attractive solution to the dilemma. Although the oscillation of electron neutrinos into other neutrino states (the µ and τ neutrinos) was suggested some time ago as a possible explanation of the solar neutrino measurements, the MSW authors showed that the theory of the weak interaction indicated resonance interactions in the presence of matter that greatly enhance the neutrino oscillation phenomenon. The theory also suggests a number of other measurements to constrain the neutrino mass and mixing angle (Dar and Mann 1987), including measuring the difference in the daytime and nighttime solar neutrino flux owing to oscillation inside the earth. Where in the past it was thought that neutrinos would pass effortlessly through light-years of lead shielding, now it is not even clear that they can pass through even the earth unscathed. However the MSW theory is not the only explanation of the neutrino measurements; it has been shown by Gilliland et al. (1986) that a sufficient number of weakly interacting massive particles (WIMPs) inside the sun could also reduce the flux of solar neutrinos.

References

Brown, T.M. and Morrow, C.A.: 1987, Ap. J. **314**, L21
Christensen-Dalsgaard, J., Gough, D., and Toomre, J.: 1985, Science **229**, 923
Christensen-Dalsgaard, J., Duvall, T.L., Gough, D.O., Harvey, J.W. and Rhodes,
    E.J.: 1985, Nature **315**, 378
Dar, A. and Mann, A.: 1987, Nature **325**, 790
Delache, P., and Scherrer, P.H.: 1983, Nature **306**, 651
Deubner, F.-L. and Gough, D.: 1984, Ann. Rev. Astron. Astrophys. **22**, 593
Duvall, T.L., Harvey, J.W. and Pomerantz, M.A.: 1986, Nature **321**, 500
Gilliland, R.L., Faulkner, J., Press, W.H., Spergel, D.N.: 1986, ap. J. **306**, 703
Gilman, P.A. and Howard, R.: 1984, Ap. J. **283**, 385
Glatzmaier, G.A.: 1985, Ap. J. **291**, 300
Goldreich, P. and Keeley, D.: 1977, Ap. J. **212**, 243
Goldreich, P. and Kumar, P.: 1986, IAU Symposium no. 123, Aarhus, Denmark
Grec, G., Fossat, E., and Pomerantz, M.: 1980 Nature **288**, 541
Leibacher, J.W., Noyes, R.W., Toomre, J., and Ulrich, R.K., Sept. 1985, Scientific
    American
Libbrecht, K.G., Popp, B.D., Kaufman, J.M. and Penn, M.J.: 1986, Nature **323**, 235
Lighthill, M.J.: 1952, Proc. Roy. Soc. London **A211**, 564
Mikheyev, S.P. and Smirnov, A.Y.: 1986, Nuovo cim. **9C**, 17

Woodard, M.F. and Noyes, R.W.: 1985, Nature **318**, 449
Wolfenstein, L.: 1979, Phys. Rev. **D20**, 2634

III. THE SOLAR DYNAMO

(M. Stix)

Traditionally the theory of the solar dynamo has been divided into two parts. The first, more difficult part, is the derivation of equations governing the mean magnetic field; the second, easier, is the solution of this equation, and the interpretation of the result in terms of observed solar magnetism. This report follows the traditional division.

1. Mean Field Equations

Mean field equations contain the effects of turbulence in form of transport coefficients, notably the turbulent diffusivity β, and the regeneration coefficient, α, for the mean poloidal field. These coefficients have often been calculated in the "approximation of second order correlations" (= "first order smoothing"). A formally complete solution has been given by Hoyng (1985). For the case of isotropic turbulence Nicklaus (1987), using an ensemble of polarized waves (Drummond et al., 1984) and the formalism of ordered cumulants, calculated corrections arising from fourth order correlations. These are proportional to $S^2 = (u\tau/1)^2$, which unfortunately is of order 1 in the solar convection zone. Moreover, not only the α-coefficient, but also the correction to the β-coefficient depends on the helicity of the turbulent flow.

In a different approach, Drummond and Horgan (1986) used the same set of polarized waves and calculated the exact Lagrangian solution of the induction equation. For the purpose of averaging they computed the paths of a large number ($\simeq 10^5$) of fluid particles. In the examples treated they obtained α and β coefficients which were surprisingly close to the results of the second order correlation approximation. The Lagrangian approach was also employed by Molchanov et al. (1984) and by Vainshtein and Kichatinov (1986) in more general investigations of a magnetic field in a turbulent medium of high conductivity.

A different derivation of an α-coefficient was given by Schmitt (1984, 1985) on the basis of dynamically unstable magnetostrophic waves (propagating in a magnetic layer at the base of the convection zone, see below).

The role of magnetic field fluctuations (on the Sun, these are large compared to the mean field!) in dynamo theory was emphasized by Hoyng (1987a,b). He derives a new equation for the tensor $\langle\mathbf{BB}\rangle$ and shows that, in addition to α and β, a third important transport coefficient, related to the mean vorticity, occurs.

2. Solar Dynamo Models

Solutions of the mean field equation in a spherical geometry have been systematically studied by Rädler (1986a), Bräuer and Rädler (1987), and Yoshimura (1984a,b,c). These studies bear on the question of mode selection, e.g. whether a mean field of odd or of even parity will be excited first, or whether the field is oscillatory or steady. Hoyng (1987b) suggests that a number of dynamo modes could be simultaneously present at any one time due to stochastic excitation, and that these modes should be compared to the modes analysed by Stenflo and Vogel (1986). The dominant mode found by these authors is a combination of odd zonal harmonics, all with the same period of 22 years, and corresponds to the leading mode predicted by most $\alpha\omega$-dynamos. Non-axisymmetric modes are strongly opposed in $\alpha\omega$-dynamos by the differential rotation (Rädler, 1986b).

Parker (1984) points out that the traditional boundary condition of vanishing toroidal field, B = 0, should be replaced by the condition $\partial B/\partial r$ = 0 because of the difficulty the field has to escape into the highly conducting corona. The new condition somewhat lowers the critical dynamo number and increases the period of the oscillatory dynamo solution (Choudhuri, 1984).

Much attention was paid to dynamos operating in an overshoot layer at the base of the Sun's convection zone. Such a dynamo would not suffer from rapid loss of magnetic flux due to instabilities. Thanks to a sign reversal of α in the lower part of the convection zone it would perhaps also avoid the poleward migration of the field which is found in dynamos based on hydrodynamic and hydromagnetic calculations (Glatzmaier, 1984, 1985a; Gilman and Miller, 1986; for the sign of α s.a. Krivodubskii, 1984b).

Inversion of p mode frequencies (Christensen-Dalsgaard et al., 1985) suggests that the base of the convection zone lies at a depth of $\simeq$ 200000 km. An overshoot layer at this depth, obtained through a non-local version of the mixing-length theory, is $\simeq$ 15000 km thick and quite capable of storing enough magnetic flux to account for the observed activity (e.g. Pidatella and Stix, 1986; for a more general approach, with "plumes", see Schmitt et al., 1984).

Unfortunately there is no indication of a concentrated shear layer at a depth of $\simeq$ 200000 km. Rotational splitting of p mode frequencies (Duvall and Harvey, 1984) yields a gradual inwards decrease of the angular velocity, i.e. $\partial \omega/\partial r > 0$, at low latitude (Duvall et al., 1984); dynamic considerations extend this functional behaviour of ω to the cylindrical isorotation surfaces known from the work of Gilman, Glatzmaier, and others (s.a. Rosner and Weiss, 1985). To concentrate magnetic flux generated in a broader shear region (and to counteract the effects of instability) one must possibly rely on mechanisms which transport flux downward into the overshoot layer. An example is the diamagnetic effect, as recently again suggested by Krivodubskii (1984a).

Independent evidence for a magnetic layer at the base of the convection zone could also come from p mode frequencies: Woodard and Noyes (1985) and Fossat et al. (1987) found a mean decrease of $\simeq$ 0.4 µHz between 1980 and 1984 for frequencies of degree 0 to 3, and attribute the change to the solar cycle. The change has been disputed by Pallé et al. (1986). Moreover, a change of 0.4 µHz would require a field strength of order 10^6 G, which is much larger than theoretically expected (Roberts and Campbell, 1986)! So the question is open. In any case it is interesting to note that a narrow magnetic layer would cause a more subtle effect: in addition to a mean term, we would expect a frequency shift which (for any given degree) is periodic in the frequency itself (Vorontsov, 1987), and perhaps detectable by this signature.

A kinematic $\alpha\omega$-dynamo model for the overshoot layer has been constructed by DeLuca (1986; s.a. DeLuca and Gilman, 1986). It employs a shear with $\partial\omega/\partial r > 0$, and $\alpha < 0$ (in the northern hemisphere; $\alpha > 0$ in the south), so that the mean field migration is equatorwards as desired. These ingredients to the kinematic model are confirmed in the full dynamic calculation of Glatzmaier (1985b). The model of Schmitt (1987) is a similar $\alpha\omega$-dynamo, he employs the α-effect arising from the magnetostrophic waves. Unfortunately all these models seem to predict the wrong phase relationship between the mean poloidal and toroidal field components (Stix, 1987).

Aperiodic behaviour, such as the Maunder minimum in the 17th century, has been attributed to the dynamic properties of the mean field equation. Non-linear interaction terms allow for the desired chaotic solutions (Weiss et al., 1984) although a different explanation, based on a stationary field in the core, has been offered by Pudovkin and Benevolenska (1985).

Reviews on the solar dynamo include those by Belvedere (1985), Gilman (1986), Moss (1986), Stix (1984, 1987), and Weiss (1986, 1987).

References

Belvedere, G.: 1985, Solar Phys. **100**, 363
Bräuer, H.-J., Rädler, K.-H.: 1987, Astron. Nachr. **308**, 27
Christensen-Dalsgaard, J., Duvall Jr., T.L., Gough, D.O., Harvey, J.W., Rhodes Jr., E.J.: 1985, Nature **315**, 378
Choudhuri, A.R.: 1984, Astrophys J. **281**, 846
DeLuca, E.E.: 1986, Thesis, University of Colorado, NCAR/CT-104
DeLuca, E.E., Gilman, P.A.: 1986, Geophys. Astrophys. Fluid Dynamics **37**, 85
Drummond, I.T., Horgan, R.R.: 1986, J. Fluid Mech. **163**, 425
Drummond, I.T., Duane, S., Horgan, R.R.: 1984, J. Fluid Mech. **138**, 75
Duvall, T.L., Harvey, J.W.: 1984, Nature **310**, 19
Duvall, T.L., Jr., Dziembowksi, W.A., Goode, P.R., Gough, D.O., Harvey, J.W., Leibacher, J.W.: 1984, Nature **310**, 22
Fossat, E., Gelly, B., Grec, G., Pomerantz, M.: 1987, Astron. Astrophys. **177**, L47
Gilman, P.A.: 1986, in "Physics of the Sun", P.A. Sturrock (ed.), Vol. 1, p. 95, Reidel
Gilman, P.A., Miller, J.: 1986, Astrophys. J. Suppl. **61**, 585
Glatzmaier, G.A.: 1984, J. Comp. Phys. **55**, 461
Glatzmaier, G.A.: 1985a, Astrophys. J. **291**, 300
Glatzmaier, G.A.: 1985b, Geophys. Astrophys. Fluid Dynamics **31**, 137
Hoyng, P.: 1987a, J. Fluid Mech. **151**, 295
Hoyng, P.: 1985, Astron. Astrophys. **171**, 348
Hoyng, P.: 1987b, Astron. Astrophys.**171**, 357
Krivodubskii, V.N.: 1984a, Astron. Zh. **61**, 354
Krivodubskii, V.N.: 1984 b, Astron. Zh. **61**, 540
Molachanov, S.A., Ruzmaikin, A.A., Sokoloff, D.D.: 1984, Geophys. Astrophys. Fluid Dynamics **30**, 242
Moss, D.: 1986, Phys. Rep. **140**, 1
Nicklaus, B.: 1987, Diplomarbeit, Univ. Freiburg
Pallé, P.L., Pérez, J.C., Régulo, C., Roca Cortés, T., Isaak, G.R., McLeod, C.P., van der Raay, H.B.: 1986, Astron. Astrophys. **170**, 114
Parker, E.N.: 1984, Astrophys. J. **281**, 839
Pidatella, R.M., Stix, M.: 1986, Astron. Astrophys. **157**, 338
Pudovkin, M.I., Benevolenska, E.E.: 1985, Solar Phys. **95**, 381
Rädler, K.-H.: 1986a, Astron. Nachr. **307**, 89
Rädler, K.-H.: 1986b, in "Plasma Astrophysics", Proc. ESA SP-251, p. 569
Roberts, B., Campbell, W.R.: 1986, Nature **323**, 603
Rosner, R., Weiss, N.O.: 1985, Nature **317**, 790
Schmitt, D.: 1984, in "The Hydromagnetics of the Sun", P. Hoyng (ed.), proc. ESA SP-220, p. 223
Schmitt, D.: 1985, Thesis, Univ. Göttingen
Schmitt, D.: 1987, Astron. Astrophys. **174**, 281
Schmitt, J.H.M.M., Rosner, R., Bohn, H.U.: 1984, Astrophys. J. **282**, 316
Stenflo, J.O., Vogel, M.: 1986, Nature **319**, 285
Stix, M: 1984, Astron. Nachr. **305**, 215
Stix, M.: 1987, in "Solar and Stellar Physics", E.H. Schröter and M. Schüssler (eds.), Springer
Vainshtein, S.I., Kichatinov, L.L.: 1986, J. Fluid Mech. **168**, 73
Vorontsov, S.V.: 1987, in "Advances in Helio- and Asteroseismology", J. Christensen-Dalsgaard (ed.), IAU Symp. 123, Reidel
Weiss, N.O.: 1986, in "Highlights of Astronomy, J.-P. Swings (ed.), Vol. **7**, p. 385
Weiss, N.O.: 1987, in "Physical Processes in Comets, Stars, and Active Galaxies", W. Hillebrandt, E. Meyer-Hofmeister and H.-C. Thomas, (eds.), Springer, p. 46
Weiss, N.O., Cattaneo, F., Jones, C.A.: 1984, Geophys. Astrophys. Fluid Dynamics **30**, 305

Woodard, M.F., Noyes, R.W.: 1985, Nature **318**, 449
Yoshimura, H., Wang, Z., Wu, F.: 1984a, Astrophys. J. **280**, 865
Yoshimura, H., Wang, Z., Wu, F.: 1984b, Astrophys. J. **283**, 870
Yoshimura, H., Wu, F., Wang, Z.: 1984c, Astrophys. J. **285**, 325

IV. SMALL SCALE MAGNETIC FIELDS

(S.K. Solanki)

Considerable progress has been made during the last four years in the theoretical and empirical investigation of the small scale solar magnetic field and associated phenomena. Although the basic outlines established in the 1970s of the structure of the field, namely small fluxtubes or magnetic elements with kilogauss fields embedded in a relatively field free medium have survived, many of the details have changed and some of the large gaps in our knowlegde of these captivating structures have been filled. In the following we briefly outline some of the highlights.

One direction in the thrust for a better understanding of magnetic elements has been towards the construction of comprehensive theoretical models, e.g. by Deinzer et al. (1984a,b) in 2-D slab geometry and by Nordlund (1986) in 3-D. Although these models still lack some essential features (too small spatial resolution in the 3-D models, no proper treatment of the radiative transfer in the 2-D models; the latter shortcoming is being remedied at the moment by a number of groups), they do illustrate the physics and give rise to the hope that within a decade models of magnetic elements of equal detail and generality as the granulation models of Nordlund will be available. However, considerable hurdles must be surmounted first, since magnetic elements are considerably more difficult to model than granulation. Waves, for example, cannot be neglected, since wave heating (perhaps involving dissipation via shocks, cf. Herbold et al., 1984) is probably quite important even in the photospheric layers and is certainly so in the chromosphere. The work of Ayres et al. (1986) actually supports the conclusion that the hot chromosphere only exists within fluxtubes. Another complexity facing 3-D fluxtube modellers is the possible presence of a boundary current sheet, which requires very fine grids for a proper treatment. Two dimensional models, like the ones of Deinzer et al. or of Steiner et al. (1986), have the advantage that they can take such boundary layers into account in detail.

A breakthrough in the radiative transfer of polarized light was achieved by Van Ballegooijen (1985). He presented a method for obtaining the formal solution of the radiative transfer equations for polarized light in the presence of a magnetic field. Besides providing deep insight into the process of solution, his method also allows contribution functions to be defined and calculated. Thus the determination of the heights of formation of the Stokes profiles has been placed on a secure theoretical footing. This advance will play an important role, not only for the proper diagnostics of observations, but also for interpreting the spectra produced by the emerging breed of comprehensive fluxtube models. The one remaining problem with his definition is that it mixes the contribution to the lines with that to the continuum. However, a remedy is already in sight.

Given the present state of theory, observational and empirical work is still indispensible. The main observational advance has come from the extension of the Fourier transform spectrometer (FTS) at the NSO McMath telescope into a spectral polarimeter, which is currently capable of registering Stokes I, V, and Q in thousands of spectral lines simultaneously at very high spectral resolution. Data from this instrument, built by J.W. Brault and converted into a polarimeter by J.W. Harvey and J.O. Stenflo, have led to a considerable fraction of the observational advances concerning small scale magnetic fields in the last three to four years. Although such data do not have high spatial or temporal resolution,

the amount of information they contain is still enormous and only a fraction of it
has so far been extracted. Some of the progress has been a result of the
derivation of an approximate form of the unpolarized line profile, Stokes I,
formed exclusively inside the magnetic elements form the observed Stokes V
(Solanki and Stenflo, 1984). This allows the rich array of diagnostic techniques
available for the unpolarized spectrum to be applied for the first time to light
arising solely from inside a magnetic element. When combined with FTS observations
this technique yields a complete atlas of a fluxtube spectrum. New empirical
models of the fluxtube temperature structure have been deduced from such profiles
(Solanki, 1986). It has also been possible to obtain information on the vertical
gradient of the magnetic field (Stenflo et al., 1987), as well as the inclination
of the fluxtubes to the vertical. Some of the other results obtained from these
data are mentioned further below.

Much effort has gone into the (mostly theoretical) investigation of dynamical
phenomena associated with a concentrated magnetic field. Of particular importance
has been the emergence of the non-linear treatment of such phenomena. For example,
the convective collapse of fluxtubes as a means of concentrating the field into
small bundles with field strengths well in excess of the value expected through
equipartition with photospheric motions, has been studied in detail numerically
(e.g. Nordlund, 1986; Hasan, 1985). One outcome of these calculations has been
that the final state is one of overstable oscillations. Also, a number of
mechanisms for exciting and amplifying the various wave modes and oscillations in
fluxtube have been recently proposed. Interesting is the result of Venkatakrishnan
(1986), who has found that very high amplitude oscillations and waves can be
excited resonantly inside fluxtubes by external pressure fluctuations. In view of
such non-linear results and the rich literature on linear calculations of fluxtube
waves (see Roberts, 1986 for a review), it appears surprising that until recently
no evidence for non-stationary mass motions in fluxtubes, except low amplitude 5-
minute oscillations (Giovanelli et al., 1978; Wiehr, 1985), existed at all. The
reason is that fluxtubes cannot be resolved, so that usually more than one fluxtube is present in the resolution element. Only the oscillations or waves which
are in phase in all of these give an oscillating Doppler shift signal. However, an
analysis of line widths by Solanki (1986) has shown that rms velocities of 2-4 km
s^{-1} are present inside the magnetic elements, from which he has inferred the
presence of non-stationary mass motions with amplitudes (in the vertical
direction) larger than in the surrounding quiet photosphere. Unfortunately, such
an analysis is not able to differentiate directly between the signatures of
different wave modes etc. Very high spatial and temporal resolution observations,
as are expected to become available from the Canary islands and perhaps later from
space, are required for this. The discovery of an asymmetry in the Stokes V
profile by Stenflo et al. (1984) is also suggestive of mass motions, although its
proper interpretation is still unclear.

Downflows inside fluxtubes and in their immediate surroundings were widely
observed in the 1960s and 1970s, but more recently Stenflo and Harvey (1985) and
Solanki (1986) among others have ruled out flows faster than 0.25 km s^{-1} in the
photospheric layers of fluxtubes. Miller et al. (1984) and Solanki and Stenflo
(1986) have shown that the previous positive detections of downflows were due to
poor spectral resolution and a neglect of the (then partly unknown) asymmetry in
the unpolarized (Stokes I) and polarized (Stokes V) line profiles. Hasan and
Schüssler (1985) have provided theoretical support for the absence of net
downflows.

A field of ever-growing importance concerns the interaction of the magnetic
field with its surroundings. Observational evidence for such interaction is still
sparse, but the investigation is gaining momentum. The classical study of line
bisectors has shown that convection is affected quite strongly by magnetic fields
(Cavallini et al., 1985, 1987) as theoretically predicted earlier. An analysis of

time sequences of white light pictures from Spacelab 2 also beautifully demonstrates that the granulation changes dramatically in character in the vicinity of magnetic elements (Title et al., 1987a). The convection pattern in magnetic regions is considerably more stable than in non-magnetic regions. Individual granules appear to live almost twice as long as on the quiet Sun. Furthermore, Title et al. (1987b) have found that magnetic elements are mainly concentrated in the intergranular lanes, as had been previously theoretically anticipated, e.g. by Nordlund's (1986) 3-D calculations of fluxtube convective collapse. His models as well as those of Deinzer et al. (1984b) take the interaction with the non-magnetic atmosphere into account in detail. Deinzer et al. have stressed the presence of a convective cell surrounding the fluxtubes, produced by the inclination of the surfaces of constant temperature in the immediate vicinity of the fluxtubes. This inclination is due to the cooling of this region by the magnetic elements through the influx of radiation and the inhibition of convection.

The Sun, being the only resolvable star, plays a central role in our understanding of stellar magnetic activity as a whole. Thus it is the only star on which the detailed structure of the magnetic field can be investigated more or less directly. Furthermore, since the solar fluxtubes are spatially unresolved, their study has often led to the development of instrumentation and techniques which can be applied to the measurement of stellar magnetic fields or stellar activity. Older examples are the Babcock magnetograph (used to measure the field on ·Ap stars) and the use of Ca II H and K flux as a measure of stellar activity. Newer examples are the use of powerful statistical techniques, originally developed for the analysis of solar spectra, to Ap stars (Mathys and Stenflo, 1986) and to active G and K main sequence stars.

References

Ayres, T.R., Testerman, L., Brault, J.W.: 1986, Astrophys. J. **304**, 542
Cavallini, F., Ceppatelli, G., Righini, A.: 1985, Astron. Astrophys. **143**, 116
Cavallini, F., Ceppatelli, G., Righini, A.: 1987, Astron. Astrophys. **173**, 155
Deinzer, W., Hensler, G., Schüssler, M., Weisshaar, E.: 1984a, Astron. Astrophys. **139**, 426
Deinzer, W., Hensler, G., Schüssler, M., Weisshaar, E.: 1984b, Astron. Astrophys. **139** 435
Giovanelli, R.G., Livingston, W.C., Harvey, J.W.: 1978, Solar Phys. **59**, 49
Hasan, S.S.: 1985, Astron. Astrophys. **143**, 39
Hasan, S.S., Schüssler, M.: 1985, Astron. Astrophys. **151**, 69
Herbold, G., Ulmschneider, P., Spruit, H.C., Rosner, R.: 1985, Astron. Astrophys. **145**, 157
Mathys, G., Stenflo, J.O.: 1986, Solar Phys. **168**, 184
Miller, P., Foukal, P., Keil, S.: 1984, Solar Phys. **92**, 33
Nordlund, Å.: 1986, in Proc. Workshop on Small Magnetic Flux Concentrations in the Solar Photosphere, W. Deinzer, M. Knölker, H.H. Voigt (eds.), Vandenhoeck & Ruprecht, Göttingen, p. 83
Roberts, B.: 1986, in "Small Scale Magnetic Flux Concentrations in the Solar Photosphere", W. Deinzer, M. Knölker, H.H. Voigt (eds.), Vandenhoeck & Ruprecht, Göttingen, p. 169
Solanki, S.K.: 1986, Astron. Astrophys. **168**, 311
Solanki, S.K., Stenflo, J.O.: 1984, Astron. Astrophys. **140**, 185
Solanki, S.K., Stenflo, J.O.: 1986, Astron. Astrophys. **170**, 120
Steiner, O., Pneuman, G.W., Stenflo, J.O.: 1986, Astron. Astrophys. **170**, 126
Stenflo, J.O., Harvey, J.W.: 1985, Solar Phys. **95**, 99
Stenflo, J.O., Harvey, J.W., Brault, J.W., Solanki, S.K.: 1984, Astron. Astrophys. **131**, 33
Stenflo, J.O., Solanki, S.K., Harvey, J.W.: 1987, Astron. Astrophys. **173**, 167
Title, A.M., Tarbell, T.D. and the SOUP Team: 1987a, in Proc. Second Workshop on

Problems of High Resolution Solar Observation, Sept. 1986, Boulder, CO
Title, A.M., Tarbell, T.D., Topka, K.P.: 1987b, Astrophys. J. **317**, 892
Van Ballegooijen, A.A.: 1985, in Measurements of Solar Vector Magnetic Fields,
    M.J. Hagyard (ed.), NASA Conf. Publ. 2374, p. 322
Venkatakrishnan, P.: 1986, Solar Phys. **104**, 347
Wiehr, E.: 1985, Astron. Astrophys. **149**, 217

V. HEATING AND DYNAMICS OF CHROMOSPHERE AND CORONA

(G.E. Brueckner)

The crucial role of magnetic fields in any mechanism to heat the outer solar
atmosphere has been generally accepted by all authors. However, there is still no
agreement about the detailed function of the magnetic field. Heating mechanisms
can be divided up into 4 classes: (I) The magnetic field plays a passive role as a
suitable medium for the propagation of Alfvén waves from the convection zone into
the corona (Ionson, 1984). (II) In closed magnetic structures the slow random
shuffling of field lines by convective motions below the surface induces electric
currents in the corona which heat it by Joule dissipation (Heyvaerts and Priest,
1984). (III) Emerging flux which is generated in the convection zone reacts with
ionized material while magnetic field lines move through the chromosphere,
transition zone and corona. Rapid field line annihilation, reconnection and drift
currents result in heating and material ejection (Brueckner, 1987; Brueckner et
al., 1987; Cook et al., 1987). (IV) Acoustic waves which could heat the corona can
be guided by magnetic fields. Temperature distribution, wave motions and shock
formation are highly dependent on the geometry of the flux tubes (Ulmschneider and
Muchmore, 1986; Ulmschneider, Muchmore and Kalkofen, 1987).

The emphasis of the literature, both theoretical and observational, is
shifting to detailed investigations of fine structures, inhomogeneities,
assymetries, singularities and the interaction of waves with changing conditions
in the surrounding media (Steinolfson et al., 1986; Lou and Rosner, 1986; Van
Ballegooijen, 1985; Einaudi and Mok, Yung, 1987; Davila, 1987). In order to
explain the increase of the smeared out emission measure distribution function
$Q(T)$ at lower temperatures, a mixture of cool and hot loops has been introduced
(Antiochos and Noci, 1986). Low lying, small scale loops (h < 5000 km) are assumed
to be the main source of the cooler emission. A possible explanation of the
dominant redshifted 100,000° K emission is based on assymetries in the loop
geometry or heating rate (McClymont and Graig, 1987). Spicules are possible
manifestations of upflows over regions of increased heating rate in a similar
model invoking inhomogeneous heating (Athay, 1984). An analysis of high resolution
C IV transition zone spectra showed that blueshifted (upward moving) material at
100,000° K cannot compensate for the observed predominant redshifted (downward
moving) material at the same temperature. The upward mass flow is 3 orders of
magnitude lower than the downward mass flux (Dere, Bartoe and Brueckner, 1986).

The role of transition zone explosive events and jets in the heating process
of the transition zone and corona has been reevaluated using much more comprehensive observations from Spacelab-2 (Cook et al., 1987). Although there are many
more events present on the sun than earlier estimates from sounding rockets
indicated, their total kinetic energy is only 2.5×10^4 ergs cm^{-2} s^{-1}, which seems
to be insufficient to compensate for the energy losses of the corona. However,
these estimates are based on an analysis of a rather narrow temperature regime,
therefore they represent only a lower limit. An analysis of intensity fluctuations
and Doppler-shifts of the NV lines (T ~ 250,000 K°) results in an upward energy
flow of 10^3 ergs cm^{-2} s^{-1} if interpreted as acoustic waves (Bruner and Polleto,
1984). This is again a lower limit because of the rather coarse spatial resolution
(3×3 arc sec^2) of the Solar Maximum Mission observations. Microwave solar
radiation at 6.3 cm displays fluctuation, which has been observed simultaneously

with two radio telescopes several 1000 kilometers apart. Most of the observed fluctuation must be associated with sources other than the sun. However, significant correlation was found in one run. A coherence length of 90 sec may indicate an association with explosive events in the transition zone (Benz and Furst, 1987).

Several investigations of the posible role of very fine structure (1 cm to 1 km) filaments in the transition zone and corona have been carried out. Filamentary electric currents in an ambient H field of 10 Gauss can dissipate enough energy in a Joule heating model of the lower transition zone (Rabin and More, 1984). Such filamentary structures can be the result of thermal instabilities in the presence of a magnetic field, the low frequency inhomogeneities are occuring along the magnetic field lines (Bodo et al., 1987). A new analysis of the "fill factor" in the transition zone supports these theoretical considerations (Dere et al., 1987) The fill factor is always smaller than 1% and can be as small as 10^{-4}. This leads to an upper limit for the dimension of the hyperfine structure of 100 km. The observations indicate, that these fine structure filaments can have a length of several thousand km and that they surround spicules. A consequence of this view of the transition zone is a revision of the amount of conductive heating in the transition zone from the corona (Bodo et al., 1987; Dere et al., 1987). The overall broadening of transition zone lines by plasma motions is sufficient to support a heating of the corona by Alfvèn waves (Brueckner, 1987).

Line profile analysis of Fe XIV in the quiet corona shows a periodic (235 s) intensity fluctuation but no periodicity of line width or Doppler velocity (Tsubaki, Saito and Suematsu, 1986). This result is in partial agreement with older measurements but it contradicts others. It confirms the intrinsic difficulty of such measurements in optically thin coronal lines. Measurements of resonantly scattered H-Ly-α radiation in the corona at solar maximum place an upper limit of 110 km s^{-1} for the combined thermal and nonthermal motions of hydrogen atoms in the corona (Withbroe, Kohl and Weiser, 1985). Observations of interplanetary scintillations with the VLA telescope show random velocities of 200–300 km s^{-1} along the bulk flow and 0–100 km s^{-1} perpendicular to it in the corona between 3 and 4.5 R$_o$. The mean speed of the solar wind seems to increase rapidly from 0 to several hundred km s^{-1} in the same region (Armstrong et al., 1986).

References

Armstrong, J.W., Coles, W.A. Kojima, M., and Rickett, B.J.: 1986, "The Sun and the Heliosphere in Three Dimensions", R.G. Marsden (ed.), D. Reidel, p. 59
Antiochos, S.K. and Noci, G.,: 1986, Ap. J. **301**, 440
Athay, R.G.: 1984, Ap. J. **287**, 412
Benz, A.O. and Furst, E.: 1987, Astr. Ap. **175**, 282
Bodo, G., Ferrari, A., Massaglia, S., and Rosner, R.: 1987, Ap. J. **313**, 432
Brueckner, G.E.: 1987, Proceedings of Eslab Symposium 21, "Small Scale Plasma Processes", ESA/ESTEC Publications, Noordwijk (Netherlands), in press
Brueckner, G.E., Bartoe, H.-D.F., Cook, J.W., Dere, K.P. and Socker, D.G., Kurokawa, H., and McCabe, M., 1987, submitted to Ap. J.
Bruner, M.E., and Polleto, G.: 1984, Mem. Soc. Astr. It. **55**, 313
Cook, J.W., Lund, P.A., Bartoe, J.-D.F., Brueckner, G.E., Dere, K.P., and Socker, D.G.: 1987, Proceedings of the 5th Cambridge Workshop on "Cool Stars, Stellar Atmospheres and the Sun", Cambrigde, MA, USA, in press
Davila, J.M.: 1987, Ap. J. **317**, 514
Dere, K.P., Bartoe, J.-D.F., and Brueckner, G.E.: 1986, Ap. J. **310**, 456
Dere, K.P., Bartoe, J.-D.F., Brueckner, G.E., Cook, J.W., and Socker, D.G.: 1987, Sol. Phys., in press
Einaudi, G. and Mok, Yung: 1987, Ap. J. **319**, 520
Heyvaerts, J. and Priest, E.R.: 1984, Astr. Ap. **137**, 63
Ionson, J.A.: 1984, Ap. J. **276**, 357

Lou, Y.Q., and Rosner R.: 1986, Ap. J. **309**, 874

McClymont, A.N. and Graig, I.J.D.: 1987, Ap. J. **312**, 402

Rabin, D. and Moore, R.: 1984, Ap. J. **285**, 359

Steinolfson, R.S., Priest, E.R., Poedts, S., Nocera, L., and Goosens, M.: 1986, Ap. J. **304**, 526

Tsubaki, T., Saito, Y., and Suematsu, Y.: 1986, Publ. Astron. Soc. Japan **38**, 251

Ulmschneider, P., Muchmore, D., and Kalkofen, W.: 1987, Astr. Ap. **177**, 292

Ulmschneider, P. and Muchmore, D.: 1986, "Small Magnetic Flux Concentration in the Solar Photosphere", W. Deinzer, M. Knölker and H.H. Voigt (eds.), Abhandl. Akad. Wissensch., Vandenhoeck and Ruprecht, Göttingen

Van Ballegooijen, A.A.: 1985, Ap. J. **298**, 421

Withbroe, G.L., Kohl, J.L. and Weiser, H.: 1985, Ap. J. **297**, 324

VI. THE SUN AS A STAR

(L.E. Cram)

Studies of the global (spatially unresolved) output from the sun are important for two main reasons: (1) the global solar output directed towards the earth plays a central role in solar-terrestrial relations, and (2) global solar observations form a link between (neccessarily) global observations of stars and the more refined spatially resolved observations which are available for the sun. This report covers both aspects (insofar as they concern the sun), using the time-scales of various phenomena as a basic distinguishing characteristic. Note that certain studies of spatially unresolved solar output have not been discussed, since they are actually directed toward the investigation of phenomena of strictly limited spatial extent [e.g. radiospectrograph observations (e.g. Wiehl et al. 1985) and studies of X-ray bursts (e.g. Thomas et al. 1985)]. Collections of relevant papers may be found in De Jager and Svestka (1985) and Labonte et al. (1984), while a review of germane stellar work is available in Baliunas and Vaughan (1985) and solar-terrestrial work in Donnelly and Heath (1985). A comprehensive summary of the subject by Hudson will appear soon in Review of Geophysics and Planetary Physics.

(a) Long-term variations (> several months)

The first 5 years (1980 to 1985) of observations of total solar irradiance by the Active Cavity Radiometer Irradiance Monitor (ACRIM) aboard the NASA Solar Maximum Mission (SMM) satellite have shown a clearly defined decline of −0.019% per year (Willson 1984; Willson et al. 1986). The trend detected by ACRIM is consistent with observations made by the NOAA Nimbus-7/ERB experiment and by sounding rockets, although there is some inconsistency in the actual values of the slopes derived from the various experiments. The observed decline could represent a dependence of the solar luminosity on the solar magnetic cycle. Pap (1986a) noted that the amplitude of variations (on time scales of days to weeks) in the irradiance is smaller at sunspot minimum than at maximum, and she also claims that the dominant period of irradiance modulation increased from 23.5 days at sunspot maximum to 28 days at minimum.

In addition to work on long-term variations in total irradiance, there have been several reports on long-term variations in narrower spectral regions. Livingston and Wallace (1987) reported that several photospheric lines exhibit long-term variations after instrumental effects are removed. In particular, a small change in the central depth of CI λ 538 nm is reminiscent of the ACRIM result described above. Cavallini et al. (1986) have shown that long-term changes occur in the asymmetry and width of photospheric lines observed at disk centre, while Snodgrass (1984) has re-examined the long-term trends in a number of absorption line properties measured as part of the Mt. Wilson magnetogram program. Deming et al. (1987) have reported what seem to be significant long-term shifts in the wavelengths of disk-averaged photospheric lines (presumably reflecting changes in the line bisector shape).

Observations of the variations in the Ca II lines over cycles 20 and 21 have been presented by Keil and Worden (1984), Sivaraman et al. (1987) and White et al. (1987). At sunspot maximum, the K-line intensity reached a maximum level about 20% above the background at sunsport minimum. To account for the total K-line emission and its waveform, Skumanich et al. (1984) found it necessary to include three separate components (active network, plage and quiet sun) in the analysis.

There have been a number of studies of intriguing long-term variations in other solar properties. Sonett and Trebisky (1986) have reported that a measure of the 11-to-22-year asymmetry of the sunspot cycle has been changing systematically over a period of ~ 2.5 Gyr (BP) with an exponential time constant of ~ 2 Gyr. Raychaudhuri (1986) has exhibited statistically significant correlations between variations in the solar neutrino flux, galactic cosmic ray flux and smoothed sunspot number which are claimed to strongly indicate that the solar activity cycle is due to the pulsating character of solar nuclear energy generation. Reiger et al. (1984) have suggested that there is a 154 d periodicity in the occurrence of hard solar flares. The suggestion has been confirmed by Bogart and Bai (1985), who conjecture that the period emerges from beating between incommensurable periods, either on rotational time scales or g-mode time scales. Delache et al. (1985) have published a power spectrum of variations of the solar diameter measured with a visual astrolabe which displays several peaks, the most prominent of which appears to be closely related to a 320 d periodicity in thè Zurich sunspot number time series.

(b) Short-term variations (irregular)

The ACRIM observations described above represent the long-term trends of a time series of irradiance data which also exhibit significant fluctuations on shorter time scales. It has been clear for some time that an important contributor to these fluctuations are due in part to sunspots which "block" the solar output to produce the irradiance deficit. This picture has been confirmed in further recent work by Hudson (1984) and Pap (1985, 1986a,b,c). However, Pap has refined this result by noting that the marked decrements in irradiance coincide with the appearance of "active" - i.e. young - spot groups, while older spot groups are less effective. She argues that the fast-developing, complex active regions may be better able to inhibit convective energy transport than are the simpler fields of older active regions.

The energy flux initially blocked by sunspots can eventually emerge in at least two quite different ways: (1) by slow diffusion over time-scales of the order of decades or millenia, or (2) over a period of a few weeks or months in association with the appearance of faculae or plages. A final choice between these alternatives cannot be made at this time. However, studies of the behaviour of specific spots and facular regions suggest that facular regions radiate over their lifetime a flux which has between 70% and 110% of the time-integrated missing flux in sunspots (Chapman et al. 1986). This implies that faculae are an important component of the global energy budget of an active region and, furthermore, that the flux blocked by spots may be "stored" for only a few weeks. Many studies of this important phenomenon have concentrated on the behaviour of individual active regions (Chapman 1984, 1987; Chapman et al. 1984, 1986; Chiang and Foukal 1985; Schatten et al. 1985; Foukal and Duvall 1985; Lawrence et al. 1985; Chapman and Meyer 1986; Chapman and Boyden 1986; Foukal and Lean 1986).

The fact that active regions change the disk-integrated solar spectrum, combined with the non-uniform disposition of active regions in longitude, implies that rotational modulation of disk-integrated spectral features could be detectable. Keil and Worden (1984), Drescher et al. (1984), Singh and Prabhu (1985), and Singh and Livingston (1987) have confirmed that rotational modulation of the Ca II flux can indeed be detected, although the intrinsic evolution of the Ca II structures on time scales of a few days often significantly alters the derived rotation rate. The relationship between rotationally modulated Ca II fluxes and other chromospheric and coronal emission has been studied by Fisher et al. (1984), who found a correlation between Ca II flux and total coronal mass over a 60 d interval. The temporal characteristics of the solar UV flux and and He I λ 1083 nm have been investigated by Donnelly et al. (1985). The possibility that photospheric lines exhibit rotational modulation of their bisector shape has been raised by Bruning and Labonte (1985). A number of workers have undertaken observational studies of spectral lines formed in solar active regions with a view towards predicting the influence of such active regions on the disk-integrated line in the spectrum of the sun or other stars (Labonte 1986a,b; Sivaraman et al. 1987).

(c) Short-term Variations (regular)

It has been known for about a decade that global "5-minute" oscillations can be detected in integrated sunlight. The modes have been identified as low degree modes (1 < 3) of high radial order (n ~ 20-30), and impressive progress has been made towards a theoretical interpretation of their observed frequencies (Fossat 1985; Osaki and Shibahashi 1986). A result of considerable interest was the announcement by Woodard and Noyes (1985) that the frequencies of the global modes appeared to decrease significantly between 1980 and 1984 (in parallel with the declining phase of the solar cycle). Pallé et al. (1986) could not confirm the decrease, but Fossat et al. (1987) did find a significant decrease in frequency between observations made in 1980 and in 1985.

The possibility that sluggish global oscillations with periods exceeding 1 month could also be present in the sun has been discussed by Wolff and Blizard (1986) and Wolff and Hickey (1987). Physically, the modes arise from Coriolis forces, and they have been variously termed r-modes, Rossby wave or inertial oscillations. Although the modes have been predicted to be damped under solar conditions, Wolff and Hickey claim that more than half of the variance of the solar irradiance in the period range 13-85 days could be due to r-modes.

(d) Temporally Invariant Global Observations

All of the work described above is related to variations in the integrated solar spectrum due mainly to activity and oscillations. There have been suprisingly few studies which seek to explore the "solar-stellar connection" independently of such variations. Published studies in this area include reviews of the formation of Fe II lines in the solar spectrum by Rutten (1987) and Rutten and Kostyk (1987), and studies on rotation, temperature and turbulence effects on photospheric lines by staff at Kiev Observatory (Sheminova 1984; Gadun and Kostyk 1985).

References

Baliunas, S.L. and Vaughan, A.H.: 1985, Ann. Rev. Astron. Astrophys. **23**, 379
Bogart, R.S. and Bai, T.: 1985, Astrophys. J. **299**, L51
Bruning, D.H. and LaBonte, B.J.: 1985, Solar Phys. **97**, 1
Cavallini, F., Ceppatelli, G. and Righini, A.: 1986, Astron. Astrophys. **158**, 275
Chapman, G.A.: 1984, Nature **308**, 252
Chapman, G.A.: 1987, J. Geophys. Res. **92**, 809
Chapman, G.A. and Boyden, J.E.: 1986, Astrophys. J. **302**, L71
Chapman, G.A. and Meyer, A.D.: 1986, Solar Phys. **103**, 21
Chapman, G.A., Herzog, A.D., Lawrence, J.K. and Shelton, J.C.: 1984, Astrophys. J.

282, L99

Chapman, G.A., Herzog, A.D. and Lawrence, J.K.: 1986, Nature **319**, 654

Chiang, W.-H. and Foukal, P.: 1985, Solar Phys. **97**, 9

de Jager, C. and Svestka, Z.: 1986 (eds.) Solar Phys. **100**

Delache, Ph., Laclare, F. and Sadsaoud, H.: 1985, Nature **317**, 416

Deming, D., Espanak, F., Jennings, D.E., Brault, J.W. and Wagner, J.: 1987, Astrophys. J. **316**, 771

Donnelly, R.F. and Heath, D.F.: 1985, Adv. Space Res. **5**, 145

Donnelly, R.F., Harvey, J.W., Heath, D.F. and Repoff, T.P.: 1985, J. Geophys. Res. **90**, 6267

Drescher, Th., Woehl, H. and Kueveler, G.: 1985, ESA SP-220, 29

Fisher, R., McCabe, M., Mickey, D., Seagraves, P. and Sime, D.G.: 1984, Astrophys. J. **280**, 873

Fossat, E.: 1985, ESA SP-235, 209

Fossat, E., Gelly, B., Grec, G. and Pomerantz, M.: 1987, Astron. Astrophys. **177**, L47

Foukal, P. and Duvall, T.: 1985, Astrophys. J. **296**, 739

Foukal, P. and Lean, J.: 1986, Astrophys. J. **302**, 826

Gadun, A.S. and Kostyk, R.I.: 1985, Kinematika Fiz. Nebesn. Tel. **1**, 24

Hudson, H.S.: 1984, Adv. Space Res. **4**, 113

Keil, S.L. and Worden, S.P.: 1984, Astrophys. J. **276**, 766

Labonte, B.J., Chapman, G.A., Hudson, H.S. and Willson, R.: 1984 (eds.) NASA CP-2310

Labonte, B.J.: 1986a,b, Astrophys. J. Suppl. **62**, 229, 241

Lawrence, J.K., Chapman, G.A., Herzog, A.D. and Shelton, J.C.: 1985, Astrophys. J. **292**, 297

Livingston, W. and Wallace, L.: 1987, Astrophys. J. **314**, 808

Osaki, Y. and Shibahashi, H.: 1986, Astrophys. Space Sci. **118**, 195

Pallé, P.L., Pérez, J.C., Regulo, C., Roca-Cortes, C., Isaak, G.R., McLeod, C.P. and Van der Raay, H.B.: 1986, Astron. Astrophys. **170**, 114

Pap, J.: 9185, Solar Phys. **97**, 21

Pap, J.: 1986a, Astrophys. Space Sci. **127**, 55

Pap, J.: 1986b, Adv. Space Res. **6**, 65

Pap, J.: 1986c, Bull. Astron. Inst. Czechosl. **37**, 202

Raychaudhuri, P.: 1986, Solar Phys. **104**, 415

Rieger, E., Share, G.H., Forrest, D.J., Kambach, G., Reppin, C. and Chupp, E.L.: 1984, Nature **312**, 623

Rutten, R.J.: 1987, in "Physics of Formation of Fe II Lines", IAU Coll. No. 94 (Reidel: Dordrecht)

Rutten, R.J. and Kostyk, R.I.: 1987, in "Physics of Formation of Fe II Lines", IAU Coll. No. 94 (Reidel: Dordrecht)

Schatten, K.H., Miller, N., Sofia, S., Endal, A.S., Chapman, G. and Hickey, J.: 1985, Astrophys. J. **294**, 689

Sheminova, V.A.: 1984, Soln. Dannye **7**, 70

Singh, J. and Livingston, W.C.: 1987

Singh and Prabhu, T.P.: 1985, Solar Phys. **97**, 203

Sivaraman, K.R., Singh, J., Bagare, S.P. and Gupta, S.S.: 1987, Astrophys. J. **313**, 456

Skumanich, A, Lean, J.L., White, O.R. and Livingston, W.C.: 1984, Astrophys. J. **282**, 776

Snodgrass, H.B.: 1984, Solar Phys. **94**, 13

Sonnett, C.P. and Trebisky, T.J.: 1986, Nature **322**, 615

Thomas, R.J., Starr, R. and Crannell, C.J.: 1985, Solar Phys. **95**, 323

Willson, R.C.: 1984, Space Sci. Rev. **38**, 222

Wilson, R.C., Hudson, H.S., Frohlich, C. and Brusa, R.W.: 1986, Science **234**, 1114

White, O.R., Livingston, W.C. and Wallace, L.: 1987, J. Geophys. Res. **92**, 823

Wiehl, H.J., Benz, A.O. and Aschwanden, M.J.: 1985, Solar Phys. **95**, 167

Wolff, C.L. and Blizard, J.B.: 1986, Solar Phys. **105**, 1

Wolff, C.L. and Hickey, J.R.: 1987, Science **235**, 1631

Woodard, M. and Noyes, R.: 1985, Nature **318**, 449

VII. INSTRUMENTATION FOR THE STUDY OF SOLAR RADIATION AND STRUCTURE

(J. Harvey)

This report complements the report on solar instrumentation presented by Commission 10 in that subjects treated there are not repeated here. Both reports taken together should give a fair, though necessarily incomplete, idea of solar instrumentation activity during the period June 1984 through June 1987.

A. Absolute Spectral Radiometry

Broad-band radiometers on the Solar Maximum Mission and Nimbus-7 satellites continue to operate and have revealed interesting changes in solar irradiance. Radiometers have been flown on Spacelabs 1 and 2 with good results (Crommelynck et al., 1986; Labs et al., 1987; Van Hoosier et al., 1986). The difficult region between 5 and 57 nm has been investigated by Carlson et al. (1984). A long term program, using modern equipment and techniques, to measure solar irradiance from the ground has started (Palmer et al. 1983).

B. Spectro- and Differential Photometry

One of the most active research areas during the past three years has been precision differential photometry, often directed toward limb observations. Most of these instruments involve rapid scanning of the image with detector arrays to reduce atmospheric noise. Seykora (1985) uses low amplitude, high frequency chopping to detect tiny intensity variations. Rösch and Yerle (1984), Hirayama et al. (1985a), Dicke et al. (1986) and Herzog et al. (1986) scan parallel to the solar limb to measure facular contrast and image geometry. Limb-to-limb scanning is employed by Chapman et al. (1986). A unique instrument to measure solar diameter fluctuations, the solar disk sextant, has recently been ground tested by Sofia and his colleagues.

Ultraviolet spectrometers and results have been described by Samain and Lemaire (1985), Hirayama et al. (1985b), Parkinson and Gabriel (1986) and Epstein et al. (1987). A 1024-element, linear array detector has been added to the spectrograph at Purple Mountain Observatory (Wang et al., 1987). At the Crimean Astrophysical Observatory, an ingenious spectrometer using a Michelson interferometer and lazer stabilization, has been used for precise Doppler shift measurements (Didkovsky et al. 1986).

C. Polarimetry

Careful design has allowed Henson and Kemp (1984) to achieve broadband circular polarization measurements to one part per million. New Stokes polarimeters are under construction by the High Altitude Observatory and the Applied Physics Laboratory of Johns Hopkins University for installation at the National Solar Observatory. The high sensitivity video magnetograph of the Big Bear Solar Observatory is described by Wang et al. (1985) and the video Solar Magnetic Field Telescope of the Beijng Observatory has been described in great detail (Beijng, 1986). Video techniques are also used in a spectromagnetograph being built at the National Solar Observatory (Jones, 1987). Markov (1985) present the characteristics of the 1024-element magnetograph operated by SibIZMIR.

D. High Angular Resolution

The flight of Spacelab 2 in the summer of 1985 at last provided extended periods of solar observations with sub-arc-second resolution and no atmospheric distortion. The visible and ultraviolet instruments and preliminary results are described by Title et al. (1986) and Brueckner et al. (1986) respectively. Rocket flights, though brief, continue to be an important method for obtaining high resolution observations. The Transition Region Camera was flown in 1985 (Damé et al., 1986) and first experimental flights of normal incidence X-ray telescopes have been conducted by groups from the Center for Astrophysics (Golub et al., 1985), Lockheed and Stanford. On the ground, speckle imaging techniques have at last been developed into a practical tool for achieving nearly diffraction-limited, white-light images of small solar structures (Laville Conde et al., 1986). Adaptive optics experiments continue at the National Solar Observatory by groups headed by R. Smithson and J. Beckers. The MSDP spectrograph at Pic du Midi has been improved and now delivers half-arc-second, two-dimensional spectroscopy (INSU), 1986).

E. Infrared Instrumentation

Using equipment designed for stellar observations, remarkable far-infrared observations of the solar limb structure, limb darkening and oscillations have been made both from high-altitude aircraft (Lindsey et al., 1984) and from the ground (Lindsey and Kaminsky, 1984). Balloons have been used to observe the thermal emission of dust near the sun (Mizutani et al., 1984), and for mapping the disk at 4.6 and 18 micrometers (Zou, 1984) and 50, 80 and 200 micrometers (Degiacomi et al., 1985). Excellent spectra were obtained by a Fourier transform spectrometer flown on Spacelab 3 and a Michelson interferometer flown on a balloon (Boreiko and Clark, 1986). Two-dimensional arrays of sensitive infrared detectors are becoming available for use in solar observations (Graves et al., 1987) and a remarkable new pair of photographic emulsions have been produced that are sensitive as far as 1.4 micrometers (Shcherbakova et al. 1985).

F. Radio Instrumentation and Techniques

Frequency agile receiving equipment has become a popular addition to solar radio telescopes (Hurford et al., 1984; White et al. 1986). Earth rotation image synthesis has been employed by Alissandrakis et al. (1985) to obtain solar images at 169 MHz and by a group at SibIZMIR to obtain microwave images (Smolkov et al., 1986). Very long baseline interferometry of solar features has been attempted by Tapping (1986) as a way of obtaining very high resolution. A fast 2.8 GHz receiver was described by Jin et al. (1986) and a sensitive 80 GHz whole sun radiometer and polarimeter have been developed by Nakajima et al. (1985). The outer corona is observed by the low frequency image synthesis telescope at Clark Lake (Kundu et al., 1987) and the magnetic field is probed using Faraday rotation of Helios transmissions (Pätzold et al., 1987).

G. Helioseismology

A new generation of instruments is under development to make observations of solar velocity oscillations. Magneto-optic filters that use resonance absorption effects in atomic vapor to reveal Doppler shifts have been developed by Cacciani and Rhodes (1984), Appourchaux (1987) and Koyama (1986). A laser-stabilized, electrically-tunable, Fabry-Perot filter has been developed by Rust et al. (1986). A birefringent filter is used by Libbrecht and Zirin (1986) to observe degrees > 4. Glenar et al. (1986) have shown that laser heterodyne spectroscopy is an effective way of measuring solar velocity oscillations. Michelson interferometers continue to be developed for Doppler shift measurements (GONG, 1987). A classical spectrograph has been equipped with three linear diode arrays to obtain phase shift measurements of oscillations (Staiger, 1987).

Equipment to measure solar intensity oscillations has been described by Duvall et al. (1986), Didkovsky and Kotov (1986), Kotov et al. (1985), Didkovsky

(1985), Jimenez et al. (1987) and Nishikawa et al. (1986).

Networks of oscillation observing equipment have been or are being established at numerous sites around the world to provide continuous observations until
suitable space experiments can be conducted. A group from Birmingham has stations
in operation in Australia, the Canary Islands and Hawaii to obtain low-degree
observations. The IRIS project based in Nice has placed one low-degree instrument
at La Silla and plans additional instruments at Stanford and Tashkent (INSU,
1986). The GONG project based in Tucson plans to install six instruments for
degrees up to about 300 at sites that are currently being investigated. A prototype instrument is under construction (GONG, 1987).

References

Alissandrakis, C.E., Lantos, P. and Nicolaides, E.: 1985, Solar Phys. **97**, 267
Appourchaux, T.: 1987, Solar Phys. **109**, 393
Beijing: 1986, Publ. Beijing Astron. Obs. **8**, 1-422
Boreiko, R.T. and Clark, T.A.: 1986, Astron. Astrophys. **157**, 353
Brueckner, G.E., Bartoe, J.-D.F., Cook, J.W., Dere, K.P. and Socker, D.G.: 1986,
    Adv. Space Res. **6**, 263
Cacciani, A. and Rhodes, E.J., Jr.: 1984, Solar Seismology from Space, 115
Carlson, R.W., Ogawa, H.S., Phillips, E. and Judge, D.L.: 1984, Appl. Opt. **23**,
    2327
Chapman, G.A., Herzog, A.D. and Templar, S.: 1985, Bull. Amer. Astron. Soc. **17**,
    896
Crommelynck, D.A., Brusa, R.W. and Domingo, V.: 1986, solar Phys. **107**, 1
Damé, L., Foing, B.H., Martic, M., Bruner, M., Brown, W., Decaudin, M. and Bonnet,
    R.M.: 1986, Adv. Space res. **6**, 273
Degiacomi, C.G., Kneubühl, F.K. and Huguenin, D.: 1985, Astrophys. J. **298**, 918
Dicke, R.H., Kuhn, J.R. and Libbrecht, K.G.: 1986, Astrophys. J. **311**, 1025
Didkovsky, L.V.: 1985, Izv. Krym. astrofiz. Obs. **72**, 217
Didkovsky, L.V. and Kotov, V.A.: 1986, Izv. Krym. astrofiz. Obs. **74**, 132
Didkovsky, L.V., Kozhevatov, I.E. and Stepanyan, N.N.: 1986, Izv. Krym. astrofiz.
    Obs. **74**, 142
Duvall, R.L., Jr., Harvey, J.W. and Pomerantz, M.A.: 1986, Nature **321**, 500
Epstein, G.L., Thomas, R.J. and Neupert, W.M.: 1987, Bull. Amer. Astron. Soc. **19**,
    in press
Glenar, D.A., Deming, D., Espenak, F., Kostiuk, T. and Mumma, M.: 1986, Appl. Opt.
    25, 58
Golub, L., Nystrom, G., Spiller, E. and Wilczynski, J.: 1985, SPIE **563**, 266
GONG: 1987, Newsletter No. 5, National Solar Obs., Tucson
Graves, B., Foukal, P., Rieke, M. and Fowler, A.: 1987, Bull. Amer. Astron. Soc.
    19, in press
Henson, G.D. and Kemp, J.C.: 1984, Solar Phys. **93**, 289
Herzog, A.D., Chapman, G.A., Lawrence J.K. and Templer, S.: 1985, Bull. Amer.
    Astron. Soc. **17**, 833
Hirayama, T., Hamana, S. and Mizugaki, K.: 1985a, Solar Phys. **99**, 43
Hirayama, R, Tanaka, K., Watanabe, T., Akita, K., Sakurai, T. and Nishi, K.:
    1985b, Solar Phys. **95**, 281
Hurford, G.J., Read, R.B. and Zirin H.: 1984, Solar Phys. **94**, 413
INSU: 1986, Ann. Rep. Inst. Nat. Sciences de L'Universe (CNRS) 1984-1985, 26
Jimenez, A., Pallé, P.L., Roca Cortes, T. and Domingo, V.: 1987, Astron.
    Astrophys. **172**, 323
Jin, S.-Z., Zhao, R.-Y. and Fu, Q.-J.: 1986, Solar Phys. **104**, 391
Jones, H.P.: 1987, Bull. Amer. Astron. Soc. **19**, in press
Kotov, V.A., Koutchmy, S., Kononovich, E.V., Ryzhykova, N.N. and Tsap, T.T.: 1985,
    Izv. Krym. astrofiz. Obs. **73**, 26
Koyama, K.: 1986, "Hydrodynamic and Magnetohydrodynamic Problems in the Sun and
    Stars", Y. Osaki (ed.), Univ. Tokyo, p. 339

Kundu, M.R., Gergely, T.E., Schmahl, E.J., Szabo, A., Loiacono, R., Wang, Z. and Howard, R.A.: 1987, Solar Phys. **108**, 113

Labs, D., Neckel, H., Simon, P.C. and Thuiller, G.: 1987, Solar Phys. **107**, 203

Laville Conde, A., von der Lühe, O. and Radick, R.R.: 1986, Bull. Amer. Astron. Soc. **18**, 933

Libbrecht, K.G. and Zirin, J.: 1986, Astrophys. J. **308**, 413

Lindsay, C. and Kaminksy, C.: 1984, Astrophys. J. **282**, L103

Lindsay, C., Becklin, E.E., Jefferies, J.T., Orrall, F.Q. and Werner, M.W.: 1984, Astrophys. J. **281**, 862

Markov, V.S.: 1985, Issled. Geomag. Aehron. Fiz. Solntsa **72**, 188

Mizutani, K., Maihara, T., Hiromoto, N. and Takami, H.: 1984, Nature **312**, 134

Nakajima, H., Sekiguchi, H., Sawa, M., Kai, K. Kawashima, S., Kosugi, R., Shibya, N., Shinohara, N. and Shiomi, Y.: 1985, Publ. Astron. Soc. Japan **37**, 163

Nishikawa, J., Hamana, S., Mizugaki, K. and Hirayama, R.: 1986, Pub. Astron. Soc. Japan **38**, 277

Palmer, J.M., Brad, L.G., Perry, D.L. and Wolfe, W.L.: 1983, J. Opt. Soc. Amer. **73**, 1964

Parkinson, J.H. and Gabriel, A.H.: 1986, Adv. Space Res. **6**, 243

Pätzold, M., Bird, M.K., Volland, H., Levy, G.F.S., Seidel, B.L. and Stelzried, C.T.: 1987, Solar Phys. **109**, 91

Rösch, J. and Yerle, R.: 1984, ESA SP-220, 217

Rust, D.M., Burton, C. and Leistner, A.J.: 1986, SPIE **627**, 39

Samain, D. and Lemaire, P.: 1985, Astrophys. Space Sci. **115**, 227

Shcherbakova, Z.A., Shcherbakov, A.G. and Shapiro, B.I.: 1985, Pis'ma Astron. Zh. **11**, 774

Seykora, E.: 1985, Solar Phys. **99**, 39

Smolkov, G.Y., Pistolkors, A.A., Treskov, T.A., Krissinel, B.B., Putilov, V.A. and Potapov, N.N.: 1986, Astrophys. Space Sci. **119**, 1

Staiger, J.: 1987, Astron. Astrophys. **175**, 263

Tapping, K.F.: 1986, Solar Phys. **104**, 199

Title, A.M. et al.: 1986, Adv. Space Res. **6**, 253

Van Hoosier, M.E., Bartoe, J.-D.F., Brueckner, G.E. and Prinz, D.K.: 1986, Bull. Amer. Astron. Soc. **18**, 675

Wang, C.-J., Lu, J., Ni, Z.-R., You, J.-Q. and Fan, C.-J.: 1987, Acta Astron. Sinica **28**, 101

Wang, J., Zirin, H. and Shi, Z.: 1985, Solar Phys. **98**, 241

White, S.M., Kundu, M.R. and Szabo, A.: 1986, Solar Phys. **107**, 135

Zou, H.-C.: 1984, Acta Astron. Sinica **25**, 1

PRESIDENT: R.W. Nicholls
VICE PRESIDENT: S. Sahal
ORGANISING COMMITTEE: A.H. Gabriel, T. Kato, F.J. Lovas, S.L. Mandel'shtam,
                        H. Nussbaumer, W.H. Parkinson, Z.R. Rudzigas, W.L. Wiese

Introduction

The Commission has, since its inception, been devoted to the continually increasing needs of astronomy and astrophysics for reliable atomic and molecular data a) for diagnostic interpretation of astronomical observations, and b) for support of theoretical modelling of astrophysical situations. At the 1985 Delhi General Assembly, the Commission reviewed the scope of its subject matter, and considered whether it should be extended to include higher energy physical processes than are commonly treated by atomic and molecular data. It was concluded that there was no strong demand for this change, which, if implemented, would make the work of the Commission too diffuse. The appropriateness of the past working group structure was also carefully reviewed in the light of contemporary needs. The following working groups, which have evolved from those of past years, together with their chairmen was approved:

1: Atomic Spectra and Wavelength Standards (excluding primary standards):
   W.C. Martin
2: Atomic Transition Probabilities: W.L Wiese
3: Collision Processes: A. Dalgarno
4: Line Broadening: N. Feautrier
5: Molecular Structure and Transition Data: W.H. Parkinson

WORKING GROUP 1: ATOMIC SPECTRA AND WAVELENGTH STANDARDS

A.   Recent Laboratory Research

Some of the references given at the end of this report are sorted according to spectra in Table 1. A Bibliography on Atomic Energy Levels and Spectra covering the period January 1984 through December 1987 is scheduled for publication in 1988 (76). Since this bibliography includes all atomic spectra for which new data have been published, Table 1 is mainly confined to a selection on the basis of astrophysical interest from work on levels, wavelengths, and line classifications for elements $Z \leq 30$.

The papers on Fe II and Fe I (Table 1) represent the most recent results of a rather extensive research program being carried out for these important spectra in several laboratories. Additional papers by S. Johansson and collaborators reviewing progress on Fe II and plans for further work are included in (77). The recent extension of the Ti I analysis (39, 40) is also based on new observations over a wide wavelength range. The extension of the V II analysis by Iglesias et al. allowed the classification of 149 lines (44).

New accurate wavelengths from laboratory measurements have been published for those lines of Cr I, II (45) and Ni I, II (62) that have also been identified in high-resolution solar spectra over the infrared wavenumber range 9000-1800 cm^{-1}. Biémont and Brault's similar observations of Mg I, II (18) and Al I, II (20) over this same infrared range allowed extensions of the analyses of these spectra. Cooksy et al. made precise far-infrared measurements of the ground term $^3P_1$-$^3P_2$ fine-structure intervals in both ^{14}N II and ^{15}N II to facilitate observation of these transitions in the interstellar medium (11).

Some of the more extensive additions to data on neutral-atom spectra not already mentioned can be cited here. Ref. (2) includes improved values and accurate series formulas for the energy levels of He I. Eriksson gives new wave-lengths for 146 N I lines in the vacuum-ultraviolet region 862-1837 Å (10). Recent data based on absorption observations in the vacuum ultraviolet include the photoionization spectrum of P I from threshold down to 765 Å (24), absorption spectra of S I (840-1220 Å) (26) and Cl I (940-1400 Å) (29), and "subvalence-shell" absorption spectra of K I (400-600 Å) (33) and Zn I (708-760 Å) (72).

B. Compilations of Laboratory Data

Sugar and Corliss have published revised and updated energy-levels compila-tions for the 235 spectra of the iron-group elements K through Ni (Z = 19-28) in a single volume (78). A compilations of levels for the P spectra was also published (79), and energy-level compilations for the spectra of S (80) and Cu (81) are underway. C.E. Moore's new compilation of levels and multiplets for O III has appeared (82). Adelman et al. have prepared a finding list from Moore's multiplet tables for spectra of H, C, N, O, and Si (83).

R.L. Kelly's most recent compilation of atomic lines below 2000 Å has been published (84). These tables, which supersede a similar 1973 compilation, cover all ionization stages of H through Kr in two sections. The first section includes energy-level classifications with observed lines sorted according to spectrum, and the second section is a finding list. Kaufman and Sugar published a compilation of forbidden (magnetic-dipole) lines for spectra of the elements Be through Mo (85). The list includes 406 observed wavelengths and gives predicted wavelengths and transition probabilities for 1660 lines over the range from 100 Å to 25.9 mm. The compilation of spectral data and Grotian diagrams for Ti V-Ti XXII by Mori et al. includes observed wavelengths with energy-level classifications (86), as does the similar compilation for Ni IX - Ni XXVIII by Shirai et al. (87). Fawcett has published calculations for the most important transition arrays in several iso-electronic sequences and included observed wavelengths in the tables. These include the O I sequence (Mg V - Ni XXI) (88), the F I sequence (Mg IV - Ni XX) (89), the Si I sequence through Ni XV (90), the P I sequence through Ni XIV (91), the S I sequence through Ni XIII (92), and the Cl I sequence (Ar II - Ni XII (93). Biémont's tables for the 3s-3p and 3p-3d transitions for V X through Ni XV (Si I sequence) also include experimental wavelengths (94).

Very complete lists of wavelengths derived from experimental energy levels comprise a part of the output of Kurucz's program of calculations for iron-group spectra (95). This work has been completed for the first and second spectra of the elements Ca through Ni and is being extended to include the first ten spectra of these elements. The accurately predicted wavelengths should allow identifica-tions of weaker lines not yet observed in the laboratory. The data can be obtained from Kurucz in computer readable form.

C. Wavelength Standards

The thorium spectrum emitted from hollow-cathode or electrodeless-lamp discharges is very useful for wavelength calibration of high-resolution spectro-meters. Palmer and Engleman's Atlas of the Thorium Spectrum (96) includes an extensive list of lines from 2277 to 13500 Å with wavenumbers accurate to ±0.002 cm^{-1} (±0.0005 Å at 5000 Å). Uranium excited in the above types of discharges also gives a very line-rich spectrum useful for wavelength calibration. The similar atlas for U lists 4928 lines (3848-9084 Å) with wavenumbers accurate to ±0.003

cm^{-1} (97). Sansonetti and Weber have measured a number of the Th and U lines (5750-6920 Å) with uncertainties of ±0.0004 cm^{-1} for Th and ±0.0003 cm^{-1} for U (98). Their measurements show the wavenumber errors of the corresponding lines in the Los Alamos atlases to be well within the quoted uncertainty estimates.

The spectrum of a Pt hollow-cathode discharge with Ne carrier gas will be used for on-board calibration of the High-Resolution Spectrograph for the Hubble Space Telescope. The accurate measurements of this spectrum by Reader et al. (99,100) and by Engleman (101) make it an excellent source of calibration wavelengths for spectroscopy over the wavelength range 1100-4100 Å. The completed measurements have yielded some 3200 wavelengths in this range with uncertainties of ±0.0005 to ±0.002 Å (100). Engleman has published accurately measured wavenumbers for about 320 Pt I lines (2200-7360 Å) and also a list of wavenumbers for Pt I lines below 2250 Å as predicted by the reevaluated energy levels (101). A list of some 560 Pt II lines together with accurate Pt II energy levels is also being published (99). A detailed atlas of the Pt hollow cathode spectrum is planned.

Eriksson has published wavelengths for 71 O II lines in the 525-4676 Å range with estimated uncertainties of about 0.001 Å (102). The vacuum-ultraviolet wavelengths (525-835 Å) calculated from the improved energy levels are probably accurate within uncertainties of 0.004 Å at 525 Å to 0.0014 Å at 796 Å. A significant addition to the available calibration wavelengths for the XUV region below 500 Å has resulted from the measurements and analysis of Y VI by Persson and Reader (103). Their complete list of some 900 lines extends from 162 to 2452 Å; the uncertainties of the more accurate Ritz-type calculated wavelengths below 500 Å vary from ±0.0003 to ±0.0020 Å.

Spectral lines of highly-ionized members of the Na I isoelectronic sequence are prominent in some laboratory and astrophysical plasmas, particularly those ionization stages centered around the iron-group elements. Reader et al. have critically reviewed and fitted the data for these spectra (Ar VIII - Sn XL) using the best available measurements and their own new observations (104). The fitted wavelengths range from 9 to 713 Å and should be especially useful for calibration in the 50-500 Å region; the uncertainties vary from about ±0.003 to ±0.007 Å.

References

1. Ito, K., Yoshino, K., Morioka, Y. and Namioka, T.: 1987, Phys. Scr. 36, p.88.
2. Martin, W.C.: 1987, Phys. Rev. A 36 (Oct. 1 issue).
3. Jannitti, E., Nicolosi, P., Tondello, G., Zheng, Y. and Mazzoni, M.: 1987, Opt. Commun. 63, p.37.
4. Jannitti, E., Pinzhong, F. and Tondello, G.: 1986, Phys. Scr. 33, p.434.
5. Dumont, P.D., Garnir, H.P. and Baudinet-Robinet, Y.: 1987, Z. Phys. D 4, p.335.
6. Baudinet-Robinet, Y., Garnir, H.P. and Dumont, P.D.: 1986, Phys. Rev. A 34, p.4722.
7. Blanke, J.H., Heckmann, P.H., Träbert, E. and Hucke, R.: 1987, Phys. Scr. 35, p.780.
8. Engström, L., Hutton, R., Reistad, N., Martinson, I., Huldt, S., Mannervik, S. and Träbert, E.: 1987, Phys. Scr. 36, p.250.
9. Dumont, P.D., Garnir, H.P., Baudinet-Robinet, Y. and Chung, K.T.: 1985, Phys. Rev. A 32, p.229.
10. Eriksson, K.B.S.: 1986, Phys. Scr. 34, p.211.
11. Cooksy, A.L., Hovde, D.C. and Saykally, R.J.: 1986, J. Chem. Phys. 84, p.6101.
12. Brown, P.R., Davies, P.B. and Johnson, S.A.: 1987, Chem. Phys. Lett. 133, p.239.

TABLE 1

Selected references on energy levels, wavelengths, and line classifications for spectra of elements Z ≤ 30. This table is supplementary to a bibliography in progress (76) and also does not include references described in Sec. B of this report.

He I 1, 2	Ar III 31
	Ar IX 32
Be I, II 3	
	K I 33
B II 4, 5	
B III 5, 6	Ca IX 34
B IV 5	Ca XI 35, 36
C IV 7, 8, 9	Sc V 37
	Sc X 34
N I 10	Sc XII 36, 38
N II 11	
N V 7	Ti I 39, 40
	Ti III, IV 41
O I 12	Ti XI 34
O II 13, 14, 102	Ti XIII 36
O VI 7	Ti XIV 42
	Ti XXI 43
F VI 15	
F VII 7, 16	V II 44
	V XII 34
Ne VIII 17	V XIV 36, 38
Mg I 18, 19	Cr I, II 45
Mg II 18	Cr XIII 34
	Cr XV 36, 38, 46, 47
Al I, II 20	
Al XI 21	Mn IV 48
	Mn XIV 34
Si II 22	Mn XVI 36, 38
Si XIV 23	
	Fe I 49, 50
P I 24	Fe II 49, 51, 52, 53
P VI–XIII 25	Fe V, VII 54
	Fe XV 34, 55
S I 26	Fe XVII 47, 56,57
SVII–XIV 27	Fe XVIII–XXIV 58
S VIII 28	Fe XXIII 59
	Fe XXIV 59, 60
Cl I 29	Fe XXV 43, 61
Cl IX 30	Zn I 72
Co XVI 34, 55	Zn IV 73
	Zn XII–XX 74
Ni I, II 62	Zn XIV, XV, XVII 68
Ni VI, VII 54	Zn XVIII 68, 75
Ni XVII 34, 55	Zn XIX 34
Ni XIX 47, 63, 64	

References (cont'd.)

13. Eriksson, K.B.S. and Wenäker, I.: 1984, Phys. Scr. 30, p.321.
14. De Robertis, M.M., Osterbrock, D.E. and McKee, C.F.: 1985, Astrophys. J. 293, p.459.
15. Engström, L.: 1985, Phys. Scr. 31, p.379.
16. Engström, L.: 1984, Phys. Scr. 29, p.113.
17. Knystautas, E.J. and Druetta, M.: 1985, Phys. Rev. A 31, p.2279.
18. Biémont, E. and Brault, J.W.: 1986, Phys. Scr. 34, p.751.
19. Chang, E.S.: 1987, Phys. Scr. 35, p.792.
20. Biémont, E. and Brault, J.W.: 1987, Phys. Scr. 35, p.286.
21. Buchet, J.P., Buchet-Poulizac, M.C., Denis, A., Desesquelles, J., Druetta, M., Martin, S., Grandin, J.P., Hennecart, D., Husson, X. and Lecler, D.: 1984, J. Phys. (Paris), Lett. 45, p.L361.
22. Artru, M.C.: 1986, Astron. Astrophys. 168, p.L5.
23. Feldman, U., Ekberg, J.O., Brown, C.M., Seely, J.F. and Richardson, M.C.: 1987, J. Opt. Soc. Am. B 4, p.103.
24. Berkowitz, J., Greene, J.P., Cho, H. and Goodman, G.L.: 1987, J. Phys. B 20, p.2647.
25. Hayes, R.W. and Fawcett, B.C.: 1986, Phys. Scr. 34, p.337.
26. Joshi, Y.N., Mazzoni, M., Nencioni, A., Parkinson, W.H. and Cantù, A.: 1987, J. Phys. B 20, p.1203.
27. Fawcett, B.C. and Hayes, R.W.: 1987, Phys. Scr. 36, p.80.
28. Trigueiros, A. and Jupén, C.: 1985, Phys. Scr. 31, p.359.
29. Cantù, A.M., Parkinson, W.H., Grisendi, T. and Tagliaferri, G.: 1985, Phys. Scr. 31, p.579.
30. Jupén, C.: 1985, Phys. Scr. 32, p.592.
31. Hansen, J.E. and Persson, W.: 1987, J. Phys. B 20, p.693.
32. Engström, L. and Berry, H.G.: 1986, Phys. Scr. 34, p.131.
33. Sommer, K., Baig, M.A., Garton, W.R.S. and Hormes, J.: 1987, Phys. Scr. 35, p.637.
34. Litzén, U. and Redfors, A.: 1987 (submitted to Phys. Scr.).
35. Jupén, C., Litzén, U., and Skogvall, B.: 1986, Phys. Scr. 33, p.69.
36. Jupén, C., Litzén, U., Kaufman, V. and Sugar, J.: 1987, Phys. Rev. A 35, p.116.
37. Smitt, R. and Ekberg, J.O.: 1985, Phys. Scr. 31, p.391.
38. Jupén, C. and Litzén, U.: 1986, Phys. Scr. 33, p.509.
39. Forsberg, P.: 1987, Thesis, University of Lund.
40. Forsberg, P., Johansson, S. and Smith, P.L.: 1986, Phys. Scr. 34, p.759.
41. Madin, M.I.: 1985, Opt. Spectrosc. (USSR) 58, p.14.
42. Jupén, C., Reistad, N., Träbert, E., Blanke, J.H., Heckmann, P.H., Hellmann, H., and Hucke, R.: 1985, Phys. Scr. 32, p.527.
43. Aglitskii, E.V. and Panin, A.M.: 1985, Opt.Spectrosc. (USSR) 58, p.453.
44. Iglesias, L., Cabeza, M.I., Garcia-Riquelme, O. and Rico, F.R.: 1987, Opt. Pura Apl. 20, p.137.
45. Biemont, E., Brualt, J.W., Delbouille, L. and Roland, G.: 1985, Astron. Astrophys., Suppl. Ser. 61, p.185.
46. Buchet-Poulizac, M.C., Buchet, J.P. and Martin, S.: 1986, J. Phys. (Paris) 47, p.407.
47. Finkenthal, M., Mandelbaum, P., Bar-Shalom, A., Klapisch, M., Schwob, J.L., Breton, C., De Michelis, C. and Mattoili, M.: 1985, J. Phys. B 18, p.L331.
48. Tchang-Brillet, W.L., Artru, M.C. and Wyart, J.F.: 1986, Phys. Scr. 33, p.390.
49. Biémont, E., Brault, J.W., Delbouille, L. and Roland, G.: 1985, Astron. Astrophys., Suppl. Ser. 61, p.107.
50. Johansson, S.: 1987, Phys. Scr. 36, p.99.
51. Johansson, S. and Cowley, C.R.: 1984, Astron. Astrophys. 139, p.243.
52. Adam, J., Baschek, B., Johansson, S., Nilsson, A.E. and Brage, T.: 1987, Astrophys. J. 312, p.337.

References (cont'd.)

53. Johansson, S. and Baschek, B.: 1987, (submitted to Nucl. Instrum. Methods Phys. Res. B).
54. Raassen, A.J.J.: 1985, Astrophys. J. 292, p.696.
55. Churilov, S.S., Kononov, E. Ya., Ryabtsev, A.N. and Zayikin, Yu. F.: 1985, Phys. Scr. 32, p.501.
56. Buchet, J.P., Buchet-Poulizac, M.C., Denis, A., Desesquelles, J., Druetta, M., Martin, S., Grandin, J.P., Husson, X. and Lesteven, I.: 1985, Phys. Scr. 31, p.364.
57. Feldman, U., Doschek, G.A. and Seely, J.F.: 1985, Mon. Not. R. Astron. Soc. 212, p.41p.
58. Seely, J.F., Feldman, U. and Safronova, U.I.: 1986, Astrophys. J. 304, p.838.
59. Lemen, J.R., Phillips. K.J.H., Cowan, R.D., Hata, J. and Grant, I.P.: 1984, Astron, Astrophys. 135, p.313.
60. Seely, J.F. and Feldman, U.: 1986, Phys. Scr. 33, p.110.
61. Briand, J.P., Tavernier, M., Marrus, R. and Desclaux, J.P.: 1984, Phys. Rev. A 29, p.3143.
62. Biémont, E., Brault, J.W., Delbouille, L. and Roland, G.: 1986, Astron. Astrophys. Suppl. Ser. 65, p.21.
63. Buchet, J.P., Buchet-Poulizac, M.C., Denis, A., Desesquelles, J., Druetta, M., Martin, S. and Wyart, J.F.: 1987, J. Phys. B 20, p.1709.
64. Haar, R.R., Curtis, L.J., Reistad, N., Jupén, C., Martinson, I., Johnson, B. M., Jones, K.W. and Meron, M.: 1987, Phys. Scr. 35, p.296.
65. Bely-Dubau, F., Faucher, P., Cornille, M. and Dubau, J.: 1986, J. Phys. (Paris), Colloq. C6, Suppl. 10, 47, p.C6-51.
66. Hsuan, H., Bitter, M., Hill, K.W., von Goeler, S., Grek, B., Johnson, D., Johnson, L.C., Sesnic, S., Bhalla, C.P., Karim, K.R., Bely-Dubau, F. and Faucher, P.: 1987, Phys. Rev. A 35, p.4280.
67. Sugar, J. and Kaufman, V.: 1986, J. Opt. Soc. Am. B 3, p.704.
68. Roberts, J.R., Pittman, T.L., Sugar, J., Kaufman, V. and Rowan, W.L.: 1987, Phys. Rev. A 35, p.2591.
69. Hutton, R., Jupén, C., Trabert, E. and Heckmann, P.H.: 1987, Nucl. Instrum. Methods Phys. Res. B 23, p.297.
70. Sugar, J. and Kaufman, V.: 1986, J. Opt. Soc. Am. B 3, p.1612.
71. Ekberg, J.O., Seely, J.F., Brown, C.M., Feldman, U., Richardson, M.C. and Behring, W.E.: 1987, J. Opt. Soc. Am. B 4, p.420.
72. Sommer, K., Baig, M.A. and Hormes, J.: 1987, Z. Phys. D 4, p.313.
73. Joshi, Y.N. and van Kleef, T.A.M.: 1987, Phys. Scr. 36, p.282.
74. Sugar, J. and Kaufman, V.: 1986, Phys. Scr. 34, p.797.
75. Hinnov, E., Boody, F., Cohen, S., Feldman, U., Hosea, J., Sato, K., Schwob, J.L., Suckewer, S. and Wouters, A.: 1986, J. Opt. Soc. Am. B 3, p.1288.
76. Musgrove, A. and Zalubas, R.: 1988, "Bibliography on Atomic Energy Levels and Spectra, January 1984 through December 1987", NBS Spec. Publ. 363, Suppl. 4 (in preparation).
77. Viotti, R., Vittone, A. and Friedjung, M., editors: 1987, Proc. IAU Colloqium 94; "Physics of Formation of Fe II Lines Outside LTE".
78. Sugar, J. and Corliss, C.: 1985, J. Phys. Chem. Ref. Data 14, Suppl. 2.
79. Martin, W.C., Zalubas, R. and Musgrove, A.: 1985, J. Phys. Chem. Ref. Data 14, p.751.
80. Martin, W.C., Zalubas, R. and Musgrove, A.: 1988, work in progress.
81. Sugar, J.: 1988, J. Phys. Chem. Ref. Data (to be submitted).
82. Moore, C.E.: 1985, Natl. Stand. Ref. Data Ser., Natl. Bur. Stand. (U.S.) 3, Sect. 11.
83. Adelman, C.J., Adelman, S.J., Fischel, D. and Warren, Jr., W.H.: 1985, Astron. Astrophys., Suppl. Ser. 60, p.339.
84. Kelly, R.L.: 1987, J. Phys. Chem. Ref. Data 16, Suppl. 1.
85. Kaufman, V. and Sugar, J.: 1986, J. Phys. Chem. Ref. Data 15, p.321.

References (cont'd.)

86. Mori, K., Wiese, W.L., Shirai, T., Nakai, Y., Ozawa, K. and Kato, T.: 1986, At. Data Nucl. Data Tables 34, p.79.

87. Shirai, T., Nakai, Y., Ozawa, K., Ishii, K., Sugar, J. and Mori, K.: 1987, J. Phys. Chem. Ref. Data 16, p.327.

88. Fawcett, B.C.: 1986, At. Data Nucl. Data Tables 34, p.215.

89. Fawcett, B.C.: 1984, At. Data Nucl. Data Tables 31, p.495.

90. Fawcett, B.C.: 1987, At. Data Nucl. Data Tables 36, p.129.

91. Fawcett, B.C.: 1986, At. Data Nucl. Data Tables 35, p.203.

92. Fawcett, B.C.: 1986, At. Data Nucl. Data Tables 35, p.185.

93. Fawcett, B.C.: 1986, At. Data Nucl. Data Tables 36, p.151.

94. Biémont, E.: 1986, Phys. Scr. 33, p.324.

95. Kurucz, R.L.: 1987, unpublished material (Harvard-Smithsonian Center for Astrophysics, Cambridge, MA 02138, U.S.A.).

96. Palmer, B.A. and Engleman, Jr., R.: 1983, Los Alamos Natl. Lab Report LA-9615, UC-4 (Los Alamos Natl. Lab., Los Alamos, NM 87545, U.S.A.).

97. Palmer, B.A., Keller, R.A. and Engleman, Jr., R.: 1980, Los Alamos Natl. Lab. Report LA-8251-MS, UC-34a (Los Alamos Natl. Lab., Los Alamos, NM 87545, U.S.A.).

98. Sansonetti, C.J. and Weber, K.H.: 1984, J. Opt. Soc. Am. B 1, p.361.

99. Reader, J., Acquista, N., Sansonetti, C.J. and Engleman, Jr., R.: 1988, J. Opt. Soc. Am. B (to be submitted).

100. Reader, J., Acquista, N. and Sansonetti, C.J.: 1988, Wavelengths of Platinum Hollow-Cathode Discharge (to be submitted for publication).

101. Engleman, Jr., R.: 1985, J. Opt. Soc. Am. B 2, p.1934.

102. Eriksson, K.B.S.: 1987, J. Opt. Soc. Am. B 4, p.1369.

103. Persson, W. and Reader, J.: 1986, J. Opt. Soc. Am. B 3, p.959.

104. Reader, J., Kaufman, V., Sugar, J., Ekberg, J.O., Feldman, U., Brown, C.M., Seely, J.F. and Rowan, W.L.: 1987, J. Opt. Soc. Am. B 4 (in press).

W.C. Martin
Chairman of the Working Group

WORKING GROUP 2: ATOMIC TRANSITION PROBABILITIES

The Data Center on Atomic Transition Probabilities at the National Bureau of Standards, Gaithersburg, Maryland, 20899, U.S.A. has continued its critical compilation work and maintains an up-to-date bibliographical data base. Work to revise and expand the existing NBS critical data compilations for the allowed and forbidden transitions in Fe-group elements, (Refs. A-D) has been completed. A single volume containing all these data for the Fe-group elements Sc to Ni is in press (Volume III of the NBS series of atomic transition probability tables) and is scheduled to be published in the near future, as a supplement to the Journal of Physical and Chemical Reference Data.

In Table 1 the important recent literature references containing atomic transition probability data are presented, which have been published since the last Working Group report of August 1984; this material is ordered according to element and stage of ionization. For brevity the references are identified there only by the running number of the general reference list given at the end of this report. In order to keep the size of this list within the allowed space, both the spectra listed here as well as the references within each spectrum had to be on a selection basis. However, the NBS Data Center will supply all-inclusive lists of references on specific spectra on request. In the general reference list supplied with this report the literature is ordered alphabetically according to principal authors. Each reference contains one or more code letters indicating the method applied by the author. These code letters are defined as follows:

THEORETICAL METHODS:

Q - quantum mechanical (including self-consistent field) calculations.
I - interpolation within isoelectronic sequences, spectral series, or homo-
    logous atoms; also, data that are presented in graphical, rather than
    tabular form.

EXPERIMENTAL METHODS:

E - measurements in emission (arc, furnace, discharge tube, shock tube, etc.).
A - measurements in absorption (King furnace, absorption tube, etc.).
L - lifetime measurements (including Hanle-effect).
H - anomalous dispersion (hook) measurements.
M - miscellaneous experimental methods (for example, Stark effect, astro-
    physical measurements, etc.).

OTHER:

C - additions or suggested revisions to data in previous articles,
    comments on particular theoretical or experimental methods, etc.
Cp- data compilations.
R - relative (non-absolute) oscillator strengths have been tabulated.
F - data on forbidden (i.e. other than electric dipole) transitions have
    been tabulated.

References for Introductory Discussion

A. Smith, M.W., Wiese, W.L.: 1973, J. Phys. Chem. Ref. Data 2, p.85.
B. Wiese, W.L., Fuhr, J.R.: 1975, J. Phys. Chem. Ref. Data 4, p.263.
C. Younger, S.M., Fuhr, J.R., Martin, G.A., Wiese, W.L.: 1978, J. Phys. Chem.
    Ref. Data 7, p.495.
D. Fuhr, J.R., Martin, G.A., Wiese, W.L., Younger, S.M.: 1981, J. Phys. Chem.
    Ref. Data 10, p.305.

TABLE 1.   Recent literature sources for atomic transition
           probability data of astrophysical interest

     This table covers the 3 year period since the publication of our last IAU
report (Reports on Astronomy, Vol. XIX A, 122 (1985); preparation date: August,
1984) to the present (September 1987). The table is arranged in alphabetical
order of element symbols, with further subdivisions according to stage of ionization (I, II, etc.). The numbers are the running numbers of the reference list
following this table.

Al I: 1	Be I: 8,11,48,74,99,104,135
Al II: 26,55,122	Be II: 38,108
Al III: 27,55	Be III: 84
Al IV: 46	
	C I: 37,57,79,105
B II: 9,46,114,115,119	C II:70,92,113
B III: 96,133	CIII: 36,45,72,102,113,114,115,119
B IV: 84	C IV: 95
Ba I: 8,11,48,74,99,104,135	Ca I: 8,10,51,54,71,80
	Ca III: 7,93
	Ca IV: 4

Co II: 116

Cr I: 20,39,130

Fe I: 21,82,112
Fe II: 85
Fe V: 73
Fe VI: 73
Fe VII: 73
Fe IX: 57, 121
Fe X: 35
Fe XI: 17
Fe XII: 124
Fe XIII: 18
Fe XIV: 49,52,134
Fe XV: 3,5,47,125,134
Fe XVII: 15
Fe XIX: 94
Fe XXI: 6,13,37
Fe XXII: 14
Fe XXIII: 24,63,91
Fe XXIV: 24,63,91
Fe XXV: 24,63

H-sequence: 106

He-sequence: 86

He I: 12,59,83,120,126,129

Li I: 44,53,97,127
Li II: 44,84,128,131

Mg I: 8,28,98
Mg II: 27,55
Mg IV: 16

Mn I: 27,75,109
Mn II: 103

N I: 25,34,56,65
N II: 37,132
N III: 72,111
N IV: 45,72,88,114,115,119

Na I: 43,55,97,101,127

Ni I: 41
Ni II: 90

O I: 2,29,40,50,76
O II: 2,37,117,140
O III: 2,30,36,37,77
O IV: 2,31,72

P I: 69
P II: 62,66,141
P III: 1
P IV: 55, 123

S I: 17,68,69
S II: 69
S III: 64,67,69
S IV: 1,78

Si I: 110,112
Si II: 1,89,136
Si III: 26,55,107
Si IV: 27,33,55

Ti I: 19
Ti II: 87

V I: 42,61,109,137
V II: 58,81,139
V III: 138

References

1. Aashamar, K., Luke, T.M. and Talman, J.D.: 1984, Phys. Scr. 30, 121. Q
2. Abramov, V.A. Zhukova, T.I., Zhidkov, A.G. and Kukushkin, A.S.: April 1984, IAEA Report INDC (CCP) - 205/GA. CP
3. Anderson, E.K. and Anderson, E.M.: 1983, Opt. Spectrosc. (USSR) 55, 500. Q, QF
4. Ansbacher, W., Inamdar, A.S. and Pinnington, E.H.: 1985, Phys. Lett. A 110, 383. L
5. Baluja, K.L. and Hibbert, A.: 1985, Nucl. Instrum. Methods Phys. Res., Sect. B 9, 477. Q
6. Baluja, K.L.: 1985, J. Phys. B 18, L413. QF
7. Beluja, K.L.: 1986, J. Phys. B 19, L551. Q
8. Barrientos, C. and Martin, I.: 1985, Can. J. Phys. 63, 1441. Q
9. Bashkin, S., McIntyre, L.C., Buttlar, H.V., Ekberg, J.O. and Martinson, I.: 1985, Nucl. Instrum. Methods Phys. Res., Sect. B 9, 593. L
10. Bauschlicher, C.W. Jr., Langhoff, S.R. and Partridge, H.: 1985, J. Phys. B 18, 1523.

11. Bauschlicher, C.W., Jr., Jaffe, R.L., Langhoff, S.R., Mascarello, F.G. and
     Partridge, H.: 1985, J. Phys. B 18, 2147. Q

12. Berrington, K.A., Burke, P.G., Freitas, L.C.G. and Kingston, A.E.: 1985,
     J. Phys. B. 18, 4135. Q

13. Bhatia, A.K., Seely, J.F. and Feldman, U.: 1987, At. Data Nucl. Data
     Tables 36,453. Q

14. Bhatia, A.K., Seely, J.F. and Feldman, U.: 1986, At. Data Nucl. Data
     Tables 35, 319. Q

15. Bhatia, A.K., Feldman, U. and Seely, J.F.: 1985, At. Data Nucl. Data
     Tables 32, 435. Q,QF

16. Biemont, E.: 1985, Phys. Scr. 31, 45. Q

17. Biemont, E. and Hansen, J.E.: 1986, Phys. Scr. 34, 116. QF

18. Biemont, E.: 1986, Phys. Scr. 33, 324. Q

19. Blackwell, D.E., Booth, A.J., Menon, S.L.R. and Petford, A.D.: 1986,
     Mon. Not. R. Astron.Soc.220, 289. A

20. Blackwell, D.E., Booth, A.J., Menon, S.L.R. and Petford, A.D.: 1986, Mon.
     Not. R. Astron. Soc. 220, 303. A

21. Blackwell, D.E., Booth, A.J., Haddock, D.J., Petford, A.D. and Leggett, S.
     K.: 1986, Mon. Not. R. Astron. Soc. 220, 549. A

22. Booth, A.J., Blackwell, D.E., Petford, A.D. and Shallis, M.J.: 1984,
     Observatory 104, 265. C

23. Bruneau, J.: 1984, J. Phys. B 17, 3009. Q

24. Buchet, J.P., Buchet-Poulizac, M.C., Denis, A., Desesquelles, J., Druetta,
     M., Grandin, J.P., Huet, M., Husson, X. and Lecler, D.: 1984, Phys. Rev.
     A 30, 309. L

25. Butler, K. and Zeippen, C.J.: 1984, Astron. Astrophys. 141, 274. QF

26. Butler, K., Mendoza, C. and Zeippen, C.J.: 1984, Mon. Not. R. Astron. Soc.
     209, 343. Q

27. Butler, K., Mendoza, C., and Zeippen, C.J.: 1984, J. Phys. B 17, 2039. Q

28. Chantepie, M., Cojan, J.L., Landais, J., Laniepce, B. and Moudden, A.:
     1984, Opt. Commun. 51, 396. L

29. Chung, S., Lin, C.C. and Lee, E.T.P.: 1986, J. Quant. Spectrosc. Radiat.
     Transfer 36, 19. Q

30. Coetzer, F.J., Kotze, T.C., Mostert, F.J. and van der Westhuizen, P.: 1986,
     Spectrochim. Acta, Part B 41, 847. L

31. Coetzer, F.J., Kotze, T.C. and van der Westhuizen, P.: 1986, Spectrochim.
     Acta, Part B 41, 243. L

32. Coetzer, F.J., Kotze, T.C., Mostert, F.J. and van der Westhuizen, P.: 1986,
     Phys. Scr. 34, 328. L

33. Cohen, M. and McEachran, R.P.: 1984, J. Phys. B 17, 2979. Q

34. Copeland, R.A., Jeffries, J.B., Hickman, A.P. and Crosley, D.R.: 1987,
     J. Chem. Phys. 86, 4876. L

35. Cowan, R.D., Bromage, G.E. and Fawcett, B.C.: 1984, Mon. Not. R. Astron.
     Soc. 210, 439. Q,C

36. Czyzak, S.J., Keyes, C.D. and Aller, L.H.: 1986, Astrophys. J., Suppl.
     Ser. 61, 159. QF

37. Czyzak, S.J. and Poirier, C.P.: 1985, Astrophys. Space Sci. 116, 21. QF

38. Davis, B.F. and Chung, K.T.: 1984, Phys. Rev. A 29, 2586. Q

39. Delibas, M., Mindreci, I. and Dorohoi, D.: 1984, Rev. Roum. Phys. 29,
     175. A

40. Doering, J.P., Gulcicek, E.E. and Vaughan, S.O.: 1985, J. Geophys. Res.,
     Space Phys. 90, 5279. M

41. Doerr, A. and Kock, M.: 1985, J. Quant. Spectrosc. Radiat. Transfer 33,
     307. E,H

42. Doerr, A., Kock, M., Kwiatkowski, M., Werner, K. and Zimmermann, P.: 1985,
     J. Quant. Spectrosc. Radiat. Transfer 33, 55. L,E

43. Engström, K., Young, L., Somerville, L.P. and Berry, H.G.: 1985, Phys.
     Rev. A 32, 1468. L.

44. Fairley, N.A. and Laughlin, C.: 1984, J. Phys. B 17, 2757. Q
45. Fawcett, B.C.: 1984, At. Data Nucl. Data Tables 30, 423. Q
46. Fawcett, B.C.: 1984, Phys. Scr. 30, 326. Q
47. Fawcett, B.C.: 1986, Phys. Scr. 34, 331. Q
48. Fisk, P.T.H., Bachor, H.-A. and Sandeman, R.J.: 1986, Phys. Rev. A 33, 2418. M
49. Froese Fischer, C. and Liu, B.: 1986, At. Data Nucl. Data Tables 34, 261. Q,QF
50. Froese Fischer, C.: 1987, J. Phys. B 20, 1193. Q
51. Froese Fischer, C. and Hansen, J.E.: 1985, J. Phys. B 18, 4031. Q
52. Frye, D. and Armstrong, L. Jr.: 1985, Phys. Rev. A 31, 2070. Q
53. Fulton, T. and Johnson, W.R.: 1986, Phys. Rev. A 34, 1686. O
54. Glass, R.: 1985, J. Phys. B 18, 4047. Q
55. Godefroid, M., Magnusson, C.E., Zetterberg, P.O. and Joelsson, I.: 1985, Phys. Scr. 32, 125. QF
56. Goldbach, C., Martin, M., Nollez, G., Plomdeur, P., Zimmermann, J.-P. and Babic, D.: 1986, Astron. Astrophys. 161, 47. E
57. Goldbach, C. and Nollez, G.: 1987, Astron. Astrophys. 181, 203. E
58. Goly, A. and Weniger, S.: 1984, J. Quant. Spectrosc. Radiat. Transfer 32, 61. E
59. Gorny, M.B., Kazantsev, S.A., Matisov, B.G. and Polezhaevs, N.T.: 1985, Z. Phys. A 322, 25. L
60. Graham, R.L. and Yeager, D.L.: 1987, Int. J. Quantum Chem. 31, 99. Q
61. Gurtovendo, E.A., Kostyk, R.I. and Orlova, T.V.: 1985, Kinemat. Fiz. Nebesn. Tel 1, No.2, 62. M
62. Harris, A.W. and Mas Hesse, J.M.: 1986, Astrophys. J. 308, 240. MR
63. Hata, J. and Grant, I.P.: 1984, Mon. Not. R. Astron. Soc. 211, 549. Q,QF
64. Hayes, M.A.: 1986, J. Phys. B 19, 1853. Q
65. Hibbert, A., Dufton, P.L. and Keenan, F.P.: 1985, Mon. Not. R. Astron. Soc. 213, 721. Q
66. Hibbert, A.: 1986, J. Phys. B 19, L455. Q
67. Ho, Y.K. and Henry, R.J.W.: 1984, Astrophys. J. 282, 816. Q
68. Ho, Y.K. and Henry, R.J.W.: 1985, Astrophys. J. 290, 424. Q
69. Ho, Y.K. and Henry, R.J.W.: 1987, Phys. Scr. 35, 831. Q,C
70. Huber, M.C.E., Sandeman, R.J. and Tozzi, G.P.: 1984, Phys. Scr. T8, 95. E
71. Hunter, L.R. and Peck, S.K.: 1986. Phys. Rev. A 33, 4452. L
72. Ishii, K., Suzuki, M. and Takahashi, J.: 1985. J. Phys. Soc. Jpn. 54, 3742. L
73. Jacques, C., Moreau, J.-P. and Knystautas, E.J.: 1984, J. Phys. (Paris) 45, 1607. L
74. Jahreiss, L. and Huber, M.C.E.: 1985. Phys. Rev. A 31, 692. C
75. Jäger, H., Neger, T. and Sperger, R.: 1987. Opt.Commun. 61, 252. ER
76. Jenkins, D.B.: 1985, J. Quant. Spectrosc. Radiat. Transfer 34, 55. A
77. Johnson, B.C., Smith, P.L. and Knight, R.D.: 1984, Astrophys. J. 281, 477. L
78. Johnson, C.T., Kingston, A.E. and Dufton, P.L.: 1986, Mon. Not. R. Astron. Soc. 220, 155. QF
79. Jones, D.W. and Wiese, W.L.: 1984. Phys. Rev. A 29, 2597. E
80. Jonsson, G., Levinson, C. and Svanberg, S.: 1984, Phys. Scr. 30, 65. L
81. Karamatskos, N., Michalak, R., Zimmermann, P., Kroll, S. and Kock, M.: 1986, Z. Phys. D 3, 391. L,E
82. Kock, M., Kroll, S. and Schnehage, S.: 1984, Phys. Scr. T8, 84. E,H
83. Kono, A. and Hattori, S.: 1984, Phys. Rev. A 29, 2981. O
84. Kono, A. and Hattori, S.: 1984, Phys. Rev. A 30, 2093. Q
85. Kroll, S. and Kock, M.: 1987, Astron. Astrophys., Suppl. Ser. 67, 225. E,H
86. Kundu, B. and Mukherjee, P.K.: 1985. Astrophys. J. 298, 844. QF
87. Kwiatkowski, M., Werner, K. and Zimmermann, P.: 1985. Phys. Rev. A 31, 2695. L

88. Lang, J., Hardcastle, R.A., McWhirter, R.W.P. and Spurrett, P.H.: 1987,
     J. Phys. B 20, 43. E
89. Lanz, T. and Artru, M.-C.: 1985, Phys. Scr. 32, 115. Q,Cp
90. Lawler, J.E. and Salih, S.: 1987, Phys. Rev. A 35, 5046. L
91. Lemen, J.R., Phillips, K.J.H., Cowan, R.D., Hata, J. and Grant, I.P.: 1984,
     Astron. Astrophys. 135, 313. Q
92. Lennon, D.J., Dufton, P.L., Hibbert, A. and Kingston, A.E.: 1985,
     Astrophys. J. 294, 200. Q
93. Loginov, A.V. and Gruzdev, P.F.: 1986, Opt. Spectrosc. (USSR) 61, 417. Q
94. Loulergue, M., Mason, H.E., Nussbaumer, H. and Storey, P.J.: 1985, Astron.
     Astrophys. 150, 246. Q,QF
95. Lunell, S., Cogordan, J.A. and Oster, P.: 1985, J. Phys. B 18, 3849. Q
96. Mannervik, S., Cederquist, H. and Martinson, I.: 1986, Phys. Rev. A 34,
     231. L
97. Martin, I. and Barrientos, C.: 1986, Can. J. Phys. 64, 867. Q
98. Mendoza, C. and Zeippen, C.J.: 1987, Astron. Astrophys. 179, 339. Q
99. Migdalek, J. and Baylis, W.E.: 1987, Phys. Rev. A 35, 3227. Q
100. Moccia, R., and Spizzo, P.: 1985, J. Phys. B 18, 3537. Q
101. Müller, W., Flesch, J. and Meyer, W.: 1984, J. Chem. Phys. 80, 3297. Q
102. Nasser, R.M. and Varshni, Y.P.: 1985, Astron. Astrophys.Suppl. Ser. 60,
     325. Q
103. Neger, T.: 1986, J. Phys. D. 19, L153. H
104. Niggli, S. and Huber, M.C.E.: 1987, Phys. Rev. A 35, 2908. E
105. Nussbaumer, H. and Storey, P.J.: 1984, Astron. Astrophys. 140, 383. Q
106. Nussbaumer, H. and Schmutz, W.: 1984, Astron. Astrophys. 138, 495. IF
107. Nussbaumer, H.: 1986, Astron. Astrophys. 155, 205. Q,QF
108. Parpia, F.A., Norcross, D.W. and da Paixao, F.J.: 1986, Phys. Rev. A 34,
     4777. Q
109. Peterkop, R.K.: 1985, Opt. Spectrosc. (USSR) 58, 7. Q
110. Peterkop, R.K.: 1985, Opt. Spectrosc. (USSR) 58, 121. Q
111. Pinnington, E.H., Ansbacher, W., Gosselin, R.N. and Kernahan, J.A.: 1986,
     Phys. Lett. A 114, 373. L
112. Pitts, R.E.: 1986, J. Quant. Spectrosc. Radiat. Transfer 35, 365. E,A
113. Reistad, N., Hutton, R., Nilsson, A.E., Martinson, I. and Mannervik, S.:
     1986, Phys. Scr. 34, 151. L
114. Reistad, N. and Martinson, I.: 1986, Phys. Rev. A 34, 2632. I
115. Rudzikas, Z.B., Szulkin, M. and Martinson, I.: 1984, Phys. Scr. T8, 141. Q
116. Salih, S., Lawler, J.E. and Whaling, W.: 1985, Phys. Rev. A 31, 744. L,E
117. Schartner, K.-H., Flaig, H.-J., Träbert, E. and Heckman, P.H.: 1985,
     Phys. Res., Sec. B 9, 642. E
118. Serrao, J.M.P.: 1985, J.Quant. Spectrosc. Radiat. Transfer 33, 219. Q
119. Serrao, J.M.P.: 1986, J.Quant. Spectrosc. Radiat. Transfer 35, 265. Q
120. Silim, H.A., El-Farrash, A.H. and Kleinpoppen, H.: 1987, Z. Phys. D 5, 61.
     L
121. Svensson, K.A., Eberg, J.O. and Edlen, B.: 1974, Sol. Phys. 34, 173. QF
122. Tayal, S.S. and Hibbert, A.: 1984, J. Phys. B 17, 3835. Q
123. Tayal, S.S.: 1985, Phys. Scr. 32, 523. Q
124. Tayal, S.S. and Henry, R.J.W.: 1986, Astrophys. J. 302, 200. Q
125. Tayal, S.S.: 1986, J. Phys. B 19, 3421. Q
126. Theodosiou, C.E.: 1984, Phys. Rev. A 30, 2910. Q
127. Theodosiou, C.E.: 1984, Phys. Rev. A 30, 2881. Q
128. Theodosiou, C.E.: 1985, Phys. Scr. 32, 129. Q
129. Theodosiou, C.E.: 1987, At. Data Nucl. Data Tables 36, 97. Q
130. Tozzi, G.P., Brunner, A.J. and Huber, M.C.E.: 1985, Mon. Not. R. Astron.
     Soc. 217, 423. E
131. Träbert, E., Blanke, J.H., Hucke, R. and Heckmann, P.H.: 1985, Phys. Scr.
     31, 130. L
132. Träbert, E., Mannervik, S. and Cederquist, H.: 1986, Phys. Scr. 33, 222. L
133. Träbert, E., Mannervik, S. and Cederquist, H.: 1986, Phys. Scr. 34, 46. L

134. Träbert, E., Hutton, R. and Martinson, I.: 1987, Z. Phys. D 5, 125. L
135. Ueda, K., Hamaguchi, Y., Fujimoto, T. and Fukuda, K.: 1984, J. Phys. Soc.
     Jpn. 53, 2501. HR
136. Van Buren, D.: 1986, Astrophys. J. 311, 400. M
137. Whaling, W., Hannaford, P., Lowe, R.M., Biemont, E. and Grevesse, N.: 1985,
     Astron. Astrophys. 153, 109. L,E
138. Wujec, T. and Musielok, J.: 1986, J. Quant. Spectrosc. Radiat. Transfer,
     35, 239. E
139. Wujec, T. and Musielok, J.: 1986, J. Quant. Spectrosc. Radiat. Transfer,
     36, 7. E
140. Zeippen, C.J.: 1987, Astron. Astrophys. 137, 410. QF
141. Zhechev, D.Z. and Koleva, I.T.: 1986, Phys. Scr. 34, 221. L

W.L. Wiese
Chairman of Working Group

WORKING GROUP 3: COLLISION PROCESSES

Of the vast array of data on electron and heavy-particle collisions that are produced each year, I select only those that have an obvious immediate bearing on astronomical research. A brief review of recent developments in atomic data for astrophysics has been published (1).

1. Electron Collisions

1.1 ELECTRON IMPACT IONIZATION

Experimental impact ionization (2) and theoretical (3,4) values of the cross sections for electron impact ionization were obtained for the magnesium-like ions S^{4+}, Cl^{5+} and Ar^{6+} and theoretical values were obtained for Al^+ (3). Experimental cross sections were published for electron-impact ionization of B^{2+} and O^{5+} (5), of N^{4+} and N^{5+} (6), of Fe^{5+}, Fe^{6+} and Fe^{9+}(7), of Fe^{11+}, Fe^{13+} and Fe^{15+} (8), of Ni^{3+}, Cu^{2+}, Cu^{3+} and Sb^{3+} L(9) and of Ti^{2+}, Fe^{2+}, Ar^{2+}, Cl^{2+} and F^{2+} (10). Double ionization cross sections were measured for Ar^+ and Ar^{4+} (11). Theoretical cross sections for the single ionization of Fe^{13+} were calculated (12). Total ionization and partial ionization cross sections of many systems have been complied by Tawara and Kato (13). A list of ionization rate coefficients for astrophysical applications was compiled by Arnaud and Rothenflug (14).

1.2 ELECTRON IMPACT EXCITATION

An evaluated compilation of data for electron-impact excitation of atomic ions was published as a JILA report (15). Many calculations of varying degrees of sophistication have appeared in the literature: excitation cross sections of transitions of He-like and Be-like ions (16) Li-like ions, outer-shell (17) and inner-shell (18), B-like ions (19), C-like ions (20) and Ne-like ions (21) and singly and multiply-charged ions of carbon and oxygen (22) have all been carried out for He^+ (23), Li^+ (24), Be^+ (25), C^+ (26), C^{2+} (27), C^{4+} (28), Ne^+ (20), Ne^{4+} and Mg^{6+} (30), Mg^{10+} (31), Mg^{3+} and Mg^{4+} (32), Al^+ (33), Si^{3+} (34), Si^{9+} (35), Si^{10+} (36), S^+ (37), S^{2+} (38), S^{7+} (30), Cl^{5+} (40), Fe^+ (41), Fe^{6+} and Fe^{22+} (42), Fe^{11+} (43), Fe^{12+} (44), Fe^{14+} (45), Fe^{16+} (46), Fe^{17+} (47), Fe^{19+} (48), Fe^{24+} (49), Ca^{14+} (50), Ca^{18} (51), and Cu^{12+} and Cu^{16+} (52). Excitation to autoionizing states and their contribution to ionization has been investigated for magnesium-like ions (2,3) and for nickel ions (53). Experiments on the excitation of Si^{2+} transitions have been carried out (54).

Electron impact excitation of neutral systems has received less attention. New cross section data are available on He (55) with a list of rate coefficients (56). Collisions with neutral C atoms (57) and with neutral S atoms (58) have been investigated theoretically and with neutral helium (59) and neutral oxygen (60) experimentally.

1.3 RECOMBINATION

An important series of papers by Nussbaumer and Storey (61) has provided theoretical estimates of dielectronic recombination rate coefficients of light elements. An updated tabulation of recombination rate coefficients has been published (14).

1.4 ELECTRON MOLECULE COLLISIONS

Experimental cross sections for electron impact excitations of H_2 are given in (64-67) and theoretical calculations are described in (68). Collisions of electrons with HCN are examined theoretically in (69).

2. Heavy-Particle Collisions

2.1 CHARGE TRANSFER

Calculations at thermal energies for Si^{4+}-He are presented in (70). A compilation of charge transfer of multiply-charged ions colliding with H and H_2 at high energies has been published in (71). Estimates of rate coefficients of charge transfer processes that may be important in HI regions are given in (72).

2.2 PROTON IMPACT

Proton impact induced transitions between fine structure levels of ions are studied in (73) and (48). Proton impact K and L shell ionization are considered in (74) and (75).

2.3 CHEMICAL REACTIONS

Lists of chemical reactions which enter into models of the chemistry of interstellar, circumstellar and shocked regions continue to grow. Listings are found in (76-79).

Many experimental and theoretical investigations of relevant ion-molecule reactions have been carried out. A few of them are described in references (80-98). Particular attention has been given to the behavior at low temperatures of reaction with neteronuclear molecules (99-101). The dependence of reaction rate coefficients on the internal rotational and vibrational level populations has been studied though rarely at the temperatures of astronomical environments. They are too many to list here.

2.4 MOLECULAR EXCITATION

Rotational excitations of astrophysically interesting molecules are investigated in (1-129). A review of calculations was presented by Green (130). Vibrational excitation was studied in (131) and fine and hyperfine structure in (132-134).

2.5 MOLECULAR DISSOCIATION

Detailed calculations have been reported on the collision-induced dissociation of H_2 appropriate to astrophysical environments (135-137).

References

1. Mendoza, C.: 1986, Publ. Astron. Soc. Pacific 98, p.999.
2. Howald, A.M., Gregory, D.C., Meyer, Phaneuf, R.A., Müller, A., Djuric, N. and Dunn, G.H.: 1986, Phys. Rev. A33, p.3779.
3. Tayal, S.S. and Henry, R.J.W.: 1986, Phys. Rev. A 33, p.3825.
4. Pindzola, M.S., Griffin, D.C. and Bottcher, C.: 1986, Phys. Rev. A 33, p.3787.
5. Crandall, D.H., Phaneuf, R.A., Gregory, D.C., Howald, A.M., Mueller, D.W., Morgan, T.J., Dunn, E.H., Griffin, D.C. and Henry, R.J.W.: 1986, Phys, Rev. A 34, p.1757.
6. Defrance, P., Chantrenne, S., Brouillard, F., Rachafi, S., Belic, D.S., Jureta, J. and Gregory, D.C.: 1985: Nucl. Meth. B 9, p.400.
7. Gregory, D.C., Meyer, F.W., Müller, A. and Defrance, P.: 1986, Phys. Rev. A 34, p.3657.
8. Gregory, D.C., Wang, L.J., Meyer, F.W. and Rinn, K.: 1987, Phys. Rev. A 35, p.3256.
9. Gregory, D.C. and Howald, A.M.: 1986, Phys. Rev. A 34, p.97.
10. Mueller, D.W., Morgan, T.J., Dunn, G.H., Gregory, D.G. and Crandall, D.H.: 1985, Phys. Rev. A 31, p.2905.
11. Müller, A., Tinshert, K., Achenbach, C., Becker, R. and Solzborn, E.: 1985, J. Phys. B 18, p.3011.
12. Butler, K. and Moores, D.L.: 1985, J. Phys. B 18, p.1247.
13. Tawara, H. and Kato, T.: 1987, Atom. Data and Nucl. Data Tables 36, p.167.
14. Arnaud, M. and Rothenflug, R.: 1985, Astron. Ap. Suppl. 60, p.425.
15. Gallagher, J.W. and Pradhan, A.K.: 1985, JILA Information Center, University of Colorado, Report No. 30.
16. Itikawa, K. and Sakimoto, K.: 1985, Phys. Rev. A 31, p.1319; Berrington, K. A., Burke, P.G., Dufton, P.L. and Kingston, A.E.: 1985, Atom. Data Nucl. Data Tables 33, p.195 and 345; Bhatia, A.K., Feldman, U. and Sealy, J.F.: 1986, Atom. Data Nucl. Data Tables 35, 473; Zhang, H. and Sampson, D.H.: 1987, Ap.J. Suppl. 63, p.487.
17. Sampson, D.H., Goelt, S.J., Petrou, G.V., Zhang, H. and Clark, R.E.H.: 1985, Atom. Data Nucl. Data Tables 32, p.343; Zhang, H., Sampson, D.H. and Clark, R.E.H.: 1986, Atom. Data Nucl. Data Tables 35, p.267.
18. Sampson, D.H., Petrou, G.V., Goett, S.J. and Clark, R.E.H.: 1985, Atom. Data Nucl. Data Tables 32, p.403.
19. Sampson, D.H., Weaver, G.M., Goett, S.J., Zhang, S.J. and Clark, R.E.H.: 1986, Atom. Data Nucl. Data Tables 35, p.223; Bhatia, A.K., Feldman, U. and Seely, J.F.: 1986, Atom. Data Nucl. Tables 35, p.319.
20. Itikawa, Y. and Sakimoto, K.: 1986, Phys. Rev. A 33, p.2320.
21. Zhang, H., Sampson, D.H., Clark and Mann, J.B.: 1987, Atom. Data and Nucl. Data Tables 37, p.17.
22. Itikawa, Y., Hara, S., Kato, T., Nakazaki, S., Pindzola, M.S., and Crandall, D.H.: 1985, Atom. Data Nucl. Data Tables 33, p.149.
23. Badnell, N.R.: 1985, J. Phys. B 18, p.955.
24. Norcross, D.W. and Christiansen, R.B.: 1985, Phys. Rev. A 31, p.142.
25. Porpia, F.A., Norcross, D.W. and De Paixo, F.J.: 1986, Phys. Rev. A 34, p.4777.
26. Lennon, D.J., Hibbert, A.L. and Kingston, A.E.: 1985, Ap.J. 294, p.200; Keenan, F.P., Lennon, D.J., Johnson, C.T. and Kingston, A.E.: 1986, MNRAS, 220, p.571.
27. Berrington, K.A.: 1985, J. Phys. B, 18, p.L395.
28. Tayal, S.S.: 1986, Phys. Rev. A 34, p.1847.
29. Bayes, F.A., Saraph, H.E., and Seaton, M.J.: 1985, MNRAS 215, 85P.
30. Aggarwal, K.M.: 1985, Ap.J. Suppl. 58, p.289; ibid 59, p.113, 1985.
31. Tayal, S.S. and Kingston, A.E.: 1985, J. Phys. B 18, p.2983; Tayal, S.S.: 1987, Phys. Rev. A 35, p.2073.
32. Mendoza, C. and Zeippen, C.J.: 1987, MNRAS 224, 7P.
33. Tayal, S.S., Burke, P.G. and Kingston, A.E.: 1985, J. Phys. B 18, p.4321.

34. Johnson, C.T., Kingston, A.E. and Dufton, P.L.: 1986, MNRAS 220, p.155.
35. Aggarwal, K.M. and Kingston, A.E.: 1986, Astron. Ap. 162, p.333.
36. Kastner, S.O. and Bhatia, A.K.: 1986, Ap.J. 309, p.883.
37. Tayal, S.S., Henry, R.J.W. and Nakazaki, S.: 1987, Ap.J. 313, p.484.
38. Hayes, M.A.: 1986, J. Phys. B 19, p.1853.
39. Mohan, M., Baluja, K.A. and Hibbert, A.: 1987, J. Phys. B 20, p.2595.
40. Baluja, K.L. and Mohan, M.: 1987, J. Phys. B. 20, p.831.
41. Baluja, K.A., Hibbert, A. and Mohan, M.: 1986, J. Phys. B19, p.3613.
42. Norrington, P.H. and Grant, L.P.: 1987, J. Phys. B 20, p.4869.
43. Tayal, S.S., Henry, R.J.W. and Pradhan, A.K.: 1987, Ap.J. 319, p.951.
44. Bhatia, A.K. and Mason, H.E.: 1986, Astron. Ap. 155, p.413.
45. Christiansen, R.B., Norcross, D.W. and Pradhan, A.K.: 1985, Phys. Rev. A 32, p.93.
46. Smith, B.W., Mann, J.B., Cowan, R.D. and Raymond, J.C.: 1985, Ap.J. 298, p.898.
47. Mohan, M., Baluja, K.L., Hibbert, A. and Berrington, K.A.: 1987, MNRAS 225, p.377.
48. Loulergue, M., Mason, H.E., Nussbaumer, H. and Storey, P.J.: 1985, Astr. Ap. 150, p.246.
49. Faucher, P. and Dubau, J.: 1985, Phys. Rev. A 31, p.3672; Pradhan, A.K.: 1985, Ap.J. Suppl. 59, p.183.
50. Bhatia, A.K. and Mason, H.E.: 1986, Astron. Ap. 155, p.417.
51. Pradhan, A.K.: 1985, Ap.J. Suppl. 59, p.183.
52. Datta, R.U., Roberts, J.R., Rowan, W.L. and Mann, J.B.: 1986, Phys. Rev. A 34, p.4751.
53. Burke, P.G., Fon, W.C. and Kingston, A.E.: 1987, J. Phys. B 20, p.2579.
54. Yu, T.L., Finkenthal, M. and Moos, H.W.: 1986, Ap.J. 305, p.880.
55. Smirnov, Iu.M.: 1984, Soviet Astronomy, 28, p.6361; Berrington, K.A., Burke, P.G., Freitas, L.C.G. and Kingston, A.E.: 1985, J. Phys. B 18, p.41351; Schneider, B.I. and Collins, L.A.: 1986, Phys. Rev. A 33, p.2982.
56. Clegg, R.E.S.: 1987, MNRAS, 229, 31P.
57. Johnson, C.T., Burke, P.G. and Kingston, A.E.: 1987, J. Phys. B 20, p.2553.
58. Ho, Y.K. and Henry, R.J.W.: 1985, Ap.J. 290, p.424.
59. Schemansky, D.E., Ajello, J.M., Hall, D.T. and Franklin, B.: 1985, Ap.J. 296, p.774; Forand, J.L., Becker, K. and McConkey, J.W.: 1985, J. Phys. B 18, p.1409.
60. Shyn, T.W. and Sharp. W.E.: 1985, Geophys. Res. Lett. 12, 1711; Doering, J.P., Gulcicek, E.E. and Vaughan, S.O.: 1985, Chem. Phys. Lett. 114, p.334.
61. Nussbaumer, H. and Storey, P.J.: 1983, Astron. Ap. 126, p.75.; 1984: Astron Ap. Suppl. 56, p.293; 1986: Astron. Ap. Suppl. 64, p.545; 1987: Astron. Ap. Suppl. 69, p.123.
62. Smith, D. and Adams, N.G.: 1984, Ap.J. Lett. 284, L13.; Adams, N.G., Smith, D. and Alge, E.J.: 1984, J. Chem. Phys. 80, p.1778.
63. Bates, D.R.: 1986, Ap.J. Letters 306, L43.
64. Shemansky, D.E., Ajello, J.M. and Hall, D.T.: 1985, Ap.J. 296, 765.
65. Nishumura, H., Danjo, A. and Suguhara, H.: 1985, J. Phys. Soc. Japan 54, p.1757.
66. Khakov, M.A., Trajumar, S., McAdams, R. and Shyn, T.W.: 1987, Phys. Rev. A 35, p.2832.
67. Nishumura, H. and Danjo, A.: 1986, J. Phys. Soc. Japan 55, p.3031.
68. Baluja, K.L., Noble, C.J. and Tennyson, J.: 1985, J. Phys. B 18, L851.
69. Jain, A. and Norcross, D.W.: 1985, Phys. Rev. A 32, p.134.
70. Opradolci, L., McCarroll, R. and Valiron, P.: 1985, Astron. Ap. 148, p.229.
71. Tawara, H., Kato, T. and Nakara, Y.: 1985, Atom. Data Nucl. Data Tables 32, p.235.
72. Péquignot, D. and Aldrovandi, S.M.V.: 1986, Astron. Ap. 161, p.169.
73. Landman, D.A.: 1985, J.Q.S.T. 34, p.365.
74. Chen, M.H. and Craseman, B.: 1985, Atom. Data Nucl. Data Tables 33, p.217.
75. Cohen, D.D. and Craseman, B.: 1985, Atom. Data Nucl. Data Tables 33, p.255.

76. Pineau des Foréts, G., Flower, D.R., Hartquist, T.W. and Dalgarno, A.: 1986, MNRAS 220, p.801.
77. Pineau des Foréts, G., Flower, D.R., Hartquist, T.W. and Millar, T.J.: 1987, MNRAS 227, p.993.
78. Viala, Y.P.: 1986, Astron. Ap. Suppl. 64, p.391.
79. Aninich, V.G. and Huntress, W.T.: 1986, Ap.J. Suppl. 62, p.553.
80. Böhringer, H. and Arnold, F.: 1986, J. Chem. Phys. 84, p.2097.
81. Lin, G.H., Maier, J. and Leone, S.R.: 1986, J. Chem. Phys. 84, p.2180.
82. Blake, G.A., Aninich, V.G. and Huntress, W.T.: 1986, Ap.J. 300, p.415.
83. Federer, W., Villinger, H., Lindinger, W. and Ferguson, E.E.: 1986, J. Chem. Phys. 123, p.12.
84. Knight, J.S., Freeman, C.G., McEwan, M.J., Smith, S.C., Adams, N.G. and Smith, D.: 1986, MNRAS, 219, p.89.
85. Daniel, R.G., Keim, E.R. and Farrar, J.M.: 1986, Ap.J. 303, p.439.
86. Bohme, D.K. and Raksit, A.B.: 1985: MNRAS 213, p.717.
87. Adams, N.G. and Smith, D.: 1985, Chem. Phys. Lett. 117, p.67.
88. Marquette, J.B., Rowe, B.R., Dupeyrat, G. and Roueff, E.: 1985, Astron. Ap. 147, p.115.
89. Herbst, E., Adams, N.G., Smith, D. and DeFrees, D.J.: 1987, Ap.J. 312, p.351.
90. Tanaka, T., Kato, T. and Koyano, T.: 1986, J. Chem. Phys. 84, p.750.
91. Adams, N.G. and Smith, D.: 1987, Ap.J. Lett. 317, L25.
92. Monson, R.J.S., Conaway, W.E. and Zare, R.N.: 1985, Chem. Phys. Lett. 113, p.435.
93. Ervin, K.M. and Armentrout, P.B.: 1986, J. Chem. Phys. 84, p.6738.
94. Gonzalez, M., Aguilar, A. and Virgili, T.: 1985, Chem. Phys. Lett. 113, p.187.
95. Adams, N.G. and Smith, D.: 1985, Ap.J. Lett. 294, p.827.
96. Smith, D. and Adams, N.G.: 1985, Ap.J. 298, p.827.
97. Borassin, J., Reynaud, C. and Borassin, A.: 1986, Chem. Phys. Lett. 123, p.191.
98. Curtiss, R.A. and Farrar, J.M.: 1986, Chem. Phys. Lett. 123, p.471.
99. Adams, N.G., Smith, D. and Clary, D.C.: 1985, Ap.J. Lett. 296, p.131.
100. Marquette, J.B., Rowe, B.R., Dupeyrat, G., Poissant, G. and Rebrion, C.: 1985, Chem. Phys. Lett. 122, p.431.
101. Morgan, W.L. and Bates, D.R.: 1987, Ap.J. 314, p.817.
102. Jones, M.E., Barlow, S.E., Ellison, G.B. and Ferguson, E.E.: 1986, Phys. Lett. 130, p.218.
103. Clary, D.C.: 1985, Mol. Phys. 54, p.605.
104. Wagner, A.F. and Graff, M.M.: 1987, Ap.J. 317, p.423.
105. Bates, D.R.: 1987, Ap.J. 312, p.363.
106. Herbst, E.: 1987, Ap.J. 313, p.867.
107. Herbst, E.: 1986, Ap.J. 306, p.667.
108. Bates, D.R.: 1985, Ap.J. 298, p.382.
109. Herbst, E.: 1985, Ap.J. 292, p.484.
110. Herbst, E.: 1985, Ap.J. 291, p.226.
111. Green, S., DeFrees, D.J. and McLean, A.D.: 1987, Ap.J. Suppl. 65, p.175.
112. Schinke, R., Engel, V., Buck, V., Meyer, H. and Diercken, G.H.F.: 1985, Ap.J. 299, p.939.
113. Flower, D.R. and Launay, T.M.: 1985, MNRAS 214, p.271.
114. Monteiro, T.S.: 1985, MNRAS 214, p.417.
115. Dewangan, D.P. and Flower, D.R.: 1985, J. Phys. B 18, L137.
116. Broquier, M., Picard-Bersellini, A., Whitaker, B.J. and Green S: 1986, J. Chem. Phys. 84, p.2104.
117. Billing, G.B. and Diercksen, G.H.F.: 1985, Chem. Phys. Lett. 121, p.94.
118. Romanowski, H., Ltt, L.-T., Bowman, J.M. and Harding, L.B.: 1986, J. Chem. Phys. 84, p.4888.
119. Meyer, H., Buck, U., Schinke, R. and Diercksen, G.H.F.: 1986, J. Chem. Phys. 84, p.4976.

120. Broquier, M., Picard-Bersellini, A. and Hall, J.: 1987, Chem. Phys. Lett. 136, p.531.
121. Danby, G., Flower, D.R., Valiron, P., Kochanski, E., Kurdi, L. and Diercksen, G.H.F.: 1987, J. Phys. B 20, p.1039.
122. Green, S.: 1986, Ap.J. 309, p.331.
123. Palma, A. and Green, S.: 1987, Ap.J. 316, p.830.
124. Danby, G., Flower, D.R. and Monteiro, T.S.: 1987, MNRAS 226, p.739.
125. Buck, V., Meyer, H., Tolle, M. and Schinke, R.: 1986, Chem. Phys. 104, p.345.
126. Billing, G.D. and Diercksen, G.H.F.: 1986, Chem. Phys. 105, p.145.
127. Palma, A.: 1987, Ap.J. Suppl. 64, p.565.
128. Danby, G., Flower, D.R., Kochanski, E., Kurdi, L., Valiron, P. and Diercksen, G.H.F.: 1986, J. Phys. B19, p.2891.
129. Danby, G., Flower, D.R. and Monteiro, T.S.: 1987, MNRAS 226, p.435.
130. Green, S.: 1985, Nuovo Cimento C 6, p.435.
131. Bacia, Z., Schinke, R. and Diercksen, G.H.F.: 1983, J. Chem. Phys. 82, p.245.
132. Monteiro, T.S. and Stutzki, J.: 1986, MNRAS 221, 33P.
133. Stutzki, J. and Winnewisser, G.: 1985, Astron. Ap. 144, p.1.
134. Monteiro, T.S. and Flower, D.R.: 1987, MNRAS 228, p.101.
135. Dove, J.E., Mandy, M.E., Sathyamurthy, N. and Joseph, T.: 1986, Chem. Phys. Lett. 127, p.1.
136. Dove, J.E. and Mandy, M.E.: 1987, Ap.J. Letts. 311, L93.
137. Dove, J.E., Rusk, A.C.M., Cribb, P.H., and Martin, P.G.: 1987, Ap.J. 318, p.379.

A. Dalgarno
Chairman of Working Group

WORKING GROUP 4: LINE BROADENING

There has been quite a renewal of interest in the field of line broadening during the last three years and the proceedings of the Eight International Conference on Spectral Line Shapes (1) showed the general state of progress of this topic. It is not the purpose of this report to be exhaustive, so we will simply give a number of useful results for astrophysical purposes and indicate some new fascinating directions in this research theme.

1. Line Broadening and Shifts in Low to Moderately Dense Plasmas

1.1 HYDROGEN OR HYDROGENIC LINES

The ionic broadening of hydrogen or hydrogenic lines in plasmas has been considered as quasistatic during the last decades. In fact, this quasistatic ion assumption is only valid at quite high densities. At low or moderate densities, ion dynamics play an important role near the center as confirmed by new experimental results (2,3,4) for Lyman and Balmer lines. Computer simulations were employed successfully for investigations of these effects (5-8). In these computations, electron broadening is accounted for using an impact operator; the ionic contribution is obtained by numerical integration of the atomic Schrödinger equation for a set of ionic mocrofield formed by superposition of Debye-screened fields from uncorrelated particles. At typical astrophysical densities (smaller than 10^{14} cm^{-3} for Ly_α) these results are in a relatively good agreement with analytic formulae giving the line halfwidth in the impact model (9). New calculations have been performed for the Ly_α profile for detunings ranging from the center to the near line wings (10) (< 10 Å). At these low densities, the fine structure splitting is not negligible and its subsequent effect on the profile has been investigated (11).

New theoretical data for the broadening of far infrared and submillimeter as well as radio (13) lines are now available.

The shift of hydrogen or hydrogenic ion lines (He II) in plasmas has become a subject of interest both from an experimental (14, 15) or a theoretical (17) point of view. This shift, important at high densities, becomes negligible in stellar atmospheres conditions.

In spite of these results, much work remains to be done in the future and particularly extensive results for Lyman, Balmer and Paschen series are required.

1.2 STARK BROADENING OF ISOLATED LINES IN MODERATELY DENSE PLASMAS

First it is necessary to present the Opacity Project in (1) page 583) which represents a major new effort to make extensive calculation of atomic data required for opacity determination. Professor M.J. Seaton assumes the responsibility of this project which is being pursued by research groups in Belfast, Boulder, Caracas, London, Munich, Paris and Urbana. The work includes accurate calculations of atomic energy levels, oscillator strengths, photoionization cross-sections and parameters for Stark broadening of spectral lines. For atoms (except H) or non hydrogenic ions, the dominant contribution to the broadening is likely to be due to electronic collisions. According to plasma line broadening theory, these electron impacts induce a Lorentzian line profile whose width given in terms of elements of the scattering matrices. The R-matrix method of computation chosen in the Opacity Project allows computation of elements of the scattering matrices. However it is not practical to use close coupling methods to calculate line broadening parameters between highly excited states. So it is proposed to make systematic studies of these parameters for transitions between states that are not too highly excited in order to obtain approximate formulae which can be used for the highly excited states. Such calculations are already under way for various isoelectronic sequences and will be published in Journal of Physics B. This work seems of particular interest for theories of stellar structure and stellar pulsations.

Another interesting and complementary direction for research concerns investigations of systematic trends of Stark broadening parameters as function of the principal quantum number and the ionization potential. Regularities of these parameters within a multiplet or a supermultiplet or within spectral series and homologous atoms have been investigated (18-23) in order to provide a simple method for critical evaluation of existing data and interpolation of new data. These regularities in Stark broadening parameters are directly related to regularities in the atomic structure, so one can expect that these studies become inadequate for complex atoms, when a large configuration mixing strongly perturbs the levels. Semi empirical methods have also been developed for ion lines (24-28) and neutral atom lines (29). Such approaches which give a good average accuracy while the accuracy for a particular line is rather poor, allow a rapid estimation of the width and may be very useful for evaluating a large amount of data.

2. Hot and Dense Plasmas

The study of spectral line profiles in hot and dense plasmas is one of the major subjects of research in the field of line broadening. In recent years, renewed interest has mostly been stimulated by inertial confinement research. Astrophysics should greatly benefit from this effort especially the physics of white dwarfs or neutron stars.

An interesting theoretical work on the so called "ion dynamics" effects on the broadening has been done by various approaches. The problem of incorporating time dependent many body ionic interactions is well taken into account by computer simulations. The field can be accurately approximated by Monte-Carlo or molecular dynamics simulations (46). It has been shown that the effects of ion dynamics may be of the order of or larger than the Doppler effects. However such computer simulation methods are computer time consuming and it appears quite important that other theoretical approaches namely the Model Microfield Method (47) or semi-analytical methods (48) should be developed at the same time. Fine structure effects (49-52) as well as coupling between Doppler and Stark broadening (46) have been investigated.

Another important problem arises in these plasma conditions as strong correlation effects appear between the internal structures of the radiating ions and the perturbing plasma. In this field ions are assumed to be static and the electric microfield produced by screened ions induces the Stark effect and causes the spectral series to merge. Correlations can be included in the microfield distribution function using a pair correlation function which yields the Debye Hückel theory at moderate densities for weakly coupled plasmas and the results obtained by Monte-Carlo method for strongly coupled plasmas (53). The ionic microfield leads to frequency shift and line broadening, and produces a potential barrier which leads to an advance of the recombination continuum and to an ionization of bound states by tunnel effect corresponding to attenuated line intensities. Actually, the physical limit between bound and free states becomes a function of the density and the temperature of the surrounding ions (54). Such studies lead to interesting improvements of the Inglis Teller formula.

Moreover there has been a controversy about the "plasma polarization shift" which must be described by an unified approach (55-57). This shift may be significant at large densities and for heavy stripped radiating ions. For densities close to the solid state densities the mean free path might be reduced to values smaller than the wavelength of the line leading to some reduction of the thermal Doppler broadening (58). But we have yet no experimental test of this assumption.

3. Line Broadening by Foreign Gases and Molecular Line Broadening

As the broadening of the sodium D lines by collisions with atomic hydrogen is very important in the studies of the chemical composition of the Sun and other stars, the determination of the width and the shift of these lines has been the subject of many experimental (59) and theoretical (60) efforts. All available theoretical and experimental results have been reviewed (61). According to the discussion given here, it appears that the latest experiment (59) agrees well with empirical determination obtained from the solar spectrum. No fully satisfactory theory exists, due essentially to inaccuracies in the intermediate range of the relevant interatomic potentials inducing the broadening effect. This particular case illustrates quite well the general situation of the atomic line broadening theory: methods for the description of detailed dynamics are efficient and accurate but depend drastically on molecular structure calculations. An important effort concerning interatomic potentials is continuing and there is now a very good agreement with experimental results for alkali-metal-atoms + rare gas systems (62) indicating large improvements on all previous results when accurate potential data are available. Therefore, one can observe that the physicists' present preoccupations are far from astrophysical applications owing to this particular difficulty. There is a great need for work in this topic.

Concerning pressure broadening of molecular lines, many important results have been obtained; for astrophysical purpose we will quote particularly the calculation of the line widths of H_2O (63), N_2 and H_2 (64) broadened CO rovibrational lines, an experimental determination of collision broadened half widths of lines of the ν_9 fundamental band of $C_2 H_6$ (12µm) (65,66). This band seen in the spectra recorded on board Voyagers 1 and 2 is an important emission feature in the thermal infrared spectrum of Titan. Measurements of pressure broadened half width have been performed for some vibration rotation lines in bands of methane (67a,b) and ammonia (68a,b). These results are of particular importance for the modelling of planetary atmospheres. The measurement of line strengths and pressure broadening coefficients of the ν_3 band R_0 and R_1 transitions of GeH_4 (69) allows the interpretation of recent observations in the atmospheres of the outer planets. Studies on the collisional broadening of H_2O (70, 71) lead to precise knowledge of water absorption required in atmosphere physics applications and for space investigations.

Owing to recent observations recorded by Voyager space craft in the infrared region, many rototranslational collision induced spectra have been obtained both experimentally and theoretically for H_2-H_2 (72), N_2-N_2 (73), N_2-H_2 (74), H_2-He (75), CH_4-H_2 (76). There is an overall good agreement for the features of free-free, free-bound and bound-bound transitions. This work is of interest for the modelling of Titan's atmosphere in the far infrared and for the determination of the vertical temperature spectra. These results will convince easily the reader that recent molecular broadening studies lead to very fruitful collaboration between physicists and astrophysicists.

4. Collisional Redistribution of Radiation and Related Topics

Collisional redistribution has been reinvigorated these last years by much experimental and theoretical work, and has been reviewed recently (79). From a theoretical point of view, the processes of absorption or emisison of radiation when collisions occur lead to different problems such as the statistics of the collision events, the correlation between photons absorbed and emitted during the same collision, the modification of atomic collision dynamics in intense fields. In view of astrophysical interest, we will focus our attention on the problem of redistribution of weak radiation at low perturber densities. It has been shown (79,80) that absorption during a collision can be thought of as populating the molecular states of a transient molecule. The observed fluorescence comes either from the same collision leading to the possibility of correlated photons or from another collision corresponding to less correlated events.

The mechanism of collisional redistribution is now understood and the theoretical description is well established (81,82). Many experimental and theoretical results have been devoted to simple systems (Sr, (83,84) or Ba (85,86) or Na (87-89) perturbed by rare gas atoms) with a particularly good agreement for redistribution of polarized radiation studies when accurate interatomic potentials exist (Na): the dependence of the far wing excited state orientation or alignment versus the detuning can be interpreted in terms of reorientation of molecular orbitals o-curing during the collision. Contrarily to some appropriate model predictions, depolarization may not be total in the far wings and depends on the details of the collision. Obviously, such work is of interest for the problem of polarized line radiation transfer for which the first experimental result has been obtained (90). Redistribution by hydrogen in plasma requires a very complicated formulation and more work is needed in this topic. It is interesting to notice that collisional redistribution and photodissociation processes are closely related problems.

Beside this field of research, one may stress the progress of theoretical work along new interesting directions such as the theory of Stark broadening in the presence of magnetic fields (91) and more generally the study of spectral lines in plasmas in the presence of external fields or turbulence (92). A workshop

on "Spectral Line Formation in Plasmas Under Extreme or Unusual Conditions" has
been devoted to this subject, invited papers will be published in J.O.S.R.T.
Finally, there has been some new work concerning the computation of the Voigt
function (93,94).

References

1. Proceedings of the Eight International Conference (Williamsburg 1986)
   "Spectral Line Shapes, Volume 4", Ed. R.J. Exton, A. Deepack Publishing
   (1987).
2. Dunzmann, K., Grützmacher, K. and Wende, B.: 1986, Phys. Rev. Lett. 57,
   p.2151.
3. Sanchez, A., Fulton, R.D. and Griem, H.R.: 1987, Phys. Rev. A 35, p.2596.
4. Baldwin, K.G., Marangos, J.P. and Burgess, D.D.: 1984, J. Phys. D: Appl.
   Phys. 17, L169.
5. Seidel, J.: 1986, Phys. Rev. Lett. 57, p.2154.
6. Seidel, J.: 1987, in "Spectral Line Shapes, Volume 4", p.57.
7. Gigosos, M.A., Cardenoso, V. and Torress, F.: 1985, Phys. Rev. A 31,p.3509.
8. Gigosos, M.A., Cardenoso, V. and Torress, F.: 1986, J. Phys. B 19, p.3027.
9. Stehlé, C. and Feautrier, N.: 1984, J. Phys. B: Atom. Mol. Phys. 17,
   p.1477.
10. Stehlé, C.: 1986, Phys. Rev. A 34, p.4153.
11. Stehlé, C. and Feautrier, N.: 1985: J. Phys. B: Atom. Mol. Phys. 18,p.1297.
12. Hoang Binh, D., Brault, P., Picart, J., Tran Minh, N. and Vallée, O.: 1987,
    Astron. Astrophys. 181, p.134.
13. Gulyavev, S.A. and Sholin, G.V.: 1986, Sov. Astrono. 30, p.31.
14. Fleurier, C. and Le Gall, P.: 1985, J. Phys. B: Atom. Mol. Phys. 18,
    p.1297.
15. Pittman, T.L. and Fleurier, C.: 1986, Phys. Rev. A 33, p.1291.
16. Vitel, Y.: 1987, J. Phys. B: Atom. Mol. Phys. 20, p.2327.
17. Griem, H.R.: 1983, Phys. Rev. A 28, p.1596.
18. Pittman, T.L. and Konjevic, N.: 1986, JQSRT 35, p.247.
19. Konjevic, N. and Pittman, T.L.: 1987, JQSRT, 37, p.311.
20. Böttcher, F., Musielok, J. and Kunze, H.J.: 1987, Phys. Rev. 36A, p.2265.
21. Dimitrijevic, M.S. and Bach, Truong: 1986, Ann. Phys. (France) 11, Suppl.
    3, p.183.
22. Dimitrijevic, M.S.: 1986, Astron. Astrophys. Suppl. Ser. 64, p.591.
23. Dimitrijevic, M.S. and Bach, Truong: 1987, Z. Naturforsch 41A, p.772.
24. Dimitrijevic, M.S., Mihajlov, A.A. and Popovic, M.M.: 1987, Astron.
    Astrophys. Suppl. Ser. 70, p.57.
25. Dimitrijevic, M.S. and Sahal-Bréchot, S.: 1986, Ann. Phys. (France) 11,
    Suppl. 3, p.181.
26. Dimitrijevic, M.S. and Krsljanin, V.: 1986, Astron. Astrophys. 165, p.269.
27. Lascicevic, I.S.: 1985, Astron. Astrophys. 151, p.457.
28. Dimitrijevic, M.S. and Konjevic, N.: 1987, Astron. Astrophys. 172, p.345.
29. Konjevic, R. and Konjevic, N.: 1986, Fizika 18, p.327.
30. Dimitrijevic, M.S. and Konjevic, N.: 1986, Astron. Astrophys. 163, p.297.
31. Purcell, S.T. and Barnard, A.J.: 1984, JQSRT, 32, p.305.
32. Arata, Y., Miyake, S. and Matsuoka, H.: 1984, JQSRT 32, p.343.
33. Goly, A. and Weniger, S.: 1986, JQSRT 36, p.147.
34. Pittman, T.L. and Konjevic, N.: 1986, JQSRT 36, p.289.
35. Kittlitz, M., Radtke, R., Spanke, R. and Hitzschke, L.: 1985, JQSRT, 34,
    p.275.
36. Nick, R.P. and Helbig, V.: 1986, Phys. Scripta 33, p.55.
37. Solakbov, M. Kh., Sorondarev, E.V. and Fishman, I.S.: 1985, Opt. Spectrosc.
    (USA) 59, p.118.
38. Uzelac, N.I. and Konjevic, N.: 1986, Phys. Rev. A 33, p.1349.
39. DiRocco, M.O., Bertuccelli, G., Almondos, J.R. and Gallardo, M.: 1986,
    JQSRT 35, p.443.

40. Konjevic, N. and Pittman, T.L.: 1986, JQSRT 35, p.473.

41. Puric, J., Cuk, M. and Kathore, B.A.: 1987, Phys. Rev. A 35, p.1132.

42. Neger, T. and Jager, H.: 1987, Z. Naturforsch A 42a, p.429.

43. n'Dolo, M. and Fabry, M.: 1987, J. Physique (France) 48, p.703.

44. Dimitrijevic, M.S. and Artru, M.C.: 1986, SPIG Sibenik, p.317.

45. Dimitrijevic, M.S. and Krsljanin, V.: 1986, SPIG Sibenik, p.321.

46. Stamm, R., Talin, B., Pollock, E.L. and Iglesias, C.A.: 1986, Phys. Rev. A 34, p.4144.

47. Boercker, D.B., Iglesias, C.A. and Dufty, J.W., to be published.

48. Oza, D.H., Greene, R.L. and Kelleher, D.E.: 1986, Phys. Rev. A 34, p.4519.

49. Calisti, A., Stamm, R. and Tallin, B.. 1987, Europhys. Lett. 4, p.1003.

50. Joyce, R.F., Woltz, L.A. and Hooper Jr., C.F.: 1987, Phys. Rev. A 35, p.2228.

51. Gaisinsky, I.M. and Oks, E.A.: 1985, J. Phys. B: Atom. Mol. Phys. 18, p.1449.

52. Stehlé, C.: 1985, J. Phys. B: Atom. Mol. Phys. 18, p.143.

53. Iglesias, C.A. and Lebowitz, J.L.: 1984, Phys. Rev. A 30, p.4.

54. d'Etat, B., Grumberg, J., Leboucher, E., Nguyen, H. and Poquerusse, A.: 1987, J. Phys. B: Atom. Mol. Phys. 20, p.1733.

55. Peach, G.: 1981, Advances in Physics 30, p.367.

56. Kelleher, D.E. and Cooper, J.: 1985, in "Spectral Line Shapes, Volume 3", p.85.

57. Nguyen, H., Koenig, M., Benredjen, D., Caby, M. and Coulaud, G.: 1986, Phys. Rev. A 33, p.1279.

58. Burgess, D.D., Everett, D. and Lee, R.W.: 1979, J. Phys. B 12, L755.

59. Lemaire, J.L., Chotin, J.L. and Rostas, F.: 1985, J. Phys. B: Atom. Mol. Phys. 18, p.95.

60. Monteiro, T.S., Dickinson, A.S. and Lewis, E.L.: 1985, J. Phys. B: Atom. Mol. Phys. 18, p.3499.

61. O'Mara, B.J.: 1986, J. Phys. B: Atom. Mol. Phys. 19, L349.

62. Nieuwesteeg, K.J., Leegnater, J.A., Hollander, Tj. and Alkemade, C. Th. J.: 1987, J. Phys. B: Atom. Mol. Phys. 20, p.487.

63. Petuchouski, S.J.: 1986, JQSRT 36, p.319.

64. Le Moal, M.F. and Severin, F.: 1986, JQSRT 35, p.145.

65. Hillman, J.J., Hasley, G.W. and Jennings, D.E.: 1985, Bull. Am. Astron. Soc. 17, p.707.

66. Chudamani, S., Varanasi, P., Giver, L.P. and Valero, F.J.P.: 1985, JQSRT 34, p.359.

67a. Ballard, J. and Johnston, W.B.: 1986, JQSRT 36, p.365.

67b. Keffer, C.E., Conner, C.P. and Smith, W.H.: 1986, JQSRT 35, p.495.

68a. Keffer, C.E., Conner, C.P. and Smith, W.H.: 1985, JQSRT 33, p.193.

68b. Keffer, C.E., Conner, C.P. and Smith, W.H.: 1986, JQSRT 35, p.487.

69. Cadot, J.: 1985, JQSRT 34, p.331.

70. Bauer, A., Godon, M. and Duterage, B.: 1985, JQSRT 33, p.167.

71. Bauer, A., Godon, M., Kheddar, H., Hartmann, J.H., Bonamy, J. and Robert, D.: 1987, JQSRT 37, p.531.

72. Borysow, J., Trafton, L., Frommhold, L. and Birnbaum, G.: 1985, Astrophys. J. 296, p.644.

73. Borysow, A., Frommhold, L.: 1986, Astrophys. J. 311, p.1043.

74. Codestefano, P., Dore, P.: 1986, JQRST 36, p.445.

75. Moraldi, M., Borysov, A., Borysov, J. and Frommhold, L.: 1986, PRA 34, p.632.

76. Codestefano, P., Dore, P. and Nencini, L.: 1986, JQSRT 36, p.239.

77. Borysov, A. and Frommhold, L.: 1986, Astrophys. J. 303, p.495; Astrophys. J. 304, p.849.

78. Burnett, K.: 1985, Physics Reports 118, No.6, p.339.

79. Van Regemorter, H. and Feautrier, N.: 1985, J. Phys. B: Atom. Mol. Phys. 18, p.2673.

80. Van Regemorter, H.: 1986, J. Phys. B: Atom. Mol. Phys. 19, p.2235.

81. Julienne, P.S. and Mies, F.H.: 1984 , Phys. Rev. A 30, p.831.
82. Cooper, J. in "Spectral Line Shapes II", Ed. K. Burnett (Walter de Gruyter,
     Berlin, 1983), p.737.
83. Julienne, P.S. and Mies, F.H.: 1986, Phys. Rev. A 34, p.3792.
84. Alford, W.J., Burnett, K. and Cooper, J.: 1983, Phys. Rev. A 27, p.1310.
85. Alford, W.J., Andersen, N., Burnett, K. and Cooper, J.: 1984, Phys. Rev.
     A 30, p.2366.
86. Alford, W.J., Andersen, N., Belsley, M., Cooper, J., Warrington, D.M. and
     and Burnett, K.: 1985, Phys. Rev. A 31, p.3012.
87. Ermers, A., Woschnik, T. and Behmenburg, W.: 1987, Z. Phys. D Molecules
     and Clusters, in press.
88. Behmenburg, W., Kroop, V. and Rebenstrost, F.: 1985, J. Phys. B: Atom.
     Mol. Phys. 18, p.2693.
89. Wahala, L.L., Julienne, P.S., and Havey, M.D.: 1986, Phys. Rev. A 34,
     p.1856.
90. Belsley, M., Streater, A., Burnett, K., Ewart, P. and Cooper, J.: 1986,
     JQSRT 36, p.163.
91. Mathys, G.: 1984, Astron. Astrophys. 139, p.196.
92. Nee, Tsu-Jye A.: 1987, JQSRT 38, p.213.
93. Claude, M.L.: 1984, JQSRT 32, p.17.
94. Drummond, J.R. and Steckner, M.: 1985, JQSRT 34, p.517.

N. Feautrier
Chairman of Working Group

WORKING GROUP 5: MOLECULAR STRUCTURE AND TRANSITION DATA

Research in molecular spectroscopy has continued to grow over the past three
years. The spectral range has expanded from the far ultraviolet to millimeter
wavelengths. The report has been limited to molecular spectroscopy of relevance
to astronomy and has been compiled from edited contributions sent to me in the
fall of 1987.

Sumner P. Davis and John G. Phillips have reported studies at the Berkeley
laboratory on the molecules C_2, CN, FeH, InI, SH, Si_2, SiC_2, TiCℓ, TiO, ZrO, ZrS
of either analysis of spectral structure or measurements of lifetimes and oscillator strengths. FeH is of special interest. A complete table of excitation
energies for its complex infrared system has been prepared, and a set of tables
for far infrared and radio wavelengths assembled to predict spectrum lines which
may exist in stellar spectra or in the spectrum of matter in interstellar space.
Underway are analysis of the blue-green system of the FeH molecule, measurements
of radiative lifetimes for FeH, radiative lifetimes of CaH, transition strength
for A-X and B-X systems of CaH, measurement of CaCℓ transition strength (collaboration with J.E. Littleton), analysis and tabulation of infrared OH and OD bands
(collaboration with R. Engleman) and analysis of ZrS in the infrared (collaboration with R. Winkel).

The bimonthly Berkeley Newsletter on Analyses of Molecular Spectra compiled
by Davis, Phillips and Eakins continues to be the invaluable, timely bibliography
of molecular spectra. There is about 150 recipients of the Newsletter.

Takeshi Oka has reported from the University of Chicago, laboratory spectra
observed of the following molecular ions, H_3^+, H_2D^+, HeH^+, NeH^+, ArH_3^+, NH_4^+,
NH_3^+, NH_2^+, H_3O^+, H_2O^+, OH^+, $HCNH^+$, CH_3^+, $HCCH^+$, $C_2H_3^+$, OH^-, C_2^-. A search for
interstellar infrared absorption spectrum of H_3^+ was conducted; the result is
inconclusive though promising.

Kurt Dressler has reported from ETH Zurich work on the electronic transition moments for the Lyman and Werner bands of H_2(1). Stephens and Dalgarno's (2) calculated radiation lifetimes of the B & C states are confirmed and its suggested that recent measurements of longer lifetimes may be affected by radiation trapping.

The Molecular Spectroscopy Division of the National Bureau of Standards, Gaithersburg, MD, carries out experimental studies in the microwave, infrared and visible regions and develops critical reviews on microwave spectra of interstellar molecules and tables of infrared absorption lines for calibration of diode laser measurements. F.J. Lovas has reported that during the past three years several critical evaluations of microwave rotational spectra have been published. The twenty-second article in the series "Microwave Spectra of Molecules of Astrophysical Interest" treats the spectrum of SO_2 (3). The 1985 Revision to "Recommended Rest Frequencies for Observed Interstellar Molecular Microwave Transitions" was published in 1986 (4). Two compilations treating vibrational spectral and properties have also been published. Pine et al. (5) report on the high temperature water vapor spectrum in the 3000 4000 cm^{-1} region and Jacox (6) has compiled the ground state vibrational energy levels for polyatomic transient molecules.

The properties and spectra of diatomic species have been examined both theoretically and experimentally. Theoretical electronic structure calculations on FeO RuO (7). AgH and AuH (8), and transition moments for excited states of NaK (9) have been reported. Jacox (10) reports on comparisons of vibrational fundamentals of diatomic molecules in the gas phase and in inert solid matrices. Experimental studies on diatomic species have treated collisional effects on the rovibrational transitions of HD (11) and infrared spectra of high temperature species LiI (12) GeO (13) LiCl (14) and PbO (15).

The effort to develop tables of accurate infrared transition frequencies continues with heterodyne measurements on CO_2 (16), N_2O (17) and OCS (18) by Wells, Maki and co-workers. Also high resolution infrared studies of the hydrocarbons, ethane (19) and allene (20) have been reported.

A.R.W. McKellar has reported from the Herzberg Institute of Astrophysics, NRC, Canada that spectroscopic studies of molecules of astronomical interest have continued with an emphasis on unstable species in general and molecular ions in particular. Electronic spectra have been studied for H_2 (21-23), He_2 (24), NO (25) XeH (26), $ArXe^+$ (27) and NF (28) with special attention to O_2 (29-33), S_2 (34), and SeS (35). Electron impact ionization studies have been made for the hydrogen halide systems HCl DCl (36) and HBr DBr (37). Infrared vibration-rotation and pure rotational studies have been made for H_2 (38), HD (39-42) SiN (43), FO (44), and HF (45).

Considerable efforts have been applied to the study of polyatomic molecular ions in the infrared, including DCO^+ (46), N_2O^+ (47-49), HOC^+ (50-52), $HOCO^+$ (53 54), D_3O^+ (55), $HCNH^+$ (56), NH_3D^+ (57), SH_3^+ (58), and especially H_3^+ and its isotopes (59-67). Other polyatomic molecules studied in the infrared include H_2S (68,69), DO_2 (70), OCS (71), H_2O (72,73), CO_2 (74), CH_2 (75,76), ND_3 (77,78), C_3O_2 (79), D_2CO (80), and cyclopropane (81-83). Theoretical studies have emphasized a combination of high accuracy ab initio force fields and sophisticated vibration-rotation calculations to predict molecular energies (76,84-89).

The technique of microwave-optical double resonance has been extensively utilized to study the molecules HNO (90,91), and H_2CS (92-95). Other polyatomic molecules studied by electronic spectroscopy include ND_2 (96), CF_2 (97), CH_2 (98), C_2H_2 (99,100), and NH_3 (101). The electron impact ionization technique has been applied also to He (102) C_2H_2 (103), and C_2H_4 (104).

R.W. Nicholls has reported from the Center for Research in Experimental Space

Science at York University, Toronto, that work has continued on experimental and theoretical aspects of astrophysically important spectra. Shock tube studies on NbO and LaO have recently been initiated. Much effort has been devoted to the realistic numerical high resolution synthesis of molecular spectra, to reduce line shepes (105), and also for diagnostic application to laboratory (106), atmospheric and astrophysical circumstances including the interstellar extinction function (107). Non-dispersive digital correlation spectrometers have been developed for diagnoses of atmospheric conditions. The behaviour of dispersive correlation spectrometers has been modelled (108). An extensive study of simple analytical representations for molecular Franck-Condon factors for astrophysical band systems whose constants are incompletely known has been extended to include bound-free and free-free transitions. An extended Birge-Mecke rule linking the vibrational and rotational constants of molecular states. which fits extensive experimental data has recently been derived (109).

Research in molecular spectroscopy at the Harvard-Smithsonian Center for Astrophysics Cambridge MA is directed towards processes of importance to atmospheric physics and astrophysics. The wavelength range studied extends from the ultraviolet to the millimeter-wave spectroscopy. W.H. Parkinson has reported that K. Yoshino and colleagues have concentrated on a number of high resolution spectroscopic studies with the 6.65 m vacuum scanning spectrometer/spectrograph. Ab initio studies of potential energy curves and dipole moments for important small molecules of astrophysical interest have been undertaken by K. Kirby and colleagues.

Absorption cross sections and band oscillator strengths of the Schumann-Runge (S-R) bands of $^{16}O_2$, $^{18}O_2$ and $^{16}O^{18}O$ at 79 K have been measured between 179.3 and 198.0 nm with an instrumental full width at half maximum (FWHM) of 0.0013 nm. Oscillator strengths of the bands have been obtained by direct integration of the measured cross sections (110).

Spectroscopic constants of the B $^3\Sigma_u^-$ state of $^{16}O_2$ (111), $^{18}O_2$ and $^{16}O^{18}O$ have been determined from the experimental data on the high resolution absorption spectrum of the S-R bands. These constants will be used in the ongoing effort to extract the predissociation linewidths of the S-R bands from the absolute cross section measurements. Level shifts in the B $^3\Sigma_u^-$ state of these isotopes due to various repulsive continuous states are being calculated. Parameters of these states are being determined and will be used to examine the vibrational and rotational dependence of the resonance linewidths.

The absorption cross section of O_2 has been measured between 195 and 241 nm at 300 K. This region contains the Herzberg I transition of O_2 and of the O_2 dimer. Our Herzberg continuum cross sections are significantly smaller than those previously used in many photochemical stratospheric modelling calculations (112).

Absolute cross sections of O_3 at the temperature of 195 K, 228 K and 293 K have been measured at several discrete wavelengths in the region 238-335 nm (113).

A high resolution vuv absorption study and vibrational analysis of the B $^1\Sigma^+$ - X $^1\Sigma^+$ system of CS was performed in the wavelength region 128-155 nm (114).

High resolution spectra and photoabsorption coefficients for CO absorption bands in the 91-112 nm region have been measured at the Photon Factory of the National Laboratory for High Energy Physics (KEK), Japan.

Potential energy curves, electronic wavefunctions and electric dipole moments for the $X^3\Sigma^-$, $A^3\Pi$, $2^3\Sigma^-$, $2^3\Pi$ and $1^5\Sigma^-$ states of NH (116). The $2^3\Sigma^-$ and $2^3\Pi$ states are repulsive and have been identified as important photodissociation pathways.

Spectroscopic constants have been obtained for the bound states. The $1^5\Sigma^-$ is shown to cross the $A^3\Pi$ at a substantially higher energy and larger internuclear separation than previously postulated and therefore is probably not the cause of observed predissociations in low-lying vibrational levels of the $A^3\Pi$.

Configuration interaction calculations have been carried out on the low-lying $^1\Delta$ and $^1\Sigma^-$ states of CO. The lifetimes of the vibrational levels of the $I^1\Sigma^-$ and $D^1\Delta$ have been computed using transition moments for $I^1\Sigma - A^1\Pi$ and $D^1\Delta - A^1\Pi$.

By using a specially-designed millimeter-wave absorption spectrometer, B.J. Connor and H.E. Radford have made laboratory measurements of pressure-broadening coefficients for several spectral lines of ozone (117) and carbon monoxide (118) at wavelengths near 3 mm. The results are useful for the analyses of ground-based observations of the same lines emitted by the earth's atmosphere.

In the period 1985-88 Patrick Thaddeus and Carl Gottlieb have completed the first laboratory millimeter-wave spectroscopy of the carbon chain radicals C_3H (119 120), C_5H (121), and C_6H (122); the CCD radical (123)· the first hydrocarbon ring identified in interstellar space, C_3H_2 (124, 125); the stable three-membered ring cyclopropene, C_3H_2 rotational spectrum led to the identification of numerous astronomical transitions including the strong ubiquitous lines at 18, 343 MHz and 85, 338 MZ. Following the measurement of its laboratory spectrum, the CCD radical was detected in Orion A with a deuterium enhancement comparable to the strongest enhancement yet found in any molecule. Measurement of the millimeter-wave spectrum of cyclopropene, obtained by attaching two H atoms to the related ring C_3H_2, allowed them to place upper limits on its abundance in Sgr B2 and TMC-1 indicating that in these sources C_3H_4 isn't much more abundant than C_3H_2. Laboratory measurements of vibrationally excited C_4H confirmed a tentative astronomical identification. The surprisingly large intensity of the vibrationally excited lines of C_4H relative to the ground state in IRC+10216 suggests the opening of a whole new subdiscipline in which laboratory spectroscopy will be crucial, since up to now little or no spectroscopy has been done on the excited state of small reactive chains and rings.

Research programs at the Observatoire de Paris (Meudon) concern new spectroscopy, photodynamics and transition probabilities of astrophysically interesting molecules. Francois Rostas has reported that a comprehensive effort has been applied to a better understanding of the CO molecule. A new analysis of the $A^1\Pi(v=0)$ perturbations extending to J = 75 has been completed (129) using absorption and emission spectra of the A-X transition. The analysis of the 0<v≤4 levels will be published shortly. The B-X transition has been reanalyzed (130) and a study of its predissociation is in progress. New data concerning the $C^1\Sigma^+$ and $E^1\Pi$ Rydberg states have been analyzed. The photodissociation cross section has been measured between 88.5 and 115 nm (131). It is shown that photodissociation occurs in discrete bands rather than in a continuum with far reaching astrophysical consequences.

The establishment of an atlas of H_2 emission lines between 78 and 168 nm is progressing steadily. This work has suggested calculation of line positions and intensities of the Lyman and Werner systems taking into account rotational coupling between the B and C states (132). These results are in agreement with observations and quantitative intensity measurements and allow a more complete analysis of the spectra.

A new emission band of the N_2^{++} ion has been observed and compared to ab initio calculations (133 134).

Emission spectra of N_2 have been obtained at very low pressure in the 85 to 108 nm region. Intensity anomalies of previous absorption spectra have been shown to be due to reabsorption effects (135).

High resolution spectra of CO_2 between 10 and 14 eV have been analyzed. nf transition bands have been characterized by sharp features in the rotational band contours. Calculated quantum defects have been used to assist in the assignments (136).

An extensive theoretical study of C_2 has been pursued in conjunction with a model of the ζ Oph. diffuse molecular cloud. Energy levels, rotational excitation by collisions with H_2 and radiative equilibrium including intercombination transitions have been determined (137 138) and included in a comprehensive chemical and radiative model of the cloud.

A new emphasis is placed on the photophysics spectroscopy and relaxation of molecular ions (139). Doubly charged ions of polycyclic aromatic molecules are studied in detail (140).

References

1. Dressler, K. and Wolneiwicz, L.: 1985, J. Chem. Phys. 82, p.4720.
2. Stephens, T.L. and Dalgarno, A.: 1972, J. Quant. Spectrosc. Radiat. Transfer 12, p.569.
3. Lovas, F.J.: 1985, J Phys. Chem. Ref. Data 14, p.395.
4. Lovas, F.J.: 1986, J. Phys. Chem. Ref. Data 15, p.251.
5. Pine, A.S., Coulombe, M.J., Camy-Peyret, C., and Flaud, J.M.: 1983, J. Phys. Chem. Ref. Data 12, p.413.
6. Jacox, M.E.· 1984, J. Phys. Chem. Ref. Data 13, p.945.
7. Krauss, M. and Stevens, W.J.· 1985, J. Chem. Phys. 82, p.5584.
8. Krauss, M., Stevens, W.J. and Basch, H.: 1985, J. Comp. Chem. 6, p.287.
9. Ratcliff, L., Konowalow, D.D. and Stevens, W.J.: 1985, J. Mol. Spectrosc. 110, p.242.
10. Jacox, M.E.: 1985, J. Mol. Spectrosc. 113, p.286.
11. Nazemi, S., Javan, A. and Pine, A.S.: 1983, J. Chem. Phys. 78, p.4797.
12. Thompson, G.A., Maki, A.G. and Weber, A.: 1986, J. Mol. Spectrosc. 118, p.540.
13. Thompson, G.A., Maki, A.G. and Weber, A.: 1986, J. Mol. Spectrosc. 116, p.136.
14. Thompson, G.A., Olson, W.B., Maki, A.G. and Weber, A.: 1987, J. Mol. Spectrosc. 124, p.130.
15. Maki, A.G. and Lovas, F.J.: 1987, J. Mol. Spectrosc. 125, p.188.
16. Petersen, F.R., Wells, J.S., Siemsen, K.J., Robinson, A.M. and Maki, A.G.: 1984, J. Mol. Spectros. 105, p.324.
17. Pollock, C.P., Peterson, F.R., Jennings, D.E., Wells, J.S. and Maki, A.G.: 1984.
18. Maki, A.G., Wells, J.S. and Hinz, A.: 1986, Int. J. Infrared and Millimeter Waves 7, p.909.
19. Maki, A.G., Pine, A.S. and Dang-Nhu, M.: 1985, J. Mol.Spectros. 112, p.459.
20. Henry, L., Valentin, A., Lafferty, W.J., Hougen, J.T., Devi Malathy, V., Das, P.P. and Rao, K. Narahari: 1983, J. Mol. Spectrosc. 100, p.260.
21. Dabrowski, I. and Herzberg, G.: 1984, Acta Physica Hungarica 55, p.219.
22. Seen, P., Quadrelli, P., Dressler, K. and Herzberg, G.: 1985, J. Chem. Phys. 83, p.962.
23. Dabrowski, I.: 1984, Can. J. Phys. 62, p.12.
24. Herzberg, G. and Jungen, Ch.: 1986, J. Chem. Phys. 84, p.1181.
25 Huber, K.P. and Sears, T.J.: 1984, Chem. Phys. Lett. 113 2, p.129.
26. Lipson, R.H.: 1986: Chem. Phys. Lett. 129, p.82.
27. Huber, K.P. and Lipson, R.H.: 1986, J. Mol. Spectrosc. 119, p.433.
28. Vervloet, M. and Watson, J.K.G.: 1986, Can. J. Phys. 64, p.1529.
29. Kerr, C.M.L. and Watson, J.K.G.: 1986, Can. J. Phys. 64, p.36.

30. Fink, E.H., Kruse, H., Ramsay, D.A. and Vervloet, M.: 1986, Can. J. Phys.
     64, p.242.
31. Borrell, P.M., Borrell, P. and Ramsay, D.A.: 1986, Can. J. Phys. 64,
     p.721.
32. Coquart, B. and Ramsay, D.A.: 1986, Can. J. Phys. 64, p.726.
33. Ramsay, D.A.: 1986, Can. J. Phys. 64, p.717.
34. Fink, E.H., Kruse, H. and Ramsay, D.A.: 1986, J. Mol. Spectrosc. 119,
     p.377.
35. Fink, E.H., Kruse, H., Ramsay, D.A. and Chang, D-C.: 1987, Mol. Phys. 60,
     p.277.
36. Nasrallah, II.K. and Marmet, P.: 1985, J. Phys. B: At.Mol. Phys. 18,
     p.2075.
37. Marmet, P. and Nasrallah, K.H.: 1985, Can. J. Phys. 63, p.1015.
38. Clouter, M.J. and McKellar, A.R.W.: 1986, J. Chem. Phys. 84, p.2466.
39. McKellar, A.R.W., Johns, J.W.C., Majewski, W. and Rich, N.H.: 1984, Can.
     J. Phys. 62, 12, p.1673.
40. McKellar, A.R.W. and Rich, N.H.: 1984, Can. J. Phys. 62, 12, p.1665.
41. McKellar, A.R.W.: 1986, Can. J. Phys. 64, p.227.
42. McKellar, A.R.W. and Clouter, M.J.: 1987, Can. J. Phys. 65, p.1.
43. Foster, S.C., Lubic, K.G. and Amano, T.: 1984, J. Chem. Phys. 82, 2,p.709.
44. Burkholder, J.B., Hammer, P.D., Howard, C.J. and McKellar, A.R.W.: 1986,
     J. Mol. Spectrosc. 118, p.471.
45. Jennings, D.A., Evenson, K.M., Zink, L.R., Demuynck, C., Destombes, J.L.,
     Lemoine, B. and Johns, J.W.C.: 1987, J. Mol. Spectrosc. 122, p.477.
46. Kawaguchi, K., McKellar, A.R.W. and Hirota, E.: 1986, J. Chem. Phys, 84,
     p.1146.
47. Amano, T.: 1986, Chem. Phys. Lett. 127, p.101.
48. Amano, T.: 1986, Chem. Phys. Lett. 130, p.154.
49. Bogey, M., Demuynck, C., Destombes, J.L. and McKellar, A.R.W.: 1986,
     Astron. Astrophys. 167, L13.
50. Beardsworth, R., Bunker, P.R., Jensen, P. and Kraemer, W.P.: 1986, J. Mol.
     Spectrosc. 118, p.40.
51. Nakanaga, T. and Amano, T.: 1987, J. Mol. Spectrosc. 121, p.502.
52. Bunker, P.R., Jensen, P., Kraemer, W.P. and Beardsworth, R.: 1987, J. Mol.
     Spectrosc. 121, p.450.
53. Amano, T. and Tanaka, K.: 1985, J. Chem. Phys. 82, 2, p.1045.
54. Amano, T. and Tanaka, K.: 1985, J. Chem. Phys. 83, p.3721.
55. Sears, T.J., Bunker, P.R., Davies, P.B., Johnson, S.A. and Spirko, V.: 1985,
     J. Chem. Phys. 83, p.2676.
56. Amano, T. and Tanaka, K.: 1986, J. Mol. Spectrosc. 116, p.112.
57. Nakanaga, T. and Amano, T.: 1986, Can. J. Phys. 64, p.1356.
58. Nakanaga, T. and Amano, T.: 1987, Chem. Phys. Lett. 134, p.195,
59. Watson, J.K.G., Foster, S.C., McKellar, A.R.W., Bernath, P., Amano, T., Pan,
     F.S., Crofton, M.W., Altman, R.S. and Oka, T.: 1984, Can. J. Phys. 62,
     p.12.
60. Lubic, K.A. and Amano, T.: 1984, Can. J. Phys. 62, 12, p.1886.
61. Amano, T.: 1985, J. Opt. Soc. Am. B 2, 5, p.79.
62. Spirko, V., Jensen, P., Bunker, P.R. and Cejchan, A.: 1985, J. Mol. Spec-
     trosc. 112, p.183.
63. Foster, S.C., McKellar, A.R.W., Peterkin, I.R. and Watson, J.K.G.: 1986,
     J. Chem. Phys. 84, p.91.
64. Jensen, P., Spirko, V. and Bunker, P.R.: 1986, J. Mol. Spectrosc. 115,
     p.269.
65. Foster, S.C., McKellar, A.R.W. and Watson, J.K.G.: 1986, J. Chem. Phys.
     85, p.664.
66. Majewski, W.A., Marshall, M.D., McKellar, A.R.W., Johns, J.W.C. and Watson,
     J.K.G.: 1987, J. Mol. Spectrosc. 122, p.341.
67. Watson, J.K.G., Foster, S.C. and McKellar, A.R.W.: 1987, Can. J. Phys.
     65, p.38.

68. Lechuga-Fossat, L., Flaud, J.-M., Camy-Peyret, C. and Johns, J.W.C.: 1984, Can. J. Phys. 62, 12, p.1889.
69. Camy-Peyret, C., Flaud, J.-M., Lechuga-Fossat, L. and Johns, J.W.C.: 1985, J. Mol. Spectrosc. 109, p.300.
70. Lubic, K.G., Amano, T., Uehara, H., Kawaguchi, K. and Hirota, E.: 1984, J. Chem. Phys. 81, 11, p.4826.
71. Hunt, N., Foster, C., Johns, J.W.C. and McKellar, A.R.S.: 1985, J. Mol. Spectrosc. 111, p.42.
72. Camy-Peyret, C., Flaud, J.-M., Mandin, Y.-Y., Chevillard, J.-P., Brault, J., Ramsay, D.A., Vervloet, M. and Cahuville, J.: 1985, J. Mol. Spectrosc. 113, p.208.
73. Johns, J.W.C.: 1985, J. Opt. Soc. Am. B 2, p.1340.
74. Clouter, M.J. and McKellar, A.R.W.: 1985, Can. J. Phys. 63, p.1559.
75. Marshall, M.D. and McKellar, A.R.: 1986, J. Chem. Phys. 85, p.3716.
76. Bunker, P.R., Jensen, P., Kraemer, W.P. and Beardsworth, R.: 1986, J. Chem. Phys. 85, p.3724.
77. Fusina, L., DiLonardo, G. and Johns, J.W.C.: 1985, J. Mol. Spectrosc. 112, p.211.
78. Fusina, L., DiLonardo, G. and Johns, J.W.C.: 1986, J. Mol. Spectrosc. 118, p.397.
79. Jensen, P. and Johns, J.W.C.: 1986, J. Mol. Spectrosc. 118, p.248.
80. Nakagawa, K., Schwendeman, R.H. and Johns, J.W.C.: 1987, J. Mol. Spectrosc. 122, p.462.
81. Pilva, J. and Johns, J.W.C.: 1984, Can. J. Phys. 62, 12, p.1369.
82. Pilva, J. and Johns, J.W.C.: 1985, J. Mol. Spectrosc. 113, p.175.
83. Pilva, J. and Johns, J.W.C.: 1986, Can. J. Phys. 64, p.1452.
84. Phillips, R.A., Buenker, R.J., Beardsworth, R., Bunker, P.R., Jensen, P. and Kraemer, W.P.: 1985, Chem. Phys. Lett. 118, 1, p.60.
85. Bunker, P.R. and Sears, T.J.: 1985, J. Chem. Phys. 83, p.4866.
86. Jensen, P. and Bunker, P.R.: 1986, J. Mol. Spectrosc. 118, p.18.
87. Beardsworth, R., Bunker, P.R., Jensen, P. and Kraemer, W.P.: 1986, J. Mol. Spectrosc. 118, p.50.
88. Sarka, K. and Bunker, P.R.: 1987, J. Mol. Spectrosc. 122, p.259.
89. Escribano, R. and Bunker, P.R.: 1987, J. Mol. Spectrosc. 122, p.325.
90. Petersen, J.C., Saito, S., Amano, T. and Ramsay, D.A.: 1984, Can. J. Phys. 62, 12, p.1731.
91. Petersen, J.C., Amano, T. and Ramsay, D.A.: 1984, J. Chem. Phys. 81, 12, p.5449.
92. Petersen, J.C. and Ramsay, D.A.: 1985, Chem. Phys. Lett. 118, 1, p.31.
93. Petersen, J.C. and Ramsay, D.A.: 1985, Chem. Phys. Lett. 118, 1, p.34.
94. Fung, K.H., Petersen, J.C. and Ramsay, D.A.: 1985, Can. J. Phys. 63, p.993.
95. Petersen, J.C. and Ramsay, D.A.: 1986, Chem. Phys. Lett. 124, p.406.
96. Muenchausen, R.E., Hills, G.W., Merienne-Lafore, M.F., Ramsay, D.A., Vervloet, M. and Birss, F.W.: 1985, J. Mol. Spectrosc. 112, p.203.
97. Comes, F.J. and Ramsay, D.A.: 1985, J. Mol. Spectrosc. 113, p.495.
98. Petek, H., Nesbitt, D.J., Moore, C.B., Birss, F.W. and Ramsay, D.A.: 1987, J. Chem. Phys. 86, p.1189.
99. Van Craen, J.C., Herman, M., Colin, R. and Watson, J.K.G.: 1985, J. Mol. Spectrosc. 111, p.185.
100. Van Craen, J.C., Herman, M., Colin, R. and Watson, J.K.G.: 1986, J. Mol. Spectrosc. 119, p.137.
101. Watson, J.K.G., Majewski, W.A. and Glownia, J.H.: 1986, J. Mol. Spectrosc. 115, p.82.
102. Marmet, P., Plessis, P. and Dutil, R.: 1987, J. Mass Spect. & Ion Processes 75, p.265.
103. Plessis, P. and Marmet, P.: 1986, Intl. J. Mass Spect. & Ion Processes 70, p.23.
104. Plessis, P. and Marmet, P.: 1987, Can. J. Phys. 65, p.165.

105. Cann, M.W.P., Nicholls, R.W., Roney, P.L., Blanchard, A. and Findlay, F.D.: 1985, Appl. Opt. 24, p.1374.

106. Nicholls, R.W., Cann, M.W.P. and Shin, J.B., Shock Tubes and Shock Waves, eds., D. Bershader and R. Hanson, (Stanford University Press, 1986) p.65.

107. Nicholls, R.W.: 1987, in press, J. Quant. Spectrosc. Rad. Transfer.

108. Nicholls, R.W.: 1985, Appl. Opt. 24, p.2046.

109. Nicholls, R.W.: 1987, in press, J. Opt. Soc. Am.

110. Yoshino, K., Freeman, D.E., Esmond, J.R. and Parkinson, W.H.: 1987, Planet. Space Sci. 35, p.1067.

111. Cheung, A.S.-C., Yoshino, K., Parkinson, W.H. and Freeman, D.E.: 1986, J. Mol. Spectrosc. 119, p.1.

112. Cheung, A.S.-C., Yoshino, K., Parkinson, W.H., Guberman, S.L. and Freeman, D.E.: 1986, Planet. Space Sci. 34, p.1007.

113. Freeman, D.E., Yoshino, K., Esmond, J.R. and Parkinson, W.H., in Atmospheric Ozone, eds. C.S. Zerefos and A. Ghazi (Dordrecht: D. Reidel, 1985) p.622.

114. Stark, G., Yoshino, K. and Smith, P.L.: 1987, J. Mol. Spectrosc. 124,p.420.

115. Cooper, D.L. and Kirby, K.: 1987, J. Chem. Phys. 87, p.424.

116. Goldfield, E.M. and Kirby, K.: 1987, in press, J. Chem. Phys.

117. Connor, B.J. and Radford, H.E.: 1986, J. Mol. Spectrosc. 117, p.15.

119. Gottlieb, C.A., Vrtilek, J.M., Thaddeus, P., Gottlieb, E.W. and Hjalmarson, A1: 1985, Astrophys. J. (Lett.) 294, L55.

120. Gottlieb, C.A., Gottlieb, E.W., Thaddeus, P. and Vrtilek, J.M.: 1986, Astrophys. J. 303, p.446.

121. Gottlieb, C.A., Gottlieb, E.W. and Thaddeus, P.: 1986, Astron. Astrophys, 164, L5.

122. Pearson, J.C., Gottlieb, C.A., Woodward, D.R. and Thaddeus, P.: submitted, Astron. Astrophys.

123. Vrtilek, J.M., Gottlieb, C.A., Langer, W.B., Thaddeus, P. and Wilson, R.W.: 1985, Astrophys. J. (Lett.) 296, L35.

124. Thaddeus, P., Vrtilek, J.M. and Gottlieb, C.A.: 1985, Astrophys. J. (Lett.) 299, L63.

125. Vrtilek, J.M., Gottlieb, C.A. and Thaddeus, P.: 1987, Astrophys. J.314,p716.

126. Vrtilek, J.M., Gottlieb, C.A., LePage, T.J. and Thaddeus, P.: 1987, Astrophys. J. 316, p.826.

127. Woodward, D.R., Pearson, J.C., Gottlieb, C.A., Thaddeus, P. and Guélin, M.: submitted, Astron. Astrophys.

128. Guélin, M., Cernicharo, J., Navarro, S., Woodward, D.R., Gottlieb, C.A. and Thaddeus, P., in press, Astron. Astrophys.

129. LeFloch, A.C., Launay, F., Rostas, J., Field, R.W., Brown, C.M. and Yoshino, K.: 1987, J. Mol. Spectrosc. 121, p.337.

130. Eidelsberg, M., Roncin, J.-Y., LeFloch, A., Launay, F., Letzelter, C. and Rostas, J.: 1987, J. Mol. Spectrosc. 121, p.309.

131. Letzelter, C., Eidelsberg, M., Rostas, F., Breton, J. and Thieblemont, B.: 1987.

132. Abgrall, H., Launay, F., Roueff, E. and Roncin, J.-Y.: 1987 (in press), Astron. Astrophys.

133. Cossart, D., Launay, F., Robbe, J.M. and Gandara, G.: 1985, J. Mol. Spectrosc. 113, p.142.

134. Cossart, D. and Launay, F.: 1985, J. Mol.Spectrosc. 113, p.159.

135. Roncin, J.-Y., Launay, F. and Yoshino, K.: 1987, Planet. Space Sci. 35, p.267.

136. Cossart-Magos, C., Jungen, M. and Launay, F.: 1987 (in press) Mol. Phys.

137. LeBourlot, J. and Roueff, E.: 1986, J. Mol. Spectrosc. 120, p.157.

138. LeBourlot, J., Roueff, E. and Viala, Y.: 1987 (in press) Astron. Astrophys.

139. Leach, S.: 1986, J. Molec. Struct. 141, p.43.

140. Leach, S., in Polycyclic Aromatic Hydrocarbons and Astrophysics, Les Houches Workshop, eds. A. Leger et al. (Dordrecht: D. Reidel, 1987) p.99.

W.H. Parkinson, Chairman of Working Group

COMMISSION 15 : PHYSICAL STUDY OF COMETS, MINOR PLANETS AND METEORITES

(L'ÉTUDE PHYSIQUE DES COMÈTES, DES PETITES PLANÈTES ET DES MÉTÉORITES)

PRESIDENT : Ľ Kresák
VICE-PRESIDENT : J Rahe
ORGANIZING COMMITTEE : M F A'Hearn, C Arpigny, C R Chapman, O V Dobrovolsky,
   H Fechtig, A W Harris, H F Haupt, L M Shulman, J T Wasson, S Wyckoff,
   V Zappalà

1 Introduction

The period covered by this report, 1984 July to 1987 June, was of extra-ordinary importance for the progress of cometary physics. For the first time in the history, special space probes were launched to comets. Vega 1, Vega 2 and Giotto encountered P/Halley, providing us with the first close-up pictures of a cometary nucleus, its surface features, and with the first in situ measurements of the matter escaping from it. ICE, Suisei and Sakigake carried out measurements relevant to P/Giacobini-Zinner and P/Halley in interplanetary space. Unprecedented worldwide campaigns of ground-based observations, with the participation of about 1000 professional and 2000 amateur astronomers, were coordinated in 8 sections of the International Halley Watch. Additional measurements were made from artificial satellites, sounding rockets, and high-flying airplanes. The wealth of data collected in this way, to a major extent thanks to an excellent international cooperation, represents a milestone in cometary astronomy. Another important step was the progress in processing the extensive 1983 IRAS observations of minor planets and comets, including the discovery of asteroid dust bands and cometary dust trails.

There were so many international meetings dealing at least partially with the subject area of our Commission, that it is only possible to quote some of them : Asteroid Missions (Graz 1984), Halley Up-Date (Graz 1984), Properties and Interactions of Interplanetary Dust (Marseille 1984), Asteroids, Comets, Meteors II (Uppsala 1985), The Evolution of the Small Bodies of the Solar System (Varenna 1985), Catastrophic Disruption of Asteroids and Satellites (Pisa 1985), Field, Particle and Wave Experiments on Cometary Missions (Graz 1985), Astrochemistry (Goa 1985), Comet Halley (New Delhi 1985), Comets Halley and Giacobini-Zinner (Toulouse 1986), The Multi-Comet Mission (Greenbelt 1986), The Comet Nucleus Sample Return Mission (Canterbury 1986), Exploration of Halley's Comet (Heidelberg 1986), Meteorites and the Early Solar System (Tucson 1987), The Diversity and Similarity of Comets (Brussels 1987), Physical Interpretation of Solar/Interplanetary and Cometary Intervals (Huntsville 1987). More information about the programs, dates, organizers, sponsors, and proceedings of these meetings can be found in the IAU Information Bulletins or in the Astronomy and Astrophysics Abstracts. The Halley meeting at Heidelberg, with 500 participants from 30 different countries and 370 papers presented, was the largest meeting on interplanetary objects in the history.

From among the numerous books, monographs, and extensive review papers, the following may be quoted : The Mystery of Comets (Whipple 1985), Asteroids (Simonenko 1985), Propriétés Physiques et Chimiques des Comètes (Arpigny 1985), The Physics of Comets (Mendis, Houpis and Marconi 1985), Cometary Dynamics (Weissman 1985), Long-term Evolution of Short-period Comets (Carusi, Kresák, Perozzi and Valsecchi 1985), Kometen (Möhlmann, Sauer and Wäsch 1986), Physics of Comets (Krishna Swamy 1986), The Origin of Comets (Bailey, Clube and Napier 1986), IRAS Asteroid and Comet Survey (ed Matson 1986), Catalogue of Short-period Comets (Belyaev, Kresák, Pittich and Pushkarev 1986), Cometary Nuclei

(Shulman 1987), Physics of Radiation of Planetary and Cometary Atmospheres (Krasnopolsky 1987), Asteroid Photometric Catalog (Lagerkvist, Barucci, Capria, Fulchignoni, Guerriero, Perozzi and Zappalà 1987). Information about the publishers or publishing journals can be found again in the Astronomy and Astrophysics Abstracts. For new editions of the catalogues of orbits, designations, absolute magnitudes and ephemerides of minor planets and comets see the reports of Commissions 6 and 20.

In spite of over 700 references included, the present report is obviously incomplete. This refers, in particular, to the broad and interdisciplinary meteorite research, from which only some results relevant to the origin of the solar system and evolutionary links to other types of objects were selected. The boom in the space exploration of comets has also made it necessary to reduce the information on asteroids. The references are given in a very succinct form, as explained by the key at the end of the report.

The preparation of this report would have been impossible without a kind cooperation of a number of colleagues. Drafts of the chapters on comets were written by B D Donn (Cometary Nuclei), M F A'Hearn (Cometary Gases), W H Ip (Cometary Plasma), E Grün and A C Levasseur-Regourd (Cometary Dust), and H Rickman (Origin and Evolution of Comets). A review of the cometary research in the USSR was provided by O V Dobrovolsky, and incorporated into the relevant chapters by C Arpigny, who also made a number of other additions and changes to avoid duplicities. The chapter on Minor Planets was prepared by A W Harris with the assistance of P Farinella and D F Lupishko. The chapter on Spacecraft Missions to Comets and Asteroids was written by J Rahe. The chapter on Meteorites was prepared by J T Wasson with the assistance of D Brownlee, J Gooding, R Grieve, R Hewins, I Hutcheon, G Kallemeyn, A S Kornacki, F Kyte, D Malvin, A E Rubin, E Scott, and P Warren. The final arrangement of the report, with a number of minor changes, was done by the Commission President, who takes the responsibility for any omissions and errors, and acknowledges with sincere thanks the kind help of all the above mentioned co-authors, especially the great share of Claude Arpigny.

2 Cometary Nuclei

Several major developments in research on the cometary nuclei have occured in the interval covered by this report. Foremost, was the Comet Halley encounter during which three space probes actually observed the nucleus. A second was the increasing acceptance of the concept that the nucleus is a non-uniform, low density aggregate of cometesimals. This picture received considerable support from the encounter images. A third is the very non-uniform surface activity of the nucleus, as demonstrated by Sekanina and Larson over several years and directly seen during the encounter imaging. These two, now generally accepted nuclear characteristics, greatly complicate theoretical analyses of the thermal behaviour and evolution of the nucleus, as the greatly simplifying assumptions of spherical symmetry and uniform angular and radial properties can no longer be made.

COMET P/HALLEY

Publications dealing with the nucleus of P/Halley during its 1986 apparition are very extensive as a result of two Vegas and the Giotto spacecraft encounters. The primary results were : (1) detection of a single nucleus; (2) direct measure of size and shape - irregular elongated nucleus about 16 km long with width 7.5 to 8 km. The surface shows features to the limit of resolution, about 50 m for the Giotto camera; (3) the surface is very non-uniform with respect to ejection of matter, with only about 20 percent being active during the encounter; (4) combination of the dimensions of the nucleus with its visual magnitude at very large heliocentric distances (> 6 AU) leads to a very low geometric albedo of approximately 4 percent; the nuclear surface is quite warm (≥ 300 K), which suggests the presence of an insulating dust layer on top of the volatile material (Sagdeev et al, Nat 321, 262; ESA 250/2, 317; Keller et al, Nat 321, 320; ESA 250/2, 347; Möhlmann et al, Bertaux and Abergel, Reitsema et al, and Wilhelm et al, ESA 250/2, 339, 341, 347, and 367; Combes et al, Nat 321, 266; Emerich et al, ESA 250/2, 381).

CHARACTERISTICS OF NUCLEUS AND NUCLEAR MODELS

Whipple (ESA 250/2, 281; NASI 156, 343), Sekanina (ASR 5, 307), and Wood (ESA 249, 123) have reviewed various aspects of nuclear models, characteristics and behaviour. Recent models emphasizing the nucleus as a compact collection of aggregates have been presented by Weissman (Nat 320, 242), and by Gombosi and Houpis (Nat 324, 43). Donn and Hughes (ESA 250/2, 523) starting with random accretion of grains and clusters have proposed a low-density porous, somewhat compressed, fractal-like structure. Greenberg (ACM II, 221) also proposed a low-density comet model. Rickman (ESA 249, 195) reported a determination of masses and densities of P/Halley and P/Kopff, and obtained densities of 0.1-0.2 g cm^{-3}.

ORIGIN

The comet formation environment has been examined by Yamamoto on the basis of equilibrium vaporization of interstellar molecules (AA 142, 31; NASI 156, 205). Clube and Napier (Ic 62, 384) and Napier and Humphries (MN 221, 105) treated comet formation in molecular clouds. A review of theories of comet origin by Bailey appeared in Vistas in Astronomy (29, 53). Implications for planet formation were discussed by Weissman (PP II, 895). The origin and dissolution was examined by Oort (Obs 106, 186). Dynamics of accreting cometesimals in the region of the outer planets was reviewed by Greenberg et al (IAUC 83, 3).

NUCLEAR ROTATION AND PRECESSION

Rotation of nuclei was discussed by Wallis (PTRS 313/A, 165). The rotational state of the nucleus of P/Halley appeared to be complex, different lines of evidence indicating periodicities of about 2.2 and 7.4 days. For the current various interpretations see Sekanina (Nat 325, 326); Julian (Nat 326, 57); Smith et al (Nat 326, 573); Wilhelm (Nat 327, 27); and references therein. For P/Arend-Rigaux, Jewitt and Meech (Ic 64, 329) derived a period of 9.58 or 6.78 hours. Precession of the nucleus and consequences for cometary evolution for several comets have been studied by Sekanina (AJ 89, 1573; AJ 90, 877 and 1370).

THERMAL MODELS AND EVOLUTION

An active area of cometary research has been modelling of the thermal behaviour and evolution of the nucleus, particularly mantle formation. Numerical simulations were carried out by Herman and Podolak (Ic 61, 252 and 267); Weissman and Kieffer (JGR 89, C235; ASR 4, 221); Weissman (ESA 250/3, 517); McKay et al (JGR 90, 1231; Ic 66, 625); Fanale and Salvail (Ic 60, 476); Ip and Rickman (ESA 249, 181); and Houpis et al (ApJ 295, 654). Kührt et al have treated thermal stresses in cometary ices (Ic 60, 512; Ic 61, 124; ASR 4, 225; ASR 5, 105; ESA 250/2, 385). The evolution of cometary ice structure was investigated by Smoluchowski et al (EMP 30, 281; NASI 156, 397) and by Klinger (NASI 156, 407). Observations of P/Kopff were interpreted by Cochran (AJ 91, 674) as evidence for a mantle development. Dust mantle formation and evolution were modelled by Rickman and Fernández (ESA 249, 185). Theoretical work of Shulman, including a theory of the dust mantle formation and evolution, and a purely analytical approach to the thermal regime problem, has been summarized in his monograph (Cometary Nuclei, Moscow 1987). The latter problem has also been treated numerically for a variety of models by Marov et al (AV 21, 47), and by Ibadinov and Aliev (KM 39, 3). The progressive dust layer growth was suggested to be a typical phenomenon of comets, as indicated by their secular brightness decrease and its dependence on perihelion distance (Dobrovolsky et al, ESA 250/2, 389). The reactive force acting on P/Halley's nucleus was evaluated by Kolesnichenko and Skorov (KIAM 84, 61).

LABORATORY STUDIES

Laboratory modelling of surface matrices continued at Leningrad and at Dushanbe. Among the items treated there were : formation of polymers under UV action on water ice with admixtures of nitriles or aminoacids (Lisunkova and Kajmakov, KTs 364 and 366); conditions of amorphous or crystalline grain formation, and structural elements of the grains (Lisunkova, KTs 366); mechanical and thermal properties of organic dust matrices (Ibadinov et al, DANT 28, 21; Ibadinov and Rakhmonov, KTs 360) - main results: densities 0.01 to 0.05 g cm^{-3}, compressive strength 0.05 to 0.5 kg cm^{-2}, thermal conductivity 10^{-1} W m^{-1} K^{-1}; creation of negative ions and clusters of both signs under simulated solar wind action (Hashimov and Shoyekubov, ESA 250/3, 189); sublimation of solid H_2O with admixture of CO_2 proceeding in the form of microbursts with ejection of icy grains up to millimeter-sized ones (Ibadinov and Aliev, KM 36, 35); these observations are in line with Kajmakov's thermobarodestruction concept (KTs 280), and with ideas by Kührt (Ic 60, 512). An experimental study of low-density residues was carried out by Saunders et al (Ic 66, 94).

3 Cometary Gases

PHOTOPROCESSES

The fluorescent emission of OD was modelled by A'Hearn et al (ApJ 297, 826), that of CO$^+$ by Magnani and A'Hearn (ApJ 302, 477), that of CN, including isotopic variants, by Zucconi and Festou (AA 150, 180; AA 158, 382), that of CO, including collisions but only for the infrared bands, by Chin and Weaver (ApJ 285, 858), and that of H_2O^+, vibrational structure only, by Lutz (ApJ 315 L147). Krishna Swamy calculated profiles of the B-X bands of CO$^+$ (EMP 34, 281) and of some Swan bands of C_2 (MN 224, 537), the latter for isotopic variants. Photodissociation lifetimes, including the Swings effect, were calculated for OH and OD by Van Dishoek and Dalgarno (Ic 59, 305) and for NH by Singh and Gruenwald (AA 178, 277). De Almeida and Singh (EMP 36, 117) calculated the photodissociation lifetimes of S_2. Pehler and Kegel (AA 155, L13) showed that formation of OH in excited states should affect the observed relative populations by no more than a few percent, whereas Bertaux (AA 160, L7) used relative brightnesses of OH on and off the nucleus to argue for the direct formation of OH in the excited A state. Crovisier (AJ 90, 670) improved his earlier models for the excitation of radio lines of H_2O, CO, HCN, and NH_3.

HYDRODYNAMIC AND PHOTO-CHEMICAL MODELS

Marconi and Mendis (ApJ 287, 445; EMP 36, 249) improved their earlier hydrodynamic models of the coma by including the effects of heating by the diffuse radiation scattered and thermally emitted by the dust in the coma. Gombosi et al (ApJ 293, 328; ApJ 311, 491) have also considered the hydrodynamic flow of a dusty gas and applied it particularly to modelling cometary outbursts. Kitamura modelled axisymmetric jets with and without dust (Ic 66, 241). Yamamoto and Ashihara (AA 152, L17) argued that the water vapor in the inner coma should recondense into icy grains. Ip (ApJ 300, 456) then calculated the heating effect on the gases of the coma due to this recondensation. Beard et al (ApJ 295, 668) argued that the pressure scale length on the sunward side should be shorter than the decay scale length for CN and C_2 and should therefore control the physical processes.

A numerical stochastic modelling of the photochemistry of cometary atmospheres was presented by Marov and Shematovich (KIAM 85, 176). On this basis the same authors gave a numerical investigation of the photochemistry of an H_2O dominated cometary atmosphere (KIAM 87, 90). Physico-chemical processes in the coma were treated in the frame of the multiphase and multicomponent radiating mixture theory with allowance for chemical reactions and heat and mass transfer (Kolesnichenko and Marov, KIAM 85, 61).

OBSERVATIONAL MODELLING

Haser model scale lengths were determined by Hu et al (ChAA 9, 86) and by Combi and Delsemme (ApJ 308, 472) using old data for CN and C_2, and by Cochran (AJ 90, 2609) using new data. There is not yet good agreement among different investigators on these scale lengths. Cochran (ApJ 289, 388) used her data to show that more complicated models with non-equilibrium photochemistry were required to fit C_2 profiles. Bockelee-Morvan and Crovisier (AA 151, 90) considered a variety of possible parents for CN and evaluated lifetimes an excitation conditions.

A generalized Haser method with account for solar light pressure and scale lengths of both parent and daughter molecules was developed and used to explain the asymmetry of coma isophotes and to improve scale lengths of molecules (Dranevich and Matveev, KM 37, 3). A method to distinguish between compact and extended sources of molecules in cometary atmospheres was developed and applied to prove the existence of extended sources (Dranevich, KTs 340). The possible role of charge transfer between N_2 and solar He^+ to excite the Meinel bands of N_2^+ was discussed by Cherednichenko (KTs 340).

Gérard (AA 146, 1) has interpreted the frequency shifts of the radio lines of OH observed in comet Austin 1982 VI as being due to Zeeman shifts which require a field of 50 gamma. Fink and Johnson (AJ 89, 1565) derived spatial profiles of [OI] emission in P/Tuttle and P/Stephan-Oterma, which agree well with models based on dissociation of water. Absolutely calibrated spectra of P/Tuttle, P/Stephan-Oterma, P/Brooks 2, and 1982 I Bowell were presented by Johnson et al (Ic 60, 351). Ip et al (ApJ 293, 609) then deduced abundances of H_2O^+ and interpreted them in terms of photoionization of H_2O. Spectrophotometry at many points in the coma of P/Stephan-Oterma was presented by Cochran and Barker (Ic 62, 72). Cochran (Ic 62, 82) then modelled these data using a variety of chemical and photochemical reactions. Cochran (AJ 92, 231) used data on many comets from three observing teams to determine the typical relative abundances of C_2, CN, and C_3. Festou (ASR 4, 165) considered the various carbonic species in a variety of comets and concluded that there are large variations in the C/OH ratio from comet to comet. He also restudied the relationship between production of OH and visual magnitude (ACM II, 299). Feldman (ACM II, 263) reviewed the observations of CO in comets. Butterworth et al (ACM II, 269) reported new emission bands of CS. Azulay and Festou (ACM II, 273) showed that optical depth effects were important for S and argued that the abundance of S could not be explained by CS_2 and CS. Cochran (AJ 91, 646) reported a major asymmetry about perihelion in her spectrophotometry of P/Kopff and suggested an interpretation in terms of mantle development.

SPECIFIC COMETS

Spectral differences between different parts of the split comet West 1976 VI, and their temporal changes were examined by Rosenbush (Ic 66, 230).

The remarkably close approach to Earth by comet IRAS-Araki-Alcock 1983 VII in May 1983 led to many unique observations, the results of which continued to appear. Cosmovici and Ortolani (Nat 310, 122) reported the existence of hitherto unobserved HCO and H_2S^+, as well as the possible presence of several other species. Temporal variations during an outburst were discussed by Feldman et al (ApJ 282, 799) in terms of a rotating model of the nucleus. They point out the value of S_2 as a tracer of activity of the nucleus. A'Hearn and Feldman (NASI 156, 487) argued that the presence of S_2 implied that the cometary ices had never been significantly heated. Bockelée-Morvan et al (AA 141, 411) and Irvine et al (Ic 60, 215) presented upper limits for HCN and showed that these were lower than the observed production of CN. Oliversen et al (Ic 63, 339) observed a "hole" in the C_2 at the center of the comet. Jockers et al interpreted the anisotropy of the coma in terms of a nuclear model with chemical differentiation (ACM II, 331). Jackson et al (ApJ 304, 515) used the spatial profiles of CS with lifetimes measured in the laboratory to infer that CS_2 must be its parent. Lutz and Wagner reported remarkable, rapid variations in the abundances of C_2, C_3, CN, and CH on May 9 (ApJ 308, 993).

Comet P/Crommelin was observed by many. Festou et al (AA 152, 170) showed that the UV spectrum was similar to that of most comets except that C was weak. Bockelée-Morvan et al (AJ 90, 2586) observed the radio lines of OH. Russell et al (MN 217, 651) obtained C_2 production rates. Wallis and Carey (MN 217, 673) determined that H was more abundant than could be explained by the observed OH. An archive of all known observations was published by Sekanina and Aronsson (JPL Publ 86-2).

Comet P/Encke was shown by Feldman et al (Ic 60, 455) to exhibit normal relative abundances but with highly anisotropic ejection. A'Hearn et al (Ic 64, 1) showed that the optical asymmetry about perihelion does not reflect the total outgassing as measured by OH. Djorgovski and Spinrad (AJ 90, 869) presented surface photometry and suggested that the shape of the isophotes was due to radiation pressure on C_2 and CN.

Comet P/Giacobini-Zinner was the first comet to be visited by spacecraft, an event which led to numerous firsts although most of these were discoveries about the plasma properties rather than the gas. The known atypical composition (C_2 and C_3 depleted) was documented by Cochran (AJ 92, 239). Ogilvie et al (Sci 232, 374) reported the detection of the long predicted H_3O^+ as well as an ion mass 23 or 24, the identity of which is still in some dispute. Na^+, Mg^+, C_2^+, and others have all been suggested in papers by Geiss et al (AA 166, L1) and by Ip and Axford (Nat 321, 682). Schleicher et al (ESA 263, 31) showed that the inversion of the lambda-doublet of OH was quenched in the inner coma as predicted by radio observers. The spatial profile of H_2O^+ was obtained by Strauss et al (GRL 13, 389) and compared with the plasma profile measured by ICE, finding good agreement (see also AJ 92, 474). They estimated an H_2O^+ abundance 1/4 that of electrons. Lyman-alpha monitoring with Pioneer Venus was discussed by Combi et al and compared with models to derive production rates (GRL 13, 385). Radio emission by OH was studied by Norris et al (PASA 6, 180). Rees et al (ASR 5, 267) monitored various neutral and ionic species for several weeks around the encounter while Perez-de-Tejada et al (ASR 5, 293) obtained spectra at the time of the encounter showing the comet steady. McFadden et al (Ic 69, 329) monitored the UV emissions both over the apparition and during the encounter, showing the comet in a very steady state at the time of the encounter and demonstrating that the production rates deduced globally with IUE agreed with those deduced from localized in situ measurements with ICE. The coma was modelled in detail by Boice et al (GRL 13, 381).

Comet P/Halley's gaseous emission was first observed by Wyckoff et al (Nat 316, 241) in the spectra at 6 AU. This was substantially confirmed by Barker and Opal (ACM II, 481) and by Meech et al (Ic 66, 561).

The spectrum of the coma was studied from the spacecraft by Combes et al (Nat 321, 266; ASR 5, 127), who discuss the abundances of H_2O and CO_2 and the detection of CO and CH-compounds; by Krasnopolsky et al (Nat 321, 269; ASR 5, 143), who discuss OH, H_2O and CN and the existence of newly created radicals in excited states; by Moreels et al (Nat 321, 271), who discuss OH, NH, CN, C_3, CH, C_2, NH_2, and H_2O^+. The principal molecules discussed from the observations were H_2O, CO_2, OH, and CN. The neutral mass spectrometers measured an inverse square law for the density and a production rate of 10^{30} per sec, as discussed by Keppler et al (Nat 321, 273), Gringauz et al (Nat 321, 282; ASR 5, 165), and Krankowsky et al (Nat 321, 326), and showed that water was the dominant gas in the inner coma. Relative abundances were also deduced for a number of other species showing, e.g., that CO_2 was a few percent of water. Eberhardt et al (ESA 250/1, 383) showed that CO (and/or N_2) was a parent but was also produced in part by a very short-lived parent. Spatial profiles of CN, C_2, and OH were also obtained by Levasseur-Regourd et al (Nat 321, 341; ASR 5, 197).

The ultraviolet emitting species were monitored with IUE by Feldman et al (ESA 263, 39). UV observations were also carried out from rockets, by McCoy et al (Nat 324, 439), and by Woods et al (Nat 324, 436) who concluded that CO was a parent molecule. Lyman-alpha was monitored by Craven et al (GRL 13, 873) from Dynamics Explorer 1. Variations in this radiation were reported by Kaneda et al (Nat 321, 297 ; GRL 13, 833). The far UV region down to 150 nm was also investigated in detail by the space station Astron (Boyarchuk et al, AZhL 12, 696; ESA 250/1, 193). Schleicher et al (ESA 263, 31) set upper limits on OD, while Eberhardt et al (ESA 250/1, 539) found the same upper limit but also a lower limit requiring D/H enhancement over the interstellar value, not by more than about 10x. The D/H ratio of cometary water molecules was found to be 0.6 to 4.8×10^{-4}, which is similar to corresponding values in terrestrial oceanic water and atmospheres of several other planetary bodies. This result is of major importance for the relationship of comets to the solar system matter.

Photometry of optical species was reported by Catalano et al (AA 168, 341). Extensive spectrophotometric observations were conducted by Spinrad et al, Cochran and Barker, and Fink et al (ESA 250/1, 437, 439; ESA 250/3, 485), while high-resolution (~ 0.09 nm) spectra were secured with the 6-m telescope: about 500 spectrograms in the region 330-660 nm with exposure times 5-10 min (Afanasev et al, KTs 352). Observations of the N_2 1P system were reported by Mamadov (ATs 1987). Sherb et al (ASR 5, 279) observed H_2O^+, H-alpha, [OI], and NH_2 at very high spectral resolution with a Fabry-Perot. Kerr et al (ASR 5, 283) made similar observations. The [OI] + NH_2 blend was also resolved with an echelle spectrometer by Arpigny et al (ESA 250/3, 81; AA, in press). Using images in CN and C_2, A'Hearn et al (Nat 324, 649; ESA 250/1, 483) discovered jets of gas uncorrelated with the previously known jets of dust.

The infrared spectrum of H_2O was studied in detail by Mumma et al (Sci 252, 1523) and by Weaver et al (Nat 324, 441) and fitted with their excitation model. Knacke et al (ApJ 310, L49) reported the 3.35-micron emission by CH compounds, first detected by Combes et al (Nat 321, 266). Emission at this wavelength was also observed by Danks et al (ESA 250/3, 103), Wickramasinghe and Allen (Nat 323, 44), and Baas et al (ApJ 311, L97), who advocate organic compounds.

In preparation for P/Halley, the cometary radio emissions were reviewed by Snyder (AJ 91, 163) and modelled by Schloerb and Gérard (AJ 90, 1117). Surveys were performed by Galt (AJ 93, 747) and by Schloerb et al, Gérard et al, Mirabel et al, and Cordes et al (ESA 250/1, 583, 589, 595; ESA 250/3, 113). De Pater et al (ApJ 304, L33) reported unusual, stable structures in the map of OH. Observations of radio emission by HCN were reported by Bockelée-Morvan et al (AA 180, 253) and by Schloerb et al (ApJ 310, L55), both of whom found outflow velocities near 1 km/s but varying with r. The HCN varied significantly from day to day.

Variability of the gaseous abundances on the time scale of hours at the time of spacecraft encounters was discussed by Festou et al (Nat 321, 361) and by Schleicher et al (ESA 250/1, 565). Leibowitz and Brosch reported a 52-hour periodicity in the molecular abundances (Ic 68, 418), whereas Millis and Schleicher found a period of 7.37 days and an amplitude near a factor of two (Nat 324, 646). This was confirmed by Williams et al (MN 226, 1p) and Festou et al (AA, in press). Feldman et al reported a sudden outburst of gases rich in CO_2 (Nat 324, 433).

4 Cometary Plasma

The years 1985 and 1986 have been particularly exciting for the study of cometary plasma physics. This was because the in situ measurements made during the spacecraft encounters with comets Giacobini-Zinner and Halley have, for the first time, enabled many physical phenomena related to comet - solar wind interaction to be revealed.

TURBULENCE AT LARGE DISTANCES
At the approaches of the spacecraft to P/Giacobini-Zinner (within 2×10^6

km) and P/Halley (within 10^7 km), high levels of magnetic and plasma wave turbulences were recorded (Scarf et al and Smith et al, Sci 232, 377 and 382; Riedler et al and Neubauer et al, Nat 321, 288 and 352). This effect is the direct result of solar wind pickup of the newly born cometary ions which were observed by plasma instruments (Mukai et al, Balsinger et al, and Johnstone et al, Nat 321, 299, 330, and 344). The magnetic field turbulence at a frequency similar to the water ion gyrofrequency (~ 0.01 Hz) has been suggested to be excited by the ion beam instability from pickup of the water-group ions (Tsurutani and Smith, GRL 13, 263; Wu and Davidson, JGR 77, 5399). Associated with the high degree of turbulence ($\Delta B/B \sim 0(1)$) is the presence of energetic heavy ions (E > 100 KeV) at large cometocentric distances (r > a few million km) which is most probably indicative of acceleration processes, such as stochastic second-order Fermi acceleration effect (Hynds et al, Sci 232, 361; Somogyi et al and McKenna-Lawlor et al, Nat 321, 285 and 347; Gribov et al, ESA 250/1, 271). Within a few 100,000 km of the cometary bow shock, which was typified by a higher degree of plasma turbulence, rapid isotropization and thermalization of cometary ions, slow down and deflection of the solar wind plasma (Tsuratani and Smith, and Richardson et al, GRL 13, 259 and 415; Balsiger et al and Johnstone et al, Nat 321, 330 and 344; Coates et al, Nat 327, 489), the hardening of the energy spectra of the cometary ions might be the combined result of diffuse shock acceleration, shock compression and shock drift acceleration - in addition to the second-order Fermi process (Amata and Formisano, PSS 33, 1243; Ip and Axford, PSS 34, 1061).

CHARACTERISTIC FEATURES OF THE COMET - SOLAR WIND INTERACTION

The ICE spacecraft went through the tail of P/Giacobini-Zinner, hence providing important information on the structure of a cometary ion tail. The draped-field model of Alfvén (Tellus 9, 92) was basically confirmed. The peak value of the magnetic field was measured to be 60 nT (Smith et al, Sci 232, 382) which was comparable to the thermal pressure exerted by the cometary pickup ions (Siscoe et al, GRL 13, 287). A central current sheet, with a projected width of 1100 km along the spacecraft trajectory, was characterized by low magnetic field (B ~ 5 nT), low electron temperature ($T_e < 13,000$ K), and high plasma density ($N_e > 670$ cm^{-3}) according to the various plasma measurements (Bame et al, Meyer-Vernet et al, and Smith et al, Sci 232, 356, 370, and 382). The scale length of the bow shock (a subsolar point distance of about 5×10^4 km), when compared to theoretical model calculations, suggested an ougassing rate of 4×10^{28} molecules/s (Mendis et al, GRL 13, 239), which is compatible with the value derived from ground-based observations (Spinrad et al, ESA 250/1, 437). The large gas production rate of P/Halley (10^{30} H_2O/s) caused a much more extended solar wind interaction effect. For instance, energetic ions presumably of cometary origin were detected at the cometocentric distance of $5-10 \times 10^6$ km (Somogyi et al, Nat 321, 285). The bow shock was encountered at about 10^6 km from the comet. In agreement with theoretical predictions (Wallis, PSS 21, 1647; Schmidt and Wegmann, CPC 19, 309), the bow shocks at P/Giacobini-Zinner and P/Halley were found to be weak, with a Mach number on the order of 2 (see Galeev, ESA 250/1, 3).

The sunward passages of the Vega and Giotto spacecraft near P/Halley have revealed the existence of several boundary structures in the subsonic flow region inside the cometary bow shock. The so-called cometopause structure, which was first detected by the Vega 2 spacecraft at a cometocentric distance of approximately 1.6×10^5 km (Gringauz et al, Nat 321, 282), has the appearance of a rather narrow layer (width 10,000 km) separating the outer region which is characterized by solar wind plasma and the inner region of slow moving plasma of cometary composition, according to the reports by the Plasmag-1 experimenters (Gringauz et al, ESA 250/1, 93). Similar transition was seen by the Giotto plasma experiments but the structure seems to be more gradual than that reported by the Vega 2 Plasmag-1 experiment (Balsiger et al, Nat 321, 330). While the formation of such compositional transition has been interpreted in terms of charge-exchange loss of the solar wind plasma, the physical configuration of the cometopause is still to be clarified.

IONIZATION IN THE INNER COMA

Except for the observation of a suprathermal electron population in the inner coma of P/Halley during the Vega flyby measurements (Gringauz et al, Nat 321, 282), no clear signatures of enhanced electron impact ionization generated by energy transfer from the cometary pickup ions to the electrons, as suggested by the critical velocity ionization effect (Formisano et al, PSS 30, 491), were found in the comas of P/Giacobini-Zinner and P/Halley. The electron ionization rate, however, can be shown to be comparable to the photoionization rate in localized regions of P/Giacobini-Zinner's coma, while the corresponding electron impact ionization appeared to be lower at P/Halley, as derived by the electron analyser on Giotto (Lin, priv. comm.). Enhanced ionization is therefore not the most likely cause of the observed ion density profile, with a maximum at cometocentric distance of approximately 13,000 km, as detected at the Giotto encounter (Krankowsky et al and Balsinger et al, Nat 321, 326 and 330). The possibility exists that such plasma structure may be of non-stationary nature, resulting from changes in the solar wind conditions or the outgassing rate of the comet itself. Under steady state, the presence of a high electron temperature gradient would be effective in reducing the ion density inside the plasma peak (Ip et al, and Sauer and Baumgärtel, AA, in press), even though the issue of the electron temperature variation in the inner coma remains ambiguous.

MAGNETIC FIELD STRUCTURE

Various magnetic directional changes were encountered by the Vega and Giotto probes during their flyby observations (Riedler et al and Neubauer et al, Nat 321, 288 and 352). These changes reflected the sweeping of the interplanetary magnetic field structures by the cometary ionosphere. A dramatic discovery was made by the magnetometer experiment on the Giotto spacecraft as it reached an inbound distance of 4600 km from the cometary nucleus: the magnetic field was observed to drop from a value of 45 nT to effectively zero within a short distance ($\Delta r \sim 1300$ km) at the boundary of the diamagnetic cavity at 4700 km (Neubauer et al, Nat 321, 352). The formation of such a magnetic field-free cavity may be explained in terms of force equilibrium between the Lorentz force of the piled-up magnetic field and the ion-neutral drag (Cravens, ESA 250/1, 241; Ip and Axford, Nat 325, 418). At the crossing of this magnetic boundary, the ion temperature was observed to drop rapidly from 2600 K to as low as 300 K (Balsiger et al, Nat 321, 330; Schwenn et al, ESA 250/1, 225). The higher ion temperature outside the contact surface has been explained in terms of frictional heating by the plasma flow in the neutral coma (Haerendel, GRL, in press). Recent work on MHD model simulation by Wegmann et al (AA, in press) also indicates a plasma speed of 1 - 3 km/s and an ion temperature up to 3000 K between the diamagnetic cavity and a distance of 10,000 km.

NEUTRAL AND IONIC COMPOSITION OF THE COMA

In the inner coma region, the electron temperature was calculated to be between 10,000 and 20,000 K by Wegmann et al (AA, in press). As both Vega and Giotto probes were not equipped with devices capable of measuring cold electrons, experimental data on electron temperature, which is vital to ion chemistry modelling, are lacking. Nevertheless, interesting results on cometary neutral and ion composition in the inner and outer comas have been obtained by several theoretical groups. For example, the spatial distributions of several ion species - O^+, OH^+, H_2O^+, H_3O^+ (the most abundant ion in the inner coma) and S^+ - published in the preliminary report of the Ion Mass Spectrometer experiment on Giotto (Balsiger et al, Nat 321, 330), can be reproduced reasonably well (Boice et al, AA, in press). Another line of approach has been to evaluate the relative abundances of CH_4 and NH_3 relative to H_2O by comparing the theoretical results from gas phase chemistry to observed count rates of the relevant ion mass channels in the inner coma (Allen et al, AA, in press). These authors report ratios of production rates $Q(NH_3)/Q(H_2O) = 0.013$, and $Q(CH_4)/Q(H_2O) = 0.02$.

A surprise in the Giotto observations has to do with the large abundance of C^+ ions measured in the outer coma (Balsiger et al, Nat 321, 330). One suggestion is that these ions might have originated from a distributed source of atomic carbon; emission from the so-called CHON dust grains could be one possibility. It should be mentioned here that the CO molecules, whose production rate was about 15 percent of the H_2O production rate, were found to originate mostly from a distributed source (Eberhardt et al, ESA 250/1, 383). Evaporating dust grains could also be a potential source of the CO molecules. On the other hand, the fact that the reported CN and C_2 jets did not correlate spatially with dust jet structures could be explained by the electromagnetic perturbation of the trajectories of electrostatically charged submicron particles (Horanyi and Mendis, ApJ 294, 357) which may be the parents of the observed radicals. Flammer et al (EMP 35, 203) had also pursued the possible correlation of the brightness variations of P/Halley at large heliocentric distances ($r > 8$ AU) with electrostatic levitation of small dust grains from the surface of the nucleus as modulated by solar wind high-speed streams.

GROUND-BASED OBSERVATIONS

Accurate work by Ivanova et al (ESA 278, in press) allowed a determination of the H_2O^+/CO^+ abundance ratio in the ion tails of P/Giacobini-Zinner and P/Halley. It was found that this ratio was 2 for P/Giacobini-Zinner, and 0.5 and 0.25 for P/Halley at heliocentric distances $r = 1.5$ and 1.0 AU, respectively. Narrow-band filter imaging observations by Jockers et al (ESA 250/1, 59) have shown the potential of such kind of measurements in tracing clearly the dynamical activity and evolution of plasma structures in cometary ion tails, as described by Brandt and Niedner (ESA 250/1, 47). One particularly interesting result, obtained by Celnik (ESA 250/1, 53), is the tracking of the motion of ion tail condensations in CO^+ emission, which revealed a periodicity of generation of these large-scale plasma structures very similar to the rotation period of the comet nucleus of 54 hours. Such a temporal effect might have been closely related to the spin-modulation of the outgassing rate, and hence to the production rate of cometary ions. Feldman et al (Nat 324, 433) analyzed an outburst of the CO_2^+ emission in the ion tail, as observed by IUE on 18 March 1986. Whether such an outburst is typical of the generation of a CO^+ condensation propagating downstream in the fashion described by Celnik (ESA 250/1, 53) is still to be investigated.

Large-scale plasma phenomena were extensively recorded by the Large-Scale Phenomena Network of the International Halley Watch during the period November 1985 - June 1986. This apparition showed a rich display of disconnection events, with 19 obvious cases and perhaps a dozen more. The 19 obvious events show a clear association with sector boundary crossings. This fact and the detailed study of the 8 March 1986 event by Niedner and Schwingenschuh (ESA 250/3, 419), utilizing both in situ and ground-based data, indicates that the sector boundary/frontside reconnection model of Niedner and Brandt (ApJ 223, 655) has survived crucial tests and, so far, is favoured. Models of disconnection events, however, are expected to be an active area of cometary plasma research Mention may also be made here of the detailed analysis published by Jockers (AAS 62, 791) on the ion tail phenomena of comet 1973 XII Kohoutek, as observed on wide-field photographs secured around the world on 17 consecutive days, 9 - 25 January 1974.

THEORETICAL WORK

Many more efforts than could be reported here have been devoted to the interpretation of the cometary plasma - solar wind interaction, and we can only mention a limited number of publications, mainly of a general character, such as the reviews by Galeev and Lipatov (ASR 4, 229); Ip et al (Advances in Space Plasma Physics, ed Buti, Singapore 1985, 1); and Galeev (ESA 250/1, 3), which contain many other references. The last mentioned survey was already prepared in the light of the results obtained from the recent cometary space missions. Several reviews of interest have also been presented by Wallis on quasi-fluid and magneto-fluid theories, Kömle and Lichtenegger on numerical modelling, Phillips et al on similarities and differences between the solar wind inter-actions with comets and with Venus, Baumgärtel et al on non-stationary models, and Sagdeev et al on plasma instabilities due to mass loading (Field, Parti-cle and Wave Experiments on Cometary Missions, eds Schwingenschuh and Riedler, Vienna 1985, 7, 19, 37, 54, and 74). The plasma processes and structures oc-curring in the interaction of the solar wind with comets and with other neu-tral atmospheres in the solar system have been reviewed and compared by Ip (ESA 235, 65), who suggests a number of related major objectives for future spacecraft missions.

The stability of the cometary ionopause has been studied extensively by Ershkovich et al (ApJ 311, 1031 and references therein), while cometopause in-stability and a possible mechanism of structure formation in ionic tails have been discussed by Ioffe (AZh 64, 145). Recent contributions on the phenomenon of tail detachment have been presented by Gerasimenko (KTs 361), Obukhov et al (AZhL 12, 942), and Russell et al (JGR 91, 1417). A chemico-physical model combining in a consistent way the description of the chemical evolution of the expanding coma with the MHD treatment of an axisymmetric solar wind inter-action has been developed and applied to P/Giacobini-Zinner (Boice et al, GRL 13, 381) and P/Halley (Wegmann et al, AA, in press).

5 Cometary Dust

Useful reviews of our pre-Halley knowledge on various topics of the study of cometary dust have been published: properties of dust particles in the coma and tail, observational evidence (Lamy, ACM II, 373); cometary polarimetry (Dobrovolsky et al, EMP 34, 189); ices in grains (Campins, NASI 156, 443); dust and neutral gas modelling of the inner coma (Gombosi et al, RG 24, 667); predictions on P/Halley dust and gas environment (Divine et al, SSR 43, 1); and predictions of the impact rate by larger dust particles (Hajduk, ACM II, 497).

Systematic studies of relatively faint comets have become possible thanks to the use of large telescopes at several observatories. Thus, for example, a comparison of the dust characteristics of ten short-period comets has been made by Hanner (ASR 4, 189). The mean size and albedo of the grains are rather similar in these objects, with the exception of P/Crommelin whose grains are larger and appreciably darker. In a systematic investigation of the wavelength distribution of the light scattered by a dozen of comets, Jewitt and Meech (ApJ 310, 937; ESA 250/2, 47) have found differences in grain properties which appear to be intrinsic to the individual objects (different size distribution or different composition ?).

A detailed review of the first analyses and results regarding the dust environment of P/Halley (S/C experiments; jets streamers, antitail, sunward spike; IR observations; effects of irradiation) has been given by Sekanina (ESA 250/2, 131). Grün et al (ESA 278, in press) have presented a critical discussion of the existing models of the dust coma, and emphasized the recent-ly discovered properties and phenomena related to the cometary dust.

INFRARED OBSERVATIONS

Routine infrared monitoring of the dust coma has been carried out from several ground-based (near IR) and airborne (far IR) observatories, on P/Crom-melin (Eaton and Zarnecki, MN 217, 659; Hanner et al, AA 152, 177), P/Churyu-mov-Gerasimenko (Hanner et al, Ic 64, 11), and comet 1982 I Bowell (Hanner and

Campins, Ic 67, 51). Further analysis of IRAS S/C data led to the discovery of dust trails behind and ahead of various comets in their orbital planes (Sykes et al, Sci 232, 1115).

P/Giacobini-Zinner and P/Halley were the first comets ever imaged from the ground with infrared arrays. The combination of nearly simultaneus visual and thermal IR maps yields the spatial distribution of the albedo, which is interpreted in terms of large, dark, fluffy particles (Hammel et al, ESA 250/2, 73). An extensive program has been developed for P/Halley and a preliminary review has been prepared by Hanner (ASR 5, 325); see also Campins et al (GRL 13, 295), Harvey and Campins (ASR 5, 335), Knacke et al (ApJ 310, L49), Tedesco et al (ApJ 310, L61), Tokunaga et al (AJ 92, 1183) and numerous reports in ESA 250/2, 81 to 130. The thermal emission was found to be dominated by relatively large particles (>1 µm); the grains exhibited a low average albedo. Emission features were detected in the 3 µm and 10 µm spectral regions, especially from the IKS spectrometer on board Vega (Combes et al, Nat 321, 266). The emission near 3.4 µm is indicative of organic matter (large molecules or small grains ?). The dust production was found to vary strongly with time.

POLARIMETRIC OBSERVATIONS

Physical properties of the dust grains can be inferred from polarimetric observations of comets, through the phase angle and wavelength dependence of the polarization. Recent analyses on this subject have been presented by Myers and Nordsieck for comet 1982 VI Austin and P/Churyumov-Gerasimenko (Ic 58, 431); by Myers for P/Kopff and P/Tempel 1 (Ic 63, 206). Telescopic linear polarization measurements have been carried out in 1985/86 over P/Halley's coma in visible light by Bastien et al (MN 223, 877), Dollfus et al (ESA 250/2, 41), Dzhapiashvili et al (ESA 250/3, 191), Kiselev et al (ESA 250/3, 29), Le Borgne et al (ESA 250/1, 571), and Mukai et al (ESA 250/2, 59). Circular polarization measurements have been reported by various authors, in particular by the observers at the Soviet-Bolivian Observatory near Taricha (Bolivia), who quote values of ~ 0.004 for April 1986 (Guralchuk et al, KFNT 3/2, 89). The same team found a gradual turning of the polarization plane in a narrow phase angle ($\propto$) around the inversion angle. This effect was interpreted as a direct result of the orientation of prolate particles (Guralchuk et al, KFNT 3/3, 93). Detailed discussion as well as analysis of a large number of linear polarization observations at various phase angles were presented by Beskrovnaya et al (ESA 278, in press). The particles appeared to be of micrometer dimensions and to have physical properties like those of graphite grains. Much larger sizes were advocated by Morozhenko et al (Ic 66, 223). The distribution function of grain sizes A was found by Dobrovolsky (KFNT 2/2, 35) to be of the form A^{-s} with s = 3.5, in accordance with Giotto measurements. Infrared data have been obtained by Brooke et al (ESA 250/2, 87); polarization measurements over the tail have been performed by Lamy et al (ESA 250/2, 69); the OPE photopolarimeter on board Giotto spacecraft has recorded in situ polarization measurements in the coma (Levasseur-Regourd et al, Nat 321, 341; ASR 5, 197).

A meeting of most of these observers was held in Paris in March 1987. There is a good agreement for the phase angle dependence of near-nucleus observations, typically with a minimum of -0.017 at $\propto$ = 10°, an inversion at $\sim$ 21°, and a value of ~ 0.2 at 73°. The question of the wavelength dependence is still disputed, however; it seems that there was a change in the colour dependence of polarization as the comet passed through different phase angles. Various results on polarization are to appear soon in AA.

IN SITU MEASUREMENTS

Direct measurements of Halley's dust have been performed by the dust instruments on board the Halley spacecraft. Significant new data on the chemical composition and the mass distribution have been obtained (Vaisberg et al, Mazets et al, Simpson et al, Kissel et al, and McDonnell et al, Nat 321, 274, 276, 278, 280, 336, and 338). The dust was found to be composed of silicates and refractory carbonaceus material in varying proportions. Many particles (referred to as CHON particles) are rich in low-Z elements: hydrogen, carbon, nitrogen and oxygen (Kissel and Krueger, Nat 326, 755). The slow release of these elements from the grains in the coma has been invoked to explain the ground-based observations of CN and C_2 jets (A'Hearn et al, ESA 250/1, 483). The average abundance of the heavier elements is chondritic within a factor of two. The composition of cometary dust grains, mainly the presence of organic matter, lends support to models emphasizing the interstellar connection of cometary materials. Although the smallest grains which are of sizes comparable to interstellar grains are by far the most numerous in the coma, they contribute little mass to the comet's total dust output, the major part of it being contained in the biggest particles recorded (McDonnell et al, ESA 250/2, 11). Over the period of one week the three Halley space probes recorded dust fluxes which varied over about one order of magnitude.

NON-UNIFORM NUCLEUS ACTIVITY AND DUST JETS IN THE COMA

Images taken by the television cameras on board the Vega 1, 2 and Giotto spacecraft (Sagdeev et al and Keller et al, Nat 321, 259 and 320) showed dramatically the irregularities of the dust emission from the nucleus. Only a small fraction (about 20 percent) of the sunlit surface actively emits dust. These active regions are the origin of jets or fans which are seen inside the coma by ground-based visual observers. Because of the inferior intensity resolution of photographic plates, only sophisticated digital image enhancement techniques reveal structures in the photographed coma (Larson and Sekanina, AJ 89, 571). From the analysis of photographs taken during the 1910 apparition of P/Halley, Sekanina et al (AJ 89, 1408; AJ 92, 462; Nat 321, 357) derived the spin vector, rotation period and distribution of active areas on the nucleus. However, recent ground-based observations (see, e.g., Grün et al, Nat 321, 144 and references on periodic variations quoted in the preceding chapters), and inclusion of the in situ data from the Halley spacecraft have shown that the rotational state of P/Halley's nucleus cannot be described as a simple spin about a fixed axis (see Wilhelm, Nat 327, 27 and references therein). The anisotropic dust emission together with the complex rotational state of the nucleus explains strong and irregular variability of the brightness, both of the nucleus at large heliocentric distances (West and Pedersen, AA 138, L9; Festou et al, AA 169, 336) and of the coma in the inner solar system (Schleicher et al, ESA 250/1, 565 and references therein; Cosmovici et al, ESA 250/2, 151). Dusty outbursts at large distances and jet development near the Sun were also reported by Dobrovolsky (KFNT 2/1, 66) and by Dobrovolsky et al (ESA 250/3, 31), respectively.

THEORETICAL INVESTIGATIONS

Different aspects of the history of solid particles from their leaving the nucleus were considered, such as possible pseudo-fluidization of the outer part of the nuclear dust envelope (Shulman, KFNT 1/5, 53); dust halo formation (Gombosi and Horányi, ApJ 311, 491); expected properties of the icy grain halo (Kajmakov and Lyzunkova, KTs 320 and 341; Andrienko and Mishchishina, AZh 63, 335; Crifo and Emerich, NASI 156, 429); properties of dirty ice grains (Mukai et al, ASR 5, 339; AA 167, 364); generalization of the Finson-Probstein dust tail theory (Chernyj and Sizonenko, KFNT 2/3, 52; Chernyj, KFNT 2/6, 64); dynamics of charged dust particles (Horányi and Mendis, JGR 91 A1, 335; ApJ 308, 800); short-wave radiation and multi-charged ion generation due to high velocity collisions of cometary and zodiacal dust particles (Ibadov, IAUC 85, 365; ESA 250/1, 377).

The problems of cometary origin and of further evolution of cometary dust particles involve relationships to the interstellar matter; meteoroids and meteor streams; and the zodiacal dust cloud. Recent progress in the relevant research areas is reviewed in the reports of IAU Commissions 34, 22, and 21, respectively.

6 Origin and Evolution of Comets

ORIGIN

Comet accretion theories and nuclear models have been discussed in the section on Cometary Nuclei. Here we shall consider mainly the progress in investigating the time scales and locations of comet nuclei formation.

Prialnik et al (ApJ 319, 993) have found a constraint on the formation time of comets from the absence of major effects of ^{26}Al heating. The evidence for very low densities (Greenberg, ACM II, 221; Rickman, ESA 249, 195; Bertaux and Abergel, BAAS 18, 794; Rickman et al, ESA 278, in press) implies that cometary nuclei are homogeneous without any major core-mantle differentiation, but also suggests a very gentle formation mechanism.

Support for an accretional origin was found by Donnison (AA 167, 359) from the cometary magnitude distribution. Modelling of cometary nuclei as accretional aggregates of planetesimals formed by gravitational instabilities was done by Yamamoto and Kozasa (ISAS 364) with the result that most comets should have formed at several 10^2 AU. Grain accretion followed by gravitational instability may also occur in wind-driven shells around protostars according to Bailey (Ic 69, 70). A new theory of cometary nuclei condensation in the protosolar nebula based on the thermodynamic compatibility concept was developed by Shulman (Cometary Nuclei, Moscow 1987). Molecular ion clusters of both signs were hypothesized to have been stored during this early stage and to serve as additional energy sources nowadays.

Icy crust eruption has been proposed as the origin mechanism from massive transplutonian planets by Radzievskij (KFNT 3/1, 66), and from satellite-like, Ganymede-type bodies by Drobyshevskij (IPTI 1132). Vorontsov-Velyaminov recognizes the possibility of different ways of nucleogenesis, but believes tha catastrophic destruction of the hypothetical planet Asteron to be the most probable one, and gives some new arguments (AZh 63, 181).

OORT'S CLOUD

The formation of the Oort cloud was reviewed by Fernández (IAUC 83, 45), its structure and observational background by Oort (Sterne 61, 270), and its evolution in more general terms by Weissman (IAUC 83, 87; SSR 41, 299; ACM II, 197) and Oort (Obs 106, 186). Survival of the cloud against the effect of GMC encounters was claimed by Hut and Tremaine (AJ 90, 1548) but appeared less certain in the work by Bailey (MN 218, 1). The treatment of stellar perturbations was improved by Remy and Mignard (Ic 63, 20). The role of the Galactic tidal field has now been recognized as important, perhaps even dominant (Heisler and Tremaine, Ic 65, 13; Heisler et al, Ic 70, 269). Injection of comets into the planetary system by Galactic tides was investigated by Torbett (MN 223, 885) and demonstrated from orbital statistics by Delsemme and Patmiou (BAAS 18, 799). New orbital computations for long-period comets were made by Everhart and Marsden (AJ 93, 753).

A dense inner core of the Oort cloud remains to some extent speculative but has found support from various arguments. Comet showers (review by Hut et al, Nat 329, 118) would depend on the existence of such a core, as explored by Fernández and Ip (Ic 71, 46). Infeed to the outer planetary region from such a core was shown by Bailey (Nat 324, 350) to yield a promising source for maintaining the short-period comet population. Observational verification of the structure of the Oort cloud, including its inner core, was considered by Baum et al (BAAS 17, 690; PASP 97, 899) and by Marochnik and Sholomitskij (IKI 942).

EVOLUTION

The problem of the lifetimes and disappearance of long-period comets was examined by Kresák (BAC 35, 129; IAUC 83, 279). Their fading was studied by Bailey (MN 211, 347) who suggested an explanation in terms of a thermal shock (IAUC 83, 311). Chemical differentiation driven by sublimation was modelled by Houpis et al (ApJ 295, 654), and effects of cosmic-ray irradiation in the Oort cloud producing severely altered material including organic refractories in meter-thick surface layers were suggested by Johnson et al (ESA 250/2, 269). The question whether P/Halley may have an irradiation-driven crust was also raised by Sekanina (ESA 250/2, 131).

Nevertheless, dust coverage appears as the main influence on short-period comet evolution (Fernández, Ic 64, 308), but may be restricted to comets that are not perturbed into too small perihelion distances (Nazarchuk and Shulman, KM 35, 27). Lifetimes and fading of short-period comets were reviewed by Kresák (IAUC 83, 279; Evolution of Small Bodies of the Solar System, Bologna 1987, 202). A typical lifetime of 300 - 500 revolutions thus appears likely; while there seem to be transient dormant phases (Kresák, ESA 250/2, 433), progressive fading can hardly be observed (Kresák and Kresáková, ESA 278, in press). Modelling work related to the photometric evolution of short-period comets was done by Dobrovolsky et al (DANT 27, 189; ESA 250/2, 389), Nazarchuk and Shulman (KM 35, 11), and Markovich (KFNT 2/1, 70). Marsden (IAUC 83, 343) reviewed the problem of nongravitational forces and indicated that "wild" and increasing forces are characteristic of comets soon to disappear. This would be indicative of a disintegration process, but Rickman et al (ESA 278, in press) found an opposite trend on the average for newly captured comets, interpreted by means of gradual dust coverage.

A continuation of the well-known Vsekhsvyatskij's cometography has been undertaken by Andrienko and Karpenko (Physical Characteristics of Comets 1976-1980, Moscow 1987). Short-term (1800-2000) orbital evolution, observing geometry at past and future perihelion passages, and various events of evolutionary significance can be found in the Catalogue of Short-period Comets (Belyaev et al, Bratislava 1986), while a longer period of orbital changes (1585-2406) is covered by the Long-term Evolution of Short-period Comets (Carusi et al, Bristol 1985).

COMETS AND ASTEROIDS

Evolution of comets into asteroids was reviewed by Rickman (IAUC 83, 149), and Hartmann et al (Ic 69, 33) pointed out that asteroids believed to be of cometary origin on dynamical grounds (Hahn and Rickman, Ic 61, 417) tend to belong to spectral classes D, P or C, indicating a relatively remote place of origin. Comet-asteroid relationships were also reviewed by Hartmann (ACM II, 191) and A'Hearn (ACM II, 187). A possible comet-asteroid connection for 3200 Phaethon was disussed by Cochran and Barker (Ic 59, 196) and Green et al (MN 214, 29p). Davies (MN 221, 19p) found a probable cometary association for the three Apollo asteroids discovered by IRAS. The low-activity comets P/Arend-Rigaux and P/Neujmin 1 were extensively observed in the visual and IR, and their nuclei proved to be big, black and hot (Tokunaga et al, ApJ 296, L13; Millis et al, BAAS 17, 688; Veeder et al, BAAS 17, 688 and AJ 94, 169; A'Hearn et al, PASP 97, 892; Birkett et al, MN 225, 285; Campins et al, ApJ 316, 847; Brooke and Knacke, Ic 67, 80).

7 Minor Planets

During the past three years, the explosion in research on asteroids has continued at a pace which precludes listing of all, or even most, papers. The reader is referred to such bibliographical sources as Astronomy and Astrophysics Abstracts (subject 098: Minor Planets) for complete listings. During the last triennium, traditional techniques (e.g., photometry, spectrophotometry, and thermal IR) have been improved, and several new techniques have been applied to a significant number of asteroids (e.g., radar, occultations, speckle interferometry, microwave and UV wavelength observations).

ASTEROID SURVEYS AND PHYSICAL OBSERVATIONS

The Infrared Astronomical Satellite mapped essentially the entire sky in four channels, from 12 to 100 μm wavelength, during 1983. Green et al (Ic 65,

517) and Davies (MN 221, 19p) discuss the results of the search for near earth
asteroids in the IRAS data. The first draft of the results of the main belt
asteroid survey (IRAS Asteroid and Comet Survey, ed Matson, JPL D-3698) has
been issued, giving observed IR fluxes for 1811 known asteroids. In addition
to these data alone, the IRAS project has stimulated progress in ground based
observations and interpretations. A new compilation of ground-based data is
being assembled in support of the IRAS project (Tedesco, ACM II, 13). The zodiacal dust bands discovered by IRAS (Low et al, ApJ 278, L19) have been interpreted by Dermott et al (Nat 312, 505) as being due to collisional debris associated with the largest Hirayama families. The ratio of the flux from these
bands to the total of the zodiacal cloud is consistent with the hypothesis
that much, if not most, of the zodiacal dust is asteroidal in origin, rather
than cometary. Sykes and Greenberg (Ic 65, 51) have offered an alternative hypothesis for the dust bands as remnants of recent collisions within the asteroid belt.

Perhaps the most dramatic asteroid discovery by IRAS was 1983 TB, now permanently named 3200 Phaethon, an asteroid which appears to be the parent body
of the Geminid meteor stream. Cochran and Barker (Ic 59, 296), Green et al (MN
214, 29p), and McFadden et al (PASP 97, 899) have looked at the visual and
near IR spectrum of Phaethon for evidence of cometary characteristics, without
apparent success. They conclude that Phaethon is more or less similar to other
near earth asteroids. Thus we are led to conclude either that comets evolve
into rather unexpected taxonomic classes of asteroids (e.g., S, Q), or that asteroids as well as comets can have associated meteor streams. The IRAS dust
bands would further argue for the latter possibility.

Brown and Morrison (Ic 59, 20) report diameters and albedos for 36 more
asteroids, based on thermal IR observations. Lebofsky et al (Ic 68, 239) have
refined the standard thermal model, based on new diameters of 1 Ceres and 2
Pallas obtained by occultation observations (see below). Brown (Ic 64, 53) has
also included the effects of ellipsoidal geometry in models used to interpret
thermal data.

NEW OBSERVING TECHNIQUES

In the past three years, the number of asteroids observed by radar has
nearly doubled, from about 20 to over 40, through the efforts of Ostro and his
co-workers (Ostro et al, Ic 60, 391 and Sci 229, 442; Ostro, PASP 97, 877).
These observations have yielded estimates for the equatorial profiles of a few
asteroids, constraints on pole positions, estimates of surface roughness and
of metal content/porosity.

Occultations of stars by at least six asteroids were reported in the last
three years (see report of Commission 20). We note here only one, by 1 Ceres,
which is of particular interest in physical studies. Millis et al (BAAS 17,
729; PASP 97, 900; Ic, in press) report a new diameter for Ceres, which has
been used in the thermal model revision mentioned above, and also derive the
flattening and mean density for Ceres. The mean density, 2.7 g/cm^3, is the same
as earlier derived for 2 Pallas. The flattening, along with the above density,
is consistent with the hypothesis that Ceres is a homogeneous body with a figure of hydrostatic equilibrium.

Another new technique which has come into use recently is speckle interferometry. This technique has been applied to search for satellites and measure
the size and shape of a few asteroids. Drummond and Hege (BAAS 16, 922; Ic 67,
251) and Drummond et al (Ic 61, 132; Ic 61, 232) report results for the asteroids 4 Vesta, 12 Victoria, 433 Eros, 511 Davida, and 532 Herculina.

Several papers report observations at new wavelengths. It is perhaps too
soon to judge the worth of these new observations, but the range of new techniques being applied is encouraging. Seidelmann et al (CM 34, 39) report observations of thermal emission of asteroids at microwave wavelengths with the VLA.
They were primarily concerned with astrometric measurements, but the observations also provide information about the physical nature and sizes of aster-

oids. Lebofsky et al (Ic 63, 192) report a first detection of an asteroid, 10 Hygiea, at the submillimeter wavelengths of 370 and 770 μm. Butterworth and Meadows (Ic 62, 305) report spectral observations of 28 asteroids in the UV range from 0.21 to 0.32 μm with the IUE. Finally, Albrecht and Schober (ACM II, 25) contemplate the range of observations which will become possible with the Hubble Space Telescope.

SPECTROPHOTOMETRY AND POLARIMETRY

Spectrophotometric observations and interpretations have proceeded at a vigorous pace. In addition to mapping out compositional classes as a function of location in the asteroid belt, two major issues currently motivating these studies are to determine the sources of near earth asteroids, and of various meteorite classes, particularly the ordinary chondrites. This most common of meteorite types seems to be spectrally distinct from all but perhaps a few asteroids. The largest project reported is the Eight Color Asteroid Survey ECAS, covering the spectral range from 0.34 to 1.04 μm, with observations of 589 objects (Zellner et al, Ic 61, 355). Tholen (PhD Dissertation, Univ. Arizona 1984) has used the ECAS data, plus radiometric albedo data, as a basis for a new taxonomic system. McFadden et al (Ic 59, 25) report spectrophotometry of 17 near earth asteroids, directed toward determining the sources of this class of bodies. Gaffey (Ic 60, 83) has looked for rotational variation of the spectrum of 8 Flora as evidence of differentiated units on the surface. Feierberg et al (Ic 63, 183) have observed 14 main-belt C class asteroids in the 2.3 - 3.3 μm spectral range for the spectral signature of water of hydration, which they find in varying degrees in 9 of the spectra. Vilas and Smith (Ic 64, 503) have observed the spectra of 19 outer main belt asteroids in the 0.5 - 1.0 μm range. Golubeva et al (AZh 63, 1179) and Golubeva (AZhL 12, 801) report spectrophotometry of the asteroids 4 Vesta and 3 Juno, respectively.

Bel'skaya et al (AZhL 11, 286 and 13, 530; KFNT 3, 19) report multicolour polarimetry of several asteroids, to search for spectral and rotational dependences of polarization.

LIGHTCURVES, SHAPES AND ROTATION

Photoelectric lightcurve observations have been reported in ever increasing numbers. Lagerkvist et al (Uppsala Obs. Report 36) present a bibliographical listing of all lightcurve publications, along with aspect data for each lightcurve published. Another publication by Lagerkvist et al (Asteroid Photometric Catalog, Rome 1987) contains the actual lightcurves as published. The authors plan to update these listings every year or two, and to make them available in magnetic tape form. Binzel (MPB 14, 39) has compiled a summary of all known unpublished lightcurve data. The reader is referred to these publications for complete listings of photoelectric observations. Harris (ACM II, 35) reviewed the current status of rotation statistics. A few lightcurve results deserve mention. Binzel (PhD Dissertation, Univ. Texas 1986) reports rotational lightcurves of over 100 asteroids, including one, 1220 Crocus (also reported in Ic 63, 99), with a lightcurve period of over 30 days. He suggests that this might be a period of spin axis precession, induced by the gravitational effects from a satellite, rather than a rotation period. Weidenschilling et al (Ic 70, 191) report a total of 257 lightcurves of 26 rapidly rotating asteroids. New lightcurves and rotation periods have been determined, among others, by Zeigler and Florence (Ic 62, 512) as a result of a high-school teaching project. Wisniewski and McMillan (AJ 93, 1264) report some of the first asteroid lightcurves obtained by CCD photometry. The technique should allow accurate photometry to perhaps 2 or 3 magnitudes fainter than previously possible, with a given telescope. Lagerkvist and Williams (AAS 68, 295) have used observations from the Carlsberg Meridian Circle on La Palma to construct composite lightcurves and phase relations for 51 asteroids.

Recent lightcurve observations have emphasized pole and shape studies, beyond just simply determining rotation periods. Magnusson (Ic 68, 1) has de-

termined 20 spin axis orientations using a combination of epoch and magnitude-
aspect methods. Zappalà and Knežević (Ic 59, 436) and Zappalà and DiMartino
(Ic 68, 40) report 14 and 10 pole positions, respectively, by the magnitude-
aspect method. Lupishko et al (VKU 278, 51; KFNT 3, 57) report senses of rotation for 8 asteroids. Smaller numbers of determinations have also been reported by Koshkin (Odessa Obs.), McCheyne et al (Univ. Leicester), Barucci et al
(IAS Rome), Binzel (Univ. Texas), Taylor et al (Univ. Arizona), and Lambert
(State Univ. New Mexico). New theoretical approaches for determining shapes
and pole orientations from lightcurves have been discussed by Ostro and Con-
nely (Ic 57, 443), Lumme et al (ACM II, 55), Pospieszalska-Surdej and Surdej
(AA 149, 186), Surdej et al (AA 170, 167), and Koshkin (KFNT 2, 44). Barucci
and Fulchignoni (ACM II, 45) review their work in laboratory simulation of asteroid lightcurves as they relate to shape studies.

COLLISIONAL EVOLUTION

Theoretical and laboratory studies of the collisional evolution of the
asteroid belt have progressed considerably in the last three years. Davis et
al (Ic 62, 30) have developed a new numerical model to simulate asteroid distributions, and using as observational constraints the properties of the major
asteroid families and of Vesta's preserved basaltic crust. The number distribution of asteroids has been investigated by Ishida et al (PASJ 36, 357),
Donnison and Sugden (MN 210, 673), and Zhou et al (PPMO 3, 22). Asteroid families have been increasingly used as natural experiments on large-scale collisions and to probe the interior composition of the presumed parent bodies
(Zappalà et al, Ic 59, 261; Farinella et al, ACM II, 109; Davis, MSAI 57, 87;
and Chapman, MSAI 57, 103). Problems in the derivation of proper orbital elements which could hamper these studies have been pointed out by Knežević (ACM
II, 129) and Carpino et al (Ic 68, 55). Binzel (PhD Dissertation, Univ. Texas
1986) discovered differences in the mean rotation rates, dispersion about the
mean, and mean lightcurve amplitudes between the Eos and Koronis families,
suggesting that the Koronis family is collisionally younger than the Eos family. Catullo et al (AA 138, 464) have estimated the shape distribution of asteroids from lightcurve amplitudes. New laboratory experiments on high-velocity impacts have been carried out and analyzed by comparing the fragment properties with those of asteroids (Cerroni, MSAI 57, 13; Fujiwara, MSAI 57, 47;
Capaccioni et al, Nat 308, 832, and Ic 66, 487; Fujiwara, Ic 70, 536). Many
difficult problems arise when laboratory results are to be scaled to asteroidal sizes, and some critical and poorly tested assumptions are still needed
for this purpose (Davis et al, Ic 62, 30; Holsapple and Housen, MSAI 57, 65).
Recently, interesting new evidence on the collisional evolution of asteroids
has been provided by meteoritical studies, whose results appear to support the
idea that many asteroids were disrupted by collisions and subsequently reassembled into gravitationally bound "rubble piles" (Taylor et al, Ic 69, 1;
Grimm, JGR 92, 2022; Scott et al, JGR 90, D137).

Regarding the origin and early evolution of the asteroid belt, Davis et
al (Ic 62, 30) and Farinella et al (MN 216, 565; ACM II, 121) independently
reach the conclusion that the collisional erosion of the asteroid belt has
been only modest, thus the original mass in the zone was only at most a few
times its present value. Interesting similarities (implying genetic connections) among C and D-type asteroids, comet nuclei, and small outer satellites
of the giant planets have been pointed out by Hartmann (ACM II, 191), and
Hartmann et al (Ic 69, 33). O'Dell (Ic 67, 71) has proposed a possible mechanism to form comet nuclei from material lying originally in the asteroid belt.

8 Spacecraft Missions to Comets and Asteroids

PAST MISSIONS

The spacecraft encounters of comets Giacobini-Zinner and Halley in 1985
and 1986, respectively, marked the beginning of a new era in cometary research.

The various activities in space were coordinated by the European Space Agency (ESA), Intercosmos of the USSR Academy of Sciences, the Japanese Institute of Space and Astronautical Science (ISAS) and the U.S. National Aeronautics and Space Administration (NASA), through the Inter-Agency Consultative Group for Space Science (IACG). Coordination of the ground-based observing activities was provided through the International Halley Watch (IHW). The single goal of both IHW and IACG was to maximize the overall scientific outcome of the un-countable efforts in all parts of the world to study Halley's comet from the ground and from space in the most comprehensive way.

Launched in 1978, the International Sun-Earth Explorer No. 3 (ISSE 3) com-pleted after four years its mission as monitor of interplanetary space about 1.5 million km upstream from the earth, and was then redirected to comet inter-cept duty. After a total of 37 propulsive and 5 lunar gravity assist maneuvers the spacecraft, then renamed International Cometary Explorer (ICE), was sent on course towards P/Giacobini-Zinner. The last maneuver was a close lunar swingby on 22 December 1983, and the comet was encountered on 11 September 1985. ICE passed through the cometary tail at a distance of 7800 km from the nucleus of P/Giacobini-Zinner before continuing its journey towards P/Halley, which it passed on 25 March 1986, upstream at a distance of 28 million km.

Launched in 1984, the two Vega spacecraft started with the exploration of Venus, which they encountered on 15 June 1985 (Vega 2) and 11 July 1985 (Vega 1), and deployed a part of their instrumentation there (see report of Commis-sion 16). In the last phases of their flight to P/Halley, the two Vegas also served as pathfinders for the Giotto spacecraft, enabling it to achieve the desired final cometary flyby distance of 600 km on the sunward side, with a deviation of just 5 km. This Pathfinder Project involved the participation of NASA in determining the exact positions of the two Vegas by the Deep Space Net-work using very-long-baseline interferometry techniques. Brief information on the Halley encounters is summarized in the following table. (The Japanese spacecraft Suisei and Sakigake were formerly called Planet-A and MS-T5, re-spectively.)

Flight parameters of the comet Halley fleet :

Spacecraft	Launched by	Launch date (UT)	Closest approach (UT)	Flyby velocity (km/s)	Flyby distance (km)
Vega 1	USSR	15 Dec 1984	6 Mar 1986	79.2	8,890
Suisei	Japan	19 Aug 1985	8 Mar 1986	73.0	151,000
Vega 2	USSR	21 Dec 1984	9 Mar 1986	76.8	8,030
Sakigake	Japan	9 Jan 1985	11 Mar 1986	75.3	6,990,000
Giotto	ESA	2 Jul 1985	14 Mar 1986	68.4	605
ICE	USA	12 Aug 1978	25 Mar 1986	60.0	28,000,000

The extensive scientific outcome of these missions is described in detail in the previous chapters of this report, arranged by topics. The highlights can be summarized as follows.

ICE provided a wealth of important data on the extent of the comet's in-fluence on the interplanetary medium and gave in-situ fields and particle data in the comet's tail. In a broad outline, the pre-encounter theories of the plasma structures were essentially confirmed, but a few surprises occurred. The effect of cometary ions on the solar wind (pickup ions) were seen as far as 28 million km from the nucleus, indicating that the influence of P/Halley on the interplanetary medium would be noticeable by the Halley probes much earlier than originally anticipated, almost a day before the actual encounter.

The Halley missions essentially confirmed Whipple's model of a comet nu-cleus. The Vega and Giotto cameras showed a peanut-shaped body, both larger (about 16x8x8 km) and darker (albedo lower than 0.04) than previously thought, making it one of the darkest known objects in the solar system. Gas and dust emanate in form of jets from only a few regions on the nucleus, while the rest is covered by a dark crust. Much of the cometary dust is organic material. Particles as small as 10^{-17} g, covering a broad variety of chemical composition, were detected in considerable quantity, all of them reportedly fluffy rather than compact. At the time of the Giotto flyby, the total gas production was 6.9×10^{29} mol/s, of which 80 percent were water molecules. As already shown by ICE, the theories about the nature of the interaction of comets with the solar wind were essentially confirmed. There is a contact surface near 5000 km, inside which the solar wind does not penetrate and cometary ions flow freely outward, and a bow wave near 400,000 km where the inflowing solar wind is slowed down substantially by cometary ions.

FUTURE MISSIONS

Following these spectacular, but still reconaissance-level cometary fly-bys, ISAS considers a cometary flyby connected with a coma sample return. NASA plans to initiate an even more comprehensive exploration by conducting a close flyby of a main-belt asteroid, followed by a multi-year rendezvous with a short-period comet (CRAF mission) in the early 1990s, to study the comet both during quiescent and active phases. and to deploy a lander/penetrator into the surface of the comet. P/Tempel 2 is currently the main candidate for the target object. A joint project of France, USSR and ESA foresees a lander/penetrator study of the asteroid 4 Vesta, combined with the flyby of another two main-belt asteroids, and possibly a short-period comet. In addition, ESA and NASA are studying a comet nucleus sample return mission (CNSR) to a short-period comet.

Although the Giotto spacecraft suffered some damage during the P/Halley encounter, it may be redirected to P/Grigg-Skjellerup at the close approach to the earth in 1990, if the instrumentation is found to be in reasonable condition. Similar plans are under consideration for the Suisei spacecraft.

9 Meteorites

METEORITES AND INTERPLANETARY DUST

Laboratory studies of micrometeorites and results from P/Halley (Kissel et al, Nat 326, 755) provided new information about interplanetary dust particles (IDP) and the relationship between IDPs, comets and the meteorites. Reviews by Mackinnon et al (RGSP 25, 1527) and Brownlee (AREP 13, 147) show that IDPs fall into anhydrous or hydrous classes (Sandford et al, ApJ 291, 383). Some hydrated particles (Tomeoka et al, Nat 314, 338; Rietmeijer et al, LPS 15, 687) are serpentinite-rich but most are dominated by a smectite-like phase not found in carbonaceous chondrites. Anhydrous particles are distinct from chondrites, but match Halley compositional data (Brownlee et al, LPS 18, 133). Solar flare tracks (Bradley et al, Sci 226, 1432) seem to link IDPs with comets (Sandford, Ic 68, 377). Raman and IR measurements were used to compare IDPs and interstellar grains (Sandford, Sci 231, 1540; Allamandola et al, Sci 237, 56). The IDP D/H ratio is 10x terrestrial (McKeegan et al, LPS 18, 627). Large amounts of cosmic dust were found in Greenland (Maurette et al, Sci 233, 869).

CHONDRITE PETROLOGY AND COMPOSITION

Yamato 82042 is compositionally similar to CM, but with textural and petrologic affinities to CI chondrites (Grady et al, MNIP 46, 162). Compositional (Kallemeyn, MNIP 46, 151) and petrographic (Scott and Taylor, LPSC 15, C699) data indicate that Karoonda, ALH 82135, PCA 82500 and ALH 84038 are grouped. Weeks and Sears (GCA 49, 1525) and Clayton et al (LPSC 15, C245) suggested differences in chalcophiles and O-isotopes between EH3 (Prinz et al, LPS 15, 653) and higher EH types. The black chondrite in the Cumberland Falls aubrite may be unique (Verkouteren and Lipschutz, GCA 47, 1625) or a metamorphosed and reduced LL chondrite (Kallemeyn and Wasson, GCA 49, 261). Equilibrated H chondrites from the Antarctic seem compositionally different from falls (Dennison et al, Sci 319, 390) suggesting a source change 300 ka ago, a time much smaller than predicted by orbital dynamics.

DIFFERENTIATED STONY METEORITES

Magma compositions represented in How, Euc and Dio (HED) polmict breccias (Delaney et al, JGR 89, C251), ferroan eucrites (Warren and Jerde, GCA 51, 713) and IE (incompatible-element)-rich eucrites (Christophe et al, BM 110, 449) were incorporated into models explaining common eucrite basalts. Much work questioned the previously accepted position that common eucrites represent primary magmas on a ferroan body. Warren (GCA 49, 577) and Ikeda and Takeda (JGR 90, C649) suggested that common eucrites formed by fractionation of IE-poor magnesium basalt. Beckett and Stolper (LPS 18, 54) showed that the Ikeda and Takeda melt cannot produce common eucrites. Longhi and Pan (LPS 18, 570) suggested polybaric formation of eucrites. Siderophile data (Newsom, JGR 90, C613) imply that eucrites were derived from a depleted source. Delaney (LPS 17, 166) discussed two-stage igneous evolution for the HED body, consistent with Fe/Mg ratios and IE concentrations. Mittlefehldt (GCA 51, 267) showed that eucrites are more depleted in volatiles than the diogenite magma. Hewins (JGR 89, C289) reported evidence of impact melting in mesosiderites, and Wasson and Rubin (Nat 318, 168) a core-onto-crust accretional model. Brett and Keil (EPSL 81, 1) showed that aubrites were not derived from enstatite chondrites. Takeda (EPSL 81, 358) suggested that ureilites formed by partial melting of carbonaceous chondritic materials.

IRON METEORITES

New iron meteorites were classified (Malvin et al, GCA 48, 785). Variations in Re/Os ratios have implications for dating (Pernicka and Wasson, GCA 51, 1717). Group IIE meteorites were reclassified (Wasson and Wang, GCA 50, 725). New Ni and P diffusion coefficients (Doan and Goldstein, MT 17A, 1131) offer improvements in cooling-rate models. Teshima et al (GCA 50, 2073) measured Ag in phases showing different shock levels. Kracher (LPSC 15, C689) modelled the formation of IAB and IIICD meteorites by partial melting and core crystallization. Kissen et al (GCA 50, 371) found sphalerite-based pressures up to 3.5 kbar for IAB.

LUNAR METEORITES

The discovery that the Antarctic meteorite ALHA 81005 is a lunar highlands regolith breccia (Bogard et al, GRL 10, 773) and the subsequent discovery of 5 more (Yanai and Kojima, SAM 12, 17) was a boon to lunar science, and enhanced the credibility of the link between Mars and SNC meteorites. Lunar meteorites manifest shock levels similar to those of Apollo samples (Bischoff et al, MNIP 46, 21). Cosmic-ray exposure ages (Nishiizumi et al, SAM 11, 58) suggest > 2 impacts produced lunar meteorites and disparities in Mg/Fe ratios also suggest distinct sites (Warren and Kallemeyn, MNIP 46, 3). Lunar meteorites have lower IE contents than typical Apollo highlands rocks.

IMPACT STUDIES

An impact origin of the Cretaceous-Tertiary (KT) extinctions gained additional credibility with the discovery of shocked quartz (Bohor et al, Sci 224, 867; Sci 236, 705). Computer modelling provided details of the consequences (O'Keefe and Ahrens; Roddy, Hypervelocity Impact Symposium, in press); a 10-km diameter impactor produces a crater 40 km deep but ejects no mantle material. An underwater impact structure was discovered (Jansa and Pe-Piper, Nat 327, 612). The ejecta from a 90-km precambrian structure (Williams, Sci 233, 200) is preserved in sediments 300 km distant (Gostin et al, Sci 233, 198). Time-series analyses of crater ages led some (Rampino and Stothers, Nat 308, 709; Alvarez and Muller, Nat 308, 718) to suggest periodic cometary showers every 30 Ma. This hypothesis was challenged (Grieve et al, EPSL 76, 1).

IMPACT RECORD IN SEDIMENTS

Although high Ir in volcanic aerosols (Zoller et al, Sci 222, 1118) encouraged proponents of a volcanic cause for the KT extinctions, new lines of research (e.g., shocked quartz) supported the impact hypothesis (Alvarez, PT 40/7, 24). In 34 Ma of sediment (Kyte and Wasson, Sci 232, 1225) the KT Ir peak is unique in magnitude and Ir accumulation rates are inconsistent with periodic comet showers. Glass et al (JGR 90, D175) showed that the North American microtektites comprise one of two late Eocene impact horizons; Keller et al (Met 22, 25) report a third horizon. There is little evidence of major impacts at other boundaries (Orth et al, LPS 16, 631); the most promising candidate is in Jurassic sediments (Rocchia et al, JGR 91, E259). Hut et al (Nat 329, 118) reviewed the possible connection between comet showers and mass extinctions.

REFRACTORY INCLUSIONS

The formation of refractory materials in CV, CO, and CM chondrites requires multistage processes under a variety of (mainly nebular) conditions. Recent advances include discussion of fluffy, type-A (MacPherson and Grossman, GCA 48, 29), plagioclase-rich (Wark, GCA 51, 221), spinel-rich (Kornacki and and Wood, GCA 49, 1219) and fremdlinge (Armstrong et al, GCA 49, 1001) inclusions. Refractory-rich materials were found in ordinary chondrites (Bischoff and Keil, GCA 48, 693) and in interplanetary dust (Zolensky, Sci 234, 1466). Laboratory studies provided constraints on phase relationships (Mysen et al, EPSL 75, 139) and cooling rates (Stolper and Paque, GCA 50, 1785).

CHONDRULES AND MATRIX

Trace element studies of chondrules indicate precursor components rich in refractory lithophiles, common siderophiles and common lithophiles (Rubin and Wasson, GCA 51, 1923). Many primitive chondrules have fine-grained silicate rims (Scott et al, GCA 48, 1741) or coarse-grained rims (Rubin, GCA 48, 1779) that formed from fine-grained rims by sintering, probably by the chondrule-forming process. Matrix materials in ordinary and CV chondrites are roughly similar to whole-rock in composition (Grossman, LPS 16, 302). McSween (Met 20, 523) found that the O-isotopic composition of barred olivine (BO) chondrules in CV chondrites lies closer to the terrestrial fractionation line than porphyritic (P) chondrules. Coarse-grained rims are poorer in ^{16}O than their associated chondrules (Clayton et al, LPS 18, 187). Chondrule crystallization experiments demonstrate that most P chondrules were incompletely melted, in contrast to BO chondrules (Radomsky and Hewins, LPS 18, 808). As expected, many relict grains have high melting temperatures (Steele, GCA 50, 1379).

NEBULAR CONDITIONS

Chondritic meteorites indicate a hot (T > 1300 K) nebula (Boynton, PP II, 772), whereas astrophysical models (Morfill et al, PP II, 493) suggest that cloud collapse produced temperatures < 1000 K. High Mo and W contents of refractory inclusions (Fegley and Palme, EPSL 72, 311) suggest O_2 pressures higher than solar, but formation by evaporation (Fegley and Kornacki, EPSL 68, 181) may generate microenvironments differing from the bulk nebula. Some chondrules have been altered in high-temperature nebular episodes (Peck and Wood, GCA 51, 1503). Refractory lithophiles are fractionated only in the highly reduced EL chondrites (Kallemeyn and Wasson, GCA 50, 2153); incomplete accretion of CaS may have produced the EL fractionation (Larimer and Ganapathy, EPSL 84, 123).

PARENT BODY PROCESSES

Ordinary chondrites may originate in S-asteroids in the vicinity of the 3:1 Kirkwood gap at 2.5 AU, and carbonaceous chondrites in C asteroids in this region (Wetherill, Met 20, 1). The association of ordinary chondrites with S asteroids is disputed by Gaffey (Ic 60, 83; Ic 66, 468). Conflicting metamor-

phism indicators and cooling rates in ordinary chondrites suggest that asteroids were fragmented and reassembled within a few days (Grimm, JGR 90, 2022).
The wide range of cooling rates in O-chondrite regolith breccias indicates
breakup and reassembly after metamorphism (Taylor et al, Ic 69, 1). Petrographic evidence also indicates mixing of the ordinary asteroids (Scott et al,
JGR 90, D137). Zoning in chondrule olivines supports metallographic cooling
rates (Miyamoto et al, JGR 91, 12804).

METEORITES FROM MARS

Considerable circumstantial evidence supports the hypothesis that the
shergottites, nakhlites, and Chassigny (SNCs) are rocks ejected from Mars
(Wood and Ashwal, LPSC 12, B1359; McSween, RG 23, 391). Trapped gases in the
shock-melted A 79001 shergottite resemble those in the Martian atmosphere
(Bogard and Johnson, Sci 221, 651; Becker and Pepin, EPSL 69, 225); its sub-
microscopic relict grains rich in Fe, S, and Cl compositionally resemble Martian dust (Gooding and Muenow, GCA 50, 1049). Formation on a hydrous planet
is indicated by amphiboles in Chassigny (Floran et al, GCA 42, 1213) and Sher-
gotty (Treiman, Met 18, 409), rust in Nakhla and Lafayette (Bunch and Reid,
Met 10, 303) and calcite and gypsum in A 79001 (Gooding et al, LPS 18, 345).
Cratering dynamics and cosmic-ray ages suggest that SNCs were ejected from a
>100 km crater $\sim$ 200 Ma ago, followed by fragmentation 0.5–10 Ma ago (Vicke-
ry and Melosh, Sci 237, 738).

ISOTOPIC ANOMALIES

Isotope anomalies in meteorites show that the solar nebula was not a homogeneous, well-mixed cloud. Unequilibrated chondrites are enriched in deuterium, $\delta D \sim 250$ percent in bulk, $\sim$ 570 percent in organic residues (Yang and
Epstein, GCA 47, 2199). Heavy carbon, $\delta^{13}C \sim 140$ percent, is found in demin-
eralized residues (Swart et al, Sci 220, 406) and extreme $\delta^{13}C$, 700 percent,
in spinels (Zinner and Epstein, EPSL 84, 359). Whole rock $\delta^{15}N$ values range
from -9 to +5 percent; except Bencubbin, 97 percent (Prombo and Clayton, Sci
230, 935). Oxygen isotopes are dominated by large variations in ^{16}O; refractory particles with $\delta^{18}O \sim -4$ percent were common at the CV, CO and CM locations
(Clayton et al, EPSL 34, 209). Large excesses of ^{26}Mg, ^{107}Ag and ^{129}Xe reflect
in situ decay of ^{26}Al, ^{107}Pd and ^{129}I, respectively (Wasserburg, PP II, 703).
Mass-dependent fractionation (F) of O, Mg and Si and unknown nucleosynthetic
(UN) anomalies of the neutron-rich isotopes of Ca, Ti and Cr are found in FUN
inclusions. Anomalies in Sr, Ba, Nd and Sm are restricted to FUN inclusions
(Papanastassiou, ApJ 308, L27; Wasserburg et al, Early Solar System, 144).
Anomalies in Mg, Ca and Ti are prominent in hibonite (Hutcheon et al, LPS 14,
339; Zinner et al, ApJ 311, L103). Ne, Kr and Xe in carbonaceous residues ex-
hibit nearly pure ^{22}Ne (Eberhardt et al, GCA 45, 1515) and excesses of the
lighter and heavier isotopes of Kr and Xe (Srinivasan et al, JGR 82, 762).
Diamonds inferred to be interstellar are major rare gas carriers in chondrites
(Lewis et al, Nat 326, 160).

Key to the References

AA   Astronomy and Astrophysics
AAS   Astronomy and Astrophysics Supplement Series
ACM   Asteroids, Comets, Meteors (Vol II: eds Lagerkvist, Lindblad, Lundstedt,
      and Rickman, Uppsala University Press, 1986)
AJ   The Astronomical Journal
ApJ   The Astrophysical Journal
AREP Annual Review of Earth and Planetary Sciences
ASR   Advances in Space Research (Vol 4: No 9; Vol 5: No 12)
ATs   Astronomicheskij Tsirkulyar
AV   Astronomicheskij Vestnik (English translations in Solar System Research)
AZh   Astronomicheskij Zhurnal (English translations in Soviet Astronomy)

AZhL Pisma v Astronomicheskij Zhurnal (English translations in Soviet Astronomy
        Letters)
BAAS Bulletin of the American Astronomical Society
BAC   Bulletin of the Astronomical Institutes of Czechoslovakia
BM   Bulletin de Minéralogie
ChAA Chinese Astronomy and Astrophysics (English translations from Acta Astro-
        nomica Sinica and Acta Astrophysica Sinica)
CM   Celestial Mechanics
CPC   Computer Physics Communications
DANT Doklady Akademii Nauk Tadzhikskoj SSR
EMP   Earth, Moon, and Planets
EPSL Earth and Planetary Science Letters
ESA   European Space Agency Special Publication (most of the papers from ESA SP-
        250 also appear in a feature issue of Astronomy and Astrophysics)
GCA   Geochimica et Cosmochimica Acta
GRL   Geophysical Research Letters
IAUC IAU Colloquium (No 83: Dynamics of Comets, Their Origin and Evolution, eds
        Carusi and Valsecchi, Reidel, Dordrecht 1985)
Ic   Icarus
IKI   Institute of Space Research, Moscow: Preprint Series
IPTI Ioffe Physico-Technical Institute, Leningrad: Preprint Series
ISAS Institute of Space and Astronautical Science, Tokyo: Research Notes
JGR   Journal of Geophysical Research
KFNT Kinematika i Fizika Nebesnykh Tel (English translations in Kinematics and
        Physics of Celestial Bodies)
KIAM Keldysh Institute of Applied Mathematics, Moscow: Preprint Series
KM   Komety i Meteory
KTs   Kometnyj Tsirkulyar
LPS   Lunar and Planetary Science Conference: Abstracts
LPSC Lunar and Planetary Science Conference: Proceedings
Met   Meteoritics
MN   Monthly Notices of the Royal Astronomical Society
MNIP Monthly Notes of the International Polar Motion Service
MPB   Minor Planet Bulletin
MSAI Memorie della Società Astronomica Italiana
MT   Metallurgical Transactions
NASI NATO Advanced Science Institutes Series C (Vol 156: Ices in the Solar Sys-
        tem, eds Klinger, Benest, Dollfus, and Smoluchowski, Reidel, Dordrecht,
        1984)
Nat   Nature
Obs   The Observatory
PASA Publications of the Astronomical Society of Australia
PASJ Publications of the Astronomical Society of Japan
PASP Publications of the Astronomical Society of the Pacific
PP   Protostars and Planets (Vol II: eds Black and Matthews, University Arizona
        Press, Tucson 1985)
PSS   Planetary and Space Science
PT   Physics Today
PTRS Philosophical Transactions of the Royal Society of London
RG   Review of Geophysics and Space Physics
SAM   Symposium on Antarctic Meteorites
Sci   Science
SSR   Space Science Reviews
VKU   Vestnik Kharkovskogo Universiteta

Ľ Kresák
President of the Commission

COMMISSION 16: PHYSICAL STUDY OF PLANETS AND SATELLITES

(ETUDE PHYSIQUE DES PLANETES ET SATELLITES)

PRESIDENT: **G. Hunt**

VICE-PRESIDENTS: **A. Brahic, D. Morrison**

ORGANIZING COMMITTEE: **J.L. Bertaux, J. Burns, Chen Dahoan, D. Cruikshank, M. Davies, T. Encrenaz, D. Gautier, M. Marov, H. Masurski, A.V. Morozhenko, T. Owen, V. Shevshenko, B. Smith, V. Tejfel.**

1. Introduction

The physical study of planets and satellites is probably one of the more active fields of research of the second half of this century. This is due to space exploration by spacecraft, but also to the use of modern detectors, of large ground-based telescopes, and of powerful computers by active researchers. Planetary research (or planetology) is a pluridisciplinary domain, which requires not only the competence of astronomers, but also of geophysicists, of mineralogists, of climatologists, of biologists, of chemists, of physicists, of "pure" mathematicians, and many other scientists. Many results are at the boundary of those of other commissions such as the 15, 20, 7, 19, 33, 40, 44, 49 and 51 ones. The study of the main results obtained during this last triennum shows a perfect complementarity between space and ground-based observations. It should be arbitrary to separate space and ground-based scientists. They have the same goal and they study the same objects. Quite often, the same individuals use both techniques, depending on the most efficient one for the problem under study. It is remarkable to see that space data collected more than ten years ago are still analysed in connection with ground-based observations. The same remarks can apply for ground-based data. In addition to that, new theoretical models, new numerical simulations and new laboratory experiments have been recently developed. They all contribute to a better understanding of planets and satellites physics.

Two major events took place at the beginning of 1986: the encounter of the Voyager 2 spacecraft with the as yet poorly known Uranus system at the end of January and the flyby of Halley comet by a set of European, Russian, and Japanese spacecrafts at the beginning of Mars. The second event, associated with an unprecedented coordinated ground-based studies, is described in the commission 15 report. Unfortunately, 1987 is the first year for 25 years where no new interplanetary spacecraft was launched in any country of the world and no new planet flyby done.

The explosive increase in the number of published papers on planetary and satellite research has made it impossible to provide an adequate summary of progress in the field over a given three year period in the few pages alloted for this purpose. Just the list of published papers should take much more room. Instead of attempting the customary abbreviated summary, it seemed more appropriate to give just a few examples arbitrarily chosen in order to show how lively, how diverse, and how developed is this field of research. This report is not an exhaustive review of the work done and published between June 1984 and July 1987 in the field of the physics of planets and satellites. It is not even a list of the most remarkable discoveries in this field. Only few results are quoted here as examples, just to show the good health of this field of research.

Most of the information can be found in journals such as Icarus, Space Science Reviews, Astronomy and Astrophysics, Soviet Astronomy, Nature, Science, and many others. More than one hundred books (technical as well as semipopular) have been published in several languages in the field of commission 16. Dozens of meetings have been organized during these past years. Among the proceedings, one can quote: The origin of the Moon (Hartmann, Phillips, and Taylor, editors; NASA Lunar and Planetary Institute, Houston, Texas 77058, 1986), Planetary rings (Greenberg and Brahic editors; The University of Arizona Press, 1984), Protostars and planets II (Black and Matthews editors; The University of Arizona Press, 1985), Saturn (Gehrels and Matthews editors; The University of Arizona Press, 1984), Planets, their origin, interior and atmosphere (Fourteenth advanced course of the Swiss society of astronomy and astrophysics, published by Geneva Observatory, Switzerland, 1984), Physico-chimie de la matière primitive du système solaire (Baglin editor, Société Française des Spécialistes d'Astronomie, Paris, 1986), Space astronomy and solar system exploration (ESA SP 268, 1987). An other source of information can be found in the annual reports of the observatories of all countries.

Every year, the Division for Planetary Sciences of the American Astronomical Society organizes a meeting where the latest results and discoveries are announced. About 400 planetologists of several countries participate to these meetings. They were held in October 1984 in Kona (Hawaii), in October 1985 in Baltimore (Maryland), in November 1986 in Paris (France), and in November 1987 in Pasadena (California). A good source of information can be found in the published abstracts (Bull. Amer. Astron. Soc., respectively vol. 16, p. 621-719, 1984; vol. 17, p. 670-750, 1985; vol. 18, p. 734-834, 1986; vol. 19, p. 795-906, 1987).

2. The origin of the solar system

Several numerical simulations of the evolution of colliding planetesimals have been studied. For example, simulations performed by Wetherill (Science, May 17, 1985) suggest that the terrestrial planets did not grow in an orderly fashion until reaching their present size, but rather are the result of violent collisions between large objects smashing into each other. Some were destroyed by the hammering while others were ejected from the solar system. At the end, only a handful remained intact. If the Wetherill conclusions are correct, this has important consequences on the primordial heating which would have melted the entire terrestrial planets, on the formation of iron-dominated cores and on the removal of primitive atmospheres.

Wetherill and Cox (Icarus 60, 40, 1984 and Icarus 63, 290, 1985) have shown that the two-body approximation is not sufficient to describe terrestrial planets accumulation. Hornung, Pellat, and Barge (Icarus 64, 295, 1985), using appropriate collision operators in a kinetic formalism, have described the thermal velocity equilibrium in the protoplanetary cloud when both gravitational encounters and inelastic impacts are present. Weidenschilling and Davis (Icarus 62, 16, 1985) have shown that orbital resonances can have important effects on accretion in the presence of the gaseous solar nebula.

Noble gases abundance should give a number of information about planets formation. As a first step, Donahue (Icarus 66, 195, 1986) has developed a model for selective loss of noble gases by thermal escape of the gases from planetesimals as they grow to form the terrestrial planets.

Lissauer (Icarus 69, 249, 1987) has presented an unified scenario of planetary accretion in which he gives timescales of formation and details on the structure of the protoplanetary disc.

Cabot, Canuto, Hubickyj, and Pollack (Icarus 69, 387 and 423, 1987) have studied the role of turbulent convection in the primitive solar nebula.

The question of the origin of the excess of ^{26}Mg in the Allende meteorite is still open. After the detection of surprisingly large amounts of radioactive ^{26}Al in the interstellar medium (Mahoney et al., Astrophys. J. 286, 578, 1984), alternatives to the explosion of a nearby supernova just before the formation of the solar system have been explored (Clayton, Astrophys. J. 280, 144, 1984; Cameron, Icarus 60, 416, 1984; Dearborn and Blake, Astrophys. J. Let. 288, 221, 1985). One can wonder if the ^{26}Mg-Al correlation observed in Allende minerals may be a manifestation of a cosmic chemical memory rather than a fossil information about the formation of the solar system.

3. Dynamical studies and chaos in the solar system

Recent discoveries of nonlinear dynamics have open a new and rapidly growing field of research. It is now well-known that the phase space of most Hamiltonian systems is divided: for some initial conditions the trajectories are quasi-periodic, for some others they are chaotic. Even if the solar system is generally perceived as evolving with clockwork regularity, it is just a dynamical system which presents sometimes a chaotic behaviour. There are several physical situations in the solar system where chaotic solutions of Newton's equations play an important role (Hyperion chaotic rotation, orbital history of irregularly shaped satellites, Kirkwood gaps, long term evolution of Pluto orbit, ...). A number of new studies have been recently developed. A review of these problems by Wisdom can be found in Icarus (72, 241, 1987).

4. The inner solar system

A. MERCURY

Observations from Mac Donald Observatory by Potter and Morgan (Science 229, 651, 1985; Icarus 67, 336, 1986) reveal that Mercury is surrounded by a faint atmosphere of sodium and potassium. Doppler shifted narrow emission lines in the spectrum of the planet correspond to 150 000 atoms per cubic centimeter at the planet's surface. This can be compared with the 4 500 helium atoms per cubic centimeter and the mere 8 of hydrogen detected by the 1974 Mariner 10 mission. The sodium comes presumably from minerals sputtered off the surface by the solar wind or from meteoritic dust formed after impacts on the surface rather from internal sources for a planet considered as having outgassed long ago. The Doppler shift of emission lines studied during several months indicates that some of the sodium is moving away from Mercury. Potter and Morgan suspect that Mercury has a tail resembling a comet's ion tail or the sodium torus of Io around Jupiter. Mac Grath, Johnson, and Lanzerotti (Nature 323, 694, 1986) and Smyth (Nature 323, 696, 1986) have discussed the nature and variability of this Mercury's sodium atmosphere.

A set of papers on the physics of Mercury have been published in a special issue of Icarus (71, 335, 1987).

B. VENUS

Astronomers are still reducing the data obtained by the pair of Soviet orbiters of Venus which arrived in October 1983 and which were, among other instruments, equipped with synthetic-aperture radar. The surface below the clouds has been probed between October 1983 and July 1984 and resolved down to about 1.5 kilometer across. The first results were presented at the 16th Lunar and Planetary Science Conference, which was held in Houston in March 1985. Venus seems a planet much more dynamic than Mars, but not more than the Earth. Some regions appear dominated by volcanic activity, others by tectonism. Most of the results have been published in Soviet Astronomy (1985 and 1986).

In mid-June 1985, two balloons were dropped in the atmosphere of Venus by the Vega spacecraft. A network of 20 radio observatories in 10 nations provided around-the-clock tracking of the balloons. Both probes encountered vertical air flows and quickly moving pockets of air. Both probes floatting at the same altitude and at similar latitudes on either side of the equator measured temperatures different by almost 10° Celsius. The results have been described in the March 21, 1986 issue of Science.

The Vega spacecraft also dropped a pair of instrumented landers onto the surface. X-ray fluorescence and gamma-ray spectrometers assessed the surface chemistry and measured relative abundances of the key radioactive isotopes of uranium, thorium, and potassium. Mineral-forming elements like silicon, aluminium, calcium, magnesium, and iron have also been identified.

Rock analyses as well as radar images tend to support the idea that a large part of Venus'surface has been dominated by frequent volcanic outpourings.

Detailed analysis of 1982 and 1985 observation seems to indicate that volcanic basalt covering Venera landing sites contains about 9% iron oxide which is in the form of FeO rather than Fe_2O_3 (Science, December 12, 1986).

Allen and Crawford (Nature 307, 222, 1984; Icarus 69, 221, 1987) have observed Venus at infrared wavelengths with the Anglo-Australian telescope. They were surprised to find rapidly evolving patterns on the dark side of Venus. A careful analysis of the data gives unique information on the temperature, the altitude and the opacity of the clouds responsible for the dark-side patterns. This observation shows that, by the use of polarimetry and spectroscopy, cloud zones can be studied remotely through ground-based observations. This is a good example of complementarity between space and ground-based observations.

An interesting review on Venus' atmospheric dynamics has been written by Golitsyn (Icarus 60, 289, 1984).

C. THE MOON

A conference was held in October 1984 in Kona (Hawaii) on the origin of the Moon. The proceedings have been published in 1986 (Hartmann, Phillips, and Taylor, editors; NASA Lunar and Planetary Institute, Houston, Texas 77058). The origin of the Moon is not yet understood. Each proposed model (intact capture, disintegrative capture, co-accretion, Earth fission or collisional ejection) should be rejected or severely modified on the basis of observed data. The pros and cons of the proposed hypotheses as well as a wealth of lunar science can be found in the proceedings.

In order to overcome most of the difficulties of the classical theories, Cameron (Icarus 62, 319, 1985) and Benz, Slattery, and Cameron (Icarus 66, 515, 1986) have developed numerical simulations on the formation of the Moon as the result of a single collision between a Mars-sized body and the proto-Earth.

D. MARS

A set of papers on the evolution of the climate and the atmosphere of Mars has been collected in a special issue of Icarus (71, 201, 1987).

New image processing techniques are now applied to the data obtained during the Viking mission and a number of new physical studies of Mars have been performed these last years. For example, Grant and Schultz (Science, August 21, 1987) explain the formation of regions marked with an array of puzzling dark, filamentary streaks by local, intense atmospheric phenomena such as strong vortex of tornadic intensity. Such tornado-like phenomena scratch the surface of Mars.

Another example is given by Lucchitta (Science 235, 565, 1987). His analysis of Viking images seems to indicate volcanic activity on Mars more recently that accepted until now. He has made a recent survey of the troughs in Valles Marineris which reveals dark patches that are interpreted to be volcanic vents. Their configuration and association with tectonic structures suggest that they are of internal origin. Their albedo and color index indicate mafic composition. Morphologic details and low albedo suggest they are young and perhaps recent.

Wind is the primary agent shaping the surface of Mars today. Greeley and colleagues have used a special wind tunnel at the NASA-Ames Research Center in order to simulate martian conditions. They have compared their results to Viking observations. It seems that Mars has a scarcity of normal sand. Sand-size particles exist on Mars, but as clumps of particles breaking apart after striking a target rather than single grains.

Owen, Maillard, de Bergh, and Lutz (Science, 1988; Bull. Amer. Astron. Soc. 19, 817, 1987) have detected from CFH Observatory HDO in the infrared spectrum of Mars and have obtained the first measurement of the D/H ratio on Mars. It is six times larger than the D/H ratio in Earth oceans. This enrichment over the telluric value implies a much more rapid escape of Hydrogen from Mars in the past, consistent with a denser and warmer atmosphere than the one we find there today. If this scenario is correct, liquid water could have existed in the past. This could explain the existence of channels cut by slowly running water.

5. The outer solar system

In 1986, the Voyager's January 24 flyby has completely changed our knowledge on the planet and its surroundings. It is interesting to compare the speculations on Uranus and its satellites as they were discussed at the A.A.S.-D.P.S. Baltimore meeting in October 1985 (Bull. Amer. Astron. Soc. 17, 685-745, 1985) with the data sent by the spacecraft. Most of the results can be found in Science (233, 1-132, 1986), in the 1986 and 1987 Icarus issues, and in the proceedings of the A.A.S.-D.P.S. meetings (Paris 1986 and Pasadena 1987) and of the A.G.U. meetings. It can be said that most of that we now know about Uranus and its satellites was transmitted to Earth during just a few days in late January 1986. The enormous quantity of data returned was such that astronomers have yet to work through it all. Analysis of Voyager 2's investigation of the Uranian system will take many years.

A. PLANETARY RINGS

(a) The discovery of Neptunian ringlike arcs

The four giant planets of the solar system are now known to possess ring systems which are all very different from the others. The discovery of Neptunian rings is a remarkable result of international cooperation and of observations organized by the universities of Paris (Brahic, Sicardy et al.) and Tucson (Hubbard et al.). Collaborating teams of French and American astronomers, observing stellar occultations by Neptune from ground-based telescopes (ESO, CFHT, IRTF, CTIO, ...) have simultaneously recorded in the close vicinity of Neptune a short reduction in the intensity of the star's light from different sites in 1984 and in 1985. Seven isolated events are also reported out of about one hundred observations. When the observation shows an event, nobody has ever seen any evidence for a second occultation, as would be expected from a uniform, complete ring around the planet. This discovery is due to Brahic, Hubbard, Nicholson, Sicardy, Vilas and colleagues (Nature 319, 636, 1985; Bull. Amer. Astron. Soc. 18, 778, 1986 and 19, 885, 1987). Neptune seems surrounded by ringlike arcs of material. More information should be obtained in August 1989 during the close flyby by Voyager 2 spacecraft. In order to understand the structure of Neptunian arcs, an intensive campaign of ground-based coordinated observations should be organized.

An upper limit of 0.006 for the normal optical depth of an hypothetical continuous disc of dark matter around Neptune has been given, from observations of ground-based stellar occultations, by Sicardy, Roques, Brahic, Bouchet, Maillard, and Perrier (Nature 320, 729, 1986).

(b) The discovery of Jupiter's gossamer ring

A reexamination of the Voyager images by Showalter, Burns, Cuzzi and Pollack (Nature 316, 526, 1985; Icarus 69, 458, 1987) has yielded a refined understanding of Jupiter's diffuse ring system. The system is composed of a relatively bright narrow ring (about 7 000 kilometer wide) and inner toroidal halo (about 20 000 kilometer full thickness) with an exterior "gossamer" ring and without the previously suspected inner ring.

(c) Uranus' rings

The Voyager 2 flyby confirmed the existence of the planet's narrow, dark ring system discovered in 1977 and continuously observed since from ground-based observatories by stellar occultation techniques. The results are reported in a series of articles in Science (233, 1-132, 1986). They confirm the picture given by ground-based observations, but they also provide a large number of new details and unexpected features. New rings, new small satellites near the rings, and previously unknown fine structure within the rings have been discovered. Images of two additional rings (1986U1R and 1986U2R) have been taken, the first one is narrow and orbits between the two outermost of the classical rings, and the second one is broad and closer to the planet than any other known ring. Stellar occultations by the rings seem to reveal possible additional rings and partial rings. One single long-exposure image obtained at very high phase angle shows a very unexpected view of the rings, the micrometer-sized particles form a complex system with the nine classical rings all discernible but not dominant in brightness. Radio occultation experiment indicates a distribution of particles in the external epsilon ring that is almost devoid of particle sizes in the one-to-ten-centimeter size range. Ring particles have an extremely low albedo, possibly as the result of methane ice bombardment by high energy protons moving in the Uranus' magnetosphere. During the Voyager crossing of the Uranus ring plane at about 4.5 Uranian radii, up to 40 micro-particle impacts per second have been recorded by the antenna of the spacecraft.

A very impressive check of the theory of ring confinement by satellites has been done by Goldreich and Porco (Astron. J. 93, 724 and 730, 1987). 1986U7 and 1986U8 are the inner and outer shepherds for the epsilon ring. 1986U7 is the outer shepherd for the delta ring. 1986U8 is the inner shepherd for the gamma ring.

(d) Saturn's rings

The data acquired by Voyager experiments such as imaging science, radio science, ultraviolet spectrometer, and photopolarimeter are still the object of intensive studies. A large diversity of phenomena has been discovered. For example the eccentric Saturnian ringlets at 1.29 Saturnian radii and at 1.45 Saturnian radii have been studied in detail by Porco, Nicholson, Borderies, Danielson, Goldreich, Holberg, and Lane (Icarus 60, 1, 1984). The kinematics of the first one seems determined solely by its interaction with Titan. The kinematics of the second one seems determined solely by Saturn's nonspherical gravity field. Porco, Danielson, Goldreich, Holberg, and Lane (Icarus 60, 17, 1984) have studied the outer edges of Saturn's A and B rings. The dynamics of the A-ring edge seems driven by the coorbital satellites through complex processes. The shape and the dynamics of the B-ring edge are primarily determined by its proximity to the Mimas 2:1 resonance. In unperturbed regions, the B-ring's vertical thickness is of the order of 10 meters.

(e) Ring dynamics. The confinement of planetary rings

Ring dynamics seem to be driven by nearby satellites. Several extensive studies of this complex phenomenon have been performed and much remain to be done. For example, Meyer-Vernet and Sicardy (Icarus 69, 157, 1987) have discussed the physics of resonant disc-satellite interaction. Petit and Hénon (Icarus 66, 536, 1986) have systematically studied satellite encounters in order to understand gravitational interactions between particles in planetary rings.

Shukman (Sov. Astron. 28, 574, 1984) has obtained a kinetic equation having a collisional term allowing for both frontal and tangential inelasticity, and taking into account the spin of the particles. He has investigated the dynamics of Saturn's rings using this equation. Lissauer (Nature 318, 544, 1985), Goldreich, Tremaine, and Borderies (Astron. J. 92, 490, 1986), Lin, Papaloizou, and Ruden (Mon. Not. R. Astron. Soc. 227, 75, 1987), Sicardy (Bull. Amer. Astron. Soc. 19, 891, 1987) have studied the confinement of planetary arcs.

B. ATMOSPHERES OF GIANT PLANETS

(a) The abundance of Hydrogen and Helium in the atmosphere of giant planets

The measurement of the relative abundance of hydrogen and helium in the atmosphere of giant planets can be considered as one of the great discoveries of solar system studies. Before the Voyager encounter with Uranus, the discrepancy of He/H ratios between Jupiter and Saturn was not well understood. Helium cannot be mixed with metallic hydrogen inside Saturn where the temperature is not high enough. As a consequence, there is a relative deficiency of helium relative to hydrogen in the atmosphere of Saturn. The relatively low value measured in 1979 for Jupiter (of the order of 24% per mass unit) corresponded to the solar value and, thus, to a still smaller value in the Big Bang if there were no separation of hydrogen and helium inside Jupiter. This was evidently a problem. The measurement of the relative abundance of helium and hydrogen in the atmosphere of Uranus gives the answer. The value which has been found lies between 26% and 27% and is in perfect agreement with new solar models (Conrath, Gautier, Hanel, Lindal, and Marten, J. Geophys. Res., 1987; Phil. Trans. Roy. Soc. London, 1987; Bull. Amer. Astron. Soc., 1986, 1987). This value suggests that helium differentiation has not occurred on Uranus. Comparisons with values previously obtained for Jupiter and Saturn imply that migration of helium toward the core began long ago on Saturn and may also have recently begun on Jupiter. This means that the relative amount of helium in the primitive solar nebula which gave birth to the solar system is given by the values measured in the Sun and in the atmosphere of Uranus. This result is in good agreement with galactic chemical models which include a substantial decrease in deuterium during the evolutionary process. About 3% of helium has been produced between the Big Bang and the formation of the solar system. In addition, the comparison of atmospheric compositions of giant planets permits a test of formation scenarios. Present available data do not definitively exclude the gas instability model, but are consistent with nucleation models.

(b) The Deuterium to Hydrogen ratio

Bézard, Drossart, Maillard, Tarrago, Lacome, Poussigue, Lévy, and Guelachvilli (Bull. Amer. Astron. Soc. 19, 849, 1987) have measured a value of 1.6×10^{-5} of the D/H ratio in Saturn's atmosphere. This confirms two other determinations of the D/H ratio from the molecule CH_3D. This result is three times larger than the determination in the visible range from HD molecules. This conflict between the two values is not yet solved.

(c) Water and sulfur abundances in the atmosphere of Jupiter

The Jupiter's atmospheric transmission windows at 2.7 microns and 5 microns have been observed from the Kuiper Airborne Observatory. Results are reported by Larson, Davis, Hofmann and Bjoraker (Icarus 60,621, 1984) and by Bjoraker, Larson and Kunde (Icarus 66, 579, 1986 and Astrophys. J., 1986).

The 2.7 micron observations suggest that photolytic reactions in Jupiter's lower troposphere may not be as significant as was previously thought. It seems that, contrary to expectations, sulfur-bearing chromophores are not present in significant amounts in Jupiter's visible clouds. The global abundance of sulfur in Jupiter may be significantly depleted. The apparent absence of hydrogen sulfide is troublesome for chemists and theorists, who had generally counted on sulfur to explain some of Jupiter's colors. Phosphorus-bearing condensates or organic polymers are now primary candidates for explaining Jupiter's visible coloration.

The 5 microns observations are a very diagnostic observational tool to probe the troposphere of Jupiter. Abundances of NH_3, PH_3, CH_4, CH_3D, CO and GeH_4 have been measured in the 1- to 6-bar pressure range in Jupiter's troposphere. The observed abundances of CO, GeH_4, and PH_3 are consistent with models of convective transport from Jupiter's deep atmosphere. There is much less water in the atmosphere of Jupiter than expected unless narrow moist convective plumes in the lower troposphere of Jupiter can reconcile the apparent depletion of water with a near-solar abundance of oxygen throughout the interior (Lunine and Hunten, Icarus 69, 566, 1987).

(d) Jupiter and Neptune ground-based images

Some of the best images ever taken of Jupiter have recently been obtained by Lecacheux, Laques and colleagues at the Pic du Midi Observatory (1987). They are particularly useful in the frame of the International Jupiter Watch before the launch of the Galileo mission.

High quality images of Neptune have been taken at Mauna Kea Observatory by Hammel and Buie during the summer 1987. The Neptune clouds rotation period lies between 17 and 18 hours and varies probably with latitude. The images reveal bright cloud features in Neptune's southern hemisphere, but no one in the northern hemisphere contrary to the images obtained by Smith and Terrile in 1983. Neptune's atmosphere seems to change with time.

(e) Chemical and dynamical studies of Jupiter's and Saturn's atmospheres

Ingersoll and Miller (Icarus 65, 370, 1986) have studied large-scale motions in the atmospheres of Jupiter and Saturn. Mac Low and Ingersoll (Icarus 65, 353, 1986) have studied the time-dependent behaviour of spots in the Jovian atmosphere. Their main result concerns the time-dependent behaviour of interacting spots. Mergings are the most frequent type of interaction. They are irreversible, and do not resemble the interaction of two solitary waves. Spots also spontaneously eject and absorb material.

Appleby and Hogan (Icarus 59, 336, 1984) have developed radiative-convective equilibrium models for Jupiter and Saturn in a study centered primarily on the stratospheric energy balance and the possible role of aerosol heating. Comparisons with Voyager data and results indicate that a dust-free model (no aerosol heating) furnishes a good mean thermal profile for the Jupiter's stratosphere, but cannot be ruled out at low latitude on Jupiter. It seems that aerosol heating played a minor role at the time of the Voyager 2 encounter in Saturn's midlatitude stratospheric energy balance. Other possibilities are discussed in the paper quoted above.

A major problem encountered in the study of the predominantly hydrogen atmospheres of the giant planets is the degree to which thermal equilibration occurs between the para and ortho states of Hydrogen molecules. Infrared spectra obtained by the Voyager spacecraft indicate that the para hydrogen fraction near the 300-mbar pressure level on Jupiter is not in thermodynamic equilibrium. The implications have been recently analysed by Conrath and Gierasch (Icarus 57, 184, 1984).

Bézard, Drossart, Maillard, Tarrago, Lacome, Poussigue, Lévy, Guelachvilli, Noll, Geballe, Knacke, and Tokunaga (Bull. Amer. Astron. Soc. 19, 849, 1987) have detected germane GeH_4 with the same concentration in the atmosphere of Jupiter and Saturn.

(f) Uranus' atmosphere

In addition to the measurements of the helium to hydrogen ratio, many fundamental data have been obtained during the flyby of Uranus by Voyager 2. Astronomers are far to have reduce all the data. We give here only few examples of the discoveries (Science 233, 1, 1986). Uranus has a methane cloud deck with a base near the 1.2-bar pressure level. Near the cloud base, the methane to hydrogen ratio is about 0.02, this corresponds to 20 times the carbon abundance seen in the Sun. Only a few discrete cloud features were seen in images, with prograde zonal wind speeds ranging from 0 near the latitude of - 20° to about 200 m/s near the latitude of - 60° and 0 again at the south pole. Near Uranus' equator, retrograde winds with speeds near 100 m/s are observed. There is a temperature inversion in the Uranus' stratosphere. The temperature rises from a minimum of about 52K at a pressure level of 0.1 bar to about 70K at 0.001 bar. In the extreme upper atmosphere, there is a very large hydrogen scale height with an associated temperature of the order of 800K and significant drag forces on orbiting ring particles. The expected auroras have not been found on Uranus, but ultraviolet emissions from sunlit portions of the very extended atmosphere show a process, which is not yet fully understood, called "electroglow".

A seasonal model of Uranus' atmosphere has been studied as a function of time, i.e. as a function of the phase angle with the Sun (Bézard and Gautier, Bull. Amer. Astron. Soc. 1987; Friedson and Ingersoll, Icarus 69, 135, 1987). The radiative transfer of an hydrogen and helium atmosphere with minor constituents such as methane ands acetylene is calculated. Thanks to the heat redistribution by the winds, the temperature suffers very little variations in the troposphere and slightly larger variations in the stratosphere. The results are in good agreement with Voyager data. The colder the planet, the longer it takes to warm it. The Uranus' dark pole "has kept the memory" of the time it was directly exposed to solar light. But the expected temperature drop at the equatorial level is not observed.

C. INTERNAL STRUCTURE OF GIANT PLANETS

The heat balance of Uranus has just been measured (Pearl, Conrath, Hanel and Pirraglia, Bull. Amer. Astron. Soc. 19, 852, 1987) from Voyager infrared spectrometer observations. Contrary to Jupiter and Saturn which have important internal sources of energy (Jupiter has still "some memory" of its primordial heat and helium is falling down in the metallic hydrogen inside Saturn), Uranus has a very weak internal source (a maximum excess of 9% has been measured) which could be explained by natural radioactivity or by a remnant of accretion heating.

The rotation period of Uranus clouds has been measured from Voyager images. It varies as a function of latitude from 14.2 hours at -70° to 16.9 hours at -27°. The rotation period near -40° is about 16.0 hours. Left-hand polarized signals from the planet show a periodicity of 17.24 (+/- 0.01) hours, which presumably corresponds to the rotation period of Uranus interior (Science 233, 1, 1986). The Uranus magnetic field is probably a consequence of the rotation of an interior ocean of ionized material. Its structure is unexpected. It is tilted 59° with respect to the rotation axis and offset from the center of Uranus by 0.3 Uranian radius. The strength of the dipole moment is intermediate between Saturn's and Earth's.

A remarkable reference book on "Planetary Interiors" has been published by Hubbard (Van Nostrand Reinhold Company, New York, 1984). A stimulating review on giant planets' interior models has been recently made by Stevenson (Icarus 62, 4, 1985). Podolak, Young, and Reynolds (Icarus 63, 266, 1985; Icarus 70, 31, 1987) have studied models in order to understand the differences between Uranus and Neptune interior structures. Internal structure of satellites has also been studied. For example, Zharkov, Leontjev, and Kozenko (Icarus 61, 92, 1985) have discussed models, figures, and gravitational moments of the Galilean satellites of Jupiter and icy satellites of Saturn.

Planetary rings can act as probes of the internal structure of the planet they surround. For example Marley, Hubbard, and Porco (Bull. Amer. Astron. Soc. 19, 889, 1987) have studied Saturnian radial p-mode oscillations and C-ring structure. Correlations between ring features and specific oscillation modes may help constrain interior models. Observation of a central flash during a stellar occultation by Neptune (Brahic, Sicardy, Roques, Mac Laren, and Hubbard, Bull. Amer. Astron. Soc. 18, 778, 1986) allows a determination of Neptune's oblateness.

D. SATELLITES OF GIANT PLANETS

(a) Io

The remarkable volcanic eruptions of Jupiter's satellite Io continue to be studied several years after their discovery in 1979. Much of the new material which has been recently published derives from the Voyager images and other data, but significant new material has been obtained from ground-based telescopic studies. In particular, Johnson, Morrison, Matson, Veeder, Brown and Nelson have observed the infrared flux simultaneously at 8.7, 10, and 20 microns (Science 226, 134, 1984). They have been able to measure the infrared emission from volcanic hotspots on Io's surface as a function of longitude. They have found that volcanic hotspots are not distributed uniformly in longitude and that most of the heat from Io's volcanic interior escapes through a relatively small number of major volcanic centers. These data suggest that the active volcanic regions observed by the Voyager spacecrafts are still active, particularly the region around the feature known as Loki. A second major emitting region corresponds probably to Pelé. The actual global average heat flow from Io has still to be measured. Any estimate depends on the assumptions introduced in a model.

(b) Ganymede

Kirk and Stevenson (Icarus 69, 91, 1987) have developed thermal models of Ganymede's interior, assuming a mostly differentiated initial state of a water ocean overlying a rock layer. They have discussed implications for surface features. Zuber and Parmentier (Icarus 60, 200, 1984) have performed a geometric analysis of Ganymede's surface deformations in order to study its tectonic evolution: while lateral motion cannot be ruled out, it may not be required to explain the geometry of the system. The tectonic features of Ganymede have also been studied by Golombek and Banerdt (Icarus 68, 252, 1986) and by Bianchi, Casacchia, Lanciano, Pozio, and Strom (Icarus 67, 237, 1986).

(c) Mutual events of Jupiter's moons

From May 1985 to Avril 1986, the Earth and Sun lied very near the Jupiter's equatorial plane. A campaign of observations of the mutual eclipses and occultations of the satellites have been organized all over the world in order to determine the orbits of Jupiter's moons with unprecedented precision (K. Aksnes and F. Franklin, Icarus 60, 180, 1984). Data reduction still continues.

(d) Titan

Titan, Saturn's largest satellite, is surrounded by a substantial atmosphere which has been probed by the infrared, the ultraviolet and the radio instruments of the Voyager 1 spacecraft. The analysis of the Voyager results continues. An interesting example is the study of the composition of the layers of clouds and haze by W.R. Thompson and C. Sagan (Icarus 60, 236, 1984). Incorporating the temperature and pressure data from Voyager observations and the absorption and emission characteristics of important atmospheric gases, they have obtained a model atmosphere of Titan which includes nitrogen, argon, methane, a cloud layer of methane droplets, and tholin hazes. They have derived quantitative estimates of the amount of liquid methane and organic solids in the clouds and the haze. Deuterium to hydrogen ratio and the origin of Titan's atmosphere have been discussed by Pinto, Lunine, Kim, and Yung (Nature 319, 388, 1986). For the first time, HCN has been detected at radio wavelengths in Titan (Coustenis, Bézard, Gautier, Marten, Samuelson, Bull. Amer; astron. Soc., 19, 873, 1987) and CO at millimeter wavelengths (Marten, Gautier, Lecacheux, Rosolen, Paubert, Bull. Amer. Astron. Soc., 19, 873, 1987). The amount of CO detected by this method differs by a factor 10 with the amount of CO detected in the infrared range. These observational results are not yet interpreted. Millimeter emission lines are formed in Titan stratosphere while infrared emission lines are formed in Titan troposphere. Thus, all photo-chemical models of Titan's atmosphere (which assume that CO is uniformly distributed) are no more valid.

(e) Hyperion

As a consequence of its out-of-round shape, its large orbital eccentricity and its tidally evolved rotation, Hyperion tumbles irregularly (Wisdom, Peale, and

Mignard, Icarus 58, 137, 1984). Its period of rotation as well as the direction of its rotation axis changes constantly on a chaotic manner.

(f) Uranus' satellites

Ten new satellites orbiting Uranus between the rings and the orbit of Miranda have been discovered by the Voyager's cameras (Science 233, 1, 1986). Their diameter ranges from 40 kilometers to 170 kilometers. The disc of 1985U1 and of the five classical satellites has been resolved and high resolution images have been obtained. The masses of the five classical satellites have been measured and thus the density are now known: Miranda (1.24 +/- 0.31), Ariel (1.55 +/- 0.23), Umbriel (1.58 +/- 0.23), Titania (1.68 +/- 0.07), and Oberon (1.64 +/- 0.06). The albedos ranges from 0.07 for 1985U1 to 0.40 for Ariel. Umbriel is particularly dark (albedo of 0.19) and the geologic activity seems to increase when the distance to Uranus decreases. There is very little evidence for geologic activity on the surface of Oberon. Titania, the largest Uranian satellite, has a surprising number of fractures across its surface. Ariel possesses a remarkably fractured surface with some indication of ice flow across parts of the surface. Miranda, the smallest of the major Uranian satellites, shows a surprising high degree and large diversity of tectonic activity. Half of its surface is relatively bland, old, cratered terrain. The remainder comprises three large regions of younger terrains, each rectangular to ovoid in plan, that display complex sets of parallel and intersecting scarps and ridges as well as numerous outcrops of bright and dark materials.

The cratering record of Uranus' satellites, studied by Strom (Icarus 70, 517, 1987), shows two different populations of different ages. In this paper, the solar system cratering record from Mercury to Uranus is rewieved. It seems very complex and any proposed origin of the impacting objects must be considered highly speculative at present time. Mobilization of cryogenic ice in outer solar system satellites has been discussed by Stevenson and Lunine (Nature 323, 46, 1986).

(g) Triton

Cruikshank (Bull. Amer. Astron. Soc. 19, 858, 1987; with Apt, Icarus 58, 306, 1984; with Brown and Clark, Icarus 58, 293, 1984) has detected liquid nitrogen on the surface of Triton. Delitsky and Thompson (Icarus 70, 354, 1987) have suggested that Triton's surface could be partially covered with a soup of liquid nitrogen and methane on which floats some organic compounds including liquid ethane C_2H_6. Lunine and Stevenson (Nature 317, 238, 1985) have discussed the physical state of volatiles on the surface of Triton and have concluded that a nitrogen ocean cannot be excluded, but requires very restrictive assumptions.

E. PLUTO - CHARON

Nearing perihelion in 1989, Pluto is actually closer than Neptune to the Sun. Consequently the planet should be near its highest temperature and astronomers have good observation conditions. It turns out that the plane of Charon's highly inclined orbit is now sweeping across the inner solar system, allowing us to see, thanks to Pluto's slow orbital motion, from 1985 to 1990, a rare series of occultations and transits of the planet and its moon. These events, which only occur every 124 years, have made it possible to determine the basic physical parameters such as the radius, the albedo and the mean density of the planet and the satellite. A major fraction of rocky material seems to lie inside Pluto. Mutual-event data have been combined with speckle interferometer measurements. Tedesco, Buratti, Binzel and Tholen have described the results in Science (1985). The total mass of the couple Pluto-Charon is no more than about two-thousandth that of the Earth, much more than previous estimates. It turns out that Charon is the largest satellite with respect to its parent planet in the solar system.

Spectroscopic evidence has suggested that Pluto possesses an atmosphere. The surface layer of solid methane gives rise to an atmosphere, but the vertical extent and the thermal properties of this gaseous layer are still unknown. From data collected by the infrared IRAS satellite, Sykes, Cutri, Lebofsky, and Binzel (Science, 1986) have developped a model in which Pluto's poles are capped by methane ice. An equatorial band extending from about - 45° to + 45° in latitude should be free of methane ice. In this model, Pluto's atmosphere, sublimed from the ice, should be about 1000 times less dense than the Earth atmosphere at sea level. Marcialis (Bull. Amer. Astron. Soc. 19, 859, 1987) has observed Charon occultation by Pluto in the near infrared range. Substraction of fluxes measured before, during and after the event has yielded individual spectral signatures for each body. Charon's surface appears extremely depleted in methane compared to Pluto and water frost has been identified at the surface of Charon.

The origin of the couple Pluto-Charon is still discussed. Lyttleton suggested in 1936 that Pluto was once a Neptunian moon ejected after a gravitational encounter with Triton which reversed the direction of Triton's orbital path. MacKinnon (Nature, September 27, 1984) has rejected this hypothesis: the presently accepted masses of Triton and the Pluto-Charon system could not have changed Triton's orbital direction and momentum conservation would have ejected Pluto-Charon not only from the Neptune system, but from the solar system as well. Mac Kinnon proposes that both Pluto and Charon began as large, independent outer solar system planetesimals rotating around the Sun.

6. Laboratory experiments

To understand the abundance of elements and the chemical reactions in planetary atmospheres, molecule spectra have been studied as a function of physical conditions: for example, one can quote laboratory studies of phosphine photolysis in the atmosphere of Jupiter and Saturn by Ferris and Khwaja (Icarus 62, 415, 1985) and laboratory studies of reactions between chlorine, sulfur dioxide, and oxygen by DeMore, Ming-Taun Leu, Smith, and Yung. To understand the initial stages of planets and satellites formation and cratering observations, collisions between various material have been studied. For example, Lange and Ahrens (Icarus 69, 506, 1987) have performed impact experiments in low-temperature ice. Fujiwara (Icarus 70, 536, 1987) has studied the energy partition into transverse and rotational motions in catastrophic disruptions by impact.

7. Other planetary systems?

Extrasolar planetary science begins to be an active (and up to now) speculative field of research. With new detectors, astronomers are intensively scanning the surroundings of nearby stars in order to discover extrasolar planets (see commission 51 report). In spite of premature announcements, no one has been still discovered (until autumn 1987), but the first observations reveal much more dark matter than expected in the immediate vicinity of a number of stars. This has may be nothing to do with any stage of planet evolution, but it is interesting to see that material is available around several stars. The answer to the question of the existence a large number of extrasolar planets is linked on to our understanding of planet formation. Anomalous long-wavelength infrared radiation has been found by the infrared satellite IRAS from Vega and other stars. Observations from the Kuiper Airborne Observatory have confirmed this discovery (Science and Nature, 1984). The best interpretation is that these stars are surrounded by a shell or a ring of solid bodies. Blocking the bright light from the star itself and using state-of-the-art imaging and processing techniques, Smith and Terrile have taken a photograph in visible light of the disc surrounding the star Beta Pictoris and seen edge-on (Science, 1985). Zuckermann and Becklin have discovered an infrared excess around a nearby white dwarf (Bull. Amer. Astron. Soc., 1987).

André BRAHIC
Vice-President of the commission

COMMISSION 19: ROTATION OF THE EARTH (ROTATION DE LA TERRE)

President: W. J. Klepczynski
Vice-Presidents: M. Feissel, B. Kolaczek
Organizing Committee: F.E. Barlier, P. Brosche, W.E. Carter,
  D.M. Djurovic, I.I. Mueller, M.G. Rochester, B.F. Schultz,
  J. Vondrak, G.A. Wilkins, Ya.S. Yatskiv, S.H. Ye, K. Yokoyama

INTRODUCTION

During the period, there have been several major events which have effected the scope and interest of Commission 19. The most significant of these has been the dissolution of the BIH and IPMS and their replacement by the International Earth Rotation Service (IERS). The correlation of higher frequency fluctuations in the Earth's rotation rate with changes in the Earth's Atmospheric Angular Momentum is also significant. Many investigators now seem to believe that the "decade variations" in the Earth's rotation rate are caused by torques between the core and mantle caused by the uneven motions at the core-mantle boundary. These events and discoveries have made this an exciting period. It seems that the future holds more in the way of discovery due to the utilization of the more accurate and precise Earth rotation data coming from the modern observing techniques.

REPORTS FROM INTERNATIONAL SERVICES

Report from the Bureau International de l'Heure

This report covers the work on Earth rotation and related reference frames. The activities on atomic time are reported to IAU Commission 31.

The activities over the reporting period have been characterized by

- the complete introduction of the results from VLBI, Lunar and Satellite Laser Ranging (LLR, SLR) in the combined Earth rotation time series,

- the implementation of a new algorithm for the BIH Terrestrial System (BTS), in cooperation with the French Institut Geographique National (IGN),

- the study of VLBI celestial reference frames, in cooperation with the Bureau des Longitudes (BDL).

Earth Rotation

The Earth Rotation Parameters (ERP: coordinates of the pole and universal time) have been continuously computed and disseminated. As in the previous years, the three main series of evaluations have been as follows:

- A scientific solution, performed yearly on the basis of long homogeneous series made available by the associated analysis centres, in conjunction with the realisation of the BIH systems. The evolution of the relative weights of the different methods of observation and the resultant precision of the series of ERP at 5-day intervals are shown in Figure I. The polhody over 1984-1986 is plotted on Figure 2.

The data introduced in the solution had been provided by the following institutes.

VLBI: National Geodetic Survey (NGS), Goddard Space Flight Center (GSFC), and Jet Propulsion Laboratory (JPL),

CERI (Connected Element Radio Interferometry): U. S. Naval Observatory (USNO),

LLR: JPL,

SLR: Center for Space Research (CSR), Deutsches Geodaetishes Forschungs-institut,
        Abteil I (DGFII), Delft Technical University (DUT), Smithsonian Astrophysical
        Observatory (SAO).

Doppler: Defense Mapping Agency (DMA)

Optical astrometry: BIH, from latitude and universal time observations of 90 stations. A
        solution based on the IAU 1980 Theory of Nutation has been recomputed from 1978
        onwards.

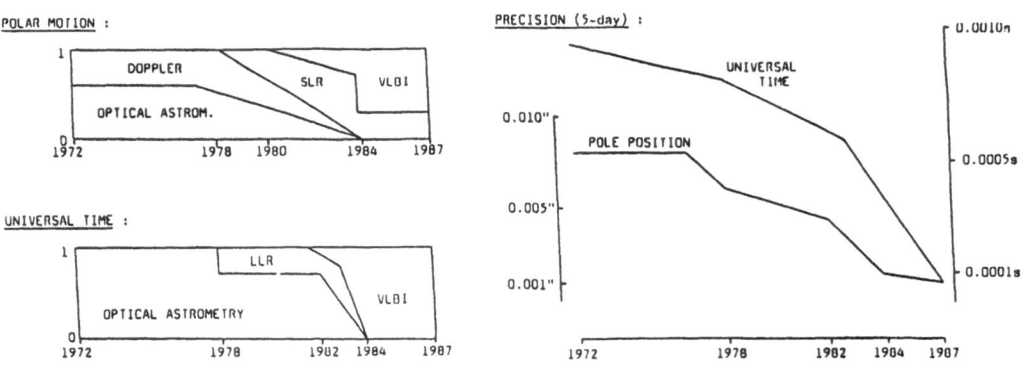

Figure 1 - Contribution of the different observing
techniques to the BIH Earth rotation series and
precision of the combined solution, 1972-1987.

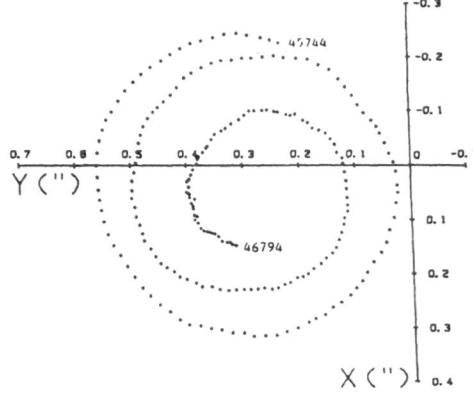

Figure 2 - Path of the pole from 1984 Feb. 14 to
1986 Dec. 30 (MJD 45744 - 46794).

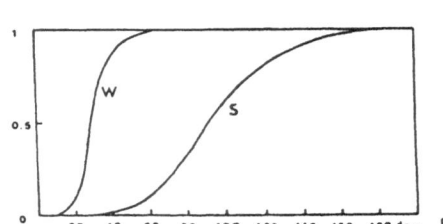

Figure 3 - Characteristics of the filters used for
the BIH monthly operational solution (Circular D),
1984-1987.
S : pole coordinates, 1984-1985
W : pole coordinates, 1986-1987
universal time, 1984-1987

- An operational solution performed monthly, based on data of all techniques available with
  a 1-2 month delay, and on preliminary evaluations of their link to the BIH system. This
  solution is made available under the form of raw and smoothed values. The degrees of
  smoothing used over 1984-1987 are given in Figure 3. The rms agreement of this solution
  with the smoothing of the scientific solution with the same filtering characteritics has
  been 0.003" for the pole coordinates and 0.0006s for UTI, while the maximum differences
  have been respectively 0.010" and 0.0020s.

- An advanced solution performed weekly, based on SLR for the pole coordinate and on optical astrometry, SLR, LLR, and since 1987, on VLBI for universal time. The maximum differences with the operational solution have been 0.01" on the pole coordinates and 0.004s on universal time. This solution was complemented by a weekly prediction issued by USNO and by a monthly prediction issued by BIH. The typical accuracy of these predictions has been 0.005" (pole) and 0.002s (UT) for a time lag of 10 days and 0.015" (pole) and 0.010s (UT) for a time lag of 40 days.

Terrestrial System

The maintenance of a terrestrial system of reference has always been an important activity of the BIH, since the series of the pole coordinates and universal time are referred to this system. The BIH system was defined in 1968 by the adoption of astronomical coordinates for a network of 68 optical instruments contributing to the monitoring of the Earth's rotation. The initial realization was concerned only with the directions of the axes: the tie to the CIO and the definition of the origin of longitudes. These directions were maintained by a stability algorithm applied to the yearly determinations of the astronomical station coordinates. Starting with 1973, the series obtained by the space geodesy and interferometric techniques were progressively introduced in the maintainance of the direction of the axes of the BIH System.

Thanks to a cooperation with the French Institut Geographique National (IGN), starting with 1984, the BIH System has been realized through a set of coordinates of sites where space geodesy stations are operated. This follows a proposal of B. Guinot (1981), "Comments on the terrestrial pole of reference, the origin of longitudes, and on the definition of UT1," Proc. 56th IAU Coll., Reidel Publ. Co. The implementation is based on a study by C. Boucher and M. Feissel (1984) "Realization of the BIH Terrestrial System," Proc. Internat. Symp. on Space Techniques for Geodynamics, J. Somogyi (ed.), Sopron, Hungary.

In its 1986 issue, the BTS is realized by the cartesian coordinates of 51 sites belonging to six of the main tectonic plates: Pacific (4), North America (23), South America (4), Eurasia (16), Africa (2), and India/Austalia (2). It is adjusted from 301 station coordinates determinations in seven networks (three VLBI, one LLR, two SLR, one Doppler), using the local ties collected by IGN. The transformation parameters which relate the seven individual systems to the BTS are evaluated in the same process. The stability of the BTS in this new realization is estimated to a few centimeters in the origin, 0.002ppm in the scale and 0.004" in the orientation.

Extragalactic Celestial Frames For Earth Rotation

During the last ten years, three groups (JPL, GSFC, and NGS) have published catalogues of precise positions of extragalactic radio sources derived from Very Long Baseline Interferometry (VLBI) observations. The catalogues provide the celestial reference frames for the series of Earth Rotation Parameters derived from the observations. It has been found that the various reference frames have their directions of axes consistent within 0.003". This is a remarkable agreement considering the diversity of observing stategies and data analysis.

The rotation angles estimated statistically with the common radio sources between catalogues have uncertainties in the range of 0.0001" to 0.0005". This range is confirmed by the stability of the estimation when the number of radio sources is changed. Consequently, the maintainance of a celestial reference system by means of VLBI coordinates of radio sources could be ensured within 0.001" by a no-rotation condition applied to the ensemble of source coordinates.

Participation In The MERIT Program

The BIH acted as the coordinator of MERIT project, during and after the main observing campaign, in 1983/1984. The different tasks accomplished in this framework were as follows.

- Collection of the precise description and geodetic ties of the sites involved in the measurement of the Earth's rotation; at IGN a Directory of MERIT Sites (DOMES) is kept up to date. Collection of the Sets of Station Coordinates (SSC) defining the terrestrial frame for each series of ERP.

- Collection of the series of ERP obtained during and after the campaign by the participating computing centres; maintainance of a data bank for storing the collected data, dissemination of copies on request.

- Issuing of results from, and information on, the campaign, in the MERIT Monthly Circular, distributed to 180 addresses. Publication of the results obtained by the Operational and Analysis Centres for Earth rotation and the related reference frames (20 Centres from 8 countries).

Other Studies

- Determination of the Earth orientation from optical astrometry: during a stay at BIH, Dr. Li Zheng-xin from Shanghai Observatory computed the ERP over 1962-1981 from optical astrometry observations; the solution is referred to the IAU 1976 System of Constants and to the IAU 1980 Theory of Nutation.

- Atmospheric excitation of irregularities in the Earth's rotation: the seasonal and higher frequency oscillations in duration of the day under the action of atmospheric currents have been studied on the basis of series of universal time and series of Angular Momentum of the Atmosphere. The "50 days" oscillations have been shown to be permanently active with unstable pseudo periods, and peak-to-peak variations from 0.2 to 0.6ms in the duration of the day. to comply with a request of the SSG 5.98 of IAG, "Atmospheric excitation of the Earth's rotation", data bank of Atmospheric Angular Momentum has been established. The associated documentation appears in the Annual Report, part. D.

Report from the International Polar Motion Service

It will be useful to start this report with the description of recent movement in Japanese astronomical community to set up a new national organization for astronomical research. This necessarily implies the reorganization of the Tokyo Astronomical Observatory (TAO) and the International Latitude Observatory of Mizusawa (ILOM), which has played a leading role in Japan for the international cooperation to monitor the Earth's rotation.

It has been agreed that the geodynamical activities, which have been conducted at the ILOM, be taken over by the Earth Rotation Branch of the new organization which will be inaugurated in 1988 fiscal year. The headquarters will be located at the same place as the present TAO, and the Earth Rotation Branch will also be at the same place as the present ILOM.

History of the ILOM

The ILOM was founded in 1899 as one on the six stations of the International Latitude Service (ILS) organized by the International Association of Geodesy (IAG). As its name indicates, the main objective of the ILOM is to continuously monitor polar motion by observing latitude variations, in collaboration with other international latitude stations. The ILOM acted as the Central Bureau of the ILS during 1922.7 - 1934.0, and also has been the Central Bureau of the IPMS since 1962.

For these ninety years, the ILOM has strived to adapt itself to the developments of science and technology relating to Earth rotation study. From the observational viewpoint, it has developed new techniques of observation, introducing the state-of-the-art techniques.

In spite of the limitation due to the prescribed objectives of the ILOM, research activities have been extended focussing on geodynamical fields, and instrumentation efforts have been stressed to construct an integrated system of geodynamical observations, composed not only of astronomical instruments but also of geophysical instruments.

In accordance with recent drastic improvement of observational accuracies of both astronomical and geophysical instruments, the ILOM has made every effort to introduce the most modern techniques, such as a very long baseline interferometer (VLBI), a super-conducting gravity meter, and a transportable absolute gravimeter.

Conclusions of the Geodetic Council of Japan for the Reorganization of the ILOM

In July 1985, the review committee, set up by the Geodetic Council of Japan, finalized a report on the reorganization of the ILOM. This report was endorsed by the Geodetic Council of Japan.

Points of the report are as follows:

1)   Although the ILOM has played an important role in the ILS and the succeeding IPMS activities, it has recently become clear that the present organizational structure of the ILOM as a branch under Ministry's jurisdiction is not suitable to meet the following requirements, which have been brought about by the recent development in science and technology:

a)   The name and the objectives of the ILOM, which are now obsolete, need overall review.
b)   In order to further promote research activities related to the Earth's rotation in Japan, it becomes necessary to form an organization, in which current scientific requirements could be effectively fulfilled.
c)   In order to introduce new high-precision measurement techniques, it becomes necessary to establish a management system, through which the equipments could be brought into effective nationwide use.
d)   In order to consolidate cooperative research activities for the scientists of various fields relevant to the Earth's rotation, establishment of a central organization is required.
e)   Establishment of the system for the common use of observational data and softwares, as well as observational facilities and equipments, is required.
f)   Establishment of a national center for international cooperation in observations and research has been required.

2)   The report concluded that in order to meet the above requirements, it would be appropriate to implement such reorganization, in conjunction with other institution(s) which is relevant to the Earth rotation research.

Recent Agreement Between the TAO and the ILOM

In 1987, the TAO and ILOM agreed to jointly establish a new national organization for astronomy in the fiscal year of 1988. The board of council for establishing the new national organization for astronomical research, set up by the Ministry of Education, Science and Culture, accepted the fundamental conception of the new organization elaborated by the TAO and the ILOM. The new organization will be composed of six scientific branches, and the present ILOM takes part as the Earth Rotation Branch and part of the Theoretical Astronomy Branch. The Earth Rotation Branch is planned to consist of five departments, which covers wide fields of geodynamics.

VLBI for the Earth Rotation Study and Astrometry (VERA) As A Tool to Participate in the New International Earth Rotation Service

In 1985, a Japanese scientist group covering positional astronomy, geodesy and geodynamics proposed to build a Japanese VLBI system for Earth rotation study and astrometry. This system, called VERA, is to be composed of two antennas spanning 2300 km. One antenna with 35 m diameter will be built near Mizusawa, and the other with 15 diameter on one of the South-Western Islands.

VERA will be dedicated to regular monitoring of the Earth's rotation in the international networks.

There are four big projects, which Japanese astronomers desire to realize in the new organization. They are:

1) Japanese National Large Telescope (JNLT) Project, to build a 7.5 m diameter optical telescope on Mt. Maunakea in Hawaii.
2) VLBI for the Earth Rotation Study and Astrometry (VERA).
3   Radio Heliograph.
4) Solar Cycle Telescope.

Role of the ILOM in the IERS

The present ILOM, or the Earth Rotation Branch in the new national organization for astronomical research, will participate in the IERS with the following roles.

a) The network center of VLBI IRIS-P network, in collaboration with the Radio Research Laboratory (RRL) in Japan.

b) A VLBI analysis center, in collaboration with relevant Japanese institutes, namely, the RRL, TAO, Japan Hydrographic Department (JHD), and Geographical Survey Institute (GSI).

c) A computing center of atmospheric excitation functions, in collaboration with the Japan Meteorological Agency (JMA).

Brief Description of VLBI IRIS-P Network

IRIS-P network became operational in April 1987. Before that, we had only one IRIS VLBI network called IRIS-A, which is composed of four regular stations (one in FRG and three POLARIS stations in USA) and several stations participating intermittently. The objectives of IRIS-P are: a) to provide Earth rotation results independently of IRIS-A for cross-check, b) to make precise time synchronization between Kashima in Japan and Richmond in USA, and b) to get experiences to design and operate Japanese VLBI system VERA.

The network center at the ILOM is responsible for smooth operation of IRIS-P, in collaboration with the VLBI coordinating center at the NGS and the correlation center at the United States Naval Observatory (USNO).

Estimation and Dissemination of the Smoothed Daily EOP's

Since September 1983, the Central Bureau of the IPMS has regularly provided optimally smoothed EOP's from optical astrometry observations using the method of Bayesian information criterion. Hyperparameters, which define smoothness of polar motion and UT1, are selected so as to minimize a Bayesian information criterion. This leads to the estimation of optimum EOP's. By the adoption of this method, precisions of the EOP's from optical astrometry were remarkably improved. Nominal rms errors of the estimates are about +/-0.003" for polar motion and +/-0.0002 s for UT1. Systematic errors in polar motion, however, are comparatively large in comparison with the results of VLBI and SLR. On the contrary, UT1 or LOD of optical astrometry has shown very close agreement with that of VLBI and SLR.

The IPMS smoothed daily EOP's have been published in the Monthly Notes of the IPMS. A whole set of the EOP series starting from 1962 was distributed on magnetic tape basis to several institutions and agencies.

Computation and Software Services

Services to compute apparent places of stars has been provided on request. Software to compute corrections of star positions and proper motions was provided by the Central Bureau to several stations on request. Also, software to evaluate the effects of ocean-tide loading developed by the ILOM has been supplied to several institutes through the Central Bureau.

Computation of the Atmospheric Excitation Functions

As one of the three computing centers under the Special Study Group 5.98 of the IAG, "Atmospheric Excitation of the Earth's Rotation", the Central Bureau has regularly reported the atmospheric excitation functions based on the data provided by the Japan Meteorological Agency, to the chairman of the SSG 5.98.

Dissemination of Data

In response to the requests of individuals, institutions and agencies relevant to Earth rotation research, various kinds of data accumulated in the IPMS data files were disseminated.

Collection of Data Related to Earth Rotation Investigation

The EOP's estimated from both astronomical observations and geophysical data have been collected and stored in the IPMS data files.

Summary

Especially from 1980 when the short campaign of Project MERIT was initiated, development in observational and analysis techniques was quite drastic in Earth rotation research. Important scientific findings have been obtained during this short period. This was never achieved by classical optical astrometry data, which have been accumulated for about century from the beginning of the ILS.

By virtue of centimeter precision and accuracy achieved by VLBI and laser ranging, scientific values of the IERS data are incomparably larger than the past data. This means that scientific aspects of the IERS activities will become more important, compared to the activities of the BIH and IPMS, which are going to be replaced by the IERS.

Most essential for the IERS is to extend current observational networks, so that they may cover the whole surface of the Earth uniformly. It is especially important for the VLBI community to establish several independent networks, each of which covers a region of the Earth's surface and is linked up with one another, so that the ensemble of the regional networks cover the whole surface of the Earth. Each of the regional network will act as an independent unit.

Cost of the new-technique facilities, however, are rather high, compared with classical instruments. Hence it may not be easy for such instruments. In addition to efforts of individual groups, international support will be very important to realize the instrumentation plans of various countries.

Report from the new International Earth Rotation Service

In accordance with resolution B2 of the IAU General Assembly at Delhi a Provisional Directing Board for the new International Earth Rotation Service (IERS) was set up with the following membership: C. D. Boucher, W. E. Carter, R. Coates, J. O. Dickey, M. Feissel, R. W. King, D. D. McCarthy, I. I. Mueller, (Vice-Chairman), P. E. C. Paquet, G. A. Wilkins (Chairman), Y. S. Yatskiv, S. - H. Ye, and K. Yokoyama. The Board held three meetings in conjunction with other conferences, as follows: at Austin, Texas, on 1986 April 29 and 30; at Coolfont, West Virginia, on 1986 October 21; and at Pasadena, California, 1987 March 20 and 21. At each meeting the Board received and discussed reports on the progress of the current activities within and related to the continuation of the MERIT/COTES programmes. The main task of the Board was, however, to prepare recommendations on the terms of reference, structure and composition of the new service.

The reports and recommendations of the MERIT and COTES Working Groups were published by the IAU in Highlights of Astronomy (7, 771-788), by the International Association of Geodesy in Bulletin Geodesique (60, 85-100), and in other publications. A letter with supplementary information about the proposed Service was sent out on behalf of the Board to all organisations that were known to be capable of participating in it. By the time of the third meeting the Board had received offers of participation from eight countries which were prepared to operate one or more technical centres, a further four countries were prepared to provide observational data, and three others indicated their intentions of setting up new stations which would contribute to the Service.

The Board discussed the submissions and decided on its recommendations which were accepted by the Executive Committee of the IAU at its meeting in 1987 September and by Councils of the IAG and of the IUGG at the General Assembly of the IUGG at Vancouver, D. C., in 1987 August. The structure and composition will be as follows:

1. The Central Bureau will be provided jointly by the Observatoire de Paris, the Institut Geographique National and the Bureau des Longitudes of France, and will be supported by two sub-bureaux.
1a. A rapid service for earth-rotation parameters to be provided by the National Earth Orientation Service (NEOS) of the USA (formed jointly by the US Naval Observatory and the National Geodetic Survey);
1b. A service for atmospheric angular-momentum data, including forecast values. The AAM sub-bureau will receive and combine data from several meteorological centres for use in the prediction and study of earth rotation parameters. It is expected that many organisations in other countries will provide data and results on reference systems, as well as on earth-rotation, for use and dissemination by the Central Bureau.
2. The Coordinating Centre for VLBI will be provided by the National Geodetic Survey (NGS) of the USA, and there will be three Network Centres: NGS for the IRIS-Atlantic network; the Jet Propulsion Laboratory (JPL), USA, for the TEMPO Deep Space Network; and the International Latitude Observatory of Mizusawa (ILOM), Japan, for the IRIS-Pacific network. Each centre will be responsible for the initial data analysis for its network, as well as for scheduling the observations at each of the radio-telescopes of the network. The Shanghai Observatory (SHO), China, the Forschungsgruppe Satellitengeoda/sie (FGS), Germany, and the three network centres will also carry out long-term analyses of the VLBI data.
3. The Coordinating Centre for SLR will be provided by the Center for Space Research (CSR) at the University of Texas at Austin, USA, and it will be supported by a Quick-look Operational Centre at the Delft University of Technology (DUT) in The Netherlands, and by Data Collection Centres provided by FGS and SHO; CSR will itself also carry out these functions. Full-rate data analysis centres are to be provided by CSR, FGS, and SHO, and it is expected that the Japanese Hydrographic Department (JHD), the University of Padova-Telespazio group (HPSN), ITALY, the Royal Greenwich Observatory (RGO), UK, and the Zentral Institut fur Physik der Erde (ZIPE), GDR, will also provide the results of their analysis of full-rate SLR data. The Intercosmos network will also contribute observational data and results to the Service. Almost all of the stations in the current global network of observing stations will provide data for the Service.

4. The Coordinating Centre for LLR will be provided by CERGA, France. It will be supported by JPL and the Astronomy Department of the University of Texas at Austin (UTEX), which will act as both Quick-look Operational Centres and as Data Analysis Centres; CERGA will also carry out these functions and SHO will act as a Data Analysis Centre. It is expected that at least three stations (Grasse, Hawaii, and Mcdonald) will provide observational data; it is hoped that Orroral will also soon contribute again from its key position in the southern hemisphere; Wettzell is expected to commence LLR observations in 1989.

All data and results submitted to the Service will be made available for general use. The recognised organisations in the Service will receive priority in their distribution. Charges may be made to cover the costs of supply of data. The Service will archive results on earth rotation and reference systems, but will not archive the observational data, although it is hoped that other organizations will do so.

The Directing Board for the new Service, which will come into operation on 1988 January 1, will consist of six members nominated by the Executive Committees of the IAU and IUGG/IAG, the Central Bureau (Paris Obs.) and the three Coordinating Centres (NGS, CSR and CERGA), respectively. The Federation of Astronomical and Geophysical Services (FAGS) is to be asked to provide financial support (comparable to that now given to BIH and IPMS) for those activities that cannot be met by the host organisations. The Board will report annually to FAGS and to each General Assembly of the IAU and IUGG/IAG.

The members of the Board are: W. E. Carter (VLBI), M. Feissel (Central Bureau), B. E. Schutz (SLR), C. Veillet (LLR), G. A. Wilkins (IAU), and K. Yokoyama (IUGG). A preliminary meeting of the Board was held in Vancouver on 1987 August 17. IA reviewed progress and agreed on actions to be taken. In particular, it was agreed that the Central Bureau (at the Paris Observatory) should issue the following series of IERS publications:

Monthly Circular (similar to BIH Circular D)
Annual Report (similar to BIH Annual Report)
Special Circulars and Technical Reports.

The Rapid Service Sub-Bureau (at the U S Naval Observatory) should issue:

Weekly Circular (with quick-look results and predictions)
Monthly Circular (with predictions for 3 months).

In general, the new Service will follow the procedures developed during and since the MERIT Main Campaign of 1983/4, and it is hoped that the transition will not cause any inconvenience to users of the services. A set of IERS Standards for using in the processing of the observational data is being prepared to replace the current MERIT Standards. During the past two years there have been further improvements in the observational results from the three principal techniques used in the service (mainly VLBI, SLR and LLR), although there is not yet a full set of LLR stations in regular operation. Forecast values, as well as synoptic values, of the angular momentum of the atmosphere are now provided by meteorological centres for use in the prediction of the short-term variations of UTI with respect to UTC. Particular attention has been given to the development of new conventional terrestrial and celestial reference systems for use by the Service, which will determine and publish estimates of the relationships between these new systems and the various reference systems now in current use.

REPORTS BY OBSERVING TECHNIQUES

Report from VLBI

The IRIS-A Network, formed by observatories in the United States, the Federal Republic of Germany and Sweden, represents the state-of-the-art Earth Orientation Monitoring System. At 5-day intervals the X and Y components of Polar Motion and the nutations in obliquity and longitude are determined with an accuracy of 1 to 2 mas, and UTI is determined with an accuracy of 0.05 to 0.10 milliseconds. At daily intervals, UTI values accurate to 0.10 milliseconds are produced. The IRIS-A observations have also been used to develop a northern hemisphere Celestial Reference Frame with a precision of a fraction of a mas, and a Terrestrial Reference Frame accurate to a few centimeters. The IRIS-A results are distributed monthly by bulletin, and are available in a more timely manner by direct access to computer files maintained at the National Geodetic Survey, the network center.

Since April 1987 a second network, the IRIS-P network, consisting of observatories in Japan and the United States had conducted monthly observing sessions to determine earth orientation. The International Latitude observatory at Mizusawa (ILOM) serves as the IRIS-P network center. As the network matures in the next few years, it will provide the redundancy needed to insure the continuity of the VLBI- based Earth Orientation Service, and contribute to the global distribution of the Terrestrial Reference Frame.

Joint VLBI projects were conducted in January and February, 1984. Over five hundred observations were successfully performed in the two experiments. The correlation processing was made by both K-3 processor at Kashima and Mark-III processor at Haystack. The two baseline-lengths between Kashima and Mojave obtained in the two experiments showed excellent repeatability. The precision of the baseline-length was higher than 0.02m in root mean square (Takahashi et al., 1985). Observations of plate motions and regional crustal deformations obtained using space technologies of Doppler observations of Navy Navigation Satellite, Satellite Laser Ranging (SLR) and VLBI were reviewed (Fujishita, 1985).

Descriptions were given for the astronomical, geophysical and geometrical models and formulae adopted in KAPRI, a program for computing the theoretical delay and delay-rate of geodetic VLBI observations and their partial derivatives with respect to physical parameters of interest. In KAPRI more sophisticated theories were adopted for the tidal and relativistic effects (Manabe et al., 1984).

A program was developed for converting a HP 1000 based Mark-III database archive to a file on MELCOM COSMO 900II. The program enables us to use VLBI observation data for various calculations and analyses on MELCOM COSMO 900II (Ishikawa, 1985).

A third network, referred to as the IRIS-S network, has conducted two short observing campaigns during January and February 1986 and 1987. In this network the IRIS-A observatories are teamed with the Hartebeesthoek Observatory, South Africa, to produce earth orientation values.

Report from Laser Ranging

The last three years have been seen an improvement in LLR data quality and the development of a Lunar Laser Ranging Network. Recent equipment and software improvements at the stations have resulted in 3-5 cm ranges (the previous data prior to 1984 had ranges of 10 cm accuracy); data are currently being acquired from three stations: CERGA (France), Maui (Hawaii-USA) and McDonald (Texas-USA). An analysis of the seventeen-year LLR data set yields a value for G/M Earth of 398 600.44 +/-0.006 km^3/s^2 in the geocentric system comparable to LAGEOS (the LLR result agrees with the LAGEOS result within one standard deviation of the error estimate). The increased accuracy of LLR should result in an improved lunar ephemeris; preparations are now under way for a new joint lunar and planetary ephemeris at JPL.

Report from Satellite Laser Ranging

During the period from 1984-1987, Satellite Laser Ranging (SLR) continued regular contributions to earth rotation that began with the short MERIT campaign in 1980. At the conclusion of the Main MERIT Campaign in October, 1984, SLR data had been collected at over 30 sites and the Operational Center had provided weekly solutions of x, y and LOD based on quick-look data. Full-rate data were used by the Designated Analysis Centers in the USA, the FRG and the GDR. Contributions were also provided by Associate Analysis Centers located in France, Italy, Japan, PRC, UK, and USSR. Comparison of results showed agreement between the SLR analyses and the VLBI technique at the 2 marc sec level for x and y.

Since the Main MERIT Campaign, SLR has continued to provide contributions to Earth rotation through both rapid (weekly) determinations and annual analyses. The rapid service began providing three-day determinations of x, y and UTI on a preliminary basis in 1985, and regular distributions began in mid-1986.

Because of the global nature of SLR and the high precision of many of the contributing SLR systems (1-5 cm), the effects of global tectonic motion have been readily apparent. Furthermore, the influence of unmodeled plate motion was shown to produce a secular drift in the pole position at the rate of about 1 marc sec/yr. Consequently, the definitions of the reference systems have become increasingly important. In some of the analyses, plate motion models have been adopted, whereas other analyses have simultaneously determined plate motion parameters.

SLR will continue its contributions in the International Earth Rotation Service. It is expected that improved measurement precision and geographic distribution of SLR systems, as well as improvements in the data analysis techniques, will take place in the next four years.

Report from Optical Astrometry

1)   Organizational Structure

During the main campaign of Project MERIT, the Central Bureau of the international Polar Motion Service at the International Latitude Observatory of Mizusawa in Japan, and Bureau International de l'Heure at the Paris Observatory acted as operational and designated analysis centers. Also the Shanghai Observatory was an associate analysis center.

2)   Stations and Instruments

The total number of stations and instruments, which participated in the main campaign are 62 and 85, respectively. Among them, 63 instruments of 52 stations provided time data, and 62 instruments of 49 stations provided latitude data.

3)   Statistics of Observation

On the average, one star was observed every 2.0 minutes for time, and every 1.6 minutes for latitude. The best instrument showed precisions in 2-hour session observation, +/- 0.0049 s for time and +/- 0.059" for latitude.

4)   Earth Orientation Parameters (EOP'S) and their Precisions

Two designated analysis centers, namely BIH and IPMS, provided optical astrometry EOP's. BIH series of EOP's are 5-day raw values with formal errors of +/- 0.007" - +/- 0.016" in Polar Motion, and +/- 0.00075 - +/- 0.00125 in UTI. IPMS Series of EOP's are smoothed daily values with formal errors of +/- 0.003" in Polar Motion, and +/- 0.00025 in UTI.

5)   Comparison with the EOP's of New Techniques

Systematic differences of optical astrometry EOP's with respect to VLBI and SLR results are: about 0.01" and 0.00035 in both annual and semi-annual components of UTI.

6)   Terrestrial Reference Systems

Both BIH and IPMS devised their own methods to estimate the variations of the station coordinates of all participating stations together with the EOP's. These methods obviously improved the EOP's.

7)   Situations After the Main Campaign

After the Main Campaign, many optical astrometry stations closed operation. Hence the optical astrometry network is now far from homogeneous in station distribution. A working group was set up in the Commission 19, chaired by Dr. Ye of Shanghai Observatory, to consider the future of optical astrometry.

8)   Rereduction of Past Optical Astrometry Observations

One of the serious error sources in optical astrometry technique is catalog errors. In consideration of the progress of Hipparcos Project, it will be extremely significant to rereduce past observations based on star catalog obtained by Hipparcos. Every station is requested to prepare machine readable data of the past records of observations.

REPORTS CONCERNING DIFFERENT AREAS OF RESEARCH

Reference Systems for Earth Rotation: Concepts, Definition, Realization

The definition and properties of the "non-rotating origin", (NRO) as introduced by Guinot (1979), have been investigated as well as the implementation of the corresponding conceptual definitions of the "stellar angle" (representing strictly the sideral rotation of the Earth) and of UTI. A ⊖/UTI relation has thus been proposed as a primary conventional definition of UTI with no appreciable step of UTI and d(UTI)dt when the model of the pole trajectory is improved.   The use of the NRO has also been shown to simplify the transformation between the terrestrial and celestial reference systems: Capitaine and Guinot (1986), Capitaine, Guinot and Souchay (1986).

The conceptual and conventional definitions of the Celestial Pole have been investigated and clarified: Capitaine, Williams, and Seidelmann (1986), Captaine (1986).

The relations between the Celestial Ephemeris Pole and several axes considered in the Earth tides have been investigated: Sevilla and Camacho (1986).

Trigonometric series to calculate the instantaneous vector of the Earth's motion with respect to the barycenter of the Solar System were derived in (Vondrak and Ron, 1986). These series, needed to calculate annual aberration, are based on Bretagnon's theory of planetary motions.

Time variations of relative positions of celestial radio sources were investigated to estimate errors on group delay observations for geodesy and geophysics with a very long baseline interferometer. It was found that systematic errors exist between the second and the third catalogues. There is no time variation exceeding 30 milli-arc-second between mean observation epoch 1975 and 1981 (Fujishita 1984B).

Practical realization of the celestial reference frame used for the reduction of the PZT results at Ondrejov was substantically improved by Ron and Vondrak, (1985).

A catalog of PZT stars was compiled by Sato et al. (1984). The calalog includes all the stars used by PZT observations at the ILOM (Mizusawa) and the United States Naval Observatory (USNO), Washington, D. C. since 1959, respectively. The stars of proposed programs for the International Latitude Service (ILS) chain by PZT s on the line of $39^{0}8'N$ latitude were also included. This catalog would give a standard for the recalculations of past observations and will be suitable for the initial observations by the ILS instruments.

A unified catalog of Washington and Mizusawa PZT stars was constructed by using recompiled and reduced data spanning from 1954 to 1983. The reduction is made based on the MERIT Standards. Computations of the right ascension and declination corrections, as well as their proper motions were based on the least squares chain method. Internal precisions of the right ascensions and declinations at the mean epochs of observations were 0.6ms and 0.008, respectively, for the best determined stars. Those of proper motions were 0.08ms/year and 0.001/years, respectively (Menabe et al., 1984).

New catalogue of declinations of latitude stars has been compiled at Kazan Observatory. Catalogue contains 2845 star declinations with mean error 0.072 arcsec (Urasina I.). It was used for reduction of 19 years of latitude observations (41.044.019). Set of coordinates of SLR stations was derived by using Kiev Geodynamics Program Complex (41.045.032., and Proc. IAU Sym. 128).

Determination of Earth Rotation Parameters

The operation of the PZT of the Deutsches Hydrographisches Institut (DHI) was discontinued on 1 July 1986. The observational data from 1962.0 to 1986.5 (new IAU constants, PZT catalogue of 1982) are available on magnetic tape in the usual BIH/IPMS format.

A comparison was made between variations of strain, which are observed at the Esashi Earth Tides Station and UT2-TAI, published by BIH, residuals of time, UTO-UTC, and latitude of the PZT and Danjon Astrolabe at Mizusawa since June 1979. Kakuta and Sato (1985) obtained a good correlation between variations of the Earth's rotational speed and UT0-UTC of the Mizusawa PZT with the east-west component of strain.

Residuals of astronomical observations of longitude and latitude in Asia might be related with the secular variations of the z-component of the internal geomagnetic field in the 1970's near Southeast Asia. Local variations for periods of several years in longitude and latitude might be attributed to thermal expansion in the fluid core near the core-mantle boundary (Kakuta, 1985).

Relative variations of UTI-TAI, Mizusawa to Washington are found to show a secular variation associated with a periodic variation during the period of 4 years. One branch of the secular variation shows a similar direction as the plate motion (Kakuta et al., 1986).

Decreases in the relative variations of UTI-TAI, Mizusawa to Washington were studied from the point of a global motion of the ocean, the El Nino events (Kakuta, 1986).

The Hydrographic Department of Japan (JHD) has been carrying out Satellite Laser Ranging (SLR) observation at Simosato Hydrographic Observatory since 1982. The mean ranging precision is about 9 cm. The numbers of return signals obtained in the years 1984 to 1986 are 483,000 of 744 passes for Lageos, 102,000 of 318 passes for Starlette and 140,000 of 361 passes for Beacon-C. Since August, 1986, the ranging to the Japanese geodetic satellite Ajisai also has been made with 139, 000 return signals of 169 passes in 1986.

A transportable SLR system is under construction at JHD, the completion being expected at the end of October, 1987.

The Royal Observatory of Belgium had maintained the activities of the Tranet Station on a daily basis. Data were provided for the monitoring of the Polar Motion and generation of precise ephemeris extensively used for the determination of the Terrestrial Reference Systems. The station will be maintained in operation till the closure of the Tranet Network.

Accuracy of Earth-rotation parameters in different frequency bands as determined by different techniques of observation was estimated by Vondrak (1984b and 1986). It was found that the difference in accuracy of different techniques are present mainly in the highest frequency band while all the techniques studied are nearly equivalent in the lowest frequencies.

New reduction of latitude observation at Pulkovo from 1904 to 1986 was made for studying secular polar motion (V. Sakharov, L. Kostina) and nonpolar variation (L. Kostina and N. Persianinova). Authors showed some arguments on the reality of secular polar motion. Six years of latitude and time observations carried out by Soviet Observatories from 1978 to 1984 have been rereduced using new system of constants and original software. The work resulted in new determination of ERP and Station Coordinates (D. Belotserovskij, A. Korsun, et al.).

Theoretical Studies of the Earth

At the University of Louvain Neuve, new theoretical nutation series have been computed by adding the inelasticity of the mantle to the Earth model adopted by IAU. It has been pointed out that mantle inelasticity increases the discrepancy between the observations and the theory. In the same group, recent researches show that an additional pressure effect at the core mantle boundary and a change of the flattening of the core could be responsible of the lag between theory and observations (Dehanv, 1987).

The rotational and tidal deformation of tri-axial elastic Earth was theoretically studied by Vondrak (1984a). It was found that the principal axes of inertia exhibit big complicated motions with respect to the mean Tisserand axes; the most pronounced motions are diurnal and they attain nearly 400 arcsec in case of equatorial axes.

The response of visco-elastic tri-axial Earth with fluid core and equilibrim oceans to atmoshperic excitation was theoretically studied by Vondrak (1987).

Motion of the deformable Earth were investigated from the point of view of a frequency-dependence of Q. Roles of statistical models were clarified to estimate parameters of the Chandler wobble and admittances of solid and ocean tides (Ooe, 1985). Then importance of Kalman filtering was pointed out in order to separate system noises and observational errors in the polar motion (Kaneko, 1985).

Polar motion for an elastic Earth Model using a canonical theory (Sevilla and Romero (1986). Variation of the inertia tensor for deformable models were studied by Camacho et al. (1986a). Global and core inertia tensor components for a classical Earth model were studied by Camacho et al.(1986).

Interpretations of the Earth's Rotation Irregularities

Long-period behaviour of polar motion between 1900.0 was studied by Vondrak (1985). The secular motion of the pole was confirmed and the stability of annual term shown in contrast to large changes of Chandler wobble both in amplitude and phase. The latter problem was further studied in a greater detail by Vondrak (1987b), where the polar motion data beak to 1860 were used. The strong correlation between the amplitude and frequency of Chandler wobble was found.

Earth rotation and Earth tides

Tidal effects and Q of the solar system were studied. Specially, tidal Q of Mars were computed by using a recent value of the secular acceleration of Phobos. Q of the solid earth also was investigated using results given by the Earth tides and the Chandler wobble. Q of the lower mantle of the Earth turned out to be almost consistent with tidal Q of other terrestrial planets and satellites (Ooe, 1986).

Complementary studies of the tidal waves of UTI confirm that the fortnightly wave M is altered and leads therefore to derived values of the Love number K which vary with time: Capitaine and Guinot (1986).

Dr. P. Brosche: Reconnaissance studies of the role of oceanic effects in the Earth rotation (with J. Sundermann, Hamburg). Search for such effects in VLBI-observations (with J. Campbell, Bonn) are so far not successful, but success may be ante portas because experiments with 1^h or 2^h resolution became available just now. Updating of reviews on the tidal friction process.

Data analysis was conducted to identify a near 50-day peak in the Solar Activity. Its possible relation with similar fluctuation in ERP and AAM has still to be identified (Djurovic and Paquet, 1987).

Crustal movement and the Earth's rotation

The variation of the seismic activity is well represented by that of the earthquake energy release. It was found that the refraction points in the pole path coincide with the active periods of earthquakes and the direction of refraction in the path was related with that of the excitation function estimated from the earth quakes in the most ackre region (Onodera, 1984).

Atmospheric and oceanic motion and excitations of the Earth's rotation

Onodera (1984) investigated the southern oscillation, one of large-scale atmospheric phenomena, in relation to variation of the rate of the Earth's rotation. The results showed that the variation of the former appears by 2-3 years in advance of the variation the latter. This fact suggested that there is some correlation between the two variations.

Fluctuations of the Earth's rotation with the period shorter than about 2 years are analyzed by using the data compiled or obtained at the BIH, the IPMS, the IRIS (International Radio Interferometric Surveying) and the CSR (Center for Space Research). The IPMS optical astrometry data show close agreement with the AAM (Atmospheric Angular Momentum) data which include the effects of both the zonal wind and the migration of the air mass (pressure term) in the period range biennial-40 days. Relatively high spectral power density is observed at the periods of around 35 days, 24 days, and in the period range of 21 to 16 days in the astronomically observed UTI by the CSR and the IRIS. Thus, the short period fluctuations of the Earth's rotation can be explained fairly well by the atmospheric excitations for the period range between biennial and 16 days (Hara and Yokoyama, 1985).

S. Molodenskij studied dynamical theory of polar tide (Physics of the Earth, Vol. 6, P. 3-18, 1985 in Russian) and M//Nutation (in book Tide, Nutation and Internal Structure of the Earth, Moscow, 1984, 213 pp.). Many new results on the investigations of ERP and tides have been reported at Second Orlov Conference in Poltava, 1986 (Proc. will be published in Kiev, 1988). V. Sakharov and L. Kostina studied Chandler Wobble for the period of 1846-1974 (Soviet Astronomy, Vol. 61, N. 6, 1984). A. Korsun studied the nature of nonpolar latitude variation (39.044.001), tidal effect (38.044.013), and correlation of secular latitude variations of Ukiah and Gaithersberg stations with Greenland ice melting (together with N. Sidorenkov). Correlations of -0.69 and -0.54 were found respectively. N. Sidorenkov proposed new interpretation of atmoshperic exitation of Polar Motion (Proc. IAG Symp. in Prague, 1986).

Longer period fluctuations in Earth rotation, the so-called "Decade Variations" can almost certainly be attributed to torques between the core and mantle. Core-mantle torques are probably largely due to dynamic pressure forces associated with core motions acting on topographic undulations of the core mantle boundary and the equatorial budge. Estimates of such torques are becoming availabe from a combination of core motion models from geomagnetism and core mantle boundary maps from seismic tomography. Geodetic torque estimates provide a strong constraint on the models and assumptions used and strongly favor the inclusion of the D" layer in the mantle and point to "bumps" on the core-mantle boundary of about 1/2 km. The magnitude of these undulations is in agreement with the findings from nutation studies.

20. POSITIONS AND MOTIONS OF MINOR PLANETS, COMETS AND SATELLITES
POSITIONS ET MOUVEMENTS DES PETITES PLANETES, DES COMETES ET DES SATELLITES

PRESIDENT: Y Kozai
VICE-PRESIDENT: Yu V Batrakov
ORGANIZING COMMITTEE: K Aksnes, J-E Arlot, M P Candy, L Kresak, L K Kristensen,
B G Marsden, R L Millis, E Roemer, D K Yeomans, J-X Zhang

I INTRODUCTION

In the past triennium members of the Commission 20 have been very active in studying positions and motions of minor planets, comets and satellites including rings by observational and theoretical investigations, as it is described in this report. In fact observers have been producing tremendous amount of astrometric data, much more than we could imagine twenty years ago, when we heard many complains and appeals to observers for needs of more observations, particularly, for minor planets and satellites. Technically, several new devices to detect faint objects and to measure their positions more effectively, have been developed. Theoreticians have faced many interesting and important problems to explain observational facts with several powerful methods developed recently.

For minor planets as observers produced over 20 000 observations each year, 300 orbits were determined by combining them with those of unnumbered minor planets and the number of the numbered minor planets is more than 3700(October 1987) as 200 minor planets were given their permanent designations annually. Although three lost minor planets were recovered at the end of 1986, three are still lost. Theoretically, many papers tried to explain the causes of Kirkwood gaps and behaviors of so-called secular commensurability regions, among others, as they are still very important problems to study dynamical stabilities and the distributions of minor bodies in the solar system.

For comets P/Halley, particularly, had attracted great attentions, as 6000 astrometric positions were obtained by 140 observatories. In the period of July 1984-June 1987 14 short-period and 18 long-period comets were discovered and 32 previously known comets were recovered. Since several papers on relations of possible periodic impact events and biological extinctions were published, many authors restudied dynamical evolutions of cometary orbits, particularly, by investigating perturbations by known stars and molecular clouds and by tidal forces of the Galaxy in the Oort clouds.

As Voyager spacecrafts discovered several faint satellites and ringlets, with very peculiar characteristics, of Saturn and Uranus, important contributions were made to explain their distributions and dynamical structures. Even for bright satellites several new theories of motion were published by using many new observations. There are still several important problems to be solved, particularly, their orbital evolutions to the present configurations of satellite systems.

Since the importance of occultations of minor planets, satellites and rings has been increased, the Commission 20 reorganized the Working Group on Occultations to try to coordinate observational activities and to channel useful informations to observers quickly. There have been several successes including determinations of the size and the shape of the minor planet, (1) Ceres.

As the activities of the Commission were increased, the number of its members was increased beyond 150 during the previous General Assembly at New Delhi. However, it is a sad news to have heard that Prof. Zhang Yu-Zhe, one of the oldest members of the Commission 20 and formerly the director of the Purple Mountain Observatory, was deceased on July 21, 1986 at the age of 83.

This report consists, besides the Introduction, of 4 Sections, namely, on Minor Planets, Comets, Satellites and Occultations, which were drafted, respectively, by B G Marsden, the director of the Minor Planet Center, and E Roemer, J-E Arlot and R L Millis, who are chairmen of Working Groups on the respective subjects, and was compiled and edited by Y Kozai, who also wrote the Introduction.

II MINOR PLANETS

a) General

Astrometric and orbital work on minor planets has been maintained at very high level of activity during the 1984–1987 triennium. The two publications sponsored by the Commission 20, the *Minor Planet Circulars*(MPC) and the *Efemeridy Malkh Planet*(EMP), continue to be the most visible indicators of this activity. The monthly batches of MPCs are issued by the Minor Planet Center at the Smithsonian Astrophysical Observatory under the direction of Marsden with assistance from Bardwell, Green and, since mid–1986, Nakano. The annual EMP volumes are published by the Institute of Theoretical Astronomy(ITA), Leningrad, under the direction of Batrakov with assistance from Ashkova, Vinogradova, Izvekov, Sumzina and Shor. Kastel' and Filippova also participate in the ITA work on minor planets.

The now rather stable annual production of about 1000 pages of MPCs has again included between 20 000 and 25 000 astrometric observations of minor planets. Further editions of the complete set of available observations on magnetic tape were issued in 1985 and 1986. The 1986 edition contained 275 636 observations of numbered and 101 179 of unnumbered minor planets; the latter included the 15 283 observations from the Palomar–Leiden survey that were made generally available for the first time in the 1985 edition of the tape and that had never appeared in print. The plan is that future editions of the complete tape will be issued less frequently, because all the observations published in the MPCs subsequent to the preparation of the 1986 tape have been made available on diskettes in MS–DOS format issued in conjunction with the printed batches of MPCs. The MPCs contain annually more than 1000 elliptical orbits computed, when there exist observations on three or more nights at a single opposition. There are also nowadays almost 300 orbits annually where observations of unnumbered minor planets can be linked at different oppositions, as well as almost 200 that refer to minor planets being newly given permanent numbers; improved orbits are also included each year for 100 already–numbered minor planets. The aforementioned MS–DOS diskettes contain, in addition to the observations in each batch of MPCs, all the pages that include orbital data.

In September 1986 the Minor Planet Center published editions of the "Catalogue of Orbits of Unnumbered Minor Planets" and the "Catalogue of Discoveries and Identifications of Minor Planets". With entries for 7095 objects the new orbit catalogue was almost three times larger than the previous 1982 edition, although almost half of the increase was due to the incorporation of all the orbits from the Palomar–Leiden survey. The number of multiple–opposition orbits, 855, was more than four times that in the 1982 edition. The discovery listing in the companion volume contained reference to 40 754 provisional designations (plus 145 P–L objects), compared to 29 157 in the 1982 edition, while the count of identifications with numbered minor planets had increased from 8550 to 12 580. The 7095 orbits were also issued on an MS–DOS diskette, as were the orbits for the 3495 minor planets numbered by that time, the latter having been updated to the standard epoch, July 24.0, 1987, and for the most part prepared by the ITA. By applying a computer program provided with the MPC diskettes users can collect in the same form all the additional orbits (for numbered and unnumbered objects) contained month by month in the diskette edition of the MPC orbit pages. Both the MPC diskettes and the orbit-catalogue diskettes also include computer programs for ephemeris computations.

The "dial–in" computer service, operated around the clock by the Minor Planet Center in association with the Central Bureau for Astronomical Telegrams, has continued to gain in popularity since its introduction in 1984, and it is now also supported by a back–up computer when the principal computer is unavailable. The user can select orbital elements and compute ephemerides, and a useful new feature, introduced in 1986, allows him to verify if particular observations can be identified with numbered minor planets. Observers are also increasingly finding the computer service convenient for submitting their observations to the Minor Planet Center, and the Center has responded by quickly providing observers with identifications or designations for their new discoveries.

The EMP volumes were published for 1986, 1987 and 1988, successive volumes increasing in size as more new minor planets are numbered. The 1988 edition contains 366 pages. The EMP pages are now prepared with the help of a computer program that utilizes a specialized computer typesetting system devised by Glejbman. In addition to the actual production of ephemerides much of the computing work involves the annual updating of the osculating elements to later epochs, and these updated elements are now also available in machine–readable form through the Stellar Data Center, Strasbourg(41.002.068). ITA staff members have computed 385 improved orbits for EMP 1986–1988 and already computed 150 for EMP 1989. The standard accuracy is now such that the positions of the great majority of the numbered minor planets can be predicted within a few seconds of arc, and less than 0.5 percent of the orbits yield ephemerides that are in error by more than 20 arcsec. ITA also prepares and distributes to participating observatories daily ephemerides for the 20 selected planets used in the ongoing program for examining systematic errors in star catalogues.

Implementation of the resolution by the Commission 20 at the 1985 IAU General Assembly concerning the new two-parameter magnitude system for minor planets was begun in January 1986. The first step consisted of the use of the single slope parameter, $G=0.25$, and stop-gap values of the visual absolute magnitude, H, derived from the old photographic value, $B(1,0)$, by $H=B(1,0)-1.0$. Tedesco of Jet Propulsion Laboratory made a preliminary analysis of the photometric data in his possession and prepared a new set of parameters (utilizing additional fixed values of $G=0.15$ and $G=0.40$ but in 237 cases making actual solutions for G) for minor planets, (1)-(3318). After some minor modifications and systematic extension (again with the single value $G=0.25$) to the minor planet (3495) by Marsden, these parameters were published in the MPCs in September 1986. The system has subsequently been extended in the same way to new minor planets when they are numbered (although for unnumbered minor planets the values of H are generally rather rough values given only to 0.5 magnitude). The values of H and G given in the EMP for the first time in 1988 are identical with those in the MPCs. It should be noted that, contrary to former practice, the predicted magnitudes for the fourth date in the standard opposition ephemerides in EMP 1988 do now include the phase effect.

b) Observations and Orbits.

Photographic observing programs that include an emphasis on the discovery of new minor planets have continued with wide-field instruments at Brorfelde(Jensen), Caussols(Heudier), Cerro El Roble(Torres), El Leoncito(Sanguin), Flagstaff(Bowell), Kiso(Kosai), Klet(Mrkos), Kvistaberg(Lagerkvist), La Silla(Debehogne, Elst and Ferreri), Nanjing(Zhang), Nauchnyj(Chernykh), Palomar(Helin, Shoemaker and Shoemaker), Perth(Candy), Smolyan(Elst and Shkodrov), Tautenburg(Borngen) and Zimmerwald(Wild), as well as by amateur astronomers, Seki, Urata and others in Japan, and Colombini, Baur and associates in Italy. New programs have been undertaken at Haute Provence(Dossin and Elst), Kavalur(Rajmohan), Pizskesteto and Piwnice(Antal) and Siding Spring(McNaught). Extensive photographic follow-up observations of unnumbered minor planets have been made as in the past, in particular, at Oak Ridge(McCrosky) and Mount John(Gilmore), and numerous photographic observations have been made at Victoria(Tatum and Balam), as well as by amateur astronomers in Japan, Australia(Herald) and the United States (Handley). Bus continued his extensive "UCAS" survey of March-April 1981 by finding and measuring images of some 80-percent of the objects in February and/or May.

Several of aforementioned programs, particularly, those at Flagstaff, Klet, La Silla, Nanjing, Nauchnyj, Perth and Zimmerwald, also yield many observations of numbered minor planets. Several USSR observatories(Abastumani, Dushanbe, Goloseevo, Kazan, Kharkov, Kitab, Moscow, Nauchnyj-Pulkovo Southern Station, Nikolaev, Pulkovo, Saratov, Tartu, Tashkent, Zvenigorod and Zelenchuk-Kazan Southern Station) concentrate on numbered minor planets, particularly, the ITA program of 20 selected objects. Other observatories involved with numbered minor planets are Asiago(Ferreri), Barcelona(Codina), Bordeaux(Rapaport), Cape Town(Churms), Cerro Calan(Wroblewski), Chorzow(Wlodarczyk), Hoher List (Geffert), La Palma(Morrison), Lick(Klemola), Pino Torinese(Zappala), Poznan(Hurnik), Quonochontaug (Penhallow), San Fernando(Quijano), Skalnate Pleso(Svoren) and Yebes(de Pascual), and there is also participation by amateur astronomers in England, France, West Germany and Italy(Balbi, Bonk, Bressler, Buczynski, Chanal, Lai, Manning, Quadri and Seiler). Quijano specifically reports that the ITA objects have been under observation with the Carte du Ciel astrograph at San Fernando, as have the 63 minor planets being observed in connection with the Hipparcos project, a total of 1062 oppositions having been obtained during the triennium from 354 plates.

CCD astrometry on minor planets, a pioneering activity at the time of the last report, has again been pursued during the current triennium in the same three pioneering programs, namely, with the Flagstaff 1.7m reflector(Bowell), the Palomar 1.5m reflector(Gibson) and the Steward Observatory's 0.9m reflector on Kitt Peak(Gehrels and Scotti). The principal problem with CCD astrometry is the expense of large CCD chips, so the available fields are even smaller than when large reflectors are used for photographic astrometry, and field plates are normally necessary for the reduction, a time-consuming process. The chip used in the Steward program, for example, yields a field only 11 arcmin across; this particular program(Spacewatch), however, utilizes the earth's rotation to make long scans in right ascension around the minor planets of interest, and it can be assured that there are enough SAOC reference stars to permit the reductions, which have frequently been completed on the same day when the observations are made. Another source of quick but reliable positions has been the 2.2m reflector at Mauna Kea, where Tholen has simply but effectively made use of the telescope's encoders to derive positions of minor planets relative to SAOC stars.

As in the past three members of the Commission 20, Bowell, Edmondson and Oterma, have continued to cooperate extensively in the measurement and sometimes remeasurement of old plates at their respective observatories(Flagstaff, Indiana and Turku). Most of the requests for measurement are channeled through the Minor Planet Center and are in response to identifications that have been suggested. West has used the facilities of the European Southern Observatory to remeasure some of the old Heidelberg plates, notably those of (473) Nolli, (719) Albert, (1026) Ingrid and (1179) Mally, lost minor planets observed only at their discovery oppositions. Following the remeasurements for (1179), Schmadel redetermined the orbit allowing the object to be found (the first recovery of a lost numbered minor planet since 1982) on a pair of plates taken by Schuster at its favorable opposition in 1986; the identity was confirmed by West's subsequent discovery and measurement of images on the Palomar Sky Survey and on exposures with the ESO and UK Schmidts. Inspired by this news in December 1986, but without the new Heidelberg measurements, Nakano was able a few days later to identify (1026) with an object observed on two nights earlier in 1986 and with single observations in 1981 and 1957. West's new remeasurements of 1901 plates of (473) had a pronounced effect on the orbit determination, and on receiving them on New Year's Eve, Marsden was quickly able to find identifications for this object in 1940, 1981 and 1986; although there was not a recognizable similarity of elements, a rather poor elliptical orbit had in fact actually been published for the identity 1981 QR. The Heidelberg discovery, (719) as well as (724) Hapag and (878) Mildred, remain lost. Gibson has remeasured the original Mt. Wilson plates of (878), but this object is evidently too faint to be identified in the existing database of observations of unnumbered minor planets. (724), one of the many visual discoveries by Palisa, remains the only numbered minor planet not known ever to have been photographed.

Multiple-opposition orbit computations have been performed by staff members of both the Minor Planet Center and ITA, by Dirkis(Riga), Hurukawa(Tokyo), Kristensen(Aarhus), Landgraf(Goettingen), Schmadel (Heidelberg) and Yeomans(Pasadena), and also by amateur astronomers in Japan(Kobayashi, Oishi and Urata), Belgium(Goffin) and Canada(Lowe). Most of the orbit computers have also been extensively involved with establishing identifications, as also have Bowell(Flagstaff) and further amateurs in Japan (Furuta), West Germany(Kippes) and the United States(Bowman). Utilizing his capability of working with survey plates taken with the large Schmidt telescopes, West has been able to confirm or deny some of more questionable identifications. Bus has also made use of Palomar Schmidt plates taken by Kowal during 1977–1979 to find images of several hundred UCAS objects at earlier oppositions.

Interest in establishing identifications continues to grow, and Marsden(41.098.017) reviewed the various methods that are used, which nowadays generally range from automatic comparison of observations of a minor planet on two nights with all the available orbits, through the more straightforward consideration of the similarity of two preliminary orbits, to the comparison of variation orbits of a particular object with all the available observations. Some researchers, notably Oishi, have recently sucessfully used as input for the last method starting orbits derived from potential multiple designations of the same object at the basic opposition. Intriguing new possibilities involves cases where there exist observations on no more than two nights at any of the oppositions concerned. In response to a condemnation by Taff(38.042.035) of the Gauss method of orbit determination from three observations, Marsden(40.042.003) demonstrated that Gauss linkages of two observations at one opposition with a series of candidate third observations at a neighboring opposition is (when there is a confirming observation on a second night at the neighboring opposition) an effective way of achieving this. Lowe and Nakano have also subsequently found some identifications of this general type from the similarity of likely two-observation "Vaisala" orbits at different oppositions or by adopting a series of Vaisala computations for the variation orbits. A paper by Laucenieks and Dirikis(*Novejshie Dostizhaniya v Teorii Komet i Dinamike Malykh Tel*, Moscow, 1986) tried to determine the maximum-likelihood region on the celestial sphere at a given moment for an object with an uncertain preliminary orbit.

van Houten reports that results are now almost complete from a new Trojan survey utilizing 85 plates taken in 1977 by Gehrels with the 1.2m Schmidt telescope at Palomar. The plates were blinked at the Leiden Observatory by van Houten–Groeneveld, who also measured the rectangular coordinates at the Catholic University in Nijmegen. Magnitudes were measured by Wisse–Schouten with an iris photometer at Leiden. Orbit computations were made by Bardwell of the Minor Planet Center. Orbital elements have been obtained for 1478 minor planets, of which 26 are new Trojans, four are Hungarias and one is a Hilda. Among the minor planets observed there were also 25 numbered objects and 17 that had previously received provisional designations. A similar survey, utilizing 78 plates taken in 1973, is in progress. Plates blinked by van Houten and van Houten–Groeneveld have been measured by the latter. Their photometry and orbit computations are expected to start soon.

Johnston, Wade, Seidelmann, Kaplan and Carroll have continued their observations of (1) Ceres, (2) Pallas, (4) Vesta and (10) Hygiea with the Very Large Array in Socorro, New Mexico. With other collaborators they made an analysis of the microwave spectrum of (1) Ceres. Of recent discoveries of minor planets 1985PA, 1986JK, PA, WA and 1987MB are Apollos, 1984QA, 1986EB and TO are Atens, and 1985JA, TB and WA, 1986DA, LA, NA and RA are Amors. 1986TO and 1987MB could be identified with 1983UH and 1959LM, respectively.

c) Theoretical Investigations
Bien and Schubart studied the long–period evolution of real and fictitious Trojan orbits by means of Labrouste's method(39.098.032), concentrating, in particular, on orbits close to secular resonances(39.098.030 and 41.098.046); they completed these studies by deriving three characteristic orbital parameters for the numbered Trojans(41.098.045 and *Astron. & Astrophys.*, 175, 292 and 299). Erdi(39.098.031) showed that, depending on the amplitude of the libration about L4, there is a critical inclination, above which the line of nodes advances. Barber(41.098.047) suggested that long–term perturbations by Saturn might explain the observed difference in the numbers of observed Trojans around L4 and L5. Continuing his analytical work on the Trojans, Garfinkel(40.042.009) investigated the kinematics of the family of homoclinic orbits that is doubly asymptotic to short–period orbits about L3. Zagretdinov(41.098.071, *Kinem. Fiz. Nebesn. Tel,* 2, 68 and 77) generalized some of Garfinkel's earlier results to the case of inclined orbits, and obtained an intermediate orbit in the circular restricted three–body probelm that could be used to determine the libration motions of 38 known Trojans(42.098.005). Agafonova and Drobyshevskij(40.098.010) discussed the origin of the Trojans in terms of explosions in icy envelopes of Jupiter's Galilean satellites and the eventual transfer of collisonal fragments of the Trojans into typical short–period comet orbits.

Following Wisdom's pioneering work, Murray and Fox(38.098.013) confirmed the qualitative behavior of chaotic trajectories that arise at the 3:1 commensurability with Jupiter. Wisdom(40.042.057) identified the principal cause of large chaotic zone. Sherbaum and Kazantsev(40.098.007) demonstrated that the eccentricity of an orbit in 3:1 resonance increases to the point of making encounters with Jupiter possible. Sidlichovsky and Melendo(41.098.009) extended the Wisdom method to 5:2 resonance and presented computations that strongly support the collisional hypothesis for gap formation. Continuing their work on the depletion of the outer part of the minor planet belt, Milani and Nobili(39.098.021 and 39.098.029) established the dominant importance of libration and secular resonance effects for stability there, and with Murray they(41.098.044) showed that there are differences in the topology of the level manifolds near the 3:2 and 2:1 resonances, an extensive chaotic region being identified in the former case and to a lesser extent in the latter(42.098.052). Ferraz–Mello(*Astron. J.*, 94, 208) analytically modeled the planar motion of high–eccentricity objects near first–order resonances and deduced approximate laws relating semimajor axis and eccentricity at the libration center and giving proper period of small–amplitude librations around the libration center and the amplitude of the forced oscillation induced by Jupiter's orbital eccentricity; these laws were found to be closely followed by the observed Hilda planets.

In a series of papers on orbital resonances Liu, Innanen and Zhang(39.042.062, 39.042.063 and 40.098.006) established the principal perturbative effects on the orbits of minor planets in resonance with Jupiter. Sergysels and Wettendorff(38.098.086) discussed the gaps in the main belt from the implication from Moser's work that circular orbits at resonances of order up to 4 evolve toward resonances, but they were unable to explain the Hilda group. Henrard and Lemaitre(39.098.028 and 39.028.034) discussed the gaps at 2:1, 3:1 and 5:2 resonances by means of the adiabatic effects of the slow dissipation of the solar nebula at the origin of the solar system. Analysis by Gerasimov(41.098.101, 42.042.085, 42.098.006 and *Astron. Zh.*, 63, 768 and 1215) of orbits near resonances of the first, second and third orders led to the development of analytical expressions for the orbital elements in terms of Weierstrass functions. Babadzhanov, Zausaev and Obrubov(40.098.012) discussed the influence of resonances on the progressive evolution of Apollo, Amor and Aten objects.

Kozai(40.098.008, 42.042.102 and 42.098.078) calculated mean values of the reciprocal of the distances of the minor planets from Jupiter and hence the secular perturbations on the assumption of circular and planar motions of disturbing planets and that the critical arguments are fixed at stable equilibrium points, and concluded that, unlike short-period comets, most minor planets can avoid close approaches to Jupiter by various kinds of mechanisms. Nakai and Kinoshita(40.098.081) examined the secular perturbations at secular resonances by a semi-analytical method with the use of numerical averaging for both non-resonant and resonant objects. Scholl and Froeschle(41.098.042, 42.042.025 and 42.042.049) extended numerical integrations for up to 1 000 000 years at the three secular resonances near the inner part of the main belt, and concluded that strong eccentricity variations can produce direct earth crossers.

Lidov and Vashkov'yak(39.098.088, 41.098.072, 41.098.090, *Kosm. Issled.*, 24, 513 and *Astron. Geodez.*, 14, 22) continued work on secular perturbations, and described a numerical-analytical method for investigating the averaged restricted circular n-body problem and calculated the extreme characteristics of the evolving orbits. Simovljevic(38.098.023) developed an approximate method for determining the mutual perturbations during close encounters between minor planets. Lazovic and Kuzmanoski(41.098.106) have examined occasions when the first four minor planets pass within 0.001AU of other numbered minor planets and found 13 cases of significant perturbations. An orbit improvement by Schmadel(42.098.057) of (29) Amphitrite over 1825-1985 yielded a value of 1047.369±0.029 for the mass ratio of the sun to Jupiter. Scholl, Schmadel and Roser(42.098.059) attempted to determine the mass of (10) Hygiea from its perturbations on (829) Academia.

With an application to the case of (126) Velleda, Arazov and Gabibov(38.098.112) showed that theories of the motions of non-resonant main-belt minor planets can be made using for an intermediate orbit the internal variant of the generalized problem of three fixed centers. Batrakov(39.098.105) developed a maximum-likelihood method for determining the orbit of a minor planet over several oppositions from orbital parameters derived at each opposition separately. Shefer(*Analiz Dvizheniya Nebesn. Tel Soln. Sistemy i Ikh Nabl*, Riga, 1986 and *Astron. Geodez.* 13, 59 and 14, 77) showed that the use of regularized equations of motion and variations can be highly efficient for orbit improvement. Shor(*Byull. Inst. Teor. Astron.*, 15, 593) described the technique used at ITA for refining predictions for occultations by minor planets when last-minute observations are available. Hahn and Rickman and also Benest and Bien(39.098.047, 40.098.028 and 41.098.048) carried out long-term integrations on several minor planets observed between 1979 and 1984 to have cometary orbits. They demonstrated, in particular, that 1983 SA and 1983 XF exhibit temporary librations about the 4:3 and 2:1 resonances with Jupiter.

Zappala, Farinella, Knezevic and Paolicchi(38.098.014) found that, although the general features of the mass distributions of members of Hirayama families can be interpreted in terms of collisional disruption of a parent body and the self-gravitational reaccumulation of the largest remnant, there remain difficulties in the way, by which families are defined(41.098.038), and the abundance of large objects with high angular momenta indicates that the original population of minor planets was several times greater than it is now(41.098.039). Differences between proper elements calculated by the Yuasa theory and the Brouwer and Williams values were discussed with regard to accuracy, particularly, for objects of high eccentricity and/or inclination(41.098.041); in collaboration also with Carpino, Gonczi and Froeschle, the accuracy was also investigated by simulating numerically the dynamical evolution of the families assumed to arise from the explosion of their parent objects(42.098.029). Davis, Chapman, Weidenschilling and Greenberg(39.098.051) simulated a collisional evolution of hypothetical initial populations of minor planets that would yield the observed number of major Hirayama families.

Halling(38.098.012) used a rough model for gravitational compaction of a porous body and the Alfven-Arrhenius spin formula to explain the relationship between the spin frequency and diameter of minor planets. This relationship, derived by Dermott and Murray, has been refined(38.098.085), and the marked rise in rotation rate as the diameter decreases below 120 km seems not to be due to observational selection effects; they felt that the minimum rotation rate at 120 km may separate primordial objects from their collision products. On examining the possible stability of binary minor planets, Whipple and White(39.042.038) found extensive regions of bounded quasi-periodic motions and regions of bounded chaotic motion. Leone, Farinella, Paolicchi and Zappala(38.098.066) established that for objects of known rotational properties there are constraints on the admissible values of mass ratio and density and that such models permit a qualitative description of the expected light curve morphology. Binzel(40.098.013) showed that a binary model can explain the appearance of two distinct periods, 30.7 days and 7.90 hours, in the light curve of (1220) Crocus.

III COMETS

a) Discoveries, Recoveries and Astrometric Observations

During the interval, July 1984 – June 1987, 18 long–period and 14 short–period comets were discovered and 32 previously known short–period comets were recovered. Eleven of the discoveries were made by amateurs, often as a result of deliberate visual searches. All of the 21 discoveries by professionals were made on wide–field photographs; seven of them on films taken by Shoemaker and Shoemaker with the Palomar 0.46m Schmidt in search of unusual minor planets. Four comets were found on plates taken with the UK Schmidt, Siding Spring. Two of the newly discovered comets had initially received minor planet designations, 1984 JD = P/Kowal–Mrkos, 1984n, and 1986 UD = P/Urata–Niijima, 1986o. Another of the discoveries was made by Skiff(*IAUC* 4250) in the course of the UK–Caltech Asteroid Survey extension back to a plate taken by Kowal in 1977. Marsden recognized the identity of Skiff's object with one observed by Kosai with the Kiso Schmidt and recorded under the minor planet designation, 1977 DV3. P/Skiff–Kosai, 1976 XVI, proved to be of short–period with an unobserved return to perihelion in 1984. In additon three more sungrazing comets were found among data from the Solwind coronagraph; Solwind 4, 1981 XXI (observed on November 3/4, 1981, *IAUC* 4129), Solwind 5, 1984 XII (July 28, 1984, *IAUC* 4129 and 4230) and Solwind 6, 1983 XX (September 24/25, 1983, *IAUC* 4229). It is probable that all three are members of the Kreutz group of sungrazers. Perhaps the most interesting and surprising of the newly discovered periodic comets was P/Machholz, 1986e, a Jupiter family member with record small perihelion distance (0.13 AU) and record high incliation (60°).

The main contributors to recoveries of periodic comets were Gibson, who now uses the 1.5m reflector at Palomar with a CCD detector (14 recoveries during the triennium), Gehrels and Scotti of the Spacewatch program, which uses the 0.91m reflector of the Steward Observatory, Kitt Peak, with a CCD detector in a scanning mode (10 recoveries), and Gilmore and Kilmartin, who use the 0.6m reflector of the Mount John Observatory in a photographic mode (5 recoveries). The last program is particularly valuable for its coverage of the Southern Hemisphere. The promise reported at New Delhi General Assembly in 1985 for the Spacewatch system has been well borne out by more than 600 observations of 47 comets since regular operations began in January 1985. A description of the system and method of operation has been given by Gehrels et al.(41.036.172).

Interest in particular comets as potential targets for planned spacecraft exploration again led to unusually early recoveries for several objects, namely, P/Grigg–Skjellerup, 1986m, by Birkle with the 3.5m telescope at Calar Alto, P/Tempel 2, 1987g, by the Spacewatch program and P/d'Arrest, 1987k, by Meech and Jewitt with the 2.1m telescope at Kitt Peak. CCD detectors were used in all three recoveries. Kresak(*IAUC* 4234 and *Bull. Astron. Inst. Czech.*, 38, 65, 1987) identified observations by Pons in February 1808 as belonging to P/Grigg–Skjellerup. The recovery and in fact the only observations of P/Honda–Mrkos–Pajdusakova at its return in 1985 were made visually by Australian amateurs, Clark, Pearce and Athanasou(*IAUC* 4055).

Comets recovered on their critical first returns included P/Boethin, 1985n, P/Russell 1, 1985b, P/Giclas, 1985g, P/Wild 3, 1987e, P/Bus, 1987f, P/Howell, 1987h, and P/du Toit–Hartley, 1986q, which was recovered on the basis of a prediction from observations only in 1982. Four comets, for which predictions for first returns were available, were missed. Perihelion times for P/Tritton and P/Kowal 2 were uncertain by many days, because observed arcs at the discovery apparitions were short and in addition the latter was close to conjunction with the sun at perihelion. P/Haneda–Campos was badly placed and P/ Schuster, which had some chance of recovery from the Northern Hemisphere in late 1985 or early 1986, also was missed. The limited field of CCD detectors presents problems for the recovery of comets subject to substantial positional uncertainty, while the current dedication of the Palomar 1.2m Schmidt to the Sky Survey II limits the availability of a valuable wide–field search capability for faint returning objects. On the other hand the Survey II is producing comet discoveries as did the first survey.

Observations of Comet Bowell, 1980b=1982 I, by Meech and Jewitt in early November 1986 and with the Spacewatch system in December 1986 set new records, as the heliocentric distance at the latest observation was 13.9AU, far exceeding that of Comet Stearns, 1927 IV, when last observed in 1931. Comet Bowell had been under observations for nearly seven years, a record long interval for a non–periodic comet. Nakano(38.103.033 and MPC 12025) identified several more observations of minor planets as actually referring to comets, including P/du Toit–Neujmin–Delporte in 1941, P/Smirnova–Chernykh in 1978 and 1981 and P/Urata–Niijima in 1986 as well as P/Whipple in 1925.

The report interval included the perihelion passages and spacecraft intercepts of both P/Halley and P/Giacobini-Zinner. Encouraged by the International Halley Watch, nearly 140 observatories participated in obtaining some 6000 astrometric observations of P/Halley. Some 28 observatories in the USSR contributed 2000 observations of P/Halley, as well as 600 observations of 22 other comets. More than 500 observations of P/Halley were made at the Perth Observatory alone. In addition to programs already referred to, substantial numbers of observations of other comets including important follow-up on discoveries as well as last observations came from Oak Ridge(McCrosky, Schwartz and Shao; more than 370 observations of some 50 comets), Lowell(Bowell, Skiff and Bus), Chamberlin(Everhart and Briggs), Klet(Mrkos and Vavrova), Geisei(Seki), as well as several other observatories in Japan, Kambah(Herald), Victoria(Balam and Tatum) and Cerro El Roble(Torres).

A new edition of the magnetic tape file of observations issued by the Minor Planet Center in June 1986 contained 25 533 observations of comets. Since October 1986 an IBM-PC-diskette version of the MPCs has been available in addition to the print version. The diskette contains the filed and sorted observations in the format of the magnetic tape as well as data and program to extract orbits in the format of the catalogue and to compute ephemerides. After a lapse Marsden (in some instances in collaboration with Green and Roemer) resumed publications of the RAS annual comet reports with issues appearing during the period of 1976-1984 inclusive(39.103.001-004, .032, 40.103.004, .047, 41.103.001 and 42.103.029). Andrienko and Karpenko continued the series of publications on physical characteristics of comets with data on comets observed in 1976-1980(39.102.005, 42.103.025, 38.103.039, .040 and 41.102.015). Some 41 numbers of the *Kiev Komet. Tsirk.*, Nos. 326-366) appeared during the triennium as Churymov became the chief editor following the death of Vsekhsvyatskij in 1984.

Ostro(40.098.089) reported on observations of four comets detected by radar; P/Encke, P/Grigg-Skjellerup and the two close-approach comets in 1983, IRAS-Araki-Alcock and Sugano-Saigusa-Fujikawa. In connection with spacecraft missions Roser(38.102.072) studied the accuracy of astrometric observations of comets over the past 200 years with special emphasis on the cases of P/Halley in 1759, 1835 and 1910. Svoren(39.103.008) and Pittich and Svoren(40.103.007) published details of positions of comets observed from the Skalnate Pleso Observatory in 1972-1975 and in 1976, respectively. De Sanctis, Ferreri and Zappala (39.103.025) likewise published positions of comets observed at Torino in October 1982 through June 1983. Zook, Fernandez and Grun(40.102.103) studied selection effects working against discovery of small comets. Radzievskij, Mamedov and Ivanov(39.102.050 and 40.102.101), by using the interval of observationability of comets as the best criterion of discovery probability, demonstrated a considerable prevalence of long-period comets with retrograde motion, and also analyzed the distribution of comets with longitude of perihelion and with perihelion distance.

b) Orbits and Ephemerides

As in the past preliminary and improved orbits and ephemerides for newly discovered comets have been determined at the Central Bureau for Astronomical Telegrams/Minor Planet Center(Marsden, Green and, from mid-1986, Nakano) and been published in the MPCs and on *IAU Circulars* as appropriate. Predictions for returning periodic comets have also been published routinely in the MPCs as well as in the *Handbook* of the British Astronomical Association, in the *Comet Handbook* of the Oriental Astronomical Association and elsewhere. Principal contributors for returning comets have been Nakano, Milbourn, Marsden and Kobayashi. Tabulations of orbital elements from new work were included as well in Marsden's RAS annual reports on comets referred above. The fifth edition of Marsden's *Catalogue of Cometary Orbits*(41.002.024), issued in January 1986 and intended to be complete for comets observed up to the end of December 1985, contains orbital data for 1187 cometary apparitions, referring to 748 individual comets, of which 135 are periodic and 85 were observed at two or more perihelion passages. For the first time this edition of the orbit catalogue was also officially issued in machine-readable form, both on magnetic tape and on a VAX/VMS or IBM-PC diskette.

Among the studies of periodic comets that link several apparitions, often with inclusion of non-gravitational forces, were the followings; P/Brooks 2 by Sekanina and Yeomans(40.103.281) and also by Evdokimov and Zaripova(42.103.122), P/Encke by Sitarski(*Acta Astron.*, 37, (1), 1987), P/Grigg-Skjellerup by Sitarski(42.103.235), P/Tsuchinshan 1 and P/Tsuchinshan 2 by Szutowicz(42.103.027), P/Harrington and P/du Toit-Neujmin-Delporte by Forti(41.103.002), and the 'intermediate' comets, P/Pons-Brooks, P/Olbers, P/Brorsen-Metcalf and P/Westphal by Yeomans(41.103.039). Festou, Morando and Rocher(39.103.101), in a study of the motion of P/Crommelin, found that the orbit has been stable over an interval of 1100 years between AD 1000 and 2100 and probably even longer.

Marsden(40.102.055) reviewed the status of recent investigations of non–gravitational effects on the motion of comets and encouraged further studies based on thermal models of the nucleus. More recently Fanale and Savail(41.103.656) have improved their model of cometary gas and dust production with application to non–gravitational force components and nuclear radii, especially for P/Halley. Froeschle and Rickman (42.103.014 and 42.103.015) continued their investigations of non–gravitational forces on short–period comets based on thermal models of the nuclei by considering both low–obliquity and high–obliquity cases.

Sekanina(39.103.183) developed a precession model for the nucleus of P/Giacobini–Zinner, which is one of the 'erratic' comets evidencing irregularities in non–gravitational perturbations due to outgassing. The motion of this comet was also studied by Evdokimov(*Komet. Tsirk.* No. 327), who found a secular retardation prior to 1972 and a secular acceleration from 1972 to 1985. Sitarski(38.102.030) included non–gravitational terms in Marsden's form in recurrent power series integrations of the equations of motion with applications to P/Kearns–Kwee, Kopff, Wolf–Harrington and Grigg–Skjellerup as well as long–period comets, Burnham(1960 II) and Rudnicki(1967 I).

Emel'yanenko(40.102.124) studied the long–term orbital evolution of several comets affected by librations of the critical arguments in orbit resonance with Jupiter and investigated the mechanisms of capture into resonance and ejection from it. Everhart and Marsden(*Astron. J.*, **93**, 753, 1987) calculated 'original' and 'future' reciprocal semimajor axes for 23 new osculating cometary orbits derived during 1982–86. Radzievskij and Tomanov(39.102.028, 40.102.107 and 41.102.056) published a statistical catalogue of the parameters of orbits of long–period comets by using the Laplacian coordinate system in connection with their studies of the origin and orbit evolution of comets. Mamedov and Radzievskij(42.002.104) also published a catalogue of orbital parameters of nearly parabolic comets.

Much effort went into improvements of the orbit of P/Halley in support of the spacecraft intercepts. Leading roles were by Yeomans on behalf of the International Halley Watch and by Batrakov, Belyaev, Medvedev and Chernetenko for the Soviet VEGA project. An independent study was made by Landgraf, who reduced a previously unused series of observations by Lamont in 1836 by extending the period of useful observations at that apparition by more than half a year(40.103.852). Landgraf also studied the effects of modeling a displacement of the center of light from the nucleus(41.103.613), which was also investigated by Sitarski(38.036.054) and was included by Yeomans and Batrakov et al. in their works.

In his comprehensive paper Landgraf(42.103.603) also considered alternate forms for presentation of non–gravitational effects. Sitarski and Ziolkowski(in *ESA Report* SP–250, 1986) looked into causes of the discordance of results obtained by different investigators in the long term motion of P/Halley, and concluded that secular variation of non–gravitational parameters has to be taken into account. Stephenson and Yau(39.103.921) published a comprehensive catalogue of Far Eastern observations of P/Halley and diagrams showing the computed path at all apparitions from 240 BC to AD 1368. Stephenson, Yau and Hunger(39.004.038) also identified observations of P/Halley at returns in both 164 BC and 87 BC among Babylonian texts in the British Museum. Yeomans, Rahe and Freitag(41.103.724) surveyed historical records of P/Halley starting from 240 BC and gave information on observations related to physical properties as well as ephemerides for each apparition.

Comprehensive results of the long–term evolution project were published by Carusi, Perozzi, Valsecchi and Kresak(41.003.008). Integrations of the orbits of 132 short–period comets, including planetary perturbations and extended over a time span of more than 800 years, permit the identification of long term processes such as temporary librations about orbital resonances and patterns at close planetary encounters. Because of limited accuracy of many reference orbits as well as omission of non–gravitational effects, results were not intended to represent accurately motions of individual comets.

In a study complementary to that of Carusi et al., Belyaev, Kresak, Pittich and Pushkarev compiled a *Calalogue of Short Period Comets*(Astron. Inst. Slovak Acad. Sci., Bratislava, 1986) based on new integrations of the motions of all known short–period comets. In addition to planetary perturbations non–gravitational effects were included when possible (25% cases). Intervals of integrations, chosen to reflect the accuracy of initial orbits, range from 200 years for objects observed repeatedly to 50 years about the time of appearance of single–apparition comets. Additional information includes geometric observing conditions for each perihelion passage of individual comets, observational histories, lists of osculating elements at uniform 25 year intervals, close planetary encounters and the like. A variety of additional information is included in appendices.

c) Theoretical Investigations; Origin and Evolution.

Dynamics of Comets(ed. by Carusi and Valsecchi, 40.012.034), the proceedings of the IAU Colloquium No. 83, contains many papers reflecting current works on various aspects of cometary dynamics. In addition a comprehensive review of current views and topics of research interest in cometary dynamics was published by Weissman(40.102.112).

Ideas related to the origin of comets were reviewed by Bailey, Clube and Napier(41.102.051). Following an historical introduction they discussed the dominant issues of the middle years of this century and then surveyed the ideas that have arisen during the past decade, particularly, those related to the realization that the outer parts of the Oort cloud are not stable in the presence of galactic influences. Clube and Napier(39.102.064) considered comet formation in molecular clouds and the possibility of repeated cycles of capture and escape of comets as the solar system orbits in the Galaxy. Tomanov(37.102.011) discussed an interstellar origin of comets including condensation of nuclei in dense dust-gas clouds, capture by the solar system and planetary perturbations of cometary orbits. Hills(42.102.070) looked into formation of comets by radiation pressure in the outer part of a collapsing protosun, while Bailey(*Icarus*, **69**, 70, 1987) concluded that a high space density of comets could arise in molecular clouds through formation of comets in wind-driven shells around protostars. Weissman(41.102.008) addressed the question of primordial vs. episodic origin of comets as related to planet formation, by noting that primordial theories have generally gained wider acceptance. Festou(42.102.063) published a paper on origin of comets, and Oort(42.102.061) gave the Halley lecture for 1986 on the subject of the origin and dissolution of comets.

Weissman(in *The Galaxy and the Solar System*, ed. by Smoluchowski, Bahcall and Matthews, U. Ariz. Press, 1986) reviewed the dynamical effects of the galactic environment on the Oort cloud. Bailey(38.102.026) also studied origin of comets on the basis of structural characteristics of the Oort cloud, and concluded that a choice between a steady-state primordial model and a recent capture event, if based on $1/a$ distribution, depends on entirely on assumptions made about comet fading. Bailey(41.102.002) examined the implications for cometary origin in a study of the mean rate of energy transfer to comets in the Oort cloud, by considering perturbations by both stars and giant molecular clouds, and concluded that the majority of primordial comets beyond 10^4 AU are removed during the age of the solar system, and that solar nebula models, that more readily generate a massive inner core of comets, need to be investigated. Lust(38.102.044) studied the distribution of aphelion directions of long-period comets in several coordinate systems, and concluded that a systematic north-south asymmetry can be at least partly explained by observational bias and that a difference in behavior is evident between dynamically 'new' and 'intermediate' comets as compared with 'old' comets with smaller semimajor axes, but no evidence was found for a concentration of aphelia around the antapex. Schreur(40.102.006) studied the production of long-period comets as a result of stellar perturbations on cometary orbits.

The primary importance of galactic tides, particularly, vertical tides caused by the galactic disk, in defining the boudary of the solar system, was identified by Smoluchowski and Torbett(38.091.005), while Harrington(39.102.004) looked into the influence of galactic tides, particularly, the vertical tide, on the distribution of orbits of long-period comets. Heisler and Tremaine(41.102.042) studied the rate of orbital evolution of comets in the Oort cloud under the influence of the galactic tides, while Hut and Tremaine (40.102.003) derived formulae for the disruption rate of the Oort cloud due to encounters with interstellar clouds and field stars. Heisler, Tremaine and Alcock(*Icarus*, **70**, 269, 1987) looked into the frequency and intensity of comet showers that might be triggered by passages of stars near the sun, and supported the conclusion that the galactic tides, rather than individual stellar perturbations, are dominant in driving the evolution of the Oort cloud.

Mignard and Remy(40.102.018 and 40.102.019) have improved their models of the evolution of the Oort cloud under the influence of random passing stars, and concluded that the cloud is stable against such disturbances for the life of the solar system, but star passages can account for the present rate of appearance of new comets in the inner solar system. Byl(42.102.028) updated his earlier studies of the effects of galactic perturbations by using a more realistic galactic model. Torbett(42.102.025) examined the capture of interstellar comets approaching with the velocity of 20km/s through three-body interactions in the planetary system, and concluded that Jupiter is the only planet capable of scattering such comets into bound orbits and that with an observational upper limit on interstellar comets of $10^{13}/\text{pc}^3$ the capture rate amounts to only one comet every 60 years. Torbett(42.102.062) also looked into the ejection of Oort cloud comets into the inner solar system by galactic tidal effects.

Bailey, McBreen and Ray(38.102.005) discussed several ways, in which limits on the mass of a possible dense inner core of comets surrounding the planetary system might be established observationally, and in which similar clouds surrounding nearby stars might be detected. Possible direct detection of the Oort cloud from submillimeter observations was discussed by Marochnik and Sholomnitskij(41.102.039). Mendis and Marconi(42.102.013) considered the implications for the total mass of comets in the solar system of possible formation of the Oort cloud by dynamical ejection from the Uranus–Neptune zone in the late phases of comet formation. If there is a substantial 'inner' Oort cloud, the total mass of comets in the solar system could approach one fiftieth of the solar mass.

Question related to possible periodic impact events and biological extinctions on the earth continued to give powerful stimulus to studies of the dynamical evolution of the Oort cloud and the ejection from it of comet showers, periodic or otherwise. The picture that has emerged is of a more massive cloud than formerly thought, smaller in extent, and possibly replenished from an inner cloud that may play a role as well in the inward diffusion and planetary capture of short–period comets. An historical perspective on extinctions and possible astronomical causes was given by Bailey, Wilkenson and Wolfendale(*Mon. Not. Roy. Astron. Soc.*, **227**, 863, 1987), who concluded, from an investigation of perturbations of the Oort cloud by known stars and molecular clouds, that comet showers induced by such processes are not likely to be the primary cause of a putative 30–Myr periodicity of impact phenomena in the terrestrial record. Fernandez and Ip(*Icarus*, **71**, 46, 1987) examined characteristics of comet showers that might be induced by close stellar encounters, and concluded that showers intense enough to be reflected in crater statistics could be produced at intervals of 80 Myr or so, provided that the Oort cloud has a massive core. In such a case the direction of approach of residual shower comets might be localized, possibly by explaining the clustering of aphelion points of long–period comets.

Stern(*Icarus*, **69**, 185, 1987) found that the influx rate of extra–solar comets, arising from encounters of the solar system with Oort–like comet clouds associated with passing stars, could have produced only a few impacts on terrestrial planets over the life of the solar system. Morris and Muller(41.102.041) studied the rate of infall of comets from the Oort cloud under the effects of the galactic tidal field, with the conclusion that the flux of comets into the inner solar system would be continuous and nearly isotropic, and would mask any possibility of determining trajectories of passing stars from analysis of the distribution of angular elements of new comets. Hills(40.107.005) concluded, from a study of the expected effects of passage of a hypothetical low–mass solar companion star(Nemesis) through the planetary system, that evident lack of damage to planetary orbits implies a low probability of any such object at the inner edge of the Oort cloud, and that limits also can be inferred on the space density of low–mass perturbers in the galactic disk.

Among investigations of the capture of comets into short–period orbits Kazantsev and Sherbaum (38.102.001 and 39.102.031) concluded that the distribution of orbits of short–period comets, according to the angular elements, can be explained by evolution under the influence of planets acting over substantial intervals of time. Bailey(42.102.069), in a study of the flux of near–parabolic comets into the inner solar system, found that the capture of short–period comets was more likely to be initiated by Uranus or possibly Saturn rather than Jupiter. Fernandez(37.102.050) studied the influence of physical processes such as sublimation and dynamical processes such as perturbations by Jupiter on the distribution of perihelion distances of short–period comets, and concluded that a combination of such effects may be adequate to account for the steep decrease in the number of short–period comets with small perihelion distances. In another study of the dynamical capture and physical decay of short–period comets Fernandez(40.102.119) concluded that the brightness behavior is not likely to be a good index to the physical lifetimes, because of build–up of a dust mantle that leads to at least temporary deactivation. Kresak(in *ESA Report*, SP–250, 433) discussed the implications of recent findings about the nucleus size and intermittent activity of P/Halley for the aging processes of other comets. Failure to observe several short–period comets at times, when they should have been bright and favorably placed, suggests that dormant phases interrupt periods of activity and that the concept of secular brightness decrease becomes meaningless in comparison with irregularities in activity on shorter time scales.

Rickman(40.102.048) reviewed various aspects of interrelations and distinctions between comets and minor planets, and concluded that there is a more substantial probability for Jupiter–family comets to develop into asteroidal objects than has previously been accepted. Hahn and Rickman(39.098.047) integrated the orbits of some 14 minor planets in cometary orbits (perihelion inside 1.7 AU and aphelion over 4 AU), ten of them discovered in the interval 1979–84, and found several kinds of orbital evolutions including stable asteroidal, stable libration about resonances and chaotic cometary. Davies(42.098.007) argued that the fewer–than–expected detections of Apollo–type minor planets by the IRAS, Infrared Astronomical Satellite, may actually have been of extinct comets, observationally selected, because of unusual characteristics. Hartmann, Tholen and Cruikshank(*Icarus*, 69, 33, 1987) examined the relationship between active comets, 'extinct' comets and dark minor planets on the basis of dynamical characteristics and spectrophotometric data. A group of ten minor planets, suggested on pure dynamical grounds to be extinct comets, have spectrophotometric properties resembling those of the most distant minor planets. They suggest that dynamically identified 'extinct comet candidates' are indeed outer solar system objects, probably of cometary origin. Nazarchuk and Shul'man(40.102.128) modeled the evolution of a cometary nucleus in the solar radiation field, and concluded that comets with perihelion distances greater than 1.5 – 2 AU tend to evolve into minor planets. Radzievskij and Tomanov(39.102.034) calculated Tisserand constants with respect to major planets for 121 comets with period shorter than 200 years under the assumption of a possible genetic relationship between short–period comets and the asteroid belt.

IV SATELLITES

a) Observations

Pascu continued photographic observations with the 26–inch telescope of the U.S. Naval Observatory for satellites of Mars, the Galilean satellites of Jupiter and Saturn I–VIII. At the Leander McCormick Observatory and the Yale–Columbia Station at Mt. Stromlo, approximately 30 photographic exposures per year for brighter satellites of Saturn and of Neptune as well as approximately 60 exposures for Uranian and Jovian satellites were made in the years of 1984–1986, and the plates were measured and the results were transmitted to the JPL, although this observation program has now ended.

At the Bordeaux Observatory about 70 plates for Saturnian satellites I–VIII were secured with the 14–inch refractor by Dourneau, Dulou and Le Campion in 1981 and 1982(39.100.001). At the ESO, La Silla, 95 plates for Saturnian satellites I–VIII were taken with the 1.50m Danish reflector in 1981 and the results were published by Dourneau, Veillet, Dulou and Le Campion(41.100.016). Observations were also made there by using GPO in 1980 by Debehogne, Freitas, Mourad and Nunes(38.099.123) and in 1981 by Debehogne(39.099.061). Observations of Saturnian satellites in 1982 were published by Debehogne(39.100.032). Observations of Saturnian satellites made by Taylor and Sinclair in 1978, 1982 and 1983 at the Royal Greenwich Observatory were published(40.100.004).

At the Pulkovo Observatory the Galilean satellites of Jupiter were observed in 1986 with the 26–inch astrograph and the normal astrograph by taking 35 plates. A party of Pulkovo observers carried out observations at Ordubad with the lunar planet telescope(D=0.7m, F= 10.0m) and 108 plates for the Galilean satellites were taken in 1984, 1985 and 1986. In 1986 60 plates for Phobos and Deimos were obtained using the same instrument. The greater part of the observations has been processed. Tolbin published results of observations for Mars and its satellites made with the 26–inch refractor in 1982 (*Izv. Glav. Astron. Obs. Pulkovo*, N 203, 44–49, 1986) and those for Saturnian satellites in 1975(40.100.006). At the Nikolaev Observatory 36 plates for the Galilean satellites and 22 plates for Saturnian satellites were taken by using the zone astrograph and Voronenko and Gorel published observations of satellites of Jupiter and Saturn in 1980–1982(*R. Zh.* 12.51.112, 1986). At the Goloseevo Observatory 60 observations of Deimos made with an astrograph in 1980 were published by Sereda(*R. Zh.*, 10.51.85, 1986). At the Engelhardt Observatory 202 observations of six satellites of Saturn obtained with the astrograph in 1980 were published by Kitkin and Chugunov(41.100.053). Kitkin published 41 positions of Saturn III, VI and VIII obtained in 1982 with the 16–inch astrograph(*R. Zh.* 10.51.91, 1986), and also results of 177 observations of Saturnian satellites made with the same astrograph and at its Southern Station at Zelenchuk by using an astrograph in 1982–1984(41.100.058).

Observations of Miranda, Uranus V, were made by Viera–Martins, Veiga and Lazzaro(41.101.061) and a paper concerning the image processing techniques for position measurements of Uranian satellites was published by Viera–Martins, Carvalho and Veiga(*Rev. Mex. Astron. Astrof.*, 12, 399, 1986). Pascu and Seidelmann continued astrometric and photometric observations with the Mark IV CCD Camera of the Space Telescope Widefield Planetary Camera Team on the 61–inch telescope at Flagstaff for Jupiter XIV, Saturn XII, XIII and XIV, Uranus I–V and Neptune I and II(39.101.021 and 41.101.033), and they with Schmidt, Santoro and Hers(*Astron. J.*, 93, 963, 1987) published observations of Miranda, Uranus I.

Aksnes and Franklin(38.099.034 and *Sky and Telescope*, **69**, 116, 1985), Arlot(38.099.016) and Arlot and Rocher(38.099.115) published predictions for some 300 mutual events of the Galilean satelites in 1985–1986 and contributed to a world–wide observing program of at least 130 events. First results were published by Arlot et al.(*Suppl. aux Ann. Phys.*, **12**, fasc. 1, 1987 and 42.099.024), by Allen and Budding(41.099.058), by Coulson(*MNASSA*, **45**, 77, 1986) and by Grigor'eva, Egorov, Tejfel' and Kharitonova, who observed at Ashkhabad in 1985(*Astron. Tsirk.*, N 1444, 1–3, 1986 and *Astron. Tsirk.* N 1447, 3–5, 1987). Aksnes, Franklin and Magnusson(42.099.068) traced the source of a puzzling longitude discrepancy between pairs of mutual eclipses and occultations occurring close in time to the phase defects on the satellites.

Mutual events of Pluto–Charon system were predicted for 1986 and 1987 by Tholen(40.101.051) and by Tholen, Buie and Swift(*Astron. J.*, **93**, 244, 1987) and were observed by Binzel, Tholen, Tedesco, Buratti and Nelson(39.101.021). A model for the eclipse events was given by Dunbar and Tedesco(42.101.054) and an asymmetry appearing between eclipses and occultations was studied by Mulholland and Gustafson (*Astron. Astrophys.*, **171**, L5, 1987).

b) Comparison of Observations with Theories

In 1984 Arlot, Morando and Thuillot(42.099.046) analyzed eclipses of Jupiter I made between 1775 and 1802, which were recently discovered and collected by Delambre. Heliometric observations made from 1891 to 1906 for the Galilean satellites were compared with modern ephemerides by Arlot(42.099.046) and observations of occultations of the Galilean satellites by Jupiter were analyzed by Fairhead, Arlot and Thuillot(42.099.048) and catalogued by Fairhead, Arlot, Jannot and Thuillot(*Astron. Astrophys. Suppl. Series*, **68**, 81, 1987). Lieske studied the evolution of the Galilean satellites (tidal perturbations and secular changes in mean motions) and showed observational evidences from a large set of data(*Astron. Astrophys.*, **176**, 146, 1987). Kiseleva(39.099.089) analyzed observations made at Pulkovo for the Galilean satellites in 1904–1910. Kiseleva, Glebova and Mal'kova(41.099.004) compared over 600 positions of the Galilean satellites obtained at the Pulkovo Observatory in 1975–1979 with Sampson's theory and with Lieske's E–1 theory, and found that the standard deviation in positions does not exceed 0."10. In two papers by Lieske(41.099.001 and *Astron. Astrophys. Suppl. Series*, **63**, 143, 1986) a collection of the Galilean satellite eclipse observations made since 1651 is presented.

Batrakov and Nikol'skaya(*Byull. Inst. Theor. Astron.*, **15**, N 9, 534–537, 1986 and N 10, 583–586, 1987 and *Kinematika Fiz. Nebesn, Tel*, **3**, N 3, 94–96) improved the orbital elements of Saturnian satellites I–VII by using modern astrographic observations. The perturbations were derived for Saturn I and II by Struve's theory, for Saturn III–V by Sinclair's theory, and numerical theories were developed for Saturn VI and VII. New values for parameters of the theories were obtained for Saturn I–V and osculating elements of Saturn VI and VII were derived for a certain epoch. Tolbin(40.100.006) compared observations of seven Saturnian satellites made in 1975 with the 26–inch refractor of the Pulkovo Observatory with the ephemerides based on the results by Batrakov and Nikol'skaya, and found that the standard deviation in position relative to the planet is close to 0".2.

Bykova, Shikhalev and Yurga(39.100.023, 40.100.009 and 41.099.005) constructed a numerical theory of Phoebe, Saturn IX, and published the collection of processed observations of Phoebe in 1898–1981, and they also refined the orbit by using modern observations made in 1940–1976, the standard deviation being 1".0. They also investigated various perturbations including relativistic effects on the motions of outer satellites of Jupiter and Saturn.

Dourneau(*These de Doctorat, Universite de Bordeau*, 1987) analyzed observations made in 1980–1985 at Bordeau, Pic du Midi, ESO and CFH and fitted them with previous theories by Kozai, Rapaport(Saturn I–VI), Taylor(Saturn VII) and Sinclair(Saturn VIII). Taylor(38.100.068) compared observations made in 1967–1982 with theories, and Sinclair and Taylor(39.100.037) generated the orbits of Titan, Hyperion and Iapetus, Saturn VI–VIII, by both numerical integrations and analytical methods and by fitting them to astrometric observations made in 1967–1982.

Laskar and Jacobson compared earth–based observations with general theories of the Uranian satellites of Laskar, derived the masses of the satellits which agree within 15% errors with the Voyager results, and provided a new ephemeris of the satellites by including Voyager data. Tholen(40.101.043) analyzed 19 speckle interferometric observations of the satellite of Pluto and observations of a partial occultation of the satellite.

c) Theoretical Studies.

Henrad(39.099.033) proposed a new method by applying Hamiltonian perturbation theories, which is valid for the large libration case of the Galilean satellites. Thuillot(39.099.032) gave a method for obtaining developments of the secular elements related to the pericenters and the nodes of the Galilean satellites with respect to variations of the dynamical parameters. Vu(*These de Doctorat, Observatoire de Paris*) developed an analytical theoy of motion of the Galilean satellites. Henrard and Lemaitre(*Icarus*, 69, 266, 1987) made a perturbation treatment of the 2:1 Jovian resonance. Greenberg(*Icarus*, 70, 334, 1987) studied evolutionary paths in deep resonance for the Galilean satellites. Campbell and Synnot(39.099.017) derived the gravity field in the Jovian system by using Pioneer and Voyager data. Goldstein and Jacobs (42.099.050) and Greenberg, Goldstein and Jacobs(42.099.052) studied the secular accelerations in the mean longitude of Io, Jupiter I, and observational evidences for the secular changes of the mean motions were investigated by Lieske(*Astron. Astrophys.*, 176, 146, 1987).

Borenenko and Schmidt(*Astron. i Geodez, Tomsk*, 13, 31–36, 1985) obtained a literal solution of the satellite case of the restricted problem of three bodies by means of averaging Lie transformations with accuracy up to the eleventh order with respect to the small parameter, which is the ratio of the mean motions of the sun and the satellite, and constructed analytical theories of Jupiter VI, VII and X. Bordovitsyna, Boronenko, Bykova and Chernitsov(41.099.002) developed numerical and analytical theories of outer satellites of Jupiter.

Salgado and Sessin(40.042.099) studied the 2:1 commensurability in the Enceladus–Dione system and Sato and Ferraz–Mello(40.042.100) studied the 2:1 resonance in the same system. Taylor, Sinclair and Message (*Astron. Astrophys.*, 181, 383–390, 1987) made a spectral analysis of the residuals derived by fitting Woltjer's theory of Hyperion to the numerical integration by Sinclair and Taylor (1985), which was extended over 50 years.

Chugunov(41.100.055) gave the root–mean–square error of representation of observations by the theory developed earlier by the author for Saturn II–VI and described a possibility of extensive application of his theory. Chugunov, Stolyarov and Stolyarov(38.100.076, 38.100.078, 39.100.006, 39.100.020, 40.091.008, 40.091.017, 40.100.005, 41.099.045, *Differenz. Uravneniya u Prikl. Zadachi, Tula*, 8–9, 1986 and *Astron. i Geodez., Tomsk*, N 14, 103–107, 1986) gave a qualitative investigation of the solution of the mathematical models describing resonant motions of some of Saturnian satellites as well as satellites in other systems.

Lazzaro, Ferraz–Mello and Veillet(38.101.018) studied the famous Laplacian resonance relation among Uranian inner satellites. Lazzaro, Veillet and Viera–Martins(40.101.062) determined the orbital parameters of Miranda, Uranus I, from Laplacian resonance analysis and Lazzaro and Viera–Martins(41.101.062) made a period analysis of Laplacian resonance of the inner satellites of Uranus. Laskar(*Astron. Astrophys.*, 166, 349, 1986) developed a general theory of the five main satellites of Uranus including the secular and short period terms.

Orlov and Chepurova(40.099.126) and Baturina(41.091.026) studied the problem of the solar perturbations, which are dominant in the motions of outer satellites. Kovalevsky(*Cel. Mech.*, 34, 243, 1984) studied the effect of resonant planetary perturbations on the evolution of the orbit of a satellite driven by tidal forces. The existence of possible satellites of Mercury and Venus was examined by Rawal(*Earth, Moon and Planets*, 36, 135, 1986).

An extensive review of satellite orbits and ephemerides was made by Ferraz–Mello(39.097.005) and a new compact representation of ephemerides was presented by Chapront and Vu(*Astron. Astrophys.*, 141, 131, 1987). By using this representation method Arlot, Chapront, Ruatti and Vu(*Astron. Astrophys. Suppl. Series*, 65, 383, 1986) published ephemerides of the main satellites of Jupiter, Saturn and Uranus.

d) Faint Satellites

Faint satellites of Saturn and Uranus, which were discovered a few years ago by ground–based and Voyager observations, have attracted much interests of astronomers, as most of them have peculiar dynamical properties. Namely, some of them are moving nearly on the same orbits with others, both faint and bright, either near equilateral–triangular points or with large libration amplitudes, and some are moving like shepherdings of nearby rings. And somebody argue that there are still fainter satellites, which could not been discovered even by Voyagers, in order to keep the shape of narrow ringlets with very sharp edges for long intervals of time.

Synnot(*Icarus*, 58, 178, 1984) reported on his estimates of orbital parameters and their uncertainties for faint satellites of Jupiter determined from Voyager imaging data. Aksnes(40.099.037) reviewed the orbital status of very faint Jovian and Saturnian satellites and their interactions with the planetary rings by using both ground-based and Voyager observations. Carussi, Roy and Valsecchi(42.100.001) discussed the stability of the Saturnian satellite system including faint satellites, which are in commensurable relations with other ones. Synnot(*Icarus*, 67, 189, 1986) showed an evidence of the existence of further small satellites of Saturn in additon to those, whose orbits had been firmly established by the time of the Voyager 2 encounter with Saturn by using Voyager imaging data.

Agafonova and Drobyshevskij(40.099.019) discussed on the origin of irregular satellites of planets by the assumption of explosions of the massive envelopes of the Galilean satellites. Burns(in *Satellites*, ed. by Burns and Matthews, 117, 1986) analyzed orbital evolutions of satellites. Sinclair(38.100.002) examined the effects of orbital resonances on satellites in tadpole or horseshoe orbits relative to Mimas, EnceLauds, Tethys and Dione, Saturn I–VI. Greenberg(38.100.143) reviewed extensively the orbital resonances among Saturnian satellites, which include fainter ones. Lissauer, Goldreich and Tremaine(41.100.009) discussed the orbital evolution of Janus–Epimetheus, which are coorbiting fainter satellites of Saturn due to torques from Saturnian rings. Henon and Petit(*Cel. Mech.*, 38, 67, 1986 and *Icarus*, 66, 536, 1986) gave series expansions for encounter type solution of Hill's problem, and described the interaction of two small satellites, which were initially on circular orbits with slightly different radii.

e) Rings

Dynamics of ring systems are one of the most exciting topics in celestial mechanics, as the Pioneer and Voyager spacecrafts discovered very peculiar properties of the rings of Jupiter, Saturn and Uranus. There are still many puzzles in them, particularly, in narrow ringlets with sharp edges and with eccentric orbits, about which much more works should be done.

Borderies(39.100.018) recalled the observations concerning the opaque rings of Saturn and Uranus, and described a model, which represents the kinematics of these rings. Brahic(40.100.003) described the dynamical processes of Saturnian rings, and reviewed in *Planetary Rings* (ed. by Greenberg and Brahic, 1984) observations, theories and prolems related to the rings. Petit and Henon(*Astron. Astrophys.*, 173, 389, 1987) made a numerical simulation for planetary rings.

Neptune arcs, whose discovery was reported by Hubbard(41.101.021) and Hubbard and Brahic(*Nature*, January, 1986), were discussed by Lissauer(40.101.055) with a shepherding model of arc rings and by Goldreich, Tremaine and Borderies(*Astron. J.*, 92, 490, 1986). Gor'kavyj(39.100.042) and Gor'kavyj and Friedman(40.091.006) studied the problem of dynamics of particles in the planetary rings and stability of rings, and they(40.101.006, 41.101.042, 41.101.043, *Pis'ma Astron. Zh.*, 13, N 3, 237–244, 1987) predicted existence of unknown satellites of Uranus on orbits close to those, which were subsequently found by using the observations made by Voyager 2, in order to explain the present ring structures.

Hill and Mendis(*Earth, Moon and Planets*, 36, 11, 1986) studied the dynamical evolution of Saturn's E-ring. Pandey and Mahra(*Earth, Moon and Planets*, 37, 147, 1987) reanalyzed the observations of occultations of MKE 31 by Neptune on September 12, 1983 and showed a possible ring system of Neptune. Goldreich and Porco(*Astron. J.*, 93, 724 and 730, 1987) studied the shepherding of the Uranian rings by 1986U7 and 1986U8. Besides a method of perturbations for particle dynamics several approaches of non-linear density waves have been made to explain detailed structures of rings; for examples, by Chu, Yuan and Lissauer(*Astrophys. J.*, 291, 356, 1985), by Chu, Dones, Lissauer, Yuan and Cuzzi(*Astrophys. J.*, 299, 542, 1985), and by Borderies, Goldreich and Tremaine(*Icarus*, 68, 522, 1985).

V OCCULTATIONS

a) Identification of Upcoming Occultations

Computerized search for future occultations are being performed independently by three sets of investigators. Wasserman, Bowell and Millis at the Lowell Observatory have published predictions for 1986 and 1987(40.096.004) and predictions for 1988 and 1989 will appear in the 1987 November issue of the *Astron. J.* In these searches the ephemerides of all minor planets, whose angular diameters are expected to reach at least 0.08 arcsec during the search years, are compared with the positions of approximately 340 000 individual stars in the SAO, AGK3, Perth 70, Pleiades Postional and Lick Voyager Catalogues. Two new Lick Voyager catalogues covering the spacecraft's encounters with Uranus (11 765 stars) and Neptune (4527 stars) were completed by Klemola and Owens during the reporting period. Dunham of Silver Spring has performed searches for about 70 of larger minor planets using a composite catalogue containing all of the above catalogues plus the U.S. Naval Observatory XZ catalogue and various astrographic catalogues. Dunham has published annual summaries of predictions in the *Sky and Telescope*(39.096.001). A third set of predictions for events involving SAO and AGK3 stars only has been prepared by Goffin of Hoboken(38.098.103 and 40.098.098).

Predictions of close appulses by P/Halley and P/Giacobinni–Zinner were published by Bowell and Wasserman(39.103.047) and Bowell et al.(39.096.008). Dunham also computed predictions for these two objects. Attempts were made to observe certain events with the IUE spacecraft and with ground–based optical and radio telescopes. Few results of these efforts have been published. It appears that most optical observations did not succeed in detecting dimming of the stars involved. Mink and Klemola (40.096.002) identified stars to be occulted by Uranus, Neptune and Pluto in the interval, 1985 through 1990, by scanning plates taken with the 0.51m Carnegie Astrograph. Events found in this search include 54 occultations by the Uranus ring system, 24 occultations by Uranus, 22 occultations by Neptune and 10 occultations by Pluto. Killian and Dalton(40.096.016) identified stars to be occulted by Saturn's rings in the years, 1985 through 1991. Photometry of occultation candidate stars was published by French et al.(42.113.016), Covault and French(41.101.058), Bosh et al.(41.101.057) and Vilas and Mink(41.096.001).

b) Prediction Refinement

Klemola continued to provide precise positions based on photography with the 0.51m astrograph for use in last–minute refinement of occultation predictions. Additionally a new 0.46m astrograph at the Lowell Observatory(39.032.007) began to contribute to refinement efforts. The Perth Observatory, the Black Birch Observatory and the Bordeau transit circle also provided needed astrometry. Basic limitations on the potential accuracy of occultation predictions made by using conventional astrometric techniques were discussed by Kristensen(41.098.001).

The Commission 20 Working Group on Occultations has been reorganized to disseminate last–minute updates more efficiently. Coordinators have been designated in various parts of the world to channel information to observers quickly. A list of more favorable occultations was compiled for concentrated prediction effort in 1987. This approach has shown promise with three of the targeted minor planet occultations observed since 1987. The International Occultation Timing Association continues to play a major role in coordinating observational activity.

c) Observations

Occultation observers of planets and their rings have been more successful than those pursuing minor planet events. The one bright spot in the minor planet area was the effort to observe the 13 November 1984 occultation of BD+8^{0}471 by (1) Ceres. This event was observed photoelectrically at 13 sites in Mexico, Florida and the Caribbean by teams from U.S. and Mexican institutions. The observations permitted a precise determination of the size, shape and bulk density of Ceres. A paper discussing the results will appear in the November 1987 issue of *Icarus*. Other minor planet occultations observed with sufficient coverage to permit a size determination since the previous report involved (47) Aglaja on September 16, 1984(*Bull. Amer. Astron. Soc.* 16, 1027, 1984) and (129) Antigone on April 11, 1985(*Bull. Amer. Astron. Soc.* 18, 797, 1986). Recent occultations by (511) Davida, (10) Hygiea and (53) Kalypso were observed in New Zealand, the United States and the Soviet Union, respectively, but in each case observations were successful at too few sites to allow a definitive diameter determination. Better luck with the weather and more participating observers are badly needed.

Since the last report at least 5 occultations by Neptune have been observed. Evidence of incomplete ring arcs in the vicinity of the planet has been reported by several investigators(40.101.049, 41.101.004, 41.101.021 and 42.101.009). The observations also have yielded new results about the size, oblateness and atmospheric properties of Neptune(40.101.050, 42.101.057, Hubbard et al. *Icarus* in press, 1987 and Hubbard et al., *Astrophys. J.*, in press, 1988). Two occultations by the Uranian system in 1985 and one in 1987 were observed at multiple sites. Data of this type continue to be of use in refining dynamical models of the Uranian rings(42.101.010).

Y Kozai
The President of the Commission

21. LIGHT OF THE NIGHT SKY (LUMIERE DU CIEL NOCTURNE)

PRESIDENT: K. Mattila
VICE PRESIDENT: A.C. Levasseur-Regourd
ORGANIZING COMMITTEE: R. Dumont, Yu. I. Galperin, R.H. Giese, M.S. Hanner,
                      P. Lamy, T. Mukai, H. Tanabe, J.L. Weinberg

I. INTRODUCTION

The different components of the light of the night sky have their origin in different formations of matter in the universe – encompassing a huge scale of distances ranging from a few kilometers in the earth's atmosphere to the most distant known galaxies and beyond. Correspondingly, the borderlines to other Commissions are not very well defined and thus material relevant to Commission 21 can also be found in the reports of other Commissions on the following topics: zodiacal light and zodiacal IR emission (Comm. 22, 44), integrated starlight (33, 25), diffuse galactic light (34), extragalactic background light (47), airglow and atmospheric scattered light (50), and space-borne observations of the LONS (44). From the Commission 21 point of view the connecting link between these various fields is the special techniques utilized in the surface photometric measurements and reductions of background radiations which extend over the entire sky. One crucial problem is the separation of the LONS into its several components. The approach for solving this task is to utilize the different spatial distributions and different broad and narrow band spectral properties of each of the LONS component. Thus the successful measurement and separation of one of the LONS components requires a knowledge of the properties of all the other components. This situation has become apparent in recent years as the infrared background radiation database, provided by the Infrared Astronomical Satellite (IRAS), has been analyzed: both the zodiacal and galactic dust emissions have to be analyzed hand in hand, and both these components must be very accurately mastered before any conclusions are possible on the extragalactic component. It is also obvious that very similar problems are encountered in the ultraviolet and infrared wavelength regions as in the more traditional optical domain. Thus the techniques developed in one of these wavelength domains are directly applicable in the others.

Knowledge of the sky brightness is also important when trying to push the detection limit for very faint objects – either extended or pointlike – to the utmost possible from ground or space. For this purpose the selection of the most suitable low-background windows, either at special positions on the sky or in special wavelength regions of the spectrum, can be selected. It has even become possible to take the measuring instrument itself to a special favorable place in the solar system: thus the orbit of the Helios sunprobe, which reached 0.3 AU in the perihel, enabled the measurement of the zodiacal light and thus the distribution of the interplanetary dust in dependence of the distance from the Sun. Equally important have been the LONS measurements from the outbound orbits of Pioneer 10 and 11. Beyond the asteroid belt ($r \geq 3.3$ AU) the zodiacal light vanished and a measurement of the sum of the pure galactic and extragalactic LONS components was possible without any foreground contamination. Unfortunately, with the cancellation of the US-spacecraft for the Out-of-Ecliptic Solar Polar Mission (ISPM) we lost the unique opportunity to obtain photopolarimetric measurements of the LONS from a position out of the ecliptic plane.

At the beginning of the triennium covered by this report the IAU Colloquium No. 85 "Physical Properties and Interactions of Interplanetary Dust" was organized by Commission 21. The proceedings of this meeting, with R.H. Giese and P. Lamy as editors, have been published in 1985 by D. Reidel, Dordrecht, as Volume 119 of the Astrophysics and Space Science Library. The proceedings include 80 reviews and contributed papers providing an up to date presentation of our knowledge of the zodiacal dust cloud.

This report was prepared with the help of many Commission members who provided information on their recent research. The task of writing the report was divided between Dr. A.C. Levasseur-Regourd (Airglow and zodiacal light) and the undersigned.

We have found it convenient to use the following very abbreviated references to which we provide the key here:

A & A = Astron. Astrophys.   ApSpaSci = Astrophys. Space Science
A & A Sup = Astron. Astrophys. Suppl.Ser.   BAAS = Bull. American Astron. Society
AdvSpaRes = Advances in Space Research   MN = Mon. Not. R. Astr. Soc.
AJ = Astron. J.   Mitt. Astron. Ges. = Mitteilungen der
AnnRevA&A = Ann.Review of Astron. Astrophys.   Astron. Gesellschaft
ApJ = Astrophys. J.   PASP = Publ. Astron. Soc. Pacific
ApJ Sup = Astrophys. J. Suppl. Ser.   JGG = J. Geomagn. Geoelectr.
                                            JGR = J. Geophys. Res.

In addition, IAU Symposium and Colloquium volumes are referred to by IAU Symp or IAU Coll and the appropriate number.

The following common abbreviations referring to the Light of the Night Sky (LONS) have also been used:

DGL = Diffuse Galactic Light
EBL = Extragalactic Background Light
ISL = Integrated Starlight
"unit" = photons cm^{-2} s^{-1} sr^{-1} $\mathring{A}^{-1}$

II. LONS, OBSERVING SITES AND FAINT SOURCE DETECTION LIMITS

A comparison of the night sky surface brightness (optical and UV) at excellent ground-based sites to the sky background in space has been presented by O'Connel (1987,AJ in press). Garstang (1986 PASP 98,364) has set up a model to permit prediction of the brightness of the night sky caused by the scattering of city light. Pilachowski et al. (1987 BAAS 19, 750) report on a systematic monitoring of the nigh sky brightness at Kitt Peak and Payne (1985 Proc. Astron. Soc. Australia 6, 182) at three observatory sites in Australia. Klimik and Shvalagin (1985 Astron. Tsirk. No. 1370, p.5) have investigated the determination of artificial sky background contamination.

III. AIRGLOW

(A.C. Levasseur-Regourd)

Numerous papers on the airglow have been published in the triennium and are likely to be included in the category of geophysics (theoretical modelling, reaction rate coefficients, lines profiles...). It should nevertheless be pointed out that airglow measurements made both from earth-based or space observatories are of major interest for astronomical observations. They provide information on the light pollution or on the fluctuations of the airglow (Meinel IR bands...) and set limits on the integration time for observations performed from space platforms in low Earth orbit.

Earthbased observations. An atlas of zenith airglow radiation (Kiso,

1979-1983) has been published by Tanabe. Spectra at the north ecliptic pole (Pic du Midi Obs.) have been analysed by Louistisserand et al. (A & A Sup 68, 539). Airglow studies in India have been presented by Agashe (Indian J. Radio & Space Phys. 16,84). Airglow variations related to seismic activity have been reported by Fishkova et al. (Ann. Geophys. 3, 689).

Space observations. Rocket airglow observations have been reported by Ogawa et al.(J. Geomagn. Geoelectr. 39, 211) and Lopez-Moreno et al.(Ann. Geophys. 2, 61 and JGR 90, 6617). UV airglow observations obtained during morning twilight and near midnight rocket flights have been analysed (Cebula & Feldman, JGR 80, 9080 ; Tennyson et al., JGR 91, 10141). Also UV nightglow has been monitored from the UVX experiment on board the space shuttle (Feldman et al., EOS Trans. Am. Geophys. Union 68, 373).

IV. ZODIACAL LIGHT

(A.C. Levasseur-Regourd)

Dramatic progress has been achieved in our knowledge of zodiacal dust, through analyses of ground-based or space zodiacal light observations, and through modelling of the zodiacal dust (Leinert & Staude, ApSpaSci 116, 415). During the last triennium, a significant number of contributions have been published in the proceedings of IAU Coll. 85, Properties and Interactions of Interplanetary Dust (= PIID) ; also major advances have been triggered by cometary dust observations and by IRAS observations of the thermal emission of the zodiacal cloud.

Zodiacal light observations

General observations. Further analyses of observations performed on board Pioneer 10 spaceprobe (Toller & Weinberg, PIID, 21), D2B spacecraft (Maucherat et al., PIID, 27), or Salyut 7 space station (Levasseur-Regourd et al., AdvSpaRes 5, No3, 27 ; Nikolsky et al., PIID, 7) have been presented. Also Dopplershifts have been measured (East & Reay, PIID, 81 ; Robley et al., PIID, 85), and surface brightness maps have been provided (Misconi & Weinberg, PIID, 11 ; Pfleiderer & Leuprecht, PIID, 17).

The thermal emission of the dust has been monitored by IRAS spacecraft ; it is indeed the most prominent component of the sky at 12 and 25 μm (see Proc. 1^{rst} IRAS Conf., Light on Dark Matter) ; it has also been observed during rocket and balloon flights (Murdock & Price, AJ 90, 375 ; Salama et al., AJ 93, 467). As emphasized in the review given by Weinberg (PIID, 1), recent results present a more complex picture of the zodiacal cloud than what was anticipated before ; mainly due to the infrared IRAS observations, it is confirmed that the cloud is neither completely smooth, nor really homogeneous.

Inner zodiacal light. A review on the F-corona and circumsolar dust has been presented by Koutchmy & Lamy (PIID, 63). The Zodiacal light experiment on board Helios B has been used to image solar mass ejection transients in the interplanetary medium (Jackson & Leinert, JGR 90, 10759).

Dopplershifts have been measured during the July 31, 1981 solar eclipse ; a prograde keplerian motion was found (compatible with grains with radii ~ 0,4 μm), as well as some retrograde motion (dust injected by Kreutz group comets ?) ; also a local stream (β meteoroids ?) was detected near the north coronal hole (Shcheglov et al., PIID, 77 ; Shcheglov et al., A & A 173, 383 ; Shestakova, A & A 175, 289). From the balloon observations performed during the June 11, 1983 solar eclipse, polarization degrees of inner zodiacal cloud were estimated, near IR photometry was performed, and an excess in infrared brightness (hot dust cloud ?) was noticed ; (Isobe et al., PIID, 49 ; Maihara et al., PIID, 55 ; Tanabe et al., AdvSpaRes 5, No1, 69).

Gegenschein. Pictures of the Gegenschein have been derived from D2B data ; temporal variations in intensity and morphology were found ; they could be associated to cometary or asteroidal debris (Maucherat et al., A & A 167, 173). Observations from Pioneer 10 also confirm the existence of small amplitude spatial fluctuations of the brightness distribution in the Gegenschein region ; a search for such structures is also made in ground based observations (Hong et al., PIID, 33).

Small scale structures. Besides the Gegenschein region, the Schuerman dust arcs region has been tentatively investigated - when observable - for small scale structures (Giovane et al., PIID, 33). From the IRAS zodiacal emission (mainly at 12 and 25 μm), it has been confirmed that the zodiacal cloud is not smooth (Low et al., ApJ 278, L 19).

Two pairs of emission bands seem to circle the ecliptic, each band being ~ 5 % of the strength of the zodiacal emission near the ecliptic ; the colour temperature implies a heliocentric distance of 2.2-3.5 AU. A disrupture collision of a ~ 10 km asteroid could produce enough particles to account for a set of bands, with a time scale for formation and dissipation of a few million years ; the EOS, Koronis and Themis families have been identified as possible progenitors (Dermott et al., PIID, 395 and Asteroids, Comets, Meteors II, 583 ; Sykes & Greenberg, Icarus 65, 51).

Symmetry surface. Annual oscillations in the brightness at the ecliptic poles have been observed by IRAS, together with asymmetries in the brightness profiles ; such features had been found in the optical domain and are also visible in rocket infrared observations (Hauser et al., PIID, 43 ; Winckler et al., A & A 143, 194 ; Murdock & Price, AJ 90, 375). Both the oscillations and the asymmetries are due to the inclination of the symmetry plane upon the ecliptic and, secondarily, to the eccentricity of the Earth's orbit (Hauser & Houck, 1^{rst} IRAS Conf., 39 ; Levasseur-Regourd & Dumont, AdvSpaRes, 6,No7, 87 ; Deul & Wolstencroft, A & A, in press).

A critical review of the different values obtained for the symmetry plane parameters has been made by Dumont & Levasseur-Regourd (1987, Prague IAU meeting). There is, between the various authors, a fairly good agreement at 1 AU on i ~ 1.5°, while Ω is found to vary between 50° and 100°. It is generally agreed that the symmetry plane is warped and that there is a systematic variation of the parameters as one moves outward from the Sun, whence the name Symmetry surface ; it is likely that the zodiacal cloud is not a simple axisymmetric volume.

Interpretation of measurements

As seen above (fluctuations at Gegenschein, oscillations at the ecliptic poles), the observations are interpreted in terms of physical properties and spatial distribution of the dust. Interpretations are made possible by fitting general models with observations or, better, by deriving local properties from inversion methods ; various methods have been developed by Hong (PIID, 215 ; A & A 146, 67), Dumont & Levasseur-Regourd (PIID, 207 ; Planet. Space Sci 33, 1) or Lamy & Perrin (A & A 163, 269). By inversion of recent infrared observations, Dumont & Levasseur-Regourd have found a heliocentric fall of the dust local temperature in $r^{-0.3}$, which conflicts with a grey-body assumption near the Earth's orbit (A & A, in press). Also scattering properties of the grains have been extensively studied from laboratory work and theory.

Albedo of the grains. Levasseur-Regourd & Dumont have applied their inversion
method (nodes of lesser uncertainty) to infrared observations and concluded that
the emissivity of the dust increases, i.e. that its albedo decreases between 0.5
and 1.5 AU (C.R.Acad.Sci., Ser II, Tom. 300, 109 ; AdvSpaRes 6,No7, 87).
Decreasing albedo was suggested by Fechtig to explain discrepancies between
optical and impact measurements of Pioneer 10 (AdvSpaRes 4, No9, 5) ; it has also
been considered by Lumme et al. (Icarus 62, 54) and recently been confirmed by
Hong. Mukai et al. have computed that sublimation and sputtering by the solar wind
produce "brighter" cometary/zodiacal grains with time, as they spiral towards the
Sun (A & A 167, 364).

Distribution of the dust in the ecliptic. The distribution has been
investigated in terms of steady state distribution under Poynting - Robertson
effect and Lorentz forces ; for grains with radii ~ 30 μm, the heliocentric
dependence is found to be ~ $r^{-1.3}$ (Mukai & Giese, A & A 131, 355). Also, it
appears that debris from disrupted meteors (cometary or asteroidal origin) would
be able to balance the mass loss due to Poynting - Robertson effect and
collisions, and could result in a $r^{-1.3}$ distribution (Leinert, PIID, 374 ; Grün et
al., Icarus 62, 244). The zodiacal dust dynamics has been studied further, taking
into account the planetary perturbations (Gustavson & Misconi, Icarus 66, 280).
Local velocities have been inferred from Dopplershift measurements ; the velocity
is nearly circular by 0.5 - 1.0 AU, while an excess over the keplerian velocity
(radial escape ?) is found by 1.5 AU (Dumont & Levasseur-Regourd, PIID, 207).

Anyhow, the heliocentric decrease of the albedo as mentioned above
invalidates the homogeneity of the cloud. The observed change of brightness $r^{-2.3}$,
interpreted up to now as a $r^{-1.3}$ change of density, could be partly due to the
fall of the albedo, while the fall of density could be in r^{-1} (Dumont &
Levasseur-Regourd, 1^{rst} IRAS Conf., 46). This result is in agreement with the
analysis made by Hong & Um, using Henyey-Greenstein functions to represent the
mean scattering phase function (Ap. J., 320, 928).

Spatial distribution of the dust. Lamy and Perrin have inverted intensity and
polarization observations, with the usual r^{-n} power law for the dust number
density ; they found that the modified fan model is compatible with the data if
the volume scattering function and the local polarization depend upon the
heliocentric distance of the observer, but are independent of the latitude (PIID,
239 ; A & A 163, 269). Comparative studies of some 3 dimensional models of the
zodiacal cloud have been made by Giese et al. (PIID, 255 ; Icarus 68, 395). The
multilobe models (Buitrago) are in strong contradiction with inner zodiacal light
observations, while the sombrero type models (Dumont) provide a satisfying fit ;
also it is found that the dust spatial density decreases by ~ 2 above the Earth's
orbit within 0.2-0.3 AU and that, beyond ~ 1.5 AU off the ecliptic, no reliable
density values have yet been obtained from zodiacal light observations.

Laboratory measurements. A review on the different methods to determine the
scattering properties of the dust has been given by Zerull, with emphasis on
future experiments needed for the interpretation of polarization and colour
effects (PIID, 197). A laser-particle dynamics laboratory has been developed in
Gainesville to study dust particles light scattering properties for roughened
surface spheres and irregular particles (Gustavson et al., Icarus, in press).
Microwave analog measurements have been continued at Bochum and Gainesville with
irregular or elongated particles (Weiss-Wrana et al., PIID, 219, 223 ; Gustavson,
PIID, 227). Also laboratory measurements have been performed in Marseilles with a
nephelometer type experimental device (Bliek et al., PIID, 231), together with
theoretical studies of the optical properties of rough grains (Perrin & Lamy,
PIID, 245).

V.GALACTIC COMPONENTS

Optical. Blue and red isophotes of the galactic background light near the celestial, ecliptic and galactic poles have been presented by Toller et al. (1987, A & A in press). The data were obtained by the Pioneer 10 imaging photopolarimeter from beyond the asteroid belt and are practically free of zodiacal light contamination. After subtraction of the ISL (star counts) an average DGL+EBL contribution of 22 $S_{10}(V)_{G2V}$ remains at the ecliptic and celestial poles. As a continuation of their p.g. UBVR surface photometry Winkler et al. (1984 A & A Sup 58, 705) present a U map of the whole Milky Way which is a compilation of a p.g. northern and a p.e. southern survey. The Pioneer 10 blue and red background starlight data have been analyzed in terms of a Galaxy model by van der Kruit (1986 A & A 157, 230). Important independent information on the photometric parameters of the old disk population has been obtained and compared with external galaxies.

A review of the diffuse galactic line emission in Hα, [N II] λ6583 and [S II] λ6716 has been presented by Reynolds (1984, IAU Coll. 81, p. 97). Observations at high galactic latitudes by Reynolds (1984, ApJ 282, 191) suggest that diffuse Hα extends over the entire sky. Another high-latitude diffuse Hα survey has been started by Münch and Pitz at the Calar Alto Observatory (1987 Mitt. Astron. Ges. 68, 168). Further Hα studies of Milky Way regions have been reported by Reynolds (1986 BAAS 18, 1023) and Brinkman et al. (1986 BAAS 18, 1036). Extended [S II] λ6716 emission has been observed and analyzed and [O III] λ5007 detected by Reynolds (1985 ApJ 294, 256; ApJ 298, L27).

Ultraviolet. A review of diffuse UV continuum and line observations has been presented by Jakobsen (1986, AdvSpaRes 6,No.7,59). The absolute intensity and spati, fluctuations of the UV background between 1450 and 2420 Å have been measured with a large (95 cm) rocket-borne telescope by Jakobsen et al. (1984, A & A 139, 481) along a 50°.5 scan path at b ≈ 50°. Significant spatial fluctuations were observed and they were found to correlate with the HI column densities. The observed UV surface brightnesses were found to be in reasonable quantitative agreement with calculations of light scattering by high-latitude dust illuminated by galactic plan. stars. Observations with the Berkeley EUV/FUV Shuttle Telescope by Hurwitz et al. (1986, AdvSpaRes 6, No. 7, 69) indicate a good correlation between UV brightness (1400-1800 Å) and HI column density, with a slope which is consistent with the dust scattering parameters obtained earlier from OAO-2 data by Witt and Lillie. Preliminary results from the Hopkins XUV Shuttle experiment by Henry et al. (1986, BAAS 18, 1023) are consistent with the previous results of this group that the UV background is ≤ 300 units (= photons cm^{-2} s^{-1} sr^{-1} $Å^{-1}$) everywhere above |b| ≥ 20°. Approximately two-thirds of the celestial sphere were scanned in the 911-1050 Å band with a 10° FOV photometer on board the STP72-1 satellite (Opal and Weller, 1984, ApJ 282, 445). The observed flux was well represented by star catalog integrations.

Discovery of diffuse extended line emission from C IV λ1550, [O III] λ1663 and O VI/Si IV λ1400 were reported by Martin and Bowyer (1986, BAAS 18, 1036; AdvSpaRes 6, No. 7, 79). Spectroscopic observations of the EUV background in the 80-650 Å band were reported by Labov and Bowyer (1986, AdvSpaRes 6, No. 7, 73).

Resonance scattering of solar radiation in the Lyα and HeI λ584 lines of the interstellar wind gaseous component has been reviewed by Bertaux (1984, IAU Coll. 81, p. 3). New results have been presented for hydrogen and helium densities based on Pioneer 10 and Voyager 2 observations in the r > 5 AU region (Shemansky et al. 1984, IAU Coll. 81, p. 24) and on Venera 11 and 12 observations in the inner Solar System (Chassefière 1986, A & A 160, 229). Holberg (1986, ApJ 311, 969) has report detection of resonance scattering in Lyγ using a very long exposure spectrum from Voyager 2.

Infrared. The galactic infrared background emission was measured by IRAS over
96 % of the sky at 12, 25, 60 and 100 μm. The emission is believed to be thermal
reradiation by dust grains (see, however, Harwit et al. 1986, Nature 319, 646 for
an alternative). Many aspects of the IR background measurements are covered by the
three IRAS conference proceedings:
Light on Dark Matter, ed. F.P. Israel, Reidel, Dordrecht 1986,
Star Formation in Galaxies, ed. C.J. Lonsdale, NASA Print. Off., Washington 1987,
Comets to Cosmology, 3rd Int. IRAS Conf., July 6-10, 1987, Queen Mary College, London,
and the review article by Beichman (1987, AnnRev A&A 25, 521).

The separation of the zodiacal and galactic emission components has been discussed by Jongeneelen et al. (1987, A & A in press), Hauser (1987, in Comets to
Cosmology), Rowan-Robinson (1987 ibid.) and Reach and Heiles (1987 ibid.).

The large scale distribution and properties of dust in the galactic disk as
derived from IRAS and other IR data and the connections to star formation have been
discussed by Burton et al. (1986, in Light on Dark Matter, p. 357), Puget (1985, in
Birth and Infancy of Stars, ed. R. Lucas et al., North-Holland, Amsterdam, p. 77),
Mezger (1985, ibid. p. 31), Cox et al. (1986, A & A 155, 380), Little and Price
(1985, AJ 90, 1812), Caux et al. (1984, A & A 137, 1; 1985, A & A 144, 137) and
Caux and Serra (1986, A & A 165, L5).

The near-IR radiation of the Galaxy is due to direct and scattered stellar
radiation. A 2.4 μm surface photometry of the galactic center area has been carried
out by Hiromoto et al. (1984, A & A 139, 309). A model for the distribution of
stellar radiation in the Galaxy at 2.4 μm and in optical (B and V) has been presented
by Oda (1985, A & A 145, 45). A similar model has been constructed by Garzon et al.
(1986, A & A 155, 63) and has been compared with 2.2 μm star counts and a 2.4 μm
surface photometry of the galactic centre area.

The infrared background emission at high galactic latitudes shows, besides a
general cosec($|b|$) dependence, large spatial fluctuations over scale lengths of tens
of degrees down to the 2' resolution limit of IRAS. The name "infrared cirrus" was
introduced by Low et al. (1984, ApJ 278, L19) for these structures. Reviews of the
infrared cirrus have been presented by Gautier (1986, in Light on Dark Matter, p. 49)
and Puget (1987, in Comets to Cosmology). Surprisingly, the cirrus was visible not
only at 60 and 100 μm but also at 12 and 25 μm, where colour temperatures of several
hundred K are observed (Leene 1985, A & A 154, 295; Boulanger et al. 1985, A & A
144, L9). Very small grains (r << 0.01 μm), transiently heated by single UV photons
(Draine and Anderson 1985, ApJ 292, 494) and macromolecules called polycyclic
aromatic hydrocarbons (PAH's) have been invoked as explanation (Puget et al., A & A
142, L19; Léger and Puget 1984, A & A 137, L5).

Infrared cirrus is closely linked to the previously observed extended bright
optical nebulae at high latitudes (see e.g. Lynds 1965, ApJ Sup 12, 163) and the
fluctuations of the UV background radiation, both of which are understood in terms
of scattering of ambient stellar radiation by high latitude dust clouds. Detailed
comparisons between infrared and optical cirrus have been presented by de Vries and
Le Poole (1985, A & A 145, L7) and Laureijs et al. (1987, A & A 184, 269). Also
the UV background fluctuations are correlated with infrared cirrus (Jakobsen et al.
1987, A & A 183, 335).

VI. EXTRAGALACTIC COMPONENTS

A review of the extragalactic background radiation, including a detailed
discussion of the visible-infrared EBL observations, has been presented by Wilkinson
(1984, Proc. 1st ESO-CERN Symp., eds. Setti and van Hove). Paresce (1985, Il Nuovo
Cim. 8, 379) has presented a review of both the observational and theoretical aspects

of the ultraviolet radiation of cosmological origin. A review of the theoretical aspects ranging from gamma rays to radio waves has been given by Narlikar (1983, 18th Int. Cosmic Ray Conf., eds. Dugaprasad et al., Vol. 12, p. 1).

Optical. Galaxy counts down to the faintest limiting magnitudes attainable today ($\sim 27^m$ in J) have been used by Tyson (1987, Proc. Vatican Conf. on Theory and Observational limits in Cosmology) to estimate an EBL lower limit of 0.43 S_{10} (J) as due to resolved galaxies.

Ultraviolet. In the ultraviolet new upper limits to the extragalactic background have been set: Bixler et al. (1984, A & A 141, 422) give 9700 units at 1040-1080 Å, Opal and Weller (1984, ApJ 282, 445) 7800 units in the 912-1050 Å region, Holberg (1986, ApJ 311, 969) gives limits of 100-200 units in the 500-1200 Å region, and Henry et al. (1986, BAAS 18, 1023) as low as 150 units in some directions at $\sim$1800 Å.

The current limits and prospects of improvement of the UV background radiation data relevant to the detection of radiative decay of fundamental particles have been reviewed by Bowyer and Malina (1986, in Inner Space/Outer Space: The Interface between Cosmology and Particle Physics, ed. Kolb et al., Chicago U.P., p. 512). A break at 1680 Å was claimed by Bochicchio et al. (1984, Proc. 4th Europ. IUE Conf., ESA SP-218, 503) in IUE background data but its statistical significance was questioned by Murthy and Henry (1987, Phys. Rev. Lett. 58, 1581). Holberg's results from Voyager UV spectrometer observations (1985, ApJ 292, 16) place lower limits on the radiative-decay lifetime of massive neutrinos.

Infrared. Boughn and Kuhn (1986, ApJ 309, 33) have searched for the EBL at 6500 Å and 2.2 μm by comparing the sky brightness in the opaque dark cloud L134 and a nearby transparent region. Upper limits on the order of $\nu I_\nu < 10^{-5}$ and $< 10^{-4}$ erg s cm^{-2}s^{-1}sr^{-1} at 6500 Å and 2.2 μm, respectively, were obtained. Boughn et al. (1986, ApJ 301, 17) have also determined upper limits to sky background fluctuations at 2.2 μm. Another near-infrared search has been performed by Matsumoto et al. (1987, preprint; IAU Symp. 124, 69; 1987, Sendai Astr. Rap. 313, 199). The authors attribute the observed isotropic residual component at 2.2 μm of $1.1\cdot10^{-4}$ ergs cm^{-2}s^{-1}sr^{-1} to be possibly of extragalactic origin.

In the far infrared domain IRAS data have been utilized by Rowan-Robinson (1986, MN 219, 737) to estimate the extraglactic component to be 5-6±2.5 MJy sr^{-1}. Differential methods for measuring the far infrared background have been discussed by Ceccarelli et al. (1984, in Clusters and Groups of Galaxies, ed. F. Mardirossian et al., D. Reidel, p. 277; 1985, Il Nuovo Cimento 8, 353).

Theoretical. A rediscussion of Olbers' Paradox has been presented by Wesson (1986, ApSpaSci) and Wesson et al. (1986, ApJ 317, 601). They conclude that the reason why EBL is low is the limited lifetime of galaxies and only to a small factor the expansion of the universe.

Calculations on the intensity, spectrum and large scale anisotropies of a cosmological IR background have been presented by Bond et al. (1986, ApJ 306, 428), de Bernardis et al. (1985, ApJ 288, 29), Fabbri et al. (1987, ApJ 315, 12), Guagzhong and Mengxian (1986, Acta Astrophys. Sin. 6, 266) and Negroponte (1986, MN 222, 19).

K. MATTILA
President of the Commission

COMMISSION 22: METEORS AND INTERPLANETARY DUST
(METEORES ET LA POUSSIERE INTERPLANETAIRE)

PRESIDENT: P.B.Babadzhanov
VICE-PRESIDENT: C.S.L.Keay
ORGANIZING COMMITTEE: W.G.Baggaley, O.I.Belkovich, W.G.Elford, H.Fechtig, M.S.Hanner, I.Hasegawa, J.A.M.McDonnell, J.Stohl, K.Tomita

I.   INTRODUCTION

As before a number of authors have contributed reviews of their own field. The contributions were editted by the President in order to avoid some overlaps, to reduce the length of the reviews and to add some publications. Reference numbers from "Astronomy and Astrophysics Abstracts" were used when available. In the end of the report a list of references not found in "Abstracts" is given. As editor, the President takes the responsibility for any shortcomings in the report.

The following Proceedings of major meetings and books were published during the last triennium:

"Properties and Interactions of Interplanetary Dust", eds. R.H.Giese and P.Lamy, Reidel, Holland, 1985.

"Asteroids, Comets, Meteors II", Proc. of a meeting held at the Astron. Obs. of Uppsala Univ., June 3-6, 1985, eds. C.I.Lagerkvist, B.A.Lindblad, H.Lundstedt, H.Rickman, 1986.

"Dynamics of Comets: Their Origin and Evolution", eds. A.Carusi and G.B.Valsecchi, Reidel, Holland, 1985.

"Meteors and Their Observations", P.B.Babadzhanov, Nauka, Moskva, 1987 (Russian).

"Meteors, Meteorites, Meteoroids", V.A.Bronshten, Nauka, Moskva, 1987 (Russian).

It is with sorrow that we record the death of a noted meteor astronomer and a member of the Commission Johannes Hoppe.

II.   PHOTOGRAPHIC AND TV METEORS   P.B.Babadzhanov

Photographic observations of meteors continued in Dushanbe, Kiev and Odessa. In Dushanbe from July 1984 to December 1986 185 meteors were photographed, 96 of which - from two stations. A catalogue of radiants, velocities and orbital elements for 70 meteors photographed in Kiev during 1967-76 was published by Sherbaum et al. (41.104.027). Kramer et al. (1986) published a catalogue of orbits of meteors photographed in 1953-83 in Odessa. Betlem et al. (40.104. 004) have published the results from 18 double-station meteors photographed in 1980-85. Simultaneous TV and radar observations of meteors continued in Dushanbe. For 13 meteors light and ionization curves are found (41.104.037). Jones and Sarma (39.104.008; 39.104. 011) published trajectories of 454 double-station meteors in the magnitude range from 0.5 to 8.5 observed with TV system. Jones, Sarma and Ceplecha (39.104.012) have presented the results of a population analysis of meteor beginning heights. Hawkes, Duffy and Jones (42.104.054) and Duffy, Hawkes and Jones (1987) have found a method of obtaining much more information from single-station meteor observations. Hawkes and Jones (1986) have carried out a comprehensive survey of electron-optical techniques used for meteor researches together with the results obtained by these methods. TV observations of the Draconid meteor shower were carried out in 1985 by Nagasawa and Kanda (41.104.066). 8 meteors were recorded in total and the shower radiant point was determined as well.

The European EN camera network was in operation during the review period. Fireball phenomena were observed in different countries. Meyer and Steiweeks (42.104.059) described a fireball observed on the 14th of November 1985 in the coast of Lower Saxony. The next day a fragment of a meteorite (chondrite) weighting 43 g was found near Salzwedel (GDR). The summer-autumn fireballs of 1908 observed at middle latitudes of Euroasia have been studied by Anfinogenov and Budaeva (40.104.013) in connection with the Tunguska event. Ognev and Maslenitsyn (42.104.019) presented astronomical and physical data on the fireball observed over the Kirov region of the USSR on 11 September, 1982. An extremely bright fireball was observed on the morning of September 23, 1981 in nearly full daylight in Belgium and also in the Netherland and West Germany during several seconds. Sauval (42.104.068) suggests that the bolid continued its flight without entering the atmosphere. Two exceptional fireballs were widely observed over Alberta in 1985. Folinsbee et al. (42.104.070) has described the Alberta fireballs. Balestrieri and Fontanelli (38.104.015) published a catalogue of fireballs including 257 observations of 1949-82. Rajchl (42.104.021) has discussed the possible causal relation between the overlights of bright fireballs and the occurence of noctilucent clouds in the Northern Hemisphere.

III. RADAR METEORS   C.S.L.Keay

Radar and theoretical studies of the Orionid and Eta-Aquarid meteor showers associated with Halley comet appear now to be in agreement. Cevolani and Hajduk (39.104.049; 40.104.040) presented results supporting the highly elliptical ribbon-like stream structure proposed by McIntosh and Hajduk in 1983 and further refined by Jones and McIntosh (1986). Hajdukova et al. (1986) suggest the presence of two populations of particles in these streams from a reduction observed in the mass index from 2.2 to 1.85 since 1984.

The 1985 return of the Draconid meteor shower associated with Giacobini-Zinner comet was observed by Lindblad (1986) who noticed $0\overset{o}{.}15$ displacement between the stream and comet orbits. The maximum zenithal echo rate of 484 per hour was an order of magnitude less than in 1933 and 1946.

Eighteen years of Geminid observations from Czechoslovakia have been employed by Šimek (40.104.039) to develope a regression method for determining equipment response as well as stream and sporadic meteor information.

Baggaley (39.104.010) has compared seasonal variations in the daily dawn maximum influx of radio-meteors with lower ionospheric irregularities occuring at dawn. Porubčan et al. (1984) report that mean heights and ranges of meteor trails exhibit minima just after sunrise, but could find no relation to the geomagnetic Kp index. Lindblad and Štohl (41.104.022) find mean mid-point heights of 102.2 93.7 and 95.7 km respectively for Perseid, Delta Aquarid and sporadic meteors.

Comparison of observations at 54, 6 and 2 MHz indicates echo height ceillings of approximately 105, 120 and 140 km respectively, with the number of echoes recorded above 105 km being an order of magnitude greater (Olsson-Steel and Elford, 1986, 1987). This confirms that conventional HF meteor radars seriously underestimate the flux of small meteoroids.

Znojil, Hollan and Šimek (39.104.009) have studied the relation between the optical brightness and ionized trail properties of meteors.

As is known the Soviet Union is a sponsor of the International project GLOBMET - Global System of Meteor Investigations - where radar observations are of great importance. Ovezgeldyev et al. (1986) have reported on the GLOBMET tasks and first scientific results. Three papers are dedicated to the method of radar meteor observations. The first paper considers some features of meteor observations when the antennae are directed to the north and south. A different character of diurnal variation of the recorded meteor rate is noted (39.104.033). The second paper (41.036.075) proposes a radar with a rotating aerial to be applied to observations of meteor showers. In the third paper Andrianov (1987) has critically considered the measurements of meteoroid deceleration in the atmosphere by the radio-method.

Physical conditions of radio-wave reflection from ionized meteor trails, diffraction, polarization and resonance phenomena are described in (39.104.045, 048; 40.104.078, 087; 42.104.005;Bayrachenko, 1985).

A great deal of attention has been drawn to the problem of optimum determination of meteor trail parameters according to the amplitude-time characteristics of meteor radio echoes (40.104.038, 075, 085; 42.104.042). From radar observations of more than 100 000 meteors up to $+12^m$ the distribution of heliocentric velocities of sporadic meteoroids is analysed by Voloshchuk et al. (40.104.022). Density distribution maps of radiants of sporadic meteoroids obtained from radar observations in Kazan and at the equator are presented by Pupishev et al. (39.104.042).

The Ottawa (Springhill) 20-kilowatt radar will contribute about 1000 hours of observations of the Eta-Aquarid and Orionid meteor showers. During the intervals in 1986 when the Earth passed through the plane of the Halley comet, the 2-megawatt radar was operated in a search for small particles outside the normal periods of the showers. Results were negative.

It is noted with regret that after thirty years of significant contributions to meteor astronomy, the Springhill Meteor Observatory in Canada was closed in 1987.

IV. METEOR SHOWERS   B.A.Lindblad

During the last triennium observations of the major meteor showers have been carried out by visual, photographic, radar and television techniques. The detection of new meteor showers and/or unexpected recurrences of old showers now largely depends on the efforts of amateur observers. The 1982 Lyrid outburst, the 1985 November Monocerotids and the 1986 Aurigid recurrence were observed by amateur groups. Visual recording by supervised terms of amateur astronomers have been carried out in many countries. For a list of amateur meteor societies and publications see Lindblad (1987a).

General studies of the formation of meteor streams and their subsequent evolution under planetary perturbations have been presented by Emelyanenko (37.104.003, 021; 38.104.006; 40.104.016), Kruchinenko (37.104.039), Karpov et al. (37.104.044), Kramer and Shestaka (38.104.018), Lebedinets (38.104.026; 40.104.076), Fox and Williams (38.104.035), Sukhotin (39.104.018), Katasev and Kulikova (39.104.040), Williams (40.104.041), Froeschle and Scholl (40.104.043), Voloshchuk and Kashcheev (40.104.073), Fox (41.104.010) and Babadzhanov and Obrubov (1986). Gravitational focusing of streams is discussed by Kramer and Shestaka (39.104.037). Šimek (1986)

compares the mass index "s" for a number of meteor showers. Lindblad (1987b) lists orbital elements and physical data for 30 major
meteoroid streams. Lebedinets (1987) derives density and fragmentation energy for meteoroids from 17 streams. Both authors conclude
that the Geminids, Virginids, Delta-Aquarids and Iota-Aquarids are
composed of denser meteoroid material than other streams.

A study of meteor streams with orbits inside the Earth's orbit
is presented by Simonenko, Terentjeva and Galibina (41.104.007).
The concept of a hollow meteor stream is discussed   by Jones (40.
104.048) and Hughes (41.104.008). Geminid observations by visual
observers appear to support this concept (41.104.076). The problem
of determining the mean radiant position of more than two meteor
trails is analysed by Steyaert (38.104.016). Various selection effects in the observations of meteors are investigated by Andreev
(38.104.032), Lebedinets and Manokhina (40.104.026) and Zabolotnikov (40.104.027). Halliday (1987) has described the first known
case of a meteorite stream, i.e. two meteorite objects moving in
essentially identical orbits. Galibina and Terentjeva (1987) suggest
that there is a stream including the Innisfree meteoroid and several fireballs.

Comet-meteor stream associations. Comet-meteor stream associations are discussed by Kulikova (38.104.034), Russell (39.104.023)
and Hughes (40.102.047; 41.104.009). Associations between ancient
comets and meteor showers are investigated by Kresakova (1986a,b;
1987). A list of theoretical meteor radiants associated with Earth-
approaching asteroids and comets is presented by Olsson-Steel (1987a)
Radar observations of meteors ascribed to Sugano-Saigusa-Fujikawa
comet (1985V) are reported by Šimek and Pečina (38.104.030; 41.104.
005). Visual observations 1977-84 of meteors associated with P/Grigg-
Skjellerup are reviewed by Lindblad (1986a). The flux of meteoroids
in the vicinity of the orbit of Halley comet is investigated by Cevolani and Hajduk (39.104.049) and Hajduk (41.103.638).

Sporadic meteors. The origin of the sporadic meteoroid complex
and its contribution to the zodiacal dust cloud is studied by Olsson-
Steel (41.104.004). The density distribution of sporadic meteor radiants is discussed by Pupyshev et al. (39.104.029) and Stohl (41.
104.019). The theory of determination of the sporadic meteoroid flux
is discussed by Pečina (37.104.006) and Bibarsov and Kolmakov (37.
104.045). Various selection effects in the observed distribution orbital elements of sporadic meteoroids and the variation of these
distributions with heliocentric distance are analysed by Zabolotnikov (37.104.041; 39.104.019; 40.104.004, 010; 41.104.032). The expected distribution of orbital elements of interstellar particles
entering the Solar system is investigated by Belkovich and Potapov
(40.131.046).

Quadrantids. Long-term radar observations are reported by McIntosh and Simek (37.104.007), Belkovich et al. (37.104.010) and Isamutdinov (40.104.089). Simultaneous radar observations from several
stations are discussed by Simek (42.104.020). The variation of the
mass index with date has been studied by Reznikov and Murav'eva (37.
104.046) and Isamutdinov (40.104.089).

Theoretical studies of Quadrantid stream evolution have been
carried out by Babadzhanov and Obrubov (38.104.012; 39.104.020, 026;
1986), Tkachuk (39.104.002), Froeschle et al. (41.042.012) and Sherbaum (42.104.057). Visual observations of Quadrantids are described
by Rendtel et al. (39.104.032).

Lyrids. Radar observations from Ottawa, Ondrejov, Budrio are
compared by Porubčan (41.104.011, 068). Budrio radar observations
1983-85 are presented by Porubčan, Cevolani and Formiggini (1986)

and the rate profile of the Lyrid shower is delineated. Porubcan and Cevolani (40.104.029) report on a short-lived outburst of Lyrid activity in 1982 as observed by meteor radars.

Extensive visual observations of Lyrid meteors are summarized by Porubčan and Štohl (37.104.019), Betlem (39.104.053), Roggemans (39.104.061) and Wood (39.104.068; 40.104.062). Telescopic observations are described by Jenniskens and Witte (40.104.035).

Eta-Aquarids. Although this meteor stream is ascribed to Halley comet and thus is of current interest surprisingly few observational results have been published. Hajduk and Vana (40.104.028) present a summary of Czechoslovakia radar observations 1969-77. These confirm a previously reported double peak of maximum activity. Australian visual observations in 1984-85 are described by Wood (39.104.067; 41.104.049). Within the framework of the International Halley Watch a number of radar groups (Onrejov, Onsala, Budrio, Ottawa, Dushanbe, Kharkov, Adelaide, Christchurch and elsewhere) have for several years monitored the main period of Eta-Aquarid activity. A preliminary list of participating stations is published (39.104.077). When these results are fully analysed it is expected that our knowledge of the stream and its activity profile will be greatly increased. One problem in studing this stream and its association with P/Halley is the almost complete lack of precise photographic meteoroid orbits on which a computer study of the orbital evolution ultimately should be based.

Delta-Aquarids. Radar visual heights for Delta-Aquarid, Perseid and sporadic meteors are discussed by Lindblad and Štohl (41.104.022).

Perseids. Long-term radar observations of the Perseid shower at Onsala (1953-78) and Ottawa (1958-74) are presented by Lindblad and Simek (41.104.013) and Simek and McIntosh (41.104.065). The activity variation with date is studied and the solar longitude of shower maximum is precisely determined. Radar observations 1982-84 in Japan are reported by Maeda (1986).

Zvolankova (39.104.004) presents Skalnate Pleso visual observations for the period 1944-53. Lindblad (41.104.012) presents Onsala visual observations 1953-81. These studies place the Perseid visual maximum slightly later than the radar maximum. The investigations indicate a surprisingly large scatter from year to year in Perseid activity at maximum. Amateur visual observations are discussed by Rendtel et al. (39.104.032), Roggemans (39.104.062; 42.104.005), Palzev (40.104.030), Betlem (40.104.031), Grishchnyuk and Bibnichenko (38.104.033) and Martynenko and Levina (42.104.017). A summary of visual and radar rate profiles is presented by Jenniskens (1986). Perseid meteor counts, heights, magnitudes and spectra are analysed by Russell (39.104.023; 41.104.067).

Taurids. A number of researches are presently undertaking extensive studies of the evolution of the Taurid meteor stream and the relation of this stream to Encke comet and sporadic complex. See papers (42.104.013, 055, 056). The relation of the Taurid stream to Apollo asteroids, the Tunguska object and the Brno meteorite has been discussed by Bailey, Clube and Napier (1986). Visual observations of Taurid meteors are reported by Veltman (1984; 39.104.050).

Draconids. Predictions for a 1985 Giacobinid (Draconid) meteor shower by Yevdokimov (39.103.190) and Yeomans and Brandt (39.103.181) indicated that shower maximum would occur on October 8 between 0.5^h and 13^h10^m. Radar and forward scatter observations showed a short-lived shower with peak activity at 09.35 hrs UT on October 8, 1985 (Lindblad 41.104.003; 1986b; Mason 1986). Radar observations 1985 are reported by Simek (1986) and Chebotarev (42.104.010). Bronshten (42.104.016) reports on the USSR visual and radar studies. Visual, photographic and TV observations in Japan are described by Koseki (1986), Nagasawa and Kawagoe (1987) and Nagasawa and Kanda (41.104. 066). Beech (41.104.061) presents a theoretical discussion of beginning and end heights of Draconid meteors based on photographic observations in 1946. An interesting summary of visual hourly rates recorded during the 1933 and 1946 displays is presented by Veltman and Jenniskens (40.104.034).

Orionids. Radar observations 1981-84 are reported by Hajduk et al. (37.104.005), Belkovich et al. (37.104.010), Cevolani and Hajduk (40.104.040; 1984), Isamutdinov and Chebotarev (41.104.036), Hajdukova et al. (1986), Hajduk et al. (1987) and Milutchenko (1987). Visual observations are presented by Porubčan and Zvolankova (39.104. 006), Veltman (39.104.050), Roggemans (39.104.063), Sisonov (41.104. 034) and Malysheva and Dubasova (1987). Telescopic observations are reported by Hajduk (42.104.050).

Leonids. Past aparitions of the Leonid shower and the relation to comet Temple-Tuttle are discussed by Kondratyeva and Reznikov (40.104.077) and Williams et al. (41.104.018). The age of the Leonid shower of 1966 is discussed by Kondratyeva (42.104.014).

Monocerotids. The possibility that some ancient fireballs are associated with the present day Monocerotid stream is investigated by Fox and Williams (40.104.047). The analysis of a number of precisely reduced photographic orbits (Lindblad 1987c) shows that the mean Monocerotid orbit agrees exactly with that of Mellish P/comet.

Geminids. Studies of the evolution of the Geminid meteor stream under planetary perturbations have been presented by Ceplecha (37. 104.036), Babadzhanov and Obrubov (38.104.012; 1986), Jones and Wheaton (39.104.007), Jones (40.104.048), Hunt, Williams and Fox (40.104.049), Kramer and Shestaka (40.104.076) and Jones and Hawkes (42.104.046). The dispersion of the Geminid stream by radiative effects is discussed by Olsson-Steel (1987b). The spatial structure and age of the stream is investigated by Belkovich (41.104.033), and Kramer and Shestaka (39.104.076; 40.104.001; 42.104.007). Radar observations 1974-76 in Kharkov are presented by Tkachuk (37.104.023) Visual and telescopic recordings are described by Wood (40.104.061), Roggemans (41.104.044) and Jenniskens (39.104.054; 41.104.076). Zvolankova (41.104.069) and Spalding (37.104.012) report on long-term visual observations.

The discovery of an asteroid (3200 Phaethon) moving in the some orbit as the Geminid stream is of great interest. The possible relation to the Geminid stream is studied by Fox, Williams and Hunt (40.104.042) and Hunt, Fox and Williams (41.104.016).

Ursids. Visual observations of the 1984 Ursids are reported by Veltman (39.104.051).

V.   METEOR ORBITS   P.B.Babadzhanov

The orbital elements for 454 TV meteors from 0.5 to 8.5 absolute magnitude observed between May 1981 and August 1982 are presented by Jones and Sarma (39.104.011). Both the cometary and asteroidal groups of meteoroids discovered in photographic survey are also given in the TV data. Many of TV meteors have retrograde - orbits with aphelia < 3 A.U. and also a greater fraction of orbits has very small perihelia when compared with Super Schmidt meteors. These features seem to suggest that meteoroid population is in a state of dynamic equilibrium with a source of TV meteoroids reple- s nishing those losses by Jovian perturbations and solar heating.

Orbital elements of 20 photographic meteors were published by Babadzhanov and Getman (40.104.091). Orbital data on about 30 multi-

station meteors photographed in the Netherlands between 1980 and
1985 are published by Betlem (40.104.005).

Jones and Hawkes (42.104.046) showed that the present duration
of the Geminid stream is the result of gravitational perturbations
combined with the initial spread in the orbits of the particles ejected from the comet. Jones (42.104.013) on the base of calculation
of the evolution of the Taurid stream complex computed the age of
the stream to be about 10^5yr.

Fox (41.104.010) calculated the orbital evolution of 53 meteoroid streams and concluded that of the 53 streams observable at present approximately half are no longer in Earth-crossing orbits either
one thousand years in the past or the future.

Froeschle and Scholl (41.104.017) investigated the dynamical
evolution of the Quadrantid-like meteoroid stream orbiting the Sun
in 2/1 mean motion resonance with Jupiter. They concluded that highly inclined streams located in resonance center may give a rise to
the formation of arcs.

Babadzhanov and Obrubov (1986) showed that planetary perturbations change strongly the stream shape due to the initial dispersion
in orbital elements of meteoroids ejected from the parent body. This
causes the streams to become much thicker and produce several coup-
les of showers active in different seasons of the year. For instance,
the Geminid stream can produce four showers, and the Quadrantid stream is predicted to produce eight showers.

VI. PHYSICAL THEORY OF METEORS AND FIREBALLS   D.O.ReVelle

The work related to meteor modelling has been carried out by
Novikov et al. (37.104.001; 38.104.003; 38.104.013; 39.104.024),
Kovshun (37.104.013), Kramer et al. (37.104.020), Kalenichenko (38.
104.020), Beech (38.104.023; 41.104.061), Jones et al. (39.104.012),
Nicol et al. (39.104.001), Novikov and Blokhin (39.104.017), Babadzhanov et al. (39.104.041), Kalenichenko (40.104.008), Bronshten et
al. (40.104.017), Lebedinets (40.104.019), Apshtejn and Shevoroshkin
(40.104.021), Andreev and Ryabova (40.104.023), Vislaya (40.104.025),
Beech (41.104.061), Sekanina (1985) and Fyfe and Hawkes (42.104.069).

Much work has continued on the detailed physical modelling of
the meteor phenomena. The efforts of Lebedinets, Babadzhanov and colleagues describe the fragmentation process as being quasi-continuous
including deceleration as well. A comparison of the theory and observations has allowed various fragmentation parameters to be derived
from the observations. Hawkes and colleagues have worked on the problem of calculating the residual mass remaining after the ablation of
very small meteoroids - the classical micrometeorites considered by
Whipple in the early 1950's. The goal of this work has been, in addition,to develope a better model of the entry process, to check the
grain model of Hawkes and Jones (1975) and to relate the residual
mass to the collections of Brownlee and others of this debris. Beech
has shown that there is a reasonably good agreement between the theory and observations with the exception that meteors are even more
fragile than assumed in the theory of Hawkes and Jones. Kovshun has
continued his work on the meteor light generation process and Kalenichenko has considered the light production from the liquid meteoroids. This could have implications for the theory of small water
comets impacting in the upper atmosphere following the work of Frank,
Sigwarth and Graven. Sekanina (39.103.182) has studied the Draconids
in relation to its parent Giacobini-Zinner comet and interpreted the
Draconids in terms of the porous meteoroid model of ReVelle rather
than the grain model of Hawkes and Jones.

The work related to fireball modelling has been carried out by Pečina and Ceplecha (37.104.009), ReVelle (37.104.026; 1985), Ceplecha (37.105.055; 39.104.056; 40.104.020; 41.104.020; 1985), Ouyang and Heusser (37.105.010), Halliday, Blackwell and Griffin (37.105.194), Gendzwill and Stauffer (38.104.021), Rietmeijer and McInnon (37.105.248), Padevet (37.105.136), Jha (39.105.105), Wetherill (39.105.212; 1986), Kyte and Brownlee (39.105.202), Padevet (39.104.030; 39.104.031), Korobeynikov et al. (39.105.233), Kalenichenko (40.104.018), Halliday (40.104.081), Bronshten (40.104.082), Pilugin and Chernova (41.104.023), Halliday (1985; 1987) and Walberg (1985).

Fireball modelling and observations have proceeded despite the loss of the MORP network from the Canadian Prairie. Data analysis is still continuing with at least one event, the Grand Prairie fireball (February 22, 1984) thought to have dropped about 12 kg to the ground despite an entry velocity of 26.5 km/s. More recently Halliday has shown that the Ridgedale fireball (February 6, 1980) has orbital characteristics identical to those of the Innisfri meteorite and searches for this object have also being planned. In the European Network Ceplecha has continued his work and searches for the Valec fireball although negative has produced some very interesting results. First, this is the deepest penetrating fireball ever photographed. Second, a precise analysis of the terminal portion of the trajectory revealed an ablation coefficient that increased with decreasing height by 20 indicating possible fragmentation. Wetherill has continued to use U.S. Prairie Network data to understand the physical origin of fireballs. In his work he has looked at the behaviour of the fireballs which are in orbits like those of the Geminids. Kalenichenko has also analysed the U.S.Prairie Network fireballs and has considered models for only two types of bodies, one chondritic and the other type which experiences intensive fragmentation. In the series of review papers Ceplecha has attempted to establish interrelations among fireballs, meteors and meteorites.He has also summarized a recent work done on the analysis of fireball trajectories using more realistic atmospheric density profiles as well as one total ablation coefficient for the entire entry. He has compared records from all the major fireball networks in terms of their future research should be done on this subject. These include detailed studies of the fragmentation process, more recoveries of bodies photographed during the entry the atmosphere etc. ReVelle has reconsidered the concept of the end point heights of fireballs in terms of the fireball energetics. He has shown that the end height can be interpreted as the point where the initial kinetic energy of the meteoroids have been reduced down to about 1 to 2 per cent.

Finally, Pilugin and Chernova have considered the optically thick approximation so that the radiative heat exchange during fireball entry can be modelled in terms of the radiant thermal conductivity. The disadvantage of this method is that it can only be used over part of the actual entry path.

Scientific debate still continues over the region and effects produced by the Tunguska event (1908). A related work has continued on this area (37.105.249; 38.105.106; 39.003.023; 39.105.230; 37.105.011; 39.105.231, 232, 234; 40.105.046, 047; 40.102.111).

The paper of Levin and Bronshten (40.105.046) is in effect a rebuttal of the work of Sekanina (34.105.010; 37.105.007) regarding with physical nature of the object that ultimately produced the famous Tunguska event, i.e. ever an asteroidal or cometary parent body to this reviewer it seems probably that the ultimate source was asteroidal (a carbonaceous chondrite) rather than cometary as supposed for many years.

Physical modelling relating to the entry the atmospheres of
planets have been carried out by Apshtejn et al. (40.104.015) and
Vislaya et al. (40.105.008).

In the work of Wang et al. (38.104.014) and Keay (39.104.025)
the so-called anomalies sounds from fireballs are considered. Bibarsov (41.104.063) has considered the production of ozone during the
interaction between meteoroids and the atmosphere. Halliday et al.
(1985) have considered the possible impacts of meteorites on people
and buildings using flux rates derived from the MORP network. Lang
and Franaszczuk (1986) have considered the fragmentation process
in terms of the fractional concept for the Lowicz meteorite and
identify two distinct stages of fragmentation during entry. Finally,
Oberst and Nakamura (1987) have reexamined the flux of meteoroids as
determined using the lunar seismic data taken during the Apollo program.

VII. METEOR SPECTRA   V.A.Bronshten

Unfortunately, meteor spectra have been poorly investigated
during the last triennium. This seems to be surprising that a serious quantitative analysis of meteor spectra should provide the abundance of information being compared with that obtained from the analysis of stellar spectra.

Halliday has studied the spectra of 17 bright meteors of the
Orionid shower, photographed within the period from 1958 to 1968.
The spectral interval from 3100 to 8700 Å was embraced. The lines
of 6 neutral atoms (NI, OI, NaI, MgI, CaI, FeI) and 5 ions (MgII,
SiII, CaII, CrII, FeII) were found. The comparison of Orionid spectra with Perseid ones showed a very small difference in the content
of the observed elements. This may be accounted for the small difference in compositions of their parent comets. Russell (41.104.067)
studied meteor spectra of the Perseids observed in 1977-83. A correlation between spectrum and magnitude of the Perseids was found.
Certain spectral characteristics appear correlated with the atmospheric density gradient along the meteor path. Inhomogeneity in the
meteoroids may play an additional role in spectral-magnitude relations.

Ovezgeldyev et al. (42.104.064) investigated spectra of meteor
trails with electron-optical equipment. Their comparison with the
spectra of meteor "heads" showed that spectra of "wakes" and trails
are poor with lines and, mainly, contain the lines of lower excitation potentials. An important feature of these spectra is a forbidden green line OI 5577 Å. Probable mechanisms of photochemical and
aeronomical reactions causing the emission of the observed lines
and bands in meteor trails were considered.

VIII.METEORS AND AERONOMY   W.J.Baggaley

Ablation. Information on the friable structure and desintegration processes of meteoroid dustballs associated with specific showers has been obtained by Beech (38.104.023) using data for characteristic of beginning and end heights, for visual meteor ablation.
Models of the residual mass produced by entry of single independent
body showed (39.104.001) a strong velocity dependent for the ablation residue and emphasized the small contribution to micrometeorites
made by major showers (because of their high velocities). This work
was extended (42.104.069) distinguishing large body ablation where
grain ejection occurs along the primary optical ionization track,
from small meteors where the majority of grains are released above
the height of grain evaporation. This dustball model permitted the pre-

diction of ablation products as a function of meteoroid parameters.

Diffusion. Estimates of the influx of meteoric dust and the mass distribution of the Solar system dust cloud are uncertain be- cause, in an important mass regime (approximately 10^{-5} to 10^{-10} kg) covered by radar meteors, a velocity dependent echo attenuation ef- fect operates. This echo ceilling - due to a large ionization column and rapid plasma diffusion - is important for radars operating in the HF band. To circumvent this observational selection, measure- ments of radar meteor fluxes have been carried out by Olsson-Steel and Elford (1987) at frequences of 2 MHz and 6 MHz.

Ionospheric effects. The redistribution of metal ions origina- ting from meteoric ablation into lower E-region ionization irregu- larities by the action of shears in the zonal wind is known to be responsible for the phenomenon of ionospheric E_s. Many factors con- trolling the processes are not understood and some measure of the time-constants involved may be gained by examining the association between E_s occurrence and the major meteor showers. In a 20-yr sur- vey (38.104.028) no correlation was found, indicating the existence of a large reservoir of meteoric ions with input from shower being only minor. Increases in ionization in the height region of 85 km observed on a VLF ionosonde and by phase anomalies on long distance propagation have been attributed (Vilas Boas et al., 1986) to the Alpha Scorpiids and Geminids. Similary long distance LF propagation effects have been associated (Sarka and De, 1985) with Leonid shower ionization yielding estimates of the effective electron attachment coefficient at 70 km.

Modelling of neutrals and ions. Progress in modelling meteoric M metals and alkalis has continued (Jegou et al. 1985; Kirchhoff 1986) with inclusion of the ablation source function, eddy diffusion, chemistry (including clustering chains), horizontal winds and E fi- elds. Such models have been applied to observed diurnal-seasonal be- haviour (Jegou et al. 1984; Granier et al. 1985). Grebowsky and Pharo (1985) demonstrated that mid-latitude nighttime metal ions can be lifted from their ablation region $\sim$100 km to F-region heights (in 30 min) by Pederson drifts produced by poleward electric fields. A review of rocket-born mass spectrometer M^+ measurements has been carried out (Swider 1984) and the data relevant to auroral conditi- ons have yielded estimates of charge transfer rates to M from NO^+. Swider (1986) has demonstrated that simple steady state chemical models are successful in describing the relative composition of neut- ral Na compounds near the 90 km sodium layer peak without recourse to 1D dynamical models. Above the layer peak ion behaviour as well as transport become important. The large seasonal changes in the sodium layer has been discussed by Swider (1985) in terms of three- body association producing NaO_2.

Measurements of ablated meteoric species. Long-lived Na clouds observed using a tilting photometer (Kirchhoff and Takahashi 1984) and resonance scattering from enhanced Li during the extended arctic twilight (Henriksen et al. 1986) have both been attributed to direct meteoric input. Lidar records of Na layers at 93-101 km associated with E_s showing growth times of only minutes, widths $\sim$ 1 km and life- times of hours have been explained by Zahn et al. (1987) in terms of the release of atoms from sodium bearing dust by auroral electrons - evidence for a reservoir of Na adsorbed onto dust. Rocket mass spectrometer identification of M^+ ions have shown (Kopp et al. 1984) layer having a sharp cut-off below 91 km while flights of concurrent with NLC on meteor shower and non-shower days have produced (Kopp et al. 1985) detailed height profiles showing dominance by Mg^+ and Fe^+ over Na^+, Al^+, Ca^+ with Si^+ behaviour reflecting its different chemistry. A variety of meteoric ion compounds MO, MO_2, MOH and hydrates was also determined as well as heavy proton hydrates $H^+(H_2O)$ n 12. Such measurements are valuable in not only ion chemistry modelling but also in the importance of the role of ions as nucleation centres for NLC development.

NLC and PMC's. Work has continued in an effort to further our understanding of aeronomy of NLC and PMC development and evolution in the summertime high latitude mesosphere. We still lack data on some of the important microphysical processes. Thomas and McKay (1985) and Roddy (1984) have considered ice growth as well as appropriate crystalline form of ice. We are still not clear as to the identity of the ice nucleation centres; either meteoric dust - coagulated meteor ablation products ('smoke') or the result of metal water cluster ions or proton hydrates growth under supersaturated conditions. Finally, are NLC's and PMC's distinct phenomena?

IX. TEKTITES   C.Koeberl

Tektite research made some progress in the years covered by this report. The close connection between tektites and impact glasses has been further quantified by a number of workers. Recent interest in tektites has focused around the question if tektites producing events are related to other terrestrial events like reversals of the Earth's magnetic field, climatic changes, and extinctions like the one at the end of the Cretaceous. The proceedings of the first International Conference on Natural Glasses (Pye et al., 1984) included a number of papers on tektites and impact glasses as well as studies on the long-term stability and leaching processes of tektite and other glass. Schreiber et al. (1984) presented a thorough study on the redox state of iron in tektites. The first data on fluorine in tektites and impact glasses were reported by Koeberl et al. (1983; 39.105.012) and Moore et al. (1984), showing that fluorine must have been lost in the tektite production process, similar to other halogens and other volatiles.

The so-called Muong Nong type tektites contain large bubbles, are inhomogeneous and layered, much larger than splash-form tektites, and not spherical symmetric or aerodynamically shaped. Muong Nong type tektite iron is present in the +3 state (Koeberl et al., 1984b). The most prominent features of Muong Nong tektites are light and dark layers, and Koeberl (1985) was able to show that the light layers are enriched in almost all elements. Dark layers have higher SiO_2 contents than light layers. Surface structures of Muong type tektites have been studied by Futrell (1986).

Storzer and Müller-Sohnius (1986) have studied the enigmatic high sodium/potassium tektites from the Australasian region. Using K/Ar dating they confirmed earlier studies like the one by Storzer (41. 105.148). The age of these tektites seems to be close to 11 million years. Recently some strange glasses similar to the high Na/K tektites have been reported from the Tikal (Guatemala) area (Essene et al., 1987). Among the most important new discoveries is the find of tektite debris in deep sea sediments at Barbados and some deep sea drilling sites (39.105.086; Thein, 1987). Previously tektites were found on land. Tektites are of some age than microtektites but are found in much younger sediments, microtektites are found in sediments of the correct (old) age. Some emphasis has been laid on trace element (Koeberl and Glass, 1986) and isotopic studies (39.105.223) of the new material, and it was shown unambiguously that the tektites found in Barbados as well as those recovered from deep sea sediments off the coast of New Jersey are part of the North American strewn

field. This result is supported by age determinations (Glass et al.,
1986). The chemistry of the tektites seems to imply that they are
slightly different than the well known bediasites and georgiaites.

The Ries crater - moldavite association has been studied by
Horn et al. (1985) using isotope studies. Possible source materials
(Middle Miocene sands) for the moldavites have been identified and
studied by Engelhardt et al. (1987) and Delano et al. (1986, 1987).
A recent review of the geochemistry of tektites and related impact
glasses was given by Koeberl (1986b).

Recent studies of the water content of tektites and impact glasses (Koeberl and Beran, 1987) have shown that there is a genetic
sequence: Muong Nong tektites contain more water than splash form
tektites, in accordance with the other volatiles. The Zhamanshin
impact crater, source of several impact glasses and tektite like
material (irghizites, Si-rich zhamanshinites) has been subject of
recent investigations, like an extensive chemical study by Koeberl
and Fredriksson (1986). Irghizites are depleted in volatiles and wate
if compared to the zhamanshinites, which in turn are very similar
to the Muong Nong type tektites and seem to constitute the lower
temperature impact glasses. Murali et al. (1987) described a similar
genetic sequence of impact materials, leading to tektite like materials at the upper end of the temperature pressure scale, from the
Lonar crater in India.

The connection between microtektites, tektites and extinctions
is currently being studied at the Eocene-Oligocene boundary, but no
conclusive results have been obtained to date (Keller et al., 1987;
D'Hondt et al., 1987). The connection between giant impacts and tektites and a soft transition between tektites and impact glasses becomes, however, more and more established.

X.   INTERPLANETARY DUST   V.N.Lebedinets

Distribution and dynamics. Voloshchuk and Kascheyev (40.104.073;
40.104.006) and Zabolotnikov (42.104.032) have carried out the mathematical simulation of the interplanetary dust distribution in the
Solar system. Kascheyev and Kolomiets (39.104.038) and Andreev et al.
(1987) found the discrepancy between the observed distribution of hy-
perbolic radio-meteors and theoretically predicted interstellar particles' orbits. Olsson-Steel (41.104.004) taking into account desintegration of meteor streams due to collisions among meteoroids
and planets as well as present concentration of dust in the zodiacal
cloud, concluded that 10^4 years ago the amount of meteor streams and
their parent comets was essentially greater than at present. From
radar observations Longo and Morris (41.081.038) have shown the absence of the Earth's natural satellites with diameters of 5cm (in
the orbits with a perigee $q \leqslant 400$ km) and 40 cm (in the orbits with
$q \leqslant 10\ 000$ km) although a great number of fragments of artificial
bodies was discovered. Barsukov and Nazarova (1985) from the cosmic
dust impact data obtained by space probes, have found that dust particles move around the Sun, Earth and Moon as separate circular systems at certain distances from the centres of gravity. Kapishinskij
(40.106.011; 41.106.070) analysing measurement data of space probes
concluded that   -meteoroids are formed as a result of impacts of
larger dust particles.

The series of works consideres the influence of different non-
gravitational effects on the motion and desintegration of dust particles. Alfonso (41.106.076) has studied the influence of the Poynting-Robertson effect and deceleration of charged particles in the
extraterrestrial plasma. Mendis (1984) calculated the interaction of

charged particles with ionosphere. Wallis (1986) and Wallis and Hassen (40.106.026) have shown that the influence of stochastic forces resulted from the fluctuations in surface-charge and interplanetary magnitic field for micron and submicron particles may exceed the influence of the Poynting-Robertson effect and cause the quicker fall of dust particles on the Sun as well as their removal from the Sun. Mukai (40.106.027) notes that the particles with radii less than 0.1 mkm resulted from heterogeneous ice meteoroids after ice sublimation can move in elliptical orbits. Kapisinski (1987) estimated that the erosion of dust particles with radii of 0.5 - 10 mkm under the action of the Solar wind and impact of smaller dust particles can essentialy reduce their lifetimes in comparison with the Poynting-Robertson effect. Steel and Elford (41.106.001) presented the Tables for calculations of the frequency of impacts of dust particles and lifetime of particles of different sizes in different orbits including orbits of 28 meteoroid streams. Grün et al. (39.106.063) calculated the balance of formation and desintegration of particles of different masses at their mutual encounters with account of mass segregation and concentration of dustballs. On the base of lunar crater statistics and data on meteor and zodiacal light observations Grün et al. (1984) calculated the equilibrium distribution of particles with masses up to meteoroids with account of their mutual collisions, the Poynting-Robertson effect and light pressure. Zook et al (40.106.055) have noted that lunar crater statistics can give overestimated values of dust influx because a part of craters is produced by the secondary dust particles ejected during impacts. Analysing the micrometeorite measurements obtained by space probes, Fechtig (39.106.028) concluded that the increase in frequency of impacts by one or two orders in the vicinity of Jupiter and Saturn can not be explained only by the gravitational focusing; this dust was ejected from the planetary satellites and crushed under the action of electrostatic forces. Alexander et al. (39.091.020, 021) estimated the flux density of micrometeorites ejected from surfaces of stone satellites of planets at meteoroid impacts. Fechtig (39.106.028) and Fechtig and Mukai (40.106.014) considered the evolution of Grinberg's particles ejected from a cometary nucleus.

Schiffer (40.106.014) noted that the observed redded zodiacal light in comparison with the solar spectrum can be explained by the roughness of surfaces of particles with diameters exceeding the light wave length. Giovane et al. (1985), Hong (39.106.036), Lamy and Perrin (42.106.001), Maucherat et al. (42.106.009), Sarma and Sommerfeld (1987) investigated the light scattering characteristics of interplanetary dust and the structure of zodiacal dust cloud.

Dermott et al. (38.106.029; 41.106.023) and Sykes and Greenberg (41.106.034) considered characteristics of intensive bands of zodiacal light near the ecliptic discovered from IRAS and showed that they could be formed due to encounters among asteroids. Sykes (41.102.077) indicates that narrow dust streams in the orbits of some comets are formed by the ejections of large dust particles from the cometary nuclei at small velocities. Walker and Aumann (1984) studied the amount of ejected dust from 9 known and 8 recently discovered comets on the base of measurement data obtained by IRAS.

Babadzhanov et al. (40.104.074; 1985) from photographic and radar observations in Dushanbe estimated the flux density of meteoroids with masses of 10^{-3} - 10^{2} g. Blinov et al. (39.106.069) from the measurements of radioactive isotope ^{26}Al in the red clay in the Pacific Ocean estimated the influx of cosmic dust onto the Earth equaled 150 ton/day. LaViolette (40.106.092) according to content of Ir and Ni in the dust filtrate from the ice kern Kemp-Century in the

layer aged 19 700- 14 200 yr estimated that the influx of cosmic
dust at those times was greater by 1-2 orders than at present.
       Physical characteristics and sources. Cosmic experiments "Vega-1
"Vega-2" and "Giotto" allowed to estimate the intensity of dust ejection from the nucleus of Halley comet: according to Sagdeev et al.
(1986a) - from 5 to 10 ton/day, according to Vaisberg et al. (41.
103.667) - 4 ton/day and McDonnel (41.103.689) - 3.1 ton/day. These
values are in satisfactory agreement with Grün's et al. ones (41.
103.649) from ground-based observations - 1.5 ton/day. McDonnel et
al.(41.103.689) from measurement data obtained by "Giotto" estima-
ted that in the mass range $10^{-17} \leqslant M \leqslant 10^{-5}$ g the observed mass distribution of dust particles is characterized by index 's' equaled
1.66+0.05 (in the coma of Halley comet). Simpson et al. (41.103.669)
from measurements at "Vega-1,2" at $M \geqslant 10^{-13}$ g obtained $s \approx 1.9$. The
absence of a cut in mass distribution at $M < 10^{-11}$ g was unexpected
for the authors of these experiments. Mazets et al. (41.103.262)
suggest that when moving away from the nucleus the fraction of the
smallest dust particles increases and this may be accounted for the
fragmentation of the larger friable particles. Fragmentation of friable particles was also noted by Storrs et al. (41.103.262) near
the nucleus of IRAS-Araci-Alcock comet from ground-based observations
       From ground-based observations of Jacobini-Zinner comet in the
visual and IR ranges in August 1985 Tedesco et al. (1986) have estimated the intensity of dust ejection from cometary nucleus equaled
0.8 ton/day at the particle ejection velocity less than 3 m/s.
       Krasnopolsky et al. (1986) from spectral observations at "Vega-2
estimated the average density of particles in the Halley comet coma
to be 0.35 g/cm^3. Reinhard's (1986) mass-spectrometric investigations at "Giotto" discovered H, C, N, Na, Mg, Si, Ca and Fe in composition of particles. Kissel et al. (41.103.688) obtained that relative to the chondrite composition the dust particles are enriched
with H, C, N and O; density of particles is very small. Sagdeev et
al.(1986b) and Kissel et al. (41.103.670) from the data of PUMA experiment at "Vega" (mass-spectrometric measurements) obtained that
particles are divided into three principle groups: (1) carbonaceous
chondrites, (2) carbonaceous chondrites enriched with carbon, nitrogen and sulphur and (3) particles, mainly consisting of hydrogen,
carbon and oxygen, and mineral nucleus may be absent in particles.
       According to measurements of IR spectra of dust comae of Cher-
nise and Bouel comets, Hoyle et al. (40.103.003) estimated that the
coma of Chernise comet contains silicate and organic carbonaceous
matter as well as about 10 per cent of ice, and the coma of Bouel comet appears not to contain free ice, but organic matter with hydroxyl
Struzzulla et al. (40.106.015) simulated the influence of solar wind
on particles of cometary origin, containing ice, and showed that over
the particle lifetime (about 10^6 yr) a strong erosion occurs; carbonaceous chondrite matter is formed from hydrocarbonates and the particle density increases. Morozhenko et al. (1986) from the data of
laboratory simulation and polarimetric and spectrophotometric observations of comets identified the ice particles near the cometary
nuclei. Frank et al (1986a,b) from UV observations of the Earth from
the board of the space probe "De-1" discovered that very small cometary nuclei consisting of snow and ice with masses up to 100 tons and
including very fragile dust cloak with masses of order of 1-10 g,
enter the atmosphere with a frequency of 20 min^{-1}. This new type of
minor bodies of interplanetary environment according to Donahue
(1987) should enrich the interplanetary space with hydrogen due to
ice sublimation. Geiss et al. (1986) according to data of mass-spectrometric measurements from the board of the space probe discovered

in the coma of Giacobini-Zinner ions of Na^+ and Mg^+ which could be resulted only from particles containing frozen solutions of Na and Mg salts. Lebedinets (1987b) from the analysis of heights and decelerations of Super Schmidt meteors noted the existence of such particles among the meteoroids of the Draconid meteoroid stream.

Struzzula (1986) concluded that composition of particles of a cometary origin significally differs from dust composition of a protoplanetary cloud.

Bradley et al. (39.106.042, 065) and Johnson and Lanzerotti (41.106.072) studied nucleous tracks in the stratospheric aerosols and estimated the lifetime of dust particles in the interplanetary space to be 10^4-10^5 yr. Bradley et al. (41.106.009), Reitmeyer (1986) and Zolensky (41.106.015) classified the interplanetary dust particles collected in the stratosphere and for the first time separated porous chondrite agrigates with porosity up to 50 per cent and pieces of glass; the main type is carbonaceous chondrites enriched with volatiles. Schramm et al. (1985) have studied the composition of interplanetary dust according to data obtained by space probe Solar Max Satellite and concluded that in most of cases the composition is close to that of chondrites. Lebedinets (1986, 1987a,b) from the analysis of decelerations and heights of Super Schmidt meteors obtained that all the main types of meteorites as well as dustballs and carbonaceous CO chondrites containing water and other volatiles up to 50 per cent are presented among the meteoroids; the main type of meteoroid is carbonaceous CI chondrites. It is shown that in meteoroid streams such as Taurid, Perseid and Alpha Capricornid (associated with Encke, 1862 III, 1954 III comets) there are iron and stone meteoroids as ordinary chondrites.

Lavrukhina and Mendybaev (1987) on the base of analysis of numerous investigations of chemical, mineral and isotopic compositions of cosmic dust from the stratosphere and oceanic sediments have tried to establish a primary composition of particles from protoplanetary cloud and pecularities of genesis of interplanetary dust.

Lumme and Bowell (39.106.049) according data of photometric and polarimetric observations of zodiacal light have estimated the optical characteristics of dust particles from zodiacal cloud and concluded that particles are resulted from encounters among asteroids. Agafonova and Drobyshevski (39.099.042; 40.098.010; 1985) investigated the possibility of the formation of minor bodies in the Solar system as a result of the explosion of ice shells of giant planets' satellites.

REFERENCES

Agafonova, I.I. and Drobyshevski, E.M. : 1985, Earth, Moon and Planets, 33, p. 1.

Andreev, G.V., Kascheev, B.L. and Kolomiets, S.V. : 1987, Meteornye issledovaniya, Moskva, p. 93.

Andrianov, N.S. : 1987, Meteorn. Rasprostr. Radiovoln, 20, p. 9-14.

Babadzhanov, P.B., Bibarsov, R.Sh., Getman, V.S. and Kolmakov, V.M.: 1985, Doklady Akad. Nauk SSSR, 284, p. 824.

Babadzhanov, P.B., Obrubov, Yu.: 1986, Doklady Akad. Nauk SSSR, p. 290.

Bailey, M.E., Clube, S.V. and Napier, W.M.: 1986, Vistas in Astron., p. 29-53.

Barsukov, V.L. and Nazarova, T.N.: 1985, Properties and Interactions of Interplanetary Dust, p. 99.

Bayrachenko, I.V.: 1985, Problemy Kosmicheskoi Fiziki, 20, p. 49-53.

Ceplecha, Z.: 1985, First GLOBMET Symposium (M.A.P.) - Dushanbe, USSR.

Cevolani, G. and Hajduk, A.: 1984, II Nuovo Cimento 7C p. 447-457.

Delano, J.W., Bouska, V. and Fernandez, M.M.: 1986, Lunar Planet. Sci. 17, p. 170-171.

Delano, J.W., Bouska, V. and Randa, Z.: 1987, Lunar Planet. Sci. 18, p. 233-234.

D'Hondt, S.L., Keller, G. and Stallard, R.F.: 1987, Meteoritics 22, p. 61-79.

Donahue, T.M.: 1987, Geophys. Res. Lett., 14, p. 213.

Duffy, A., Hawkes, R.L., Jones, J.: 1987, Mon. Not. Roy. Astr. Soc. (submitted).

Engelhardt, W.V., Luft, E., Arndt, J., Schock, H. and Weiskirchner, W.: 1987, Geohim. Cosmohim Acta 51.

Essene, E.J., Moholy-Nagy and Nelson,F.W.: 1987, EOS-Trans. Am. Geophys. Union 68, p. 462.

Frank, L.A., Sigwarth, J.B. and Graven,J.D.: 1986a, Geophys. Res. Lett. 13, p. 303.

Frank, L.A., Sigwarth, J.B. and Graven,J.D.: 1986b, Geophys. Res. Lett. 13, p. 307.

Futrell, D.S.: 1986, J. Non-cryst. Solids 86, p. 213-218.

Galibina, I.V. and Terentjeva, A.K.: 1987, Astron. Vestnik, 21, p. 251-261.

Geiss, J., Bochsler, P., Ogjlriee, K.W. and Coplan, M.A.: 1986, Astron. and Astrophys., 166, p. LI.

Glass, P.B., Hall, C.M. and York, D.: 1986, Chem. Geology 59, p. 181-186.

Giovana, F., Weinberg, J.I., Mann, H.M. and Oliver, J.P.: 1985, Properties and Interactions of Interplanetary Dust, p. 39.

Granier, C., Jegou, J.P., Chanin, M.L. and Megie, G.: 1985, Annales Geophysicae, 3, p. 445-450.

Grebowsky, J.M. and Pharo, M.W.: 1985, planet. Space Sci., 33, p. 807-816.

Grün, E., Fechtig, H., Zook, H.A. and Giese, R.H.: 1984, Lunar and Plan. Sci., 15 Pt. 1, p.333.

Halliday, I.: 1987, Icarus, 69, p. 550.

Halliday, I.: 1985, J. Roy. Astron. Soc. Can. 79, p. 197.

Halliday, I., Blackwell, A.T. and Griffin, A.A.: 1985, Nature 318, p. 317.

Hajduk, A., Hajdukova, M., Cevolani, G. and Formiggini, C.: 1987, Bull. Astron. Inst. Chech. 38, p. 129-131.

Halliday, I.: 1986, Proc. 20-th ESLAB Symp., ESA SP-250, p. 245-248.

Hajdukova, M., Halduk, A., Cevolani, G. and Formiggini, C.: 1986, Proc. 2o-th ESLAB Symp., ESA SP-250, p. 371-373.

Hawkes, R.L. and Jones, J.: 1975, Month. Not. Roy. Astron. Soc.,173, p. 339.

Hawkes, R.L. and Jones, J.: 1986, Quart. J. Roy. Astron. Soc. 27, p. 569-589.

Horn, P., Müller-Sohnius, D., Kohler, H. and Graup, G.: 1985, Earth Planet. Sci. Lett. 75, p. 384-392.

Jegou, J.P., Granier, C., Chanin, M.L. and Megie, G.: 1985, Annales Geophysicae 3, p. 163-176.

Jegou,J.P., Granier, C., Chanin, M.L. and Megie, G.: 1985, Annales Geophysicae 3, p. 299-312.

Jones, J. and McIntosh,B.A.:1986, Proc. 20th ESLAB Symp., ESA SP-250.

Kapisinsky, I.: 1987, Bull. Astron. Inst. Czech. 38, p. 7.

Keller, G. et al.: 1987, Meteoritics 22, p. 25-60.

Kirchhoff, V.W.J.H. and Takahashi, H.: 1984, Planet. Space Sci. 32, p. 831-836.

Kirchhoff, V.W.J.H.: 1986, IAGA 5th Scientific Assembly, Prague.

Koeberl, C.: 1985 Lunar Planet. Sci. 16, p. 449-450.

Koeberl, C.: 1986, Ann. Rev. Earth Planet. Sci 14, p. 323-350.
Koeberl, C. and Beran, A.: 1987, Lunar Planet. Sci. 18, p. 497-498.
Koeberl, C. and Fredriksson, K.: 1986, Earth and Planet. Sci. Lett. 78, p. 80-88.
Koeberl, C. and Glass, B.P.: 1986, Meteoritics 21, p. 421-422.
Koeberl, C., Kiesl, W., Kluger, F. and Weinke, H.H.: 1983, Lunar Planet. Sci. 14, p. 381-382.
Koeberl, C., Kluger, F. and Kiesl, W.: 1984, Meteoritics, 19, p. 253-254.
Kopp, E., Eberhardt, P. and Herrmann, U.: 1985, J. Geophys. Res. 90, p. 13041-13053.
Kopp, E., Ramseyer, H. and Bjorn, L.G.: 1984, Adv. Space Res. 4, p. 157-161.
Koseki, M.: 1987, J. Meteor Obs. (Nippon Meteor Soc.), No. 2, p. 31-47 (Japanese).
Kramer, E.N., Markina, A.K., Shestaka, I.S.: Meteor Orbits from Photograph. Obs. in 1957-1983 yr., Moscow, VINITI, 1986, 19.
Krasnopolsky, V.A., Morel's,G., Gogoshev, M. et al.: 1986, Pis'ma v Astron. J., 12, p. 616.
Kresakova, M.: 1986a, Bull. Astron. Inst. Czechosl., 37, p. 339-344.
Kresakova, M.: 1986b, Proc. 20th ESLAB Symp., ESA SP-250.
Kresakova, M.: 1987, Bull. Astron. Inst. Czechosl., 38, p. 75-80.
Lang, B. and Franaszczuk, K.: 1986, Meteoritics 21, p. 428.
Lavrukhina, A.K. and Medybaev, R.A.: 1987, Thesisy Dokl. XX Meteorit. Konferenc. USSR, Moskva, p. 72.
Lebedinets, V.N.: 1986, Dokl. Akad. Nauk USSR, 291, p. 313.
Lebedinets, V.N.: 1987a, Astron. vestnik, 21, p. 65.
Lebedinets, V.N.: 1987b, Astron. vestnik, 21, p. 262.
Lindblad, B.A.: 1986, Proc. ESLAB Symp., ESA SP-250, p.229-231 and p. 399-400.
Lindblad,B.A.: 1986, in "Asteroids, Comets, Meteors II" (eds. Lagerkvist et al.), p. 531-535, p. 545-548 and p. 615-620.
Lindblad,B.A.: 1986a,b Proc. 20th ESLAB Symp., ESA SP-250.
Lindblad, B.A.: 1987a, Meteor studies, Proc. IAU Coll. No 98, Paris, 1987 (in press).
Lindblad B.A.: 1987b, Proc. Int. School of Physics "Erico Fermi", (eds. Fulchioni and Kresak).
Tedesco, C.M., Decher, R., Baugher, C. et al.: 1986, Astrophys. J. 310, p. 61.
Thein, J.: 1987, Init. Rept. Deep Sea Drilling Proj. 95, p. 565-579.
Thomas, G.E. and McKay,C.P.: 1985, Planet. Space Sci. 33, p. 1209.
Veltman R.: 1984, Radiant, 6, p. 5-7.
Vilas-Boas, J.W.S., Paes Leme, N.M., Rizzo Piazza, L. and Da Costa, A.M.:1986.
von Zahn,U., von der Gathen, P. and Hansen, G.: 1987, Geophys. Res. Lett. 14, p. 76-79.
Walberg, G.D.: 1985, J. Spacecraft 22, Special Section: AOTV, p.3-67.
Walker, R.G. and Aumann, H.H.: 1984, Adv. Space Res. 4, p. 197.
Wallis, M.K.: 1986, Nature, 320, p. 146.
Wetherill, G.W.: 1986, Meteoritics 21, p. 537.

P.B.Babadzhanov

President of the Comission

24. PHOTOGRAPHIC ASTROMETRY (ASTROMETRIE PHOTOGRAPHIQUE)

PRESIDENT: Arthur R. Upgren
VICE-PRESIDENT: William F. van Altena
ORGANIZING COMMITTEE: A. N. Argue, T. E. Corbin, Ch. de Vegt,
                      W. Gliese, I. I. Kanaev, T. E. Lutz, J. Stock

I. Introduction

One of the continuing problems facing Commission 24 is its name. It has been known and reported for some years that the present title is not suitable for reasons that are all too obvious. This problem has continued during the last three years, despite the growing development and use of non-photographic methods and techniques.

Most of the recent meeting activity in which the Commission and its membership has been involved, has centered on its shared interests with Commission 8 and with the Working Group on reference frames formed at the last General Assembly. Since its inception at that time, C. A. Murray has been appointed to represent the Commission on that group.

Meetings of interest to the Commission and its membership held during this period include the following:

IAU Symposium No. 128, "Earth's Rotation and Reference Frames for Geodesy and Geodynamics" was held at Washington in October 1986. The SOC was chaired by D. D. McCarthy and W. E. Carter was the chairman of the LOC.

IAU Symposium No. 133, "Mapping the Sky - Past Heritage and Future Directions" was held at Paris in June 1987. The meeting celebrated the centennial of the Carte du Ciel. J. A. Eddy and A. R. Upgren served as co-chairmen of the SOC. They, S. Débarbat and H. K. Eichhorn are the editors of the proceedings. The chairman of the LOC was P. Charvin.

IAU Colloquium No. 96, "The Few Body Problem" was held at Turku in June 1987. The chairman of the SOC was V. A. Brumberg and the chairman of the LOC was M. Valtonen.

IAU Colloquium No. 97, "Wide Components in Double and Multiple Stars" was held at Brussels in June 1987. The meeting was organized by J. Dommanget who also served as chairman of the SOC. The chairman of the LOC was E. Van Dessel. The meeting was dedicated to W. J. Luyten and the proceedings are being edited by Z. Kopal.

IAU Colloquium No. 100, Fundamentals of Astrometry" was held at Belgrade in September 1987 to celebrate the centennial of the Astronomical Observatory there. The meeting had been organized by G. Teleki. Following his very unfortunate death in February 1987, Eichhorn accepted an invitation to succeed Dr. Teleki as chairman of the SOC and I. Pakvor chaired the LOC. Eichhorn, Murray and Upgren are editing the proceedings.

"Astrophysics of Brown Dwarfs" is the title of a meeting and its published proceedings, edited by M. C. Kafatos, R. S. Harrington and S. P. Maran. The meeting was held at Fairfax, Virginia in October, 1985.

Meetings with papers of intrerest held prior to the period covered by this report but published during this period include IAU Symposia Nos. 109, 111, 114 and 118 and Highlights of Astronomy 7, 1986. Each contains a number of contributions dealing with astrometric subjects.

The astrometric work on HIPPARCOS and other satellite projects is covered in detail in the report from Commission 8 and consequently will not be repeated extensively here. In brief, A. N. Argue, Chairman of HIPPARCOS Input Catalogue Subgroup 2130, reports on the link to the extragalactic reference frame. The final catalogue has an average density of 2.5 stars per square degree. A detailed account of it is being prepared for publication, edited by C. Turon and M. A. C. Perryman. E. Hog, J. Kovalevsky and Turon report on the progress since the last General Assembly. The satellite is expected to be finished in early 1988, but the launch is scheduled for 1989, for a 2.5-year mission.

Kovalevsky has reviewed the present achievements of ground-based astrometry and the prospects of stellar astrometry from space. The methods used to construct a quasi-inertial reference frame are discussed along with HIPPARCOS and its payload and detector. The establishment of an absolute reference frame and the use of VLBI observations are also described (38.041.031, 38.041.042).

D. Hoffleit is conducting research on the history of astronomy at the Yale Observatory through the directorship of Dirk Brouwer, much of which involves the history of astrometry. She is also updating the tape version of the Bright Star Catalogue with W. Warren. Several thousand corrections and additions have been made since the most recent printed version of 1982.

II. Trigonometric Parallaxes and Nearby Stars

The fourth edition of the General Catalogue of Trigonometric Stellar Parallaxes is nearing completion. W. F. van Altena gave a status report on it at IAU Symposium No. 133. R. B. Hanson, T. E. Lutz and van Altena are analyzing the contents to refine the luminosity calibration of field subdwarf stars.

Lists of newly determined trigonometric parallaxes were reported from several observatories. Among them are Lick, where Hanson, A. R. Klemola and S. Vasilevskis have redetermined the Lick parallaxes for about 125 stars, McCormick, where P. A. Ianna is continuing observations on Vyssotsky stars, subdwarfs and binaries (40.111.005) and the U. S. Naval where Harrington and his colleagues have published a seventh list of faint stars (39.111.005). At Sproul, W. D. Heintz has measured parallaxes and mass ratios (38.111.001, 42.041.016) and at Van Vleck, Upgren and his colleagues have published a seventeenth list, the first to include PDS measures of subdwarfs among other stars (39.111.016). From Yale, van Altena reports that about 5100 stars brighter than the 21st magnitude at regions at a galactic latitude of $16°$ have been determined using the 4m reflector at Kitt Peak. Similar observations are being obtained near the South Galactic Pole with the Cerro Tololo 4m reflector. Two papers were published of Yerkes parallaxes, one by van Altena et al. (41.111.018) and the other by E. U. Vilkki et al. (42.111.018). An experimental parallax program with the 0.4m refractor of the Shanghai Observatory is now in progress. Results have been published by J.-j. Wang et al. (40.111.022).

Parallaxes of individual objects have been obtained by Heintz at the Sproul Observatory of HR 4550 (38.111.014) and by D. Monet and C. C. Dahn at the U. S. Naval Observatory of LHS 2924 (39.111.027).

Murray and his colleagues have conducted a survey of parallaxes, proper motions and photometry of several thousand stars near the South Galactic Pole, using the UK Schmidt telescope. The observed parallactic motion is used to

estimate the proper motion zero point and to calibrate the trigonometric
parallaxes. A number of other kinematical results have been derived from the data
(42.111.014, 42.111.015).

At the Van Vleck Observatory, Upgren and his associates completed two sets of
measures of stars in 13 fields in the central region of the Hyades cluster. The
measures of 25 members and 260 field stars were made by hand and on the Yale PDS
Microdensitometer. The parallaxes of members agree with the currently accepted
cluster distance. The external mean errors from the PDS measures are only about
half those from the hand measures. This implies that about 3/4 of the total
variance of the hand measures lies in the measuring uncertainties.

During the period under review, W. Gliese and H. Jahreiss continued the
compilation for the Third Catalogue of Nearby Stars, which will be extended to 25
pc from the Sun. Trigonometric Parallaxes are adopted from the new edition of
the Yale Parallax Catalogue, and spectral class-luminosity and color-luminosity
relations are calibrated from them. At least 3000 systems with about 3500
individual stars are expected to be included in the catalogue. The Malmquist and
Lutz-Kelker corrections are being applied to the absolute magnitudes determined
from B-V, R-I and other colors (38.115.014, 39.115.021, 40.041.013, 40.115.012,
41.041.071).

E. W. Weis completed his photometric survey of all NLTT stars in the Northern
Hemisphere brighter than about 14.5 with color class m and no previous parallax
(41.111.014). Of about 2000 stars covered, some 300 appear to be closer than 25
pc, but only about ten appear to be within ten pc and none within five pc of the
Sun.

III. Proper Motions and Positions

W. J. Luyten, in collaboration with G. Hill and S. Morris of the Dominion
Astrophysical Observatory, has published two further catalogues of proper motions
and approximate colors for 13700 faint stars in the Pleiades and Hyades regions
with mean errors of 0."007/yr. This raises the total number of published proper
motions in these regions to 24000 stars. The data are determined from plates
taken with the Mt. Palomar 1.2m Schmidt telescope (39.111.025, 41.111.027).

Together with H. Hughes, Luyten has continued the analysis of proper motions
from Schmidt plates and they have published motions for 350 additional faint stars
found on the Mt. Palomar Schmidt plates and 270 stars found on the ESO Schmidt
plates. Luyten has also published two lists giving data for 927 more double stars
with common proper motion.

The second-epoch observations at El Leoncito, Argentina, for the Southern
Proper Motion Survey was started by T. Girard, C. E. Lopez and van Altena. The
results are referred to faint galaxies and the average epoch difference is 20
years, which will yield an accuracy of 0."005/yr for the average star.

At the Yale Observatory, van Altena and his colleagues obtained proper
motions of stars in open clusters. Together with L. A. Marschall he completed a
proper motion membership study of Tr 37. He and J.-F. Lee, J. T. Lee, P. K. Lu and
Upgren completed an investigation of the internal motions of the Orion Nebula
Cluster. They found the internal velocity distribution for stars brighter than
magnitude 12.5 to be 1.46 ± 0.13 km/sec, with no evidence for anisotropy in the
velocities or equipartition of energy, as is to be expected for a very young
cluster. Probabilities of membership for stars in NGC 188 are being derived from
the proper motions by Y.-W. Lee and van Altena. Astrometric calibration fields
have been established for the field of this cluster and for Omega Centauri for the
focal plane of the Space Telescope. Both fields of view have flat field

rectangular positions accurate to about 0".01.

At the Lick Observatory, B. F. Jones and M. Walker continued a program of measuring proper motions for membership and internal motions in young clusters. They have completed a study of the Orion cluster region, which includes more than 1000 stars within 15' of the cluster center. They also find isotropic velocity dispersions of about 2 km/sec in each coordinate. Klemola continues an astrometric survey of the SMC halo field around NGC 121 and has begun a similar study of the LMC around NGC 2257 in order to isolate red giants for further observation. A major objective is the determination of the absolute motion of the Magellanic Clouds. Klemola has also started an astrometric and photometric survey of the extended Pleiades field in order to isolate faint K and M type members for subsequent astrophysical study.

Second-epoch photography for the Lick Northern Proper Motion (NPM) program is now complete for 97% of the 1246 fields north of declination -23°. Plate measurement is complete north of -3° for the sky outside the Milky Way and reductions are complete for the 600 fields between -3° and +68°.

Proper motions and photometry have been determined by K. Cudworth and his associates at the Yerkes Observatory for the globular clusters M 71 (39.154.003), M 22 (42.154.019) and M 2 and the Ursa Minor dwarf spheroidal galaxy. In each of the globular clusters, the internal proper motion dispersion and the membership has been derived, and for the latter two, the distance as well, from a combination of the data with radial velocities.

At the Bonn Observatory, Relative proper motions were derived in the regions of the globular cluster M 12 by H.-J. Tucholke and M. Geffert (38.154.073) and more recently in the open clusters M 16 (42.153.021) by Tucholke, Geffert and P.S. The and the Pleiades by Geffert. The absolute space motions of the globular clusters NGC 4147 and NGC 6218 have also been derived.

Optical positions and identifications of QSO's and other radio sources were undertaken at Bonn (38.041.048), Yerkes (41.159.086) and Hamburg where Ch. de Vegt reports that work on optical positions of radio sources has continued with precise positions for about 100 sources in the Northern Hemisphere determined. H. G. Walter and R. M. West (41.041.009) also determined optical positions from Schmidt plates relative to reference stars in the Perth 70 Catalogue.

Position and proper motion studies also continued at Bonn by P. Brosche and H. Schwan (38.111.019, 41.043.018) who studied motions of FK4 stars. They found that the motions of young stars from fundamental catalogues reveal a breaking wave in the velocity field along the local spiral arm and they have discussed the consequences of this result.

At Hamburg, de Vegt et al. continued work on the CPC2, and at Brussels, Dommanget has improved the Catalogue of the Components of Double and Multiple Stars (CCDM). Some 20000 accurate positions of components of systems have been introduced into the CCDM.

At Strasbourg, A. Fresneau compiled a survey of the Zones from +1° to +31° of the Astrographic Catalogue for about a million stars. Stars brighter than magnitude 11.5 can readily be detected with precision +0".006/yr (39.111.020). Klemola and his colleagues at Lick constructed star positions in the fields of Uranus and Neptune for the Voyager spacecraft encounters with these planets, and along part of the inbound trajectory of Comet Halley.

Measurement was completed of plates taken for the Zodiacal Zone Catalogue at Yale and at the U.S. Naval Observatory with the 0.2m astrograph. This project includes all SAO stars within 16° of the Ecliptic with epochs around 1980 but with proper motions. The catalogue should be complete by the end of 1987. Observations of the northern section of their astrographic catalogue were completed and the telescope was relocated at the site on Black Birch Mountain near Blenheim, New Zealand. Measurement of the northern plates has commenced on the Starscan measuring machine in Washington.

At the Kazan Observatory, N. Rizvanov reports on a photographic survey of northern stars and an astrographic catalogue of stars in both hemispheres.

A precise position was determined by Girard, Lopez and van Altena for SN1987A from plates of the 0.5m double astrograph at El Leoncito. It is found to lie very close to Star 1 in the complex image of Sk -69*202.

IV. Instrumentation, Techniques and Reduction Methods

Brosche and his associates at Bonn have investigated the accuracies of measuring machines and of plates of the Astrographic Catalogue.

First astrometric tests at Hamburg have been performed with the new PDS 1010G by de Vegt et al. A HP dual axis laser interferometer has been obtained and is being adapted to it. The visual F422 Mann-Comparator was converted to a semi-automatic measuring machine using a Hamamatsu CCD-camera. First results have shown that 0.5-micron accuracy will be possible for plate measurements. Design studies are begun for a new granite-based measuring machine with several camera detectors. Also at Hamburg, optical design studies for a new type of astrometric telescope of aperture 1.5m are in progress. It is expected that the system will be capable of positioning stars to magnitude 17 to a few hundredths of an arcsecond. Reductions are also being improved. A new block adjustment program is being finished and work on transformations of the principal photographic catalogues to the FK5 system has been started.

D. Jones reports that the 1.0m Kapteyn telescope at La Palma, in the Canary Islands is in routine use for astrometry. The highly corrected Harmer-Wynne f/8 arangement with a 1.5° field is used on most nights. A search for common proper motion companions to nearby stars and precise positions of radio sources are among the programs in progress. A Cassegrain focus of f/15 is also used for a parallax program with the CCD camera. Jones also reports that astrometry at Herstmonceux has ceased but old plates are repeated if they show stars on one of the La Palma programs.

A new tailpiece/autoguider was installed on the Mt. Stromlo telescope and a similar system is being installed on the 1.0m reflector of the McCormick Observatory.

Dommanget has pursued his research into Schmidt plate reduction methods, taking into account the bending of a glass plate in a plate holder. Measures have been made of a network registered on a plate in bended and unbended positions.

Rizvanov reports on improvements in methods of observation, measurement and reduction (40.002.133).

At the Shanghai Observatory, Wang has published (42.036.144) an iteration procedure using Murray and Corben's method for the reduction of trigonometric parallax data.

Eichhorn has developed formulae which derive corrections to spherical coordinates of stars directly, without the use of the standard coordinates explicitly other than as auxiliary quantities. The formulae are needed to perform an overlap solution in an extended area (40.041.024).

An expression has been developed by H. Smith for the probability density of the true parallax of a star, given its apparent magnitude and measured trigonometric parallax and the absolute magnitude distribution of its type. This allows the calculation of the most probable parallax which may differ from the weighted mean. This has implications for the compilation of a catalogue of nearby stars (40.111.020).

V. Acknowledgment

I am very grateful to all who responded to my request for information on which this report was based.

Arthur R. Upgren
President of the Commission

COMMISSION 25. STELLAR PHOTOMETRY AND POLARIMETRY
PHOTOMETRIE ET POLARIMETRIE STELLAIRES

PRESIDENT: F.G. Rufener
VICE-PRESIDENT: I.S. McLean
ORGANISING COMMITTEE: R. Buser, J. Dachs, P.J. Edwards, I.S. Glass,
D. Kilkenny, J.S. Miller, E.F. Milone,
A.J. Penny, V.E. Piirola, N.M. Shakhovskoj,
J. Tinbergen, F.J. Vrba, R. Wielebinski,
A.T. Young.

1. INTRODUCTION.

The outstanding feature of the last triennium was most
certainly the abrupt generalisation of the use of array detectors,
particularly CCDs (charge coupled devices). The latter pervade all
subdivisions of instrumental astronomy. The gains achieved by
their high quantum efficiency, their stability, their capability of
delivering immediately recordable signals which can be processed by
appropriate computational means, have been the cause of spectacular
progress regarding the photometric precision of weak signal
measurements.

The applications made in stellar- or surface photometry and
polarimetry, high- and low resolution spectroscopy, astrometry, are
innumerable. Every aspect of experimental research is concerned by
these new detectors (at least 60% of the IAU commissions are
concerned!). New developments have been made regarding either the
number of pixels (chips having 2000 x 2000 pixels have been
announced) or lower noise levels at signal readout which thus
improve the signal to noise ratio for weak sources and shorten
integration times.

Since the evolution of the characteristics of these detectors
is not guided by the sole needs of astronomers, we encounter
discontinuities regarding the availability of chips having already
excellent properties. The varying specifications adopted by the
commercial manufacturing companies force renewals which are not
always beneficial.

Apart from this infatuation for CCDs, one should not lose
sight of the performances already attained by other types of bi-
dimensional detectors which are particularly adapted for very weak
extended sources. Devices using combinations of elements such as
image intensifiers, multichannel plates, CCDs, resistive anodes,
etc.. allow the counting of incident photons and their X-Y
localisation in the focal plane. The quantum efficiency, resolution
and limiting acquisition frequency of these IPCSs (image photo-
counting system) have also been improved.

A difficulty which is still in want of a satisfactory solution
is the storage of these data. Astronomy makes much use of data
centres where classified collections of photographic spectra and
stellar fields are kept in view of being available for purposes
different to those originally intended. Setting up and coordinating
the storage of these "frames" and their instrumental
characteristics is a challenge which must be met in regard of
future generations.

243

The continued use of more traditional detectors such as the photographic plate and the photomultiplier (PM), and the quest for improvements of their performances have been but slightly affected by the new techniques. If the disappearance of the AAS photo-bulletin is regrettable it is, on the other hand, satisfying to note that several large scale projects based on the use of hypersensitised photographic plates are still been either continued or newly started. We may point out the following:
- The forthcoming completion of the blue and red atlases of the southern sky made jointly by ESO and SRC.
- The starting of a second version of the POSS in blue, red and IR making use of modern emulsions.

Among the greater efforts undertaken in photoelectric photometry (UBVRI, uvby, Walraven, Geneva) one must mention the acquisition of photometric data necessary for the HIPPARCOS mission. This astrometric satellite conceived by ESA will measure the proper motions of about 100 000 stars and a large number of parallaxes.

The present report has no intent of being exhaustive; it does not represent a summary of Astron & Astrophys abstracts, which is so useful for our community (we take the opportunity here to thank its promoter, prof. W. Fricke, and wish him a profitable retirement). The editors, I.S. McLean for sections 2.1.3. and 3. and F. Rufener for the rest, hope that they have commented on the prominent events with sufficient fairness. To know more about those who have marked the "good old days", one should read (40.004.005) where the early stages of photoelectric photometry are related.

2. ADVANCES IN PHOTOMETRY TECHNIQUES.

2.1 Instrumentation.

2.1.1 Few channel photometers.

The analysis of the properties of PMs used as precision instruments has led Rosen & Chromey to trace the fatigue effects on pulse-counting photometry (39.034.006), Hildebrandt & Lange to determine counting losses (39.036.034), Ebbets to characterise signal to noise rate from pulse-counting devices in presence of various noises (41.034.039), Lamarre to investigate the photon noise in the far IR domain (41.034.090) and Raouff & al to revisit the noise power spectrum (42.036.104). Stecklum investigates stellar scintillation using photon counting statistics (39.082.059). The consultation of the proceedings of the San Diego workshop (40.012.062) is recommended.

We notice an increasing interest for photometers capable of high frequency sampling of signals from rapidly varying objects such as cataclysmic variables or stellar occultations (40.034.057; Bouchet & Guttierrez, Messenger, 45; Pfau at Jena university observatory). This aim is achieved by the complex photometer built in München by Barwig & al which also records 5 colours (UBVRI) on 3 channels (2 stars and one sky-background) simultaneously (38.034.119; Astron & Astrophys 175, 327). The performances of this instrument (MCCP) are unique for the monitoring of weak rapidly variable objects. It uses the transmission of light by fibre optics in a most astute manner. Other photometers also make use of this device as well as differential measurement techniques (39.034.021; 42.034.066; 42.034.218 to 223). An interesting solution is the use

of photo-diodes as detectors (37.034.056; 40.012.062; 42.036.085).
Provided that the necessary precautions are taken to conserve the
passbands, such a technique could be applied to the measurement and
monitoring of the brightest stars, as proposed by Crawford
(42.113.022).

Much effort has been devoted to the development of photometers
for the near- and far infrared (38.034.012; 38.034.166-167;
39.034.030; 39.034.062; 40.034.175-176).

The following references concern new equipment, several of
which have been described by our Russian colleagues (39.034.43;
39.034.168; 40.034.058; 40.034.071; 40.034.196-197; 42.034.058).
Particular care is taken regarding the conservation of the
passbands and long term stability in Young´s photometer
(42.034.069). Mégevand (Thesis 2190, Geneva University) has
presented the concept of a photometer which does rapid sampling of
the 7 Geneva colours simultaneously in the star-field and in the 6
background-fields surrounding it. The latter two instruments use
the statistical tests presented by Bartholdi & al (37.036.111).

2.1.2 Two dimensional or array photometry.

This rapidly evolving field has several orientations. To
gather information on this subject one must consult the main
conference proceedings and the following review articles:
- Proceedings of SPIE (The International Society of Optical
  Engineering) vols: 445, 501 and 627 of the conferences held at
  London, San Diego and Tucson (38.012.098; 40.012.047;
  42.012.092).
- CCD in Astronomy by C.D. Mackay in Annu. Rev. of A & A vol. 24,
  255 (42.034.051).
- An enthusiastic review by J.A. Tyson: Low-light-level CCD
  imaging in astronomy (42.034.162).
The use of CCDs is extensively described by the references
given in the above mentioned reviews. I would nevertheless like to
point out some works concerning sensitisation in the near UV
(40.034.081; 41.034.030) and the applications of CCD cameras using
rapid-sampling (37.036.142; 41.034.010; 41.034.041; 42.034.227;
42.036.100). This technique allows the time-resolution of flux
variations. The descriptions of CCD cameras are often concise and
not detailed enough for the reproduction of similar equipment. One
must however welcome the presentation by Gunn & al (PASP 99, 618)
who give us a detailed description of the state of development of
the Palomar Observatory camera, thereby allowing the astronomical
community to profit by their 10 years of experience. Walker
(38.113.002) has shown that an accuracy of 1 mmag is possible for
the repeated photometry of standard stars.

The processing of information obtained via array detectors has
been the object of numerous publications, several of which are in
the proceedings of an international workshop held at Erice on data
analysis in astronomy (40.012.096), or in the ESO-OHP workshop held
at Saint-Michel (June 1986): The optimisation of the use of CCD
detectors in Astronomy. Software packages have been developed and
distributed abroad. We may mention IRAF (KPNO, Tucson), DAOPHOT
(DAO, Victoria) and MIDAS (ESO, München). These systems contain
specialised software, such as that necessary to deconvolve crowded
fields (FOCAS, ROMAFOT). The analysis of performances achieved has
been discussed (41.036.045; K. Mighell in Messenger 47). A review
of the methods based on maximum entropy image restoration in
astronomy is to be found in (42.036.060).

A.N. Argue reports as follows: Argue et al have made CCD
observations with the 1.0 m Kapteyn Telescope at la Palma of over
3000 binary stars, mostly with separations less than 5 arcsec,
selected by J. Dommanget (Obs. Royal de Belgique) as part of the
HIPPARCOS project. Reductions are being carried out using M.J.
Irwin´s "Automatic Analysis of crowded fields" program
(40.036.256) .

Several types of image photo-counting systems (IPCS) have been
improved and tested. Their intrinsic limitation remains the
frequency that each pixel can sustain for a reliable detection to
be made (number of photons and X-Y coordinates). These devices are
best adapted for weak and extended sources. We may mention
Mepsicron (37.034.068), the Multi-Anode Multichannel Array or MAMA
(38.034.088; 41.034.161), the proposal by Shectman (42.034.121) or
that of Hickson (PASP, 98, 622). The problems related to noise
reduction (41.034.168) and the statistical properties of these
detectors have also been tackled (42.034.164).

The ability of electronography to achieve high resolution, the
large number of pixels and the easy storage of the information
continue to interest French (CFHT, Hawaï) and Russian (41.034.047)
groups. A good example of the performances that can be attained was
presented by Blecha (37.036.143).

Improvements regarding the hypersensitising of photographic
plates have extended their possibilities of application. The IAU
colloquium 78: Astronomy with Schmidt-type telescopes (37.012.077)
and the Edinborough workshop (38.012.067) have examined their
potentiality. The photometric-reading of glass-based emulsions has
been reviewed by Pierre (41.036.171). The atlases have been
discussed in the introduction.

2.1.3 Near Infrared - New Techniques (section by I.S. McLean)

In the past three years near infrared astronomy has undergone
a major revolution of staggering importance to the subject. The
cause has been the advent of two-dimensional solid-state detector
arrays; the infrared analogue of silicon CCDs. A number of major
observatories have established "common-user" or "facility"
instruments, i.e. cameras and/or array spectrometers including:

- the U.K. Infrared Telescope Unit of the Royal Observatory,
  Edinburgh,
- the National Optical Astronomy Observatories at Kitt Peak
  and Cerro Tololo,
- the European Southern Observatory,
- many independent institutes (such as the Institute of
  Astronomy, University of Hawaii, the Steward Observatory,
  University of Arizona, Rochester University and many others)
  have constructed their own instruments.

Of course, it is early days yet, the array formats are still
relatively small (64 x 64 approximately), but the sensitivity per
pixel is at least as good as the best single-element IR sensors. In
the near infrared the current front-runner is the 62 x 58 Indium
Antimonide array from SBRC, but recent arrays of Mercury-Cadmium-
Telluride (HgCdTe) from Rockwell International also appear very
promising. Long-wavelength systems employing extrinsic Silicon
photoconductor arrays from Aerojet Electrosystems, SBRC and others
have been used by groups at Berkeley, Goddard and UCSD. Depending
on the dopant, these arrays are sensitive to approximately $30\mu m$.

European manufacturers are also participating in these developments. For example, SAT and LETI-LIR (France) and Battelle (Germany) companies are involved in providing array detectors for the ESA Infrared Satellite Observatory (ISO) and Mullards (UK) are making HgCdTe arrays with 3.5 μm cut-offs. For reference to the latest developments see SPIE volumes 627, 782 and "Infrared Astronomy with Arrays" eds. C.G. Wynn-Williams and E.E. Becklin, University of Hawaii, Institute of Astronomy (1987).

It seems likely that IR photometry with panoramic detectors will now encounter problems similar to those already faced by optical astronomers using CCDs. The discussions on standardisation planned for the Baltimore General Assembly should therefore include photometry with infrared arrays. Some effort by this commission to stimulate discussions on standardisation would therefore be very valuable.

2.1.4 Automatic photoelectric telescopes (APT).

The dynamic activity of the IAPPP (International Amateur Professional Photometry) has brought forth an interesting series of communications where Boyd, Genet and Hall (37.036.026) have presented a new and low-priced application of the APT concept which had already been conceived and tried out in the sixties. The successive improvements and the demonstration of the efficiency of the new concept have been described in (38.032.029; 40.032.005; 42.036.082). The scientific motivations and the operation of several APT have been the object of a meeting of which the proceedings have already been published (IAPPP commun. 25). This effort which tends to orient the use of small telescopes to the systematic monitoring of well defined objects is full of promise.

2.2. Extinction.

Atmospheric extinction, its measurement and the limitations it imposes on the photometric accuracy of ground-based observations have been discussed during specialised workshops in Paris (38.012.078) and San Diego (40.012.062). Our Russian colleagues have re-examined the optimisation of methods of reduction outside the atmosphere for wide- and intermediate band photometries (39.036.083; 39.113.026; 39.113.034; 42.036.071). For ground-based observations in the far-IR we can mention (42.082.085). Several retrospective analyses have given a description of the short-, mean- and long-term variations of atmospheric extinction in the visible range and have evaluated the role of volcanic eruptions, particularly that of El Chichón (42.082.029; 42.082.044; 42.082.045; 42.082.083).

2.3. Photometric systems (some references limited to basic aspects).

2.3.1 Broad band systems.

South African colleagues, Cousins first and Menzies & Lang later have collected large pieces of work for standard stars of UBV(RI)c (38.113.049; 38.113.050; 40.113.060). Stobie proposes faint galactic reference stars in 8 fields (39.113.038). Many transformation equations for VRI have been gathered by Taylor (41.036.021). In the USSR an effort to establish an instrumentally well defined and homogeneous WBVR system is related in (39.113.050)

by Khaliullin & al. Lanz gives a general catalogue of the Johnson
UBVRI (42.002.036). The basics of JHKLM is reviewed by Glass
(39.113.022). Standard stars for the L and M filter bands are
proposed (38.113.022) and absolute calibrations are given in this
range (39.113.027; 39.113.028).

2.3.2 uvby & β.

Several publications gather lists of standard stars. Perry and
Crawford for the bright stars (42.002.048), Kilkenny, Menzies and
Cousins for the E regions (42.113.015; 42.113.019; 41.113.061).
Secondary standards near the south galactic pole are proposed in
(37.113.040). Philip (39.113.011) has collected faint secondary
standard stars for use with large telescopes. A preliminary uvby
calibration for G & K type stars is given by Olsen (38.113.014).
Manfroid and Heck analyse the accuracy of reduction procedures
(37.036.036). Using numerical simulations, Manfroid computes
synthetic observations filtered through standard passbands and
reduced via the usual transformations. He compares these with those
obtained for mismatched passbands. Large deviations are found for
objects outside the zone where $0 < b-y < .4$ (38.113.034). More
recently, Manfroid and Sterken have strikingly confirmed these
effects due to a mismatch by comparing observations of a same
sample of stars made with non-identical instrumental systems at the
same observatory (Astron & Astrophys, in press). Similar work has
also been undertaken by Reimann & Boehm (Astron. Nachr. submitted).
Hauck & Mermilliod have compiled a new edition of the uvby,β
photoelectric catalogue with 40 484 entries (39.002.016).

2.3.3 Others.

Relationships between stellar magnitudes and photon fluxes are
given by Ralys for the Vilnius photometric system (38.113.061). For
the same system, an automated two-dimensional classification has
been issued from photographic plates (42.113.018). Cramer
investigates relations between the β-index and Geneva photometry
for B-type stars (38.113.036). A new description of the passbands
of the Geneva system, calibrated from recent spectrophotometric
data for α Lyr, allows the latest edition of the Geneva photometric
catalogue (≈ 28 000 stars) to be used as a low resolution absolute
spectrophotometry (Nicolet and Rufener, Astron. & Astrophys. suppl.
in press). Luminosity classification with the Washington system is
re-examined by Geisler (38.113.026).

2.4. Synthetic photometry and energy distributions.

In the same sense as the points of view expressed during the
joint meeting at New-Delhi concerning this subject (42.012.020),
synthetic analyses of photometric systems had already been carried
out by the Vilnius group (37.113.001-003). The properties of the U
filter of UBV and those of the uvby,β system have been reviewed by
Bessel, Howell and Lester & al respectively (41.113.034;
41.113.039; 42.113.002). As a by-product from the Geneva passbands
evaluation (Nicolet & Rufener, Astron. & Astrophys., in press) a
small correction is proposed for the absolute calibration of the
Gunn & Stricker spectrophotometric catalogue (33.002.042). Among
the new absolute spectrophotometric data we may mention those of
Kiehling for stars of type F and later (39.113.063; Astron. &
Astrophys. Suppl. 69, 465), those of Taylor (37.113.026) and

Filipenko & Greenstein for faint spectrophotometric standards
useful with large optical telescopes (38.113.017). Calibrations of
Vega in the IR (40.113.048; 40.113.057) from 5 to 100 μm have
served to fix the instrumental responses of IRAS. Synthetic colours
of the Sun (37.113.020) and the optimisation of an ideal
photometric system have been discussed this way (41.113.059).

2.5. Photometric faint star sequences.

Apart from the Catalogue of Photometric Sequences one can
mention some other sources where sequences of very faint stars are
established up to mv ≈ 22 (38.154.038; 38.154.040; 39.113.039;
39.113.061; 40.113.006; 40.113.061; 41.113.048). The method
proposed by Jelley (42.034.146) can be a useful alternative.

2.6. Data Centres and Catalogues.

The steadily increasing activities of Data Centres(Strasbourg,
Washington, Moscow) which are now as much appreciated as they are
essential no longer need to be emphasised here. We may nevertheless
mention the major contribution of the CDS in Strasbourg to the
preparation of the Input Catalogue by the INCA consortium. This
catalogue gathers precise data for ≈ 120 000 stars which can thus
be observed during the astrometric mission of the HIPPARCOS
satellite. The intersection of this catalogue with the available
photometric data produces a list of ≈ 15000 stars without known
magnitudes or colours. They are currently being observed with all
the available photometric systems. This effort which is already 3/4
completed is being carried out by numerous observers.
        From W.H. Warren Jr. (NASA Centre) we get a Report on Machine-
Readable Photometric Catalogs Received or Modified in the Period 1
July 1984 through 30 June 1987. Numbers are in the international
astronomical data centre numbering system used by Strasbourg and
NASA data centres:

Num.   Description (limited to the title)

2092   Revised S201 Catalog of Far-Ultraviolet Objects.
2094   The Revised AFGL Infrared Sky Survey Catalog and Supplement.
2097   ANS Ultraviolet Photometry Catalogue of Point Sources.
2098   Catalog of Infrared Observations.
2101   A Catalogue of Photoelectric Magnitudes and Colours of
        Visual Double and Multiple Systems.
2106   A Search for Ultraviolet-Excess Objects.
2108   The Two-Micron Sky Survey: Nearest SAO Star and Locations on
        Palomar Sky Survey Prints.
2114   A Catalog of Ultraviolet Interstellar Extinction Excesses
        for 1415 Stars.
2115   Faint Blue Objects at High Galactic Latitude. Catalog of
        Objects in SA 28, 29, 55, 57, and 94.
2116   Photoelectric Photometric Catalogue in the Johnson UBVRI
        System.
2117   A Catalogue of Concentric Aperture UBVRI Photoelectric
        Photometry of Globular Clusters.
2118   UBVRI Photometric Standard Stars Around the Celestial
        Equator.
2119   Stellar Distribution Near the South Galactic Pole,
        Photographic V Magnitudes and B-V Colours of 640 Stars in
        Region I.

2120   A Photometric and Spectrophotometric Investigation in a
        Region at the South Galactic Pole.
2121   Stellar Distribution Near the Selected Areas 127, 141 and
        189.
2122   UBV Photoelectric Photometry Catalogue.
2123   UBV Photometry of Faint Stars (V > 14.5) in the Open Cluster
        M 67.
2124   Catalogue of UBV Photometry and MK Spectral Types in Open
        Clusters.
2125   IRAS Catalog of Point Sources.
2126   IRAS Serendipitous Survey Catalog (SSC).

2.7. Books and Proceedings (related to the field but not already mentioned).

37.003.082   Solar System Photometry Handbook (Genet).
38.003.002   Advances in photoelectric photometry, Vol. 2
              (eds.: Wolpert & Genet).
38.012.037   Rapid variability of early-type stars. Proc. of the
              Hvar workshop.
40.012.038   Calibration of fundamental stellar quantities. Proc.
              IAU symposium 111, Como.
41.003.035   Photometric & polarimetric investigations of celestial
              bodies (ed. Morozhenko).
42.003.048   Méthodes physiques de l'observation (Léna).
42.012.006   Spectroscopic & photometric classification of
              population II stars (meeting at New Delhi).
42.012.053   Instrumentation and research programs for small
              telescopes. Proc. IAU symposium 118, Christchurch.

3. POLARIMETRY (by I.S. McLean).

3.1. Introduction.

The field of optical and infrared polarimetry (including
photopolarimetry, spectropolarimetry and imaging polarimetry) has
continued to grow and blossom over the current triennium. At our
meeting in New Delhi we held a special session on polarimetry. One
of the highlights of that meeting was the description of
polarization facilities for the Hubble Space Telescope by Dr.
Olivia Lupie. The tragic loss of the Space Shuttle Challenger, of
course, has delayed the launch of the HST but much more work has
now gone into the HST instrumentation including polarimetry modes.
Similarly, more thought and more work has gone into understanding
the effects and consequences on the measurement of polarization
implied by the optical designs of the next generation of very large
telescopes. Steve West of the University of Arizona has analysed
the polarization effects associated with the Multiple-Mirror
Telescope (MMT) and demonstrated that the problem is more benign
than previously imagined at least for some applications. An
analysis by Jaap Tinbergen shows that high precision photometry on
VLTs can be affected by polarization effects, especially for full
moon.
      Much of the real progress in optical polarimetry over the past
three years has been in the range and depth of astrophysical
applications rather than in new instrumentation. This was evident
in Delhi and the trend has strengthened. In the near infrared,
however, there has been a colossal impact from the recent advent of
solid-state imaging devices; several groups have lost no time in

producing "infrared array polarimeters" (see below).

When considering the preparation of this report, it became evident to me that "astronomical polarimetry" is now such a fully-fledged subject, that to properly describe progress in this field one has to refer to the progress in a wide range of astrophysical topics, from interstellar dust to BL Lacs, from solar system studies to Seyfert galaxies, and many more. Alternatively, in describing the innovations in polarization instrumentation one finds that these largely stem from detector development, rather than from development of polarization optics. In many ways therefore, there is a huge overlap with the work of other commissions and, in particular, with Commission 9.

3.2. Instrument Developments.

Several new optical polarimeters have come into use in the past three years. One of these is the "Multi-Purpose-Fotometer (MPF)" at the La Palma Observatory. This instrument, designed by Jaap Tinbergen, is a 12-channel photopolarimeter operating with 6 beams. It is a superb, high precision instrument. Another multi-channel photopolarimeter has been built by V. Piirola. The CCD imaging photopolarimeter developed by Ian McLean at the Royal Observatory Edinburgh has been emulated by other groups. Dr. Joe Miller at Lick Observatory reports that he has converted the IDS spectropolarimeter to a CCD instrument. At the Anglo-Australian Observatory, Jeremy Bailey has upgraded the Pockels cell + IPCS spectropolarimeter (originally developed by Ian McLean) to work with a CCD camera. This innovation was beautifully timed to coincide with the discovery of SN1987a! Dr. James Hough, of Hatfield Polytechnic, England, together with Jeremy Bailey, have completed and commissioned a dual optical/IR photopolarimeter system capable of working from $0.3 - 2.5$ μm using rotating superachromatic waveplates. In Arizona, Roger Angel and Steve West have developed a new form of retarder based on stress birefringence which, unlike the piezo-optic modulators invented by Jim Kent, operates at a low frequency. Steve has applied this technique to a near infrared array polarimeter. The new 62 x 58 infrared camera on UKIRT has been "adapted" to perform imaging polarimetry by Ian McLean. Many other optical and IR photopolarimeters have come into use or have been busily working e.g. the "Vatpol" developed by George Coyne, Director of the Vatican Observatory, is a robust, modern version of the classical "Minipol" originally pioneered by Kris Serkowski and Tom Gehrels at LPL (Tucson). A remarkable polarimeter-cum-telescope has been designed by Jim Kemp at Oregon to observe the Sun! Jim reports the astonishing polarization levels of 10^6 (i.e. 0.0001%)!

3.3. Astronomical Progress.

The only conference or workshop dealing specifically with polarimetry in this triennium was hosted by George Coyne at the Vatican Observatory, Castel Gandolfo in June 1987. This 2 week meeting included sessions of several days on the polarization properties and modelling of AM Her magnetic binaries, cool long period variables, symbiotic stars, early-type binaries (Wolf-Rayet, T Tauri etc.) and future prospects for polarimetry in astrophysics. The proceedings for this workshop are currently in preparation with a publication date of mid-88 being aimed for. The book will represent one of the most important contributions to the field of

astronomical polarimetry in recent years since it will be a
combination of review articles and new results covering the whole
field of polarization from the circumstellar environment of stellar
sources. It will turn out to be an excellent source book, the first
for many years.

                              F.G. Rufener

                              President of the Commission

COMMISSION 26: DOUBLE AND MULTIPLE STARS

PRESIDENT: K. D. Rakos

VICE PRESIDENT: H. A. McAlister

ORGANIZING COMMITTEE: H. A. Abt, P. Couteau, J. Dommanget,
    M. G. Fracastoro, R. S. Harrington, A. A. Kiselyov

Preface:

    In the recent years new techniques are expanding visual binary star
astronomy into previously inaccessible regimes of angular resolution and accuracy of measurements regarding the astrometric and astrophysical parameters of
their components. New techniques and algorithms are providing enhanced sensitivity to low mass companions and the determination of photometric properties
of the components of close visual binaries. The potential now exists for significantly narrowing, if not eliminating entirely, the historic gap separating
spectroscopic from visual binaries. Advances in closely related fields as in
precision radial velocity measurement and enhanced accuracy of parallax determination from ground and space based observatories will place demands upon double
star astronomy that have been absent for half a century. Today double and multiple star astronomy is perhaps closer to fulfilling its true potential than
it has ever been, entering a fundamentally important area of modern astrophysics.
    Visual and photographic methods are rapidly being replaced by more power-
ful electronic methods of measurement by astronomers from all over the world.
Access for double star observers using modern techniques is finally being routinely granted at the largest telescopes located at the best observing sites,
an often quoted dream of such leading figures in our field as van den Bos and
Finsen. Groups in the U.S., France, Australia and Germany are constructing
or planning long baseline interferometers that will turn systems of stars having
orbital periods of a few days into "visual" doubles. Other interferometric
techniques using single aperture telescopes in the visible and infrared spectral
regions are currently being used by French, German, British, Soviet, U.S. and
other astronomers to obtain diffraction-limited measurements of double star
geometry and to measure magnitudes and colors of the individual components
of these objects. This most classic of all subfields in astronomy, the last
to which direct measurement with the eye has made a major contribution, is
truly in the midst of a revolution.

I. The Hubble Space Telescope - Double and Multiple Stars:

    Research in binary star astronomy is the principal activity of the Center
for High Angular Resolution Astronomy at Georgia State University. Research
activities center around the capabilities of a speckle camera developed and
operated by CHARA. The camera system incorporates a microchannel plate intensified CCD array, which is read-out at standard video rates and supplies speckle
pictures for video recording and subsequent processing or for real time processing by means of a hardwired vector-autocorrelator. The camera has been
used an average of 15 nights per year on the 4-meter telescope on Kitt Peak
and for 50 nights per year on the 1.8-meter Shane telescope at Lick Observatory,

the 2.5-meter Hooker telescope at Mt. Wilson and the 3.8-meter CHF telescope on Mauna Kea. These last three telescopes were used in a survey for duplicity on Hubble Space Telescope guide stars.

The primary program, as carried out on Kitt Peak, involves the attempted resulution of spectroscopic binaries, close visual binaries and other stars. Limiting resolution and magnitude for the routine program is 0.030 arc-second and V = +11.0. The ICCD speckle camera has produced some 2700 measurements published in early 1987. Numerous revisions of orbits are underway at CHARA. Observations taken at the Perkins telescope (Lowell Observatory) are aimed at detecting low mass companions, including Jupiter-sized objects, by means of submotions revealed through the high accuracy of speckle interferometry. A concerted effort toward the recovery of intensity ratios of the components of close binaries is underway. The feasibility of an optical array of telescopes capable of sub-milliarcsecond resolution and applicable to the direct angular resolution of short period spectroscopic binaries is currently under discussion. Participants in these efforts include or have included: H.A. McAlister, D.J. Hutter, J.R. Sowell and W.G. Bagnuolo, Jr. of CHARA, O.G. Franz of Lowell Observatory and M.M. Shara of Space Telescope Science Institute.

The second large and very productive team working in the field of speckle interferometry using the 2m telescope at Pic du Midi, Canada-France-Hawaii telescope at Mauna Kea and 1.93m telescope at Haute-Provence is guided by A. Labeyrie, D. Bonneau, A. Blazit, R. Foy, L. Koechlin, R.V. Stachnik, D.Y. Gezari, Y.Y. Balega, J.M. Carquillat, J.L. Vidal, F. Vakili and M. Faucherre. Beside the Observation of very close binary stars, the measurements of star diameters, color dependent limb darkening, determinations of inclination and masses for the system Algol AB-C, high resolution of Pluto-Charon system, Uranus and Neptune and 3C273 have been published.

The third large groupe using speckle technique and ESO telescopes at La Silla is working on similar observations. There are G. Weigelt, K.H. Hofmann, T. Reinheimer, G. Baier, R. Ladebeck, U. Bastian, E. Keller, R. Mundt, J. Ebersberger, M. Walter, M. Mueller, R.B. Orellana and others.

II. Hipparcos - Double and Multiple Stars

During the last three years J. Dommanget, with the collaboration of O. Nys, has carried on his contribution to the ESA Hipparchos Astrometric Mission preparation as a member of the Input Catalog Consortium, where he is in charge since 1981 of problems concerning double and multiple stars.

In order to solve these problems it appeared essential to establish a specific "Catalogue of the Components of Double and Multiple Stars (CCDM)" giving accurate astrometric and photometric data on each component. To collect the necessary information, an international working group was established. Its activity has mainly been directed to accurate astrometric observations and identifications. The members are presently: J.P.Anosova (Leningrad), A.N. Argue (Cambridge), P. Bacchus (Lille), P. Brosche (Bonn), U. Bastian (Heidelberg), P. Candy (Perth), J. Guibert (Paris), L. Louys (Uccle), E. Oblac (Besancon), V.V. Orlov (Leningrad), R. Pannunzio (Torino), R. Perdomo (La PLata), F. Schmeidler (München), D. Sinachopoulos (Bonn), J. Torra (Barcelona) and G. Soulie (Bordeaux). The work of this group may be considered from the following points of view:

a) The Catalog of the Components of Double and Multiple Stars:
The basic reference of this catalogue is the Index (1976.5) whose format has been fundamentaly modified in such a way, that each component is described by one record. This enables primarily the introduction of information missing

in the Index as: DM numbers for the secondary components or other identifications. It presently consits of 62.300 systems split into four files of 56.600 double stars, 4.400 triple stars and 1.000 quadruple stars and 300 multiple stars of more than four components.

More than 2.000 newly published AGK2/3 and SAO identifications of components have been introduced in the CCDM. Hundreds of double stars having separations less than 3", that cannot be identified on photographic plates, are observed by amateur astronomers in West Germany and in France using finding charts, under leadership of U. Bastian and P. Bacchus. Simultaneously, this work has led to the discovery of hundreds of errors in the Index: a manuscript is being prepared for publication.

The format of the CCDM also permits the introduction of accurate astrometric data. Today, the completion of this catalogue is under way. Some 20.000 accurate positions of components of systems have been extracted from a new general astrometric catalogue prepared at Heidelberg especially for the Hipparcos mission and have already been introduced in the CCDM. A few hundreds of meridians observations made at Berlin-Babelsberg, more than three thousand photographic positions obtained at Leningrad, Perth, Uccle, Torino, La Plata, and Bordeaux, nearly two thousand relative positions of secondary components computed at Bonn and some 2.500 CCD photometric and astrometric data of double stars recently obtained at La Palma (Cambridge group) as well as micrometric observations made at München, are being introduced in the CCDM. Also some 800 "semiaccurate" positions (+- 0.1 arc-minute) published by P. Bacchus and useful for identifications have also been considered.

b) The double and multiple stars in the Hipparcos Input Catalogue:

The intercomparison of the CCDM with the INCA DATA BASE (210.000 stars) has allowed the identification of some 20.000 systems. One may thus estimate by 10.000 the presently identified known double and multiple systems in the HIPPARCUS INPUT CATALOGUE that will consist of some 100.000 stars.

In view of their possible observation by Hipparcos, all these systems had to be studied from the point of view of the number of entries to consider in the Input Catalogue as a function of their components separation and their differences in magnitudes. Research has also been made on the observability problems of astrometric and spectroscopic pairs and on the possibility of orbit computations based on the sole Hipparcos observations.

As a by-product of the identification work done for the CCDM and for the Hipparcos Input Catalogue, a full list of the BD stars having a suffix "a" or "b" has been established for publication in view to bring the attention to the fact that these suffixes do not mean anything concerning components A or B.

Research has been made on Schmidt plate reduction taking into account the curvature introduced in the plate exposed to the sky.

In connection with a program called "Contribution of the Stars of Various Masses and Compositions to the Chemical Evolution of the Galaxy", a new study has been started on the orientation of the orbital planes of visual double stars in the surrounding of the sun.

In the course of preparation of Hipparcos data analysis, members of the Fundamental Astronomy by Space Techniques Consortium (FAST) and of the Northern Data Analyses Consortium (NDAC) have studied algorithms and worked out s/w programs in order to detect astrometric binaries and multiple stars among Hipparcos program stars and to derive their astrometric parameters. The state of the art of these studies is represented by several papers. Detection should occure for binaries brighter than 13 magnitudes with separation larger than 0.07 arc-second and magnitude difference between the components smaller than three. These studies have been carried out by J. Kovalevsky, P.L. Bernacca, W. Delaney, L. Borriello, A. Farilla, M. Froeschle, J.L. Falin, F. Mignard,

S. Soederhjelm, F. D'Alessandro, L. Lindegren, G. Prezioso, E. Canuto, B. Fassino and H.G. Walter.

III. Visual and Photographic Observations:

Visual and photographic observations have been continued at different observatories. U.S. Naval Observatory, Observatoire de Nice, Bosscha Observatory, Osservatorio Astronomico di Torino, Observatorio Astronomico Universidad de Santiago de Compostella and other observatories have published a large number of observations of double stars.

Beside the classical visual and photographic methods, modern measuring techniques as CCD cameras, solid state area scanner, speckle cameras, interferometric methods, multicolor photometry, radial velocity measurements, have been used on regular basis. Here we have to mentione the strong trend toward the complete analysis of physical parameters of selected double stars instead of compilation of large number single observations of orbital parameters only. Also the increasing collaboration between different observatories and individual astronomers in the double star research should be pointed out.

IV. IAU Colloquium No.97 :

The colloquium "Wide components in double and multiple stars" was held at the Brussels Congress Center, Belgium, June 8-13, 1987. It was co-sponsored by comissiones 26 and 24 with the collaboration of comissions 30, 33 and 42. The colloquium was organized by the Royal Observatory of Belgium and dedicated to W.J. Luyten. The meeting was devided into seven sessions including one poster session that were devoted to the following specific subjects: Cataloguing wide pairs; Wide nearby binaries; Wide components in multiple systems; Observation of wide pairs and their results; Statistical studies on wide pairs and origin and evolution of wide pairs. Sixty astronomers from 17 countries registered and delivered 48 papers and posters including 6 introductory lectures. The aim of this colloquium was to show how much double star research may be important for better knowledge in many astronomical domains including star formation and evolution as well as galactic structure and dynamics. It drew attention on wide pairs when so much is done today in the field of close binaries. It has shown that astrometric, spectroscopic and photometric investigation of wide pairs appear today as an indispensible counterpart to observation of close systems. This was the second colloquium organized on double stars by the Royal Observatory of Belgium and the seventh by IAU Comission 26. The proceedings of the colloquium will be published in Astrophysics and Space Science.

I am most grateful to those who have supplied information for this report.

                                        Karl D. Rakos
                                   President of the Commission

27. VARIABLE STARS (ETOILES VARIABLES)

PRESIDENT: B. Szeidl
VICE PRESIDENT: M. Breger
ORGANIZING COMMITTEE: A.N. Cox, R.E. Gershberg, M. Jerzykiewicz, L.N. Mavridis,
L.N. Mirzoyan, J.R. Percy, M.A. Smith, A.M. van Genderen,
B. Warner

1. Introduction

The field of variable star research has become so broad and the amount of research to be reported on has grown so rapidly that it is a vain hope that a report of this kind, in a very limited space, could cover the whole field of research and could mention all the papers that have been published in the last three years. It is only hoped that this report presents the significant results achieved in the field of the most important aspects of variable star research. Some important subjects (e.g. cataclysmic variables) relevant to the variable star research are reviewed in the reports of other commissions. This is a consequence of the fact that the research has become very complex and the phenomena producing light variability belong to the field of interest of other commissions, too.

Before reporting the scientific works, three short reports about ongoing activities of the Commission are given. In general the abbreviations for identifying the cited publications are the same as used in the previous triennial report.

The commission president is very grateful to the authors for their individual contributions.

2. Commission Activities

A. ARCHIVES OF UNPUBLISHED OBSERVATIONS: 1984-1987 (M. Breger)
The Archives of Unpublished Photoelectric Observations of Variable Stars was created to provide permanent archives in different parts of the world. The Archives can replace lengthy and expensive tables in scientific publications by a single reference to the archival file number. Furthermore, many valuable observations are never used for scientific publications, and the Archives makes such observations available to other astronomers at a time when they might become very important.

The number of completed files with data sets has grown to 174, and considerably more files have been assigned to data not yet received. The time period has been one of rapid growth. Rapid retrieval of past files, often at no cost, has been reported from the depositories.

The Publications of the Astronomical Society of the Pacific has kindly agreed to publish, on a regular basis, the summaries of recent files. Detailed reports can be found in PASP 91.408; 93.528 and 97.85. Other summaries and announcements can be found in the Information Bulletin on Variable Stars. At present, astronomers who wish to obtain unpublished photoelectric measurements on variable stars may do so by requesting whole files (not partial files) from one of the three archives:

P.D. Hingley, Librarian,	Dr. C. Jaschek,	Dr. Yu.S. Romanov,
Royal Astronomical Society,	Centre de Données Stellaires,	Odessa Astronomical
Burlington House,	Observatoire de Strasbourg,	Observatory,
London, W1V ONL,	11, Rue de l'Université,	Shevchenko Park,
Great Britain	F-67000 Strasbourg,	Odessa 270014,
	France	U.S.S.R.

The requested file number should be specified. There is no charge for short files.
Astronomers who wish to submit new files, should submit three copies of the data and five copies of the cover sheet to the Coordinator, who will forward the data with the cover sheet to the three archives. Please consult IBVS, 2853, 1986 for helpful details and formats. If a new file number is required for scientific publications (in place of extensive tables of measurements), the file number can be assigned by the Coordinator before receipt of the actual measurements. The address is:

Prof. Dr. Michel Breger,
Institut für Astronomie,
Türkenschanzstr. 17,
A-1180 Wien,
Austria

The main purpose of the Archives has in the past been to preserve and make available data in paper form. Through the Centre de Données Stellaires in Strasbourg it is possible to submit new data by computer tape (ASCII), diskettes, or computer networks.
Since our previous report to IAU Commission 27, the following files have been assigned and completed:

53.	HD 219989	– D.S. Hall	183.	λ And	– E.S. Guinan + S.W. Walker
54.	HR 454	– D.S. Hall	184.	EG Cep	– J. Kaluzny + I. Semeniuk
90.	SV Cam	– A. Cellino	185.	VY Lac	– J. Kaluzny + I. Semeniuk
139.	V396 Per	– D.S. Hall	186.	λ Eri	– J.R. Percy
142.	λ And	– A. Cellino	187.	27 Cyg	– J.R. Percy
143.	AR Dra	– P. Broglia	188.	YY Gem	– E.H. Geyer + B.D. Kämper
144.	RS Cep	– E.C. Olson	189.	TZ CrB	– A. Giménez
145.	Φ Cas	– A. Arellano-Ferro	190.	U Sag	– E.C. Olson
146.	AC And	– R. Robb	191.	CW Cep, Y Cyg, AG Per	
147.	22 late-type stars				– A. Giménez
		– J.A. Eaton	192.	BP Mus	– Z. Kviz
148.	RS CVn	– J.A. Eaton	193.	Cluster Stars	
149.	V444 Cyg	– J.A. Eaton			– J.R. Sowell
150.	HR 6469	– D.S. Hall	194.	YZ Cas	– R. Diethelm
151.	HD 8358	– D.S. Hall	195.	RU Cnc, VV Mon, TY Pyx	
152.	CZ Aqr	– J.R. Burton			– F. Scaltriti
153.	σ Gem	– K. Strassmeier +	196.	SAO 77615, AZ Vir	
		D.S. Hall			– R.M. Campbell
179.	Bright Southern Be Stars		197.	RS Ind	– M.A. Cerruti
		– C. Stagg	198.	V677 Cen	– R.H. Koch
181.	UV Psc, UX Ari		199.	BW Vul	– A.P. Odell
		– F. Scaltriti	203.	RZ Oph	– E.C. Olson
182.	FO Vir	– J.D. Fernie	204.	CG Cyg	– J.R. Sowell

B. GENERAL CATALOGUE OF VARIABLE STARS (N.N. Samus)

During the last three-year cycle the compilation and publication of the 4th edition of the General Catalogue of Variable Stars (GCVS) has been continued. Volume I (Andromeda - Crux), and Volume II (Cygnus-Orion) of the GCVS appeared in 1985, Volume III (Pavo-Vulpecula) was published by "Nauka" Publishers in 1987 (Editor-in-chief: Prof. P.N. Kholopov). Thus, the publication of the main body of the Catalogue, comprising data for 28435 variable stars of our Galaxy, has been completed. The magnetic tape version of the main table of the GCVS for astronomical data centres has been prepared. The 67th Name-List of Variable Stars was published in 1985; the 68th Name-List was compiled in 1987. These two Name-Lists contain final GCVS designations for 1310 new variable stars not yet entered in the GCVS IV. The Name-Lists have been prepared in a new form providing users not only with identification data, but also with basic information on the character of variability of the stars.
The preparation of Volume IV of the GCVS is now under way. It will contain

information on approximately 10000 variable stars in external galaxies (information of this kind is being included in the GCVS for the first time), as well as on
optically variable quasars, galactic nuclei,and BL Lacertae objects. In the course
of this work numerous inconsistencies in the coordinates of Magellanic Cloud variable stars, as well as unknown cross-identifications among these variables and of
these variables with stars of the main body of the GCVS or with stars of the NSV
catalogue have been found. The work on Volume V has been commenced; this volume
along with the lists of variables in the order of their right ascensions and of
variable stars arranged according to their types of variability, will contain
detailed cross-identification tables with principal astronomical catalogues. Such
cross-identification tables have not been published since 1958. The compilation
of the Astrometric Supplement to GCVS has also been initiated; it will present
accurate coordinates for variable stars with available astrometric data.
    The members of Commission 27 will have their complimentary copies of the GCVS
Volume III forwarded to them in due course. The 68th Name-List has been published
in IBVS 3058.

C. INFORMATION BULLETIN ON VARIABLE STARS (L. Szabados)
    During the period covered by this report, the number of issues of the IBVS
progressed from No.2544 (July 1984) to No.3041 (June 1987). The yearly average of
published issues is somewhat lower than during the previous three years. This
decrease is mainly due to the fact that the editors now impose stricter requirements when receiving the submitted manuscripts. This procedure cannot be avoided
because Konkoly Observatory is unable to publish more than 200 issues per year,
nor have the editors the intention of allowing the scientific level of the IBVS to
decrease. If a system of independent referees were to be introduced, this would
jeopardize the promptness of publication. All correspondence concerning the IBVS
should be addressed to:

                        The Editors, IBVS
                        Konkoly Observatory,
                        H-1525 Budapest, P.O. Box 67,
                        Hungary

                        3. Early-Type Variable Stars
                           (John R. Percy)

INTRODUCTION
    Review papers on early-type variables appear in Highlights of Astronomy vol.
7 p221; 229; 247; 255; 265 and 273; and in proceedings of other conferences on
stellar pulsation and on UV astronomy. For a complete bibliography, see Astronomy
and Astrophysics Abstracts.
    Information about early-type variables has been obtained not only by conven-
tional techniques but also by strategies such as: UV spectroscopy and photometry
from space (IUE, Voyager and Pioneer), sometimes during multi-technique and multi-
longitude "campaigns", collaborative monitoring of slow variables (such as by the
ESO observers led by Sterken /IAPPP Comm 22.42/), and high S/N spectroscopy of
line-profile variations. To monitor southern variables continuously in winter,
LeContel and Valtier proposed a photometric telescope at the south pole, and other
astronomers have proposed a network of automated (or non-automated) telescopes at
different longitudes. Automated photometric telescopes are now in routine
operation.

NORMAL B STARS
    Searches for β Cep stars have been carried out in NGC 6231 by Balona &
Engelbrecht (MN 212.889), in the Scorpius complex by Waelkens & Cuypers (AA 152.
15), in NGC 3766 by Balona & Engelbrecht (MN 219.131), and in IC 4996 by Delgado
et al. (AA Supp 61.89). Chapellier et al. (AA 143.466) observed αVir in 1983 and

found marginal evidence for pulsation; Sterken et al. (AA 169.166) found no pulsation on 13 nights in 1985. Smith (ApJ 297.206 and 224) found commensurable l=m=-8
and -16 modes and a quasitoroidal mode in α Vir. Goossens et al. (AA 140.223) reanalyzed van Hoof's photometry of σ Sco. Sterken (AA Supp 58.657) and Jerzykiewicz
& Sterken (MN 211.297) obtained new observations of this star, and derived four
periods. Costa & Ringuelet (Rev Mex AA 10.293) discussed the van Hoof effect in
this star, and Dunham (Occultation Newsletter 3.16.346) discussed its close binary
nature. Kubiak and Seggewiss (Acta Ast 34.41) obtained spectroscopy and photometry
of HR 6684. Cuypers (AA Supp 69.445) analyzed new and existing photometry of
τ' Lup, finding a single period with dP/dt < 0.5 s/century. Sterken et al. (AA
Supp 66.11; AA 177.150) obtained photometry of BW Vul from 13 sites, and found a
stable light curve. BW Vul was observed by Herrera et al. (IBVS 2597) and Jung
& Lee (J Korean Ast Soc 18.1). Barry et al. (ApJ 281.766) observed BW Vul with
Voyager 2, and measured the temperature and UV and radius amplitude. Peters et
al. (ApJ 314.261) observed γ Peg and δ Cet with Pioneer 10. Van der Linden &
Sterken (AA 150.76) identified HR 5488 as a β Cep star, and found a triplet of
frequencies. Valtier et al. (IBVS 2843) observed three β Cep stars. Chapellier et
al. (AA 176.255) observed small variations in ι Her, Wolf (IBVS 3003) found an increased amplitude in 53 Psc. Sterken (AA, in press) found no variability in 53 Ari.
Clarke (AA 161.412) found no pulsation-related polarization variations in three
β Cep stars.

Shobbrook (MN 214.33) used an improved luminosity calibration to derive
Q-values for β Cep stars, which suggest that most of the stars are pulsating in
the first overtone. Balona (preprint) noted the occurrence of β Cep stars near
the ZAMS in clusters. Engelbrecht & Balona (MN 219.449) discovered that HD 92024
in NGC 3293 is an 8-day eclipsing binary β Cep star. Engelbrecht (MN 223.189) used
the β Cep stars in NGC 3293 to derive Q-values and period ratios for mode identification. Other studies involving β Cep stars in clusters are noted above.

Chapellier (AA 147.135) proposed that all long-term period changes in β Cep
stars are abrupt. He also noted slow cyclic changes in amplitude (AA Supp 64.275),
and in phase lag between light and velocity (AA 163.329) in several β Cep stars.
However Jerzykiewicz (AA, in press) cautioned against inferring amplitude changes
in such stars from inadequate data. Balona (MN 217.17p) suggested that the amplitude changes in α Vir and 16 Lac might be due to precession in a binary system.
Jerzykiewicz (Acta Ast 34.409) has commented on the absence of β Cep stars in
close binaries. Sato & Hayasaka (PAS Japan 38.47) confirmed the slow amplitude
change of one period in 16 Lac. Sareyan et al. (AA Supp 65.419) detected slow
amplitude changes in δ Cet from multi-longitude photometry. Cuypers (AA 167.282)
used a new period analysis method to derive a dP/dt of +0.14 s/century in δ Cet.
Bloome & Hensberge (AA 148.97) determined mass-loss rates for BW Vul and σ Sco
from UV spectra.

53 Per stars are normal OB stars which undergo pure non-radial pulsation
(NRP). Until recently, NRP has been studied by comparing theoretical and observed
line profiles. Balona used the moments of the line profile to derive NRP periods
and modes (MN 219.111) and developed a quantitative algorithm for applying it (MN
220.647). He also investigated the effects of temperature variations on the line
profiles (MN 224.41). Baade (IAU Symp 132) and Gies & Kullavanijaya (preprint)
carried out period analyses of time series of line profiles of B stars, deconvolved with the window function by means of the CLEAN algorithm. This enabled them to
derive NRP periods and modes in these stars. The bright B0.7III star ε Per was
studied by Smith (ApJ 288.266: identification of three low-order prograde modes
from line-profile variability; ApJ 307.213: investigation of NRP and vertical
shocks as a means of driving mass loss), by Smith et al. (ApJ, in press: results
of a coordinated spectroscopic-photometric study of the star in 1984) and by Gies
& Kullavanijaya (preprint: detection of four low-order modes by the method
mentioned above). Smith (ApJ 304.728) identified six commensurable NRP modes in
the B0.3IV star δ Sco. Photometric studies of 53 Per stars were carried out by
Balona & Engelbrecht (MN 214.559); Waelkens & Rufener (AA 152.6) discovered
several probable 53 Per stars. Balona (preprint) reinterpreted existing photometry

of 53 Per in terms of a double-wave light curve with a period of 3.45 days. Balona
& Laing (MN 223.621) suggested that two suspected 53 Per stars may actually be
ellipsoidal variables. Gry et al. (AA 137.29) and Prinja & Howarth (ApJ Supp 61.
357) discussed the narrow absorption components which occur in the UV spectra of
ε Per and many luminous OB stars.

The "hypothetical" Maia variables are a group of mid- to late-B stars, suspected by Otto Struve and others of light and velocity variations. Photometric
surveys of Pleiades stars by McNamara (ApJ 289.213) and of mid-B field stars by
Waelkens & Rufener (AA 152.6) yielded several new low-amplitude variables with
periods about a day. This suggests that the Maia variables are the same as 53 Per
stars - probably non-radial pulsators. The Maia star 23 Tau is an X-ray source
(Micela et al. ApJ 292.172).

Be STARS

Most aspects of Be stars were reviewed in "The Physics of Be Stars" (ed.
Slettebak, Cambridge, 1987); see also the report of Commission 29 in this volume.
That Commission's Be Star Newsletter (ed. Peters) contains a complete bibliography
on Be stars. Baade (AA 148.59) reviewed concepts of modelling Be stars.

Long-term photometric variations of Be stars continue to be monitored by
Harmanec and colleagues (Ondrejov), Percy (Toronto), Halbedel (Las Cruces) and
others. Spectroscopic variations are monitored by Barker (London, Canada). Doazan
and colleagues studied long-term visual/UV variability of several stars: 59 Cyg
(AA 152.182), Θ CrB (AA 158.1; 170.77; 173.L8), 88 Her (AA 159.65 and 75) and
γ Cas (AA, in press). The polarimetric monitoring program of Hayes at Columbia U.
has unfortunately been terminated. Short-term variability of southern Be stars
was surveyed by Stagg (MN 227.213). Balona studied Be stars in NGC 3766 (MN 219.
131), and in the field (IAU Coll 92), and found that the majority have double-wave
light curves. Balona et al. (MN 227.123) found 1.26 day photometric variability in
α Eri - the brightest Be star.

There are two main models for the short-term variability of Be stars: rotation (spots or circumstellar matter) and NRP. Evidence for the former includes the
time scales, the double wave light curves, and the analogy to magnetic oblique rotators such as σ Ori E (Harmanec, BAI Czech 35.193). Harmanec et al. (IAU Coll 92)
fitted a spot model to light curves of o And. Clarke & McGale (AA 169.251;178.294)
modelled the polarization curves of Be stars in terms of inhomogeneities in an extended atmosphere. The non-radial pulsation model is supported most strongly by
the extensive spectroscopic studies of Baade and of Penrod (IAU Coll 92), who find
low-order (l=2) NRP modes in all Be stars observed. Photometric variations with
the same period, and (variable) amplitudes up to 0.1 occur in the same stars.

Changes in the NRP amplitude may produce mass-loss episodes; see the proceedings of "The Connection between Non-Radial Pulsation and Stellar Winds in Massive
Stars" (PASP 98,29) for a discussion. It is difficult to test this theory observationally, but Guinan & Hayes (ApJ 287.L39) and Peters (ApJ 301.L61) observed the
rapid onset of emission episodes in ω Ori and μ Cen. A major campaign on o And,
KY And, ω Ori and λ Eri was carried out in 1987 to obtain a "multi-wavelength
understanding" of the short-term variability. As in the case of other pulsating B
stars, the cause of the variability is unclear. Osaki's (ApJ 189.469) mechanism is
still promising, and Ando (AA 108.7; 163.97) explored the wave-rotation interaction as a way of feeding pulsation energy into the wind.

O STARS AND WOLF-RAYET STARS

The variability of O and WR stars in the visible and UV has been reviewed by
Baade and by Henrichs in "O, Of and WR Stars" (ed. Conti & Underhill, NASA 1987).
Many O stars are photometric variables, though no thorough survey has been done.
Fullerton et al. (J.P. Cox Symp) have surveyed line-profile variability in 46 O
stars, and found it in about a third. O stars previously known to be line-profile
variable include 10 Lac, υ Ori, ζ Oph and ζ Pup. The variability is thought to be
due to non-radial pulsation, but the driving mechanism is unknown.

Vreux et al. proposed that the velocity variations in HD 192163 (AA 149.337) and other WR stars (PASP 97.274) may have periods < one day (rather than several days as previously believed) and may therefore be due to pulsation. Both the time scale and the complex spectra of WR stars make it difficult to confirm this suggestion. Maeder (AA 147.300) and Cox & Cahn (preprint) have found a radial pulsation instability in some model WR stars, driven by nuclear processes; the period is < one hour. Lamontagne & Moffat (preprint) found no brightness variations in one WR star on this scale. Scuflaire & Noels (AA 169.185) found that NRP modes may be trapped and amplified in H-burning shells in WR stars.

OB SUPERGIANT VARIABLES

S Dor or Hubble-Sandage variables were reviewed in three papers in "Luminous Stars and Associations in Galaxies" (ed. de Loore et al., D. Reidel Co.) p139,p151 and 157, and at an April 1986 conference on "Instabilities in Luminous Early Type Stars" (ed. Lamers & de Loore, to be published by D. Reidel Co.). Van Genderen and his colleagues continued to search for and observe OB supergiant variables (AA 151.349; 153.163; 157.163; AA Supp 61.213; 62.291). They have irregular light curves with amplitudes of up to 0.1 and time scales of days to weeks, depending on their Mbol and Te. Stahl & Wolf (AA 154.243) studied R127, an erupting S Dor star. Percy et al. (AA, in press) investigated the light curve of P Cyg over two years, and Markova (AA 162.L3; Ap Sp Sc 123.5) investigated the spectrum, and concluded that there are both pulsations and expanding shells in the atmosphere of this star. Both van Gent & Lamers (AA 158.335) and Percy et al. have concluded that P Cyg is not periodic, but has a quasi-period of about 40 days. Baade & Ferlet (AA 140.72) observed line-profile variability in γ Ara (B1Ib). This and other studies suggests that the variability of OB supergiants is due to nonradial pulsation, but Harmanec (BAI Czech 38.52) proposed that it may be due to duplicity or rotation.

HOT DEGENERATE VARIABLES

The PV Tel variables (as they are now called in the GCVS) are hot hydrogen-deficient stars with temperatures and luminosities similar to those of the β Cep stars. They are part of a sequence which includes the R CrB stars. Various papers on these stars appear in "Hydrogen-Deficient Stars and Related Objects"(ed. Hunger et al., D. Reidel Co., 1986), including a comprehensive review by Landolt. Hill and colleagues at St. Andrews (Scotland) published several papers on individual stars: BD+10°2179 (MN 207.823), BD+13°3224 (MN 209.387; 210.731; 221.975), BD+1°4381 (MN 213.61P; 224.1083), HD 168476 (AA Supp 61.303), BD-9°4395 (MN 217.701), CPD-58°2721 (IAU Circ 4086; 4097; MN 225.1005), υ Sgr (MN 222.543), LSIV-1°2 (MN 224.1083), V348 Sgr (IAU Circ 4399), KS Per and LSII+33°5 (MN, in press) and HD 160641 (MN, in press). Virtually all the hot hydrogen-deficient stars show small, complex (except in the case of BD+13°3224) and often sporadic variability, in some cases radial, and in other cases, non-radial. BD+13°3224 appears to be contracting to the helium main sequence (Jeffery, MN 210.731). Ando (MN 221.1P) constructed models of BD+13°3224; the radial periods agreed with the observed period, but the radial modes were all stable.

The DB variables are hot helium white dwarfs, pulsating non-radially with periods of minutes. According to Liebert et al. (ApJ 309.241), 23,000 < Te < 28,000 K approximately; the prototype GD358 has Te = 24,000 K and log g = 8.0 (Koester et al., AA 149.423). Winget et al. (ApJ 316.305) observed PG1115+158, which has a complex mixture of periods, and PG1351+489, which has only a single period (489.5 s), unlike the other three members of this class. Hansen et al. (ApJ 297.544) and Cox et al. (ApJ 317.303) studied the pulsational stability of models of these stars.

The PG1159-035 (GW Vir) stars are hot (80,000 < Te < 160,000) degenerate (6 < log g < 8) stars which show complex, non-radial pulsation. Five such variables are known: PG0122+200, PG1159-035, PG1707+427, PG2131+066 and the nucleus of PN K1-16. Grauer et al. (preprint) surveyed 15 candidate stars and found only one new variable. This seems to indicate that not every star in the PG1159-035 instability strip is variable. The pulsations are thought to be due to the effect of the C and O ionization zones, or possibly to helium shell burning instabilities (Kawaler et al., ApJ 306.L41). Barstow et al. (ApJ 306.L25) reported X-ray pulsations in PG1159-035 itself. These stars were reviewed by Winget and by Cox (Highlights of Astronomy vol. 7, p221 and 229). (A more detailed review on these objects is given by D.E. Winget in this report.)

4. Cepheids
(E.G. Schmidt)

INTRODUCTION

During the interval surveyed, the proceedings of two conferences on pulsating stars have appeared, "Cepheids: Theory and Observations", IAU Colloquium No. 82, and "Stellar Pulsation", a conference held as a memorial to John P. Cox (Lecture Notes in Physics No. 274, eds. Cox, Sparks & Starrfield, Springer-Verlag, referred to below as St Puls). Both contain reviews and contributed papers which summarize recent work on Cepheids. Böhm-Vitense and Querci (in Exploring the Universe With the IUE Satellite, ed. Kondo, Reidel, p. 233) reviewed ultraviolet observations of Cepheids and other pulsating stars.

LIGHT AND VELOCITY DATA

A considerable amount of new, high quality, well-sampled light and velocity data has been published during the past few years. Notable studies include those of Moffett & Barnes (photometry of 112 Cepheids: ApJ Supp 55.389; summary of mean parameters: ApJ Supp 58.843), Barnes & Moffett (velocity curves for 88 stars from the same sample: IAU Coll 82.32), Berdnikov (photometry of 77 stars: Per Zv 22. 69), Caldwell & Coulson (photometry of SMC Cepheids: SAAO Circ 8.1), Coulson et al. (photometry and velocities for six stars: ApJ Supp 57.595), Coulson & Caldwell (photometry and velocities for 27 stars: SAAO Circ 9.5), Eggen (intermediate band photometry of 50 Cepheids: AJ 90.1297), Evans & Lyons (velocities for three stars: AJ 92.436) and Imbert et al. (velocities for six LMC Cepheids: AA Supp 61.259).

PULSATIONAL PROPERTIES

Fourier decomposition has been employed extensively to study the pulsational properties of Cepheids. Topics explored in this way include trends of light curve properties with period (Simon & Moffett, PASP 97.1078), the properties of type II Cepheids (Simon, ApJ 311.305), the phase lag between velocity and light (Simon, ApJ 284.278) and the comparison of LMC Cepheids with those in the Galaxy (Andreasen & Petersen, St Puls, p195). Petersen (AA 170.59) has developed methods of estimating the errors in this technique.

Fernie & Chan (ApJ 303.766) showed that stars with periods between 7.2 and 11 days exhibit a remarkably small range of amplitude while Tsvetkov (Ap Sp Sc 116. 43) derived relations among various amplitudes. Moffett (IAU Symp 118.305) pointed out that light curve dips are potentially important to pulsation.

Arellano Ferro (MN 209.481) and Antonello & Poretti (AA 169.149) have studied the properties of the low amplitude short period type S (also called type C) Cepheids and conclude that they are almost all fundamental pulsators. Why they appear distinct from the larger amplitude Cepheids (as shown by Eggen, AJ 90.1278) is not yet clear. Eggen's (AJ 90.1260) suggestion that the S Cepheids are on the first crossing of the instability strip deserves further consideration.

Double mode Cepheids seem to have attracted less interest from observers than in the past but Gieren (IAU Coll 82.98) presented evidence for double mode behavior for EU Tau and new periods were derived for EW Sct (Cuypers, AA 145.283) and CO Aur (Antonello et al., AA 159.269). The latter star has the unusual period ratio of 0.8007 which the authors attribute to the presence of the first and second overtones. Balona & Engelbrecht (IBVS 2758) showed that BQ Ser is near the cool edge of the instability strip unlike other known double mode Cepheids.

RADII

There is continued interest in the radii of Cepheids. Fernie (ApJ 282.641) compared the period-radius relations derived by various methods and found serious disagreements. A number of investigators have used versions of the Baade-Wesselink method to try to improve this situation. Noteworthy are studies by Burki (IAU Coll 82.34), Caccin et al. (IAU Coll 82.43), Welch et al. (IAU Coll 82.51), Barnes & Moffett (IAU Coll 82.53), Gieren (ApJ 282.650) and Coulson et al. (ApJ 303.273). A recent discussion of the period-radius relation by Moffett & Barnes (St Puls p169) shows that progress has been made.

THE TEMPERATURE SCALE

Teays & Schmidt (St Puls p173) derived a new effective temperature scale from Cepheid energy distributions. It agrees with the cooler among previous scales.

CEPHEIDS IN CLUSTERS AND ASSOCIATIONS

The search for additional calibrating stars for the PLC relation has been an active area. Cluster or association membership has been discussed for SU Cas (Turner & Evans, ApJ 283.254), BB Sgr (Turner & Pedreros, AJ 90.1231), three Cepheids in NGC 6067 (Walker, MN 214.45; Coulson & Caldwell, MN 216.671; Moffett & Barnes, MN 219.45P), V 378 Cen (Turner, ApJ 292.148), TW Nor (Anderson et al., IAU Coll 82.203), SZ Tau (Gieren, AA 148.138) and S Vul (Turner, JRAS Can 79.175). Attempts to locate associations near southern Cepheids have been mostly unsuccessful (van den Bergh et al., ApJ Supp 57.743). Walker (SAAO Circ 11.131) summarized work on known and suspected cluster Cepheids.

BINARY CEPHEIDS

A number of investigators have searched for companions to Cepheids using radial velocities and photometry (Gieren, ApJ 295.507; Coulson et al., ApJ 303.273), CaII H and K line profiles (Evans, IAU Coll 82.79), ultraviolet energy distributions (Böhm-Vitense & Profitt, ApJ 296.175; Arellano Ferro & Madore, Obs 105.207) or collections of available data (Szabados, IAU Coll 82.75; Gieren, AA 148.138). While the various methods sometimes yield contradictory results, the binarity of individual stars and the nature of the companions (e.g. Leonard & Turner, JRAS Can 80.240) are becoming more and more certain.

Orbits for binary Cepheids are important to resolving the long-standing mass question. Imbert (AA Supp 58.529) and Evans & Bolton (St Puls p163) derived spectroscopic orbits for SU Cyg. The latter authors then used IUE spectra to measure the velocity curve of the companion and obtained a mass for the Cepheid in agreement with evolution theory. Unfortunately, the companion is also a spectroscopic binary; its total mass and thus the Cepheid mass depends on the inferred types of the two components. Jacobsen et al. (PASP 96.630) were unable to determine an orbit for W Sge even with additional new data.

LUMINOSITIES AND THE PERIOD-LUMINOSITY-COLOR RELATION

There have been many studies aimed at addressing the question of Cepheid luminosities. The much-cited Sandage & Tammann (ApJ 157.683) PLC relation corrected to a Hyades distance modulus of 3.3 (referred to as S&T) will be used as a basis to compare the results. It should be realized that the comparison is complicated by various assumptions regarding such things as reddening, metallicity corrections, etc. so the values given below are only approximate.

Some investigators have redetermined distances of clusters with Cepheid members. H-Beta photometry of cluster B stars (Schmidt, ApJ 285.501; revised by Balona & Shobbrook, MN 211.375) implied Cepheid luminosities 0.4 mag fainter than S&T. Pel's (IAU Coll 82.1) Walraven observations of main sequence stars in two clusters, M 25 and NGC 6087, implied luminosities about 0.2 mag fainter than S&T. Additional UBV observations of individual clusters were made by Pedreros et al. (NGC 7790, implies luminosities 0.1 mag brighter than S&T, ApJ 286.563), Walker (Ly 6, 0.3 mag fainter than S&T, MN 213.889), Walker & Laney (NGC 6649, 0.2 mag fainter than S&T, MN 224.61), Turner (NGC 6087, 0.15 mag fainter than S&T, AJ

92.111) and Walker (NGC 6067, 0.4 mag fainter than S&T, MN 214.45). A study of blue companions of five Cepheids by Böhm-Vitense (ApJ 296.169) produced magnitudes which were about 0.8 mag fainter than S&T while Evans & Arellano Ferro (St Puls p183) obtained luminosities by the same method which were about 0.3 mag fainter than S&T. Gieren (ApJ 306.25) applied the surface brightness technique and obtained luminosities about 0.3 mag fainter than S&T. Although there is still a range, it appears that the various determinations are converging on values a few tenths of a magnitude fainter than was accepted several years ago.

Welch et al. (ApJ Supp 54.547; ApJ 292.217) derived an improved PL relation for the J, H and K bands. Pritchet & van den Bergh (ApJ 316.517) found that the resulting distance of M 31 was in good agreement with what they obtained from the RR Lyrae stars. Cester & Marsi (Ap Sp Sc 107.167) rediscussed the color term in the PLC relation while Tsvetkov (Ap Sp Sc 117.227) derived separate PL relations for the long and short period Cepheids and for the S Cepheids.

Eggen (AJ 90.1278) used narrow band indices to argue that at a particular metallicity no color term is needed to account for the scatter in the Period-Luminosity relation. He then derived a Period-Luminosity-Abundance relation.

MASSES AND EVOLUTION

The fainter luminosity scale implied by many new observations again causes the pulsational masses to disagree with the evolutionary and theoretical masses of Cepheids (see Schmidt, ApJ 285.501, for example). When masses implied by Wesselink radii are compared with theoretical masses, various investigators have found conflicting results (Gieren, MN 222.251; Burki, IAU Coll 82.71).

Böhm-Vitense (ApJ 303.262 and St Puls p159) used binarity to infer the masses of Cepheids. However her conclusion that the Cepheids are under-massive compared with evolutionary theory, conflicts with the mass of SU Cyg found by Evans & Bolton (discussed above). While the important question of Cepheid masses is still unresolved, these new studies of binaries promise to finally put it to rest.

Deasy & Wayman (MN 212.395) discussed period changes of Magellanic Cloud Cepheids and concluded that evolution caused some but not all of the effect.

McAlary & Welch (AJ 91.1209) and Deasy & Butler (Nature 320.726) searched the IRAS catalogue for Cepheids. The majority of detected classical Cepheids exhibited only a photospheric continuum in the infrared while most of the type II Cepheids had an infrared excess. Szabados (IBVS 2910) suggested that the presence of a companion might affect the infrared behavior. Deasy & Wayman (Irish Ast J 17.286) derived a very low mass loss rate for Zeta Gem from MgII h and k line profiles. The suggestion that pulsational mass loss plays an important role in classical Cepheid evolution (Willson & Bowen, Nature 312.429) seems unlikely in view of these results.

ATMOSPHERES AND ELEMENTAL ABUNDANCES

Harris & Pilachowski (ApJ 282.655), Luck & Lambert (ApJ 298.782), Luck & Bond (PASP 98.442), Wallerstein et al. (PASP 96.613), Giridhar (IAU Coll 82.100) and Sanwal et al. (Ap Sp Sc 123.183) carried out high dispersion spectral studies of Cepheids. The abundance gradient in the galactic disk found previously from photometry was confirmed. On the other hand, within the limits of present day model atmospheres and spectroscopy, there is no compelling reason to postulate a dredge up of processed material in these stars.

TYPE II CEPHEIDS

Harris (IAU Coll 82.232 and St Puls p274) and Wallerstein & Cox (PASP 96.677) have reviewed the properties of type II Cepheids and Harris (AJ 90.756) has compiled a catalogue. Petersen & Diethelm (AA 156.337), Simon (ApJ 311.305), Petersen & Andreasen (AA 176.183) and Carson & Lawrence (St Puls p293) applied Fourier decomposition to light curves of these stars. While some evidence for resonant behavior was claimed, type II Cepheids are a very heterogeneous group and more photometric and abundance studies like that of Diethelm (AA Supp 64.261) are needed.

EXTRAGALACTIC CEPHEIDS

Studies of extragalactic Cepheids have continued to be primarily concerned with distance determination (as reviewed by Aaronson & Mould, ApJ 303.1 and Walker, SAAO Circ 11.125). Older photometry of Cepheids in M 33 has been recalibrated using modern techniques (Christian & Schommer, AJ 93.557) while Walker (MN 225.627) redetermined the LMC distance with new CCD photometry of Cepheids. Following a suggestion by Madore and collaborators near infrared photometry of Cepheids has been used to determine the distances of the Magellanic Clouds (Visvanathan, ApJ 288.182), M 31 (Welch et al., ApJ 305.583), M 33 (Madore et al., ApJ 294.560) and NGC 2403 (McAlary & Madore, ApJ 282.101). On the other hand, Welch (ApJ 317.672) was unable to use JHK photometry of type II Cepheids to obtain the distance of the Magellanic Clouds due to infrared excesses. Freedman et al. (ApJ Supp 59.311) pointed out advantages offered by R band photometry with CCD's. Madore (ApJ 298.340) and Madore & Freedman (AJ 90.1104) discussed ways to optimize observations of extragalactic Cepheids for distance determination. However, one suggestion, the "Feinheit" function, was shown by Moffett & Barnes (ApJ 304.607) to be inappropriate to extragalactic studies.

Discoveries of new extragalactic Cepheids were reported for the LMC (van Genderen et al., IBVS 3026), M 31 (Ivanov, Ap Sp Sc 115.409), M 33 (Kinman et al., AJ 93.833), M 101 (Cook et al., ApJ 301.L45), NGC 205 (Richer et al., ApJ 287.138), NGC 300 (Graham, AJ 89.1332), Sextans B (Sandage & Carlson, AJ 90.1019) and WLM (Sandage & Carlson, AJ 90.1464).

Studies of the properties of individual extragalactic Cepheids have also been reported. Wallerstein (AJ 89.1705) inferred the metallicity of the SMC from the spectra of seven Cepheids while Caldwell et al. (MN 220.671) and Imbert (AA 175.30) obtained radii of Magellanic Cloud Cepheids from Baade-Wesselink analyses. Caldwell & Coulson (MN 212.879) obtained reddening measurements of individual Magellanic Cloud Cepheids. Schmidt & Simon (St Puls p180) obtained light curves for stars in NGC 6822 which were of sufficient accuracy to study their pulsational properties. Ivanov & Sharov (Ap Sp Sc 124.329; 125.201) compared the period-amplitude relation in various galaxies.

STARS RELATED TO THE CEPHEIDS

There is considerable interest in luminous variable stars located in an extension of the Cepheid strip. Photometric observations were published for such stars in the Magellanic Clouds (Grieve et al., ApJ 294.513; Grieve & Madore, ApJ Supp 62.451), M31 (Ivanov, IBVS 2729) and the galaxy (Arellano Ferro, MN 216.571; Eggen, AJ 91.890) and Arellano Ferro (Rev Mex AA 11.113; PASP 96.641) made spectroscopic observations. The pulsation of Rho Cas was investigated by Sheffer & Lambert (PASP 98.914) and Percy & Kieth (IAU Coll 82.89). Sasselov (IAU Coll 82.85) discussed the status of such stars in the galactic halo.

Arellano Ferro & Madore (ApJ 302.767) surveyed F and G supergiants near the instability strip for companions using IUE. Similarly to the Cepheids, about 26% of such stars are binaries and this seems to suggest that binarity cannot inhibit pulsation and account for non-variables inside the instability strip. However, in a study of yellow giants in clusters, Schmidt (ApJ Supp 55.455; ApJ 287.261) was able to find few within the instability strip and concluded that they might be rare or non-existent contrary to previous findings (for example, Bidelman, IAU Coll 82.83).

PECULIAR CEPHEIDS

The peculiar Cepheid V473 Lyr (HR 7308; see Burki, IAU Symp 105.453, for a summary of its properties) continued to attract interest although no convincing explanation for the long term amplitude modulation was found. Burki et al. (AA 168.139) carried out an extensive observing campaign and were able to estimate the radius. Second or higher overtone pulsation was suggested. The metallicity appeared from intermediate band photometry to be solar.

Teays & Simon (ApJ 290.683) studied the peculiar variable XZ Cet. They concluded that this star is possibly an anomalous Cepheid. If so, it is only the second found in the galaxy.

MISCELLANEOUS

Jacobsen & Wallerstein (PASP 99.138) studied atmospheric level effects and the period of Eta Aql. Caldwell & Coulson (AJ 93.1090) applied Cepheids to the determination of galactic rotation and the distance to the galactic center. They obtained a galactocentric distance of 7.8 kpc in accord with other recent determinations. Giridhar (J Ap Ast India 7.83) used Cepheids to study local chemical inhomogeneities in the galactic disk.

Cepheids were used to probe the structure of the Magellanic Clouds by Mathewson et al. (ApJ 301.664), Caldwell & Coulson (MN 218.223) and Laney & Stobie (SAAO Circ 10.51; MN 222.449).

5. Delta Scuti Stars
(Michel Breger)

LARGE-AMPLITUDE DELTA SCUTI STARS

This group of pulsators with $Av \geq 0.30$ mag contains both Population I and II stars. Burki & Meylan (AA 159.261) determined Wesselink radii using CORAVEL radial velocities and Geneva photometry for two stars. They found 3.2 $R_\odot$ for BS Aqr and a smaller value of 1.4 $R_\odot$ for DY Peg, which has an evolutionary status similar to the other SX Phe-like variables. The connection between slow rotation and large amplitude was examined by McNamara (PASP 97.715), who found v sin i values less than 20 km/s for a group of large-amplitude variables. He also found these stars to occupy a smaller region in the HRD than the δ Scuti stars with smaller amplitudes. Cox et al. (ApJ 284.250) considered the observed period ratio of 0.80 for VZ Cnc in terms of atmospheric depletion of helium as well as the position in the HRD. Period decreases were reported for DY Peg (Pena & Peniche, AA 166.211), CY Aqr (Rolland et al., AA 168.125; Kämper, IBVS 2802) and BS Aqr (Kozerska & Stepien, Acta Ast 34.377).

Other studies include: HD 200295 = V1719 Cyg (Johnson & Joner, PASP 98.581; Mantegazza & Poretti, AA 158.389), HD 38882 = RY Lep (Diethelm, AA 149.465), EH Lib (Joner, PASP 98.651; Hamdy, IBVS 2810), YZ Boo (Peniche et al., PASP 97.1172), SZ Lyn (Bardin & Imbert, AA Supp 57.249), AI Vel (Bates & Halliwell, AA 151,403), HD 94033 (McNamara & Budge, PASP 97.322; Hobart et al., Rev Mex AA 11.19).

FREQUENCY AND MODE DETERMINATIONS

The difficulty of extracting the multiple frequencies present from a limited amount of photometry is demonstrated by the large number of studies with inconclusive results of frequency analysis. We can only repeat the comments of Kurtz made in these pages three years ago: "Observationally, extensive observation sets of selected multiperiodic δ Scuti stars are needed now, rather than short studies...". The progress of the last years shows that the most promising approach for the future consists of extensive high-quality measurements of one observatory or, better, at several observatories. A multisite campaign at the Beijing (China), McDonald (USA), and Merate (Italy) observatories led to the discovery of four close (nonradial) frequencies in Θ^2 Tau (Breger et al., AA 175.117) which were confirmed in a subsequent campaign. Despite the complexity of some variables, BD +43°1894 was found to have only one pulsation frequency (Costa et al., AA Supp 57. 233). 28 And = HR 114 is also monoperiodic, but has a variable amplitude from season to season (Garrido et al. AA 144.211).

Other studies include: HR 151 and HR 239 = AZ Phe (Kreidl, MN 216.1017), HD 37819 (Padalia & Gupta, Acta Ast 34.303), HD 96008 (Lampens, AA 172.173), HD 101158 (Lampens, Ap Sp Sc 127.27), HR 4684 = FM Com (Antonello et al., AA 146.11; Paparó & Kovács, Ap Sp Sc 105.357), 73 Vir (Sterken & Jerzykiewicz, AA 169.164), HD 126859 (Vander Linden & Sterken, AA 168.155), HR 7222 = LT Vul (Lopez de Coca et al., AA Supp 58.441), V 1719 Cyg (Poretti, AA Supp 57.435), BD-7°1108 (Lampens,

IBVS 2794), BD+28°1494 (Broglia & Conconi, AA 149.15), Y Cam (Broglia & Conconi, AA 138.443).

LINE PROFILE VARIATIONS AND RADIAL VELOCITIES

The examination of line-profile variations is a powerful tool to examine nonradial oscillations. In a series of papers (see MN 224.41), Balona discusses the problem of mode identification from line-profile variations by introducing moments. He also points out that the temperature variations during a pulsation cycle can significantly affect the line profiles, if the projected rotational velocity is large. Yang & Walker (PASP 98.1156) detected absorption features moving across spectral line profiles in the star HR 1298 = 38 Eri. They offer an explanation for the different radial velocities found for different spectral lines and point out that the velocity/light amplitude ratio may not be an appropriate criterion for mode identifications. For HR 5017 = AO CVn a radial velocity amplitude of just over 1 km/s has been reported (Yang & Walker, PASP 98.862) while HR 21 = β Cas shows a variable radial-velocity amplitude associated with its (presumably) single frequency of oscillation (Weiss et al., PASP 99.303).

MISCELLANEOUS

The search for δ Scuti variables in clusters has received some attention: NGC 2516 (Antonello & Mantegazza, AA 164.40), NGC 6871 (Delgado et al., AA Supp 58.447), and NGC 6405 (Schneider, IBVS 2626).

Nonlinear mode coupling is a promising explanation for the complex frequency combinations found for many (but not all!) low-amplitude pulsators. Dziembowski & Krolikowska (Acta Ast 35.5) theoretically examined acoustic and gravity mode instabilities. Moskalik (Acta Ast 35.229) reported that resonant mode coupling can lead the periodic amplitude modulation with a time scale of years.

Tsvetkov (see Ap Sp Sc 128.319) compared the observed period lengths of δ Scuti stars with theoretical evolutionary tracks. Antonello et al. (see AA 171.131) examined the Fourier decomposition of light curves for various subgroups of δ Scuti stars. Hauck et al. (AA 149.167) published a detailed abundance analysis of HR 5017 = 20 CVn and determined a [Fe/H] ratio relative to the Sun of +0.49. Eggen (AJ 90.1046) discussed the known δ Scuti variables in the context of the Hyades Supercluster.

The "mythical" Maia variables comprise a group of stars outside the hot border of the δ Scuti instability strip in the HRD, which may or may not be really variable with δ Scuti-like periods. In the latest installment of the continuing saga, Philip & Hayes (PASP 96.546) report that for the star 109 Vir, no evidence for photometric or spectroscopic variations larger than the rms errors of observations seems to exist.

PULSATION OF Ap STARS

Cool magnetic Ap stars have rapid oscillations with periods in the range of 4 to 15 minutes (e.g., HD 60435, Matthews et al., ApJ 300.348). The light amplitudes are on the order of millimags with the largest known value being 15.7 millimag (HD 60435, Kurtz, MN 209.841). The frequency splitting can be understood in terms of Kurtz's oblique pulsator model (Dziembowski & Goode, ApJ 296.L27; Kurtz & Shibahashi, MN 223.557; Gabriel et al., AA 143.206). The mode identification of the pulsation frequencies requires extensive observational data of the highest accuracy. One of many excellent recent examples is the study of α Cir = HD 128898 (Kurtz & Balona, MN 210.779), in which observations covering 38 nights showed two close high-overtone frequencies near 6.8 minutes. Kurtz (Ap Sp Sc 125.311) searched for rapid oscillations in FG Sge, since the spectrum of FG Sge is starting to resemble that of Przybylski's star. However, over 25 hours, no variations were found.

Further observational studies to determine which Ap stars have normal, longer-period δ Scuti-like variations were also reported. Generally, no further variability was found (see Kreidl, MN 212.337; 216.1013). However, for 49 Cam (Matthews & Wehlau, PASP 97.841) and HD 92664 (Megessier, AA Supp 59.485) longer periods were suspected, which need to be reconfirmed in further studies.

6. RR Lyrae Stars
(B. Szeidl)

This section is devoted to field RR Lyrae stars. The studies carried out on RR Lyrae stars in clusters and related systems are described in the following section.

Great efforts have been made to obtain new accurate photometric and spectral photometric data on RR Lyr stars. Poretti (AA Supp 57.435) performed UBV observations of V1719 Cyg. From its photometric characteristics the star is classified as an RRc variable. The star may be a double mode pulsator in the first and second radial overtones (Mantegazza & Poretti, AA 158.389). The unusual pulsating variable XZ Cet was observed by Teays & Simon (ApJ 290.683). Their new photometry has confirmed the star's long period of 0.823 day. Energy distributions obtained from spectrum scans were used to derive the temperature and surface gravity of this star. Pulsation models suggest the possibility that XZ Cet is an anomalous Cepheid or an overtone BL Her star. Tan (Acta Ast Sin 26.301; IBVS 2533) obtained UBV photometry for a new RR Lyr star in Leo discovered by Huruhata (IBVS 2402) and determined its correct period. From photometric and spectroscopic observations of UY Phe – previously classified as a dwarf nova – Warner & Barrett (Ap Sp Sc 124. 199) found that it is a type RR Lyr halo variable with a period of 0.512 day. Grauer (PASP 96.84) investigated a new variable (15h 20.9m, $+52^{\circ}$ 39'; 1950) in the UBV system and recognized that the new variable is a halo RRc star. Saha (ApJ 283.580) reported on searches for distant halo RR Lyr stars. Photoelectric photometry and spectroscopy of the faint RR Lyr stars found in this survey were used to derive the chemical abundances and radial velocities of these objects (Saha & Oke, ApJ 285.688).

Nikolov et al. (Publ Astron Dept Univ Sofia) compiled a catalog of mean light and colour curves of 210 RR Lyr type stars.

Observations of RR Lyr and X Ari were obtained in the ultraviolet spectral range with the IUE satellite (Bonnell & Bell, NASA CP-2349 p334; PASP 97.236). Excellent agreement between observed and calculated fluxes was found for X Ari, but the analysis of the IUE observations of RR Lyr encountered some difficulties which are believed to be due to the star's secondary cycle.

Butler et al. (AJ 91.570) investigated the oxygen abundances for a large number of RR Lyr stars and found some connection between the metal abundance and the [O/Fe] values. Alania (Ap Sp Sc 132.313) used $uvby\beta$ photometry to accurately determine the metallicity index ml and Δs for eight RR Lyr stars. Grenon & Waelkens (AA 155.24) paid special attention to the long period, small amplitude RR Lyr star HD 47147 which is extremely metal poor. The evolutionary stage of this star appears to be rather exceptional. New photometric observations and ephemerides were obtained for the variables in the field RRI(MWF 361A) (Kinman et al., AJ 89.1200) and the metal abundances of RR Lyr variables in this field were determined and discussed (Kinman et al., AJ 90.95).

A new search for RR Lyr stars in Baade's Window was carried out and the relationship among the different parameters of light curves were considered (Blanco, AJ 89.1836); the distance to the galactic centre, R_0, has been determined using the newly obtained data (Blanco & Blanco, Mem S A It 56.15). Using CCD measurements of eleven RR Lyr variables in Baade's Window Walker & Mack (MN 220.69) have given a new estimate of R_0. Their data tend to deny the existence of an absolute magnitude-metallicity relation for RR Lyr stars. Gratton (MN 224.175) obtained radial velocities for seventeen RR Lyr stars in Baade's Window. New mean radial velocities were determined for a further 46 RR Lyr stars (Hawley & Barnes, PASP 97.551). Later, the same authors (Barnes & Hawley, ApJ Lett 307.L9) rediscussed these data.

The Baade-Wesselink method (or a modification of this) was utilized to obtain the physical parameters and the distances of a number of RR Lyr stars (SW And, X Ari, YZ Cap, RR Cet, DX Del, SU Dra, SW Dra, SS For, RR Lyr, RV Phe, V440 Sgr, VY Ser) and the results have been discussed in detail (Burki & Meylan, AA 156. 131, 159.255; Meylan et al., AA Supp 64.25; Jones et al., ApJ 312.254, 314.605; Longmore et al., MN 216.873; Buonaura et al., Mem S A It 56.153; Cacciari et al. ibid 57.345; Jameson, Vistas Astron 29.17).

The kinematic properties and absolute magnitudes of RR Lyr stars have been investigated by Wan et al. (Publ Beijing Astr Obs No.6.191), Saha (ApJ 289.310), Hawley et al. (ApJ 302.626) and Strugnell et al. (MN 220.413). A thorough investigation using earlier data and new infrared observations, was carried out on the dependence of absolute magnitude on metallicity and the distance to the Galactic Center (Fernley et al., MN 226.927).

Fourier decomposition parameters were determined for RR Lyr variables (Hansen & Petersen, IAU Coll 82.272) and employed to compare the light curves with those emerging from hydrodynamic models (Simon, BAAS 17.559; ApJ 299.723).

Elements have been derived, revised or refined for a great number of known RR Lyr stars and for some accidentally discovered ones. The results have mostly been published in MVS, IBVS, Per Zv, Per Zv Supp, JAAVSO, J Br Astron Assoc, Astr Tsirk, BAV Rundbrief and GEOS Circ. Ephemerides of RR Lyr variables have been compiled by Tsessevich, Firmanyuk and Kreiner for the years 1985, 1986 and 1987 (Rocznik Astron Obs Krakow).

The Blazhko-effect was investigated by Tsessevich & Mandel (Per Zv 22.237) in V421 Her and by Firmanyuk et al. (Astr Tsirk 1395.6) in TT Cnc.

7. Variable Stars in Globular Clusters and Related Systems
(Amelia Wehlau)

STUDIES OF PULSATING VARIABLES IN GALACTIC GLOBULAR CLUSTERS

A brief review of the properties of RR Lyrae stars and of current problems in RR Lyrae research in globular clusters is given by Hazen (JAAVSO 15.201).

C0911-646 (NGC 2808) Clement et al. (JRAS Can 79.235) report the discovery of 3 more variables.

C1323-472 (NGC 5139) Spectra of 18 RR Lyrae variables in ω Cen have been used by Gratton et al. (AA 169.111) to derive values of ΔS which do not correlate with period shift within the cluster in contrast to 17 variables within Baade's window which do show such a correlation. The authors suggest RR Lyrae stars in the cluster are not ZAHB stars but show the effects of post-HB evolution. In a paper on blue stragglers in the cluster, Da Costa et al. (ApJ 308.743) point out that further observations of the dwarf Cepheids in ω Cen would yield masses which would constrain competing theories for the blue stragglers. An analysis by Nemec et al. (AJ 92.358) of Martin's photometry for 55 RR Lyrae stars in the cluster has found no double-mode variables. Walker & Mack (SAAO Circ 11) present CCD light curves for V84 and V85.

C1339+286 (NGC 5272) Meinunger presents observations of 12 variables in M3 (MVS 10.89; 10.134).

C1514-208 (NGC 5897) Spectra taken by Smith (AJ 90.1242) of a possible non-variable in the RR Lyrae gap of the cluster show that the star has a metallicity and radial velocity consistent with cluster membership.

C1516+022 (NGC 5904) Cohen & Gordon (ApJ 318.215) report on a determination of the distance to M5 obtained by applying a modification of the Baade-Wesselink method to four RR Lyrae variables in the cluster.

C1620-264 (NGC 6121) Yao (Ap Sp Sc 119.41) reports on four possible red variables in M4.

C1639+365 (NGC 6205) Ruseva & Rusev (Per Zv 22.49) present observations for seven variables in M13 and discuss period changes as well as cluster period-luminosity, period-amplitude and period-color relations. Most of the red variables in the cluster are included in a radial velocity study by Lupton et al. (AJ 93.111).

C1654-040 (NGC 6254) Fifty-two years of observations of three slow variables in M10 are discussed by Clement et al. (AJ 90.1238).

C1715+432 (NGC 6341) Several new variables have been found by Kadla et al. (Per Zv 21.827) in the central region of M92. Data are given for 11 stars.

C1732-447 (NGC 6388) Hazen & Hesser (AJ 92.1094) present photometry for four newly found field variables and for 14 new and three previously known variables within the tidal radius of this cluster, the cluster with the highest metallicity thought to contain RR Lyrae variables.

C1800-300 (NGC 6522) Walker & Mack (MN 220.69) report that four of the RR Lyrae variables in Baade's window appear to be cluster members on the basis of CCD photometry of the cluster and of nearby variables. Blanco (AJ 89.1836) also discusses the possible cluster membership of other variables in Baade's window. Also see Gratton et al. (AA 169.111).

C1810-318 (NGC 6569) Hazen-Liller (AJ 90.1807) presents photometry of 17 new and 6 previously known variables within the tidal radius of this cluster, which has an unusually blue horizontal branch for its metallicity.

C1821-249 (NGC 6626) Wehlau et al. (AJ 91.1340) present period changes for variables in M28 derived from observations from 1939 through 1985. Margon & Anderson (PASP 97.962) report that a spectrum of V7 shows it to be a normal Mira variable. Low resolution spectra of V7, one red variable and 4 RR Lyrae stars in the cluster are discussed by Smith & Wehlau (AJ 298.572).

C1827-255 (NGC 6638) Spectroscopy of six RR Lyrae stars, confirming their designation as c-type, is reported on by Smith & Stryker (AJ 98.453) who attribute the large ratio of c-type stars to ab-type stars in the cluster to a gap in the color distribution of HB stars.

C1828-323 (NGC 6637) A possible cataclysmic variable in M69 (IAUC 4247) has been shown to be a late type Mira-type variable by Charles et al. (IAUC 4285).

C1832-330 (NGC 6652) A search by Hazen has failed to find any variables (IAU Symp 118.287).

C1833-239 (NGC 6656) Marinchev (Per Zv 22.65) presents period changes for 12 RR Lyrae stars and observations of 8 long period variables in M22. Of the 13 variables included in a proper motion study of the cluster by Cudworth (AJ 92.348), 11 are confirmed as cluster members. V17 is shown to be a non-member, and V16 only a possible member.

C1850-087 (NGC 6712) Infrared photometry of 15 red giants including 5 variables is presented by Frogel (ApJ 291.581).

C1914+300 (NGC 6779) Wehlau et al. discuss 50 years of observations of V1 and V6 in M56 (ed. Madore, Cepheids:Theory and Observations, 284). Photometry of all 12 known variables (including 5 probable non-members of the cluster) are presented by Wehlau & Sawyer-Hogg (AJ 90.2514).

C1915+186 (NGC 6838) Pogossiantz (Per Zv 22.85) presents observations of the semiregular variable Z Sagittae, V1 in M71. Welty (AJ 90.2555) confirms the variability of 1-29 and S142. All the known variables are included in the proper motion study of the cluster by Cudworth (AJ 90.65).

C2127+119 (NGC 7078) Kadla et al. (Astron Tsirk 1314.1; 1342.8) report the discovery of 8 new variables in M15 while Gordenko et al. (Prob Kosm Fiz Vyp 19.93) have studied period changes in 35 of the variables.

C2130-010 (NGC 7098) Several of the known variables in M2 are included in the dynamical studies by Pryor et al. (AJ 91.546) and Cudworth & Rouscher (AJ 93. 856).

VARIABLES IN DISTANT GALACTIC CLUSTERS, LMC CLUSTERS AND DWARF SPHEROIDAL GALAXIES
     A review of variables in 23 Palomar-like clusters is given by Rosino & Ortolani (Mem S A It 56.113). CCD photometry by Ortolani (AA 137.269) of AM-1 (C0353-497), one of the most distant galactic globular clusters known, includes 5 stars in the variable star gap, three of which show suspected variability, although a search during a similar CCD study of the cluster by Aaronson et al. (ApJ 276.221) turned up no variables.
     Using CCD photometry of 9 RR Lyrae stars in the LMC cluster NGC 2210, Walker

(MN 212.395; SAAO Circ 9.111) has obtained a distance modulus of 18.42 ± 0.10 using a value of MvRR = 0.6. In this cluster a few additional variables have been found by Hazen. In another LMC cluster, NGC 1786, Graham (PASP 97.676) has identified 12 variables. Periods obtained for 10 of them show an Oosterhoff type II distribution. One star may be an anomalous Cepheid. Nemec et al. (ApJ Supp 57.287) studied 41 RR Lyrae stars in the LMC cluster NGC 2257 and 47 field variables in the region while Nemec et al. (ApJ Supp 57.329) have derived period changes for 38 of the cluster variables.

A review of the properties of dwarf spheroidal galaxies by Zinn (Mem S A It 56.223) includes a summary of the properties of the variables found in these galaxies and their associated globular clusters. Light et al. (BAAS 17.883) report on CCD photometry of the two brighter variables discovered along with several probable RR Lyrae stars by Buonanno et al. (AA 152.65) in the fields of clusters of the Fornax dwarf galaxy. One star appears to be an anomalous Cepheid and the other a Pop II Cepheid, the first to be found in a dwarf spheroidal galaxy. Demers & Irwin (MN 226.943) have identified 30 long-period variables in Fornax and determined periods for 26 of them. Smith & Stryker (AJ 92.328) present metal abundances obtained from low resolution spectra for three anomalous Cepheids in the Sculptor dwarf galaxy. Saha et al. (AJ 92.302) report on a search for variables in the Carina dwarf galaxy. Light curves and periods were obtained for 73 variables, most of which seem to be RR Lyrae stars. Of these, 58 appear to be members of the galaxy with a mean period for the ab-type stars of 0.62, intermediate between Oosterhoff types I and II. Using new data, Nemec & Wehlau have determined periods for 49 more RR Lyrae stars in Ursa Minor. Wehlau & Demers (IBVS 2914) present recent observations of bright variables in Draco.

COLOUR STUDIES OF RR LYRAE STARS AND THE OOSTERHOFF DICHOTOMY

Longmore et al. (MN 220.279) present near-infrared photometry of RR Lyrae stars in three clusters, yielding a period-absolute K magnitude relation with remarkably small scatter which could be a useful tool in determining distances. Scaria (Ap Sp Sc 103.207) derives a possible P-L relation for cluster RR Lyraes, luminosity decreasing with period. Magnitudes and colors of RR Lyrae stars belonging to 10 globular clusters are discussed by Cacciari et al. (Mem S A It 56.97) and relations among periods, amplitudes and metallicity are shown. Caputo and her collaborators (AA Supp 55.463) present synthetic horizontal branch computations leading to constraints on the expected properties of RR Lyrae variables in Pop II systems and in a further series of papers (AA 138.457; 143.8; 172.67; Mem S A It 57.437) use the properties of the RR Lyrae stars in M4, M15 and M3 to derive reddenings, distance moduli and ages of 16 billion years for each cluster. VBLUW photometry of RR Lyrae stars in M4 and ω Cen by de Bruijn & Lub (eds. Cox et al., Stellar Pulsation, 233) has been used to derive reddening, blanketing, Teff and g for these stars.

Stellar evolution calculations which suggest possible explanations of the Oosterhoff dichotomy, the Sandage effect, and the second parameter problem are discussed by Caputo (Mem S A It 56.73), Torambe (Mem S A It 56.85) and Torambe & Gratton (Mem S A It 57.361). Using a new grid of ZAHB models, Sweigart et al. (ApJ 312.762) fail to find any satisfactory explanation for the Sandage effect while Jones et al. (ApJ 314.605), using the Baade-Wesselink method, find no evidence of a difference in luminosity between two field RR Lyraes which differ greatly in metallicity. Using their derived absolute magnitudes they find the metal-poor cluster M92 to be considerably older than the metal-rich cluster 47 Tuc. Cacciari & Clementini give a preliminary report on a similar study (Mem S A It 57.345). In order to investigate the Oosterhoff dichotomy, Castellani & Quarta (AA Supp 71.1) present an updated catalogue of RR Lyrae period distributions in graphical form with the clusters ranked according to metallicity. Caloi et al. (AA Supp 67.181) point out the importance of the RR Lyrae rich cluster, M62 (NGC 6211, C1658-300), in the context of globular cluster evolutionary status. Kadla & Gerashchenko (Izv Glav Ast Obs Pul No. 202) discuss the numbers and period distribution of RR Lyrae variables for 48 clusters.

POPULATION II CEPHEIDS AND RED VARIABLES

Two reviews by Harris (ed. Madore, Cepheids:Theory and Observations, 232; eds. Cox et al., Stellar Pulsation, 274) present statistics for Pop II pulsating variables (excluding RR Lyrae stars) in globular clusters and in field. Included are the 8 Pop II Cepheids recently identified in globular clusters as discussed by Clement et al. (ed. Madore, Cepheids:Theory and Observations, 262). The evolutionary status of Type II Cepheids in clusters and in the field is reviewed by Gingold (Mem S A It 56.169) and the population distribution of these stars discussed. Three of the cluster Cepheids are included in the low mass variables whose light and color curves are discussed by Eggen (AJ 91.890).

Results from infrared photometry of red variables in globular clusters are summarized by Frogel (Mem S A It 56.193) and Frogel & Elias (ApJ, 1988). Mass loss rates are discussed and evidence is presented for circumstellar dust shells about these stars. Menzies & Whitelock (MN 212.783) present JHKL photometry for 31 Mira variables in 15 galactic globular clusters and use it to obtain a period-luminosity relation which differs from that found for LMC Miras. Whitelock (MN 219.525) derives a P-L relationship which includes cluster variables with periods ranging from 1 to 300 days. Observations of 7 red variables are included in a paper on spectroscopic data for giants in ω Cen by Lloyd Evans (SAAO Circ 10.1). A search for giant and asymptotic branch variables in 6 clusters by Welty (AJ 90.2555) has found no variables below the tip of the giant branch although Yao (Ap Sp Sc 119. 41) presents evidence for the variability of 4 such stars in M4.

PULSATION MODES AND FOURIER ANALYSIS OF LIGHT CURVES

A review of double-mode RR Lyrae stars by Cox (IAU Coll 95 on Faint Blue Stars) includes work by Ostlie showing that double-mode behaviour may require time-dependent convection with a high helium content. Nemec (AJ 90.240) has reanalyzed the double-mode variables in M15 using simulated photometry to determine the effects of random scatter on the derived periods and showing that the secondary oscillations are probably real. Nemec (AJ 90.204) has also studied 10 double-mode stars in the Draco dwarf galaxy and compared their derived physical characteristics to those of double-mode stars in galactic globular clusters. In the Oosterhoff type I cluster IC 4499 (C1452-820), Clement et al. (AJ 92.825) have identified 13 double-mode RR Lyrae stars and obtained a mean mass of 0.54, about 0.11 smaller than that found for Oosterhoff type II systems. However, thorough searches by Nemec and his collaborators have failed to find any double-mode RR Lyraes in ω Cen or M5.

Cox & Proffit (ApJ, 1988) present detailed pulsation studies of the anomalous Cepheids in the Draco galaxy and galactic globular clusters, but find they are not able to distinguish between fundamental and first overtone pulsation for any of the stars. The question of second-overtone pulsation in RR Lyrae stars has recently been discussed by Stothers (ApJ 319.260) who concludes that such pulsators probably do not exist among RR Lyrae stars although Stellingwerf et al. (ApJ 313.L75) feel such stars might exist.

Peterson reports on Fourier analysis of the RR Lyrae light curves in ω Cen (AA 139.496) and M15 (AA 170.59). Although the ω Cen light curves exhibit a Cepheid-like progression for both the overtone variables and the fundamental mode variables with periods from 0.5 to 1.5 day, confirmation or disproof of such a sequence in M15 requires better observations than currently available. Kovács et al. (ApJ 307.593) compare their Fourier decomposition parameters for variables in the two Oosterhoff I clusters, NGC 6171 (M107, C1629-129) and NGC 6723 (C1856-367), to those in the Oosterhoff II cluster M15 and find there may be small systematic differences between the two Oosterhoff classes. In a study of Fourier coefficients of RR Lyrae stars in NGC 6171 Stellingwerf & Dickens (ApJ, 1988) point out that the stellar light curves are not sensitive to composition or interior structure of the star. Several globular cluster stars are included in a study of Fourier phases of Type II Cepheids by Petersen & Andreasen (AA 176.183).

CATACLYSMIC BINARIES AND X-RAY VARIABLES
     Shara and his collaborators present observations of an outburst of the dwarf
nova V101 in M5 (AJ 94.357) and report that a search for cataclysmic binaries in
M3   found   no   bright   emission-line sources between 4 and 30 core radii from the
center (IAU Symp 113.103). Margon & Bolte (ApJ Lett 321) found no optical evidence
of cataclysmic binaries in ω Cen despite a thorough search. Another such search by
Shara et al. (AJ, 1988) in two fields in ω Cen and one in 47 Tuc (NGC 104,
C0021-723) was equally unsuccessful, indicating a density of contact binaries at
least 8 times lower than in the solar neighborhood. On the basis of CCD photometry
Shara et al. (ApJ 311.796) suggest an optical candidate for nova 1938 in M14 (NGC
6402, C1735-032). A search for millisecond pulsars by Hamilton et al. (AJ 90.606)
turned up one source in M28 within one core radius of the center of the cluster,
later confirmed by Erickson et al. (ApJ 314.L45). Lyne et al. (Nature 328.399)
find a period of 3054 μs and no evidence for a binary period. CCD U-band photometry by Ilovaisky et al. (AA 179.L1) indicates an orbital period of 8.537 hours for
the X-ray binary in M15 in agreement with the spectroscopic period found by Naylor
et al. (IAUC 4263) and the X-ray period of Hertz (IAUC 4272). Verbunt (ApJ 312.
L23) presents mechanisms for cluster binaries containing a white dwarf and a
neutron star as is indicated by the discovery by Stella et al. (ApJ 312.L17) of a
685 sec orbital period for the X-ray source in NGC 6624 (C1820-303). Data on this
source are also used by van Paradijs & Lewin (AA 172.L20) to derive constraints on
the mass-radius relation for the neutron star.

8. Mira Variables and Related Objects
(M. W. Feast)

     A knowledge of the distances of Mira variables is basic to practically all
studies of these objects. The infrared period-luminosity relation, first found in
the LMC has been further refined (Glass et al., Late Stages of Stellar Evolution,
ed. Kwok & Pottasch, Reidel 1987 (= LSSE) p51) and now shows a scatter of only σ =
0.13 mag. Further LMC Miras have been discovered and studied by Reid, Glass and
Catchpole (in preparation). The P-L zero point was rediscussed by Feast (eds.
Gilmore & Carswell, The Galaxy, Reidel 1987 p1). An important new result is that
SMC Miras follow an infrared P-L relation of closely the same slope and zero point
to that in the LMC (Lloyd Evans et al., MN in press). This, together with the
close similarity of P-L slope and zero point for Miras in globular clusters, the
galactic bulge and the LMC (Feast & Whitelock, LSSE p33) strengthens considerably
the value of Miras as distance indicators in stellar systems which may differ in
age-metallicity relations. Herman & Habing (Phys Rep 124.255) give a general review of distance determinations for OH/IR Miras by the phase lag method. Herman
et al. (AA 143.122) use distances for six objects derived in this way, together
with kinematic distances, to infer a distance to the galactic centre (9.2 ± 1.2
kpc). The empirical relation established between the OH luminosity and the shell
radius, has been used by Herman et al. (AA 167.247) to deduce distances of a number of OH/IR sources. Luminosities based on kinematics and membership of the galactic bulge are discussed for some OH/IR sources by Baud et al. (ApJ 292.628 see
also Herman et al. AA 139.171). Empirical relations involving VRI colours were
used by Celis (AJ 89.1343) to estimate Mira distances.
     Whitelock (MN 219.525) studied the period-luminosity-temperature relationship
for red and yellow variables (including Miras) in globular clusters and found that
the higher temperature, more metal-deficient cluster variables pulsate in the fundamental mode whilst the lower temperature more metal-rich variables pulsate in
the first overtone. On the theoretical side, a new linear survey of the Mira instability region was carried out (Ostlie & Cox, ApJ 311.864). A detailed review of
the properties of Miras in symbiotic systems and their relation to single Miras
was prepared by Whitelock (PASP in press).
     The recognition (Habing; ed. Israel, Light on Dark Matter, Reidel 1986 (=LDM)

p329; Feast, LDM p339; Whitelock et al., MN 222.1; Glass, MN 221.879) that at least a substantial fraction, if not all, IRAS sources in the galactic bulge are Miras or Mira-like objects, has stimulated considerable work on these stars. Ground-based studies suggest an upper limit to the luminosity of these stars of $M(Bol) \sim -4.7$ corresponding to an upper limit to their periods of ~ 400 days and a lower limit to the age of the bulge population of ~ 5 Gyr (Feast & Whitelock, LSSE p33). This is consistent with recent work on masses of planetary nebulae in the bulge (Kinman, Feast, Lasker, in preparation). There is considerable other work on the identification of IRAS sources as OH/IR (Mira) stars (Whitelock et al., MN 210.25p; 213.51p; Lewis et al., Nature 313.200; Engels et al., AA 140.L9; Hrivnak et al., ApJ 294.L113; Zuckerman & Lo, AA 173.263; Le Bertre & Epchtein, AA 171.116; Sivagnanam & Le Squeren, AA 168.374). Problems in identifying IRAS sources were noted by Craine et al. (PASP 97.303). Persi & Ferrari-Toniolo (AA Supp 55.165) have made ground-based infrared searches for AFGL sources. The periods of some Miras in the galactic bulge (Plaut fields) have been revised (T. Wesselink, Thesis, Nijmegen, 1987) and the completeness of discovery discussed.

Detailed high dispersion spectroscopy of emission and absorption lines in Miras including variation with phase is now yielding results which may be meaningfully compared with shock-wave models, enabling a beginning to be made in the quantitative understanding of the complex structure and kinematics of these atmospheres (Gillett et al., AA 148.155; 150.89; Fox et al., ApJ 286.337; 297.455; Bertschinger & Chevalier, ApJ 299.167). Hinkle et al. (ApJ Supp 56.1) continued their work on time series infrared spectroscopy of Miras. Data on CO gives evidence for an outward propagating shock-wave driven by stellar pulsation. The appearance of the silicate and other infrared spectral features in Miras is related to the degree of asymmetry of the visual light curves and this indicates that the nature of the dust condensates depends on the strengths of the atmospheric shock-waves (Vardya et al., ApJ 304.L29). The results of spectropolarimetry of Miras is complex and no general pattern has yet been established. Both photospheric and circumstellar polarization are, in general, important (Boyle et al., AA 164.310). In the case of L2 Pup there are secular changes in polarization, indicating the evolution of a circumstellar cloud on a time scale of several years (Magalhaes et al., AA 154.1). The time dependent spectropolarimetry of the SRa variable, V CVn (Magalhaes et al. AJ 91.919) can be interpreted in terms of a scattering layer at an intermediate level in the stellar atmosphere. Dominy & Wallerstein (ApJ 310. 371) derive s-process element abundances in some Miras and interpret the diverse results in terms of differing neutron exposure events (e.g. an s-process event $\leqslant 1.5 \times 10.0E5$ yr ago in X Cyg). Wallerstein et al. (MN 215.67) interpret the spectrum of R Cyg at an unusually low maximum as indicating enhanced limb brightening.

Mennessier (AA 144.463) has attempted to classify the various types of visual and infrared light curves of Miras and Hoeppe (AA Supp 68.419) has listed maxima and minima of R Leo since 1757. The IRC catalogue, late type, long period (~ 500 day) Miras are an important intermediate group between the "optical" Miras and the very long period OH/IR stars. Lockwood (ApJ Supp 58.167) has given near infrared photometry for stars in this class. Amongst other basic data published is the spectral classification of 72 southern Miras round their cycles (Crowe, JRAS Can 78.103), UBVRI of Miras and discussions of their light and colour variations (Celis, ApJ Supp 60.879; AJ 91.405).

Mass loss at the Mira stage (that is, at the top of the AGB) is a crucial factor in the evolution of low mass stars. Consequently a large amount of effort continues to be devoted to the study of the circumstellar molecules and dust of Miras. This work is carried out principally at infrared, mm and cm wavelengths and includes studies of circumstellar masing. Mass loss rates from Miras and other stars can be calculated from infrared data and models (for the dust component) and from mm CO observations from the gaseous component (Knapp et al., ApJ 292.640; 293.273; 293.281). The results suggest a ratio of gas to dust mass in the envelopes of ~ 160 for O-rich stars and ~ 400 for C-rich stars. Knapp suggests that the carbon stars in her sample all have masses $\sim 2M_\odot$. The CO observations also suggest that $^{12}C/^{13}C$ is in the range 5-20 for M type Miras and 30-100 for C stars.

Further data on mass loss from CO observations are given by Wannier & Sahai (ApJ 311.335) and Zuckerman et al. (ApJ 304.394; 304.401). The latter workers find a class of (probably massive) carbon stars with large (~34 km/s) outflow velocities. In addition they suggest that the well known object IRC +10216 (a C-type Mira with a thick dust and molecular shell) is only 100-150 pc away. Extensive studies continue of the circumstellar chemistry and shell structure of this object (Thaddeus et al., ApJ 294.L49; Guelin et al., AA 157.L17; Lucas et al., AA 154.L12; Glassgold et al., AA 157.35; Rengarajan et al., ApJ 289.630; Lester et al., ApJ 304.623; Huggins & Healy, ApJ 304.418; Sahai et al., ApJ 284.144). The four known carbon Miras with thick dust shells all have very long periods (495-650 days), (as do the thick-shelled O-rich Miras, OH/IR stars), showing that these objects are not simply normal C-type Miras undergoing episodic mass ejection (Feast et al., MN 215.63p). Zuckerman & Dyck (ApJ 311.345) use IRAS data to estimate dust grain emissivities for C- and O-rich late type stars. They also find V Hya (a C star) to be the first known CO maser. Willems (Thesis, Amsterdam 1987) has given an extended discussion of data on carbon stars and their circumstellar shells from IRAS data. Quasi-simultaneous JHKL and IRAS observations of 18 Miras were analysed by Whitelock et al. (LSSE p269). There is a relationship, with very little scatter, between the period (or luminosity), the pulsation amplitude and the mass of the dust shell. This shows rather clearly that pulsation is important in facilitating mass loss and that Miras behave in a rather predictable manner. Infrared data on OH/IR sources were discussed by Le Bertre et al. (AA 132.75) and Keping et al. (Ap Sp Sc 107.373). Infrared speckle techniques are being increasingly used to study the circumstellar emission from cool stars, including Miras and related objects. A number of objects seem to have bipolar rather than spherical symmetry (Dyck et al., ApJ 287.801; Cobb & Fix, ApJ 315.325; Ridgway et al., ApJ 302.662). Cobb & Fix interpret their data on several OH/IR sources at 5 and 10 μm, as indicating partially resolved circumstellar shells together with dominant unresolved cores. Several discussions of circumstellar dust models in late type stars are of particular relevance for Miras (Rose, ApJ 312.284; Papoular & Pegourie, AA 156.199; Rowan-Robinson et al., MN 222.273; Gail & Sedlmayr, AA 161.201). Contrary to earlier work, Volk & Kwok (ApJ 315.654) find that the interstellar component of the 10 μm dust feature in OH/IR stars is small. Studies of circumstellar dust at submillimeter wavelengths has the advantage that the shells may be assumed to be optically thin and the emissivity is less sensitive to temperature than at shorter wavelengths. Observations at 400 μm of a range of evolved objects yield gas/dust ratios of ~100 in most cases (Sopka et al. ApJ 294. 242).

Maser and thermal emission from SiO in Miras and other objects has been intensively studied in recent years. One reason for this is that the SiO masers are believed to reside close to the central star and hence are important in studying mass loss mechanisms. Several studies have been devoted to detecting new SiO masers (Jewell et al., ApJ 298.L55; Nyman et al., AA 160.352; Bujarrabal et al., AA 162.157; Barcia et al., AA 142.L9). The possibility of determining the SiO maser luminosity function of OH/IR stars is opened up by the detection of SiO in some of these objects in the galactic centre region (Lindquist et al., AA 172.L3). SiO-infrared phase lag observations (Clark et al., ApJ 283.174) suggest that the SiO is excited by shocks rather than by radiation. Polarization studies yield much detailed information on the circumstellar shells (Barvainis & Predmore, ApJ 288. 694; Miller et al., ApJ 287.892; Clark et al., ApJ 289.756; Olofsson et al., AA 150.169). These and other studies (Nyman & Olofsson, AA 147.309; Gomez et al., AA 159.166; Snyder et al., AJ 92.416) suggest that the variations are in general related to the periodicity of the stellar pulsation. The line structure is complex, but the time-averaged centre of the weak emission pedestal gives the stellar radial velocity. To explain thermal SiO emission in Miras, the existence is suggested of an extended inner envelope (~10.0E15 cm) in which grains have not completely formed (Bujarrabal et al., AA 162.157). In the supergiant variable VX Sgr, the SiO masers exist in dense cloudlets rather than in a spherically expanding wind (Alcock & Ross, ApJ 310.838) and Langer & Watson (ApJ 284.751) have indeed shown that simple wind models fail to produce sufficient maser power. The importance of SiO formation, not only as a first step in dust production, but as an efficient radiative cooling agent has been emphasized by Muchmore et al. (ApJ 315.L141).

Search for H_2O masers in OH/IR stars has continued. Both the H_2O luminosity and the ratio of H_2O/OH maser luminosities appears to depend on mass loss rate (Bowers & Hagen, ApJ 285.637). The increase of H_2O intensity with mass loss rate is expected theoretically (Cooke & Elitzur, ApJ 295.175). The H_2O masers are within the OH maser shell and vary with larger amplitude and less regularity than the OH emission (Engels et al., AA 167.129). Periodicities in H_2O variations equal to multiples of the optical period as well as aperiodic behaviour have been reported (Gomez et al., AA 159.166). Aperiodic outbursts can occur in OH maser emission (e.g. R Leo, Le Squeren & Sivagnanam, AA 152.85). VLA observations show that the structure of H_2O masers round Miras consists of small (unresolved) knots distributed over a region of size $\sim$8x10.0E14 cm (larger by a factor $\sim$10 than some earlier estimates). Observations at two epochs indicate considerable changes in structure and velocity with time (Johnston et al., ApJ 290.660). An unsuccessful search for 183 GHz H_2O emission from Miras and other objects was carried out (Kuiper et al., ApJ 286.310). Diamond et al. (AA 174.95) have carried out detailed mapping of the H_2O and OH masers surrounding the supergiant red variable S Per. The masers around the supergiant red variable VX Sgr were mapped by Chapman & Cohen (MN 220.513) and the structures, magnetic field and driving force on the shell, discussed. Lopez (IBVS 2905) suggests an appreciable proper motion in declination for VX Sgr which is surprising for a supergiant.

The problem of the symmetry of the OH masing shell in Miras and OH/IR stars is important both for physical questions of shell formation, structure and excitation, and for the use of the phase-lag method of distance determination. Continuing work suggests at least rough spherical symmetry but with a good deal of complexity, including incomplete shells (Diamond et al., MN 121.1; 216.1p) and fluctuating velocity fields (Ukita & Le Squeren, AA 138.343). High resolution 18 cm spectra of Miras (Fix, AJ 93.433) as well as other data, are consistent with a model in which mass loss is by the ejection of blobs of material which evolve into plate-like structures in the OH masing region (Alcock & Ross, ApJ 305.837). Alcock & Ross (ApJ 290.433; 299.763; 306.649) have also made a detailed theoretical study of saturation and beaming in astrophysical masers. Dickinson (ApJ 313.408) finds that the far infrared flux is sufficient to pump all types of OH masing stars. For OH/IR stars the pump efficiency is $\sim$8% and for optical Miras $\sim$4%. The possibility of H_2O photodissociation as a pump mechanism for OH masers was discussed by Andresen (AA 154.42) in the light of a revision of the assignments in the Λ-doublet states of OH (see also Field, MN 217.1). Herman et al. (AA 144.514) failed to detect 6 cm continuum emission from OH/IR sources. Such emission might be expected if any of the objects were in the transition phase to planetary nebulae (with an inner ionized region developing). This suggests that the transition time must be very short. Infrared and OH data on a radio complete sample of OH/IR stars is given by Willems & de Jong (AA Supp, in press). Studies of circumstellar masing have now been extended to the shorter period, hotter Miras and SRd variables and semiregular variables. R Cet (P = 166 days) has OH main line emission (Dickinson, AJ 92.627) and three short period Miras are found with SiO emission. Fix & Claussen (ApJ 287.L35) find OH masers in two SRd variables (other work on SRd variables includes an analysis of DDO colours (Mantegazza, AA 135.300) and the removal of IS Gem from this class (Crimi & Mantegazza, Ap Sp Sc 100.255).

Studies of the remarkable object OH 231 + 4.2 (= OH 0739 −14), which contains a red long period variable component (possibly a Mira), continue. Jura & Morris (ApJ 292.487) note that it is in the cluster M46 and deduce an initial mass of 3$M_\odot$. They and Knapp (ApJ 311.731) discuss the high mass loss rate. Beautiful CCD images of the object have been obtained by Reipurth (Nature 325.787). He and Cohen et al. (ApJ 297.702) discuss this object which has shocked bipolar bubbles expanding supersonically (200 km/s) at right angles to the dust disc. The dynamic age is 1500 yr and there are Herbig-Haro objects at the front of the bubbles. The

object shows HCN emission (Deguchi & Goldsmith, Nature 317.336) as does IRC +10420
(Jewell et al., Nature 323.311) a probable post AGB object. OH 231.8 + 4.2 was
originally unique in showing a 12 μm feature attributed to relatively pure water
ice but another OH/IR source has been found to show a similar feature (OH 32.8
-0.3, Roche & Aitken, MN 209.33p).

9. Observations of Pulsating Compact Stars
(Don E. Winget)

The pulsating compact objects were last reviewed in these pages by Robinson
(Trans IAU vol.19A p303). Here, we will discuss only the important developments
since that review; other recent reviews have been published by Kawaler (IAU Coll
95, in press), Starrfield (IAU Coll 95, in press), Cox (ed. J.-P. Swings,
Highlights of Astron vol.7 p229), and Winget (ed. J.-P. Swings, Highlights of
Astron vol.7 p221; IAU Symp 123 in press). The compact pulsators divide into at
least three distinct classes of pulsating variable stars, which we will refer to
according to the classification scheme of Sion et al. (ApJ 269.253): the DOV
stars, the DBV stars and the DAV stars. The three classes are almost uniformly
distributed in log Te, and lay along white dwarf evolutionary tracks in the H-R
diagram.

In spite of the wide separation of the three classes in the H-R diagram, they
do seem to have remarkably similar pulsation properties. All are multiperiodic,
with periods in the range from about 100 s to 2000 s (although this upper limit
may be an observational selection effect imposed by limits of the observing technique). All are low-amplitude pulsators in the optical, with typical semi-amplitudes of individual modes of one percent or less. All seem to be pulsating in non-
radial gravity-modes with very low spherical harmonic index. It is interesting to
note that the radial overtone number required to match the observed periods decreases dramatically with the decreasing effective temperatures of the respective
classes.

The hottest group, the DOV stars, have effective temperatures in excess of
100,000 K. This group includes the pulsating PG 1159-035 (GW Vir) stars, and possibly the pulsating planetary nebula nucleus K1-16 - although this may be a different sort of beast representing a possible fourth class. The total number of
known pulsators in this group has risen to five with the discovery of pulsations
in PG 0122+200 by Grauer et al. (preprint).

The survey for additional pulsating DOVs has also turned up 15 null results
(Grauer et al., preprint). These candidates were selected because of their similarity to the other DOV stars. Three were spectroscopically identical with PG 1159-
035 and four were similar to K1-16. The absence of pulsations in these objects
presents a serious challenge to our understanding of the driving mechanism in
these stars.

Although the amplitude of the DOV pulsation modes is relatively small in the
optical, the high effective temperatures of the DOVs suggest that it is very interesting to look at shorter wavelength. Indeed, Barstow et al. (ApJ 306.L25) have
reported the first observation of pulsations in the X-ray band. Their observations
in the soft X-ray (44-150 A) demonstrated large amplitude pulsations (up to 17%
for individual mode semi-amplitudes) at the same frequencies detected in the
optical.

The power spectra of the light curves of the DOV stars separate into distinct
bands of power, with most bands exhibiting some fine structure. In a series of
recent papers Kawaler (cf. Kawaler, IAU Coll 95, in press and references therein)
has shown, using the observations of PG 1159-035, that the period spacing between
these bands can be used to determine the mass of the DOV to two significant figures, as well as to determine the l values of the observed modes. Observations of
the new DOV, PG 0112+200, by Hill et al. (IAU Coll 95, in press) have resolved the
period bands present in it. They find the same sort of regular period spacing ex-

pected by Kawaler - thereby demonstrating the usefulness of this new technique to extract seismological information from the DOV stars. For this reason the resolution of the band structure in the other DOV stars should be of highest priority.

Recent observations of the PG1159-035 by Koupelis & Winget (IAU Coll 95, in press), have revealed that at least four new bands of power were found in the 1987 data which could not be detected in the considerable body of archival data from 1979-1984. Most puzzling is that the eight bands of power previously reported in the star (Winget et al., ApJ 292.606) are still present at amplitudes consistent with the previous data - indicating that the new period bands grew in leaving the old unchanged. This view is lent further support by the analysis of the 1987 data for the phase of the 516 s peak (a single isolated peak) by Winget & Kepler (Workshop on Multiperiodic Variable Stars, Comm Konkoly Obs, in press). The new data are consistent with the ephemeris of Winget et al. (ApJ 292.606), and suggest that the 516 s peak has not only maintained its amplitude and frequency but its phase as well (including the slow secular evolutionary change), even as the new bands appeared. These observations will severely challenge models for the mode-selection mechanism in these stars, and also possibly provide unprecedented information about their nonlinear behavior.

The pulsating compact stars of intermediate temperature are the DBV stars. Since the review of Robinson, three new DBV stars have been found: PG 1351+489, PG 1115+158 (Winget et al., ApJ 316.305), and PG 1456+103 (Grauer et al., preprint), bringing the total known to five. Studies of the temperature range of these stars indicate that all the variables fall into a very narrow range near the highest temperatures of the DB stars. The exact values of this temperature range remain somewhat uncertain due to difficulties in reconciling optical and IUE temperature estimates (cf. Liebert et al., ApJ 309.291 and Koester et al. AA 149.423). The IUE results (Liebert et al., ApJ 309.291) imply a blue edge of 28,000 +/- 2,000 K, and a red edge of 24,000 +/- 2,000 K. However, if the optical temperature scale is adopted the blue edge may be up to 3,500 degrees cooler, and the red edge about 2,000 K cooler. The work of Liebert et al. serves to define an empirical instability strip, and suggests that most or all of the stars in the temperature strip pulsate and those outside do not. This suggests that similar to the DAV stars the pulsations are strictly an evolutionary effect, and that the DBV stars are otherwise normal DB stars. These conclusions must be regarded as tentative however, because the sample of stars, variable and nonvariable, upon which they are based is perilously small.

Only the light curve of PG 1351+489 can be considered resolved, and its simplicity and similarity to some DAV light curves suggest that it may be a special case (cf. Winget, IAU Symp, 123 in press). Attempts to resolve the light curve of GD 358 by Hill (IAU Coll 95, in press), succeeded in demonstrating, surprisingly, that the pattern of sets of five regularly spaced (in frequency) modes are not stable, and that this spacing appears to change, and on occasion the star appears to be nearly a mono-periodic pulsator. Hill points out that this sort of behavior casts serious doubt on the rotational splitting explanation for the equally spaced modes, and also indicates that the pulsations are not stable, since beating is not a plausible explanation for the dramatic changes in the character of the light curve.

Rotational splitting still seems to be the explanation of choice for at least one of the coolest class of compact pulsators, the DAV stars. The work of O'Donoghue & Warner (MN, in press) on L19-2, has demonstrated that rotational splitting is very succesful in explaining not only the generally equally-spaced structure of the power spectrum, but also that the slight deviations from equal spacing can be accounted for by the next highest order terms due to rotational splitting.

O'Donoghue and Warner are also monitoring the phase of the pulsation in this star attempting to measure a rates of period change. The same is being done for R 548 (Tomaney, preprint), and for G117-B15A (Kepler & Winget, IAU Symp 123, in press). Currently, the limits on all three stars are rapidly approaching the values expected from theoretical evolutionary calculations, and the best limit comes from G117-B15A: $dP/dt < |9.9 \times 10^{-15}|$, at the 68% confidence level.

Observations of a DA star (PG 2303+243) from the Palomar Green survey by Vauclair et al. (AA 175.L13), indicate that it is the 20th DAV, and the only new one discovered since the review of Robinson (Trans IAU vol. 19A p303). New variables may be found at a somewhat higher rate in the future, however, after the work of Fontaine et al. (AJ 90.1094). They showed that the well defined temperature instability strip for the DAV stars based on G-R colors (cf. Robinson's Review), can be almost as sharply defined using the much more readily available Stromgren colors. This result should increase the ease of identification of candidate stars.

10. Theory of Stellar Pulsation
(J. Christensen-Dalsgaard)

INTRODUCTION

There has been considerable activity within the area of stellar pulsation theory. Thus, rather than a comprehensive summary, the following is largely the reviewer's biased selection from a considerable larger number of relevant publications.

The understanding of non-linear stellar oscillation is advancing rapidly. Simple models have been analysed which reproduce important features of observed light curves. Resonant mode interaction may be able to explain the low amplitudes of main sequence pulsators, and amplitude variations in certain classes of pulsators. A very interesting development is the emerging connection between chaotic dynamics and irregular stellar variability.

An important trend has been the growing interest in using observed frequencies, for stars where many modes of oscillation are excited simultaneously, to carry out asteroseismic investigations of stellar interiors. This has so far concentrated on solar-like stars, rapidly oscillating Ap stars and white dwarfs; however an analysis of δ Scuti variables with rich spectra of oscillations would also be rewarding. The related subject of helioseismology is not touched upon here.

In the period considered a number of relevant conferences have been held. These include NATO Advanced Research Workshop on "Chaos in Astrophysics", Palm Coast, Florida, April 1984 (eds.: J.R. Buchler et al., Reidel, 1985; in the following referred to as FLORIDA); IAU Colloquium No 82 "Cepheids: Theory and Observations", Toronto, May 1984 (ed.: B.F. Madore, Cambridge University Press, 1985); NATO Advanced Research Workshop on "Seismology of the Sun and the Distant Stars", Cambridge, June 1985 (ed.: D.O.Gough, Reidel, 1986, in the following referred to as CAMBRIDGE); IAU Symposium No 123 "Advances in Helio- and Asteroseismology", Aarhus, July 1986 (eds.:J. Christensen-Dalsgaard & S. Frandsen, Reidel, in press); and "Stellar Pulsation-a Memorial to J.P. Cox", Los Alamos, August 1986. The proceedings of the latter two conferences were not available at the time of writing. In addition the proceedings of the conference on "Theoretical Problems in Stellar Stability and Oscillations", Liege, July 1984 (Institut d'Astrophysique, Liege; eds.: A. Noels & M. Gabriel, in the following referred to as LIEGE) have been published.

PHYSICS OF STELLAR PULSATION

Stellar stability is largely determined by the interactions between the radiation field and the motion. The fundamental aspects of these interactions were described by Mihalas & Mihalas in "Foundations of Radiation Hydrodynamics" (Oxford University Press, 1984), which will undoubtedly provide the basis for a more consistent treatment of effects of radiative transfer in stellar stability. The specific application to stellar oscillations was considered by Mihalas (ApJ 284.299).

The difficulty in treating convection remains a major impediment to the theory of most pulsating stars. Gonczi (AA 157.133) calculated stellar stability with a linearly perturbed non-local mixing length theory. The effects of Reynolds stress perturbations on solar oscillations were considered by Gabriel (AA 175. 125). Stellingwerf applied his, essentially phenomenological, model of convection to RR Lyr stars (ApJ 284.712); he also presented a simple one-zone model (ApJ 303. 119) which may be useful for understanding the effects of convection. A diagnostic of the behaviour of convection, as a function of oscillation phase, may eventually be obtained from measurements of turbulent velocities in the stellar atmosphere (Benz & Stellingwerf, ApJ 297.686).

Asymptotic techniques have proved very useful for the understanding of the properties of stellar oscillations, particularly for the high-order acoustic modes observed in, e.g., the Sun. Gough (Hydrodynamic and Magnetohydrodynamic Problems in the Sun and Stars, p117, ed.: Y. Osaki, University of Tokyo; 1986) presented an illuminating derivation of the asymptotic relations on the basis of a ray analysis.

NONLINEAR DYNAMICS OF STELLAR PULSATION
Analysis of observed light curves

Detailed hydrodynamical nonlinear calculations have generally been used to interpret the observed light curves. However it is becoming clear that a more fundamental understanding can be obtained from expansions of the non-linear equations in terms of amplitude equations. They were reviewed by Buchler (FLORIDA, p137). Klapp et al. (ApJ 296.514) compared the results of using the amplitude equations with detailed hydrodynamical calculations. They also applied the results to the interpretation of Fourier transform analysis of observed light curves (e.g. Simon, ApJ 284.278). This was investigated in more detail by Buchler & Kovács (ApJ 303.749; 318.232).

Stellingwerf & Donohue (ApJ 306.183; 314.252) computed Fourier transform coefficients from one-zone models, with results that were remarkable similar to the observations.

Resonant mode coupling

Dziembowski & Krolikowska (Acta Ast 35.5) showed that in a model of a ZAMS δ Scuti star parametric resonance between an unstable acoustic mode and a pair of stable gravity modes limits the acoustic mode amplitude to less than about 0.02 mag. This accounts naturally for the low observed amplitude of these stars, and explains why it is not reproduced by non-linear radial calculations. Dziembowski (LIEGE, p346) suggested that the observed decrease in amplitude with increasing rotation rate may be explained by the increasing probability of resonance due to non-uniform rotational splitting. Amplitude modulation caused by resonant nonlinear interactions was investigated by Moskalik (Acta Ast 35.229; 36.333): he suggested that with a reasonable choice of parameters this can reproduce the observed Blazhko-effect for RR Lyr stars.

Irregular stellar variability

Irregular variability is an extreme manifestation of nonlinearities. An extensive review of both observational and theoretical aspects was given by Perdang (FLORIDA, p11). Nonlinearly coupled sets of adiabatic modes exhibit chaotic phenomena similar to those found in simpler dynamical systems (Perdang & Blacher, AA 136.263; Däppen & Perdang, AA 151.174; Däppen, FLORIDA, p273). It remains to be seen whether these phenomena set in at sufficiently low amplitudes to be relevant to, e.g., solar oscillations. Auvergne & Baglin (AA 142.388) and Regev & Buchler (FLORIDA, p285) obtained chaotic behaviour in simple one-zone models. Very recently Buchler & Kovács (ApJ Lett 320.L57) found that hydrodynamical models of W Vir variables display period-doubling bifurcations and transition to chaos.

Chaos of a different nature, caused by departures from sphericity, was investigated by Perdang (CAMBRIDGE, p141); it manifests itself as irregularities in the frequency spectrum of the pulsating star.

Auvergne & Baglin (AA 168.118) investigated methods for determining parameters characterizing irregular variability from observed time series.

ASTEROSEISMOLOGY
Solar-like stars
     The potential, and in a few cases actual tentative, observations of stellar
analogies to the solar five min oscillations have prompted investigations of the
properties of such oscillations in stellar models. Christensen-Dalsgaard (Space
Research Prospects in Stellar Activity and Variability, p11, ed.: F. Praderie,
Paris Observatory; 1984) showed that a two-dimensional classification, separating
stellar mass and evolutionary state, can be established on the basis of frequency
separations among low-degree high-order acoustic modes. This was discussed further
by Ulrich (ApJ Lett 306.L37). As emphasized by Gough (Nature 326.257) the utility
of such a classification depends on its sensitivity to the other parameters characterizing the star.
     Observations of apparent 5 min oscillations in α Cen A are very difficult to
reconcile with normal models (Demarque et al., ApJ 300.773). The interpretation
of observations of such oscillations in ε Eri was discussed by Guenther &
Demarque (ApJ 301.207) and Guenther (ApJ 312.211).

Rapidly oscillating Ap stars
     The modes observed in these stars are similar to the solar five min oscilla-
tions. However, in contrast to the solar case, the oscillations maintain phase
over very extended periods of time; furthermore the modes are closely linked to
the large-scale magnetic field of the stars.
     Oscillation frequencies for models of Ap stars were computed by Shibahashi &
Saio (PAS Japan 37.245) and Gabriel et al. (AA 143.206). The results were largely
consistent with the observed frequencies, although problems remained with the
details of the frequency separations. Furthermore Shibahashi & Saio found that
some observed frequencies exceeded the critical frequencies for the assumed, simple atmospheric models.
     Dziembowski & Goode (ApJ Lett 296.L27) showed that the association between
the oscillations and the magnetic field can be understood in terms of the rotational and magnetic splitting of the oscillation frequencies. An alternative explanation, based on the surface inhomogeneities of the stars, was proposed by
Mathys (AA 151.315), but not developed in detail.

Compact pulsators
     The study of the pulsating compact stars has undergone a rapid observational
(see the section by Winget) and theoretical development. Pulsations are observed
in central stars of planetary nebulae (PNN), and oxygen (DO), helium (DB) and
hydrogen (DA) white dwarfs.
     Cox et al. (ApJ 317.303) made a detailed analysis of the stability of DA and
DB white dwarfs. To match the observed blue edge of the instability strip, convection had to be very efficient, as first suggested by Fontaine et al. (LIEGE,
p328). Instability in PNN stars was studied by Starrfield et al. (ApJ Lett 293.
L23); stellar models consisting solely of carbon and oxygen were found to be unstable due to partial ionization of these elements. Kawaler et al. (ApJ Lett 306.
L41) found efficient driving from the perturbation in the nuclear energy generation rate, in PNN stars with a helium-burning shell, but at periods considerably
shorter than those observed. The absence of oscillations at the periods found to
be unstable might indicate that He burning is not active in these stars.
     Period changes have been measured for the pulsating DO star PG 1159-35, and
the observed upper limits on period changes for DB and DA variables are approaching the theoretically expected values. Kawaler et al. (ApJ 295.547; 302.
530) calculated period changes; for PG 1159-35 a period <u>increase</u> was found, of
the same order of magnitude, but opposite sign, as the observed period <u>decrease</u>.
This discrepancy can be explained by variations in the rotational <u>splitting</u>
(Kawaler et al., ApJ 298.752), if the observed mode is prograde; however it is not
obvious why such a mode should preferentially be excited.

11. Flare Stars
(R.E. Gershberg and N.I. Shakhovskaya)

The symposium "Flare Stars and Related Objects" was held in Byurakan, 1984, the Proceedings volume was edited by Mirzoyan and published by the Publishing House of the Armenian Academy of Sciences, Erevan, 1986; hereafter it is referred to as 'M'. Activity in red dwarf stars has been considered in conjuction with solar activity at many meetings: the 4th Cambridge Workshop "Cool Stars, Stellar Systems and the Sun", Santa Fe, 1985, the Proceedings being edited by Zeilik and Gibson and published in Lecture Notes in Physics Vol. 254, Springer Verlag and referred to as 'ZG'; the meeting "Solar and Stellar Flares", Rutherford-Appleton Laboratory, Chilton, 1986, Gondhalekar edited the Proceedings, it is referred to as 'G'; the 12th symposium "Solar and Stellar Activity" at the 26th COSPAR General Assembly, Toulouse, 1986 (Adv Sp Res, in press); the meeting "Solar and Stellar Physics", Titisee, 1987 (Lecture Notes in Physics, in press). Many reviews on different aspects of flare stars (FSs) have been published: by Mirzoyan & Chavushian on photographic observations of FSs (Comm Byurakan Obs 40.27); by Mirzoyan on FSs in clusters and on the Byurakan conception of stellar flares (Vistas in Astron 27.77; Comm Konkoly Obs 86.409); by Mullan on energy dissipation mechanisms in FSs (Unstable Current Systems and Plasma Instabilities in Astrophysics, eds.: Kundu & Holman, Reidel, 1985, p245) and on non-thermal radio emission from FSs ("Radio Stars", eds.: Hjellming & Gibson, Reidel, 1985, p173); by Gary on quiescent microwave emission (ibid, p185); by Rodono on coordinated ground-based and space observations of FSs (M, p19) and on starspots and plages (Highlights in Astronomy, ed.; Swings, Reidel, 1986, p429); by Oláh on starspots (Comm Konkoly Obs 86.393); by Baliunas & Vaughan on stellar activity cycles (Ann Rev AA 23.379); by Butler on solar activity phenomena among dKe-dMe stars (G, in press); by Haisch on stellar coronae (Irish AJ 17.200); by Pallavicini on solar-stellar relationships (G, in press) and on solar and stellar coronae (Lecture Notes in Physics, in press); by Pettersen on atmospheric activity of red dwarf stars (Vistas in Astron, in press).

For the last three years new, important results have been obtained due to the realization of methods for photospheric magnetic field measurements on FSs, to the X-ray observations with the EXOSAT, to microwave observations with the VLA, and to fast photometry of 3x10.0E-7 s resolution with the 6 m telescope.

In the solar vicinity a number of new FSs were found and several objects are suspected to be of the same kind: Doyle & Byrne (AA 154.370 registered 3 faint flares during 11.7 h in Gliese 812; Kovalchuk & Pugach (IBVS 2557) registered a flare with an amplitude of 0.25 mag and duration of about 1 min in the G5 star HD 282773; Good (IBVS 2581) found a flare on an anonymous star of 16.5 mag; Yang & Liu (IBVS 2705) registered a flare of 0.11 mag amplitude in the contact binary CN And; Pettersen & Hawley (AA, in press) discovered a flare activity in the dMe star BD $+3°4138B$; Wenzel (IBVS 2740) confirmed that SVS 2559 is a FS; Donahue et al. (ZG) registered a flare in the dM star HD 95735 during photoelectric observations in Ca II H and K lines and the continuum between them; Pesch & Sanduleak (IBVS 2989) found a FS and two probable late type emission line variables on an objective prism plate; Bopp et al. (IBVS 2604) found spottedness on the dK0 star HD 91816; Udalski & Geyer (IBVS 2691 and 2525) and Bopp et al. (in press) found a variability probably due to starspots on HD 102077, 119285, 127535, 139084,155555, 174429 and 202134; Bopp (in press) found a faint H-alpha emission in spectra of dwarf stars Gliese 256, 425A, 900 and 907.1. Barden & Nations (BAAS 17.879) discovered variable H-alpha emission and strong Ca II IR triplet reversals for the binary HD 80715 consisting of two K dwarfs, and they estimated a period of 3.8025 days. Oskanian (M, p11) has given 8 cases of photometric observations of the UV Cet type flares in stars of different spectral classes and luminosities. Kovalchuk (Comm Konkoly Obs 86.443) gave light curves of fast flares in 4 antiflare stars. Pettersen et al. (AA, in press) studied the fast rotating dwarf Gliese 890, estimated its period of 0.4312 day and rotational velocity of 60 km/s

and found its flare activity. Van Leeuwen (M, p289) found spottedness and fast
rotation for the brightest FSs in Pleiades.

Results of patrol observations of known FSs were published by Tsvetkov et al.
(IBVS 2618, 2954 and 2972), by Melikian (IBVS 2630), by Malcolm (IBVS 2647), by
Reglero et al. (IBVS 2752), by Orchiston et al. (IBVS 2785), by Geyer & Kämper
(IBVS 2819), by Panov et al. (IBVS 2826), by Ilyin (IBVS 2985), by Oláh et al.
(IBVS 2889), by Avgoloupis et al. (IBVS 2997, 2998 and 3016), and by Chugainov
(Bull Crimean Ap Obs 68.114).

New FSs in stellar clusters were discovered by Rosino et al. (IBVS 2620 and
2981), by Melikian (IBVS 2621), by MacConnell & Mermilliod (IBVS 2633), by Hojaev
(IBVS 2635 and 2636), by Santangelo (IBVS 2665), by Tsvetkova et al. (IBVS 2730),
by Yao et al. (Publ Beijing Ast Obs 6.226), by Chavushian & Jankovics (IBVS 2814),
by Melikian & Della Valle (IBVS 2929), by Parsamian & Pogosian (M, p130; Astrofizika 24.239), by Gasparian & Parsamian (Astrofizika, in press), by Hambarian
(M, p120), by Parsamian et al. (Astrofizika 22.87, 22.315; M, p79), by Tsvetkova
(M, p84), by Tsvetkov et al. (Comm Konkoly Obs 86.429) and by Tsvetkova & Tsvetkov
(ibid. p431).

Hojaev (M, p91; Astrofizika 24.65) compared the statistical properties of FSs
in dark clouds in Taurus and in other clusters. Chavushian (M, p125) tested the
track method as a means of registering flares in stellar clusters. Parsamian
(Comm Byurakan Obs 57.79) and Kelemen (Comm Konkoly Obs 86.433) analysed distortions of light curves of flares due to low time resolution. Szécsényi-Nagy
concluded that in the Pleiades the most FSs with a high frequency of flares are
known, but there exist many unknown FSs with lower flare frequency (M, p101); he
suggested an activity variation in FS HII 2411 with a characteristic period of
10-15 years (Comm Konkoly Obs 86.425). Mirzoyan & Ohanian (M, p68) suggested
that the flare activity of FSs in clusters have a cyclic character. Mnatsakanian
(Astrofizika 24.621) represented the time distribution of flares in the Pleiades
by two Poisson groups with different flare frequencies and estimated the total FS
number of 750 for this cluster, 1250 for the Orion cluster and 350 for the Taurus
clouds; Natsvlishvili (Comm Konkoly Obs 86.427) found the total number of FSs of
1850 for Orion. Parsamian (Astrofizika 22.633) concluded that the fraction of
visual binaries among FSs in the Pleiades is 3 times higher than the field stars.
Gasparian (Comm Konkoly Obs 86.439) showed that in the Orion cluster the FS and
H-alpha star densities decrease whereas the distance from the cluster centre increases and for FSs this decrease is smaller. Mirzoyan (in press) found a clear
dependence of a fraction of FSs among stars of a luminosity fixed on the age of
stars under consideration.

Kaluzny (IBVS 2627) published BV observations of a spotted star HD 175742.
Zsoldos (IBVS 2860) improved the orbital period of YY Gem. Dahn et al. (IBVS 2796)
specified the spectral type as M6.5e and $M(V)= 16.6$ mag for the FS AZ Cnc.
Melikian (IBVS 2622) described a long - about 4 h - flare in AZ Ori with unusual
colours: $\Delta U < \Delta B < \Delta V$.

Byrne et al. (Armagh Obs Preprint 43; AA, in press) and Byrne & Doyle (AA,
in press) improved the rotational period of a FS Gliese 867A (4.39 days), found
non-synchronous rotation of the binary components, did not find any significant
seasonal variations in total flare energy output and noted the advantages of the
Walraven photometric system for investigating stellar flares. Doyle et al. (AA
156.283) registered about a dozen flares in Gliese 867B during 6.4 h patrol observations, did not find noticeable variations of flare activity level in 1977-81 and
discovered brightness variations with an amplitude $\Delta V = 0.15$ mag and a period of
1.95 days. Rojzman (Ast Tsirk 1484) confirmed periodic brightness variations for
EV Lac (P = 4.38 days, $\Delta V = 0.44$ mag), discovered such variations for V1396 Cyg
(P = 3.34 days, $\Delta V = 0.044$ mag) and V1285 Aql (P $\sim$ 12-13 days, $\Delta V = 0.03$ mag) and
found - for Gliese 48, 687, 699, 752A, 908, GX And and GQ And - that periodic variations, if they exist, have amplitudes less than 0.025 mag.

Pettersen et al. (AA Supp 66.235; ZG, p91) carried out a statistical analysis
of more than a hundred of the AD Leo flares and found that their mean frequency
showed a weak periodicity with P $\sim$ 8 years, confirmed the random time distribution

of flares, and did not find noticeable variations of energy flare spectra from
season to season. They found that flare fading times are distributed at random and
there is a maximum near 30-50 s in a flaring up time distribution.

Doyle (AA 177.201) analysed observations of EV Lac in 1973-82 and found that
in 1973-76 there was a modulation of flare frequency by stellar rotation but later
the modulation disappeared. Rojzman & Kabichev (Sov Ast 62.1095) did not find
the modulation of flare frequency in EV Lac by stellar rotation in 1980-81, they
registered 3 very long flares and found that fast preflare dips occur; the minima
of preflare dips are not lower than the normal stellar brightness. Panov &
Korhonen (in press) registered intricate drifts of the EV Lac flare radiation
in the U-B,B-V diagram. Gershberg & Petrov (M, p38) registered a flare in EV Lac
with a total duration of 2.4 s. Tsvetkov et al. (Comm Konkoly Obs 86.423) registered fast flares in this star with total duration of 1-2 s and amplitudes till
$\Delta U = 1.8$ mag. Zalinian & Tovmassian (ibid, p435 and IBVS 2992) tested a new method
for flare detection with a photoelectric photometer programmed to wait for a
brightness increase exceeding the fixed level and registered flares in EV Lac
with a duration of 0.1 - 0.2 s. Mavridis & Avgoloupis (AA 154.171 and in press)
found small (0.3 mag) and slow (P $\sim$5 years) fluctuations of the mean brightness
level of EV Lac and a clear correlation of this level with the stellar flare activity level. At different phases of fluctuations the relative number of fast and
slow flares is different, the energy spectra of flares vary and the time-averaged
energy of flares changes by at least double. Avgoloupis (AA 162.151) considered
183 light curves of the EV Lac flares and concluded that statistical correlations
become stronger if one considers flares of different Oskanian's type separately.

Melikian & Grandpierre (IBVS 2638) analysed the time distribution of about
a hundred flares in UV Cet and found essential deviations from random distributions.

Mirzoyan & Melikian (M, p153) concluded that flares of larger energy have
longer flaring up stages, the majority of such flares have complicated light
curves, and flares with slow flaring up have "redder" colour indices. Melikian
(IBVS 2661) found a similarity of the mean amplitudes of flares for 8 FSs with
different luminosities. Gershberg (Astrofizika 22.531) showed that the mean amplitude of flares in a FS is mainly determined by the amplitude of the faintest
detectable flare in the star and therefore the mean amplitude is not a measure of
the flare activity level. The intrinsic frequency of flares is higher in absolutely brighter FSs although due to the observational selection in absolutely
fainter FSs the flares are registered more often. The constructed luminosity
function for FSs in the Pleiades permitted one to conclude that the total number
of FSs in the cluster exceeds 1100 and they have mean frequencies of flares photographically registered from 10.0E-4 to 10.0E-2 /hour.

Shevchenko (M, p135) found that relative portions of fast and slow flares are
the same for FSs in the solar vicinity and in stellar clusters.

Beskin et al. (M, p60; Bull Crimean Ap Obs 79, in press) registered more than
a hundred flares during patrol observations for 8 FSs with a time resolution of
3 x 10.0E-7 s at the 6 m telescope. Data analysis permitted the conclusion that
in all flares there is no fine structure within the range 10.0E-6 - 10.0E-1 s, the
lowest lifetime of significant variations are 0.3-0.8 s at phases of fast flaring
up and these data fit the prediction of the gasodynamical model of flares. Flares
with total durations of several seconds are registered in detail, their fading
phases are found to be determined by the relaxation time of gas heated in the
flare. The energetics of faintest registered flares is compatible with the solar
subflare energetics. The upper limit for the total power of stellar microflares
does not exceed the time-averaged power of individually registered flares, i.e.
there is no reason to believe that microflares dominate in stellar coronae
heating.

Chugainov & Lovkaya (M, p52) suggested global oscillations in BY Dra with the
periods 160 and 82 min.

Rodono et al. (AA 165.135) mapped the spotted surfaces of FSs BY Dra and AU
Mic using multicolour observations and two-spot approximation. Poe & Eaton (ApJ

289.644) proposed a computer program for generating light curves for rotating
spotted stars and used it to solve multicolour observations of BY Dra; they found
that the BY Dra photometry can be represented by the single cold spot model, and
the wavelength dependence of limb darkening contributes a surprisingly large a-
mount of the variation in the visible colours. Melkonian (Astrofizika, in press)
found variations of the mean annual brightness of Gliese 5, 836.7, 841.1, 908,
BD $+14^{o}4637$ and BD $+22^{o}3208$ probably due to activity cycles.

Pettersen & Hawley (Publ Inst Theor Astrophys Oslo 2) published the spectral atlas of 26 dwarf G9-M6 FSs in the solar vicinity in the quiet state.
Pettersen & Tsvetkov (IBVS 2660) did not find emission in either the Balmer lines
or the Ca II lines in the spectra of 4 FSs, one of them -V654 Her- shows spectral
features inherent in giants. Pettersen et al. (AJ 90.2296) found two faint M-stars
without emission lines and concluded that the effect of emission lines on the B-V
index is large enough to separate dMe from dM stars.

Byrne (Irish Ast J 17.N3) found unusually strong emission lines in the spectrum of FS HDE 319139. Tomkin & Pettersen (AJ 92.1424) found that Gliese 268 is
a spectroscopic binary with the period of 10.4 days, the eccentricity of the orbit
is 0.34 and it consists of two dKe components with masses of 0.16 and 0.19 solar
mass.

Bopp (ApJ Supp 54.387) determined fluxes in CaII H and K lines for 14 spotted
stars. Kodaira (M, p43) analysed H-beta emission in the system YY Gem in 1978-84
and interpreted the variations found in terms of the rotation of components and
changes in a sector structure of the main component. Using coordinated observations Baliunas et al. (ESA SP-263 p181) found noticeable surface inhomogeneity of
the YY Gem components and they registered flares in H-beta and brightening in
ultraviolet.

The main success in FS studies over the last 3 years is the direct registra-
tion of photospheric magnetic fields on late dwarfs. Analysing the IR region of
the AD Leo spectrum with a Fourier spectrometer, Saar & Linsky (ApJ 299.L47) found
strong magnetic fields (B = 3.8 kGs) overlapping about 73% of the stellar surface.
Then Saar et al. (ApJ 302.777) studied line profiles in the optical range of the
EQ Vir spectrum and found a magnetic field of 2.5 kGs over 80% of the stellar surface. At the Santa Fe meeting Saar & Linsky (ZG) reported magnetometric results
for 5 additional chromospheric active stars: HD 17433 (1.9 kGs, 50%), HD 28099
(1.7 kGs, 30%), PZ Mon (2.2 kGs, 80%), HD 45088 AB (2.4 kGs, 50% and 3.0 kGs, 65%)
and noted a correlation between stellar rotational velocity and the portion of the
stellar surface occupied by magnetic fields.

Important results have been obtained in the course of simultaneous observations of FSs at different wavelengths.

Butler et al. (AA 174.139; ESA SP-263) found a weak anticorrelation between
UV line intensities and optical brightness for BY Dra that can be interpreted as
the effect of bright plage above photospheric spots; but such a correlation is absent for AU Mic; for both FSs there exist dependences of differential emission
measures on temperature similar to the solar ones.

Andersen et al. (ESA SP-263 p87) found anticorrelation between UV chromospheric line intensities and the visual brightness for EV Lac, but no correlation for
transition zone UV lines that can be due to the large size of the transition zone.
Ambruster et al. (ESA SP-263) registered a two-fold drop of the whole UV spectrum
of EV Lac for 1.5 h that may be due to an episode of mass expulsion during a noticeable flare that occurred just before the UV observations; no modulation of Mg II
and C IV line intensities by stellar rotation was found.

Doyle (MN 224.1p) studied the relation between the intensity of Mg II lines
and the rotational period for dMe stars and the correlation found follows the
known correlation for hotter stars. Doyle et al. (MN 220.223) when carrying out
photometric investigations of Gliese 1, 461, 825, 899 and 908, found neither
flares nor starspots; this fits the known correlation between X-ray luminosity and
optical flare activity level.

De Jager et al. (AA 156.95) registered - in the course of coordinated observations of BY Dra in optical, radio and X-ray ranges - one strong and two weak

flares; in the strong one the brightness temperature reached 25000 K at a surface of about 2 x 10.0E7 km² for 5 min, and the heated gas (with a density of 2 x 10.0E11 cm⁻³ and area of 10.0E10 km²) was responsible for one hour X-ray emission. Using the Einstein and EXOSAT data, Smale et al. (MN 221.77) discovered a flaring X-ray source that was identified with a 13 mag dMe star at a distance of 13 pc from the Sun; they showed that the quiet stellar corona can be represented by a symmetrical homogeneous model but not by a loop structure.

Harris & Johnson (ApJ 294.649) observed 4 nearby binaries containing M dwarfs and found X-ray flares with durations of dozens of minutes in Gliese 34A, 338B and 669B. Johnson (ApJ 316.458) observed the multiple system Wolf 630 in soft X-ray and found one of the components - VB 8 - an anomalous cold (0.6 x 10.0E6 K) corona and hotter radiation during a flare.

Connors et al. (ApJ 303.769) systematically revised X-ray data from the A-2 HEAO-1 experiment, discovered 5 flaring sources with flare durations from 1 to 30 minutes, and concluded that most of such sources are dKe-dMe stars; one may expect about 2 x 10.0E4 such X-ray flares per year over the whole sky.

Doyle et al. (MN 223.1p; AA in press) carried out photometric, spectral, radio and X-ray observations for YZ CMi and, during optical flaring with the amplitude ΔU = 1.2 mag, did not register any X-ray flux variations. During faint flares no correlation was found between the X-ray flux and U band brightness, and a weak correlation between X-ray flux and the Balmer line emissions was suggested. In the maximum of the strong flare the essential broadening of Balmer lines was registered, the H-gamma and H-delta lines were symmetric, and the higher members of the series had an enhanced red wing. During one of the long flares the whole X-ray energy was 5 times greater than the optical radiation, and Balmer lines' emission was about 10% of the U band radiation.

Elgaroy et al. (ESA SP-263; AA, in press) carried out optical and UV observations for the binary AT Mic and found that the mean frequency of flares in the U band is about 1.3 per hour, UV line flux densities are similar to the solar active region ones, variations of Mg II lines are not well correlated with the U band brightness, "hot" UV lines - C IV and Si IV - are more broadened and blue shifted compared with "cold" UV lines - Fe II and Mg II; the high flare activity of the star casts doubt on a simple model of magnetic energy dissipation within the stellar atmosphere as the flare energy source.

Pettersen (ESA SP-263 p157) carried out optical and UV observations of a strong flare in AD Leo and found that near flare maximum coronal plasma radiation (10.0E7 K) can well represent optical radiation but it gives too high values for the UV region; the upper limit of gas density in the flare is estimated as 6 x 10.0E13 cm⁻³ using the Inglis-Teller formula.

Haisch et al. (Armagh Obs Preprint 42; AA, in press) performed UV and X-ray observations for EQ Peg AB and found the Mg II lines' enhancement during a flare; X-ray data gave a coronal temperature of 26 x 10.0E6 K at the flare maximum and 14 x 10.0E6 K at the fading stage thereby permitting one to note a similarity between this stellar flare and two-ribbon solar flares. An unusual feature of the stellar flare is that the soft X-ray flaring up time exceeds the duration of the fading stage although in moderate X-ray the light curve is of usual shape.

Doyle et al. (Armagh Obs Preprint 50; AA in press) registered in Gliese 644AB a strong flare in H-alpha and X-ray; flaring ups in soft and moderate X-ray and in H-alpha occurred simultaneously but in soft X-ray the maximum took place slightly later and the fading stage was longer than for moderate X-ray. The X-ray flare can be represented by 2-3 loops with a height of about 10.0E9 cm and an electron density of about 10.0E12 cm⁻³ . De Jager et al. (in preparation) registered a strong flare in UV Cet whose parameters differ essentially from a mean solar flare: a factor of 30 in vertical scale, a factor of 40 in optical thickness, a factor of 4 in area. The temperature of the coronal plasma was (1-4) x 10.0E6 K, EM = 10.0E51 cm⁻³ and n_e = 10.0E12 cm⁻³ .

Pallavicini et al. (ZG, p225) registered with EXOSAT and VLA a strong flare in EQ Peg with a duration of about 2 hours, peak luminosity L(X) ~ 2 x 10.0E30 erg/s, and a total X-ray energy E(X) ~ 10.0E34 erg. Pallavicini (BAAS 18.962) reported that in the course of patrol observations with EXOSAT for 9 hours for each of 4 FSs - UV Cet, EQ Peg, YZ CMi, and AD Leo - very different X-ray activity and different levels of quiet states were found.

Agrawal et al. (MN 219.225) estimated a mean level of X-ray emission from 7 FSs in quiet states as 2 x 10.0E28 erg/s, found a coronal temperature of about 3 x 10.0E6 K. They registered gradual variations of radiation from Gliese 735 and a flare in Gliese 729. For all FSs they found a close correlation between bolometric and X-ray luminosities but neither L(X) nor L(X)/L(bol) correlate with rotational periods or stellar equatorial velocities.

Bookbinder et al. (ZG) found that X-ray emission from K-M dwarfs is strongly dependent on the stellar age and effective temperature, and distributions of L(X) (or L(X)/L(bol)) along T(eff) for each kinematic population change markedly at the point where theory predicts the onset of full convection in the stellar interior.

Ambruster et al. (ApJ 284.270) carried out one week patrol observations for EV Lac with the Einstein Observatory and registered two strong flares with the amplitudes 7 and 50, and durations of 1-4 hours. Variations of the X-ray emission in the quiet state were discovered. Time-averaged X-ray flare luminosity is found to be larger than X-ray luminosity of the star in the quiet state and than time-averaged optical flare luminosity. Using the Einstein Observatory data, Ambruster et al. (ApJ Supp, in press) studied the stability of the X-ray flux levels in the quiet state for FSs: in the 16 of 19 cases they found variability in the scale from 10.0E2 to 10.0E3 s with the amplitude of about 30% independently of the spectral classes. They concluded that microflares cannot be responsible for stellar corona heating.

Doyle & Butler (Nature 313.378) and Byrne (Irish Ast J 17.216) noted a similarity between total X-ray luminosities of stellar coronae of FSs and time-averaged powers of stellar flares and suggested that coronal heating is due to flares. Butler & Rodono (Irish Ast J 17.131) and Butler et al. (Nature 321.679) carried out ground-based and space patrol for FSs UV Cet, EQ Peg and Prox Cen and discovered fast - less than a minute - soft X-ray variability and close correlation between X-ray radiation and the H-gamma intensity for UV Cet that gave grounds for the hypothesis that quiet X-ray stellar radiation is due to superposition of many microflares but not to stellar corona. Skumanich (Austr J Phys 38.971 and ApJ 309.858) found that the Sun fits the relation which exists between time-averaged flare luminosity and X-ray coronal luminosity for FSs and proposed an evolutionary scenario for flare activity where the total flare rate remains more or less constant but the mean flare yield decreases linearly with coronal X-ray flux.

Pallavicini et al. (AA 149.95) confirmed the existence of quiet emission at 6 cm for UV Cet, EQ Peg and YZ CMi and estimated the upper limits of such emission from YY Gem and EQ Vir. They registered impulse (10 s) and gradual (10-20 min) radio flares for YZ CMi with the increase in the degree of circular polarization.

Lang (Solar Phys 104.227) patrolled YZ CMi at 3 frequencies and found flux variability at 1415 and 1515 MHz with a characteristic time of about an hour and non-coincident maxima and flux constancy at 4885 MHz. Within a framework of cyclotron maser mechanism these data lead to a magnetic field strength of several hundred Gs within the stellar corona. Gary (in "Radio Stars", p385) registered the AD Leo radio flare at 6 cm with nearly 100% circular polarization. Lang et al. (ibid, 267) studied 31 stars with active chromospheres and coronae at 6 cm with VLA, detected radio emission from 6 stars but not from stars of G-K spectral classes. Twenty hours of observations of AD Leo enabled Lang & Willson (ApJ 305. 363) to register two flares at 1415 MHz of 50 and 25 s durations. For the longer flare quasi-periodic oscillations with a period of 3.2 s were observed during 25 s and peaks of such pulsations had a fine structure of 5 ms with nearly 100% circular polarization. Data analysis led to the conclusion that the size of the radiating region was about 10.0E8 cm; the high brightness temperature and high circular polarization were due to coherent maser emission, and quasi-periodic pulsations that modulate the maser action were due to magnetoacoustic oscillations in a coronal loop.

Kundu & Shevgaokar (ApJ 297.644) observed YZ CMi, L 726-8A and UV Cet at 6
and 20 cm simultaneously and concluded that quiet radio emission of FSs is due to
gyrosynchrotron radiation of electrons having a power-law energy distribution and
this emission originates from sources whose sizes are of several stellar radii.
From the lifetime of 1 hour for non-thermal particles they estimated stellar photospheric magnetic fields to be a few kGs. Jackson et al. (ApJ 316.L85) registered
the UV Cet flare of about 10 min duration at 4 frequencies from 1385 to 1652 MHz
and found a rather complex dynamic spectrum involving both positive and negative
frequency drifts.

Nelson et al. (MN 220.91) made simultaneous observations for AT Mic in optical, infrared and microwave ranges and found essential disagreement between the
observable features of flares and the expected ones within the framework of
Gurzadian's model. The data obtained can best be accounted for in terms of a hot
(10.0E5 K) plasma model; the ratio of optical to microwave flare luminosities
shows a variation of at least several orders of magnitude which means that various
mechanisms are responsible for these radiations.

Stewart et al. (Proc Astr Soc Australia, in press) found a correlation between peak flare luminosity at 8.4 GHz and stellar surface velocity. Vaughan &
Large (MN 223.399) found variable emission at 8.4 GHz for 2 of 8 studied FSs; the
emission mechanism is not known (Proc Astr Soc Australia 6.319 and in press).

A set of theoretical studies on FSs have been carried out.

Zarro & Zirin (AA 148.240) demonstrated that the series merging of Stark
broadened line profiles can produce a significant continuum enhancement that may
be responsible for the "blue" continuum in solar and stellar flares, Gurzadian
(Ap Sp Sc 106.1) calculated postflare relaxation of different chromospheric lines.
Cram & Giampapa (ApJ, in press) developed a theory of the formation of H-alpha
and Ca II lines in dM stars. Poletto et al. (BAAS 18.962; Adv Sp Res, in press)
showed that the solar two-ribbon flare model can be used to interpret long-lived
stellar flares. Gurzadian (Ap Sp Sc 125.127) concluded that a usual fast flare
that occurred on the opposite side of a star can be seen as a very slow event due
to the diffusion of flare radiation.

Grandpierre (M, p176) proposed a convective theory of flares. On concluding
that the theory of current sheets cannot energize the most powerful solar and
stellar flares Gershberg et al. (IZMIRAN Preprint No 41a(655); Kinematics and Physics of Celestial Bodies 3.N5) proposed that a soliton gas is formed within subphotospheric convective layers as a flare energy source.

Badalyan (Sov Ast 63.762) found a general analytical formula for EM of isothermal hydrostatic stellar coronae; using this formula and X-ray observations,
Katsova et al. (Sov Ast 64.N5) estimated the EM of FS coronae and n_e in their
lower layers, found a systematic increase of the coronal densities for FSs from
G to M3 spectral classes and a decrease for later stars. Katsova & Livshits (M,
p183; Comm Konkoly Obs 86.437) estimated the least flaring up times within the
framework of their gasodynamical model of flares and found these times for optical
flares to be several tenths of a second.

Gurzadian (Ap Sp Sc 113.213) proposed stellar aggregates containing many FSs
as sources of galactic cosmic rays. Caillault et al. (ZG) concluded that the integrated contribution to the diffuse soft X-ray background from M dwarfs is less
than 10 %.

12. T Tauri Stars
(Claude Bertout)

Major progress has been achieved in our understanding of the T Tauri
phenomenon since the previous Commission 27 report and this overview concentrates
on these advances. In the picture which is now emerging, a typical T Tauri
"system" is made up of a late-type, active star, surrounded by and interacting

with the dusty equatorial disk which has been formed together with the star during
the protostellar collapse of a dense molecular core.

Recent reviews of several theoretical and observational aspects of the
formation of low-mass stellar objects are found in "Protostars and Planets II",
(eds. M.S. Matthews & D.C. Black, Tucson, University of Arizona Press), in
"Nearby Molecular Clouds", (ed. G. Serra, Lecture Notes in Physics No. 237,
Heidelberg, Springer), in "Circumstellar Matter", (IAU Symp 122, eds. I.
Appenzeller & C. Jordan, D. Reidel) and in "Protostars and Molecular Clouds",
(eds. T. Montmerle & C. Bertout, Saclay, Commissariat a l'Energie Atomique).
Current theoretical ideas about low-mass star formation are summarized by Shu et
al. in Ann Rev AA 25.23.

A great deal of work has been devoted to the study of the spectral appearance
of protostar models by Adams & Shu (ApJ 296.655; 308.836), who show that observed
spectra of several low-luminosity infrared sources can be reproduced by models
involving a central protostar surrounded by the nebular disk expected to be formed
at the same time as the star during the protostellar collapse of a slowly-rotating
molecular cloud core (e.g. Terebey et al., ApJ 286.529). Wilking & Lada (ApJ
287.610) classified the ρ Ophiuchi infrared sources in different categories
according to the shape of their infrared spectra, and Adams et al. (ApJ 312.788)
proposed an interpretation of these classes according to their evolutionary
status. In this theory, stars with mid-infrared spectra rising toward long
wavelenghts (Class I) are still surrounded by a disk and an infalling envelope.
Stars with power-law infrared spectra whose flux decreases toward longer
wavelengths (Class II) have accreted their envelope but are still surrounded by
optically thick disks, and stars with approximate blackbody spectra (Class III)
have lost both their envelope and disk. Many T Tauri stars have Class II energy
distributions.

The observational search for protostellar disks has been quite active, and
many of these efforts were aimed at HL Tauri, a star long suspected to have an
equatorial disk (Cohen, ApJ 270.L69). The presence of a scattering disk around
this object was inferred by Grasdalen et al. (ApJ 283.L57) from direct infrared
imaging followed by maximum-entropy image reconstruction and by Beckwith et al.
(ApJ 287.793), who used speckle interferometric techniques. Beckwith et al. (ApJ
309.755) observed an elongated structure in ^{12}CO with extension about 4000 AU, and
Sargent & Beckwith (ApJ, in press) found that this molecular structure may be
rotating around the star. At this point, this is the only direct evidence for a
disk around a low-mass young stellar object.

There are, however, several lines of indirect evidence for disks surrounding
T Tauri stars (TTS). Rucinski (AJ 90.2321) searched the IRAS Point Source Catalog
for TTS and found that a number of them display far-infrared power-law spectra
reminiscent of those expected from optically thick accretion disks ($\lambda F_\lambda \propto \lambda^{-3/4}$).
Appenzeller et al. (AA 141.108) were first to suggest that the lack of red
emission often noted in the forbidden lines of TTS optical spectra is best
understood if a large (typically 100 AU), optically thick disk hides from our view
the receding parts of the stellar wind. Detailed profile computations by Edwards
et al. (ApJ, in press) confirm this and demonstrate that latitude-dependent wind
best accounts for the observed profiles. Rydgren & Cohen (in Protostars and
Planets II) discuss additional evidence suggesting that dust is anisotropically
distributed around TTS. Bertout (IAU Symp 122) suggested that the interaction
between a fast-rotating, Keplerian accretion disk and the slow-rotating TTS
photosphere should give rise to a boundary layer emitting mostly in the ultraviolet, so that infrared excess and the strong ultraviolet excess characteristic
of TTS (Herbig & Goodrich, ApJ 309.294) may be correlated as is often the case.

There is some controversy about the exact nature of the disks surrounding
TTS. Adams et al. (ApJ 312.788) favor geometrically thin but optically thick
"passive" disks, which merely intercept stellar photons and re-distribute them at
longer wavelenghts. They show that the system luminosity can then be up to 1.25
times the stellar luminosity and that the expected spectral slope in the IR is the
same as that of an accretion disk. Kenyon & Hartmann (ApJ, in press) also favor

passive disks, but argue that they need not be geometrically thin ("flaring" disk). In that case, the outer parts of the disk can intercept more stellar photons, with the results that the system luminosity can be up to 1.5 times the stellar luminosity and that more energy can be emitted at long IR wavelenghts. There are several TTS whose "flat" infrared spectra cannot be reproduced when assuming a passive, flat disk, and a "flaring" disk represents one possibility to account for them in the framework of the passive disk model. Other possibilities are discussed by Adams et al. (ApJ, in press), who propose instead that TTS with far-IR spectra are surrounded by "active" disks (i.e., intrinsically luminous disks) with temperature distributions different from those of passive disks or classical accretion disks. A recent detailed study of the continuous energy distribution from the IUE range to the far-IR range of several TTS by Bertout et al. (ApJ, in press) confirms that active accretion disks with boundary layers between the disk and the star are probably present around a number of TTS. In this case, the system luminosity can range from 1 to 3 times the stellar luminosity depending on the mass-accretion rate, which typically ranges from 5x10.0E-8 to 5x10E-7 M$_\odot$ /yr. It is clear, however, that the TTS with flat IR spectra discussed above cannot be understood in such a simple framework.

Progress has also been made in recent years by studying TTS with low emission characteristics, which differ little from other active late-type stars. In the framework outlined above, TTS with small UV and IR excesses and low-level activity are probably more evolved than more active TTS. Their energy distribution is not dominated by the disk, which either optically thin or has been accreted or dissipated. Thus, we expect their activity to be stellar rather than circumstellar.

Rotational velocities of a large sample of TTS with low to moderate emission characteristics were measured by Hartmann et al. (ApJ 309.275) using Fourier techniques and by Bouvier et al. (AA 165.110) using mainly CORAVEL. Both groups find that v sini ranges typically from 5 to 25 km/s. The detection limit of CORAVEL is about 2 km/s, but Bouvier et al. did not find TTS rotating slower than about 5 km/s in their sample of 30 stars. If confirmed by further work, this result would have far-reaching consequences. Studies of correlations between various activity diagnostics and rotation rates demonstrate that TTS, together with RS CVn systems, continue upwards the relationship found for main-sequence late-type stars between v sini and X-ray flux. This suggests that X-ray activity is dynamo-driven in TTS just as it is in RS CVn stars. Furthermore, the anticorrelation between X-ray flux and rotation period observed in RS CVn systems also seems present in TTS, although the small number of TTS for which both X-ray flux and rotation period are known makes this conclusion somewhat controversial at this point (Bouvier in Protostars and Molecular Clouds).

Intensive photometric and spectroscopic monitoring of various TTS was performed at several observatories around the world. The large number of publications that resulted over the last few years cannot be detailed here because of limited space, but the interested reader can consult the Astronomy and Astrophysics Abstracts for works about specific objects.

One highlight of this intensive monitoring was the discovery of periodic components in the light-curves of more than a dozen TTS with usually low or moderate emission characteristics, including T Tauri itself (cf. Vrba et al., ApJ 306.199; Bouvier et al., AA 158.149; Bouvier in Protostars and Molecular Clouds and Herbst et al., ApJ 310.L71). Bouvier reproduces the periodic light-curves by assuming that the modulation is rotational and is caused by an inhomogeneous surface temperature distribution. He shows that the derived properties of the spots on the stellar surface are similar to those of RS CVn spots. Thus the intrinsic properties of TTS - as opposed to properties caused by their circumstellar environment - appear similar to those of other late-type active stars.

Optical line emission, a characteristic property of TTS, has also been the subject of many investigations, but only a few results can be mentioned here. No correlations were found between the Hα or CaII K line fluxes and the rotation rate (e.g. Hartmann et al., ApJ 309.275), ruling out dynamo mechanisms for the production of these lines. Detailed models of the formation of emission lines in deep-lying chromospheres by Calvet et al. (ApJ 277.725) confirmed that Hα cannot be purely chromospheric. However, Calvet et al. (ApJ 293.575) report a strong correlation between MgII k and CaII K line fluxes, and Bouvier (op. cit.) finds that Hα and CaII K line fluxes are also strongly correlated at all emission levels. This suggests that these three lines are all formed in the same atmospheric region and the same physical mechanism is responsible for their formation even in stars with different activity levels. Bertout et al. (ApJ, in press) suggest that these strong emission lines originate partly from the atmosphere of the boundary layer between disk and stellar photosphere.

The long-standing problem of the apparent lack of post-T Tauri stars (i.e., stars more evolved than TTS but still in their pre-main-sequence evolutionary phase) may also be on the way to a solution. Walter (ApJ 306.573) and Feigelson et al. (ApJ, in press) identified a number of pre-main-sequence objects with low-emission characteristics through follow-up observations of X-ray surveys in star-forming regions (see also the review by Feigelson in Protostars and Molecular Clouds). While some of these objects appear older than classical TTS, many of them fall in the same region of the H-R diagram as TTS and are denoted "naked" TTS by Walter because their circumstellar environment appears empty. The calcium-emission stars found in a survey of the Taurus-Auriga region by Herbig et al. (AJ 91.575) fall in the same category; they are not genuine post-T Tauri stars, but their study brings useful insight into the structure of their atmospheric layers. Finkenzeller & Basri (ApJ 318.823) performed high-resolution spectroscopy of a number of stars with low emission characteristics. They are able to show that their metallic emission spectra are very similar to the solar chromospheric spectrum.

Walter (ApJ 306.573) also finds that a significant fraction of pre-main-sequence objects found from X-ray surveys appear older than TTS, and that there is little besides their X-ray properties to distinguish them from field stars.
Another way to detect objects in the post-T Tauri phase might be through surveys at radio wavelenghts. Andre et al. (AJ 93.1182) used VLA to survey the ρ Ophiuchus cloud and discovered there a number of sources apparently associated with Class III stars that are as close to the molecular cloud core as are typical TTS. The evolutionary status of these objects is however unclear at this point, as is the nature of their radio emission (see Andre in Protostars and Molecular Clouds).

Finally, yet another long-standing question - the apparent deficit of binary systems among TTS - may be finding an answer with the discovery via Moon occultation techniques of close companions to several TTS (Simon et al., ApJ 320.344).

B. Szeidl
President of the Commission

COMMISSION 28: GALAXIES

PRESIDENT: P.C. van der Kruit
VICE-PRESIDENT: G.A. Tammann
ORGANIZING COMMITTEE: S. d'Odorico, J. Einasto, I.D. Karchentsev, E.Ye. Khachikian, J. Lequeux, Li Qi-Bin, D. Lynden-Bell, H. Quintana, V.C. Rubin, A. Toomre, K.-I. Wakamatsu

This report covers the period 1 July 1984 to 30 June 1987 and - as usual - has been prepared by the President, Vice-President, members of the Organizing Committee and the chairpersons of the Working Groups. There has been emphasis on highlights rather than on completeness of references. In comparison with earlier reports one section (Galactic Dynamics) has been added, but otherwise the same line of reporting has been used. For readability and quick reference, in general abbreviated notation for literature references to major journals have been used; more complicated ones usually follow the code of Astronomy and Astrophysics Abstracts. The abbreviation et al. has been replaced by the symbol +.

1. Highlights since Delhi
(V.C. Rubin)

The arrival of optical photons and neutrinos from a supernova in the LMC has surely made the present time one of the most exciting period for extragalactic astronomers. Moreover, other galaxy studies during the last 3 years have led to many significant advances;some are of such fundamental importance that they will will alter the way in which we look at the universe. While my predecessors have used this space to detail survey work, catalogues, and observations in various spectral ranges, I will use this opportunity to make some general comments on three exciting topics which I view as the highlights during the past few years.In successive years, such reports could constitute an interesting historical document of which accomplishments we viewed as most important.

I. SN 1987a

On february 23, 1987, the 15-29 solar mass star SK-69.202 "went supernova"; in less than a few seconds most of the released gravitational energy ($\sim 10^{53}$ ergs) was emitted as neutrinos, of which one in 10^{47} traveled through the earth. Of those wich interacted with matter, 19 were detected by sensitive detectors at Kamioka, Japan and outside of Cleveland, Ohio, USA. Within these few seconds, the sience of observing neutrinos from supernovae was born. Surely these seconds rank among the most important in the history of astronomy. Present analysis of the observations sets a likely upper limit to the neutrino mass at about <25 eV.

At the time of this writing, June 1987, SN 1987a has peaked at all observed wavelenghts, and its luminosity is slowly decreasing. While it seems likely that its luminosity in the X-ray, the γ-ray, and the UV spectral regions will increase in the coming months or years, a significant dust component may blot out our optical view. Still to be explained is the speckle observation of a bright companion (now the second brightest object in the LMC) only 0.06 arcsec away from the SN. Reflection? Object? Its properties are presently unknown.

While the theorists rejoice with the observational confirmation of theories of core collapse and SN production, observers question why the SN is less bright than expected, and the role of the companion object. We can anticipate an enormous

increase in our knowledge of SN within the next decade, due to our very great
fortune in having appropriate detectors available when the event occured. We can
hope for a second naked-eye SN before 2019, thus to better the Brahe/Kepler record
of SN 1572a and 1604a.

II. REDSHIFT SURVEYS, LARGE-SCALE STRUCTURE, AND LARGE-SCALE MOTIONS

Conclusions from studies of the large-scale structure and of large-scale
motions within hundreds of megaparsecs have produced intreging results; this is
probably the most exciting time to be studying the large-scale universe since
Zwicky and Smith noted the high velocity dispersion of galaxies in the Coma and the
Virgo clusters. The distribution of galaxies in the universe remains clumpy on
scales extending to the distances of the deepest surveys. This result in each of
the many redshift surveys.

The total number of redshifts is now
about 24,000; up from 10,000 in 1979
(Fig. 1), and it is increasing at a rate
of several thousand velocities per year.
We should easily have 100,000 by the
year 2000; it is difficult to guess if
will have observed 1,000,000 by the year
2010. But we are well on the way to the
10^6 redshifts that Ambartsumian said we
will require in order to understand the
universe.

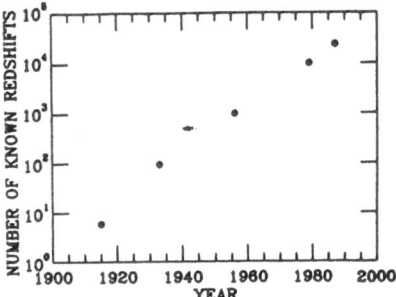

Large area surveys have determined velocities of all IRAS galaxies with intensities
greater then 2 Jy (Davis, Huchra, Tonry, Yahill, and others); velocities of 2000
optically selected ESO galaxies (Da Costa, Davis, Fairal, Sargent and others); and
velocities of 4000 galaxies complete to m=15·5 contained in four slices (Geller and
Huchra). Velocity surveys in clusters of galaxies, independently made in the USSR
(Koplov, Karachentsev, and Klypin) and in the US (Postman, Huchra, and Henry;
Gionavelli and Haynes) have added thousand of more radial velocities. And pencil-
beam surveys (Ellis, Efstathiou, and Peterson), those narrow regions extending as
far as m=22 (Koo, Kron, and Szalay), have confirmed that even at these large
distances, the distribution of galaxies in redshift space is not smooth. Where is
the smooth, isotropic universe we were all taught to expect?

In a universe in wich the random motions of galaxies are small, three-dimensional plots of redshifts and positions will map the distribution of galaxies. But
if random motions are not small, or if systematic streaming motions exist on scales
wich are large then the transformation from position-redshift to position-distance
is non-trivial.

Major observational programs to map large scale bulk motions on scales of
100´s of Mpc have produced equally surprising results. Distances to galaxies,
independent of their observed velocities, are required. The set of residual velocities from a smooth Hubble flow are then analysed for systematic patterns, with
the cosmic microwave background defining the rest-frame.

Almost without exception, analyses for a bulk motion all imply a motion of the
nearby galaxies with a velocity in the 500-700 km/sec range, in a direction not far
from the apex of the microwave dipole. Aaronson and colleagues, who have studied
galaxies in clusters, and de Vaucouleurs and Peters have used the Tully-Fisher
relation to obtain absolute magnitudes, and hence distances; Faber and her six
colleages have used the Faber-Jackson relation with a modified magnitude criterion.
Even the 1976 Rubin-Ford ScI galaxies, as well as a new sample of rotation curve
spirals, determine a motion whose apex is within 20° of the MWB apex. In regions

where the various studies overlap, there is a good agreement among the residual velocities. Many questions still remain. Is a model based on the attraction of a single dominant mass a better fit to the observations than a model based on a bulk flow? How significant are the differences between the solutions; are they merely all optical approximations to the microwave dipole?

Motions this large put serious constraints on the properties of the early universe. We have much to learn concerning the large-scale distribution of matter of motions, but progress has been rapid in the last few years; we can anticipate exciting new discoveries in the next three years.

III. GALAXIES IN THE INFRARED

From January until November, 1983, the Infrared Astronomical Satellite (IRAS) scanned the sky at wavelengths of 12, 25, 60, and 100 μm, ultimately viewing 98% of the sky at wavelengths inaccesible from the ground. Over 20,000 galaxies were detected, primarily at 60 and 100 μm, and revealed a view of the extragalactic universe only dimly perceived previously. New insights into galaxy evolution, active galactic nuclei, and QSOs have been obtained, and still offer much to be learned. Advances in our knowledge of external galaxies, surely as great as those which came from the initial 21-cm observations, have once again come from opening up a new spectral band.

"An exciting and surprising revelation". That is what G. Neugebauer has called the observation that some galaxies emit more than 50% in the infrared and as much as 30^0 to 300^0 K is sampled; in galaxies this radiation corresponds to emission from large dust comlexes which have been heated by starlight. The surprise that came with this discovery that many peculiar, disturbed, or tidally interacting galaxies emit 98% of their energy in the IR compared with galaxies of undisturbed morphology, where the fraction emitted in the IR is less than half.

Follow-up ground based studies of IRAS galaxies indicate short phases of active massive star formation, probably induced by the supersonic collision of gas clouds in gravitational interacting galaxies. Star formation rates ten times normal are experienced over relatively short time scales, generally within a few kilopasecs of the nucleus. Heavy-element enrichment at small radii is thus a natural consequence of such models. Extension of these ideas to even larger scales may account for QSO activity. Overall, IRAS observations have offered a new approach to the study of galaxy evolution, and have taught us how important is the role of gas in the evolution of galaxies.

IV. IMPORTANT CATALOGUES AND ATLASES
(prepared by J. Lequeux)

1984 – Rosa +		IUE UV spectra of extragalactic HII regions I the catalogue and the atlas, AA Suppl. 57, 361
	Sandage +	(Atlas of dwarf galaxies in the Virgo Cluster) AJ 89, 919
	Helou +	(HI observations of magnitude-limited sample of Virgo spiral galaxies) ApJ Suppl. 55, 43
1985 – Verter +		Catalog of CO observations in galaxies ApJ Suppl. 57, 261
	Takase +	An atlas of selected galaxies with illustrations of photometric analyses U. Tokyo press
	Karachentseva +	Atlas of dwarf galaxies in the region of the M81 group AA Suppl. 60, 213

Grosbol +	(Catalogue of disk parameters for 605 galaxies) AA Suppl. 60, 261
Lonsdale +	Catalogued galaxies and quasars observed in the IRAS survey, JPL
Ballick +	Catalogue of dusty elliptical galaxies A.J. 90, 183
Sandage +	(Atlas of Virgo Cluster spiral galaxies) AJ 90, 395
Binggeli +	A catalog of 2096 galaxies in the Vigo cluster area AJ 90, 1681
Bothun +	A catalog of radio, optical and infrared observations of spiral galaxies in clusters ApJ Suppl. 57, 423
1986 - Gallagher +	UBV colors of Virgo cluster irregular galaxies AJ 92, 557
Mazzarella +	A catalog of Markarian galaxies ApJ Suppl. 62,
Huchtmeier +	HI observations of galaxies in the Kraan Korteweg-Tammann Catalogue AA Suppl. 63, 323
Wevers +	The palomar-Westerbork survey of northern spiral galaxies AA Suppl. 6, 505
1987 - Hoffman +	(HI survey of dwarf irregular galaxies in Virgo) ApJ Suppl. 63, 2476
Bettoni +	A catalogue of early type galaxies with emission lines AA Suppl. 67, 341
Gioia +	Radio continuum observations of early and late-type spiral galaxies ApJ Suppl. 63, 4
Deutsch +	Far-IR luminosities of Markarian starburst galaxies ApJ Suppl. 63, 803
Lewis +	HI survey of face-on galaxies: the frequency of distortions in HI disks ApJ Suppl. 63, 515

2. Structure and Evolution of Galaxies
(J. Lequeux)

Many recent meetings have been devoted totally or partly to this field: 1984: "Spectal evolution of galaxies", ed. Gonghalekar, RAL report 84-008; 1985: "Theoretical aspects on structure, activity and evolution of galaxies", ed. Aoki +, Tokyo Astr.; "New aspects of galaxy photometry", ed. Nieto, Springer, Verlag; "The Virgo cluster of galaxies", ed. Richter, ESO; "Extragalactic infrared astronomy", ed. Gondhalekar, RAL report 85-086; "Production and distribution of CNO elements", ed. Danzinger +, ESO; "Birth and evolution of massive stars and stellar groups", ed. Boland +, Reidel; "Spectral evolution of galaxies", ed. Chiosi +, Reidel; "Star forming dwarf galaxies", ed. Kunth +, Editions frontières; "Luminous stars and associations in galaxies", ed. de Loore + (IAU Symp. 116), Reidel; "Light on dark matter", ed. Israel (first IRAS conf.), Reidel; "Stellar populations", ed. Norman +, Cambridge U. Press; "Gaseous halos of galaxies" ed. Bregman +, NRAO. To be published: "Nearly normal galaxies", ed. Faber (Santa Cruz), Springer Verslag; "Starbursts and galaxy evolution", ed. Montmerle + (moriond), editions frontières; "Stellar evolution and dynamics in the outer halo of the galaxy", ed. Azzopardi +, ESO; etc. One should mention the book by Hodge "Galaxies" Haward U. Press.

I. ELLIPTICAL AND LENTICULAR GALAXIES

1. Structure and Evolution

Kormendy (ApJ 292, L9; 295, 73) shows that the cores of ellipticals (and bulges) are generally not isothermal, and that dwarf E and SO galaxies are not connected genetically to giant E, a conclusion confirmed by van den Berg (ApJ 91,

271) and Ichikawa (ApJ Suppl 60, 475) although Wirth + (ApJ 282, 85) do not agree.
The presence of disks in ellipticals is indicated by Kodaira + (ApJ Suppl 62, 703),
Carter (ApJ 312, 514) and Bender + (AA 177, 71) while Michard (AA 140, L39) also
finds similarities between flattened E and S0. Color and metallicity gradients in E
have been found by Estathiou + (MN 215, 37P), Baum + (ApJ 309, 572), Carter + (ApJ
315, 451) while Glass (MN 211, 461) finds no IR color gradients and Boroson + (AJ
92, 33) find color gradients only in the most lumious E Shells have now been
discovered around many (Fort + ApJ 306, 110; hence ApJ 310, 597) and are usualy
considered as resulting from merging (Dupraz + AA 166, 53; Huang + AA 174, 13). On
the other hand, the galaxy NGC 6166 is no longer considered as cannibalizing
smaller galaxies (Lachieze-Rey + AA 150, 62; Lauwer ApJ 311, 34). Globular clusters
have been found and studied in a large sample of E up to the Coma Cluster (Haris +
ApJ 287, 175 and 185; 291, 147; 315, L29; AJ 91, 822; Prichett + AJ 90, 2027; Hanes
+ ApJ 300, 279; 304, 599; 309, 564; Grilmair + ApJ 91, 1328, Thomson + ApJ 315,
L35); although they differ from those in spirals (van den Berg + AJ 90, 535; Mould
+ AJ 92, 53) they might be usefull as standard candles. Variations of parameters
with luminosity amongst E have been studied by Mould (PASP 96, 273), Pickles (ApJ
294, 134; 296, 340), Rose (AJ 90, 1927), Vader (ApJ 306, 390) and others, with the
main result that metallicity is correlated with luminosity. Djorgovski + (ApJ 313,
59) finds that E form a 2-parameter family (surface brightness and velocity dispersion). Several authors have tried to explain these statistical results in terms
of formation and evolution. Weeraghavan + (ApJ 296, 336) find no objection for E
resulting from merged S, but this possibility is dismised by Kent (ApJ Suppl 59,
115) and Lake + (ApJ 310, 605). Vader (ApJ 306, 390) finds little difference
between cluster and field E while Bica + (preprint) claim that field giant E and S0
have fainter metallic lines. Dwarf E have been suggested to be stripped dwarf
irregulars, but the studies of Bothun + (AJ 92, 1007), Thuan (ApJ 299, 881) and
Binggeli (in Star-Forming Dwarf Gal. 53) show that this is unlikely in the Virgo
Cluster. Alternative models for the formation of E invoke galactic winds at early
stages (Vader ApJ 305, 669; Arimoto + AA 133, 23; Yoshii + AA, in press).

The problem of whether E have an intermediate-age population is still controversial. While spectral synthesis or colors suggest to Burnstein + (ApJ 287, 586),
Rose (AJ 90, 1927), O'Connell (in Stellar Populations, 167), Thuan (ApJ 299, 881)
and Rocca-Volmerage + (AA 175, 15), the presence of a large amount of stars 5-10
Gyr old, this conclusion is disputed by Renzini (in Stellar Populations, 213).
Hamilton (ApJ 297, 371) finds all E up to $z=0.8$ to be much more than 8 Gyr old. The
nature of the UV excess might well be due to old evolved objects (Nesci + AA 145,
296; Mochkovitch AA 157, 311; O'Connell + ApJ 303, L37). However some authors favor
a young population (Kjaergaard AA 176, 210; Bertola + ApJ 303, 624) in some giant
E.

2. Gas and Young Stars

It is now clear that many E and S0 galaxies contain gas and dust and presently
form stars, although massive stars only in small quantities. Neutral hydrogen has
been detected in many (and even CO in NGC 185: Wiklind + AA 164, L22). In E its
morphology and other properties suggest that it has been accreted (Knapp + AJ 90,
454) and it is often very extended (Wilkinson MN 217, 779; van Gorkom + AJ 91, 791;
Lake + ApJ 314, 57). An HI bridge has been found between a dwarf galaxy and the
giant E NGC 4472 (Sancisi + ApJ 315, L39). The situation for S0 seems more complex.
Many have inclined HI disks (e.g. Sancisi + MN 210, 497; Krumm + AA 144, 202; Knapp
+ AA 142, 1; van Gorkom + ApJ 314, 457) and the gas is very extended (Chamareaux +
AA Suppl 69, 263; Shostak AA 175, 4), suggesting accretion; the culprit may have
been found for NGC 1023 (Cappaccioli + AA 169, 54). NGC 5084 may be a counter-
example to accretion (Gottesman + MN 219, 759). Dust is present in 25 (Lauer MN
126, 429) to 40 per cent (Sadler + MN 214, 177) of E. NGC 5266 is an E where
coinciding dust and HI rings have been seen (Varnas + ApJ 313, 69). More than 50
percent of E show far-IR-emission (Jura + ApJ 312, L11) from dust heated by the

50 per cent of E show far-IR-emission (Jura + ApJ 312, L11) from dust heated by the general radiation field (Jura ApJ 306, 483; Wrobel + ApJ 311, L11) although star formation may contribute (Brosch MN 225, 257). More than half of the E and SO also contain ionized gas (Demoulin-Ulrich + ApJ 285, 527; Phillips + AJ 91, 1062; Wodge + AJ 92, 291) which may or may not coincide with HI dust (e.g. Phillips + Nature 310, 733; Davies + ApJ 302, 234). The ionization may be through non-thermal UV radiation (Rose + ApJ 285, 55; Wilson + ApJ 291, 627; Tadhunter + Nature 325, 504; Robinson + MN in press) but there are counter-opinions (Diaz + MN 214, 41P). Hot gas giving X-ray emission is common place around giant E in clusters (Fabian + Nature 310, 733; Steward + ApJ 285, 1; Forman + ApJ 293, 102; Stanger + MN 229, 363; Trinchieri + ApJ 310, 637; Canizares + ApJ 312, 503; Maccagni + ApJ 316, 132); note that old low-mass stars contribute to the X-ray emission (Trinchieri + ApJ 296, 447). This gas cools and settles in the central region and star formation is expected, as well as from the cold gas. Indeed recent star-formation is seen either as blue nuclei (Vigroux + AA 139, L9) or spectral features (Véron + AA 145, 433; Johnstone + MN 224, 71) or perhaps distortions in the optical appearance (Nulsen MN 225, 939). Star formation is not very active and, in the case of SO, not correlated with the amount of gas (Balkowski + AA 167, 223). Where has the cooled gas gone? Perhaps it forms only low-mass stars (Romanishin ApJ 301, 675; Fabian + ApJ 305, 9; Silk + ApJ 307, 415; Nulsen MN 221, 377).

3. Dwarf Spheroidal Galaxies

Intensive activity has been deployed on the "seven dwarfs" around our Galaxy: L-M diagrams (da Costa ApJ 285, 483; Olszewski + AJ 90, 2221; Buonnano + AA 152, 65; Stetson + PASP 97, 908; Carney + AJ 9, 23; Cudworth + AJ 92, 766), surveys of carbon stars (Azzopardi + AA 144, 38; 161? 232 and in prep.; Westerlund + AA 178, 41; Demers + in prep.; Aaronson + ApJ. 296, L7), infrared stellar photometry (Aaronson + ApJ 290, 191), abundance determinations (Bell PASP 97, 219; Suntzeff + AJ 91, 1091). The presence of carbon stars, the diagrams and the spread in metallicities indicate star formation over a long period, with the last episode a few billion years ago. M81B might be a dwarf spheroidal with present star formation (Schmidt + ANachr 306, 257). Morphologically the dwarf spheriodicals are more related to the dwarf E (Kormendy ApJ 295, 73) but the existence of a generic relationship is not established.

II. SPIRAL AND IRREGULAR GALAXIES

1. Structure of Normal Galaxies

Separation and statistical properties of bulges and disks have been discussed by Kent (ApJ Suppl 59, 115), Simien + (ApJ 302, 564) and Kodaira + (ApJ Suppl 62, 703) (see also Schombert + AJ 91, 60): only the bulge/disk ratio is well correlated with the Hubble type and there seems to be a discontinuity between E, SO and S, dismissing the possibility of type-to-type evolution.

Population syntheses in nuclei indicates a composite population (Frogel ApJ 298, 528; Ciani + AA 137, 223), a possible metallicity gradient in M31 (Martinez-Roger + AA 161, 237) and variations in metallicity and fraction of young stars from nucleus to nucleus (Bica + AA in press). The properties of X-ray emission indicate a population of low-mass binaries in the M31 bulge (Fabbiano + ApJ 316, 127). Many bulges contain gas (Mathews + ApJ 312, 66) often ionized; contamination by stellar lines complicates its study (Rubin + ApJ 305, L35). The bulge of M31 contains dust (it is a strong far-IR source: Soifer + ApJ 304, 651) and ionized gas with a complex distribution (Jacoby + ApJ 290, 136) and kinematics (Boulesteix + AA 178, 97), wich appears connected with a strange radio-continuum structure (Walterbos + AA 150, L1). However there are no massive stars in this region, and the UV excess is due to evolved stars (Bohlin + ApJ 298, L37).

Surface photometry of disks confirms the absence of color gradients (Carignan ApJ Suppl 58, 107) including in the IR (Glass MN 211, 461; Beckman + AA 161, 70; Walterbos + AA Suppl 69, 311); local regions of star formation are somewhat bluer. Old-star spiral arms can be seen in some galaxies (Kennicut + ApJ 300, 132; Adamson + MN 224, 367). The constancy of the general brightness, expotential decrease and cut offs at about 4.5 scalelengths are confirmed in an unbiased sample of isolated galaxies by van der Kruit (AA 173, 59) who suggests that this is naturally explain-ed by the momentum-conserving collapse of uniform rotating spheres. Grand design S are somewhat larger than the flocculent ones (Elmegreen + ApJ 314, 3) but their differences in color are marginal (Romanishin). Irregular galaxies, dwarf I and some blue compact galaxies have normal disks (Schild Aplett 24, 85; Carigan ApJ 299, 59; Loose + in Star Form. in Dwarf Gal. p 73; Comte +, in prep.) although weaker (Manousoyannaki + AA 160, 331) and their JHK colors are rather uniform except in extreme cases (Hunter + AJ 90, 1457). Loose + ApJ 309, 59 claim that the blue compact Ho2 has an underlying E. The apparent flattening distribution of the I is simular to that of the spirals, but the dwarf I might be triaxial (Feitzinger + AA 167, 215). HI has been mapped in many S galaxies in particular M31 (Brinks AA 141, 195; 169, 14) M33 (Deul + AA Suppl 67, 509) and NGC 6946 (Tacconi + ApJ 308, 600); one of the most interesting features is the presente of holes probably dug by stellar winds and supernova explosions. Strong warps are frequent (e.g. Appleton + MN 212, 393). A statistical study shows that HI disks as extended as the optical disks can produce the observed CaII absorption in extragalactic sources (Morton + ApJ 302, 272). Dwarf I and blue compact galaxies often have a relatively large HI extent (Krumm + AJ 89, 1319; Huchtmeier + AA 143, 216; Hunter + AJ 90, 1464) but their morphology varies from a regular disk to a clumpy irregular gas distribution (Sargent +, Skilman +, Comte +, Brinks +, in Star Form. in Dwarf Gal., resp. pp 253, 263, 273, 281; Briggs ApJ 300, 613). Bothun + (AJ 90, 697) show that dwarf E in the Virgo Cluster have no HI, and cannot be quiescent dwarf I.

CO has been detected in many galaxies (Verter ApJ Suppl 57, 261; Young + ApJ 288, 487; Odenwald ApJ 310, 86, etc.) and even ^{13}CO (Rickard + ApJ 292, L57; Young + ApJ 302, 680). Dickman + (ApJ 309, 326) state that the amount of H_2 can be derived from the line intensity. The latter is proportional to the blue Luminosity both in Virgo Cluster and field galaxies (Young + ApJ 288, 487). The distribution of CO differs from that of HI; this is probably related to the formation/disruption of molecular clouds (Allen + Nature 319, 296; Wyse ApJ 311, L41). Detailed CO maps of portions of M31 have been made, showing individual giant molecular clouds (Boulanger + AA 140, L5; Ryden + ApJ 305, 823; Casoli + AA 173, 43; Ichikawa + PASJ 37, 439). M33 is more difficult (Blitz ApJ 296, 481) but Viallefond + (in prep.) have detected and mapped CO in many places. The Caltech interferometer is producing detailed CO maps (IC 342: Lo + ApJ 282, L59; NGC 6946: Ball + ApJ 298, L21, ect): these observations and others show that CO is an excellent tracer of bars (Ohta + PASJ 38, 677) and rings (Garman + AA 154, 8). No CO has been found in two clumpy I galaxies (Sofue + PASJ 38, 161), only weak emission in several I (Tacconi + ApJ 290, 602) and none in dwarf I and blue compacts (Israel + AA 168, 369; Combes + in prep.). This is not too surpising given the low metallicity (also explaining their low far-IR flux: Gondhalekhar + MN 219, 505). Crawford + (ApJ 91, 755) have detect-ed using the Kuiper Airborne Observatory the C^+ 158 μm line in 6 galaxies. Its intensity is proportional to that of CO and its space distribution is very similar, while in relation with HI. A large amount of data on dust is coming from IRAS observations (Becklin in light on Dark Matter p. 415). Studies in our galaxy (Boulanger + ApJ in press; Pérault + in prep.), in M31 (Walterbos, thesis) and in other galaxies (Lonsdale-Persson + ApJ 314, 513) show that in normal galaxies the larger fraction of the dust is heated only by the general radiation field (the "cirrus" component) and thus that the far-IR emission is not a good indicator of star formation. A variable construction of star-forming regions explains the observed anticorrelation between the 12/24 μm and the 60/100 μm flux ratio (Helou ApJ 311, L33). The effect of optical thickness may not be negligible even at 100 μm since face-on galaxies are statistically stronger 100 μm emitters than the edge-on ones (Burstein + ApJ 301, 693).

radio-continuum studies have confirmed that the non-thermal radiation of S has an
exponential distribution similar to the blue light (Harnett MN 210, 13, etc.).
However M81 has a ring structure (Beck AA 152, 237). The radio brightness is correlated with the B-H color, an indicator of recent star formation (Burnstein + ApJ
315, L99). Higher-resolution observations show a bar and arms in M83 (Ondrechen AJ
90, 1474) and arms in M81 (Bash + ApJ 310, 621). Walterbos + (AA Suppl 61, 451) and
Viallefond + (AA Suppl 64, 237) find terminal sources in M31 sometimes coinciding
with optical HII regions and nonthermal sources. Dwarf I and blue compact galaxies
are weak radio emitters as expected (Altschuler + AA 177, 22) with a flatter
spectrum than S (Klein AA 141, 241; 161, 155; 168, 65; Sramek + ApJ 302, 640). The
structure of the magnetic field has been studied in several S through radio polarization (Sofue + AA 144, 257; Ap. Sp. Sci. 19, 191; Scarrott + MN 224, 299; Beck +
AA 152, 237; Vallée AJ 91, 541) or optical polarization (King MN 220, 485); it
generally follows the arms. Baryshnikova + (AA 177, 22) make dynamo models to explain this. Hummel (AA 160, L4) finds that deviations from the radio continuum/ 100
μm mean relation (de Jong + AA 147, L6) have the same distribution as $B^{1.9}$ if B is
the equipartition magnetic field, a strong argument in favor of equipartition.

Further studies have been made of the z-distribution of radio radiation in
edge-on galaxies (Schlickheiser + AA 140, 277; Sukumar + MN 212, 367; Werner AA
144, 502; Harnett + MN 215, 247; Broeils + AA 153, 281). Extended radio emission is
always present, with often a steepening of the spectral slope; several authors find
agreement with the dynamic halo model. Attempts to find hot-gas emission in the
halo of NGC 4244 have given marginal results (Deharveng + AA 154, 119). Bothun (AJ
90, 1982) has discussed the constants on HI in halos. New surveys of globular
clusters in M31 have been performed by Crampton + (ApJ 288, 494), Battistini +(AA
Suppl 67, 447) and Wirth + (ApJ 290, 140), the latter concentrating on the
difficult central regions. Globular clusters have also been found around NGC 2683
(Harris + AJ 90, 2495) and NGC 253 (Blecha 154, 321). IR photometry has been
performed by Sitko + (ApJ 286, 209) for globular clusters in M31 and Andromeda
dwarf E: a M_k, J-K relation offers hopes for using globular clusters as distance
indicators. Comparisons of globular clusters in different galaxies have been made
by Burnstein + (ApJ 287, 586) and Zinn (in Stellar Populations p. 73).

2. Individual Stars, Color Magnitude Diagrams

Local-group galaxies are sufficiently nearby for CM diagrams to be built and
even spectra to be taken of individual stars. CM diagrams in outer parts of M31
have been obtained by Crotts (AJ 92, 292): the disk shows a mixture of ages and
populations, the outer bulge (or halo) has a large spread in metallicities but the
authors seem to disagree as to the mean of metallicity. The halo of M33 is of lower
metallicity (Mould + ibid.). Massey + (AJ 92, 1303) and Odewahn (AJ 92, 310) have
obtained CM diagrams in star-forming regions of M31 and the former made detailed
studies of the stellar population. Upper CM diagrams have been obtained for many
local-group irregulars (Hoessel + ApJ 286, 159; ApJ 286, 159; ApJ Suppl 60, 507;
Demers + ApJ 89, 1160; 90, 1967; Sandage + ApJ 90, 1019 and 1964; 91, 496; Walker
MN 224, 935) and more distant galaxies (Freedman ApJ 299, 74; Pierre + AA, in
press). The upper luminosity function (OB stars) is simular in all the galaxies
exept perhaps M81 and indicate similar initial mass functions (Hoessel IAU Symp
116, p 439; Scalo, Fund. Cosm. Phys. 11, 1). Surveys have been performed in several
S and I: variables (e.g. Kinman + AJ 93, 833); carbon stars and M giants (Richer +
ApJ 298, 240; 298, L31; Cook + ApJ 305, 634; Aaronson, in Stellar Populations p.
45) (their ratio depend on metallicity); S stars (Aaronson + ApJ 291, L41). Planetary nebulae (Nolthenius + ApJ, preprint; Lequeux + AA Suppl 67, 1); Wolf-Rayet
stars (Bohannan + AJ 90, 600; Armanhof + ApJ 291, 685; Lequeux +, ibid.; Azzopardi
+ AA, in press.): according to the latter authors, the relative abundance of Wolf-
Rayet stars depends critically on the galaxy metallicity. Wolf-Rayet features are
commonly seen in the spectra of extragalactic HII regions (Kunth + AA 142, 411;

169, 71; Keel AA 172, 43; Campbell + and Rosa IAU Symp 116). Photometry and spectroscopy of individual stars have been made for many candidates in the surveys above and for supergiants (Humphreys + AJ 89, 1155; 91, 522 and 808; Elias + ApJ 289, 141); IUE spectra of a few OB stars in M31 and M33 surprisingly show winds like in the Magellanic Clouds (Massey + AJ 90, 2239).

3. "Quiet" Massive-Star Formation (SF)

Tracers of SF have been summarized by Lequeux (Spectral Evol. Gal., 57) and comprise direct star counts, colors, far-UV radiation, Balmer line emission, far-IR radiation, radio continuum flux, and even soft X-rays. There are good correlations between these quantities, e.g. far-IR/Hα (Dennefeld in Star-Forming Dwarf Gal. p. 351), far-IR/radio continuum (Helou + ApJ 298 L7; Gavazzi + ApJ 305, L13; Kunth + in Star Forming Dwarf Gal. p 331), soft X-rays/radio continuum althrough part of the X-rays can come from an old population (Fabbiano + ApJ 296, 430; see also Trinchieri + ApJ 290, 96; Palumbo ApJ 298, 259; Cox + ApJ 304, 657; Fabbiano + ApJ 315, 46). Strangely, Turner + (ApJ 313, 644) find too much Balmer line emission in the nucleus of M83 for the observed radio flux. Other correlations are less straightforward to interpret e.g. far-IR/blue light (Iyengar + AA 148, 43; Thronson + ApJ 311, 98), Hα/blue light (Phillips + MN 217, 435) or FIR/2 μm light (other tight correlation between SF (measured by the far-UV flux) and HI has been studied by Donas + (AA 140, 325 and in prep.) but Kennicutt + (AJ 89, 1279) find only a loose correlation between Hα and HI while Rengarajan +(AA 165, 300) shows a good correlation far-IR/HI. SF seems to be proportional to HI mass in M33 (Nakai + PASJ 36, 313). There appears to be a HI surface density threshold for SF (Skillman + AA 165, 45; Guideroni AA 172, 27), and the SF/HI ratio seems somewhat type-dependent, decreasing from Sb to I (Mochkovitch + AA 137, 298; Comte, in Spectral Evol. Gal. p. 81). SF seems also rather tightly correlatedto molecular gas as expected (Rickard + AJ 89, 1520; 90, 1175; Israel + ApJ 283, 81; Sanders + Ap.J. 298, L31; Young + ApJ 304, 443). Chini +(AA 166, L8) find a good correlation between warm dust emission in the far-IR and the total amount of interstellar matter by the 1.3 mm dust emission.

Relations between SF and morphology have been examined by various authors. McCall + (ApJ 311, 548) and Elmegreen + (ApJ 311, 554) find no difference between flocculent and grand-design galaxies. Only strong bars and/or rings enhance star formation (Hawarden + MN 221, 41P; Prieto + AA 146, 297; Hummel + AA 172, 32). The time history of SF in galaxies, mainly I, has been touched upon by e.g. Gallagher + (Ann. Rev. AA 22, 37; ApJ Suppl. 58, 533) and Sandage (AA 161, 89) who confirm that it is does not decrease appreciably with time in the I contrary to the S (see e.g. Isserstedt + AA 167, 11). The proportionality of SF rate to the mass of gas favors self-regulated models like those of Dopita (ApJ 295, L51 and preprint), Shore + (ApJ 316, 663), or Tanaka + (Ap. Sp. Sci. 119, 207); see also Scalo + (ApJ 301, 77) and Cox (ApJ 288, 465).

Extensive surveys of HII regions have been made in several spirals by Viallefond + (AA 154, 357), Courtès + (AA 174, 28), Bonnarel + (AA Suppl. 66, 149) and Kaufman + (ApJ in press) while Hodge (PASP 98, 1095; IAU Symp 116) reports on stellar associations. Detailed properties of individual HII regions are discussed by Viallefond + (ibid.), de Gioia Eastwood (ApJ 288, 175) and Skilmann (ApJ 290, 449); see also Kennicutt (ApJ 287, 116) and Kennicutt + (PASP 96, 944). Their internal velocity dispersion is studied by Roy + (ApJ 300, 264), Hippelein (AA 160, 374), Chu (ApJ 311, 85) and Arsenault + (AJ 92, 576).

4. Chemical Abundances and Evolution

Spectrophotometric data on HII regions and planetary nebulae have been gather-ed by Stauffer (AJ 89, 1702), Dufour + (NASA CP 2349 - 107), McCall (ApJ Suppl. 57, 1), Stasinska + (AA 154, 352, Peimbert + (AA 158, 266), Dufour (PASP 98, 1025),

Fierro + (PASP 98, 1032), Garnett (PASP 98, 1041), Campbell + (MN 223, 811),
Zamorano + (AA 170, 31), Jacoby + (ApJ 304, 490), etc. and their interpretation in
terms of abundances discussed by McCall (ibid) and Dopita + (ApJ 307, 431; application to M101 in Evans + ApJ 309, 544). The very low abundances in IZw 18 are
still unique (discussion in Kunth + ApJ 300, 496). Correlations between elemental
abundances are: excellent for Ne/O (Vigroux + AA 172, 15); poorer for S/O, a
difficult elements, probably positive but loose for He/O, hence one difficulty in
determining the primeval He abundance (Pagel in Paris Conf. on Nucleosynthesis;
Peimbert +, preprint; see also Davidson + ApJ Suppl 58, 321); N independent on O at
low metallicities with fluctuations (primary?) and secondary at low metallicities;
C more or less like N. Models based on recent nucleosynthesis recipes (see Truran,
in Stellar Populations p. 149) account reasonably well for all these features (Wyse
+ ApJ 296, L1; Matteucci + MN 217, 391; 221, 911; Tosi + MN 217, 571; Díaz + AA
158, 60) provided that infall or galactic winds are included (see also Arimoto + AA
164, 260 and Clayton + ApJ 307, 411).

5. Environmental Influences

The influence of the environment on the HI content of galaxies is well established: for early work see Haynes +(Ann. Rev. AA 22,445): S and S0 galaxies in the
center of dense clusters are statistically HI-deficient, probably due to stripping
by the intergalactic medium (Giovanelli + ApJ 292, 404; several papers in The Virgo
Cluster of Gal.; Huchtmeier + AA 149, 118; Guideroni + AA 151, 108; Chincarini + AA
153, 218; Vigroux + AJ 91, 70; Haynes + ApJ 306, 466; Chamaraux + AA 165, 15;
Giraud + AA 167, 25; Warmels, in prep.) However the molecular component is not
affected (Kenney + ApJ 301, L13). Gavazzi + (ApJ 294, L89; 310, 53) show that
galaxies have a higher SF rate in clusters, even HI-poor S. Moss (Starbursts and
Evol.) find Hα-emitting galaxies more frequent in rich clusters and Petrosian + (AA
163, 39) show that Markarian galaxies are preferably in dense clusters. All this
may be the effect of encounters. Clusters also contain an excess of low-surface
brightness or red galaxies (Giuricin + AA Suppl 62, 157; Galagher + AJ 92, 557; van
der Hulst + AA 177, 63): this may by the final result of encounters. Burnstein +
(ApJ 305, L11) find that the mass distribution in S is a function of environment.
Bosma (AA 149, 482) however warns about attributing problems with the IR surface
brightness/HI line width to environmental effects.

6. Starbursts

The discovery of very strong starbursts in galaxies is one of the major events
in recent years. Although some cases were known before, far-IR observations with
IRAS have shown how frequent and strong these starbursts can be; the energy output
can be 100 times larger than normally (Emmerson + Nature 211, 237; Houk + ApJ 290,
L5; Antonucci + AJ 90, 2203 and 91, 56; Soifer + ApJ 283, L1 and 303, L41;
Fairclough, MN 219, 1P; Hill + ApJ 316, L11). As expected the dust that re-radiates
the energy is hotter than usual and the flux is very strong from a few μm to a few
hundreds of μm (Lawrence + ApJ 291, 117; Sekiguchi ApJ 316, 145, etc.). All SF
tracers described previously can be used to find starburst galaxies: colors
(Belfort + AA 76, 1; many Markarian and KISO galaxies are starburst, see e.g.
Deutch ApJ 306, L11); line emission, including Lyman α (Djorgovski in Starbursts
and Gal. Evolution; Deharveng in Starforming Dwarf Gal. p 431); far-UV (Hartman +
ApJ 287, 487; Walsh + MN 220, 453); radio continuum (Maehara + PASJ 37, 451; Karoji
+ AA 155, 43), of course IR radiation, and X-rays (Fabbiano + ApJ 286, 491). Star-
burst galaxies have strong CO emission, with good CO/far-IR and CO/radio relations
(Young + ApJ 287, L65). Very often the starburst is confined to the central regions
and offers interesting morphology (rings, etc. also in CO): eg. NGC 3310 (Telesco +
ApJ 284,544), NGC 2903 (Wynn-Williams + ApJ 290, 108) Arp 220=IC 4553 and NGC 6240
(Rieke + ApJ 290, 116; Neugebauer + AJ 93, 1057; Scoville + ApJ 311, L47), NGC 1068
(Myers + ApJ 312, L39; Telesco ApJ 282, 427; Wynn-Williams + ApJ 297, 607; Evans +
ApJ 310, L15; Planesas + AA, in press); see also Blitz + ApJ 311, 142.

Most far-IR emitting galaxies are believed to be starburst (Elston + ApJ 296, 106; Moorwood + AA 160, 39) but the distinction with Seyferts is not always easy. Starburst blue compact dwarfs can be Seyfert (Kunth + AJ 93, 29) and Seyferts can also show starbursts (Rodriguez-Espinosa + 309, 76 and 555; Wilson + ApJ 310, 121; Heckman in starbusters and Gal. Evol.; Weedman ApJ 291, 72 for NGC 1068). Arp 220 is definitley an edge-on Seyfert (Norris MN 216, 701; de Poy + ApJ 307, 116 and 316, L63) and NGC 4418 might be similar (Roche + MN 218, 19P). Merging galaxies (mergers) can produce starbursts, Seyfert nuclei or both (Sanders + ApJ 312, L5; Byrd + AA 166, 75 and 171, 16; Netzer + AA 171, 41; Hutchings + AJ 92, 6 and 14; Joseph in light on dark matter p. 47 and in starbursts and Gal. Evol.). The presence of shocks yielding excited H_2 is not a good distinction criterion (Joseph + Nature 311, 132). Moorwood (AA 166, 4) notices that the 3.28 μm feature is rare in Seyferts, while Rieke + (ApJ 304, 326) and Rowan-Robinson (Light on dark matter p. 421 and Starbursts and Gal. Evol.) propose interesting far-IR criteria.

It is now clear that galaxy interactions trigger starbursts. This can be seen statistically (Altschuler + AJ 89, 1531; Joseph + MN 209, 111; Lonsdale + ApJ 287, 95; Keel + AJ 90, 708; Sanders + ApJ 305, L45; Young + ApJ 311, L17; Kennicutt + AJ 93, 1011; Brink in Spectral Evol. of Gal, although Lawrence + (MN 219, 687) disagree), or in individual cases (van der Hulst + AA 150, L7; Urbanik + AA 152, 291; Hummel + AA 155, 151 and 161; Lester + ApJ 302, 280; Telesco + ApJ 302, 632; Boissé + AA 173, 229). Ring galaxies fall in the same category (Bonoli AA 174, 57; Appleton + ApJ 21, 566; Wakamatsu + ApJ 315, L23); star formation often occurs in both members of a pair of interacting galaxies (Shaver + AA 148, 143; Kollatschny + AA 163, 31; Madore, in Spectral Evol. of Gal. p. 97). Extreme cases are mergers (Joseph + MN 214, 87) wich may become radiogalaxies (Heckman + ApJ 311, 526); Arp 220 and NGC 6240 might be mergers (Fried + AA 152, L14; Joy + ApJ 307, 110). Various cases are described by Graham + (Nature 310, 213), Bergvall + (AA 149, 475), van Breugel + (ApJ 311, 58) and Joy + (ApJ 315, 480). Mk 171 has been particulary well studied (Augarde + AA 147, 273; Telesco + ApJ 299, 896; Sargent + ApJ 312, L35) and there is a suggestion that only massive stars are formed (see also Olofsson + AA 137, 327). Harwit + (ApJ 315, 28) and Larson (in Starbursts and Gal. Evol.) propose simple models for starbursts in collisions. Jets may also trigger starbursts (Tresch-Fienberg + ApJ 312, 542; Reay + MN 218, 13P) van Breugel (ApJ 293, 83).

Blue compact dwarfs (BCD), clumpy I and Amorphous galaxies are special cases of starbursts. BCDs have been discussed above but see also Gonghalekar + (MN 209, 59), Bergvall (AA 146, 269) and Wynn-Williams + (ApJ 308, 620). Clumpy I are discussed by Heidman (IAU Symp. 116, p 599) see also Klein + (AA 154, 373) and Yin (Sci. Sin. A. 28, 1090). Amorphous galaxies show SF over their whole surface (Hunter + ApJ 303, 171; Arp + AJ 90, 1163; Bothun AJ 91, 507; Melnick + AA 149, L24; Lamb + ApJ 291, 63; Noreau + AJ 93, 1045).

There is abundant literature concerning the two best-studied pure starburst galaxies: M82 (Jaffe + ApJ 285, L31; Weliachew + AA 137, 335; Olofsson + AA 136, 17; Young + ApJ 287, 153; Unger + MN 211, 783; Jones + ApJ 285, 580; Watson + ApJ 286, 144; Houck + ApJ 287, L11; Kronberg + ApJ 291, 693; Seaquist + ApJ 294, 546; Dietz + AJ 91, 758; Lutgen + ApJ 311, L51; Lo + ApJ 312,574; Le Van + ApJ 312, 592; Duffy + ApJ 315, 68; Heckman + AJ 92, 276; Mc Carthy + AJ 92, 264) and NGC 253 (Mc Carthy + ibid.; Hummel + AA 137, 138; Young + ApJ 289, 129; Turner ApJ 299, 312). These galaxies and the most powerful starburst galaxies are favorite places to detect galactic-scale winds (models by Chevalier + Nature 317, 44 and Schiano ApJ 299, 24) and molecules other than CO (Henkel + AA 150, L25; Seaquist + ApJ 303, L67). Strong starburst galaxies often show HI, OH and other molecular absorptions due to the strong concentration of gas and the intense radio continuum, and also OH and H_2O megamasers (at least for OH) from IR pumping by the intense radio field (Schmeltz + ApJ 315, 492; AJ 92, 1291; Hashick + Nature 314, 144; Bann + Nature 315, 26; ApJ 293, 394; 298, L51; 305, 830; 313, 102; Norris + MN 213, 821; 221,

51P; Bottinelli + AA 151, L7; Kazès + AA 152, L9 and in prep.; Unger + MN 220, 1P; Martin + ApJ 308, L7; Claussen + ApJ 308, 592; Gardner + MN 221, 537; Nakai + PASJ 38, 603; Henkel + AA 168, L13).

3. Galactic Dynamics
(P.C. van der Kruit)

Much progress has been made in the area of structure and dynamics of elliptical galaxies. A number of papers have been published on Schwarzschild's method of constructing model galaxies by populating stellar orbits and finding self-consistent models through linear programming techniques. The method itself and results are discussed and presented by Vandervoort (ApJ 287, 475) and Richstone and Tremaine (ApJ 286, 27). Characteristics of orbits are being discussed by Robe (AA 142, 351), Davoust (AA 156, 152) and Levison + (ApJ 295, 349).

A major new development is constituted by the work of de Zeeuw (MN 216, 273) who showed that potentials are separable in Stäckel potentials and that all orbits then have three isolating integrals. This has been followed up by de Zeeuw (MN 216, 599; MN 215, 731) and the Zeeuw + (MN 221, 1001; 215, 713; ApJ 317, 607).

Construction of models for elliptical galaxies by and studies of appropriate distribution functions can be found in Newton (MN 210, 711), Bertin + (AA 137, 26) and Stiawell + (MN 217, 735). Dynamical instabilities and perturbed galaxy models have been studied by Barner + (ApJ 300, 112), Aguilar + (ApJ 307, 97), Merritt + (MN 217, 787) Gerhard (AA 151, 279) and Martinet + (AA 173, 81).

A major issue is the traxiality of elliptical galaxies and the anisotropy of the velocity distributions. Models and distribution functions can be found in Merrit (MN 214, 258; AJ 90, 1029) and for boxshaped ellipticals in Petrou (MN 226, 111). The effects of a black hole in the centre have been investigated in Gerhard and Binney (MN 216, 467) and Norman + (ApJ 296, 20). Tests for triaxiallity are reported by Barney (MN 216, 767) and Bacon (AA 143, 84). The corresponding question of mass-to-light ratio M/L is addressed by Levison + (ApJ 295, 340) and Richstone + (AJ 92, 72). Dissipative formation of elliptical galaxies has been discussed by Carlberg (ApJ 286, 403; 286, 416; 300, L1), including effects of dark halos. Constraints on dark halos from shells around elliptical galaxies (for a review see Anthanassoula + Ann. Rev. AA 23, 147) have been derived by Hernquist + (ApJ 312, 1).

The dynamics of bulges of spiral galaxies is also increasingly better understood now that detailed models have been constructed and tested against observation (Jaris and Freeman ApJ 295, 314 and 324). Binney and Petrou (MN 214, 449) derive distribution functions appropriate to box-shaped bulges. The effects of disks on spheroidicals and formation scenarios for disk galaxies have been discussed by Binney + (MN 218, 734) and Barnes + (MN 211, 753). There is also significant progress in the dynamics of galactic disks now that stellar velocity dispersion can be measured and stability studied observationally (v.d. Kruit and Freeman, ApJ 303, 556). Distribution functions and their evolution have been studied by Villumsen (ApJ 290, 75; 295, 388). The effects of massive black holes in the halos on disk kinematics is investigated by Lacey and Ostriker (ApJ 299, 633), while dynamical evolution due to transient spirals is studied in the models of Caldberg and Sellwood (ApJ 292, 79). Stability questions for our galaxy are adressed by Sellwood (MN 217, 127). Numerical three-dimensional simulation of disks in massive halos are presented by Miller (Celest. Mech. 37, 307). Carlberg + (ApJ 298, 486) describe simulations of dissipative spiral galaxy formation. Three-dimensional orbits and consequences have been studied by Contopoulos + (Celest. Mech. 37, 387; 38, 1; AA 161, 244; 153, 44). Recently Athanasoulla + (AA 179, 23) used disk stability and kinematics considerations in constructing mass models for spiral galaxies.

Lubow + (ApJ 309, 496) studied the dynamics of both stars and gas in spiral density waves, while Hausman + (ApJ 282, 106) presented models for a cloud-derivated gas components in disks with density waves. Stellar 4/1 resonances in disks and the effects on spiral depaumis are discussed by Contopoulos + (AA 155, 11; see also comm. Astrophys. 11, 1). The origin of spiral instabilities was investigated from the point of view of gas accretion (Sellwood + ApJ 282, 61) and of bar potentials (Schwarz, MN 212, 677; MN 209, 93). Athanassoula and Sellwood (MN 221, 195; MN 221, 213) have described instabilities in model disks with a view at stability and bar-,or spiral responses.

Schwarzschild (ApJ 311, 511) has studied the orbit families in a particular model bar and found them restricted to box orbits. Michalodinitrakis + (AA 150, 83) also studied orbits in three dimensions and Pfenniger (AA 134, 373; 141, 171; 150, 97; 150, 112) has addressed the question of resulting velocity fields, instabilities and bulge dynamics. Teuben and Sanders (MN 212, 257) define dynamical rules to which realistic barred galaxy models conform. Weinberg (MN 213, 451) studied the angular momentum transfer between bar and halo.

Dynamical studies of galactic warps are producing constraints on their possible origin. Sparke (MN 211, 911) has produced models that seem to enclude bars and suggests weakly barred retrograde figure-rotating dark halos.

Interest in polar rings is also increasing. Whitmore + (ApJ 314, 439) find weak constraints on dark halo flattening. Coupled orbital characteristics can give simular constraints (Katz + AJ 89, 975). The stability appears related to its self-gravity in models by Sparke (MN 219, 657). Binney + (MN 226, 144) also discuss the flattening of dark halos.

There have been a number of papers addressing the question of dynamical friction on galaxy satelites and the corresponding orbit decay and capture (Weinberg, ApJ 300, 93; Tremaine + MN 209, 729; Quinn + ApJ 309, 472; Byrd + MN 220, 619; Bontekoe + MN 224, 349).

There is major interest in possible revisions of Newtonian gravity to explain flat galaxy rotation curves without dark matter (Milgrom, ApJ 287, 571; Sanders, AA 136, L21; AA 154, 135). Futher discussion has been given by Kuhn + (ApJ 313, 1), while Christodoulou + (ApJ 307, 449) and Hernquist + (ApJ 312, 17) show apparent inconsistencies with observed warps in spirals and shells in ellipticals.

4. Groups and Clusters of Galaxies
(G.A. Tammann)

Deeper and deeper galaxy surveys reveal larger structures. The scale length over which galaxies are distributed uniformly has not been reached yet. There is no question that the clustering of galaxies on all scales contains still the most decisive clues for galaxy information. Clusters are also the best sites for studying galaxy evolution, i.e. evolution in the universe in general. The clustering of galaxies and its effect on the Hubble expansion field offer also a unique chance for large amounts of hypothetical non-baryonic mass wich might close the universe (cf. IAU Symp. 117: Dark Matter in the Universe). The cosmological aspects are covered in the Report of Commission 47. Yet the detailed understanding of nearer groups and clusters lies at the basis of the understanding of the large-scale structure as well as of the cosmogony of individual galaxies.

The following Conference Proceedings have appeared since the last report: Clusters and Groups of Galaxies, Trieste, 1983 (cited here: Clusters and Groups); The Virgo Cluster, Garching, 1984; Galaxy Distances and Deviations from Universal Expansion, Kona-Hawaii, 1986/cited here: Galaxy Distances). Not available at the time of writing are the Proceedings of: IAU Symp. 130: Evolution of Large-Scale Structures in the Universe, Balatonfüred, 1987; 3rd IAP Astrophysics Meeting: High Redshifty and Primeval Galaxies, Paris, 1987: as well as the lectures (by A. Fabian, M. Geller, and A. Szalay) of the Saas-Fee Course: Large-Scale Structure of the Universe.

I. THE LOCAL GROUP

Distances to LG galaxies have been considerably improved by various methods, i.e. RR Lyr stars (van den Berg and Pritchet) and novae (Cohen) in M31, population II giants in M31 and M33 (Mould and Kristian), and most importantly infrared photometry of Cepheids (Weech +, McAlary +, Madore +, Freedman +, Visvanatahan, - for detailed references see IAU Symp. 124, 151), the latter method being exceptionally sensitive to absorption and metallicity effects. The distances to LMC and SMC are reviewed by Feast (Galaxy Distances p 7). A mass of the LG of 3.10^{12} $M_\odot$ and hence - $M/L_B = 13$-20 has been determined by Sandage (ApJ 317, 557) from its decelerating effect on very nearby field galaxies and from the Kahn-Woltjer paradox; the value is compatible with Lynden-Bell (IAU Symp 106, 461). Mishra (MN 212, 163) argues that the mass of the Milky Way is at least half of that of M31.

The motion of the Sun relative to the centroid of the LG has been redetermined by Richter + (AA 171, 33). An X-ray survey of the LG was discussed by Helf and (PASP 96, 913).

II. GROUPS

Galaxy groups were reviewed by Geller (Clusters and Groups p 353) and Vennik (AN 307, 157). Their M/L_B ratios of 220 (see also Heisler + ApJ 298, 8; Mezzetti + AA 143, 188) and 65, respectively, show a clear discrepancy with the LG (cf. Trimble, Nearly Normal Galaxies p 313), perhaps because of the inclusion of unbound groups (the values here are reduced to $H_0 = 50$). Schneider + (AJ 92, 742) conclude that the mass determined from the galaxies'rotation is sufficient to bind small groups; this is clearly the case for compact groups (Williams + ApJ Suppl. 63, 265). Byrd + (ApJ 289, 535) take unbound group members as the explanation of apparent non-Doppler redshifts in groups (cf. Arp + ApJ 291, 88). Spiral galaxies in groups seem to have normal HI content (Huchtmeier, Clusters and Groups p 221; Giuricin + AA 146, 317). Optical properties of group members may depend somewhat on compactness (Guiricin + AA Suppl 62, 157). X-ray emitting gas has been detected in the group elliptical NGC 5846 (Biermann + Clusters and Groups p 395); some of the hot gas found in three groups by Bahcall + (ApJ Lett 284, L29) may lie in the ultragroup region. The discovery of a large, intergalactic HI cloud in the Leo group by Schneider + (1983) ·has drawn considerable interest; it is now interpreted as a ring surrounding two galaxies (Schneider ApJ Lett 288, L33). Dickel + (Clusters and Groups p 389) suggest an HI cloud also in Seyfert's group. Narrow-angle tailed radio galaxies in poor groups are discussed by Burns + (BAAS 18, 707). ^{12}CO emission in the UMa group is taken as evidence for galaxy interaction (Odenwald ApJ 310, 86). Group rotation (Williams ApJ 311, 25) and preferential orientation of group members (Rosino + Astrofizika 19, 834) are suggested in individual cases; for a ring-shaped group see Danks + (AA 139, 455). Karachentseva + (AA Suppl. 60, 213) published an atlas of dwarf galaxies in the M81 group. Cepheid distances have become available for two group members, NGC 300 (Graham AJ 89, 1332; Freedman, Galaxy Distances p 21) and tentatively M101 (Cook + ApJ Lett 301, L45). A catalogue of 100 compact groups (Hickson + Clusters and Groups p 367) may contain ~50% of chance superpositions (Mamon ApJ 307, 426). The effect of gravitational lensing is discussed by Hammer + (AA 155, 420). The time scale of dynamical evolution of (compact) groups seems amazingly short (Williams ApJ 290, 462; Williams + ApJ Suppl. 63, 265; Barnes MN 215, 517; Navarro + MN 228, 501). Fricke + (Mitt Astr. Ges. 67, 383) find Seyfert and AGN galaxies to lie preferentially in compact groups.

III. THE VIRGO CLUSTER

Much work has been devoted to the Virgo cluster. A complete catalogue of morphologically selected members to B $<$ 18^m has been published, the majority of them being dEs (Bingeli + AJ 90, 1681). A large body of redshifts is available for the cluster members, mainly from the work of Huchra (cf. The Virgo Cluster p.181). Surface distribution, type segregation (cf. also Salpeter, Galaxy Distances p.159) with a pronounced lack of Im's towards the cluster center, and subclustering are discussed; even the E core of the cluster seems dynamically young (Bingeli + AJ 94, 251). In the outerpart spirals are still falling in (Tully + ApJ 281, 31). As compared to the field, cluster spirals are HI deficient (e.g. Balkowski + The Virgo Cluster p.37; Giovanelli ibid.p.67; Huchtmeier ibid.p.23; Warmels ibid.p.51), ram pressure (van Gorkum + ibid. p.61) and turbulent viscous stripping (Haynes + ApJ 306, 466) being proposed as the cause. Both these mechanisms are compatible with the normal CO content of these galaxies (Kenny + ApJ Lett 301, L13). HI deficiency is accompanied by color (Kennicutt + The Virgo Cluster p.91 and 227; Guideroni AA 151, 108) and geometrical (MacGillivray + The Virgo Cluster p.217) pecularities of the cluster members; their present star formation rate from IRAS data is lower than in their field counterparts (de Jong, The Virgo Cluster p.111). The radio continuum is field-like except in the very core; M87 may be a wide-angle tail source (Kotanyi The Virgo Cluster p.13). Also the X-ray emission of Virgo and field galaxies is comparable, except for M87 wich contains ~10^{12} $M_\odot$ of hot gas, comparable to NGC 4696 in the Centaurus cluster (Forman + The Virgo Cluster p. 323). The center of very extended, hard X-ray component may possibly coincide with the cluster center of M87 (Smith + ibid. p.345).

While Bothun + (AJ 90.697) suggest Virgo dEs to be stripped dwarf irregulars, Binggeli (Star-Forming Dwarf Galaxies p.53) finds this transaction to be improbable. Hoffman + (preprint) find no evidence for dIm´s being distributed more uniformly than spirals; this would be expected from biased galaxy formation (Dekel + ApJ 303, 39). Even in the field the number ratio low/high surface brightness galaxies seems to remain constant (Bothun + ApJ 308, 510; Binggeli, Sandage, and Tarenghi, work in progress; however Davis + ApJ 299, 15).

The luminosity function (LF) of E and S galaxies is roughly Gaussian, while the dE's have an flatter LF (Sandage + AJ 90, 1759). The shape of the LFs, separated according to Hubble types, is nearly the same in the Coma cluster and in the field, however the LF over all types changes with the type mixture, i.e. with the surrounding galaxy density (Binggeli, Nearly Normal Galaxies p.195). Several independent distance indicators seem to converge towards a Virgo cluster distance near $(m-M)^0$ = 31.6, i.e. globular clusters (van den Berg + AJ 90, 595), novae (Prichet + ApJ 318, 507), the L/σ relation of spiral bulges (Dressler ApJ 317, 1), and the Tully-Fischer relation of a nearby complete sample of 81 Virgo spirals (Kraan-Korteweg + Basel Preprint No. 26), while supernovae Ia (Tammann, IAU 124.151) may indicate a somewhat higher value; cf. however de Vaucouleurs (The Virgo Cluster p.413).

IV. OTHER CLUSTERS OF GALAXIES

The Fornax cluster´s population was studied by Caldwell (AJ in press); a survey on a large-scale plates is underway (Sandage and Ferguson). Surface photometry is carried out by Phillips (Cluster and Groups p.183). Photometry of SO members is provided by de Carvalho + (AA 149, 449). Redshifts were determined by Richter + (AA Suppl 59, 433).

The structure and dynamics of the Coma cluster were reviewed (Peach, Clusters and Groups p.89; Schipper PhD thesis). Millington + (MN 221, 15) adopt a model of constant velocity anisotropy to derive a M/L ratio of 240. The cluster mass is a few times 10^{15} $M_\odot$ and M/L roughly 100 (Gerbal + Clusters and Groups p.147; The + AJ

92, 1248). This is a good agreement with the mass inferred from X-ray data (Kriss, Clusters and Groups p.313; Cowie + ApJ 317, 593). For three groups within the Coma-A1367 supercluster Williams (Clusters and Groups p.375) derives M/L_B = 30-40. Fichett + (preprint) suggest the cluster to be dynamically young. The specific energy of the intracluster medium is studied by Gerball + (AA 158, 177). The radial X-ray profile of the cluster is analyzed by Chanan + (ApJ 287, 89). Branduardi + (Space Sci Rev 40, 647) find an excess of X-ray sources in the cluster background. Upper limits on the far-UV emission from the cluster center are set by Holberg + (ApJ 292, 16). Cordey (MN 215, 437) discusses two extended radio sources in the cluster. Gavazzi + (ApJ 310, 53) find that E/SO galaxies in the cluster are more likely to be radio sources than in the field; the effect is much more pronounced for spirals; the authors conclude that the same mechanism wich causes the HI deficiency of spirals triggers also bursts of star formation. Blue disk galaxies in Coma, similar to those observed in high-redshift clusters, require a mechanism for enhanced stare formation (Bothun ApJ 301, 57). Several determinations of the distance modulus difference Coma-Virgo yield values of $\Delta(m-M)^0$ = 3.6-4.0 (Dressler ApJ 281, 512 and 317, 1; Vader ApJ 306, 390; Lucey MN 222, 417; Kraan-Korteweg + Basel Preprint No 26).

The Centaurus cluster may form together with the Hydra I cluster and clustering in Atlas a large supercluster (Hopp + AA 61, 93; da Costa + AJ 91, 6). This is of great interest because the "big attractor" is to be expected in this general direction in order to explain the MWB dipole in conjunction with our Virgocentric flow (see below); of course the motion towards the MWB pole could also be due to more than one attractor. Lucey + (MN 221, 453 and 222, 427) find in Centaurus two galaxy concentrations at 3000 and 4600 km s^{-1}, but believe them to be at the same distance, contrary to Lynden-Bell + (preprint). In either case the finding requires large streaming velocities.

The Hydra cluster is in several respects similar to Virgo, but apparently more evolved, perhaps it lies in a void and infall of spirals has ceased (Richter, The Virgo Cluster p.427). The structure of the Perseus (super-) cluster has been investigated by Focardi + (AA 136, 178), Egikyan + (Astrofizica 23, 5), Tanaka (Publ Astr Soc Jpn 37, 481), and Giovanelli (BAAS 17, 581), and that of the Cancer cluster by Bicay + (BAAS 17, 578) and Perea + (MN 222, 49); for A149 see Nemiroff + (AJ 90, 163) and Mazure (AA 157, 159). Studies of non-random orientations of spiral galaxies in the Virgo complex remain inconclusive (MacGillivray + AA 15, 269; Anderson + Clusters and Groups p.63; Flin + ibid.p.65). Dodd + (Ap Space Sci 123, 145) find galaxy alignment in the Shapley-Centaurus cluster. Flin (Clusters and Groups p.163) suggests preferred ellipticities in individual clusters. Large-scale alignment of neighboring clusters, i.e. the Binggeli effect, was questioned by Stubble + (AJ 90, 582), see however Rhee + (AA 183, 217). Luminosity segregation in 0004.8-3450 is interpreted as the effect of cannibalism (Capelato + Ap Space Sci 108, 363). Wakamatsu + (AJ 92, 700) have discussed the interacting ring galaxies in the Hercules cluster. The fraction of emission line galaxies in A634 is surprisingly high (Stepanyan Astrofizica 21, 245), while Dressler + (ApJ 288, 481) find from a large sample that the emission line fequency is much higher in field galaxies

Photometry was provided in A1213 (Egikyan + Astrofizika 21, 21) and A2052 (Couture + J R Astr Soc Can 78, 211); for other clusters see Le Fèvre + (AA Suppl 66, 1), Yamagata + (Publ Astr Soc Jpn 38, 661; Ann Tokyo Obs Ser II 21, 31; Ap Space Sci 118, 459), Melnick + (AJ 89, 1288), and Couch + (ApJ Suppl 56, 143); for 175 brightest cluster galaxies see Hoessel + (AJ 90, 1648). The luminosity functions in several Abell clusters were derived by Oemler + (AJ 93, 519) and Lugger (ApJ 303, 535). HI and radio continuum mapping of the Hercules cluster was carried out by Salpeter + (ApJ 292, 426) and Dichey (ApJ 284, 461). Nonthermal radio sources in clusters were reviewed by Fanti (Cluster and Groups p.185). Faraday rotation measures of background radio sources in A2319 show a large-scale magnetic field in the cluster (Valée + AA 156, 386).

Of great interest is the discovery of gigantic luminous, somewhat knotty arcs in A2218 and 2242-02 (Lynds + BAAS 18, 1014) and in A370 (Soucail + AA 172, L14).

A distance of the Hercules cluster has been published by the Vaucouleurs + (ApJ 297, 23) and Buta + (ApJ Suppl 62, 283). The distances to ten clusters at $4000 < v < 11000$ km s^{-1} (Aaronson + ApJ 302, 536) have statistically been improved by Bottinelli + (AA 181, 1) and Kraan-Korteweg + (ApJ in press).

A compilation of cluster redshifts is given by Schmidt (AN 307, 69), see also Postman + (AJ 90, 1400). 418 high-redshift clusters have been found in a systematic survey by Gunn + (ApJ 306, 30). The automated detection of clusters is described by Dodd + (AJ 92, 706). A much needed test of the reliability of the Abell catalogue has been provided from clusters in the Shane-Wirtanen catalogue (Shectman ApJ Suppl 57, 77) with somewhat alarming results. Viral masses of clusters, in addition to those already mentioned, have become available for A2197 and A2199 with M/L ~ 100-250, but the double cluster method indicates M/L ~ 50 (Gregory + ApJ 286, 422 and 305, 580). If the Perseus cluster is a flat system seen nearly edge-on its mass corresponds to M/L ~ 50 (Tanaka Publ Astr Soc Jpn 37, 481). A value of M/L = 68 is found for the poor cluster MKW 4 from the viral theorem and X-ray data (Malamuth + ApJ 308, 10). Valtonen + (ApJ 303, 523) find groups and clusters largely unbound and hence still smaller M/L values.

An illuminating review on the evolution of cluster galaxies has been given by Dressler (Ann Rev Astr Ap 22, 185; Clusters and Groups p.117). More recent investigations have concentrated mainly on three questions: 1) Are cluster galaxies HI-deficient? The answer is positive (Giovanelli + ApJ 292, 404), positive for early-type spirals only (Dressler ApJ 301, 35), respectively negative (Bothun + ApJ 291, 586). Besides gas stripping enhanced star formation (Gavazzi + ApJ Lett 294, L89; Tammann, Star-Forming Dwarf Galaxies p.41) may be a cause of gas deficienty. 2) Have galaxies in very distant clusters higher star formation rates than locally? The answer is positive (Butcher + ApJ 285, 426 and Nature 310, 31; Schild ApJ 286, 450; Lavery ApJ Lett 304, L5; Ellis + MN 217, 239), occasionally (Dressler + Space Telsc Sci Inst Prepr 130, 65), marginally (Thompson ApJ 300, 639 and 306, 384, Sharples + MN 212, 687; Couch + MN 213, 215; Dressler + ApJ 294, 70), restricted to a few galaxies (Lilly + MN 217, 551), and negative (Laurikainen + New Aspects of Galaxy Photometry p.309). Tyson (AJ 92, 691) proposes enhanced star formation in galaxies near to QSOs. 3) How do first-ranked cluster galaxies evolve? Brightest cluster galaxies are not the extreme members of a statistical population (Bhavsar + MN 213, 857; however Morley PASP 96, 874). If they are cDs they are ~0^{m}5 brighter than "normal" first-ranked galaxies (Lugger ApJ 286, 106). Hoessel + (AJ 90, 1648) find a strong correlation between the structure and luminosity of the brightest cluster galaxies. Their specific structural properties suggest environmental effects (Schombert ApJ Suppl 60, 603) like bound satellites (Cowie + ApJ Lett 305, L39), accretion flows (Lindblad + New Aspects of Galaxy Photometry p.337) or mergers (Tonry AJ 90, 2431; Merrit ApJ 289, 18; Malamuth ApJ 291, 8; Hoessel + ApJ 293, 94). But globular clusters contradict the merging of spirals to form ellipticals (van den Berg, Cluster and Groups p.139; cf. also Muzzio + ApJ 285, 7).

The intracluster medium has been investigated by means of radio tail galaxies (O'Dea + AJ 90, 927 and 954) and distorted radio sources (Hanisch + AJ 90, 1407). extended Hα feature around NGC 4438 is interpreted as due to the bow-shoch caused by the motion of the galaxy in the medium (Chincarini + AA 153, 218). The X-ray emission of clusters requires cooling flows (Arnaud + MN 21, 981; Jones + Clusters and Groups p.319; Bertschinger + ApJ Lett 306, L1; for reviews see Fabian + Nature 310, 733; Fabian, Mitt Astr Ges 65, 123; Steward + ApJ 285, 1) Hendriksen + (ApJ 292, 441) find the original isothermal model to be non-physical (see however Gerbal + AA 146, 119; Miller MN 220, 713). The X-ray properties of individual clusters are discussed by many authors, e.g. of the clusters in Perseus (Branduardi-Raymont + Adv Space Res 5, 133) and Pegasus (Canizares + ApJ 304, 312); typical X-ray properties of clusters are discussed by Kowalski + (ApJ Suppl 56, 403), Ulmer + (Clusters and Groups p.307), Beers + (ApJ 283, 33), Kalinkov (New Aspects of Galaxy Photmetry p.331), and Quintana + (AJ 90, 410). Star formation is an accreting cluster is found only in the Perseus cluster (Romanishin ApJ 301, 675; cf. als Silk + ApJ 307, 415). Thermal instabilities in cooling flows are the cause for the optical emission filaments in and around a number of galaxies (Fabian + ApJ 305, 9). A radio tail may be identified with an X-ray source (Burns + ApJ 291, 611). From high-excitation Fe lines in the X-ray gas an abundance of 0.4-0.5 times the solar value is derived (Rothenpflug + AA 144, 431; Singh + ApJ Lett 308, L51); this seems high if the acreting gas is assumed to be primordial. Theoretical models of cluster evolution has been presented by several authors (e.g. Peebles, Clusters and Groups p.405; Bhavsar + ibid.p.415; Salpeter Ann NY Acad Sci 422, 95; Yabushita + MN 213, 117; Allen + MN 216, 155; Cavaliere + ApJ 305, 651; Kaiser + MN 222,323).

Superclustering has been reviewed by Oort (Clusters and Groups p.1; Large-Scale Structure of the Universe p.209). He has proposed to detect super-pancakes by matching absorbtion lines in neighboring QSOs (AA 139, 211). Very large-scale structure is revealed in the redshift distribution of galaxies at $B \sim 22^m$(Koo + IAU Symp 124, 383). No clear filamentary structure has been found in a deep galaxy sample (MacGillivray + J Astr Ap 7, 293; see also Fry ApJ 306, 366). But filamentary structure around the Coma cluster was pointed out by Fontanelli (AA 138, 85). "Bubble" structures of diameters of 5000km.s^{-1} in a volume containing the Coma cluster are detected in the extended CfA redshift survey (Geller + IAU Symp 124, 301). Haynes + (ApJ Lett 306, L59) find a connection between the Pisces-Perseus supercluster and the Virgo complex (cf. also Haynes + Galaxy Distances p.117). The Hercules and Perseus superclusters were studied by Moles + (MN 213, 365). Superclusters in Zwicky's classical sence, i.e. clustering of clusters, were searched for and detected in the distribution of Abell clusters by Bahcall + (ApJ 270, 20; also Burns IAU Symp 124, 319). The superclusters are elongated in the redshift direction, possibly indicating large-scale deviations from a quiet Hubble flow (Bahcall + ApJ 311, 15; Bahcall IAU Symp 124, 335).

The infall velocity of the LG towards the Virgo cluster was determined to be 220 ± 50 km s^{-1} (Tammann + ASpj 294, 81); more recent determinations have found similar values (de Freitas Pacheco AJ 90, 107; Kraan-Korteweg, The Virgo Cluster p.397; Visvanathan, Galaxy Distances p.99). A more complex flow pattern has been suggested by de Vaucouleurs + (ApJ 297, 27). A quadrupole moment of the velocity field in the Virgo complex could be due to the "great attractor" in the Hydra-Centaurus direction (Lilje + ApJ 307, 96); this apex direction was originally proposed to explain the MWB dipole (Shaya ApJ 280, 470; Sandage + Large-Scale Structure of the Universe p.127). A corresponding dipole moment has been suggested for the distribution of IRAS galaxies (Meiksin + AJ 91, 191; Yahil + ApJ Lett 301, Distances of nearly 400 ellipticals indicate a large-scale flow towards an apex at 4500 km s^{-1} in the direction of the Centaurus cluster with a peculiar motion of the Sun of $\sim$570 km s^{-1}; the superimposed random velocities are <245 km s^{-1}; the flow, however, if combined with our Virgocentric infall velocity does not explain the MWB dipole (Lynden-Bell ApJ in press), particularly for the last two paragraphs see also the Report of Commision 47.

5. Quasars and Related Topics
(S. D'Odorico)

It is not possible in this limited space to discuss all of the many interesting results which have appeared in this widely defined subject. The aim of this report is just to provide to a newcomer to the field a reasoned list of the most important contributions in the various subtopics. For a complete view see also the progress reports on galaxies, radiostrouomy and cosmology in this volume.

I. CONFERENCES, WORKSHOPS AND CATALOGUES

Three IAU Symposia were closely related to QSOs: IAU No. 119 "Quasars", IAU No. 124 "Observational cosmology" and IAU No. 130 "Evolution of Large Scale Structure in the Universe". Proceedings of other meetings on the subject include the 24th Liege Meeting "QSOs and Gravitational Lenses", the workshop in Manchester "Active galactic nuclei", the 7th Santa Cruz Workshop "Astrophysics of active galaxies and QSOs", the Tata Institute school on "Extragalactic energetic sources", the NRAO workshop "Physics of energy transport in extragalactic radio sources",the Trieste Meeting "Structure and evolution of active galactic nuclei" and the NOAO workshop "Continuum emission in active galactic nuclei". No references to articles published in the Proceedings of these conferences are made in this report. Three new editions of QSO catalogues have appeared in 1987: in ApJ Suppl. Ser. 62, 751; as ESO SR No. 5 and as an Asiago-Padova Observatory Contribution. Other related catalogues are those of Markarian galaxies (ApJ Suppl. Ser. 62, 751) and of BL Lac objects (AJ 93, 1).

II. SURVEYS AND SYSTEMATIC INVESTIGATIONS

In the last three years the sample of QSOs available for statistical studies has significantly expanded with the results of several new surveys in selected areas of the sky being published. Both color and grism -based techniques have been employed (ApJ 283, 50; ApJ 287, L3; MN 213, 485; ApJ Suppl 57, 523; AJ 89, 1658; AJ 90, 987; Nature 314, 238; ApJ 295, 94; MN 216, 589 and 623; MN 218, 445; MN 220, 1; AA Suppl 63, 1; PASP 98, 285; ApJ 310, 518; MN 223, 87; AA Suppl 67, 551; AJ 92, 203; ApJ 314, 129; ApJ 316, L1).

The new results have been used to discuss the problems of the density and luminosity distribution of QSOs at different epochs and that of their evolution. These topics have been also addressed in ApJ 298, 448; ApJ 299, 109 and 799; MN 213, 389; AA 170, 37; PASP 38, 611; ApJ 300, 224; ApJ 311, 156; ApJ 312, 589; AA Suppl 67, 267; Nature 325, 131; MN 227, 717 and ApJ 316, L5.

It is worth noting that the discovery of several QSOs with z > 4 has been reported while this review was being written. This suggests that the techniques which have been employed so far have not been very effective in the detection of objects at high redshifts. We can now expect further progress in this area, and it remains to be seen whether the generally accepted notion of a decrease of QSO in number density at large z will eventually be confirmed.

The QSO data base has also benefitted from other systematic investigations (Act. Ast. 34, 117; MN 211, 105; ApJ 285, 584; AA Suppl 61, 191 and 225).

The research mentioned above has been carried out mostly at optical and radio wavelengths, but important results on QSOs have been also obtained in the X-ray band (ApJ 284, 491; AJ 89, 1658; ApJ 292, 357; J AA 6, 49; MN 220, 51; Bologna X-Ray Astr.'84, 419 and 463; ApJ 297, 177; ApJ 299, 814; ApJ 303, 614; ApJ 308, 53; ApJ 310, 291; ApJ 314, 111; ApJ 318, 188) and in the infrared (MN 212, 631; ApJ 308, 815 and L1; ApJ 309, L69).

III. QSOs NEAR GALAXIES AND GALAXIES NEAR QSOs

Special QSO surveys have been conducted in the vicinity of bright galaxies and in clusters either to prove an association (with success, according to the authors of some of these investigations) or simply to obtain background sources to be used to study the interstellar and intergalactic medium (MN 226, 58; AJ 89, 958; MN 210, 373; AA 138, 408; ApJ 283, 59; MN 211, 443; ApJ 285, 44; Chin AA 8, 238; ApJ 285, 355; AA 151, 264; ApJ 288, 82; ApJ 288, 201 PASP 97, 1149; MN 221,897; MN 222, 787; ApJ 319, 687 and 693).

The opposite approach, that is to search for faint galaxies near QSOs, has tempted more observers. Many studies have dealt with the possible presence of QSO "host" galaxies, and many detections have been reported (AA 138, 337; ApJ Suppl 55, 533; ApJ 283, 64; ApJ 287, 555; ApJ 293, 120; ApJ 295, L27; MN 214, 241; ApJ 291, L37; AJ 90, 1642; ApJ 298, 275; ApJ 306, 64; ApJ 311, L1; ApJ 312, 518; ApJ 316, 584; ApJ Suppl 62, 681; AJ 93, 255). Most of these studies have been confined to objects at z < 1 and have found that different galaxies can host a QSO, the morphological type being possibly correlated with the radio properties of the QSO.

To the study of faint galaxies in the direction of QSOs belong also deep photometry of galaxies in QSO fields (PASP 97, 684; MN 223, 173; ApJ 319, 28; ApJ 62, 681 and ApJ 139, L39) and the successfull searches for galaxies close to the line of sight to the QSO and at the redshift of metal absorption systems seen in the spectrum of the QSO (AA 155, L8; MN 210, 873; MN 223, 173; AA 161, 206; AA 175, L1; AA 180, 1 and ApJ 319, 683).

IV. CLUSTERING

A number of authors have addressed the question of clustering of QSOs at various scales and at various epochs, starting from the results of a survey or from a statistical analysis of existing data. Weak or no clustering was generally found, but with larger,more complete samples becoming available, the case for strong clustering at high redshifts is building up (AA 136, 69; MN 214, 905; MN 218, 139 and 587; ApJ 311, 578; MN 227, 1; MN2 27, 739; MN 227, 921; Nature 326, 773). Discussions on QSO pairs in particular can be found in AA 143, 451; Nature 317, 413 and Nature 323, 185 and 186.

V. THE EMISSION SPECTRUM

The emission lines in the spectrum have been used to investigate the physical conditions and the composition of the matter near to the cores of QSOs. Note that the studies of Seyfert and other active galaxies, not referenced here in detail, are strictly related to this topic as these objects are likely to represent QSOs of low luminosity at small redshifts. QSO spectra, their modelling and interpretation, are treated in NASA CP-2349, 133; ApJ 283, 70; ApJ 284, 497; ApJ 288, 94; ApJ 290, 394; ApJ 291, 128; PASP 97, 966; MN 218, 331;AJ 91, 226; ApJ 295, 394; AA 145, 324; ApJ 298, 114; AA 156, 121; MN 225, 55; ApJ 302, 56; ApJ 308, 805; MN 226, 629; ApJ 310, 40 and ApJ 314, 145.

VI. THE BROAD ABSORPTION LINE QSOs (BALs)

A proportion of QSOs possibly as high as 10% show broad absorption lines in their spectra indicative of an outflow of matter at velocities which can reach a significant fraction of the velocity of light. Detailed studies on these lines share light on the mechanism and energetics of the central power source while studies on the frequency of these objects versus "normal" QSOs have been used to try to understand which evolutionary stage they represent or under which conditions they develop, both questions being far from being fully resolved (ApJ 282, 33; MN 211, 813; ApJ 294, L1 and L73; ApJ 296, 416; ApJ 302, 64; ESA SP-263, 627; AA 177, 42; ApJ 310, L1; ApJ 317, 450 and 460).

VII. ENERGY DISTRIBUTION ANS MODELS OF THE CENTRAL POWER SOURCE

Studies on the overall energy distribution in the continuum and on the central engine have been presented in AJ 89, 1275; JAA 5, 495; MN 213, 97; MN 216, 63; AJ 90, 405; ApJ 288, 32; ApJ 289, 451; AJ 90, 998; ApJ 296, 423; AZh 62, 662; ApJ 300, 216; MN 224, 257; MN 226, 601; ESA-SP 263, 601; ApJ 313, 164 and 171.

III. GRAVITATIONAL LENSES

More than 60 papers have been published on gravitational lenses in the last three years, testifying the amount of work which went into this subject which is 8 years old only. Apart from the interest of studying in the real world an effect predicted by the theory of gravity, GL are potential sources of information on the lensing objects themselves, galaxies or clusters at high redshifts, provide a confirmation of the cosmological distance of QSOs and may affect the statistics of QSO counts. Reports on the actual discoveries, new data and discussions of individual cases can be found in ApJ 282, L1; MN 210, L1; AA 138, L19; ApJ 283, 512; AJ 90, 691; Nature 316, 102; AJ 90, 1399; ApJ 294, 66; AA 149, L13; ApJ 300, 209; AA 158, L5; Nature 321, 139; ApJ 303, 605; AA 166, 119; ApJ 312, 45; Nature 3 16, 268 and Nature 329, 696.

The rise and fall of the GL identification 1146+111 A,B (most likely a close QSO pair) is told in as much as 10 published papers (see eg Nature 321, 142 and 569; Nature 323, 784; ApJ 313, 28) and provides instructive reading for anyone dealing with the interpretation of observational data.

Theory of GL for different lensing objects and different configurations is discussed in Nature 310, 112; ApJ 287, 26; 4th Grossman Meeting p1549; ApJ 310, 568; ApJ 312, 22; MN 224, 283; ApJ Suppl 68, 223; ApJ 313, 13; AA 174, 361; ApJ 315, 283; ApJ 317, 11; ApJ 319, 9. The effect of GL on the luminosity function and in general on the density of QSOs is treated in MN 215, 639; ApJ 300, 68; AA 179, 71 and 80; ApJ 316, L7; ApJ 318, L1.

IX. QSOs AS PROBES OF MATTER AT HIGH REDSHIFTS

With more large optical telescopes and efficient high dispersion spectro graphs now in operation, it has become possible to observe a relatively large number of QSOs at high dispersion, and then to study in detail through the absorption lines in the spectrum the distribution and the metal content of intervening matter at high redshifts. This technique has been very efficient and we may expect that it will be one of the main areas of application for telescopes of 8m in diameter or larger now under construction. We can distinguish between studies centered on absorption lines originating in HI clouds at redshifts smaller than the emission redshift of the QSOs (the so-called Lalfa forest) and studies centered on the metal lines of low and high ionization. In the first category fall the papers in AA 144, L17; ApJ 292, 58; MN 218, 25P; Astrofiz. 24, 321; MN 220, 1; ApJ 309, 19; ApJ 310, 583; MN 224, 13; ESA-SP 263, 435; ApJ 311, 610; AJ 92, 247; MN 224, 675; PASP 98, 1140; MN 224, 13P; MN 225, 1P; ApJ 316, L59; ApJ 319, 14 and 709.

To the metals systems, their ionization state, abundances and nature are dedicated the following references: AA 145, 59; ApJ 292, 362; ApJ 293, 387; ApJ 301, 116; MN 220, 429; AA 168, 6; AA 169, 1; ApJ 303, L27; ApJ 307, 504; ApJ 310, 40; ApJ 311, 610; ApJ 312, 50; ApJ 315, L5; ApJ 320, L75. For metal systems where the intervening galaxy has been identified, see also the work referred to in Chap. III.

X. 3C 273

We do not discuss in this report studies of single QSOs except for 3C 273. The first QSO to be discovered still attracts considerable attention because it allows very detailed studies due to its high brightness and to the small distance. For example the jet structure, which is related to the central power source, can be investigated in 3C 273 in much more detail than in any other object (Nature 314, 425; MN 216, 679; Nature 318, 343; AA 154,15; JAA 7,225; ApJ 289, 109; ApJ 313, 136). Studies at different wavelengths have been reported in AA 136, 351 and AA 182, L1 (X-rays), ApJ 283, 329, PASP 97,118, PASP 97,395, Nature 316,524, AJ 90, 2474, AJ 92, 1030 (UV and optical) and ApJ 298, 114 (radio). The overall spectrum is discussed in AA 140, 341 and Nature 323, 134. Models of 3C 273 are presented in AA 139, 289 (as the effect of a GL), in ApJ 285, 64 (as the effect of a continuous beam) and in ApJ 298, 114 (to explain radio outbursts).

XI. QSOs AS STANDARD CANDLES

Accurate optical and radio positions of a QSO have been obtained as a step to establish an extragalactic reference frame (AJ 93, 261). A study on aberration in QSOs has confirmed the constancy of c on a scale comparable to the size of the universe (ApJ 295, 24).

6. Galaxy Redshifts
(J.P. Huchra)

Surveys of galaxy cluster redshifts have been spurred onwards by the growth of interest in the large scale structure of the universe. In the last few years it has become increasingly obvious that the topology of large scale structures places crucial constraints of theories of galaxy formation, cluster formation and our basic cosmological models.

At the time of this writing, the number of galaxies with measured redshifts is approaching 25,000. The majority of these can be found in either of two large compilations; one maintained at the Center for Astrophysics by J. Huchra and collaborators and one maintained at Bologna by G. Palumbo and G. Vettolani. Versions of these are obtainable from the authors or from the Natinal Space Science Data Center in the U.S. or the Strasbourg Astronomical Data Center in Europe. Figure 1 is a plot of the surface distribution of galaxies with redshift in J. Huchra's ZCAT as of June, 1987. The Nearby Galaxy and Catalog of 2367 galaxies with redshifts less than 3000 km s^{-1} has been published by R.B. Tully and R. Fisher.

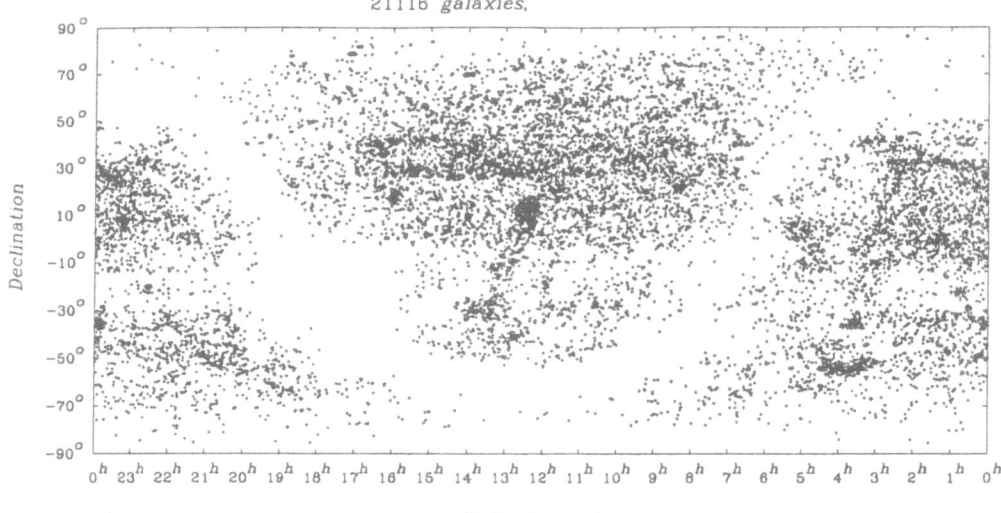

Figure 1. The Surface Distribution of Galaxies in ZCAT as of June 1987.

Similarly, the number of galaxy clusters from the lists of Abell, Zwicky, Corwin, Ortwin and Abell and deeper surveys (eg. Gunn and Oke, and Sandage et al.) with measured redshift is nearing 1000. Catalogs of cluster redshift are maintained by H. Rood and M. Struble, and by J. Huchra in the U.S. and by M. Kalinkov and I. Kuneva in Bulgaria.

As digital detectors have become commonplace on small telescopes and as radio receivers became more sensitive, the number of galaxy redshift surveys has increased. Major large area surveys of optically selected galaxies (from the catalogs of Zwicky et al., Vorontsov-Velaminov, Nilson, and Lauberts et al.) are being carried out by Giovanelli and Haynes at Arecibo, by Huchra and Keller at Mt. Hopkins, by da Costa et al. in Brazil, by Davis, Tonry and Sargent at Las Campanas and by Fairal and Menzies and Davies in South Africa. Deeper redshifts surveys, usually over small areas where galaxies are selected by scanning photographic plates, are being done by Kirshner, Oemler, Schechter and Shectman at Las Campanas and McGraw Hill (the Bootes Void Survey), by Koo, Kron and Szalay at Kitt Peak, by Ellis, Shanks and Broadhurst at the AAT, by Geller et al. at the MMT and McGraw Hill. Particular succes has been obtained at the AAT with a multiple object fiber-coupled spectrograph. These redshift surveys have confirmed the complexity of the galaxy distribution that was hinted at in earlier surveys and uncovered large voids, superclusters, and shell-like structures.

Studies of the internal dynamics of nearby galaxy clusters have also benefited greatly from the proliferation of advanced detectors and multiple object spectrographs. A decade ago, the number of galaxy clusters with more than 20 measured redshifts was only of order a dozen. The number now is nearly 100. Major contributions have come from the work of Melnick and Quintana, Dressler, Geller and Huchra and collaborators, Richter and and Huchtmeier and collaborators, Salpeter and Hoffmann (primarily the Virgo cluster), and from other groups using EFOSC and OPTOPUS at ESO, the fiber-fed spectrograph at the AAT, and the multislit spectrograph at the CFHT. Coupled with X-ray observations from the Einstein satellite, the emerging picture is that clusters are now in a variety of states with many far from dynamical relaxation.

Two major surveys of Abell cluster redshifts are now underway, prompted by the initial analyses of a 100 cluster sample by Bahcall and Soniera. These are the surveys of Karachentsev, Klypin and Kopylov using the SAO 6-m, and of Huchra, Henry, Postman and Geller using the MMT and CFHT. Although the results are only preliminary (because the Abell cluster catalog suffers from a variety of difficult selection effects) these surveys are confirming the large amplitude of the cluster-cluster correlation function found earlier. That degree of clustering had not been matched by any of the extant theories of large scale structure.

Observations of objects at large redshifts continued to be pioneered by Spinrad, Djorgovski and collaborators who have discovered several galaxies with strong Lyα emission at redshifts in excess of 1. Also noteworthy are observations by A. Wolfe and collaborators of high redshift damped Lyα absorbers that may be galactic disks.

In addition to optically selected galaxies, the survey done with the Infrared Astronomy Satellite (Neugebauer et al.) has provided extragalactic astronomers with tens of thousands of high latitude sources. Not only is that catalog a new finding list of active nuclei, but it is also a nearly whole sky and uniform survey of star forming galaxies. Radial velocity and spectroscopic identification work has been done by de Grijp, Miley, Lub and de Jong, by Lawrence et al., and by Kleinmann et al., and a large survey of all 2600 objects brighter than 1.95 Jy at 60 μm and further than 10° from the galactic plane has just been completed by Davis, Huchra, Strauss, Tonry and Yahil. This last survey is being used to compare the anisotropy of the galaxy distribution with our observed motion w.r.t the μ-wave background to derive the value of Ω.

Since the last report, progress on the derivation of a more accurate value for the Hubble constant through improvements in primary and secondary calibrators (eg. the Infrared Cepheid observations of McAlary, Madore, Freedman and others) has been offset both by the discovery of large scale flows within 6000 km s^{-1} by Burnstein, Davies, Dressler, Faber, Lynden-Bell, Terlevitch and Wegner and by the untimely death of Marc Aaronson, a good friend and a leader in distance scale research.

I would like to close by stressing to all astronomers the need to publish heliocentric velocities (cz) or redshifts (as z = $\Delta\lambda/\lambda$ in the optical convention), and also accurate 1950 positions for the galaxies they observe.

7. Extragalactic Research in the U.S.S.R.
(E.Ye. Khachikian)

Abreviations:

Kaz	– Trudy Kazan Astron Obs.	Ye U	– Yerevan State University
Af	– Asrofizika, Erevan	Crim	– Izvestia Crimean Astrophys. Obs.
AZ	– Astron. Zhurnal	LGU	– Trudy Astron. Obs. Leningrad Univ.
LAZ	– Letters to AZ (Pis'ma)	SAO	– Communic. Special Astrophys. Obs.
AZ	– Astron. Circ.	BAO	– Communic. Byurakan Astrophys. Obs.
Tartu	– Publ. Tartu Obs.	SAO Iz	– Izvestia Special Astrophys. Obs.
SA	– Superassociation		

I. SURVEYS AND LISTS OF GALAXIES

Markarian + (Af 20, 513; 23, 439; 25, 345) continue the second Buyrakan Spectral Survey (SBS); The number of UV excess objects reaches 500. Stepanian + (SAO, 50, 21) described the method of observations and selection of SBS's objects. The history and perspectives of the first Byukarian Survey are discussed by Lipovetsky (SAO 50, 12). Afanas'ev + described the deep spectral survey (up to 23m) on the 6-meter telescope using the slitness spectroscopy method (SAO 50, 31). Poljakova obtained objective prism spectra of about 100 galaxies with dispersion 660 A/mm by Hγ (AC No. 1368). Markarian + (Af 26, 15) showed that FBS and SBS more effectively select QSO than other surveys. Khachikian made a survey of UV-galaxies (Highlights of Astronomy, 6, 459), Markarian galaxies and star formation regions (IAU Collo-quium No. 78; SAO 50, 39) and studied morphology and spectro-photometry of the central regions of Markarian galaxies (IAU Symposium No. 121). I. Pronik made a survey of variability in the spectrum of the nuclei of 37 Seyfert galaxies (IAU Symposium No. 121). It is shown that the spectra are changing on a time scale from hours to days. Kazarian and Khachikian made a survey of morphology, spectroscopy and forms of activity of UV-galaxies (Ye U, Publication, Physics 5, 149; Problems of theory of superdence bodies, Ye U press, p.195, 1984). 15 new dwarf galaxies in the IC 342 complex of galaxies was found by Borngen and Karachentseva (Astron. Nachr 306, 301). A catalogue of 1051 isolated galaxies with m < 15.7 was published by Karachentseva + (Bull. inf. Cent Donnees stellairie No. 30, 125).

II. ACTIVE GALACTIC NUCLEI

Spectral observations of galaxies with UV excess from Bukarian surveys was conti-nued by Markarian + (Af 21, 35; 21, 419; 22, 215), Denisyuk + (Af 20, 525; SAO 50, 88). Spectral and optical variability of active galactic nuclei is observed and discussed by Merkulova + (Crim 68, 85; 71, 160), Pronik (Crim 68, 75; SAO 50, 64),

Lipovetsky + (Af 24,437), Chuvaev (SAO 50, 73; LAZ 11,803), Merkulova (Crim 75, 175), Neizvestnij (SAO 50, 74), Ljutij + (LAZ 10, 803), Bocharev (LAZ 10, 239), Shevchenko (LAZ 10, 896; 11, 83; 11, 432), Gorbatskij (Af 22, 267), Ljutij + (LAZ, 12, 187), Doroshenko + (Crim 73, 143; AA 163, 321). Spectrophotometric investi- gation of active galaxies is carried out by Lipovetskij + (Af 21, 5; 24, 437; SAO 50, 43), Andreassian and Khachikian (Af 24, 17), Burenkov (Af 24, 349), Kazarian + (Af 22, 431), Amirkhanian + (Af 22, 239). UBV photometry of galaxies with UV excess has been made by Tamax zian (Af 20, 43), Kazarian + (LAZ 10, 815); B,V photometry by Kostjuk + (SAO 20, 68); Hα-electrophotometry by Asadulaev + (AC No. 1415)). Polarimetric and electrophotometric study of BL LAC and BL Lac type objects are carried out by Gagen-Torn + (AZ 61, 925;Af 22, 5; 25, 485) and Marchenko (Af 22, 15). Surface brighness distribution in Seyfert galaxies is studied by Afanas´ev + (Af 24, 333; 24, 425; 25, 5; SAO 50, 60) and Metik + (Af 23, 451) for NGC 1275. Dependence between luminosities of Seyfert galaxies and their active nuclei is mentioned by Dibaj + (AZ 62, 468). A method of determination of luminosity function was proposed and applied to Seyfert galaxies by Arakelian (Ap J 301, 92). The lumi- nosity function of a sample of 219 Seyfert galaxies is determined by Reshetnikov (Af 24, 33); for faint galaxies with UV continuum by Stepanian (Af 21, 445) and Terebizh (SAO 50, 34). Gas dynamics in active galactic nuclei in terms of emission line properties is discussed by Pronik (Crim 72, 135; SAO 50, 64), Mikhailov (SAO 50, 84), Bochkarev (SAO 50, 77), Vilkoviskij (SAO 50, 100), Shevchenko (LAZ 11, 181). Models of active galactic nuclei are discused by Schklovskij (AZ 61, 833), Ljutij, Cherepaschuk (AZ 63, 897), Romanova (SAO 50, 80), Dibaj (AZ 61, 417), Silchenco + (A Sp Sci 117, 293), Afanas'ev (SAO 50, 57), Suchkov (SAO 50, 91). Some theoretical aspects of galactic nuclei are discussed by Zentsova (AZ 62, 1227), Galev (LAZ 11, 181), Lavrushkina (Acad USSR, 48, 2080). Burenkov and Khachikian measured the velocity field of UV-galaxies with starbirth regions Mark 7 (Af 19, 619) and Mark 297 (IAU Symposium No 121). Petrosian completed the statistical investigation of SA in UV-galaxies (Af 21, 57). Andreassian and Khachikian (Af 22, 441) carried out spectrophotometrical investigation of SA in NGC 2820 A. Detailed spectrophotometry with 6 m telescope and long slit have carried out by Burenkov + for Mark 111 (Af 21, 433), by Khachikian + for Mark 306 (Af 24, 5), by Burenkov and Khachikian for Mark 3353 (Af 24, 349), by Andreassian + for Mark 71 (Af 25, 507), Petrosian + discovered four UV-galaxies with jets (Af 22, 229). Kalinkov + applied numerical methods of the photometry of Markarian galaxies (Af 26, 29). Joeveer found a deficiency of normal elliptical galaxies among Markarian galaxies (Af 24, 25). Yegiazarian carried out detailed spectral observations of 12 UV-galaxies from Kazarian lists. The redshifts and some physical properties are measured. No 199 shows Sy characteristics (Af 25, 425; BAO 57, 8; 58, 68).

III. QUASARS AND COMPACT GALAXIES

Markarian + (AC No 1346, 1381; 1384, LAZ 13, 3) observed new QSO from his lists of UV excess galxies. Levshakov + (SAO 50, 48) studied three quasars from SBS with 6 m telescope. Varshalovich + (Adv space Res 3, 187) studies absorption-line spectra of quasars Levshakov and Varshalovich (MN 212, 517) found molecular hydrogen in z=2.811 absorbing material towards the quasar PKS 0528-250. Properties of optical variability of quasars and correlation of optical and radio activity of 3C 345 is studied by Babadjanyanz + (Af 21, 217; 22, 247; 23, 459; AZ 62, 627). A method of the separation of radiation components of variable extragalactic sources was pro- posed by Gagen-Torn (Af 22, 449). Photographic photometry of compact extragalactic objects was continued by Skulova (LGU 39, 43). Classification method of compact galxies is discussed by Byorngen and Kalloghlian (Astr Nachr 306, 81). Multiple light scattering as a possible cause of visual polarization is discussed by Loskutov + (Af 23, 307). Some indications of the symmetry of emitting regions of quasars with broad absorption lines are discussed by Grinin (LAZ 10, 643). The possible contribution of dwarf galaxies and globular clusters to Lyα and 21 cm lines in the spectrum of distant QSO´s is discussed by Komberg (Af 24, 321). Pro- perties of hot gas in the nuclei of quasars is discussed by Zentsova (AC No 1417).

Lebedev and Lebedeva (SAO Izv 19, 16) found no significant periodicity in the space distribution of quasars in a search for "antipodal" images of quasars (SAO Izv 19, 12). Komberg (Af 20, 351) discussed double quasi-stellar object Q 0957-561 A, B as a possible pair of galaxies.

IV. STRUCTURE AND KINEMATICS OF GALAXIES

New stellar associations in M31 were reported by Efremov + (IAU Symposium 116). The resolution of blue star-like objects in Arp's ring around M81 was made by Efremov + (LAZ 12, 434). Ten candidates per luminous red supergiants were found in M31 by Efremov + (A Sp Sci, 129, 39). Investigation of colour-luminosity diagrams of 12 globular clusters in the SMC & 22 in the LMC showed that star formation histories are different in LMC and SMC (Frantzman, LAZ 12, 281). A singular pronounced aniso-tropy in the apparent major axis orientations of Uppsala and ESO/Uppsala catalogue galxies was shown by Mandzhos + (LAZ 11, 495). From a study of the cross-section of a spiral arm of Andromeda galaxy Efremov showed, that the structure of S4 arm between OB 75 and OB 82 agrees with the density wave theory (LAZ 11, 169). Photo-electric photometry of globular clusters in Andromeda nebulae was made by Sharov + (LAZ 10, 583; AZ 61, 245). BV photometry and spectroscopy of "Garland"- an unusual object near NGC 3077- was made by Karachentsev + MN, 217, 731). The integrated colours of some very red spiral galaxies is discussed by Silchenko (A Sp Sci, 117, 83). Optical variability of the nucleus of M33 was discovered by Ljutii + (LAZ 12, 187). Photometry of nuclear region of NGC 1569 was made by Merkulova + (Kaz, 50, 83). A catalogue of rotation curves of normal galaxies was compilled by Kyazumov (AZ 61, 846). Rotation curves were studied and masses of galaxies were estimated by Mineva (LAZ 11, 811). The direction of rotation of spirals in 109 galaxies was studied by Pascha (LAZ 11, 3). Estimates of masses of disks and haloes of galaxies on the basis of local stability criterion for a disk was made by Zasov (LAZ 11, 730).

Photometric colour profiles of spiral arms of galaxies was dicussed by in Traat (Tartu 51, 181). Effects of galaxy inclination in surface photometry are discussed by Kaazik (Taru, 98, 100), and the problem of hidden mass in galaxies by Gurzadian (AZ 63, 812). Gas/dust ratio in M33 is determined by Sharov (LAZ 11, 313). Halo/disk mass ratio's were discussed by Zasov (LAZ 11, 307). The magnetic field distribution in spiral galaxies was investigated by Ruzmaikin + (AA 148, 335). Problem determination of masses of elliptical galaxies and their X-ray radia-tion was discussed by Krol (Af 23, 227). Gas dynamics in galaxies is studied by Volkov (Af 24,57; 24, 477) and Sotnikova (Af 25, 139). Ionization equilibrium con-dition was dicussed by Sidorov (Af 21, 353). Some aspects of galaxy modelling was discussed by Silchenko (AZ 61.634), (SAO 50, 91). Kalloghlian and Kandalian showed that radio emission of SB galaxies more often is connected with the central regions of galaxies (Af 24,47). Malumian (Af 22, 25) discussed the question of colours and Byurakan classification of nuclei of galaxies.

V. SYSTEMS OF GALAXIES.

Burenkov + described four pairs of galaxies one of which is a UV galaxy (LAZ 10, 403). Vennik (A.N. 307, 157) determined the visual parameters of groups of galaxies. Mahtessian studied the connection of some physical parameters of galaxies with density and morphological type (BAO 17, 13, 21) of the groups in which they are located. Shakhbazian and Shapovalova (Af 20, 179) studied morphlogical struc-ture of two compact groups of compact galaxies. Plaksina (Probl. Cosm. Physics, Kiev 19, 118) studied the orientation of double galaxies on the sky. Djomin + (AZ 61, 625) studied CI of galaxies in double systems and confirmed the Holmberg effect. Bisnovatii-Kogan (Af 21, 87) derived the kinematical characteristics of pairs of galaxies. Karachentsev considered the relation of surface brightness of isolated and double galaxies with some of their parameters (SAO Izv 19,3). Konukov has shown that the distribution of the galaxy component of a cluster can be determined from a boundary problem for the gravitational potential of a self-consistent field (Af 22, 273). Vinokurov + discussed the probable mechanism of formation of breaks in the luminosity function of galaxy clusters (AC No.1390,5). Arkhipova + (AZ 64, 233; 63, 16; AC No. 1329,1) studied the kinematics of gas in close pairs, chains and groups of galaxies. Their M/L ratio is small and testifies to their stability. Stephanian (Af 21, 445) discovered 35 emission galaxies in A634. Egikian + made two colour photometry of about 420 galaxis in the clusters A 1213 (Af 21, 21) and A 426 (Af 23, 5). Petrossian and Turatto considered the relation of Markarian galaxies with Zwicky clusters (AA 163, 26) and determined redshifts of 12 of them (Af 24, 205). Petrossian has shown that physical character-istics of Sy do not depend on the multiplicity of the system of which they are members (BAO 57, 3). Some aspects of dynamical evolution of triple systems was discussed by Anosova + (LGU 40, 66), by Anosova (A Sp Sci 124, 217. Gorbatsky (Af 20, 61) discussed a possible mechanism of heating in intergalactic gas in clusters. Gas equiluilibrium conditions in gravitational fields of massive galaxies in clusters was discussed by Volkov (Af 62, 450).

Spectral observations of low surface brightness galaxies in M81 group region was made by Karachentsev + (Af 21, 641). Vennik and Kaazik listed new redshift for 31 galaxies, selected in the vicinity of nearby groups of galaxies (Af 23, 213). Kopylov + presented the redshift measurements of rich clusters of galaxies (AC No. 1393). Karachentsev + measured individual masses of galaxies in pairs (AZ 62, 417; LAZ 10, 563); Malykh and Orlov proposed a method for the statistical study of multiple system configuration to judge their dynamic properties (Af 24, 445). Barausov and Chernin showed that interaction of shock waves with interstellar medium gives a good chance for formation of the group of spirals (LAZ 11, 883). Aliakbarov + (SAO 8, 81) found no evidence for the Sunyaev-Zeldovich effect in X-ray clusters of galaxies.

VI. LARGE SCALE STRUCTURE OF THE UNIVERSE

Fesenko has shown that the irregular extinction provides the observed depend-ence of covariance function on distance (Af 24, 453), and that there are no evi-dences for the existence of the local supercluster from the distribution of radial velocities and angular diameters of galaxies (Af 25, 161). Stepanian (LAZ 11, 575) using the space distribution of SBS emission galaxies found a large scale structure with D=50 Mpc. Shandarin and Klipin (AZ 61, 837), in the frame of adiabatic theory have shown that rich galaxy clusters are formed as a result of long-scale schemes inside of a supercluster. Gilfanov and Syunjaev have shown that the diffusion of elements in intergalactic gas can increase the abundance of Deuterium, Helium and Lithium in the central parts of rich clusters (LAZ 10, 329). Sotnikova considered the evolution of interstellar clouds initially out of pressure balance with extern-al hot intergalactic gas (Af 25, 139). The prediction of the global rotation of the universe is discussed and the problems of cosmic magnetic fields and energy problem in active galaxies are considered by Muradian (IAU Symposium No 121). Ozernoy + (LAZ 12, 325) found arguments for an explosional origin of cosmic void in boots. Tago + (MN 218, 177) discussed a prominent 50 Mpc long string of galaxies in Bootes. Einasto + (MN 219, 457) continued his investigation of structure of super-clusters and formation. Agekyan and Yakimov proposed a method for determining the presence of structure in a field of objects and for estimates of the parameters of a structure if it exists (AZ 63, 214).

8. Working Group on the Magellanic Clouds
(M.W. Feast)

Some 200 papers on, or related to, the Magellanic Clouds (MC) are published each year. This level of activity reflects the crucial importance of MC for extra-galactic and galactic distance scales; for many areas of stellar evolution, stellar structure and pulsation theory; as well as for the chemical evolution, structure, kinematics and interaction of galaxies. In the following, references are frequently omitted when they can be traced from papers cited.

I. DISTANCE, STRUCTURE, MAGELLANIC STREAM

Evidence on MC distances has been summarized (Galaxy Distances and Deviations from Universal Expansion (GD) Reidel 1986 p 7 and Feast, Walker An. Rev. AA in press). True moduli of 18.47 (LMC) and 18.78 (SMC) were derived, based primarily on Cepheids. Main sequence fitting to MC intermediate age clusters may yield slightly lower moduli (~18.2 for the LMC), but the effects of convective overshooting (wich remains contentious GD p 15) and other effects (see above references) may have led to these distances being underestimated. In more recent work, Mateo, Hodge (ApJ in press) find for the LMC intermediate age cluster LW 79, 18.4±0.2. Strömgren photometry of LMC B supergiants (MN 219, 495) gave 18.8±0.3 whilst spectral types of OB stars gave 18.3±0.3 (AJ 92, 48). A direct Baade-Wesselink analysis using BV photometry and Kraft-Parsons coefficients gives for 6 LMC Cepheids, 18.4 (internal error ~±0.1) (Imbert, Marseilles preprint 44). This is in accord with the result that MC Baade-Wesselink radii of Cepheids agree with the galactic period-radius relation (MN 220, 671).

The Gascoigne sequence on wich Grahams LMC RR magnitudes depend has been strengthened (PASP 98, 1162) whilst 10 RR Lyraes have been found in NGC 1786 (LMC) (PASP 97, 676). There is as yet no compelling evidence for different MC distances for objects of different ages. Nevertheless given the known great depth of the SMC such differences might well arise. Whilst there is general agreement that the SMC is very extended in the line of the sight, there is still considerable uncertainty as to its structure and kinematics. In particular the depth deduced from Cepheids agree with extensive phase coverage (MN 218, 223; 222, 449) is considerably less than infered from less extensively observed Cepheids (ApJ 301, 664). Welch et al. (ApJ in press) suggest that this is due to the greater uncertainties in the adopted distances of the latter. To obtain a clear picture of the kinematics it will probably be necessary to combine accurate Cepheid distances from multi-colour, multi-observational work with extensive radial velocities (E.G. Stobie and Hatzidimitriou, in progress). The structure may be quite complex since interstellar line velocities suggest the two 21 cm components, the lower velocity one is closest, whilst two of the most distant Cepheids in the sample with accurate distances belong to the low velocity component (cf above references). Also, ultraviolet work suggests additional high and low velocity components in front of the main body of the SMC (ApJ Suppl 59, 77, ApJ 292, 122).

There still remains some problem in assigning gas to the LMC or the halo of our galaxy (cf. the complex interstellar spectrum of SN 1987A Vidal-Madjar + AA in press) some components of which may be due to shells expanding from 30Dor (see also ApJ 310, 700). It has been suggested (ApJ 303, 1987) that the gaseous components of both MC are very extended in the line of sight and it remains to be seen whether this can be reconciled with the evidence (spatial, kinematic) that young LMC stars are confined to a plane. Further evidence for optical nebulosity, and for young stars, in the bridge between the MC has been obtained (Nature 316, 705; 318, 160; MN 223, 317). A new 21 cm study of the LMC (AA 137, 343) shows considerable noncircular motions superimposed on the differentially rotating disk. Extensive Coravel radial velocities of F-M stars in the MC (AA Suppl 62, 23; 67, 423) wil enable more detailed kinematic modelling. The Marseille group find that super-associations in the LMC have a very small velocity dispersion (5 km s^{-1}). A study of the radio continuum morphology of the MC (1.49 Hz) has been published (AA 159, 22). Amongst basic data produced are BVR data on 2600 MC foreground stars (AA Suppl 68, 63). Much still remains to be learned about the Magellanic Stream, a high sensitive radio survey shows the highly complex gas distribution in the stream (AJ 90, 1801). Spicker and Feitzinger (AA in press) have made a statistical turbulence analysis of the LMC, the first complete stellar system in wich this has been done. The largest eddies are 1-2 kpc in size. Feitzinger, Braunsfurth also attempt a quantitative analysis of the overall (spiral) structure of the LMC (AA 139, 104).

II. ABUNDANCES, CLUSTERS

A variety of arguments suggest metal deficiencies (Fe peak) for young objects in the LMC, SMC of 1.4 times and 4.0 times respectively and some recent work agrees with this general result (cf. AA 155, 72; 155, 145; AJ 89, 1705). Light elements (CNO) in HII regions appear more deficient than this (cf. ApJ 292, 155 and earlier papers). It would be very important to obtain more information on the abundances of different elements and their variation with stellar age. A particulary interesting result concerns the prototype young globular cluster NGC 330 (SMC). Various lines of evidence (AA 168, 197) (Spite +) ApJ 312, 195 and references there) suggest that stars in this cluster have $[Fe/H]$ as low as -1.5. Some of the results depend on the $[CNO/Fe]$ value (but not Spite +). Further work on this and other clusters is desirable since, taken at its face value, the result suggests that the origin of the young globular cluster stars is quite different from that of the young field (which have $[Fe/H] \sim -0.6$). A number of elements show a depletion factor of ~2 in the interstellar medium of the LMC compared to our galaxy (MN 217, 115). Low abundances are amongst the factors responsible for low CO emission in the MC (ApJ 303, 186). Old red giants in the SMC halo have $[Fe/H] = -1.6$ with a real spread, $\Delta[Fe/H] \sim 0.3$ comparable to the abundance spread shown by globular clusters in the halo of our galaxy (AJ 91, 275).

A very large amount of work continues on clusters of all ages in the MC. Their relevance to distance scales is mentioned above. Other primary concerns are ages, metallicities and comparisons with evolutionary calculations. The clusters can be roughly classed in three groups; (1) Intermediate age (1-2 Gyr) with $[Fe/H] \sim -0.5$ (LMC) or -0.9 (SMC) (ApJ Suppl 60. 893; ApJ 305, 214; 297, 582; 304, 265; 284, 108; 283, 552; AJ 92, 1334; AA Suppl 67, 373; 64, 189; PASP 97, 753; Mateo, Hodge in press). NGC 1831 sometimes considered in this class is apparently younger (0.3 Gyr, PASP 96, 947); (2) some old globular clusters of age ~10 Gyr (ApJ 298, 544; 311, 113; Olszewski +, Steward Preprint 701) Kron 3 is 5-8 Gyr old (ApJ 286, 517); (3) Young clusters (e.g. ApJ 292, 130; 304, 617; Obs 104, 161; AA 134, L1; PASP 98, 1133). Clusters younger than ~8 Gyr have C stars at the tips of their giant branches, Younger than ~0.8 Gyr the giant branches are populated by the M stars (ApJ 288, 551). It has been argued that convective overshooting in intermediate mass stars affects very considerably the interpretation of the C-M diagrams of MC clusters (AA 150, 33 cf. also Brocato, Castellani, preprint). The age-metallicity and age-frequency distributions of MC clusters are important indicators of evolution in the MC and from them Elson, Fall (ApJ 299, 211) find no evidence for bursts or star formation though this is not the conclusion of other workers (AA 165, 84; 156, 261; see also ApJ 285, 595; and Wielen I.A.U. Symposium 126 in press). The luminosity function of LMC clusters is simular to that of open clusters in our galaxy (PASP 97, 692). A mass to light ratio of 0.42 is found for the old LMC cluster NGC 1835 (ApJ 288, 521). Contrary to earlier suggestions there now seems no correlation between LMC cluster ellipticities and age (ApJ 283, 598; AA 146, 293). Various parameters of MC clusters, are discussed by Kontizas and co-workers (AA 131, 58; 159, 305; AA Suppl 65, 207; 65, 283; 67, 147, 68, 357; 68, 493). 213 new clusters and a list of SMC associations were published (PASP 98, 1113; 97, 530).

III. HII REGIONS, PLANETARY NEBULAE, SNR, INTERSTELLAR MATTER

30 Dor is the prototype supergiant HII region and its stellar content, particularly its concentration of WR stars remains a challenge to theoretical models (ApJ 295, 109; 312, 612; AA 153, 235; MN 224, 435). The possibility that R136 contained a supermassive star was finally laid to rest by Weigert, Baier (AA 150, 218) who resolved R136a into a cluster of 8 objects (cf. also PASP 96, 999; ApJ 283,

560). It has been concluded that R136a alone can energize the observed kinematics and structure of the 30 Dor Nebula (MN 211, 867 cf. also MN 211, 521; AA 153, 235). Kinematic studies of ionized shells in the MC place restrictions on models though as yet these features are not fully understood (AA 137, 512; 138, 57). The nature of LMC ring shaped nebula has been discussed (AA 160, 21). There is considerable interest in compact (high excitation, high density) HII regions in MC (AA 139, 330; 144, 98; 145, 170; 162, 180). These may represent a distinct class of HII region. Other compact HII regions are of low excitation (MN 209, 241). Molecular hydrogen has been detected in the compact HII region N81 (SMC) (ApJ 291, 156). A study of MC HII regions shows that the Hα luminosity function follows a power law and that the frequency distribution of nebular diameters is exponential with a scale length of 80 pc (ApJ 306, 130). Hα, Hβ fluxes of LMC HII regions have been used to model the extinction and reddening, some at least of which appears to be associated with the HII region (AA Suppl 62, 63; AA 155, 297).

Because MC PN are at known distance the resolution of their shells by speckle interferometry opens up important opportunities for nebular mass and age determinations. It is however a cause of some concern that for the one object (SMC N2) wich has been independently studied by two groups, quite different results have been obtained (ApJ 311, 632; MN 223, 151). Earlier, high masses of the central stars of three bright MC PN have been revised downwards and are now similar to estimated masses of galactic PN (ApJ 313, 268; MN 223, 151; AA 138, 317). Analysis of new PN radial velocities implied 9.10^8 $M_\odot$ within 3 kpc of SMC centroid. There is an excellent correlation between expansion velocity and excitation class for SMC PN wich may be traced to a correlation of these quantities with the mass of the central star (ApJ 296, 390). Ten new SMC PN were discovered (MN 213, 491). Some extremely energetic LMC PN are found to be bipolar. They may represent the upper end of the mass range of PN precursor stars and are perhaps related to symbiotics and such galactic objects as OH 0739-14 (ApJ 297, 593). The MC are crucial to studies of SNR and much work in this field continues; new SNR have been found (ApJ Suppl 58, 197) and detailed studies made of some of these (AA 164, 26; ApJ 314, 103). Observations (ApJ Suppl 51, 345; MN 216, 365) suggest that MC SNR are in free, rather than in adiabatic (Sedov), expansion but this result has been attributed to selection effects (MN 209, 449; AA 157, 6). New and revised chemical abundances have been obtained for LMC SNR (MN 216, 365; AA 174, 5). An earlier result (Obs 104, 193) that A_V/E_{B-V} in the SMC is close to the galactic value, has been confirmed (AA 149, 330; MN 211, 895). The ultraviolet extinction curve in the 30 Dor region of the LMC is distinctly different from that in our galaxy. Differences, though present, are less significant for the LMC outside 30 Dor (AJ 92, 1068; ApJ 299, 219; 288, 558). The diffuse interstellar band at 4430A is present in reddened SMC stars both with and without the 2200A feature suggesting that 4430A is not directly associated with graphite (MN 215, 5p). IRAS cirrus near the LMC coincides with galactic nebulosity previously identified by the Vaucouleurs and Freeman (MN 221, 543).

IV. STELLAR CONTENT AND EVOLUTION

The outburst by a known precursor (a B3 supergiant) at a known distance with intense optical observations beginning within 18 hours of the initial rise and preceeded by a neutrino bursts makes SN 1987A in the LMC an object of major importance. The behaviour has been in many respects unexpected and the results will have a major impact on SN theories. Many observational and theoretical papers are in press dealing with a variety of matters, from the extragalactic distance scale to the mass of the neutrino. It has been suggested that this star evolved to a SN without being a red supergiant and that energy fed into the system from a millisecond pulsar could explain the gradual rise in bolometric luminosity over a period of ~100 days. During the period under review one nova was discovered (SMC) (I.A.U. Circ. 4283, 4290, 4299).

The MC are the most favourable place for the study of the evolution of massive stars. The S Dor variables and other hot emission line objects of high luminosity (including B[e] stars) are of particular interest in this connection and the intensive study of these objects has continued, mainly by the Heidelberg group (AA 140, 459; 143, 421; 153, 168; 154, 243; AA 158, 371; 164, 321; 153, 163; 164, 435; AA Suppl 61, 237). The observations suggest a two-component stellar wind model for the B[e] stars. In a number of LMC emission line stars (including S Dor variables) [NII] and other nebular emission lines have been found, suggesting the classification of these objects as "supergiant PN" (similar to the galactic object AG Car). Further evidence has been obtained (ApJ 293, 407; AA 148, 379) for lower mass loss rates from MC blue supergiants than from those in our galaxy. Such results are explicable in terms of a lower metallicity (AA 173, 293). A comparison of Kurucz models with parameters of LMC B supergiants suggest that the late B stars are considerably less massive than expected from evolutionary theory. This may indicate mass loss during a long-lived hot supergiant phase (ApJ 312, 596). Equivalent widths of lines in galactic and LMC B supergiants have been compared. HeI lines appear weaker at a given class in the LMC (MN 210, 131). A very hot, massive O star in an LMC HeIII region may be evolving to the left of the main sequence (AA 170, L4). Extreme B stars range up to $M_V \sim -6$, brighter than anticipated from galactic work (AJ 90, 2009). There have been analyses of a number of spectroscopic binaries (PASP 96, 81; ApJ 310, 715). WR star research depends heavily on MC work and activity continues in this field including the discovery of new members of the class, analysis of binaries, and the discussion of individual stars including a possible WR runaway star in the LMC and the presence of a WN star in an old association implying a low (~ 20 $M_\odot$) mass (AA 173, 405; 149, 213; ApJ 300, 379; 292, 511; 309, 714; PASP 96, 968; MN 216, 459 see also 30 Dor above). Some general aspects of the WR problem are reviewed by (PASP 97, 5).

Reid, Glass and Catchpole (preprint) have discovered more Miras in the LMC and studied them in the infrared. They discuss the possible age distribution of these objects. Extensive SAAO infrared work on MC Miras is being analysed. The Mira P-L relation in the LMC is very narrow (Glass + Calgary Workshop. Ast. Sp. Sci. Library 132 p 51). The 200-day LMC Miras have a velocity dispersion of 30 km s^{-1} (ApJ 310, 710). They therefore constitute a population wich is less flattened to the LMC plane than young objects, and probably also less flattened than the LMC PN wich must be primarily old objects (note that in our own galaxy, 200-day Miras also constitute a flattened, rather than halo, population). The red giant population (AGB etc.) in the LMC has been discussed in relation to evolutionary models and dredge-up processes (ApJ 299, 236; 284, 98; 294, L7). Objects similar to galactic supergiant OH/IR sources have been identified from IRAS and ground based infrared observations and one found to show OH masing. In general the OH intensity of these objects appears lower than in galactic supergiant OH/IR stars (ApJ 302, 675; 306, L81). Amongst basic data published are colours, spectral types and luminosities of MC M supergiants (ApJ Suppl 57, 91; ApJ 289, 141) and the first part of a GRISM survey for C stars in the SMC (AA Suppl 65, 79).

Superluminous giants (SLG) were reported some years ago in young MC globular clusters and have been discussed in the past as post-AGB stars. It now appears that most, or all, of these are either non-MC members or blended images of more than one star (AJ 91, 80; 91, 1136). The research for, and study of protostars and related objects has continued (MN 219, 603; 215, 103; AA 140, 67). Considerable reference to star forming regions in MC is made in I.A.U. Symposium 115 whilst luminous stars are disscussed in I.A.U. Symposium 116. Some stars in the nebulosity N70 (LMC) may have circumstellar dust shells (AA 148, 397). Cepheid period changes have been discused and found to be primarily non-evolutionary in nature (MN 212, 395). Short period Cepheids and RR Lyrae stars in the direction of the LMC were also studied (ApJ 299, 728) some of these are probably foreground objects. An attempt has been made to systematize the nomenclature of MC objects (AA Suppl 64, 303; Bischoff, Strasbourg Circ.).

Helf and (PASP 96, 913) has given a general review of XR results in the MC. An 0.25 KeV survey of the LMC shows the emission to arise in two regions, Shapley III and the Bar (ApJ 313, 185). Optical pulses have been detected from the (Crab-like) 50 ms pulsar 0540-69.3 (ApJ 315, 142). Amongst other results are a suggested 164 day periodicity in the γ-ray burst source, GBS 0526-66 and the detection of an XR ionized He III region round the black hole candidate LMC X-1 (Nature 307, 41; 312, 737).

The initial mass function (IMF for MC massive stars is consistent with the Miller-Scalo IMF for the solar neighbourhood (ApJ 284, 565). Amongst relevant reviews are Garmany on the evolution of massive MC stars (PASP 96, 779), Frogel on some aspects of MC stellar populations (PASP 96, 856) and Wood on MC and stellar evolution theory (Calgary Workshop, Astrophys. Sp. Sci. Library 132). A number of workers have published population studies in specific MC areas. In the Wing of the SMC the IMF is relatively flat with an upper mass limit of 30 $M_\odot$ (AA 154, 249). In an LMC field the presence of disk and halo components has been studied and the disk found to be of constant thickness for stars ranging from M_V = -3 to +1 (AA 148, 263). A field at the periphery of the SMC shows two populations one of age 3.10^9 year and one (halo?) of age at least 5.10^9 year but younger than galactic globular clusters (MN 216, 165). The general ideas of self propagating star formation seem to satisfy results from Shaply III (LMC) (ApJ 297, 599; see also AA 150, 151) though it has been suggested that the overall distribution of LMC stars, dust and gas is not compatible with such a theory (AA 139, 115) and whilst interstellar matter and stars are well correlated in normal spirals, in the LMC they are not. BV magnitudes and positions for 1300 stars brighter than V=18.3 in a 0:02 sq. degree area of the LMC were tabulated (AA Suppl 61, 473). Contrary to previous work there now seems no significant evidence for runaway supergiants in the LMC. This places constraints on mechanisms for runaway star production (AA 152, 243)

9. Working Group on Galaxy Photometry and Spectrophotometry
(J.L. Nieto)

The field of photometry and spectrometry has benefited in the past three years from major technological advances in CCD technology, wich allow fast aquisition of more reliable and more accurate data. They yield automatic and accurate analyses of large galaxy samples and carefull detailed photometry of individual galaxies. This flow of CCD data has also forced photographic observers to produce higher accuracy data, a requirement that the development of sophisticated algorithms of image analysis has made possible. Targets were as usual, luminosity standards and other bright galaxies, but evidently also galaxies and components of galaxies, the analysis of wich has been facilitated by these improvements.

Several meetings took place during this triennum summarizing our present knowledge of the field: "The Virgo +cluster of Galaxies" (39.012.069), "New Aspects of Galaxy Photometry" (39.012.114), "Dark Matter in the Universe" (IAU Symposium 117), "Structure and Evolution of Elliptical Galaxies" (IAU Symposium 127, in press), "Nearly normal galaxies: From Planck Time to Present" (Eight Summer Santa-Cruz workshop, in press), etc.... A supplement to the bibliography on surface photometry of galaxies has been published by Pence and Davoust (39.002.044) (Further updates will be circulated through a mailing list). A paper reviewing technical aspects and results is in preparation (Okamura).

Kent: 167 galaxies of all morphological types (ApJ Suppl 56, 105) 37 Sb or Sc galaxies with optical rotation curves (AJ 91, 1301) and 16 galaxies with HI rotation curves (AJ 93, 816); van der Kruit (AA 173, 59): 51 disk galaxies; Michard (AA Suppl 39, 205), Lauer (ApJ Suppl 57, 473), Djorgovski (Thesis, Univ. of Calif.), Jedrzejewski (MN 226, 747): respectively 39, 42, 262, and 49 early-type objects; Schombert (ApJ Suppl 60, 603): 261 brightest cluster members; Ichikawa (ApJ Suppl

60, 475): 69 dwarf ellipticals in Virgo; Pierce and Tully (1987 in prep.): 300
nearby galaxies. Their aim is essentially to discuss light profile decomposition,
notably in terms of ($r^{1/4}$) spheriod + (expotential) disk models, even for E's
(Kent, ApJ Suppl 53, 115; Kodaira +, ApJ Suppl 62, 703; see also Schombert + AJ 92,
60). Analysis of published and new data led Simien and the Vaucouleurs (ApJ 302,
564) to discuss systematics of bulge-to-disk rotations. However severe difficulties
regarding the laws used in these decomposions were raised by Simien and Michard
(39.157.291) and Capaccioli + (preprint).

Very carefull (often multicolor) studies brought new insights on galaxies in
very small samples or considered individually. These comprehensive analyses were
made either from photographic (often Schmidt) material (e.g. Pence and de Vaucouleurs, ApJ 298, 560; Baumgart and Peterson, 41.157.051; Walterbos and Kennicut, AA
Suppl 68, 311; Carigan, cited below; Prugniel + AA 173, 49; Hamable and Wakamatsu
ApJ Suppl 56, 283; Duval and Monnet, ApJ Suppl 61, 141; Buta, ApJ Suppl 61, 609; de
Carvalho and da Costa, AA 171, 66; Forte, AJ 92, 301; Wevers + AA Suppl 66, 505) or
with CCD data (e.g. Daly + AA Suppl 68, 33; Davis + AJ 90, 169; Boroson and
Thompson AJ 92, 33) or, notably for large galaxies, with both CCD and photographic
data (Rampazzo, Thesis, University of Padova). In particular, Capaccioli + (preprint) have analysed in great detail the luminosity standard NGC 3115 relying upon
a large colection of photographic (including Schmidt) plates and CCD frames,
permitting them to cover a very wilde dynamic range.

It is especially in the field of elliptical galaxies that these new accurate
data have brought new insights. indeed not only were new photometric properties
discovered, but also very informative structural details on their origin and evolution were unraveled, notably dustlanes, disks, shells, boxy isophotes. Further
discoveries after IAU Symposium 127 (mid-1986) (Carter ApJ 312, 514; Fort + ApJ
306, 110; Prieur, ApJ preprint; Rampazzo, cited above; Dettmar and Wielebinski, AA
167, L21; Mollenhoff and Bender, AA 174, 63; Bender and L Mollenhoff, AA 177, 71;
Jarvis, ApJ in press) deserve attention as samples become increasingly larger and
allow searches for correlations with other quantities. The origin of such features
is mainly discussed in terms of formation of Es through mergers. Catalogues of
dusty E's are being updated by Ebneter and Ballick (AJ 90, 183) and Zeilinger
(Thesis, University of Vienna). A compilation of E's with emission lines was made
by Bettoni and Buson (AA Suppl 67, 341). New properties brought out by the geometry
of E's are discussed by Lauer (MN 216, 429), Jarvis (AJ 91, 65), etc... Notably
Michard (AA 140, L39) stressed the analogy between the ellipticity profiles of
elongated Es and SOs. The two-dimensional light decomposition of NGC 3115
(Capaccioli +, cited above) prompted the study of disks in SOs (Capaccioli +, preprint): they show an increase of thickness with distance unlike spirals that exhibit a constant scale height.

The correlation of L with color and metallicity for E's has been studied by
Pickels and Visvanthan (ApJ 294, 134) and Pickels (ApJ 296, 340) with the population synthesis technique. The stellar content of E's and SO's was investigated
notably by Carter + (ApJ 311, 637), Rocca-Volmerange and Guideroni (AA 175, 15) and
Kjaeraard (AA 176, 210). The earlier discovery of a correlated scatter in the L-σ
and L-Mg_2 relations has stimulated a series of papers whose aims are to investigate
the nature of the second parameter discribing Es (Burnstein +, preprint and accompanying papers; see also Djorgovski cited above). Metallic linestrength profiles
themselves have been investigated by Efstathiou and Gorgas (MN 215, 37p), Gorgas
and Efstathiou (1987, in press), Baum + (ApJ 301, 83). The question of possible
physical relations existing between normal Es and E-like systems has been addressed
by several authors. Lachièze-Ray + (AA 150, 62) and Lauer (ApJ 311, 34) found no
evidence for multiple nuclei to be merging with the cD galaxy NGC 6166. Different
observational arguments suggest that dwarf spheriodical systems are not the low-
mass end of the luminosity distribution of Es (Wirth and Gallagher, ApJ 282, 85;
Kormendy, ApJ 295, 73); Bothun, AJ 92, 1007; see also Ichikawa, cited above).

Probably because spirals are better understood than early-type objects, they were less studied - observationally at least - during this triennum. However crucial problems deserve continuous attention, namely Freeman's constant (van der Kruit, AA 173, 59) and dark matter especially for edge-on and elongated (Meisels, AA 145, 138; Skrutskie +, ApJ 299, 303) and pure-disk spirals (Carignan, ApJ Suppl 58, 107; ApJ 299, 59; Carignan and Freeman, ApJ 294, 494).

Among irregulars, blue compacts stil have special place (Zamorano and Rego, AA 170, 31; Loose and Thuan, ApJ 309, 59), as well as those objects called UV-excess (Maehara +, PASJ, in press; Kodaira +, PASJ, in press) or starburst (Johansson, AA, in press), but this triennum has seen a special interest for ring-galaxies (Wakamatsu +, AJ 92, 700; Bonoli, AA 174, 57; Brosch, AA 153, 19; Schweizer +, ApJin press; Buta, ApJ Suppl, 61, 609; 61, 631 and preprints; sealso a review on rings and shells by Athanassoula and Bosma, Ann. AA 23, 147). Other intersting studies of peculiar galaxies were carried out by Noreau and Kronberg (AJ, 92, 1048) and Karachensev + (MN 217, 743), etc....

The photometric profiles of the central regions have received special attention. A black hole has been suggested in M32 by Tonry (ApJ 283, L27). Lauer (ApJ 292,104) from deconvolution procedures and Kormendy (ApJ 292, L9) from high resolution data have found that (early-type) galaxies have non-isothermal cores, either due to velocity anisotropies or black holes. The nucleus of M31 has been also discussed in these terms (Nieto, ApJ 108, 111; Nieto +, AA 165, 189; Kormendy, IAU Symp. 127, in press) as well as the central regions of NGC 3379 and M81 (Bendinelli +, AA 140, 174; Parmeggiani, 39.157.125).

Photometry and spectrophotometry have yielded quite interesting results in the UV (e.g. Ciani +, AA 137, 223; Bohlin +, ApJ 298, L37; Donas and Deharveg, AA 140, 325; Israel +, AA Suppl 66, 117) or in the (near) IR (e.g. Frogel, ApJ 298, 528; Hunter and Galagher, AJ 90, 1457; Martinez +, AA 161, 237; Burkhead +, AJ 91, 777; Giles +, MN 218, 615; Adamson +, MN 224, 364). IRAS observations have brought new insights on star forming regions in galaxies (e.g. AA 154, 373; MN 218, 19P; ApJ 303, 171; ApJ 304, 651) and stimulated further IR investigations from the ground on stellar content and formation (e.g. Morwood +, AA 160, 39; Neugebauer +, AJ 93, 1057; Carico +, AJ 92, 1254) and on galactic evolution and activity (see Schweizer, Science, 231, 227).

Several papers have brought still usefull photoelectric photometry data (Lauberts, AA Suppl 58, 249; 68, 15; Véron-Cetty, AA Suppl 58, 665; Poulain, AA Suppl 64, 225; Bergvall and Olfsson, AA Suppl 64, 469; Peterson and Baumgart, AJ 91, 530; Gallagher and Hunter, AJ 92, 557; Burnstein +, cited above; see also Lucey, MN 222, 417 for a discussion on the L-σ test). A large copilation of such measurements has been made by Lauberts and Saddler(39.002.054).

Improvements in CCD technology should allow progress in the near future in several respects. A higher accuracy in photometric measurements and colors has already been reached with scanning CCD data (Boroson and Thompson, cited above) in the continuation of a previous study in this mode; only sky fluctuations seem to limit the accuracy of such data, and no longer the pixel-to-pixel variations. Field limitations are starting to be overcome with CCD mosaics, allowing larger field-studies (Cohen, AJ 92, 1039).

This report is intensionally limited to a) non-active galaxies (Active galaxies are discussed in another report), b) bright galaxies. However, new frontiers have been reached as far as very deep photometric measurements are concerned. Faint galaxies have been measured with photographic plates (e.g. Koo, ApJ 311, 651) and with CCDs (e.g. Schneider +, AJ 92, 523). The latter however have brought very spectacular results: Djorgovski and Spinrad (ApJ 1987, preprint) have extended the Hubble diagram in the B, V, R, up to z=1.82, while objects as faint as the 27th magnitude in the V band have been detected with chopping techniques cancelling low-level systematics (Tyson, JOSA, 3, 2131). Note also an attempt of measuring redshifts of 22th mag galaxies from multicolor CCD photometry (Loh and Spillar, ApJ 303, 154) and the very powerful technique of multi-aperture spectophotometry (Soucail +, AA, preprint) applied to derive magnitude and redshifts of a large number of faint galaxies in clusters through metallic masks (Mellier +, AA, preprint). There are all the reasons to believe that these permanent technological improvements wil continue in the next years and yield fundamental discoveries on the structure, the formation and the evolution of galaxies.

10. Working Group on International Motions in Galaxies
(C.J. Peterson)

Over the last three years, the major portion of the work of several hundred publications on international motions in galaxies and their interpretation has continued to be the accumulation of data from optical and radio studies, for individual galaxies as well as in more extensive surveys of larger samples of galaxies selected for a wide variety of reasons. Major stimuli for continued study have been the desires to clarify the relationship between kinematical and morphological properties of galaxies (Whitmore ApJ 278, 61), the use of such relationships (Tully-Fischer, Faber- Jackson) for distance determination (for E galaxies, see Dressler + ApJ 313, 42; Djorgovski + ApJ 313, 59; for spirals, see Giraud AA 155, 283; ApJ 301, 7 and ApJ 301, 7), and especially the improvement of observational constraints on dark versus visible matter (e.g. Kent AJ 91, 1301).

A major highlight of the past three years is the progress that has been achieved in obtaining high quality data to ascertain the range of galaxy kinematical properties and their relations to other physical parameters. As part of a long-term study on the kinematical properties of disk galaxies, Burnstein and Rubin (ApJ 297, 423) have analyzed rotation curves of 60 Sa, Sb, and Sc galaxies to conclude a) that galaxies of very different optical morphology and luminosity can have simular rotation curves, and b) that derived mass distributions for most galaxies fall into three well-defined integral mass types. The form of the mass distribution does not correlate with Hubble type or with Bulge-to-disk ratio, mass or mass density, size or any other global property of the galaxies. Burnstein + (ApJ 305, L11) also have shown that rotation curves of 20 galaxies in large clusters provide evidence for dependence of the distribution of mass types not on environment, but on Hubble type. Rubin and Graham (ApJ 1987) are now studying high-resolution rotation curve data to probe the distribution of matter at small radial distances from the nuclei of spiral galaxies.

Analysis of kinematical data for the shape of the gravitational potential in most disk galaxies is model dependent, but polar-ring galaxies provide a rare circumstance for assessing the distribution of dark halo perpendicular to the galaxy disk. Whitmore + (ApJ 314, 439), comparing disk circular velocities to the rotational motions in the polar rings of three galaxies, show that the gravitational well is essentially spherical. The origin of polar rings has been addressed by Whitmore + and also by van Gorkom + (ApJ 314, 457).

Barred galaxies continue to be of interest. Davoust (37.151.092) has produced a comprehensive review of kinematical and dynamical models for barred systems and Pfenniger (AA 141, 171) has produced new models. Teuben + (MN 212, 257), from numerical integration of 2-dimensional orbits, have constructed dynamical rules wich barred spirals theoretically obey. Schwarz (MN 209, 93) has studied how galactic disks are affected by bar strength and pattern speed. Observational studies have appeared for NGC 1097 and 1365 (van der Hulst 37.157.043), NGC 6221 (Pence + MN 207, 9), NGC 7496 and 289 (Pence + MN 210, 547), NGC 1365 (Jorsater + AA Suppl 58, 507 and AA 140, 288), NGC 3359 (Ball 39.157.246), NGC 7741, 3359, and 7479

(Duval + AA Suppl 61, 141), NGC 1566 (Beckman + AA 157, 49), and NGC 3992 and 4731 (Gottesman + ApJ 286, 471). Tremaine + (ApJ 282, L5) have proposed a combined kinematical/photometrical technique for measuring the pattern speed in barred galaxies; this has been applied to an SB0 galaxy by Kent (AJ 93, 1062).

Elliptical galaxies also have continued to be of interest as their true shape is still not known; at least three tests have been proposed wich might indicate the true shapê of ellipticals. Biney (MN 212, 767) has suggested a test for triaxiality by comparison of rotational velocities along the major and minor axes; the test has been applied to observational data for 10 galaxies. Capaccioli + (MN 209, 317) have tested ellipticals by consideration of the correlation of velocity dispersion with ellipticity. Wyse + (ApJ 286, 88) have proposed consideration of mean surface brightness versus rotation as means of obtaining information on the true shape of a galaxy. Davies + (ApJ 303, L45) have studied the E2 galaxy NGC 4261 for which the rotation implies a prolate object. From study of extended gaseous emission in the E4 NGC 7097, Caldwell + (ApJ 305, 136) have referred the existence of a dark matter halo. More interest has been shown for peculiar elliptical galaxies. The kinematics of NGC 5128 have been studied by Wilkinson + (MN 218, 297) who have determined a relatively slow rotation of the stars perpendicular to the dust lane and by Hesser + (ApJ 303, L51) who have obtained data on the globular cluster Davies + (40.158.205) and van Gorkom (AJ 91, 791) have studied NGC 1052; the misalignment of the stellar and gas kinematical axes implies a recent capture for the gaseous material. Mollenhoff + (AA 154, 219) have surveyed dust lane elliptical galaxies.

An increasing amount of data is being obtained via conventional Fabry-Perot observations applied to the hydrogen emission lines, for example, the work of Buta (cited below), and other studies on NGC 300 (Marcelin + AA 151, 144), NGC 3109 (Carignan, ApJ 299, 59), and the Vela ring galaxy (Taylor + MN 208, 601). Fabry-Perot line reconstructions techniques (TAURUS) will be of increasing importance in the future. Studies have already been done, for example, in NGC 5642 and 7582 by Morris + (MN 216, 193) and for M83 by Allen + (39.157.019, 40.157.166, and Nature 319, 296). The use of emission lines from CO is also beginning to be exploited for kinematical purposes, as, for example, in studies of M51 by Rydbeck + (AA 144, 282) and the SBc galaxy NGC 5383 by Ohta + (P. Japan 38, 677).

Of the many other studies made on other galaxies, a number deserve specific attention. Giovanelli + (ApJ 301, L7) have studied UGC 12591, an S0/Sa galaxy with the largest rotation velocity (500 km/s) yet observed in a disk galaxy. Buta (ApJ Suppl 61, 609; 61, 631; 64, 1; and in press) has observed the rings in normal galaxies and argued that the strong case exists for a link between rings and orbital resonances in a bar. Jarvis (ApJ, in press) has investigated the box-shaped S0 galaxy IC 3370, wich shows cylindrical rotation to a height of $8(50/H_0)$ Kpc above the plane in the agreement with predictions of theoretical models (May + MN 214, 131 and Binney + MN 214, 449). The motions of the nuclear gas in M31 have been considered by Boulesteix + (AA, in press) and Goas + (ApJ 297, 98), respectively. A comprehensive high resolution 21 cm survey of M31 has been presented by Brinks + (AA Suppl 55, 179 and AA 141, 195). Tonry (ApJ 283, L27) has given kinematical evidence for a central mass concentration in M32. Jarvis + (ApJ 295, 324) have investigated the dynamics of the bulge components of two spiral systems, NGC 7814 and 4594; the bulge of NGC 7814 has been similary studied by Bacon (AA 147, L16).

While the emphasis of the Working Group has traditionally concentrated on observations, advances in theory and theoretical interpretation of data cannot be ignored and a few items must be mentioned here. Of especial interest are the self-consistent N-body experiments of Smith and Miller (ApJ 309, 522) which produce model galaxies with flat totation curves. The linear programming technique of Schwarzschild for producing self-consistent galaxy models is being exploited by Richstone and collaborators (ApJ 281, 100, ApJ 286, 27, ApJ 296, 331, ApJ 295, 340 and 349, ApJ 296, 370). This method also has been applied by Schwarzschild (ApJ 31,

511) to galactic bars and by Vandervoort (ApJ 287, 475) to spherically symmetric galaxies. In other work, Martinet (AA 1987) has investigated the instabillity of radial periodic orbits along the rotation axes of bulges, spheroids, and rotating triaxial ellipsoids by numerical experiments in realistic gravitational potentials.

Several useful reviews appeared in this last triennial period. Athanassoula + (Ann Rev AA 23, 147) have reviewed rings and shells in galaxies. Pismis (41.157.263) has discussed the general kinematical properties of spiral galaxies. Efstathiou + (37.157.004) have reviewed rotational properties and the origin of rotation via tidal torques. The kinematics of the Magellanic Clouds and the LMC-SMC Galaxy interaction was reviewed by Freeman (37.156.030). A comprehensive discussion of the historical development and current understanding of galactic properties, concentrating on work by Carnegie astronomers has appeared as How Galaxies Rotate: Clues to Their Past? (Carnegie Inst. Perspectives in Science #2, 1986).

Finally we mention several catalogues that will serve as valuable guides to the available literature. Two compilations of velocity dispersion data have appeared: Whitmore + (ApJ Suppl 59, 1) have compiled 1096 central velocity dispersion measuremants in 725 galaxies; of these, 51 with at least three concordant measures are defined as standard galaxies. The Davoust + (AA Suppl 61, 273) catalogue tabulates 880 measures in 546 galaxies; systematic corrections are applied to yield a homogenous set of data. A catalogue of rotation curves for 116 galaxies has been prepared by Kyazumov (38.157.097).

This review of the literature of the past three years shows that study of internal motions in galaxies remains a very active endeavor at this time. The field wil remain active as the investigators move to refine and exploid the use of kinematical properties (both rotation curves as well as internal velocity dispersions) as an alternative method for distance determination to explore the local distribution and motions of galaxies and to improve the determination of the Hubble constant. Other questions of interest are the nature and distribution of dark matter, the true shapes of elliptical galaxies, and galaxy interactions and mergers. A significant increase in activity is promised in the near future as both the next generation of larger telescopes and the Hubble Space Telescope come into use

11. Working Group on Space Schmidt Surveys
(J. Lequeux)

The activity of this working group formed in 1976 has been very limited. In spite of the persisting need for a large-aperture, wide-field survey space telescope in the UV the chances for flying a long-duration mission of this type in a reasonable future look very small, and the technical studies have slowed accordingly. The existing ASTRO equipment contains a 38 cm imaging telescope with a 40 arc minute field and a 1.8 arc second resolution: this is a short-duration Space Shuttle mission wich unfortunatly will not fly before mid-89 but will give a first hint as to how the far-UV sky looks like at deep levels. For the moment, balloon flights are conducted succesfully by the Laboratoire d'Astronomie Spatiale at Marseilles at 2050A with a 39 cm telescope (2.3° field, 20 arcsecond resolution, intensified photographic detector). A 60 cm telescope with a 1° field and 5 arcsecond resolution using an existing 40 mm photon-counting detector has been proposed by a European collaboration to fly on the ESA EURECA platform (Astronomy version, flight duration 6 months); but the fate of this project is linked to the uncertain fate of the platform itself, which is related to the Space Station project. There is also a project in USSR for a 40 cm Ritchey-Chrétien telescope. Thus the present trend is towards more modest projects than the original 1 m Schmidt concept, but even those have an unclear future.

12. Working Group on Supernovae
(V. Trimble)

The supernova event of the triennium was, of course, 1987A. It will have aged by about a factor five between the writing and reading of these words, and many ideas will have changed. At the moment, we can say with some confidence that it occured (IAU Circ 4316) and showed the hydrogen lines of a type II (IAUC 4317, 4318). The optical light curve (IAUC 4316, 4320, 4329, 4330, 4333, 4348, 4377, 4388) was un-precedented (IAUC 4359), with very rapid rise, brief dip, and smooth linear in-crease lasting about 60 days before leveling off. The optical light was variable polarized (IAUC 4319, 4328, 4337, 4339, 4340, 4358, 4361) and reddened very quickly (IAUC 4320, 4325, 4326, 4328, 4332, 4338), energy shifting out of the UV (IAUC 4317, 4320, 4320, 4327, 4330, 4348, 4367, AApLett 1 May) and into the IR (IAUC 4347, 4351, 4353, 4354, 4368, 4370, 4374,) at the same time. Optical color plateau-ed and the UV began to cover as the ligh curve flattened (IAUC 4341, 4369, 4377). A radio outburst was much fainter and shorter-lived than those associated with more distant SNe (Nature 327, 38). The spectrum has evolved in complex ways and shows much structure due to absorbtion in interstellar gas in both galaxies (IAUC 4320, 4323, ESO Messenger No 47, AAp Lett 1 May). The gas velocity was initially at least 17,000 km/s but dropped rapidly (IAUC 4320, 4326, 4331, 4336, 4342, 4352, 4361). SN 1987A emitted neurinos in one or two bursts (IAUC 4323; PRL 58, 1490 & 1494) whose temporal pattern constrains the rest mass of the electron neutrino to <10-15 eV (PRLJ 58, 1906; Nature 326, 476). X-ray and γ-ray detectors had yield only upper limits (e.g. 0.4 mCrab and 0.3 γ/cm^2.s) up to day 90 (IAUC 4336, 4365, 4367, 4387). The progenitor was almost certainly a B3Ia star previously catalogued as Sk-69^0 202 (IAUC 4317, 4366, 4325, 4333, 4349, 4356). There is a compact HII region nearby (IAUC 4376), and early speckle work shows the SN itself unresolved at 0".015 (IAUC 4369, 4389), but there seems to be a 7 m object of uncertain nature 0".06 away i PA 194° (IAUC 4382, 4391). Models of the progenitor's evolution and the explosion process are exceedingly numerous, but largely still in preprint form.

On other fronts, the Berkeley automated supernova search found its first three events, 1986 I, N, and O (IAUC 4219, 4287, 4298), two of them early enough for important follow-up work to be done. Robert Evans continued to find SNe by visual methods, including 1986G in Cen A (IAUC 4208) which will surely be the most under-appreciated supernova in history, 1987B! (IAUC 4321). The triennium also saw a probable SN in a QSO host galaxy (ApJ 291,L37) and one in a Seyfert (ApJ 293, L77)

Theorists continued to try to understand how energy from core collapse manages to eject Type II envelopes and to identify suitable progenitors for the nuclear detonations/deflagrations that are thought to make Type I's (Ann. Rev. AAp 24, 205). An unassisted core-bounce shock has great difficulty in getting out (Prog. Theor. Phys. 71, 524; Comm. 10, 149; ApJ Supp 58, 711; ApJ 299, L19). Among possi-ble SNI progenitors, there are problems both with cataclismic variables (ApJ 279, 166; 283, 241) and with merging white dwarf pairs (AAp 150, L21; ApJ 297, 53; ApJ 308, 161; MN 223, 319; ApJ 315, 229; ApJ 316, 733). The neutrino cooling of a new-born neutron star has been calculated (ApJ 307, 178), and the possibility that Type II with linear light curves might be powered by nuclear rather than gravitational energy has been explored (AJ 90, 2303). Shklovskii's suggestion (Sov AJ Lett 10, 302) that we see no SNII's in irregular galaxies because the low metallicity does not permit development of the extended envelope needed to make the light curves looks remarkably prescient given the properties of 1987A.

Meanwhile, back at the telescope, radio detections became sufficiently numerous for subtypes to be categorized (ApJ 285, L85; Sci. 321, 1251; ApJ 293, 400). M82 revealed a population of compact, variable radio sources that are argu-ably young SNe and SNR's (MN 211, 783; ApJ 291, 693; Sci. 227, 28); and a simular population may have been seen in NGC 3448 (AJ 93, 1045). The curious proto-type V, 1961V, is apparently still present at radio and optical wavelengths (ApJ 297, L29 & L33; PASP 98, 467)

A new subtype, generaly called Ib, was defined (ApJ 294, L17; AA 149, L7; ApJ 306, L77; AJ 91, 691; MN 200, 27p; ApJ 313, L69; ApJ 317, 1 June). Its members differ from the class Ia's in lacking spectral feature at 6115 and 6855 Å, in being more likely to emit detectable radio fluxes while being $1^m.5$ fainter and more likely to be associated with young stellar populations, and, in one case, in showing spectroscopic evidence for about 0.3 M_0 of Fe (MN 218, 93). They have been modeled as comming from massive Wolf-Rayet progenitors (ApJ 302, L59; AA 167, 265 & 274; ApJ 316, 231), from Helium white dwarfs (Sov AJ Lett 12, 152), from WD mergers (ApJ 304, 201; ApJ 305, 225), WD + RG binaries (AA 164, L13), and Helium star binairies (ApJ 310, L35)

The relative rarity of associations between SNR's and neutron stars continues to puzzle (Nature 307, 215; Comm. 11, 15). Evidence has accumulated for significant gas ejection just prior to SN explosions, presumably in a superwind (PASP 96, 789; ApJ 288, L17). And supernovae are proving a useful tool for probing the interstellar medium of other galaxies (ApJ 281, 585).

Finally, a number of lines of both observational and theoretical evidence have begun to suggest that standard supernova rates are overestimated by a factor of two or so, in M31 (AA 169, 14), the LMC (ApJ 281, L25), and elliptical galaxies (ApJ 213, 503), as well as in the Milky Way (Sov AJ 28, 137; AA 140, 431; MN 216, 691; ApJ 304, 657; ApJ 315, 555) for which the correct rate is probably about two per century, divided among Ia, Ib and II in the ratio 3:4:11 (van den Berg PASP in press 1987).

About 300 archival journal papers, many fine reviews, and half a dozen conference proceedings concerning supernovae appeared in English over the past three years. Only a very few of these could be mentioned, and many important topics are not addressed here at all. Please read the literature!

PRESIDENT: G. Cayrel de Strobel
VICE-PRESIDENT: P.S. Conti
ORGANIZING COMMITTEE: M.S. Bessell, A.M. Boesgaard, J. Jugaku, D.L. Lambert, O.H. Levato, M.A. Smith, J. Smolinski, M. Spite, S.C. Wolff.

In the triennium under review, from the second half of 1984 to the first half of 1987, Commission 29 has sponsored or cosponsored the following IAU Conferences: Symp. 120, "Astrochemistry", GOA, India, December 1985; Symp. 122, "Circumstellar Matter", Heidelberg FRG, June 1986; Symp. 132, "The impact of very high S/N Spectroscopy on Stellar Physics", Paris, France, June–July 1987; Symp.123, "Advances in Helio and Asteroseismology", Aarhus, Denmark, July 1986; Coll. 87, "Hydrogen Deficient Stars and Related Objects", Bangclore, India, December 1985; Coll. 88, "Stellar Radial Velocities", Schenectady, N.Y., U.S.A., October 1984; Coll. 90, "Upper Main Sequence Stars with Anomalous Abundances", Crimea, May 1985; Coll. 92, "Physics of Be Stars", Boukder, CO, U.S.A., August 1986; Coll. 94, "Physics of Formation of FeII Lines Outside LTE", Capry, Italy, July 1986; Coll. 95, "Second Conference on Faint Blue Stars", Tucson, AZ, U.S.A., June 1987; Coll. 102, "UV and X Ray Spectroscopy of Astrophysical and Laboratory Plasmas", Beaulieu-sur-Mer, France, September 1987; Coll. 108, "Atmospheric Diagnostic of Stellar Evolution: Chemical Peculiarities, Mass Loss and Explosion", Tokyo, Japan, September 1987.

This report is a common effort by experts in various fields of stellar spectroscopy.

Short reports on the activity of the four working groups (WG) sponsored by Com. 29 and prepared by their chairmen are also included.

A list of abbreviations is given at the end of this report.

1. O, Of, and WOLF-Rayet stars. (by P.S. Conti)

Several detailed reviews concerning these stars have appeared recently: a monograph viewing the entire field has been completed by Conti and Underhill (NASA SP – in press). Chiosi and Maeder (Ann Rev 24, 329) have discussed the evolution of massive stars with mass loss; the problems of WOLF-Rayet stars have been considered by Abbott and Conti (Ann Rev 25, 113), and by Underhill, summarizing her rather different views (Pub Asp 98, 897). An IAU Symposium on luminous stars and associations in galaxies has also appeared (De Loore, Willis and Laskarides, eds.; Reidel 1986). I will give here a personalized overview of what I consider the most important developments and issues at the present time. The reviews listed above contain more details and very complete references.
The stellar parameters of luminosity, effective temperature, mass and composition are now relatively well understood for the O and Of stars in broad outline, but additional data on the masses from well studied binary systems would be most welcome. Spectroscopic studies with high signal to noise of selected O and Of stars are now possible, given the increased detector quantum efficiencies, and groups in Boulder and Munich have been taking the lead in these efforts. For example, Bohannan et al. (ApJ 308, 728) have analysed the atmosphere of ζ Pup, a well known Of star, and have found an enhanced helium abundance. Their continuing work (in progress) has also revealed enhanced helium for αCam (but not for either of the O supergiants in the belt of Orion). This careful work suggests that at least some O stars, including (all?) those already identified as Of, are already turning back

towards the hotter part of the HR diagram, as predicted by the newest models of stellar evolution.

The effective temperature of W-R stars remains controversial although most investigators now feel the values are somewhat higher than had been inferred in the past from continuum studies. Hummer et al. (ApJ in press) have shown that such methods are inadequate to determine stellar temperatures since the emergent continua above the Lyman limit are all similar for these very hot stars. They are able to reproduce the continuum of ζ Pup with a range of model effective temperatures, although only one of which will also give the observed line spectrum.

W-R stars have abundance anomalies as seen in their emission line spectra. These represent stars with their upper layers stripped away so the results of nuclear fusion in the stellar cores are visible. Hillier (ApJ in press) has completed a detailed analysis of the WN5 star HD 50896 in which the inferred abundances of carbon and nitrogen are close to that predicted by highly evolved stellar models. Using detailed models for the stellar wind, he also finds a relatively hot effective temperature for this object, similar to what would be expected for a helium burning core (ApJ Supp 63, 948 and 965).

Pollock (ApJ 320, 283) has discussed the X-ray emission detected with the IPC on the Einstein satellite for 48 W-R stars. Their X-ray luminosities cover a range of 2 ordres of magnitude; some of the X-ray brightest objects also have evidence of non-thermal radio emission. The source of the X-ray flux is probably similar to that in the Ob stars, namely shocks in their winds (e.g. Owocki and Rybicki ApJ 299, 265). Infrared photometry of 41 W-R stars have been given by Williams et al. (AA in press). Circumstellar dust emission is inferred about half the sample, primarily those of late WC subtype.

Detailed models of the winds of O type stars in the Magellanic Clouds has been presented by Kudritzki et al. (AA 173, 293). These incorporate the lower abundances believed to obtain for the stars; they predict lower terminal velocities and mass loss rates. The former prediction seems to be confirmed by observations but the mass loss rates do not seem to scale this way (Garmany and Fitzpatrick ApJ 1988 submitted). The stellar winds of the hot stars of M31 seem to be inordinately weak according to preliminary optical and IUE data presented by Hutchings et al. (ApJ 1987 in press). The normally strong CIV and NV Resonance lines are weak in the handful of stars so far observed in this galaxy. It is not clear whether this is a result of sampling only the brightest stars, which are often pathological, or whether it represents some fundamental new result concerning hot stars in SB type galaxies.

Discovery of new W-R candidates in M31 has been carried out by Massey et al. (AJ. 92, 1303) and Moffat and Shara (ApJ 320, 266). Massey and associates also discussed the Ob star population of several associations in M31 using CCD photometry, finding relatively massive stars in those groups that also contained W-R stars. Spectroscopy of W-R stars in this and other local group galaxies has been carried out by Massey et al. (AJ in press). The WN stars in the solar vicinity, the LMC and M33 seem similar, as do the WC stars. Those W-R stars found in the SMC, NGC 6822 and M31 seem to have relatively weak emission lines as if the winds were less energetic but these results are still provisional and will need more data. Late type WC stars have not been found in any locales other than inward from the Sun in our Galaxy. The explanation for this is not settled. Kunth and Schild (AA 169, 71) have identified some additional more distant galaxies with W-R emission features (HeII). There seems to be a correlation between the HeII emission measure and the luminosity of the parent galaxy, which suggests the mass of newly formed stars scales as the mass of the galaxy.

Moffat and Associates (3 papers in press in the ApJ) have carried out a series of polarization studies of W-R stars. The polarization is due to electron scattering in the wind and is variable in a number of cases: in those known W-R plus O type binaries it is periodic and due to the orbital motion and rotation of the star; in some apparently single stars the changes in polarization are random and caused by inhomogeneities in the wind. They also give an ingenious method for deriving the mass loss rate in the binary W-R stars. Questions of variability in the emission line spectrum and in the continua of W-R stars are still unsettled issues but progress has been reported by Lamontagne and Moffat (AJ 94, 1008). They find variability in just about all the stars they have observed. The entire topic of variability is one in which a concerted observational effort will help our understanding of complicated physical phenomena.

Improvements in modelling of high mass stars, with the effects of mass loss and overshooting are continuing, (Maeder AA 173, 247 and 178, 159; and Maeder and Meynet AA 182, 243). Although main sequence mass loss is reasonably well known (to a factor two or so) the question of overshooting is unsettled theoretically. It is estimated empirically by attempting to fit the observed distribution of hot luminous stars in the HRD by adjusting the mixing outside of the convective core. This seems a reasonable approach and does lead to W-R stars with predicted properties in agreement with those observed. The luminous blue variables (LBV) may play a key role in the evolution of the most massive O stars to the W-R phase. Langer and El Eid (AA 167, 265 and 274) and Langer (AA 171, L1) have considered the evolution of very luminous stars and the origin of the different W-R Sub-types. The stellar models still differ in the details but seem to be converging towards consistency in interpretation of the W-R stars as being the helium burning end products of massive star evolution.

B and Be stars   (by D. Baade)

The primary track left behind by B-type stars in the literature of the report period once more was one of mass loss and variability – even without the case of Sk –69 202. Percy's contribution to the report of Comm. 27 and the progress reported by Comm. 36 in the treatment of stellar winds therefore contain essential complements of this summary. The available X-ray observatories were, for various reasons, not suited for the study of the weak X-ray flux from single B stars. Most of their B-type targets were therefore in close binaries so that the field is left to Comms. 42 and 44. This approach is partly extended to binaries in general since there is no new evidence that binarity is more important for B stars than for other stars. (References to proceedings, *etc.* are by *Astronomy and Astrophysics Abstracts* numbers where available. Papers that reference other recent work by their author(s) on closely related subjects, are marked by a plus sign "_+". Mentioning of review papers has purposely been kept to a minimum; the relevant ones are easy to spot.)

   Be star envelopes. For IRAS, Be stars not unexpectedly proved the most conspicuous among the B-type stars and inspired an ongoing series of papers (*AA*, **176**, 93 _+) by Waters, Coté, and Lamers. The density decreases as $r^{-2.5\pm0.5}$, indicating a slightly accelerated outflow. Mass loss rates are with a simple disk model up to two orders of magnitude higher than inferred from UV resonance lines but agree roughly with earlier analyses of Hα emission lines and ground-based IR data. While the IRAS data *per se* do not constrain the envelope geometry, the existing evidence for equatorial disks was considerably extended by Hayes's (*ApJ (Letters)* **287**, L39 _+) finding that the intrinsic polarization of ω Ori varies at *constant* polarization angle. From the analysis of a major body of Hα profiles Dachs *et al.* (*AA* **159**, 276 _+) also infer flattened enevelopes. A first, though only one-dimensional, interferometric resolution in Hα of the envelope of γ Cas was achieved by Thom *et al.* (*AA* **165**, L13). With the lifetime of *IUE* now approaching 10 years, important long-term studies become possible. Doazan *et al.* (*AA* **182**, L25 _+) discovered in 2 Be stars an association between the still unexplained long-term V/R variability of the Hα emission (but for which Okazaki, 42.064.067, has renewed the interesting model of one-armed oscillations of a cool disk) and the strength of the C IV UV resonance lines. Similarly, *IUE* (Barylak & Doazan *AA* **159**, 65) and *Voyager* (Peters & Polidan in Proc. IAU Coll. **92** *Physics of Be stars*, eds. A. Slettebak and T. Snow, p. 278) observations indicate long-term variations in the far-UV flux of Be stars; a correlation with the appearance of envelope features is possible.

Winds and discrete UV components. The discrete components of UV resonance lines continue to annoy theorists and to challenge observers. Henrichs (in *O, Of and Wolf-Rayet Stars*, eds. P. Conti & A. Underhill, *NASA/CNRS Monograph*) meticulously discusses in particular the variability patterns and the numerous attempts of an explanation. In non-supergiant B stars, discrete components are closely coupled to the Be phenomenon (Henrichs, *op. cit.*) and, there, intimately related to the so-called superions (Marlborough & Peters *ApJ Suppl.* **62**, 875_+, Grady *et al.* *ApJ* **320**, 376) which in O-stars Pauldrach (Ph. D. thesis, Munich) finds to be a non-LTE effect. Hubený *et al.* (*BAI Czech.* **36**, 214) caution against confusions of HVC's with shell lines; but the observed variability and an extension of the analysis beyond a range of a few Å should provide a clear discrimination.

High-quality observations are now also possible of luminous stars in other galaxies. The Heidelberg group (Wolf, Stahl, Leitherer, Zickgraf, Appenzeller) continued their work on luminous blue variables (LBV's) in the LMC and, *e.g.*, observed an outburst in R 127 (*AA* **127**, 49_+), detected extended circumstellar emission line regions in long-slit spectra (*AA* **158**, 371), and discovered R 81 (*AA*, **184**, 193) as the first eclipsing LBV. Kenyon & Gallagher III (*ApJ* **290**, 542) observed massive winds also in LBV's in M31 and M33; in the same galaxies Massey *et al.* (*AJ* **90**, 2239) found low outflow velocities but no P Cygni profiles in several seemingly normal OB supergiants. Garmany and Humphreys (*AJ* **90**, 2009) identified a population of unusually luminous Be stars in the Magellanic Clouds. Polarimetry (Hayes, *ApJ* **289**, 726_+) and optical (Markova *AA* **162**, L3_+) and UV (Lamers *et al.*, *AA* **149**, 29) spectroscopy saw variations in the mass loss of the galactic LBV P Cygni which appear to be due to semi-regularly (∼ 50-200 days) ejected shells; van Gent & Lamers (*AA* **158**, 335) attribute the shell ejection to possible nonradial pulsations, and Lamers (*AA* **159**, 90) argues that numerous weak metal lines in the Balmer continuum are the main driving agent of the very dense wind. Finally, Leitherer and Zickgraf (*AA* **174**, 103) report a visually resolved shell around P Cygni.

Pulsations and rapidly variable mass loss. In most OB stars UV mass loss rates associated with non-saturated resonance lines are fairly constant (Prinja & Howarth, *ApJ Suppl.* **61**, 357). However, this does not apply to LBV's and Be stars; and in normal supergiants considerable variability of HVC's, continuum polarization (Hayes *ApJ* **302**, 403) and Hα emission (Baade, in *O, Of and W-R Stars*) signals at least some structural variations. Two conferences (*PASP* **98**, 29-55, and *Instabilities in Luminous Early-type Stars*, eds. H. Lamers & C. de Loore, Reidel) have been devoted to this subject. Polarimetry by Hayes (*ApJ (Letters)* **287**, L39_+) and spectroscopy by Peters (*ApJ (Letters)* **301**, L61) and others of a few Be stars as well as the work on LBVs further strengthen the evidence for the existence of mass loss events. Penrod (41.122.026_+) found new examples of Be outbursts being related to amplitude variations of low-order ($m \approx 2$) nonradial pulsation (NRP) modes. The important clue is probably the ubiquity of such modes in Be stars and their absence in Bn stars (see also 40.112.126, 42.116.004, 42.122.025, and *Physics of Be Stars*). M.A. Smith (*ApJ* **304**, 728_+) continued his work on other nonradially pulsating B stars; at amplitudes near the soundspeed considerable nonlinearities seem to develop. Application of more 'objective' spectroscopic period search techniques to ε Per by Gies & Kullavanijaya (preprint) suggests more modes to be present than found before. It remains to be seen if this will reduce some of the former 'oddities'. The same method indepently found multi-periodicity also in a Be star (Baade, in IAU Symp. **132** *The Impact of very High S/N Spectroscopy on Stellar Physics*, eds. G. Cayrel and M. Spite); accordingly the NRP-versus-rotational-modulation discussion (Harmanec *BAI Czech.* **35**, 193; Balona *et al.* *MNRAS* **227**, 123_+) could come closer to a (partial) conclusion. Higher-order ($m \approx 10$) NRP's have also been detected in a few broad-lined B supergiants (Baade, in *O, Of and Wolf-Rayet Stars* _+).

Rotation, binarity and runaways. Abt & Cardona (*ApJ* **285**, 190) and Kogure & Suzuki (*Astr. Sp. Sc.* **127**, 143_+) note a near-complete lack of Be stars with periods below ∼ 50 days. If the rapid rotation of Be stars is not a surface phenomenon, this might explain their high specific angular momentum. Slettebak (*ApJ Suppl.* **59**, 769) gives v sin i's and spectral types for Be stars in 12 open clusters; color-magnitude diagrams do not furnish important evolutionary differences between B and Be stars. Rotation velocities of 53 Be stars have also been determined by Gao & Cao (40.116.058). Ruusalepp (42.116.012_+) and Stoeckley & Buscombe (*MNRAS* **227**, 801) have attempted to separate v and i by the comparison of different lines. With their shape-distorted gravity-darkened models, Carpenter *et al.* (*ApJ* **286**, 741) achieve consistency between the v sin i's derived from UV and optical lines. – In a thorough study of 36 proposed OB runaway stars Gies & Bolton (*ApJ Suppl.* **61**, 419) find only two binaries with certainty and discuss the origin of the singles.

Magnetic fields and abundances. Except for rather strong diople fields (and the possible first quadrupole: Thompson & Landstreet *ApJ (Letters)* **289**, L9) in stars which furthermore prove to have anomalous surface abundances (*e.g.* Bohlender & Landstreet in IAU Symp. 132) attempts (*e.g.* Barker *et al. ApJ* **288**, 741 _+) to detect ordered large–scale magnetic fields have been insuccessful down to the ~100 gauss level which, unfortunately, is still too high to settle the question of major magnetic atmospheric effects. Likewise, abundance analyses of most 'normal' Pop. I stars not known to be magnetic usually did not yield significant deviations from the average (*e.g.*, Ptitsyn & Ryabchikova *Astron Zh.* **63**, 527; Klochkova & Panchuk *Pis'ma Astron. Zh.* **12**, 928; Adelman *AA Suppl.* **64**, 173 _+); interestingly this also includes several stars probably formed above the galactic disk (Keenan *et al. AA* **178**, 194 _+). Similarly (but contrasting with analyses of HII regions), from B stars in clusters Gehren *et al.* (in *Production and Distribution of C, N, O Elements*, eds. Danziger *et al.*, ESO, p. 171) find no convincing evidence for a galactic abundance gradient. Wolff & Heasly (*ApJ* **292**, 589) confirm that in comparison with field stars B stars in h and χ Per are helium deficient by one third. CNO processed material is probably seen *in situ* at the surface of supergiants (Lennon & Dufton *AA* **155**, 79) while Michaud *et al.* (*ApJ* **299**, 741 _+) required an interplay between diffusion and mass loss to explain the abundances patterns of some OB subdwarfs. Gerbaldi *et al.* (*AA* **146**, 341) investigated correlations between chemical pecularities on the one hand and variable radial velocity (binarity, but possibly also pulsation), eccentricity and period on the other; they note a lack of peculiar stars with short periods and/or low eccentricity. Massa *et al.* (*ApJ* **287**, 814) attribute anomalously strong winds in some B V stars of NGC 6231 to a suspected carbon overabundance. Kilogauss magnetic fields do have dramatic effects on the wind (*e.g.*, S.N. Shore *et al. AJ* **94**, 737 _+) and may in connection with the wind also explain the probably non–thermal radio flux detected in two He-r stars with the VLA (S.A. Drake *et al.*, 40.116.082).

Unclassified. A most curious group of B stars was discovered by Downes (*ApJ* **316**, 763) which generally resemble sdB's but may switch from phases of normal Hα strength to no Hα at all and *vice versa* without other variations having been detected at low spectral resolution.

Line identifications and spectral and photometric calibrations. Rogerson *et al.* (*ApJ Suppl.* **58**, 265 _+) have listed identifications for 2200 UV features in the narrow–lined B0 V star τ Sco. Numerous spectral energy distributions have been published, *e.g.* by Fitzpatrick (*ApJ* **312**, 596) for LMC supergiants and by Goraya (*MNRAS* **222**, 121 _+) and M. Singh (*Astr. Spac. Sci* **120**, 133 _+) for Be stars. Millward & Walker (*ApJ Suppl* **57**, 63;) and Hill *et al.* (*PASP* **98**, 1186) have established empirical Hγ luminosity calibrations for luminosity classes V–III and supergiants, respectively. New calibrations of various photometric systems have been given by Schuster (*Rev Mex. A. A.* **9**, 53), Balona (*MNRAS* **211**, 973 _+), Kilkenny and Whittet (*MNRAS* **216**, 127), Cramer (*AA* **141**, 215), and Shulov (*Astron. Zh.* **63**, 734). Sekiguchi & Anderson (*AJ* **94**, 129) have presented an improved equivalent width/spectral type/luminosity class relation for the Si IV and C IV UV resonance lines.

A survey of Be stars in the λλ7500–8800 region was published (Andrillat, Jaschek and Jaschek, AA Suppl, in press). About one hundred stars were observed. IRAS excess radiation in Be stars and the behavior of the CaII infrared triplet was analysed (M. Jaschek, Andrillat, C. Jaschek and Egret, AA in press).

3a A, Ap, Am and CP Stars (by J. Jugaku)

The current state of knowledge concerning A, Am, Ap, and related stars was extensively reviewed at IAU Colloquium N°90 held at the Crimean Astrophysical Observatory, USSR, May 13–19, 1985. The proceedings of this meeting were edited by C.R. Cowley, M.M. Dworetsky and C. Megessier and published by Reidel in 1986. Some of the conclusions reached in the Colloquium may be summarized as follows.

The recent availability of ultraviolet spectra of many stars has had a major impact on the study of these objects. Simultaneously, traditional studies of optical spectra have been advanced by data obtained at very high spectral resolution and with high signal–to–noise detectors. It is now clear that the unusual chemistry

and magnetic structure of these objects have relevance across the broad domain in
the H—R diagram from the upper main sequence to horizontal branch stars and white
dwarfs. It may be that the majority of A and B stars have non—solar abundances, at
least for some elements. (Two of the brightest early A-type stars in the sky,
Sirius and Vega, show signatures of chemical peculiarity).

Much of the chemistry of Ap and Am stars is explicable in terms of the diffusive
fractionation. However, it has also become clear that the detailed chemistry
cannot be understood without consideration of several hydrodynamic and
hydromagnetic processes. For example, high mass—loss rates are capable of
explaining the λ Bootis stars, while somewhat lower rates may produce the
anomalies that appear in Am stars. Additional physical factors that may be
important include meridional circulation, magnetic fields, and turbulence. The
chemical differentiation is the simplest and most promising mechanism for
explaining the peculiarities of CP stars. However, several other mechanisms can
cause abundance anomalies in main sequence A-stars, and it may be reasonable to
consider additional processes in order to explain the observational complexity and
variety of the chemically peculiar stars. We may ask, for example, what abundance
patterns to expect in the case of mass transfer in binary systems? Theoretical
discussions of stellar magnetism focused on a global approach in which fossil and
dynamo—generated fields were intercompared. General problems with stellar
magnetism are not restricted to Ap stars and much has to be learned about
hydromagnetic processes that are active during star formation and pre—main
sequence phases, as well as during the hydrogen—burning phase. Although many Ap
stars have strong magnetic fields, recent advances in Zeeman—effect techniques
have demonstrated that only few Hg—Mn stars could have detectable magnetic fields.
The standard AOV star Vega has been shown not to have a detectable field (B=—
9±19 G). The Hg—Mn stars may become favorable objects in which to study the
distinct abundance patterns that manifest themselves in non—magnetic chemically
peculiar stars.

Andrillat, Jaschek and Jaschek studied the variability of Ae and A shell stars
(Harvard Obs. Publ.7, 193). They also studied the behavior of the Hα line (AA
Suppl. 65, 1) in these stars. A survey of 28 northern stars in the photographic
region was carried out by M. Jaschek, C. Jaschek and Andrillat (AA Suppl., in
press). The infrared behavior of these stars was analysed by means of the IRAS
satellite (Jaschek, Jaschek and Egret – AA 158, 325) and some βPic like objects
were detected.

A copy of the detailed reference list of papers in the period from the second half
of 1984 through the first half of 1987 is available on request from J. Jugaku.

3b Horizontal Branch Stars (HBS) (by J. Jugaku)

The book by Adelman and Leckrone on "Horizontal-branch and UV-bright stars" (ed.
A.G.D. Philip, Davis press inc., Schenectady, N.Y.), contains a very comprehensive
review of this subject and will be referred as HBUVBS, hereafter). In it, (p.75)
the authors made an abundance analysis of the prototype HBS star: HD 109995, using
UV and optical spectral regions. Adelman and Hill (MN, 226, 581) also made fine
analyses of HD 109995, HD 16817, and HD 64488 based on coadded high—dispersion
spectrograms. Using infrared lines Adelman et al. (PASP 98,783) determined oxygen
and nitrogen abundances of the above three stars as well as those of the A stars,
Θ Leo and HR 6559. Klochkova and Panchuk (Astr ZH, 64, 74) determined chemical
abundances of 6 stars (HD 2857, 64488, 93329, 105262, HDE 281679, and BD +20 5009)
using high—dispersion spectra obtained with the USSR 6—M telescope. Spectra of
three proto—type HBS (HD 86986, 109995, 161817) were also obtained with the 6—M
telescope by Philip and Lee (HBUVBS, p.57). An LTE analysis of HD 214080 was made

by Keenan and Lennon (AA 130, 179) but this star turned out to be an ordinary Pop I star. An analysis of five high-latitude blue stars were made by Kilkenny and Lydon (MN 218, 279). They used 30 A/MM Reticon spectra and UVBY photometry to determine T_{eff} and log g and checked the results by fitting Kurucz models. Hayes (News Lett Astr Soc NY 2, No9, 13) discussed a method of determining T_{eff} and log g for HBS. The energy distributions of 16 HBS have been measured by Hayes and Philip (IAU Symp 111, p.469). Similarly IUE ultraviolet energy distributions of four field HBS were obtained by Philip et al. (PASP 99, 54). Jaschek et al. (AA152, 439) pointed out that the features at 1600 and 3040 Å in IUE low-resolution spectra are similar to those previously found in λ Boo stars. Peterson (HBUVBS, p.85) summarized the current status of the presence of primordial stellar rotation among HB stars in globular clusters. Philip (HBUVBS p.41) identified a number of field horizontal branch (FHB) A-type stars in areas at high galactic latitude. Radial velocities and line widths are measured for 7 blue HB stars in the globular cluster NGC 288 by Peterson (Ap J 294, L35). The data support the idea that rotation increases the proportion of blue stars on the HB of some intermediate-metallicity clusters.

4. F, G and K stars (by M. Spite)

The F, G, K stars, characterized by temperatures between about 4000 and 7000K, include a large variety of stars at different stages of evolution. Some supergiants are massive stars crossing rapidly this region , but the majority of the F, G, K stars are low mass stars evolving slowly.
Many of these stars are young but some of them are as old as the Galaxy . Therefore, this type of stars is particularly suitable, to study the metal enrichment of the Galaxy. Definite progresses have been made over the last few years in determining the chemical evolution of the Galaxy in particular because efficient detectors and spectrographs became available at the large telescopes. Because of their low rotation, G and K stars are also very suitable for line-profile studies, evidencing effects of convection, stellar oscillations and magnetic fields.

FIELD STARS (abundances)

Large progresses have been made in the determination of the abundances of the light elements in the old stars.
- Carbon and nitrogen abundances have been determined in a large sample of disk and halo stars from intermediate dispersion spectra by Laird (ApJ.289,556). He found that C/Fe is constant over a wide range of metallicities; the N/Fe ratio is also constant for the majority of the stars indicating that in the halo the nitrogen production has been largely primary. Moreover four halo stars are found with a large N/Fe ratio .
Independently carbon abundance have been measured in 32 halo dwarfs by Tomkin and Lambert (ApJ.302,415). They show that indeed carbon follows iron down to [Fe/H]=-1.8 , but that at lower metallicities it seems that there is a positive trend in [C/Fe].
-Gratton and Ortolani (AA.169,201) have derived the [O/Fe] ratio in 18 stars of different chemical composition. They confirm a constant overabundance of oxygen in metal-poor stars.
-The ratios of the light metals Na, Mg, Al, Si ... to iron have been studied in particular by Tomkin et al. (ApJ.290,289), Laird (ApJ.303,718), François (AA.160,264 and AA.165,189), Luck and Bond (ApJ.292,559).
The abundance of all these elements relative to iron is constant in the halo stars. But the even elements are overabundant in the halo and the odd elements (at least down to about [Fe/H]≈-2) have a solar ratio relative to iron. As a consequence the odd/even effect is enhanced in the halo stars.

An isotopic abundance analysis for Mg in disk and halo stars is also reported by Barbuy (AA.151,189) and Lambert (ApJ.304,436). They confirm an enhancement of the odd/even effect in the metal deficient stars.

An important point is also the discovery of a metal poor star with "r" process overabundances by Sneden and Pilachowski (ApJ.288,L55). In this star the distribution of the heavy elements abundances is consistent with pure r-process neutron synthesis reactions.

FIELD STARS (atmospheric structure)

The fine structure of stellar photospheric convection: the stellar equivalent of solar granulation can be analyzed with the help of high resolution spectroscopy. The study of stellar granulation and the ensuing inhomogeneities on stellar surfaces has considerable potential, not only for analysis of stellar atmospheres, but also in applications for other astrophysical problems. A fair number of stars has now being observed for photospheric line-asymmetries, revealing significant asymmetry changes across the HR-diagram. Detailed line shape observations for several G and K stars are in Dravins (AA 172, 200 and AA 172, 211).

Observations of stellar granulations coupled with two-stream numerical simulations (D.F. Gray and C.G. Toner, PASP 97; 543) have helped us to understand some of the factors, such as the Expanding-Star Effect and the Rotation Effect (D.F. Gray, PASP 98, 319) that mold the line asymmetries of F, G, and K stars. A reversal of the sense of the line asymmetries was discovered in Ib stars (D.F. Gray and C.G. Toner, PASP 98, 499). Conventional granulation is seen for stars cooler than G0 Ib, but for hooter stars, a remarkable reversal is found, possibly indicative of upward velocities ≈ 25 km/s in the photosphere. Preliminary evidence points to a granulation boundary in the HR diagram at which these sudden alterations in granulation occur (D.F. Gray 1986 Advances in Space Research 6, 161).

The important development of helio-seismology has yield very important results in these last years, both in Doppler-shifts and broad-band photometry. Fossat tried to detect the same kind of small amplitudes p-modes on solar-type stars. He presented at the IAU Symp.132 results concerning seismology of the stellar cores of Procyon, α Centauri and ε Eridani.

The Zeeman effect has been used to detect magnetic fields in F, G and K stars (Marcy ApJ 276, 286). An interesting review paper (L.W. Hartmann and R.W. Noyes, Ann. Rev. Astr. Astrophys. Vol 25) parallels the stellar and solar characteristics of magnetic fields and associated phenomena.

The structure of the FGK giant and supergiant stars is still not completely understood, but some progresses have been made :

Boyarchuk and Lyubimkov (Astrofizika 20,85) have studied the atmosphere of some F supergiants. The analysis of the FeI lines led to lower values of the microturbulent velocities than the lines of FeII or TiII. This discrepancy is explained by deviations from LTE in the degree of ionisation. In the upper layers of a few supergiant stars, supersonic microturbulent velocity is observed.

On the other hand Tsuji (Astrophys. and Sp. Sc.118,227) has shown that in red giant stars radial velocity gradients are larger for stars with larger turbulent velocities. The stellar turbulence may thus have something to do with the differential velocity field in stellar atmosphere.

OPEN CLUSTERS

High resolution spectra of twelve Hyades dwarfs have been analysed by Cayrel et al. (ApJ.146,249) to determine accurate abundances for metals. The temperatures of the stars have been derived from Ha profile. They obtain [Fe/H]=0.12±0.03. For two stars with high chromospheric activity they deduce however lower abundances. Diagnostic indices of active chromospheres have been also given by Rose (AJ.89,1238) for 35 Hyades and 31 Pleiades dwarfs from low resolution spectra.

Rose also proposes that active regions affect seriously many of the spectral lines
of the stars.

GLOBULAR CLUSTERS

The main problem with the globular clusters is to understand the reason of their
inhomogeneity. In M3, Norris and Smith (ApJ.281,255) have shown that variations in
the strength of the CN band persist down to the luminosity at which the theory
(meridional circulation) predicts that CN variations should originate. Moreover
Hesser et al. (ApJ.295,437) have shown that the abundance spread observed in wCen
originates near the main sequence. Norris and Pilachowski (ApJ.299,295) from a
spectroscopic survey of 5 globular clusters , found a positive correlation between
the behaviour of the sodium lines and the abundance of nitrogen.
On the other hand Leep et al. (AJ.91,117) found that some stars in M13 are very
oxygen poor.
Spite et al.(AA.168,197) have derived the distribution of the abundances in a star
of a globular cluster of the Small Magellanic Cloud. For the first time, lithium
has been detected outside of our Galaxy.

THE ABUNDANCE OF LITHIUM IN OUR GALAXY

The high efficiency of the new detectors in the red has allowed a precise study of
the abundance of lithium in different objects. Lithium is one of the few elements
synthetized during the Big-Bang , but it is very fragile and can be destroyed at
high temperature in the deep layers of the atmosphere of the stars.It is supposed
to be formed also in novae , in some giant stars, and by cosmic rays (spallation).
The aims of the precise determination of the abundance of lithium is, on one
hand the determination of the canonical abundance, and on the other hand, a check
of the structure of the atmosphere of the stars.
A/ -A decisive step has been made in our understanding of the destruction of
lithium in the atmospheres of the stars owing to the observation of the behaviour
of the lithium in the Hyades cluster.
First , Cayrel et al. (ApJ.283,205) studied the decline of the abundance of lithium
in the G and K dwarfs. They found from high quality spectra that the abundance of
lithium is decreasing more rapidly with decreasing temperature than heretofore
realized. The simple models of lithium burning in the convective zone (with or
without overshooting) does not agree with the observations.
Second , Boesgaard and Tripicco (ApJ.302,L49) have obtained spectra of F stars in
the Hyades showing the existence of an abundance gap of lithium (the Li-chasm) in
the F dwarfs of that cluster. There is a remarkable variation of Li/H with stellar
surface temperature : stars with Teff near 6600K show Li depletions by factors
>100 relative to stars 300K hotter and cooler. An explanation of this "chasm" by
diffusion in the F stars was proposed by G.Michaud (ApJ.302,650).
-The abundance of lithium has been observed in some other galactic clusters
spanned in age, but younger than the Sun.
The abundance of lithium has been measured in Praesepe by Soderblom and Stauffer
(AJ.282,L7),   NGC752 by Hobbs and Pilachowski (ApJ.309,L17), M67 by Hobbs and
Pilachowski (ApJ.311,L37) and Spite et al. (AA.171,L8). All the authors deduce
that in these clusters the initial abundance of lithium is about the same as in the
Hyades.
On the other hand in NGC7789 Sneden and Pilachowski (ApJ.301,860) have shown that
some of the "lithium rich" giants have a very low 12C/13C ratio (a signature of
mixing). Therefore, the surface lithium in these giants has not survived simply,
through a lack of convective envelope mixing.
B/In the field stars the abundance of Li has been measured in some more stars of
different types:
Pop.I dwarfs (Soderblom, PASP.97,54; Boesgaard and Tripicco (ApJ.303,724), Pop.I

giants (Smith and Lambert ApJ.303,226), Pop.II dwarfs (Spite et al. AA.141,56 and
Spite and Spite AA163,140). Moreover the ratio $^6Li/^7Li$ could be obtained for
several F and G stars (Andersen et al. AA, 136, 65; Hobbs, ApJ, 290, 284; Rebolo et
al. AA, 166, 195).

5   M, S, C Stars. (by D.L. Lambert)

Warning! What follows is a personal selection of the many results that deserve the
label "recent highlights of spectroscopic investigations of giants and dwarfs of
spectral types M, S, and C and the intermediate classes SC and CS; this
contribution is NOT a comprehensive summary of work reported since 1984. The
reader is urged to consult *Astronomy and Astrophysics Abstracts* and to read some of
the many review papers on the atmospheres, internal structure and evolution of
these stars.

5.1   Chemical Composition of Giants.   At last, abundance analyses are being
reported for M, S and C stars with a completeness approaching that attained for G
and K stars. Three factors might be identified as responsible for this happy
development:
(i) Fourier spectroscopy at KPNO and CFHT has provided access to diatomic
molecules (e.g., C_2, CH, CN, CO, OH, NH, H_2) in the relatively uncrowded infrared
windows. These molecules provide the elemental and isotopic C, N, and O abundances
that are tracers of H and He burning in interior layers. (ii) Near-infrared
spectroscopy with silicon diode arrays provides atomic lines in windows between
the strong molecular bands such as TiO in the M giants. These lines provide the
metallicity [Fe/H] as well as abundances of s-process products of He-burning.
(iii) Analyses of these spectra and others of crowded regions containing specific
lines (e.g. the TcI resonance lines near 4260 Å and the Li I doublet at 6707 Å)
are greatly facilitated by development of model atmospheres and spectrum synthesis
techniques.
Investigations of elemental and isotopic C, N, and O abundances shows that, as
predicted, the first dredge-up has rearranged the abundances: C is depleted and N
enhanced (Smith and Lambert, ApJ 294, 326 and ApJ 311, 843, here SL; Tsuji, ApJ
118, 227, and Astr. Ap., 156, 8). The oxygen isotopes ^{17}O and ^{18}O were studied by
Tsuji (above, and 1985 in *Cool Stars with Excesses of Heavy Elements*, p. 295) and
Harris and colleagues (ApJ 285, 674; ApJ 299, 375, here HLS): the ^{17}O abundances
appears to depend on the stellar mass and reconciliation of the observed and
predicted ^{18}O abundances was possible when the adopted ^{18}O (p,α) rate was reduced
on remeasurement in the laboratory.
The MS and S stars have been shown, as long expected from convincing arguments
involving the dissociation equilibrium of TiO, ZrO, etc. for mixtures with
C/O$\lesssim$1, a higher C/O ratio than the M giants. Abundances of s-process products are
enhanced. With the use of line blanketed model atmospheres, the M, MS, S stars are
shown to have near solar metallicities (see SL). The reported compositions suggest
that products of the He-burning shell (nearly pure ^{12}C plus s-process elements)
are mixed to the surface of these low mass (M $\leqslant$ 3 M_0) of MS, S, (and C) stars.
These observations challenge the theoreticians to produce satisfactory models of
low-mass thermally-pulsing AGB stars.
Using Fourier spectra, the CNO analyses were extended to a sample of cool carbon
stars (Lambert et al., ApJ Suppl., 62, 373; Harris et al., ApJ 316, 294). A
majority of the C stars show a surprisingly modest C enrichment (C/O $\leqslant$ 1.1). The
$^{12}C/O$ and the high $^{12}C/^{13}C$ ratios are consistent with the addition of pure ^{12}C to
the envelope. Fourier spectra of SC stars were analysed by Dominy and colleagues
(ApJ 300, 325; ApJ 317, 810). The CNO abundances generally resemble those of the S
and C stars. A key remaining problem is provided by the high $^{16}O/^{17}O$ and $^{16}O/^{18}O$
ratios in the M, S, and C stars (see HLS).

5.2 Atmospheric Structure Fourier spectra have been used also to investigate atmospheric structure and velocity fields. Hinkle, Scharlach and Hall (ApJ Suppl., **56**, 1) extended their earlier work on time series analysis of Miras to nine additional stars. These observations provide extensive information on the Mira atmosphere throughout the pulsation cycle. Wallerstein et al. (P.A.S.P., **96**, 222) report similar data for the S star R Cyg. Dominy, Wallerstein, and Suntzeff (M.N.R.A.S., **212**, 671) discuss the line doubling seen in infrared CO and visual and infrared atomic lines of V Cnc about 1 month after maximum light. Tsuji (1987, in *Circumstellar Matter*, in press) finds evidence in high resolution CO line profiles for a quasi-static turbulent layer above the stellar photosphere.

5.3 Winds and Shells Stellar spectroscopy across the spectrum is providing new insights into the structure of stellar winds and circumstellar shells of cool giants and giants. The selected highlights illustrate this claim. Mutilation of the 2800 Å Mg II chromospheric emission lines by Mn I and Fe I absorption provides evidence for a turbulent layer between the photosphere and the circumstellar expanding shell of the cool carbon star TX Psc (Eriksson et al., Astr. Ap., **161**, 305). The circumstellar gas around cool evolved stars is detectable by resonance line scattering of photospheric radiation. Mauron et al. (Astr. Ap., **165**, L9) provide an image of the extensive shell around the M supergiant μ Cep. Sahai and Wannier (ApJ, **299**, 424) used a Fourier spectrometer to detect resonantly scattered CO vibration-rotation emission lines at 4.6 μm from different regions of the shell around the shrouded carbon star IRC+10216, and constructed a model of the inner envelope. As 2D arrays in the visible and infrared are exploited more fully, spatially resolved spectroscopy is sure to reveal more of the secrets of stellar winds, shells, and the mass loss experienced by AGB stars. With new telescopes and more sensitive arrays, the triennium has seen new observations at millimeter and radio wavelengths of the molecular line emission from shells (e.g., Knapp, ApJ, **293**, 273; ApJ, **311**, 731). The maximum mass loss rates are close to the maximum predicted for a wind driven by radiation pressure on the mass in the wind (here, the dust) – see Jura (ApJ, **275** 681), and Knapp and Morris (ApJ, **292**, 640). With the implementation of mm-wave interferometers, the spatial structure of the line emitting shell is being mapped (Masson 1987, in *Late Stages od Stellar Evolution*, p. 119).

5.4 Magellanic Clouds No report on the spectroscopy of M, S, and C AGB stars would be a fair reflection of recent progress without a mention of the Magellanic Clouds and other distant ensembles of stars (e.g., the Galactic bulge). Through low resolution spectroscopy and photometry of LMC and SMC stars, many clues have been provided to the evolution and structure of AGB stars –see, for example, reviews by Wood (1987, in *Late Stages of Stellar Evolution*, p. 197; and 1987, in *Stellar Pulsation*, p. 250). Results include one period-luminosity law for low mass LPVs and another for high mass/supergiant LPV, and a proposal that stars above the mass M_{up} ($<3M_\odot$, perhaps) may shed their envelopes before the thermal pulses convert them from O-rich to C-rich. I expect these pioneering studies to continue and to be supplemented by high resolution spectroscopy of the brightest stars in the Magellanic Clouds and the Galactic bulge where metallicities as high as [Fe/H]$\approx$1 have been proposed from analyses of low resolution spectra of K giants (Whitford and Rich, ApJ, **274**, 723).

5.5 Cool dwarfs To conclude, I offer two comments on the spectroscopy of the cool dwarfs. One exciting development has been the observation of Zeeman split line profiles in infrared spectra of AD Leo, an M dwarf flare star (Saar and Linsky, ApJ, **299**, L47). The clear resolution of the Zeeman components of Ti I lines is indubitable evidence for a strong field (H$\approx$3000 to 4000 G). The interested onlooker may not have been convinced by earlier evidence for fields in various dwarfs and giants that was based on rather subtle broadenings of line profiles in the visible and near-infrared.

Since the tools have just become available, one expects to see M dwarfs subjected to detailed abundance and structural analyses. Analyses of pop. II dwarfs may yield novel data not readily provided by the warmer dwarfs; the molecular lines may provide critical isotopic ratios. Undoubtedly, peculiar stars will be discovered. However, few, if any, will rival the dwarf carbon star G77-61. Recent work shows that it is binary with an unseen companion (Dearborn et al., ApJ, 300, 314) and that the star is very C-rich but extremely metal-poor: [Fe/H]≥-5.6 is proposed by Gass, Liebert, and Wehrse (1987, preprint).

6. Pre-main-sequence Stars (by J. Krautter and R. Mundt)

In the period covered by this report (1984-1987) the number of outflow sources found increased enormously. This threw new light on our understanding of the star formation process, since mass outflow is now considered to be a very common and important phase in the evolution of low- and intermediate-mass stars. This development was - at least in part - only possible because of further improvements of spectrographs and detectors. Another sign of this period is the increasing number of publications based on simultaneous observations in different wavelength regimes.

The proceedings of the following meetings contain reviews and many original papers on the spectra of PMS stars: "Nearby Molecular Clouds", Toulouse 1984, Lecture Notes in Physics 237; "Protostars and Planets II", Tucson 1984; IAU Symposium no.115, "Star Forming Regions", Tokyo 1985; and IAU Symposium no.122, "Circumstellar Matter", Heidelberg 1986.

6.1 T Tauri Stars.

Recent reviews concerning spectroscopic aspects of T Tauri stars were given by Cohen (Phys Rep 116, 173), Herbig (Birth and infancy of stars, 535), Appenzeller (Phys Scr T11, 76), and Herbst (PASP 98, 1088).

Spectroscopic surveys in order to find new T Tauri stars were carried out by Herbig et al. (AJ 91, 575), Wiramihardja et al. (Stellar activities and observational techniques, 137), and Wilking et al. (AJ 94, 106).

Petterson (AA 171, 101) presented many detailed observations of hitherto unknown T Tauri stars in the Gum nebula, Whittet et al. (MNRAS 224, 497) studied the Chamaeleon T-association. Spectroscopic observations of a larger sample of T Tauri stars were carried out by Rydgren et al. (Publ US Naval Obs, 2nd ser., Vol. 25, 114) and Sun et al. (AA Supp. 62, 309).

Grinin et al. (1985, Astroph 22; 1985, Bull Crim Aph Obs 71, 111), Ismailov and Rustamov (1987, Sov Astron Lett 13), and Walker (PASP 99, 392) studied the variability of emission lines and continuum in several stars in Taurus and Auriga. High resolution spectra (R 2 x 10^4) were used by Appenzeller et al., (AA Supp 64, 65) to derive the basic properties of the different types of emission lines and of the photospheric absorption spectrum in the stars S CrA and GQ Lup. They explain the differences in the two spectra with the same physical model seen under different inclination angles. Kolitov (Sov Astron Lett 12) found strong Hα emission in LHα 324, contrary to earlier publications, where no Hα emission had been reported. Emission line widths were used by Saal. (MNRAS 222, 213) to study the flow velocities in the winds of T Tauri stars. A list of 520 absorption lines in the spectrum of T Tauri is given by Korutin and Krasnobatsev (1986, Bull Crimean Ap Obs 75). Physical properties of AS 353 A were determined by Böhm and Raga (PASP 99, 265). Boesgaard (AJ 89, 1635) found from high-resolution spectroscopy of RU Lup significant changes of emission line profiles and intensities within a few days. Giovanelli et al. (ESA SP 263, 95) report on X-ray, ultraviolet, optical, infrared and ratio simultaneous observations of RU Lup from 1983-1986. Strong variations on different time scales were detected in all the wavelength regions, both in the line intensities and in continuum.

Many papers deal with UV observations of T Tauri carried out with the IUE satellite. Review papers were given by Imhoff (NASA CP 2349, 81) and Giampapa and Imhoff (Protostars and Planets II, 386). The UV properties of PMS objects in general were reviewed by Imhoff and Appenzeller (Exploring the Universe with the IUE Sat, 295). Brown et al. (1984, NASA CP 2349, 338) present high- and low-resolution IUE spectra of RU Lup. Strong P Cygni profiles are observed. They derive a transition region density of 3×10^{10} cm^{-3} and discuss the status of the atmospheric modeling.

Further UV results are published by Simon (1984, NASA-CP2349, 183), Calvet et al. (ApJ 293, 575), Brown et al. (1986, ESA-SP 263, 177), de la Reza et al. (1986, ESA-SP 263, 107), and Herbig and Goodrich (ApJ 309, 294). A more theoretical interpretation of the most relevant IUE observational results is given by Lago et al. (1984, ESA SP-218, 233). The main topics are the presence of high temperature regions around T Tauri stars, stellar winds, and the origin of the continuum and molecular emission.

Finkenzeller and Basri (IAU Symp 122, 103; ApJ, in press) obtained calibrated optical high-resolution and high S/N spectra of 7 T Tauri stars of low to intermediate activity level. They conclude that all important features are clearly chromospheric, and that the physical processes do not differ qualitatively from the ones found in extremely active T Tauri stars. Upper limits to coronal line emission in X-ray detected T Tauri stars were measured and discussed by Lago et al. (MNRAS 212, 151) and Lamzin (1985, Sov Astron 29).

High-resolution spectra do now allow to very accurately measure rotational velocities in T Tauri stars. Hartmann et al. (ApJ 309, 275) confirm that T Tauri stars are generally slow rotators. They do not find a correlation between Hα emission and rotation. Bouvier et al. (AA165, 110) present rotational velocities of 28 T Tauri stars down to a resolution limit of a few km/sec. The lack of very slow rotations (vsini ≤ 6 km sec^{-1}) suggests that a minimum rotational velocity may be necessary for the T Tauri phenomenon to turn-on.

Several papers deal with IR spectroscopy. Persson et al. (ApJ 286, 289) studied Bracket-alpha line profiles with a velocity resolution of 45 km/sec of several T Tauri stars and embedded sources. Simon and Cassar (Ap 7283;, 179) use high-resolution spectra in the 2 - 2.4 μm region to study the conditions in the envelope of LHα 101. Cohen and Witteborn (ApJ 294, 345) show that silicon emission features around 10 μm are common in T Tauri stars. Thompson (ApJ 312, 784) finds an excess of hydrogen line emission.

Papers concerning more theoretical aspects of the T Tauri spectra and the line formation in the atmospheres and circumstellar envelopes are the review paper on "Theories of mass loss" by Hartmann (Fund Cosmic Phys 11, 279), the model for XUV ionization of He I in the atmospheres of T Tauri stars by Calvet (1984, Rev Mex Astron Astrofis) and Lago's wind model for RU Lup (MNRAS 210, 323).

6.2 Post-T Tauri stars (PTT). An increasing number of studies have been carried on these objects since the last report. A large number of potential PTT have been found by analysis of Einstein X-ray data (Walter, ApJ306, 573; Feigelson et al., AJ, in press) and by searches for CaII H+K emission through objective prism surveys (Herbig et al., AJ 91, 575). Spectroscopic studies of these low activity stars have discussed by Walter et al. (ApJ 314, 297). It has been argued by Walter that many of these stars are not PTT but so called "naked" TTS, which are coeval with the TTS; he argues that for some reasons (e.g. lack of an accretion disk) they show a much lower activity then normal TTS. The rotational velocities and kinematic properties of some of these stars has been studied by Hartmann et al. (AJ 93, 907). Several of these PTT show period light variations (Rydgren et al. AJ 89, 1015), which is attributed to spots on their surface.

Spectroscopic and photometric observations of the spotted star RY Lup are reported by Liseau et al. (AA 183, 274).

6.3 Herbig Ae/Be stars (HBeS). High quality spectra of 27 HBeS are discussed by Finkenzeller and Jankovics (AA Supp 57, 285). The line profiles are discussed and several important quantities (e.g. radial velocities) are derived from the data. Rotational velocities (v sin i), spectral types and the properties of their forbidden lines are derived by Finkenzeller (AA 151, 340). Intermediate mass pre-main sequence stars (Herbig Ae stars) were studied to establish further their properties as a class (Catala et al., AA 154, 103). The star AB Aur revealed short-term variability, with a rotational modulation of the wind velocity at various depths (Praderie et al. Ap. J. 303, 311, Catala et al. Ap. J. 308, 791, Catala et al., AA, 182, 115). Spectroscopic and photometric observations of the star Z CMa and Lk Hα 198 are reported by Covino et al. (AJ 89, 1868) and Chavarria (AA 148, 317), respectively.

6.4 FU Ori object (FUORS). The recent years brought decisive progress in the understanding of these unusual pre-main sequence stars. A number of observational data have been obtained, which strongly suggest that FUORS are surrounded by luminous accretion disks. The outburst is explained by very high mass accretion rate, being caused by an accretion disk instability. This model is based on new optical and IR spectra and is discussed in a series of papers by Hartmann and Kenyon (ApJ 299, 462; ApJ 312, 243; ApJ 322). The outflow source L1551-IRS5 was recognized as a FUOR by Mundt et al. (ApJ Lett 297, L41) on the basis of its optical spectrum. The Hα and NaD line profiles of the brightest FUORS have been discussed by Bastian and Mundt (AA 144, 57) and Crosswell et al. (ApJ 312, 227). A detailed study of the FUOR HH57-IRS has been carried out by Reipurth (AA 143, 435). The ring-shaped nebula of FUORS are discussed by Goodrich (PASP 99, 116). The photometric behaviour and light curve of FU ori and V1057 Cyg in the optical and IR are described by Kolotilov and Petrov (1985, Sov Astr Lett 11(6), 358) and by Kapatskaya (1984, Astrofisika 20, 138).

6.5 Herbig-Haro objects (HHO), optical jets. A strongly increasing number of papers has been published about this topic the last three years. Through intensive CCD imaging many more jets and also HHO have been discovered. It is now generally accepted that HHO and jets are highly related outflow phenomena in star formation regions. Both phenomena are tracing the outflowing matter in these regions with the highest degree of collimation (a few degrees) and with the highest velocity (200-400 km/s). On deep CCD images long-known HHO often turn out to be the brightest parts of a jet or collimated outflow. The following reviews have been published the last three years on this topic: Mundt, Schwartz (1985, in Protostars and Planets II, eds. D.C. Black and M.S. Matthews); Canto, Mundt (1985, in "Nearby Molecular Clouds", Lect. Notes Physics 237), Mundt, Schwartz (1986, Can. J. Phys. 64); Canto (1986, in "Cosmical Gas Dynamics", ed. F.D. Kahn); Staude (Ap Sp Sc 128, 179); Cohen, Dyson, Mundt, Norman (1987, in IAU Symp. 122) Mundt et al. (ApJ 319, 275). A review on molecular outflows and related topics has recently been given by Lada (Ann Rev AA 23, 267).

CCD imaging data and for spectroscopic data on individual HHO and jets have been collected by Axon and Taylor (MN 207, 241),Krautter et al. (AA 133, 169), Lenzen et al. (AA 137, 202), Mundt et al. (AA 140, 17), Brugel et al. (ApJ Lett 287, L93); Bohm and Solf (ApJ 294, 533), Hartigan and Lada (ApJ Supp 59, 383), Walsh and Malin (MN 217, 31), Schwartz et al. (AJ 90, 1820), Bührke et al. (AA 163, 83), Hartigan et al. (AJ 92, 1155), Krautter (AA 161, 195), Strom et al. (ApJ Supp 62, 39), Ray (AA 171, 145), Meaburn and Dyson (MN 225, 863), Solf and Bohm (AJ 93, 1172), Neckel et al. (AA 175, 231).

Recently several HHO have been discussed in terms of radiating bow shocks, created by a rapidly propagating jet or a bullet. Several HHO have been discovered recently, which have indeed a shape very similar to that of a bow shock. The most striking example is HH34 (Reipurth et al. 1986), several other examples are discussed by Mundt et al. (ApJ 319, 275). HHO models based on radiative bow shock are discussed by Raga (AJ 92, 637), Raga and Böhm (ApJ 308, 829), and Hartigan et al. (ApJ 316, 323).

7. The Spectra of White Dwarfs (by J. Liebert)

Observational advances in the last several years may be divided into the broad wavelength categories of (1) ultraviolet spectrophotometry, at both low and high resolutions, (2) high precision optical spectrophotometry, as discussed below. Some of the complementary advances in model atmospheres and envelope studies are also mentioned, as they pertain to the observational results. Wehrse, R. (1985, in Reports on Astronomy, Transactions of the IAU, Vol. XIXA, p. 513) recently reviewed the studies of white dwarfs using model atmospheres for Commission 36. A general discussion of the evolution of surface abundances in white dwarfs is included in the review by Sion, E.M. (1986, Publ. Astron. Soc. Pacific, 98, 821). This paper also includes a discussion of the observations of white dwarf primary stars of cataclysmic variables, an exciting subject which is not included here due to space limitations. A comprehensive review of colors and luminosities is given by Eggen, O.J. (1985, Publ. Astron. Soc. Pacific., 97, 1029). Trimble, V. (1986, Q.J.R. Astron. Soc., 27, 38) summarizes the diverse results on low luminosity stars presented at a conference in honor of Prof. J.L. Greenstein. Also worth noting in a more general context is the publication of model atmospheres and emergent fluxes for cool DA and DQ white dwarfs (Galdikas, A. 1985, Vilniaus Astron. Obs. Biul., NR. 72 and 73).

Ultraviolet spectroscopy. The international ultraviolet explorer (IUE) Observatory has permitted white dwarfs to be studied in the ultraviolet (1150-3100 Å) over the last decade, and an impressive list of discoveries and related advances are reviewed in Liebert, J. (1984, "White dwarfs at ultraviolet wavelengths") and in Vauclair and Liebert (1987, in "Exploring the universe with the IUE Satellite", eds. Y. Kondo et al., D. Reidel publishing Co., Dordrecht, p. 355). The unexpected spectroscopic features discovered by the IUE in a variety of white dwarf stars represent the most important contributions. On the other hand, the limitations in both the Sec Vidicon and the aperture have limited the accuracy of the absolute spectrophotometry for estimating the effective temperatures of hot stars. The comparison of IUE temperature determinations with those derived from optical measures have been quite good for several hot DA stars (Holberg, D., Wesemael, V. and Basile, J. ApJ, 306, 629). Wesemael, F., Lamontagne, R. and Fontaine, G. Astron. J. 91, 1376, used IUE data to refine the temperature range of the instability strip of the pulsating ZZ Ceti variables. Less encouraging was the agreement between optical and ultraviolet data fits for the hot DB stars (Liebert et al., ApJ, 309, 241).

The identification of strong, broad absorption features near 1400Å and 1600Å in cool DA (hydrogen atmosphere) white dwarfs was one of the unexpected IUE results. Koester et al. (Astr. Ap., 142, L5) and Nelan, E.P. and Wegner, G. (ApJ Let., 289, L31) independently and simultaneously proposed the explanation that the absorptions are caused by quasi-molecules of H_2 and H_2^+. IUE observations indicate that the features appear in all DA white dwarfs cooler than about 20,000 K. A similarly pervasive set of features in cool white dwarfs with helium-rich atmospheres had also been discovered earlier with the IUE; these are attribute to atomic and molecular carbon. Time-dependent calculations of the dredge-up of trace material from the carbon core offer quantitative agreement with the derived carbon abundances, provided that the helium envelope has a rather small mass (Pelletier et al., ApJ, 307, 242).

Earlier in the decade, metallic line features were identified in high dispersion
IUE Echelle spectra of hot DA white dwarfs, principally by F. Bruhweiler, Y. Kondo
and collaborators. While the low ionization transitions were generally
attributable to the interstellar medium, lines of higher ionization were found
which appeared to come from the stellar photospheres, while still others showed
modest blueshifts indicative of a selective stellar wind. During the more recent
period, theorists have made progress in evaluating these observations. Line
profiles and equivalent widths for a large range of heavy element abundances have
been presented for a grid of hydrogen-rich atmospheres by Henry, R.B.C., Shipman,
H.L. and Wesemael, F., ApJ, Suppl., 57, 145). The study of Morvan, E., Vauclair, G.
and Vauclair, S., Astr. Ap., 163, 145, indicates that the theory of radiative
acceleration offers a qualitative, if not yet a fully quantitative explanation.

Similar calculations predict more and stronger features in hot, helium-rich white
dwarfs, and a nitrogen abundance greater than solar has in fact been found in the
80,000 K do star PG1034+001 (Sion, E.M., Liebert, J. and Wesemael, F, ApJ, 292,
477). More spectacular still are the species found in ultraviolet and optical
spectra of the pulsating PG1159-035 stars (Wesemael,F., Green, R.F. and Liebert,
J., ApJ Suppl., 58, 379). The O VI, C IV and He IV transitions appear to link this
group to the helium-rich planetary nebula nuclei of the "O VI" and WC types (Sion,
E.M., Liebert, J. and Starrfield, S.G., ApJ292, 471). Surface abundances have not
yet been determined, however.

Optical spectrophotometry. CCD detectors have been used to obtain spectra of
unprecedented signal-to-noise ratio within the last several years. Greenstein,
J.L., ApJ, 304, 334) found numerous weak features in cool white dwarfs, and
demonstrated conclusively that the fraction of stars having hydrogen-rich
atmospheres drops decisively at $T_{eff} < 10,000$ K. In the optical ultraviolet, image
tube spectra especially using photon-counting detectors are still competitive. A
somewhat complementary study of a complete sample of hot DA white dwarfs (Fleming,
T.A., Liebert, J. and Green, R., ApJ, 308, 176) resulted in a revised determination
of the local birthrate of white dwarfs which is now less than half of the commonly
accepted formation rates for planetary nebulae nuclei.

Important leaps in the understanding of magnetic white dwarf stars have been
possible because of concurrent advances in the quality of theory and observations.
The first such star to show both circularly and linearly-polarized light,
GRW+70 8247, also exhibits a set of absorption features whose cause has been
unclear untill the last few years. That these are due to Zeeman-shifted
transitions of hydrogen in the presence of a magnetic field well in excess of 10^8
Gauss had been suggested in earlier papers by J.R.P. Angel and J.L. Greenstein.
However, this explanation is now firmly accepted following the precise matching of
the absorption features at optical and near-infrared wavelengths with the detailed
predictions from two independent sets of theoretical calculations (Greenstein,
J.L., Henry, R.J.W. and O'Connell, R.F., ApJ Let., 289, L25; Angel, J.R.P.,
Liebert, J. and Stockman, H.S., ApJ, 292, 260; Wunner et al., Astr. Ap. 149, 102).
Even more spectacular is the analysis of another case, the rotating magnetic
degenerate PG1034+234 which appears to have a surface field approaching 10^9 Gauss
(Schmidt et al., ApJ309, 218). Also worthy of special note is the discovery of
Zeeman-Split emission lines in the apparently single white dwarf GD356
(Greenstein, J.L. and McCarthy, ApJ, 289, 732).

8. Symbiotic stars (by M. Friedjung)

Nearly all who work in the field now agree on a basic binary model for these stars.
A compact mass gainer (main sequence star, white dwarf or perhaps even sometimes a

neutron star) accretes mass from a cool giant companion either as a result of Roche lobe overflow of the latter, or if this latter star does not fill the lobe by accretion from its wind. This conception is described in two books (Kenyon 1986, and the proceedings of IAU Colloquium n°103, Mikolajewska, Friedjung, Kenyon and Viotti, eds., in press).

Orbits are becoming much better known, especially through the measurements of the absorption lines of the cool component (Garcia 1986, Astron. J. 91, 1400, Garcia and Kenyon, Proc. IAU coll. 103). The radial velocity curve so determined agrees at least in the case of AG Dra with the orbit found from the reflection effect (Leibowitz and Formiginni, Proc. coll. 103).

Symbiotic stars can generally be divided into two classes depending on whether the cool giant is or is not a Mira variable (Whitelock 1987, PASP 99, 573 and Proc. IAU coll. 103). First results by Bensammar et al. (1987 preprint) indicate the power of Fourier Transform spectroscopy in studying the cool component and circumstellar material on the line of sight between the cool component and the observer. Spectral classification of this component is becoming more precise (Kenyon and Fernandez-Castro 1987 preprint). However metal underabundance and other abundance anomalies may disturb classification criteria as for AG Dra (Iijima et al. AA 178, 203). Most symbiotics seem nevertheless to have abundance ratios characteristic of those of normal M giants (Nussbaumer, Proc. IAU coll. 103). IRAS observations studied by Kenyion et al. (AJ 92, 1118) suggest that in some cases the cool but not the hot components are heavily reddened.

Many individual stars have been studied. A detailed investigation of PU Vul, which is probably a symbiotic nova, by Belyakina et al. (1985, Iz. Krym. Astrofiz. Obs. 72, 3) indicates that the active component near maximum had a spectrum indistinguishable from a normal F supergiant, while the cool component had an effective temperature near 2400 K. Much work has been done on R Agr which possesses what appears to be a jet. In a detailed study Kafatos et al. (AJ Suppl. 62, 853) showed that high ionization ultraviolet HeII and NV emission observed with IUE came from the jet; they interpreted their results in the framework of a thick accretion disk model. Viotti et al. (AJ 319 L7) confirmed the last observation; the same region seems also to emit X rays, detected with EXOSAT. Solf and Ulrich (AA 148, 274) studied the nebula of R Agr at high angular and spectral resolution. They interpreted their informations in terms of a double hour-glass structure, the jet being part of the inner hour-glass. High spatial and spectral resolution observations of CH Cyg by Solf (AA 150, 207) showed the presence of very compact nubulosity in 1986 about 1" from the star, and was identified by him with the at least apparently expanding radio jet observed by Taylor et al. (Nature 319, 38) and perhaps also associated with the X-ray emission detected by Leahy and Taylor (AA 176, 262).

Some recurrent novae are now considered to be symbiotic or closely related; they satisfy the definition given above. The 1985 outburst of RS Oph was studied in detail and early results are collected in the proceedings "RS Ophiuchi and the Recurrent Nova phenomenon", ed. M.F. Bode (1987). Bode and Kahn (MNRAS Soc. 217, 205) constructed a model involving the interaction of high velocity ejecta with a slow pre-outburst wind. Models of recurrent novae in general as well as RS Oph in particular have been considered by Livio et al. (AJ 308, 736) and by Webbink et al. (AJ 314, 653). Some, such as RS Oph and T Cor Bor, have according to these authors outbursts produced by accretion events.

As far as symbiotic stars in general are concerned, modelling continues. New more rigorous work on colliding wind models (winds coming from each stellar component) by Girard and Willson (AA 183, 247) should be particularly mentioned in this connection.

9. **The importance of stellar spectra libraries for studies of galactic evolution**
   (by B. Rocca-Volmerange)

Models of spectral synthesis or galactic evolution need significant samples of
stellar spectra, well representative of the stellar population preferentially
emitting in the considered wavelength range. Ideal conditions will be obtained
when stellar and galactic spectra are observed with the same instrument. When it is
possible, the best way is to use publications in litterature. One of the first
samples published in visible is the catalogue by Straizys and Sviderskiene (1972,
Bull. Vilnius Astr. Obs. 35, 1) from 3000Å to 10000Å with a resolution 50Å,
followed more recently, by Gunn and Stryker (ApJ Sup Ser 52, 121) from 3130Å to
10800Å with a resolution 20 and 40Å and Jacoby et al (ApJ Sup Ser 56, 257) from
3510Å to 7427Å with a better resolution of ≃4.5Å.

Recent improvements in spectral synthesis of galaxies as well as in the large field
of evolution of distant galaxies are essentially due to the extension of these
libraries to far-UV range, in particular the far-UV spectra atlases published from
IUE observations (Wu et al, NASA News 22, Heck et al, 1984, ESA SP-1052) are very
useful to build libraries of stars connected with the visible. A library of O to M
spectral types and two classes of luminosity, in the 1200Å to ≃1μm wavelength
range has been built by Rocca-Volmerange et al, AA 104, 177, and Guiderdoni and
Rocca-Volmerange, in press, and introduced in evolutionary models of galaxies.

For a long time, performent minimisation procedures fitted observational spectra
of galaxies (essentially ellipticals in visible and near-infrared (from Stebbins
and Whitford, 1948, ApJ, 108, 413, followed by many authors) with a synthetic
spectrum, giving detailed solutions of stellar populations. The strongest default
of this method is the lack of unicity (see Peck, ApJ, 238, 79 and references
therein). Such a disadvantage is less severe for evolutionary synthesis since the
number of solutions is constrained by stellar evolutionary models but it still
exists. One way to limit the number of solutions is to extensively obtain a good
fit on a larger wavelength range, including UV (and in a next future, IR) stellar
emission. Far-UV satellites observed massive young stars (OB stars) as well as hot
evolved stars (Horizontal Branch stars, blue stragglers, planetary nebulae,
etc...). That was a rich source of data which could be used to analyze stellar
populations of HII regions (Rosa et al, AA Sup Ser, 57, 361 and Lequeux et al, AA
103, 305), the far-UV emission of early-type galaxies or many other specific
problems. One of the most questionable problem concerning early-type (elliptical
or SO) galaxies is the origin of the "so-called UV-excess". It is not the place
here to detail the various episodes of this question: It was successively
attributed to massive stars or evolved hot stars according to the epoch and the
authors. In a recent paper, Rocca-Volmerange and Guiderdoni (AA 15, 22) used
stellar spectra of these far-UV libraries to give an interpretation of the
emission of early-type galaxies. The comparison of a spectrum of a F8 star with the
galactic spectrum of the SO galaxy NGC3115, treated with the same software (IHAP,
ESO) as the stellar spectra, concludes to a dominant population of F stars at the
assumed turn-off of the galaxy. With an estimated age 3 to 8 Gyrs, the turn-off
appears to be younger than assumed by the classical theories of galaxy formation
for ellipticals (Larson, MNRAS, 164, 585). Another conclusion attributing the UV-
excess to a star formation process, and not to horizontal branch stars, is also
proposed from this analysis. This result is confirmed from the synthetic spectrum
of Rocca-Volmerange and Guiderdoni, compared to the observed spectrum of M87
(Bertola et al, ApJ Let 237, L65). Unfortunately spectral resolution in galactic
spectra is not sufficient to detect spectral features of hot stars; only the Hubble
Space Telescope will give us the opportunity to identify these stars in nuclei of
early-type galaxies.

A second (and not the least!) domain of analysis which is also strongly depending on far-UV stellar spectra, is the large field of distant galaxies, in rapid progress since performances of the present and future large telescopes allow to fix-high-redshift and possibly primeval galaxies as preferential objectives. One way to study formation and evolution of such distant galaxies from observations is to stimulate synthetic galaxies, assumed to be template ones, from a model of evolution including stellar population and nebular component and to apply cosmological effects to the synthetic galaxy. The redshifted resulting spectrum will be compared in spectroscopy as well as in photometry (through filters) with observations. For distant galaxies, optical or infrared observations will correspond to far-UV emissivities from stars in the rest frame. This justifies importance and need of far-UV stellar spectra for such studies. A library of 30 stars has been used in the spectrophotometric model of galaxies (Guiderdoni and Rocca-Volmerange, 1987, AA, in press) to analyze and predict magnitudes and colors of galaxies at different redshifts. A comparison of very distant radiogalaxies observed by Djorgovski et al, (1987, ApJ, preprint) with the evolutionary models of Guiderdoni and Rocca-Volmerange, shows, among other results, that a large factor of evolution has to be taken into account justifying to pursue this kind of studies.

In conclusion, we may subline that any model of galactic evolution (spectrophotometric as well as chemical) is essentially based on stellar data. The interest of the far-UV wavelength range for spectral studies as well as extensive developments of detailed stellar spectra in infrared wavelength range has been sublined.

Working Group on Standard Stars (by A. Batten)

During the General Assembly held in Delhi, the Group discussed the question of finding a suitable solar analogue amongst the stars. No completely satisfactory analogue is yet identified.

The principal activity of the group between Assemblies has been the continued distribution twice a year, of the <u>Standard Star Newsletter</u> under the editorship of Prof. Pasinetti. As well as providing a means of keeping members informed about group activities, the <u>Newsletter</u> also carries brief contributions drawing attention to areas in which standard stars are needed or should be further studied. A running bibliography of papers relevant to standard stars recorded 165 items in the six issues since the last General Assembly.

Members of the Group who also belong to Commission 30 have been active in the re-examination of radial-velocity standards undertaken by that Commission. A full account of that work will be found in the report of Commission 30.

Faraggiana reports that a detailed analysis of the spectrum of Procyon is in progress at Trieste Observatory. An atlas of the spectrum from 2030Å to 2371Å, based on observations with IUE and BUSS VIII, has been published (ApJSup 61, 719, 1986). A further atlas covering the region 2600Å - 3200Å is in preparation. Gerbaldi and Faraggiana are examining suitable photographic spectra in the search for a catalogue of Vega.

Working Group on Be Stars (by D. Baade)

For the working group "Be stars" the most noteworthy event of the report period

was IAU Colloquium n° 92. *Physics of Be stars* in August 1986 which was co-sponsored
by IAU Commissions 29 and 45 and kindly hosted by the Joint Institute of Laboratory
Astrophysics at the University of Colorado in Boulder IAU Coll N°92, brought
together 101 participants from 17 IAU member countries. The proceedings were
edited by A. Slettebak and T.P. Snow and give a fairly complete and up-to-date
account of the status of the knowledge and speculations about Be stars. (Another
very brief summary has been incorporated into the Section on Be stars of Commission
29's triennal report.) New instruments and long time base lines accumulated with
others have further substantiated the evidence that in addition to the defining
emission lines there are at least four more domains in which Be stars differ
substantially from B stars : UV resonance lines (*i.e.*, *mass loss*), IR excess (*i.e.*,
amount and extent of circumstellar matter), polarization (*i.e.*, *shape of
circumstellar mass distribution*), and photospheric line profile variations (*i.e.*,
nonradial pulsations). The latter seem to provide the first direct observational
evidence that *Be stars* do in fact differ from *B stars* (which in principle of course
is a trivial statement because somehow it must be the *stars that do or do not eject
an envelope*).

The insights gained from these new observations, intriguing as they may be, have on
the other hand also made clear that real progress probably will come only from a)
long-term, b) multi-wavelength and c) multi-site observations, ideally a
combination of all three. Some attempts in this direction were already undertaken
and proved quite successful, but it appears desirable to put them on a somewhat
broader basis. The biannual *Be star newsletter* (edited by G.J. Peters and
distributed to nearly 125 different places [≠ from number of recipients by the
European Southern Observatory) provides the forum for organizing such campaigns
and keeping the community updated on the progress made. An important experience
with the campaigns carried out has been that applications for observing time seem
to be perceived by time allocation committees rather positively when the proposals
are put on a broad (but qualified) basis. An important role of the working group
will, therefore, be to further back such efforts.

Like in New Delhi in 1985, the working group and by then newly elected Organizing
Committee (currently consisting of the undersigned, P.K. Barker, V. Doazan, J.M.
Marlborough, G.J. Peters, A. Slettebak, and T.P. Snow) will meet also during the
General Assembly in Baltimore. There, the intended brief confrontation of the
implications derived from various types of data is hoped to provide another
incentive to coordinate also observational efforts from which so far rather
conflicting conclusions have been drawn.

Working Group on Ap Stars (by C. Cowley)

The Observatoire de Lausanne hosted a workshop on Elemental Abundance Analyses on
7 - 11 September 1987. The workshop was organized by Adelman (chairman, SOC),
Boyarchuk, and Cowley. The hosts were Hauck, Lanz, and North (LOC). The 21
participants from 9 countries compared results of the analyses of digitized
spectra of Omicron Peg and Phi Her. The proceedings of the workshop are being
prepared for publication by Adelman and Lanz (October 1987).

A. V. Tutukov (Moscow) and G. Michaud (Montreal) are currently planning a meeting
on stellar evolution and surface chemistry. It is anticipated that some fraction
of this meeting will be devoted to problems of CP stars.
[A meeting of the Ap Working Group is planned for the Baltimore IAU, at which time
plans for a future meeting devoted entirely to CP stars will be discussed. Please
check the Commission 29 Bulletin Board for information concerning the time and
place of this meeting].

Working Group on Peculiar Red Giants (by C. Jaschek and P.C. Keenan)

This working group has published a "Newsletter of chemically peculiar late-type stars" of which N°5 is being published at Strasbourg, and which is distributed in 300 exemplars. IAU Coll. N°106 is being organized at Bloomington, Indiana, in July 1988, on the subject "Evolution of peculiar Red Giant stars".

Abbreviations

AA .... Astronomy and Astrophysics
AA Supp .... Astronomy and Astrophysics Supplement Series
AJ .... Astronomical Journal
Ann Rev Astr Astroph .... Annual Review Astronomy and Astrophysics
ApJ .... Astrophysical Journal
ApJ Suppl .... Astrophysical Journal Supplement Series
Astr Tsirk .... Astronomicheskij Tsirkulyar
Astr Zh .... Astronomicheskij zhurnal
Astrofiz .... Astrofizika
Astrometr Astrofiz ...... Astrometriya Astrofizika
Astrophys Space Sci ..... Astrophysics and Space Science
Bull Inf CDS .... Bulletin d'Information du Centre de Données
                         Stellaires
Fund Cosmic Phys .... Fundamentals of Cosmic Physics
IBVS .... Information Bulletin on Variable Stars
Izv Crim .... Izvestiya Ordena Trudovogo Krasnogo Znameni Krymskoj
                         Astrofizicheskoj Observatorii
J Astrophys Astron ...... Journal of Astrophysics and Astronomy of India
Mem Soc Astr It .... Memorie della Società Astronomica Italiana
MNRAS .... Monthly Notices of the Royal Astronomical Society
PAS Japan .... Publications of the Astronomical Society of Japan
PASP .... Publications of the Astronomical Society of the
                         Pacific
Perem Zvesdy .... Peremennye Zvezdy
Phys Scr .... Physica Scripta
Pisma Astr Zh .... Pis'ma v Astronomicheskij Zhurnal
QJRAS .... Quarterly Journal of the Royal Astronomical Society
Rev Mex Astr Astrofis ... Revista Maxicana de Astronomia y Astrofisica
Sov Astr .... Soviet Astronomy

                                    G. Cayrel de Strobel
                                    President of Com. 29

30. RADIAL VELOCITIES (VITESSES RADIALES)

PRESIDENT: J. Andersen
VICE-PRESIDENT: D.W. Latham
ORGANIZING COMMITTEE: A. Florsch, E. Maurice, M. Mayor, R.D. McClure, A.G.D. Philip

1. Introduction

The present report on the activities of IAU Commission 30, covering the triennium June 1, 1984 through June 1, 1987, will be somewhat different from its recent predecessors in both content and style. Over the preceding decade or so, the reports mainly emphasized the dramatic improvements in observing efficiency, achieved primarily through the general adoption of cross-correlation techniques, combined with modern detectors attached to either specialized spectrometers or to existing, more conventional instruments. A great surge of observational activity followed, directed towards a variety of astrophysical problems, some of which are of a more classical nature, but many of which are in entirely new classes of research. At the time of the previous reports, most of the major observational projects were still underway, even if some preliminary results were emerging. The proceedings of IAU Colloquium No. 88, *Stellar Radial Velocities* (L. Davis Press, 1985) contains a collection of papers on instrumentation and reduction techniques as well as on ongoing observing programs which remains a very useful source of references to this developmental phase as well as to the current state of the art.

In contrast, the past triennium has been a period rather of maturity and refinement on the technical side; no major new developments have been reported. The main contributors of accurate photoelectric cross-correlation determinations of radial velocities continue to be the instruments at the Center for Astrophysics (CfA), the Dominion Astrophysical Observatory (DAO), and the Franco-Swiss CORAVEL spectrometers operating at Observatoire de Haute-Provence, France, and at ESO in Chile; typically, the same instrument/detector/software systems are used to observe widely different types of object, both galactic and extragalactic.

With the instrumentation now working in a routine fashion, the astronomers have concentrated their attention on the astrophysical problems. As an appetizer for the more detailed account below, we shall mention just a few of the new, exciting scientific results which have been based on radial-velocity measurements: The recent revolution in our ideas concerning the large-scale structure of the Universe, with the unexpected distribution of the (luminous) matter in bubble-like structures surrounding huge voids, is due to the systematic and painstaking collection of accurate velocity (redshift) measurements for thousands of faint galaxies. The secular stability of dwarf spheroidal galaxies in the Local Group, and the possible presence of dark matter also in these old, distant systems, has been investigated by means of accurate radial-velocity data for individual, very faint ($V \approx 17$) member stars. Finally, the first accurate radial-velocity curves for extragalactic Cepheid variables (in the Magellanic Clouds) have been obtained.

Turning to objects in our own Galaxy, radial velocities for numerous faint, individual member stars in several globular clusters have contributed significantly to the new, much more profound understanding of the structure and dynamical evolution of these clusters. In open clusters, the central concentration of the spectroscopic binaries, along with the blue stragglers and other massive stellar constituents, has been demonstrated observationally. The dynamics of the Galaxy itself - both disk and halo components - is being studied from data sets which exceed earlier samples by orders of magnitude in precision and homogeneity as well as in num-

ber. Similar improvements in the observational basis for Baade-Wesselink analyses
of the radii, luminosities, and pulsation properties of RR Lyrae stars are providing independent, fundamental data for these important distance indicators and are
being brought to bear on other types of pulsating variables; with high-precision
techniques, pulsations with amplitudes of a few tens of m s^{-1} have been discovered
in Arcturus and other bright stars. Long-accepted results on the frequency of
(spectroscopic) binaries amongst various types of stars are being challenged as new
observations become available and quality standards stricter. And, last but not
least, ultra-precise radial-velocity measurements over nearly a decade have produced the first substantial (if not indisputable) evidence for the existence of
planet-size companion(s) to at least one other solar-type star.

An exhaustive account of all published and ongoing research involving determination of (stellar or other) radial velocities during the past triennium, as has
traditionally been the goal, would now far exceed the allotted space for this report and, moreover, presumably be of limited interest to the readership. It will
not be attempted. Instead, the intention is to highlight, primarily for the benefit
of readers from *outside* the Commission, what appears to be the more significant developments during this period in a number of selected subjects. References to some
of the key papers are given, with no pretention of completeness, while the reader
is referred to the appropriate sections of *Astron. Astrophys. Abstracts* for a more
complete bibliography on subjects of her/his specific interest. The main weight of
the report will be on programs in stellar astronomy, with a certain bias towards
the contributions of those colleagues who supplied material in time for its
inclusion.

This approach has some foundation in the composition of the Commission membership, currently at 64 after a 25% increase at the 1985 General Assembly. This is
far fewer than the number of IAU members actively engaged in research based on
radial-velocity data, especially in non-stellar fields. Hence, this report should
be considered, in a sense, as the "tip of the iceberg"; many results have been
included in the reports of other Commissions (*e.g.* 9, 26, 27, 28, 29, 33, 40, 42).
Similarly, while IAU Colloquium No. 88, *Stellar Radial Velocities* (see above) was
the only meeting during the triennium entirely devoted to the subject of radial
velocities, Commission 30 co-sponsored several other meetings: IAU Symposium No.
132, *The Impact of High S/N Spectroscopy on Stellar Physics* (Meudon, June-July
1987), IAU Colloquium 95, *Second Conference on Faint Blue Stars* (Tucson, May-June
1987), and IAU Colloquium No. 97, *Wide Components in Double and Multiple Stars*
(Brussels, June 1987), as well as the future IAU Colloquium No. 107, *Algols* (Victoria, August 1988) and the Joint Discussion on *Formation and Evolution of Stars in
Binary Systems* at the upcoming IAU General Assembly. Radial velocity work featured
prominently at many other meetings.

Commission 30 itself continues to fill an important role of coordination and
standardization, as exemplified by the initiative to publish a Commission *Newsletter* (by Dr. A. Florsch, Observatoire de Strasbourg), and by the current effort to
refurbish the IAU system of radial-velocity standard stars to meet present-day
demands (see 8, below). In is also appropriate to record here the valuable contribution made by the publication of the *Bibliographic Catalogue of Stellar Radial
Velocities*, by Barbier-Brossat and Petit (*Astron. Astrophys. Suppl.* 65, 59), and
the imminent publication, by the same authors, of the corresponding *Catalogue of
Mean Stellar Radial Velocities*.

2. Large-Scale Distribution of Matter in the Universe

As noted above, an extensive account of extragalactic radial-velocity work is
outside the scope of this report. It is, however, appropriate to review the redshift survey projects which have recently made a major impact and received considerable public attention. The following section was kindly prepared by D.W. Latham:

Mapping out the distribution of galaxies is a first step towards understanding the origin and evolution of large-scale structure in the universe. Over the past few years, surveys of the redshifts (radial velocities) of galaxies have uncovered unexpectedly large structures. The 21-cm survey by Giovanelli and Haynes (*Astron. J.* **90**, 2445) delineated the Perseus-Pisces chain. Optical redshift surveys were used to probe the enormous void in Boötes (Kirshner, Oemler, Shectman, and Schechter, *Astrophys. J. Lett.* **248**, L57) and the remarkably similar structure in the adjacent Corona Borealis region (Postman, Huchra, and Geller, *Astron. J.* **92**, 1238). The completion of the first slice of the extension to the CfA Redshift Survey (de Lapparent, Geller, and Huchra, *Astrophys. J. Lett.* **202**, L1) suggested that galaxies are distributed on thin sheets which surround or nearly surround vast regions nearly devoid of bright galaxies in a bubble-like structure. The completion of a second slice adjacent to the first (Geller, Huchra, and de Lapparent, *IAU Symp.* **124**, p. 301) supports this picture. Further slices are now being observed at the CfA.

Surveys in the south reach less deep than those in the north, but a recently completed survey once again reveals enormous voids and sheet-like structures (da Costa, Pellegrini, Sargent, Tonry, Davis, Meiksin, and Latham, *Astrophys. J.* in press). Deeper surveys in the south are needed to help investigate whether a fair sampling of the universe has been achieved, to help determine the average properties of the large-scale structure, and to help analyze the large-scale flows.

3. Stars in Local Group Galaxies

Once the possibility of determining accurate radial velocities for very faint stars was established, the brightest giants in some of the nearby dwarf spheroidal galaxies became targets of high priority. Not only could accurate systemic velocities for these objects help to estimate the total mass of our own Galaxy, but their velocity dispersions would indicate whether they could be stable over the long periods indicated by their colour-magnitude diagrams, and could be used to estimate their total mass as well as the presence (and nature) of any dark matter within them. The pioneering attempt in Draco and Ursa Minor by Aaronson (*Astrophys. J.* **266**, L11) was followed up by work on the Carina, Fornax, and Sculptor systems by Seitzer and Frogel (*Astron. J.* **90**, 1796), Armandroff and da Costa (*Astron. J.* **92**, 777) and Aaronson and Olszewski (*Astron. J.* **92**, 580; **94**, 657). The latter authors conclude that conventional forms of dark matter (low-mass stars) are adequate to explain the observations.

In the nearest Local Group galaxies, the Magellanic Clouds, hundreds of individual stars are accessible to existing instruments. Adding to previously existing data of lower precision, mostly on earlier-type supergiants, catalogues of CORAVEL radial velocities have been published for some 400 late-type supergiants in the LMC (Prévot et al., *Astron. Astrophys. Suppl.* **62**, 37) and for about 230 such stars in the SMC (Maurice et al., *Astron. Astrophys. Suppl.* **67**, 423). The results are currently being studied for determination of the rotation parameters and mass of (especially) the LMC. A similar study has been carried out from accurate radial velocity measurements of planetary nebulae by Dopita and collaborators (*ESO Workshop on Galactic Halos*, Garching, April 1987). With the best current and future instruments, these observations can be extended to the fainter giant stars (young and old), and lead to a better understanding of the dynamical history of both field and cluster population in the MC.

Finally, accurate radial-velocity curves have been observed for several Cepheid variables in the LMC (Imbert et al., *Astron. Astrophys. Suppl.* **61**, 259) - the first such data for extragalactic Cepheids. These data have been used in a Baade-Wesselink analysis by Imbert (*Astron. Astrophys.* **175**, 30) to determine radii and luminosities for the variables, and hence a value for the LMC distance. No doubt, with more good radial velocity curves combined with photometry in suitable colours and better understanding of the pulsation phenomenon, this approach will supply

important independent data on the distance to the Clouds and thereby on the extragalactic distance scale.

4. Structure and Dynamics of the Milky Way

The study of the structure and dynamics of our own Galaxy is another field
which has received new vitality with the relatively easy access to obtaining accurate radial velocities for large numbers of faint stars. Here, however, typical observing programs comprise hundreds or thousands of stars, and the resulting radial
velocities require supplementary, homogeneous photometric, spectroscopic, and
astrometric data of matching quality for their interpretation in terms of the past
evolution and present structure of the Galactic disk and halo. Hence, results are
presently available mostly as progress reports, while full exploitation of the information inherent in the data will take considerable time. A few programs are
designed to simply provide velocity data for a complete sample of stars for future
analysis. Among these, we mention the large-scale Marseille-Haute-Provence objective-prism and spectrographic program to supply radial velocities for the HIPPARCOS
stars, where some 17 000 spectra have already been measured to give radial velocities with a precision of about ±4 km s^{-1} for some 5000 stars (3-4 plates per star);
the publication by Griffin (*MNRAS* 219, 95) of velocities for more than 400 faint
stars in the Clube selected areas; and the completion of radial velocity data for
all stars in the *Bright Star Catalogue* by Nordström et al. (*Astron. Astrophys.
Suppl.* 59, 15; 61, 53).

Several large-scale programs are designed to explore the information contained
in the correlations between the kinematics, the metal abundances, and the ages of
stars in the Galactic disk and halo. The criteria used to define the sample under
study, and the various biases they introduce into the correlations to be examined,
are then of crucial importance. Different approaches are being followed simultaneously, the main sample definition criteria being either proper motion, which
introduces a kinematical selection but in principle no metallicity bias, or photometric or spectroscopic metallicity or age indices applied to a magnitude-limited
parent sample, which in principle avoids both kinematic and chemical bias.

The two main projects of the first type are the surveys of proper-motion stars
by Sandage and Fouts (*Astron. J.* 93, 74) and by Carney and Latham (*Astron. J.* 93,
116), both of which discuss radial velocities and *UBV* photometry for some 900 stars
from the Lowell Proper Motion Survey, the overlap between the two programs being
30-40%. Sandage and Fouts conclude that their data support the classical Eggen,
Lynden-Bell, and Sandage view that the galaxy collapsed from a spherical halo to a
flat rotating disk in essentially a single free fall time of about 10^8 years, although with the addition of the Gilmore and Wyse (*Astron. J.* 90, 2015) "thick
disk". The Carney-Latham data seem to support the alternate view (Norris, *Astrophys. J. Suppl.* 61, 667) that the collapse was more chaotic and lasted many free
fall times. The Carney-Latham photometric metallicity indices are being supplemented by abundance measurements on the low-S/N spectra obtained for the radial-velocity determinations (Carney et al., *Astron. J.* 94, 1066) as well as by photometry
in other bands to improve the accuracy of their distance determinations. Both surveys give preliminary values for the Galactic escape velocity of 500-550 km s^{-1}.

Surveys based on essentially magnitude-limited samples defined by objective-
prism and/or photometric data on spectral types, metal abundances, and ages are
being carried out by several groups. Halo red giants were studied by Carney and
Latham (*Astron. J.* 92, 60). Ardeberg and Lindgren are carrying out a massive survey
of *uvbyβ* photometry and radial-velocity determination for some 5000 metal-poor F,
G, and K stars; a detailed description and progress report is given in *IAU Colloq.
88*, p. 151. Andersen, Mayor, Nordström, and Olsen have completed radial-velocity
observations for some 5000 nearby F dwarfs in the whole sky, selected from the
complete, magnitude-limited catalogue of homogeneous *uvbyβ* photometry by Olsen

(*Astron. Astrophys. Suppl.* 54, 55); these include all stars with metal abundances less than about half of that of the Sun as well as all those for which individual photometric *ages* - the salient characteristic of this program - can be computed. Extension of the observations to some 1200 fainter F stars in the direction of the North Galactic Pole, and to an all-sky sample of about 2000 G dwarfs, is under way. An outline of the program is given in *IAU Colloq.* 88, p. 171, and a preliminary analysis in the (perhaps last) paper by B. Strömgren in *The Galaxy* (Ed. G.F. Gilmore and R.F. Carswell, in press). Finally, Sandage and Fouts (*Astron. J.* 93, 592, 610) use radial velocities of magnitude-limited samples of stars in three directions to estimate that the "thick disk" contributes about 10%, and the halo 0.5%, of the stars near the Sun, much higher than earlier estimates.

The new possibilities for obtaining accurate radial velocities for very faint stars are being used to study the halo population *in situ*, rather than inferring its properties from data for samples near the Sun. Extensive programs are being conducted by G. Gilmore and others; Ratnatunga and Freeman (*Astrophys. J.* 291, 260) found an unexpectedly low velocity dispersion for a sample of metal-poor giants toward the South Galactic Pole.

5. Star Clusters

The pioneering work by Gunn and Griffin in 1979 (*Astron. J.* 84, 752) of obtaining and analyzing radial velocities for a large number of individual stars in the globular cluster M3 with measuring errors much smaller than the expected internal velocity dispersion of the cluster produced a breakthrough in the theoretical modelling of globular cluster dynamics. The additional constraints on the models posed by the detailed knowledge of the velocity dispersion in the centre of the cluster as well as its radial variation showed the inadequacy of simple one-component models, and multi-mass King models with anisotropic velocity dispersions in the outer parts of the cluster were introduced.

During the past three years, this observational material has been greatly extended and refined, and similar data and analyses are now available for five additional clusters: M2 (Pryor et al., *Astron. J.* 91, 546), M13 (Lupton et al., *Astron. J.* 93, 1114), M92 (Lupton et al., *IAU Symp.* 113, p. 19), ω Cen (Meylan, *Astron. Astrophys.* 184, 144), and 47 Tuc (Meylan and Mayor, *Astron. Astrophys.* 166, 122). In all except M2, velocity anisotropy was found to be important, strongest in ω Cen. Significant rotation has been found in three clusters (M13, 47 Tuc, and ω Cen), with a maximum of 6-8 km s^{-1} at a distance of a few core radii from the centre. Masses derived for the clusters are in the range 0.4-0.9 10^6 $M_\odot$, with the exception of the unique giant cluster ω Cen ($\sim$4 10^6 $M_\odot$). The mass fraction locked up in unobserved heavy remnants (presumed to be white dwarfs) seems to vary from some 2% (M3) through $\sim$10% (M2) and $\sim$30% (ω Cen, 47 Tuc) to perhaps 50% (M13); the origin of these is not yet fully understood. Extensive radial velocity data are becoming available for many further clusters, from observations by both Griffin et al. (Palomar), Mayor et al. (ESO), and Pryor et al. (DAO).

Galactic *open clusters* are, in fact, even more challenging from an observational standpoint than globular clusters, due to their far smaller internal velocity dispersions of typically 1 km s^{-1} or even less. Several current instruments can achieve a precision substantially better than this figure. However, any binary systems present in the clusters with periods of decades, or even centuries, can have velocities - even when observed with infinite precision - which differ from the γ-velocity of the system by amounts similar to the velocity dispersion of the cluster. For the derivation of the true velocity dispersion of the cluster, orbital motion in binaries is a major factor and must be considered carefully - yet another example of a field where observational uncertainty has ceased to be the main limitation.

Precise radial velocity data are, first, an effective means of identifying member and non-member stars in the cluster field - especially important if the field contains a single star of singular interest, e.g. a Cepheid (Mermilliod et al., *Astron. Astrophys. Suppl.* **70**, 389). Moreover, identification not only of cluster members, but also of cluster binaries, is crucial for a correct interpretation of the details of a cluster colour-magnitude diagram, as shown beautifully by Nissen et al. (*Astron. J.* **93**, 634), who used the accurate radial velocity data for M67 (and M11) by Mathieu et al. (*Astron. J.* **92**, 1100). The same data enabled Mathieu and Latham (*Astron. J.* **92**, 1364) to conclude that the spectroscopic binaries and blue stragglers in M67 were (equally) concentrated toward the cluster centre relative to the single stars, an effect readily explained as the result of dynamical relaxation (reviewed by Mathieu, *IAU Symp. 113*, p. 427); the same result was found for NGC 188 by Harris and McClure (*IAU Colloq. 88*, p. 257). Finally, we mention the alternative method for age determination of open clusters from the cutoff period for tidal circularization of cluster spectroscopic binary orbits proposed by Mathieu and Mazeh (*Astrophys. J.*, in press), a further development of the observation by Mayor and Mermilliod (*IAU Symp. 105*, p. 411).

6. Spectroscopic Binaries

While detection and orbit determination of spectroscopic binaries are within the domain of Commission 30, progress in the study of binary frequencies in various stellar types and determination of masses and other physical properties from observations of binaries are thoroughly reviewed in the reports of Commissions 42 (close binaries) and 26 (visual binaries) elsewhere in this volume. Hence, the present report will be limited to a few highlights:

Much progress has been made on the detection and frequency determination of spectroscopic binaries in various types of stars during the triennium under review. The earlier belief that spectroscopic binaries are rare or absent in Population II has been proved wrong by Stryker et al., (*PASP* **97**, 247), Carney and Latham (*Astron. J.* **93**, 116), and Ardeberg and Lindgren (*IAU Colloq. 88*, p. 371); one example is the spectroscopic orbit for the extremely metal-deficient star BD+13°3683 ([Fe/H]≈-2.3) published by Jasniewicz and Mayor (*Astron. Astrophys.* **170**, 55). Progress in identifying and measuring orbits for the spectroscopic binaries in the Hyades has been made by Griffin et al. (*Astron. J.* **90**, 609). Griffin is also making steady progress in defining the long-period part of the spectroscopic binary population in his long series of papers in *Observatory* and elsewhere, which includes the first binary with $K < 1$ km s^{-1} (McClure et al., *PASP* **97**, 740) and the first S-type binary (*Obs.* **104**, 224). Similar systematic work, on a smaller scale, is done by the Toulouse group, who have also published their 14th *Catalogue Complémentaire* (Pédoussaut et al., *Astron. Astrophys. Suppl.* **58**, 601). From some eight years of data on Ba II, CH, and R-type carbon stars, McClure has determined complete or partial orbits for 17 out of 20 Ba II stars; both the Ba II and the CH stars appear to have significantly lower orbital eccentricities than normal red giants.

The binary Cepheid DL Cas was studied by Harris et al. (*Astron. J.* **94**, 403) and Mermilliod et al., (*Astron. Astrophys. Suppl.* **70**, 389); both groups are continuing work on binary Cepheids. Also the frequency and properties of triple and multiple systems are being studied in greater detail than before, e.g. by Mayor and Mazeh (*Astron. Astrophys.* **171**, 151) and Duquennoy (*Astron. Astrophys.* **178**, 114); see also the report of Commission 26. At the limit of very large separations, Mazeh and Latham (*IAU Colloq. 97*, in press) find a real cutoff in the number of physical pairs at separations greater than about 0.1 pc. Finally, on this subject, the reality of some published orbits has been called into doubt by Morbey and Griffin (*Astrophys. J.* **317**, 343); while it is not clear that future, more definitive results on binary frequencies will differ substantially from earlier estimates, the uncertainties are clearly larger than previously believed, and standards for accepting orbital solutions will be stricter in the future than before.

A special case of a spectroscopic binary survey is that by Campbell and Walker (*IAU Symp. 132*, in press), who from six years of data with a precision of about ±15 m s^{-1} find a few stars to exhibit velocity variations which are plausibly (but not unequivocally) explained as due to companions of a few Jupiter masses. Companions of masses corresponding to those expected for brown dwarfs were not detected, which suggests that brown dwarfs - should they exist at all - are generally not found as companions to solar-type stars.

Finally, a few examples to illustrate how application of the modern techniques allows new types of results on some individual binary systems: High-resolution, high S/N observations in the far red allowed Tomkin and Popper (*Astron. J.* 91, 1428) to measure the orbit of the very faint secondary of α CrB; such systems are especially informative in comparisons with stellar evolution models. The ability of the cross-correlation method to "see" only the late-type system in a composite spectrum has been utilized by, e.g. Knee et al. (*Astron. Astrophys.* 168, 72) and Griffin and Griffin (*J. Astrophys. Astron.* 7, 195). Finally, the mere improvement in precision possible with the new instruments has been used by Andersen et al. (*Astron. Astrophys.*, in press) in the hitherto most successful comparison of stellar evolution models with observations of a star (AI Phe) other than the Sun.

7. Pulsating stars

Measurement and analysis of precise radial-velocity curves of pulsating variables is another field which has seen enormous progress in the period covered by this report. Not only can observations be obtained of fainter and more distant stars than before - even of stars in the Magellanic Clouds (Imbert et al., *Astron. Astrophys. Suppl.* 61, 261) - but the new level of observational accuracy has stimulated the theoretical work on their interpretation. This is especially true as regards the application of the Baade-Wesselink method to the determination of radii, luminosities, and distances to such stars as RR Lyrae and Cepheid variables. These stars are of prime importance as extragalactic (and Galactic) distance indicators, but their absolute-magnitude calibrations are still affected by considerable uncertainties.

Jones et al. (*Astrophys. J.* 312, 254, and further papers in press) have shown that self-consistent radius variations (from photometry and spectroscopy) for RR Lyrae variables can be obtained if light curves in the *K* band are used and the *V-K* colour index is used to infer the temperature variations over the cycle; a systematic program is under way for field and globular cluster variables which should place our knowledge of the absolute magnitudes of RR Lyrae stars, and their dependence or otherwise on metallicity, on a new observational basis. Similar work, utilizing photometry in the Geneva system, is carried out by Burki et al. (*Astron. Astrophys.* 156, 139; 159, 255, 261; 181, 34) on RR Lyrae, δ Scuti, and other types of variables. At the CfA, observations are under way under the coordination of J.P. Huchra to obtain radial-velocity curves for all the calibrating Cepheids in Galactic clusters, as part of one of the Key Programs for the Hubble Space Telescope. Similar programs are also under way by the CORAVEL group and elsewhere. Work is also in progress to examine to what extent the methods developed for Baade-Wesselink analysis of RR Lyrae stars can be applied to the Cepheids, which have much more extended atmospheres; in the Cepheids, the cross-correlation profiles show a decidedly non-static behavior of the photosphere during the expansion phase.

Finally, the techniques developed for ultra-precise radial-velocity measurements have also been applied to the detection and measurement of pulsating stars. The Arizona group has recently discovered that Arcturus is a pulsating star with a peak-to-peak range of 160 m s^{-1} and a period near 2 days, while Pollux, and perhaps Aldebaran, pulsate with much smaller velocity ranges (~10 m s^{-1}) and periods near 3 hours (Smith et al., *Astrophys. J.* 317, L79 and *IAU Symp. 132*).

8. The IAU System of Standard Stars

At its 1985 meeting in New Delhi, Commission 30 decided to appoint a Working Group, consisting of J. Andersen (chairman), W.I. Beavers, B.Campbell, D.W. Latham, M. Mayor, and R.D. McClure, with the task of proposing a new system of radial-velocity standard stars, the existing (1955) system having been rendered hopelessly inadequate by the recent great improvements in observational accuracy (see, e.g. Batten, *IAU Colloq. 88*, p. 325). The Working Group was to report to the Commission at its meeting in Baltimore in 1988; meanwhile, intensive observation was recommended of those of the former standard stars with $|\delta| < 20°$ which had proved to be most stable (Mayor and Maurice, *IAU Colloq. 88*, p. 308), as well as of minor planets for the purpose of establishing the velocity zero-point. The radial velocities of these minor planets can be calculated from their known orbital elements to a precision better than 10 m s^{-1}.

These observations have been pursued during the period of this report, the main weight of the data originating again from the three photoelectric instruments at the CfA, the DAO, and the southern CORAVEL in Chile. At the time of writing, about 250 minor planet observations have already been collected (nearly 200 from the CfA alone, primarily through the efforts of R.P. Stefanik), and typically 100-200 observations over a 7-10 year baseline are available for each of the stars under consideration as future primary standards. These observations are currently being collected for analysis by the Working group.

Preliminarily, the minor planet observations by each group appear to confirm the validity of their individual zero-points to about ± 100 m s^{-1}. It remains to be seen whether their observations of the standard star candidates can be reconciled after being adjusted to a common zero-point. However, our original goal of proposing, in 1988, a revised set of about 20 bright, late-type, primary standard stars for which the individual velocities and the common zero-point are known to about ± 100 m s^{-1}, still appears to be within reach. Moreover, results from the Canadian programme of high-precision radial velocity observations by the HF absorption-cell technique indicate that a half-dozen or so of these stars may have velocities which are constant to about ± 10 m s^{-1} over almost a decade; after independent verification, these stars would be candidates for later adoption as "standards of relative radial velocity". Note that, due to uncertainties in the effects of gravitational redshifts and convective blueshifts in stars other than the Sun, the actual *center-of-mass* velocities of these stars will not be known to similarly high accuracy; however, for most or all known applications, this limitation should be of no practical consequence.

The Working Group is also considering the situation as regards secondary and early-type standards. Velocity standards for use with large telescopes and modern detectors may have to be as faint as $V=13-15$ in order to be easily observable. Such stars can be readily measured relative to the primary standards with existing instruments; stars in open clusters whose velocities agree with the cluster mean appear to offer the best prospects for selecting long-term constant-velocity stars from a relatively short series of observations. For the early-type stars, cross-correlation techniques are also being developed to greatly increase the observational accuracy. However, the difficulties of having to deal with spectra suffering from appreciable rotational broadening, and of establishing the long-term stability of prospective standards to the new, higher level of accuracy remain; with luck, the Commission may perhaps be able to celebrate the start of the next millennium by adopting a set of radial-velocity standards covering all major spectral types.

Johannes Andersen
President of the Commission

31. TIME (L'HEURE)

PRESIDENT: D. D. McCarthy   VICE PRESIDENT: P. Paquet

ORGANIZING COMMITTEE: S. Aoki, J. Benavente, N. Blinov,
B. Guinot, G. Hemmleb, J. Kovalevsky, Y. Miao, I. Mueller,
P. Paquet, J. Pilkington, E. Proverbio, S. Ye

INTRODUCTION

The work of IAU Commission 31 members is contained mainly in
the reports contributed by members and listed below. Unfortunately because of limitations in space, journal references to the
work could not be listed. Readers interested in any of the topics mentioned in the reports are urged to contact those who have
submitted that particular report for further information.

During the period from 1984 to 1987 the membership list of
Commission 31 was revised to reflect new addresses and telephone
numbers. Those listed as members previous to this period who
were no longer concerned with Commission matters were deleted
from membership. Members were also asked to submit their recommendations for future research in the area of time. The result
of this poll was the formation of two working groups.

The first group, formed with D. Allen as chairman, is entitled "The Use of Millisecond Pulsars and Timing of Pulsars". The
second is entitled "Time Transfer with Modern Techniques" with H.
Fliegel serving as chairman. Both of these groups will present
their findings and recommendations to the 1988 Baltimore General
Assembly. In the period covered by this report the Commission
sponsored IAU Symposium 128 at Coolfont, West Virginia (USA)
entitled "The Earth's Rotation and Reference Frames for Geodesy
and Geodynamics". The papers presented at this meeting are contained in the Symposium proceedings published by D. Reidel and
edited by Wilkins and Babcock.

Many members of the Commission took an active part in the
success of Project MERIT which resulted in the creation of the
International Earth Rotation Service which began operation on 1
January 1988 providing information on UT1-UTC as well as the
motion of the pole of rotation. During this period the responsibility for the formation of the UTC time scale was transferred
from the Bureau International de l'Heure to the Bureau International des Poids et Mesures. The importance of the proper implementation of the theory of relativity to timekeeping remained an
issue for research and discussion. Important developments in the
transfer of precise time also took place in this period. These
are mainly concerned with the development of the Global Positioning System, but other space techniques were also shown to be
useful in the area of time transfer. These topics as well as
many more important developments are contained in the individual
observatory and laboratory reports which follow. Because of the
severe space limitation for this report, the reader is urged to
consult these reports for more detailed information.

BUREAU INTERNATIONAL DE L'HEURE
(BUREAU INTERNATIONAL DES POIDS ET MESURES), FRANCE
(Reported by B. Guinot)

Formation of TAI and UTC

The transfer of BIH activities on atomic time took place in April, 1985. A team of five persons, staff members of BIPM, is in charge of TAI. The BIH has continued to produce TAI and UTC under the form of corrections to the master clocks of timekeeping laboratories, given at 10-day intervals. These corrections are disseminated monthly by Circular D and the General Electric Mark 3 system.

The first step consists in establishing a free atomic time scale EAL with stable, but not necessarily accurate, clocks. The number of clocks processed by the BIH stability algorithm has reached 173 in 1986, among which are 6 primary clocks (at NRC and PTB) and about 10 hydrogen masers. Some of the latter rival the best primary clocks for the long term stability (τ = 2 months) and have no apparent frequency drift. The frequency of EAL is estimated from the calibrations performed by 10 frequency standards located in Canada, Fed. Rep. of Germany, Japan, USA, and USSR.

The relationship between TAI and EAL includes a frequency offset which is modified by steps of 2×10^{-14} in order to maintain the accuracy of the TAI frequency. However, this steering has not been applied since February 29, 1984. Since that date TAI runs as a free time scale, its frequency remaining nevertheless in agreement with the data of the best primary standards, within a few units of 10^{-14}.

Development of the Time Links

In March 1987, 38 laboratories or national consortiums of laboratories participated in the establishment of TAI and UTC. Twenty of them operated GPS time receivers. In addition, four countries are soon expected to join the TAI club, Argentina, India, Israel, South Africa.

In 1986, the responsibility for establishing the schedules for simultaneous tracking of GPS satellites has been transferred from NBS to BIH/BIPM. The principles orginated by NBS, according to which each area of the world should be linked to a maximum of other areas, have been kept. However, with the increase in the number of areas and the future deployment of the GPS system, a different organization of the common views might keep the number of daily tracks at a reasonable level, while increasing the accuracy. One could have a small network of highly accurate long distance links complemented by regional links. To investigate this possibility, the BIH has undertaken several studies: improvement of the models for ionospheric refraction by using dual frequency observations of other systems; improvement of the models of orbits; and experiments in cooperation with other laboratories on ultra-accurate, short-distance links.

Jointly with the NBS, the BIPM has measured the GPS receiver and antenna delays of laboratories in Europe and the USA by carrying a GPS receiver taken as a reference. Biases up to 100 ns have been found. It is intended to pursue these calibrations. Most of the GPS time comparisons used in establishing TAI are processed by the BIH. The agreement with computations made by other countries is normally within a few nanoseconds. The Loran-C time comparisons have been entirely reorganized in order to link the laboratories by simultaneous tracking of the same station. Daily measurements are then used in a similar way as GPS data. A substantial improvement has been obtained, but the comparison with GPS has confirmed a seasonal variation in the Loran-C time comparisons which may reach several tenths of a microsecond. To illustrate the performances of the time comparison methods, we have considered the scattering of the ten-day averages of the TAI-TA (lab), after removing a trend by using a high-pass filter with cut-off frequency of 60 days^{-1}. The quadratic mean of the residuals, for the years 1985 and 1986 ranges

for GPS, from 8 to 17 ns (average = 12 ns),
for Loran-C, from 31 to 54 ns (average = 44 ns).

Other Studies

(a) Tests on real data have shown that the linear production of the clocks rates is optimum with the rhythm used in the BIH algorithm. But with the advent of more precise time comparisons, new studies of the predictions are needed and have been undertaken. A major difficulty remains in the existence of seasonal variations of the clock rates, the period involved being a large fraction of the life of the clocks.

(b) An improved atomic time scale for pulsar studies has been established and made available. Since 1977 its maximum departure from TAI has been 9 μs.

(c) The definition of TAI in general relativity has been considered. The relationship between TAI and the time-like argument of dynamical theories has been discussed by Seidelmann and Guinot.

(d) A detailed study of the non-rotating orgin and its application to the definition of UT1 has been completed in cooperation with Paris Observatory.

SHANGHAI OBSERVATORY, SHAANXI ASTRONOMICAL OBSERVATORY
BEIJING ASTRONOMICAL OBSERVATORY, BEIJING INSTITUTE OF RADIO
METROLOGY AND MEASUREMENT, NATIONAL INSTITUTE OF METROLOGY,
INSTITUTE OF GEOPHYSICS, CHINA.
(Reported by Y. R. Miao)

There are 6 institutes which are engaged in time service in China. They are: Shanghai Observatory; Shaanxi Astronomical Ovservatory; Beijing Astronomical Observatory; Beijing Institute of Radio Metrology and Measurement; National Institute of Metrology; Institute of Geodesy and Geophysics.

Shanghai Observatory's major purpose in time research activities is to serve the VLBI system at Shanghai Observatory and the Chinese VLBI network which is being developed. The Hydrogen masers which have been used at Shanghai Observatory were improved again in 1986. The frequency stability for periods of several hours is 5×10^{-15}. A transportable Hydrogen maser which has been developed at Shanghai Observatory for the Chinese VLBI network was successful in early 1987. Time comparisons using the Japanese GMS satellete ranging signals between Shanghai Observatory and RRL are being made routinely. The precision of the time comparision is about ±15ns. During recent years, Shanghai Observatory's atomic time scale which is established by using 3 HP commercial Cesium clocks has reached much better performance.

In reference frame reseach, Shanghai Observatory has discussed the definition, application, research, and transformations of the solar system barycenter and Earth-centered frames in detail, and obtained some good results in satellite laser ranging processing.

Shaanxi Astronomical Observatory has two stations for transmission of time signals, a short-wave transmitting station (BPM), and a long-wave transmitting station (BPL). An intensive investigation with ground-wave propagation of long-wave signals was made in recent years. The synchronization precision within a 2000-km range for the complex condition of propagation path has been able to reach ±0.5 μs. The synchronization precision using the TV signal of the Russian Screen Satellite is ±3 μs which has been achieved within China. Meanwhile, time synchronization research through a Chinese satellite is also being made to obtain higher precision. In addition to keeping the local atomic time scale, Shaanxi Astronomical Observatory is responsible for the Joint Atomic Time System which is composed of the HP commercial Cesium clocks of five institutes mentioned above except the National Institute of Metrology. By using ionosphere D-Layer data observed for 6 years, Shaanxi Astronomical Observatory has determined the ionospheric parameters and the pattern of activities for ionosphere and radio wave absorption, etc. With these investigations, the timing precision for the sky-wave of long-wave signals has been able to reach the order of 10^{-12}.

The Institute of Geodesy and Geophysics joined the Joint Atomic Time System with its 3 HP commercial Cesium clocks. Using the reception of short-wave time signals, research of near-distance progagation characteristics of short-wave signals has obtained very good results.

Beijing Astronomical Observatory has been using HP commercial Cesium clocks to control a Chinese national radio broadcasting station which has a timing program for civil purposes. At present, it is planned to put time code into the time signal broadcast. Also, HP commercial Cesium clocks of Beijing Astronomical Observatory have joined the Joint Atomic Time System.

The National Institute of Metrology has also established a local atomic time scale with its HP Cesium clock. In addition, the HP commercial clocks are being used to control frame pulse and line pulse signal of TV pictures for the national central TV brodcasting station, so as to be used for the purpose of TV syn-

chronization. The primary Cesium standards which were developed
at the National Institute of Metrology have an accuracy of 3x10⁻
13.

The Beijing Institute of Radio Metrology and Measurement is
engaged in time metrology with several HP commerical Cesium
clocks and two hydrogen masers which were made in China. Meanwhile, Bejing Institute of Radio Metrology and Measurement has
developed crystal circuits for many years, and achieved better
performance.

In China, investigations of the relationship between time
definition, time coordinates, relativity, and time metrology have
been made in recent years. The determination of more accurate
time periods by using pulsars is also being investigated in China. We have started to use the GPS timing receiver for international time comparison. Currently VLBI, Solar-Earth investigations, and satellite geodesy as well as space techniques and
earthquake prediction have begun to use more stable and accurate
time and frequency standards in China.

ZENTRALINSTITUT FUER PHYSIK DER ERDE (ZIPE), POSTSDAM, DDR
(Reported by G. Hemmleb)

The time scale UTC (ZIPE) is based upon one HP Cesium clock
5061 A. UTC (ZIPE) is compared daily with UTC (ASMW) using the
TV method. UTC (ASMW) is based on two atomic clocks (one since
1987.0). For time comparison against clocks of PTB, DHI, TP and
AOS the TV method is also used. Furthermore, Loran-C signals
from Sylt were received for time comparison. The ZIPE was included in the regular portable clock trips of SU.

Throughout 1985, 1986, and 1987 the Amt fuer Standardisie-
rung, Messwesen und Warenpruefung (ASMW) has continued to form
the independent atomic time scale TA (DDR) from the readings of 3
(two since 1987.0) commercially produced Cesium clocks of ZIPE
and ASMW. Time and latitude observations with PZT 2 were continued, whereas observations with the Danjon astrolabe (OPL 10) -
made from 1957.7 onwards without interruption - were discontinued
since January 1986. A Photo-eletric Zenith Tube (PEZR) was constructed. Preliminary observations are planned still in 1987.
The results of time observations and comparisons are communicated
weekly to BIH, IPMS, and to the time service of USSR. They were
published in quarterly series bulletins.

DEUTSCHE FORSCHUNGS-UND VERSUCHSANSTALT FUER LUFT-UND
RAUMFAHRT (DFVLR), Oberpfaffenhofen, F.R. GERMANY
(Reported by S. Starker)

From Oct. 30 to Nov. 6, 1985 in the first German Spacelab-
Mission D1 a Cesium and a Rubidium clock were flown in an orbit
of 326 km height. During this experiment, called NAVEX, the
onboard clocks were compared with ground clocks using a two-way
method with PN-code signals at 1.5 GHz. For the data transfer
spread spectrum techniques were used. The slope between onboard

and ground clocks could be measured with an uncertainty of ±1.6 x 10^{-14} during 3 days. The expected relativistic effects (velocity and gravitational effect) were $\Delta f/f$ = (-3.3009 + 0.3538) x 10^{-10} = 2.9471 x 10^{-10}. The measurement result agreed with this value within 0.1%.

In connection with this experiment 4 clock transportations were accomplished between PTB-Braunschweig and DFVLR-Munich with parallel GPS-time comparisons. The deviations between both methods were in every case smaller than 5 ns.

Before and after the D1-mission, the flight clocks at Cape Kennedy were compared with the Cesium clocks at DFVLR-Munich via GPS with a frequency uncertainty of ±1 x 10^{14} during one week.

DEUTSCHES HYDROGRAPHISCHES INSTITUT, FEDERAL REPUBLIC OF GERMANY
(Reported by H. Enslin)

The time signal transmissions of the Deutsches Hydrographisches Institut (DHI) were terminated on October 13, 1985, and the high-precision time comparisons on December 31, 1985. The Cesium clocks were switched off on January 3, 1986. One of the clocks (HP 5061 Ser. No. 11050, Opt. 004) was given to Wettzell where it contributes to TA1; the other clock is in operation in Munich for research work on GPS.

ASTRONOMICAL OBSERVATORY, CAGLIARI, ITALY
(Reported by E. Proverbio)

The Time Service of the Cagliari Observatory is based on 2 commercial cesium standard and 2 quartz clocks. During the period 1984-1987 the local reference time scale has been compared continuously by VLF and Loran-C techniques. The UTC (CAO) scale was also compared via television pulses with IEN (Turin) and ISPT (Rome).

The accuracy of the UTC (CAO) scale versus UTC is about 1-3×10^{-13} The results of time and frequency comparisons are published in the Monthly Bulletin of Cagliari Astronomical Observatory and in Circular D of the BIH.

Particular techniques for time dissemination and synchronization were investigated. The Time Service actively participated in the MERIT Campaign and is participating in the new LASSO enterprise.

ISTITUTO ELETTROTECNICO NAZIONALE, ITALY
(Reported By P. G. Galliano)

Since May 1985, Istituto Elettrotecnico Nazionale (IEN) has been using GPS satellite reception to relate UTC (IEN) to the international time scales. Nevertheless, the reception of two Loran-C stations of the Mediterranean Sea Chain has been continued.

Thanks to the daily synchronization link by means of television measurements, an investigation of the long-term behavior
of the commercial Cesium standards kept at IEN and in the ISPT
Time and Frequency Laboratory in Rome, with reference to the
enviromental conditions was carried out in 1986/1987. The results were presented at the First Time and Frequency European
Forum held in Besancon, France in March 1987.

Regarding time transfer techniques, the IEN performed the
following experiments:
1. A two way time synchronization experiment between the IEN and
the Shaanxi Observatory time scales via Sirio 1 geostationary
satellete. The experiment lasted two weeks in May/June 1984 and
the precision of the time comparisons was 30 ns (one sigma). In
the same period, the time scales of other Chinese observatories
(Peking, Shanghai, and Wuhan) were also compared by different
synchronization techniques with that of the Shanghai
Observatory. A paper dealing with this work was published.
2. A two way time synchronization experiment via the Sirio 1
satellite using Mitrex modems, developed by Prof. Hartl of the
Stuttgart University, to evaluate the capabilities of PRN codes
and using at one site an Earth station equipped with a small
antenna. This experiment lasting one week in March, 1985, was
performed by IEN with Politecnico of Turin in Italy, the Shannxi
and Beijng Astronomical Observtories in China, and, in a one way
mode, the Technische Universitaet of Graz in Austria. The standard deviation of the measurements was of the order of 300 ps.
Finally I wish to point out that a general review of the activity
of the IEN Time and Frequency laboratory can be found in the
paper presented at the Eighteenth PTTI in 1986.

INTERNATIONAL LATITUDE OBSERVATORY OF MIZUSAWA, JAPAN
(Reported by C. Kakuta)

1. SPACE TECHNIQUES

Time variations of relative positions of celestial radio
sources were investigated to estimate errors of group delay observations for geodesy and geophysics with a very long baseline
interferometer. It was found that systematic errors exist between the second and third catalogues. There is no time variation exceeding 30 milli-arc-seconds between mean observation
epoch 1975 and 1981.

The system-level experiments on the Japan-U.S. joint VLBI
project were conducted in January and February, 1984 Over five
hundred observations were successfully performed in the experiments. The correlation processing was made by both the K-3 processor at Kashima and the Mark-III processor at Haystack. The
two baseline-lengths between Kashima and Mojave obtained in the
two experiments showed excellent repeatability. The precision of
the baseline length was higher than 0.02 m in root mean square.
Observations of plate motions and regional crustal deformations
using space technologies, Doppler observation of Navy Navigation
Satellite, Satellite Laser Ranging (SLR), and VLBI were reviewed.

Descriptions were given for the astronomical, geophysical, and geometrical models and formulae adopted in KAPRI, a program for computing the theoretical delay and delay-rate of geodetic VLBI observations and their partial derivatives with respect to physical parameters of interest. In KAPRI more sophisticated therories were adopted for the tidal and relativistic effects. A program was developed for converting an HP1000 based Mark-III database archive to a file on MELCOM COSMO 900II. The program enables us to use VLBI observation data for various calculations and analyses on MELCOM COSMO 900II.

2. POLAR MOTION AND THE ROTATION OF THE EARTH

During the main campaign of Project MERIT (a program of international cooperation to monitor Earth rotation and intercompare the techniques of analysis and observation) from September 1983 through October 1984, the Central Bureau of the International Polar Motion Service (IPMS) at the International Latitude Observatory of Mizusawa (ILOM) acted as an analysis center for the optical astrometry technique.

Prior to the main campaign, the ILOM in collaboration with the Tokyo Astronomical Observatory (TAO) and the Japan Hydrographic Department (JHD) developed software to compute star positions based on the new system of astronomical constants. The method developed by the Japanese group was adopted as the standard method and given in the MERIT Standards compiled by the MERIT group. The computer programs were distributed to all countries where optical astrometry techniques were in operation.

Since the beginning of the main campaign, the Central Bureau of the IPMS (ILOM) has published daily smoothed Earth orientation parameters. This method developed at the ILOM proved to be a very efficient and reliable method to analyze the optical astrometry data through the comparison of the results with those of the new techniques.

At the IAG/IAU Symposium No. 128 held at West Virginia, USA, the results of the daily smoothed EOPs based on a revised method of computation were presented. This method made it possible to estimate the station coordinates simultaneously with the EOPs. As a result of comparison with the VLBI results, agreement is known to be very close. Recomputation of the EOPs based on the past observations will bring about results more reliable than any other past series of EOPs determined using optical astrometry.

The Central Bureau of the IPMS started computation of the atmospheric excitation functions of the Earth's rotation from September 1984, using the data computed at the Japan Meteorological Agency (JMA). JMA's data as one of the three data sources, will contribute very much to understanding atmospheric effects on Earth rotation. The ILOM will continue this work in the coming new International Earth Rotation Service (IERS). The global analysis data of the JMA and the computing method of the effective Atmospheric Angular Momentum (AAM) functions were explained in the Annual Report of the BIH.

In accordance with the establishment of the IERS, the IPMS, as well as the Bureau International de l'Heure (BIH), will be closed. The ILOM is preparing to act as analysis center for the

VLBI technique in the IERS. As the coordinator of optical astrometry in Project MERIT, K. Yokoyama, the Director of the IPMS,
has made various activity reports.

The ILOM, in order to continue Earth rotation activities
using the techniques which fulfill precision requirements of the
IERS, is planning to build a Japanese VLBI system, VERA (VLBI for
the Earth Rotation Study and Astrometry). VERA will be composed
of two antennas spanning about 2300 km and a correlator of multi-
station capability. The diameters of the antennas are 35 m and
15 m.

A catalog of PZT stars was compiled by Sato et al. The
catalog includes all the stars used in PZT observations at the
ILOM (Mizusawa) and the United States Naval Observatory Washington, D.C. since 1959 and 1915, respectively. The stars of proposed programs for the International Latitude Service (ILS) chain
of PZT's on the line of 39 8'N latitude were also included. This
catalog would give a standard for the recalculations of past
observations by the ILS instruments. A unified catalog of Washington and Mizusawa PZT stars was constructed by using re-compil-
ed and re-reduced data from 1954 to 1983. The reduction is made
based on the MERIT Standards. Computations of the right ascension and declination corrections, as well as their proper motions
were based on the least squares chain method. Internal precisions of the right ascensions and declinations at the mean epochs
of observations were 0.6 ms and 0".008, respectively, for the best
determined stars. Those of proper motions were 0.08 ms/year and
0".001/year, respectively.

Estimations for the effects of the nutation errors on the
determination of the Earth orientation parameters and the corrections to stellar positions in source catalogs, which were determined internally by the chain method using the observed latitude
and UTO-UTC, were made. Internal corrections were found to be
related to the adopted nutation series. It was required that
every station re-reduce all the past observations in the new
system and re-estimate internal corrections based on observed
data.

Li's expression for the gravitational deflection of light in
the case of astrolabe observations was corrected by deriving a
rigorous and simpler expression based on elementary methods. The
derived corrections are to be added to group values of time and
latitude.

The Mizuaswa PZT (Photographic Zenith Tube) was moved from
its semi-basement location (the second PZT observation room) to a
location 6.65 m above ground level (the third PZT observation
room) in order to avoid the effects of temperature variation
near the ground during observations. The PZT base was constructed on a stable layer of silt 14.8 m in depth. Larger tremors
were expected for the tall base than lower ones. Comparison of
measured micro-seisms confirmed that the amplitude of horizontal
displacement of the new tall base for PZT 2 is even smaller than
the rest of the bases, which are used for astronomical observations at the ILOM with the old VZT, automatic astrolabe and the
semi-basement type observing house.

near Southeast Asia. Local variations for periods of several years in longitude and latitude might be attributed to thermal expansion in the fluid core near the core-mantle boundary.

Relative variations of UT1-TAI in Mizusawa and Washington were found to show a secular variation associated with a periodic variation during the period of 4 years. One branch of the secular variation shows a similar direction as the plate motion. Decreases in the relative variations of UT1-TAI, in Mizusawa and Washington, were studied from the point of a global motion of the ocean and the El Nino events.

HYDROGRAPHIC DEPARTMENT OF JAPAN
(Reported by Kubo)

For the purpose of monitoring the relation between dynamical time reduced from the orbital longitude of the Moon TDT_M and TAI, the observation of occulations of stars by the Moon have been continued at the head office of the Hydrographic Department of Japan (JHD) in Toyko and three branch Observatories, Sirahama, Simosato, and Bisei. About 900 timing data including 600 photo-electric data were obtained each year. TDT_M-TAI obtained from the occulation observations for the epochs 1984.5 and 1986.5 were 32.98s, 32.93s and 32.78s, respectively, with mean errors of 0.04s. Details are published in Data Report of Hydrographic Observations, Series of Astronomy and Geodesy as well as in the Japanese Ephemeris.

The services of the International Lunar Occultation Center have been continued since 1981. The number of data reported to the Center in the years 1984 to 1986 amounts to 31,370.

Satellite Laser Ranging (SLR) observations have been carried out at Simosato Hydrographic Observatory since March, 1982. The mean ranging precision of the SLR system is about 9 cm. Total return signals obtained from Lageos, Starlette and Beacon-C satellites in the years 1984 to 1986 amount to 725,000. Since August 1986, ranging to the Japanese geodetic satellite Ajisai also has been done with 139,000 return signals in 1986. A transportable SLR system is under construction at JHD, the completion being expected at the end of October, 1987.

TOKYO ASTRONOMICAL OBSERVATORY, JAPAN
(Reported By S. Aoki)

Astronomical observations for time and latitude have been made regularly with the PZT, using the star system (α_{85}, δ_{85}) since January 1, 1985. PZT plates have been measured with an automated-measurement system equipped with a linear CCD since January 1, 1985.

UTC(TAO) has been kept with a master clock, selected out of five HP Cesium clocks, controlled with a phase-microstepper. Three of five Cesium clocks, have been in operation in a shielded room, which has been intended to reduce the effect of eletromagnetic waves on the Cesium clocks, and to keep the appropriate

A Tsubokawa type astrolabe has been developed at the ILOM. To investigate effects of the meteorological enviroment around observational sites upon anomolous refractions for astrometry at the ILOM, measurements of horizontal and vertical profiles of temperature were made during periods before and after thinning out trees around the observational site. Results showed that an inhomogeneous distribution of the environments in the site mainly produces inhomogeneous distributions of temperature, particularly for stable nights, due to a radiative effect of the enviroment. Refraction corrections for VZT observation have been made traditionally by using the mean values of temperature and pressure during the observations of a pair of stars. Because temperature and pressure variations exist during the observations of a pair of stars it is necessary to employ the temperature and pressure measured during the observation of a pair of stars to eliminate refraction error completely in VZT observations.

Refraction effects currently remain on the order of 0."01 in the VZT and astrolabe observations. In order to make such errors as small as possible we must not only improve the meteorological measurements but also keep the meteorological enviroment, such as the roughness height homogeneous in all directions around the observing room. Even the new techniques such as VLBI and laser ranging have errors which originate from the meteorological environment near the observing sites.

Onodera investigated the southern oscillation, one of large-scale atomospheric phenomena, in relation to variations of the rate of the Earth's rotation. The results showed that the variation of the former appears by 2-3 years in advance of the variation of the latter. This fact suggested that there is some correlation between the two variations.

Fluctuations of the Earth's rotation with periods shorter than about 2 years were analyzed using the data compiled or obtained at the BIH, the IPMS, the IRIS (International Radio Interferometeric Surveying), and the CSR (Center for Space Research).- The IPMS optical astrometry data show close agreement with the AAM (Atmospheric Angular Momentum) data which include the effects of both the zonal wind and the migration of the air mass (pressure term) for periods between two years and forty days. Relatively high spectral power density is observed at periods around 35 days, 24 days, and in the period range of 21 to 16 days in the astronomically observed UT1 by the CSR and the IRIS. Thus, the short period fluctuations of the Earth's rotation can be explained fairly well by the atmospheric excitations for the periods between two years and sixteen days.

A comparison was made between variations of strain, which are observed at the Esashi Earth Tides Station and UT2-TAI, published by BIH, residuals of UT0-UTC and latitude of the PZT and Danjon Astrolabe at Mizuaswa since June 1979. Kakuta and K-S. Sato obtained a good correlation between variations of the Earth's rotational speed and UT0-UTC of the Miszusawa PZT with the east-west component of strain.

Residuals of astronomical observations of longitude and latitude in Asia might be related with the secular variations of the z-component of the internal geomagnetic field in the 1970's

temperature and humidity. Domestic time comparison of UTC clocks
has been continued by using a Cesium portable clock of TAO twice
a year with those of ILOM, NRLM, RRL, GSI,(Geodetic Survey Institute) and KGO (Kanozan Geodetic Observatory). TAO clocks have
also been linked with TV-signals.

     For international time comparisons, the receptions of Loran-
C signals from the Iwo Jima Master station (9970-M) and Okinawa
station (9970-Y) have been continued. By means of the reception
of GPS timing signals the TAO has continued time comparisons
among the TAO and six institutes: NBS, NPL, OP, PTB, RRL, and
USNO with an accuracy of the order of a few tens of nanoseconds.-
Clock comparison between the TAO and the USNO with a flying clock
of the USNO has been performed regularly once a year by courtesy
of the USNO.

JET PROPULSION LABORATORY, USA
(Reported by J.O. Dickey)

     The Jet Propulsion Laboratory has been actively involved in
the following areas key to IAU Commission 31:
     * the development of constants, models, and ephemerides to
be used as standards by the community and by the MERIT analysis
centers;
     * reference frame studies: 1) determination of the JPL Radio
Frame; 2) the determination of the Dynamical Reference Frame of
the Lunar/Planetary Ephemerides; 3) determination of ties
between the various reference systems; and 4) development of the
concept of the dynamical equinox as a reference point for the
modern ephemerides and the unification of coordinate systems;
     * acquistion, reduction analysis of VLBI data;
     * intercomparisons of Earth rotation results from the vari-
ous techniques;
     * combination of data types via a Kalman filter;
     * analysis of the scientific implications of these measure-
ments.
During both the Main MERIT and the continuing campaign, Earth
rotation and polar motion studies have demonstrated the unprecedented accuracy (~5 cm) achieved by modern space geodetic techniques, permitting us to determine tectonic plate motion at
various stations and observatories located on separate plates;
these results are in general agreement with those of Minster and
Jordan. In fact, the combination of data sets from several sites
and different techniques demands the inclusion of tectonics;
omission of such considerations would result in a combined series
that would be systematically corrupted and degraded.
     High-quality estimates of the atmospheric excitation of
Earth rotation and polar motion, provided by the routine analyses
of global weather data for operational weather forecasting, together with the modern Earth orientation measurements have allowed new insight into atmospheric and non-atmospheric excitation of
Earth rotation and polar motion variations. We have shown that
changes in the Length of Day (LOD) at seasonal and higher frequencies are dominated by the exchange of angular momentum be-

tween the atmosphere and the solid Earth. The correlation is significant at the 99% level indicating that atmospheric angular momentum (AAM) analysis and forecast fields may be useful in providing near-real time estimates and prediction of Earth rotation changes. Our studies using the forecasts of AAM from large numerical models of the major forecast centers indicate that these models can forecast the AAM better than the currently used statistical predictors (done in collaboration with Rosen, Salstein (AER), Miller and McCalla (NMC)). The dominance of the atmosphere in short-period variations and the simple power law spectrum of the AAM made possible the development of a Kalman filter for smoothing and predicting Earth orientation. We have established the existence of rapid polar motion; by comparison with AAM data, we found that it is produced at least in part by atmospheric pressure changes. The atmospheric excitation of the Chandler wobble was investigated using both modern polar motion determinations and meteorological estimates of atmospheric forcing; some correlation was found, but the nature of the Chandler wobble forcing remains unclear. The relationship between the Southern Oscillation and the Length of Day (LOD) was established; the observed correlation at lead times of 2.6 years was used to derive a statistical predictor of the Southern Oscillation Index based on the LOD.

Turning to longer period fluctuations in Earth rotation, torques are probably largely due to dynamic pressure forces associated with core motions acting on topographic undulations of the core mantle boundary and the equatorial budge. Estimates of such torques are becoming available from a combination of core motion models from geomagnetism and core mantle boundary maps from seismic tomography. Our geodetic torques estimates will provide a strong constraint on the models and assumptions used (done in collaboration with R. Hide, Met Office (UK)) and strongly favor the inclusion of the D" layer in the mantle and point to bumps on the core-mantle boundary of about 1/2 km. The magnitude of these undulations is in agreement with the findings from nutation studies. Corrections at the 2 milliarcsecond level required for the annual term of the Standard IAU Nutation Model as inferred by the analysis of both the IRIS and the JPL Deep Space Network VLBI data are in agreement with those of Herring et al. at CfA. Analyses of seventeen years of lunar laser ranging (LLR) data allowed calculation of corrections to the IAU accepted values of the 18.6-year nutation terms and the precession; results indicate agreement with optical astrometry.

The last three years have seen an improvement in LLR data quality and the development of a Lunar Laser Ranging Network. Recent equipment and software improvements at the stations have resulted in 3-5 cm ranges (the previous data prior to 1984 had ranges of more than 10 cm accuracy); data are currently being acquired from three stations: CERGA (France), Maui (Hawaii-USA) and McDonald (Texas-USA). An analysis of the seventeen year LLR data set yields a value for GM_{EARTH} of 398 600.443 ± 0.006 km^3/s^2 in the geocentric system comparable to LAGEOS (the LLR result agrees with the LAGEOS result within one standard deviation of the error estimate). The increased accuracy of LLR should result in an improved lunar ephemeris; preparations are now underway for a new joint lunar and planetary ephemeris at JPL.

Reference frame studies have included the determination of the JPL Radio Frame, the Dynamical Reference Frame of the Lunar/Planetary Ephemerides, ties between the various reference systems, and the development of the concept of the dynamical equinox and the unification of coordinate systems. Ties between VLBI and optical frames have been established via the observations of radio stars (in collaboration with our French colleagues); a link has been determined between the VLBI and the Ephemeris frame through the use of differential VLBI. Very Large Array measurements of Jupiter, Saturn, Uranus and Neptune provide both a tie between the outer planet ephemerides and the radio frame and a means of improving the ephemerides themselves.

U.S. NAVAL OBSERVATORY (USA)
(Reported by W.J. Klepczynski)

Optical observations for time were made with Photographic Zenith Tubes in Washington, D.C. (PZT7, 65cm) and at the Naval Observatory Time Service Alternate Station (NOTAS) in Richmond, near Miami, Florida (PZT6, 20cm). Observations with PZT2 (20cm) were discontinued in Richmond on June 24, 1987. Daily radio observations with the Connected Element Interferometer located in Green Bank, West Virginia were obtained throughout the period. Measurements were made of three 35-km baseline vectors to determine Earth orientation parameters. The Richmond POLARIS Observatory, located at NOTAS participated in the VLBI Observations of the International Radio Interferometric Survey (IRIS).

The Washington Mark IIIA VLBI Correlator was dedicated at the U.S. Naval Observatory on August 25, 1986 in a ceremony attended by representatives of the four sponsoring agencies (National Geodetic Survey, National Aeronautics and Space Administration, Naval Research Laboratory and U.S. Naval Observatory).-

After several months, the Washington Mark IIIA VLBI Correlator expanded from a 3-station, 3-baseline correlator to a 5-station, 8-baseline facility. The addition of the final two baselines to make a 5-station, 10-baseline correlator is in progress.

The primary reference time system, designated UTC(USNO,MC), is now derived from the hydrogen maser based master clock. Two Smithsonian Astrophysical Observatory VLG 11 Hydrogen Maser Frequency Standards are phase-locked together and are steered to UTC(USNO,MEAN) by adjustments to the frequency synthesizer of the lead hydrogen maser. Steering of the master clock system is done in steps no larger than 7 parts in 10 to the 15th (0.6 ns/day).

Evaluation of the Hewlett-Packard Mercury Ion Frequency Standards (MIFS), delivered in July 1986 and May 1987, for long-term performance has begun. The difference between the MIFS's and the Hydrogen maser based master clock is being studied for use in extrapolating UTC(BIH)-UTC(USNO). Preliminary analysis indicates that the MIFS agrees with UTC(BIH) in frequency to better than a part in 10 to the 14th.

33. STRUCTURE AND DYNAMICS OF THE GALACTIC SYSTEM (STRUCTURE ET DYNAMIQUE DU SYSTEME GALACTIQUE)

PRESIDENT: W.B. Burton
VICE-PRESIDENT: M. Mayor
ORGANIZING COMMITTEE: J. Bahcall, L. Balasz, J. Binney, L. Blitz, J. Einasto, G. Lynga, M. Tosa, R. Wielen

1. Introduction

Commission 33 has the long-standing tradition of producing two versions of the report covering its activities in the general field of Galactic Structure during the past triennium. The short version of the report is the one presented here. A longer, more complete version, is being printed by the President for distribution to all members of the Commission.

Compilation of the triennial summary is one of the major tasks of Commission officers. Section 2 of this report was prepared by M. Major; section 3 by L. Balasz; sections 4 and 8 by L. Blitz; section 5 by G. Lynga; section 6 by D. Hartmann; section 7 by J. Bloemen; sections 9 and 10 by U. Haud and J. Einasto; section 11 by R. Wielen; section 12 by B. Fuchs; section 13 by J. Binney; and section 14 by V. Malyuto and J. Einasto.

2. Basic Data and Calibrations

2.1 Basic Data

Basic data useful for Galactic astronomy are more and more numerous but with the generalization of the machine-readable catalogues and data bases they become very easily accessible. More than 450 machine catalogues are presently available to observatories upon request. This short report merely summarizes work done in the field of the acquisition of basic data; reading of the different issues (numbers 27 to 32) of the "Bulletin d'Information du Centre de Données Stellaires" of Strasbourg (France) is recommended, or of the NSSDC-WDC-A-RS publication no. 83–84 and Astrosoviet preprint **3–86** of the Soviet Center of Astronomical data. You will find, for example, detailed documentation for machine-readable catalogues done by Warren and Roman for the majority of large catalogues (for ex. 37.002.008 to .015, 37.002.096, 38.002.039 to .042). Procedures for obtaining data from the Astronomical Data Center and policies for distribution have been described by Warren (42.002.049) or can be found in the Bull. Inf. CDS **32** (1987), p. 127. Star catalogues and files available at the Stellar Data Center are listed in (41.002.075). The various data retrieval capabilities of the new *SINBAD* data base of the Centre de Données Stellaires have been described by Wenger and Ochsenbein (37.002.031/032). Among the new catalogues let us mention *IRAS* catalogues and atlases: Atlas of low-resolution spectra by Olnon *et al.* (42.002.038); the Bibliographic Catalogue of stellar radial velocities (1970–1980) by Barbier-Brossat and Petit (42.002.035); the Catalogue of 900 faint star UV spectra based on observed data of the Space Observatory Orion-2 by Gurzadyan et al. (41.002.027); the *HEAO A-1* X-ray source catalogue by Wood et al. (38.002.048) including positions and intensities for 842 X-ray sources; the General Catalogue of variable stars discovered until 1982, by Kholopov (39.002.023, 40.002.068): a Supplement to the Bright Star Catalogue containing data for stars with $V = 7.1$ and brighter that are not listed in the Bright Star Catalogue, Hoffleit *et al.* (37.002.007). For metal deficient stars, see Bartkevicius (38.002.005, 37.002.39 and 38.002.075), for $uvby\beta$, see Hauck and Mermilliod (39.002.016), for Vilnius photometry (38.002.052) and for *UBVRI* see Lanz (42.002.036), for MK classification (39.002.108) and for $[Fe/H]$ determinations (39.002.001). A new and extensive catalogue of *UBV* photoelectric photometry has been compiled by Mermilliod (1987) who collected all data published between 1953 and 1985, which led to a final catalogue of 136,319 entries concerning 87,267 stars. A new version of the Geneva photometric catalogue is being prepared by Rufener and should include about 4×10^4 stars.

A large set of basic data for galactic structure is expected from the future *HIPPARCOS-TYCHO* space mission: star positions, parallaxes, proper motions, magnitudes. Intensive preparation work has been done in relation with the Input Catalogue (INCA). Ground-based astrometric and photometric observations are in progress. A first possible Input Catalogue is described by C. Turon (1986), Scientific Aspects of the Input Catalogue preparation have been discussed in an ESA colloquium (40.012.019).

Becker and collaborators have published volume ten of the Basel Photometric Catalogue (*RGU* system) for fields in Anticenter, Carina, Centaurus, Aquila and Cassiopeia (39.002.010).

During the last triennium Gliese and Jahreiss continued the compilation of the "Third Catalogue of Nearby Stars" which will be extended to 25 pc from the Sun. At least 3000 objects (singles and systems) are expected to be included in this catalogue.

A Catalogue of Infrared Observations by Gezari *et al.* (37.002.079) summarizes all infrared astronomical observations published between 1965 and 1982.

A new edition of the Yale parallax catalogue is in preparation by van Altena (37.002.086, 38.002.029). New parallaxes have been determined by Upgren et al. (39.111.016), Vilteki *et al.* (42.111.018), Ianna (42.111.009), Scales and Zhao (37.111.011) and Harrington *et al.* (39.111.005).

From the proper motion survey with the 48-inch Schmidt telescope, Luyten and collaborators have published a list of double stars with common proper motions (37.111.020), the proper motions of 6056 stars in the Pleiades-Hyades region (39.111.025) and for 7698 stars in the Hyades region (41.111.027).

A survey of trigonometric parallaxes and proper motions with the UK Schmidt telescope yields astrometric and photometric data for a complete sample of 6125 stars brighter than $B = 17.5$ in the South galactic cap. Murray *et al.* (42.111.014, 42.111.012).

A search has been made by Halbwachs to find common proper motion systems among AGK3 stars (42.111.012). Part of the Second Cape Photographic Catalogue 1950.0 containing provisional positions of stars in the Cape zone −40° to −52° has been published by Nicholson *et al.* (37.002.024); new proper motions have been determined for 22,731 stars.

Absolute proper motions of stars relative to galaxies have been determined by Zhu and Wang (38.111.016) and by Baltahaev, Rakhimov and Umarova (39.002.086/087/088/089). An objective prism survey carried out by Stephenson provided a sample of more than 2000 candidates for K and M dwarf status, generally having proper motions that are either unknown or less than 0.2 per annum (41.114.085) and a sample of high proper motion K and M stars (42.111.007).

A very large number of radial velocity measurements was published during the last three years. Many investigations have been directed towards halo kinematics. Fouts and Sandage (41.111.016) have obtained new radial velocities for 889 high-proper-motion stars and have rediscussed previously known relations between space motion and chemical composition. Carney and Latham (42.155.035) have examined the kinematics from new radial velocities of 85 metal-poor field red giants. Carney, Latham (1987) have also published *UBV* photometry and radial velocities for about 900 G-stars selected from the Lowell Proper Motion Survey without any metallicity bias. Their radial- velocity data indicate a binary fraction of the high velocity stars probably exceeding 25%. Radial velocities of 57 RR Lyrae stars have been obtained by Hawley and Barnes (40.111.008). Kinematic investigations of the outer galactic halo have been published by Sommer-Larsen and Christensen (39.111.001, 41.111.004) using a sample of blue horizontal branch field stars and by Ratnatunga and Freeman (39.155.100) using faint K giants. Radial velocities of these giants show that the outer halo is at most slowly rotating and that the line-of-sight velocity dispersion is approximately constant with increasing distance from the Sun. Stock's catalogue of radial velocities from objective prism plates of (40.002.083) has been used to select southern stars of high radial velocities, Stock *et al.* (38.111.021, 38.111.009).

Radial velocities of 790 late-type bright stars, Andersen *et al.* (39.111.002) and 551 A and F type stars, Nordstroem and Andersen (40.111.001) complete the V_r data of the Bright star catalogue. Part of a large survey of the population II stars radial velocities has been acquired for a preliminary list of 146 F stars, by Andersen and Nordstroem (40.111.001).

Photoelectric radial velocities of 406 ninth-magnitude KO stars located in ten small regions at galactic latitudes ±35° have been acquired by Griffin (41.111.003). Jones and Fisher (37.111.012) have published the velocities of 116 southern red stars.

The first eight years of radial velocity studies at Fick Observatory have yielded 16,000 observations of over 2000 late-type stars; Beavers and Eitter (42.111.010) and (1986).

Radial velocities of standard stars have been discussed by Barnes *et al.* (41.111.010), Maurice *et al.* (38.111.003), Mayor and Maurice (39.111.040) and Stefanik *et al.* (1985).

Fehrenbach and collaborators (42.111.020; 42.111.021) published prism objective radial velocities. Part of these velocities were obtained for stars of the *HIPPARCOS* astrometric mission.

We also wish to mention the Bibliographic Catalogue of Stellar Radial Velocities done by Barbier (39.002.097) and Barbier & Petit (42.002.035).

Numerous photographic photometry surveys are done in various directions in order to determine reddening density and luminosity function, but also to test galactic models. Due to the developments of an automatic measuring device, star counts to faint magnitudes are more and more recognized as an important information on galactic structure, Morton (41.155.082), Bahcall (42.155.031).

In *RGU* three-colour photometry, Fenkart and collaborators have done investigations in the directions of the anticenter (38.113.005, 39.113.004) as well as the anticenter-northern galactic meridian (38.113.012). Other starfields were investigated in the direction of VelaII by Marsoglu (38.113.052), around NGC 2158 by Topaktas (38.113.051), near IC 2581 by Alfaro and Garcia-Pelayo (38.113.019), in the direction of Praesepe by Karaali (38.113.004) or in the direction of Scutum (39.155.073). The galactic structure was examined in a faint object survey in a field in Aquarius by Tritton, Morton and collaborators (38.155.008, 39.114.015, 41.155.083). A photometric survey of two high-galactic-latitude fields has been carried out by Friel and Cudworth (41.155.088). Observed magnitudes and colour distributions have been compared with star-count predictions from the Bahcall-Soneira galaxy model.

Three fields of the galactic bulge have been analyzed by Rodgers *et al.* (42.155.038) using star counts, colour distributions and colour-luminosity arrays. For these star counts more than 120,000 photographic stellar images were used.

Reed and Fitzgerald have determined the distribution of B5 to M5 stars to $V \simeq 12.5$ in the direction of Puppis (38.115.034/035). Dubyago (37.113.053) carried out *UBV* photographic measurements of 2200 stars in Cygnus (37.113.053) and Frogel *et al.* (38.113.007) *RIJHK* photometry of K and early M-giants in Baade's window.

Late-type dwarfs have been the subject of different studies. Robertson has carried out an objective prism survey for late M dwarfs (38.114.013); Upgren and Lu have presented broadband *BVRI* photometry for nearby K and M stars (42.113.031). Hartwick *et al.* (38.114.121) have been able to separate a true halo population of M stars. The authors have obtained an upper limit to the relative density of halo to disk stars at $M_v \simeq 14$ and argue that the faint halo M stars are unlikely candidates for the solution of the "missing mass" problem in the Galaxy. Reid and Gilmore (37.115.004) have presented new optical and infrared photometry of a large sample of very low mass dwarfs.

UBV photometry was carried out by Sandage and Kowal (41.113.053) for 1690 high-proper-motion stars which provided a finding list for potential high-velocity stars of various metallicity values. Using the variation of the tangential velocity maximum with galactic longitude, the authors estimate the rotation of the subdwarf system to be V_{rot} (halo) $\sim 0 \pm 50$ km.s^{-1}. On the same subject, see the work done by Carney and Latham (1987) on some 900 G stars, including *UBV* photometry and radial velocities. *UBV* photoelectric photometry has been carried out by Oja (40.113.003) for about 700 stars near the North Galactic Pole. Vilnius photometric measurements of stars near the North Galactic Pole has been published by Bartasiūtė and H_β photometry of A and F stars in the direction of the South Galactic Pole by Andersen and Jensen (39.113.005). New spectral types have been determined by Lee (37.114.105) for high-proper-motion stars. Nikolaev *et al.* (40.002.139) have published mean light curves and $B - V$ and $U - B$ colours for 210 RR Lyrae-type stars of the galactic plane.

In order to study the distribution of interstellar matter in the solar neighbourhood and to investigate ages, abundances and kinematics of F type stars, Olsen and Perry have done H_β photometry of 2699 A5 to G0 population I stars. Large and kinematically unbiased samples of F-type stars (38.113.028) and G and K type stars (42.113.032) have been studied by Eggen; four colours, H_β and *RI* photometry have also been published. Golay et al. (39.155.059, 41.155.068) did a first analysis on the UV survey of the galactic plane (2000 Å) in a balloon-borne experiment. The second catalogue of stellar UV-excess objects searched with the Kiso Schmidt Telescope has been presented by Kondo *et al.* (38.113.055).

Positions and infrared photometry of 338 sources have been published from the Valinhos 2.2 μm survey of the South Galactic Plane by Epchtein *et al.* (40.002.003).

2.2 Intrinsic Colours, Interstellar Reddening and Absolute Magnitudes

IAU Symposium no. 111 has been devoted to "Calibration of fundamental stellar quantities" (40.114.092). Equally of interest for stellar classification are "Spectroscopic and photometric classification of population II" (42.012.106) and "The MK process and stellar classification" (42.012.033).

Among detailed studies we can mention that on H_γ luminosity calibration for spectral types O to early A of luminosity classes V–III by Millward and Walker (39.115.003) and for A and B supergiants (39.115.006) and (42.114.160). Effective temperatures and angular diameters of non-supergiant O9–G8 stars have been determined by Moressi and Malaguini (39.114.090); see also (42.115.001). Absolute magnitude and T_{eff} of B-type stars have been derived from 13-colour photometry by Conconi and Mantegazya (40.113.043). Balona and Shobbrook have carried out re-calibration of the luminosities of early-type stars (38.115.011) (see also 38.115.016); they discussed in particular the effect of the calibration on the zero-point of the cepheid luminosity scale. Effective temperature scales for B-type stars in relation with UBV and Geneva photometry is discussed by Cramer (37.113.017). The Effective Temperature Scale of O to F stars has been analysed by Theodossiou (39.114.089). Pastori and Malaspina have applied the visual surface brightness scale for B5–F5 main sequence stars, (38.115.001) and Shulov the luminosity of OB stars from photometry in the $uvby\beta$ system (42.115.004).

F-type stars have been calibrated by McNamara and Powell (41.114.008) and by Saxner and Hammarbäck (40.113.047). Moon and Dworetsky have published grids for the determination of effective temperature and surface opacity of B, A and F stars using $uvby\beta$ photometry. The infra-red flux method has been used by Leggett et al. (41.115.007) to obtain effective temperatures, diameters and luminosities of 22 bright stars.

Calibrations of G and K type stars have been discussed by Olsen (38.113.014), Frisk and Bell (40.114.088), Wing et al. (40.114.092) and Ardeberg and Lindgren (40.113.036); population II giants by Geisler (42.113.027); M supergiants by Elias et al. (39.115.004) and by Abramyan (38.115.005); M giants by Mennessier and Grenon (40.114.090); OH/IR sources by Feast (39.115.014); supergiants in open clusters by Keenan and Pitts (39.115.022); normal stars by Grenier et al. (39.115.009); red dwarfs by Gliese (39.115.021) and by Thé et al. (37.113.019); the Barnes-Evans relation by Eaton and Poe (38.115.012); stars in the $uvgr$ system by Kent (39.113.013); absolute magnitudes of K and M giants by Mikami (41.115.003).

The existence of tight linear correlation between the stellar absolute magnitude M_v and the MgII k-line emission has been confirmed using IUE high-resolution spectra, by Parthasarathy (40.115.014).

Buser and Kurucz have treated some of the basic problems involved in the synthetic colour calculations and discussed the theoretical calibration of UBV photometry (40.113.037).

The Malmquist correction has been derived using different shapes of the luminosity function by Jaschek and Gomez (1985). Space density of B5 to A5 stars and extinction was analyzed by Burns et al. (37.155.012) in a direction $(l, b) = (253°, -7°)$. The distribution of OB stars and absorbing matter in the region around P Cyg (38.155.026) as well as in the direction of Per OB1 (38.114.130) has been discussed by Garibdzhanyan et al. Kunde, (41.113.003) using $uvby\beta$ photometry, has obtained the local dust distribution in the direction $(l, b) = (359°, +24.5°)$.

The catalogue of Savage et al. (1985) containing UV interstellar extinction excesses for 1415 stars exists in machine-readable version (42.002.063).

2.3 Stellar Luminosity Function

The search for brown dwarfs in relation with the missing mass in the solar neighbourhood is always a topic in the forefront of astrophysics. The proceedings of the workshop on "Astrophysics of brown dwarfs" (42.012.099) give a general idea of the state of current research in this field. The crucial role played by the luminosity function in galactic structure modelling using star counts has been analyzed by Bahcall (42.155.031). This author gives in particular a list of the more important questions which can be answered by future observations.

A RI survey for low-luminosity M-dwarfs from deep UK Schmidt plates has been analyzed by Hawkins (42.155.-081). The resulting luminosity function shows a decrease in space density in the range $M_R = 12 - 15$ after which the space density rises again. The author explains this subsequent rise as the reappearance of a population of non-hydrogen burning "brown dwarfs" in degenerate cooling phase. This result gives new hope that the missing mass in the solar neighbourhood resides in faint stars after all. On the contrary, the analysis done by Boeshaar and Tyson (39.155.096) on faint red star counts does not support the hypothesis that the missing mass of the disk is constituted of very low mass stars near the hydrogen-burning limit. Studies of the disk main-sequence luminosity function using photometric and kinematic absolute magnitude calibration methods result in significantly different estimates of the number density of very faint $(M_v > +14)$ dwarfs. Reid (37.115.003) has investigated biases affecting the latter method. The stellar distribution in apparent V magnitude and $B - V$ colour has been determined for complete samples toward $(l, b) = (0°, -9°)$ and $(37°, -51°)$. From these data Gilmore et al. (39.155.062) derive the form of the stellar luminosity function in the solar neighbourhood. The low-mass stellar luminosity function obtained shows a broad maximum near $M_v \sim +13$ and a decline for less luminous stars. The conclusion of the authors being that missing mass in the solar neighbourhood cannot be explained by unusual low-mass luminous stars. Luyten has continued this survey for proper motions of faint stars on Schmidt plates and finds evidence that the frequency of such proper motions decreases with galactic latitude and this appears to become more prominent for fainter stars. This may mean that eventually we may have to use different luminosity functions for different galactic latitudes.

From the luminosity function of main-sequence nearby stars, D'Antona and Mazzitelli have derived the Initial Mass Function between 0.1 and 1 $M_\odot$ (42.155.001). The local mass density of halo stars has been discussed by Lee (41.155.110) and Dawson (41.155.092/.106). The initial mass function for massive stars in the Galaxy and the Magellanic Clouds has been determined by Humphreys and McElroy. The authors find no observational evidence that the slope of the massive IMF differs from the normal IMF for the solar neighbourhood or that it varies with galactocentric distance. Observational constraints on the form of the high-mass stellar IMF have been reviewed by Scalo (41.155.121). There is no convincing evidence for any systematic variations of the shape of the high-mass IMF.

References

Beavers, Eitter, *Astrophys. J.*, **62**, 147

Carney, B.W., Latham, D.W., 1987, *Astrophys. J.*, **92**, 116

Jaschek, Gomez, *Astron. Astrophys.*, **146**, 387

Mermilliod, J.C., 1987, *Astron. Astrophys. Suppl.*, in press.

Turon, C., 1986, *A first possible Input Catalogue*, 3rd Fast Thinkshop, Bari (Italy)

3. Stellar Studies of Local Galactic Structure

This subsection reports results on the existence and nature of the possible unseen matter near the Sun, the local density of disc and halo stars, the respective luminosity functions and the local distributions of some special type objects. It summarizes work relating to the space density of stars of different types in the plane of the Galaxy and dealing with photometric stellar studies in the Milky Way fields of strong interstellar extinction. It also reviews galactic models relevant to the local structure.

3.1 The Volume Closest to the Sun

The existence and nature of unseen matter in the solar neighbourhood received much attention in the past triennium; the subject remains controversial.

Bahcall (38.155.084) determined the total amount of matter in the vicinity of the Sun by comparing the observed distributions of K giants with the predictions of detailed Galaxy models. Hartwick *et al.* (38.114.121) studied a sample of 65 faint cool stars from the LHS catalog. Strömgren (39.155.020) discussed the determination of the galactic force K_z and the local mass density. Boeshaar and Tyson (39.155.096) searched for faint red stars out to 40–100 pc from the Sun. A survey of low luminosity M-dwarfs from deep UK Schmidt plates has been described by Hawkins (42.155.081).

The local density of disc and halo stars, and the respective luminosity functions were investigated on different kinematically selected new samples. Eggen (42.113.032) discussed the distributions of abundances and space motions based on four-colour, H_β, and (R, I) photometry for some 5000 stars in four kinematically unbiased samples. He computed (1987) the luminosity and abundance distributions based on a catalog of VRI photometry of stars with $V > 15.1$ and annual proper motion $> 0.5''$. Sandage and Kowal (41.113.053) gave UBV photometry for 1690 high-proper-motion stars, providing a finding list for potential high-velocity stars of various metallicities. Sandage (1987) determined the disk and halo densities at the galactic plane from star counts at the galactic poles. From a kinematically unbiased sample of halo stars, S.G. Lee (41.155.165) estimated the local mass density of halo dwarfs. The luminosity function of the main-sequence galactic halo population was derived from the LHS catalog by Dawson (42.155.106).

Westin (38.155.063) studied the distribution of early type stars and the orientation and extent of the Gould's Belt. A population of super-metal-rich stars was searched by Grenon (39.155.165). He estimated the age and space density of the SMR subpopulation.

3.2 Objects at Low Galactic Latitudes

Efforts to determine the space density of stars of different types in the plane of the Galaxy continue. Results of a spectrophotometric study of 277 OB stars in a region around P Cyg was published by Garibdzhanyan et al. (38.155.026). Reed and FitzGerald (38.155.034) presented spectroscopic and photometric parallaxes for 108 OB stars in Puppis. These authors (38.155.035) also classified 3339 B5–M5 stars in this field. Vega et al. (42.155.-016) studied the spiral structure in the Vela. Spectral classification of carbon stars with the Kiso Schmidt telescope was carried out by Maehara (40.114.053) in the Cassiopea region. The distribution of late-type giant stars in the galactic plane were studied by Melik-Alaverdyan and Tovmassyan (42.155.011) based on the "SAO and Atrophysical Data" catalog. Downes (42.126.007) determined the space densities of white dwarfs, subdwarfs, and cataclysmic variables based on UV excess objects with $B > 15.3$ mag. and $b < 11°$. Winkler et al. (38.155.060) published a complete isophote map of the Galaxy in the ultraviolet spectral region ($\lambda = 356$ nm).

RGU photometry was extensively applied to study the interstellar reddening and space density in Milky Way fields of strong interstellar extinction. Fenkart and Karaali (38.113.005) measured colours for 1700 stars down to $G = 18$ mag. in the anticentre direction in order to determine the reddening, density and luminosity function. Alfaro and Garcia-Pelayo (38.113.019) studied the reddening and space density of 2099 stars in the Carina region. Karaali et al. (39.155.073) completed RGU photometry of 2647 stars in a field of Scutum. Becker and Steppe (39.155.075) reported application of the method of 3-colour photometry on Milky Way fields with strong interstellar extinction. Fenkart et al. (39.113.004) gave RGU colours for 1362 stars down to $G = 18$ in the anticentre direction (A6). RGU photometry in a complexly reddened Milky Way field in the direction to SA 193 was given by Fenkart and Topaktas (1987).

3.3 Objects at Intermediate and High Galactic Latitudes

Experimental testing of current galactic models has motivated a number of attempts to prove, or exclude, the existence of a "thick disc" component. Gilmore et al. (39.155.062) determined V and $B-V$ for 10,000 stars in 11.5 sq. deg. towards $(l, b) = (0°, -90°)$ and for 28,000 stars in 17 sq.deg. towards $(l, b) = (37°, -51°)$. They derived the parameters of the galactic spheroid, and the form of the stellar luminosity function. Stobie et al. (41.155.017) obtained UBV colours of stars in the $12 < V < 18$ range covering 24 sq.deg. in the NGP region. Yoshii et al. (1987) obtained UBV data of 18,303 stars of $V > 19$ mag. over 21.46 sq.deg. towards the NGP. Del Rio and Fenkart (1987) have attempted to compare observed density gradients in a field near the galactic centre using the models of Bahcall and Soneira and of Gilmore and Wyse. Fenkart and Karaali (1987) measured 1806 stars down to $G = 19.5$, and made comparisons with current galactic models. A study was performed on 993 stars of $V < 12.5$ in a 19 sq.deg. field centered on NGC 7686 by Balazs et al. (41.155.095). The best fit to the space densities was obtained by an isothermal model.

The Basel group has continued study of star fields at different galactic latitudes in order to get more detailed distribution of stars in the galactic disc and halo. Fenkart and Esin-Yilmaz (38.155.058) applied the Basel Halo Program methods (BHP) in UBV to SA 82, already treated in RGU. Photometry in the RGU system was carried out by Karaali (38.133.004) in a field of 3.56 sq.deg. containing a large fraction of the Praesepe cluster with 1500 stars down to $G = 16.2$ mag. Fenkart and Karaali (38.113.012) derived space density functions of 759 stars in a 1.70 sq.deg. starfield near M67, based on RGU photometry.

McNeil (42.155.036) investigated the distribution of G5–M stars in a south galactic pole region, studying the density distribution as a function of distance from the galactic plane. Opal and Weller (38.155.025) scanned two-thirds of the celestial sphere in the 911–1050 Å band. The correlation between the stars and the continuous 240 MHz radiation in the Loop I main ridge region were calculated by Anisimova (38.155.073). Tritton and Morton (38.155.008) identified and classified all objects with $B < 20.0$ in a 0.31 sq.deg. region centered on $l = 36°.5, b = -51°.1$. They determined the B and V magnitudes of 601 normal stars and 2 white dwarfs to this magnitude limit. An objective prism survey was completed by Stephenson (41.155.040) of 583 late-type stars more then 10° from the galactic plane and north of −25° declination.

References

del Rio, G., Fenkart, R.P., 1987, *Astron. Astrophys. Suppl.*, **68**, 397
Eggen, O.J., 1987, *Astron. J.*, **93**, 393
Fenkart, R.P., Karaali, S., 1987, *Astron. Astrophys. Suppl.*, **69**, 33
Fenkart, R.P., Topaktas, L., 1987, *Astron. Astrophys. Suppl.*, **69**, 281
Sandage, A., 1987, *Astron. J.*, **93**, 610
Toshii, Y., Ishida, K., Stobie, R.S., 1987, *Astron. J.*, **93**, 323

4. Studies of the Interstellar Medium Near the Sun

The local interstellar medium was the subject of a meeting held in Madison Wisconsin in June 1984. Much of the discussion centered around the structure of the local interstellar medium (LISM). Results presented at the meeting included observations of the very local interstellar medium within a parsec of the sun probed by spacecraft using backscattered solar radiation (38.131.240, 38.131.241). The diffuse warm gas in the LISM has been explored by a variety of means including various UV and optical absorption lines (38.131.243), chromospheric Lyman α (38.131.245), Mg I and Mg II (38.131.246, 38.131.247, 38.131.248, 38.131.249, 39.131.004), H_α emission (38.131.254) and Na (38.131.255, 38.131.256, 38.131.257). Other probes of the diffuse gas come from hot DA dwarfs (38.131.253), EUV and continuum absorption (38.131.263), observations of β Canis Majoris (38.131.250), and HI and radiocontinuum observations (38.131.273). The surprising existence of a cold component is inferred from CO observations (38.131.267) and possibly from extinction maps (38.131.258, 38.131.260, 38.131.266). The hot component is seen by means of soft X-rays (38.131.265, 38.131.280), Fe XIV (38.131.264), and OIV lines (38.131.262). An observational overview was presented by York and Frisch (38.131.244) and Cowie presented a theoretical model of the LISM as a supernova remnant (38.131.279).

One of the new results regarding the structure and composition of the LISM this triennium has been that of the clear identification of a cold component seen in both molecular line emission and infrared continuum detected by *IRAS*. The discovery of the infrared 'cirrus' (37.133.021) and the high latitude molecular clouds (HLCs; 38.131.021) was followed by work showing that the two are intimately related (42.131.049), and that the distances to them puts one cloud clearly within the local hot interstellar cavity (42.131.051), and the remainder at distances of about 100 pc (42.131.161). Other studies give the properties of the HLCs, and show that the Sun is about 30 pc above the galactic midplane defined by the clouds (40.155.015, 40.131.291, 40.131.263). Keto and Myers made an independent study of the southern hemisphere HLCs (41.131.276), and Magnani *et al.* estimated the completeness of the surveys (41.131.104, 41.131.391). Other reviews of the cirrus were done by Gautier (41.131.348), and de Vries (42.131.345); southern hemisphere cirrus was also studied (42.131.063).

Other work on the local dark cloud and molecular cloud components included two new large scale surveys of obscuring material (41.131.022, 41.131.390, 42.155.092). CO surveys have shown that half of the emission from the Cygnus Rift can be decomposed into 10 molecular clouds at distances of 200–2300 pc (40.155.049), and that an extended complex of molecular clouds close to the Sun can be linked to the Aquila Rift, ρ Ophiuchus, and Gould's Belt (37.131.321).

Work on the warmer components of the LISM examined the vertical distribution and extent of the local HI halo (41.131.103), and the source of ionization for the warm local gas. (42.131.173). The warm LISM was probed by a variety of optical and UV absorption line studies, some using high resolution spectroscopy (42.131.352, 42.131.353), Ca II (42.131.157), Mg I (42.131.360), Mg II (42.131.068, 42.131.348), CIV and Si IV (42.131.359), as well as the hot gas sampled by Fe X (37.131.228), and Fe XIV (37.131.322).

An already frequently cited work by Paresce gives a local N(HI) map from *EUV* observations of local stars that clearly shows the low density hole in which the Sun is embedded (38.131.007). Both Soviet and American space probes as well as European rocket flights have sent back data allowing the analysis of the heliospheric bow shock caused by the LISM (37.131.318, 38.131.226, 40.131.154). UV, HI, and soft X-ray data were analyzed to constrain the nature and location of the hot bubble (39.131.119, 42.131.346, 42.131.349). Theoretical investigations suggested an origin for the hot gas (38.131.002, 42.131.355) and examined absorption line mechanisms to constrain the relationship of HI to HII in the bubble (40.131.154). Although not necessarily related to the bubble, one work suggested that some of the high velocity clouds are less than 150 pc from the Sun (38.131.079).

Finally, an important review of galactic constants by Kerr and Lynden-Bell (42.155.028) provided the basis for the revision of the rotation constants recommended by Commission 33 at the previous IAU.

5. Stellar Studies of the Overall Galactic Structure

During the triennium, there have been relevant symposia on "Luminous Stars and Associations in Galaxies" (41.012.079) and "Star Forming Regions" (Peimbert and Jugaku, 1987). The first *IRAS* conference "Light on Dark Matter" was held in Noordwijk (41.012.080) and the second conference on "Faint Blue Stars" took place in Tucson during June 1987. A meeting on "Stellar Populations" (42.012.082) was held at in Baltimore, May 1986.

The initial mass function (IMF) for luminous stars was discussed by Humphreys and McElroy (38.155.041) and by Garmany (40.155.067, 41.155.119). The issue was reviewed by Scalo (41.155.121), who found no convincing evidence of systematic variations of the IMF over the Galaxy. Oda (39.155.071) constructed a model distribution of stellar radiation accounting for the galactic light in the B and V bands. Van der Kruit (41.155.032) analysed the surface brightness of the galactic background in terms of models of stellar distribution.

Francois (41.155.047) discussed the chemical evolution of the Galaxy on the basis of light-metal abundances relative to iron. Gilmore and Wyse (42.155.005) showed that consideration of the chemical properties of the thick-disk population of the Galaxy results in a self-consistent model for galactic chemical evolution. Clayton (39.155.076) studied models of chemical evolution with star formation rate proportional to the square of the mass of gas. Lacey and Fall (39.155.088) presented models with radial inflows as well as infall from outside the Galaxy. Rana and Wilkinson (41.155.008, 41.155.009) developed a model of chemical evolution showing that the star formation rate is correlated with the surface density of molecular hydrogen and has remained practically constant over the lifetime of the disk. Guesten (39.155.118, 41.155.051) has reviewed the framework of chemical evolution models and has given some special solutions. Giridhar (41.155.117) used 23 Cepheids to analyse chemical inhomogeneities. A scenario involving supernova-induced star formation is invoked. Olive *et al.* (1987) identified constraints on the history of the galactic star formation rate, and found that reliable predictions of stellar and supernova nucleosynthesis are required for optimal use of the methods.

Bahcall and Casertano (42.155.033) reanalysed Eggen's proper-motion and radial-velocity data for high transverse velocities. Bahcall *et al.* (40.155.086) discussed the 12 Basel fields that are situated above $b = 20°$. Observations require the presence of a thin as well as of a thick disk. The latter may satisfy the model of Bahcall and Soneira (37.155.078) as well as that of Gilmore (37.155.016).

Tritton and Morton (38.155.008, 41.155.083) made an exhaustive survey down to $B = 20$ in a field at $l = 36°.5$, $b = -51°.1$. Comparisons with the Bahcall-Soneira model revealed considerable discrepancies. Gilmore *et al.* (39.155.062) derived structural parameters of the galactic spheroid from about 28,000 stars in 17 sq. deg. towards $l = 37°, b = -51°$. Bahcall and Ratnatunga (39.155.063) found these results in good agreement with the Bahcall-Soneira standard model of the Galaxy. Inante (41.155.066) studied stars near the south galactic pole down to $J = 22$. The data agree well with the model of Bahcall and Soneira; they are not consistent with the Gilmore-Reid normalisation of the thick disk component, although they do not rule it out.

From the metallicity distribution perpendicular to the plane to distances of the order of 2 kpc, Gilmore and Wyse (40.155.023) divided the stars into components with quite different spatial density distributions. Sandage (1987) and Sandage and Fouts (1987) used star counts and radial velocities to derive density ratios of 1:0.11:0.005 for the thin disk, the thick disk, and the halo. Friel and Cudworth (41.155.088), from photometry in two high-latitude fields, find good agreement with the Bahcall and Soneira model, although they do not find as many blue stars.

Whitford (39.155.103) reported a considerable spread of metallicity among bulge giants, the dominating component being super metal rich. Drilling (42.155.103) concluded that extreme helium stars and cool, hydrogen-deficient stars belong to the bulge population. Habing *et al.* (40.155.051) reported more than 2500 *IRAS* stars with flux densities above 1 Jy at 12 μm. According to Habing (41.155.122), the faintest and reddest of the *IRAS* sources show the bulge of the Galaxy. Feast (41.155.076, 41.155.123) suggested that the majority of the *IRAS* sources in the bulge are Mira variables. Frogel (41.155.124) found that the M giants of the bulge have significantly different colours and luminosities than the solar neighbourhood giants. Feast (41.155.123), Feast and Whitelock (preprint), and Feast and Spencer Jones (preprint) identified the majority of *IRAS* sources in the bulge with Mira variables.

Henry, DePoy and Becklin (38.155.047) identified, from a deep CCD image, the galactic center non-thermal radio source with IRS 16. A visual extinction of 38 magnitudes was extrapolated from the infrared measurements of IRS 14. Sandqvist (preprint) demonstrated the great resolving power of lunar occultation observations of the galactic centre. Adams *et al.* (preprint) resolved IRS 16 into 3 sources. Winnberg *et al.* (39.155.102) found 33 OH/IR stars strongly concentrated towards Sgr A West. Shklovskij (39.155.081) explained a number of features of the galactic centre by a permanent star formation process, while Maddox (39.155.092) concluded from the measurements of the size of the galactic core that it contains a black hole. Lo *et al.* (39.155.093) found from *VLBI* measurements an upper limit of 20 AU to the diameter of Sgr A, also revealing an elongated structure at 3.6 cm. Crawford *et al.* (39.155.143) studied the mass distribution from spectroscopic measurements in the central 10 pc of the Galaxy. Data are consistent with a large point mass. Serabyn and Lacy (39.155.149) observed $[NeII]$ emission from Sgr A West and found a mass distribution suggestive of a black hole.

Tobin (41.114.017 and preprint) analysed arguments concerning the nature of faint high-latitude B stars and suggested relevant observations. Carney (38.155.076) discussed stellar systems beyond 25 kpc. Saha (39.155.078) studied the space density of RR Lyrae stars out to 33 kpc from the galactic centre. Ratnatunga and Bahcall (40.155.033) predicted the field star densities towards some globular star clusters and local group galaxies as a guide for planning ground-based and *HST* observations of objects at intermediate and high latitudes.

Forward Look: The chemical evolution of the galactic disk has received much attention but no single, convincing model has yet emerged. One would wish for more reliable data pertaining to galactic gradients, the IMF, and other observable features which might form the basis for a unique model. The thick disk has been confirmed by several, independent observations. Its structure and its dynamical properties will be subject of continuing discussions. The precise nature of objects near the galactic centre is still a controversial topic, on which new observations are needed. Observations planned for the Hubble Space Telescope promise considerable advances in all aspects of stellar studies in our Galaxy.

References

Olive, K.A., Thielmann, F.K., Truran, J.W., 1987, *Astrophys. J.*, **313**, 813

Peimbert, M., Jugaku, J. (eds.), 1987, *Star Forming Regions*. IAU Symposium No. 115. D. Reidel, Dordrecht, Boston, Lancaster, Tokyo

Sandage, A., 1987, *Astron. J.*, **93**, 610

Sandage, A., Fouts, G., 1987, *Astron. J.*, **93**, 592

6. Large-Scale Aspects of the Distribution of Interstellar Matter

6.1 General

The previous report on structure and dynamics of the galactic system was given by Wielen (41.155.100). The recently recommended values for solar distance to the galactic center (8.5 kpc) and our rotation speed around it (220 km.s^{-1}) were discussed by Trimble (42.155.043).

The distance scale of the Galaxy was reviewed by Barkhatova *et al.* (40.155.088). A discussion of typical corrugation scales in the Galaxy was given by Spicker and Feitsinger (42.155.003), who concluded that three distinct scales seem to exist: 1–2 kpc, 4–8 kpc, and > 13 kpc. These corrugations are reflected in the distribution of O and B-stars and HII regions, and to a lesser extent in the HI distribution. Feitsinger and Spicker (39.155.026) investigated the corrugation phenomenon for the (heliocentric) longitude range $10° \leq l \leq 240°$ as derived from HI studies.

The proceedings of the 106th IAU Symposium : "The Milky Way Galaxy", were edited by van Woerden, Allen and Burton (39.012.007).

6.2 Observations and their Interpretations

6.2.1 Neutral Hydrogen

An HI survey of the southern Milky Way for $240° \leq l \leq 350°$ and $|b| \leq 10°$ using the 18-m Parkes antenna, was presented by Kerr *et al.* (42.155.040). The HPBW is 48′, and the velocity resolution is 2.1 km.s^{-1}.

The Leiden-Green Bank survey of atomic hydrogen was published by Burton (40.155.056) and Burton and te Lintel Hekkert (40.155.057). Observations with the *NRAO* 140-foot telescope cover the complete longitude range accessible at $\delta > -46°$ and the latitude interval $|b| \leq 20°$. The sampling interval is 1° in both l and b. The kinematic resolution is 1 km.s^{-1}. The observations are displayed as cross sections through the data cube in longitude, velocity coordinates at constant latitudes, and as cross sections in latitude, velocity coordinates at constant longitudes. Maps of integrated velocity intervals, and of HI column densities are also presented. Burton and te Lintel Hekkert (39.155.024) discuss the HI survey north of $\delta = -40°$ in the latitude range $|b| \leq 20°$. An atlas of cuts through the galactocentric data cube of HI observations at constant R-values in Θ, z coordinates, at constant Θ-values in R, Θ coordinates was published by Burton and te Lintel Hekkert (42.155.025). Maps of the galactic arrangement in R, Θ coordinates of projected HI surface densities, z-height of the gas-layers centroids, maximum volume density, as well as measures of the layer thickness are presented.

Pöppel and Viera (40.155.060) published an HI survey of the region $240° \leq l \leq 359°$, $+3° \leq b \leq +17°$ with a velocity resolution of 2 km.s^{-1} and an extent from -100 km.s^{-1} to $+100$ km.s^{-1}. The l and b separation was 1°. The region $-3° \leq l \leq 21°$ and $-4° \leq b \leq +3°$ was surveyed with the Effelsberg 100-m telescope by Braunsfurth and Rohlfs (38.131.009). The velocity range was -300 km.s^{-1} $\leq v_{lrs} \leq +300$ km.s^{-1}, with sampling intervals of 0.1 and 2 km.s^{-1}. The sensitivity was about 0.5K. An HI survey in the region $120° \leq l \leq 142°$ and $-5° \leq b \leq +5°$

with the Effelsberg 100-m telescope was presented by Braunsfurth and Reif (38.155.059). The velocity range of this survey was -180 km.s$^{-1} \leq v_{lrs} \leq +80$ km.s^{-1}, and the sensitivity was about 0.3 K.

HI observations made by Olano (40.131.094) of the region $290° \leq l \leq 320°$ and $+3° \leq b \leq +17°$ reveal the presence of HI features with filamentary characterisctics, and a close correlation of radio-continuum emission and HI gas at large height above the galactic plane. A discussion of the vertical distribution of galactic HI was given by Bania and Lockman (39.155.025), based on data from the Arecibo-Green Bank survey.

Dickey and Garwood (42.155.053) used the *VLA* to measure 21-cm absorption in directions with $|b| < 1°, |l| < 25°$ to probe the cool atomic gas in the inner Galaxy. Most of the absorbing gas is associated with molecular cloud complexes. A 21-cm study of 9 areas that have the smallest known amount of HI in the northern hemisphere was carried out by Lockman *et al.* (41.131.137). The data indicate that the HI column density never drops significantly below 4.5×10^{19} cm^{-2} anywhere in the sky. Observations of HI profiles in 34 directions in the region ($172° \leq l \leq 97°$) at high galactic latitude and in 59 directions towards the LMC were published by McGee and Newton (42.131.311). Using the HI column densities as references, depletions in the abundances of calcium and sodium in the halo, spiral arms and the LMC disk were estimated.

Lockman *et al.* (41.131.103) mapped the extent of the local HI halo. The principal result is that the total column density of HI at $|z| > 1$ kpc is, on the average, $5 \pm 3 \times 10^{19}$ cm^{-2}, or 15 % of the total N_{HI}. The HI halo in the inner Galaxy was discussed by Lockman (38.155.027). There is 21-cm emission from corotating HI in the inner Galaxy to $|z| \geq 1000$ pc from the plane. Over most of the inner Galaxy more than 10% of the HI emission at the subcentral points comes from $|z| > 500$ pc. High-z HI is not present ≤ 3 kpc from the galactic center, but appears suddenly near R = 3.5 kpc.

Heiles (38.155.014) used combined existing HI surveys to derive lists of new shell-like objects that cross survey boundaries at $b = 10°$. Spatial filtering revealed wormlike structures in the inner Galaxy. Sodroski *et al.* (39.131.-040) studied the structure and kinematics of the HI associated with Gould's Belt, using data of high velocity resolution and large latitude extent ($10° \leq l \leq 350°$).

Observations in the 21 cm line made with the *NRAO* 43-m telescope of 20 randomly selected intermediate and high-galactic-latitude regions were published by Jahoda *et al.* (39.131.119). The data were examined for evidence of the neutral-gas clumping which is required by models in which a substantial fraction of the diffuse soft X-ray background originates outside the galactic disk and is absorbed by interstellar gas. No such evidence was found. Shull and van Steenberg (40.131.042) presented high resolution (0.1 Å) spectra obtained with the *IUE* satellite. From these spectra, HI column densities toward 244 early-type stars were derived.

The kinematics and distribution of HI in the Galaxy was reviewed by Petrovskaya (40.155.080). Topics like non-circular motions and the application of the wave-theory of spiral structure to investigations of HI in the Galaxy were treated. The large-scale distribution of HI in the Galaxy was discussed by Petrovskaya (41.155.094). The distribution of atomic hydrogen and diffuse gas in the Galaxy was discussed by Heiles (41.155.103).

Feitzinger and Spicker (39.155.148) showed on the basis of recently published HI data and new radial velocity fields that the so-called rolling motion phenomenon is only partly explained by geometric effects. An improved method and a new warp model to correct the observed velocity gradients for these apparent rolling motions were used. The HI at the outer edge of the Galaxy was discussed by Jackson (39.155.027).

An investigation of small-scale HI structure at high galactic latitude was made by Jahoda *et al.* (38.131.274). The number density and random motions of interstellar HI clouds have been studied by Anantharaimaiah *et al.* (41.131.-241) and (38.131.072) using a method which involves comparison of thermal velocities of HI absorption spectra in the direction of HII regions with their recombination line velocities. Using HI data, Kulkarni and Fich (39.155.084) find that a detectable amount of higher than normal velocity dispersion HI exists. The "fast" HI probably constitues of $\leq 20\%$ of the total mass of galactic HI, but contains most of the kinetic energy of the ISM.

Results of 21 cm HI observations in the region $220° \leq l \leq 260°, |b| \leq 15°$ with the *RATAN*-600 radio telescope were presented by Yudaeva (40.155.079). Observations were made at constant declination scans on right ascension with 5° steps in declination. The results are shown as antenna temperature $(\alpha - v_{lrs})$ maps of gas distribution in the vicinity of the outer spiral arm of the Galaxy.

Mirabel and Morras (39.131.044) reported the results of a search for high-velocity hydrogen around the direction of the galactic center. About 2000 positions were surveyed with the 43-m *NRAO* telescope with an *rms* of 0.03 K on a velocity interval of $-1000 \leq v \leq +1000$ km.s^{-1}. A deep Dwingeloo survey for high-velocity clouds was reported by Hulsbosch (39.131.043). Preliminary results were presented for $0° \leq l \leq 200°, -70° \leq b \leq +70°$.

6.2.2 Carbon Monoxide

The Massachusetts-Stony Brook galactic plane CO $(J = 1 \rightarrow 0)$ survey was discussed in several publications. Clemens (41.155.090) presented the observations, and gave some discussion on galactic structure, and of cloud identification. The (b, v) maps of the first galactic quadrant were given by Sanders et al. (41.155.001). The data from the *FCRAO* 14-m telescope consists of 40,551 spectra in the longitude range $8° \leq l \leq 90°$ and $-1°.05 \leq b \leq +1°.05$. The spectral coverage was 300 km.s^{-1} at 1 km.s^{-1}. The small grid spacing enables detection of all clouds larger than ~ 15 pc diameter in the $R = 0.4 - 0.8R_o$ molecular cloud ring. The (l, v) maps were given by Clemens (41.155.002) in (l, v) format as seven gray-scale maps, and in (l, b) format as 17 contour- and gray-scale maps of integrated CO intensity. The presentations are useful for comparison with other first-quadrant surveys. The disk and spiral-arm molecular cloud population derived from the survey was discussed by Solomon et al. (39.155.112). Molecular clouds and cloud components in the inner Galaxy with a size larger than 10 pc could be detected. The total number of emission centers is seen to be distributed as 75% cold molecular cores in the disk, and 25% warm molecular cores in the spiral arms.

Determination of the dependence of CO radial velocity on galactic longitude along loci of tangent points in the inner Galaxy was made by Clemens (40.155.016). The measurements were combined with published data for HI in the nuclear region, outer Galaxy CO–HII regions, and globular clusters, to yield a rotation curve. Rivolo et al. (41.155.035) discussed the statistical clustering properties of the ~ 2000 molecular cloud cores that were identified in the survey between $20° \leq l \leq 50°$. Evidence was presented that the warmest cores are strongly clustered into groups with a characteristic size of $\sim 50 - 150$ pc.

The Columbia CO survey of molecular clouds in the galactic quadrant was announced by Cohen et al. (39.155.-029), and published by Cohen et al. (41.131.141) The galactic disk was surveyed from $12° \leq l \leq 60°$ and $|b| \leq 1°$ with a sampling of $0°.125$ for $|b| \leq 1°$, and $0°.25$ elsewhere. The entire collection of spectra as well as spatial and (l, v) maps were presented.

Maps of ^{12}CO emission in the first galactic quadrant were published by Knapp et al. (39.155.058). Their survey consists of strip maps in latitude at 38 galactic longitudes ($4° \leq l \leq 90°$), spaced at equal intervals of sin l. The latitude sampling is $2'$. The velocity resolution is 0.65 or 2.6 km.s^{-1}. The data are presented as (l, v) maps, as latitude-averaged maps, and as a latitude-averaged $(v, \sin l)$ map.

A CO $(J = 2 \rightarrow 1)$ survey of the southern Milky Way was discussed by van der Stadt et al. (39.155.031). The survey consisted of three parts: the galactic plane ($b = 0°$) in the range $270° \leq l \leq 355°$, 88 dark clouds with and without associated nebulosity, and 47 bright HII-region complexes.

A latitude survey of CO $(J = 1 \rightarrow 0)$ emission near the galactic center was presented by Bania (39.155.032, 42.155.044). The region surveyed covered $350° \leq l \leq 25°$ at latitude points $b = 0', \pm 10'$ and $\pm 20'$. The bulk of the ^{12}CO emission in the inner Galaxy could be produced by three massive objects: the nuclear disk/bar, the "3 kpc arm", and the "+135 km.s^{-1} feature". A wide latitude CO survey of molecular clouds in the northern Milky Way was made by Dame and Thaddeus (38.131.268). The area $12° \leq l \leq 100°$ and $-5° \leq b \leq +6°$ was fully sampled with the Columbia 1.2-meter telescope.

A wide-latitude CO survey of molecular clouds in the third quadrant was published by May et al. (40.131.293). The region surveyed was $180° \leq l \leq 280°$ and $|b| \leq 5°$, at a angular resolution $0°.5$. Distances and masses of more that 30 molecular clouds related to the Perseus arm and Cygnus arm were calculated. McCutcheon et al. (39.155.030) discussed the distribution of CO in the southern Milky Way from a survey of the $(J = 1 \rightarrow 0)$ line in the area $294° \leq l \leq 358°$ and $-0°.075 \leq b \leq +0°.075$ with a sampling interval of $3'$. Robinson et al. (38.155.032) presented the distribution of CO $(J = 1 \rightarrow 0)$ for the longitude interval $294° \leq l \leq 86°$. A global average radial distribution of molecular gas out to R_o was derived.

The distribution of ^{13}CO in the inner galactic plane was discussed by Liszt et al. (38.155.055). The longitude range was $20°.5 \leq l \leq 40°.0$. The emissivity ratio of ^{12}CO over ^{13}CO, the cloud mean free path and cloud-cloud random velocity dispersion, and the effect of cold HI in molecular clouds on 21-cm HI profiles were discussed.

CO $(1 \rightarrow 0)$ and CS $(2 \rightarrow 1)$ observations of the neutral disk around the galactic center were presented by Serabyn et al. (42.155.034). A $2' \times 6'$ region at the center of the Galaxy was mapped in CO with a $21''$ resolution using the *IRAM* 30-meter telescope. Additional spectra were measured in the CS line. The observations are consistent with an inclined disk orbiting about the galactic center in an almost circular orbit. Harris et al. (40.155.009) mapped the central 10 pc of the Galaxy in the CO $(J = 7 \rightarrow 6)$ line. The emission comes from a dense clumpy disk of temperature 300 K. Possible heating mechanisms were discussed. The data show that the rotational velocities drop by a factor of 1.4 to 2 between 2 and 6 pc from the center.

The merits of HI and CO observations as tracers of spiral structure were discussed by Kerr (41.155.072). Individual structural features are often easier to identify in CO. Robinson et al. (41.155.019) presented a geometrical framework provided by CO and HI observations of six directions where gas is seen tangentially along a spiral feature, and four directions where extended structures cross the $R = R_o$ circle.

A comparative analysis of ^{13}CO with ^{12}CO and HI emission in the galactic plane was made by Xiang (41.131.313). Statistics indicate that most peaks of ^{13}CO integrated emission anticorrelate with the corresponding HI integrated intensities.

6.2.3 Sub-millimeter and Infrared

A survey of the galactic plane in the first quadrant ($-5° \leq l \leq +62°$) at wavelengths 150, 250, and 300 μm with a $10' \times 10'$ beam, was published by Hauser et al. (38.155.046). The emission detected arises mostly from sources known to have 5 GHz or CO emission. A total of 80 prominent discrete sources were identified and characterized. Campbell et al. (38.155.030) presented a far-infrared and submillimeter survey of the the galactic plane ($11°.5 \leq l \leq 17°.5$) with $11'$ resolution at wavelengths of 93 μm, 154 μm, and 190 μm. The maps were interpreted in terms of the temperature and spatial structure of diffuse far-infrared and submillimeter sources associated with evolved HII regions and a continuous ridge of galactic emission.

Caux and Serra (42.155.023) reported galactic disk observations ($-150° \leq l \leq 82°$) at $\lambda_{eff} = 380$ μm made with the *AGLAE* 83 balloon-borne instrument. The longitude profile exhibits diffuse emission all along the disc with bright peaks associated with resolved sources. A far-infrared (FIR) survey of the galactic disc ($250° \leq l \leq 20°$) in the southern hemisphere was presented by Caux et al. (38.155.012). The observations were made with a baloon-borne instrument in the wavelength range $114 - 196$ μm. The FIR emission could be due to very large complexes of HII regions, giant molecular clouds, and lower density gas, which are distributed along galactic spiral arms. The complete data set of this survey, presented in the form of brightness contour maps, was published by Caux et al. (39.133.003). A comparison was made with the radio continuum data at 5 GHz.

Campbell et al. (39.155.106) presented a far-infrared and submillimeter survey of the galactic center and nearby galactic plane in the range $359° \leq l \leq 5°$. The data were obtained with a balloon-borne telescope from channels filtered for a bandpass of 70 μm $\leq \lambda \leq$ 110 μm and for a longpass of $\lambda < 80$ μm. Continuous emission was mapped along the galactic plane, and discussed in detail.

Mid-infrared observations at 4, 11, 20, and 27 μm of the galactic center region were discussed by Little and Price (40.155.007). The diffuse emission around the galactic center can be separated into three components: foreground emission from the 4–5 kpc ring, a spheroidal component surrounding the center, and an elliptical component immediately surrounding the galactic center. Catchpole et al. (39.155.014) presented infrared scanning observations of the galactic bulge. Interstellar absorption was derived from J, H, and K bands of a $7' \times 200'$ strip of the sky from Sgr I to the galactic center. The visual absorption (excluding dark clouds) ranges from 3 to 30 mag. in that region.

The large-scale mapping of the Galaxy by *IRAS* was discussed by Gautier and Hauser (39.155.036). The high sensitivity of the *IRAS* instrument for detection of interstellar matter in the survey mode was illustrated in terms of visual extinction and dust and gas column densities.

Tereby and Fich (42.131.308) found a strong, apparently linear correlation between the IR cirrus at 100 or 60 μm and HI near the galactic plane. *IRAS* sky brightness images were compared with the Weaver-Williams HI survey in two regions near $l = 125°$ and $l = 215°$. The dust temperature inferred is nearly uniform, and in reasonable agreement with theoretical predictions.

6.2.4 Molecular Clouds

The large scale distribution of molecular clouds as a function of galactic radius and azimuth was discussed by Scoville et al. (38.155.006). Particular emphasize was given to the 5–8 kpc molecular cloud ring and to the issue of CO spiral structure. A summary of properties of the emission regions was provided.

The number and distribution of molecular clouds in the inner Galaxy was discussed by Thaddeus and Dame (38.155.007). Two CO (l, v) diagrams, covering of much of the first and second quadrants, are given. Huang and Thaddeus (42.131.307) carried out a CO survey toward every confirmed outer Galaxy supernova remnant in the range $70° \leq l \leq 210°$, and found that half of them revealed spatial coincidence with large molecular cloud complexes.

A comparison of CO (2.6 mm), HII ($H100\alpha : 6$ cm), and far-infrared (150 μm, 250 μm) surveys over $-1° \leq b \leq +1°$ and $12° \leq l \leq 60°$ was published by Myers et al. (41.131.105). Some 54 molecular cloud complexes with mean mass of 10^6 $M_\odot$ were identified. The estimated star formation efficiency for the entire sample lies near 0.02.

An estimate of the radial distribution of molecular hydrogen from star formation rates was given by Rana and Wilkinson (41.155.009). The derivation was made independently of the CO surveys, and a comparison is made with the controversial estimates of Σ_{H_2} based on various CO surveys.

Gatley et al. (42.155.029) presented the first detailed maps of the surface brightness and velocity field made in the $(\nu = 1 \rightarrow 0\ S(1)\ )$ line of molecular hydrogen, with a spatial resolution of $18''$ and a velocity resolution of 130 km.s^{-1}. The molecular ring that surrounds the nucleus of the Galaxy was confirmed to be tilted $\sim 20°$ out of the plane of the Galaxy. Detailed far-infrared observations of several atomic and ionic fine-structure lines and molecular rotational lines toward the galactic center were discussed by Genzel et al. (40.155.050). The dominant motion of the observed neutral gas disk is rotation about an axis similar to the rotation axis of the Galaxy. The dynamics of interstellar clouds in the galactic center as determined by the nuclear mass distribution was also examined.

A CO survey of high-latitude molecular gas was carried out by Magnani et al. (40.155.015). About 57 clouds were found in 35 complexes at $|b| \leq 25°$. The clouds are distributed asymmetricaly with respect to $b = 0°$; the distribution is consistent with a displacement of the sun of 30 pc above midplane.

The face-on distribution of molecular gas in the first quadrant, derived from the Massachusetts-Stony Brook galactic plane CO survey, was compared to the galactic distribution of giant HII regions by Clemens et al. (42.155.055). The HII regions were found to preferentially select gas regions of higher than average density, and showed a strong correlation with the second power of the gas density. Burton et al. (40.131.292) gave a preliminary analysis of the distribution of dust in the Galaxy and a comparison of the dust distribution with that of the gaseous components.

The distribution of CH in the Galaxy was investigated by Johansson (39.155.034). Observations with $2°.5$ spacing in the range $(10° \leq l \leq 60°)$ and $(310° \leq l \leq 350°)$ were made in the main-line transition in the $^2\pi_{1/2}, J = 1/2$ ground state Λ-doublet at 3335 MHz. Maurice et al. (39.131.035) studied absorption lines of interstellar sodium, covering a substantial part of the Galaxy at high spectral resolution.

A survey of the galactic center region in the CS $(J = 2 \rightarrow 1)$ line was published by Stark et al. (42.155.051). Güsten et al. (42.155.052) reported 88 GHz observations of HCN $(J = 1 \rightarrow 0)$ emission and absorption in the central 5 pc of the Galaxy.

6.2.5 Radio Continuum

A complete VLA survey in the outer Galaxy was published by Fich (42.155.057). All continuum sources stronger than 0.3 Jy at 21 cm, smaller than $2'$, and in the area defined by $93° \leq l \leq 163°, -4° \leq b \leq +4°$ were observed at 6 cm with a resolution of $4''$ and a sensitivity of 1 mJy. The unresolved objects were also observed at 2 cm with the same resolution and sensitivity. The purpose of the study was to identify objects within the disk of the outer Galaxy. Beuermann (40.155.054) presented a three dimensional model of the galactic radio emission at 408 MHz, based on the all-sky survey of Haslam et al. In this model, the Galaxy consists of a thick non-thermal radio disk in which a thin disk is embedded. Both disks exhibit spiral structure. The thick disk emits about 90% of the total power at 408 MHz.

The first part of a survey of the southern sky at 2.3 GHz was presented by Jonas et al. (40.155.026). The surveyed area was $-63° \leq \delta \leq -24°$, $12^h00 \leq \alpha \leq 22^h00$. The angular resolution was $20'$, the sensitivity better than 16 mK.

Reich et al. (38.155.057) presented a radio continuum survey at 11 cm for the area $357°.4 \leq l \leq 76°$, $-1°.5 \leq b \leq +1°.5$. The angular resolution is about $4'.3$, and the sensitivity is 20 mJy/beam area. A catalogue of 1212 small-diameter radio sources was compiled. A very deep survey of the Galaxy at 7.6 cm was published by Berlin et al. (38.155.020).

A description of recent surveys of galactic continuum radiation was given by Reich (40.155.022). The surveys dicussed include the Bonn 408 MHz all-sky survey, the Bonn 1420 MHz survey which will be extended by observations with the Argentinian 30-m dish, the Effelsberg 1420 MHz and 4875 MHz surveys, and the Effelsberg 2695 MHz survey.

6.2.6 Kinematics and Spiral Structure

The Milky Way halo gas kinematics were discussed by Danly (42.155.054). A distinction was made between higher column density material in the form of condensed clouds, and low column density diffuse material.

Feitzinger and Spicker (42.155.007) investigated the vertical velocity asymmetries observed in HI spiral arms, the so-called rolling motion phenomenon. A descriptive energy model of spiral arm regions with pronounced vertical velocity gradients, so-called VAR's (Velocity Active Regions) was presented, and compared with the energetics of star forming regions. The spiral structure of the Galaxy for the region $38° \leq l \leq 70°$ was discussed by Jacq et al. (40.155.081). The Bordeaux ^{13}CO survey and the Arecibo HI survey have been used to explain most of the features observed in the (l, v) diagrams, and consequences of small kinematical deviations from the mean rotation curve on the shapes of predicted spirals in the (l, v) plane were examined.

The galactic nucleus was reviewed by Oort (39.155.051). The mini-spiral and the possibility of a central black hole were discussed, as were the supernova remnants, HII regions, molecular clouds and other phenomena in the central region. Possible expanding features, the asymmetry and low rotation of the bulk of the molecular gas, and the tilt of the gas layer were reviewed.

Lisst (39.155.046) considered some results of attempts to trace the spiral structure in HII regions, HI, and CO. Deriving galactic structure in CO seems to be recapitulating the history laid down by HI observers.

A model of a two dimensional quasi-steady solution of the gas-dynamical equations in the gravitational potential of a weakly barred galaxy was presented by Mulder and Liem (41.155.031). From the solution, (l, v) diagrams were constructed and compared with HI observations in our Galaxy. Gorbatskij and Usovich (42.131.036) presented results of computations that show that ringlike structures consisting of clouds must be formed in spiral galaxies due to viscosity. The origin of GMCs is discussed.

6.2.7 Rotation Curve

Gerhard and Vietri (42.155.046) showed that the narrow peak ($v_{max} \simeq 250$–260 km.s^{-1} at $r \simeq 500$ pc) and the steep decline for 600 pc $\leq r \leq 1.5$ kpc down to a rather broad minimum ($v_{min} \simeq 195$ km.s^{-1}) around 2.8 kpc as seen in the rotation curve as inferred from HI and CO terminal velocities, can be reconciled with the bulge distribution determined by infrared observations and the local density of spheroid stars only if the bulge of our Galaxy is non-axisymmetric and the resulting potential triaxial.

A determination of the galactic rotation curve from selected HII regions was made by Rohlfs et al. (41.155.036). For $R < R_o$ the rotation curve relies mainly on radio data, and in the innermost 4 kpc non-circular gas motion was taken into account. For $R > R_o$ the curve is based on HII region data. The resulting rotation curve is flat on a large scale.

The rotation curve up to 16 kpc was determined by Kolesnik and Yurevich (39.155.195) using the relation between parameters of OH molecular features of clouds and the distance to the clouds. From the position of the central peak in the rotation curve, the sun's galactocentric distance was estimated to be ~ 8.5 kpc. The mass of the Galaxy interior to 100 kpc from the galactic center is $\sim 3 \times 10^{12}$ $M_\odot$, as was derived from the rotational velocity distribution.

Haud (38.155.024) discussed the rotation curve of the Galaxy for $R > R_o$. It was argued that the apparent rise of rotational velocity in the outer regions of the Galaxy may be the result of a too-simple method of processing the observational data. It was also shown that the wavy form of the rotation curve may be evoked by the spiral density waves in the Galaxy.

The rotation curve of the Galaxy in the distance range $1 < R/R_o \leq 1.6$ was derived by a method which utilizes the whole HI 21-cm line profile by Petrovskaya and Teerikorpi (42.155.002). No evidence was found for a fall-off below a flat rotation curve in the distance range covered by the method. The rotation curve of the outer parts of the Galaxy was determined by Petrovskaya and Teerikorpi (41.155.102) from neutral hydrogen 21-cm line profiles.

A graphic way of presenting the HI data and determining the galactocentric distance R/R_0 in the method of Agekyan et al. (1964) was introduced by Teerikorpi (40.155.075).

Brand (*"The Velocity Field of the Outer Galaxy"*, Ph.D. Thesis, Leiden, 1986) analyzed photometric and CO data of Nebulous objects in the outer Galaxy, and calculated the rotation curve. He finds that $R_o = 8 \pm 0.5$ kpc, for a flat rotation curve and $\Theta_o = 220$ km.s^{-1}. Part of this work has already been published (Brand, Blitz, & Wouterloot, *Astr. Astrophys. Suppl.* **65**, 537; Brand, Blitz, Wouterloot, & Kerr, *Astr. Astrophys. Suppl.* **68**, 1; **69**, 343).

7. Galactic X- and Gamma-Radiation, Magnetic Fields and Pulsars

7.1 Diffuse Galactic X-ray Emission

The past triennium saw considerable activity on the interpretation of the diffuse X-ray background. Marshall and Clark (38.142.091) analyzed the *SAS-3* survey of the soft X-ray sky in the *C*-band ($\sim 0.1 - 0.28$ keV) and concluded that the counting rates in this band and HI column densities are generally anticorrelated down to the angular resolution of their detector ($\sim 3°$). They showed that the data can be fitted by a two-component model: a local hot plasma and a galactic halo extending beyond most of the absorbing clouds. Knude (39.131.186) has modeled such a scenario. Burrows *et al.* (38.155.081) studied the Wisconsin survey and HI observations of sky areas near the galactic poles and concluded that most of the observed X-ray flux must originate on the near side of the most distant neutral gas. In any case, the observed anticorrelation is weaker than expected if the obscuration would be due to a uniform absorbing layer. Jacobsen and Kahn (42.142.034) showed that this weakness of the anticorrelations can be understood if the absorbing material is highly clumped. Jahoda *et al.* (39.131.119; 42.131.436) searched for evidence of this clumping, studying HI column density variations at medium- and high-latitude regions to $10'$ resolution. No such evidence was found.

Hirth *et al.* (40.131.219) found high-negative-velocity HI clouds in close positional coincidence with enhancements in the soft X-ray surveys. They suggested that the X-rays are caused by conversion of the collective motion of the high-velocity clouds into thermal energy during a deceleration process in the ambient gas.

Rocchia *et al.* (37.131.007) performed spectral observations of the soft X-ray background and detected CV, CVI, and OVII lines, which confirms the temperature of about 10^6 K of the hot medium in the solar vicinity. They found evidence for a weak component at a higher temperature which, they suggested, could be produced by a hot halo. Bloch *et al.* (42.142.018) reported the results of very soft X-ray observations (Be band, $\sim 0.078 - 0.111$ keV) of a section of the northern galactic hemisphere; the rates are consistent with the expectations from a plasma of $\sim 10^6$ K.

Kahn and Caillault (41.142.090) combined several *EINSTEIN* IPC fields and found an excess of diffuse M-band emission (0.1 –2 keV) at low latitudes which exhibits a resolved, double-peaked profile that is roughly symmetric about $b = 0°$ with a half-width of $\sim 1° - 2°$. A similar galactic ridge (without the absorption dip) was found at higher energies and was recently mapped by *EXOSAT* (40.155.037). Modelling of the *EINSTEIN* and *EXOSAT* data indicates that the scale height of the emitting region is of the order of 100 – 200 pc and that the radial scale length is roughly 5 kpc. The physical origin of the emission is uncertain – it is probably at least partly due to discrete sources. Several possibilities have been considered (40.155.037; 41.142.090; 40.142.064; 42.131.434). Koyama *et al.* (41.155.042) reported the detection of an intense emission feature in the emission along the galactic ridge at about 6.7 keV attributed to iron.

7.2 Diffuse Galactic Gamma-Ray Emission

No new γ-ray experiments that are of direct interest for studies of the structure of our Galaxy were flown during the last three years. There are, however, interesting new developments in the studies of low-energy γ-ray lines being detected from the general direction of the galactic centre and in studies of very high energy γ-rays (> 1000 GeV), which are reported elsewhere in this volume. Most of the new results that are relevant for this report were obtained from analyses of the *COS-B* data base (~ 50 MeV – 5 GeV) which became publicly available in 1985. Discussions of the diffuse galactic γ-ray emission at lower energies ($\sim 1 - 30$ MeV), using balloon observations, are given by Sacher and Schönfelder (37.143.039) and Lavigne *et al.* (42.143.017). Two text books on γ-ray astronomy, by Hillier (38.003.026) and Ramana Murthy & Wolfendale (41.003.030) appeared.

The galactic γ-ray emission in the *COS-B* energy range seems to originate largely from cosmic-ray/matter interactions (through $\pi°$-decay and bremsstrahlung). Dermer (41.143.005) reconsidered the production of $\pi°$-mesons in these interactions using accelerator data. Bloemen (39.131.128) re-evaluated the γ-ray contribution of a weak third component, namely the inverse-Compton emission originating from the interaction of cosmic-ray (CR) electrons with the interstellar photon field.

Significant progress in the interpretation of the *COS-B* data could be made mainly because large-scale CO surveys of the Galaxy became available, particularly the one by the Columbia group (Dame *et al.* 1987, *Ap. J.* **322**, 706). These have been used together with various HI surveys in γ-ray /gas correlation studies, which gave insight in the galacto-centric distribution of CR particles, the CO-H_2 conversion factor (on a galactic scale as well as for some local molecular clouds), and the amount of molecular gas in the Galaxy. Several authors presented these type of analyses, although with various different approaches: Bloemen *et al.* (37.155.072; 38.143.012; 41.155.003), Harding & Stecker

(39.155.101), and the Durham group (37.143.013; 38.143.025; 38.144.045; 39.155.091; 42.155.059; 39.143.061; Bhat et al. 1987, *J. Phys. G*, **10**, 1087; Mayer et al. 1987, *Astron. Astrophys.* **180**, 73). There are discrepancies between the results, but these can to a large extent be understood and the findings are converging in recent work. The main conclusions that can be drawn are that the radial CR gradient in the Galaxy is very weak (a radial exponential scale length of roughly 10 – 15 kpc), the ratio $N(H_2)/W_{CO}$ is $< 3 \times 10^{20}$ mol.cm^{-2}.K^{-1}.km^{-1}.s and thus near the lower edge of independent prevous estimates, and the H_2 mass inside the solar circle is $\lesssim 1.0 \times 10^9$ $M_\odot$.

Blitz et al. (39.143.022) found that the γ-ray flux from the central few hundred parsecs of the Galaxy is nearly an order of magnitude smaller than the value expected from the H_2 masses generally estimated to be present in the center and from the γ-ray emissivity measured for the galactic disk.

Pollock et al. (39.143.040; 40.143.093) searched for γ-ray excesses in the *COS-B* data which cannot be explained by the predicted γ-ray emission from HI and CO observations (with a smooth CR distribution). They found indeed some point-like γ-ray sources which do not have counterparts in the gas data and which may be due to localised enhancements of the CR density or, alternatively, genuine point sources.

Strong (39.131.089), Strong et al. (40.143.088), and Lebrun and Paul (40.143.086) continued their studies of the observed γ-ray emission at intermediate latitudes using galaxy counts as a gas tracer. Lebrun and Paul showed that the well-known discrepancy between the observations and predictions toward the inner Galaxy at medium latitudes (the observed intensities are larger than expected) is larger than found previously. This is the result of an improvement in the calibration of the galaxy counts after an observational bias was detected (Lebrun 1986, *Ap. J.* **306**, 16). Lebrun and Paul (40.143.086) and Bhat et al. (1985, *Nature* **314**, 515) suggested that this excess may be attributed to Loop I. Bloemen et al. (1987, *Astron. Astrophys.*, in press) suggested recently that it may originate from CR-matter interactions in the ionized medium with a large scale height, as traced by pulsar Dispersion-Measure data (which was not taken into account in the model).

Bloemen (1987, *Astrophys. J. Lett.* **317**, L15) and Bloemen et al. (1987, *Astron. Astrophys.*, in press) used the high-energy part of the *COS-B* data base to study CR spectral variations throughout the Galaxy. They found that the γ-ray spectrum shows a flattening with increasing latitude toward the outer Galaxy. This effect is not seen toward the inner Galaxy. They compared their findings with similar results from a recent study of the galactic radio-continuum emission at 408 and 1420 MHz (summarized in the following section) and conclude that significant CR spectral variations exist in the Galaxy. They argued that this spectral behaviour can be explained by the galactic-wind model of cosmic-ray propagation proposed by Lerche and Schlickeiser (31.063.020; 32.143.056), but some other possibilities cannot be excluded.

7.3 Magnetic Fields

Two text books on magnetic fields in astrophysics, including discussions on galactic fields, appeared in this triennium (Bochkarev - 39.003.047; Seymour - 42.003.061). Sofue et al. (42.157.056) published a review of the global structure of magnetic fields in galaxies. Heiles (1987) gave an extensive overview of interstellar magnetic fields and Zweibel (1987) reviewed the theoretical aspects (both in *Interstellar processes*, eds. Hollenbach & Thronson, Reidel, Dordrecht).

The relationship between the interstellar magnetic field strength and the gas density is still poorly understood. Troland and Heiles (41.131.100) presented a compilation of observations. The field strengths show no clear evidence of increase for $n = 0.1 - \sim 100$ cm^{-3}. At higher densities, a modest increase in field strength is observed in some regions. Several new measurements of the Zeeman splitting of OH lines were presented (42.131.025; 41.131.286; 41.131.098). Aperture synthesis observations of the 21-cm Zeeman effect toward Cas A (41.131.098) indicate considerable spatial structure of the magnetic-field strength; peaks often coincide with clumps of molecular gas. In general, the findings are not inconsistent with theoretical expectations for selfgravitating clouds, but questions still exist about how the magnetic field strength remains rather constant for densities up to ~ 100 cm^{-3}.

Vallée and collaborators continued their analyses of interstellar "magnetic bubbles" with sizes of 100 – 300 pc (37.131.179) and found a relation between the observed magnetic field strength in a shell and the degree of compression of the shell (39.131.300). Simonetti et al. (38.155.037) used rotation measures of extragalactic sources to investigate variations in the interstellar magnetic field on length scales of $\sim 0.01 - 100$ pc.

There is increasing observational evidence that the field is often morphologically related to the interstellar gas, e.g. parallel or perpendicular to filaments, and systematically oriented in large shells (38.131.134; 38.131.098; Heiles 1987). Theoretical aspects of these phenomena in clouds and the implications for star formation are discussed by Mestel and Paris (38.131.011). Several studies of the role of magnetic fields in star formation have been carried out, but these are reported elsewhere in this volume.

Wielebinski (39.141.024) discussed the available radio-continuum surveys of the sky and presented the likely survey developments in the future. Beuermann *et al.* (40.155.054) constructed a three-dimensional model of the radio emission at 408 MHz based on the all-sky survey of Haslam *et al.* (31.141.036). In this model, the Galaxy consists of a thick (several kpc) non-thermal disk and a thin disk which contributes only about 10% to the total power at 408 MHz. The authors suggested that the magnetic field and relativistic particles in the thick disk are dynamically coupled to the hot halo gas. Reich and Reich (41.141.002) have completed the Stockert 1420 MHz continuum survey of the northern sky and used this map and the 408 MHz survey of Haslam *et al.* to calculate a spectral-index map of the northern sky (Reich and Reich 1987a, *Astron. Astrophys. Suppl.*, in press). This map shows a flattening of the spectra with increasing latitude, particularly toward the outer Galaxy. At lower frequencies, the spectra show, if any large-scale variation, a steepening with increasing latitude (Lawson *et al.* 1987 reviewed these observations – *Mon. Not. R. Astron. Soc.* **225**, 307). The comparison with gamma-ray observations was discussed in the previous section.

Seymour (41.155.027) considered the coupling between the dynamics of the interstellar gas and the galactic magnetic field. Several authors have reanalysed the role of magnetic fields and cosmic rays in the (quasi) hydro-static equilibrium and stability of the galactic disk (40.144.021; 40.144.207; 41.131.264; Bloemen 1987). Chernoff *et al.* (41.155.096) studied the stability of the galactic magnetic field in the presence of a magnetic monopole halo. Kulsrud (42.155.104) argued that the wind-up problem of the magnetic field of the Galaxy can be solved and that a primordial origin of the galactic field is therefore possible.

7.4 Pulsars

This section reports on statistical studies of the galactic pulsar population and on the impacts of pulsar observations on studies of the ISM. Taylor and Stinebring (42.126.009) and Radhakrishnan (42.126.010) reviewed the developments of pulsar research. Stokes *et al.* (40.126.027; 42.126.078) and Clifton and Lyne (41.126.019) presented the results of recent pulsar surveys at Green Bank, Arecibo, and Jodrell Bank, which extend the presently known sample to over 440 pulsars.

Studies of the galactic distribution of pulsars indicate a scale height of about 400 pc and a smooth increase of the number density toward the galactic centre (Guseinov and Yusifov - 39.126.050; Lyne *et al.* - 39.126.038). Including in this work the 32 new pulsars discovered by Clifton and Lyne (41.126.019) shows convincingly that the pulsar density falls within about 5 kpc (Clifton, priv.comm.).

Allakhverdiyev *et al.* (40.126.021) and Trimble (40.125.105) discussed the connection between the distribution of pulsars and SN remnants. Amnuel *et al.* (41.126.017) studied the proper motions of a pulsar sample and concluded that the birthplaces of pulsars are located within OB-associations and/or in spiral arms. Blaauw (40.126.067) evaluated the local evaporation of massive stars during the past 50 Myr and showed that the estimated evaporation may well account for the local replenishment of pulsars. The author argued that the high space velocities of pulsars cannot be explained as runaway velocities of the progenitors due to binary disintegration following a SN explosion, contrary to the conclusion reached by Radhakrishnan and Shukre (41.126.005). Chevalier and Emmering (41.126.058) modeled the observed properties of pulsars and used their model to calculate the variation of pulsar number, period, and characteristic time with z. Huang *et al.* (40.126.042) studied these parameters for the two types of pulsars they proposed in earlier work.

Using dispersion measures and independent distance estimates for a subset of pulsars, Lyne *et al.* (39.126.038) estimated the electron density distribution in the Galaxy (within several kpc from the Sun). They found that the electron density model that is most consistent with the data consists of a layer of much greater scale height than that of the pulsars (with a density of ~ 0.03 cm^{-3} at $z = 0$) and a thin layer with $n(0) \simeq 0.015$ cm^{-3} and a scale height of ~ 70 pc, both components having a weak density increase toward the galactic centre. Several studies on fine scale electron density fluctuations in the ISM have been reported during the past few years, using measurements of scintillations and temporal broadening of pulsar signals (and angular broadening of galactic and extragalactic sources) (37.131.182; 37.131.190; 37.131.251; 38.131.162; 39.126.040; 39.131.292; 42.131.052; 39.131.265; 42.131.376; 42.131.128; 41.131.182).

8. Galactic Kinematics

8.1 Stars

A substantial body of new radial velocity and proper motion measurements have been made during the past triennium.

8.1.1 Radial Velocities

Radial velocities for standard stars have been obtained for standard stars by at least three groups (41.111.010, 39.111.040, 39.111.044). The *CORAVEL* spectrometers have been an important instrument for these as well as many of the studies mentioned here. The southern sky has also been the target of several extensive surveys (40.111.001, 39.111.002, 38.111.009, 42.111.019, 39.111.036, Preprint 1). Measurements of particular spectral types have been as follows: OB stars including those in clusters and proposed runaways (41.111.020, 39.111.022, 40.155.047), F dwarfs (39.155.153), K0 stars (41.111.003), and RR Lyraes (40.111.008). Other classes include barium stars (37.111.019, 39.111.030), Mira variables (Preprint 2), blue horizontal branch field stars (39.111.001, 41.111.004), high proper motion subdwarfs (41.111.016), population II stars (39.111.032) and stars in globular clusters (41.154.019, 37.154.047, 42.111.003). Mayor reports that the *CORAVEL* spectrometers have already made 25,000 measurements of 10,000 stars for the *HIPPARCOS* mission (40.111.010), and preliminary results of a survey of O–F8 stars within 15° of the North Galactic Pole was also published (39.111.045).

8.1.2 Proper Motions

The following proper motion studies were published during the past triennium: Pleiades-Hyades region stars (39.111.025), R Coronae Borealis stars (39.111.017), stars within 20° of the South Galactic Pole to 17 mag (42.111.014, 42.111.015), dwarf K and M stars (42.111.007), halo stars (39.155.152), and selected areas of the Pulkovo zone (39.111.055). Many regions and stellar types were extensively investigated, with several studies containing ∼ 5000 stars.

8.1.3 The Disk

In the following two sections we report studies which are primarily analytic in nature which make extensive use of previously published measurements. Early type stars were the subject of several kinematic investigations which investigated the O star velocity ellipsoid (37.111.002), kinematics as a function of age (38.155.063), corrections to the precession constant (38.111.025) and the motions of open clusters (39.155.016). Brosche and Schwan (38.111.019) present evidence for a breaking wave in the velocity field of nearby young stars. Palous has low values of the Oort A constant from analysis of B and A stars (40.111.011), and Balasz argues that the kinematics of A stars may be the result of periodic star formation with a characteristic time that is near a galactic rotation time (37.155.100).

The solar motion and velocity ellipsoid has been the subject of several new studies (39.111.014, 37.155.041, 39.155.082, 41.155.113, Bassino, *et al.*) stellar velocity dispersions from the Yale bright star catalogue were derived (39.155.164), and a random walk analysis has been made to investigate the effect of giant molecular clouds on stellar velocity dispersions (37.151.061). Phase mixing in the distribution of stars was investigated by Fuchs (38.151.110), and Clube (39.155.017) reanalyzed the concept of star streams first proposed by Kapteyn.

Other kinematic studies investigate constraints on the past star formation history of the Milky Way (39.155.144), evolutionary phases of peculiar red giants (39.155.162), relativistic and perspective effects in radial velocity and proper motion measurements (41.111.001), the linkage of solar neighbourhood kinematics to large-scale galactic structure (42.155.027), and the old stellar population (39.155.012). *IRAS* sources in the bulge are argued to be Mira variables by Feast. Allen, *et al.* (Preprint 3) study an unusual nearby wide binary which they argue has an apogalacticon of 115 kpc.

8.1.4 The Halo

An important study of halo kinematics was published by Ratnatunga and Freeman (39.155.100) who found that the outer halo is, at most, slowly rotating and that the line of sight velocity dispersion is independent of distance from the Sun. From the velocity dispersion of carbon stars at the north galactic pole, Mould, *et al.* (39.111.012) find a higher velocity dispersion than predicted from the Ratnatunga and Freeman model which was modeled by White

(40.155.010) with a spherical potential and flat rotation curve. The density and abundances of halo stars were measured (42.155.033, 42.155.061, 42.155.021). The usefulness of the Stock radial velocity survey was evaluated (38.111.021, 39.111.037), as were the halo red giants (42.155.035, 39.155.154), and A and B stars (37.155.092). Carney also presented a useful review of stellar systems with distances greater than 25 kpc from the galactic center (38.155.076). The kinematics and metal abundances of globular clusters were the subjects of two new studies (38.154.013, 38.154.029). Fall and Rees also presented a new theory for the origin of the globular clusters.

8.2 Interstellar Matter

8.2.1 The Galactic Center

Kinematic studies of the nucleus of the Milky Way have brought about the discovery of a dense, clumpy CO disk 10 pc from the galactic center (40.155.009, 42.155.034). Other kinematic studies in the 63 μm fine structure line of oxygen (37.155.025), and 158 μm line of [CII] (42.155.015) have probed the mass distribution suggesting a thin azimuthally symmetric disk is Keplerian rotation around a central mass point of 7×10^6 $M_\odot$. These studies are confirmed by observations of S(1) hydrogen line emission from the same region (42.155.029), and complemented by hydrogen recombination line synthesis observations (39.155.052). Vietri (42.151.004) examined the dynamical consequences of the gas ring and concluded that stability requires a triaxial galactic bulge rotating in a direction opposite to the disk material. The kinematic consequences of the ring have also been studied (40.155.020). An unusual, wide feature in the CO spectra of the galactic center is shown to consist of a number of gravitationally bound clouds (42.131.073). Two reviews of the radio and infrared observations of the galactic center were published by Oort (39.155.051) and by Hyland (41.155.013).

8.2.2 Galactic Rotation

A sizable number of papers on galactic rotation were published during the past triennium. New observations of HII region/molecular cloud complexes in the outer Galaxy (39.155.008, 38.155.022, 39.155.123, 39.155.195), as well as newly catalogued objects beyond the solar circle (42.155.026) have provided important data to determine the rotation curve in the disk to large R. An interesting new twist in deriving the rotation curve beyond the solar circle from 21 cm HI observations has also been published (42.155.002, 40.155.080). An unusual B supergiant was also used to determine the distant rotation law of the Galaxy (37.113.045).

Existing observations were used to reanalyze the inner and outer Galaxy rotation curves (40.155.016, 40.155.085, 41.155.036). An analysis by Haud (38.155.024) argues that the rise in the outer Galaxy rotation curve may be due to radial gas motions. A dynamical analysis of the inner Galaxy rotation curve argues forcefully that the data require a triaxial bulge potential (42.155.046). Caldwell and Coulson have completed a major new survey of distances and velocity measurements of Cepheid variables from which they derive the distance to the galactic center and Oort's A constant (Preprint 4). Nelson finds that the influence of magnetic fields on the rotation curves in the outer disks of galaxies can be quite significant (preprint 5).

8.2.3 Disk and Solar Vicinity

The velocity dispersion of HI clouds was subject to a number of new studies. The differences between HI and HII terminal velocities was confirmed (38.131.072), but the fraction of the high velocity dispersion HI gas was shown to be an order of magnitude smaller than originally proposed (39.155.084). The motions of some anomalous velocity HI clouds are argued to be from the envelopes of giant molecular clouds (39.131.091). An analysis of the motions of local clouds seen in absorption is found to be in good agreement with the standard solar motion (38.131.016), and the spectrum of turbulence appears to closely resemble a Kolmogoroff law (39.131.032).

The kinematics of molecular clouds has been the subject of some disagreement. Some investigators find a dispersion of $\sim$ 7 km.s^{-1} (37.131.320), while others find a value closer to 4 km.s^{-1} (38.155.055, 40.155.016). The velocity dispersion of the small, high-latitude molecular clouds shows an intermediate value of 5.4 km.s^{-1}, however (38.131.021, 38.131.267, 40.155.015). The random motions of molecular clouds have been argued to originate from the differential rotation of the Milky Way (38.131.229, 37.155.057). HI motions resulting from effects of the spiral arms was also studied by two different groups (37.155.083, 39.155.148, 42.155.007, Preprint 6).

The kinematics of various components of the solar vicinity were studied by Goulet and Shuter (38.155.079) who found strong deviations from circular rotation. Palous (42.155.060) has investigated how giant molecular clouds effect the stellar kinematics in the solar neighbourhood.

8.2.4 High Velocity Clouds and the Halo

A number of new observations of high velocity clouds has been made (39.131.042, 39.131.043, 37.155.063 39.131.044, 39.131.045), the last of which is interpreted as high velocity inflow of HI toward the Galaxy. Other more local interpretations were also presented (38.131.079, 38.155.015), the first of which shows components of the high velocity clouds in the spectra of nine stars. This important observation, if confirmed, would imply that a large fraction of the high velocity clouds are within 200 pc of the Sun. Another study (42.155.054) suggests distances of 1–3 kpc. Clearly, the problem of the high velocity clouds requires more observations. A theoretical investigation by Lacey and Fall suggests that the star formation history of the Milky Way requires either radial infall or inflow from the outer regions of the galactic disk.

References

M. W. Feast and J. H. Spencer Jones, Preprint 2.

J. Denoyelle, Preprint 1.

L. P. Bassino, V. H. Dessaunet, and J. C. Muzzio, 1986, *Rev. Mez. Astron. Astrof.*, **13**, 9–14.

M. W. Feast, 1986, *Light on Dark Matter*, F. P. Israel, Ed., Reidel, Dordrecht, p. 339–348.

C. Allen, M. A. Martos, and A. Poveda, Preprint 3.

S. M. Fall, and M. J. Rees, 1985, *Astrophys. J.,* **298**, 18–26.

C. G. Lacey, and S. M. Fall, 1985, *Astrophys. J.,* **290**, 154–170.

J. A. R. Caldwell, and I. M. Coulson, Preprint 4.

A. H. Nelson, Preprint 5.

J. V. Feitzinger and J. Spicker, Preprint 6.

9. The Outer Galactic Environment

One of the highlights of the study of the galactic environment is the demonstration that companions and hydrogen clouds surrounding our Galaxy form a ring-like structure similar to that surrounding external polar-ring galaxies. Probably this feature is common in giant galaxies. Dwarf galaxies may possess their own dark coronas, which fact, if confirmed, puts severe constraints on the nature of dark matter. Available evidence confirms earlier suggestions that our Galaxy with its massive corona, companions, and surrounding gas forms a single system with many mutual interactions. Most companions of our Galaxy as well as the main hydrogen streams are located in a narrow strip inclined 70° to the galactic plane.

(042.155.108) suggested that nearby companions and hydrogen streams probably form a polar ring around the Galaxy. There are evidently two types of high-velocity clouds (HVC's) of neutral hydrogen: relatively nearby features, and clouds at large distances from the Sun.

The polar ring of the Milky Way is composed of high-velocity gas of the second type. It is rotating with a velocity of ~ 200 km.s^{-1}, approximately equal to the rotation speed of the Galaxy in its main plane. A similar match is observed in external galaxies with polar rings (043.151.002). These results demonstrate that coronas are triaxial with axial ratios of their equipotentials $c/a \sim 0.96$. The mass of the dark corona was derived using a number of various test particles. RR Lyrae stars and globular clusters yield the mass within 20 – 25 kpc; the results lie in an interval $2.6 - 2.9 \times 10^{11} M_\odot$ (039.155.078; 040.122.159; 039.155.104); 039.155.113; 041.154.019; 040.154.015). The outer radius of the ring system is estimated to be ~ 90 kpc. HVC's of the first type can be considered a consequence of interaction between the polar ring and galactic gas. Near the anticenter region accretion of the intergalactic gas to the Galaxy takes place. HVC streams, beginning there and smoothly merging to the polar ring structure may be the infalling hydrogen clouds. The infall of ring clouds into the galactic disk may give rise to the bending of the plane of the Milky Way, and may also trigger the formation of spiral structure, and thus explain a number of features in the kinematics of the population of young stars.

(039.151.016), (038.151.020), (038.151.070) and (038.157.165) discussed the stability of the galactic disk and the developing of its warp under the influence of the heavy halo. The infalling gas can also explain the constant scale height of the disk of Galaxy and the formation of HI loop structures. (039.155.054), (039.155.117), (040.131.283) and (042.155.087) reviewed the observations and theories involving the gaseous corona of the Milky Way.

(039.131.044), (039.131.045), (039.131.296), (040.156.008), (042.131.053), (039.131.043) and (039.131.042) describe recent observations of HVC's. (042.131.189) used *IRAS* observations to look for infrared emission from dust in HVCs. None of these clouds is detected. (043.155.002) detected no associated *IRAS* 100 μm flux in the Magellanic Stream, indicating that the Stream has a different dust content from that of the gas in the Galaxy. (039.156.006) and (041.156.003) suggested that the SMC approached the LMC as close as 3 to 7 kpc about 200 million years ago, and that the Magellanic Stream is due to the gravitational interaction among the triple system of the Galaxy, LMC and SMC. (042.151.055) present two-dimensional hydrodynamic simulations for the interaction of HVC's with a galactic disk. The calculations show the build-up of massive structures, able to retain for a significant time (043.155.003). General trends support the hypothesis that the mechanism of cloud-Galaxy interactions may be responsible for some of the most energetic structures in the Galaxy, such as supershells.

(039.155.100) located, from an automated objective-prism survey, a sample of over 150 K giants in the outer galactic halo. Radial velocities show that the outer halo is at most slowly rotating and the line-of-sight velocity dispersion, 101 km.s^{-1}, is approximately constant with distance from the Sun (039.111.012; 041.111.004).

(040.114.149) reviewed the determination of C, N, and O abundances in old stars of the halo field, globular clusters, and the seven dwarf spheroidal satellites of the Galaxy. Of crucial importance for the understanding of the nature of the dark matter is the possible presence of dark halos around dwarf galaxies. Observed and derived structure parameters are tabulated for 154 galactic globular clusters, 7 dwarf spheroidal satellites, and 6 globular clusters in the Fornax dwarf spheroidal by (039.154.069). Dwarf spheroidal companions can be divided into nearby (Carina, Draco, Sculptor and Ursa Minor) and distant ones (Fornax and Leo). In Carina accurate radial velocities have been obtained for six carbon stars (040.157.025). The observed rms velocity dispersion is ∼ 6 km.s^{-1}. The derived M/L is then 9.7. Accurate radial velocities have been obtained for 3 carbon stars in Sculptor by (040.157.025). (042.157.128) presented radial velocities for 16 K giants. The observed *rms* velocity dispersion is ∼ 6 km.s^{-1}, consistent with an M/L ratio of 6.0. Precise $(1-2$ km.s$^{-1})$ velocities have been obtained for three Fornax globulars (042.157.095). The observed *rms* velocity dispersion of five carbon stars is ∼ 6 km.s^{-1}(040.157.025). The derived M/L is then 0.5. These values indicate that both the Carina and Sculptor dwarfs contain a substantial amount of additional mass not found in globular clusters, but Fornax does not. Another possible explanation for the large velocity dispersion in nearby companions is that these galaxies are undergoing tidal disruption. *BV* and near-infrared photometry was done (041.156.012) for 161 Cepheids in the Small Magellanic Cloud to derive their relative distances. The line-of-sight distribution of the younger Cepheids splits into two components, each of depth of about 6 kpc and with centers 12 kpc apart.

Recent radial velocity measurements of stars and their interstellar CaII absorption lines show convincingly that the near and far components should be identified with the low and high-velocity portions in the SMC HI distribution respectively. The results demonstrate that SMC is in a stage of tidal disruption. (043.154.001) argued that the distribution of the globular clusters with respect to the galactocentric distance shows a gap around 30 – 40 kpc. The outer halo clusters are intrinsically faint ($M_V = -4$ to –6 against –7 to –8), with a very large core radius (5–20 kpc against 1–2 kpc).

10. Dark Matter in the Galaxy

10.1 Dark Matter

Following an early suggestion by Kahn and Woltjer (1959) and Oort (1970), Gunn (1974), Einasto *et al.* (1974) and Ostriker *et al.* (1974) suggested that our Galaxy as well as other giant galaxies are surrounded by a corona. These initial suggestions on the presence of massive coronas around galaxies were based on the observed flat rotation curves of galaxies. The present state of the evidence for dark matter galactic halos derived from rotation curves of spiral galaxies is summarized by Sancisi (1985). Properties of the "dark" matter component of the galactic halos have been inferred from the constancy of the rotation curves by Malagoli and Ruffini (1985). In order to detect the gravitational effect of the dark corona component of disk galaxies, it is necessary to have surface photometry and rotation data that extend well beyond three disk scale-lengths. The new observational data for such analysis were discussed by Freeman (1986), Kent (1986) and Athanassoula *et al.* (1987). From the analysis of the observations Carignan and Freeman (1985) concluded that the mean halo-to-disk mass ratio at the Holmberg radius is 1.0. Bahcall and Casertano (1985) mentioned that the unseen matter in a sample of spiral galaxies exhibits simple regularities and characteristic numerical values. Athanassoula *et al.* (1987) have made an analysis of the rotation curves of a sample of spiral galaxies for which both photometric and kinematical data of reasonable quality are available in the literature, assuming constant mass-to-light ratios for bulge and disk separately. They suggested that all galaxies need a halo to fit their rotation curve. The velocity dispersion of the isothermal sphere best fitting the halo correlates well with the maximum velocity of the disk component. The central density of the isothermal sphere correlates with the central disk surface density divided by a disk characteristic length. They considered both as manifestations of a disk halo conspiracy and noted that the halos of early type spiral galaxies are more concentrated than those of later types. Fabian et al. (1986) have shown that the average total binding mass associated with early-type galaxies is large, $M_T \geq 5 \times 10^{12} M_\odot$. This implies that the average mass-to-light ratio $(M/L_B) > 74$ and significant amounts of dark matter are present in early-type galaxies. The radial distribution of matter in three elliptical galaxies with extensive shell systems was investigated by Hernquist and Quinn (1987a). They reported that the form of the galactic potential can be constrained by several independent observable quantities: 1) the number of shells between two fixed radii; 2) the radial distribution of shells; 3) the location of the innermost shell; and 4) the velocity dispersion of the underlying elliptical galaxy. A simple argument was used to show that the luminous material in NGC 3923 cannot account for the number of shells surrounding this galaxy and that the potential must be dominated by an extended and massive dark component. The total mass-to-light ratio was $\sim 100 - 200$. In their next paper (Hernquist and Quinn, 1987b) the shell method was used to study the modified Newtonian dynamics introduced by Milgrom as an alternative to the existence of dark matter in galaxies. The results are in disagreement with both the observed number and radial distribution of shells around NGC 3923. The modified dynamics have been suggested as an alternative to the hidden mass hypothesis. Kuhn and Kruglyak (1987) mentioned that from an empirical perspective we could conclude that there are no significant constraints on possible spatial variations in Newton's constant at large distances. They considered a correction term in analogy to a power-law expansion. Milgrom (1984) considered self-gravitating isothermal spheres using dynamics that differ from the Newtonian in the limit of small accelerations. He found (Milgrom, 1986) that if the mass discrepancy in galactic systems is due to a break-down of Newtonian dynamics, it may be possible to find configurations in which the required "hidden mass" is negative. Sanders (1984, 1986ab) demonstrated that the modification of Newtonian gravitational attraction which arises in the context of modern attempts to unify gravity with other forces in nature can produce rotation curves for spiral galaxies which are nearly flat from 10 to 100 kpc, bind clusters of galaxies, and close the universe with the density of baryonic matter consistent with primordial nucleosynthesis. This is possible if a very low mass vector boson carries an effective anti-gravity force which on scales smaller than that of galaxies almost balances the normal attractive gravity force. At the same time Goldman (1986) showed that the strength of antigravity of a range $\gg 1$ AU is severely constrained by terrestrial and solar system experiments. The Eotvos-Dicke experiments impose $|\alpha| \leq 10^{-9}$; the gravitational redshift experiment yields $|\alpha| \leq 7 \times 10^{-5}$ and the Mercury perihelion shift measurement implies $|\alpha| \leq 10^{-2}$. These constraints rule out antigravity as a possible cause for the flat rotation curves of spiral galaxies. A brief outline of the conflicting determinations of galaxy masses based on observable luminous matter and those based on dynamical arguments is presented by Gallagher III (1986). Upper limits have been set by Skrutskie et al. (1985) to the luminosity from the massive halos of three late-type edge-on spiral galaxies: NGC 2683 (Sb), NGC 4244 (Scd), and NGC 5907 (Sc). The limits resulted from simultaneous photometry in the visual (V) and 2.2 μm (K) photometric bands. Valtonen and Byrd (1985) discussed the evidence for dark matter in different scales in the universe, from our Galaxy to large clusters of galaxies. They find that in spiral galaxies the mass-to-light ratio $M/L \leq 15$ in solar units $(H_o = 72$ km.s^{-1}.Mpc^{-1} is assumed). They have presented evidence pointing to the possibility that groups and large clusters of galaxies are not gravitationally bound units. Frenk and White (1985) reported on a symposium "Dark matter in the Universe", held at Princeton, 24–28 June 1985 and on a related meeting "Galaxy formation", held at Toronto, 19–21 June 1985. Some aspects of this question were reviewed by Miyamoto (1986). Carney (1984) discussed recent and continuing studies of the outer halo of our Galaxy. A brief inventory has been conducted of the stellar systems lying at distances exceeding 25 kpc from the galactic center. The spatial distributions of such systems and the field stars have been reviewed, as well as the galactic mass estimates that follow from considerations of their kinematics. The question of a gradient in the halo's metallicity has been addressed, plus the scant information available on the chemical abundance pattern of outer halo systems has been discussed.

10.2 The Nature of Dark Matter

The principal particle candidates for galactic dark matter were discussed, and the detection methods available for each were summarized by Smith (1986). A review was given of the present status of two general classes of dark matter experiment: 1) the detection of light bosons (e.g. axions) by conversion of photons, and 2) the detection of new heavy particles (e.g. photinos) by measurement of nuclear recoil energy using low temperature calorimetric or photon detection techniques. The distinction between "hot" and "cold" varieties of dark matter was reviewed by Primack (1986), and the evidence against hot dark matter was briefly summarized. The hypothesis of cold dark matter with a Zel'dovich spectrum of primordial Gaussian fluctuations gives a picture of galaxy and cluster formation that is in reasonably good agreement with the available observations. However, this model appears to lead to less structure on very large scales than is observed. Possible remedies were discussed, including: 1) decaying dark matter, 2) an additional feature in the fluctuation spectrum on large scales, such as can arise in a hybrid model with more than one kind of dark matter, and 3) non-Gaussian fluctuations, for example those that arise from cosmic strings. The particle physics of the most popular cold dark matter candidates was reviewed, Sciama (1984b) illustrated recent development by brief discussions of the possible roles of massive neutrinos, photinos and gravitinos in providing the "missing" matter in the universe as a whole, in galaxy clusters and in individual galaxies. If these particles have non-zero rest-mass they might dominate the universe, providing it with the critical density, and also individual galaxies, providing them with their missing mass Sciama (1984a). This hypothesis might be tested by searching for the photons which these particles (except gravitinos) would be expected to emit. Srednicki et al. (1986) pointed out that if the galactic halo is composed heavy, weakly interacting particles then the pair annihilation can produce potentially observable sharp peaks in the diffuse cosmic γ-ray background. The possibility of detecting of heavy neutral fermions in the Galaxy was discussed by Wasserman (1986). The decaying dark matter cosmology postulates that a heavy elementary-particle species X first drives the formation of galaxies and clusters, and then decays non-radiatively, providing a smooth, undetected background of relativistic particles. It has been found (Flores et al. , 1986) that the observed flat rotation curves cannot be obtained in these decaying dark matter models. Thus, a relativistic, weakly interacting decay product cannot be dominant. Krauss (1985) demonstrated that dark matter consisting of any type or types of stable weakly interacting elementary particles is incompatible with the minimal predictions of inflation, based on present observations of galaxy clustering, and assuming galaxies are good traces of mass in the Universe. Datta et al. (1985) pointed out that several independent considerations rule out the hypothesis that the missing mass in galactic halos is dominated by massive neutral fermions such as neutrinos, gravitinos or photinos. The analysis of the data on the small-scale anisotropy of the relic electromagnetic radiation leads to a conclusion on the advantages of the models of the Universe with super-massive carriers of its "hidden mass" (Zabotin and Nasel'skij, 1985). The relations between micro-wave background anisotropy and decaying cold particle scenarios are analyzed by Kolb et al. (1986). Ruffini and Song (1987) introduce a general theoretical framework which imposes constraints upon the spin, masses, and phase space densities of the cosmological "inos" forming the dark matter component of the Universe. Solar system constraints and signatures for dark matter candidates were analysed by Krauss et al. (1986b). A note on a lower limit to the rest mass of ions in the halo of our Galaxy was published by Fang and Gao (1984). Melnick and Terlevich (1986) discussed the nature of dark matter in dwarf galaxies. Fermions whose masses exceed $\sim$ 500 eV may cluster in these objects (Melott and Schramm, 1985), but they cannot provide the missing mass, as long as such dwarf galaxy halos constitute a small fraction of the dark matter in the Universe. Nasel'skij and Polnarev (1984, 1985) reviewed the possible forms of hidden mass in inflationary Universe models and Zee (1986) discussed the relations between fractional statistics, exceptional preons, scalar dark matter, lepton number violation, neutrino masses, and hidden gauge structure. Paczynski (1986) pointed out that if the halo is made of objects more massive than $\sim 10^{-8} M_{\odot}$, then any star in a nearby galaxy has a probability of 10^{-6} to be strongly microlensed at any time. Monitoring the brightness of a few million stars in the Magellanic Clouds over a time scale between 2 hr and 2 yr may lead to the discovery of "dark halo" objects in the mass range $10^{-6} - 10^2 M_{\odot}$ or it may put strong upper limits on the number of such objects. Cremonesi (1986) presented limits on dark matter candidates calculated using the data of an experiment on double beta decay of ^{76}Ge carried out by the Milan group in the Mont Blanc laboratory using two big Ge(Li) detectors. The detectability of certain dark matter candidates was discussed also by Goodman and Witten (1985) and Drukier et al. (1986). Possible dark-matter candidates are discussed below.

10.3 Baryonic Halos

Skrutskie et al. (1985) set upper limits to the luminosities from the massive halos of three late-type edge-on spiral galaxies. The limits resulted from simultaneous photometry in the visual (V) and 2.2 μm (K) photometric bands. The results virtually eliminate the possibility that hydrogen-burning stars comprise more than a fraction of the halo masses. Gurzadyan (1986) obtained an upper limit on the mass of the objects the hidden mass consists of, which seems to exclude the possibility that the hidden mass is constituted of stars being on the late stages of their evolution. Theoretical considerations would predict that sub-stellar masses have formed more frequently under the metal-poor conditions in the early Galaxy (Zinnecker, 1986). Thus the missing mass in the galactic halo and in the dark halos around other spirals may well reside in these metal-poor Population II brown dwarfs. Some of the major observational and theoretical issues in the study of brown dwarfs were reviewed by Bahcall (1986). It was concluded that all of the unseen local disk matter could be in the form of brown dwarfs without conflicting with any available observations. Nelson et al. (1985) presented the results of the first numerical evolutionary calculations for very low-mass stars (masses in the range of $0.01 - 0.1 M_{\odot}$). Tayler (1985) commented on new stellar evolution calculations for low-mass sub-luminous stars. The study of such objects may be directly relevant to the development of an understanding of the dark-matter problem in our Galaxy and many extragalactic systems. A survey for low-luminosity M-dwarfs from deep UK Schmidt plates was described by Hawkins (1986). The resulting luminosity function shows a decrease in space density in the range $M_R = 12 - 15$; thereafter the space density rises again. On the basis of stellar evolution models for low-mass stars, the turnover in the luminosity function is associated with the end of hydrogen burning, and its subsequent rise with the appearance of a population of degenerate brown dwarfs. An astrophysically plausible brown dwarf population was defined by Probst (1986) and yields a dark mass density ~ 0.5 times the observed density. Near infrared imaging of 60 nearby stars and 8 stars in the young Pleiades cluster revealed no substellar companions down to a limit corresponding to a mass of ~ 0.04 $\mathcal{M}_\odot$ (Skrutskie et al., 1986). The authors concluded that the dark matter in the galactic disk cannot reside solely in sub-stellar companions unless it is largely in objects less massive than the survey limit. Hills (1986) ruled out any appreciable fraction of the mass in the disk being in solid objects with masses less than about 10^{22} g. The corresponding limit for the galactic halo is about 10^{21} g. Halos composed of snowballs, dust and rocks, planets, stars, dead stellar remnants, and hot and cold gas were considered by Hegyi and Olive (1986). The serious problems that would arise for each of these types of matter lead to the conclusion that halos cannot plausibly contain substantial amounts of such matter. Olive and Hegyi (1986) reviewed a number of arguments which indicate that it is very unlikely that galactic halos contain substantial amounts of baryonic matter.

10.4 Black Holes

Lacey and Ostriker (1985) considered the idea that galaxy halos are composed of massive black holes as a possible solution of two problems: the composition of dark halos, and the heating of stellar disks. It is found that in order to account for the disk heating, the black holes must have masses $\sim 10^6 \mathcal{M}_\odot$. This heating mechanism makes predictions for the dependence of the velocity ellipsoid, that are in good agreement with observations. Kamahori and Fujimoto (1986) found that when the whole halo mass is attributed to black holes, the mean mass of the halo black holes must be smaller than $2 \times 10^6 \mathcal{M}_\odot$. McDowell (1985) provided independent constraints on black holes, by the requirement that their radiation due to accretion from the ISM should not make the nearest ones directly observable as optical objects. He showed that halo black holes must be less massive than about $10^3 \mathcal{M}_\odot$, and that the dark matter in the galactic disc cannot be made up of black holes of mass more than 10 $\mathcal{M}_\odot$. Carr (1985) considered the various constraints on the form of the dark matter and concluded that black holes could have a significant cosmological density only if they were of primordial origin or remnants of a population of pre-galactic stars.

10.5 Massive Neutrinos

Cowsik and Vasanthi (1986) concluded that neutrino condensates with $M \sim 10^{16} \mathcal{M}_\odot$ were the first objects to be formed in the Universe at a redshift of $\sim 10^4$. Subsequent to formation they expanded much slower than the rest of the Universe and fluctuations in the density of baryonic matter grew effectively with the formation of the galaxies. Assuming that neutrinos have a rest mass of ~ 10 c^{-2} eV details of the dynamical motions of galaxies and stars can be understood quantitatively. Ruffini and Stella (1983) investigated the properties of self-gravitating massive neutrino halos in structures ranging from galaxies to clusters of galaxies and attempted to obtain testable predictions. Assuming that massive neutrinos dominate the dark matter in galactic halos, Paganini et al. (1986, 1987) calculated the mass-radius relations and rotational curves for an isothermal neutrino distribution as well as for a distribution with a spatial cutoff (King model). For $M_\nu \geq 40$ eV, galaxies between 10 and 100 kpc radius can be reproduced with satisfactory (flat) rotational curves in the isothermal model. The authors also included baryons in the galactic core. The King model fails to give acceptable rotational curves for all baryon distributions considered. Chau and Stone (1985) reported on attempts to fit the observed rotation curves of spiral galaxies out to ~ 60 kpc assuming the required dark matter to be made up of neutrinos. The results indicate that the isothermal assumption made has to be relaxed in this "inner" region of a galaxy for a really good fit. Lower limits for the neutrino mass were obtained by Madsen and Epstein (1985), if one flavor of neutrino is predominantly responsible for the dark matter in galaxies. From rotation curves of spiral galaxies, it is found that the neutrino mass must exceed 35 $h^{1/2}$ eV (where h is the Hubble constant in units of 100 km.s^{-1}.Mpc^{-1}) if the velocity distribution of halo neutrinos is isotropic. If radial dispersion dominates, limits are slightly weakened. Madsen and Epstein (1984) found from the well studied galaxies M87 and M31 that the neutrino mass must exceed 8 eV. The preliminary reports of dark matter in dwarf galaxies would imply that the neutrino mass exceeds 125 eV. Phase-space considerations would require $M_\nu > 50$ eV if neutrinos dominate the missing mass in halos of large spiral galaxies and moreover $M_\nu > 200$ eV is implied in the case of dwarf spheroidals (Sivaram, 1985). These larger neutrino masses would be in conflict with observed constraints on the age of the Universe unless a cosmological constant is invoked. In order to investigate the mass-to-light ratio in irregular Magellanic galaxies, a sample of 21 Sdm, Sm, Im objects has been selected by Comte (1985). Implications for the particle mass of hypothetical massive neutrinos have been discussed. Neutral hydrogen observations were used to measure the total mass of dwarf galaxies by Davies (1984a). The dimensions and velocity fields of dwarfs would require a neutrino mass of ~ 150 eV. The observed luminosity profiles of dwarf spheroidals imply densities for the dark matter in the range 10^{-26} to 10^{-25} g.cm^{-3}, and mass-to-luminosity ratios which are typically an order of magnitude greater than those of globular clusters (Cowsik and Ghosh, 1986). Neutrinos of mass ~ 10 eV and $<V> \sim 1000$ km.s^{-1}can provide this requisite density for the background. The theoretical expectation of the high mass of ≥ 400 eV for the particles constituting the dark matter in dwarf-spheroidals is an artifact of the implicit assumption that the density of particles vanishes at the visible edge (Cowsik, 1986). On the contrary, if dwarf-spheroidals are embedded in a neutrino condensation of the cluster, then $M_\nu = 10$ eV can accommodate all the observations. The determinations of the neutrino mass were reviewed also by Huang et $al.$ (1983), Ho (1984), Sciulli (1986), Madsen and Epstein (1986) and Press (1986). Some results of astrophysical importance on neutrino physics have been published in physical journals (Fukugita and Yanagida, 1984; Ching and Ho, 1984; Möβbauer, 1985; Galeotti and Gallino, 1985; Freese, 1986; Grifols et $al.$, 1986; Takahara and Sato, 1986).

10.6 Photinos

Observational tests of the hypothesis that the Universe is flat and dominated by dark matter in the form of massive photinos include the production of significant fluxes of cosmic rays and gamma rays in the galactic halo (Silk and Srednicki, 1984). Specification of the cosmological photino density and the masses of scalar quarks and leptons determines the present annihilation rate. The predicted number of low-energy cosmic-ray antiprotons is comparable to the observed flux. So, the stable photinos can explain both the "missing mass" in galactic halos and the cosmic-ray antiproton spectrum up to the highest energies observed so far. This requires a photino mass around 15 GeV (Stecker et $al.$, 1985a). As a consequence, the observed cosmic-ray antiproton-to-proton ratio is predicted to decrease abruptly just above the measured energy range, at $E = M_{\overline{\gamma}}$. Stecker et $al.$ (1985b) considered the physics of the annihilation of photinos ($\overline{\gamma}$) as a function of mass, in order to obtain the energy spectra of the cosmic-ray $\overline{p}$'s produced under the assumption that $\overline{\gamma}$'s make up the missing mass in the galactic halo. The authors then compared the modulated spectrum at 1 AU with the cosmic-ray $\overline{p}$ data. A very intriguing fit is obtained to all of the present $\overline{p}$ up to 13.4 GeV data for $M_{\overline{\gamma}} \sim 15$ GeV. A cutoff in the $\overline{p}$ spectrum is predicted at $E = M_{\overline{\gamma}}$ above which only a small flux from secondary production should remain. Silk et $al.$ (1985) reported that if the Universe contains a nearly critical density of photinos, then gravitational trapping by the Sun and ensuing annihilation in the solar core yields a significant flux of ~ 250 MeV neutrinos.

10.7 Magnetic Monopoles

The lifetime of monopolonium was used by Stein and Schabes (1985) to put limits on the mass of the monopole. The author found that in order to be in accord with observations of the energy density of the Universe, the isotropy of the radiation backgrounds and the abundance of primordial light elements, the mass of the monopolonium cannot be greater than 10^{16} GeV, so making it very difficult to accommodate super-heavy monopoles in our observable Universe. The author investigated also the possibility of identifying monopolonium with the heavy particle recently proposed to solve the Ω-problem i.e. how to reconcile Universe with $\Omega = 1$ and a cold dark matter scenario capable of predicting the right large-scale structure of the Universe. It has been found that by choosing the radius of monopolonium, it is possible to solve the Ω-problem. The halo models are consistent with monopole masses $M_M \leq 7 \times 10^{19}$ GeV, and monopole fluxes in the range $F_M \geq 3 \times 10^{-13}$ cm^{-2}.s^{-1} (Farouki et $al.$, 1984). Some aspects of monopoles in astrophysics were discussed by Turner (1983).

10.8 Other Particles

Recent work on superstring theories has prompted interest in "shadow matter", exotic matter which interacts only gravitationally with normal matter (Krauss et $al.$, 1986a). Such a theory could result, at low energies, in the existence of two sectors: an 'observed' sector associated with all familiar particles and interactions, and another Universe. Of the particles whose mass may account for the missing mass of the Universe, those known as axions are the most shadowy (Maddox, 1986). What is known of stellar evolution helps to define their properties. Some effects of the axion halo on bound electrons were discussed by Slonczewski (1985). It has been recently proposed by Witten (1984) that dark matter in the Universe might consist of nuggets of quarks which could populate the "nuclear desert" between nucleons and neutron star matter. Audouze et $al.$ (1985) examined a consequence of Witten's proposal and showed that the production of relativistic quark nuggets was accompanied by a substantial flux of potentially observable high energy neutrinos. The gravitinos as the cold dark matter in an $\Omega = 1$ Universe has been discussed by Olive et $al.$ (1985).

References

Athanassoula, E., Bosma, A., and Papaioannou, S., 1987, $Astron.$ $Astrophys.$, $\mathbf{179}$, 23

Audouze, J., Schaeffer, R., and Silk, J., 1985, in: *19th International Cosmic Ray Conference*, Vol. 8, HE session: High Energy phenomena, ed. F.C. Jones, J. Adams, G.M. Mason, NASA Conf. Publ., NASA CP 2376, pp. 290

Bahcall, J.N., 1986, in: *Astrophysics of brown dwarfs*, ed. M.C. Kafatos, R.S. Harrington, S.P. Maran, Cambridge University Press, Cambridge-London-New York-New Rochelle-Melbourne-Sydney, pp. 233

Bahcall, J.N. and Casertano, S., 1985, *Astrophys. J. Lett., 293*, L7

Carignan, C. and Freeman, K.C., 1985, *Astrophys. J., 294*, 494

Carney, B. W., 1984, *Publ. Astron. Soc. Pac., 96*, 841

Carr, B. J., 1985, in: *Observational and theoretical aspects of relativistic astrophysics and cosmology*, ed. J.L. Sanz, L.J. Goicoechea, World Scientific Publishing Co. Pte. Ltd., Singapore, pp. 1. (Proc. international course).

Chau, W. Y. and Stone, J. M., 1985, *Astrophys. J., 297*, 76

Ching, C.R. and Ho, T.H., 1984, *Phys. Rep. 112*, 1

Comte, G., 1985, in: *Serendipitous discoveries in radio astronomy*, ed. K. Kellermann, B. Sheets, National Radio Astronomy Observatory, Green Bank, pp. 169. (Proc. NRAO Workshop No. 7).

Cowsik, R., 1986, *J. Astrophys. Astron., 7*, 1

Cowsik, R. and Ghosh, P., 1986, *J. Astrophys. Astron., 7*, 7

Cowsik, R. and Vasanthi, M., 1986, *J. Astrophys. Astron., 7*, 29

Cremonesi, O., 1986, in: *Cosmology, astronomy and fundamental physics*, ed. G. Setti, L. Van Hove, pp. 265. (ESO Conf. Workshop Proc., No 23, Proc. Second ESO-CERN Symp.)

Datta, B., Sivaram, C., and Ghosh, S. K., 1985, *Astrophys. Space Sci., 111*, 413

Davies, R. D., 1984a, in: *Large-scale structure of the universe, cosmology and fundamental physics*, ed. G. Setti, L. Van Hove, ESO, Garching bei Munchen, pp. 206. (Proc. First ESO-CERN Symp.)

Drukier, A. K., Freese, K., and Spergel, D. N., 1986, *Phys. Rev. D, 33*, 3495

Einasto, J., Kaasik, A., and Saar, E., 1974, *Nature, 250*, 309

Fabian, A. C., Thomas, P. A., Fall, S. M., and White III, R. E., 1986, *Mon. Not. R. Astron. Soc., 221*, 1049

Fang, L. Z. and Gao, J. G., 1984, *Phys. Lett. B, 139B*, 351

Farouki, R., Shapiro, S. L., and Wasserman, I., 1984, *Astrophys. J., 284*, 282

Flores, R. A., Blumenthal, G. R., Dekel, A., and Primack, J. R., 1986, *Nature, 323*, 781

Freeman, K. C., 1986, in: *Third Asian-Pacific Regional Meeting of the IAU. Part 1*, ed. M. Kitamura, E. Budding, pp. 337. (*Astrophys. Space Sci., 118*, No. 1/2).

Freese, K., 1986, *Phys. Lett. B, 167B*, 295

Frenk, C. and White, S., 1985, *Nature, 317*, 670

Fukugita, M. and Yanagida, T., 1984, *Phys. Lett. B, 144B*, 386

Galeotti, P. and Gallino, R., 1985, *G. Fis., 26*, 25 (In Italian)

Gallagher III, J. S., 1986, in: *Inner space / outer space. The interface between cosmology and particle physics*, ed. E. W. Kolb, M. S. Turner, D. Lindley, K. Olive, D. Seckel, The University of Chicago Press, Chicago-London, pp. 199.

Goldman, I., 1986, *Astron. Astrophys., 170*, L1

Goodman, M. W. and Witten, E., 1985, *Phys. Rev. D, 31*, 3059

Grifols, J. A., Mendez, A., and Ruiz-Altaba, M., 1986, *Phys. Lett. B, 171*, 303

Gunn, J. E., 1974, *Comments Astrophys. Space Phys., 6*, 7

Gurzadyan, V. G., 1986, *Astron. Zh.*, Tom 63, Vyp., 4, 812 (In Russian)

Hawkins, M. R. S., 1986, *Mon. Not. R. Astron. Soc., 223*, 845

Hegyi, D. J. and Olive, K. A., 1986, *Astrophys. J., 303*, 56

Hernquist, L. and Quinn, P. J., 1987a, *Astrophys. J., 312*, 1

Hernquist, L. and Quinn, P. J., 1987b, *Astrophys. J., 312*, 17

Hills, J. G., 1986, *Astron. J., 92*, 595

Ho, T.H., 1984, in: *Flavor mixing in weak interactions*, ed. L.L. Chau, Plenum Publishing Corporation, New York, USA, pp. 163. (Proc. Europhysics Topical Conference).

Huang, W., Xu, C., Qing, C., and He, Z., 1983, in: *High energy astrophysics and cosmology*, ed. J. Yang, C. Zhu, Science Press, Beijing, and Gordon and Breach Science Publishers Inc., New York-London Paris-Montreux-Tokyo, pp. 467.

Kahn, F. D. and Woltjer, L., 1959, *Astrophys. J., 130*, 705

Kamahori, H. and Fujimoto, M., 1986, *Publ. Astron. Soc. Jpn., 38*, 151

Kent, S. M., 1986, *Astron. J., 91*, 1301

Kolb, E. W., Olive, K. A., and Vittorio, N., 1986, *Phys. Rev. D, 34*, 940

Krauss, L. M., 1985, *Gen. Relativ. Gravitation, 17*, 89

Krauss, L. M., Guth, A. H., Spergel, D. N., Field, G. B., and Press, W. H., 1986a, *Nature*, **319**, 748

Krauss, L. M., Srednicki, M., and Wilczek, F., 1986b, *Phys. Rev. D*, **33**, 2079

Kuhn, J. R. and Kruglyak, L., 1987, *Astrophys. J.*, **313**, 1

Lacey, C. G. and Ostriker, J. P., 1985, *Astrophys. J.*, **299**, 633

Maddox, J., 1986, *Nature*, **319**, 717

Madsen, J. and Epstein, R. I., 1984, *Astrophys. J.*, **282**, 11

Madsen, J. and Epstein, R. I., 1985, *Phys. Rev. Lett.*, **54**, 2720

Madsen, J. and Epstein, R. I., 1986, in: *Inner space / outer space. The interface between cosmology and particle physics*, ed. E. W. Kolb, M. S. Turner, D. Lindley, K. Olive, D. Seckel, The University of Chicago Press, Chicago-London, pp. 510.

Malagoli, A. and Ruffini, R., 1985, in: *Galaxies, quasars and cosmology*, ed. L. Z. Fang, R. Ruffini, Wold Scientific Publishing Co. Pte. Ltd., Singapore, pp. 11.

McDowell, J., 1985, *Mon. Not. R. Astron. Soc.*, **217**, 77

Melnick, J. and Terlevich, R., 1986, *Observatory*, **106**, 69

Melott, A. L. and Schramm, D. N., 1985, *Astrophys. J.*, **298**, 1

Milgrom, M., 1984, *Astrophys. J.*, **287**, 571

Milgrom, M., 1986, *Astrophys. J.*, **306**, 9

Miyamoto, M., 1986, *Astron. Her.* **79**, 234 (In Japanese)

Mößbauer, R. L., 1985, *Phys. Bl.*, **41**, 391

Nasel'skij, P. D. and Polnarev, A. G., 1984, *Inst. kosm. issled. Akad. Nauk SSSR*. Prepr., 20 pp. (In Russian.)

Nasel'skij, P. D. and Polnarev, A. G., 1985, *Astron. Zh.*, **62**, 833 (In Russian)

Nelson, L. A., Rappaport, S. A., and Joss, P. C., 1985, *Nature*, **316**, 2

Olive, K. A. and Hegyi, D. J., 1986, in: *Inner space / outer space. The interface between cosmology and particle physics*, ed. E. W. Kolb, M. S. Turner, D. Lindley, K. Olive, D. Seckel, The University of Chicago Press, Chicago-London, pp. 112.

Olive, K. A., Schramm, D. N., and Srednicki, M., 1985, *Nucl. Phys. B, Part. Phys.* **255**, 495

Oort, J. H., 1970, *Astron. Astrophys.*, **7**, 381

Ostriker, J. P., Peebles, P. J. E., and Yahil, A., 1974, *Astrophys. J.*, **193**, L1

Paczynski, B., 1986, *Astrophys. J.*, **304**, 1

Paganini, R., Straumann, N., and Wyler, D., 1986, *MPA Rep.* (21 pp)

Paganini, R., Straumann, N., and Wyler, D., 1987, *Astron. Astrophys.*, **177**, 84

Press, W. H., 1986, in: *Cosmogonical processes*, ed. W. D. Arnett, C. Hausen, J. W. Truran, S. Tsuruta, VNU Science Press, Utrecht, The Netherlands, pp. 35. (Proc. symp.)

Primack, J. R., 1986, in: *Cosmology, astronomy and fundamental physics*, ed. G. Setti, L. Van Hove, pp. 193. (ESO Conf. Workshop Proc., No. 23, Proc. Second ESO-CERN Symp.)

Probst, R. G., 1986, in: *Astrophysics of brown dwarfs*, ed. M. C. Kafatos, R. S. Harrington, S. P. Maran, Cambridge University Press, Cambridge-London-New York-New Rochelle-Melbourne-Sydney, pp. 22

Ruffini, R. and Song, D. J., 1987, *Astron. Astrophys.*, **179**, 3

Ruffini, R. and Stella, L., 1983, in: *Proceedings of the Third Marcel Grossmann Meeting on General Relativity*. Part A: Session Papers, ed. N. Hu, Science Press, Beijing and North-Holland Publishing Company, Amsterdam, pp. 545

Sancisi, R., 1985, in: *Italian astronomy and the scientific potential of Space Telescope*, ed. F. Bertola, pp. 709. (*Mem. Soc. Astron. Ital.*, **56**, No. 4, Proc. workshop, held in honor of Riccardo Giacconi)

Sanders, R. H., 1984, *Astron. Astrophys.*, **136**, L21

Sanders, R. H., 1986a, *Astron. Astrophys.*, **154**, 135

Sanders, R. H., 1986b, *Mon. Not. R. Astron. Soc.*, **223**, 539

Sciama, D. W., 1984a, in: *The Big Bang and Georges Lemaitre*, ed. A. Berger, D. Reidel Publishing Company, Dordrecht-Boston-Lancaster, pp. 31. (Proc. symp. in honor of G. Lemaitre fifty years after his initiation of Big-Bang cosmology)

Sciama, D. W., 1984b, in: *Plasma astrophysics*, ed. T. D. Guyenne J. J. Hunt, ESA Spec. Publ., ESA SP-207, pp. 171. (International School & Workshop on Plasma Astrophysics)

Sciulli, F., 1986, in: *Inner space / outer space. The interface between cosmology and particle physics*, ed. E. W. Kolb, M. S. Turner, D. Lindley, K. Olive, D. Seckel, The University of Chicago Press, Chicago-London, pp. 495

Silk, J. and Srednicki, M., 1984, *Phys. Rev. Lett.*, **53**, 624

Silk, J., Olive, K., and Srednicki, M., 1985, *Phys. Rev. Lett.*, **55**, 257

Sivaram, C., 1985, *Astrophys. Space Sci.* 116, 39

Skrutskie, M. F., Shure, M. A., and Beckwith, S., 1985, *Astrophys. J., * **299**, 303

Skrutskie, M. F., Forrest, W. J., and Shure, M. A., 1986, in: *Astrophysics of brown dwarfs*, ed. M. C. Kafatos, R. S. Harrington, S. P. Maran, Cambridge University Press, Cambridge-London-New York-New Rochelle-Melbourne-Sydney, pp. 82

Slonczewski, J. C., 1985, *Phys. Rev. D* **32**, 3338

Smith, P. F., 1986, in: *Cosmology, astronomy and fundamental physics*, ed. A G. Setti, L. Van Hove, pp. 237 (ESO Conf. Workshop Proc., No. 23, Proc. Second ESO-CERN Symp.)

Srednicki, M., Theisen, S., and Silk, J., 1986, *Phys. Rev. Lett.,* **56**, 263

Stecker, F. W., Rudaz, S., and Walsh, T. F., 1985a, *Phys. Rev. Lett.,* **55**, 2622

Stecker, F. W., Walsh, T., and Rudaz, S., 1985b, in: *19th International Cosmic Ray Conference Vol. 2*, OG session: Cosmic ray and gamma-ray Origin and Galactic phenomena, ed. F. C. Jones, J. Adams, G. M. Mason, NASA Conf. Publ., NASA CP-2376, pp. 358

Stein-Schabes, J., 1985, *Mon. Not. R. Astron. Soc.,* **215**, 659

Takahara, M. and Sato, K., 1986, *Phys. Lett. B,* **174**, 373

Tayler, R. J., 1985, *Nature,* **316**, 19

Turner, M. S., 1983, *Magnetic monopoles*, 127

Valtonen, M. J. and Byrd, G. G., 1985, *Rep. Ser.*, Dep. Phys. Sci., Univ. Turku (Turku Univ. Obs., Informo No. 91, 31 pp)

Wasserman, I., 1986, *Phys. Rev. D,* **33**, 2071

Witten, E., 1984, *Phys. Rev. D.,* **30**, 272

Zabotin, N. A. and Nasel'skij, P. D., 1985, *Astron. Zh.,* **62**, 410 (In Russian)

Zee, A., 1986, in: *Particles and the universe*, ed. G. Lazarides, Q. Shafi, North-Holland Physics Publishing, Amsterdam-Oxford-New York-Tokyo, pp. 257 (Proc. international symp.)

Zinnecker, H., 1986, in: *Astrophysics of brown dwarfs*, ed. M. C. Kafatos, R. S. Harrington, S. P. Maran, Cambridge University Press, Cambridge-London-New York-New Rochelle-Melbourne-Sydney, pp. 212

11. Galactic Dynamics : Stellar Orbits

11.1 General Problems

The interest in stellar orbits focussed on two areas: Firstly, on the existence of non-classical integrals of motion ("third integral") and on the occurence of stochastic or chaotic motions in systems with two or three degrees of freedom. While the case of three degrees of freedom is the more realistic one, results on systems with only two degrees of freedom can be applied to situations such as the motions of stars in the co-moving meridional plane of axisymmetric galaxies or in the equatorial plane of spiral or barred galaxies. Secondly, there is continuing interest in the orbital motions of stars in triaxial systems, which may represent either triaxial elliptical galaxies or galactic bars. The ultimate aim of many of the studies on stellar orbits is to build self-consistent models of stellar systems on the basis of the individual orbits of the stars.

A review of stellar dynamics has been given by Dejonghe (41.151.112). Orbital theory and the existence of non-classical integrals of motion have been reviewed or generally discussed by Antonov (40.042.118), Binney (38.151.018), Cleary (43.151.068), and Contopoulos (38.151.075).

Best approximations for quadratic integrals have been discussed by de Zeeuw and Lynden-Bell (40.151.158). The number of effective integrals in galactic models was studied by Magnenat (39.151.158). Models of stellar systems with third integrals have been constructed by Petrou (40.151.061), Vandervoort (38.151.106), and Villumsen and Binney (40.151.033). Non-isolating integrals were studied by Genkin and Genkina (43.151.048).

Periodic orbits in systems with two or three degrees of freedom have been investigated by Barbanis (40.151.063), Caranicolas, Diplas and Varvoglis (38.151.064, 41.151.105, 42.151.104), Cartigny, Desolneux and Hayli (38.042.028), and Hadjidemetriou (39.042.028). Resonant orbits were studied especially by Andrle (39.151.046), Caranicolas (38.151.042, 39.151.099, 40.151.050), and Contopoulos and Barbanis (40.151.078).

The consequences of bifurcations of families of periodic orbits, the collisions of bifurcations and complex instability in systems with three degrees of freedom have been studied by Contopoulos (39.151,073, 40.151.098, 41.151.087,

41.151.104), Contopoulos and Magnenat (41.151.086), Martinet and Pfenniger (43.151.034, 43.151.100), and Pfen-niger (40.151.025, 43.151.092).

The transition from integrable orbits to chaotic motions was studied by Contopoulos (40.151.053, 1987a), Contopoulos and Polymilis (1987), Contopoulos, Varvoglis and Barbanis (43.151.027), Evangelidis and Neethling (38.042.007) and Innanen (39.151.156). Stochastic orbits in galaxies have been discussed by Barbanis (38.151.071, 43.151.073) and Gerhard (39.151.179, 40.151.065, 42.151.037).

The effect of dynamical friction on stellar orbits was investigated by Casertano, Phinney and Villumsen (41.151.-060), Hoffer (39.151.045), and Pfenniger (42.151.031). A general definition of orbital excentricity was given by Ninkovich (41.151.048.)

11.2 Spiral and Barred Galaxies

The orbits of stars in spiral or barred galaxies were investigated under various aspects. The occurence of Lindblad resonances in general were discussed by Dzigvashvili and Malsidze (43.151.091). Contopoulos (42.151.053) and Contopoulos and Grosbol (41.151.048, 41.151.019) investigated the orbits near the 4/1 resonance in spiral galaxies. Contopoulos (1987b) reviewed non-linear phenomena in spiral galaxies.

Periodic orbits in barred galaxies have been studied by Michalodimitrakis and Terzides (40.151.023, 40.151.052, 42.151.007), especially for explaining inner rings in barred galaxies (39.151.026, 40.151.003). The implications of the 1/1 resonance for barred galaxies were investigated by Petrou and Papayannopoulos (41.151.018), and the response density of irregular orbits by Petrou (38.151.039). Pfenniger (38.151.063) derived the velocity field in barred galaxies on the basis of orbit calculations.

11.3 Oblate Elliptical Galaxies

Orbits in classical oblate models of elliptical galaxies have been investigated by Andrie (42.151.076), Caranicolas (39.151.076), Caranicolas (39.151.009), and Caranicolas and Vozikis (42.151.054).

11.4 Triaxial Systems

De Zeeuw (40.151.014, 40.151.035) and de Zeeuw, Peletier and Franx (42.151.036) studied mass models of elliptical galaxies with separable potentials, especially of Stäckel form in ellipsoidal coordinates. The effect of a nucleus at the center of a triaxial galaxy on the stellar orbits has been investigated by Gerhard and Binney (40.151.036) and by Spyrou and Varvoglis (43.151.119). Periodic orbits in triaxial ellipticals have been calculated by Davoust (41.151.020) and Robe (39.151.006, 42.151.001). Preferred orbital planes in triaxial systems have been studied by David, Steiman-Cameron and Durisen (38.151.081, 40.151.017). Habe and Ikeuchi (39.151.047) investigated gas orbits in prolate triaxial galaxies. Schwarzschild (42.151.120) discussed the perfect ellipsoid and derived a truncated perfect elliptic disk as a model for galactic bars.

References

Contopoulos, G.: 1987a, in: *Chaotic Phenomena in Astrophysics*, Eds. J.R. Buchler and H. Eichhorn, New York Acad. Sciences (in press)

Contopoulos, G.: 1987b, in: *The Galaxy*, Eds. G. Gilmore and R. Carswell, D. Reidel Publ. Comp., Dordrecht, p.199

Contopoulos, G., Polymilis, C.: 1987, *Physica*, **24D**, 328

12. Galactic Dynamics : Computer Simulations

Computer simulations have become a standard tool for investigating the structure and evolution of gravitating systems.

12.1 General Problems of Stellar Dynamics and New Methods

Reviews on numerical methodes for simulating gravitating have been given by Aarseth (42.151.078), Saslaw (40.003.011), Anosova (40.042.119), Tajima, Clark, Craddock, Gilden, Leung, Li, Robertson, and Saltzman

(39.014.043), Hut (42.151.057), and Efstathiou (42.011.002). Bettwieser and Sugimoto (39.151.001) have considered the validity of the gas mdoel for gravitational N-body systems.

New integration schemes and methods for the constuction of model galaxies have been presented by Aarseth and Bettwieser (38.042.083), Vandervoort (38.151.106), Richstone and Tremaine (38.151.079), Marciniak (39.042.088), Monaghan and Lattansio (40.021.004), Mikkola (40.042.013), White (42.151.081), and Chau et al. (42.151.115).

The dynamics of one-dimensional systems has been investigated by Severne, Luwel, and Rousseuw (38.151.049, 40.151.049, 40.151.071, 41.151.088) and Miller et al. (38.151.031).

Shapiro and Teukolsky (40.151.075, 40151.076) developed methods to realise relativistic stellar dynamics on the computer. Johns and Nelson (40.151.089) described particle simulations of three-dimensional galactic hydrodynamics. Tremaine and Weinberg (38.151.012) studied dynamical friction in spherical systems and Inagaki and Wiyanto (41.151.077) considered the effect of gravitational focusing on the dynamical evolution of stellar systems.

12.2 Clustering of Galaxies

Numerous N-body simulations of the clustering of galaxies have been reported and analysed under various aspects, especially with regard to the dynamical role of dark matter and to global effects due to tidal interactions between galaxies, by Yabushita (38.151.011), Schwekendiek and Wielen (38.151.085, 42.161.351), Shandarin and Klypin (38.160.030), Smith (38.160.047, 39.160.024), Saarinen and Valtonen (38.160.070, 40.151.079), Saarinen, Dekel, and Carr (39.151.151), Melott (38.160.075, 42.161.050), Cavaliere, Santangelo, Tarquini, and Vittorio (38.160.131, 41.160.121), Ryden and Turner (38.160.148), Miller (39.151.083), Mamon (41.160.082), Yabushita and Allen (39.160.011), Barnes (40.151.011), Barnes, Dekel, Efstathiou, and Frenk (40.161.100), Evrad and Yahil (40.161.11, 40.161.112, 42.151.050, 42.151.100, 43.151.095), Saslaw (40.161.130), Ishizawa (41.151.012, 43.151.038), Muzzio (41.151.028, 42.151.003, 43.151.067, 1987, in press), Muzzio et al. (38.160.046, 43.151.044), Navarro et al. (41.161.172), White (42.161.048), Villumsen and Davis (42.160.055, 42.161.176), and Hoffman (42.160.048).

12.3 Interacting Galaxies

Tidal interactions between galaxies have been studied in detail by Aguilar and White (40.151.032, 42.151.027), Miller (42.151.035), Borne (38.151.107), Lukkari and Salo (38.042.010), Capaccioli and Malvasi (39.151.062), Nishida and Wakamatsu (39.151.169), Song (39.151.171), Song and Stewart (42.151.047), Martel (42.151.005), and Byrd et al. (42.1541.033). Specifically the dynamics of satellites of galaxies and the shell phenomenon have been treated by Byrd et al. (41.151.111), Bontekoe and van Albada (43.151.003), Dupraz and Combes (39.151.175), Quinn and Goodman (42.151.094), Huang and Stewart (43.151.035), and Quinn and Hernquist (42.151.083). Merging of galaxies has been simulated by Duncan (38.151.061) and Mezzetti et al. (38.160.126). The gas dynamics in interacting galaxies has been modelled by Foster (38.151.098), Gaetz (39.151.043, 39.151.044), Noguchi and Ishibashi (39.131.366, 41.151.023, 43.151.018, 43.151.104), Appleton and Struck-Marcell (41.151.054), Tenorio-Tagle and Bodenheimer (42.151.055), and Gaetz et al. (43.151.116). The generation of spiral structure by tidal encounters of disk galaxies has been investigated by Sorensen (39.151.004), Icke (39.151.040), Korchagin and Prokhovnik (40.151.105), and Undelius et al. (43.151.036).

12.4 Collapse of Protogalaxies

Numerical collapse experiments to study the formation of galaxies and their halos have been described by May and van Albada (38.151.002), Villumsen (38.151.037), Lada et al. (38.151.048), White (38.151.080), Carlberg (38.151.103, 38.151.104), Carlberg et al. (41.151.003), Polyachenko (41.151.069), Moreno and Pismis (41.151.115), Smith and Miller (42.151.096), Arbolino (43.151.009), Burkert and Hensler (43.151.016), Quinn et al. (43.151.081), Kim et al. (43.151.083), and Nakamura and Tosa (43.151.103).

12.5 Evolution of Galactic Structures

The dynamics and evolution of elliptical galaxies have been simulated numerically by Gerhard (39.151.177), May et al. (39.151.033, 39.151.048, 40.151.034), Habe and Ikeuchi (39.151.047, 39.151.080), Gerhard and Binney (39.151.074), Barnes (39.151.111), Merritt and Aguilar (40.151.091), Tohline et al. (40.062.085), Levison and Richstone (40.151.030, 40.151.031, 43.151.064), Huang and Stewart (40.151.080), Katz and Richstone (40.157.064), Barnes et al. (41.151.080), Katz and Richstone (40.151.106), Dupraz and Combes (42.151.002), Madejsky and Mollenhoff (42.151.069), Sparke (43.151.042), and van Albada (43.151.080).

Numerical models of disk galaxies and their varous properties, in particular their stability, have been studied by Hayes and Comins (38.151.062), Whitmore *et al.* (38.157.231), Vandervoort *et al.* (39.151.086), Bishop (39.151.087), Zasov and Morozov (39.151.144), Fujiwara and Hozumi (39.151.163), Sellwood (40.151.078), Athanassoula and Sellwood (42.151.012, 42.151.013, 43.151.040), Carlberg and Freedman (40.151.087), Zhou and Zheng (40.151.103), Casertano *et al.* (41.151.060), Miller (41.151.061, 42.151.075), and Nishida *et al.* (41.151.063, 41.151.076). The evolution of galactic disks due to internal relaxation has been treated by Villumsen (39.151.013, 39.151.049), Mishurov (38.151.093, 39.151.159), Balasz (42.151.073), Villumsen and Binney (40.151.033), Palous (41.151.084), Kamahori and Fujimoto (43.151.072), and Zotav and Morozov (43.151.023).

The appearance and evolution of spiral structures in disk galaxies have been investigated by numerical simulations by Sellwood and Carlberg (38.151.007), Hausman and Roberts (38.151.008), Inagaki *et al.* (38.151.028), Freedman *et al.* (38.151.032), Leisawitz and Bash (38.151.047), Byrd *et al.* (38.151.070), Korchagin (42.151.011, 43.151.086), Morosov *et al.* (42.151.015, 42.151.016, 42.151.017), and Bakrunov *et al.* (42.151.049).

Numerical models of galactic bars and barred spiral galaxies have been presented by Schwarzschild (38.151.056, 42.151.120), Schwarz (38.151.003, 39.151.003, 39.151.056, 41.151.085), Combes and Gerin (40.151.048), Pfenniger (38.151.063, 39.151.180), Barnes and White (38.151.069), Sparke (38.151.070), Sparke and Sellwood (43.151.025), Weinberg (39.151.037), and Liu (43.151.113).

A review on galactic gas dynamics has been given by Roberts (41.151.015). Extensive numerical simulations of gas flow in disk galaxies have been reported by Johns and Nelson (39.151.017, 41.151.062, 43.151.013), Nelson *et al.* (38.151.021, 39.151.021), Kritsuk (38.131.045), Mikkola *et al.* (38.131.093), van Albada (39.151.008), Roberts and Hausman (39.151.031, 39.151.032), Roberts and Stewart (39.151.147, 39.151.148, 43.151.060), Carlberg (39.151.032), Fukunaga and Tosa (39.151.167, 41.151.118, 41.131.027, 43.151.017), Mulder (41.151.021), Tomisaka (41.151.037), and Varnas (43.151.069).

The conjecture of spiral structure as being due to large-scale stochastic self-organization of galaxies was developed further by Seiden (39.151.029), Feitzinger (39.151.030), and Comins and Balser (39.151.085). Spurzem and Langbein (39.151.079) and Miller and Smith (42.151.116, 43.151.124) treated the evolution of the centers of galaxies.

12.6 Evolution of Star Clusters

The dynamics of star clusters has received great interest in recent years. Various aspects of cluster evolution, including multicomponent structures, energy input by environmental effects, and the role of binaries, have been studied by numerical methods by Tomley (38.151.005), Inagaki and Wiyanto (38.151.094), Bettwieser and Fritze (38.151.095), Bettwieser *et al.* (38.151.090, 39.151.112, 41.151.0720, Spurzem *et al.* (38.151.092, 39.151.105), Giannone and Molteni (39.151.039, 39.151.116), Inagaki and Saslaw (39.151.077), Ostriker *et al.* (39.151.091), Duncan (39.151.126), Cohn (39.151.102), Aarseth (39.151.108), Lightman and McMillan (39.151.109), Jernigan (39.151.110), McMillan (39.151.119), Stodolkiewicz (39.151.124), Shapiro (39.151.125), Casertano and Hut (40.151.077), Wiyanto *et al.* (40.151.093, 41.151.075), Giannone *et al.* (40.151094), Danilov and Beshenov (40.151.107, 43.151.029), Fritze and Fricke (41.151.120), Heggie (43.151.079), and Statler *et al.* (43.151.117).

A review on the dynamics of open star clusters has been given by Wielen (39.151.129). Further N-body simulations of open star clusters have been reported by Terlevich (39.151.131, 43.151.002), Dearborn *et al.* (39.151.092, 39.151.130), Danilov (40.153.011), and Inagaki (41.151.007).

Spitzer has given reviews on the dynamical evolution of globular clusters (38.151.1013, 39.151.100, 43.151.077). Simulations of globular cluster evolution, including core collapse, gravothermal oscillations, and effects due to compact massive objects, either inside or passing the clusters, have been presented by McMillan and Lightman (38.151.034, 38.151.035), Duncan (39.151.090), Heggie (39.151.101), Bettwieser (39.151.106), van Albada and Bontekoe (39.151.128), Inagaki (39.151.168, 42.151.107), McMillan (42.151.028), Cohn and Hut (39.154.087), Chernoff *et al.* (42.151.080), Makino *et al.* (42.151.108), Murphy and Cohn (42.151.114), Cohn *et al.* (43.151.082), and Aarseth *et al.* (1987, in press).

Binary and triple star systems are known to be of particular importance for the evolution of star clusters as well as of galaxies in general. The evolution of such star systems, as driven by encounters with further stars, has been investigated by Hills (38.151.041), Hut and Paczynski (38.151.043), Mikkola (38.151.077, 39.151.120, 42.151.098), Hut (39.151.057, 39.151.107) Ostriker (39.151.123), Baranov (39.151.173), Weinberg *et al.* (41.151.050, 43.151.020), Anosova and Orlov (42.042.010, 42.151.010, 43.151.051, 43.151.089), Alexander (42.042.043), Mikkola and Valtonen (42.151.064), and McMillan (42.154.005).

13. Galactic Dynamics : Stability and Evolution

13.1 Spiral Structure

Our knowledge of spiral structure at the beginning of the period under review has been surveyed by Athanassoula (1984). It appears that a zeroeth-order understanding of spiral structure can be obtained by treating galactic disks as resonant cavities within which almost stationary patterns of Lin-Shu-Kalnajs (LSK) density waves resonate. Losses due to Landau damping at the Lindblad resonances are made good by energy input at corotation, probably by Goldreich-Lynden-Bell-Toomre swing amplification.

Since the WKBJ approximation that lies at the heart of the LSK dispersion relation is in practice marginal, there has been continued interest in understanding disk modes without appealing to density waves. The most promising techniques for this work are N-body simulation and analysis of gaseous disks. Sellwood and Athanassoula (1986) have shown how N-body simulations may best be analysed in terms of modes, while Iye (1984) has emphasised the close connection between modes in gaseous disks and the modes of stars. Ueda *et al.* (1985) have shown how inferences about the mass distributions of galaxies may be drawn from stability analyses of gaseous disks.

Clearly there is much more to spiral structure than can be captured even by mode analyses of realistic stellar disks. In particular, Korchagin & Korchagin (1984) and Tagger (1987) have emphasised the importance of taking into account the effects of non-linearity and it is essential to consider the response of the interstellar medium to a spiral in the stellar disk. There are two aspects to this response: (i) the passive response to the non-axisymmetric components of the gravitational field; (ii) detonation waves that will arise whenever large quantities of gas are induced to fragment into young, massive stars. The paper of Roberts & Hausman (1984) explores both aspects. Another important question is whether spiral structure is responsible for the observed correlation between the ages and velocity dispersions of disk populations (e.g. Palous & Piskunov 1985; Knude *et al.* 1987; Wielen & Fuchs 1987). Sellwood & Carlberg (1984) and Carlberg & Sellwood (1985) have used N-body models and analytic treatments to argue that spiral structure can account for the dispersions of stars within the plane, but is probably unable to account for the vertical dispersions. Binney & Lacey (1987) take a slightly less optimistic view of the ability of spiral structure to produce the hottest disk populations. Since several studies have concluded that giant molecular clouds cannot unaided heat the disk (but see Kamahori & Fujimoto 1986), they suggest that we should seriously consider the possibility that massive objects in the near or far halo constitute the responsible agency (Rogers *et al.* 1981; Lacey & Ostriker 1985; Ipser & Semenzato 1985).

13.2 Bars

The prevalence of large "forbidden" velocities in the (l, v) plane at $|l| < 20°$ strongly suggests that ours is a barred galaxy. Schwarz (1984) and Combes & Gerin (1985) have considered the generation of spiral structure by such a bar. Depending on the bar's pattern speed either outer or inner rings of gas and stars can be formed. If the 3.5 kpc ring is, from the bar's point of view, an outer ring, the bar's pattern speed must be relatively large. Mulder & Liem (1986) and Gerhard and Vietri (1986) have argued that the apparent peak near $R = 0.7$ kpc in rotation curves derived from tangent velocities is an artifact of the large non-circular velocities generated by the bar.

Robust rotating bars commonly form in N-body simulations of initially cold stellar systems. Hence there is, in a sense, no mystery about the prevalence of bars in galaxies like ours. Yet we are still very far from understanding how bars work at the stellar level. Until recently nearly all studies were of strictly planar bars, which are already remarkably intricate structures (e.g. Teuben & Sanders 1985; Petrou & Papayanopoulos 1986). During the period under review we have begun to get a glimpse of the vastly greater complexity of three-dimensional bars. This complexity derives from the large number of resonances introduced when the extra degree of freedom is unfrozen (Mulder 1983; Pfenniger 1984), and from qualitative changes in the structure of orbits in phase space when the latter moves from four to six dimensions (Martinet & Pfenniger 1987). Stochasticity generated by the interaction of these many resonances probably precludes the possibility of a truly steady-state rotating bar; we should expect all rotating bars to evolve secularly as stars diffuse through the stochastic parts of phase space. The development of a framework for the calculation of this sort of evolution is an important task for the future.

Pending the development of such general machinery, Weinberg (1985) has discussed the secular loss of angular momentum from a bar to the surrounding halo by treating the bar as a rigid object and following the response of individual halo stars.

13.3 Warps and Corrugations

Mathur (1984) and Sparke (1984) have explored the possibility that warps are forced by either an off-axis or a triaxial halo. Sparke (1986) has emphasized the importance of self-gravity for the longevity of even a light inclined ring. A large body of data on the distribution of molecular gas within the Galaxy (Dame & Thadeus 1985; Sanders *et al.* 1986) is now available for comparison with calculations of dynamics of corrugation waves in this layer by Johns & Nelson (1987). Spicker & Feitzinger (1986) have studied corrugations in the HI layer.

13.4 Dynamics of the Halo Populations

Though no satisfactory working model of any halo population has yet been constructed, there has been progress in understanding how we might accomplish that task. De Zeeuw's (1986) reexamination of a class of potentials first examined by Jacobi and Stäckel has clarified the connection between distribution functions that employ non-classical integrals and the morphology of the substems such distribution functions generate (May & Binney 1986a; Binney 1987). In particular, these developments establish for the first time a clear connection between models with spherical potentials, such as those employed by White (1985) to explain the observations of Ratnatunga & Freeman (1985), and flattened and even triaxial systems. Also, if we choose to express the distribution functions as functions of action integrals, the adiabatic invariance of these integrals can be exploited to discover how each population responded to slow changes in the Galaxy's gravitational potential, such as that associated with a slowly growing disk (Barnes & White 1984; Binney & May 1986; Blumenthal *et al.* 1986; Ryden & Gunn 1987).

An important development during the period under review has been the realization that spherical stellar systems with large numbers of stars on highly eccentric orbits are prone to non-axisymmetric instabilities (Antonov 1973; Polyachenko & Shukhman 1981; Merritt & Aguilar 1985; Barnes *et al.* 1986; May & Binney 1986b). Even if the halo components on their own would be spherically symmetric, in the presence of a massive disk velocity dispersion ratios $\sigma_r/\sigma_\theta \approx 0.6$ of the order frequently derived for extreme population II objects indicate that these components are significantly flattened (Binney *et al.* 1987).

It is clear that the Galaxy owes some of its present substance to cannibalism of smaller stellar systems. It has even been suggested (Peebles & Dicke 1968) that all population II field stars were formed in clusters that have since been disrupted: the range of clusters capable of avoiding either internal disruption (of light, tight clusters) or tidal shredding (of massive, diffuse clusters) is small (Fall & Rees 1977; Goodman & Hut 1985). Fall & Rees (1985) reanalyze this problem and suggest that metal-poor proto-galactic material would have preferentially formed clusters of sizes well suited survival in the galactic halo. Terlevich (1987) has modelled the evolution of the open cluster population as clusters are dissolved by tidal interaction with giant molecular clouds.

Whatever the Galaxy's gains at the expense of the cluster populations, the Magellanic Stream constitutes vivid evidence that cannibalism of dwarf galaxies is probably an important process. Tremaine & Weinberg (1984), Quinn & Goodman (1986) and Bontekoe & van Albada (1987) have studied the action of dynamical friction on the orbits of such systems. Lance (1987) has accumulated an impressive body of evidence that the young stars observed 1 kpc and more above the plane are the debris of disrupted late-type systems. These studies raise an important question: are the Galaxy's stars strongly clumped into groups associated with different accretion events rather than being smoothly distributed through phase space as we have tended to assume in the past? From the work of Malin & Cater (1983), Quinn (1984) and others we suspect clumping in phase space is important for many elliptical galaxies. Perhaps it is not too fanciful to imagine a field of galactic archaeology opening up, in which painstaking sifting of the contents of each element of phase space will enable us to piece together a fairly complete picture of how our Galaxy grew to its present grandeur and prosperity.

References

Antonov, V. A. 1973, In: *The Dynamics of Galaxies and Star Clusters* ed. G. B. Omarov, (Alma Ata: Nauka), p. 139

Athanassoula, E. 1984, *Phys. Reports*, **114**, 319

Barnes, J., Goodman, J. & Hut, P. 1986, *Astrophys. J.*, **300**, 112

Barnes, J. & White, S. D. M. 1984, *Mon. Not. R. Astron. Soc.*, **211**, 753

Binney, J. J. 1987, In: *The Galaxy* ed. G. Gilmore & R. Carswell, p. 399. Dordrecht: Reidel

Binney, J. J. & May, A. 1986, *Mon. Not. R. Astron. Soc.*, **218**, 743

Binney, J. J. & Lacey C. G. 1987, *Mon. Not. R. Astron. Soc.*, in press.

Binney, J. J., May, A. & Ostriker, J. P. 1987, *Mon. Not. R. Astron. Soc.*, **226**, 149

Blumenthal, G. R., Faber, S. M., Flores, R. & Primack, J. R. 1986, *Astrophys. J.*, **301**, 27

Bontekoe, Tj. R. & van Albada, T. S. 1987, *M.N.R.A.S.*, **224**, 349

Carlberg, R. G. 1987, *Astrophys. J.*, submitted

Carlberg, R. G. & Sellwood, J. A. 1985, *Astrophys. J.*, **292**, 79

Combes, F. & Gerin, M. 1985, *Astron. Astrophys.*, **150**, 327

Dame, T. M. & Thadeus, P. 1985, *Astrophys. J.*, **297**, 751

de Zeeuw, T. 1986, *Mon. Not. R. Astron. Soc.*, **216**, 273

Fall, S. M. & Rees, M. J. 1977, *Mon. Not. R. Astron. Soc.*, **181**, 37

Fall, S. M. & Rees, M. J. 1985, *Astrophys. J.*, **298**, 18

Gerhard, O. E. & Vietri, M. 1986, *Mon. Not. R. Astron. Soc.*, **223**, 377

Goodman, J., & Hut, P., ed. 1985, *Dynamics of Star Clusters*, IAU Symposium No. 113. Dordrecht: Reidel

Ipser, J. R. & Semenzato, R. 1985, *Astron. Astrophys.*, **149**, 408

Iye, M. 1984, *Mon. Not. R. Astron. Soc.*, **207**, 491

Johns, T. C. & Nelson, A. H. 1987, *Mon. Not. R. Astron. Soc.*, **224**, 863

Kamahori, H. & Fujimoto, M. 1986, *Publ. Astron. Soc. Japan*, **38**, 77

Korchagin, V. I. & Korchagin, P. I. 1984, *Astr. Zh.*, **61**, 814 (translated in *Sov. Astron.*, **28**, 476)

Knude, J., Schnedler Nielsen, H. & Winther, M. 1987, *Astron. Astrophys.*, **179**, 115

Lacey, C. G. & Ostriker, J. P. 1985, *Astrophys. J.*, **299**, 633

Lance, K. 1987, *Astrophys. J.*, submitted

Malin, D. F. & Carter, D. 1983, *Astrophys. J.*, **274**, 534

Martinet, L. & Pfenniger, D. 1987, *Astron. Astrophys.*, **173**, 81

Mathur, S. D. 1984, *Mon. Not. R. Astron. Soc.*, **211**, 901

May, A. & Binney, J. J. 1986a, *Mon. Not. R. Astron. Soc.*, **221**, 857

May, A. & Binney, J. J. 1986b, *Mon. Not. R. Astron. Soc.*, **221**, 13

Merritt, D. & Aguilar, L. A. 1985, *Mon. Not. R. Astron. Soc.*, **217**, 787

Mulder, W. A. 1983, *Astron. Astrophys.*, **121**, 91

Mulder, W. A. & Liem, B. T. 1986, *Astron. Astrophys.*, **157**, 148

Palous, J. & Piskunov, A. E. 1985, *Astron. Astrophys.*, **143**, 102

Peebles, P. J. E. & Dicke, R. 1968, *Astrophys. J.*, **154**, 891

Petrou, M. & Papayannopoulos, T. 1986, *Mon. Not. R. Astron. Soc.*, **219**, 157

Pfenniger, D. 1984, *Astron. Astrophys.*, **134**, 373

Polyachenko, V. L. & Shukhman, I. G. 1981, *Astr. Zh.*, **58**, 933 (translated in *Sov. Astr.*, **25**, 533)

Quinn, P. J. 1984, *Astrophys. J.*, **279**, 596

Quinn, P. J. & Goodman, J. 1986, *Astrophys. J.*, **309**, 472

Ratnatunga, K. U. & Freeman, K. C. 1985, *Astrophys. J.*, **291**, 260

Roberts, W. W. & Hausman, M. A. 1984, *Astrophys. J.*, **277**, 744

Rogers, A., Harding, P. & Sadler, E. 1981, *Astrophys. J.*, **244**, 912

Ryden, B. S. & Gunn, J. E. 1987, *Astrophys. J.*, **318**, 15

Sanders, D. B., Clemens, D. P., Scoville, N. Z. & Solomon, P. M. 1986, *Astrophys. J. Suppl.*, **60**, 1 & 297

Schwarz, M. P. 1984, *Mon. Not. R. Astron. Soc.*, **209**, 93

Sellwood, J. A. & Athanassoula, E. 1986, *Mon. Not. R. Astron. Soc.*, **221**, 195

Sellwood, J. A. & Carlberg, R. G. 1984, *Astrophys. J.*, **282**, 61

Sparke, L. S. 1984, *Mon. Not. R. Astron. Soc.*, **211**, 911

Sparke, L. S. 1986, *Mon. Not. R. Astron. Soc.*, **219**, 657

Spicker, J. & Feitzinger, J. V. 1986, *Astron. Astrophys.*, **163**, 43

Tagger, M., Sygnet, J. F., Athanassoula, E. & Pellat, R. 1987, *Astrophys. J. Lett.*, **318**, L43

Teuben, P. & Sanders, R. 1985, *Mon. Not. R. Astron. Soc.*, **212**, 257

Terlevich, E. 1987, *Mon. Not. R. Astron. Soc.*, **224**, 193

Tremaine, S. D. & Weinberg, M. D. 1984, *Mon. Not. R. Astron. Soc.*, **209**, 729

Ueda, T., Noguchi, M. Iye, M. Aoki, S. 1985, *Astrophys. J.*, **288**, 196

Weinberg, M. 1985, *Mon. Not. R. Astron. Soc.*, **213**, 451

Wielen, R. & Fuchs, B. 1986, In: *The Milky Way Galaxy* IAU Symposium No. 106, eds. H. van Woerden, R. J. Allen & W. B. Burton (Dordrecht: Reidel) p. 481

White, S. D. M. 1985, *Astrophys. J. Lett.*, **294**, L99

14. Supplement: Galactic Structure Studies Recently Carried Out in the USSR

14.1 Introduction

During the triennium under review some proceedings and books have been published in the USSR concerning the field of galactic research. Among them there are the proceedings of the international colloquium "Stellar Catalogues: Data Compilation, Analysis, Scientific Results", held in Tbilisi, USSR, 10–15 September, 1984 (Kharadze and Kogoshvili, 1985), and "Problems of Astrometry" (Podobed, 1984). A number of books were published: "The Galaxy" (Marochnik and Suchkov, 1984), "Physics of Gravitating Systems", volumes I and II (Fridman and Polyachenko, 1984ab), "Stellar Astronomy" (Agekyan, 1985), "Magnetic Fields in Space" (Bochkarev, 1985), "Methods in the Qualitative Theory of Dynamical Systems in Astrophysics and Dynamics" (Bogyavlensky, 1985), "Supernovae and Stellar Wind: Interaction with the Gas of the Galaxy" (Lozinskaya, 1986).

14.2 Basic Data and Calibrations

Glushneva (1985ab) published a spectrophotometric star catalog in which effective temperatures, angular diameters, bolometric corrections, radii, luminosities and gravities were presented for different groups of stars. The scale of effective temperatures has been constructed. A catalog of metal-deficient F–M stars was compiled by Bartkevichius (1984ab, 1985, 1986). Alksnis (1985) presented a catalog of carbon stars based on the data obtained with a Schmidt telescope. Salukvadze (1985) presented a catalog of trapezium-type multiple systems. Lebedev and Lebedeva (1985) and Lebedev (1986ab) compiled a catalog of chemically peculiar stars and studied their spatial distribution and kinematics.

Catalogs of stellar magnitudes and colour indices of stars in different regions of the sky were compiled by Kazanasmas et al. (1984, 1985), Kazanasmas et al. (1985), Tokhtas'ev (1985), and Baltabaev et al. (1985). Zdanavicius and Cerniene (1985) published new, accurate values of photoelectric magnitudes and colour indices of 33 stars in the Cygnus standard region in the Vilnius photometric system. Eglitis (1986) obtained the absolute energy distribution shown in the spectra of 22 carbon stars. Kazanasmas et al. (1986) published stellar magnitudes and $B-V$ colour indices in the BV system for eight photometric standards in the direction of Wirtanen-Vyssotsky areas.

Analysis of spectra of four red giants in the Hyades by the model atmosphere method, performed by Mishenina et al. (1986) showed metallicity excess in the atmospheres of Hyades giants of $\Delta[Fe/H] = 0.08$. Mishenina and Panchuk (1986) found from analysis of homogeneous spectroscopic observations of a sample of K giants that the metallicity dispersion is equal to 11 with an error of $\Delta[Fe/H] = 0.1$. Sil'chenko (1984) confirmed a new metallicity scale of globular clusters on the basis of calculations of model integrated $B-V$ colours. Kopylov (1985) constructed a $(U-B)_0$ vs. $(B-V)_0$ diagram for O5–K5 stars of the luminosity class V with solar chemical composition.

Avedisova (1985) cataloged star formation regions in our Galaxy. Sleivyte (1986) calculated the ratios of colour excesses and the R ratios in the UBV and Vilnius photometric systems for carbon and barium stars. Straizys (1985) summarized the calibration of a number of important photometric systems. A program for two-dimensional quantitative spectral classification of stars in the Vilnius photometric system was developed by Jasevichius (1986). Smriglio et al. (1986) described an automatic method of two-dimensional stellar classification from photographically determined magnitudes using the seven-colour Vilnius photometric system. Malyuto and Shvelidze (1985), Kuzmin et al. (1986) and Malyuto (1986) developed a method of deducing external weights of catalogs from data residuals and applied this method to $[Fe/H]$ and T_{eff} catalogues.

14.2.1 Photometry

Bartasiute (1984, 1985ab) published results of photoelectric photometry of stars near the galactic poles in the Vilnius photometric system. Results of photoelectric photometry in the same system in the region of open cluster M29 were presented by Kazlauskas and Jasevichius (1986). Straizys et al. . (1986) published photometric data for about 100 metal-deficient stars. Sleivyte (1986b) presented results of photoelectric photometry in the Vilnius photometric system for 27 low-temperature carbon stars. Cernins (1986) performed the Vilnius photometry in Kapteyn areas 92, 108 and 112, and Janulis (1986) did the same in the direction of globular cluster M56. Zdanavicius (1986) measured about 30 metal-deficient G–K giants in the $UBVR$ system . Kizla and Paupers (1985) presented observations of 73 stars also in the $UBVR$ system. Dzervitis and Paupers (1985) published results of photoelectric photometry in the Vilnius system for 12 stars in the immediate neighbourhood of WCMa. A photometric classification of these stars has been performed. Andruk and Kharchenko (1987) performed observations (photoelectric and photographic) for the determination of $UBVR$ magnitudes.

14.2.2 Spectra and Luminosity

Karimova and Pavlovskaya (1985) compiled a list of high-luminosity stars and Cepheids for meridian observations. It contains stars with radial velocities, photometry and a two-dimensional spectral classification. Garibdzhanyan *et al.* (1984) presented results of a spectro-photometric study of 277 OB stars in the region around P Cyg. Zakomanova (1984) identified faint carbon stars in the direction of the galactic anticenter on infrared-sensitive spectral plates. Nikolashvili (1987) detected 180 carbon stars on the basis of the low dispersion spectral material. A spectral classification of 89 faint O–B starts around the cluster IC 1805 has been carried out by Kuznetzov (1984) with the use of unwidened objective prism spectra. Catalogs of *BV* magnitudes and spectral classes of 6000 stars have been compiled by Pugach *et al.* (1985).

14.3 Local Galactic Structure

14.3.1 Low Galactic Latitudes

On the basis of catalogs of photographic *BV*-magnitudes and spectral classes of stars Guseva *et al.* (1984) investigated the distribution of absorbing matter and B0–B2 stars in $5° \times 5°$ region toward the Rosette nebula. In the same direction Guseva (1986) identified the interstellar dust clouds of various densities. Guseva and Metreveli (1986) studied the distribution of the interstellar absorbing matter in the direction $l = 207°$, $b \approx -3°$. Levina (1985) investigated the distribution of the dust material in the region centered at $\alpha_{1950} = 4^h34^m$ and $\delta_{1950} = +26''$. Kolesnik and Metreveli (1985ab) investigated the structure of galactic spiral arms and star formation regions. Kolesnik (1986) studied the distribution of dust clouds along the line of sight in the direction of the CO molecular cloud in the W3 region of star formation. Kalandadze *et al.* (1986) studied absorbing matter in NGC 2264 and its relation to the star formation.

On the basis of photoelectric absorption curves Uranova (1985) determined the form and location of the dust-arm axis up to the distance of 4 kpc from the Sun. Pavlovskaya and Suchkov (1984) performed a statistical analysis of the space distribution of bright stars and open clusters in the fourth quadrant of the Galaxy. It was shown that the Sagittarius-Carina spiral feature cannot be a chance density fluctuation in an actually homogeneous distribution of stars. Basharina *et al.* (1985) studied the stellar structure of the Orion arms by numerical experiments. Melik-Alaverdyan and Tovmasyan (1986) studied the distribution of late-type giant stars in the galactic plane. Avedisova and Kondratenko (1984) derived the distribution of diffuse nebulae in the galactic plane. Avedisova (1985) found some parameters of spiral arms of the Galaxy traced by 255 emission nebulae with known photometric distances. Petrovskaya (1986) studied the large-scale distribution of neutral hydrogen in the Galaxy. Yudaeva (1985) discussed the fine structure of the gas layer in the region $220° < l < 260°$, $|b| \leq 15°$ obtained on the basis of the 21-cm radio line observations with the *RATAN*-600 radio telescope. Amnuel *et al.* (1986) showed that the birthplaces of pulsars are located within OB-associations and/or in the galactic arms.

14.3.2 High Latitude Optical Studies

Bartasiute (1987) estimated the variation of metal abundance perpendicular to the galactic plane, using data on metallicities and distances of 190 F–K stars in the direction of the North Galactic Pole. The resulting gradient amounts to -0.7 ± 0.1 kpc^{-1}. Einasto *et al.* (1985) described an observational program of the main meridional section of the Galaxy. Complex (astrometric, photometric, spectral) observations of stars in selected areas have been continued. The aim of this program is study of the spatial and kinematic structure of physically homogeneous subsystems of stars. Gradients and other parameters of subsystems will be determined. In the frame of this program Kharchenko (1984, 1987) outlined the data to be used and compiled a catalog of proper motions of 14,100 stars with respect to 206 galaxies. Rybka (1985a) recommended reference stars at high galactic latitudes with relatively small proper motions to be used to decrease the cosmic error of stellar proper motions.

14.4 Overall Galactic Structure

14.4.1 Galactic Disk

Kharadze *et al.* (1985) obtained stellar density values in the solar neighbourhood, and the β parameter defining stellar distribution along the z-coordinate. The distribution of B stars in the projection on the plane of the Galaxy was investigated. Mdzinarishvili (1985) obtained information on the stellar density function $D_s(r)$. Catalogs of *BV* photometry and MK classification were the source of this information.

Statistical investigation of the field-star luminosity function was carried out by Vereshchagin and Piskunov (1984) on the basis of some photometric and spectral machine-readable catalogs. Myakutin (1984) estimated the influence of small systematic differences in the definition of luminosity and effective temperature in the 4-colour Strömgren photometric system and MK spectral classification on the inclination of the initial mass function. Dlushnevskaya et al. (1985) found mass distributions on the basis of photo-electric UBV data for 12 open clusters. Malkov (1987) showed that the uncertainties of the BC-scale and effects of unresolved binaries could essentially influence the slope of resulting IMF.

Suchkov (1986) analyzed the mass-diameter relation for stellar systems. Gvaramadze and Lominadze (1986) modelled the galactic disk by an oblate spheroid with confocal isodensity surfaces. Kasak (1986) applied the accretion theory to the generation of spiral structure of galaxies and derived formulae to calculate brightness profiles of spiral arms. Gusejnov and Yusifov (1986ab) found the luminosity function of pulsars and evolutionary ages of most known pulsars ($5 - 10 \times 10^6$ years). Vladimirskij (1985) studied the spatial distribution of pulsars and found that their mean distance to the galactic plane is $\sim$ 50–70 pc.

14.4.2 Galactic Center

Zakhozhaj (1984) analyzed stellar populations within 10 pc. Gurzadyan (1985) studied dynamical structure of the central region of the Galaxy. Ozernoj (1986) proposed a hybrid model for the active source at the center of our Galaxy, containing a very massive star coupled with a black hole. Kardashev (1985) developed a phenomenological model of the galactic core. Stern (1984) collected data for and against the black hole hypothesis. Ozernoj (1984a) argued that a symbiotic object consisting of a superstar and a moderate-mass black hole seems to be able to explain the principal features of the galactic centre emission. Ozernoj (1984b) found that recent observations of broad HeI and HI lines from IRS 16 located at the galactic centre indicate an outflow of matter from it. Seitnepesov and Khanberdiev (1984, 1985) proposed a mechanism for the activity of the non-thermal radio source in the centre of the Galaxy. Different models of the activity of the non-thermal radio source in the galactic centre were analyzed.

14.4.3 Galactic Rotation Curve

Haud (1984) demonstrated that the existing observational data do not conflict with the hypothesis of the flat rotation curve of our Galaxy. Kolesnik and Yurevich (1985) determined the galactic rotation curve up to 16 kpc using the relation between parameters of OH molecular absorption features of clouds and distances to the clouds. Yurevich (1985) determined the distance of the Sun to the galactic centre from the rotation curve with the use of OH clouds. The rotational curve of the Galaxy was constructed by Avedisova (1985) with the aid of 178 diffuse nebulae with known radial velocities and photometric distances obtained in a uniform way. The rotation curve of the outer parts of our Galaxy from neutral hydrogen 21-cm line profiles was analyzed by Petrovskaya and Teerikorpi (1986).

14.4.4 Integrated Galactic Spectrum for UV and Optical Data

Zavarzin (1984) presented results of surface photometry of the northern Milky Way in the V system. Zvereva et al. (1985) presented observations of the far-UV spectrum (1300–1800 Å) of the sky background at different galactic latitudes.

14.4.5 Evolution of the Galaxy

Marsakov and Suchkov (1985) discussed the chemical pattern of the Hertzsprung-Russel diagram of red giants and the age of the galactic disk. They used photometric data for about 1400 disk-population red giants. The data support the view that the disk is $\sim 6 \times 10^9$ years younger than the halo, the age of which is apparently not less than 13×10^9 years. Marsakov and Suchkov (1984, 1985b) discussed also the distribution of metallicity $[Fe/H]$ and found that it cannot be described by a Gaussian. The center of the distribution of giants at the galactic poles is found to demonstrate a statistically significant gap similar to the one found earlier for the main sequence stars. Suchkov et al. (1987) found some discrepancies in characteristics of UV excesses and metallicity distributions of F-G-K dwarfs. They can be eliminated if we assume that UV excess of K and late G dwarfs and $[Fe/H]$ values from detailed analysis of F dwarfs were underestimated. Bartkevichius (1984) also discussed the metallicity distribution of metal-deficient field stars. Shchekinov (1985) discussed the abundance of deuterated molecules in the Galaxy. Shklovskij (1985) concluded that specific features of the galactic center, observed at different regions, can be explained by a permanent star formation process. Shatsova (1984) discovered traces of the torus-like structure of the Kapteyn group, which permitted an evolutionary interpretation. In the paper of Traat

(1986) spiral arms of galaxies have been modeled by a young stellar component with prolonged and symmetrical star formation rate.

14.5 Kinematics

14.5.1 Stars

Loktin (1984) discussed kinematics of red giants in the solar neighbourhood. Karimova and Pavlovskaya (1984b) studied kinematics of young objects in the Galaxy. Kharchenko (1984) dealt with kinematics of stars in the galactic plane. Barkhatova et al. (1984) discussed the distance scale of the Galaxy. Barkhatova et al. (1985) estimated the angular velocity gradient and scaling galactic parameter for open cluster systems.

Palous and Piskunov (1985) investigated the mean velocity and the velocity dispersion versus age relations for B and A stars. Rybka (1985b) used the maximum likelihood method to divide 2463 faint stars in 30 areas with galaxies into two groups at different distances. Kinematics of distant stars was shown to have departures from the Oort-Lindblad model. Khrutskaya (1984) analysed galactic rotation and the correction to the constant of precession and coordinates of the solar apex according to proper motions of bright stars. Petrovskaya (1987) discussed the structure of the neutral hydrogen subsystems in the Galaxy. Allakhverdiyev et al. (1985) investigated space kinematic characteristics of pulsars and their connection with supernova remnants.

14.5.2 Interstellar Matter

Gorbatskij and Usovich (1986) found that ringlike structures consisting of clouds must be formed in spiral galaxies due to viscosity. Shapirovskaya and Bocharov (1986ab) performed an analysis of observed and theoretically calculated characteristics of pulsar radiation scattering. The results show that hot ionized plasma may be responsible for the scattering properties of the diffuse interstellar medium. They constructed the spatial distribution of the mean electron density for the diffuse interstellar medium in the galactic disk on the basis of dispersion measures of pulsars with known distances. Dogiel et al. (1986) investigated processes of distribution of accretion and energy losses of cosmic rays in the vicinity of molecular clouds.

Eelsalu (1986) discussed formulae underlying the use of intensity profiles of the interstellar gas medium for large-scale galactic studies. Gurevich et al. (1985) investigated the mechanism of generation of fluctuating electromagnetic fields affected by neutral gas turbulence of giant molecular clouds. Dogiel et al. (1985) made an attempt to determine the mechanism which could lead to the increase in the cosmic-ray density in clouds. Suchkov and Shchekinov (1985) discussed ionization processes in the interstellar gas. Kissel'man and Frolov (1986) found a numerical solution for the size distribution function of solid charge particles in interstellar clouds. Mirzoyan and Ambaramyan (1986) discussed the connection of optical HII regions with molecular clouds of the Galaxy.

14.6 Radio Studies

Berlin et al. (1984) presented the results of the very deep cross-cut of the Galaxy at 7.6 cm wavelength. Berlin et al. (1985) also presented the results of observations of the giant HII regions situated at the longitudes $4° - 10°$, also at 7.6 cm wavelength. Pyatunina (1984, 1985, 1986) performed a survey of the galactic plane in the region of some associations carried out at 7.6 cm wavelength at the *RATAN*-600 radio telescope, and discussed the outer Galaxy and different statistics of radio sources. Vitkovskij et al. (1985) performed a radio survey in the Orion Loop region, a giant ring-like feature in the radio-continuum emission. Anisimova (1984) calculated the correlation between the stars and the continuous 240 MHz radio emission intensity distribution in the Loop I system. Abramenkov (1985) carried out decameter observations of HII regions in the galactic disk, $147° < l < 153°$. Strukov and Skulachev (1987) presented the results of an atmospheric survey of the galactic plane at the frequency of 37 GHz. Gulyaev and Sorochenko (1985) reported data on the catalog of radio recombination lines. For the radio source at the centre of the Galaxy $V_{exp} = 2 - 1 \pm 7$ km.s^{-1}. Bystrova (1985) made an additional representation of the Pulkovo sky survey results for the HII radio line.

14.7 Dynamics of Our Galaxy

14.7.1 Dynamics

The basic theory for the galactic, spatial and kinematical structure as a whole is provided by dynamics starting with the theory of stellar orbits dictated by gravitational potential and the integrals of motion (Genkin and Genkina, 1984, 1985; Ivannikova and Maksumov, 1985). The galactic gravitational potential has been modeled by Kosenko (1986), Abramyan (1986), Kutuzov and Osipkov (1986), Kondrat'ev (1984) and Danilov (1984). Various aspects of the large-scale dynamics have been discussed by Osipkov (1985) and Omarov (1985). The equations of dynamics have been analysed by Osipkov (1985), while Batt (1986) has attempted to introduce the Vlasov-Poisson equations into stellar dynamics. A special problem related to phase dynamics has been treated by Genkin and Genkina (1984). Dissipational and interactive mechanisms have been treated by a number of theoreticians: Gursadyan and Savvidin (1984, 1986), Fridman et al. (1985), and Sagintaev and Chumak (1984). Vinokurov et al. (1985) have worked at collisionless relaxation. Numerical experiments have been conducted by Zotov and Morozov (1987).

14.7.2 Disk Dynamics

If the presence of a spiral structure is postulated, its study becomes a matter of disk dynamics. Spiral density waves are consistent with the solutions of gas-dynamical equations. Because of the small dynamical role of the galactic gaseous medium, the two kinds of theories need not be mutually exclusive. The interaction of the gaseous spiral waves with stars has been studied by Korchagin (1985) and Korchagin and Ryabtsev (1987). The following authors have attacked the theory of spiral waves: Abramyan and Mikhajlova (1986), Litvintsev (1985), Korchagin and Korchagin (1984, 1985), Abramyan and Arutyunyan (1985), Moroz et al. (1985), and Korchagin and Ryabtsev (1986). The stability of disks has been studied by Zasov et al. (1985), Abramyan (1985), Morozov et al. (1985), Morozov (1985), and Bisnovatyi-Kogan and Seidov (1985). The latter two authors have studied magnetized disks, too (Bisnovatyi-Kogan and Seidov, 1985ab). Disk dynamics have been interpreted in terms of observational data by Zasov and Morozov (1985ab), Fridman (1986) and Abramyan et al. (1986). Mishurov (1984), Korchagin (1986), Grivnev (1985) and Morozov et al. (1985) have carried out numerical computations, while Nezlin et al. (1986) have made laboratory experiments. Korchagin and Prokhovnik (1985) have simulated numerically a situation where a wave pattern is imposed by a companion stellar system. Global properties of disks have been discussed by Zasov and Osipova (1987) and Morozov and Khoperskov (1984).

14.7.3 Violent Dynamics

Violent dynamics are concerned with explosive point sources such as the galactic nucleus or supernovae, but also with the nature of shock waves occurring in disk dynamics, as well as with early stages of the galactic evolutionary scenario, where the gravity is inefficient. Violent phenomena can be either local or global. The galactic nucleus as an energy source has been discussed by Illarionov and Romanova (1986), Dokuchaev and Ozernoj (1985) and Baranov (1986). The influence of a supernova explosion upon the interstellar gas medium has been described by Kovalenko and Shchekinov (1985). The effects of galactic shock waves to radio source scintillations have been described by Pimenov (1984), and their effects upon cosmic rays by Ptuskin (1986), Galeev et al. (1986) and Berezhko (1986).

Galaxy formation scenarios include studies of collapsing systems. Collapsing disks have been studied by Nuritdinov (1985) and Kolykhalov and Shandarin (1984); collapsing spheres have been studied by Malkov (1987) and Nuritdinov (1985). Osipkov (1985) and Nezhinski and Osipkov (1987) have discussed violent relaxation. Concepts of thermodynamic stabilization have been developed by Suchkov et al. (1985), Terletskij (1984) and Tsitsin and Simentsov (1984). A pulsation model for stellar systems has been outlined by Malkov (1984). Korchagin and Ryabtsev (1986) have modeled the star formation process.

14.8 X-rays, Cosmic-rays, Gamma-rays, Magnetic Fields

Bochkarev (1987) considered the structure of the local interstellar medium and sources of the soft X-ray background radiation. Ginzburg and Ptuskin (1986) reviewed the cosmic-ray origin problem. Kuznetsov (1986) discussed the influence of cosmic rays on the stability and large-scale dynamics of the interstellar medium. According to Gursadyan (1985), stellar aggregates may be powerful sources of cosmic rays. Fomin et al. (1985) searched super-high-energy γ-rays from various objects and regions of the Galaxy. Agaronyan et al. (1985) analyzed the ultra-high energy γ-ray absorption on the microwave background radiation. Mukhanov and Fomin (1986) observed the galactic disk at ultra-high energies. Dogiel and Uryson (1986) calculated the distribution of relativistic protons in the Galaxy. Shapirovskaya and Bochkarev (1985) discussed the distribution of electron density and scattering inhomogeneities over Galaxy. Seitnepesov and Khanberdiev (1985) discussed the structure of the large-scale magnetic field of the Galaxy. Radio background intensity variations and the structure of the galactic magnetic field has been studied by Dogkemansky and Shoutenkov (1987).

References

Abramenkov, E. A., 1985, *Sov. Astron.*, **29**, 616

Abramyan, M. G., Arutyunyan, S. V., 1985, *Astron. Zh.*, **62**, 871

Abramyan, M. G., Mikhajlova, E. A., Morozov, A. G., 1986, *Astrophysics*, **24**, 99

Abramyan, M. G., Mikhajlova, E. A., Morozov, A. G., 1986, *Astrofizika*, **24**, 167

Abramyan, M. G., 1969, *Astrofizika*, **25**, 173

Abramyan, M. G., 1985, *Astrofizika*, **22**, 487

Agaronyan, F. A., Mamidzhanyan, E. A., Nikol'skij, S. I., 1985, *Astrofizika*, **23**, 55

Agekyan, T. A. (ed.), 1985, in: *Stellar Astronomy*, Vsesoyuznyj Institut Nauchnoj i Tekhnicheskoj Informatsii, Moskva. 156 pp. (Itogi Nauki i Tekhniki, Seriya Astronomiya. Tom 26)

Alksnis, A., 1985, in: *Stellar catalogues: data compilation, analysis, scientific results*, ed. E. K. Kharadze, N. G. Kogoshvili, 81 (Proc. International Colloq., Abastumanskaya Astrofiz. Obs. Byull. No. 59)

Allakhverdiyev, A. O., Guseinov, O. H., Kasumov, F. K., 1985, *Astrophys. Space Sci.*, **115**, 1

Amnuel, P. R., Guseinov, O. H., Sustamov, Yu., 1986, *Astrophys. Space Sci.*, **121**, 1

Andruk, V. N., Kharchenko, N. V., 1987, *Kinematika Fiz. Nebesn. Tel*, **3**, 76

Anisimova, G. B., 1984, *Astron. Zh.*, **61**, 1226

Avedisova, V. S., Kondratenko, G. I., 1984, *Nauchn. Inf. Astron. Sov. AN SSSR*, **56**, 59

Avedisova, V. S., 1985, *Pis'ma Astron. Zh.*, **11**, 898

Avedisova, V. S., 1985, *Bull. Inf. Cent. Donnees Stellaires*, **29**, 79

Avedisova, V. S., 1985, *Pis'ma Astron. Zh.*, **11**, 448

Bakrunov, A. O., Sladkov, O. S., Shchukin, I. V., 1986, *Astrometriya*, **2**, 90

Baranov, A. S., 1986, *Astron. Zh.*, **63**, 220

Barkhatova, K. A., Gerasimenko, T. P., Blum, M. E., Lukhanov, K. B., 1985, in: *Stellar catalogues: data compilation, analysis, scientific results*, ed. E. K. Kharadze, N. G. Kogoshvili, 169 (Proc. International Colloq., Abastumanskaya Astrofiz. Obs. Byull. No. 59)

Barkhatova, K. A., Pyl'skaya, O. P., Seleznev, A. F., 1984, in: *Problems of Astrometry*, ed. V. V. Podobed, Izdatel'stvo Moskovskogo Universiteta, Moskva, 275 (Proc. of the 22nd Astrometric Conf. of the USSR)

Bartasiute, S., 1984, *Vilniaus Astron. Obs. Biul.*, **68**, 33

Bartasiute, S., 1985, *Vilniaus Astron. Obs. Biul.*, **69**, 24

Bartasiute, S., 1986, *Vilniaus Astron. Obs. Biul.*, **74**, 15

Bartasiute, S., 1987, *Pis'ma Astron. Zh.*, **13**, 393

Bartkevichius, A., 1986, *Vilniaus Astron. Obs. Biul.*, **74**, 55

Bartkevichius, A., 1984, *Vilniaus Astron. Obs. Biul.*, **66**, 89

Bartkevichius, A., 1985, in: *Stellar catalogues: data compilation, analysis, scientific results*, ed. E. K. Kharadze, N. G. Kogoshvili, 75 (Proc. International Colloq., Abastumanskaya Astrofiz. Obs. Byull. No. 59)

Bartkevichius, A., 1983, *Astron. Tsirk.*, **1289**, 5

Bartkevichius, A., 1984, *Vilniaus Astron. Obs. Biul.*, **8**, 3

Basharina, T. S., Pavlovskaya, E. D., Filippova, A. A., 1985, *Sov. Astron.*, **29**, 17

Batt, Yu., 1986, in: Nonlinear system of Vlasov-Poisson equations with partial derivatives in: *stellar dynamics Novosibirsk*, **47** (Tr. Mezhdunar. konf. po differ. uravneniyam schast. proizvodnymi).

Berezhko, E. G., 1986, *Pis'ma Astron. Zh.*, **12**, 842

Berlin, A. B., Bulaenko, E. V., Vitkovskij, V. V., Kononov, V. K., Korol'kov, D. V., Pariiskij, Yu.N., Trushkin, S. A., 1984, *Soobch. Spets. Astrofiz. Obs.*, **43**, 43

Berlin, A. B., Golnev, V. Ya., Lipovka, N. M., Nizhel'skij, N. A., Spangenberg, E. E., 1985, *Astron. Zh.*, **62**, 229

Bisnovatyi-Kogan, G. S., Seidov, Z. F., 1985, *Astrophys. Space Sci.*, **115**, 275

Bisnovatyi-Kogan, G. S., Seidov, Z. F., 1985, *Inst. Kosm. Issled. Akad. Nauk SSSR, Prepr.*, **994** 31 pp.

Bisnovatyi-Kogan, G. S., Seidov, Z. F., 1985, *Pis'ma Astron. Zh.*, **11**, 395

Bochkarev, N. G., 1985, Magnetic fields in space, Glavnaya Redaktsya Fiziko-Matematichskoj Literatury, Nauka, Moskva, 208 pp. (Seriya "Problemy Nauki i Tekhnicheskogo Progressa")

Bochkarev, N. G., 1987, *Astron. Zh.*, **64**, 38

Bogyavlensky, O. I., 1985, Methods in the qualitative theory of dynamical systems in: *astrophysics and dynamics Springer Verlag*, Berlin-Heidelberg-New York-Tokyo, 301 pp. (Springer Series in Soviet Mathematics)

Bultabaev, Yu., Rakhimov, A. G., Umarova, K.: 1985, *The Catalogue of photometric and spectral data of 4221 stars obtained from an identification of 17 selected Pulkovo areas with the Bergedorf Survey*, Astron. inst. Akad. Nauk USSR, Tashkent, 163 pp.

Bystrova, N. V.L: 1985, *Bull. Spec. Astrophys. Obs.*, **18**, 78

Cernins, K.: 1986, *Vilniaus Astron. Obs. Biul.*, **75**, 31

Dagkesmansky, R. D., Shoutenkov, V. R., 1987, *Pis'ma Astron. Zh.*, **13**, 182

Danilov, V. M., 1984, *in Vopr. astrofiz. Saransk*, **41**

Dluzhnevskaya, O. B., Myakutin, V. I., 1985, in: *Stellar catalogues: data compilation, analysis, scientific results*, ed. E. K. Kharadze, N. G. Kogoshvili, 155 (Proc. International Colloq., Abastumanskaya Astrofiz. Obs. Byull. No. 59)

Dogel', V. A., Uryson, A. V., 1986, *Pis'ma Astron. Zh.*, **12**, 831

Dogiel, V. A., Gurevich, A. V., Istomin, Ya.N., Sharov, G. S., Zybin, K. P.: 1986, in: *Plasma Astrophysics* ed. T. D. Guyenne, L. M. Zeleny, 287 A Spec. Publ., ESA SP–251)

Dogiel, V. A., Gurevich, V. V., Istomin, Ya. N., Zybin, K. P., 1985, *19th International Cosmic Ray Conference. Vol. 9, OG sessions: Cosmic ray and Gamma-ray Origin and Galactic phenomena*, ed. F. C. Jones, J. Adams, G. M. Mason, 195 (NASA Conf. Publ., NASA CP–2376)

Dokuchaev, V. I., Ozernoj, L. M., 1985, *Pis'ma Astron. Zh.*, **11**, 335

Dzervitis, U., Paupers, O., 1985, *Issled. Solntsa Krasnykh Zvezd*, **23**, 43

Eelsalu, H., 1986, *Tartu Astrofuus. Obs. Pub.*, **51**, 91

Eglitis,I., 1986, *Issled. Solntsa Krasnykh Zvezd*, **24**, 39

Einasto,J. E., Malyuto, V. D., Kharchenko, N. V., 1985, *Astron. Tsirk.*, **1394**, 1

Fomin, Yu. A., Khristiansen, G. B., Kulikov, G. V., Nazarov, V. L., Silaev, A. A., Solovyeva, V. I., Trubitsyn, A. V., 1985, *19th International Cosmic Ray Conference. Vol. 1, OG sessions: Cosmic ray and gamma-ray Origin and Galactic phenomena*, ed. F. C. Jones, J. Adams, G. M. Mason, 259 (NASA Conf. Publ., NASA CP–2376)

Fridman, A. M., Morozov, A. G., Palous, J., Piskunov, A., Polyachenko, V. L., 1985, in: *The Milky Way Galaxy* ed. H. van Woerden, R. J. Allen, W. B. Burton, D. Reidel Publishing Company, Dordrecht-Boston-Lancaster, 509 (Proc. IAU Symp. No. 106)

Fridman, A. M., Polyachenko, V. L., 1984a, *Physics of gravitating systems* , I. Equilibrium and stability Springer Verlag, New York-Heidelberg-Tokyo, 468 pp.

Fridman, A. M., Polyachenko, V. L., 1984b, *Physics of gravitating systems* , II. Nonlinear collective processes: nonlinear waves, solitons, collisionless shocks, turbulence. Astrophysical application Springer Verlag, New York-Heidelberg-Tokyo, 358 pp.

Fridman, A. M., 1986, *Astron. Zh.*, **63**, 884

Galeev, A. A., Sagdeev, R. Z., Shapiro, V. D., 1986, *Plasma astrophysics*, ed. T. D. Guyenne, L. M. Zeleny, 297, (ESA Spec. Publ., ESA SP–251)

Garibdzhanyan, A. T., Gasparyan, K. G., Oganesyan, R. Kh., 1984, *Astrofizika*, **20**, 245

Genkin, I. L., Genkina, L. M., 1984, *Tr. Astrofiz. Inst. Alma-Ata*, **43**, 82

Genkin, I. L., Genkina, L. M., 1984/85a, in: *Problems in field theory Kazakh.Univ.*, Alma-Ata, 42 (Collected papers)

Genkin, I. L., Genkina, L. M., 1984/85b, in: *Problems in field theory Kazakh. Univ.*, Alma-Ata, 53 (Collected papers)

Ginzburg, V. L., Ptuskin, V. S., 1986, in: *Plasma astrophysics* ed. T. D. Guyenne, L. M. Zeleny, 317 (ESA Spec. Publ., ESA SP–251)

Glushneva, I. N., 1985, *Sov. Astron.*, bf 29, 659

Glushneva, I. N., 1985, in: *Stellar catalogues data compilation, analysis, scientific results* ed. E. K. Kharadze, N. G. Kogoshvili, 61 (Proc. International Colloq., Abastumans-kaya Astrofiz. Obs. Byull. No. 59)

Gorbatskij, V. G., Usovich, I., 1986, *Astrofizika*, **25**, 125

Grivnev, E. M., 1985, *Astron. Zh.*, bf 62, 681

Gulyaev,S. A., Sorochenko, R. L., 1985, in: *Stellar catalogues data compilation, analysis, scientific results*, ed. E. K. Kharadze, N. G. Kogoshvili, 135 (Proc. International Colloq., Abastumanskaya Astrofiz. Obs. Byull. No. 59)

Gurevich, A. V., Dogiel, V. A., Zybin, K. P., Istomin, Ya.N., 1985, *Sov. Astron. Lett.*, **11**, 284

Gurzadyan, G. A., 1985, *Astrophys. Space Sci.*, **113**, 213

Gurzadyan, V. G., Savvidin, G. K., 1984, *Dokl. Akad. Nauk SSSR. Ser. Mat. Fiz.*, bf 277, 69

Gurzadyan, V. G., Savvidin, G. K., 1986, *Astron. Astrophys.*, **160**, 203

Gurzadyan, V. G., 1984, in: *Particles and cosmology. Part 2. Grand unified theories. Gauge theories.* Institut yadernykh issledovanij Akademii Nauk SSSR. Moskva. 43

Gusejnov, O. Kh., Yusifov, I. M., 1986a, *Astron. Zh.*, bf 63, 78

Gusejnov, O. Kh., Yusifov, I. M., 1986b, *Astron. Zh.*, bf 63, 265

Guseva, N. G., Metreveli, M. D., 1986, *Kinematika Fiz. Nebesn. Tel*, **2**, 3

Guseva, N. G., Kolesnik, G., Kravchuk, S. G., 1984, *Pis'ma Astron. Zh.*, **10**, 741

Guseva, N. G., 1986, *Kinematika Fiz. Nebesn. Tel*, **2**, 22

Gvaramadze, V. V., Lominadze, J. G., 1986, in: *Plasma astrophysics* ed. T. D. Guyenne, L. M. Zeleny, 551 (ESA Spec. Publ., ESA SP-251)

Haud, U., 1984, *Astrophys. Space Sci.*, **104**, 337

Illarionov, A. F., Romanova, M. M., 1986, *Inst. Kosm. Issled. Akad.* Nauk SSSR, Prepr. 1088, 46 pp.

Ivannikova, E. I., Maksumov, M. N., 1985, *Dokl. Akad. Nauk Tadz. SSR*, **28**, 631

Janulis, R., 1986, *Vilniaus Astron. Obs. Biul.*, **75**, 8

Jasevichius, V., 1986, *Vilniaus Astron. Obs. Biul.*, **74**, 41

Korchagin, V. I., 1986, *Astrofizika*, **25**, 149

Kalandadze, N. B., Kuznetsov, V. I., Voroshilov, V. I., 1986, *Kinematika Fiz. Nebesn. Tel.*, **2**, 27

Kardashev, N. S., 1985, *Astrophys. Space Phys. Rev.* **4**, 287

Karimova, D. K., Pavlovskaya, E. D., 1985a, *Tr. Gos. Astron. Inst. Shternberg*, **57**, 245

Karimova, D. K., Pavlovskaya, E. D., 1984b, in: *Problems of astrometry*, ed. V. V. Podobed, Izdatel'stvo Moskovskogo Universiteta, Moskva, 280, (Proc. of the 22nd Astrometric Conference of the USSR)

Kasak, E., 1986, *Tartu Astrofuus. Obs. Publ.*, **51**, 111

Kazanasmas, M. S., Mis'kin, N. A., Tomak, L. F., Zavershneva, L. A., 1984, Catalogue of stellar magnitudes and colour indices of stars in: *the Kapteyn 77 area*, Odes. univ. Odessa, 20 pp.

Kazanasmas, M. S., Mis'kin, N. A., Tomak, L. F., Zavershneva, L. A., 1984, Spectra and Luminosities of stars in: *the Andromeda-Perseus region*, Odes. univ. Odessa, 28 pp.

Kazanasmas, M. S., Zavershneva, L. A., Tomak, L. F., 1985, Catalogue of Stellar magnitudes and colour indices of stars in: *the direction of six selected Kapteyn areas*, Odes. univ. Odessa, 18 pp.

Kazanasmas, M. S., Zavershneva, Tomak, L. F., 1986, *Perem. Zvezdy*, **22**, 443

Kazlauskas, A., Jasevichius, V., 1986, *Vilniaus Astron. Obs. Biul.*, **75**, 18

Kessel'man, V. S., Frolov, A. B., 1986, *Astron. Zh.*, **63**, 476

Kharadze, E. K., Bartaya, A., 1985, in: *Stellar catalogues data compilation, analysis, scientific results* ed. E. K. Kharadze, N. G. Kogoshvili, **91**, (Proc. International Colloq., Abastumanskaya Astrofiz. Obs. Byull. No. 59)

Kharadze, Kogoshvili (ed.), 1985, *Stellar catalogues data compilation, analysis, scientific results*, 260 pp. (Proc. International Colloq., Abastumanskaya Astrofiz. Obs. Byull. No. 59)

Kharchenko, N. V., 1987, *Kinematika Fiz. Nebesn. Tel*, **3**, 63

Kharchenko, N. V., 1984, in: *Problems of astrometry* ed. V. V. Podobed, Izdatel'stvo Moskovskogo Universiteta, Moskva, 284, (Proc. of the 22nd Astrometric Conf. of the USSR)

Khrutskaya, E. V., 1984, Galactic rotation, correction to the constant of precession and coordinates of the solar apex according to proper motions of bright stars" Glav. Astron. Obs. Akad. Nauk SSSR, Leningrad, 26 pp.

Kizla, J., Paupers, O., 1985, *Issled. Solntsa Krasnykh Zvezd*, **23**, 38

Kolesnik, I. G., Yurevich, L. V., 1985, *Astrofizika*, **22**, 461

Kolesnik, L. N., Metreveli, M. D., 1985, in: *Stellar catalogues* data compilation, analysis, scientific results ed. E. K. Kharadze, N. G. Kogoshvili, 203 (Proc. International Colloq., Abastumanskaya Astrofiz. Obs. Byull. No. 59)

Kolesnik, L. N., Metreveli, M. D., 1985b, *Kinematika Fiz. Nebesn. Tel*, **1**, 53

Kolesnik, L. N., 1986, *Astron. Astrophys.*, **169**, 268

Kolykhalov, P. I., Shandarin, S. F., 1984, *Pis'ma Astron. Zh.*, **10**, 820

Kondrat'ev, B. P., 1984, *Astrofizika*, **21**, 499

Kopylov, I. M., 1985, *Sov. Astron.*, **29**, 200

Korchagin, V. I., Korchagin, P. I., 1984, *Astron. Zh.*, **61**, 814

Korchagin, V. I., Korchagin, P. I., 1985, *Astron. Zh.*, **62**, 202

Korchagin, V. I., Prokhovnik, N. A., 1985, *Astrofizika*, **23**, 237

Korchagin, V. I., Ryabtsev, A. D., 1987, *Astron. Zh.*, **64**, 31

Korchagin, V. I., Ryabtsev, 1986, *Astrophys. Space Sci.*, **126**, 1

Korchagin, V. I., 1985, *Astrofizika*, **23**, 91

Kosenko, I. I., 1986, *Prikl. mat. i mekh.*, **50**, 194

Kovalenko, I. G., Sechekinov, Yu. A., 1985, *Astrofizika*, **23**, 363

Kutuzov, S. A., Osipkov, L. P., 1986, *Astrofizika*, **25**, 545

Kuzmin, G., Malyuto, V., Eelsalu, H., 1986, *Tartu Teated*, **83**, 4

Kuznetsov, V. I., 1984, *Astrometr. Astrofiz.*, **53**, 34

Kuznetsov, V. D., 1986, *Astron. Zh.*, **63**, 446

Lebedev, V. S., Lebedeva, I. A., 1985, in: *Stellar catalogues data compilation, analysis, scientific results*, ed. E. K. Kharadze, N. G. Kogoshvili, 225 (Proc. International Colloq., Abastumanskaya Astrofiz. Obs. Byull. No. 59)

Lebedev, V. S., 1986, *Astrofiz. Issled. Izv. Spets. Astrofiz. Obs.*, **21**, 21

Lebedev, V. S., 1986, *Bull Spec. Astrophys. Obs.*, bf 21, 28

Levina, N. F., 1985, *Kinematika Fiz. Nebesn. Tel 1*, 80

Litvintsev, S. I., 1985, *Byull. Inst. Astrofiz.*, 8

Loktin, A. V., 1984, in: *Problems of astrometry* ed. V. V. Podobed, Izdatel'stvo Moskovskogo Universiteta, Moskva, 280 (Proc. of the 22nd Astrometric Conf. of the USSR)

Lozinskaya, T. A., 1986, *Supernova and stellar gas of the Galaxy*, Glavnaya redaktsiya fiziko-matematicheskoj literatury, Nauka, Moskva, 304 pp.

Malkov, E. A., 1986, *Astrofizika*, **24**, 416

Malkov, E. A., 1986, *Astrofizika*, **24**, 377

Malkov, O. Yu., 1987, *Nauchnye Inf. Astron. Sov.*, AN SSSR **63**, 19

Malyuto, V., 1986, *Tartu Astrofuus. Obs. Teated.*, **83**, 11

Malyuto, V. D., Shvelidze, T., 1985, in: *Stellar catalogues data compilation, analysis, scientific results* ed. E. K. Kharadze, N. G. Kogoshvili, **217** (Proc. International Colloq., Abastumanskaya Astrofiz. Obs. Byull. No. 59)

Marochnik, L. S., Suchkov, A. A., 1984, *The Galaxy Nauka*, Moskva, 392 pp.

Marsakov, V. A., Suchkov, A. A., 1984, *Astron. Tsirk.*, **1297**, 1

Marsakov, V. A., Suchkov, A. A., 1985a, *Sov. Astron.*, **29**, 403

Marsakov, V. A., Suchkov, A. A., 1985b, *Astron. Zh.*, **62**, 847

Melik-Alaverdyan, Yu.K., Tovmasyan, G. G., 1986, *Astrofizika*, **25**, 73

Mdzinarishvili, T. G., 1985, in: *Stellar catalogues data compilation, analysis, scientific results*, ed. E. K. Kharadze, N. G. Kogoshvili, **117** (Proc. International Colloq., Abastumanskaya Astrofiz. Obs. Byull. No. 59)

Mirzoyan, L. V., Ambaryan, V. V., 1986, *Astrofizika*, **24**, 257

Mishenina, T. V., Panchuk, V. E., Komarov, N. S., 1986, *Astrofiz. Issled. Izv. Spets. Astrofiz. Obs.*, **22**, 13

Mishenina, T. V., Panchuk, V. E., 1986, *Astrofiz. Issled. Izv. Spets. Astrofiz. Obs.*, **21**, 12

Mishurov, Yu.N., 1984, *Astron. Zh.*, **61**, 1074

Moroz, A. G., Nezlin, M. V., Shezhkin, E. N., Torgashin, Yu.M., Fridman, A. M., 1985, *Astron. Tsirk.*, **1414**, 7

Morozov, A. G., Khoperskov, A. V., 1986, *Astrofizika*, **24**, 467

Morozov, A. G., Nezlin, M. V., Snezhkin, E. N., Torgashin, Yu.M., Fridman, A. M., 1985, *Astron. Tsirk.*, **1414**, 1

Morozov, A. G., Nezlin, M. V., Snezhkin, E. N., Torgashin, Yu.M., Fridman, A. M., 1985, *Astron. Tsirk.*, **1414**, 4

Morozov, A. G., 1985, *Astron. Zh.*, **62**, 209

Mukanov, D. B., Fomin, V. P., 1986, *Bull. Crimean Astrophys. Obs.*, **69**, 60

Myakutin, V. I., 1984, *Nauchn. Inf.*, **56**, 52

Nezhinskij, E. M., Osipkov, L. P., 1987, *Astron. Tsirk.*, **1474**, 1

Nezlin, M. V., Polyachenko, V. L., Snezhkin, E. N., Trubnikov, A. S., Fridman, A. M., 1986, *Pis'ma Astron. Zh.*, **12**, 504

Nikolashvili, M. G., 1987, *Astrofizika*, **26**, 209

Nuritdinov, S. N., 1985, *Astron. Zh.*, **62**, 506

Nuritdinov, S. N., 1985, *Pis'ma Astron. Zh.*, **11**, 89

Omarov, T. B., 1985, *Astrofizika*, **23**, 77

Osipkov, L. P., 1985, *Astron. Tsirk.*, **1399**, 1

Osipkov, L. P., 1985, *Astron. Tsirk.*, **1399**, 3

Osipkov, L. P., 1985, *Astron. Tsirk.*, **1359**, 7

Osipkov, L. P., 1985, in: *Astronomical-geodetical investigations Sverdlovsk*, 9, (Collection of papers of the Ural University)

Ozernoj, L. M., 1984a, *Astron. Tsirk.*, **1349**, 1

Ozernoj, L. M., 1984b, *Astron. Tsirk.*, 1

Ozernoj, L. M., 1986, *Ann. N. Y. Acad. Sci.*, **470**, 385

Palous, J, Piskunov, A. E., 1985, *Astron. Astrophys.*, **143**, 102

Pavlovskaya, E. D., Suchkov, A. A., 1984, *Astron. Zh.*, **61**, 665

Petrovskaya, I. V., Teerikorpi, P., 1986, *Prepr. Ser. Dep. Phys. Sci.* Univ. Turku N. O. FTL-R94, 13 pp.

Petrovskaya, I. V., 1985, in: *Itogi Nauki i Tekhniki. Seriya Astronomiya.* Tom 26. Stellar astronomy. ed. T. A. Agekyan, Vsesoyuznyj Institut Nauchnoj i Tekhnicheskoj Informatsii, Moskva, 113

Petrovskaya, I. V., 1986, *Vestn. Leningr. Univ. Mat. Mekh. Astron.*, **1**, 126

Petrovskaya, I. V., 1987, *Pis'ma Astron. Zh.*, **13**, 474

Pimenov, S. F., 1984, *Pis'ma Astron. Zh.*, **10**, 523

PodobedV. V. (ed.), 1984, *Problems of astrometry* Izdatel'stvo Moskovskogo Universiteta, Moskva, 341 pp. (Proc. 22nd Astrometric Conference of the USSR)

Polyachenko, V. L., 1985, *Astron. Tsirk.*, **1405**, 4

Ptuskin, V. S., 1986, in: *Plasma astrophysics* ed. T. D. Guyenne, L. M. Zeleny, 279 (ESA Spec. Publ., ESA SP–251).

Pugach, A. F., Voroshilov, V. I., Guseva, N. G., Kalandadze, N. G., Kolesnik, L. N., Kuznetsov, V. I., Metreveli, M. D., Shapotalov, A. N., 1985, *Catalogue of BV-magnitudes and spectral classes of 6000 stars*, Naukova Dumka, Kiev, 140 pp.

Pyatunina, T. B., 1984, *Soobch. Spets. Astrofiz. Obs.*, **43**, 63

Pyatunina, T. B., 1985, *Astron. Zh.*, **62**, 218

Pyatunina, T. B., 1986, it *Pis'ma Astron. Zh.*, **12**, 353

Rybka, S. P., 1985a, *Kinematika Fiz. Nebesn. Tel*, **1**, 37

Rybka, S. P., 1985b, *Kinematika Fiz. Nebesn. Tel*, **1**, 88

Sagintaev, B. S., Chumak, V., 1984, *Tr. Astrofiz. Inst. Alma-Ata*, **43**, 51

Salukvadze, G. N., 1985, in: *Stellar catalogues data compilation, analysis, scientific results*, ed. E. K. Kharadze, N. G. Kogoshvili, 149 (Proc. International Colloq., Abastumanskaya Astrofiz. Obs. Byull. No. 59)

Seitnepesov, Ch. N., Khanberdiev, A. Kh., 1985a, *Astrofizika*, **22**, 293

Seitnepesov, Ch., Khanberdiev, A. Kh., 1985b, *Izv. Akad. Nauk TSSR, Ser. Fiz.tekhn.*, geol. nauk, **3**, 14

Seitnepesov, Ch., Khanberdiev, A., 1984, *Izv. Akad. Nauk TSSR, Ser. fiz. tekh. i geol. nauk*, **5**, 27

Shapirovskaya, N. Ya., Bocharov, A. A., 1984, Inst. Kosm. Issled. Akad. Nauk SSSR. Prepr., **936**, 44

Shapirovskaya, N. Ya., Bocharov, A. A., 1986a, *Astron. Zh.*, **63**, 1111

Shapirovskaya, N. Ya., Bocharov, A. A., 1986b, *Astron. Zh.*, **63**, 666

Shatsova, R. B., 1984, Nauchn. Inf. *Astron. Sov. AN SSSR*, **56**, 104

Shchekinov, Yu.A., 1985, *Astron. Zh.*, **62**, 182

Shklovskij, I. S., 1985, *Pis'ma Astron. Zh.*, **11**, 163

Shtern, B. E., 1984, in: *Particles and cosmology. Part 2. Grand unified theories. Gauge theory* Institut yadernykh issledovanij Akademii Nauk SSSR, Moskva, 32

Sil'chenko, O. K., 1984, *Sov. Astron. Lett.*, **10**, 209

Sleivyte, J., 1986, *Vilniaus Astron. Obs. Biul.*, **75**, 36

Sleivyte, J., 1986, *Vilniaus Astron. Obs. Biul.*, **74**, 24

Smriglio, F., Boyle, R. P., Straizys, V., Janulis, R., Nandy, K., MacGillivray, H. T., McLachlan, A., Coluzzi, R., Segato, C., 1986, *Astron. Astrophys. Suppl. Ser.*, **66**, 181

Straizys, V., Cernins, K., Vansevicius, V., Janulis, R., Jasevichius, V., Tautvaisiene, G., Zdanavicius, K., 1986, *Vilniaus Astron. Obs. Biul.*, **75**, 3

Straizys, V., 1985, in: *Calibration of fundamental stellar quantities* ed. D. S. Hayes, L. E. Pasinetti, A. G.D. Philip, D. Reidel Publishing Comany, Dordrecht-Boston-Lancaster-Tokyo, **285**, (Proc. IAU Symp. No. 111).

Strukov,,I. A., Skulachev, 1987, *Pis'ma Astron. Zh.*, **13**, 469

Suchkov, A. A., 1986, *Astron. Tsirk.*, **1421**, 1

Suchkov, A. A., Kovalenko, I. G., Shchekinov, Yu.A., 1985, *Astron. Tsirk.*, **1390**, 1

Suchkov, A. A., Marsakov, V. A., Shevelev, Yu.G., 1987, *Astron. Zh.*, **64**, 586

Suchkov, A. A., Shchekinov, Yu.A., 1985, *Izv. Sev.Kavkaz. nauchn. tsentra vyssh. shk. Estestv. nauk*, **1**, 41

Terletskij, Ya.P., 1984, in: *Probl. kvant. i stat. fiz.*, 3

Tokhtas'ev, S. S., 1985, *Izv. Astron. Ehngel'dordt*, **51**, 60

Traat, P., 1986, *Tartu Astrofuus. Obs. Publ.*, **51**, 181

Tsitsin, F. A., Simentsov, V. A., 1984, *Astron. Tsirk.*, **6**

Uranova, T. A., 1985, *Pis'ma Astron. Zh.*, **11**, 251

Vereshchagin, S. V., Piskunov, A. Eh., 1984, Nauchn. Inf. *Astron. Sov. AN SSSR*, **57**, 76

Vinokurov, I., Kats, V., Kontorovich, V. M., 1985, *J. Stat. Phys.*, **38**, 217

Vitkovskij, V. V., Nachel'skij, N. A., Trushkin, S. A., 1985, *Soobch. Spets. Astrofiz. Obs.*, **48**, 61

Vladimirskij, B. M., 1985, *Izv. Krymskoj Astrofiz. Obs.*, **72**, 125

Yudaeva, N. A., 1985, *Soobch. Spets. Astrofiz. Obs.* **46**, 77

Yurevich, L. V., 1985, Astrofizika, **23**, 265

Zakhozhaj, V. A., 1984, in: *Problems of astrometry* ed. V. V. Podobed, Izdatel'stvo Moskovskogo Universiteta, Moskva, 278, (Proc. of the 22nd Astrometric Conf. of the USSR)

Zasov, A. V., Osipova, T. A., 1987, *Pis'ma Astron. Zh.*, **13**, 174

Zasov, A. V., Morozov, A. G., 1985, it Astron. Tsirk. **1356**, 1

Zasov, A. V., Morozov, A. G., 1985, *Sov. Astron.*, **29**, 277

Zavarzin, Yu.M., 1984, *Astron. Tsirk.*, **1313**, 4

Zdanavicius, K., Cerniene, E., 1985, *Vilniaus Astron. Obs. Biul.*, **69**, 3

Zdanavicius, K., 1986, *Vilniaus Astron. Obs. Biul.*, **74**, 3

Zlakomanova, I., 1985, *Issled. Solntsa Krasnykh Zvezd*, **23**, 19

Zotov, V. M., Morozov, A. G., 1987, *Pis'ma Astron. Zh.*, **13**, 333

Zvereva, A., Severny, A., Granitsky, L., Courtes, G., Cruvelier, P., Hua, C. T., 1985, *The Milky Way Galaxy*, ed. H. van Woerden, R. J. Allen, W. B. Burton, D. Reidel Publishing Company, Dordrecht-Boston-Lancaster, **237**, (Proc. IAU Symp. No. 106)

34. INTERSTELLAR MATTER

(MATIERE INTERSTELLAIRE)

PRESIDENT: J. Lequeux
VICE-PRESIDENT: J.S. Mathis
ORGANIZING COMMITTEE: K.S. de Boer, S. D'Odorico, B.G. Elmegreen, D. Flower,
H. Habing, M. Peimbert, P.A. Shaver, B. Shustov, P.G. Wannier, D.G. York.

I. Introduction
(J. Lequeux)

The previous report started with optimistic remarks about the increasing importance of the study of interstellar matter in astronomy. This trend has largely
been confirmed in the 1985-87 period and it is clear that the subject of our Commission is one of the most active fields of astronomical research. This is also
shown by the rapidly growing number of members and by the constitution of new
working groups. The major new event in the period has undoubtly been the availability of IRAS data.

The present report covers the period mid-84 to mid-87 and is divided in self-
contained sections - A new section concerns the intergalactic medium - References
are given by commonly used abbreviations (see the previous reports) and in order
to save place only the name of the first author is cited, followed by a + sign if
several authors are involved. It should be noted that due to the space allotted
for this report it was not possible to cover all the papers relevant to the subject of our Commission during this period, and to do full justice even to the most
important papers. A detailed list of articles on the physics of the interstellar
medium carried out in the Soviet Union has been prepared by B. Shustov. Only a
fraction is discussed in the present report; copies of the list are available from
the president of Commission 34.

The presentation of this report was made possible by the reviewing efforts
of its writers, mostly members of the Organizing Committee or presidents of Working Groups, and by the cooperation of those of the members of Commission 34 who
submitted relevant information. I wish to thank especially Dr. T. Landecker who
accepted to write on short notice the chapter on Supernova Remnants. Dr. D. Flower
wishes to acknowledge the hospitality of the Bordeaux Observatory where he could
write his part. We list below a selection of books, conference proceedings, catalogues and review articles published in the period covered. When cited again in
the report, these books, conference proceedings or reviews are only referred to
by the year and the underlined name of the first author or editor. Many others,
in particular the more specialized conference proceedings and review papers, are
cited at the beginning of the various sections of the report. More information can
also be found in reports by Commissions whose fields overlap with that of Commission
34, e.g. Comm. 14, 28, 40 and 47.

1. MONOGRAPHS AND OTHER BOOKS

Aller,L.H.: 1984,"Physics of Thermal Gaseous Nebulae", Reidel,Dordrecht
Bowers,R.L.+: 1984,"Astrophysics.Vol.II. Interstellar Matter and Galaxies",
    Jones and Bartlet, Boston
Gurevich,L.E., Chernin,l.D.: 1987,"Origin of Galaxies and Stars". Nauka, Moscow,
    2d edition (in Russian)
Khromov,G.S.: 1985,"Planetary Nebulae: Physics, Evolution, Cosmogony". Nauka,
    Moscow (in Russian)

Kitchin,C.R.: 1987,"Stars,Nebulae and the Interstellar Medium",Adam Hilger,Bristol
Kondo,Y.+,ed.: 1987,"Exploring the Universe with the IUE Satellite", Reidel,
    Dordrecht (many review papers)
Lozinskaya,T.A.: 1986,"Supernovae and Stellar Winds: Interaction with the Galactic
    Gas". Nauka, Moscow.(in Russian)
Marochnik,L.S.+: 1984,"The Galaxy", Nauka, Moscow.(in Russian)
Pottasch,S.R.: 1984,"Planetary Nebulae. A Study of Late Stages of Stellar
    Evolution", Reidel, Dordrecht

2. SYMPOSIUM REPORTS AND CONFERENCE PROCEEDINGS

Black,D.C.+,ed.: 1985,"Protostars and Planets II", Univ. of Arizona Press, Tucson
Bochkarev,N.G.+: 1985,"Physics of ISM and Nebulae", Astr.Zh.$\underline{62}$,1234
Bochkarev,N.G.: 1986,"Systematization of Atomic Data for Astrophysics", Astr.Zh.
    $\underline{63}$, 823
Boland,W.+,ed.: 1985,"Birth and Evolution of Massive Stars and Stellar Systems",
    Reidel, Dordrecht
Bregman,J.N.+,ed.: 1986,"Gaseous Halos of Galaxies", NRAO, Charlottesville
Chiosi,C.+,ed.: 1986,"Spectral Evolution of Galaxies", Reidel, Dordrecht
Danziger,J.,ed.: 1985,"Production and Distribution of CNO Elements",ESO, Garching
Diercksen,G.H.F.,ed.: 1985,"Molecular Astrophysics. State of the Art and Future
    Directions", Reidel, Dordrecht
Faber,S.M.,ed.: 1987,"Nearly Normal Galaxies", Springer-Verlag, Berlin
Gahm,G.,ed.: 1986,"Astrophysical Aspects of the Interstellar Medium and Star
    Formation", Phys.Scripta topical issue
Gondhalekar,P.M.,ed.: 1984,"Gas in the Interstellar Medium", RAL-84-101, Chilton
Henning,T.+,ed.: 1986,"The Role of Dust in Dense Regions of Interstellar Matter",
    Ap Sp Sci. $\underline{128}$, 1
Hollenbach,D.J.+,ed.: 1987,"Interstellar Processes", Reidel, Dordrecht
Israel,F.P.,ed.: 1986,"Light on Dark Matter", Reidel, Dordrecht
Kahn,F.D.,ed.: 1985,"Cosmical Gas Dynamics", VNU Science Press", Utrecht
Kessler,M.F.+,ed.: 1984,"Galactic and Extragalactic Infrared Spectroscopy",
    Reidel, Dordrecht
Kondo,Y.+,ed.: 1984,"Local Interstellar Medium", NASA-CP 2345, Washington
Lucas,R.+,ed.: 1985,"Birth and Infancy of Stars", North-Holland, Amsterdam
Longdon,N.,ed.: 1986,"Space-Borne Sub-Millimeter Astronomy Mission",ESA SP-260,Paris
Mead,J.M.+,ed.: 1984,"Future of Ultraviolet Astronomy based on Six Years of IUE
    Research", NASA CP-2349, Washington
Morfill,G.+,ed.: 1987,"Physical Processes in Interstellar Clouds",Reidel,Dordrecht
Peimbert,M.+,ed.: 1985,"Star Forming Regions", Reidel, Dordrecht
Péquignot,D.,ed.: 1986,"Model Nebulae", Obs. de Paris-Meudon, Meudon
Seggewiss,W.+ : 1985,"Interstellar Materie" Mitt.Astron.Ges. 63, available from
    H.U. Keller, Planetarium Stuttgart
Serra,G.,ed.: 1985,"Nearby Molecular Clouds", Springer-Verlag, Berlin
Shaver,P.A.+,ed.: 1985,"(Sub)millimeter Astronomy", ESO Conf. Workshops Proc. 22,
    ESO, Garching
Shields,G.A.,ed.: 1986,"The Texas-Mexico Conference on Nebulae and Abundances",
    PASP, $\underline{98}$, 956
van Woerden,H.+,ed.: 1985,"The Milky Way Galaxy", Reidel, Dordrecht
Wilson,R.,ed.: 1986,"New Insights in Astrophysics: 8 Years of UV Astronomy with
    IUE", ESA SP-263, Paris

3. CATALOGUES AND SURVEYS

Avedisova,V.S.: 1985,"Catalogue of Star Formation Regions in our Galaxy", Bull.
    Inf.Cent. Données Stellaires, $\underline{29}$, 78
Brand,J.+: 1987,"The Velocity Field of the Outer Galaxy in the Southern Hemisphere:
    II CO Observations of Galactic Nebulae", AA Suppl. $\underline{68}$, 1 and $\underline{69}$, 343

Caux,E.+: 1985,"Far-Infrared Survey of the Southern Galactic Plane", AA,144, 37
Cernicharo,J.+: 1984,"A Catalogue of Visual Extinction in Taurus and Perseus",
     AA Suppl. 58, 327
Dorschner,J.+: 1984,"A Catalogue of Equivalent Widths of the Interstellar 2200 A
     Band", Bull.Inf.Cent.Données Stellaires, 27, 141
Feitzinger,J.A.+: 1984,"Catalogue of Dark Nebulae and Globules for Galactic
     Longitudes 240 to 360 Degrees", AA Suppl. 58, 365
Feitzinger,J.A.+: 1986,"A Large-Scale Atlas of the Milky Way Dark Clouds",
     Vistas A., 29, 291
Friedemann,C.: 1987,"A Second Catalogue of Equivalent Widths of the Interstellar
     2170 A Band", Bull.Inf.Cent.Données Stellaires, 32, 77
Hartley,M.+: 1986,"A Catalogue of Southern Dark Clouds", AA Suppl. 63, 27
Helou,G.+: 1986,"IRAS Small Scale Structure Catalog", US Gov't Printing Office,
     Washington; also distributed by Centre de Données Stellaires, Strasbourg
IRAS Science Working Team: 1985,"Point Source Catalog", US Gov't Printing Office,
     Washington; cf. Bull.Inf.Cent.Données Stellaires, 29, 107
Klinglesmith,D.A.+: 1987,"A Palomar Observatory Sky Survey Atlas of Selected
     Molecular Clouds", ApJ.Suppl. 64, 127
Neckel,Th.+: 1985,"Atlas of Galactic Nebulae", Treugesel-Verlag, Düsseldorf
Rosa,M.+: 1984,"IUE UV Spectra of Extragalactic HII Regions. I.The Catalogue and
     the Atlas", AA Suppl. 57, 361
Sabbadin,F.: 1984,"A Catalogue of Expansion Velocities in Planetary Nebulae",
     AA Suppl. 58, 273
Savage,B.D.: 1985,"A Catalog of Ultraviolet Interstellar Extinction Excesses
     for 1415 Stars", ApJ.Suppl. 59, 397
Scoville,N.Z.+: 1987,"Molecular Clouds and Cloud Cores in the Inner Galaxy",
     ApJ.Suppl. 63, 821
Ungerechts,H.+: 1987,"A CO Survey of the Dark Nebulae in Perseus, Taurus and
     Aurifa", ApJ.Suppl. 63, 645
Whiteoak,J.B.: 1985,"3.3 GHz Ground-State Transitions of CH towards Southern HII
     Regions. I.An Atlas of CH Profiles", Proc.A.Soc.Aust. 6, 6

4. REVIEW ARTICLES

Blinnikov,S.I.+: 1987,"Supernovae and Supernova Remnants: Observations and
     Theory", AdvSpace Phys. Sov. Sci. Rev. Sect. E, in press
Cowie,L.L.+: 1986,"High-Resolution Optical and Ultraviolet Absorption-line
     Studies of Interstellar Gas", Ann.Rev.AA., 24, 499
Kaler,J.B.: 1985,"Planetary Nebulae and their Central Stars", Ann.Rev.AA., 23, 89
Lada,C.J.: 1985,"Cold Outflows, Energetic Winds and Enigmatic Jets around Young
     Stellar Objects", Ann.Rev.AA., 23, 267
Raymond,J.C.: 1984,"Observations of Supernova Remnants", Ann.Rev.AA., 22, 75
Scalo,J.M.: 1986,"The Stellar Initial Mass Function", Fund.Cosm.Phys. 11, 1
Voshchinnikov,N.V.: 1986,"Interstellar Dust", Summaries in Science and Technology,
     Ser. Space Res. 25, 98
Yorke,H.W.: 1986,"The Dynamical Evolution of HII Regions - Recent Theoretical
     Developments", Ann.Rev.AA., 24, 49

II. Diffuse Interstellar Medium
(K. S. de Boer)

Emphasis in this report is put on observations since these are vital to further our understanding. The period covered showed a surge in investigations dealing with the local interstellar medium based on a large variety of (mainly) space observations. Also interstellar depletion received a lot of attention due to the large amount of good spectroscopic data available today. At the end of the reporting period SN 1987a came which boosted studies of the ISM in the LMC and of the galactic halo. In the following, mainly papers of the refereed litterature are mentioned; for the relevant conferences and reviews the reader is referred to the listings above. Some reviews pertaining to our topic which appeared in places one would not immediately look for are: Blades (1984,ESA SP-218,11); Savage (1984,NASA CP-2349,3); Salpeter (1985,Seggewiss+,11); de Boer (1985,Seggewiss+,63); Pettini (1985,Danziger,355) and many in (1987,Kondo book).

1. THE DIFFUSE DISK ISM

Neutral hydrogen surveys continued using the 21-cm emission line and the Lyman 121.6nm absorption line. Braunsfurth+(1984,AA.Suppl,57,189) presented 21-cm data for the Milky Way disk at longitudes between -3 and +21 deg from Effelsberg. Crovisier et al.(1985,AA,146,223) investigated small-scale structure of HI using absorption spectra towards background double sources, and showed that column densities may vary by factors larger than 3 over 1/4pc. Greisen+(1986,ApJ,303,702), using the VLA, found that structural changes are smooth on a scale of 1 arcmin. A catalog of HI 21-cm absorption line spectra of galactic radio sources was published in (1987,Izv Spec Ap Obs,24,93). Anomalous velocities for HI gas were analysed by Bash+ (1985,AA,145,127). Shaver (1984,AA,138,131) found that most deviations of observed velocities from galactic rotation are due to random cloud motions. The average density along interstellar lines of sight was investigated by Spitzer (1985,ApJ,290, L21), who found two types of clouds in more uniform warm HI gas. A survey of the southern sky in the H166α recombination line was published by Hart+(AA Abs,40.131. 195). Carbon recombination lines were seen in absorption towards CasA (Ershov+ 1984, Pis'ma Astron Zh,10,846; 1987,ibid.,13,19) which led Sorochenko+(1987,Pis'ma Astron Zh,13,191) to estimate the CR intensity needed to ionize carbon. Shull+(1985,ApJ, 294,599) published a large survey of HI seen in absorption in IUE spectra of 205 early type stars up to 8.5kpc and with 68 stars in the halo. The gas to dust ratio as derived from Copernicus data is confirmed but now based on a much larger sample (the Rho Oph N(HI) was revised, see below). The distribution of highly ionized gas (SiIV, CIV) was analysed by Kool+(1985,AA,149,151), while Savage+(1985,ApJ,295,L9; 1987,ApJ,314,380) observed these ions and detected NV as well in the general direction of the galactic centre. Hobbs (1984,ApJ,284,L47) presented evidence for very weak absorption by high-velocity FeX from the hot phase of the ISM but showed later (1985,ApJ,298,357) that the absorption feature more likely is due to some agent in cool diffuse clouds. Using CCD echelle data, also Pettini+(1986,ApJ,310,700) failed to detect FeX absorption in the galactic halo on the line of sight to LMC stars.

The pervasiveness of dust in the diffuse medium was highlighted by the discovery of the "infrared cirrus" in the IRAS measurements. Boulanger+ (1985,AA,144,L9) succeeded in demonstrating the connection of individual cirrus features with HI clouds at high latitude north, while McGee+ (1986,MN,221,543) identified similar structures in the south (see further with local ISM). General surveys of extinction, using wide-angle photographs (1984,AA,137,287), photometry (1985,AJ,90,301), and galaxy counts (1986,AA,154,181) were presented, and the spectral structure in the UV was analysed (Savage+,1985,ApJ.Suppl.,59,397; Carnochan, 1986,MN,219,903; 1986, Fitzpatrick+,ApJ,307,286). Diffuse bands were measured (Isobe+,1986,PAS Japan,38, 511; Federman +,1984,ApJ,282,485). Obviously the amount of dust has to match the extinction and the infrared emission; however, there is a large amount of freedom in the models and consistency is often found (de Muizon+,1985,AA,143,160; Dall'Oglio

ApJ,289,609; Cox+,1986,AA,155,380; Rowan-Robinson,1986,MN,219,737). Doubt on the
dust nature of the IRAS cirrus came from Harwit et al.(1986,Nature,319,646) who
claim that most if not all of the 60 and 100 μm flux may be due to fine-structure
emission by OI and OIII.

The ionization structure of the interstellar medium determines which absorption and emission features can be observed. An up to data compilation of the pertinent atomic data with ionization balance calculations for the diffuse medium became
available from Pequignot and Aldrovandi(1986,AA,161,169). Mathis(1986,ApJ,301,423)
aimed largely at the emissivity of the ionized phase. His calculations were triggered by the continuing observations of diffuse line emission (Hα, and the newly
detected OIII and SII) of Reynolds (1984,ApJ,282,191; 1985,ApJ,294,256; 1985,ApJ,
298,L27), Ogden+(1985,ApJ,290,238), and of Sivan+(1986,AA,158,279). Combining such
measurements with hints at substantial ionization in the very local gas (see next
section) Reynolds (1986,AJ,92,653) argues for an excess local EUV flux or a transient recent energetic event. Bixler+(1984,AA,141,422) achieved a new determination
of the interstellar radiation field at 100 nm, and Bloemen+(1985,AA,145,391) considered the effect of the diffuse radiation on the production of inverse-compton gamma rays in the galaxy. Fujimoto+(1984,PAS Japan,36,319) presented models for the
interaction network of interstellar gas and its diffusive energy. McKee summarized
the processes due to injection of radiative, wind, and explosive energy by stars
(1986,ApSpSci,118,383), and Kegel+(1986,AA,161,23; 1986,AA,164,337) investigated
the effects of friction between the ISM and the system of stars. Hartquist+(1984,
ApJ,287,194) adressed Alfven waves and energy dissipation at cloud boundaries, while
Shull and Woods (1985,ApJ,288,50) discussed the formation of clouds from thermally
instable intercloud gas.

Scintillation of radiation from pulsars was further investigated. At cm wavelengths flicker of 3% was seen over periods of days (Blandford+1986,ApJ,301,L53)
but at meter waves with periods of months (Rickett,1986,ApJ,307,564), the latter
explaining variations observed in extragalactic radiosources. Shapirovskaya+(1985,
Pis'ma Astron.Zh.11,686; Sov Astron,30,no.4) find that low frequency variations do
not correlate with galactic direction and they propose that hot ionized plasma is
responsible. Also Balasubramanian+(1985,J.AA,6,35) favour a homogeneous medium to
explain scintillation seen in pulsars, but Cordes+(1985,ApJ,288,221) required a two
phase medium, adding a clumped component with scale height of less than 100pc.
Krishnamohan(1986,MN,220,119) then postulated a high density (6 cm^{-3}pc) layer below
the galactic plane to explain the anomalous pulsar distribution on the sky. Alurkar+
(1986,Aust J Phys,39,433) added new pulse broadening measurements of 33 pulsars
with Parkes and Cordes+(1986,ApJ,307,L27) predicted that multiple pulsar images due
to scintillation might be visible with VLBI. Further theoretical analyses by
Goodman+(1985,MN,214,519), by Romani+(1986,MN,220,19) and by Cordes+(1986,ApJ,310,
737) showed the effects of diffractive scintillations and of refraction due to a
multicomponent interstellar medium on the radiation of pulsars and other radiosources.

More detailed studies of the abundance of elements in the ISM became possible
with the increasing size of the IUE data base and a new and independent body of
data from BUSS. In addition some newly analysed Copernicus observations were published. In the visual new equipment boosted observational activities. LiI was observed by Ferlet+(1984,AA,138,303), by Hobbs (1984,ApJ,286,252) and in particular
by White (1986,ApJ,306,777) who found good correlations with Na, K, C and n(e).
Boesgaard (1985,PASP,97,587) further pursued Be with new instrumentation and now
finds a factor of 2 depletion. Based on new determinations of f-values Hibbert+
(1985,MN,214,721) and Keenan+(1985,AA,147,89) reanalysed resp. NI and OI data and
confirmed the earlier less than 50% depletion, and about 60% depletion of Mg (1984,
ApJ,282,481). ClI was investigated by Harris+(1984,ESA SP-218,157) and Federman
succeeded in identifying the line found at 108.805nm in Copernicus spectra as also
due to ClI. The remaining heavy elements were investigated in several papers, from

Copernicus (Mg P Cl Mn Fe Cu Ni) by Jenkins+(1986,ApJ,301,355), from BUSS (Mg Mn Fe Cr Zn) by de Boer+(1986,AA,157,119), and from IUE (Si S P Fe Zn) by Harris+(1986,MN,220, 271;1986,ApJ,308,240), Gondhalekar (1985,MN,217,585) and Shull+(1987,ApJ Supp, preprint). All authors agree on increased depletion with increased column density, possibly through gas density but definitely correlated with reddening (1984,ApJ,284, 157;1984,ApJ,287,238;1985,ApJ,293,230;1986,MN,222,143;1986,AA,157,119;1986,ApJ,309, 771). An important discovery was that both S (1985,MN,217,585;1986,ApJ,308,240) and Zn (1986,MN,220,271;1986,AA,157,119) are depleted, in contrast to earlier beliefs, by up to a factor of 2 on heavily reddened lines of sight. A summary of current abundance and depletion values is given by de Boer+(1987,Kondo+ book,p.485). Also molecules were seen in absorption such as C_2 (Gredel+,1986,AA,154,336), CN(Federman+ 1984,ApJ,287,219;Snow,1984,ApJ,287,238), CH(Lien,1984,ApJ,284,578; Jura+,1985,ApJ, 294,238) and CH^+(Lambert+,1986,ApJ,303,401). CH was detected for the first time in emission from a rotationally excited lever (Ziurys+,1985,ApJ,292,L25). Absorption of CS^+ was reported but later turned out to be misidentified (1986,AA,168,259). For data on other molecules see the section on molecular clouds.

Rho Oph gave one of its depletion mysteries away: both new IUE spectra (Shull+, 1985,ApJ,271,408) and a reanalysis of the Copernicus Lyman profile (de Boer+,1986, AA,157,119) showed that log N(HI) is only 20.4, thus reverting this line of sight to the normal category in all respects. The star HD147889 in the Rho Oph cloud shows heavy depletion, with Si possibly behaving like Ca, and many lines of molecules (1985,ApJ,288,277;1985,ApJ,290,251;1985,ApJ,296,213;1986,PASP,98,857;1986,ApJ,302, 492;1986,ApJ,303,433;1986,ApJ,309,771). Meyers+(1985,ApJ,288,148) analysed absorption components and showed that a weak shock processed the gas in the Rho Oph cloud. Klose (1986,Ap Sp Sci,128,135) discussed the Rho Oph cloud with respect to gamma rays and Young+,1986,ApJ,304,L45) presented IRAS far-infrared maps and identified 13 of the 18 sources seen at 12 μm.

Zeta Oph served as prime star in investigations by Federman+(1985,ApJ,290,L55 on Rb), by Hawkins+(1985,ApJ,294,L131 on C isotopes), by van Dishoeck+(1986,ApJ,307, 332 on C_2) and by Pwa+(1986,AA,164,116), while Keene+(1987,ApJ,313,396) found emission in the sub-mm finestructure transitions of CI, which were in agreement with earlier analyses of the CI resonance line strengths. Crutcher+(1987,ApJ,316,L71) resolved CO radio emission towards Zeta Oph into two narrow components. Draine(1986, ApJ,310,408) constructed a shock model for the molecular gas seen towards Zeta Oph and claims that the HII region drives the shock.

The Pleiades were intensively observed both for absorption lines (White,1984, ApJ,284,685 and 695; Younan+,1984,MN,209,123) and radio emission lines (Federman+, 1984,ApJ,283,626). The reflection nebulae were reobserved by Witt+(1985,ApJ,294,216; 1986,ApJ,302,421) and they conclude that the reflection nebulae are within 0.1pc from the stars. IRAS maps of the Pleiades were presented by Castelaz+(1987,ApJ,313, 853) who argue for 50K and very small grains with large UV absorption efficiency to match the Witt+ UV surface brightness.

Shells were seen with various sizes, near OB associations (Nichols-Bohlin+, 1986,AJ,92,642;Silich,1985,Astrofizika,22,563), very thin ones possibly outside the disk (Kulkarni+,1985,ApJ,291,716;Dubner+,1985,Rev Mex AA,10,151) or due to supernovae (Jones+,1985,MN,213,711;Fich,1986,ApJ,303,465). Of particular interest were the analyses of the Cygnus region, where Heske+(1985,AA,148,439) measured radio recombination lines. Bochkarev+(1985,Ap Sp Sci,108,237) summarized all information for the Cygnus superbubble involving OB associations, but they required also 100pc diameter regions of coronal gas at distances of 0.5 to 2.5kpc from the sun in the Carina-Cygnus spiral arm. Broten+(1985,Ap Lett,24,165) reviewed the evidence for a correlation between the compression and the magnetic field strength in interstellar bubbles. The dynamics and the observability of shells and cavities were investigated by Silich (1985,Astrofizika,22,563) and Bocharev+(1985,Ast.Zh,62,103 and 875). Shocks received a lot of attention, in particular in connection with molecular clouds and grain disruption (e.g. 1985,ApJ,288,148; 1985,MN,215,125; 1985,AA,151,121; 1986,MN,218, 729; 1987,Astrofizika,26,113).

2. LOCAL INTERSTELLAR MEDIUM (LISM)

The reporting period started with (1984, Kondo+) "Local Interstellar Medium". The earlier descriptions of Frisch+ (1983,ApJ,271,L59) and Paresce (1984,AJ,89, 1022) were supplemented by contributions dealing with new and old observations of Lyman alpha and of various metals, such as MgII, CII, FeII and SiII, which also result in estimates of local gas columns. Three independent UV data bases are available, each with different instrumental characteristics (Copernicus, BUSS, IUE). Landsman+ (1984,ApJ,285,801; 1986,ApJ,303,791; 1987,ApJ,315,675) used in particular late type stars, but the analysis of the stellar/IS Lyman alpha profile is difficult. Molaro+ worked on MgII stellar emission with superimposed IS absorption (1985 AA,144,81; 1986,AA,161,339), de Boer+ included nearby A-F stars (1986,AA,157,119), and fast rotating stars were used (1984,ESA SP-218,133; 1987,AA,177,228). York+ worked mostly with early type stars and detailed profile fitting (1985,ApJ,296,593; 1986,ApJ,308,232), and Vidal-Madjar and Ferlet et al. concentrated on NaI and CaII with very high dispersion (1986,AA,155,407; 1986,AA,163,204; 1986,AA,168,225). Kondo+ (1984, Kondo+,200) and de Boer et al.(1985,Mitt.AG,63,155; 1986,AA,157,119) analysed the physical conditions of the LISM using NaI, MgI and MgII. The LISM can be described as hot and of low density, in particular in the 3rd galactic quadrant, with large concentrations of gas towards $l=130°$ and $l=290°$, and with a very close (6pc?) cloud towards $l=40°$ (1986,AA,157,119; 1986,AA,163,204). Discussions on the diffuse soft X-ray emission centered on spatial models for the source function of the hot gas (see also next section).

EUV-resonance emission from the LISM in the interplanetary space was investigated by Clarke+ (1984,AA,139,389), Shemansky+ (1984,IAU Coll.81,24), Suess+ (1985,Nature,317,702), and Baranov (1986,Sov.Astron.Lett.,12). The interactions of the LISM with the interplanetary medium and the perturbed plasma interface of the heliosphere were further analysed by Fahr+ (1984,AA,139,551; 1985,AA,142,476; 1986, Adv.Sp.Res,6,13). The systematics of the motions of the local gas and the interplanetary medium were analysed by e.g. Bertaux+ (1984,AA,140,230; 1985,AA,150,1 and 21) and Lallemant+ (1986,AA,168,225) and it appears that there is a systematic flow of interstellar gas from about l=300 deg and at about v=25 km/s through the solar neighbourhood, including the planetary system. However, the exact values differ between the analyses from interstellar lines and the backscattering profiles. The density of the interplanetary medium just inside the Heliopause derived from the Venera 11 and 12 data (Chassefiere+ 1986,AA,160,229) compared favourably with results from other satellites at n(H)= 0.06 cm^{-3}. Further contributions can be found in the proceedings of the COSPAR meeting in Toulouse, 1986.

An individual cloud was recognized by Hobbs+ (1986,ApJ,306,L109) at $b=-34°$ and $l=159°$ with d=65pc. This and other clouds, such as those found in CO by Magnani + (1985,ApJ,295,402) may be related to the so called "cirrus" as seen in IRAS far-infrared emission. These structures apparently have small radial velocities and, although at high latitude, do not belong to the halo (see next section).

3. GAS AT HIGH LATITUDES, IN THE HALO, AT HIGH VELOCITIES

The vaguely defined "halo" of the Milky Way is best seen at high latitudes and may contain high-velocity gas. However, low-velocity gas may exist well outside the Milky Way disk, and gas at high latitudes may be very local. This ambiguity played its part in the research carried out over the past few years. Studies of gas outside the heavily populated galactic disk region continued with searches for CO emission. Both in the northern hemisphere (Blitz and Magnani et al. 1984,ApJ,282,L9; 1985,ApJ,295,402; 1986,ApJ,301,395; Mattila+, in 1985, Serra,15) and in the southern hemisphere (Keto+,1986,ApJ,304,466) CO clouds were found. The discovery from the IRAS data of the existence of veils of dusty gas at high galactic latitudes continued to arouse great interest. Comparisons with HI, CO and extinction data were made, of which the latter were least successful due to overall small reddening values. However, the cirrus could be recognized on PSS plates as faint blue whisps, due to the reflection of diffuse galactic light by cirrus dust (de Vries+,1985,AA,145,L7). The comparisons with CO showed very productive and Weiland+ (1986,ApJ,306,L101) established that all previously discovered high latitude CO clouds coincided with the peaks of IRAS 100 micron cirrus emission. Wakker + (1986,AA,170,84) then looked for coincidences of the IRAS cirrus with halo high velocity clouds, but found none. For a few cirrus-CO clouds distance estimates could be mase based on star counts or absorption lines: Magnani+ (1986,AA,168,271) found 75 pc in one case, Weiland+ (1986,ApJ,306,L101) arrive at about 100pc, and Hobbs+ (1986, ApJ,306,L109) find 65 pc for a cloud at $l=159°$, $b=-34°$.

The high latitude Draco cloud was intensely investigated by Mebold+. High velocity gas is seen stretching along the sky and ending near "the" Draco cloud, which in itself shows substantial velocity structure. The vicinity shows enhanced soft X-ray emission (Hirth+ 1985,AA,153,249), which might be due to a collision of high velocity gas with more local gas at rest. Further observations then revealed emission from molecules such as CO, H_2CO, and NH_3 (1985,AA,151,427; 1987,AA,180, 213). Using various methods, Goerigk+ (1986,AA,162,279) estimate the distance of the cloud at between 0.8 and 2.5 kpc. Johnson (1986, ApJ,309,321) catalogued all 100 micron sources in the Draco field.

The soft X-ray intensities (McCammon+,1983,ApJ,269,107) show an anticorrelation with HI which was intensively investigated. Marshall+ (1984,ApJ,287,633) presented similar results and discussed a two component model for the radiation source. Knude (1985,AA,147,155) looked for cloud shadows and Jahoda+ (1985,ApJ,290, 229; 1986,ApJ,311,L57) made a statistical analysis; both required complex X-ray source models, and it appears likely that little of the detected soft X-ray flux stems from the galactic corona. Rather a hot stratum interspersed with neutral clouds is producing the observed anticorrelation (Mebold+, preprint).

All the work on classical high velocity clouds (HVCs) was reviewed by van Woerden+ (1985,van Woerden+,387). Further reports may be found in that symposium. McGee+ (PAS Austr. 1986,6,358, and 471) discussed new 21-cm data of halo gas in the direction of the Magellanic Clouds and of the bridge between LMC and SMC. Observational results from absorption line studies (mostly in the UV) have been presented by Savage (NASA CP-2349, 3) and by de Boer (1985,Mitt.A.G,63,21). York+ (1986, AJ,91,354) presented a list of RR Lyr stars near HVCs to plan absorption observations. New absorption data in the direction of M3 (galactic north pole) show inflow of gas (de Boer+ 1984,AA,136,I.7) while Songaila+ (1985,ApJ,293,L15) report on possible absorption towards an RR Lyr star at 2 kpc. Observations of extragalactic objects added little hard data on clouds in the Milky Way halo, except maybe for the direction to NGC 3783 (West+, 1985,MN,215,481), where an absorbing cloud was seen at +241 km/s, the same speed as a HVC; the distance to this cloud remains elusive. QSO observations showed many absorption line systems as usual (see further Comm. 28 and 47), but Meyer+ (1987,ApJ,315,L5) argued that at least 15% of the lines classified as Lyman alpha forest are rather misidentified metal lines. Analysing the little data available in the literature, Morton+ (1986, MN,220,927) suggest that the CaII absorption arises in a thick disk. The metallicity of the HVCs seen earlier towards the LMC are near solar (see 1985, Danziger, p.355; Proc.Astron.Soc.Aust.6,358 and 42.131.311). Fine structure in HVCs was investigated by Arzamasova (1985, Soobshch.Spets.Astrofiz.Obs.46,69). Saar+ (1984, Tartu Obs.Pub.50,280) analysed the possible distances of three clouds, finding less than 1 kpc or more than 30 kpc; on the other hand, Kaelble+ (1985,AA,143,408), who analysed the older Giovanelli HVC sample and included IUE observations, found consistency for distances between 2 and 5 kpc, with typical motions of 100 km/s towards the disk, parallel to galactic rotation and towards the rotation axis of the

Milky Way. Infalling clouds may plunge into the disk and Tenorio-Tagle (1986,AA,
170,107; 1987,AA,179,219) mode models for the shape of the shocks in such collisions. Most papers considered the ionizations structure of the gas in the halo, in
particular SiIV vs CIV. Savage+ presented absorption data for long sight lines
reaching outside the disk in the direction of the galactic centre and they found
pronounced NV in addition to the obvious SiIV and CIV (1985,ApJ,295,L9). If the
halo is hot it might contain FeX, but Pettini+ (1986,ApJ,310,700) failed to detect
it on the line of sight to the LMC. Chevalier+ (1984,ApJ,279,L43; 1985,ApJ,296,35)
tested models with cosmic ray ionization. Bregman+ (1986, ApJ,309,833) showed that
plain photoionization plays a very important role with for each wavelength domain
a prolific contributor, such as disk OB stars, halo PN and pAGB stars (also 1986,
AA,142,321), energetic galaxies, and so on.

4. MAGELLANIC CLOUDS

The diffuse interstellar medium in the Magellanic Clouds has been probed by
means of interstellar absorption lines in the visual and the UV and through surveys
of HI. However, a pronounced imbalance persists between the set of stars studied
in the visual and in the ultraviolet (de Boer, 1984,ESA SP-218, p.179). Cohen (1984
AJ,89,1779) published CaII data for 31 SMC stars and Ferlet + observed NaI lines
in the LMC (1985,AA,152,151). From UV absorption of weak lines (O Mg S Cr Mn Ni Zn)
de Boer+ (1985,MN,207,115) found a metal abundance in the LMC in front of R136 of
a factor 2-3 below galactic. A new radio 21-cm survey was produced by Rohlfs+(1984,
AA,137,343), which is the follow up of the early 21-cm work of the 1960s. McGee+
(1986,PAS.Austr.6,358 and 471) presented new 21-cm measurements towards the Magellanic Clouds and the Bridge region. The SN1987a allowed very accurate measurements
of interstellar lines in all wavelength domains (1987,AA,177,L17 and L37), which
showed to be similar to those towards R136. An analysis of all information of gas
in Shapley III indicated that its edges expand with 35 km/s and that the origins
of the structure lie 15Myr ago (Dopita+, 1985,ApJ,297,599).

III. Molecules and Molecular Clouds

(D. Flower, H. Habing, P.G. Wannier)

1. CHEMISTRY

There has been considerable activity in the field of interstellar chemistry
during the period under review, including several more or less directly relevant
conferences and workshops. Those not cited at the beginning of this report are :
Doschek G.A.,ed.: 1984,"UV and X-ray Spectroscopy of Astrophysical and Laboratory
Plasmas", NRL, Washington;Haschick,A.D.,ed.: 1986,"Masers, molecules and mass outflows in star-forming regions", Haystack Obs., Westford, Mass.; Vardya,M.S.+,ed.:
1987, "Astrochemistry", Reidel, Dordrecht. The fundamental question "Is interstellar chemistry useful?" was addressed by Dalgarno (1986: QJRAS,27,83). Much work has
been done on the assumption that the answer to this question is positive, including
a survey of the important category of bimolecular ion-molecule reactions (Anicich+:
1986, ApJ.Suppl,62,553). References to more specialized work are given below.

1.1. Molecular Processes

Photo-dissociation is an important process, at least in diffuse molecular
clouds. Photodissociation rates for OH, OD, and CN have been reported by Nee+(1985,
ApJ,291,202), who measure the OH photodissociation cross-section to be 2 to 3 times
higher than calculated by van Dishoek+ (1983,JCP,79,873). The VUV predissociating
states of CO have been studied spectroscopically in the laboratory by Eidelsberg+
(1987,J.Mol.Spectrosc,121,309),and CO self-shielding in the interstellar medium by
Glassgold+ (1985,ApJ,290,615). Cross-sections for the photoionization of C_2 in its

ground state have been computed by Padial+(1985,ApJ,298,369). At high kinetic temperatures, the collisional dissociation of H_2 by H may become important (Dove+,1986, ApJ,311,L93).

Much work has been devoted to ion-neutral reactions, particularly with hydrogen molecules. Adams+(1984,MN,211,857) measured the dependence on the centre-of-mass kinetic energy of the rates of the reactions with H_2 of C^+, N^+, $C_2H_2^+$, $C_2H_3^+$, C_3H^+ and $C_3H_2^+$. The experimental data relating to the $C^+(H_2,H)CH^+$ reaction were reevaluated by Chesnavich+(1984,ApJ,287,676). Such studies are relevant to, for example, CH^+ formation in MHD shocks. The rate of the $N^+(H_2,H)NH^+$ reaction was measured at low temperature ($8 \leq T \leq 70$ K) by Marquette+(1985,AA,147,115) and Luine+, 1985,ApJ,299,L67. Reactions leading to the chemical fractionation of deuterium have been studied in the laboratory by Adams+(1985,ApJ,294,L63).

Clary (1985,Mol Phys, 54,605) suggested that reactions of ions with polar molecules would proceed very much faster than the Langevin rate at interstellar temperatures, and this has been confirmed experimentally (Adams+,1985,ApJ,296,L31). Rate constants for reactions of ions with the cyanopoly-ynes have been measured (Bohme+,1985,MN,213,717;Daniel+,1986,ApJ,303,439;Knight+,1986,MN,219,89). Rates of reactions relevant to the formation and destruction of HCl have also been measured (Smith+,1985,ApJ,298,827), and this chlorine-bearing molecule has been detected (see below).

Rate coefficients for radiative association reactions have been revised by Herbst (1985,ApJ,291,226). The radiative association of H_2 and CH_3^+ has been studied by Herbst(1985,AA,153,151) and Bates(1985,ApJ,298,382;1987,ApJ,312,363), and of H_2 and C_3H^+ by Herbst+(1984,ApJ,285,618). The rates of dissociative recombination and of electron collisional dissociation have been compared by Zhdanov (1986,Sov Astron, 30,278), and the vexed question of the products of the dissociative recombination of polyatomic molecular ions addressed by Bates(1986,ApJ,306,L45). Protonation reactions have been discussed by Pauzat+(1986,AA,159,246), and the specific case of the protonation of CO by Dixon+(1984,JCP,81,3603).

1.2. Identification of Molecules

The molecular ion H_3^+ is believed to play a key role in interstellar chemistry but is difficult to detect because it has no permanent dipole moment. Its (forbidden) rotational spectrum has been computed by Pan+(1986,ApJ,305,518). The $1_{10}-1_{11}$ sub-mm line of the isotope H_2D^+, which does possess a dipole moment, has been measured in the laboratory by Bogey+(1984,AA,137,L15) and possibly detected by Phillips· (1985,ApJ,294,L45). Another molecular ion, H_3O^+, has been sought (Wootten+,1986,AA, 166,L15) and perhaps detected (Hollis+,1986,Nature, 322,524). The deuterated molecule CCD has been detected at a level which is believed to be consistent with ion-molecule formation schemes (Combes+,1985,AA,147,L25). Other heavy hydrocarbon molecules have also been observed and identified: C_3H (Thaddeus+,1985,ApJ,294,L49; Gottlieb+,1985,ApJ,294,L55;Gottlieb+,1986,ApJ,303,446), cyclic-C_3H_2 (Thaddeus+,1985, ApJ,299,L63;Matthews+,1985,ApJ,298,L61;Matthews+,1986,ApJ,307,L69), C_5H (Cernicharo+ 1986,AA,164,L1;Gottlieb+,1986,AA,164,L5) and C_6H (Suzuki+,1986,PAS Japan,38,911; Guélin+,1987,AA,175,L5). The deuterated form of C_3H_2 has also been detected (Gerin+, 1987,AA,173,L1), as well as the ^{13}C-substituted form (Gomez-Gonzalez+,1986,AA,168, L11;Madden+,1986,ApJ,311,L27) following a laboratory study (Bogey+,1986,AA,159,L8). A chlorine-bearing molecule,HCl, has been observed in OMC-1 (Blake+,1985,ApJ,295, 501). The structure and spectrum of the cyclic molecule SiC_2 have been studied (Oddershede+,1985,JCP,83,1702). Protonated HCN has been detected (Ziurys+,1986,ApJ, 302,L31).

1.3. Theoretical Models

Much work continues to be devoted to developing and refining chemical models of both cold and hot (shocked) molecular gas in the interstellar medium. The chemistry in dynamically evolving clouds has been studied by Tarafdar+(1985,ApJ,289, 220). The time-dependence of the chemistry in dense clouds has been investigated by Mitchell (1984,ApJ,287,665), Watt (1985,MN,212,93), Watt+(1985,MN,213,157), Millar+(1985,MN,217,507), Williams (1986,QJRAS,27,64), Duley (1986,QJRAS,27,403) and Brown+(1986,MN,223,405). Models have been constructed by Mann+(1984,MN,209,33), d'Hendecourt+(1985,AA,152,130) which take into account grain-surface reactions. Diffuse cloud models are discussed by Mann+(1985,MN,214,279) and van Dishoeck+(1986, ApJ Suppl,62,109). Steady-state chemical models of both diffuse and dense clouds have been reported by Viala(1986,AA Suppl,64,391) and Viala+(1986,AA,160,301). Models of photodissociation regions have been evolved by Tielens+(1985,ApJ,291,722 and 747). Theoretical studies of specific aspects of the interstellar chemistry have been undertaken: the chemistry of OD (Croswell+: 1985,ApJ.289,618) and of chlorine-bearing molecules (Blake+: 1986,ApJ.300,145), the CH_3NC/CH_3CN isomer ratio (DeFrees+: 1985,ApJ.293,236), and the high HCS^+/CS ratio in TMC-1 (Millar+: 1985, MN,216,1025). The effects on models of dense clouds of the very large low-temperature rate coefficients for reactions between ions and polar molecules have been investigated by Herbst+(1986,ApJ,310,378). The physics and chemistry of polycyclic aromatic hydrocarbon (PAH) molecules are discussed by Omont (1986,AA,164,159) and the destruction of these molecules in reactions with atoms and ions by Duley+(1986, MN,219,859). No chemical model has yet given a satisfactory explanation of the high C/CO ratio observed by Keene+(1985,ApJ,299,967). Elaborate models of chemical processes in interstellar shocks have been developed. These models have been used to study molecule formation, in particular CH^+ formation, in diffuse interstellar clouds (Mitchell+: 1985,AA,151,121;Pineau des Forêts+: 1986,MN,220,801;Draine+: 1986,ApJ,310,392;Draine: 1986,ApJ,310,408). A correlation between the column densities of CH^+ and rotationally excited H_2 has been observed (Lambert+: 1986,ApJ, 303,401), suggestive of a common shock origin. The formation of SH^+ in shocks has also been studied (Millar+: 1986,MN,221,673;Pineau des Forêts+: 1986,MN,223,743).

2. ISOTOPE ABUNDANCES

Molecular transitions have continued to provide valuable information about the interstellar C, N and O isotopes. In this regard, a reference of special usefulness is Danziger,ed.: 1985. A more general set of results is presented in the proceedings from the Texas-Mexico Conference on Nebulae and Abundances (1986,PASP, 98,956). A very active area is that of abundances in the winds from red giant stars. Also, a major effort has been to understand interstellar deuterium.

2.1. Interstellar C, N and O Isotopes

A survey of the inner galaxy in the lines of ^{13}CO and $C^{18}O$ has extended work previously confined to the region from 4-12 kpc (Taylor+: 1986,BAAS,18,1026). The results are generally consistent with earlier conclusions: there is a modest enhancement of ^{13}C in the galactic disc with a large enhancement in the Galactic Center. A galactic survey of the $^{14}N/^{15}N$ ratio in NH^3 shows a similar result: a slight, constant, enrichment in the disc and a large enhancement in the Galactic Center (Güsten+: 1985,AA,145,241). The generally large ^{13}C abundance inferred from CO observations in GMC's has found support from observations of CH^+ in the local gas (Hawkins+: 1985,ApJ.294,131;Hawkins+: 1987,ApJ.317,926) which yield $^{12}C/^{13}C$ = 43±4 and from observations of isotopic forms of C_3H_2 and CH_3OH in GMC's (Gomez-Gonzalez+: 1986,AA,168,L11;Blake+: 1984,ApJ.286,586).

2.2. Interstellar Deuterium

Interstellar D/H has the lure of providing cosmological information, but many molecules suffer severe fractionation and contamination from processed stellar material. That also makes the subject interesting to interstellar chemists. HI observations are used in connection with H_2 and HD observations and a recent measurement is consistent with previous estimates of $D/H > 10^{-5}$ in the local gas (Landsman+,1984, ApJ,285,801). In dense clouds, several new deuterated species have been detected: C_3HD (Bell+,1986,ApJ,311,L89), CCD (Vrtilek+,1985,ApJ,296,L35), and C_3DH (Gérin+, 1987,AA,173,L1) and one species predicted (Croswell+,1985,ApJ,289,618). Such results have been interpreted in finely tuned chemical models generally consistent with a D/H ratio of 10^{-5} (Dalgarno+,1984,ApJ,287,L47; Herbst+,1987,ApJ,312,351; Brown+, 1986,MN,223,429). Such results have been input to models of galactic nucleosynthesis which indicate a need for an early generation of stars (Vangioni-Flam+,1987,AA, submitted).

2.3. Circumstellar C, N, O and Si Isotopes

The understanding of interstellar isotopic ratios is intimately connected with red giant stars, which are responsible for the production and injection of several of the CNO isotopes. In carbon stars with the largest mass-loss rates, millimeter emission lines were used to measure $^{17}O/^{18}O$, $^{18}O/^{16}O$ and $^{12}C/^{13}C$ (Wannier+,1987,ApJ, 319) and to survey $^{12}C/^{13}C$ (Knapp+,1985,ApJ,293,281). In stars with less opaque envelopes, IR absorption provided C and O isotopes in 21 C-stars (Harris,1987,ApJ, 316,294); four SC stars (Dominy+,1986,ApJ,300,325); nine MS, S and SC stars (Dominy+,1987,ApJ,317,810) and $^{12}C/^{13}C$ in 15 others (Sneden+,1986,ApJ,311,826). ^{17}O is always enriched though highly variable, leading to a considerable puzzle when compared to the very constant and modest enrichment in the Galactic disc (Wannier, Danziger,ed.,1985). ^{13}C is always enriched, with values from 4 to 100 and with some indication that C-rich objects have less ^{13}C enrichment, consistent with a ^{12}C dredge-up from the stellar core. Observations of $^{29}Si/^{30}Si$ are, as expected, consistent with the terrestrial abundance ratios (Cernicharo+,1986,AA,167,L9;Fox+,1984, BAAS,16,491).

3. CLOUD MORPHOLOGY, DYNAMICS AND EVOLUTION

3.1. Mainly theoretical considerations on individual clouds

Mass distribution in clouds, fragmentation, clumpiness have been dealt with in general terms by Bhatt+(1984,MN,209,69), Arquila +(1985, ApJ,297,436), Mundy+ (1986,ApJ,306,670), Kwan+(1986,ApJ,309,783), Pérault+(1985,AA,152,371; 1986,AA,157, 139), Falgarone+(1985,AA,142,157; 1986,AA,162,235), Blitz+(1986,ApJ.Lett,300,L89), Evans+(1987,ApJ,312,344), Chièze (1987,AA,171,225). Thermal instabilities have been discussed by Gilden (1984,ApJ,283,679), and (equilibrium) structures of rotating clouds by Hachisu+(1985,AA,143,435), Arquila+(1986,ApJ,303,356), Kiguchi+(1987, preprint). Boss (1985,ApJ.Lett,292,L71) argues that velocity information may be misinterpreted in case of binary formation. Turbulence in clouds was discussed by Stenholm (1984,AA,137,133), by Silk (1985,ApJ.Lett,292,L71) and by Canuto+(1985, ApJ.Lett,294,L125); on the line formation in turbulent fields see Albrecht+(1987, AA,176,317). Van de Hulst (1987,AA,173,115) started a series of papers on multiple scattering in spherical dust clouds. Tielens+(1985,ApJ,291,722,747) discuss photo dissociation regions; for evaporation of clouds see Balbus (1985,ApJ,291,518); for collisions: Lattanzio+(1985,MN,215,125). A simulation of life cycles of molecular clouds going through periods of star formation has been made by Bodifée+(1985,AA, 142,297).

3.2. Systems of molecular clouds

Clouds in the Inner Galaxy: see Myers+(1986,ApJ,301,398), Dame+(1986,ApJ,305, 892); in the Carina Arm: Cohen+(1985,ApJ.Lett,290,L15); Grabelsky+(1987,ApJ,315, 122). There is also some interest in the Outer Galaxy: Terebey+(1986,ApJ,308,357), Huang+(1986,ApJ,309,804), Mead+(1986,ApJ,311,321). For clouds in Gould's belt see Taylor+(1987,ApJ,315,104). Results of large surveys from the northern hemisphere have been reported by the Columbia group (1985,ApJ,297,751) and the Massachusetts-Stony Brook group (1985,ApJ.Lett,292,L19; 1986,ApJ.Lett,301,L19; 1985,ApJ,289,373).

See further Feitzinger+(1986,ApJ,305,534) and Schlosser+(1984,AA,137,287), Drapatz +(1984,MN,210,11P), Kwan+(1987,ApJ,315,92) and Peters+(1987,ApJ,317,646).

3.3. Shocks, magnetic fields, collapse

Chernoff (1987,ApJ,312,143) and Draine+(1984,ApJ,282,491) discuss shocks. Collapse and fragmentation is discussed by Hachisu+(1984,AA,140,259), Larson (1985, MN,214,379), Rengarajan (1984,ApJ,287,671). Bonazzola+(1987,AA,172,293) find a smaller tendency toward collapse in a turbulent medium. The collapse of rotating clouds has been calculated by Boss (1987,ApJ,316,721). For effects on dense clouds by supernova remnants see Oettl+(1985,AA,151,33), Tenorio-Tagle+(1986,AA,155,120; 1987,AA,176,329), Tenorio-Tagle+(1985,AA,148,52), Odenwald+(1985,ApJ,292,460), White+(1987,AA,173,337), and Pollock (1985,AA,150,339). Magnetic fields in dense clouds are discussed by Heiles+(1986,ApJ,301,339); the dissipation of magnetic structure is discussed by Nakano+(1986,MN,218,663; 1986,MN,221,319), Elitzur+(1985, ApJ,298,170), Elmegreen (1985,ApJ,299,196). The structure of magnetic gas clouds has been calculated by Benz (1984,AA,139,378); the polarization of molecular lines by Deguchi+(1984,ApJ,285,126); OH Zeeman Splitting is reported by Kazès+(1986,AA, 164,328).

3.4. Globules

The formation of globules is calculated by Sandford+(1984,ApJ,282,178); their thermal emission by Lee+(1987,ApJ,317,197). Casali (1986,MN,223,341) reports near-IR observations, Jones+(1984,ApJ,282,675), Bachiller+(1984,AA,140,414) and Stenholm (1985,AA,144,179) discuss the inner structure, and Williams+(1985,MN,212, 181) and Joshi+(1985,MN,215,275) report polarization measurements of background (!) stars. Reipurth+(1984,AA,137,L1) report about the formation of stars of low mass. See also Menten+(1984,AA,137,108) and Turner+(1986,AA,167,157).

3.5. High-latitude clouds

Two independent detections of molecular clouds at high galactic latitudes have been reported: a cloud in Draco (Mebold+: 1985,AA,151,427; 1986,AA,180,213; 1986,AA,162,279). Several high-latitude clouds have been detected and studied by Blitz and others (1984,ApJ.Lett,282,L9; 1985,ApJ,295,402; 1986,ApJ,301,395; 1986, AA,168,271; 1986,ApJ.Lett,306,L109); Halpern+(1987,ApJ,12,L31) find an X-ray sour-ce in one such cloud and suggest it to be a very young star. Further papers of in-terest are Keto+(1986,ApJ,304,466), Heithausen+(1987,AA,179,263), Sandell+(1987, AA,179,255), de Vries+(1985,AA.Lett,145,L7), Weiland+(1986,ApJ.Lett,306,L101). Wakker+(1986,AA,170,84) searched the IRAS data for far infrared emission from dust in high-velocity clouds and found none; this probably indicates a significant dif-ference between these clouds and the high latitude molecular "cirrus" clouds.

3.6. Individual clouds

Many papers have appeared dealing with individual clouds or regions;although most of these have relevance for some of the topics discussed above, they are sum-marized here per region. The most studied region remains the Orion molecular cloud; two other regions of prominent interest remain the Taurus dark cloud complex and the cloud complex containing ρ Oph. Orion: Bloemen (1984,AA,139,37), Omodaka+ (1984,ApJ.Lett,282,L77), Werner+(1984,ApJ.Lett,282,L81), Hasegawa+(1984,ApJ,283, 117), Vogel+(1984,ApJ,283,655), Jaffe+(1984,ApJ,284,637), Loren+(1984,ApJ,286,232), Hasegawa+(1984,ApJ.Lett,287,L91), Lester+(1985,AJ,90,2331), Padman+(1985,MN,214, 251), Bastien+(1985,AA,146,86), Hermsen+(1985,AA,146,134), Heske+(1985,AA,149,199), Mason+(1985,ApJ Lett,295,L47), Vogel+(1985,ApJ,296,600), Wright+(1985,ApJ Lett,297, L11), Watson+(1985,ApJ,298,316), Goldsmith+(1985,ApJ,299,405), Zivrys+(1986,ApJ. Lett,300,L19), Maddalena+(1986,ApJ,303,375), Sugitani+(1986,ApJ,303,667), Crawford +(1986,ApJ Lett,303,L57), Davis+(1986,ApJ,304,481), Mundy+(1986,ApJ Lett,304,L51),

Crutcher+(1986,ApJ,307,302), Dragoran+(1986,ApJ,308,270), Goldsmith+(1986,ApJ,310, 383), Pendleton+(1986,ApJ,311,360), Wilson+(1986,AA,158,L1), White+(1986,AA,162, 253), Takaba+(1986,AA,166,276), Nakajima+(1986,MN,221,483), Zeng+(1987,AA,172,299), Walmsley+(1987,AA,172,311), Bally+(1987,ApJ Lett, 312,L45), Blake+(1987,ApJ,315, 621), Plambeck+(1987,ApJ Lett,317,L101), Geballe+(1987,ApJ Lett,317,L107).

Taurus: Kleiner+(1984,ApJ,286,255; 1985,ApJ,295,466), Gaida+(1984,AA,137,17), Cerni- charo+(1984,AA,138,371), Moneti+(1984,ApJ,282,508), Irvine+(1984,ApJ,282,516), Schloerb+(1984,ApJ,283,129), Goldsmith+(1984,ApJ,283,140), Mollar+(1985,MN,216, 1025), Colgan+(1986,AJ,91,107), Takano+(1985,AA,144,363), Ungerer+(1985,AA,146,123) Cernicharo+(1985,AA,149,273), Crutcher(1985,ApJ,288,604), Blake+(1985,ApJ,295,501), Brown+(1985,ApJ,297,302), Murphy+(1985,ApJ,298,818), Lebrun+(1986,AA,154,181), Cernicharo+(1986,AA,160,181), Duvert+(1986,AA,164,349), Matthews+(1986,ApJ,300,766) Tamura+(1986,MN,224,413), Cernicharo+(1987,AA,176,299), Olano+(1987,AA,179,202), Kleiner+(1987,ApJ,312,837).

ρ Ophiuchi: Zeng+(1984,AA,141,127), Lada+(1984,ApJ,287,610), Meyers+(1985,ApJ,288, 148), Snow+(1985,ApJ,288,277), Feigelson+(1985,ApJ Lett,289,L19), Crutcher+(1985, ApJ,290,251), Castilaz+(1985,ApJ,290,261), Wilking+(1985,ApJ,293,165), Wadiak+ (1985,ApJ Lett,295,L43), Cardelli+(1986,ApJ,302,492), Snow+(1986,ApJ,303,433), Young+(1986,ApJ Lett, 304,L45), Loren+(1986,ApJ,306,142).

Two isolated clouds that merit further studies are a very cold and quiet cloud in Monoceros (Maddalena+,1985,ApJ,294,231) and one in Norma, that produces new star (Alvarez+,1986,ApJ,300,756).

Other papers concerning individual clouds are the following:
M17: Schultz+(1985,AA,142,363), Keene+(1985,ApJ,299,867), Rainey+(1986,AA,171,252) Snell+(1986,ApJ,304,780), Schultz+(1987,AA,171,297). W3: Zeng+(1984,AA,140,169), Kolesnik (1986,AA,169,268), Thronson+(1984,ApJ,284,597), Claussen+(1984,ApJ Lett, 285,L79), Turner+(1984,ApJ Lett,287,L81), Thronson+(1985,ApJ,297,662), Mauersberge +(1986,AA,166,L26), Melnick+(1986,ApJ,303,638), Thronson (1986,ApJ,306,160), Reid+ (1987,ApJ,312,830), Gordon (1987,ApJ,316,258). CasA: Barcla+(1984,AA,136,127), Goss+(1984,AA,139,317), Troland+(1985,ApJ,298,808). DR21: Johnston+(1984,ApJ Lett, 285,L85), Dickel+(1985,ApJ,290,256), Dickel+(1986,AA,162,221). W49: Dreher+(1984, ApJ,283,632), Goss+(1985,MN,215,197), Miyawaki+(1986,ApJ,305,353). Chamelopardalis Torisera+(1985,AA,153,207), Jones+(1985,AJ,90,1191), Whittet+(1987,MN,224,497).

Other clouds have been described in the papers by Gardner+(1984,MN,210,23), Federman+(1984,ApJ,283,626), Joy+(1984,ApJ,284,161), Chini+(1984,AA,137,117), Eiroa (1984,AA,141,263), Richardson+(1985,MN,216,713), Rossano+(1985,AJ,90,308), Martin-Pintado+(1985,AA,142,131), Arnal+(1985,AA,145,369), Olofsson+(1985,AA,146, 337), Mauersberger+(1985,AA,146,168), Kahane+(1985,AA,146,325), Despois+(1985,AA, 148,83), Crovisier+(1985,AA,149,209), Fulkerson+(1985,ApJ,287,723), Hayashi+(1985, ApJ,288,170), Smith+(1985,ApJ,291,571), Haschick+(1985,ApJ,292,200), Zheng+(1985, ApJ,293,522), Makinen+(1985,ApJ,299,341), Kutner+(1985,ApJ,299,351), Martin-Pintad +(1985,ApJ,299,386), Churchwell+(1986,ApJ,300,729), Graham (1986,ApJ,302,352), Redeman+(1986,ApJ,303,300), Loughran+(1986,ApJ,303,629), Lebrun (1986,ApJ,306,16) Mundy+(1986,ApJ Lett,311,L75), Chini+(1986,AA,154,L8), Matthews+(1986,AA,155,99), Sorochenko (1986,AA,155,237), Ungerechts+(1986,AA,157,207), Chini+(1986,AA,157,L1 White+(1986,AA,159,309), Mattila (1986,AA,160,157), Celnik (1986,AA,160,287), Stenholm+(1986,AA,161,150), Murphy+(1986,AA,167,234), Andersson+(1986,AA,167,L1), Menten+(1986,AA,169,271), Casoli+(1986,AA,169,281), Wilking+(1986,AJ,92,103), Kuiper+(1987,MN,227,1013), Richard+(1987,MN,228,43), Menten+(1987,AA,177,L57), Walmsley+(1987,AA,179,231), Clark (1987,AA,180,L1), Avery+(1987,ApJ,312,848), Straw+(1987,ApJ,314,283), Jaffe+(1987,ApJ,316,231), Vallée (1987,ApJ,317,693), Henkel+(1984,ApJ,282,L93), Moreno+(1986,AA,161,130), Cunninghamm+(1984,MN,210,891 Liszt+(1985,AA,142,237).

The galactic center remains of interest: Hiromoto+(1984,AA,139,309), Liszt+ (1985,AA,142,245), Güsten+(1985,AA,142,381), Sandqvist+(1985,AA,152,L25), Armstrong +(1985,ApJ,288,159), Ho+(1985,ApJ,288,575), Harris+(1985,ApJ Lett,294,L93), Gezari +(1985,ApJ,299,1007), Gatley+(1986,MN,222,299), Walmsley+(1986,AA,155,129), Mezger +(1986,AA,160,324), Serabyn+(1986,AA,161,334), Serabyn+(1986,AA,169,85), Thronson+ (1986,ApJ,300,396), Churchwell+(1986,ApJ,305,405), Bania (1986,ApJ,308,868), Goldsmith+(1987,ApJ Lett,313,L5), Goldsmith+(1987,ApJ,314,525), Heiligman (1987, ApJ,314,747), Vogl+(1987,ApJ,316,243).

On the Magellanic clouds only 3 papers appear in this list: Morgan (1984,MN, 209,241), Epchtein+(1984,AA,140,67), Israel+(1986,ApJ,303,186). On M31/M33 : Boulanger+(1984,AA,140,L5), Blitz (1985,ApJ,296,481), Ryden+(1986,ApJ,305,823), Casoli+(1987,AA,173,43), Vogel+(1987,ApJ,321,L145).

On other galaxies: Sadler+(1985,MN,214,177), Bhatt+(1986,MN,219,217), Rydbeck+ (1985,AA,144,282), Stark+(1986,ApJ,310,660), Blitz+(1986,ApJ,311,142), Young+ (1986,ApJ Lett,311,L17), Wyse (1986,ApJ Lett,311,L41), Sanders+(1987,ApJ Lett,312, L5), Sargent+(1987,ApJ Lett,312,L35), Myers+(1987,ApJ Lett,312,L39), Lo+(1987,ApJ, 312,574), Lo+(1987,ApJ Lett,317,L63).

4. HIGH VELOCITY OUTFLOWS

Flows of matter at high speed from young stellar objects have been reviewed by Lada (1985,ARAA,23,267); the review contains a catalogue of 68 high-velocity molecular flows. The flows are noticed by a variety of indicators: high velocity H_2O masers; broad emission lines of thermally excited molecules (e.g. CO); emission lines of vibrationally excited H_2 ; Herbig-Haro objects; "FU Orionis" type objects (very rare). A new detection method is probably through the absorption lines of CO at 4.8 micron: in M8E the lines show the presence of warm CO near the star (Larson+:1986,ApJ,307,295). An IR polarization study by Lenzen (1987,AA,173, 124) uncovered several reflection nebulae very close to the central objects. Near-infrared cameras have shown the existence of loops or arcs (Forrest+:1986,ApJ Lett, 311,L81). Observations of emission lines of CS by Heyer+(1986,ApJ,308,134) show no symmetries associated with the bipolar axis and they conclude that the wind is not focussed by large interstellar objects. Mundt+(1984,AA,140,17) discovered three more optical jets, and Krauter (1986,AA,161,195) discovered a jet in Th28. A systematic mapping program in a CO line led Fukui+(1986,ApJ Lett,311,L81) to the discovery of 7 new outflow sources. Masses and energetics in six well-studied flows have been estimated by Margulis+(1985,ApJ,299,925).

From IRAS data luminosities have been derived of 45 sources associated with molecular outflows by Morzukewich+(1986,ApJ,311,371); the range is from a few solar luminosities for e.g. TTau to several times 10^6 solar luminosities for NGC 6334. Low-luminosity sources have been studied in infrared lines by Smith+(1987, ApJ,316,265), who conclude that low-luminosity sources are likely to be qualitatively different from high-luminosity sources.

The mechanisms that produces these huge flows are not yet known; explanations have been proposed by Cameron (1985,ApJ Lett,299,L83), by Kwok+(1985,ApJ,299,191), and by Choe+(1986,ApJ,305,131). More earthly approaches to theoretical support have been taken by Reynolds (radio continuum spectra; 1986,ApJ,304,713), by Liseau +(geometrical information from the shapes of CO line profiles; 1986,ApJ,304,459, see also Cabrit+:1986,ApJ,307,313); and in papers on the bow shocks associated with HH regions (Choe+:1985,ApJ,288,388; Raga, 1986,AJ,92,637; Hartigan+:1987,ApJ, 316,323; Raga+:1986,AJ,92,119).

The most surprising area (Orion/Monoceros) contains at least three different active regions: Orion A (M42), Orion B (NGC2024), and the Monoceros Area (Mon R2).

In the cloud L1641, that contains M42, there are now 7 high outflow sources providing enough momentum to stabilize the cloud; 4 of these 7 have been discovered
via a systematic CO survey for high velocity wings (Fukui+,1986,ApJ Lett,311,L85).

    Other papers concerning the flows in this general area are: Orion A:
Kuiper+,1984,ApJ,283,106; Jones+,1985,AJ,90,1320; Schwarz+,AJ,90,1820; Thronson+,
1986,AJ,91,1350; Meaburn, 1986,AA,164,358; Wilson+,1986,AA,167,L17; Taylor+,1986,
MN,221,155; Geballe+,1986,ApJ,302,693; White+,1986,ApJ,302,701; Ziurys+,1987,ApJ
Lett,314,L49. Several papers deal with Herbig Haro objects in the Orion nebula
(HH1 and HH2): UV and optical spectrum: (Meaburn+,1984,AA,138,36; Brugel+,1985,
ApJ Lett,292,L75; Boehm+,1985,ApJ,294,533; Boehm+,1987,ApJ,316,349); infrared:
(Strom+,1985,AJ,90,2281; Pravdo+,1987,ApJ,314,308; Harvey+,1986,ApJ,301,346);
radio continuum emission: (Pravdo+,1985,ApJ Lett,294,L117); Martin-Pintado+,1987,
AA,176,L27).

    Orion B: Sanders+(1985,ApJ Lett,293,L39); Chalabaev+(1986,AA,168,L7); Russel+
(1987,MN,226,287).
    Mon R: Brugel+(1984,ApJ Lett,298,L73); Aspin+(1985,AA,149,158); Hughes+(1985,
ApJ,289,238).
Papers concerned mainly with specific regions are:
L1551 IRS5: radio sources: (Bieging+,1985,ApJ Lett,289,L5; Snell+,1985,ApJ,290,587;
Rodriguez+,1986,ApJ Lett,301,L25); optical obs.: (Mundt+,1985,ApJ Lett,297,L41;
Sarcander+,1985,ApJ Lett,288,L51); infrared: (Edwards+,1986,ApJ Lett,306,L65; Clark
+,1986,AA,154,L25; Clark+,1986,AA,158,L1); velocity gradients in CO flow (Fridlund
+,1984,AA,137,L17); no rotating disk (NH3: Menten+,1985,AA,146,369; CS: Moriarty-
Schievan+,1987,ApJ Lett,317,L95); high velocity OH: (Mirabel+,1985,ApJ Lett,294,
L39); rotation measures in background galaxies: (Simonetti+,1986,ApJ,303,659); a
thick disk: (Strom+,1985,AJ,90,2575).
    AFGL2591: 12CO observations indicating the flow and IR observations by Lada+
(1984,ApJ,286,302); a compact rotating disk from NH3 (Takano+,1986,AA,158,14) and
some faint radio sources (Cambell, 1984,ApJ,287,334).
    Cepheus A: source of flow (Cohen+,1984,MN,210,425; Guesten+,1984,AA,138,205;
Torelles+,1986,ApJ,305,721); optical and IR observations (Linzen+,1984,AA,137,202;
Hartigan+,1986,AJ,92,1155); CO(3-2) line: (Richardson+,1987,AA,174,197); HCO+:
(Loren+,1984,ApJ,287,707); rotation measures of background galaxies: (Simonetti+,
1986,ApJ,303,659).
    AFGL 490: (Kawabe+,1984,ApJ Lett,282,L73); Gear+(1986,MN,219,835). R Cor Aus:
Ward-Thompson+(1985,MN,215,537); Castelaz+(1987,ApJ,314,317); Hartigan+(1987,AJ,93,
913). G35.2-0.74: Matthews+(1984,AA,136,282); Dent+(1985,AA,136,282). NGC 2071 :
Takano+(1984,ApJ Lett,282,L69); Takano+(1986,AA,167,333); Takano (1986,ApJ Lett,
300,L85); Scoville+(1986,ApJ,303,416) (high resolution mapping of radio line).

    Various other papers: Zealey+(1984,AA,140,L31); Campbell (1984,ApJ Lett,282,
L27); Torrelles+(1985,ApJ,288,595); Harvey+(1985,ApJ,288,725); Richardson+(1985,
ApJ,290,637); Sato+(1985,ApJ,291,708); Schwarz+(1985,ApJ,295,89); Cohen+(1985,ApJ,
296,620); Cohen+(1985,ApJ,296,633); Little+(1985,AA,142,378); Aspin+(1985,AA,144,
220); Lightfood+(1986,MN,221,993); Castelaz+(1986,ApJ,300,406); Zealey+(1986,AA,
158,L9); Buehrke+(1986,AA,163,83); Guesten+(1986,AA,164,342).

5. CIRCUMSTELLAR MOLECULAR ENVELOPES (CSE's)

The study of dense winds from red giants has evolved dramatically due to progress in the techniques for observing millimeter-wave, infrared and maser transitions of circumstellar molecules. A reference of special use is from a conference on Mass Loss from Red Giants (1985, Morris and Zuckerman, eds., Reidel, Dordrecht) which includes reviews of nearly all aspects of CSE's up to 1984. Isotopic abundances in CSE's are reported above, in the section on Isotope Abundances.

5.1. Millimeter-wave Observations

Surveys of millimeter-wave CO emission have been carried out by several groups using infrared luminosities as guides to the source selection process (Knapp+: 1985,ApJ.292,640; Zuckerman+: 1986,ApJ,304,394; Zuckerman+: 1986,ApJ,304, 401; Wannier+: 1986,ApJ,311,335; Likkel+: 1987,AA,173,L11; Wannier+,ApJ, in press). The total number of known CSE's is now about 100, of which half are C-rich and half O-rich. With the improved statistics, there has been progress in understanding the interrelations of mass-loss rate, gas-to-dust ratios, IR luminosities and abundances, but not without considerable scatter and notable exceptions to each rule (Zuckerman+: 1986,ApJ.311,345; Knapp: 1985,ApJ.293,273; Knapp: 1986,ApJ.311, 731). In this regard, two objects of special note are 1) a particularly hot and photo-dissociated outflow of very modest proportions and displaying HI emission (Huggins: 1987,ApJ,313,400; Bowers+: 1987,ApJ,315,305), and 2) IRAS 09371+1212, a particularly cold and dense outflow of large proportions and displaying strong circumstellar ice bands (Forveille+: 1987,AA,176,L13). Also, with additional observations, the division of stars into clear C/O abundance classes has been muddied by the detection of HCN in O-rich sources (Deguchi+: 1985, Nature,317,336; Jewell+: 1986,Nature,323,311).

Great progress has been made on determining the structure and chemistry of CSE's, subjects which are intimately intertwined. Many relevant articles appear in the proceedings of an IAU symposium on Astrochemistry (Vardya+,1985). Direct observations of CSE structure have been made by new arrays, and by decreasing antenna beamsizes. CRL 2688, in particular, has been mapped at cm wavelengths in NH_3 and HC_7N with the VLA (Nguyen-Q-Rieu+: 1986,AA,165,204), at mm wavelengths in CO with the OVRO interferometer and the Nobeyama 45m telescope (Heiligman+: 1986,ApJ,308,306; Kawabe+: 1987,ApJ,314,322) and in the near IR in H_2 (Beckwith+: 1984,ApJ,280,648). These complementary techniques lead to a detailed picture of a rapid bipolar outflow and a dense molecular toroid surrounded by a cool, extended envelope. The greatest attention has been focused on IRC+10216 where interferometry of HCN (Bieging+: 1984,ApJ,285,656) and maps of C_2H and CN (Truong-Bach+: 1987, AA,176,285) confirm predictions based on models which include photodissociation chemistry. Several new CSE chemical models have appeared focusing on molecule formation, especially near the stellar surface (Tsuji: 1986,Ann.rev.AA,24,89; Shmeld+ : 1985,Sov.Astron.Lett,11; Slavsky: 1984,Proc.Southwest Reg.Conf. for Astron.Astrophys.,9,43). One model, including photo-dissociation effects, predicts the spatial extent of circumstellar molecular ions (Glassgold+: 1986,AA,157,35) and models of ion-molecule chemistry have been made for both C-rich and O-rich stars (Nejad+: 1987,AA,in press; Nejad+: 1987,MN, submitted).

The steady march of new molecular lines in CSE's is also led by IRC+10216. Of particular interest is a series of transitions of vibrationally excited states of well-known molecules, namely: 1) HCN (v=1) (Ziurys+: 1986,ApJ,300,L19), 2) CS (V=1) (Turner: 1987,ApJ, submitted, and 3) SiS (v=1) (Turner: 1987,ApJ, submitted). These transitions probe small radii and reflect intersting envelope kinematics. Other new lines are of SiCC (Thaddeus+: 1984,ApJ,283,L45; Snyder+: 1985,ApJ, 290,L29), SO_2 (Lucas+: 1986,AA,154,L12), C_8H (Cernicharo+: 1986,AA,167,L5), CO (6-5) (Sahai: 1987,ApJ,318,in press), C_6H (Guelin+: 1987,AA,175,L5; Saito+: 1987, PASJ,39,193), C_2H_4 (Goldhaber+: 1987,ApJ,314,356), HSiCC or HSCC (Guelin+: 1986,

AA,157,L17), a new line of NH₃ (Nguyen-Q-Rieu+: 1984,AA,138,L5), and a line survey
of IRC+10216 and CIT6 (Henkel+: 1985,AA,147,143). Each of these lines, along with
high signal-to-noise spectra of CO in IRC+10216 (Huggins+: 1986,ApJ.304,418)pro-
vides grist for models which use radiative transfer and self-consistent molecular
excitation to yield a detailed radial molecular structure in IRC+10216 (Sahai:1987,
ApJ.318 in press; Morris+: 1985,AA,142,107). Models of O-rich and non-spherically
symmetric objects are less advanced, but will be called for with such new observations as those of SO and SO2 in O-rich stars (Guilloteau+: 1986,AA,165,L1) and HCN
in C-rich bipolar reflection nebulae (Deguchi+: 1986,ApJ,303,810). Sought, but not
found, were CCℓ, CℓO, MgO and TiO (Millar, 1987,AA,in press), while Cernicharo+
(1987,AA,183,L10) have found NaCℓ, AℓCℓ, KCℓ and perhaps AℓF.

5.2. Infrared Observations

Infrared spectroscopy has yielded important information about the inner envelopes, where mass-loss is initiated, and about the outer photospheres, where there
is interaction with the Mira shocks. Most of the work has been from CO bands,
though a 2400-2778 cm⁻¹ atlas has revealed lines of OH, CH, SiO, CS and HCl in observations of K,M,C and S stars (Ridgway+: 1984,ApJ,54,177). The CO bands themselves provide a lot of flexibility, and the hotter, denser, regions are probed by
successively larger vibrational overtone numbers and by higher rotational levels.
The fundamental 4.6 micron band has been used, with annular observing apertures,
to isolate weak CO emission features separated by up to 6arcsec in IRC+10216, revealing an unexpectedly hot rotation temperature (Sahai+: 1985,ApJ,299,424). The
firstovertone band at 2.3 microns has been observed in 18 M-type stars to measure
circumstellar C/H (Tsuji: 1986,AA,156,8). Second overtone bands, formed near the
stellar photosphere, have yielded time-sequence observations which follow the passage of regular shocks in eight Mira variables and one SRa variable star (Hinkle+:
1984,ApJ.Suppl,56,1). Finally, several authors have used the CO bands to derive C
and O isotope abundances in CSE's and these results are discussed above in the section on "Isotope Abundances" (above). In addition to the CO vibration/rotation
bands, significant attention has been focused on the formation of very large molecules (PAH's) and the initiation of dust formation. That subject is treated more
extensively under the section on "Interstellar Dust" (below). In circumstellar environments, laboratory experiments have demonstrated a possible path to form long-
chain hydrocarbons and small PAH's (Heath+: 1987,ApJ,314,352) and several authors
have focused on the problem of initiating grain growth formation (Kozasa+: 1984,
Ap.Space Sci,98,61; Dojkov: 1985,Ref.Zh,51, Astron.5.51.431; Jura+: 1985,ApJ.292,
487; Shmeld: 1985,Astron.Zh.62,1229). The problem of distinguishing C-rich from
O-rich chemistries has been further muddied by observations of silicate features
around C-rich stars, (Little-Marenin: 1986,ApJ,307,L15; Willems+: 1986,ApJ,309,
L39).

5.3. OH Maser Observations

A general survey of galactic OH masers confirms a population peak at R≈7.5kpc
(Tong+: 1984, Chinese AA,8,343). A survey near the galactic center with the VLA
has revealed 33 stars, strongly concentrated at the Sgr A West source and with
unusually large expansion velocities (Winnberg+: 1985,ApJ,291,L45). A survey at the
galactic tangent point has been used to derive intrinsic physical parameters of
OH/IR stars (Baud+: 1985,ApJ,292,628). In addition, an IRAS source selection criterion has been proven effective (Lewis+: 1985,Nature,313,200) and a general survey has turned up seven new CSE sources (Slootmaker+: 1985, AA Suppl,59,465). A
search for OH emission from symbiotic stars was unsuccessful (Norris+: 1984,Proc.
Astron.Soc.Aust.,5,562) and a search for emission from short-period Mira's partially successful (Dickenson+: 1987,AJ,92,416).

With the regular spectral line operations of the VLA and MERLIN have come
many continuing studies of the spatial structure of OH emission. VLA observations

of 11 objects near the tangent point reveals general circular symmetry, in agreement with an earlier survey and places the Sun at 9.2 ± 1.2 kpc from the galactic center (Herman+ 1985,AA,143,122). Detailed VLA studies have been made of notable objects IRC+10420 (Bowers: 1984,ApJ,279,350) and OH231.8+4.2 (Bowers: 1984,ApJ,276, 646). MERLIN observations have also continued apace, with maps of OH127.8-00 (Diamond+: 1985,MN,216,1), U Ori (Chapman+: 1985,MN,212,375) and five other objects, including a review of prior maps (Diamond+: 1985, MN,212,1).

Other techniques include studies of the regular (Herman+: 1985,AA Suppl,59, 523; Bowers+: 1984,ApJ,276,646) and sudden (Lewis+: 1986,ApJ,302,L23; Yudaeva:1986, Pis'ma Astron.Zh.12,No5,361; Le Squeren+: 1985,AA,152,85) time variabilities of the maser features, allowing details of structure to be inferred and sometimes simply leading to puzzlement. Studies using high-resolution spectrometers reveal the presence of many (thousands) of small masing elements (Fix: 1986,AJ,93,433; Fix: 1987,AJ,92,433) and of unexpected magnetic fields (Cohen+: 1987,MN,225,491). Maser polarization has been used to infer time variation in the velocity field (Ukita+: 1984,AA,138,343). General models of OH masers have been made (Alcock+: 1986,ApJ, 305,837; Dickinson: 1987,ApJ,313,408) as well as ones applicable to specific cases (Grinin: 1985,Izv.Krymskoj Astrofiz.70,139).

5.4. Other Masers

A new strong maser, of HCN, has revealed itself in CIT6 (Guilloteau+: 1987,AA, 176,L24), being the first strong maser ever seen in a C-rich envelope. Otherwise, H_2O and SiO masers, in O-rich envelopes have provided most information. SiO maser emission has been studied using the v=1, J=1-0, J=2-1, J=3-2 and J=5-4 lines as well as v=2, J=2-1. The most extensive SiO survey used the NRO antenna, simultaneously observing six SiO transitions. SiO masers were detected in 83 stars, with a v=3 maser in eight stars and a suggestion of ^{29}SiO (v=0) masing in six stars (Cho+: 1986,Astr.Space Sci,118,237). Using OH masers to guide source selection an NRO survey also detected SiO masers (v=1 and 2) near the galactic center (Lindqvist +: 1987,AA,172,L3). Other surveys have also revealed a significant number of new SiO maser sources (Jewell+: 1985,ApJ,298,L55; Barcia+: 1985,AA,142,L9; Nyman+:1986, AA,160,352; Bujarrabal+: 1987,AA,175,164). Time variability and polarization of SiO v=1 maser sources has been undertaken by several groups (Nyman+: 1985,AA,147, 309; Nyman+: 1986,AA,158,67; Miller+: 1984,ApJ,287,892; Clark+: 1985,ApJ,289,756; Snyder: 1987,AJ,92,416), suggesting a correlation with the Mira pulsation and a pump located close to the star. The pedestal features, not the intense spikes, are revealed to be the most useful in deriving general properties of the envelopes. One star, χ Cyg, has also been monitored in the v=2, J=2-1 line (Olofsson+: 1985, AA,150,169). A multi-transitional polarization study of SiO masers reveals that features arising from the same rotational state, but different vibrational states originate in the same volumes of gas, but different rotational states are from different volumes (Barvainis+: 1985,ApJ,288,694). A collisional model of SiO masers now includes all observed transitions and predicts relative maser strengths including non-observed transitions (Bolgova: 1984, Nauchn.Inf.57,39). The polarization of SiO masers has been modeled with the use of radiative transfer in the presence of magnetic fields (Deguchi+: 1986,ApJ,302,108) and a detailed model of the SiO maser in VY CMa includes a rotating disc and two independent gas streams(Zhou+ : 1984,AA,138,359).

A number of surveys of the 22 GHz H_2O maser have been made (Engels+: 1986,AA, 167,129; Zuckerman+: 1987,AA,173,263; Engels+: 1984,AA,140,L9; Bowers+: 1984,ApJ, 285,637). These surveys indicate emission generally from about 10^{15} cm, with amplitudes correlated with mass-loss rate and OH and IR variability. A suggested phase lag with respect to OH implies a collisional pump for the H_2O maser. One oddball maser source, associated with the carbon star EU and is shown to originate in a binary pair including an M giant (Benson+: 1987, ApJ,316,L37). The spatial structures of H_2O masers have been studied in VX Sgr with MERLIN (Chapman+: 1986,MN,220,

513) and in RX Boo, R Aql, RR Aql and NML Cyg with the VLA (Johnston+: 1985,ApJ, 290,660) revealing details of the spatial, velocity and magnetic field structure. Time variability of H_2O emission reveals aperiodic behaviour, suggesting the passage of individual Mira shocks far out into the CSE (Gomez Balboa+: 1986,AA,159, 166). A new model of H_2O maser emission, using current collision cross-sections, satisfactorily reproduces the 22 GHz maser and predicts other strong maser lines in the 227-789 micron region (Cooke+: 1985,ApJ,295,175).

6. INTERSTELLAR MASERS

The observational and theoretical study of masers in both the interstellar gas and stellar envelopes continues to provide valuable information on the physical state of the emitting regions. Of the various masers which have been detected, OH and H_2O continue to be the most extensively observed, followed by NH_3, SiO, H_2CO and CH_3OH (methanol). General references not cited at the beginning which are relevant to the period under review are: Burke,ed.:1984,"Quasat - a VLBI Observatory",ESA-SP213; Haschick, ed.1986:"Masers, Molecules and mass outflows in starforming regions", Haystack, Westford.

6.1. OH masers

Observations of W51 (main) using the VLBI technique (Benson+: 1984,AJ,89,1391) display Zeeman splitting indicative of mG magnetic fields. There have been VLA observations of NGC 7538(IRS1) by Palmer+,(1984,MN,211,41P) and of Sgr B2 (Gardner+: 1987,MN,225,469). MERLIN observations of 1665 MHz emission from W75N were reported by Baart+ (1986,MN,219,145). Other observations relate to OH megamasers (Baan+: 1985,ApJ,298,L51; Bottinelli+: 1985,AA,151,L7), a maser outburst in Ceph A (Cohen +: 1985,MN,216,51P), late-type stars (see Section 6.3 above), and to Mira variables, infrared stars, and molecular clouds (Slootmaker+: 1985,AA Suppl,59,465).The peculiar OH maser G24.3+0.1 has been observed by Paschenko (1984,Sov.Astron.Lett., 10,303). The collisional pumping of the main lines has been discussed by Andresen+ (1984,AA,138,L17), on the basis of laboratory data. The problem of radiation transport has been considered by Field+(1984,MN,211,799, erratum in 1985, MN,213,495) and Field (1985,MN,217,1), and also by Deguchi+: (1986,ApJ,300,L15).

6.2. H_2O masers

The H_2O masers in star-forming regions have been discussed by Downes (Lucas+, 1985,557). A young stellar object has been discovered near the water masers in W3 (OH) (Turner+: 1984,ApJ,287,L81). Flares have been observed in the 8 km/s H_2O maser source in Orion (Abraham+: 1986,AA,167,311), in W75N (Lekht+: 1984,Sov.Astron.Lett. 10,307) and H_2O maser outbursts which may be connected with protoplanetary rings by Matveenko (1986,Sov.Astron,30). Maser emission from stars in the IRAS point source catalogue is reported by Zuckerman+: (1987,AA,173,263), and in nearby galaxies by Whiteoak+ (1986,MN,222,513). The related questions of radiative transfer are discussed by Chandra+ (1984,AA,140,295), of maser pumping by Kylafis+ (1986,ApJ, 300,L73), and of the influence of magnetic fields by Deguchi+ (1986,ApJ,302,750). A cloud-cloud collision model for H_2O maser excitation has been presented by Tarter + (1986,ApJ,305,467).

6.3. Other masers

New interstellar masers in non-metastable ammonia have been reported by Madden +(1986,ApJ.300,L79), and an interstellar $^{15}NH_3$ maser by Mauersberger+ (1986,AA,160, L13). Both maser and thermal emission from ammonia have been observed towards the star-forming region W51 (IRS2) by Mauersberger+ (1987,AA,173,352). Observations have been made with the VLA of both methanol (Menten+:1985,ApJ,293,L83) and formaldehyde (Gardner+: 1986,MN,218,385). New maser lines of methanol are reported by Morimoto+(1985,ApJ,288,L11). The masers of silicon oxide have been observed in

red-supergiants and in the vicinity of molecular clouds by Ukita+(1984,AA,138,194),
and in Mira variables, short-period variables, and OH/IR stars by Jewell+(1985,ApJ,
298,L55).

IV. Interstellar Dust
(J.S. Mathis)

In addition to the references listed at the beginning, several specialized
symposia are: Wolstencroft R.D.+,ed.,1984, "Laboratory and Observational Infrared
Spectra of Interstellar Dust", Occas.Rep.Roy.Obs.Edinburgh No 12;Nuth,J.A.+,ed.,
1985,"Interrelationships among Circumstellar, Interstellar and Interplanetary Dust",
NASA CP-2403;Léger,A.+,ed.,1987, "Polycyclic Aromatic Hydrocarbons and Astrophysics"
Reidel,Dordrecht;Geogenthal workshop: 1986, "The Role of Dust in Dense Regions of
Interstellar Matter", ApSpSci,128.

Emission from dust: Many objects, all of which contain dust, (Willner,1984,
Kessler+,84) show "Unidentified Infrared Bands" (hereafter "UIBs", but also called
"UIR bands" by many authors). They range from 3.28 to 11.4 μm, with a plateau of
emission extending out to 13 μm and probably extending to wavelengths beyond this
value (Cohen+,1986,ApJ,302,737). They are strong in the reflection nebulae NGC 2023
and NGC 7023 (Sellgren+,1985,ApJ,299,416). Witt+(1984,ApJ,281,708;1986,ApJ,294,216
and 225) found that the emission extends to wavelengths of about 0.6 μm and has
a component of broad bands and a continuum. Castelaz+(1986,ApJ,313,853) determined
the IRAS surface brightness maps of the reflection nebulae in the Pleiades and
found a high 12/25 μm color temperature at both large and small distances from the
various stars. As regards interstellar dust, probably the most important findings
of the IRAS satellite were the existence of streaks of diffuse material called
"cirrus" (Low+,1984,ApJ Lett,278,L19). This material is primarily visible in the
60 and 100 μm IRAS filters, is quite faint at 25 μm, but is surprisingly strong at
12 μm, with a high color temperature (typically 250 K) between the 12 and 25 μm
band ratios (Hauser+,1984,ApJ Lett,278,L15). The high 12/25 μm color temperature
showed that the emission is a non-equilibrium process. High-latitude molecular
clouds and cirrus are closely related in space (Weiland+,1986,ApJ Lett, 306,L101),
suggesting that the cirrus is only about 100pc from the plane of the Galaxy. However, cirrus is not always bright where the CO emission is strong, and conversely.
There are streams of cirrus seen near the LMC but of Galactic origin (McGee+,1986,
MN,221,571) which have narrow-line 21-cm emission.

Interpretation of the excess emission: The excess emission at wavelengths of a
few micrometers can arise from one or both of two related processes: (a) It might
be radiative cascading from high vibrational levels of a molecule excited after
absorption of a photon. This process tends to concentrate the emitted energy into
bands, but there might be a real continuum if the molecule is excited enough. (b)
The emission may arise from the thermal emission of a tiny grain, heated to hundreds of K by the absorption of a single photon and cooling. This grains might be
thought of as a disordered molecule, so large that its bands are replaced by a
continuum. Temperature fluctuations in small grains have been modelled by Draine+
(1985,ApJ.292,494) and by Désert+(1986,AA,160,295), treating the grain classically.
Fluctuations caused by a grain being struck by a single energetic electron in hot
gas have been modelled by Dwek (1986,ApJ,302,363).

There is at present little doubt that some form of hydrogenated carbon is
responsible for the UIBs and associated continuum, but precisely what form is controversial. Léger (1984,AA,137,L5) and Puget+(1985,AA,142,L19) suggested that the
UIBs can be accounted for by about 6% of the cosmic carbon being in the form of
polycyclic aromatic hydrocarbons (benzene-ring structures), hereafter "PAHs", of up
to about 60 carbons, and that the spectrum of an excited PAH should correspond to
the UIBs. Independently, Allamandola+(1985,ApJ.Lett.290,L25) pointed out that the

Raman spectra of auto soot, containing PAHs and amorphous hydrogenated carbon particles, strongly resembles the 5 - 10 μm emission spectrum of the Orion Nebula.
Barker+(1987,ApJ Lett,315,L61) use the width of the observed UIBs to suggest that
no more than 20 or 30 C atoms are in the PAHs. Reasons for associating PAHs with
the UIBs and possibly the NIR emission continuum are discussed extensively in Léger
and d'Hendecourt's article in (1987,Léger+,223), where there are many other important papers about these molecules. The properties of PAHs, such as their ionization
and recombination cross-sections, response to photons, and the like have been also
discussed by Omont (1986,AA,164,159). PAHs in space are likely to be ionized, either
negatively in low radiation fields or positively near stars (as in reflection nebulae). There are problems with understanding the UIBs completely on the basis of
PAHs. Duley+(1986,MN,219,859) have shown that chemical reactions in the interstellar
medium might destroy PAHs rapidly. Donn+(1987,BAAS,18,1030) have objected to the
idea on the basis of a lack of agreement between the emission of the UIBs and PAHs
obtained in the laboratory, and the likely production of fluorescence from a molecule in a high energy state. Such fluorescence is not seen in the spectrum of many
reflection nebulae, nor in the diffuse galactic light of the night sky.

A highly ordered form of carbon which is produced when solid carbon is irra-
diated with a laser is a hollow molecule of sixty regularly spaced carbon atoms
called "buckminsterfullerene" after the architect who designed houses with a similar structure (Kroto+, 1986,Nature,318,162). Zhang+(1986,J Phys Chem,90,525) discuss
spherical forms of carbon. These forms are very stable against photodissociation.
They may be the cause of the λ 2175 bump (Hoyle+,1986,Ap Sp Sci,122,181). There is
another form of carbon which might well be responsible for some or most of the IR
emission: hydrogenated "amorphous" carbon (HAC). Such material should be called
"disordered", because there are domains of graphite-like structure within it, as
well as tetrahedral bonds (diamond) and randomly placed carbons. There can well be
H atoms both on the surface and within the interior since H is readily intercalated
in such material. Laboratory studies of HACs (Borghesi+,1987,ApJ,314,422) show many,
but not all, of the UIBs. Duley (1984,ApJ,287,694) measured the indices of refraction of amorphous carbon in the optical through the UV. His material contains tetrahedral (diamond) bonding which may affect the far-UV absorption properties. The
optical properties of HACs and many other material are discussed in an excellent
article by Tielens and Allamandola in (1987,Hollenbach+,p.397). Tiny (50 Å) diamonds
are actually found in some meteorites (Lewis+,1987,Nature,326,160). Materials which
produce the UIBs in the laboratory have been produced by discharges through methane
gas (Sakata+,1984,ApJ Lett,287,L51). This material is probably some sort of disordered aromatic hydrocarbon.

The "Red Rectangle" (HD 44179) shows that the red/IR emission can be quite
strong in some cases. It is a dense nebula surrounding HD 44179 (AO III).
d'Hendecourt+(1986,AA,170,91) showed that its broad emission bands can be accounted
for by PAHs even if they are isolated molecules. They require about 10% of cosmic
C in PAHs. The PAHs in the Red Rectangle might be smaller and simpler than in the
general ISM because the radiation field there is more benign (Cohen+,1986,ApJ,302,
737). Duley(1985,MN,215,259) showed that the emission from the Red Rectangle could
also be caused by HACs. Wdowiak(1985,Nuth+,A41) suggested that PAHs condensed onto
solid grains would produce broadband emission.

The properties of Ultraviolet (UV) Extinction have been reviewed by Friedemann
(1986,Ap.Sp.Sci.128,71) and by Mathis (1987,Kondo+ book,517). It has become increasingly clear that the UV extinction (from λ = 0.3 μm to 0.12 μm) can be separated
into three components:(a) a smooth (assumed linear) background increasing with λ^{-1},
(b) the λ2175 "bump", and (c) the "far-UV rise", which is a rapidly increasing extinction for λ < 0.16 μm. Witt+(1984,ApJ.279,698), Carnochan (1986,MN,219,903) and
Nicolet (1987,AA,177,233) have determined the variations of UV extinction in many
objects. Massa+(1986,ApJ.Suppl,60,305) and Fitzpatrick+(1986,ApJ,307,286) have done
careful analyses of the extinction laws of five clusters and of the best-determined

early-type stars. They find that the width of the bump varies significantly (up to
a factor of two) from region to region, with dust in dense clouds having a wider
bump than in the diffuse interstellar medium. A very interesting result is that the
central wavelength of the bump, λ_0, is very constant within the sample of stars they
studied: a maximum variation from the mean of only 17 Å, and a mean variation of 9
Å. However, the variations among lines of sight from one cluster to another show
that the variation in λ_0 are real.

By comparison of the spectrum of the central star in a reflection nebula with
that of the nebula, Witt+(1986,ApJ.Lett,305,L23) conclude that there is a scattering
feature on the long-wavelength side of the bump. Most theories regarding the origin
of the bump attribute it to very small particles which do not scatter efficiently.
There is an anticorrelation between the 60/100 μm ratio in diffuse clouds and the
strength of the bump (Leene+: 1987,AA,174,L1). The origin of the bump is still controversial. Small graphite particles are probably the most favored explanation for
it (Hecht: 1986,ApJ,305,817; 1987,ApJ,314,429, and other authors). It can possibly
be fitted by oxides and glassy carbon (Duley+: 1983,Ap.Sp.Sci,95,187). It might also
be absorption of OH⁻ ions in tiny silicates (Steel+: 1987,ApJ,315,337) or be of
biological origin (Hoover+: 1986,Earth, Moon and Planets,35,19).

The "Far-UV Rise": There is a steep rise in extinction with λ^{-1} below λ = 0.16
μm which seems uncorrelated with optical properties of the extinction (Greenberg+:
1983,ApJ.272,563;Franco+: 1985,AA,147,191). The shape of the extinction is similar
from one line of sight to another, so there is probably a single grain component
responsible for this extinction. The spectra of reflection nebulae (Sellgren+: 1985,
ApJ,299,416) strongly suggest that this material is not small silicate particles
(Désert+: 1986,AA,159,328) because there is a deficiency of emission at the 9.7 μm
band of silicates, while the emission is caused by the absorption of energetic
photons by small grains or large molecules. Clayton+(1986,AJ,93,157) find that there
is a very strong far-UV rise in stars in the cluster Trumpler 37.

Distribution of dust: The locations of clouds and local regions of excess
extinction have been studied by several workers. Knude published uvbyβ photometry
of stars in Selected Area 132 (1986,AA Suppl,63,313). Urasin+(1987,Astr.Nach,308,5)
determined the distribution of dust in the interval 7°-222° and obtained a model
of a two-arm spiral structure. Uranova determined the distribution of dust in the
local spiral arm in Cygnus (1985,Pis'ma AZh,11,251). Star counts from various regions
are used to determine the distribution of dust (Feitzinger+: 1984,AA Suppl,58,365;
1986,ApJ,305,534).Bochkarev (1984,Pis'ma AZh,10,184;Sov.Astr.Lett,10) discusses the
formation of arclike dust structures from the action of stellar winds and supernovae on the ISM. The distribution of absorption in the Rosette Nebula and others was
determined by Guseva+(1984,Pis'ma AZh,10,741; 1985,Astrofiz,22,505), in the W3 complex by Kolesnik (1986,AA,169,268), and in NGC 2516 by Clocchiatti+(1986,AJ,92,1130).

The dust along the line of sight to the galactic center is similar to that in
the usual diffuse interstellar medium rather than that in molecular clouds (Roche+:
1985,MN,215,425). Butchart+(1986,AA,154,L5) have determined an excellent profile
of the 3.4 μm absorption band of IRS 7 near the Galactic Center. The band is important because it is caused by the C-H stretch of materials and coatings on grains.
The 1.3 mm emission of dust near the galactic center has been determined by Mezger+
(1986,AA,160,324).

Extinction at optical wavelengths is modulated by a Very Broad Band structure
(Krelowski+: 1986,AA,166,271) which has been reported in the past. For wavelengths
of a few micrometers, extinction is closely proportional to λ^{-1} (Rieke+: 1985,ApJ.
288,618). Extinction has been studied from the spectroscopy of molecular H by Davis+
(1986,ApJ.304,481) with similar results.

Theories of dust: Many current ideas have some form of amorphous carbon (e.g.,

Rowanta-Robinson,1986,MN,219,737) in a bare silicate-carbon mixture, or hydrogenated amorphous carbon (Hecht,1986,ApJ,305,817;1987,ApJ,314,429). The λ 2175 bump is
contributed by graphite on these theories. There is also the strong possibility
that PAHs are present. The biological origin of grains has been advocated (Jabir+,
1986,ApSpSci,123,351) but criticized by Duley (1984,QJRAS,25,109,1984).

Extensive calculations concerning the extinction and polarization properties
of dust were reported by the Leningrad group (Voshchinnikov+,1986,Astrofizika,24,
307 and 523; 25,197). The spectral properties of silicates in the infrared were
modelled by Pavlova (1984,Trudy Ap Inst Alma-Ata,44,20,1984).

Far-infrared (FIR) radiation is primarily a diagnostic of the energy sources
of the dust, but it can be interpreted by various grain models because the optical
properties of the grains do play a role in the spectrum of the emerging radiation.
Unfortunately, the size distribution of the particles also enters. The "standard"
means of interpreting FIR observations, and the quantities which can be derived
from them, has been explained very clearly by Hildebrand(1983,QJRAS,24,267). The
determination of the temperature distribution of a mixture of particles has been
given by Pajot+(1986,AA,157,393).

Observations of the far-infrared/sub-mm radiation from the Galaxy have been
reported by Hauser+(1984,ApJ Lett,278,L15; 150,250,and 300 μm), Pajot+(1986,AA,154,
55), and de Bernardis+(1984,ApJ,278,150; 150-400 and 350-3000 μm). The FIR observations were combined with previous measurements of the radiation field from 2 μm
through 3 mm to construct models of the heating and cooling of the dust in the
Galaxy (Cox+,1986,AA,155,380). The FIR continuum in several other galaxies has been
measured (Thronson+,1987,ApJ,318,645). Helou (1986,ApJ,311,L33) has discussed the
IRAS colors for normal galaxies. The IRAS measurements have been discussed, in connection with other FIR data, by Persson+(1987,ApJ,314,513). Unfortunately, 100 μm
is somewhat too short a wavelength to detect most of the luminosity from most
galaxies.

The time-dependent chemistry of dense clouds as regards expulsion of icy mantles from dust grains has been discussed by d'Hendecourt+(1986,AA,158,119), Grim+
(1986,AA,167,161), and Greenberg (1984,Origins of Life,14,25; 1983,Wolstencroft+,
1). This process might be in competition with the "standard" gas-phase chemistry
picture of interstellar chemistry. The sticking probability of gas atoms sticking
to grains was calculated by Leitch-Devlin+(1985,MN,213,295), and the changes of the
grain surface caused by accretion and the growth of mantles by grains in dark cloud:
have been discussed by Jones+(1985,MN,217,413,1985).

The formation of grains in circumstellar shells has been reviewed by Jura(1985,
Nuth+,3) and Draine (ibid.,19). It is safe to say that condensation theory is one
of the poorest-understood aspects of the interstellar medium. The condensation and
grain growth in circumstellar shells was discussed by Pearce (1986,AA,157,335);Gail+
(1986,AA,166,225; 1987,AA,171,197;177,186), and Muchmore+(1987,ApJ Lett,315,L141).
The radiation pressure on the grains can be the dominant force in the stellar wind,
and the available material puts a limit for the rate of mass loss from the late-
type stars. Nucleation has been studied in the laboratory by Nuth+(1986,J Chem Phys,
85,1116), and theories do not fit the results well. The far-infrared energy distribution of both C-rich and O-rich stars (Sopka+,1985,ApJ,294,242) indicates a rather
slow decrease in the absorption at very long wavelengths, which is characteristic
of an amorphous rather than a crystalline material. The destruction of dust, mainly
by shocks arising from supernovae arising within the galaxy, has been modelled by
McKee+(1987,ApJ,318,674). The theoretical lifetime for grains is disturbingly low
(< 1 Gyr).

The Diffuse Interstellar bands (DIBs) are very likely caused by some material
coated onto dust grains. They seem to fall in three groups, within which the

correlation of strengths is quite good (Chlewicki+: 1986,ApJ,305,455; 1987,AA,173, 131; Krelowki+: 1987,ApJ,312,860). The groups are distinguished by the breadth of the feature. Nos diffuse interstellar bands have been discovered in the ultraviolet portion of the spectrum, although careful searches have been made. The correlation of the DIB at 4430 Å with polarization and UV extinction was studied by Krelowski+ (1987,ApJ,316,449). Shapiro+ (1986,ApJ,310,872) consider theoretical profiles of DIBs associated with resonant impurities in grains. Measurements of three of the DIBs for stars with low reddening (Federman+: 1984,ApJ,282,485) show a correlation with the column density of molecular hydrogen. PAHs may account for the DIBs (Léger +: 1985,AA,146,81; van der Zwet+: 1985,AA,146,76; Crawford+: 1985,ApJ.Lett.293,L45).

Depletions of gas-phase elements have been discussed in Part II of this com- mission 34 report. For our purpose, one should note that carbon has poorly-deter- mined depletions (Welty+: 1986,PASP,98,857) and that depletions increase in general with the mean gas density along the line of sight.

The UV extinction in the LMC (see Nandy,IAU Symp.108,341) was studied by Clayton+(1985,ApJ,288,558) and Fitzpatrick (1986,AJ,92,1068). The extinction is especially different from Galactic in the 30 Doradus region, where the 2175 Å "bump" is weak and the FUV rise is very strong. In other parts of the LMC, though, the ex- tinction is fairly similar to Galactic. The gas-to-dust ratio is about four times larger than in the Galaxy (Koornneef, IAU Symp.108,333), which is consistent with the O/H ratio in the LMC relative to the Galaxy. In the SMC, Bouchet+(1985,AA,149, 330) find a normal visible-IR extinction curve although the far-UV is very diffe- rent from galactic, and a gas-to-dust ratio about 8 times larger than in the Galaxy.

Polarization of the light from the Pleiades was studied by Breger (1986,ApJ, 309,311) and in three stars by Clarke (1986,AA,156,213). A very interesting obser- vation is the polarization of emitted thermal radiation from two positions deep within the Orion molecular cloud (Hildebrand+: 1984,ApJ.Lett.284,L51; Dragovan: 1986,ApJ.308,270). This observation shows that grains are aligned even in very dense regions of space, where there should be thermal equilibrium between the grains and gas. Mathis (1986,ApJ.308,281) explained the wavelength dependence of polari- zation by assuming that there are superparamagnetic inclusions within grains, and only those grains which have one or more inclusions are aligned. Polarization by scattering has been modelled by Matsumura+(1986,Ap.Sc.Sci.126,155). Lee+(1985,ApJ. 290,211) used the polarization of the 0.7 μm feature in the BN object can lead to estimates of the band strength of the feature and the shape of the grains. They suggest that the grains are oblate.
Polarization in the dark lane in Cen A shows a perfectly normal wavelength dependence (Hough+: 1987,MN,227,1P) but peaks at 0.43 μm, which is considerably smaller than the average (0.55 μm) for our Galaxy. This may mean the grains are smaller; it could also be affected by alignment mechanisms.

V. Star Formation
(Bruce G. Elmegreen)

Research on star formation published between July 1984 and June 1987 is summa- rized here. The topics considered are: the clumpy structure of molecular clouds, magnetic fields, the collapse and stability of clouds, Herbig-Haro objects and jets, pre-main-sequence stellar winds, pre-main-sequence circumstellar disks, the initial mass function and efficiency of star formation, scenarios for star formation, and general properties of particular regions of star formation. Because of a lack of space, there are no references to extragalactic star formation, starburst galaxies, and models of the early solar system. There is also no discussion of the general literature on T Tauri stars and molecular clouds, unless a direct reference to im- plications for star formation is made by the authors. Presumably these neglected topics will be covered elsewhere in these IAU Reports (see Reports of Commissions 28,29,33,36,37,40 and the subsections on interstellar molecules in Commission 34).

Conference proceedings that include discussions of star formation are apart from those cited at the beginning of the report: "Workshop on Star Formation"(1984) ed. Wolstencroft (Occas.Rep.Royal Obs. Edinb.);"Frontiers of Astronomy and Astrophysics" (1984) ed. Pallavicini (Soc.Astron.Ital.); "Second Asian-Pacific Regional Meeting on Astronomy" (1984)ed. Hidayat+ (Pustaka); "Theoretical Aspects on Structure,Activity and Evolution of Galaxies III" (1985)ed. Aoki+ (Tokyo Astron.Obs.); "Tercera Reunion Regional Latinoamericana de Astronomia" (1985)ed. Sahade+ (Rev. Mex.Astron.Ap.10); "Third Asian-Pacific Regional Meeting of the IAU" ed. Kitamura+ (Ap.Sp.Sci.118,119); "Luminous Stars and Associations in Galaxies" (1986)ed. de Loore+ (Reidel). Individual contributions to conferences are not referenced here.

Reviews of T Tauri stars were written by Imhoff (1984,NASA Publ CP-2349,81) and Herbst (1986:PASP,98,1088). L1551 was reviewed by Emerson (1985:Nature,318, 604). Welch+ (1985:Science,228,1389) reviewed pre-main-sequence jets and Pudritz (1986:PASP,98,709) reviewed pre-main-sequence disks. Stahler (1986:PASP,98,1081) reviewed primordial star formation. Other reviews of star formation, aside from those in conference proceedings, were published by Turner (1984:Vistas Astron.27, 303) and York (1985:Mitt,Astron,Ges,63,89).

1. CLUMPY STRUCTURE IN MOLECULAR CLOUDS

Clumps with masses between 1 and 1000 solar masses and sizes on the order of 0.01 to 0.1 pc have been observed with NH_3, H_2CO or CS in molecular clouds associated with Cas A (Batrla+: 1984,AA,136,127), NGC 7538 (Henkel+: 1984,ApJ,282,L93), ρOph (Zeng+: 1984,AA,141,127; Wadiak+: 1985,ApJ,295,L43), W3(OH) (Turner+:1984,ApJ, 287,L81), W51A (Arnal+: 1985,AA,145,369), ON1 (Zheng+: 1985,ApJ,293,522),the Orion ridge (Mundy+: 1986,ApJ,304,L51), W49A (Goss+: 1985,MN,215,197), CepA (Torrelles+: 1986,ApJ,305,721), M17 (Snell+: 1986,ApJ,304,780), M17, S140 and NGC 2024 (Mundy+: 1986,ApJ,306,670), G34.3+0.2 (Andersson+: 1986,AA,167,L1), DR21 (Dickel+: 1986,AA, 162,221; Richardson+: 1986,MNRAS,219,167), NGC 7129 and GGD12-15 (Güsten+:1986,AA, 164,342), in Perseus globules (Bachiller+: 1986,AA,168,262), IC 348 (Bachiller+: 1987,AA,173,324) and in several other clouds (Mauersberger+: 1986,AA,162,199). Ammonia clumps have also been seen in the envelope of W3(OH)(Reid,Myers and Bieging 1987,ApJ,312,830). Hot OH clumps were observed by Walmsley+ (1986,AA,167,151).

Clumps have been observed with CO in several clouds by Pérault+ (1985,AA,152, 371; 1986,AA,157,139),in Orion by Bally+ (1987,ApJ,312,L45),and in IRAS star forming regions by Casoli+ (1986,AA,169,281). Jaffe+ (1987,ApJ,316,231) found dense cores with the 7-6 transition of CO. A CO interclump medium in the Rosette cloud was discussed by Blitz and Stark (1986,ApJ,300,L89).

Clumps have also been observed at FIR or sub-mm wavelengths in S255, W3 and OMC1 (Jaffe+: 1984,ApJ,284,637), W51 and DR21 (Harvey+: 1986,ApJ,300,737), and W3, M42, W49A and W51A (Gordon: 1987,ApJ,316,258).

Absorption of background starlight reveals small cloud clumps too, as shown by Rossano (1985,AJ.90,308), Cernicharo+ (1985,AA,149,273) and Casali (1986,MNRAS, 223,341).

Research on the physical structure and dynamics of clumps and cloud turbulence was done by Falgarone and Puget (1985,AA,142,157; 1986,AA,162,235), Stenholm (1985,AA,144,179), Kessel'man+ (1985,SovAstron,29,417), Yoo+ (1986,J.Korean Astr. Soc,19,33), Kleiner+ (1984,ApJ,286,255; 1985,ApJ,295,466; 1987,ApJ,312,837) and Dickman+ (1985,ApJ,295,479). Spectra of clumpy clouds were calculated by Kwan+: (1986,ApJ,309,783)and the energy dissipation rate in clumpy clouds was evaluated by Elmegreen (1985,ApJ,299,196).

Several formation mechanisms for clumps were considered: self-gravitational cloud fragmentation (Larson,1985,MNRAS,214,379;Tohline,1985,ApJ,292,181;Schloerb and Snell,1984,ApJ,283,129), stellar winds (Silk,1985,ApJ,292,L71), fragmentation in an ionization-front shock (Rainey+,1987,AA,171,252) or supernova shock (White+,1987, AA,173,337), and thermal instabilities (Gilden,1984,ApJ,283,679;Grazini+,1987,Ap Lett,25,235).

Clump or globule compression by surrounding ionization was discussed by Ho+ (1986,ApJ,305,714) and King (1987,MNRAS,226,473).

2. MAGNETIC FIELD OBSERVATIONS

Magnetic field strengths in cloud clumps were measured from HI (Schwarz+,1986, ApJ,301,320) and OH lines (Heiles+,1986,ApJ,301,331) on the line of sight to CasA. The magnetic field was found to be perpendicular to the Taurus filaments, sugges- ting a compression of gas along the mean field direction (Moneti+,1984,ApJ,282,508; Tamura+,1987,MNRAS,224,413). The same orientation occurs in L204 (McCutcheon+,1986, ApJ,309,619). The field is parallel to filaments in Cygnus (McDavid,1984,ApJ,284, 141) and in Can Maj OB1 (Vrba+,1987,ApJ,317,207). In the Horsehead nebula, the field is parallel to the surrounding fields, suggesting that the grains may not be align- ed in dense regions (Zaritsky+,1987,AJ,93,1514). The field orientation in B5 was studied by Joshi+,(1985,MNRAS,215,275). A summary of observed field strengths is in Troland+(1986,ApJ,301,339).

3. COLLAPSE AND STABILITY

Infall or collapse onto a protostar was observed by Gee+(1985,MNRAS,215,15p), Garay+(1985,ApJ,289,681), Zinchenko+(1985,AZ,62,860), Walker+(1986,ApJ,309,L47), and Menten+(1987,AA,177,L57). This interpretation of the observations for W3 (OH) was questioned by Welch+(1987,ApJ,317,L21). Collapse was inferred for the globule B5 by Boss(1985,ApJ,288,L25).

The globule B335 was said to be in rotational equilibrium by Frerking+(1985, Icarus,61,22), but most globules are not rotating fast enough for such support (Casali+,1987,MNRAS,225,481). The density gradient in molecular clouds was determi- ned to be inverse square by Fulkerson+(1984,ApJ,287,723).

Theoretical work on cloud stability criteria was done by Kiguchi+(1987,ApJ,317, 830), and Schmitz(1986,AA,169,171). Stability in the presence of turbulence was dis- cussed by Bonazzola+(1987,AA,172,293). The stability of magnetic disks was discus- sed by Mestel+(1985,MNRAS,212,275). The equilibrium structure of a magnetic cloud around a new star, and the tendency for this structure to help collimate a bipolar flow, was discussed by Nagasawa+(1985,PASJ,37,369).

Hachisu+(1984,AA,140,259) determined a criterion for fragmentation of a rota- ting collapsing cloud. Monaghan+(1984,PAS Australia,5,493) and Boss (1986,ApJ Suppl, 62,519) studied fragmentation in collapse models and Phillips (1986,PAS Australia, 6,205) discussed fragmentation in colliding magnetic clouds. Vanajakshi+(1985,ApJ, 294,502) considered collapse with turbulent viscosity. Liu (1984,Chin AA,8,310) simulated ring formation and fragmentation. Boss (1987,ApJ,316,721) calculated the collapse of a rotating cloud further than had been done before and got a disk and an evacuated polar cavity into which a wind might be channeled.

Analytical calculations of cloud collapse were made by Terebey+(1984,ApJ,286, 529), Whitworth+(1985,MNRAS,214,1) and Hunter (1986,MNRAS,223,391).

Analytical studies of the emergent radiation from collapsing clouds or proto- stars were made by Adams+(1985,ApJ,296,655;1986,ApJ,308,836), Kolesnik (1985,AZ,62, 518), Wolfire (1987,ApJ,315,315), Adams+(1987,ApJ,312,788) and Crawford+ (1986,MNRAS,

221,923). Beall (1987,ApJ316,227) calculated the appearance of a protostellar disk.

Mestel+ (1984,AA,136,98) calculated some general properties of magnetic cloud collapse. Three-dimensional magnetic collapse models were made by Benz (1984,AA,139, 378) and Phillips (1986,MNRAS,221,571; 1986,MNRAS,222,111). Collapse with ambipolar diffusion was studied by Mouschovias+(1985,ApJ,291,772) and Nakano (1986,MNRAS,218, 663; 1986,MNRAS,221,319), and the effect of diffusion on angular momentum transport was discussed by Mouschovias+(1986,ApJ,308,781). Angular momentum transport between clumps in a cloud was studied by Mouschovias+(1985,ApJ,298,190-205). Angular momentum transport by gravitational torques in a disk was calculated by Boss (1984,MNRAS, 209,543).

4. HERBIG-HARO OBJECTS

Several new HH objects were found by Magakyan (1984, Pis'ma AZ,10,661), Reipurth (1985,AA,Suppl,61,319) and Krautter (1986,AA,161,195).

Proper motions of HH objects show expansion away from a central source (Gyul'budagyan,1984,Astrof,20,115;Jones+,1984,AJ,89,1404;Schwartz+,1984,AJ,89,1735, and Jones+,1985,AJ,90,1320). Time variability was noted by Walsh (1986,Ap Sp Sci, 118,439) and Reipurth+(1986,Nature,320,336).

The exciting sources for some HH objects were determined to be T Tauri stars, some of which show radio emission and extended infrared emission (Bieging,Cohen+, 1984,ApJ,282,699;Reipurth+,1985,AA,150,307;Rodriguez+,MNRAS,214,9P;Pravdo+,1985, ApJ,293,L35;Vrba+,1985,AJ,90,2074;Cohen+,1986,ApJ,302,L55,1986,ApJ,307,L21; Cohen+, 1987,ApJ,316,311; Bührke+,1986,AA,163,83;Goodrich,1986,AJ,92,885; Tapia+,1987,MNRAS, 224,587; Böhm+,1987,PASP,99,265). A study of bare T Tauri stars and a discussion of the implications for star formation were made by Walter (1987,PASP,99,31).

The possibility that FU ORI type stars could power the associated outflow was discussed by Reipurth (1985,AA,143,435), Graham+(1985,ApJ,289,331), Silvestro+(1984, ESA.Pub.SP-207 p.235), Mundt+(1985,ApJ,297,L41), Hartmann+(1985,ApJ,299,462) and Goodrich(1987,PASP,99,116). Periodic ejections were discussed by Meaburn+(1985, MNRAS,215,761).

Excited emission from molecular hydrogen was found or mapped around HH objects by Zealey+(1984,AA,140,L31), Zealey+(1986,AA,158,L9), Harvey+(1986,ApJ,301,346) and Lightfoot+(1986,MNRAS,221,993).

Other properties of HH objects determined from emission lines, such as velocities, densities, time-variability and shock structure, were discussed by Lenzen, Hodapp and Solf (1984,AA,137,202), Meaburn+(1984,AA,138,36), Pettersson (1984,AA, 139,135), Brugel+(1985,ApJ,292,L75), Brugel+(1984,NASA.Publ.CO-2349,p.171),Cohen+ (1985,ApJ,296,620), Schwartz+(1985,AJ,90,1820),Hartigan+(1986,AJ,91,1357), Meaburn (1986,AA,164,358),Böhm+(1985,ApJ,294,533), Solf+(1986,ApJ,305,795), Taylor+(1986, MNRAS,221,155), Lighfoot+(1986,MNRAS,221,47p), Raga+(1986,AJ,92,119), Böhm+(1987, ApJ,316,349), Hartigan+(1986,AJ,92,1155), Meaburn+(1987,MNRAS,225,863) and Solf+ (1987,AJ,93,1172). Low excitation emission around embedded stars was mapped by Morgan+(1984,ApJ,285,L71) and Br α emission was surveyed by Persson+(1984,ApJ,286, 289).

These observations imply that many HH objects are bow shocks. Some form around bullets or jets emitted by the exciting stars, others form around stationary or slowly moving clumps that are exposed to winds from the exciting stars.

Deep photographic surveys of the regions around HH objects show jets, knots and loops (Walsh+,1985,MNRAS,217,31; Hartigan+,1985,ApJ Supp,59,383; Bohigas+,1985 Rev Mex AA,11,149; Strom+,1986,ApJ Suppl,62; see also section below on jets). An

IRAS survey of the region around HH 1 and 2 found several faint embedded stars (Pravdo+: 1987,ApJ,314,308). Cardelli+ (1984,NASA.Publ.CO-2349,p.175) studied the dust near HH objects. NH_3 observations of HH 46 and 47 show a disk at the IR source which is too low in mass to confine the outflow (Kuiper+: 1987,PASP,99,107). NH_3 observations of HH 1 and 2 show dense clumps close to the exciting star, and they show the compressed wall of the cavity associated with the high velocity flow (Martin+: 1987,AA,176,L27). High resolution CO observations of Orion, NGC 2071, GL490 and S140 also show a shell at the edge of the wind cavity (Snell+: 1984,ApJ, 284,176).

Theoretical aspects of shock models for HH objects were discussed by several of the above authors, in addition to Hartigan+:(1987,ApJ,316,323), Tenorio-Tagle+ (1984,AA,137,276; 1984,AA,141,351), Choe+ (1985,ApJ,288,338), Liseau+ (1986,ApJ, 304,459) and Raga (1986,AJ,92,637). The possibility that some HH objects are shocks around old jets was discussed by Lightfoot+ (1986,MNRAS,221,993; 1986,MNRAS,221, 47p).

5. JETS

Optical jets are occasionally observed near embedded windy pre-main-sequence stars. In addition to the jets associated with HH objects discussed above, new jets were discovered in HH 6-5B, HH 33/40 and HH 19 (Mundt+: 1984,AA,140,17),R Mon (Brugel+:1984,ApJ,287,L73), Orion B (Sanders+:1985,ApJ,293,L39), DG Tau (Jones+: 1986,ApJ,311,L23), HH34 (Reipurth+:1986,AA,164,51) and GGD34, RNO43, HH3,5 (Ray: 1987,AA,171,145). The jet in RNO43 extends for 1.4pc, which the longest found so far. The jet in DG Tau was observed to turn on (Cohen+:1986,AJ,92,1396). A jet near Orion IRC9 was suggested by the observation of fingers of excited H_2 emission (Taylor+:1984,Nature,311,236). The jet in L1551 IRS5 is occulted on one side by a circumstellar disk (Snell+:1985,ApJ,290,587). Another jet in V645 Cyg was studied by Goodrich (1986,ApJ,311,882), and a spectrum of the jet in L1551 was studied by Sarcander+:(1985,ApJ,288,L51).

The theory of pre-main-sequence jets was discussed by Fukue+(1986,Nature,321, 841), who considered precession, Shibata+(1986,ApSpSci,118,443; 1986,PASJ 38,631), who modeled the acceleration of disk gas by a twisting polar magnetic field, Kaburaki+ (1987,AA,172,191), who consider jet alignment by a toroidal magnetic field in an accretion disk, and Sakashita+ (1986,PASJ,38,879), who model jet formation by the interaction between a spherical wind and a plane parallel gas layer. Fujue+ (1986,PASJ,38,895) studied the effect of gravity from a stellar torus on the jet flow, and the shock structure inside a jet was studied by Falle+ (1987,MNRAS, 225,741), A model of wind acceleration from a massive disk was made by Pudritz (1985,ApJ.293,216).

6. PRE-MAIN-SEQUENCE STELLAR WINDS AND BIPOLAR FLOWS

One of the most active fields in star formation research today is pre-main-sequence winds and bipolar flows near embedded stars. Molecular outflows were studied for the following sources: AFGL 2591 (Lada+:1984,ApJ.286,302), G327.3-0.6 (Brand+:1984,AA,139,181), B335, L723 and L1455 (Goldsmith+:1984,ApJ,286,599), NGC 6334 IRS V-1 (Simon+:1985,MNRAS 212,21), L379 (Hilton+:1986,AA,154,274), an IRAS source in Orion (Wolstencroft+:1986,MNRAS,218,1), and a high resolution study of Orion (Wilson+:1986,AA,167,L17), DR21 (Richardson+:1876,MNRAS 219,167), 3 IRAS sources in B5 (Goldsmith+:1986,ApJ,303,L11), IRAS 1827-145 (La Bertre,Epchtein+: 1984,AA,138,353), B335 (Langer+:1986,ApJ,306,L29), Ori A, L1641, NGC 2071 and the Oph dark cloud (Fukui+:1986,ApJ,311,L85), and another study of NGC 2071 (Takano+: 1986,AA,167,333), Mon OB1 (Margulis+:1986,ApJ,309,L87), 9 globules (Avery+:1987, ApJ,312,848), NGC 7023 (Watt+:1986,AA,163,194), the Boomerang nebula and PV Ceph (Neckel+:1987,AA,175,231), V645 Cyg (Torrelles+:1987,AA,177,171), R Cr Aus(Hartigan +:1987,AJ,93,913), and S235B (Nakano+:1986,PASJ,38,531). High velocity OH absorption was found in L1551 (Mirabel+:1985,ApJ,294,L39). The energies and masses of outflow sources were discussed by Margulis+:(1985,ApJ,299,925).

High density gas or clumps were observed in the outflows associated with Ori A (Masson+:1984,ApJ,283,L37), Ori B (Russell+:1987,MNRAS,226,237), Cep A (Loren+: 1984,ApJ,287,707; Richardson+:1987,AA,174,197), L1551 (Walmsley+:1987,AA,179,231), NGC 2071 (Takano:1986,ApJ,300,L85), NGC 2071, NGC 7538 and W49 (Scoville+: 1986, ApJ,303,416), M16 and the Rosette nebula (Meaburn+:1986,MNRAS,220,745), NGC 7538 (Kameya+:1986,PASJ,38,793), and in several other sources (Thronson+:1984,ApJ,284, 135; Richardson+:1985,ApJ,290,637). The possibility of a wind near the Horsehead nebula in Orion was discussed by Reipurth+ (1984,AA,137,L1), Warren-Smith+ (1985, MNRAS,215,75), and Neckel+ (1985,AA,147,L1).

Shells surrounding outflows were observed in L1551 (Snell+:1985,ApJ,295,490; Rainey+:1987,AA,179,237), and Taurus (Murphy+:1985,ApJ,298,818), and the shocked cloud/wind interface was seen in M8E (Larson+: 1986,ApJ,307,295). Images of cones or bubbles around wind sources were obtained by Campbell+ (1986,ApJ,305,336) and Forrest+ (1986,ApJ,311,L81). The M8 hourglass region was imaged in the infrared by Allen(1986,MNRAS,219,35).

Winds were observed very close to the stars in W3 IRS5 (Claussen+:1984,ApJ. 285,L79), GL2591 (Geballe+:1985,ApJ.291,L55) and NGC 2071, W49 and NGC 7538 (Scoville+:1986,ApJ.303,416).

Radio continuum radiation from the central stars was observed by Schwartz+: (1985,ApJ,295,89) and Snell+:(1986,ApJ,303,683). The mass loss rates found in this latter study are too low to explain the observed molecular flows. Other observations of ionized winds from embedded stars were by Tanaka+(1985,PASP,97,1112), Schwartz+ (1986,ApJ,303,233), Chalabaev+ (1986,AA,168,L7), and Smith+ (1987,ApJ, 316,265). Other observations of ionized structures in wind regions were made by White+ (1986,AA,156,301). Winds near Herbig Ae-Be stars were discussed by Canto+ (1984,ApJ,282,631) and Scarrott+ (1986,MNRAS,223,505).

Excited molecular hydrogen emission was observed in the vicinity of outflow sources by Phillips+ (1985,AA,145,118), Matsumoto+ (1985,PASJapan,37,129), Garden+ (1986,MNRAS,220,203), Lane+ (1986,ApJ,310,820), Oliva+ (1986,AA,164,104), Geballe+ (1986,ApJ,302,693), and Longmore+ (1986,MNRAS,221,589).

Confinement of deflection of the flow by cloud clumps was discussed by Fridlund+ (1984,AA,137,L17), Weliachew+ (1985,AA,153,139), Wootten+ (1987,ApJ,317, 220), and Torelles+ (1985,ApJ,288,595).

Flow orientations are often found to be close to that of the surrounding magnetic field, or to that of another flow source (Cohen+:1984,MNRAS,210,425; Vrba+: 1986,AJ,92,633; Langer+:1986,ApJ,306,L29). Magnetic field strengths in flow regions were estimated by Simonetti+ (1986,ApJ,303,659).

Optical and infrared polarization studies of winds often show the locations of the exciting stars, and sometimes suggest that the line of sight from the star to the wind is obscured less than the line of sight from the star to the observer, as if an obscuring disk surrounds the star (Beckwith+:1984,ApJ,287,793; Draper+: 1985,MNRAS,216,7; Hodapp:1984,AA,141,255; Cohen+:1985,ApJ,296,620; King+:1985,MN RAS,213,11; Castelaz+:1986,ApJ.300,406; McLean+:1987,MNRAS,225,393; Harvey+:1987, ApJ,317,173, and Lenzen:1987,AA,173,124). A polarization map of excited H_2 emission in Orion was made by Hough+ (1986,MNRAS,222,629).

The dust around wind sources may be heated by uv radiation from the shock (Clark+:1986,AA,168,L1), and it may be located at the wind-cloud interface (Clark+: 1986,AA,154,L26). The dust and gas is also heated by the embedded stars (Takano:

1986,ApJ,303,349). The energy of the flow in L1551 was determined by Edwards+(1986, ApJ,307,L65). Without a large amount of light scattering, radiation pressure is too small to drive the winds (Mozurkewich+,1986,ApJ,311,371).

Theoretical discussions on the observable properties of winds were made by Dyson (1984,ApSpSci, 106,181), Bastien (1987,ApJ,317,231), Okuda+(1986,PASJ,38,199), and Cabrit+(1986,ApJ,307,313). Polarization models for scattered light were made by Heckert+(1985,AJ,90,2291). An energy conserving wind model was made by Kwok+(1985, ApJ,299,191).

7. PRE-MAIN-SEQUENCE DISKS

Disks or elongated structures have been observed in the vicinity of embedded pre-main-sequence stars using emission from scattered light in HL Tau (Grasdalen+, 1984,ApJ,283,L57), from NH_3 in Cep A (Güsten+,1984,AA,138,205; Torrelles+,1986,ApJ, 305,721), although Menten+(1985,AA,146,375) found no NH_3 disk in L1551,from CS in NGC 2071 (Takano+,1984,ApJ,282,L69), GL 490(Kawabe+,1984,ApJ,282,L73), the Orion KL region (Hasegawa+,1984,ApJ,283,117), and in several other sources (Heyer+,1986, ApJ,308,134), although Moriarty-Schieven+(1987,ApJ,317,L95) found no CS disk in L1551, from HCN in S106 (Bieging,1984,ApJ,286,591), from H_2CO in L1551 (Duncan+,1987, MNRAS,224,721), from CO in IRAS 16293-2422 (Mundy Wilking+,1986,ApJ,311,L75), from IR polarization in several sources (Sato+,1985,ApJ,291,708), and from the near and far-infrared and sub-mm wavelength range in L1551 (Strom+,1985,AJ,90,2575), GL 490 (Gear+,1986,MNRAS,219,835), M8E (Simon+,1985,ApJ,298,328), S106 (Harvey+,1987,ApJ, 316,L75), and a number of other sources (Cohen+,1985,ApJ,296,633). Orientations of the disks perpendicular to the magnetic field were noted in the above papers by Sato+ and Mundy+.

The orientations of disks or elongated structures perpendicular to bipolar outflows or jets have been discussed for the sources L1551 (Bieging+,1985,ApJ,289, L5; Rodriguez+,1986,ApJ,301,L25), although no velocity gradient similar to that expected from a disk was found in L1551 (Batrla+,ApJ,298,L19), for G34.3+0.1 (Heaton+ 1985,MNRAS,217,485),G35.2-0.74 (Matthews+,1984,AA,136,282; Dent+,1984,MNRAS,210, 173),G35.2N (Dent+,1985,AA,146,375), Orion (Vogel+,1984,ApJ,283,655), HH24-26 (Little+,1985,AA,142,378), GGD12-15 (Harvey+,1985,ApJ,288,725), R Mon(Aspin+,1985, AA,149,158), HH1,2 (Torrelles+,1985,ApJ,294,L117; Strom+,1985,AJ,90,2281), CRL 2591 (Takano+,1986,AA,158,14), NGC 7538 (Scoville+,1986,ApJ,303,416;Campbell,1984,ApJ, 282,L27), L723 (Torrelles+,1986,ApJ,307,787), MWC 349 (Hamann+,1986,ApJ,311,909; White+,1985,ApJ,297,677), R Cor Aust (Castelaz+,1987,ApJ,314,317), and HH46-47 (Kuiper+,1987,PASP,99,107).

8. THE INITIAL MASS FUNCTION AND STAR FORMATION EFFICIENCY

The initial mass function (IMF) was modelled by cloud fragmentation and coalescence by Yoshii+(1985,ApJ,295,521), who considered that the stars that already formed heat the remaining gas. A time dependence of the IMF was also discussed by Di Fazio (1986,AA,159,49) and Brown (1986,ApSpSci,122,287). The influence of hierarchical fragmentation was studied by Zinnecker (1984,MNRAS,210,43). Smith (1985,ApJ, 293,251) studied gas accretion and a possible turnover in the IMF at low mass.

De Gioia-Eastwood (1984,PASP,96,582) devised a method to determine the IMF in an embedded cluster. Cudlip+(1984,MNRAS,211,563) suggested that low mass stars in Ophiuchus formed quiescently and that a high mass star formed in an interacting region. Van Buren (1985,ApJ,294,567) suggested that the IMF is shallower than the Salpeter function after correcting for local dust on the line-of-sight to stars of various masses. Low-mass stars near the Trapezium cluster were observed by Herbig+ (1986,ApJ,307,609). A model of bimodal star formation was compared to observations of the solar neighborhood by Wyse+(1987,ApJ,313,L11). Wouterloot+(1986,AA,168,237) found that the luminosity function of IRAS sources agrees with the standard IMF.

Reid (1987,MNRAS,225,873) studied the IMF in low mass stars.

Rengarajan(1984,ApJ,287,671) determined that the star formation efficiency is constant in molecular clouds because the CO flux is proportional to the FIR flux. Myers+(1986,ApJ,301,398) also determined the efficiency of star formation by comparison of IR, radio continuum and CO fluxes. The efficiency in regions of cluster formation was discussed by Elmegreen+(1985,ApJ,294,523).

9. SCENARIOS FOR STAR FORMATION

Self-regulating models for star formation on various scales have been discussed by Franco(1984,AA,137, 85), Fujimoto+(1984,PASJ,36,319), Bodifiée+(1985,AA,142, 297), Chiang+(1985,ApJ,297,507), Brosche+(1985,AA,153,157), Bodifée(1086,ApSpSci, 122,41), Korcgagin+(1986,Kinematika Fiz.Nebesn.Tel2,22), Pudritz+(1987,ApJ,316,213), and Nepveu(1987,AA,175,91).

Various theories of propagating star formation were discussed by Sandford+ (1984,ApJ,282,178), McCray+(1987,ApJ,317,190) and Kimura+(1987,ApSpSci,129,261). Applications to specific sources were made by Dopita+(1985,ApJ,297,599), Thronson+ (1985,ApJ,297,662), Ho+(1986,ApJ,305,714), Handa+(1986,PASJ,38,361), Arnal+(1987, AA,174,78), Felli+(1987,AA,175,193), Olano+(1987,AA,179,202), and Kun+(1987,ApSpSci, 134,211). A model of chemical evolution including propagation was made by Shore+ (1987,ApJ,316,663). Cameron (1984,Icarus,60,416) discussed extinct radioactivity and propagating star formation.

Star formation in globular clusters was modelled by Tenorio-Tagle+(1984,MNRAS, 221,635) and Smith+(1987,ApJ,316,206). Pre-main-sequence evolution of primordial stars was modelled by Stahler+(1986,ApJ,308,697).

Star formation by cloud-cloud collisions was discussed by Scoville+(1986,ApJ, 310,L77). Lattanzio+(1985,PAS Australia,5,495;1985,MNRAS,215,125) suggest that such collisions will only destroy the clouds. Star formation triggered by colliding gas flows or crashing turbulent eddies was discussed by Sabano+(1985,ApSpSci,115,85) and Hunter+(1986,ApJ,305,309). Observations of collisionally-triggered star formation were suggested by Haschick+(1985,ApJ,292,200).

10. OTHER STUDIES OF REGIONS OF STAR FORMATION

Large-scale maps and analyses of the distribution of molecules, dust, embedded stellar sources, etc., were made for the following sources: G35.2-0.74 (Tapia+,1985, MNRAS,213,833), G30.8-0.0 and G25.4-0.2 (Lester+,1985,ApJ,296,565), GL961 (Lenzen+, 1984,AA,137,365; Castelaz+,1985,AJ,90,1113), GL4176 and GL4182 (Persi+,1986,AA,157, 29),L1641 (Nakajima+,1986,MNRAS,221,483; Takaba+,1986,AA,166,276),L810(Neckel+,1985, AA,153,253; Turner,1986,AA,167,157), M8 (Woodward+,1986,AJ,91,870), M17 (Schulz+, 1987,AA,171,297), NGC 1333 (Jennings+,1987,MNRAS,226,461), NGC 2244(Guseva+,1984, Pis'ma AZ,10,741), NGC 2264 (Kalandadze+,1986,Kinematika Fiz Nebesn Tel,2,27),NGC 3603(Baier+,1985,AA,151,61), NGC 6334(Laughran+,1986,ApJ,303,629), NGC 6334, NGC 6193 and IC 4628(Phillips+,1986,AA Suppl,65,465), NGC 6357(Persi+,1986,AA,170,97), NGC 7129 (Draper+,1985,MNRAS,212,5p),NGC 7538(Lynds+,1986,ApJ,306,532), IC 443 (Odenwald+,1985,ApJ,292,460), Serpens MC 2(Churchwell+,1986,ApJ,300,729),Chamaeleon (Thé+,1986,AA,155,347; Jones+,1985,AJ,90,1191;Toriseva+,1985,AA,153,207; Whittet+, 1987,MNRAS,224,497), the Gum nebula (Graham,1986,ApJ,302,352; Pettersson,1987,AA, 171,101), Carina (Whiteoak+,1985,PAS Australia,5,552), a cloud in Monocerous (Maddalena+,1985,ApJ,294,231), the Cas OB2 region (Lozinskaya+,1986,ApSpSci,121, 357), Taurus (Duvert+,1986,AA,164,349), Lupus (Murphy+,1986,AA,167,234), Ophiuchus (Young+:1986,ApJ,304,L45; Loren+,1986,ApJ,306,142; André+,1987,AJ,93,1182), Draco (Johnson+;1986,ApJ,309,321), Orion (Thronson+,1986,AJ,91,1350; Lester+,1985,AJ,90, 2331; Garay Moran+,1987,ApJ,314,535; Plambeck+,1987,ApJ,317,L101; Geballe+,1987, ApJ,317,L107), OMC-1 (Masson+,1985,ApJ,295,L47), OMC-2 (Pendleton+,1986,ApJ,311,360) the KL nebula in Orion (Viscuso+:1985,ApJ,296,142), Ori B (Crutcher+:1986,ApJ,307, 302), ON1 (Matthews+:1986,AA,155,99), Orion and Cepheus (Wouterloot+:1986,AA,168, 237), the magellanic bridge (Meaburn:1986,MNRAS,223,317), the LMC (Epchtein+:1984, AA,140,67), R136a in 30 Dor (Weigelt+:1985,AA,150,L18), N160 in the LMC (Heydari-Malayeri+:1986,AA,162,180), W3 (Thronson+:1984,ApJ,284,597; Thronson:1986,ApJ,306, 160), W3(OH) (Mauersberger+:1986,AA,166,L26), W33 (Stier+:1984,ApJ,283,573), W40 (Smith+:1985,ApJ,291,571), W49A (Miyawaki+:1986,ApJ,305,353), W49N (Dreher+:1984, ApJ,283,632), W51 (Cunningham+:1984,MNRAS,210,891; Ohishi+:1984,PASJ,36,505; Rengarajan+:1984,ApJ,286,573; Cox+:1987,MNRAS,226,703; Mauersberger+:1987,AA,173, 352), W51 and DR21 (Harvey+:1986,ApJ,300,737), W3, M42, W49A and W51A (Gordon:1987, ApJ,316,258), DR6, 7, and 22 (Odenwald+:1986,ApJ,306,122), Mon R2 (Hugues+:1985, ApJ,289,238; Hodapp:1987,AA,172,304), S140 (Lester+:1986,ApJ,309,80), S156 (Joy+: 1984,ApJ,284,161), S128 (Haschick+:1985,ApJ,292,200), S255 (Richardson+:1985,MNRAS 216,713), S254-S257 (Henkel+:1986,AA,165,197), Sandqvist 187 (Alvarez+:1986,ApJ. 300,756), LkHα101 (Redman+:1986,ApJ,303,300; Dewdney+:1986,ApJ,307,275), IRAS 04238+5336 (Wynn-Williams+:1986,ApJ,304,409), IRAS 19520+2759 and IRAS 01133+6434 (Arquilla+:1987,AA,173,271), Cep A (Hugues:1985,ApJ,298,830), K3-50, W51-IRS 2E and G333.6-0.2 (Hofmann+:1986,AA,160,18), B62 (Reipurth+:1986,AA,166,148), B335 (Frerking+:1987,ApJ,313,320), SgrB2 (Goldsmith+:1987,ApJ,313,L5; 1987,ApJ,314,525; Vogel+:1987,ApJ,316,243), RCW 108 (Straw+:1987,ApJ,314,283), Sco OB2 (Cappa de Nicolau+:1986,AA,164,274), and Serpens SVS20 (Eiroa+:1987,AA,179,171).

Surveys of dust, IR, HII and CO emission, embedded stars, etc. in regions of star formation were made by Waller+:(1987,ApJ,314,397), Chini+:1986,AA,167,315), Braz+:1987,AA,176,245), Heckert+:(1984,AJ,89,1379), McGregor+:(1984,ApJ,286,609), Chini+(1986,AA,157,L1), Arquilla+:(1986,MNRAS,220,125), Beichman+:(1986,ApJ,307, 337) and Clark:(1987,AA,180,L1).

VI. HII Regions
(M. Peimbert)

1. INTRODUCTION

H II regions are paramount for the study of: the formation and evolution of massive stars, the energy input to the interstellar medium and the evolution of the chemical abundances in our galaxy and other galaxies. Hundreds of papers based on observations in almost all domains of the electromagnetic spectrum were published in the 1984-1987 period covered by this review. The following discussion will be restricted to representative papers in this area.

Apart from the books, proceedings, atlas and catalogues cited at the beginning of the report, the general relevant literature contains: Kunth, ed.: 1986, "Star Forming Dwarf Galaxies", ed. Frontières, Paris, hereafter SFDG; Backer, ed.: 1987, "The Galactic Center" AIP, Chicago, hereafter GC. The conference proceeding edited by Pequignot, 1986, is designated hereafter as MNP.

2. STRUCTURE

A large number of papers based on radioobservations has been devoted to determine the structure of galactic H II regions. High resolution studies have been carried out with the VLA, the RATAN-600, and other radiotelescopes (e.g.: 1984, SoobSpecApObs, 43, 56; 1984, MN, 209, 209; 1984, AA, 136, 53; 1984, ApJ, 282, L27; 1984, MN, 210, 173; 1985, AJ, 90, 59; 1985, AJ, 90, 310; 1985, ApJ, 288, L17; 1985, PASJapan, 37, 123; 1985, AA, 147, 84; 1985, AZh, 62, 229; 1985, AZh, 62, 482; 1985, AJ, 90, 2061; 1985, AA, 152, 387; 1986, MN, 219, 39P; 1986, ApJ, 307, 275; 1986, ApJ, 308, 288; 1986, AJ, 92, 75). From radio observations: the stellar types of the ionizing stars have been estimated (e.g.: 1984, MN, 209, 209; 1985, AJ, 90, 2061), compact or supercompact H II regions have been studied (e.g.: 1984, SoobSpecApObs, 43, 56; 1985, AZh, 62, 218; 1985, AJ, 90, 310; 1985, ApJ, 288, L17; 1985, AZh, 62, 482; 1985, Pis'maAZh, 11, 27; 1986, Pis'maAZh, 12, 353; 1986, IzvSpecApObs, 23, 3; 1987, IzvSpecObs, 24), and blister type H II regions have been analyzed (e.g.: 1985, AA, 152, 387; 1986, ApJ, 309, 553; 1986, AJ, 92, 75). Radio and optical data have been combined to stablish the structure of many H II regions (e.g.: 1984, Astrofisika, 20, 199; 1984, MN, 211, 149; 1984, MN, 211, 155). Radio images of ionized bright rim structures show discrete condensations that may be the result of Rayleigh-Taylor instabilities at the interface between the ionized and neutral phases of the interstellar medium (1985, ApJ, 298, 292). Ershov considered the possibility of detection of hydrogen fine-structure radio lines in H II regions (1987, Pis'maAZh, 13, 285).

Many papers on infrared mapping of H II regions are presented in the literature, the main results derived from these observations are: the dust temperature, density and optical depth distributions, the stellar infrared luminosities and the presence of stellar winds (e.g.: 1984, MN, 211, 15; 1984, ApJ, 283, 573;1985,AJ,90, 88; 1985, AA, 144, 275; 1985, ApJ, 291, 571; 1985; AA, 146, 337; 1985, ApJ, 296, 565; 1985, NASA CP-2353, 164; 1986, AA, 154, L8; 1986, ApJ, 300, 737; 1986, AA, 157, L1; 1986, AA, 158, 143; 1986, ApJ, 303, 629; 1986, ApJ, 303, L57; 1986, ApJ, 306, 122; 1986, AA, 170, 97).

Optical studies of emission line intensities have been used to determine the density structure and, in some cases, the ionization and velocity structures (e.g.: 1984, AstronTsirk, 1309, 1338, 1339; 1984, AA, 140, 24; 1984, Astrofizika, 20,199; 1985, AstronTsirk, 1369; 1985, AJ, 90, 92; 1985, ApJ, 294, 578; 1985, MN, 216, 761; 1985, AA, 152, 254; 1986, AstronTsirk, 1444, 1449,1457; 1986, ApSpSc, 119, 131; 1986, AZh, 63, 246; 1986, AA, 162, 265). By means of the ASTRON satellite UV energy distributions (λλ1500-3500) for Orion, M8 and M17 were obtained by Pronik and Petrov (1986, AZh, 63, 1016).

The interface between the H II and H I regions has been studied by observations of neutral hydrogen associated with H II regions (1984, ApSpSc, 107, 271; 1985, JKoreanAS, 18, 100; 1986, PASJapan, 38, 347); models have been presented to explain the dynamical interaction between the H II and H I regions (1984, ApSpSc, 107, 289; 1986, PASJapan, 38, 347).

OH and H_2O masers associated with compact or ultracompact H II regions have been detected in many sources, in some cases the H_2O masers are closer to the central star than the OH masers, moreover while some authors argue that the OH masers form in expanding shells others srgue that they are part of a remnant envelope which is still collapsing toward the newly formed star (e.g.: 1984, ApJ, 283, 632; 1984, MN, 211, 41P; 1985, AA, 153, 179; 1985, ApJ, 289, 681; 1985, MN, 213, 641; 1985, ApJ, 298, 830).

The relationship between H II regions and molecular clouds has been studied by several groups. The genetic connection between H II regions and molecular clouds has been stressed, many molecular and atomic transitions have been mapped across the interface regions, models of the H II-molecular cloud structure, including the velocity field, have also been presented (e.g.: 1984, ApJ, 282, L81; 1985, ApJ,290, L59; 1985, ApJ, 292, 200; 1985, ApJ, 293, 522; 1985, ApJ, 299, 341; 1985, ApJ, 299, 351; 1986, AA, 160, 287; 1986, Astrofisika, 24, 257; 1986, ApJ, 303, 638; 1986, ApJ, 303, 683; 1986, PASJapan, 38, 361; 1986, ApJ, 307, 302).

3. PHYSICAL CONDITIONS

X-ray observations and predictions have been made for H II regions. The extended X-ray emission from the Rosette nebula is coincident with the central radio minimum suggesting that hot gas originating from the central O stars is responsible for the diffuse X-rays and the thick hollow shell appearance (1985, MN, 217, 69). High temperature plasma with $T \simeq 4 \times 10^7 K$ has been detected in the Orion nebula as well as an X-ray emission feature at 6.69±0.06 keV (1986, PASJapan, 38, 723). The X-ray emission from ring nebulae around WR and Of stars, due to stellar winds, has been predicted (1985, AZh, 62, 103; 1986, MNP, 246).

The local electron densities were determined from: optical observations of the [S II] and [Cℓ III] lines, far-infrared observations of the [O III] and [N III] lines, Stark broadening of radio recombination lines, and from radio observations of two Hnα lines and continuum (e.g.: 1983, AstronTsirk, 1301; 1984, AstronTsirk, 1320; 1985, AA, 145, 347; 1985, Pis'maAZh, 11, 17; 1985, ApJSS, 57,

571; 1986, ApJ, <u>306</u>, 532).

The mean electron temperature has been determined for a large number of galactic H II regions based on radio observations, in many cases the accuracy of the
determination depends on the evaluation of deviations from LTE (e.g.: 1984, AA,
<u>138</u>, 225; 1984, MN, <u>211</u>, 149; 1984, MN, <u>211</u>, 155; 1984, MN, <u>211</u>, 339; 1985, AA,<u>146</u>,
<u>L19</u>; 1985, AZh, <u>62</u>, <u>1057</u>; 1985, ApJSS, <u>57</u>, 571; 1986, ApJ, <u>301</u>, 813; 1986, AA, <u>160</u>,
129; 1986, RMexAA, <u>13</u>, 15).

Two reviews of the effects of dust in H II regions were presented by Mathis
and Münch (1986, PubASP, <u>98</u>, 995; 1985, MittAGes, <u>63</u>, 65). One millimeter continuum observations of galactic H II regions imply that most of the dust emission appears to come from dust within the H II region (1984, AA, <u>137</u>, 117). The effect of
dust scattering internal to H II regions has been studied based on simple analytical
models (1984, AA, <u>139</u>, 30). Measurements of dust scattered light from M20 indicate that grains have high albedo and a strongly forward-throwing phase function
(1985, ApJ, <u>288</u>, 164). Observations of M8 favor models with small refractory grains
(1986, AJ, <u>91</u>, 870). The heating of H II regions by electrons ejected from grains
has been considered (1986, RMexAA, <u>12</u>, 257; 1986, ApSpSc, <u>126</u>, 211). The broad
Balmer profiles present over the face of Car II are shown to be due to dust scattered Balmer emission from η Car (1986, RMexAA, <u>13</u>, 27). A 4 mm excess over the
free-free emission from W3 has been found and has been ascribed to large-size
grains (1986, PASJapan, <u>38</u>, 775). The stellar polarization in the direction of S252
is similar to the values observed for the nearby field stars (1986, ApSpSc, <u>128</u>,
125).

Important improvements in computer capabilities have made possible more
sophisticated treatments of photoionized models of H II regions (e.g.: 1986, MNP,
235). Different problems have been studied in the literature: ionization correction
factors for nebulae of low degree of ionization (1985, ApJ, <u>291</u>, 247), the helium
ionization correction factor for H II region complexes (1986, PubASP, <u>98</u>, 1061),
the importance of the ionization parameter (1986, MNP, 225; 1986, PubASP, <u>98</u>, 1072),
different geometries like blister nebulae (1984, ApJ, <u>287</u>, 653), O and N infrared
emission from blister nebulae (1986, AA, <u>161</u>, 347), homogeneous grids of models
for different heavy element abundances (1984, ApJSS, <u>57</u>, 349; 1985, ApJSS, <u>58</u>,125),
the effect of dust on radiative transfer (1986, MN, <u>221</u>, 923), the effect of Lyman
line pumping on model H II regions (1986, AA, <u>156</u>, 393), star in motion relative to
the surrounding medium (1986, ApJ, <u>300</u>, 745), the importance of ionization from the
n = 2 level of hydrogen for regions with densities higher than 10^8 cm^{-3} (1984, ApJ,
<u>283</u>, 165).

Optical and near-infrared emission lines have been observed by means of high,
intermediate and low dispersion spectra as well as Fabry-Pérot interferometry, from
these observations the structure and dynamics of many H II regions have been studied (e.g.: 1984, AA, <u>137</u>, 245; 1984, AA, <u>138</u>, 451; 1984, ApJ, <u>283</u>, 640; 1984,
RMexAA, <u>9</u>, 119; 1984, MN, <u>211</u>, 267; 1984, MittAGes, 62, 302; 1985, ApJ, <u>288</u>, 142;
1986, AA, <u>155</u>, 6; 1986, MN, <u>219</u>, 895; 1986, ApJ, <u>304</u>, 767; 1986, MN, <u>220</u>, 745; 1986,
ApJ, <u>307</u>, 649). Observational constraints on dynamical models have been discussed
by Meaburn (1986, MNP, 167). The mean velocity dispersion as a function of size
has been studied for a few H II regions in terms of turbulent motions (e.g.: 1984,
ApJ, <u>285</u>, 109; 1985, ApJ, <u>288</u>, 142; 1986, ApJ, <u>304</u>, 767; 1986, ApJ, <u>307</u>, 649; 1986,
PubASP, <u>98</u>, 1002). Kelvin-Helmholtz instabilities have been suggested for the generation and maintenance of nebular turbulence for Sharpless 142 (1985, ApJ, <u>288</u>,
142). An explanation of the observed turbulent motions based on Kolmogorov's
theory seems indicated to a first approximation, but there is poor agreement in
many major details (1986, ApJ, <u>304</u>, 767; 1986, PubASP, <u>98</u>, 1002). A method for determining large scale expansion (or contraction) velocities based on observations
of radio recombination lines was developed by Gulyaev and Sorochenko (1985, Bull
AbastumaniApObs, <u>59</u>, 135).

4. EVOLUTION

High dynamic range radio continuum maps show that ionized gas appears to flow around dense neutral condensations, this observation supports a model where radiation-driven ionization shock fronts lead to: the efficient implosion of near-by neutral condensations and the formation of second generation stars (1986, ApJ, 305, 714).

Models to study the evolution of H II regions emphasizing different aspects have been computed. The evolution of a neutral envelope surrounding an H II region ionized by an O5 star was calculated (1985, AstrofizIIzvS, 20, 95). The effect of a strong stellar wind on the structure of an H II region was considered (1983, Astrofizika, 19, 559). The kinetic efficiencies of stellar wind bubbles were estimated (1986, ApJ, 306, 538). Comparison of models with observations suggests that, for hollow H II regions, the shock at the edge of the wind cavity is not adiabatic but strongly dissipative (1986, AA, 160, 1). A model is presented for the dynamical evolution of an H II region in a cloudy medium (1986, MNP, 211). Models of line formation have been computed for the champagne phase (1984, AA, 138, 325). A self-consistent model of the evolution of a spherical nebula including the ionization structure, the energy balance and a hydrodynamics code has been computed (1986, MNP, 215). The time evolution of the Hβ equivalent width and the [O III]/Hβ line ratio for models with a single burst of star formation has been computed, it is found that these two ratios decrease monotonically and that they can be used as age indicators (1986, AA, 156, 111).

5. ABUNDANCES AND GALACTIC GRADIENTS

Aller gave a review paper on fifty years of nebular chemical compositions (1986, PubASP, 98, 957). The potential of infrared and far-infrared emission lines for studying nebular abundances has been reviewed by Dinerstein (1986, PubASP, 98, 979). The abundances of Ne^+, Ar^+, Ar^{++}, S^{++} and S^{+3} have been determined for five compact H II regions based on infrared line emissions (1984, ApJ, 285, 174). The He^+/H^+ abundance ratio for three positions of the Rosette nebula has been determined (1985, AA, 144, 171). The nitrogen and helium enrichment for four ring nebulae has been obtained (1984, ApJ, 287, 840). Variations of the relative abundances of N, O and S in NGC 6164-4 have been determined and are discussed (1986, ApSpSc, 120, 17). The Si^+/H^+ abundance ratio has been detected near θ^1 Ori C (1986, ApJ, 301, L57). The N/H, O/H and Ne/H values have been determined for the Orion nebula based on far-infrared observations, the N/O ratio is about a factor of two higher than that derived from optical lines (1986, ApJ, 311, 895).

A galactic electron temperature gradient of 310 K kpc^{-1} has been derived from H I 166α observations (1985, RMexAA, 10, 179).

6. GALACTIC CENTER

In addition to the Symposium on the galactic center, a review on Saggitarius A and its environment was presented by Brown and Liszt (1984, AnnRevAA, 22, 223). The ionized gas in the galactic center forms a unique H II region in the Galaxy; the inner 1.0 pc, centered at SgrA west or at IRS 16, is dust defficient and it is not yet clear if a central source (central engine) or if star formation is responsible for the observed structure (1987, GC, 1,8,30,39,79). There are several observations that indicate that almost certainly star formation in the inner 10 pc of the galaxy is going on at present and that O and B stars are being formed (e.g.: 1985, MN, 215, 69P; 1987, GC, 79).

7. EXTRAGALACTIC H II REGIONS

The study of extragalactic H II regions has increased at a faster rate than

that of galactic H II regions. Supergiant H II regions, that are almost absent
in the Galaxy, have been used for the study of: star formation, the chemical evolution of galaxies, preliminary estimates of the value of H_0 and the determination
of the pregalactic or primordial helium abundance.

The structure, luminosity and distribution of giant and supergiant H II regions has been studied by several groups (e.g.: 1984, ApJ, 287, 116; 1984, PubASP,
96, 944; 1985, ApJ, 293, 400; 1985, AASS, 62, 63; 1985, AA, 145, 170; 1986, AA,154,
357; 1987, ApJ, 319, 61). The electron density has been determined for a few objects
based on the [O II] 3726/3729 ratio (1984, ApJ, 283, 158). Extragalactic H II regions have been used as tracers of the stellar content and of starburst galaxies
(e.g.: 1985, ApJSS, 58, 533; 1985, AA, 143, 347; 1985, ApJ, 288, 175; 1985, AJ,
90, 80; 1985, AA, 148, 443; 1985, AA, 152, 427; 1986, IAUSymp,116, 355; 1986,
SFDG, 395; 1986, MN, 223, 811; 1986, PASJapan, 38, 571; 1986, PubASP, 98, 1032;
1987, RMexAA, 14, 144; 1987, Exploring the Universe with the IUE Satellite, ed. Y.
Kondo, Reidel, hereafter EUIUE, p. 605; 1987, EUIUE, 623; 1987, AA, 180, 12).
Molecular hydrogen emission has been detected in the direction of giant H II regions in the SMC, the LMC and M33 (1985, ApJ, 291, 156; Israel, F.P. preprint). The
velocity field at various scales has been determined for 30 Doradus and NGC 604
(1984, MN, 211, 521; 1984, AA, 141, 49). The relationship between the linear diameter and the velocity width for 47 H II regions and between the Hα luminosity and
the velocity dispersion for 43 objects has been analyzed (1986, ApJ, 300, 624;
1986, AA, 160, 374), furthermore the integrated Hα velocity profiles of 47 extragalactic H II regions have been discussed (1986, AJ, 92, 567).

Dufour (1987, EUIUE, 577) presented a review of IUE observations of galactic
and extragalactic H II regions with an emphasis on the CNO abundances. Pagel (1986,
PubASP, 98, 1009) presented a review of current problems related to abundances in
extragalactic H II regions. Several reviews on the chemical evolution of galaxies
based on H II observations are present in the literature (e.g.: 1985, MN, 217,391;
1986, Highlights of Astronomy, ed. J.P. Swings, Reidel, p. 377; 1986, SFDG, 403;
1986, PubASP, 98, 973; 1986, PubASP, 98, 1057; 1987, AA, 172, 15; 1987, Interstellar
Processes, ed. D.J. Hollenbach and H.A. Thronson, Reidel, p. 667). Abundances of
many extragalactic H II regions have been reported in the literature (e.g.: 1985,
ApJ, 290, 449; 1985, ApJ, 292, 155; 1986, AA, 154, 352; 1986, AA, 158, 266; 1986,
PubASP, 98, 1025; 1986, AA, 162, 180; 1987, ApJ, 317, 163; 1987, MN, 226, 19; 1987,
RMexAA, 14, 178). An extensive study of abundance gradients of extragalactic H II
regions reveals a good fit between models and observations if nitrogen is a product
of secondary nucleosynthesis (1985, ApJSS, 57, 1), a similar result is obtained by
an independent study (1986, ApJ, 307, 431). Many papers on abundance gradients of
spiral galaxies are present in the literature (e.g.: 1984, MN, 211, 507; 1984, AJ,
89, 1702; 1986, PubASP, 98, 1032; 1986, ApJ,309, 544; 1987, ApJ, 317, 82).

A review on the pregalactic helium abundance and its relationship with cosmology was presented by Boesgaard and Steigman (1985, AnnRevAA, 23, 319). Many observations of the He/H ratio and many determinations of the pregalactic helium
abundance, Y_p, from extragalactic H II regions are present in the literature (e.g.:
1985, AA, 146, 269; 1985, ApJSS, 58, 321; 1986, SFDG, 183; 1986, SFDG, 197; 1986,
PubASP, 98, 984; 1986, PubASP, 98, 1005; 1986, ApJ, 311, 45; 1986, AA, 158, 266;
1986, AA, 154, 352). The best observations indicate that Y_p = 0.232±0.005 (1σ) and
that $\Delta Y/\Delta Z$ = 3.5±0.7 (1σ). Ferland (1986, ApJ, 310, L67) has suggested that the effect of collisional excitation from the He I 2^3S state reduces Y_p to about 0.207,
on the other hand Peimbert and Torres-Peimbert (1987, RMexAA, 14, 540) argue that
the collisional excitation effect has been overestimated and that Y_p should be
reduced only from about 0.232 to about 0.228.

Giant and supergiant H II regions have also been used as distance indicators
and in principle they provide an independent value of H_0 (1985, AA, 143, 469; 1987,
MN, 226, 849; 1987, RMexAA, 14, 158). Finally a 100 kpc cloud of ionized gas at a

redshift of 1.825 has been detected through observations of Lyman α emission (1987, ApJ.319,L39), these type of objects might be indicative of young or recently forming galaxies.

VII. Supernova Remnants
(T. Landecker)

Supernova remnant (SNR) research at the beginning of the review period is well summarized by Lozinskaya (1985,ApSpPhysRev,3,35) and IAU Colloquium 101, held at the end of the period, provides an overview of recent developments. Progress in instrumentation has in the usual way led to new understanding. The IRAS data have become available, revealing shock-heated dust in SNRs. IR and X-ray spectroscopy have advanced allowing study of the composition of ejecta. Better detectors have led to improved optical images and have permitted spectroscopy of faint features. Oxygen-rich ejecta in SNRs have been studied. Many more extragalactic SNRs have been studied at all wavelengths. Improved radio telescope sensitivity has revealed low-brightness features of known SNRs and allowed the detection of many new ones. Good data have been obtained on temporal changes in young SNRs at radio wavelengths. SNRs resembling the Crab are being found in increasing number; in particular, a very close facsimile to the Crab Nebula has been discovered in the LMC. Composite SNRs with Crab-like and shell SNR features, have been recognized. Advances have been made in associating various types of SNRs with the various types of super-novae and statistical studies of SNR radio properties continue, but not without controversy over their interpretation. Theoretical work has continued on SNR evolution in an inhomogeneous ISM; the modification of its environment by the progenitor star has been recognized as a very important influence on SNR evolution. Relevant conference publications include: Bartel,ed.: 1985,"Supernovae as Distance Indicators",Springer-Verlag,Berlin; Durgaprasad+,ed.: 1985,"18th International Conference on Cosmic Rays",Tata Inst.,Bombay; Kafatos+,ed.: 1985,"The Crab Nebula and Related Supernova Remnants",Cambridge U.Press,London; Jones+,ed.: 1985,"19th International Cosmic Ray Conference",NASA CP-2376,Washington; Peacock,ed.: 1985, "X-ray Astronomy in the EXOSAT Era",SpSciRev,40,1; Srinivasan+,ed.: 1986,"Super-novae, their Progenitors and Remnants",Suppl. to JAA; Proc. IAU Colloquium 101, to be published.

The theory of SNR evolution in homogeneous and cloudy media remains the subject of much work. Cox (1986,ApJ,304,771) considers SNR evolution in isotropic media, and shows that SNR properties depend on present density and post-shock temperature only. Kundt (1985,Kafatos+,151) argues that SNR shells consist of rampresure confined filaments, not Sedov-Taylor blast waves. Raymond (1986,Pequignot+, 145) models radiative and non-radiative shocks and predicts optical and UV properties. Hester (1987,ApJ,314,187) has developed a sheet description of shocks in middle-aged SNRs like the Cygnus Loop. Shull+(1985,MN,212,799 and IAU Coll.101) present analytical models of SNR evolution in the environment created by its progenitor. Cioffi+(IAU Coll.101) develop an analytical model of adiabatic and radiative SNRs. Chieze (1986,AdvSpRes,6,129) discusses evaporative SNRs. Innes+(1986, Pequignot,153; 1987,MNRAS,224,179), Bertschinger (1986,ApJ,304,154) and Gaetz (IAU Coll.101) consider instabilities of radiative shocks and Bychkov (1985,Pis'ma AZh, 11,911; 1986,AstroSpetsAstroObs,21,58; 1986,AZh,63,939) discuss their optical and radio emission. Hamilton (1985,ApJ,291,523) develops similarity solutions for ejecta driven blast waves. Hugues+(1985,ApJ,291,544) have modelled X-ray emission from Kepler's SNR including non-equilibrium ionization (NEI) in a hydrodynamic (HD) calculation. Jerius+(IAU Coll.101) discuss effects of NEI on X-ray emission from adiabatic SNRs. Pimenov (1985,Pis'ma AZh,11,265) interprets radio features of SNRs as magnetic field perturbations.

Preite-martinez (1986,Pequignot,204) uses a HD code to study interaction of a young SNR with an interstellar (IS) cloud. Tenorio-Tagle+ use a 2-dimensionel HD

code to study:-SNR-SNR (1984,AA,138,215) and SNR-Molecular cloud interaction (1985,
AA,145,70), applying the latter model to the Cygnus Loop (1986,AA,148,52); cloud
crushing (1986,AA,155,120) and condensation and ejection of clouds (1987,AA,176,329)
by a SNR; shock reflection (1986,AA,167,120); and sequential SN explosions in an OB
association (1987,AA,182,120). Rumyantsev (1985,Astrofiz,22,157) considers filament
formation by shock collisions.

Dopita+(1984,ApJ,282,142) discuss models for optical emission from oxygen-rich
ejecta in young SNRs. Kafatos (1985,Kafatos+,13) models forbidden line emission in
the Crab Nebula. Sunyaev+(1984,Pis'ma AZh,10,483) predict radio lines from heavy
elements in SNRs. Gondhalekar (1985,MN,216,57P)has studied depletion of elements in
SNR filaments. Shull (IAU Coll.101) reviews line emission processes in atomic and
molecular shocks.

Injection of energy into the ISM by stars before and after they become SNe is
reviewed by McKee (1986,ApSpSci,118,383). SNRs expand in an ISM modified by the
progenitor; Lozinskaya (IAU Coll.101) discusses the influence of WR and O-f stars,
including those in associations, on their surroundings. Clark (1984,Gondhalekar,169)
critically reviews the use of SNRs as probes of the ISM. Braun (IAU Coll.101) discusses ISM interactions which most influence SNR dynamics and brightness. McKee
(IAU Coll.101) reviews SNR shocks in an inhomogeneous ISM. Clifford (1984,MN,211,
125) models SNR interaction with a two-phase ISM. Cowie (1984,Kondo+,287) shows
that the local ISM may be described as the interior of an SNR.

Stellar winds and SNe in OB associations act together to generate supershells
in the ISM. Examples are found in Cygnus (Bochkarev+: 1985,ApSpSci,108,237) and in
the Perseus Arm (Fich: 1986,ApJ,303,465) and perhaps in the local ISM (Innes+: 1984,
MN,209,7; Arnaud+: 1984,Kondo+,301 and 1986,AdvSpRes,6,119). Supershell structure
is reviewed by McCray (IAU Coll.101) and modelled by Wolff (1987,MN,224,701),
Tomisaka+(1986,PASJ,38,697), Silich (1985,Astro,22,563), and MacLow+(IAU Coll.101).
Collective effects of SNRs and supershells on galaxies have been considered by
Kafatos and McCray (1986,de Loore+,ed.:IAU Symp.116,Reidel), MacLow+(1986,PASP,98,
1104), Dopita (IAU Coll.101), and by Tomisaka and Ikeuchi (IAU Coll.101).

Chevalier (1985,Kafatos+,63) reviews evolution of the Crab Nebula. Davidson+
(1985,AnnRevAA,23,119) review knowledge of the ejecta and the progenitor. Kennel+
(1984,ApJ,283,694) study energization by the central pulsar and Craig+(1985,AA,149,
171) the resultant synchrotron source. The shock wave expected around the Crab, has
been searched for (1985,Kafatos+,81,89,115;1986,AA,157,335). Lundquist+(1986,AA,162,
L6) predict a fast shell should be detectable in the UV. X-ray observations give
information on injection of energy in the nebula (1985,Nature,313,662;1985,Kafatos+,
197;Ogowara+,1986,in "X-Ray Astronomy 1984",ed.Oda+,ISAS,Tokyo). Abundances have
been studied by X-ray (Schattenburg+:1986,ApJ,301,759) and optical spectroscopy (1986,
Pis'ma AZh,12,440;1986,PASP,98,1044). Van den Bergh+(1986,Nature,321,46) have resolved filaments into small knots. Henry+(IAU Coll.101) have studied Ni emission.
Radio emission from the Crab has been mapped (1984,Pis'ma AZh,10,730;1985,MN,212,
359;Kafatos+,115,133;1986,Pis'ma AZh,12,275). The jet seen in the Crab has been well
studied (1984,ApJLett,285,L75;1985,Nature,313,661;Kafatos+,115,127;ApJ Lett,294,
L121;AA,151,101;1986,ApJ,306,259). A significant discovery is the close resemblance
of the SNR LMC 0540-693 to the Crab Nebula. It contains an X-ray and optical pulsar
(1984,ApJ Lett,287,L19;1985,Nature,313,659) and is a synchrotron nebula (1984,ApJ
Lett,287,L23;1986,Srinivasan,119). Reynolds (1985,ApJ,291,152) considers its evolution and age. A synchrotron shell has been sought around 3C58 with no success
(1985,AJ,90,2312;1986,MN,218,533;Landecker+: 1987,AJ,94,111). Fesen+(IAU Coll.101)
have measured expansion velocities in 3C58 and proper motion of filamants. CTB80,
an unusual SNR with a Crab-like core, has been studied optically (1985,Kafatos+,
257; Observatory,105,7), in X-ray (1984,ApJ,285,607) and in the radio (1984,AA,139,
43). Strom (preprint) finds pulsar-like properties in its central source. The Vela-X
region has been observed in X-rays (1985,AdvSpRes,5,53;Kafatos+,203;SpSciRev,40,487;

ApJ.299,821). Milne+(1987,AA,167,117) have shown that Vela-X does not differ in
radio spectral index from the rest of the Vela SNR.

Srinivasan+(1984,JAA,5,403) and Lozinskaya (1986,AZh,63,914) consider the evolution of pulsar-driven SNRs. Radio observations have been made of Crab-like SNRs
G20.0-0.2 (1985,ApJ Lett,297,L25), and G291.0-0.1 (1986,MN,219,815). X-ray observations have been made of G21.5-0.9, 3C58 and Vela-X (1985,Kafatos+,219;SpSciRev,
40,513; 1986,ApJ Lett,300,L59), G291.1-0.1, G308.7+0.0 and G328.4+0.2 (1986,ApJ,
302,718). Composite remnants with a Crab-like component and an outer shell are receiving increasing attention. Radio and X-ray observaitons have been made of G0.9+
0.1 (Helfand+: 1987,ApJ,314,203), G18.94-1.06 (Barnes+: IAU Coll.101), G24.7+0.6
(Becker+: IAU Coll.101), G27.4+0.0 (1985,ApJ,288,703), G29.7-0.3 (1984,ApJ,283,154;
1985,ApJ,295,456).

Considerable effort continues to be expended on observations and theoretical
investigations of Cas A. Van den Bergh+(1986,ApJ,307,723) put new limits on any
stellar remnant of the SN. Fesen+(1986,ApJ,306,248) have observed nitrogen-rich
ejecta. Koyama+(1986,Oda+,ed."X-Ray Astronomy 84",ISAS,Tokyo,289), Tsunemi+(1986,
ApJ,306,248) and Markert+(IAU Coll.101) have used X-ray spectroscopy to study
abundances. Jansen+(1985,AdvSpRes,5,49) have mapped X-ray temperature structure.
Morfill+(1984,Nature,311,358) have observed an X-ray halo beyond the shock front.
Dwek+(1987,ApJ,315,571) present IRAS observations; the emission is largely from
shock-heated dust. Dinerstein+(1987,ApJ,312,314) find IR evidence for dust within
the ejecta. Kenneys+(1985,ApJ,298,644) have mapped the 86 GHz brightness and
polarization.

Recent work on Tycho's SNR has concentrated on X-ray data. Davelaar+(1985,SpSci
Rev,40,467) have measured abundances spectroscopically. Hamilton+(1986,ApJ,300,713)
model the X-ray spectrum. NEI models are reported by Brinkmann+(IAU Coll.101). Itoh+
(IAU Coll.101) model the X-ray spectrum with a carbon deflagration Type Ia SN.
Smith+(IAU Coll.101) predict radio and X-ray emission from a simple HD model. X-ray
spectra of Kepler's SNR have been modelled by Ballet+(IAU Coll.101) using NEI HD
models. Matsui+(1984,ApJ,287,295) have compared radio and X-ray images. Optical
structure has been mapped by Bandeira+(1986,MemASocItal,56,773); a catalog of features is presented by D'Odorico+(1986,AJ,91,1382). Bandiera (preprint) considers
the origin of Kepler's SNR. The memnant of SN1006 has been mapped in the radio by
Reynolds+(1986,AJ,92,1138) and by Roger+(preprint) and in the X-ray by Jones+(1986,
Oda+,ed. "X-Ray Astronomy 84",ISAS,Tokyo,305). Vartanian+(1985,ApJ Lett,288,L5) have
detected strong X-ray emission lines. Hamilton+(1986,ApJ,300,698) analyze the X-ray
spectrum using NEI models.

The dividing line between SNe and SNRs is indistinct. Weiler+(1986,ApJ,301,
790; Science,231,1251) review observations of radio SNe. Graham+(1986,MN,218,93)
have made a significant detection of iron in a SN using IR spectroscopy. Bandiera+
(1984,ApJ,285,134) suggest that radio SNe evolve into Crab-like SNRs. Chevalier
(1984,ApJ Lett,285,L65; 1985,Gondhalekar+,128,IAU Coll.101) Fransson (1986,in Mihalas+,
ed. "Radiation hydrodynamics of Stars and Compact Objects", Springer-Verlag,Berlin,
141) and Dickel+(1986,NASA TM-88342,67 and IAU Coll.101) discuss interaction of
young SNRs with the circumstellar medium (CSM). Liang+(1984,AnnNYAcadSci,422,233)
study interaction of ejecta with the CSM and ISM. Band+(IAU Coll.101) extend this
work to adiabatic SNRs. Bedogni+(IAU Coll.101) discuss instabilities due to electron heat conduction. Itoh (1984,ApJ,285,601) discusses effects of temperature relaxation on X-ray properties. Canizares (1986,Oda+,ed."X-Ray Astronomy 84",ISAS,
Tokyo,275) reviews X-ray results on ejecta dominated SNRs. Glushak (1986,Sov.A.Lett
11,350) discusses evolution of the radio spectrum of young shell SNRs. Reynolds
(IAU Coll.101) has studied radio structure in 3C58 and SN1006 and discusses particle acceleration and transport. Bohigas (1984,RevMexAA,9,13) deduces the mass ejected in SN events from studies of young SNRs. Mezger+(1986,AA,167,145) have observed
Cas A and the Crab Nebula at 1.2mm; the IR luminosity of the Crab is dominated by

synchrotron emission but dust emission prevails in Cas A.

Changes with time of the radio appearance of a number of young SNRs have been detected. Tuffs (1986,MN,219,13)has compared 5 GHz images of Cas A over a 4-year period and Green (IAU Coll.101) has used 151 MHz data with a 2.3-year baseline. Braun+(1987,Nature,327,395) have used 3 images spread over 4-years to show clumps of ejecta overtaking the radio shell; resultant bow shocks cause the observed structure. Temporal changes in total radio flux density of Cas A are recorded by Barabanov+(1986,AZh,63,926) and Walczowski+(1985,MN,212,27P). Van den Bergh+(1985, ApJ,293,537) review optical studies of Cas A from 1951 to 1983. Aller+(1985,ApJ, Lett,293,L73) have measured a decrease in the 8 GHz flux density of the Crab Nebula from 1968 to 1984. However, Green (1987,MN,225,11P) shows that the flux density of 3C58 has increased over a comparable period.

Theory indicates that some SNe form a condensed object and an expanding SNR. Seinivasan (1986,Srinivasan+,105), Qadir+(1986,ChinPhysLett,3,189) and Akujor (1987, ApSpSci,135,187) discuss pulsar-SNR correlations. Nomoto+(1986,ApJ Lett,305,L19; 1986,Comments Ap, 11,151)discuss neutron star cooling and its effect on X-ray emission from young SNRs. Seward (1986, Comments Ap, 11,15) summarizes X-ray observations of five SNRs containing neutron stars. Manchester (1987,AA,171,205) models radio emission from most SNRs as the result of biconical outflow from a central object. Feigelson (1986,CanJPhys,64,474) and Becker (1986,CanJPhys,64,482) discuss production of outflows from degenerate stars in SNRs. Haynes+(1986,Nature,324,233) have observed a nebula associated with Cir X-1 which may have been ejected from an SNR. Van Gorkom+(1986,Srinivasan+,93) have made a radio search for compact sources in SNRs. Compact sources have been shown to be NOT associated with the SNRs G74.9+ 1.2 (1984,ApJ,286,284), G127.1+0.5 (1984,JAA,5,425), G109.1-1.0 (1986,CanJPhys,64, 479) and Tycho (Green+: 1987,MN,224,1055).

The Cygnus Loop and IC443 are the canonical old SNRs and attract observers at all wavelengths. Hester+ have studied filaments of the Cygnus Loop (1986,ApJ,300, 675,698; IAU Coll.101; Raymond+, preprint) using optical, UV and X-ray data. Based on this data Hester (1987,ApJ,314,187) has proposed a model of SNR filaments as sheets. Fesen+(1985,ApJ,295,43) have investigated a non-radiative filament in the Cygnus Loop. Teske+(1985,ApJ,292,22) present images in a coronal line of iron. Greidanus+(IAU Coll.101) have mapped filaments using an imaging Fabry Perot interferometer. Kritsuk (1986,Astrofiz,26,45) suggests that thermal instabilities play a part in filament formation. Ballet+ and Charles+(1985,SpSciRev,40,481;ApJ,295,456) have mapped the Cygnus Loop in X-rays, and Vedder+(1986,ApJ,307,269) have found evidence of NEI from X-ray lines. Straka+(1986,ApJ,306,266) have compared high-resolution radio and optical images. Green presents a detailed image at 408 MHz. Braun+ (1986,AA,164,208) analyze IRAS observations of the Cygnus Loop and conclude that the expanding SNR has encountered a pre-existing high density shell.

Braun+(1986,AA,164,193) have also analyzed IRAS data for IC443 and conclude that this SNR too has expanded into a pre-existing shell, part of a network of stellar wind bubbles. Mufson+(1986,AJ,92,1349; IAU Coll.101) have investigated IC443 at radio, IR,optical, UV and X-ray wavelengths. Gensheimer+ (1986,PASP,98, 1147) and Ballet (IAU Coll.101) have mapped IC443 in a coronal line of iron. Graham+ (1987,ApJ,313,847) and Wright+ (IAU Coll.101) present IR spectroscopy of a number of iron lines. Molecular gas and IRAS sources associated with IC443 have been studied by Huang+ (1986,NASATM-88342,69). White+ (1987,AA,173,337) and Burton (IAU Coll.101) have observed a number of molecules in these clouds, and Mitchell (IAU Coll.101) models their chemistry. Green (1986,MN,221,473) has mapped IC443 at 151 and 1419 MHz and has detected spectral index variation across the source, but Erickson+(1985,ApJ,290,596) show that the overall spectrum follows a power law from 20 MHz to 11 GHz.

In the review period, the increasing sensitivity of radio telescopes has led

to discovery of new SNRs and to improved maps of others. G357.7-0.1 and G5.3-1.0
have been mapped by Becker+(1985,Nature,313,115,118) who consider them a new type
of non-thermal radio source, powered by outflow from a binary system. G357.7 -0.1
was also observed by Shaver+(1985,Nature,313,113). On the other hand Caswell+(1987,
MN,252,329) conclude that G5.3-1.0 is part of a larger shell SNR. Papers which present data on many SNRs include Milne+(1985, ProcASocAust,6,78) -50 southern SNRs;
Fürst+(1985,MittAGes,63,149) -14 new SNRs;Reich+(IAU Coll.101) -32 new SNRs;
Tateyama+(IAU Coll.101) -flux densities of 15 SNRs at 22 GHz; Reich+(1986,AA,155,
185) -2 new SNRs; Milne (IAU Coll.101 and preprint) radio polarization catalogue
for 70 SNRs; Trushkin+(1986,preprint) SNRs between longitudes 85 and 135 degrees.
Green (1985,MN,216,69) has searched for very young SNRs and lists 42 sources thought
NOT to be SNRs.

New radio data on the following SNRs or possible SNRs has been published: -
G2.4+1.4 (Green+,1987,MN,225,221), G7.7-3.7 (1986,MN,223,487), G8.7-0.1 (1986,AJ,
92,1372,G11.2-0.3 (1984,MN,210,845;1985,ApJ,296,461), G18.9-1.1 (1985,Nature,314,
720;1986;AA,92,1372;Fürst+,IAU Coll.101), G19.96-0.18(1986,AA,92,1372), G24.7+0.6
(Becker+,1987,ApJ,316,660),G41.1-0.3(1985,ApJ,296,461),W50(1986,MN,218,393), G50.29-
0.4 and G54.09+0.26 (1985,AA,151,L10), G54.5-0.3 (1985,AJ,90,1224: 1986,JAA,7,105),
G70.68+1.2 (1985,AA,151,L10; 1986,MN,219,39P), G71.23+1.47 (1985,AA,151,L10), G73.9+
0.9(Chastenay+,IAU Coll.101), G93.7-0.3 and G94.0+1.0 (1984,AA,138,469;1985,AJ,90,
1082), G109.1+1.0 (1984,ApJ,283,147), G132.6+1.5 (Landecker+,1987,AJ,94,111),G160.9+
2.6 (Leahy+,IAU Coll.101), G166.2+2.5 (1986,MN,221,809; Kim+,IAU Coll.101), G166.0+
4.3 (Pineault+,1987,ApJ,315,580), G179.0+2.7 (1986,AA,154,303), S147 (1986,AA,163,
185), G192.8-1.1 (1985,AJ,90,1076), G206.9+2.3 (1986,ApJ,301,813), G292.0+1.8 (1986,
AA,162,269), G312.4-0.4 (1985,MN,216,753), G348.5+0.1 and G348.7+0.3(1984,MN,210,
845), G349.7+0.2(1985,Nature,313,113), G350.1-0.3(1986,AA,162,217). Radio studies of
SNRs at the end of the review period were summarized by Caswell(IAU Coll.101).

Evidence continues to accumulate suggesting that the galactic loops are old
SNRs. Parkinson+(1985,SpSciRev,40,503) have studied soft X-ray emission from Loops I
and III. Bhat+(1985,Jones+,342) have detected a gamma-ray excess from Loop I. Arnal+
(1986,RevMexAA,12,298) discuss possible star formation induced by the Lupus Loop.
Radio spectroscopy of HI and CO lines has provides powerful probes of SNR/ISM interaction. Gosachinskij+(1985,ApSpSci, 108,303) present observations of HI shells near
12 SNRs. Braun and Strom(1986,AA Suppl, 63,345) have found post-shock HI in four
SNRs (G78.2+1.8, G109.1-0.1, G166.0+4.3, and IC443). HI has been observed near: -
G296.5+10.0(1986,AJ,91,343), G166.2+2.5(1986,MN,221,209), W50(1986,Astrofiz,25,287),
G166.0+4.3 (Landecker+,IAU Coll.101); G18.95-1.1(Fürst+, IAU Coll.101). CO has been
observed near W28 and W44 (Velusamy,IAU Coll.101); G109.1-1.0 (Tatematsu+,IAU,Coll.
101). Fukui+(IAU Coll.101) present observations of CO in the vicinity of five SNRs.
Dubner+(IAU Coll.101) have detected HI and CO near Puppis A. Huang+(1986,ApJ,309,
804) list molecular clouds which they associate with SNRs between longitudes 70
and 210°. They use these associations to derive a new surface brightness-diameter
(sigma-D) relationship (1985,ApJ Lett,295,L13).

Studies of the sigma-D relationship, based on the radio properties of SNRs,
have continued. Green (1984,MN,209,449) has critically reviewed the available distance estimates. He concludes that sigma-D is of little value. Allakhverdiyev+(1986,
ApSpSci,121,21; Astrofiz,24,97,397) conclude that sigma-D is relevant for shell SNRs
in dense environments but not for low-brightness SNRs. Berkhuijsen(1986,AA,166,257;
preprint;IAU Coll.101) concludes that the sigma-D relationship results from dependence of SNR properties on ISM density, not from an evolutionary sequence. Duric+
(1986,ApJ,301,308)have theoretically reproduced the observed sigma-D relationship.

Preite-Martinez+(1986,AA,157,6) have studied energy and diameter for LMC SNRs.
Berkhuijsen(1984,AA,140,431) has studied SN rates in M31, M33 and the Galaxy and
concludes that only a small fraction of SNRs have been detected. Clark (1984,Gondha-
lekar,ed."Mass Loss from Astronomical Objects",RL-82-075) reviews work on SN rates,

explosion energy, and on progenitors. Weiler(1985,Kafatos+,227) reviews evolution
of Crab-like SNRs. Sakhibov+(1985,ByulInstAstrofiz,75,3) have used SNR statistics
to deduce explosion energy. Xinji+(IAU Coll.101) have studied the distribution and
birthrate of galactic SNRs. Li+(IAU Coll.101) have examined the correlation of
galactic SNRs and spiral arms. Van den Bergh+(preprint) have studied the SN rate in
Shapley-Ames galaxies. Van den Bergh (preprint) relates the various types of SNRs
to the various types of SNe.

Improved instrumentation has permitted observation of faint optical features
in SNRs. Fesen+(1985,ApJ,292,29) have studied abundance gradients in the Galaxy
from spectrophotometry of 7 SNRs. Pineault+(1985,AA,151,52) have mapped G166.0+4.3.
Winkler+(1985,ApJ,299,981) have detected oxygen in Puppis A. Van den Bergh and
Pritchet(1986,PASP,98,448) have used CCDs to map faint features in Cas A, the Crab
Nebula and CTB80. Meaburn+(1986,MN,222,593) have used echelle spectrograms to detect
blast wave-cloud interactions in RCW103. Itoh(1986,PASJ,38,717) has studied OI emission from Puppis A and Cas A. Kirshner+(1987,ApJ,315,L135) have studied high velocity emission in SN1006 and Tycho. Teske+(1987,ApJ,318,370) have observed coronal
iron lines in Puppis A.

The data from IRAS has had a substantial impact on SNR studies. This field is
reviewed by Dwek+(1986,NASA-TM-88342) and by Dwek(IAU Coll.101). Most IR emission
from SNRs arises from shock heated dust, and IR emission appears to be an important
cooling mechanism for SNRs. Braun (1987,AA,171,233) has prepared IR images of Cas A,
Tycho and Kepler from IRAS data. The emission arises from IS or CS material, not
from ejecta. Graham+(1986,Israël,397;1987,ApJ,319,126)have examined IRAS data for
LMC SNRs. Arendt(IAU Coll.101) shows that one third of known galactic SNRs are detectable in the IRAS data. Moorwood+(IAU Coll.101) have studied the IR spectra of
galactic and LMC SNRs. Dennefeld (1986,AA,157,267) has studied abundances using IR
spectroscopy. Rengarajan+(IAU Symp.120) show that IRAS point sources are found preferentially on SNR shells, and interpret these as dust knots heated by X-ray emissio

X-ray studies have been reviewed by Aschenbach (1985,SpSciRev,40,447; 1986,
Highlights A,7,649) and by Strom (1986,ESA,SP-239,43). Hamilton+(1984,ApJ,284,601)
interpret X-ray spectra including NEI, and Bleeker (1985,MN,220,501) and Markert+
(1986,IAU Coll.86,NRL,Washington,76) discuss observational evidence of NEI. Seward
(IAU Coll.101) has reprocessed X-ray images of 44 SNRs from Einstein data. Smith
(IAU Coll.101) presents Exosat spectral data for 8 SNRs. X-ray data have been pu-
blished for the following SNRs -G29.7-0.3 (1985,Jones+,394), W28 and 3C400.02 (1985,
Kafatos+,211), W44 (1985,MN,217,99), W49B (1985,ApJ,296,469), HB3 (1985,ApJ,294,183),
Monoceros (1985,MN,213,15P; 1986,MN,220,501), Puppis A (1986,ESA SP-239,137;Fisch-
bach+,IAU Coll.101), RCW86 (1986,ESA SP-239,107), RCW103 and MSH 14-63 (1984,ApJ,
284,612), PKS 1209-52 (1985,SpSciRev,40,475;Kellett+: 1987,MN,225,199;Matsui+,IAU
Coll.101), 1E1149.4-6209 in Crux (1986,ApJ,302,606).

Extensive bodies of new data on Magellanic Cloud SNRs have been published by
Mills+(1984),AustJPhys,37,321-radio data), Mathewson+(1985,ApJ.Suppl,58,197-optical
data), Danziger+(1985,MN,216,365-spectrophotometry), Graham+(IAU Coll. 101-infrared
data), Dopita+(1984,ApJ,282,135-spectrophotometry of young oxygen-rich SNRs). Con-
tini (1987,AA,174,5) has fitted models to optical line ratios for LMC SNRs. Other
studies of Magellanic Cloud SNRs are:-N157B (1984,AA,140,390), N63A (1986,AA,164,
26), N132D (Hughes: 1987,ApJ,314,103; Lasker,IAU Coll.101) and 1E0102-7219 (Hughes,
Blair+, Lasker, IAU Coll.101).

Long (1985,SpSciRev,40,531) has reviewed observations of extragalactic SNRs
and Dickel+(1985,Bartel,100) have reviewed techniques for detecting them. SNRs have
been observed in:-M33 (1985,ApJ,289,582; 1985,JAA,6,145; Blair+,Duric,Long+,IAU Coll.
101), M82 (1985,Bartel,88), NGC 185, IC 1613 and NGC 6822 (1985,AJ,90,414), M101(1986,
ApJ,311,85). Cox+(1986,ApJ,304,657) estimate the SNR rate in M101 from X-ray data.
Artyukh+(1986,Pis'ma AZh,12,739)deduce the SNR density in M33 from radio data.

Cowan+(1985,Bartel,75) discuss radio observations of historical SNe.

The subject of cosmic ray (CR) acceleration in SNRs is an active field; only a few papers can be referred to here. Conferences Durgaprasad+, Jones+ and 1986, Shapiro,ed. "Cosmic Radiation in Contemporary Astrophysics", Reidel, include many relevant references. Wolfendale (1986,Shapiro (ibid),135) reviews CR origin and propagation. CR acceleration by SNR shocks is treated by Bulanov+(1984,Pis'ma AZh, 10,594), Jokipii+(1985,ApJLett,290,L1), Dorfi+(1985,Jones+,136), Allakhverdiyev+ (1986,ApSpSci,123,237), and Axford (1986,Kahn,119). Cowsik+(1985,Durgaprasad+,306) consider electron acceleration in SNRs and effects on SNR spectra and evolution. Volk (1985,Audouze+,ed. "High Energy Astrophysics",Ed. Frontières,Paris) treats the non-linear theory of CR acceleration. The effects of particle acceleration on SNR evolution are discussed by Heavens (1984,MN,211,195), Glushak (1986,Pis'ma AZh, 11,825), Volk+(1985,Jones+,148; 1986,Kahn,101) and Droge+(1987,AA,178,252). Jokipii (1987,ApJ,313,842) examines diffusive shock acceleration, and shows that orienta- tion of the magnetic field affects acceleration efficiency. Blandford (IAU Coll.101) reviews particle acceleration in SNR shocks and the confrontation of theory with observations.

VIII. Planetary Nebulae
(Y. Terzian)

1. GENERAL STUDIES

A number of relevant books, conference reports, catalogues or surveys and review papers are cited at the beginning of this report. Moreover, IAU is scheduled to con- duct Symposium 131 on Planetary Nebulae (PN) in Mexico City during the first week of October 1987. Acker edited "Les Nébuleuses Planétaires: Comptes Rendus Sur Les Journées De Strasbourg" in 1986. Statistical surveys of PN and their central stars were discussed by Amnuel+(1984,ApSpSci,107,19;1985,ApSpSci,113,59).

Acker+ worked on producing a new general catalog of PN, which should be avail- able by the IAU General Assembly in 1988. Acker+(1987,AASuppl,in press), and Sabbadin+(1987,AASuppl,in press) examined in detail a group of misclassified nebulae. Saurer+(1987,AASuppl,in press), and Hartl+(1987,AASuppl,in press) reported on the identification of new PN, many of them of very low brightness. Maehara+(1987,AA, 178,221) reported on the identification of NGC 2242 as a PN rather than as a galaxy. Shaw+(1985,PASP,97,1071) have identified seven new PN in Baade's Window.

Méndez+(1987,AA,in press) presented high resolution spectroscopic observations of many central stars of PN. Ishida+(1987,AA,178,227) have identified two extremely faint and old PN, and present a list of 31 nebulae within a distance of 500 pc from the Sun. They estimate a large birthrate of 8×10^{-3} kpc^{-3} yr^{-1}, and conclude that the total number of PN in our galaxy is larger than 10^5. Gathier (1987,AA,in press) derived a very different birthrate of 2.4×10^3 kpc^{-3} yr^{-1} from a group of 30 PN with individually determined distances. Stenholm (1986,Acker,25) described a spec- troscopic program to observe 1500 PN of which 476 objects have already been obser- ved. Pottasch+(1986,AA,161,363) described measurements of IRAS spectra of PN.

2. DISTANCES

The problem of distance determinations of PN has remained very difficult, inspite of several studies that have examined this issue carefully (Gathier,1987, AA,in press; Kwok,1985,ApJ,290,568). Kaler+(1985,PASP,97,594) measured accurate extinctions towards 8 nebulae and derived dust-distances, however they conclude that typical distance uncertainties are ±55%. Gathier+(1986,AA,157,171) derived distances to 12 nebulae using the reddening-distance method and conclude distance accuracies from 10 to 40%. Gathier+(1986,AA,157,191) use HI absorption observations in the direction of PN to derive kinematical distances. For 12 nebulae the derived distances have an uncertainty of less than ∿ 50%. Kaler+(1985,ApJ,288,305) have suggested deriving wind distances from the P Cygni profiles, where one can infer the gas escape velocity. Masson (1985,ApJ,302,L27) reported the measurement of the angular expansion of NGC 7027 with VLA observations and derived a distance to this object of 940 ± 200 pc. Terzian (1987, Sky and Telescope, p.128) described the nebular angular expansion method of determining distances with VLA observations for the nearby nebulae.

3. MORPHOLOGY

Substantial progress has been made in observations of the morphology of PN. Balick (1987,Sky and Telescope, p.125) presented a comprehensive study of the shaping of PN. Chu+(1987,preprint) have studied the multiple shell nebulae and conclude that the frequency of multiple shell events is as high as ≥ 0.5. Kwok (1985,AJ,90,49) reported on VLA maps of 10 compact nebulae, and Basart+(1987,ApJ, 317,412) reported on additional VLA maps of BD 30°3639,NGC 6572, NGC 6590 and NGC 7027. In addition Balick+(1987,preprint) reported on optical and radio (VLA) images of NGC 6543, IC 3568 and NGC 40. Sabbadin+(1985,MN,217,538) discussed in detail the structure of NGC 3587 from optical spectroscopic observations. Hua+(1986,AJ,92,853) reported high resolution optical observations of IC 351. Moreno+(1987,AA,178,319) presented very deep photographic, narrow band images of NGC 6720 and have revealed the presence of extended filanetary structure in its halo. Jewitt+(1986,ApJ,302, 727) showed results of a CCD survey to detect outer halos around 44 PN, and have reported that 2/3 possess extensive outer halos. Zuckerman+(1986,ApJ,301,772) have examined a large sample of PN and concluded that among 108 nonstellar objects ∿ 50% show bipolar morphological symmetry, and ∿ 30% display elliptical symmetry. Pascoli (1987,AA,180,191) discussed in detail the origin of bipolar structures and has computed models assuming intra-nebular magnetic fields.

4. MOLECULAR AND NEUTRAL HYDROGEN OBSERVATIONS

Exciting new developments have taken place during the last few years in our understanding of the transition from red giant stars to PN. A recent review on this topic has been given by Kwok(1987,"Late Stages of Stellar Evolution", Reidel 321). Knapp (1987, MittAstronGes, in press) has discussed the molecular line observations and the mass loss from red giants, and has concluded that the mass returned to the Galaxy from red giants is about $1M_\odot$/yr. Ground based and IRAS observations of pre-planetary nebulae have revealed dust grains in the circumstellar envelopes of luminous red giant stars (Zuckerman+,1986,ApJ,311,345; Kwok+,1986,ApJ,303,451; Likkel+, 1987,AA,173,L11).

Thronson+(1986,ApJ,300,749) have searched for molecular emission from NGC 7027, and other than CO, they detected CN. Phillips+(1985,AA,151,421) have presented CO maps of NGC 7027. Huggins+(1986,ApJ,305,L29; 1986,MN,220,33P) have detected significant CO emission associated with NGC 7293, NGC 2346 and NGC 6720, implying that a significant fraction of these nebulae have not yet been ionized. Pottasch+(1987,AA, 177,L49) reported the detection of radio continuum radiation from two OH/IR stars, and Payne+(1987,ApJ, in press) reported the detection of OH from NGC 6302 and Vy 2-2; they also present a detailed VLA map of the 1612 MHz OH maser emission of NGC 6302. Rodriguez+(1985,MN,215,353) reported the detection of λ 21 cm neutral hydrogen in the PN NGC 6302, and Taylor+(1987,AA,176,L5) also reported HI from IC 418. HI has also been observed by Altschuler+(1986,ApJ,305,L85) from IC 4997.

5. SPECIAL OBJECTS AND STUDIES

Atherton+(1986:Nature,230,423) detected the central star of NGC 2440 and derived a Zanstra temperature of 350,000K. Heap(1987:Nature,326,571) reported IUE observations of the same star derived a stellar temperature of 200,000K.

Feibelman (1987,PASP,99,270) reported a stellar temperature of 94,000K for the high galactic latitude star of the nebula 75 + 35°1, and Bianchi+(1987,AA,181,85) using IUE observations derived a temperature of 90,000K, log $L/L_\odot$ = 4.4, R = 0.66 $R_\odot$, and M = 1.1 $M_\odot$ for the central star of NGC 40. Kaler+(1985,ApJ,297,724) studied the IUE spectra of 32 central stars and derived Zanstra temperatures ≥ 70,000K. Aller+(1985,ApJ,296,492) derived a temperature of 160,000K for the central star of NGC 6741. de Korte+(1985,AdvSpaceRes,5,57) reported the detection of X-ray emission from the PN NGC 1360. Pronik+(1986,Astron Zh, 63,1016) made UV observations of NGC 2352 and NGC 6853 with the ASTRON satellite. Kostyakova +(1985-6,Astron Tsirk, 1380-1460) studied the variability of PN including HM and FG Sge. YM29 was studied by Lozinskaya+(1986,Astron Zh, 63,255). NGC 7293 was studied by Leene+(1987,AA,173,145), and by Walch+(1987,MN,224,885). [OII] studies of PN were made by O'Dell+(1984,ApJ, 283,158). Studies of galactic abundance and electron temperature gradients were made using PN by Faundez-Abans+(1986,AA,158,228), and by Maciel+(1985,AA,149,365) respectively. Sabbadin+(1986,AASuppl,65,259) studied the internal motions 32 PN, and Sabbadin+(1985,MN,213,563) reported on the expansion velocities of 22 nebulae.

6. PN IN THE MAGELLANIC CLOUDS, M31 AND M33

Barlow (1987,MN,227,161) studied 32 nebulae in the Magellanic Clouds and deduced a mean ionized mass of 0.27 $M_\odot$. Small nebulae were observed using speckle interferometry by Wood+(1985,ProcAstronSocAust,6,54); these results show a mean ionized mass of 0.08 $M_\odot$. Barlow+(1986,MN,223,151) have also made speckle interferometric observations of the SMC N2 nebula. Morgan+(1985,MN,213,491) have detected 10 additional PN in the SMC, raising the number to 44. Gull+(1986,Wilson,295) presented IUE results on 15 nebulae in the LMC and the SMC. Dopita+(1985,ApJ,297,593) discussed the identification of highly energetic nebulae in the LMC. A series of preprints by Dopita+, to be published in ApJ, discuss various aspects of the LMC PN. Nolthenius+(1987,ApJ,317,62) have surveyed PN in the outer parts of M31, and Lequeux+(1987,AASuppl, 67,169) have made a new survey of PN in M33.

7. NEBULAR EVOLUTION

The dynamical evolution of PN was considered under the Interacting-Stellar-Winds model by Volk+(1985,AA,153,79). Schmidt-Voight+(1987,AA,174,211 and 223) have studied the influence of stellar evolution on the evolution of the nebulae. Balick (1987,AJ,in press) discussed the evolution of PN with respect to structure, ionization and morphological sequences. O'Dell+(1985,ApJ,289,526) formulated a detailed model for NGC 2392.

Theoretical models of PN nuclei were considered by Tylenda (1986,AA,156,217), in an attempt to explain the appearance of double-envelopes in the nebulae.

IX. Intergalactic Medium
(P. Shaver)

This new section in the Commission 34 Report is devoted to the diffuse intergalactic medium (IGM). Thus, halos of galaxies, for example, are not included, and even quasar absorption lines, which may have more to do with galaxies, are referred to here only insofar as they have a bearing on the more diffuse IGM. Symposium or conference proceedings where relevant information can be found are: Mardirossian+, ed.: 1984,"Clusters and Groups of Galaxies",Reidel; Richter+,ed.: 1985,"The Virgo Cluster of Galaxies",ESO,Garching, plus a few cited later.

For the most part the IGM has proved difficult to detect, let alone study in detail, so it is understandable that most of the work has concentrated on hot gas in clusters of galaxies, for which there is a relative abundance of observations. Indirect information about this intracluster medium has been obtained by studying its apparent effect on ejecta from active galaxies, e.g. radio jets and lobes (Spangler+: 1984,AJ,89,1478; Smith: 1984,MN,211,767; Arnaud+: 1984,MN,211,981; Jaffe+: 1984, Mardirossian+,293; O'Dea+: 1985,AJ,90,929; Hanisch+: 1985,AJ,90,1407; Stocke+: 1985,ApJ,299,799; Miller+: 1985,MN,215,799; Jaffe+: 1986,AJ,91,199; Fedorenko+: 1986,Sov.Astron.30,No.1; Feigelson+: 1987,ApJ,312,101). The effects of a dense, hot intracluster medium on gas in galaxies (Giovanelli+: 1985,ApJ,292,404; Balkowski+: 1985, Richter+,37; Haynes: 1985,ibid.45; Warmels: 1985,ibid.51; van Gorkom+: 1985,ibid.61; Giovanelli: 1985,ibid.67; Kennedy+: 1985,ibid.165) and intergalactic clouds (Rephaeli+: 1985,MN,215,453; Sotnikova: 1986,Astrofiz,25,139) have been considered, also the influence of galaxy winds and ejecta on the IGM (Silk: 1984, in Mass Loss from Astronomical Objects,ed. Gondhalekar,128; Kundt+: 1985,AA,142,150).

X-ray emission from intracluster gas has made possible relatively detailed studies - morphology, spectroscopy, inferences regarding density, temperature, and abundances, theoretical models, etc. (Feretti+: 1984,AA,139,50; Gorbatskij: 1984, Astrofiz,20,42; Sotnikova: 1984,Astrofiz,21,415; Bahcall+: 1984,ApJ,284,L29; Stewart+: 1984,ApJ,285,1; Fabricant+: 1984,ApJ,286,186; Chanan & Abramopoulos:1984, ApJ,287,89; Gil'fanov & Sunyaev: 1984,Sov.Astr.Lett.10,137; Ulmer+: 1984, Mardirossian+,307; Kriss+: 1984,ibid.313; Jones+: 1984,ibid.319; Porter: 1984,ibid. 351; Branduardi-Raymont+: 1985,AdvSpRes,5,133; Rothenflug+: 1985,AA,144,431; Gerbal+ 1985,AA,146,119; Konyukov: 1985,Astrofiz,22,163; Matilsky+: 1985,ApJ,291,621; Henrikson+: 1985,ApJ,292,441; Hu+: 1985,ApJSuppl,59,447; Caganoff+: 1985,ProcASA, 6,151; Fabian: 1985,SpSciRev,40,653; Smith+: 1985,ibid.661; Norgaard-Nielsen+: 1985,ibid.669; Henrikson+: 1985,ibid.681; Jones+: 1985,"X-ray Astronomy '84",ed. Oda ISAS,355; Fabricant+: 1985,ibid.381; Rothenflug+: 1985,ibid.391; Gerbal+: 1985, ibid.395; Gerbal+: 1986,AA,158,177; Friaca: 1986,AA,164,6; Henry+: 1986,ApJ,301, 689; Henrikson+: 1986,ApJ,302,287; Ulmer+: 1986,ApJ,303,162; Canizares+: 1986,ApJ, 304,312; Bertschinger+: 1986,ApJ,306,L1; Miller: 1986,MN,220,713; Sarazin: 1986, RevModPhys,58,1). Kowalski+(1984,ApJSuppl,56,403) have analyzed X-ray data on 3600 clusters.

Much theoretical work has concentrated on the possibility of cooling flows (e.g. Pallister: 1985,MN,215,335; Fabian+: 1985,MN,216,923; Fabian+: 1986,ApJ,305, 9). Radio emission (Kotanyi: 1985,Richter+,13; Dennison: 1986,AA,159,251), UV emission (Holberg+: 1985,ApJ,292,16) and γ-rays (Houston+: 1984,J Phys,G10,L147) from intracluster gas have been considered. Vallée+(1987,Ap Lett,25,181) discuss the determination of the intracluster magnetic field by combining rotation measures and X-ray emission. Work has continued on searches for and possible detections of the Sunyaev-Zeldovich effect (Davies+: 1984,Mardirossian+,267; Aliakberov+: 1985, S.S.Astrofiz.Obs.Vyp,48,81; Andernach+: 1986,AA,169,78; Radford+: 1986,ApJ,300,159; Birkinshaw+: 1986,Highlights Astron,7,321; Partridge+: 1987,ApJ,317,112; Chase+: 1987,MN,225,171).

Searches for intergalactic HI clouds have generally been negative (e.g. Altschuler+: 1987,AA,178,16). The discovery of one such cloud in Leo was therefore of great interest, stimulating VLA observations (Schneider+: 1986,AJ,91,13), detection of smaller nearby HI clouds (Schneider: 1985,ApJ,288,L33), infrared and optical searches (Skrutskie+: 1984,ApJ,282,L65; Pierce+: 1985,AJ,90,450; Kibblewhite+: 1985,MN,213,111) with a possible detection in Hα (Reynolds+: 1986,ApJ,309,L9), and interpretations in terms of galaxy interactions (Rood+: 1984,ApJ,285,L5; 1985,ApJ, 288,535).

The possibility of intergalactic dust has received further attention (Voshchinnikov+: 1984,Adv Sp Sci,3,443; 1984,Astrofiz,21,401; 1984,Ap Sp Sci,103, 301; Shaver: 1985,AA,143,451; de Bernardis+: 1985,ApJ,288,29; Khersonskij+: 1985,Ap Lett,24,217; 1985,Ap Sp Sci,117,179; Evans+: 1985,MN,213,1p; Rudnicki+: 1985,Nuovo Cimento C,8C,368; Tanaka: 1985,PASJ,37,481; Evans+: 1985,Sp Sc Rev,40,701; Reaves: 1985,Richter+,433; Horstmann: 1986,Mitt Astron Ges,65,237; Greenberg+: 1987,Nature, 327,214). There is some evidence for extinction of quasars by foreground clusters (Shanks+: 1986,in Quasars,ed. Swarup+,Reidel,37; Phillipps: 1986,Ap Lett,25,19), but no evidence of reddening in the spectra of distant quasars (Steidel+: 1987,ApJ, 313,171; Wright+: 1987,preprint).

A new, more stringent limit on HI in the IGM has been set using the Gunn-Peterson test (Sargent: 1987,"Observational Cosmology",ed. Hewitt+,Reidel,777). The photoionization of the high-redshift IGM has been studied (e.g. Shapiro: 1986,PASP, 98,1014; Bechtold+: 1987,ApJ,315,180; Bajtlik+: 1987,preprint; Donahue+: 1987,preprint; Barcons+: 1987,preprint), and the possibilities of detecting it through dispersion (Wiita+: 1984,Observatory,104,270; Barcons+: 1985,ApJ,289,33) or refraction (Burke: 1984,ESA-SP,213,153) effects were considered. A 0.6 keV OVIII Lyα absorption feature has been observed in the spectrum of a z = 0.1 BL Lac object (Canizares+: 1984,ApJ,278,99), which could be intrinsic (Krolik+: 1985,ApJ,295,104), or possibly from the distributed IGM (Shapiro: 1986,"Galaxy Distances and Deviations from Universal Expansion",ed. Madore+,Reidel,203). Other papers dealt with heating and evolution (e.g. Couchman: 1985,MN,214,137; Barausov+: 1985,Sov Astron Lett,11,372; Ikeuchi+: 1986,ApJ,301,522), molecules (Shchekinov+: 1984,Astr Zh,61,460; MacLow+: 1984,BAAS,16,962; Shchekinov+: 1985,Astr Zh,62,841; Couchman: 1985,MN,214,137; Nakai+: 1986,PASJ,38,603; Shchekinov+: 1986,Pis'ma Astr Zh,12,499), and magnetic fields (Beech: 1985,Ap Sp Sci,116,207; Andreasyan: 1986,Astrofiz,24,363).

It is still unclear whether the IGM contributes significantly to the X-ray background, but many recent papers have dealt with this important question (Marshall+: 1984,ApJ,283,50; Elvis+: 1984,ApJ,283,479; Maccacaro+: 1984,ApJ,283,486; Kowalski+: 1984,ApJ Suppl,56,403; Setti: 1985,"Non-Thermal and Very High Temperature Phenomena in X-Ray Astronomy",ed. Perola+,U. Roma,159; Fabian: 1985,"Observational and Theoretical Aspects of Relativistic Astrophysics and Cosmology",103; Zamorani: 1985,"X-Ray Astronomy '84",ed. Oda+,ISAS,419; Schmidt+: 1986,ApJ,305,68; Tucker+: 1986,ApJ,308,53; Guilbert+: 1986,MN,220,439; Setti: 1986,Mitt Astron Ges Nr,67,133; Anderson+: 1986,"Quasars",ed. Swarup+,Reidel,247; Schmidt: 1986,"Structure and Evolution of AGN",ed. Giuricin+,Reidel,3; Giacconi+: 1987,ApJ,313,20; Segal: 1987, ApJ,313,543; Barcons: 1987,ApJ,313,547).

35. STELLAR CONSTITUTION (CONSTITUTION DES ÉTOILES)

PRESIDENT: D. Sugimoto (Japan)
VICE-PRESIDENT: A. Maeder (Switzerland)
ORGANIZING COMMITTEE: P.H. Bodenheimer (USA), C.S. Chiosi (Italy), A.N. Cox (USA), D.O. Gough (UK), R. Kippenhahn (FRG), Y. Osaki (Japan), J.L. Tassoul (Canada), J.W. Truran (USA), V. Weidemann (FRG), and J.C. Wheeler (USA).

I.   INTRODUCTION   (D. Sugimoto, University of Tokyo)

This report of Commission 35, as in past reports, consists of some details of only a few selected topics. This is necessary because a survey of the entire field of stellar formation, structure, stability, evolution, explosion, and nucleosynthesis for the three year period from mid-1984 to mid-1987 would be excessively long. Our topics here, in order from early to late evolutionary pahses are: Convective Overshooting (N.H. Baker), Mass Loss (I. Appenzeller), Novae (M. Livio), Presupernova Models and SN1987A (K. Nomoto), and Structure of X-Ray Bursting Neutron Stars (M.Y. Fujimoto and D. Sugimoto). In addition, Asteroseismology (H. Shibahashi) is reported briefly as one of the new disciplines now being developed. About two decades ago, Professor Martin Schwarzschild suggested convection, mass loss, and calculation of models through supernova stage as ones of the most important problems to attack. Though great progress has been achieved in these topics, we still have some fundamental questions concerning physical mechanisms involved. This is the reason why these topics are selected for this report.

One of the biggest events of interest of Commission 35 during this period was the occurrence of SN1987A in the Large Magellanic Cloud. This delayed event of nearby supernova from statistical point of view was lucky in the sense that astronomers have been prepared not only with different observational facilities through different wave-length bands even including neutrino detectors but also detailed model calculations through the presupernova stages and some models of explosion into the interstellar space. Nomoto's report herein is somewhat more concentrated on this supernova rather than is exhaustive in the presupernova models in general. Though such a report on the current topics seems premature, it is a topics of urgent concern. Full report could be given in the future.

References are given as their numbers in the Astronomy and Astrophysics Abstracts when possible. In many cases, the conventional references needed to be given because the 1987 volumes are not available at the moment of preparing this report. Because our reports for the preceding or succeeding periods do not necessarily contain the same topics covered here, some essential papers published even before 1984, and some important papers now in press or in preprints are also included.

General interest in stars continues at high level, because the stars are fundamental constituents of the universe and provide a variety of physical phenomena inherent to self-gravitating systems. Some people, or rather onlookers say that interests in astronomy is being shifted from stellar physics to galaxies and cosmology. This is true, in a sense, because the galaxies and observational cosmology are relatively new and expanding fields. Nevertheless, we will be able to see in the following sections that we still have wealth of interesting topics in the field of Stellar Constitution.

During the reporting period seven IAU Symposia and ten IAU Colloquia were held on topics of interest to Commission 35. They are in order of dates: Symposium 116 Luminous Stars and Associations in Galaxies, Porto Heli, Greece, May 26-31, 1985;

Symposium 115 Star Forming Regions, Tokyo, Japan, November 11-15, 1985; Symposium
125 The Origin and Evolution of Neutron Stars, Nanjing, China, May 26-29, 1986;
Symposium 122 Circumstellar Matter, Heidelberg, FRG, June 23-27, 1986; Symposium 123
Advances in Helio- and Asteroseismology, Aarhus, Denmark, July 7-11, 1986; Symposium
126 Globular Cluster Systems in Galaxies, Cambridge, MA, USA, August 25-29, 1986;
Symposium 131 Planetary Nebulae, Mexico City, October 5-9, 1987; Colloquium 90 Upper
Main Sequence Stars with Anomalous Abundances, Crimea, May 14-17, 1985; Colloquium
89 Radiation Hydrodynamics in Stars and Compact Objects, Copenhagen, Denmark, June
11-20, 1985; Colloquium 87 Hydrogen Deficient Stars and Related Objects, Mysore,
India, November 10-15, 1985; Colloquium 93 Cataclysmic Variables, Bamberg, FRG, June
18-20, 1986; Colloquium 92 Physics of Be Stars, Boulder CO, USA, August 18-22, 1986;
Colloquium 95 Second Conference on Faint Blue Stars, Tucson, AZ, USA, May 31 - June
3, 1987; Colloquium 97 Wide Components in Double and Multiple Stars: Problems of
Observation and Interpretation, Brussels, Belgium, June 8-13, 1987; Colloquium 101
Interaction of Supernova Remnants with the Interstellar Medium, Penticton, BC,
Canada, June 9-12, 1987; Colloquium 103 The Symbiotic Phenomenon, Torun, Poland,
August 18-21, 1987; and Colloquium 108 Atmospheric Diagnostics of Stellar Evolution:
Chemical Peculiarity, Mass Loss and Explosion, Tokyo, Japan, September 1-4, 1987.

II. CONVECTIVE OVERSHOOTING (N.H. Baker, Columbia University)

The structure of stellar convection zones is usually calculated using some form
of mixing-length theory, according to which it is postulated that the velocity of
convective elements is nonzero only in those regions of the star that are con-
vectively unstable according to a linear stability analysis. (Such a theory is
called a "local" theory.) In fact, there is no reason to expect that the velocity
should vanish as soon as the acceleration is no longer positive. The true boundaries of a convection zone will thus lie outside the boundaries of the unstable
("superadiabatic") region, and it is this phenomenon that is called "overshooting".

It has long been recognized that this may have detectable consequences for
stellar structure and evolution. Energy transport by the overshooting elements may
change the structure of the star. In chemically inhomogeneous regions there may be
additional mixing of species, which can alter the nuclear processes in subsequent
evolution. There may be dynamical effects also, such as momentum transport and wave
generation.

The problem that has received most attention is overshooting above the convective cores of massive stars, which may alter the luminosity, the temperature, and
even the surface composition of evolving models. In the sun, overshooting from the
top of the subsurface convection zone has a strong effect on the velocity fields
observed at the surface, and the overshooting region at the bottom may be important
for magnetic flux storage. In main-sequence stars overshooting at the bottom of
such a zone may have a bearing on the lithium-depletion problem. Mixing by the over
shooting elements may be important also in other stars like red giants, novae, and
supernovae.

Several approaches to the problem are currently being used. The most realistic
are the large two- and three-dimensional simulations now possible on large computers. These are too complex to be used in stellar evolution calculations, and so
the search has continued for simpler formulations, usually nonlocal generalizations
of mixing-length theory. In evolution studies these are often used, but a purely
empirical approach, in which a free parameter is introduced which is proportional to
the overshooting distance, has also proved fruitful. In both main-sequence and
later stages there are consequences that can be compared with observation.

1. Numerical Simulations

Several types of numerical simulation have been used. Modal expansions with one or two horizontal planforms, as used recently by Massaguer et al. (38.064.041) allow good vertical resolution, but the horizontal structure is distorted by the planform truncation. Two-dimensional simulations like those of Hurlburt et al. (38.065.037 and Ap.J., 311, 563, 1986) also restrict the form of the horizontal flow, but not nearly so severely as the modal approach, and they allow better spatial resolution than is possible in three dimensions. Three-dimensional simulations have also been performed, most with application especially to the solar convection zone: Nordlund (40.080.105), Glatzmeier (39.065.045), Gilman and Miller (Ap.J. Suppl., 61, 585, 1986), and Chan and Sofia (Ap.J., 307, 222, 1986 and Science, 235, 465, 1987) and others have contributed. In most of this work the convective cells are found to extend vertically over a number of pressure scale heights, suggesting that the mixing-length treatment, which postulates a characteristic distance of the order of a scale height, is inapplicable to this type of convection. On the other hand Chan and Sofia (Science, 235, 465, 1987) find a correlation length of the order of one scale height. While there are differences in the results obtained with the various techniques, there are many points of agreement, and certainly we are now beginning to get an idea of what stellar convection zones must really look like.

The work most directly applicable to the overshooting problem is that of Massaguer et al. (38.064.041) (modal) and of Hurlburt et al. (Ap.J., 311, 563, 1986) (two-dimensional). In these papers three polytropic layers are studied, and the mutual interactions of the middle unstable layer and the surrounding stable ones is investigated. Strong pressure fluctuations can counteract the temperature fluctuations and reverse the sign of density fluctuations, leading to a reversal of buoyancy in the upper part of the unstable zone. This produces marked up-down asymmetries in the flow. There is strong penetration into the lower stable region, which induces internal gravity waves. Overshooting into the upper stable zone is much less pronounced, having very little effect on the heat transport there; however, relatively weak velocity fields, which could give rise to mixing, do penetrate to a distance of the order of a scale height. On the whole, the picture appears to be at variance with what any of the mixing-length type theories predict.

The two-dimensional and especially the three-dimensional simulations must be run on rather coarse grids, so only large-scale motions can be explicitly represented. The smaller scales, including the smallest ones in which viscous dissipation occurs, are usually modeled crudely by introducing some kind of eddy diffusivity. In this sense there is a link to the nonlocal mixing-length theories, though the simulations of course take full account of compressibility and have more realistic horizontal structure. The virtues and pitfalls of the simulation approach are summarized in a review of solar and stellar convection by Zahn (in "Solar and Stellar Physics", eds. Schroeter and Schluessler, Springer-Verlag, 1987, in press).

2. Mixing-length Theories

Overshooting is excluded in standard mixing-length theories, but a number of authors have attempted to fix up the mixing-length formalism to take account of penetration. Some authors have used a scheme proposed several years ago by Shaviv and Salpeter, in which the dynamics of the overshooting elements is modeled in a simple way. The method has recently been criticized by Langer (A & A 164, 45, 1986), who finds that the solution is not unique. In any case there are difficulties in matching the overshooting to the solution in the unstable region. Another scheme is that of Roxburgh, which is based on a conservation theorem; this has been criticized by several authors, most recently Baker and Kuhfuss (A & A 1987, in press). Others have used their own recipes. One of the more sophisticated treatments is due to Xiong (40.065.008, 40.065.030), who obtained a nonlocal statistical theory by approximating higher-order correlations, which arise from the nonlinear terms in the equations of motion, with diffusion terms. Mixing of species is also included in a consistent way. Such a theory requires the specification of

at least one length scale other than the mixing length, namely some kind of diffusion length. A rather similar approach was discussed by Kuhfuss (A & A, 160, 116,
1986). A much simplified treatment that also invokes diffusion has been proposed by
Doom (A & A, 1987, submitted). Another statistical theory, incorporating a length
scale partly based on the scale height of potential temperature, has been proposed
very recently by Cloutman (Ap.J. 313, 699, 1987). Nonlocal mixing-length theories
and other theories have been discussed by Unno et al. (40.064.057, 41.065.007).
These authors propose improved ways of constructing stellar models with convection
zones.

3. Applications

Unno et al. (40.080.033) applied Xiong's nonlocal mixing-length theory to the
solar convection zone. They obtained a rather steep temperature gradient in the
overshooting region at the top, in good agreement with empirical data. Photospheric
velocities were comparable with those observed in granulations. Schmitt et al.
(Ap.J., 282, 316, 1984) applied plume theory (used in atmospheric physics and
distantly related to the Shaviv-Salpeter formulation) to the question of overshooting at the bottom of the solar convection zone. They found the overshooting
region to be adiabatic, and only a few tenths of a scale height in thickness.

Most of the other applications have been to the evolution of massive stars. A
number of papers on this subject are found in the proceedings of IAU Symposium No.
116, "Luminous Stars and Associations in Galaxies" (41.012.079). The evolution of
these stars is also affected by mass loss, which is the main subject of a review by
Chiosi and Maeder (Ann. Rev. Astron. Astrophys., 24, 329, 1986).

Overshooting from the convective core can be significant in the evolution of
stars with initial masses of as little as a few solar masses, and studies have
included stars as massive as 120 solar masses and more. The usual technique is to
construct sequences of stellar models from which theoretical isochrones can be
obtained. These are then the basis of a comparison with observed HR diagrams,
especially those of young clusters and associations. (See particularly Stothers
(40.065.087) and Mermilliod and Maeder (41.153.005).) Most authors include mass loss
(according to an empirical formula) in addition to overshooting, but the effect of
overshooting alone can be seen in the paper of Stothers and Chin (39.065.054), who
compute tracks with and without mass loss. They model the effect of overshooting by
a parameter which is the ratio of the "overshooting distance" to the local pressure
scale height at the top of the unstable region, and find the best fit with a parameter equal to 0.35. This is in good general agreement with Maeder and Meynet (A & A,
1987, in press), who use a similar procedure. Another evolution study in which the
effect of overshooting alone can be seen is that of Xiong (A & A 167, 239, 1986),
who uses his own nonlocal theory, with special attention to details of the overshooting.

Models having a convective core enlarged by overshooting are more luminous at
the main-sequence turnoff and have longer H-burning lifetimes than standard models.
In general this means that the main-sequence band is wider, which improves greatly
the agreement with observational data. The exception to this is that Maeder and
Meynet (A & A, 1987, in press) note that their models, including mass loss and
overshooting, produce a narrowed main sequence for O-type stars, which they find to
be in good agreement with observations for the youngest clusters and associations.
Overshooting also limits the extent of the blueward loops in the HR diagram during
He burning, thus affecting the statistics of red giants in the diagram. In addition
to the works already mentioned, there has been extensive work by other authors,
notably by Bertelli et al. (40.065.029, 40,065.107, 41.065.171, 41.065.174), who use
an overshooting model something like that of Shaviv & Salpeter, by Doom and collaborators (39.065.004, 41.065.076, 40.065.109), using Roxburgh's overshooting
criterion, and by Langer & El Eid (A & A, 167, 265, 1986). It is also to be

expected that the effects mentioned will influence mass transfer in close binary systems containing massive stars, the consequences of which have been studied by Sybesma (39.117.033, 41.065.034, A & A 168, 147, 1986). On the whole there seems to be a consensus that models with overshooting produce better agreement with observations, but recently Vanbeveren (A & A, 182, 1987) has questioned this.

In the more massive stars, large convective cores produced by overshooting together with rapid mass loss may lead to the uncovering of nuclear-processed material. This provides the opportunity for confrontation of current ideas of nucleosynthesis with observational data, especially the peculiar abundances seen in various classes of Wolf-Rayet stars. Recent papers that have been especially concerned with such questions are those of Prantzos and collaborators (41.065.099 and Ap.J., 315, 209, 1987), Doom et al. (41.065.076), Langer (A & A 171, L1, 1987), and Maeder & Meynet (A & A, 1987, in press).

4. Future Developments

The stellar evolution studies appear to have given a good idea of the amount of overshooting needed to make contact with observations, and they thus give convection theorists something to aim at. They can be used to calibrate theories having free parameters, but they clearly do not yet provide any sensitive test of the theories. Indeed, the simple parameterizations, which do not pretend to any physical content, seem to be about as good as anything. Evolution studies will surely continue, and may at least help to exclude some of the theories.

In the end, however, one wishes to have a serviceable theory of convection and overshooting, simple enough to be used in stellar evolution calculations and agreeing in at least essential details with more sophisticated models. Possibly the generalized mixing-length theories that are increasingly being developed and applied will be satisfactory; perhaps quite different approaches will have to be tried. One can only be sure that this will continue to be an active field for astrophysical fluid dynamicists. One can also expect that large-scale numerical simulations of compressible convection will become increasingly sophisticated and realistic, providing a body of data which will furnish a guide as well as a challenge to those who wish to devise simpler convection models for use in stellar-structure calculations.

III. MASS LOSS (I. Appenzeller, Landessternwarte, Heidelbrg-Konigstuhl)

1. General Remarks, Meetings, Reviews

During the past decade it became clear that mass loss occurs in practically all stars and at all evolutionary phases and that mass loss may profoundly influence a star's evolution. Hence, the number of papers concerning various aspects of stellar mass loss has been increasing steadily. In this report we shall briefly discuss those new results on stellar mass loss which are directly related to the internal constitution and evolution of stars. Concerning new investigations of the spectroscopic manifestation of stellar mass loss, atmospheric effects and the interaction of stellar mass loss with the interstellar medium we refer to the reports of Commissions 29, 36, and 34.

Among the recent scientific meetings the IAU Symposia Nos. 116 (Luminous Stars and Associations in Galaxies, May 1985, Porto Heli, Greece, 41.12.79) and 122 (Circumstellar Matter, June 1986, Heidelberg, FRG, D. Reidel 1987) and the IAU Colloquia Nos. 89 (Radiation Hydrodynamics, June 1985, Copenhagen, Denmark, 42.12. 102) and 92 (Physics of Be Stars, August 1986, Boulder, USA) were to a large part devoted to problems related to stellar mass loss. Many new results concerning cool stars of high and low luminosity were reported at the conference on "Mass Loss from Red Giants" held in June 1984 in Los Angeles (40.12.28) and at the Dunsink Bicente-

nary Colloquium in 1986 (42.12.24). Mass loss from hot stars was discussed at the
NASA-GSFC workshop in June 1984 (39.12.23) and at the workshop on "Instabilities in
Luminous Early Type Stars" which took place in April 1986 at Lunteren (Netherlands),
honouring C. de Jager. The relation between chromospheric heating and mass loss was
the topic of the 1984 "Trieste Workshop" which was held at the Sacramento Peak
Observatory, USA (40.12.97). The proceedings of all these meetings contain, in
addition to exciting new results, excellent reviews on many details of the mass loss
problem. Two other highly useful and comprehensive reviews which must be mentioned
here have been published in Volume 24 of the Annual Review of Astronomy and Astrophysics: Chiosi and Maeder (42.65.44) summarized our knowledge about mass loss from
hot stars, while Dupree (42.112.43) treated mass loss from cool stars.

2. Mass Loss Rates

An important input parameter of modern stellar evolution computations of
massive stars and evolved stars are accurate mass loss rates as a function of
a star's position in the HR diagram. Due to more sensitive IR and Radio measurements (for hot stars) and the introduction of molecular microwave data (for cool
stars) the accuracy of empirical mass loss rate derivations could be significantly
improved during the past few years. New or better mass loss rates of individual hot
stars have been reported by Lamers and Waters (38.64.01, 38.64. 06), by Kenyon and
Galagher (39.112.73) and by Schmutz and Hamann (42.64.27). De Jager, Nieuwenhuijzen
and van der Hucht (41.112.126) compiled a list of 189 individual stellar mass loss
derivations taken from the literature and covering spectral types from O to M. For
cool stars many new mass loss rates derived from CO observations were publised by
Knapp and Morris (39.112.93), Knapp (42.112.86, 42.112.185), and Wannier and Sahai
(42.112.69). Valuable new data on mass loss rates derived from UV spectroscopy has
been listed in several contributions to the new IUE book ("Exploring the Universe
with the IUE Satellite", Kondo et al. eds. D. Reidel 1987).

3. Mechanisms

Although there are still many open questions, the period covered by this report
brought real progress in our theoretical understanding of the physical mechanisms
leading to stellar mass loss. Greatly improved atmospheric models of hot stars
developed by Abbott and Hummer (40.64.06) and by Kudritzki, Pauldrach, and Puls
(40.64.03, 42.64.04, A.A. 173, 293, 1987) resulted in much better agreement between
theory and observations of radiation-driven winds of O stars. These new results
strongly support the basic ideas of the Castor-Abbott-Klein theory. Lamers (41.112.
30., 41.112.127) and Appenzeller (41.65.163) showed that line radiation pressure may
also explain the highly non-stationary and extreme mass loss from S Dor and P Cyg
stars. Friend and MacGregor (38.64.25) and Uchida (39.64.20) investigated possible
magnetic acceleration effects in the wind of hot stars. The stability of line-
driven winds was studied by Owocki and Rybicki (38.64.33) and by Lucy (38.64.34).
Optically thick radiative wind acceleration at luminosities near or above the
Eddington luminosity was modelled by Quinn and Paczynski (39.64.35) and Kato
(39.64.55). Stellar pulsations as a mechanism driving mass loss was studied by
Dziembowski (39.65.118) and by Willson an Bowen (38.65.65). De Jager further
developed his theory of turbulent-pressure driven winds (38.64.16). Mass loss due
to radiation pressure on molecules in cool stellar atmospheres was reinvestigated by
Stelnitskij et al. (41.64.24). Gail and Sedlmayr (40.64.05, 41.64.37), Drinkwater
and Wood (40.65.50) and Yorke (42.64.38) developed new detailed models of the dust-
driven mass loss from cool stars. Using better opacities (which included isotopic
variants of diatomic molecules) Lucy, Robertson, and Sharp calculated improved
models of carbon stars. Compared to earlier results the new evolutionary tracks
reach lower effective temperatures and therefore are now consistent with the onset
of strong dust-driven mass loss during the evolution of carbon rich stars.

CCD imaging and spectroscopic observations confirmed the highly non-isotropic character of the mass loss from PMS stars (see e.g. 38.121.35, 40.121.55, 41.121. 14, 42.121.04). New theoretical models explaining the collimated wind flows were developed by Sakashita, Hanami, and Umemura (39.64.76) and by Pudritz and Norman (41.64.20). Bastian and Mundt (39.121.76) and Croswell, Hartmann, and Avrett (Ap.J. 312, 227,1987) used high resolution spectroscopy to analyse the winds from FU Ori objects. They found very low temperatures in the wind acceleration zones, which rules out a thermal acceleration mechanism and supports the earlier suggestion of a magnetic driving mechanism of PMS stellar mass loss.

4. Evolutionary Computations with Mass Loss

Mitalas and Falk (38.65.33), Pylyser, Doom, and de Loore (40.112.65), Sreenivasan and Wilson (39.65.39, 39.65.58), Bertelli, Bressan, and Chiosi (40.65. 29), and Prantzos, Doom, Arnould, and de Loore (41.65.99, 41.65.136, 41.65.137) carried out new theoretical model computations of the structure and evolution of mass-losing stars.

Although differing in various model assumptions and physical details, these new calculations give qualitatively similar results. General discussions of the consequences of mass loss on evolutionary tracks have been presented by Maeder (42.65.15) and by Iben (40.64.26). Maeder also used his evolutionary computations of massive stars with mass loss to study the vibrational stability behaviour during the evolution to the WR stage (39.65.74). Schild and Maeder (38.65.02) compared the results of evolutionary computations with mass loss with the Wolf-Rayet star content of open clusters. In another very interesting comparison of computed theoretical isochrones with observed CM diagrams of 25 galactic clusters, Mermilliod and Maeder (41.153.05) found discrepancies which cannot be explained by the uncertainties of the mass loss rates alone. Brunish, Gallagher, and Truran (41.65.111) re-examined the number ratio of blue and red supergiants as predicted from the evolutionary computations which include mass loss in a self-consitent way. They find satisfactory agreement between the present theory and the observations within the expected errors.

Taken together the improved evolutionary computations show reasonable agreement with the basic observational results. However, in detailed comparisons with galactic clusters there still remain some discrepancies which must be overcome either by even better theoretical models, or perhaps by more accurate empirical mass loss rate derivaitons.

IV. NOVAE (Mario Livio, University of Illinois, Urbana and Technion, Haifa)

1. Introduction

Our understanding of classical novae has benefited in recent years especially from simultaneous multi-frequency observations. In particular, observations in the UV, infra-red, and X-rays have added new dimensions to our coverage of nova energetics. Consequently, much of the present summary of highlights in nova research in the past three years, concentrates on such observations. Some theoretical developments are also outlined. Two books, "Interacting Binary Stars" (39.003.030) and "Interacting Binaries" (1985, Dordrecht: D. Reidel), contain material on novae and related objects. Ritter's fourth edition of the Catalogue of Cataclysmic Binaries, Low Mass X-Ray Binaries and Related Objects (1987, Astron. Ap. Suppl., in press) provides up to date information on nova systems. The most impressive compilation of data on novae and related systems in recent years, is provided by Duerbeck's Reference Catalogue and Atlas of Galactic Novae (1987, Space Science Reviews, 45, 1). The catalogue and atlas contain positions, finding charts, apparent magnitudes, light curve types and principal bibliographical references, and will surely prove immensely useful to all nova researchers.

2. Oxygen-Neon-Magnesium Novae

A recent compilation of abundances observed in nova ejecta revealed the fact that all novae for which reasonable abundance data exist show enrichments in heavy elements or helium or both (Truran and Livio 1986, Ap. J., 308, 721; Truran 40.124. 010, Williams 40.124.011, de Freitas Pacheco and Codina 39.124.322). Furthermore, recent abundance determinations established the existence of a class of novae showing enrichment in Oxygen, Neon, and Magnesium (V693 Cr A 1981, Williams et al. 39.124.102, V1370 Aql 1982, Snijders et al. 38.124.201, Nova Vul 1984II, Andrillat and Houzioux 40.124.135, Gehrz et al. 40.124.133). While it has been shown, that breakout of the CNO cycle can yield neon and heavier nuclei (Wiescher et al. 41.065.129), the temperatures required for such a breakout ($T > 4 \times 10^8$ K) are not achieved in nova outbursts. It has therefore been suggested that these enrichments originate from the underlying white dwarf, suggesting the existence of a population of ONeMg white dwarfs in nova systems (Delbourgo-Salvador et al. 40.124.017, Truran and Livio 1986, Ap. J., 308, 721). The white dwarf material is probably brought into the envelope by shear mixing (Fujimoto 1987, Astron. Ap., 176, 53, Livio and Truran 1987, Ap. J., 318, 316) or to some extent by diffusion (Kovetz and Prialnik 39.065.047). It has been shown (Truran and Livio 1986, Ap. J., 308, 721) that selection effects associated with the fact that ONeMg white dwarfs, being more massive, experience their nova outbursts more frequently, act to ensure that the relatively small fraction of the systems in which ONeMg white dwarfs do occur, can account for about one third of the observed outbursts. A simulation of a thermo-nuclear runaway, when the envelope contains oxygen, neon, and magnesium has been performed by Starrfield et al. (41.065.077), resulting in an extremely violent outburst.

3. X-Ray Observations of Novae

Several EXOSAT X-ray observations of classical novae were reported. First, soft X-rays (0.04-2 keV) were detected from Nova Muscae 1983, about 4×10^7 sec after the outburst maximum (Gelman et al. 38.124.144). This was followed by a second set of observations which revealed that Nova Muscae 1983 continued to emit X-rays at the same intensity in the interval 460-700 days and then started to decay on a timescale of $\sim$ 300 days (Gelman et al. 40.124.014 and 1987, Astron. Ap., 177, 110). Observations of Nova Vulpeculae 1984I and Nova Vulpeculae 1984II have shown the emission arising from the outburst to about 300 days (Gelman et al. 1987, Astron. Ap., 177, 110). Based on the fact that a decline has been observed in $\sim$ 2 years in Nova Muscae 1983 (compared to a calculated plasma cooling time of 30-60 years) and on the absence of coronal [Fe XIV] lines, Gelman et al. argued against an interpretation of emission from a shocked circumstellar gas. Instead, they suggested that the observations are consistent with a constant bolometric luminosity model of a hot white dwarf, evolving like the central star of a planetary nebula.

EXOSAT observations of GK Per during its 1983 optical outburst (Bianchini et al. 41.124.162) detected hard X-ray emission showing a strong coherent modulation at 351 seconds, thus identifying the white dwarf as an intermediate polar (Watson et al. 39.124.201). In addition, an aperiodic modulation on a timescale of $\sim$ 3000 sec has also been detected. Both of these modulations were also identified by high speed photometry (Mazeh et al. 40.124.201). It has been suggested that transition fronts, moving in a small inner disk, may be responsible for the 3000 sec variation (Duschl et al. 40.124.202). Intense soft X-ray emission has also been detected from the recurrent nova RS Oph about two months after its optical outburst. The X-ray flux declined rapidly between 60-90 days (Mason et al. 1986, to be published in "RS Ophiuchi and the Recurrent Nova Phenomenon").

4. Dust in Nova Shells

Considerable progress has been achieved in observations and to a lesser extent in theory of dust formation in nova shells. On the theoretical side, the effects of hydrogen and carbon ionizations on grain growth and destruction have been investigated by Mitchell and Evans (38.124.001). Observations of FH Ser have shown that grain growth occurred at 60-111 days after the outburst, while the grains have undergone a significant reduction in size at 111-129 days. The bolometric luminosity was found to stay constant until 200 days and to decrease as t^{-1} afterwards. In Nova Aquilae 1982, the ejecta was found to contain comparable amounts of gas and dust (Snijders et al. 38.124.201, also Roche et al. 38.124.202, and Catchpole et al. 40.124.008). The general noted trend that slow novae contain larger amounts of dust than fast novae has been confirmed by the finding that five classical novae have counterparts in the IRAS Point Source Catalog (Dinerstein and Robinson 41.124.018).

Somewhat surprisingly perhaps, an excess at 10 micrometers, consistent with the heating of dust, was found for the "dustless" nova V1500 Cyg (Bode and Evans 40.124.163). Even more surprising was the fact that infrared thermal dust emission did not occur (Whitelock et al. 38.124.142) in Nova Mus 1983 (which had a decay rate of ~ 0.06 mag day^{-1}).

5. The "Hibernation" Scenario of Classical Novae

An interesting development in nova theory has been provided by the suggestion of the "hibernation" scenario for classical nova (CN) systems (Shara et al. 1986, Ap. J., 311, 163, Livio and Shara 1987, Ap. J., in press). In this scenario, it is suggested that the mass transfer rate in some CN systems decreases some 50-300 years following the outburst, to values which allow the system to undergo dwarf nova eruptions. The mass transfer rate is then assumed to return slowly to its pre-outburst high value. The cause for a behavior of this type could be an increase in the binary separation as a result of mass loss during the nova outburst, followed by a reduction of the separation due to magnetic braking. The basic motivation for this scenario has been provided by the following facts: (a) The accretion rates deduced from observations (Patterson 37.117.200, Warner 1987, M.N.R.A.S., in press), are too high to produce strong outbursts (Prialnik and Shara 1986, Ap. J., 311, 172, Livio et al. 1987, Ap. J., in press). (b) Observations of the two oldest recovered nova systems CK Vul (1670) and WY Sge (1783) have shown that these systems are in a state of very low mass (Shara et al. 40.124.141, Shara et al. 38.124.181) and RR Pic (1925) was observed to undergo a decrease in its luminosity (Warner, 41.124.221). (c) Some CN systems were observed to undergo dwarf nova eruptions (e.g. Q Cyg, Shugarov 39.124.281, BV Cen, Menzies et al. 41.117.154, V1017 Sgr, Webbink et al. 1987, Ap. J., 131, 493). (d) There exists a known discrepancy between the space density of CNe as deduced from sky surveys and that deduced from conventional nova theory (Patterson 37.117.200).

The most appealing aspect of the hibernation scenario is the fact that it provides a picture of cyclic evolution in which cataclysmic variable systems metamorphose between classical novae and dwarf novae. Some difficulties with this scenario were outlined by Livio (1987, Comments on Astrophysics, in press), as well as crucial observations.

6. Recurrent Novae

Recurrent novae have received much attention in recent years, especially with the outburst of RS Oph in 1985 (39.124.014, 39.124.016, 39.124.018, 39.124.019). The outburst has been observed also in the UV, by Cassatella et al. (40.124.018), in the radio, revealing an unusual radio source (Hjelming et al. 41.124.013) and in X-rays, where intense soft X-ray emission (Mason et al. 1986, to be published in "RS Ophiuchi and the Recurrent Nova Phenomenon") has been detected. New radial velocity measurements for T CrB confirmed that an M3 giant is transferring mass onto a main sequence star in this system (Kenyon and Garcia 41.124.007, also Peel 41.124.007). Observations of U Sco at quiescence revealed He II emission lines on a nearly featureless continuum, no hydrogen lines were detected (Hanes 39.124.004). U Sco was detected in outburst again in 1987, the previous outburst being in 1979 (IAU Cir., 4395-4397, 4399, 4405). In a very comprehensive work, Webbink et al. (1987, Ap. J., $\underline{314}$, 653) examined all of the available observational data on recurrent novae, in an attempt to suggest an appropriate model for each system. They concluded that WZ Sge, VY Aqr, RZ Leo, and V1195 Oph are dwarf novae and V1017 Sgr is a classical nova that has undergone two dwarf nova eruptions. Of the remaining systems, the outbursts of T CrB and RS Oph represent very probably accretion events from a red giant to a main sequence companion (as discussed also by Livio et al. 1986, Ap. J., $\underline{308}$, 736). The X-ray emission in RS Oph is probably caused by high velocity ejecta interacting with a slow moving pre-outburst wind (Bode and Kahn 40.124.001). Webbink et al. (1987, Ap. J., 314, 653) suggested that the outbursts of T Pyx result from thermonuclear runaways (TNR) on a massive accreting white dwarf. Based on a successful simulation of Starrfield et al. (39.124.001), they also suggested that the outbursts of U Sco may be thermonuclear. However, Truran et al. (1987, Ap. J., in press) have shown that if the helium to hydrogen ratio in the accreted material is indeed 2:1 (by number), as suggested by the observations of Hanes (39.124.004), then it is extremely difficult to produce a visually bright nova by a TNR. Remembering in addition, that the time interval between the last two outbursts of U Sco is only eight years, an accretion event interpretation has to be reexamined (Truran et al. 1987, Ap. J., in press).

In an interesting development, recent spectroscopy of V616 Mon has revealed surprisingly large radial velocity variations in the spectrum of the K-star in this system, indicating a massive compact companion (McClintock and Remillard 1986, Ap. J., 308, 110). V616 thus became a very good black hole candidate (Watson 41.117.10!

7. Final Note on Novae as Distance Indicators

Finally, we would like to mention that it has been suggested that the M_V (max) to rate-of-decline relationship is potentially a very powerful tool for the calibration of the extragalactic distance scale (Cohen 39.124.009, van den Bergh and Pritchet 41.161.057).

V. PRESUPERNOVA MODELS AND SN 1987A (Ken'ichi Nomoto, University of Tokyo)

The occurrence of SN 1987A in the Large Magellanic Cloud had a big impact on theoretical works on stellar constitution and a lot of work has been done since then. Spectral observations have shown that SN 1987A is a Type II supernova and thus an explosion of a massive star. Indeed a massive blue supergiant star, Sk-69 202, has been identified as the most likely progenitor of SN 1987A. This is the first identification of a supernova progenitor. The historic observations of neutrino burst from SN 1987A (Hirata et al. Phys. Rev. Lett., $\underline{58}$, 1490, 1987; Bionta et al. Phys. Rev. Lett., $\underline{58}$, 1494, 1987) have confirmed the basic validity of the current theory of Type II supernovae and opened a new era of neutrino astronomy. Further observations in all wave bands are providing us with very interesting materials to test the theory of massive star evolution, nucleosynthesis, and supernova explosion. This opportunity is important because massive stars have been considered to be one of the most important sites of nucleosynthesis, yet their presupernova structure and explosion mechanism still involve considerable uncertainties.

SN 1987A is certainly one of the highlights in the field of stellar constitution, so that I will focus on the problems of presupernova models that is directly related to SN 1987A. Related meetings are Supernova Workshop in Honor of Hans Bethe (Phys. Rep., 1987, ed. G.E. Brown), ESO Workshop on SN 1987A (ESO, 1987, ed. J. Danziger) and IAU Colloquium No. 108, Atmospheric Diagnostics of Stellar Evolution (Lecture Note in Phys., 1987, ed. K. Nomoto). Comprehensive reviews on the current understanding of presupernova evolution are seen in Woosley and Weaver (42.125.057)

Nomoto (42.065.126), and Woosley (42.065.136).

1.   Evolution of the Progenitor of SN 1987A

Sk-69 202 is the most likely progenitor of SN 1987A (West et al. A&A, $\underline{177}$, L1, 1987; Kirshner et al. Ap.J. , 1987, in press; Gilmozzi et al. Nature, $\underline{328}$, 318, 1987). It is a B3 blue supergiant and its luminosity is estimated to be about 1.3 x 10^8 $L_{\odot}$ which corresponds to the presupernova luminosity of a helium core of mass about 6 $M_{\odot}$ (Woosley et al. Ap.J., $\underline{318}$, 664, 1987; Nomoto, Shigeyama, and Hashimoto in $\underline{\text{SN 1987A, ESO}}$, 1987). Its main-sequence mass is estimated to be M_{ms} = 17 - 20 $M_{\odot}$ which depends on the helium-core mass to M_{ms} relation and thus on the convective overshooting during hydrogen burning (Hillebrandt et al. Nature, $\underline{327}$, 597, 1987; Maeder in $\underline{\text{SN 1987A, ESO}}$, 1987).

From the luminosity and the spectral type of Sk-69 202, the progenitor's radius is estimated to be about 3 x 10^{12} cm. The occurrence of a Type II supernova from such a blue supergiant progenitor is not known before. Previously Type II super-novae are subclassified into II-P (plateau) and II-L (linear) according to their light curve shape (Doggett and Branch 40.125.080). Type II-P supernovae are well modeled by the explosions of red supergiants (Woosley and Weaver 42.125.057). The early light curve of SN 1987A reached plateau but its luminosity is about 20 times lower than the typical plateau luminosity of SN II-P. Another important observation is that it took only 3 hr from the neutrino burst to the optical brightening at 6.4 mag. This requires the progenitor's radius of as small as the order of 10^{12} cm in order for the shock wave to propagate through the envelope in sufficiently short time (Shigeyama et al. Nature, $\underline{328}$, 320, 1987). The low luminous plateau is well modeled with a progenitor whose radius is much smaller than that of red supergiants (Arnett Ap.J., $\underline{319}$, 136, 1987; Shigeyama et al. 1987; Woosley, Pinto, and Ensman Ap.J., 1987, in press). All these observations point to the blue supergiant progenitor.

Then the question is why the progenitor of SN 1987A evolved to become a blue supergiant at the explosion. Basically two reasons have been argued. One is the low metallicity of the LMC, which is 1/3 - 1/4 of our Galaxy. With such low metallicity and without mass loss, Hillebrandt et al. (1987) and Arnett (1987) found that their 15 - 20 $M_{\odot}$ stars remain blue throughout the evolution without becoming red. Woosley, Pinto, and Ensman (1987) found in a similar calculation that 15 - 20 $M_{\odot}$ stars without mass loss once evolve to the red supergiant and return to the blue if metallicity is low and the Ledoux criterion is applied to convective stability. The other reason is mass loss during the evolution. Maeder (in $\underline{\text{SN 1987A, ESO}}$, 1987a) argues that mass loss leads to disappearance of the intermediate convective shell and drives the expansion of the star to the red even with low metallicity. The stars more massive than about 20 $M_{\odot}$ return to the blue when the mass of the hydrogen-rich envelope is reduced below 1 $M_{\odot}$. Maeder (1987a) and Wood and Faulkner (Proc. Astr. Soc. Australia, 1987, in press) have shown that the low metallicity plays a roll to reduce the mass loss rate by a factor of 2 - 3 so that the return from the red to blue is delayed to the beginning of carbon burning stage and the star ends up as a blue supergiant before evolving to a Wolf-Rayet star. If the low metallicity is essential, the envelope mass could be large, while it is smaller than 1 $M_{\odot}$ if the latter is the case.

The exact reason for such numerical behavior has not been well analyzed yet. The stellar radius is very sensitive to the temperature and density gradient near the bottom of the hydrogen-rich envelope and thus to the luminosity at the core edge, opacities and the appearance of intermediate convective shell (e.g., Renzini in $\underline{\text{SN 1987A, ESO}}$, in press). It has been known that stars around 15 - 20 $M_{\odot}$ show complex evolutionary behavior between blue and red supergiants in the H-R diagram (e.g., Maeder and Meynet A&A, $\underline{182}$, 243, 1987). The low metallicity would affect the structure near the core-envelope boundary through the hydrogen shell burning the opacities.

Because of the complexity of the theoretical models, observational constraints
are useful. One is the number ratio of the bright red supergiant to blue supergiant
in the SMC, LMC, and Galaxy, which is larger for smaller metallicity (Chiosi and
Maeder 42.065.014). Another is the UV observations of SN 1987A that show the
existence of circumstellar materials where nitrogen is overabundant relative to
carbon and oxygen (Kirshner et al. IAU Cir. 4435). These two facts are consistent
with the scenario that the progenitor evolved first to become red and then came back
to the blue. Moreover, the progenitor should have undergone significant mass loss
during the red supergiant phase to expose the nitrogen-rich layer (Maeder A&A, <u>173</u>,
247, 1987b). How much hydrogen-rich envelope was left at the explosion may depend
on the metallicity and the treatment of convection (Woosley, Pinto, and Ensman 1987)

Another constraint on the mass of the hydrogen-rich envelope, M_{env}, at the
explosion has been obtained from the light curve model (Woosley, Pinto, and Ensman
1987; Nomoto, Shigeyama, and Hashimoto 1987). The slow increase of the light curve
up to the peak requires a relatively slow expansion of the core which requires the
existence of at least 3 $M_\odot$ envelope. Whether such constraints are consistent with
each other is an open question. This may not be an easy question to answer because
it is deeply related to the mass loss and convection that have been long-standing
problems.

2.   Relation to Other Types of Supernovae

Suppose that uniqueness of SN 1987A is mainly due to the progenitor's mass loss
history rather than the low metallicity. Then the relation of SN 1987A to other
types of supernovae, namely, Type II-P, II-L, and Ib may be as follows: II-Ps are
the explosion of red supergiants, so that their progenitor must have suffered from
less mass loss than SN 1987A. This implies that the progenitors of II-P are less
massive than 17 - 20 $M_\odot$. For II-L, progenitor may either be AGB stars (Wheeler,
Harkness, and Cappellaro in <u>13th Texas Symp.</u>, 402, 1987) or stars somewhat more
massive than SN 1987A. Because of larger amount of ^{56}Ni production or smaller
amount of envelope mass, radioactive decays may dominate the light curve. Type Ib
supenovae may be either a kind of accreting white dwarfs (Branch and Nomoto 42.125.
014) or Wolf-Rayet stars (Wheeler and Levreault 40.125.261; Begelman and Sarazin
41.125.281; Fillipenko et al. 42.125.223; Gaskel et al. 42.125.028; Wheeler et al.
Ap.J., <u>313</u>, L69, 1987; Schaeffer, Casse, and Cahen, Ap.J., <u>316</u>, L31, 1987). If the
latter is the case, the progenitor of Ib is even more massive than those of II-L and
SN 1987A, although the low occurrence frequency of very massive star explosion and
too broad a light curve (Woosley 42.065.136) are problems with this Ib model. (The
Ib progenitors might be helium stars in close binaries. Even so, the progenitor's
main-sequence mass must be larger than SN 1987A because the amount of ^{56}Ni required
from the Ib light curve is about 0.15 $M_\odot$ (Wheeler and Levreault 40.125.261) which is
roughly twice as much as in SN 1987A.) In the above sequence, SN 1987A corresponds
to a rather narrow mass range, say, 17 - 20 $M_\odot$ which depends on the metallicity
(Maeder 1987a). It is interesting to note that the progenitor of Cas A is very
likely a WN star (Fesen et al. Ap.J., 1987, in press) and thus might be closely
related to SN 1987A.

The idea that the low metallicity is essential in making the presupernova blue
might be a good explanation of the fact that Type II supernovae have not been
observed in irregular galaxies (Shklovski 38.125.030). In this view, SN 1987A-like
supernovae are not rare in metal-poor young galaxies but rare in metal-rich galaxies.

3.   Presupernova Core Models and Explosion Mechanism of SN 1987A

The evolution of the core of massive stars is relatively straightforward up to
the beginning of the iron core collapse, which is almost independent of the com-

plexity of the envelope. The observation of neutrino burst has dramatically
confirmed that the gravitational collapse actually occurred in SN 1987A. Most of
the energy released by the collapse has been predicted to be emitted as neutrinos
during the proto-neutron star contraction. Because of high density, diffusion of
neutrino takes place for over 10 sec (Burrows and Lattimer 42.067.031). The
duration of observed neutrino burst over 12 seconds is entirely consistent with this
prediction. The total energy of neutrinos emitted from SN 1987A is estimated to be
$1 - 3 \times 10^{53}$ erg which corresponds to the binding energy of a neutron star with mass
of 1.2 - 1.7 $M_\odot$ (Burrows and Lattimer Ap.J., 318, L63, 1987; Sato and Suzuki, Phys.
Let., 1987 in press). Unless the equation of state is very soft, a black hole
formation seems to be unlikely.

One of the central problems in Type II supernova modeling has been to find
a mechanism that transforms gravitational collapse into explosion. Two mechanisms
have recently demonstrated. One is that the hydrodynamical shock wave ejects
overlying materials in a "prompt" manner in mill-seconds. This occurs if the iron
core mass is smaller than about 1.6 $M_\odot$ and the equation of state is sufficiently
soft (Baron, Cooperstein, and Kahana 40.066.123). The other is "delayed" heating of
the material behind the stalled shock by neutrinos. The neutrinos are emitted from
a hot neutron star that undergoes mass accretion. In a second, the shock wave
revives and ejects the material (Wilson et al. 42.065.125). Unfortunately, statistics of neutrinos is too poor to tell us which mechanism did work in SN 1987A.

The hydrodynamical behavior of collapse and explosion depends sensitively on
the iron core mass and a density structure in the surrounding heavy element mantle.
The major uncertainties in determining the iron core mass include the carbon
abundance that is closely related to the treatment of convection. Recent increase
in the $^{12}C(\alpha,\gamma)^{16}O$ rate by a factor of about three (Fowler 37.061.073) results in
the lower carbon abundance (Thielemann and Arnett 40.065.033). Overshooting near
the end of core helium burning mixes fresh helium into the central region and
decreases carbon abundance (Bertelli, Bressan, and Chiosi 40.065.029; Habets
42.065.032). Moreover, the mass of the evolved core is close to the Chandrasekhar
mass, so that the core structure is sensitive to the equation of state of strongly
coupled plasmas (Ichimaru Rev. Mod. Phys., 54, 1017, 1982) and electron mole number
Y_e. For the latter, treatment of silicon burning and associated electron capture
involves large uncertainties (Thielemann and Arnett 40.065.033).

In previously published models by Woosley and Weaver (42.125.057), overshooting
was included but Coulomb interaction was neglected in the equation of state. For
their 25 $M_\odot$ star carbon abundance is so small that no convective carbon burning
shell forms. The absence of active burning shell results in the formation of an
iron core as large as 2.1 $M_\odot$ and a relatively high density silicon and oxygen-rich
layer. For such a large iron core, the "prompt" mechanism certainly fails to work
(Burrows and Lattimer 40.065.132). If overshooting is suppressed, active carbon
burning shell forms even with the enhanced $^{12}C(\alpha,\gamma)^{16}O$ rate and the resulting iron
core masses are as small as 1.4 $M_\odot$ for 20 - 25 $M_\odot$ stars (Nomoto and Hashimoto Prog.
Part. Nucl. Phys. 17, 267, 1986); also densities in the heavy element mantle are
lower and thus smaller amount of ^{56}Ni would be produced. If this is the case,
"prompt" mechanism could operate. Since the progenitor of SN 1987A is about 20 $M_\odot$
star, it is interesting whether the observations could provide us with any indication of the deepest core structure.

For stars as small as 13 $M_\odot$, the iron core mass is as small as 1.2 $M_\odot$ because
inclusions of Coulomb interaction term and electron capture from earlier stages are
more effective to reduce the effective Chandrasekhar mass than in more massive stars
(Nomoto and Hashimoto, Sci. Rep. Tohoku U. 7, 259, 1987). This indicates that
presupernova models still have a large uncertainties. Hence, it is still crucial to
improve presupernova models in solving the long-standing problem of supernova
mechanism.

In SN 1987A, though the mechanism is not yet clear, the shock wave generated at bounce propagated through the star and should have synthesized heavy elements. To find a direct evidence of nucleosynthesis in the deep cores of SN 1987A, we have to wait until emission lines are observed in super-nebula phase (Fransson and Chevalier Ap.J. , 1987, in press). Indirect but good evidence of ^{56}Ni production are seen in the exponential decline of the light curve of SN 1987A, because its decline rate exactly coincides with the ^{56}Co decay rate (Catchpole et al. M.N., 1987, in press; Hamuy et al. Ap.J., 1987, in press). The mass of ^{56}Ni initially synthesized is estimated to be about 0.07 $M_\odot$ (Nomoto, Shigeyama, and Hashimoto 1987). This amount reflects the explosion energy, masses of the resulting neutron star and ejecta, and composition and density structures in the deepest layers.

If the explosion energy is known, it is useful to discriminate the explosion mechanism. For this purpose, quite unique optical light curve of SN 1987A may be useful. From the initial sharp rise, relatively low luminous plateau, and slow increase to the peak, explosion energy of SN 1987A is estimated to be $1 - 1.5 \times 10^{51}$ erg (Woosley, Pinto, Ensman 1987; Nomoto, Shigeyama, and Hashimoto 1987). Consistent calculation of collapse, explosion, nucleosynthesis, and light curve will provide a valuable material to probe the interior of SN 1987A and to test the theory of massive star evolution. In other words, the comparison with the observations of SN 1987A will make it possible to determine some previously unknown quantities and could lead us to deeper understanding of the presupernova configuration, mass loss processes, and convection, which are three major problems in the current stellar evolution theory.

VI.   STRUCTURE OF X-RAY BURSTING NEUTRON STARS   (M.Y. Fujimoto, Niigata University,
                                                   and D. Sugimoto, Univeristy of Tokyo)

1. Introduction

    X-ray bursts were discovered in 1975. More than thirty X-ray burst sources have been observed to date with a number of satellites. Characteristics of X-ray bursts are: risetime of 1 - 10 sec, decay time of several to several tens of seconds, and duration of $10 - 10^5$ sec (Lewin and Joss 34.142.63; Tanaka 41.142.42; Matsuoka 41.142.68). This phenomenon has greatly stimulated theoretical studies on nuclear burnings near the surface of neutron stars. Soon after the discovery, a model of thermonuclear flashes on accreting neutron stars was proposed as a mechanism to produce X-ray bursts (Maraschi and Cavaliere 20.142.63; Woosley and Taam 18.142.64). This model has been studied extensively in these ten years (for the earlier works, see reviews by Taam 39.67.13; Joss and Rappaport 38.117.125). The validity of this model had been generally established. In the reporting years, the X-ray bursts have been studied furthermore as a powerful tool of probing neutron stars. This side is somewhat more stressed in this report rather than the X-ray bursts themselves. Since this topics has never been reported from Commission 35 in the preceeding volumes, we briefly refer also to some former works for easy understanding of the background.

2. Characteristics of Shell Burnings

    Properties of the nuclear shell-burnings near the the surface of neutron stars are rather simple and have now been understood fairy well: The pressure of the burning shell is practically indifferent to its thermal state because the thickness of the burning shell is negligible as compared with the radius of the neutron star, and because the nuclear energy release is small as compared with the potential energy of the shell. Therefore, the nature of the shell burning is determined solely by the properties of nuclear reactions. Unless the accretion rate is very low, the accreted hydrogen burns stably, because the rate of CNO cycle is limited by intervening β decays. On the contrary, helium shell-burning, once ignited, grows to a flash and brings about high temperature of the order of 10^9 K where proceeds the formation of heavy elements (Ergma and Tutukov 27.66.511; Taam 29.66.511; Fujimoto et al. 29.66.522). Of particular interest was the nucleosynthesis through successive proton captures and β decays of heavy elements in the hydrogen-rich material, which is termed rp-process (Wallace and Woosley 29.65.105; Hanawa et al. 34.67.81; Wallace and Woosley 39.67.17). On the other hand, Miyaji and Nomoto (40.67.72) discussed the influence of the non-resonant 3α reactions in the low temperature regime on the evolution leading to the shell flash.

In earlier models, only one cycle of thermonuclear flashes were calculated, and only the outer layers were took into computation with the interior replaced by appropriate inner boundary conditions. Models of recurrent shell flashes were constructed later by Woosley and Weaver (39.67.14) with the inner boundary conditions. Hanawa and Fujimoto (38.67.63) computed a recurrence of shell flashes, taking also into account the whole structure of a neutron star and effects of general relativity in spherical symmetry. They demonstrated that a limit-cycle behavior was readily reached after several cycles of flashes because of a small heat capacity of the neutron star envelope.

As for the interaction with the interior of neutron stars, Fujimoto et al. (37.67.70) found that the surface layers and the interior were thermally well decoupled at the shell where heat transports by photon diffusion and by electron conduction are both ineffective during the timescale of accretion. Therefore, the evolution of shell burnings has little to do with the thermal state of the interior for accretion rates of the order as usually observed. For lower accretion rates, however, Hanawa and Fujimoto (41.67.31), and Fujimoto et al. (1987a, Ap.J., 278, 813) discussed a finite effect which could discriminate between different internal thermal states due to possible existence or non-existence of pion condensation, for instance.

One of the interesting findings was the expansion of the envelope and the resulting mass loss from the neutron star during relatively strong X-ray bursts (Hanawa and Sugimoto 31.65.508; Taam 32.66.505; Wallace et al. 32.66.502; Starrfield et al. 32.66.501; Paczynski 33.67.25). They could sound strange because the nuclear energy is much less than the gravitational potential energy. However, there acts the mechanism that concentrates the released nuclear energy into only a small fraction of the accreted surface layer. The expansion is understood as an effect of the decrease of the electron scattering opacity at higher temperatures; more opaque surface layer is pushed out by the photon flux coming from the interior at the rate close to the Eddington luminosity (Hanawa and Sugimoto 31.65.508; Paczynski 33.67.25). Paczynski and Anderson (41.67.33) constructed models of static extended envelope with the general relativity taken into account, and asserted that the effect of general relativity was also crucial to the expansion.

Further expansion leads to the mass loss which was modeled in terms of steady stellar wind (Ebisuzaki, Hanawa, and Sugimoto 33.67.18; Kato 33.67.19). It was elaborated by Melia and Joss (39.67.19), and Quinn and Paczynski (40.64.35), and earlier results were confirmed. Paczynski and Proszynski (41.67.33) also studied the effects of general relativity to observe an influence on the photospheric radius. Helium envelope was assumed in all these models, but the hydrogen-rich layer should be more easily blown off because of higher opacity. Taniguchi and Hanawa (40.67.106) followed the expansion of hydrogen-rich envelope with general relativity included, and Kato (41.67.30) constructed models of mass loss wind from upper hydrogen-rich layer. Observations by Tenma X-ray satellite were interpreted in terms of these mass losses from neutron star (Sugimoto, Ebisuzaki, and Hanawa 38.142.80). The models mentioned above were the stellar wind in steady state. Though Yahel et al. (38.67.93) constructed time-dependent models, they neglected the temperature-dependence of the electron-scattering opacity which was the essential mechnism of driving the mass loss.

3. Problems of Super-Eddington Luminosity

During a burst, the X-ray luminosity and its color change in time, and, accordingly, the star draws a track on a plane similar to the HR diagram. Though this traverse along the track finishes in minutes, it corresponds to the whole evolutionary tracks followed by novae in years. When analyzed peoperly, it provides us with information on the neutron star. Van Paradijs (22.142.5) discussed average radius of neutron stars assuming that the burst peak-flux was equal to the Eddington luminosity. Goldman (26.142.42) and van Paradijs (26.66.517) discussed the general relativistic effects on it.

If the peak-flux is really equal to the Eddington luminosity, the X-ray bursts can play a role of a standard candle. Later, however, this assumption were questioned by the following two independent aruguments. One concerns the temperatures; when the spectra were fitted with the Planckian function, many bursts showed color temperatures much higher than the maximum effective temperature that was shown, from the first principles, to be attained at. The other concerns the flux; for some burst sources, for which independent estimates of their distances were available, the observed luminosity indicated the super-Eddington luminosity for plausible masses of neutron stars.

For the spectra, van Paradijs (31.66.504) suggested a modification of continuous spectra due to suppression of emissivity in the scattering-dominant envelope. Including the Comptonization effect, Czerny and Sztajno (34.67.107), Ebisuzaki, Hanawa, and Sugimoto (38.142.66), and Hoshi (39.67.18) showed that the observed color temperature could be higher than the effective temperature. Self-consistent model atmospheres were constructed by London, Taam, and Howard (38.67.147; 42.67.8), Ebisuzaki and Nomoto (41.67.60), and Ebisuzaki (1987, Publ. Astron. Soc. Japan, 39, 287), who solved the radiative transfer together with the hydrostatic equilibrium in the envelope of neutron star. Foster, Ross, and Fabian (42.67.28) computed also the radiative transfer but not hydrostatic equilibrium. These models gave the relation of the color to the effective temperatures, and enabled detailed comparison of theoretical models with observations.

As to the burst luminosities, no theoretical models produced luminosities appreciably exceeding the Eddington limit. From observational side, there were strong indications for saturation of the flux at a certain level. They were interpreted in relation with expansion of the neutron star envelope which took place at the luminosity very close to the Eddington luminosity. Paczynski (33.67.15) and Ebisuzaki et al. (38.142.66) discussed the expansion of the envelope in relation with observed double peaks in hard X-ray band. Tawara et al. (37.142.9), and Lewin, Vacca, and Basinska (37.142.30) showed that long-lasting bursts with a precursor could also be due to the expansion in much larger scales. Sugimoto et al. (38.142. 80) interpreted the observed gap in the peak luminosities in terms of the expansion and mass loss from the distinctly different surface chemical compositions, i.e., from hydrogen-rich envelope and from helium envelope. Though the concept of the Eddington luminosity was essential in these arguments, observations could not give any definite absolute value for the saturation level. Concerning this point, Ebisuzaki et al. (38.142.66) discussed that errors might exist in the estimates of the distance to the burst sources because the Eddington limit was the plausible upper limit for X-ray bursts which proceeded much more slowly as compared with dynamical timescale in the surface layers of the neutron star.

The luminositiy exceeding the Eddington limit may, however, be expected if the the emitting regions deviates from spherical symmetry and the X-ray emission is anisotropic. Lapidus and Sunyaev (40.67.59) argued that the scattering and reflection of photons incident upon the associated accretion disk could distort the angular distribution of the fluxes to produce the super-Eddington luminosity up to by a factor of 1.5 if the source was seen pole on. Melia and Joss (42.142.38) suggested the interaction between the relativistic stellar wind and the accretion disk, but the origin of such high power wind was obscure.

4. Mass and Radius of Neutron Stars

If the observed upper limit to the luminosities is fitted to the Eddington luminosity corresponding to specified chemical compositions, a mass-radius relation results from the observed fluxes and color temperatures where the spectral hardening, i.e., the deviation from the Planckian, is taken into account. This relation is also free from possible anisotropies in the burst fluxes. It was applied to some sources [4U 1636-53 by Fujimoto and Taam (41.67.113); 4U 1636-53 and 4U 1608-52 by Ebisuzaki (1987); 4U 1746-47 by Sztajno et al. (1987, M.N.R.A.S., 226, 39); 4U 1820-30 by van Paradijs and Lewin (1987, A&A, 172, L20); MXB 1728-34 by Foster et al. (42.67.28) and Kaminker et al. (preprint)]. The derived mass-radius relations turn out to be compatible with current models of neutron stars. Some argued that these results could rule out very soft and very hard equations of state. The results, however, depend strongly on the derived color temperatures and the hardening ratios, and thus require further elaborations. In this relation, Jaroszynski (1986, Acta Astron. 36, 335), and Asaoka and Hoshi (1987, Publ. Astron. Soc. Japan, 39, 475) discussed the importance of the effect of rotation and deviation from the Schwarzschild geometry on the distortion of the black body spectra.

For burst sources in globular clusters, the distance estimate is available from optical observations. If we use it, further constraints can be given. Thus, Sztajno et al. (1987) derived a value lower than one solar mass for 4U 1746-37 in NGC 6641 with reservation for the anisotropy, while van Paradijs and Lewin (1987) obtained a mass range for 4U 1820-30 in NGC 6624 which was compatible with those theoretically believed. On the other hand, Horne, Verbunt, and Schneider (41.117.9) suggested a small mass, i.e., 0.6 ± 0.2 $M_\odot$ for 4U 2129+47 (V 1727 Cyg) from spectroscopic observation of emission lines.

An absorption feature was observed in the bursts of 4U 1636-53 and 4U 1608-52 (Waki et al. 38.172.78; Nakamura et al. 1987, Publ. Astron. Soc. Japan, submitted). When it is combined with the mass-radius relation discussed above, the mass and the radius of neutron stars could be determined separately (Fujimoto and Taam 41.67.113; Ebisuzaki 1987). The observed redshift yielded the ratio of the stellar radius to the gravitational radius of $R/R_g = 1.6$ if the line was identified with Kα line of Fe XXV; it seemed too small when compared with theoretical models of neutron star. If we assigned it with other elements, larger radii were possible ($R/R_g = 2.1$ and 3.8, respectively, for Cr XXIII and Ti XXI; Waki et al. 38.172.78), but difficulty lies in the origin of these elements. Fujimoto (39.67.122) presented another senario in which the transverse Doppler effect due to rotation of the accreted gas played a role, and gave the value of $R/R_g = 2.4$ for Fe XXV. This interpretation is, however, subject to a strong condition on the aspect angle of observation. Proper understanding of the absorption feature requires further investigations not only on its identification but also on the line formation in the electron-scattering dominated atmosphere, since an absorption line may suffer from severe smearing due to the Comptonization effect as dissussed by Foster, Ross, and Fabian (1987, M.N.R.A.S., in press).

Additional information can be obtained from the ratio of the persistent flux to the time-averaged burst flux (usually termed as α-value), which is thought to be equal to the ratio of the surface potential to the nuclear energy released in the burst. Lapidus and Sunyaev (40.67.59) pointed out that the α-value was largely affected by the anisotropy both in the burst and persistent emissions. Allowing for the possible anisotropy, Fujimoto (1987b, Ap.J. in press) derived the radius in the range of $R/R_g = 2.2 - 3.4$ for 4U 1636-53 and EXO 0748-676 from the EXOSAT data (Lewin et al. 1987, Ap.J., 319, 893; Gottwald et al. 1986, Ap.J., 308, 213).

5. Limitations in Spherically Symmetric Models

Most of the models, which were constructed to date, assumed spherical symmetry. Moreover, the accreted material was assumed to pile up layer by layer, successively. On the other hand, there are growing evidences which can hardly be accommodated within such models. Among them are: large variations in the burst properties (such as the peak fluxes, the time-integrated burst fluxes and the recurrence intervals) without any apparent correlations to the accretion rate; too rapid succession of bursts (in less than 10 min) to replenish nuclear fuel by accretion; and absence of burst activity for very bright sources with very high accretion rate.

Various attempts have been continued during the reporting period. Livio and Regev (40.67.10) discussed variability due to the influence of burst emissions to the accretion process. Woosley and Weaver (39.67.14) presented a model of bursts in rapid successions where utilized was residual fuel left unburnt in the preceding shell flash. Nozakura, Ikeuchi, and Fujimoto (38.67.133) studied lateral propagation of the burning front in relation to the rapid succession of bursts due to partial burnings. They were not necessarily granted as successful, and yet assumed the simple accumulation of accreted fuel. Another possible solution was proposed on the basis of detailed analysis of observations. According to EXOSAT data (Lewin et al. 1987), the majority of bursts show much smaller energies and recur at much shorter intervals than predicted by existing models. Fujimoto et al. (1987c, Ap.J. 319, 902) pointed out that an extra mechanism for mixing and heating was necessary to explain them, and suggested hydrodynamical instabilities invoked by the inflow of angular momentum of accreted gas. Fujimoto (1987, A&A, in press) studied the efficiency of elemental mixing and redistribution of angular momentum due to the instabilities, Fujimoto and Hoshi (39.67.121) discussed the effect of dissipative heating on the evolution leading to the ignition of the flashes. Fujimoto et al. (1987c) also argued that this mixing mechanism and the smallnesss of the burst energies could resolve the problem of bursts in rapid succession, since the energy supply was readily explained by the fuel stored only in the upper layers where nuclear burning was inactive.

Effect of magnetic field has so far been ignored or not well explored bacause the magnetic field has been believed to be very weak or even absent in such old neutron stars as the burst sources. However, Kulkarni (42.126.25) asserted, by analysing the nature of optical counterpart of binary pulsar systems, that the decay of the field could be slowed down or might even be stopped when the field strength became of the order of 10^9 G (see van den Heuvel 41.65.158 and references therein). In some models explaining the origin of quasi-periodic oscillations, which were found in some burst sources, the presence of magnetic field of such strength was postulated (Lamb 41.67.159, and references therein). Further studies seem to be necessary to explore the existence of such field and its possible effects on the accreting neutron stars.

VII. ASTEROSEISMOLOGY (H. Shibahashi, University of Tokyo)

Stimulated by the success of helioseismology, a research field called astero-seismology, in which one probes the internal structure of stars by means of obser-vations of their oscillations, is steadily developing. During the past three years, mid 1984 - mid 1987, a couple of conferences whose titles include the term of asteroseismology or its equivalence were held; the NATO Advanced Research Workshop on "Seismology of the Sun and the Distant Stars" in Cambridge in 1985 (41.012.026, ed. D.O. Gough), and the IAU Symposium No.123 on "Advances in Helio- and Astero-seismology" in Aarhus in 1986 (in press by Reidel, ed. J. Christensen-Dalsgaard). The titles of these two conferences are representative of the present status of asteroseismology; the main topics of these conferences were helioseismology, but asteroseismology has been very much activated by the solar oscillation study and is evidently to follow. Some useful theoretical guides have been given and aspiring

observations have also been attempted. Good reviews on the subject are available in
the proceedings of these conferences (Christensen-Dalsgaard, 41.080.051; Dziembowski
and Däppen, in press), and reference may also be made to other reviews by Christensen-Dalsgaard (38.065.129; 42.080.023), de Jager (38.065.130), Dziembowski (39.065.
013), and Shibahashi (1986, Proc. Workshop: Hydrodynamic and Magnetohydrodynamic
Problems in the Sun and Stars, U. Tokyo, p.195). The following conferences are also
somehow related to the subject: the Joint Discussion at the last IAU General
Assembly "Solar and Stellar Nonradial Oscillations" (42.012.011, ed. A.N. Cox) and
the Los Alamos conference on "Stellar Pulsation" in 1986 (published by Springer
1987, ed. A.N. Cox, W.M. Sparks, and S.G. Starrfield).

1. Solar-like Oscillations

The observable modes in stars in general are limited only to modes with low
degrees $\underline{l}$, where $\underline{l}$ denotes the degree of the spherical harmonic function. Since
high order p-modes with such low degrees penetrate into the stellar deep interior,
they are useful to the seismological approach. By assuming Goldreich and Keeley's
(19.080.007) mechanism for the stochastic wave excitation by convective motion in
late type stars, Christensen-Dalsgaard and Frandsen (33.065.044) estimated the
amplitudes of high order p-modes with low degrees of those stars as $\delta I/I \sim 10^{-5}$. Some
aspiring attempts to detect such small amplitude oscillations have been made and two
groups (Gelly et al. 42.116.03; Noyes et al. 38.116.024) have so far reported
possible detection of them in stars αCen (G2V) and αCMi(F5IV), and εEri (K2V),
respectively, and Frandsen (1987, A&Ap., $\underline{181}$, 289) reported an upper limit on p-mode
amplitudes in βHyi (G2IV).

The asymptotic theory for high order p-modes with low degree predicts that the
eigenfrequencies of p-modes with even and odd $\underline{l}$ alternate, to first order, with an
equal spacing (Tassoul 28.065.044). Since the periodic oscillation is not conspicuous in their raw data, those two groups first calculated the power spectra of the
data on the variability and then searched for expected regular patterns on them. By
using the frequency spacing thus obtained, we can impose some constraints on the
physical parameters of those stars. As for αCMi, the radius thus inferred seems
reasonable (Gelly et al.). However, as for αCen, the radius inferred from the
oscillation data is inconsistent with that independently determined in terms of the
precise measurements of the mass, temperature, and distance of the star (Demarque et
al. 41.115.004; Gelly et al. 42.116.03). Guenter and Demarque (41.116.007) compared
the calculated oscillation spectrum of models of εEri with the observation by parame-
trizing the mass, the metallicity, and the mixing length. But the best fit model is
so old that the high chromospheric activity and rapid rotation rate of the star seem
to contradict with their conclusion. Guenter (1987, Ap.J., $\underline{312}$, 211) discussed this
apparent contradiction, and Däppen and Soderblom (1986, IAU Symp. 23) have proposed
another possibility of young stellar model.

The quantity of spacing between eigenfrequencies of p-modes with low, even and
odd degrees $\underline{l}$ is, roughly speaking, proportional to the mean density of the star.
On the other hand, the small departure from the true equi-distance of the power
spectrum is significantly sensitive to physical conditions near the center of the
star and then it depends on the evolutionary stage of the star. Christensen-
Dalsgaard (38.065.129; 41.080.051; 1986, IAU Symp.123) argues, based on this
consideration, that we can distinguish various models differing in masses and ages
on a two-dimensional diagram, on which the quantities of the spacing and the
departure from the real equal spacing is plotted. Ulrich (42.065.004) argues we can
determine the radius and age of isolated stars by means of frequencies of pairs of
modes differing by two in the degree $\underline{l}$ and by one in the radial order $\underline{n}$. Precise
measurements of eigenfrequencies of p-modes will become a powerful tool to determine
these fundamental quantities of stars.

2. Rapidly Oscillating Ap Stars

The rapidly oscillating Ap stars are another group of stars which reveal high
order p-mode oscillations with low degrees $\underline{l}$. The detailed power spectrum analysis
(e.g., Kurtz and Seeman 34.122.052; Matthews et al. 41.122.001) has revealed that
some of these stars are pulsating in several modes with uniformly spaced frequencies
as theoretically expected for low $\underline{l}$ high order p-modes (Shibahashi and Saio
40.065.047; Gabriel et al. 39.065.018). The amplitudes of light variation are
modulated with the same period and phase as the magnetic strength variation. In
order to explain this character, Kurtz (32.122.022) has proposed a model called the
oblique pulsator model, in which the observed oscillations are interpreted as very
high order p-modes of zonal ($\underline{m}=0$) nonradial oscillations of low degree whose
symmetry axis is aligned to the magnetic axis but oblique to the rotation axis of
the star. This model was generalized by Dziembowski and Goode (40.065.044; 41.065.
069), who took account of both the oblique magnetic field and advection to formulate
the pulsation as an eigenvalue problem. According to their formula and that of
Kurtz and Shibahashi (42.116.081), ($2\underline{l}+1$)-frequency components are to be observed as
a fine structure in the power spectrum even if only a single eigenmode is excited,
and their relative amplitudes are not equal each other but dependent on the rotation
and the magnetic field of the star. This leads to a possibility of using the fine
structure of oscillation frequencies as diagnosis of rotation and the internal
magnetic field of Ap stars. In this sense, the observation of the rapidly oscillating Ap stars will open a new aspect of asteroseismology.

A nice review on the observational aspect of these stars was given by Kurtz
(41.122.096) in Cambridge, and the theoretical aspect was reviewed by Shibahashi
(1987, Stellar Pulsation, p.112).

3. Pulsating White Dwarfs

So far eighteen DA white dwarfs have been found to be pulsating variables, and
several pulsating DB white dwarfs and pre-white dwarfs have been discovered. Since
the number of modes of an individual pulsating white dwarf is larger than those of
other-type pulsating stars but for the Sun, asteroseismology of white dwarf may be
the most promising. The oscillations in white dwarfs are regarded as nonradial
g-modes. According to the asymptotic theory for high order g-modes with low degrees
$\underline{l}$, there is a simple relation among the periods, the degree $\underline{l}$, and the radial order
$\underline{n}$ of the g-modes (e.g. Tassoul 28.065.044) as in the case of p-modes. Kawaler
(1987, Stellar Pulsation, p.367; 1986, IAU Symp.123; 1987, IAU Colloq. 95) applied
this relation to get some constraints on individual pulsating white dwarfs.

Cooling rates of white dwarfs are thought to be rapid enough to enable us to
detect the resultant period change of pulsations (McGraw et al. 27.126.070; Winget
et al. 33.065.063). In fact, such a measurement of period changes has been realized
by Winget et al. (39.126.069). Kawaler et al. (40.065.031) showed that the transi-
sition from negative to positive dP/dt occurs at nealy 1000 solar luminosity for
a wide variety of the planetary nebula nuclei and pre-white dwarf models and that
dP/dt is near zero only for a very short time. Measurement of period change will
give us some clues to understand the evolutionary stage of pulsating hot pre-white
dwarfs.

The eigenfrequencies of g-modes are mainly governed by the Brunt-Väisälä
frequency distribution in the star. Therefore, we can, in principle, obtain some
information about the Brunt-Väisälä frequency by means of a seismological approach.
The mathematical procedure for the inversion to get the Brunt-Väisälä frequency
distribution in a white dwraf is outlined by Shibahashi(1986). A comprehensive
review on the seismological investigations of compact stars was given by Winget
(1986, IAU Symp.123) in Aarhus.

36. THEORY OF STELLAR ATMOSPHERES (THEORIE DES ATMOSPHERES STELLAIRES)

PRESIDENT: K. Kodaira
VICE PRESIDENT: D.F. Gray
ORGANIZING COMMITTEE: J.P. Cassinelli, L.E. Cram, B. Gustafsson, A.G. Hearn,
 I. Hubeny, W. Kalkofen, P.-P. Kudritzki, D. Mihalas, M.J. Seaton, A.B. Underhill

I. COMMISSION ACTIVITIES

Commission 36 acted or acts as a sponsor or a cosponsor of the following Symposia and Colloquia.

(1)Symposium No. 122: "Circumstellar Matter", Heidelberg, FRG (June 1986), (2)Symposium No. 132: "The Impact of Very High S/N Spectroscopy on Stellar Physics", Paris, France (June 1987), (3)Colloquium No. 95: "Second Conference on Faint Blue Stars", Tucson, AZ, USA (May 1987), (4)Colloquium No. 102: "UV and X-ray Spectroscopy of Astrophysical and Laboratory Plasmas", Beaulieu-sur-Mer, France (September 1987), (5)Colloquium No. 106: "Evolution of Peculiar Red Giant Stars", Bloomington, IN, USA (July 1988), (6)Colloquium No. 108: "Atmospheric Diagnostics of Stellar Evolution: Chemical Peculiarity, Mass Loss and Explosion", Tokyo, Japan (September 1987).

II. RECENT PROGRESS IN THE THEORY OF STELLAR ATMOSPHERES

The theoreticians on stellar atmospheres have been continuously being challenged by new observational results obtained with ever increasing signal to noise ratio and over a wide range of wavelength. The high S/N ratio spectroscopy demands us to take subordinate physical processes into account such as realistic velocity fields and the gravity-settling effects in the quasi-hydrostatic stellar atmospheres (cf. Review A). The observations in UV and X brought up the theory of hot stellar wind, and those in IR and mm wavelength now stimulate the development of the theory of cool stellar wind (cf. Review B). The theory of expanding atmospheres have been elaborated for stationary cases as well as for dynamical cases (cf. Review C, E), and the community requires more comprehensive treatments in radiative transfer than before (cf. Review D, E). The LMC supernova 1987A allowed us for the first time to investigate the details of the dynamical structure and the development of the expanding SN atmosphere (cf. Review C, E).

As the previous reports of Commission 36, this report presents a few reviews focused on the selected topics which shall characterize the recent progress in the theory of stellar atmospheres.

A. Velocity Fields in Stellar Atmospheres (D. F. Gray)

This portion of the report draws attention to only certain selected topics of current interest regarding velocity fields in stellar atmospheres. It is not intended to be a comprehensive review. Items limited to the solar atmosphere appear in the reports of Commissions 10 and 12.

1. Convection and Granulation.
Convective motions below the photosheres of F, G, and K stars interact with the stars' rotation to generate dynamo activity and they push upward into the

visible layers producing granulation and turbulence, and setting off various oscillatory motions. Rotation of main sequence stars, its decline with time, and its relation to chromospheric emission is summarized by Soderblom (1985). Futher evidence for dynamos and rotational braking is accumulating (Gray and Nagar 1985; Gilliland 1985). Rotational decline may be regulated by convection (the Rotostat Hypothesis, Gray 1986a, 1986b, Gondoin et al. 1987).

Granulation in M giants have been detected by Nadeau and Maillard (1987) using the well-known technique of differential radial velocity shifts with excitation potential. Higher spectral resolution data, which allows the line asymmetries to be measured (Dravins 1987a, 1987b; Gray 1988) have been used to show that G and K dwarfs do have C-shaped bisectors, in agreement with the solar case and in contrast to higher luminosity stars and F dwarfs which show only the upper half of the C. These results, taken with earlier studies (Gray 1982) effective temperature and decreasing luminosity. Macroturbulence dispersion follows this same pattern and suggests that classical stellar macroturbulence is largely the motion of granulation.

Physical modeling of granulation (Nordlund and Dravins in prep.; Nordlund 1985; Steffen 1987) has shown that observed bisectors, with typically ~200 m/s excursions, reflect only about one tenth of the actual velocity displacements going on in the photosphere. While there is little hope for the spatial resolution of stellar granules, and disk-average values for line asymmetries make it hard to separate the real factor from the specific-intensity factor, the situation is made even worse by complications from large-scale non-uniformities on stellar surfaces that cause distortions of the line bisectors which vary as the star rotates (Gray and Toner 1988 in prep.). Secular variations of granulation associated, for example, with cycles needs to be studied (Wallace et al. 1987).

Dramatic differences in granulation patterns of Ib supergiants are seen in stars hotter than G0-2 Ib (Gray and Toner 1986). Studies of even higher luminosity stars show further evidence for vertical motions (Boer and deJager 1988 in prep.). Supersonic and near-supersonic macroturbulence velocities are indicated in high-luminosity supergiants (de Jager et al. 1988, in prep.).

2. Oscillation, Rotation and Outflow.

Non-radial oscillations have been identified in twelve A stars (Kurtz and Martinez 1987; Shibahashi 1987). Mode identification is progressing mainly using photometry, but phase-averaged spectroscopy has been done on HR 1217 (Matthews et al. 1987).

The observational techniques for detecting non-radial ossilations in cooler stars are reviewed by Harvey (1987), and spectroscopic methods have a 40-fold predicted advantage over photometric ones. Radial velocity variations have been detected in Arcturus (~2 days; 160 m/s) and in Pollux (~3 hr; 10 m/s) by P. H. Smith et al. (1987).

Velocity fields in the outer regions of stellar atmospheres also show complex behavior. Interesting (radial) differential rotation has been detected in AB Aur (A0ep) by Catala et al. (1986) in which Ca II shows a 32 hour period while Mg II show a 45 hour period. Modeling of stellar co-rotating interacting regions (Mullan 1984a, 1984b, 1986) has shed light on the UV emission anomalies of hybrid stars.

Modeling of outflows from bright giants have been successful in reproducing observed line widths as well as satisfying radio data and density-sensitive line ratios (Brosius and Mullan 1986). Mass loss and dust formation for long period variables are greatly enhanced by shock waves, according to recent calculations of Bowen (1988).

B. Impact of IR and mm-Wave Observations on the Theory of Stellar Envelopes of Red Giants(T. Tsuji)

The infrared and mm-wave observations, after the pioneering attempts in 1960's and 1970's, respectively, are now being matured to be used as standard tools in stellar astrophysics. In this report, we focus our attention to observations (mostly published in 1984-1987) of stellar envelopes that may be related to our understanding on stellar mass-loss, which is one of the major unresolved problems in stellar astronomy today.

1. Infrared Observations.

The completion of IRAS survey undoubtedly gave great impact on our subject as on other fields. The unbiased survey is especially useful in providing some statistics on dust formation in evolved stars (Hacking et al., 1985; Habing, 1987). IRAS photometric data provided important constraints on the physical structure of the dust envelope. For example, it was shown that the opacity of dust grains is nearly proportional to λ^{-1} and hence dust grains should be highly amorphous, both in oxygen-rich and carbon-rich envelopes (Rowan-Robinson et al., 1986; Jura, 1986; Zuckerman, Dyck, 1986b). This confirmed the conclusion based on submillimeter observations of limited sample of evolved stars by Spoka et al. (1985). The database of low resolution spectra (LRS) by IRAS is a rich source of new materials to study the chemistry of dust envelopes. For example, some carbon stars by optical spectral classification appeared to have silicate emission features in IRAS RLS: Such carbon stars may be binaries with M-giant stars (Little-Marenin, 1987) or may be just evolved from M-giant stars with remnant of oxygen-rich envelopes (Willems, De Jong, 1987). The later idea has further been detailed (Willems, 1987). Also, formation of dust in Mira variables was found to be correlated with the asymmetry of light curve that may depend on the shock strength (Vardya, De Jong, Willems, 1986).

Another important information in understanding the nature of dust around red giants is the spatial structure of dust envelopes and the speckle interferometry in the near infrared has successfully been applied to the envelopes of red giant stars: Dyck et al. (1984) have shown that the dust shell of many evolved stars are pretty large and this fact suggests the presence of dust free zone extending out to a few stellar radii around the central stars. Ridgway et al. (1986) have shown that the infrared interferometry, combined with photometric data, could provide further details of the dust shell (extinction law, optical depth, dust temperature etc). Dyck et al. (1987) indicated a possibility of highly asymmetric structure of dust envelope of IRC+10216 and suggested it to have a bipolar structure.

High resolution infrared spectra by Fourier transform spectroscopy (FTS) provided unique information on the inner circumstellar envelope: An interesting attempt is to isolate the contribution of the gas in the envelope from that of the central infrared continuum source with the use of annular entrance apertures by Sahai and Wannier (1985), who showed that resonantly scattered lines of the CO fundamental bands could be well separated and be used as probes of the physical structure of the inner part of the expanding envelope of IRC+10216. The separation of different layers by spectroscopic observations is generally difficult when velocity structures of the layers are not very different. From the detailed analysis of CO overtone bands, however, a possible presence of the quasi-static molecular dissociation zone in the transition region between the warm chromosphere and the cool wind in several normal red giants has been suggested (Tsuji, 1987). This new component is characterized by high turbulent velocity but shows little relative motion to the photosphere. For this reason, such a layer has been difficult to be recognized in normal red giant stars, but has been recognized more clearly in Mira variable stars because of the Doppler-shift of the photospheric lines due to photosheric pulsation (Hinkle,

Hall, Ridgway, 1982). Thus, the presence of such a quasi-static layer may be a basic characteristics of cool luminous stars in general. The dynamics of the envelope of Mira variables is best explored by infrared spectroscopy and it has clearly been shown that the shock is developed by the outwardly propagating waves generated by stellar oscillation (Hinkle, Scharlach, Hall, 1984; and references cited therein). It is, however, not likely that the shock wave is the direct driving force of mass-outflow, in view of the presence of the static layer noted above. These works also showed that SiO masers may be not originating from the pulsating envelope but rather they may be connected with the quasi-static turbulent layer noted above.

2. Millimeter-wave Observations.

In recent years, molecular spectroscopy in the millimeter wavelength region has made important progress: For example, CO pure rotational transitions have been observed not only in very cool stars with massive winds but also in many ordinary red giant stars (e.g., Wannier, Sahai, 1986; Olofsson, Eriksson, Gustafsson, 1987). From these observations, stellar mass loss rate has been determined with better accuracy than by any other method for a large sample of red giant stars (e.g., Knapp and Morris, 1985). Also, termianl flow velocities, which are confirmed to be rather small in most red giant stars, have been determined with high accuracy from the CO spectra (e.g., Zuckerman, Dyck, Claussen, 1986), together with the stellar radial velocity. These results enable a comparison of the momentum in the outflow gas $\dot{M}v$ and momentum in the stellar radiation field L/c; a necessary condition for the winds to be accelerated by radiation pressure on dust seems to be fulfilled for a large number of stars, but not for all the stars (Zuckerman, Dyck, 1986a, Knapp, 1986). The answer to the question of whether the dust formation initiates mass-loss or vice versa seems to be not very clear yet. Molecules other than CO are also observed: For example, HCN was found not only in carbon-rich stars (e.g., Rieu et al., 1987) but also in oxygen-rich stars (Deguchi, Claussen, Goldsmith, 1986). While rich chemistry has been known in envelopes of carbon stars such as IRC+10216, it has recently been shown that the chemistry in oxygen-rich envelope such as of OH231.8+4.2 (M-giant star with bipolar nebulosity) is similarly rich (Morris et al., 1987). Such information on gaseous component of the circumstellar matter should be most useful in our understanding on physics and chemistry of stellar envelopes when they are combined with the new information on dust component revealed by IRAS.

So far the three dimensional (i.e., two spatial dimensions plus radial velocity) mapping of the circumstellar envelopes at sub-arc second resolution has been done by radio interferometry with SiO, H_2O, and OH masers: For example, VLBI observations of SiO masers showed that the masers occur within a few stellar radii but do not show any systematic expanding patern (Lane, 1984). Thus, SiO masers may originate in cloudlets of rather high density in the turbulent region of the inner circumstellar envelope (Alcock, Rose, 1986). MERLIN observations (e.g., Chapman, Cohen, 1986; and references cited therein) clearly showed that OH maser at 1612MHz originates in a thin expanding shell at some 100 stellar radii while H_2O and OH main line masers in accelerating thick shell at some 20 stellar radii. The velocity extents of the maser emissions increase systematically with angular extent and, if this fact be interpreted as showing the continuous increase of expansion velocity, the driving force per unit mass must be increasing with the distance from the star. Thus, maser emissions are very useful in probing the inner circumstellar envelope where the acceleration may take place, but they are so far limited to oxygen-rich stars. However, recent discovery of HCN masers in carbon-rich stars (Guilloteau, Omont, Lucas, 1987; Izumiura et al., 1987) has opened a new possibility of probing inner part of the circumstellar envelope of carbon stars. Also, a possible weak maser of CO has recently been found in a classical carbon star V Hydrae by Zuckerman and Dyck (1986b). A spatially resolved time series observation of CO in V Hya revealed that anomalous excitation, which may occasionally induce maser, should be closely

related to the inhomogenous radiation field due to bipolar geometry, and thus V
Hya may be a carbon star already at the initial phase to planetary nebula via
bipolar object (Tsuji *et al.*, 1987). Such observations of masers and related
phenomena in carbon stars will also provide new impacts on our understanding
of molecular excitations and masers in stellar envelopes.

3. Envelopes of Red Giant Stars.
    Now, radio interferometry revealed that the circumstellar matters are
continuously accelerated until at least some 50 stellar radii before they gain
the terminal flow velocity that is still observed at some 1000 stellar radii by
radio thermal emissions. On the other hand, infrared interferometry revealed
the presence of dust-free zone in the inner part of the envelope. Then, even
if dust may play some role in accelerating the outflow, it could not provide the
initial acceleration in the inner envelope. Thus, inner turbulent region revealed
by SiO maser and infrared CO lines may have important key in our understanding
of the origin of mass-outflow. Although little theoretical work has yet been done
to have unified understanding on all these observations now available, one
interesting suggestion in this regard is that the thermal instability due to
molecular cooling plays an important role in determining the physical structure
of the outer atmosphere of cool giant stars (Muchmore, Nuth, Stencel, 1987). Also,
evidences of bipolar structure for the envelopes of red giant stars are
increasing, and bipolar stage may represent an important phase of the evolution
of mass-losing stars. Then, how to understand the formation of such a bipolar
structure should be another important problem in the theory of stellar envelopes
of red giant stars.

C. Supernovae Spectra (R. Harkness)

    The computation of supernovae spectra is still a relatively young field.
Initial modelling by means of LTE spectrum syntheses has been very successful
in explaining the complex spectra of both Type Ia and Ib supernovae, while
progress has been made in attempting non-LTE calculations for Type II
supernovae. Of course, the explosion of SN 1987A in the Large Magellanic Cloud
has intensified interest in this area, not only because SN 1987A has been
observed in unprecedented detail, but because its spectral evolution and light
curve appear to be unique. IUE observations of supernovae of all types have
provided a new avenue of investigation. Supernovae represent some of the most
extreme conditions for the theory of stellar atmospheres: rapid expansion,
extended spherical geometry, non-thermal radioactive decay energy sources,
extreme departures from LTE, unusual and radially stratified abundances.

1. Type Ia Supernovae
    The spectra of 'normal' Type Ia supernovae are now relatively well
understood. There seems little doubt that the explosion mechanism is the
thermonuclear incineration of a white dwarf. Theoretical spectra of a carbon
deflagration supernova model by Nomoto *et al.* (1984) show extremely good
agreement with observations at the time near maximum light (Branch *et al.* 1985,
Harkness 1985). The spectrum at this time is dominated by intermediate mass
elements, and the models require a substantial fraction of the white dwarf be
converted to these elements. Unfortunately, current theory suggests that the
central ignition will lead to a supersonic combustion, or detonation, rather than
a subsonic deflagration. The net result is that the entire white dwarf is burned
to Ni^{56}, leaving no intermediate mass elements to account for the maximum light
spectrum. It remains a puzzle as to how a detonation may be damped to allow
production of the intermediate mass elements in the outer layers.

    IUE observations show that all Type I supernovae have an 'ultraviolet
deficit', or paucity of ultraviolet flux compared to that which one would expect

on the basis of the optical colour temperature. It now seems, on the basis of
theoretical atmosphere models, that the deficit arises quite naturally in
moderately extended atmospheres and is due mainly to the strong near-uv
resonance lines of Fe II, Ni II and perhaps other iron-peak elements such as Cr
II and Mn II (Branch and Venkatakrishna 1986, Harkness 1986). In SN Ia there is
no shortage of these elements, and indeed one problem is to prevent them from
swamping the optical spectrum at maximum light. The strong abundance gradients
in the partially burned outer layers require some mixing to take place, to
homogenise the intermediate mass elements. If this mixing involves the almost
pure Fe-Co-Ni core, the spectrum deteriorates. It may be that the dying
combustion wave promotes this homogenisation as it disintegrates into small
burning regions or blobs and this mechanism will require future study.

Most spectrum calculations to date have been confined to conditions near
maximum light or very late phases (more than 100 days past maximum). Fu and
Arnett (1986) consider the effects of flux dilution on the UBVRK colours of SN
Ia from 20-50 days after maximum light. They find evidence for wavelength
dependent flux dilution, the dilution increasing with increasing photon energy.
Consequently, SN Ia may be dimmer than their infra-red colours would indicate,
but brighter than their UBV colours would indicate.

Just a few years ago it seemed that all Type Ia supernovae shared a
practically identical spectral and photometric development. However,
counter-examples are beginning to show up (Branch 1987) and hopes of using these
events as standard candles may begin to fade. Matching individual supernovae
may shed some light on the role of mixing in the outer layers. Wide spectral
coverage on a daily basis for the first four weeks of a typical SN Ia would allow
very detailed models to be constructed. Observations of the first few days would
be of the greatest interest, but obviously the hardest to obtain. It may be
possible to detect the presence of a small amount of hydrogen at this time,
lending further suppout to the theoretical model on an accreting white dwarf
undergoing a runaway as it approaches the Chandrasekhar mass.

2. Type Ib Supernovae

Apart from the explosion of SN 1987A, the most exciting developments of
the last few years concern a subtype of SN I now denoted as SN Ib. The prototype
was observed in the summer of 1983 in NGC5236 (M83), and a second, very similar
example SN 1984L was observed in NGC991 in August 1984 (Wheeler and Levreault
1985). Meanwhile, unrelated observations of SN 1985F showed a unique late-time
spectrum of an unknown kind of supernova (Filippenko and Sargent 1986). The
spectrum was dominated by [O I] and [Ca II], with no evidence for Hα which is
usually the dominant feature of Type II events. The spectrum was also quite
unlike the late-time spectrum of SN Ia, which is due principally to [Fe II] lines.
When observations of SN 1983N made in February 1984 were reduced, it was found
that the spectra were a close match to the Filippenko-Sargent object (Gaskell
et al. 1986). Furthermore, analysis of SN 1984L also shows evidence of the
emergence of a strong [O I] emission. The early-time spectra of 83N and 84L have
also been shown to be dominated by He I lines, so this type of SN appears to
be most consistent with the explosion of a Wolf-Rayet star (Harkness et al. 1987).

Atmosphere models of SN Ib are in a very early stage of development. It
seems clear, however, that large departures from LTE are necessary for the
formation of the helium lines and this poses a major computational challenge.
From the light curve, one requires the synthesis of a substantial mass of Ni^{56} (but
less than an SN Ia), which provides most of the energy input from maximum light
onwards. The ultraviolet deficit at maximum light appears to be formed in the
same manner as SN Ia.

Other 'peculiar' SN I such as SN 1983I and 1983V may be related to the apparently helium rich SN Ib. Atmosphere models with power-law density profiles consisting of helium/oxygen mixtures provide a fair match with the observations of these two events (Wheeler *et al.* 1987). It seems possible that these are directly related to normal SN Ib, but defferentiated by the extent to which the helium envelope has been removed through mass loss.

3. Type II*l* and IIp Supernovae

Type II supernovae are so defined by the presence of hydrogen lines in their spectra. The Type IIp (plateau light curve) events are consistent with explosions of red supergiants. Shock energy originating in a core collapse is deposited in a massive extended envelope. At maximum light the spectra tend to be somewhat featureless, with weak Balmer lines. SN II*l* (linear light curve) may represent lower envelope masses, but Doggett and Branch (1985) note that the light curves resemble Type I events and tend to be almost as bright, leading them to suggest that SN II*l* may be due to some kind of thermonuclear explosion, as opposed to the conventional core collapse model. The spectral development of the two subclasses is distinctly different, the metal lines being more dramatic in SN II*l*.

Non-LTE analysis of SN II is a much more practical proposition than for the $Z = 1$ composition of SN Ia. All supernova atmospheres are expected to be strongly scattering dominated. If true, distance estimates using the Baade method could be seriously in error. Shaviv *et al.* (1985), Hershkowitz *et al.* (1986a,b) and Hoflich *et al.* (1986) address the problem of non-LTE spectrum formation for low density hydrogen atmospheres. They find that deviations in level populations from LTE values generally increase with radius through the atmosphere. Near the surface NLTE effects occur because of the non-local nature of the radiation field and are enhanced by dilution of the radiation field due to sphericity. However, Hoflich *et al.* (1986) also find that electron scattering can act to restore LTE if a sufficiently extended scattering atmosphere exists above a low density photosphere. Furthermore, they note that for power law dinsity profiles, NLTE effects become more important with steeper dinsity gradients.

Hempe (1985, 1986) has calculated line profiles for Hα, Hβ formed in SN II model atmospheres with power-law density profiles.

4. Peculiar Supernovae and SN 1987A

Excellent spectral coverage of the LMC supernova 1987A has given us a new opportunity to unravel supernova spectrum formation. SN 1987A was not, however, a 'normal' Type II event. It was very subluminous and apparently originated in a blue supergiant. Spectra were obtained only 24 hours after the explosion, and were observed to undergo extremely rapid changes over the first few days as the shock heated photosphere cooled rapidly. Unlike normal Type IIs, the progenitor of SN 1987A is known to have had a small radius of about 2×10^{12}cm because the supernova was acquired visually just three hours after detection of a neutrino pulse presumably due to the core collapse. Adiabatic expansion losses were therefore very significant for this supernova. IUE data show the ultraviolet flux dropping hourly over the first two to three days. Initial ultraviolet spectra were unusual for a Type II, and then passed through a phase similar to the maximum light UV spectra of SN I. The optical spectra displayed distinct Balmer lines which became almost saturated before fading at about 10 days. He I 5876Å was visible for a couple of days, but no He lines have re-emerged. The spectrum after 10 days remained remarkably constant for the next three months, and most closely resembled that of a SN II*l* after maximum light.

Modelling the rapid spectral evolution of SN 1987A over the first 10 days presents a whole new challenge. It is likely that the full time dependence of the radiation field will be required to simulate this event, including non-LTE effects.

As more and more high quality supernovae spectra are obtained it is becoming clear that many have some unique feature. Adapting conventional atmosphere theory to supernovae should continue to be a rewarding endeavour.

D. Numerical Methods in Radiative Transfer(W. Kalkofen)

This report surveys perturbation methods for the numerical solution of radiative transfer problems in unpolarized light, as well as solution methods for polarized radiation.

1. Perturbation Methods.

Recent theoretical and numerical developments in radiative transfer have been towards more efficient, faster methods for the solution of radiative transfer problems, such as line transfer in media in statistical equilibrium and the construction of model atmospheres in radiative equilibrium. The speed of these new methods is due to the use of approximate matrix operators that are simpler in structure and often lower in order than the corresponding exact operators, leading to equations that can be solved much more rapidly than the equations of conventional, direct methods. An apparent drawback is that even linear equations must be solved iteratively. But the timing advantage can be several orders of magnitude. In addition, a given problem can usually be separated into simple parts that can be solved still more efficiently on parallel processors, leading to further substantial savings in computer time. Problems that can be attacked with these methods are, for example, radiation in multi-dimensional media, time-dependent radiative transfer coupled with gas dynamics, and the radiation emerging from stochastic media. In the following we describe some of the recent methods, emphasizing the physical principles on which the approximate operators are built, and mention some of the applications.

A typical line transfer problem contains many scale lengths, which range from the monochromatic photon mean free paths in the core of a line to the thermalization length, *i.e.*, the distance over which information about the structure of a medium is communicated to the source function. These lengths must be represented in the operators used in the numerical solution. Operator perturbation methods based on integral equations contain these scales in different operators: the individual, monochromatic photon mean free paths are expressed mainly in the exact intergral operator; the thermalization distance, in the approximate operator. The exact operator is used in the solution of uncoupled, or low-order coupled, equations that determine the error made by a provisional solution in a conservation equation; the approximate operator is used in the calculation of corrections to a provisional solution.

The approximate operator of Scharmer's (1981, 1984) method, which has inspired many of the recent operator perturbation methods, simplifies the description of the transfer along a ray by equating the specific intensity at a given point to the source function at unit optical distance in the upstream direction. This construction leads to a nearly triangular matrix operator (in a semi-infinite atmosphere) and thus a system of equations that can be solved very rapidly, with timing that scales more nearly as the square of the order of the system than as its cube. -- An ingenious application of this efficient method is by Carlsson & Scharmer (1985; cf. also Scharmer & Carlsson 1985, Carlsson 1985) to line formation in stochastic velocity fields in the solar atmosphere.

The simplification of approximate integral operators has been carried to an extreme with a purely *diagonal* operator, in which the intensity at a given point is proportional to the source function. The proportionality factor is equal to unity for a source function in a saturated line, *e.g.*, deep inside an atmosphere, but near the surface it is reduced to account for escaping radiation. Olson *et al.* (1986) base this factor on the escape probability whereas Werner & Husfeld (1985), Werner (1986, 1987) and Hamann (1985, 1986, 1987) base it on the core fraction of the Λ−operator. —— In the applications showing the features of the method, Olson *et al.* solve a two−level atom problem, Werner constructs line−blanketed hydrogen model atmospheres in radiative, hydrostatic, and statistical equilibrium, and Hamann solves the multi−level equations for a spherically expanding pure helium atmosphere of a Wolf−Rayet star of given structure.

The distinctive and significant feature of *diagonal* operator for the desciption of the radiative transfer is that the system of equations for N depth points and K unknown functions, whose order would ordinarily be $N \times K$, is split into two systems, one of order N, the other of order K. This feature is of enormous advantage for model atoms with many levels since the computation time scales as the cube of the order of the system of equations. —— Avrett & Loeser (1987) reduce the size of the matrices in another way, by writing the equations for a multilevel atom as a series of (so−called equivalent) two−level atom problems. This also separates a problem into parts of order N and order K, but insteasd of solving the transfer equation *approximately* and the equations of statistical equilibrium *exactly*, Avrett & Loeser do the opposite, solving the transfer equation *exactly* and the equations of statistical equilibrium *approximately*. Of course, the converged solution in either case is the (numerically) exact solution of the problem.

One potential drawback of Scharmer's method is that for atoms with many levels the timing may approach that of th complete linearization method (Auer & Mihalas 1969). The reason is that the solution time for such problems is dominated by the time for solving a system of equations of high order. Thus for large atomic systems, the computation time of both methods scales the same way with the "size" of the problem (*i.e.*, as $(N \times K)^3$) . Diagonal operators avoid this difficulty. On the other hand, they may give slowly converging equations since the maximal eigenvalue of a critical matrix appearing in the expansion for the solution (cf. Kalkofen 1985) need not be small compared to unity. This is a problem studied by Olson *et al.* (1986) whose remedy is to extrapolate the solution based on several consecutive iterates (cf. also Auer 1987); this cuts the number of iterations necessary to reach a required accuracy considerably although, as with operator perturbation methods generally, convergence remains linear.

An operator perturbation method that solves only Feautrier (*i.e.*, second−order differential) equations, both scalar and low−order coupled equations, is offered by Kalkofen (1987) who derives the differential operator for the perturbed system of equations from an integral equationm formulation. His demonstration is for grey, scattering, radiative equilibrium atmosphere.
Nordlund (1985) describes a method in which he perturbs the transfer equation for a two−level atom along individual rays. This approach is suitable also for problems in more than one dimension. He applies it to the formation of an iron line in an atmosphere obtained from a three−dimensional hydrodynamic simulation of the solar granulation.

A differential equation method not based on the perturbation of an operator is proposed by Anderson (1985, 1987). His approach is analogous to that in the variable Eddington factor technique, applied here to frequencies. Anderson groups photons with similar interaction probabilities into blocks, defining block average coefficeints, and performs a Newton−Raphson iteration on the block equations. With an improved solution he updates the frequency averages. He has used this method to construct line−blanketed model atmospheres in which the fine−grained frequency set has 2000 grid points and the coarse−grained set, about 100 points. Thus the order of the system of equations is reduced by a factor of 20 (not counting angle points), implying a very significant reduction in the solution time.

2. Polarized Radiation.

Polarization plays a role in the Zeeman effect, i.e., in line formation in a magnetic field, and in electron scattering near the surface of the solar atmosphere, for example. Solution methods for the transfer equation of polarized radiation must deal with the complication of an opacity that is a matrix and an intensity that forms a vector, given by the Stokes parameters. When departures from LTE combine with polarization the equations become still more complicated. Fortunately, it is often possible to make the weak-field approximation and thus to separate a transfer problem into two parts. In the first, the magnetic field is completely ignored and the source function of a line is computed as in standard non-LTE theory; in the second, the source function is given and the polarized radiation field is computed for given magnitude and orientation of the magnetic field. This approach is frequently used, e.g., by Rees and Murphy (1987) who discuss several methods of solving the transfer equation for polarized radiation. In one method they transform the first-order differential equation for the Stokes vector into a second-order differential equation for the even part of the radiation field, i.e., the analogue of the Feautrier equation. The difference equation form of this equation is block tridiagonal; thus the equation is solved by standard techniques. In another approach, they separate out the diagonal part of the absorption matrix to arrive at an equation with a diagonal transfer operator but with off-diagonal elements. For the formal integration of this equation, the off-diagonal terms are treated as known source terms. The result is an integral equation for the unknown Stokes vector. By using a quadrature scheme, such as piecewise linear expansion of the Stokes vector, a system of linear equations involving a block triangular matrix is derived from which the Stokes vector can be determined by means of a simple forward elimination.

A solution method that does not make the weak-field approximation is due to van Ballegooijen (1987). He supposes that the magnetic field is sufficiently strong to permit the assumption that the magnetic sublevels of the atom do not interfere with each other. Then the problem of this Zeeman effect with departures from LTE is completely described by a scalar line source function that depends only on optical depth. For this source function van Ballegooijen derives an integral equation which he solves by expanding the source function in piecewise linear and quadratic segments. -- This method is a welcome addition to methods for polarized radiation; it is also very useful in allowing an assessment of the limits of applicability of the weak-field approximation.

A completely different method for solving the transfer equation of polarized radiation in the presence of departures from LTE is described by Schmidt & Wehrse (1987). Their method, which is well suited for homogeneous media, uses the eigenvalues and eigenvectors of the matrix multiplying the intensity, i.e., the matrix that describes the opacity and the redistribution properties of the transfer. After diagonalizing with the modal matrix, the transfer equation together with the statistical equilibrium constraint can be solved analytically. If only the emergent intensity is wanted, the intensities elsewhere in the medium need not be calculated, a significant advantage of the method. However, for a multi-level atom in an inhomogeneous atmosphere this method is expensive since it requires the calculation of the eigenvalues of the interaction matrix at many grid points, a calculation that scales as the cube of the order of the matrix. An approach described by Peraiah (1987) achieves the same end, albeit after more matrix manipulations that are needed for insuring the stability of this doubling method. -- Note that both Schmidt & Wehrse's and Peraiah's methods are very general; they can be used for transfer of polarized radiation with complete or partial redistribution in atmospheres with plane or spherical symmetry.

Several recent conference proceedings and monographs contain papers on the subject of this review or on related topics. In particular:

Chromosheric Diagnostics and Modelling, 1985, B.W.Lites ed.,
  National Solar Observatory/Sacramento Peak, Sunspot, New Mexico
  (ref. as *Sac. Peak*)
Measurements of Solar Vector Magnetic Fields, 1985, M.J.Hagyard ed.,
  NASA Conference Publication 2374.
Numerical Radiative Transfer, 1987, W.Kalkofen ed.,
  Cambridge University Press (ref. as *NRT*).
Progress in Stellar Spectral Line Formation Theory, 1985, J.E.Beckman & L. Crivellari ed.,
  Reidel Publ. Co., Boston (ref. as *Trieste*).
Theoretical Problems in High Resolution Solar Physics, 1985, H.U.Schmidt ed.,
  MPA Garching bei *München*
  (ref. as *München*).

E. Line—blanketing in Expanding Atmospheres(D. C. Abbott)

1. Introduction.
   "Line–blanketing" is the term for the effect of bound–bound transitions from spectral lines on the global properties of the atmosphere, such as the temperature, density, and emergent spectrum. It is well known from the study of radiative equilibrium, plane–parallel, static atmospheres that line–blanketing leads to two main results (*e.g.* Mihalas 1978): (i) flux blocking and redistribution, and (ii) heating and backwarming. The strength of these effects depends strongly on the fraction of the spectrum that is blocked, which means that the cumulative effect of many, relatively weak lines may dominate the process, as the strongest lines occupy only a small frequency interval of the spectrum.

   In an expanding atmosphere, line–blanketing is enormously enhanced if two conditions are met: (i) the expansion velocities are large compared to the velocity width of the line profiles, and (ii) the expanding atmosphere is dense enough that the stronger lines are still optically thick. Two examples where these conditions are met are stellar winds from hot stars and the photospheric phase of novae and supernovae. Each optically thick line may cover a frequency interval that is a thousand times larger than in a static atmosphere. Large regions of the spectrum may contain no line–free continuum, as the overlapping lines form a "pseudo–continuum". Examples demonstrating this have been given by Abbott (1982, figure 7) for hot star winds, and by Karp *et al.* (1977, figure 3) for expanding atmospheres from supernovae. As there are no gaps in frequency between the strong lines under these circumstances, the importance of the cumulative contribution of many weak lines is greatly diminished.

   A unique feature of expanding atmospheres is that the Doppler–shifted frequencies of distinct spectral lines can coincide, so that a given photon can interact with many lines before escaping from the atmosphere. The presence of this "multiply scattered" radiation field also means that the radiative excitation rates of distinct lines from distinct elements are coupled, so the statistical eqilibrium of all the species are interlocked. Line–blanketing in an expanding atmosphere also creates a very favorable environment for the transfer of momentum from the radiation to the gas, which is the driving agent in hot star winds. Finally, the interpretation of features in a region of a line pseudo–continuum can be very difficult. For example, an apparent emission feature may indicate an absence of lines, where the radiation leaks out.

2. Optically–Thin Expanding Atmospheres.
   In many cases of interest, such as the OB stars, expansion is confined to

the outer layers of the atmosphere, which are optically-thin to continuum opacity. These atmospheres can be modeled by a "core-halo" approach. The continuum radiation is formed in a "core" represented by a standard radiative-equilibrium, plane-parallel, hydrostatic atmosphere. The expanding part of the atmosphere forms an extended "halo" about this core, which contains many optically-thick lines, but negligible continuum absorption or emission processes. Models which self-consistently solve for the radiation hydrodynamics of the expanding atmosphere were first developed by Castor, Abbott, and Klein (1975). Recent refinements to the model include correctly accounting for multiple scattering of radiation (Abbott and Lucy 1985, Puls 1987), and solving for the non-LTE statistical equilibrium of the thousands of transitions formed in the wind (Pauldrach 1987, Puls 1987). These models provide a detailed description of the radiation that is reflected back onto the core by line scattering and electron scattering in the wind. This reflected radiation field can greatly modify the structure of the hydrostatic atmosphere where the photospheric spectrum originates (e.g. Hummer 1982, Abbott and Hummer 1985), a process called "wind-blanketing".

Wind-blanketing mainly heats the atmosphere, while changing the shape of the emergent spectrum only somewhat. A wind-blanketed model atmosphere which fits the photospheric spectrum will have a lower effective temperature than a standard unblanketed model. In the case of the star Zeta Puppis (O4f), wind-blanketing decreased the inferred effective temperature by 4500 K or about 10% (Bohannan et al. 1986), while in Alpha Cam (O9.5I) the effective temperature was reduced by roughly 7% (Voels et al. 1986). For the more luminous OB stars, spectral classification is truly three-dimensional, with the mass loss rate, the gravity, and the effective temperature all playing important roles in determining the emergent photospheric spectrum.

3. Optically-Thick Expanding Atmospheres.
In many cases of interest, such as Wolf-Rayet stars, Luminous Blue Variable stars, as well as Novae and Supernovae, the expanding atmosphere is optically thick to continuum opacity. The photosphere is formed out in the wind. Line-blanketing will be of great inportance in these objects from the outer atmosphere all the way through the depth of continuum formation. Models which can accurately treat the radiation-hydrodynamics of expanding, optically-thick winds are still under development.

Models of optically-thick expanding atmospheres were calculated with a Monte Carlo code which extends the method of Abbott and Lucy (1985) to account for continuum absorption opacity under the assumption of LTE (see also Lucy 1987). The preliminary result from a supernova model calculation (Abbott 1987) shows tremendous flux-blocking in the UV, where it is impossible to ascribe the observed features to any one element, because there are overlapping, optically thick lines from many different elements at virtually all frequencies. The flux-blocking is much more severe in these models than in hot stars because the atmosphere is less extended, so the scattered photons tend to diffuse back down in the atmosphere to be thermalized, rather than to escape. This in turn leads to significant backwarming, and the temperature at depth is roughly 15% higher in the line-blanketed model, as reflected in the heightened continuum at optical wavelengths. The development of models which can correctly account for line-blanketing in optically thick winds is clearly a fruitful area of research for the coming years.

References by Sections of the Report

A. Velocity Fields in Stellar Atmospheres

Bowen,G.H.: 1988, Astrophys.J. submitted.

Brosius,J.W. and Mullan,D.J.: 1986, Astrophys.J. 301, 650.
Catala,C. *et al.*: 1986, Astrophys.J. 308, 791.
Dravins,D.: 1987a, Astron.Astrophys. 172, 200.
Dravins,D.: 1987b, Astron.Astrophys. 172, 211.
Gilliland,R.L.: 1985, Astrophys.J. 299, 286.
Gondoin,P. *et al.*: 1987, Astron.Astrophys. 174, 187.
Gray,D.F.: 1982, Astrophys.J. 255, 200.
Gray,D.F.: 1986a, Highlights in Astronomy 7, 411.
Gray,D.F.: 1886b, Adva. Space Res. 6, 161.
Gray,D.F.: 1988, in The Impact of very High Signal-to-Noise Spectroscopy
          on Stellar Physics, IAU Symp No.132.
Gray,D.F. and Nagar,P.: 1985, Astrophys.J. 298, 756.
Gray,D.F. and Toner.C.G.: 1986, PASP 98, 499.
Harvey,J.: 1987, in Advances in Helio- and Asteroseismology, IAU Symp No.123.
Kurtz,D.W. and Martinez,M.: 1987, MNRAS 226, 187.
Matthews,T.A. *et al.*: 1987, Astrophys.J. accepted.
Mullan,D.J.: 1984a, Astrophys.J. 283, 303.
Mullan,D.J.: 1984b, Astrophys.J. 284, 769.
Mullan,D.J.: 1986, Astron.Astrophys. 165, 157.
Nadeau,D. and Maillard,J.-P.: 1987, Astrophys.J. submitted.
Nordlund,A.: 1985, Small-Scale Dynamical Processes in Quiet Stellar
          Atmospheres, S.L.Keid, ed., NSO/Sac Peak.
Shibahashi,H.: 1987, Lecture notes in Phys. 274, 112.
Smith,P.H. *et al.*: 1987, Astrophys.J.Lett. in press.
Soderblom, D.R.: 1985, Astron.J. 90, 2103.
Steffen,M.: 1987, The Role of Fine-Scale Magnetic Fields
          on the Structure of the Solar atmosphere, La Laguna, Tenerife.
Wallace,L.V. *et al.*: 1987, Astrophys.J. submitted.

B. Inpact of IR and mm-Wave Observations on the Theory of Stellar Envelopes of Red Giants

Alcock,C. and Rose,R.R.: 1986, Astrophys.J., 310, 838.
Chapman,J.M. and Cohen,R.J.: 1986, MNRAS, 220, 513.
Deguchi,S., Claussen,M.J. and Goldsmith,P.F.: 1986, Astrophys.J., 303, 810.
Dyck,H.M., Zuckerman,B., Leinert,Ch. and Beckwith,S.: 1984, Astrophys.J., 287, 801.
Dyck,H.M., Zuckerman,B., Howell,R.R. and Beckwith,S.: 1987, PASP, 99, 99.
Guilloteau,S., Omont,A. and Lucas,R.: 1987, Astron.Astrophys. 176, L24.
Habing,H.J.: 1987, in Circumstellar matter ed. I.Appenzeller and C.Jordan,
          IAU Symp No.122, Dordrecht;Reidel, p.197.
Hacking,P. *et al.*: 1985, PASP, 97, 616.
Hinkle,K.H., Hall,D.N.B. and Ridgway,S.T.: 1982, Astrophys.J., 252, 697.
Hinkle,K.H., Scharlace,W.W.G. and Hall,D.N.B.: 1984, Astrophys.J.Suppl., 56, 1.
Izumiura,H., Ukita,N., Kawabe,R., Kaifu,N., Tsuji,T., Unno,W. and Koyama.K.:
          1987, NRO Report.
Jura,M.: 1986, Astrophys.J., 303, 327.
Knapp,G.R.: 1986, Astrophys.J., 311, 731.
Knapp,G.R. and Morris,M.: 1985, Astrophs.J., 292, 640.
Lane,A.P.: 1984, in VLBI and Compact Radio Sources
          eds. R.Fanti, K.Kellerman and G.Setti, IAU Symp No.110, Dordrecht;Reidel,
          p.329.
Little-Marenin,I.R.: 1986, Astrophys.J.Letters, 315, L141.
Morris,M., Guilloteau,S., Lucas,R. and Omont,A.: 1987, IRAM Preprint.
Muchmore,D., Nuth III,J.A. and Stencel,R.E.: 1987, Astrophys.J.Letters, 315, L141.
Olofsson,H., Eriksson,K. and Gustafsson,B.: 1987, Uppsala Preprints in Astronomy,
          No.18.
Ridgway,S.T., Joyce,R.R., Connors,D., Pipler,J.L. and Dainty,C.:
          1986, Astrophys.J., 302, 662.

Rieu,N.Q., Epchtein,N., Bach,T. and CohenM.: 1987, Astron.Astrophys., 180, 117.
Rowan-Robinson,M., Lock,T.D., Walker,D.W. and Harris,S.: 1986, MNRAS, 222, 273.
Sahai,R. and Wannier,P.G.: 1985, Astrophys.J., 299, 424.
Spoka,R.J., Hildebrtand,R., Jaffe,D.T., Gatley,I., Roelling,T, Werner,M., Jura,M. and
Zuckerman,B.:
         1985, Astrophys.J., 294, 242.
Tsuji,T.: 1987, in Circumstellar Matter ed. I.Appenzeller and C.Jordan,
         IAU Symp No.122. Dordrecht;Reidel, p.377.
Tsuji,T., Unno,W., Kaifu,N., Izumiura,H., Ukita,N., Cho,S. and Koyama,K.:
         1987, NRO Report.
Vardya,M.S., DeJong,T, and Willems,F.J.: 1986, Astrophys.J.Letters, 304, L29.
Wannier,P.G. and Sahai,R.: 1986, Astrophys.J., 311, 335.
Willems,F.J.: 1987, Ph.D.Thesis, Univ.Amsterdam.
Willems,F.J. and De Jong,T.: 1986, Astrophys.J.Letters, 309, L39.
Zuckerman,B., Dyck,H.M.: 1986a, Astrophys.J., 304, 394.
Zuckerman,B., Dyck,H.M.: 1986b, Astrophys.J., 311, 345.
Zuckerman,B., Dyck,H.M. and Claussen,M.J.: 1986, Astrophys.J., 304, 401.

C. Supernovae Spectra

Branch,D.: 1987, Astrophys.J., 316, L81.
Branch,D. and Venkatakrishna,K.L.: 1986, Astrophys.J., 306, L21.
Branch,D, Doggett,J.B., Nomoto,K. and Thielemann,F.-K.: 1985, Astrophys.J., 294, 619.
Doggett,J.B. and Branch,D.: 1985, Astron.J., 90, 2303.
Filippenko,A.V. and Sargent,W.L.W.: 1986, Astron.J., 91, 691.
Fu,A.J. and Arnett,W.D.: 1986, Astrophys.J., 307, 726.
Gaskell,C.M.,   Cappellaro,E.,   Dinerstein,H.L.,   Garnett,D.R.,   Harkness,R.P.   and
Wheeler,J.C.
         1986. Astrophys.J., 306, L77
Harkness,R.P.: 1985, in Supernovae as Distance Indicators,
         ed. N.Bartel, Berlin:Springer-Verlag, p.183.
Harkness,R.P.: 1986, in Radiation Hydrodynamics in Stars and Compact Objects,
         eds. D.Mihalas and K.-H.A.Winkler, Berlin:Springer-Verlag, p.166.
Harkness,R.P.: 1987, in 13th Texas Symp. on Relativistic Astrophysics,
         ed. M.P.Ulmer, World Scientific Publishing, p.413.
Harkness,R.P., Wheeler,J.C., Margon,B., Downes,R.A., Kirshner,R.P., Uomoto,A., Barker,E.S.,
Cochran,A.L., Dinerstein,H.L., Garnett,D.R., and Levreault,R.:
         1987, Astrophys.J., 317, 355.
Hempe,K.: 1985, in Supernovae as Distance Indicators,
         ed. N.Bartel, Berlin:Springer-Verlag, p.192.
Hempe,K.: 1986, Astron.Astrophys., 158, 329.
Hershkowitz,S., Linder,E. and Wagoner,R.V.: 1986, Astrophys.J., 301, 220.
Hershkowitz,S., Linder,E. and Wagoner,R.V.: 1986, Astrophys.J., 303, 800.
Hoflich.P., Wehrse,R. and Shaviv,G.: 1986, Astron.Astrophys., 163, 105.
Nomoto,K., Thielemann,F.-K. and Yokoi,K.: 1984, Astrophys.J., 286, 644.
Shaviv,G., Wehrse,R. and Wagoner,R.V.: 1985, Astrophys.J., 289, 198.
Wheeler,J.C., Harkness,R.P., Barker,E.S., Cochran,A.L. and Wils,D.:
         1987, Astrophys.J., 313, L69.
Wheeler,J.C. and Levreault.R.: 1985, Astrophys.J., 294, L17.

D. Numerical Methods in Radiative Transfer

See the end of Review D for the keys, NRT, München, Sac. Peak, and Trieste
Anderson,L.S.: 1985, Astrophys.J., 298, 848.
Anderson,L.S.: 1987, in NRT, Ch.7.
Auer,L.H.: 1987, in NRT, Ch.4.
Auer,L.H. and Mihalas,D.: 1969, Astrophys.J., 158, 641.
Avrett,E.H. and Loueser,R.: 1987, in NRT, Ch.6.
Carlsson,M.: 1985, in München, 67.

Carlsson,M. and Scharmer,G.: 1985, in *Sac. Peak*, p137.
Hamann,W.-R.: 1985, Astron.Astrophys., **148**, 364.
Hamann,W.-R.: 1986, Astron.Astrophys., **160**, 347.
Hamann,W.-R.: 1987, in *NRT*, Ch.2.
Kalkofen,W.: 1985, in *Trieste*, 153.
Kalkofen,W.: 1987, in *NRT*, Ch.8.
Nordlund,A.: 1985, in *Trieste*, 215.
Olson,G.L., Auer.L.H.: and Buchler,J.R.: 1986,
          J.Quant.Spectrosc.Rad.Transfer, **35**, 431.
Peraiah,A.: 1987, in *NRT*, Ch.13.
Rees,D.E. and Murphy,G.A.: 1987, in *NRT*, Ch.10.
Scharmer,R.: 1981, Astrophys.J., **249**, 720.
Scharmer,R.: 1984, Methods in Radiative Transfer,
          W.Kalkofen ed., Cambridge University Press, Cambridge, 173.
Scharmer,R.: 1987, in *NRT*, Ch.8.
Scharmer,G. and Carlsson,M.: 1985, J.Comp.Phys., **59**, 56.
Schmidt,M. and Wehrse,R.: 1987, in *NRT*, Ch.14.
van Ballegooijen,A.: 1987, in *NRT*, Ch.12.
Werner,K.: 1986, Astron.Astrophys., **161**, 177.
Werner,K.: 1987, in *NRT*, Ch.3.
Werner,K. and Husfeld,D.: 1985, Astron.Astrophys., **148**, 417.

E.   Line-Blanketing in Expanding Atmospheres

Abbott,D.C.: 1982, Astrophys.J., **259**, 282.
Abbott,D.C.: 1987, Bull.AAS, **19**, 705.
Abbott,D.C. and Lucy,L.B.: 1985, Astrophys.J., **288**, 679.
Abbott,D.C. and Hummer,D.G.: 1985, Astrophys.J., **294**, 286.
Bohannan,B., Abbott,D.C, Voels,S.A. and Hummer,D.G.: 1986, Astrophys.J., **308**, 728.
Castor,J.I., Abbott,D.C. and Klein,R.I.: 1975, Astrophys.J., **195**, 157.
Hummer,D.G.: 1982, Astrophys.J., **257**, 724.
Karp,A.H., Lasher,G. and Chan,K.L.: 1977, Astrophys.J., **214**, 161.
Lucy,L.B.: 1987, Astron.Astrophys.Letters, in press.
Mihalas,D.: 1978, Stellar Atmospheres, San Francisco:Freeman.
Pauldrach,A.: 1987, Astron.Astrophys., in press.
Puls,J.: 1987, Astron.Astrophys., in press.
Voels,S.A., Bohannan,B., Abbott,D.C. and Hummer,D.G.: 1986, Bull.AAS, **18**, 953.

COMMISSION 37: STAR CLUSTERS AND ASSOCIATIONS
(AMAS STELLAIRES ET ASSOCIATIONS)

PRESIDENT: D.C. Heggie
VICE-PRESIDENT: G.L.H. Harris
ORGANISING COMMITTEE: K.C. Freeman, J.E. Hesser, P.E. Nissen, C. Pilachowski, G.N. Salukvadze

1. Introduction

The last three years have been very productive for cluster research, especially now that observations with CCDs have become so routine. After a section on recent meetings, and one on data catalogues (by J.-C. Mermilliod), subsequent sections of this report go into details on individual topics: associations (P.E. Nissen), open clusters (G.L.H. Harris), globular clusters and cluster systems (R.E. White) and dynamical theory (D.C. Heggie).

The rapid growth in the volume of observational data has meant regrettably that the information has had to be tabulated in an even more compressed form than in earlier Reports. As in previous Reports, three-figure references are to Astronomy and Astrophysics Abstracts, and two-figure references to Physics Abstracts. Most entries cover publications from 1 July 1984 to 30 June 1987, but some exceptions are noted under the individual sections. For abbreviations see §6.

2. Symposia and Colloquia

The highlight of the past triennium was IAU Symposium 126 on *Globular Cluster Systems in Galaxies* (August 25-9, 1986), on which there is a short report in the IAU *Information Bulletin* no.57, p.27. Proceedings were published from two earlier IAU meetings bearing on the work of the Commission: IAU Symposium no.113 on *Dynamics of Star Clusters* (39.012.070) and no.116 on *Luminous Stars and Associations in Galaxies* (41.012.079). Another publication of interest is the proceedings of a meeting on *The Age of Stellar Systems*, held in 1986 (42.012.104).

3. Catalogs of Cluster Data

Several new catalogues or updates of old catalogues have been announced or are available through the data centres in Strasbourg and at NASA:

Mermilliod, J.-C. (38.002.054, CDS 2-104) - Preliminary Catalogue of Photographic UBV photometry for stars

White, R.E. & Philip, A.G.D. (38.002.071) - Catalog of Cluster CM diagrams

Zhao, T., *et.al.* (40.002.081) - Special Issue for Tables of Membership to 42 Open Clusters

Popova, M., & Kraicheva, Z. (40.153.049) - Catalog of eclipsing and spectroscopic binary stars in the region of Open Clusters

Mermilliod, J.-C. (41.002.005, CDS 4-017) - Catalogue of DM, HD(E) and other cross-identifications for stars in Open Clusters

Lyngå, G. (CDS 7-066) - Catalog of Open Cluster Data (4th Edition)

Mermilliod, J.-C. (CDS 2-124) - Catalogue of UBV photometry and MK spectral types in open clusters (3rd edition)

G. Lyngå reports that he has sent to the Strasbourg Data Centre a copy of the fifth version of his Catalogue of Open Cluster Data.

4. Stellar Associations

INTRODUCTION

This part of the report covers the full three-year period January 1984 - December 1986. Papers on associations in general, and on OB, T and R associations are listed in the following four tables. For abbreviations see §6.

TABLE 1. ASSOCIATIONS IN GENERAL

(37.113.008) var Ap and He-weak stars
(37.115.014) HR-diagram, stf
(37.131.113) IR obs of R ass
(37.144.018) WR stars, cosmic and γ rays
(38.116.007) rot, stf
(38.155.064) stf
(39.112.044) stellar winds
(40.152.011) dynamics, stf
(40.152.013) ages from phot
(40.065.087) evolution of massive stars
(40.131.265) stf in mol clouds
(40.155.047) rv of OB stars
(40.157.135) ass in M31
(41.126.017) birthplaces of pulsars
(41.156.014) age and stf in LMC
(41.157.218) ass in the galaxy NGC4449
(42.122.059) flare stars
(42.157.167) massive stars in M31

(37.113.014) ex law, IR excess
(37.121.029) X-ray, T-Tauri stars
(37.131.255) far-UV ex
(38.065.002) evolution of WR stars
(38.153.012) frequency of WR stars
(38.156.007) ass in LMC
(40.152.005) ass in M31
(40.152.012) frequency of binaries
(40.152.014) HI supershells
(40.131.262) stf, ages
(40.153.047) stellar abund
(40.156.016) ass in SMC
(41.152.010) ass in galaxies
(41.153.039) stellar abund
(41.157.013) ass in the galaxy NGC2403
(42.121.028) multiple stars in T-ass
(42.157.099) phot of ass in M31

TABLE 2. OB ASSOCIATIONS

Ara OB1	Whiteoak, Gardner (41.131.268)	ist NaI
Cam OB1	Wesselius *et al.* (37.112.049)	IR obs, massive stars
Cas OB2	Lozinskaya *et al.* (41.131.112)	ring nebula, shells, etc
Cen OB1	Doom *et al.* (39.152.003)	time-dependent stf
Cep OB2	Gyul'budagyan *et al.* (41.152.001)	dark globules
	Kun (42.113.006)	Hα em stars
	Kun (42.123.019)	var stars
Cep OB3	Massa, Savage (37.131.174)	UV obs of ist ex
CMa OB1	Gaylard, Kemball (38.132.028)	radio obs, HII regions
	Pyatunina, Taraskin (42.131.328)	radio obs, HII distr
Cyg OB1	Philips *et al.* (37.131.015)	high-vel ist gas
Cyg OB2	Giovannelli *et al.* (38.112.020)	IUE spectra
	Giovannelli *et al.* (42.112.149)	mass loss
Gem OB1	Braun, Strom (42.125.012)	IRAS obs, SNR shock
Mon OB1	Ogura, Hasegawa (37.131.188)	Hα em stars, Bok globules
	Rosino *et al.* (38.122.174)	flare stars
	Parsamyan *et al.* (42.122.060)	flare stars
Mon OB2	Leahy (40.142.044)	X-ray em
Ori OB1	Rydgren, Vrba (37.113.038)	IR excess of G stars
	Rössiger (37.122.147)	young var stars
	Mermilliod, Mayor (39.152.008)	rot vel
	Gasparyan (39.113.030)	phot of Hα stars
	Klochkova (40.114.001)	Bp stars
	Gasparyan (41.122.123)	distr of Hα and flare stars
	Brown, Shore (42.114.124)	spphot of λ1400 ist feature
Per OB1	Doom *et al.* (37.152.004)	age
	Garibdzhanyan (38.114.130)	spphot, d, ist ex
	McLachlan, Nandy (40.152.001)	shock-heated gas
	Schild, Berthet (42.152.001)	blue stragglers
Per OB2	Klochkova, Kopylov (40.152.002)	d, age
	Klochkova, Panchuk (42.114.053)	helium abund of B stars
Sco OB1	van Genderen (38.152.007)	phot, T_{eff}, g
	Heske, Wendker (38.153.003)	phot, relation to Tr24
	Heske, Wendker (40.152.008)	phot, sp, relation to Tr24
Sco OB2	Lipovka *et al.* (37.152.006)	radio em
	Cappa de Nicolau, Pöppel (40.131.194)	HI distr

Klochkova, Panchuk (42.114.053)   helium abund of B stars
Cappa de Nicolau, Pöppel (42.131.024) ist gas distr

TABLE 3. T ASSOCIATIONS

Cyg T4	Satyvoldiev (37.123.003)	var stars
	Satyvoldiev (39.123.004)	var stars
	Satyvoldiev (40.123.019)	var stars
Tau T1	Nurmanova (39.152.002)	phot, memb
Tau T3	Nurmanova (37.123.004)	var stars
	Nurmanova (39.152.002)	phot, memb

TABLE 4. R ASSOCIATIONS

CMa R1	Nakano *et al.* (38.152.009)	radio obs
	Pyatunina (39.152.001)	radio sources, stf
	Pavolva, Rspaev (42.116.114)	pol
Mon R1	Ogura (37.152.001)	Hα em stars
	Pyatunina (30.152.001)	radio sources, stf
	Pavlova, Rspaev (42.116.114)	pol
Mon R2	Pyatunina (39.152.001)	radio sources, stf
Per R1	Pavlova, Rspaev (42.116.114)	pol

5. Open Clusters

INTRODUCTION

Some papers have discussed a large number of open clusters, general properties of open clusters, overall interpretations of large data samples or models of cluster properties. Those papers are referred to below in a separate tabulation. Extragalactic cluster observations have been excluded, primarily due to the length of this report. However, given the increasing interest in this area, some means of handling such data in future will need to be decided.

Papers and projects which refer to individual open clusters are listed first, where the clusters are ordered according to IAU number. For abbreviations see §6.

CLUSTER-BY-CLUSTER LITERATURE CITATIONS

C0017-302 (Bl 1)
(40.153.005) BV pg phot; E; d; ms gaps.
C0027+599 (NGC 129)
(38.153.001) DDO, uvby, VRI phot; memb.   (39.115.022) sp cl; abs mag.
(41.113.019) Vilnius phot; bright st; RG.
C0039+850 (NGC 188)
(38.153.028) W UMa st; bin; age.   (40.065.053) th isochrones; age.
(40.153.021) W UMa systems.   (40.153.034) sp; ms turnoff st; abund spread.
(41.153.007) R phot; faint st.   (39.116.038) FK Comae-like st.
(39.153.056) age.   (40.117.020) W UMa st; origin.
Kaluzny and Shara (STSI) phot; short-per var. (42.153.016) Washington phot; RG; abund; CN.
Lee & van Altena (Yale, in press) pm; astrom calibration field.
C0040+615 (NGC 225)
(38.153.001) DDO, uvby, VRI phot; memb.
C0049+563 (NGC 281)
(39.153.055) reddening law.
C0115+580 (NGC 457)
(42.153.005) IUE sp; extinction law.   (39.115.022) sp cl; abs mag.
C0126+630 (NGC 559)
(39.153.035) rv; blue stragglers.
C0129+604 (NGC 581)
(39.115.022) sp cl; abs mag.   (40.153.010) semi-automatic phot.
(41.153.025) memb; mass, age distributions.

C0132+610 (Tr 1)
(41.153.025) memb; mass, age distributions.
C0140+616 (NGC 654)
(38.153.001) DDO, uvby, VRI phot; memb.   (41.153.025) memb; mass, age distributions.
C0140+604 (NGC 659)
(38.153.001) DDO, uvby, VRI phot; memb.
C0142+610 (NGC 663)
(40.153.010) semi-automatic phot.   (40.115.034) rot; sp types; B st.
C0154+374 (NGC 752)
(38.153.001) DDO, uvby, VRI phot; memb.   (41.153.026) positions; V mag.
(42.153.029) sp; Li abund.
Schiller and Milone (Calgary) UBVRI phot; ecl bin;d.
(42.153.016) Washington phot; RG; abund; CN.
C0215+569 (NGC 869, h Per)
(38.153.001) DDO, uvby, VRI phot; memb.   (38.153.008) JIIK(LM) phot.
(39.115.022) sp cl; abs mag.   (40.115.034) rot; sp types; B st.
(40.118.018) bin.
C0218+568 (NGC 884, χ Per)
(38.153.001) DDO, uvby, VRI phot; memb.   (38.153.008) JIIK(LM) phot.
(39.115.022) sp cl; abs mag.   (40.115.034) rot; sp types; B st.
(40.118.018) bin.   (42.114.158) SG memb; abund.
C0228+612 (IC 1805)
(38.114.087) sp cl.   (40.153.019) rv.
(41.153.025) memb; mass, age distributions.
C0233+557 (Tr 2)
(41.113.019) Vilnius phot; bright st; RG.
C0238+613 (NGC 1027)
(39.153.035) rv, blue stragglers
C0238+425 (NGC 1039)
(42.153.010) δa phot; CP2 st.
C0318+484 (Mel 20, α Per)
(38.113.025) ms gap.   (39.153.015) sp; pm; memb; rot; evol low mass st.
(42.153.044) sp types; rot vel; pec st.   (40.115.034) rot; sp types; B st.
(42.153.019) space vel; d; memb; vel gradient; expansion.
(42.114.053) He abund; Be st.
Stauffer et al. (NASA Ames,Lick,Harvard) phot; light curves; rapid rot K st.
(42.116.052) rot vel; rapid rot.; evolution.
Fehrenbach et al. (Haute-Provence) rv 258 st; space vel.
C0328+371 (NGC 1342)
(39.153.035) rv, blue stragglers
C0344+239 (Pleiades)
(38.153.004) sp; UBV phot; E.   (38.153.009) sp; Ca II em; pec.
(38.153.011) E; pol.   (38.153.014) rot;var; bin.
(38.153.010) pm; flare st; sd.   (38.153.024) energy distr; Be st;
(38.153.033) th isochrones; age.   (38.041.017) astrometry
(38.113.025) ms gap.   (39.153.012) E; pol; pre-ms st.
(38/9.253.016) X-ray survey; B, A st.   (39.153.028) X-ray survey; LF.
(39.153.047) pm.   (40.153.040) turnon point; age; pre-ms.
(42.153.009) VLBUW phot; memb; pre-ms st.   (39.116.015) Hα em; periodicity.
(40.114.100) pec st; sp; age.   (39.036.132) phot detection of bin
(40.115.034) rot; sp types; B st.   (41.114.088) Fe abund.
(41.116.041) X-ray lum; convection.   (41.116.057) BVRIJHK phot.
(42.153.022) lower ms st abund; d.   (42.153.028) pol; E; molecular cloud.
(42.114.053) He abund; Be st.   (42.122.064) flare st.
(42.122.065) flare st.   (42.112.133) CaII, MgII em; late ms st.
(42.122.068) UV-Ceti st in clusters.   (42.122.073) fast-rotating flare st.
(42.122.100) radio observations of flare st.
van Leeuwen et al. (RGO,Leiden)VBLUW phot; light curves; G and K dwarfs
McNamara (New Mexico State) uvby phot; variability; B st; periods.

Stauffer *et al.* (NASA Ames,Lick,Harvard) phot; light curves; rapid rot K st.

C0403+622 (NGC 1502)
(38.153.001) DDO, uvby, VRI phot; memb.

(39.153.035) rv; blue stragglers.

C0411+511 (NGC 1528)
(38.153.001) DDO, uvby, VRI phot; memb.

C0417+501 (NGC 1545)
(38.153.001) DDO, uvby, VRI phot; memb.

C0417+368 (NGC 1548)
(38.153.001) DDO, uvby, VRI phot; memb.

C0424+157 (Hyades)
(38.153.002) rv; memb; d.
(38.153.014) rot;var; bin.
(38.153.018) color anomalies; starspots.
(38.153.018) astrom and kin par.
(38.113.025) phot; var; CaII em.
(38.113.025) ms gap.
(39.153.008) memb search; ms; supercluster.
(39.153.013) Cousins VRI phot.
(39.153.025) coude sp;abund.
(39.153.049) α Tauri CD, pm, parallax; memb.
(39.153.053) bin population analysis
(40.153.029) abund; d
(40.153.015) BVRI phot; memb.
(41.153.015) X-ray var survey.
(41.153.018) Li abund.
(41.153.043) Hα, OI phot.
(39.120.008) sp orbits; bin.
(41.153.035) bright memb
(41.114.076) sp; abund.
(41.116.052) coronal X-ray em.
(41.119.060) sp; phot; bin orbit.
(42.065.033) star spots; model comparison.
(42.113.025) by phot; var; dust.
Peterson & Solensky (SUNY) sp; bin; d.
(42.116.052) rot vel; rapid rot.; evolution.

(38.153.009) sp; Ca II em; pec.
(38.153.017) Li abund.
(38.153.019) astrom r anomalies; starspots.
(38.153.020) search for memb.
(38.116.023) CaII em; rot.
(38.117.023) IUE sp; bin; model atm.
(39.153.009) memb search; RG; supercluster.
(39.153.023) MgII em; mass and age dependence.
(39.153.039) metallic-line st; ultrashort period Cep
(39.153.051) rv; convergent point; bin.
(41.153.041) sp vel; memb.
(40.153.036) plages.
(41.153.014) EXOSAT phot; coronal t.
(41.153.016) Li abund gap.
(41.153.023) positions.
(39.112.109) CaII em; star spots.
(40.153.013) sp; F st; Fe abund.
(41.113.036) uvby phot; interpretation.
(41.116.041) X-ray lum; convection.
(41.119.041) bin analysis.
(41.120.001) speckle interferometry bin st.
(42.112.133) CaII, MgII em; late ms st.
(42.116.064) X-ray em; modeling.
(42.153.016) Washington phot; RG; abund; CN.

C0443+189 (NGC 1647)
(38.153.001) DDO, uvby, VRI phot; memb.

(40.122.011) Cep; light curves; memb

C0447+436 (NGC 1664)
(38.153.001) DDO, uvby, VRI phot; memb.

C0504+369 (NGC 1778)
(41.153.025) memb; mass, age distributions.

C0509+166 (NGC 1817)
(38.111.015) pm; memb.

C0519+333 (NGC 1893)
(38.153.001) DDO, uvby, VRI phot; memb.

C0524+352 (NGC 1907)
(38.153.001) DDO, uvby, VRI phot; memb.

C0520+295 (Be 19)
(38.153.047) scanner sp; memb; abund.

C0528+342 (NGC 1931)
(41.153.004) UBV phot; var E; d; age.

C0530+341 (NGC 1960, M36)
(41.153.031) diam; E; d; LF.

(40.115.034) rot; sp types; B st.

C0532+099 (Orion)
Lee *et al.* (Yale) internal motions; vel distr.

C0532-054 (Trapezium)
(42.153.012) V,I$_c$ phot; age; st density.

Herbig: low mass st formation; molecular clouds.

C0546+336 (King 8)
(38.153.033) th isochrones; age.

(38.153.047) scanner sp; memb; abund.

C0548+217 (Be 21)
(39.153.058) phot; d
C0551+003 (NGC 2112)
(39.153.007) uvby phot; E; d; age.
C0604+241 (NGC 2158)
(40.153.050) BV phot; sp; age; d.   (41.153.042) erratum
(41.154.003) CMD; ft st.
C0605+243 (NGC 2168, M35)
(38.153.001) DDO, uvby, VRI phot; memb.   (41.153.033) pm; memb; pg phot.
(42.153.043) internal motion; mass segregation
C0609+054 (NGC 2186)
(38.153.001) DDO, uvby, VRI phot; memb.
C0613-186 (NGC 2204)
Cameron and Reed (Cambridge,RGO) sp; rv; memb; blue stragglers; bin st.
Claria and Lapasset (Cordoba) UBV, CMT$_1$T$_2$ phot; RG; abund.
C0624-047 (NGC 2232)
(39.153.005) UBV, DDO phot; no RG memb.   (42.153.045) Δa phot; search for CP2 st.
C0627-312 (NGC 2243)
Cameron and Reed (Cambridge,RGO) sp; rv; memb; blue stragglers; bin st.
C0629+049 (NGC 2244)
(39.153.065) dust shell; abs.   (41.153.008) phot; sp; P-Q method; memb.
(40.114.173) UV ext,colors; O st.   (41.131.005) UV E; d; memb.
C0632+084 (NGC 2251)
(38.153.001) DDO,uvby,VRI phot; memb.
C0638+099 (NGC 2264)
(39.153.029) star formation.   (39.153.043) star formation
(39.153.050) IUE sp; X-ray em; chromospheric activity.
(39.153.005) UBV, DDO phot; no RG memb.   (39.121.005) pre-ms st; Hα em.
(40.112.021) EXOSAT obs; st coronae.   (41.153.025) memb; mass, age distributions.
(41.131.130) BV phot; abs.   (42.116.052) rot vel; rapid rot.; evolution.
C0644-206 (NGC 2287)
(38.153.001) DDO, uvby, VRI phot; memb.   (38.153.056) rot vel.
(40.153.019) rv.   Ianna et al. (Virginia) pm; UBV phot; memb; d.
FitzGerald and Harris (Waterloo) UBV phot; st types; E; d; st counts; memb.
C0649+005 (NGC 2301)
(41.153.045) pm; memb
Claria and Lapasset (Cordoba) UBV, DDO phot; RG; abund.
C0652-245 (Cr 121)
(38.065.002) memb; initial mass; WR st   (39.115.022) sp cl; abs mag.
C0701-207 (NGC 2324)
(41.153.046) pm; memb   (41.123.035) var st.
C0704-100 (NGC 2335)
(39.153.005) UBV, DDO phot; probable RG memb.
C0705-105 (NGC 2343)
(39.153.005) UBV, DDO phot; probable RG memb. (42.153.045) Δa phot; search for CP2 st.
C0712-102 (NGC 2353)
FitzGerald and Harris (Waterloo) UBV phot; st types; E; d; st counts; memb.
C0712-256 (NGC 2354)
(42.153.046) BV phot; E; d.
C0716-248 (NGC 2362)
(41.153.039) abund.
C0721-131 (NGC 2374)
(39.153.040) pg, pe phot; E; d; age.
C0721-122 (Haff 8)
Claria and Lapasset (Cordoba) UBV, DDO, CMT$_1$T$_2$ phot; RG; abund.
C0722-321 (Cr 140)
(42.153.045) Δa phot; search for CP2 st.
C0722-261 (Ru 18)
(39.153.005) UBV, DDO phot; possible RG memb.

Claria and Lapasset (Cordoba) CMT_1T_2 phot; RG; abund.

<u>C0724-287 (Ru 20)</u>
Claria and Lapasset (Cordoba) UBV, DDO, CMT_1T_2 phot; RG; abund.

<u>C0724-476 (Mel 66)</u>
(38.153.045) Washington phot; abund.   (42.153.016) Washington phot; RG; abund; CN.
Cameron and Reed (Cambridge, RGO) sp; rv; memb; blue stragglers; bin st.
Claria and Lapasset (Cordoba) UBV, DDO phot; RG; abund.

<u>C0734-205 (NGC 2421)</u>
(38.153.001) DDO, uvby, VRI phot; memb.   (40.115.034) rot; sp types; B st.

<u>C0734-143 (NGC 2422)</u>
(38.153.001) DDO, uvby, VRI phot; memb.   (38.153.034) uvby $H\beta$ phot; age; d.

<u>C0734-137 (NGC 2423)</u>
Claria and Lapasset (Cordoba) UBV, DDO, CMT_1T_2 phot; RG; abund.

<u>C0735+216 (NGC 2420)</u>
(40.065.053) th isochrones; age.   (39.153.035) rv; blue stragglers.
(42.153.033) st counts.   (42.153.016) Washington phot; RG; abund; CN.
Fenkart (Basel) RGU phot; space densities.
Smith and Suntzeff (Texas, Mt.Wilson-Las Campanas) sp; rv; abund; giants.
Cameron and Reed (Cambridge,RGO) sp; rv; memb; blue stragglers; bin st.

<u>C0735-119 (Mel 71)</u>
(41.153.020) BV phot; E; d; age.

<u>C0738-315 (NGC 2439)</u>
(39.115.022) sp cl; abs mag.   (40.115.034) rot; sp types; B st.

<u>C0743-378 (NGC 2451)</u>
(39.153.005) UBV, DDO phot; no RG memb.   (40.153.033) wd search; 4 memb.
(42.153.013) δa phot; CP2 st; d.   (39.115.022) sp cl; abs mag.
(42.153.035) uvby $H\beta$ phot; E; B8 to A0 st; age.

<u>C0745-271 (NGC 2453)</u>
(41.134.006) VBLUW phot; E; d.

<u>C0746-261 (Ru 36)</u>
(39.153.020) UBV phot; E; d.

<u>C0750-263 (NGC 2467)</u>
(38.153.001) DDO, uvby, VRI phot; memb.

<u>C0750-384 (NGC 2477)</u>
Claria and Lapasset (Cordoba) UBV, DDO phot; RG; abund.

<u>C0753-278 (NGC 2483)</u>
(38.153.001) DDO, uvby, VRI phot; memb.

<u>C0754-299 (NGC 2489)</u>
Claria and Lapasset (Cordoba) UBV, DDO, CMT_1T_2 phot; RG; abund.

<u>C0757-284 (Ru 44)</u>
(38.065.002) memb; initial mass; WR st

<u>C0757-106 (NGC 2506)</u>
Cameron and Reed (Cambridge,RGO) sp; rv; memb; blue stragglers; bin st.

<u>C0757-607 (NGC 2516)</u>
(42.153.002) var st; phot.   (40.115.034) rot; sp types; B st.
Claria and Lapasset (Cordoba) UBV, DDO phot; RG; abund.

<u>C0759-193 (Ru 46)</u>
(39.153.005) UBV, DDO phot; probable RG memb.
Claria and Lapasset (Cordoba) UBV, DDO phot; RG; abund.

<u>C0802-461 (Cr 173)</u>
(42.153.035) uvby $H\beta$ phot; E; B8 to A0 st; age.

<u>C0803-280 (NGC 2527)</u>
(39.153.005) UBV, DDO phot; probable RG memb.
Claria and Lapasset (Cordoba) UBV, DDO, CMT_1T_2 phot; RG; abund

<u>C0805-297 (NGC 2533)</u>
(38.153.001) DDO, uvby, VRI phot; memb.

<u>C0808-126 (NGC 2539)</u>
(41.153.017) DDO, CMT_1T_2 phot; RG; E; d; age; abund.
(42.153.008) UBV phot; memb; E; d; age.   (41.153.025) memb; mass, age distributions.

C0809-491 (NGC 2547)
(40.153.030) UBV Hβ, DDO phot; sp; rv; pre-ms.   (41.153.038) uvby Hβ phot; d.
(42.153.035) uvby Hβ phot; E; B8 to A0 st; age.
C0810-374 (NGC 2546)
(38.153.001) DDO, uvby, VRI phot; memb.   (39.153.005) UBV, DDO phot; probable RG memb.
C0811-056 (NGC 2548)
(39.153.005) UBV, DDO phot; probable RG memb.   (40.153.019) rv.
C0816+304 (NGC 2567)
(39.153.005) UBV, DDO phot; probable RG memb.
(41.153.027) UBV, DDO, CMT₁T₂ phot; memb; E; d; age.
C0832-441 (Pi 4)
Claria and Lapasset (Cordoba) UBV, DDO, CMT₁T₂ phot; RG; abund.
C0837+201 (NGC 2632, Praesepe)
(38.153.014) rot;var; bin.   (38.153.025) Li abund.
(38.153.033) th isochrones; age.   (38.113.025) ms gap.
(41.153.022) sp; MK types; Am st; Ap st.   (39.036.132) phot detection of bin
(40.117.282) contact bin; phot; ms   (42.153.017) MK sp cl; faint st; d.
(42.122.068) UV-Ceti st in clusters.   (42.153.016) Washington phot; RG; abund; CN.
C0838-528 (IC 2391)
(40.153.033) wd search; memb.   (42.153.013) δa phot; CP2 st; d.
C0839-480 (IC 2395)
(42.153.045) Δa phot; search for CP2 st.
Claria and Lapasset (Cordoba) UBV, DDO, CMT₁T₂ phot; RG; abund.
C0843-486 (NGC 2670)
(38.153.001) DDO, uvby, VRI phot; memb.
C0846-423 (Tr 10)
(38.153.039) positions; pm; mag.   (42.153.035) uvby Hβ phot; E; B8 to A0 st; age.
(42.153.045) Δa phot; search for CP2 st.
C0846+120 (NGC 2682, M67)
(40.065.053) th isochrones; age.   (38.114.076) CNO abund; giants.
(39.153.013) Cousins VRI phot.
(39.153.064) UBV phot; pm; memb; RG; blue stragglers
(40.153.023) sp; RG; CNO abund.   (40.153.027) phot; sp; abund; d.
(41.153.034) sp; metallicities.   (39.153.057) uvby phot; ms st.
(40.113.061) RI phot.   (42.153.051) Li abund
(42.153.036) precise rv; memb.   (42.153.037) sp bin; blus stragglers; spatial distr.
(42.153.042) MgII,CaII em; RG   (42.153.016) Washington phot; RG; abund; CN.
Brown (Texas) sp; CN abund; evolution.
Nissen et al. (Aarhus,Kansas,KPNO) uvby Hβ phot; E; d; age; abund.
Anthony-Twarog (Kansas) uvby phot; E; d; abund.
Claria and Lapasset (Cordoba) CMT₁T₂ phot; RG; abund.
C0915-495 (Pi 11)
(41.113.001) VBLUW phot; hypergiant membs
C0926-567 (IC 2488)
Claria and Lapasset (Cordoba) UBV, DDO phot; RG; abund.
C0939-536 (Ru 79)
(38.153.001) DDO, uvby, VRI phot; memb.
C0940-438 (Ru 80)
(39.115.022) sp cl; abs mag.
C1001-598 (NGC 3114)
(38.153.001) DDO, uvby, VRI phot; memb.
Claria and Lapasset (Cordoba) UBV, DDO, CMT₁T₂ phot; RG; abund
C1025-573 (IC 2581)
(39.117.061) UBV light curves; per; mass; age.   (40.115.034) rot; sp types; B st.
C1033-579 (NGC 3293)
(39.115.022) sp cl; abs mag.   (41.153.039) abund.
(41.122.018) β Cep var; ecl bin.   (42.122.085) β Cep st; periods; modes.
(41.131.005) UV E; d; memb.
C1041-597 (Cr 228)

(38.065.002) memb; initial mass; WR st;
C1041-593 (Tr 14)
(40.114.173) UV ext,colors; O st.   (41.131.005) UV E; d; memb.
C1043-594 (Tr 16)
(38.065.002) memb; initial mass; WR st; evol   (40.153.031) sp cl.
C1057-900 (NGC 3496)
Claria and Lapasset (Cordoba) CMT_1T_2 phot; RG; abund.
C1104-584 (NGC 3532)
Claria and Lapasset (Cordoba) UBV, DDO phot; RG; abund.
C1112-609 (NGC 3603)
(38.153.026) WR st; sp; rv; bin.   (38.065.002) memb; initial mass; WR st
C1123-429 (NGC 3680)
(42.153.016) Washington phot; RG; abund; CN.
C1133-613 (NGC 3766)
(39.153.002) uvby $H\beta$ phot; Be st; E; st.   (39.153.046) H II regions; bin.
(39.115.022) sp cl; abs mag.   (40.115.034) rot; sp types; B st.
(41.112.012) Be st; var.
C1134-627 (IC 2944)
(41.153.039) abund.   Walborn (STSI) sp types
C1141-622 (Stock 14)
(38.153.001) DDO, uvby, VRI phot; memb.   (41.119.018) uvby light curves; ecl bin.
Peterson and FitzGerald (DAO) UBV phot; E; d
C1148-554 (NGC 3960)
Claria and Lapasset (Cordoba) CMT_1T_2 phot; RG; abund.
C1154-623 (Ru 97)
Claria and Lapasset (Cordoba) UBV, CMT_1T_2 phot; RG; abund.
C1222+263 (Mel 111)
(38.153.014) rot;var; bin.   (39.153.013) Cousins VRI phot.
(42.153.044) sp types; rot vel; pec st.   (41.153.043) $H\alpha$, OI phot.
(39.036.132) phot detection of bin   (41.114.088) Fe abund.
(42.002.047) coordinate catalogue   (42.112.133) CaII,MgII em; late ms st.
C1239-627 (NGC 4609)
Claria and Lapasset (Cordoba) UBV, DDO, CMT_1T_2 phot; RG; abund.
C1250-600 (NGC 4755)
(38.153.035) UBV phot;d; age; E(var).   (40.115.034) rot; sp types; B st.
(41.153.039) abund.
C1315-623 (Stock 16)
(39.153.027) UBV phot; E; d; age.
C1317-646 (Ru 107)
Claria and Lapasset (Cordoba) UBV, DDO, CMT_1T_2 phot; RG; abund.
C1324-587 (NGC 5138)
Claria and Lapasset (Cordoba) UBV, DDO, CMT_1T_2 phot; RG; abund.
C1328-625 (Tr 21)
Peterson and FitzGerald (DAO) UBV phot; E; d
C1350-616 (NGC 5316)
Claria and Lapasset (Cordoba) UBV, DDO, CMT_1T_2 phot; RG; abund.
C1356-619 (Ly 1)
Peterson and FitzGerald (DAO) UBV phot; E; d
C1404-480 (NGC 5460)
(40.153.016) δa phot; CP2 st;
Claria and Lapasset (Cordoba) UBV, DDO, CMT_1T_2 phot; RG; abund
C1426-605 (NGC 5617)
Claria and Lapasset (Cordoba) UBV, DDO, CMT_1T_2 phot; RG; abund.
C1431-563 (NGC 5662)
Claria and Lapasset (Cordoba) UBV, DDO, CMT_1T_2 phot; RG; abund.
C1440+697 (Cr 285, U Ma)
(41.153.035) memb.   (42.153.001) kinematics; origin; evolution.
C1445-543 (NGC 5749)
Claria and Lapasset (Cordoba) UBV, DDO phot; RG; abund.

<u>C1501-541 (NGC 5822)</u>
(39.153.021) UBV, DDO, CMT$_1$T$_2$ phot; RG; memb; E; d; age; abund.
<u>C1511-588 (Pi 20)</u>
Peterson and FitzGerald (DAO) UBV phot; E; d
<u>C1601-517 (Ly 6)</u>
(39.153.014) d.   (39.153.018) UBV phot; E; d; memb TW Nor.
<u>C1609-540 (NGC 6067)</u>
(38.153.001) DDO, uvby, VRI phot; memb.   (39.153.019) UBV phot; E; d; Ceph memb.
(40.153.016) δa phot; CP2 st   (41.122.020) Cep in cluster.
Claria and Lapasset (Cordoba) DDO, CMT$_1$T$_2$ phot; RG; abund.
<u>C1614-577 (NGC 6087)</u>
(42.153.018) UBV phot; Cep; E; d.
Claria and Lapasset (Cordoba) UBV,DDO phot; RG; abund.
<u>C1624-490 (NGC 6134)</u>
Claria and Lapasset (Cordoba) UBV, CMT$_1$T$_2$ phot; RG; abund.
<u>C1642-469 (NGC 6204)</u>
(38.153.001) DDO, uvby, VRI phot; memb.
<u>C1650-417 (NGC 6231)</u>
(38.153.054) IUE sp; mass loss.   (38.065.002) memb; initial mass; WR st
(38.152.007) VBLUW phot; E; d; evol.   (39.153.003) phot; β Cep instability strip
(39.112.006) B st; st winds.   (40.114.173) UV ext,colors; O st.
(41.131.005) UV E; d; memb.
<u>C1652-394 (NGC 6242)</u>
Claria and Lapasset (Cordoba) UBV, DDO, CMT$_1$T$_2$ phot; RG; abund.
<u>C1653-405 (Tr 24)</u>
(38.153.003) UBV phot; d; E; age   (40.152.008) phot; sp; memb; age; pre-ms st.
<u>C1654-447 (NGC 6249)</u>
Claria and Lapasset (Cordoba) UBV, DDO phot; RG; abund.
<u>C1675-446 (NGC 6259)</u>
(42.153.046) BV phot; E; d.
Claria and Lapasset (Cordoba) CMT$_1$T$_2$ phot; RG; abund.
<u>C1701-378 (NGC 6281)</u>
(38.153.015) Δa phot; CP2 st.   (40.153.001) uvby phot; E; memb.
<u>C1720-499 (IC 4651)</u>
(42.153.047) ms turnoff, structure; E; d.   (42.153.048) uvby phot; ms turnoff; age.
<u>C1722-343 (Pi 24)</u>
(38.065.002) memb; initial mass; WR st;
<u>C1731-325 (NGC 6383)</u>
(40.153.028) Walraven phot; E; d; sp; IR excess.
<u>C1732-334 (Tr 27)</u>
(38.065.002) memb; initial mass; WR st
<u>C1734-349 (NGC 6396)</u>
Claria and Lapasset (Cordoba) CMT$_1$T$_2$ phot; RG; abund.
<u>C1736-321 (NGC 6405)</u>
(38.153.015) Δa phot; CP2 st.
(40.153.001) uvby phot; E; memb; possible δ Sct st, CP1 st.
(39.115.022) sp cl; abs mag.
<u>C1741-323 (NGC 6416)</u>
(38.153.001) DDO, uvby, VRI phot; memb.
<u>C1743+057 (IC 4665)</u>
(38.153.015) Δa phot; CP2 st.   (40.153.019) rv.
<u>C1750-348 (NGC 6475)</u>
(38.153.001) DDO, uvby, VRI phot; memb.   (38.153.037) UBV phot; sp; bin; memb.
(38.113.025) ms gap.   (40.153.002) rv; memb; sp bin orbit; high vel st.
<u>C1753-190 (NGC 6494)</u>
(41.113.023) Vilnius phot; E; d.
Claria and Lapasset (Cordoba) UBV, DDO, CMT$_1$T$_2$ phot; RG; abund
<u>C1758+029 (Cr 359)</u>
Rucinski (DAO) UBVRI phot; W UMa st; reality of cluster.

C1801-225 (NGC 6531)
(41.153.039) abund.
C1801-243 (NGC 6530)
(38.153.055) UV sp; age; mass loss.   (39.153.043) star formation
(39.112.005) sp; line profiles; st winds.   (41.153.025) memb; mass, age distributions.
C1816-138 (NGC 6611)
(41.153.003) uvby phot; sp; abund.   (41.153.044) polarimetry
(41.153.025) memb; mass, age distributions.   (42.153.021) pm; memb.
C1825+065 (NGC 6633)
(40.153.019) rv.   (39.153.035) rv; blue stragglers.
(41.113.020) Vilnius phot; E; memb; RG.   (42.153.016) Washington phot; RG; abund; CN.
C1828-192 (IC 4725, M 25)
(38.153.001) DDO, uvby, VRI phot; memb.   (40.153.016) δa phot; CP2 st;
C1830-104 (NGC 6649)
Walker and Laney (SAAO) UBV phot; Cep; d; E; var st.
C1834-082 (NGC 6664)
(38.153.001) DDO, uvby, VRI phot; memb.
C1836+054 (IC 4756)
(40.153.020) pm technique; memb.
Claria and Lapasset (Cordoba) CMT$_1$T$_2$ phot; RG; abund.
C1842-094 (NGC 6695)
(38.153.001) DDO, uvby, VRI phot; memb.
C1848+102 (NGC 6705, M 11)
(38.153.027) pm; lf; str; dyn.   (42.153.036) precise rv; memb.
(42.153.016) Washington phot; RG; abund; CN.
C1850-204 (Cr 394)
(40.153.006) UBV, VRI phot; E; d; Cep memb.
C1851-199 (NGC 6716)
(40.153.006) UBV, VRI phot; E; d.
C1905+041 (NGC 6755)
(38.153.001) DDO, uvby, VRI phot; memb.   (39.153.035) rv; blue stragglers.
C1909+129 (Be 82)
(41.153.022) UBV phot; memb; E; d; RG; age.
C1919+377 (NGC 6791)
(38.153.050) DDO, UBV phot; RG; d; abund; E.   (39.153.023) phot; ms; d.
(42.153.016) Washington phot; RG; abund; CN.
C1923+200 (Cr 399)
(41.113.019) Vilnius phot; bright st; RG.
C1939+400 (NGC 6819)
(42.153.016) Washington phot; RG; abund; CN.
C1941+231 (NGC 6823)
(41.153.025) memb; mass, age distributions.
C1947+272 (S Vul)
(40.153.012) new cluster; st counts; d; S Vul; Cep.   (42.153.015) UBV phot; E; d; age; Cep memb.
C2004+356 (NGC 6871)
(38.153.036) uvby Hβ phot, var.   (38.065.002) memb; initial mass; WR st
C2009+264 (NGC 6882)
(38.153.006) E; d.   (39.153.035) rv, blue stragglers
(41.113.019) Vilnius phot; bright st; RG.
C2009+263 (NGC 6885)
(38.153.006) E; d.   (39.153.035) rv, blue stragglers
(41.113.019) Vilnius phot; bright st; RG.
C2014+374 (IC 4996)
(39.153.006) uvby Hβ phot; E; d; age.   (40.153.003) var st search; bin; β Cep st.
C2018+392 (Har 5)
Claria and Lapasset (Cordoba) UBV, DDO phot; RG; abund.
C2022+383 (NGC 6913)
(41.153.025) memb; mass, age distributions.
C2032+281 (NGC 6940)

(40.153.019) rv.
C2109+454 (NGC 7039)
Schneider (Göttingen) uvby Hβ phot; E; d; reality of cluster.
C2121+461 (NGC 7062)
(38.153.029) rv, blue stragglers
C2122+362 (NGC 7063)
(38.153.001) DDO, uvby, VRI phot; memb.   Schneider (Göttingen) uvby Hβ phot; E; d.
C2128+487 (Anon.)
(42.153.031) reality of cluster.
C2130+482 (NGC 7092, M 39)
(38.153.013) memb.   (39.153.011) UBV phot; E; d; age.
(40.153.019) rv.   (42.153.044) sp types; rot vel; pec st.
(42.153.010) δa phot; CP2 st.
C2137+573 (Tr 37)
(39.153.037) UV E; B st; dust.
Clayton and Fitzpatrick (Wisconsin, JILA) UV sp; E; dust.
Marschall and van Altena (Gettysburg, Yale, in press) pm.
C2151+470 (IC 5146)
(38.153.022) UBVRI phot; n st; age.
C2210+570 (NGC 7235)
(38.153.001) DDO, uvby, VRI phot; memb.
C2245+578 (NGC 7380)
(40.132.038) 21 cm em, cont obs; evolutionary history
C2306+602 (Ba 2)
(39.153.004) UBV phot; E; A; d; memb; earliest sp
C2309+603 (NGC 7510)
(38.153.001) DDO, uvby, VRI phot; memb.
(39.153.004) UBV phot; E; A; d; memb; earliest sp
(41.131.112) OIII, NII, SII phot.
C2313+601 (BA 3)
(39.153.004) UBV phot; E; A; d; memb; earliest sp (41.131.112) OIII, NII, SII phot.
C2313+602 (MA 50)
(38.065.002) memb; initial mass; WR st
C2322+613 (NGC 7654)
(38.153.001) DDO, uvby, VRI phot; memb.
(39.153.004) UBV phot; E; A; d; memb; earliest sp
C2343+619 (Stock 2)
(42.153.006) UBV phot; E; d; age.
C2347+624 (King 21)
(38.153.041) phot; E; d; age.
C2350+616 (King 12)
(38.153.023) phot; E; d; age.
C2354+564 (NGC 7789)
(38.153.007) Li abund.   (40.153.007) uvby phot; blue stragglers.
(40.153.037) sp; giants; met abund.   (41.153.006) C isotope ratio; wk G-band st.
(41.153.001) Li abund.   (42.153.016) Washington phot; RG; abund; CN.
C2355+609 (NGC 7790)
(38.153.049) UBV phot; E; d; bin.

SURVEY PAPERS

Abt (39.012.033) pec st in clusters.
Abt (40.153.009) sp; ages; blue stragglers.
Ambartsumian (39.151.132) dynamics of open clusters (republication).
Anthony-Twarog (42.153.020) uvby phot with CCD.
Balazs (42.153.025) integrated phot.
Balona and Shobbrook (39.122.070) early st; lum.
Barbaro and Pigatto (38.153.005) RG; evolution; age.
Barkhatova et al. (40.153.025) kin par of clusters.

van den Bergh and Lafontaine (38.153.046) F and G st search
Bertelli *et al.* (41.153.036) problems in evol.
Bica and Aloin (42.153.014) grid for population synthesis
Cameron (39.153.030) d; abund; 38 clusters.
Cameron (39.153.031) abund; age; 38 clustrs.
Campbell (42.114.050) strong CN st; binary coalescence.
Casertano and Hut (40.151.077) core radius; density; N-body models.
Catalano and Marilli (41.153.037) rot; ang mom; MgII, CIV em.
Collier and Jenkins (38.117.121) blue stragglers; models; bin.
Cudworth (41.111.026) astrometry in st clusters; review.
Dabrowski (39.153.061) bin st in clusters, frequencies and masses.
Danilov *et al.* (40.153.008) star counts in 22 clusters.
Danilov (41.153.024) dyn; disintegration of clusters.
Danilov *et al.* (41.153.040) diameters; n of st in clusters.
Danilov and Beshenov (41.153.019) dyn par multiple clusters
Dearborn *et al.* (39.151.092) N-body calculations; st formation; cluster formation.
DeGioia-Eastwood (38.131.120) IMF; cluster formation.
Dluzhnevskaya *et al.* (40.153.024) IMF from young clusters.
Duran and Graziati (42.114.130) IUE sp; blue stragglers.
Efremov (40.153.048) stellar complexes, review.
Ehjgenson and Yatsyk (42.153.007) young clusters.
Elmegreen and Clemens (40.131.039) formation rate; models; cluster formation
Elmegreen (40.131.008) IMF; models; cluster formation.
Feinstein *et al.* (41.115.008) Of st in clusters; ages.
Gehren *et al.* (40.153.047) sp; B st in clusters; Gal abund gradient.
Geyer and Nelles (38.153.043) rv.
Gotz (42.153.027) metal abund of clusters.
Hensberger and van Rensbergen (42.115.005) systematics; CP st in clusters.
Hill *et al.* (42.114.160) M_V-$H\gamma$ calibration; O6 to A3 SG.
Hirschfeld and Sinnott (39.002.019) catalogue; double st; var st
Hron *et al.* (39.111.022) rv; OB st.
Iben (42.153.041) bin origin of blue stragglers
Illarionov (42.153.023) angular momentum of gas in clusters.
Janes (42.153.038) cluster ages; applications.
Jaschek and Mermilliod (38.115.006) sp; defined ms width.
Jones *et al.* (39.153.035) synthetic CMD's.
Jura (UCLA) mass loss; RG.
Klochkova and Panchuk (40.114.036) sp; abund; young clusters
Kutuzov (41.153.028) Lindblad diagram
Lada and Wilking (38.153.053) star formation; cluster formation.
van Leeuwen (39.153.048) pm in clusters, compilation of studies.
Leisawitz (42.153.026) molecular clouds and young clusters.
Leonard (41.153.012) blue stragglers; field st capture?
Lotkin (38.153.042) method of d determination.
Lundstrom and Stenholm (38.153.012) WR; bin.
Lyngå (39.002.002) computer-based catalogue of open cluster data
Lyngå (40.002.051) use of the cluster database.
Maitzen (41.153.047) CP2 st in clusters.
Mathieu (39.153.045+052+062, 42.153.003) structure, kinematics, dynamics of clusters.
Melnick (41.131.351) IMF in violent st forming regions
Mermilliod (38.002.054) catalogue UBV pg phot.
Mermilliod (40.153.017) clusters for HIPPARCOS.
Mermilliod and Maeder (41.153.005) theoretical isochrones; mass loss; clusters.
Mermilliod (40.002.011) DM,HD(E) etc. cross-identifications.
Millward and Walker (39.115.003) $H\gamma$ lum calibration; class V-III st
Pandey and Mahra (42.153.011) age distribution; lifetimes of clusters.
Pandey *et al.* (Uttar Pradesh) age distr; correlation with diameter, cluster mass.
Parsamian (40.122.033) flare st in clusters; evolution.
Pedreros (41.153.029) new approach to ZAMS fitting.

Pigatto (42.153.039) ages; overshooting models.
Popova (40.153.049) catalogue; ecl bin, sp bin in clusters.
Roth-Hoppner (42.153.026) age determinaton of clusters.
Roth-Hoppner et al. (42.002.060) remarks on (42.153.026)
Sagar (40.153.004) effect of ms st on age determination.
Schmidt (38.153.049) yellow giants; evolution.
Singh et al. (40.151.070) tidal effects of Galaxy on st clusters.
Smith (39.131.278) accretion; IMF for clusters.
Sowell (Michigan) sp; rv; memb; yellow giants.
Spassova (39.153.041) integrated mag, col - 50 clusters.
Stecklum (39.153.010) IMF; age correlation.
Stothers (40.065.087) WR st; evolution; young clusters.
Terlevich (39.151.131) N-body simulations.
The and Perez (42.153.049) obs of intermediate-mass pre-ms st in very young clusters
Wielen (40.153.018) pm of clusters from HIPPARCOS.
Wielen (39.151.129) dynamics of clusters.
Zhao and Tian (40.153.041) method of pm memb in clusters.
Zhao et al. (40.002.081) pm; memb catalogue; 42 clusters.
Zhao and Jing (42.111.022) pm algorithm; cluster memb.
Zinnecker (42.153.004) IMF; young clusters.

REFERENCES TO TABLE

Anthony-Twarog, B.J. 1987, *Astron. J.* **93**, 641.
Brown, J.A. 1987, *Astrophys. J.* **317**, 701.
Cameron, A.C. and Reed, N. 1987, *Mon. Not. Roy. Astron. Soc.* **224**, 821.
Clayton, G.C. and Fitzpatrick, E.L. 1987, *Astron. J.* **93**, 157.
Fehrenbach, Ch., Burnage, R., Figuiere, J. Traversa, G. and Agnid, Cl. 1987, *Astron. and Astrophys. Suppl.* **68**, 515.
Claria and Lapasset: private communication.
Fenkart, P., Topaktas, L., Boydag, S. and Kandemus, G. 1987, *Astron. and Astrophys. Suppl.* **67**, 245.
FitzGerald and Harris: private communication.
Ianna, P.A., Adler, D.S. and Faudree, E.F. 1987, *Astron. J.* **93**, 347.
Jura, M. 1987, *Astrophys. J.* **313**, 743.
Kaluzny, J. and Shara, M.M. 1987, *Astrophys. J.* **314**, 585.
van Leeuwen, F., Alphenaar, P. and Meys, J.J.M., 1987, *Astron. and Astrophys. J.* **67**, 483.
McNamara, B.J. 1987, *Astrophys. J.* **312**, 778.
Nissen, P.E., Twarog, B.A. and Crawford, D.L. 1987, *Astron. J.* **93**, 634.
Pandey, A.J., Bhatt, B.C. and Mahra, H.S. 1987, *Astrophys. and Sp. Sci.* **129**, 293.
Peterson, C.J. and FitzGerald, M.P.: private communication.
Peterson, D.M. and Solensky, R. 1987, *Astrophys. J.* **315**, 286.
Rucinski, S.M. 1987, *Pub. Astron. Soc. Pacific* **99**, 487.
Schiller, S.J. and Milone, E.F. 1987, *Astron. J.* **93**, 1471.
Schneider, H. 1987, *Astron. and Astrophys. Suppl.* **67**, 545.
Shobbrook, R.R. 1987, *Mon. Not. Roy. Astron. Soc.* **225**, 999.
Smith, V.V. and Suntzeff, N.B., 1987, *Astron. J.* **93**, 359.
Sowell, J.R. 1987, *Astrophys. J. Suppl.* **64**, 241.
Stauffer, J.R., Schild, R.A., Baliunas, S.L. and Africano, J.L. 1987, *Pub. Astron. Soc. Pacific* **99**, 471.
Terlevich, E. 1987, *Mon. Not. Roy. Astron. Soc.* **224**, 193.
Walborn, N.R. 1987, *Astron. J.* **93**, 868.
Walker, A. R. and Laney, C.D. 1987, *Mon. Not. Roy. Astron. Soc.* **224**, 61.

6. Globular Star Cluster Research

INTRODUCTION

The present compilation has been prepared from the literature cited in *Astronomy and Astrophysics Abstracts*, **37-42** inclusive, representing the contributions from 1984 through 1986. What is presented below is a condensed version of the original report, which is Preprint No.765 of the Steward Observatory. As much as possible, only those papers which present or analyze observational data have been included here. Review articles, syntheses of data, theoretical and/or numerical modelling papers, and citations of abstracts have not been incorporated into this Report; otherwise, its length would have been doubled, easily. As has become a standard format, the literature citations are arranged under each relevant cluster. Papers which refer to data from more than ten clusters, or to a specific and complete set thereof (e.g. the eight X-ray globulars), may be found in the 'Catalogues' section, at the beginning of the tabular information; multiple listings occur under the relevant clusters' IAU designation for those papers in which ten or fewer clusters are analyzed. Finally, all of the known clusters are listed in order of their IAU 'C' designation, ignoring the resultant scrambling of NGC numbers. The purpose of the listing is two-fold: 1) to provide the community with an up-to-date reference, in correct IAU 'C' order, of the known clusters; and, 2) to see how many important clusters are being ignored in favor of the 'biggest and brightest'. The large telescope/CCD combinations should be put to use on these fainter, more sparse, and extremely important members of the cluster sub-system of our Galaxy.

Following the request to this writer by the Commission President three years ago, and repeated by the incumbent this year, the Report also includes the relevant citations for the globular star clusters in both Magellanic Clouds, and in external galaxies. The length of their tables is an accurate statement of the ease with which the CCD, coupled to a large telescope, permits reliable work at formerly heroic levels of apparent faintness.

GLOSSARY OF ABBREVIATIONS

1-D = one-dimension(al)	abund. = abundance(s)	AGB = asymptotic giant branch
ass = association(s)	astrom = astrometr-y,-ic	BHB = blue horizontal branch
bin. = binar-y,-ies	br. = bright, -er, -ness	c.c. = chemical composition
CCD = charge-coupled device	chem. = chemical	cl. = cluster
comp. = com-pare(d), -parison	csc = cosecant	cts. = counts
d = distance(s)	determ. = determin-e(d),-ation	discov. = dis-covery, -covered
disp. = dispersion	dist. = distance	distr,-ib. = distribut-ed,-ion
dwf. = dwarf	dyn. = dynamic,-al,-s	E = colour excess, reddening
electronogr. = electronographic	ellip. = elliptical	em = emission
est. = estimate,-ed	evol = evolution,-ary	ex = extinction
gal. = galaxy, galactic	Gyr. = 10^9 yrs.	HB = horizontal branch
IMF = initial mass function	I/S = interstellar	integr. = integrated
interp. = inter-pret(ed),-pretation	IR = infra-red	ist = interstellar
I.T. = image tube	IUE = International UV Explorer	kin. = kinemati-c,-cal,-cs
lf,LF = luminosity function	m.-p. = metal-poor	m.-r. = metal-rich
meas. = measure(d)	memb. = member, -ship	metal. = metall-ic , -icity
modif. = modi-fy, -fication	mol = molecular	ms = main sequence
obj. = object(s), objective	obs. = observ-ation(s), -ed	PAGB = post-AGB
par = parameter(s)	p.e. = photo-electric	pec = peculiar
pg = photographic	phot. = photomet-er,-ric,-ry	photog. = photogra-phy, -phic
pm, PM = proper motion	PN = planetary nebula(e)	pol = polarization
prelim. = preliminary	resol. = resolve(d), resolution	red. = reddening
RG(B) = red giant(s) (branch)	rot. = rotation,-al	rv,RV = radial velocity
SB = spectroscopic binary	SG = supergiants	SN = supernova
sp. = spectr-a,-al,-oscopic	spphot = spectrophotometry	st. = star(s), stellar
stf = star formation	str. = strength(s)	subgt. = subgiant
surf. = surface	temp. = temperature	theo. = theoretical
UV = ultraviolet	v. = very	var. = variable
vel. = velo-city,-cities		

CATALOGUES OF CLUSTER DATA

Bica, E. & D. Alloin (42.154.001) - integr. sp. of 63 cl.; grids of st. cl. properties.

Brodie, J.P. & D.A. Hanes (41.154.001) - abs. line indices from low-disp. sp. lead to [Fe/H]-values for 36 cl.

Burstein, D. *et al.* (38.154.081) - integr. sp. of 17 cl. in the Galaxy comp. to the same for 19 cl. in M31; sp. indices comp. to ellip. gal.

Caloi, V. *et al.* (38.154.030) - IUE sp. of cl. central regions.

Djorgovski, S. & I.R. King (41.154.066) - surf. phot. of 100+ cl.; 20% show signs of collapsed cores.

Ehjegenson, A.M. & O.S. Yatsyk (42.154.002) - generalized characteristics for 97 cl. from 'principal component method'.

Gratton, R.G. (39.154.027) - phot. data of MS st. in 26 cl.

Gratton, R.G. & S. Ortolani (40.154.045) - dist. and ages for 26 cl. with MS phot.

Grindlay, J.E. et al. (38.142.009) - *Einstein* obs. of X-ray cl.; tidal-capture hard binary X-ray sources have $0.9 \leq M/M_\odot \leq 1.9$.

Hamuy, M. (38.154.003) - integr., half-light, UBVRI p.e. photom. for 72 cl.

Hamuy, M. (41.154.009) - multi-aperture, integr., UBVRI phot. of 79 southern cl.; incorporates data in (38.154.003).

Hanes, D.A. (38.154.020) - integr. UBV p.e. photom. for 31 cl. with $\delta < +12°$.

Hanes, D.A. & J.P. Brodie (39.154.070) - multi-aperture UBVRI phot. for 71 cl.

Hesser, J.E. & S.J. Shawl (40.154.009) - I.T., integr., 1-D, sp. types for 90 cl.

Hesser, J.E. et al. (41.154.057) - RV investigation from I.T. sp. of 90 cl.

Kron, G.E. et al. (37.154.042) - partial surf. br. profiles for 69 cl. with use of electronic camera.

Kron, G.E. & K.C. Gordon (41.154.068) - br. distrib. from multi-aperture BVRI obs. of 21 cl.

Lugger, P.M. et al. (39.154.059) - CCD UBVR surf. phot. of 11 cl. sample to test for presence of central cusps in surf. br. profile.

Lee, S.-W. (41.154.062) - 16 well-obs. cl. with [Fe/H] used to determine age and Y.

Peterson, C.J. (41.154.022) - BV surf. phot., concentric apertures, of 101 cl.

Peterson, R.C. et al. (42.111.003) - high precision RV for RG st. in eight nearby and four remote cl.

Petrovskaya, I.V. & A.M. Ehjgenson (37.154.035) - age correlations of 51 cl. with metal., mean density.

Reed, B.C. (39.154.008) - integr. UBVRI phot. of 114 cl. reduced to van den Bergh UBV and Hamuy VRI systems.

Rose, J.A. (40.157.056) - integr. sp. of 10 cl. (and one open cl.: M 67) comp. to 12 ell. gal.

Shawl, S.J. & R.E. White (41.154.031) - accurate ($\pm 1''$) optical positions of 109 cl. centers.

Smith, H.A. (37.154.055) - ΔS-index correlated with RGB $(V - K)_0$; Zinn's Q_{39} metal. index recalibrated in terms of [Fe/H] via ΔS-index.

Webbink, R.F. (39.154.069) - obs. and derived structural parameters tabulated for 154 cl. in Galaxy, 6 cl. in Fornax, and 7 dwf. spheroidals.

White, R.E. & S.J. Shawl (43.154.xxx) - axial ratios and Galactic orientation for 105 cl.

van den Bergh, S. & C.L. Morbey (38.154.075) - cl. radii correlated more strongly with Galactocentric distance than are metal.

Zasov, A.V. et al. (40.154.008) - E(B-V) dependence upon gal. lat., b; csc-law.

Zinn, R. (39.154.072) - spatial distrib., kin.s, and metal. for 121 cl.

Zinn, R. & M.J. West (37.154.047) - RV and metal. meas. for 60 cl.; compilation of metal. for 121 cl.

CLUSTER-BY-CLUSTER LITERATURE CITATIONS

C0021-723 (NGC 104, 47 Tuc)

(39.154.025) - UBVRI p.e. seq. (24 st.).

(38.154.016) - computed synthetic sp. comp. to low-disp. sp. obs.

(39.154.056) - RV for cl. memb.; dynam. study.   (38.154.021) - low-disp. IUE sp. of UV-br. st.

(37.154.034) - photog. UV surf. phot.; search for UV-excess similar to that shown by NGC 7078 (M 15).

(39.114.021) - high-disp. sp. analysis of 13.5-mag. RG st.

(39.154.052) - chem. and kin. prop.; chem gradient.(39.154.074) - importance of st. RV in dyn. studies.

(41.154.043) - TiO absorption features in high-resol. sp. of RG st.

(42.154.018) - high-disp. sp. comp. to Arcturus for [Fe/H].

(37.154.012) - BV CMD to MS.   (38.154.012) - review of cl. parameters.

(39.154.058) - CCD BV phot. to obtain CMD and LF to $B_{lim} \approx 25$.

(40.154.056) - photog. phot. of 1200 st., $13.5 < V < 17.5$; BV CMD and LF.

(38.154.011) - RG st. in cl. sp. comp. to RG st. in NGC 6723.

(42.154.031) - central depths for CH- and CN-features and prominent abs. lines of Fe I, Ca I, Sr II, Ba II, and H for 22 RG st.

(37.154.040) - kin. studies of 169 st.; no sp. bin. detected.

(39.154.078) - RV of 169 RG st.; cl. rot. and vel. disp.

(39.154.060) - CORAVEL RV study; virial mass, M/L.

(41.154.010) - kin. and dyn. from 272 memb. st. RV.

(38.154.024) - Walraven (VBLUW) phot. of RG st.; [Fe/H]-values determined.

(37.154.032) - CN and CH anticorrelated on RGB.

(40.154.039) - Sabbatier technique to produce isodensity contours; ellipticity.

(40.154.014) - theo. isochrone fits to cl. with UBVRI and Strömgren uvby p.e. phot.

C0050-268 (NGC 288)

(37.154.063) - photom. to cl. MS.

(38.154.043) - cl. CMD shows gap in subgt. branch; interp., modif. to theory.

(38.154.016) - computed synthetic sp. comp. to low-disp. sp. obs.

(37.154.030) - BV CMD from 3000 st. images; $Y = 0.23 \pm 0.03$.

(38.154.002) - BV CMD from 2350 st. images; [Fe/H] ≈ -1.4.

(42.154.031) - central depths for CH- and CN-features and prominent abs. lines of Fe I, Ca I, Sr II, Ba II, and H for 1 RG st.

(37.154.056) - BV CMD to V= +21.

(37.154.044) - electronogr. BV CMD to V= +21; age determ.

(37.154.064) - electronogr. and CCD photom.

(40.116.004) - $< RV > = -43.2 \pm 1.5 kms^{-1}$ from 7 st.; line-broadening obs. give $v \sin i = 22 \pm 3 kms^{-1}$.

(37.154.012) - BV CMD to MS.

(38.154.024) - Walraven (VBLUW) phot. of RG st.; [Fe/H]-values determined.

C0100-711 (NGC 362)

(38.154.016) - computed synthetic sp. comp. to low-disp. sp. obs.

(40.154.051) - CO and CN band str. among RG st.

(42.154.024) - Washington phot.; temp., abund., memb.

(42.154.031) - central depths for CH- and CN-features and prominent abs. lines of Fe I, Ca I, Sr II, Ba II, and H for 2 RG st.

(38.154.024) - Walraven (VBLUW) phot. of RG st.; [Fe/H]-values determined.

(38.154.037) - CN variations among RG st.

C0234+209 (DEJC 1)

C0310-554 (NGC 1261)

(38.154.040) - UBVRI p.e. seq.

C0325+794 (Pal 1)

(40.154.046) - prelim. BV CMD.

C0345-718 (NGC 1466)

C0354-498 (AM 1, E 1)

(37.154.005) - BV CMD, est. [Fe/H]= -1.8.

(37.154.070, 38.154.018) - CCD-derived BV CMD to $V_{lim} = +23$.

(40.154.003) - sp. of two RG st.; [Fe/H] $= -1.7 \pm 0.2$, $RV = +116 kms^{-1}$.

C0422-213 (Erid 1)

(39.154.022) - CCD BV CMD; m-M= 19.95.

(40.154.015) - RV ($\pm 15 kms^{-1}$) to est. Galaxy's mass.

(41.154.007) - CCD BV CMD, st. cts., to $V_{lim} \approx 23.5$.

(42.154.042) - cl. age from CMD.

C0433+313 (Pal 2)

C0444-840 (NGC 1841)

C0512-400 (NGC 1851, 4U 0512-401)

(37.154.034) - photog. UV surf. phot.; search for UV-excess similar to that shown by NGC 7078 (M 15).

(41.142.033) - 6h obs. baseline of soft- and medium- energy X-ray light-curves.

(41.154.004) - BV and DDO p.e. phot.; radial color gradients.

C0522-245 (NGC 1904, M79)

(42.154.008) - CCD BV CMD to below MS turnoff. (37.154.012) - BV CMD to MS.

(41.154.053) - CCD BV CMD below MS turnoff.

(38.154.024) - Walraven (VBLUW) phot. of RG st.; [Fe/H]-values determined.

C0536-618 (E 2)

(42.154.037) - BV CMD; comp. with VandenBerg models gives age of 1.5 Gyr; same dist. as LMC; LMC cl.?

C0647-359 (NGC 2298)

(41.154.024) - CCD BVRI CMD.   (42.154.008) - CCD BV CMD to below MS turnoff.

C0734+390 (NGC 2419)

C0737-337 (AM 2)

C0911-646 (NGC 2808)

(41.154.029) - p.e. UBVRI seq. in field.   (37.154.039) - BV CMD to $V_{lim} = +21$.

(37.154.045) - BVRI Reticon profiles.   (42.154.008) - CCD BV CMD to below MS turnoff.

(41.154.004) - BV and DDO p.e. phot.; radial color gradients.

C0921-770 (E 3)
(37.154.072) - BV CMD.   (40.154.017) - CCD BV CMD.

C0923-545 (UKS 2)

C1003+003 (Pal 3)
(38.154.044) - CCD photom. to V about 23.2; BV CMD.
(42.154.042) - cl. age from CMD.

C1015-461 (NGC 3201)
(37.154.063) - photom. to cl. MS.   (38.154.040) - UBVRI p.e. seq.
(40.154.051) - CO and CN band str. among RG st. (37.154.064) - electronogr. and CCD photom.

C1052+407 (SAHB 1)

C1126+292 (Pal 4)
(41.154.026) - CCD BV CMD to $V_{lim} \approx 25$.   (41.154.019) - st. RV; $M_{Galaxy} = (5 \pm 2) \times 10^{11} M_\odot$.
(41.154.028) - CCD BV CMD to $V_{lim} \approx 24.8$.

C1207+188 (NGC 4147)
(40.154.012) - space motion from PM of 39 st.

C1223-724 (NGC 4372)
(41.154.029) - p.e. UBVRI seq. in field.

C1236-264 (NGC 4590, M68)
(38.154.040) - UBVRI p.e. seq.   (37.154.064) - electronogr. and CCD photom.

C1256-706 (NGC 4833)
42.154.035) - JHK phot. for 21 RG st.

C1310+184 (NGC 5024, M53)

C1313+179 (NGC 5053)

C1323-472 (NGC 5139, ω Cen)
(38.154.040) - UBVRI p.e. seq.
(38.154.043) - cl. CMD shows gap in subgt. branch; interp., modif. to theory.
(38.114.069) - IUE sp. of st. ROA 5701; T_{eff} and M_{bol}.
(40.114.151) - c.c. and abund. of RG st.
(41.154.065) - CNO abund. from sp. and narrow-band phot. of 72 st.
(40.114.128) - ROA 577, RV and PM memb. of cl., is CH st.
(42.154.025) - sp. of six candidate blue stragglers; RV; st. E39 is dwf. Cepheid (AI Vel) var.
(39.154.052) - chem. and kin. prop.; chem. gradient.
(39.154.074) - importance of st. RV in dyn. studies.(40.154.011) - SIT Vidicon sp. of subgt. st.
(39.154.032; Engl. transl.: 40.154.049) - scanner obs. of 2 HB st. (1403 and 1853).
(42.154.031) - central depths for CH- and CN-features and prominent abs. lines of Fe I, Ca I, Sr II, Ba
II, and H for 217 RG st.
(39.154.078) - RV of 298 RG st.; cl. rot. and vel. disp.
(39.154.060) - CORAVEL RV study; virial mass, M/L.
(41.154.010) - kin. and dyn. from 318 memb. st. RV.
(38.154.063) - large cl. mass deduced from virial theorem and tidal radius determination.
(41.154.004) - BV and DDO p.e. phot.; radial color gradients.
(37.154.014) - BV CMD to MS.
(40.154.039) - Sabbatier technique to produce isodensity contours; ellipticity.
(41.154.027) - BHB st. radial distrib.
(42.154.044) - sp. of RR Lyr var. to derive state of HB evol.; Sandage period-shift effect half of Sandage's
result.
(42.154.017) - EXOSAT obs. of 2 soft X-ray sources discov. by Einstein satellite.

C1339+286 (NGC 5272, M3)
(39.154.001) - low-disp. IUE sp of UV-br. st.
(42.154.046) - BV CMD from high precision phot. of 10000 st. with internal $\delta m \approx \pm 0.03$ and $V_{lim} \approx 22$.
(38.154.021) - low-disp. IUE sp. of UV-br. st.   (39.154.074) - importance of st. RV in dyn. studies.
(40.154.051) - CO and CN band str. among RG st.
(40.154.010; Engl.transl.: 41.154.049) - cl. rot. using Griffin/Gunn RV.
(39.154.080) - bi-modal CN distrib. interpreted by mixing scenario.
(40.154.054) - sp. obs. of 9 st. pairs (one st. N-rich, the other N-poor); comp. to NGC 104 (47 Tuc) and
NGC 6752 results.
(38.154.077) - CN variations of st. on lower RGB.
(39.154.062) - RV for 112 RG st.; no convincing evidence for SB systems.

(39.154.063) - deep Hβ narrow- and wideband images of cl. to search for cataclysmic bin.; none found brighter than $M_B = 6$.

(37.116.011) - discovery of intrinsic polarization in some RG st.

<u>C1343-511 (NGC 5286)</u>

(40.154.022) - BV CMD.

<u>C1354-269 (AM 4)</u>

<u>C1403+287 (NGC 5466)</u>

(37.154.029) - BV CMD for stars br. than B = +19(39.154.009) - photog. phot.; BV CMD.

(40.154.019) - BV CMD to fainter than MS turnoff.

(39.154.028) - galactic orbit highly inclined (80° ± 10°) to plane and v. eccentric ($e > 0.7$).

(41.154.064) - RV and vel. disp.

<u>C1427-057 (NGC 5634)</u>

<u>C1436-263 (NGC 5694)</u>

<u>C1452-820 (IC 4499)</u>

<u>C1500-328 (NGC 5824)</u>

(37.154.034) - photog. UV surf. phot.; search for UV-excess similar to that shown by NGC 7078 (M 15).

<u>C1513+000 (Pal 5)</u>

(40.154.025) - br. RG st. sp.; [Fe/H] = -1.4, CN strength varies.

(41.154.052) - CCD BV CMD.

<u>C1514-208 (NGC 5897)</u>

(40.154.002) - sp. of non-var. st. (SK 120) in RR Lyr gap.

<u>C1516+022 (NGC 5904, M5)</u>

(39.154.001) - low-disp. IUE sp of UV-br. st.　　　(37.154.051) - PM for var. and other unusual st.

(39.154.029) - UV-images of cl. at 1540Åand 2360Å.

(41.153.034) - $< \Delta Fe >$ defined and meas. for individual st.

(40.154.051) - CO and CN band str. among RG st. (38.154.021) - low-disp. IUE sp. of UV-br. st.

(42.154.018) - high-disp. sp. comp. to Arcturus for [Fe/H].

(39.154.032; Engl. transl.: 40.154.049) - scanner obs. of HB st. (I-6).

(39.154.080) - bi-modal CN distrib. interpreted by mixing scenario.

(39.154.089) - synth. sp. fit to low-resol. sp. of 3 CN-strong and 3 CN-weak RG st.; all have similar lum. and T_{eff}.

(40.154.054) - sp. obs. of nine st. pairs (one st. N-rich, the other N-poor); comp. to NGC 104 (47 Tuc) and NGC 6752 results.

<u>C1524-505 (NGC 5927)</u>

(38.154.052) - narrow-band p.e. phot. of TiO strength, leading to $[m/H] = -0.4 \pm 0.2$.

<u>C1531-504 (NGC 5946)</u>

<u>C1542-376 (NGC 5986)</u>

(38.113.030) - p.e. UBV seq. of 22 st. around cl.

<u>C1608+150 (Pal 14)</u>

(38.154.062) - st. cts., BV CMD.

(40.154.015) - RV ($\pm 15 km s^{-1}$) to est. Galaxy's mass.

(41.154.019) - st. RV; $M_{Galaxy} = (5 \pm 2) \times 10^{11} M_\odot$.

<u>C1614-228 (NGC 6093, M80)</u>

(37.154.034) - photog. UV surf. phot.; search for UV-excess similar to that shown by NGC 7078 (M 15).

<u>C1620-720 (NGC 6101)</u>

<u>C1620-264 (NGC 6121, M4)</u>

(37.154.063) - photom. to cl. MS.

(38.154.036) - BVRI p.e. and photog. photom.; BV CMD.

(39.154.004) - red., m-M, and age from RR Lyr properties.

(41.154.010) - search for UV-excess st. to identify wd component; UBV phot. of 6 objects consistent with hot wd.

(42.154.024) - Washington phot.; temp., abund., memb.

(42.154.018) - high-disp. sp. comp. to Arcturus for [Fe/H].

(37.154.038) - BHB st. [Fe/H] = -0.4.

(42.154.031) - central depths for CH- and CN-features and prominent abs. lines of Fe I, Ca I, Sr II, Ba II, and H for 2 RG st.

(40.154.054) - sp. obs. of 9 st. pairs (1 st. N-rich, the other N-poor); comp. to NGC 104 (47 Tuc) and NGC 6752 results.

(39.116.021) - rot. of HB st.; $< RV >= 72.3 \pm 0.9 kms^{-1}$ (9 st.); no rot. larger than $v \sin i = 15 kms^{-1}$ for any of the 9 st.

(41.154.064) - RV and vel. disp.   (41.154.050) - JHK phot. of RG and HB st.

(37.154.031) - UBV CMD from CCD photom.   (37.154.062) - CCD-derived CMD; isochrones.

(40.154.014) - theo. isochrone fits to cl. with UBVRI and Strömgren p.e. phot.

C1624-259 (NGC 6144)

C1624-387 (NGC 6139)

(38.154.038) - p.e. BV seq.

C1625-325 (Ter 3)

C1629-129 (NGC 6171, M107)

(38.154.008) - BV CMD to MS turnoff $\Rightarrow [Fe/H] = -0.9$.

(42.154.031) - central depths for CH- and CN-features and prominent abs. lines of Fe I, Ca I, Sr II, Ba II, and H for 2 RG st.

(38.154.010) - BV CMD to $V_{lim} = 21$ and $B_{lim} = 21.5$.

(42.154.023) - DDO phot. of RG st.

C1639+365 (NGC 6205, M13)

(39.154.001) - low-disp. IUE sp of UV-br. st.

(40.154.051) - CO and CN band str. among RG st. (39.154.055) - dyn. modelling from PM of indiv. st.

(38.154.021) - low-disp. IUE sp. of UV-br. st.   (41.154.005) - p.e. BVRI seq. (20 st.) in field.

(39.154.032; Engl. transl.: 40.154.049) - scanner obs. of 2 HB st. (II-25, AJ-63) show cl. He-abund. is nearly normal.

(39.154.080) - bi-modal CN distrib. interpreted by mixing scenario.

(41.154.054) - high-res. CCD sp. of 4 st.; O-, Sc-, Fe-, and La-abund.

(39.154.051) - RV and PM techniques for indiv. st. (41.154.032) - UBV CMD, mass function.

(38.154.017) - PM used to determine cl. mass.

(40.154.054) - sp. obs. of 9 st. pairs (1 st. N-rich, the other N-poor); comp. to NGC 104 (47 Tuc) and NGC 6752 results.

(41.154.037) - CCD UBV CMD fainter than V = 25.

(37.116.011) - discovery of intrinsic polarization in some RG st.

C1644-018 (NGC 6218, M12)

(38.154.069) - p.e. photom. of faint st.

C1645+476 (NGC 6229)

(40.154.019) - BV CMD to fainter than MS turnoff.

C1650-220 (NGC 6235)

C1654-040 (NGC 6254, M10)

(39.154.001) - low-disp. IUE sp of UV-br. st.   (38.154.021) - low-disp. IUE sp. of UV-br. st.

(38.154.069) - p.e. photom. of faint st.   (37.154.071) - BV CMD.

C1656-370 (NGC 6256)

C1657-004 (Pal 15)

(38.154.062) - st. cts., BV CMD.   (41.154.019) - st. RV; $M_{Galaxy} = (5 \pm 2) \times 10^{11} M_{\odot}$.

C1658-300 (NGC 6266, M62)

(39.154.017) - IUE obs. of cental region; brightest cl. in UV so far.

C1701-246 (NGC 6284)

C1702-226 (NGC 6287)

C1707-265 (NGC 6293)

C1711-294 (NGC 6304)

C1713-280 (NGC 6316)

C1714-237 (NGC 6325)

C1715+432 (NGC 6341, M92)

(39.154.001) - low-disp. IUE sp of UV-br. st.   (39.154.009) - photog. phot.; BV CMD.

(40.154.051) - CO and CN band str. among RG st. (42.154.013) - CCD BV CMD.

(39.154.032; Engl. transl.: 40.154.049) - scanner obs. of 2 HB st. (II-40 and II-27).

(41.154.059) - subgt. st. C-abund.   (39.154.051) - RV and PM techniques for indiv. st.

(38.154.017) - PM used to determine cl. mass.

(40.154.054) - sp. obs. of 9 st. pairs (1 st. N-rich, the other N-poor); comp. to NGC 104 (47 Tuc) and NGC 6752.

 C1716-184 (NGC 6333, M9)

 C1718-195 (NGC 6342)

 C1720-177 (NGC 6356)

C1720-263 (NGC 6355)
(38.154.038) - p.e. BV seq.
C1721-484 (NGC 6352)
(38.154.082) - [Fe/H]-value from Washington phot. and echelle sp.
(42.154.024) - Washington phot.; temp., abund., memb.
(38.154.052) - narrow-band p.e. phot. of TiO strength, leading to $[m/H] = -0.5 \pm 0.2$.
C1724-307 (Ter 2, 4U 1722-30)
(40.142.093, 41.142.112) - EXOSAT obs. over 11h baseline; sp., lum. consistent with Einstein obs.; quasi-periodic oscillations energy range 2-8 kev.
C1725-050 (NGC 6366)
(40.154.004) - structure.
C1726-670 (NGC 6362)
(41.154.030) - CCD BVRI CMD.
C1727-315 (Ter 4)
C1727-299 (HP 1)
C1729-338 (Grindlay 1)
C1730-333 (Liller 1)
C1731-390 (NGC 6380)
C1732-304 (Ter 1)
C1732-385 (Pismis 26)
C1732-447 (NGC 6388)
(37.154.045) - BVRI Reticon profiles.
(41.154.004) - BV and DDO p.e. phot.; radial color gradients.
C1733-390 (Ton 2)
C1735-032 (NGC 6402, M14)
C1735-238 (NGC 6401)
C1736-536 (NGC 6397)
(37.154.063) - photom. to cl. MS.   (41.154.029) - p.e. UBVRI seq. in field.
(38.154.021) - low-disp. IUE sp. of UV-br. st.   (42.154.064) - sp. of 4 BHB st.
(42.114.056) - sdO st. ROB 162 has $T(3)_{eff} = 51 \pm 2°K$; resembles m.-p. central st. of PN; is a PAGB st.
(37.154.038) - BHB st. [Fe/H] = -1.4.
C1740-262 (Pal 6)
C1742+031 (NGC 6426)
C1745-247 (Ter 5)
C1746-203 (NGC 6440)
(37.154.034) - photog. UV surf. phot.; search for UV-excess similar to that shown by NGC 7078 (M 15).
C1746-370 (NGC 6441)
(37.154.034) - photog. UV surf. phot.; search for UV-excess similar to that shown by NGC 7078 (M 15).
(41.154.004) - BV and DDO p.e. phot.; radial color gradients.
C1747-312 (Ter 6)
C1748-346 (NGC 6453)
(38.154.038) - p.e. BV seq.
C1751-241 (UKS 1)
C1755-442 (NGC 6496)
(38.154.038) - p.e. BV seq.
C1758-268 (Ter 9)
C1759-089 (NGC 6517)
C1800-260 (Ter 10)
C1800-300 (NGC 6522)
(40.154.048) - BV CMD.
C1801-003 (NGC 6535)
C1801-300 (NGC 6528)
C1802-075 (NGC 6539)
C1804-437 (NGC 6541)
(41.154.004) - BV and DDO p.e. phot.; radial color gradients.
C1804-250 (NGC 6544)
C1806-259 (NGC 6553)
C1807-317 (NGC 6558)
(38.154.038) - p.e. BV seq.

C1808-072 (IC 1276)
C1809-227 (Ter 12)
C1810-318 (NGC 6569)
(38.154.038) - p.e. BV seq.
 C1812-121 (Kodaira 1)
 C1814-522 (NGC 6584)
(40.154.013) - p.e. UBVRI seq. in field.
 C1820-303 (NGC 6624, 4U/MXB 1820-30)
(40.142.067) - HEAO A-2 database searched for abs. events: 1 found in this cl.
(37.154.034) - photog. UV surf. phot.; search for UV-excess similar to that shown by NGC 7078 (M 15).
(41.142.035) - analysis of 6 X-ray bursts from cl.
 C1821-249 (NGC 6626, M28)
(40.141.011) - X-ray source obs. at 30.9 and 50.5 MHz has sp. index -2.44; source is fast pulsar?
(40.154.033) - met. abund. and occurrence of Cepheids.
 C1827-255 (NGC 6638)
 C1828-235 (NGC 6642)
(38.154.038) - p.e. BV seq.
 C1828-323 (NGC 6637, M69)
(38.154.082) - [Fe/H]-value from Washington phot. and echelle sp.
 C1832-330 (NGC 6652)
(38.154.038) - p.e. BV seq.
 C1833-239 (NGC 6656, M22)
(37.154.063) - photom. to cl. MS.
(42.154.019) - PM, photog. phot. for 672 st.; BV CMD.
(41.133.003) - IRAS 18333-2357: cooler than all OH/IR st., hotter than most gal.
(39.114.040) - est. sp. type of st. II-81 is O6; $RV = -140 km s^{-1}$.
(42.154.031) - central depths for CH- and CN-features and prominent abs. lines of Fe I, Ca I, Sr II, Ba
II, and H for 23 RG st.
(41.154.064) - RV and vel. disp.
 C1838-198 (Pal 8)
 C1840-323 (NGC 6681, M70)
(39.154.001) - low-disp. IUE sp of UV-br. st.   (38.154.023) - low-disp. IUE sp. of cl. nucleus.
(37.154.034) - photog. UV surf. phot.; search for UV-excess similar to that shown by NGC 7078 (M 15).
 C1846-652 (NGC 6684)
 C1850-087 (NGC 6712)
(39.154.023) - IR phot. for 15 RG st., incl. 5 var.
 C1851-305 (NGC 6715, M54)
(38.154.023) - low-disp. IUE sp. of cl. nucleus.
 C1852-227 (NGC 6717)
(38.154.038) - p.e. BV seq.
 C1856-367 (NGC 6723)
(42.154.024) - Washington phot.; temp., abund., memb.
(38.154.011) - RG st. sp. comp. to NGC 104 (47 Tuc) RG st.
(42.154.031) - central depths for CH- and CN-features and prominent abs. lines of Fe I, Ca I, Sr II, Ba
II, and H for 2 RG st.
(41.154.004) - BV and DDO p.e. phot.; radial color gradients.
(42.154.023) - DDO phot. of RG st.
 C1902+017 (NGC 6749)
 C1906-600 (NGC 6752)
(39.154.001) - low-disp. IUE sp. of UV-br. st.
(38.154.043) - cl. CMD shows gap in subgt. branch; interp., modif. to theory.
(38.154.034) - SIT Vidicon sp. of 26 st. over $\Delta m = 5$: from tip of RGB to MS turnoff.
(37.036.143) - CCD BV CMD.   (41.154.006) - CCD BV CMD to MS.
(42.154.015) - photog. image reconstruction techniques of 5000+ st.; BV CMD.
(37.154.054) - optical and UV spectr. of BHB st.   (41.154.011) - more optical and UV spectr. of BHB s
(42.154.029) - two HB st. with strong He-lines (λ4026).
(42.154.018) - high-disp. sp. comp. to Arcturus for [Fe/H].
(37.154.066) - BV CMD.   (42.114.001) - 9 HB st., $9.3 \leq T(3)_{eff} \leq 40.0°K$.
(38.154.024) - Walraven (VBLUW) phot. of RG st.; [Fe/H]-values determined.

(41.154.063) - CCD BV CMD.

(39.154.064) - are cl. properties due to SN enrichment?

C1908+009 (NGC 6760)

C1914+300 (NGC 6779, M56)

(41.154.034) - spatial structure as function of limiting mag. of st. cts.

C1914-347 (Ter 7)

C1916+184 (Pal 10)

C1925-304 (Arp 2)

C1936-310 (NGC 6809, M55)

(42.154.024) - Washington phot.; temp., abund., memb.

(37.154.006) - automated star cts.   (37.154.064) - electronogr. and CCD photom.

(38.154.078) - c.c. from echelle sp. of five RG st.

C1938-341 (Ter 8)

C1942-081 (Pal 11)

C1951+186 (NGC 6838, M71)

(41.153.034) - $< \Delta Fe >$ defined and meas. for individual st.

(40.154.036) - IR obs. of CO-band of 21 RG st.; comp. to 2 open cl. [NGC 2158 and 2682 (M 67)].

(39.154.003) - photom., PM for 350+ st. with $V < 16$; BV CMD.

(38.154.082) - [Fe/H]-value from Washington phot. and echelle sp.

(42.154.024) - Washington phot.; temp., abund., memb.

(42.154.018) - high-disp. sp. comp. to Arcturus for [Fe/H].

(37.131.148, 38.131.001) - I/S red. in direction of cl.; Vilnius p.e. system.

(41.154.064) - RV and vel. disp.

C2003-220 (NGC 6864, M75)

C2031+072 (NGC 6934)

(41.154.055) - sp. of RG st. interp. by mixing hypoth.

C2033-048 (NGC 6941)

C2050-127 (NGC 6981, M72)

C2059+160 (NGC 7006)

(40.154.019) - BV CMD to fainter than MS turnoff.

C2127+119 (NGC 7078, M15)

(37.134.018) - IUE obs. of PN, K648; central st. T_{eff}; mass loss est. from C IV abs. feature.

(39.154.001) - low-disp. IUE sp of UV-br. st.   (39.154.054) - high spatial resol. of cl. core.

(38.142.021) - optical candidate (Auriere-Cordoni Catalog No.211) for cl. X-ray xource.

(40.154.007) - photog. phot. of BHB using GALAXY; 300 st.

(38.154.016) - computed synthetic sp. comp. to low-disp. sp. obs.

(39.154.009) - photog. phot.; BV CMD.

(38.114.069) - IUE sp. of st. ROA 5701; T_{eff} and M_{bol}.

(38.154.022) - self-consistent age determination from a variety of techniques.

(40.154.051) - CO and CN band str. among RG st. (38.154.021) - low-disp. IUE sp. of UV-br. st.

(39.154.073) - CCD UBV CMD; red., m-M, LF determined.

(39.154.032; Engl. transl.: 40.154.049) - scanner obs. of HB st. (I-70).

(38.154.024) - Walraven (VBLUW) phot. of RG st.; [Fe/H]-values determined.

(38.154.017) - PM used to determine cl. mass.   (40.154.052) - C and N abund. among RG st.

(37.154.062) - CCD-derived CMD; isochrones.

(40.154.014) - theo. isochrone fits to cl. with UBVRI and Strömgren p.e. phot.

C2130-010 (NGC 7089, M2)

(39.154.001) - low-disp. IUE sp. of UV-br. st.   (37.154.051) - PM for var. and other unusual st.

(38.154.016) - computed synthetic sp. comp. to low-disp. sp. obs.

(41.154.051) - RV of 69 st.; internal cl. dynamics.

C2137-234 (NGC 7099, M30)

(37.154.063) - photom. to cl. MS.   (39.154.001) - low-disp. IUE sp of UV-br. st.

(39.154.054) - high spatial resol. of cl. core.   (38.154.023) - low-disp. IUE sp. of cl. nucleus.

(38.154.058) - BV CMD.

(37.154.034) - photog. UV surf. phot.; search for UV-excess similar to that shown by NGC 7078 (M 15).

(41.154.004) - BV and DDO p.e. phot.; radial color gradients.

C2143-214 (Pal 12)

(38.154.016) - computed synthetic sp. comp. to low-disp. sp. obs.

C2304+124 (Pal 13)

(39.154.005) - BV CMD to upper MS; 122 st.
<u>C2305-159 (NGC 7492)</u>
<u>C2346-732 (AM 3)</u>

CLUSTER SYSTEMS

<u>LMC clusters</u> (see also 'Combined Studies' below)
(37.156.013, 42.154.053) - BVRI phot. of cl. in Bok Region: **NGC 1847, 1850, 1854,** and **1856** are most conspicuous.
(37.154.009, 38.154.055) - electronogr. photom. BV CMD of **Hodge 11** (= **SL 868**).
(41.156.004) - electronogr. photom. BV CMD of **NGC 2210**.
(37.036.143) - CCD BV CMD of **Hodge 11** (= **SL 868**).
(37.156.097, 38.154.068) - IUE integr. sp. of 17 cl.
(40.154.021) - CCD BV CMD for **NGC 2213** to MS turnoff; age.
(40.156.027) - LF for cl. br. than $M_B = -4.5$.   (40.156.042) - integr. UBV colors; empirical age calib
(39.154.011) - lum. profiles for **NGC 1835, 2210,** and **2257**; M/L ratios.
(37.154.058, 37.156.083) - age estimates (all < 1 Gyr) using time-dep't of model RG st. interior lum. for **NGC 1783, 1868, 1978, 2121, 2209,** and **2231**.
(37.156.007) - prelim. results of SIT surf. phot. of LMC cl.: UBV and RGU phot. for **NGC 1806** (red cl.) and **NGC 1818** (blue cl.)
(37.154.013) - SIT Vidicon BV photom. and CMD for **NGC 1466, 2203, 2210,** and **2257**.
(37.156.012) - CCD and SIT BV CMD for **NGC 1466, 2203, 2210,** and **2257; H11; LW 4**.
(38.154.079) - BV CMD of **NGC 1831,** to V = 21.5.
(37.154.017) - br., bl., rich cl. **NGC 1856** age est. from MS-turnoff to be $80(\pm 30)Myr$.
(37.154.029) - BV CMD of **NGC 2133** and **2134**. (41.156.038) - st. cts. for tidal radii and masses.
(37.156.082) - sp. obs. of 16 O-rich AGB st.   (41.156.021) - CCD UBVR CMD of **Hodge 4**.
(41.154.036) - phot. to MS turnoff, age estimate of **NGC 1651**.
(42.154.028) - age of **NGC 2173** est. to be $1.8 \pm 0.7Gyr$.
(37.154.010) - CCD BV CMD to V = 23.5 of **Hodge 11** (= **SL 868**).
(38.154.061) - CCD BV CMD of **NGC 2210,** to V = 21.5.
<u>SMC clusters</u> (see also 'Combined Studies' below)
(38.154.043) - **Kron 3** CMD shows gap in subgt. branch; interp., modif. to theory.
(41.156.013) - C st. candidates in **NGC 419**.   (40.156.002) - phot., sp. of **NGC 330** st.
(41.154.048) - BV CMD below MS turnoff and age est. for **NGC 411**.
(42.156.007) - finder charts and sp. class for br. st. in 15 cl.
(37.156.095, 39.154.024) - ellipticities of 24 cl. using BV isodensity contours.
(41.156.011, 41.156.037) - Schmidt obj. prism sp. class. for st. br. than B about 18.5.
(41.156.039) - st. cts. in 25 disk, 3 intermed., and 18 halo cl.
(41.156.040) - sp. class of br. st. in **NGC 1806, 1818, 1831, 2098,** and **2157**.
(42.156.007) - finder charts and sp. class for br. st. in 17 cl.
(37.156.093) - BV CMD of 20 cl.; Schmidt photog. phot.
(37.156.093) - LF for 10 old cl.
(37.154.059) - LF of **NGC 152** and of **Kron 3** in B and V.
(42.156.006) - st. cts. in 24 cl.; masses, concentration parameter.
(42.156.026) - CCD BV CMD of cl. **ESO121-SCO3**; aperture phot.
(37.154.052) - depending on SMC dist. mod., age of cl. **Lindsay 113** is 4 or 5 Gyr; est. [m/H]= -1.4.
(37.154.011, 38.154.076) - BV CMD to MS of **Kron 3**.
(42.114.054) - supergt. A7 st. has metal. < field gt. st. in SMC.
(40.154.032) - MS isochrone fits to BV CMD for **NGC 121**.
<u>Combined Studies of LMC and SMC</u>
(40.156.039, 41.156.007) -integr. G-band and Hb phot. for 41 LMC and 10 SMC cl.
(37.156.017) - separating field st. contrib. from IUE sp. of LMC cl. **NGC 1786** and SMC cl. **NGC 121** and **330**.
(37.156.015) - IUE results summarized.
(39.156.025) - linear correlation for ellip. vs. mass of disk cl.
<u>Fornax Group</u> (of galaxies)
(42.157.095) - MMT-plus-echelle RV for cl. 3, 4, 5; systemic $RV = 55 \pm 5 km s^{-1}$.
(42.157.156) - cl. population found around five ell. gal. (NGC 1374, 1379, 1387, 1399, and1404).
<u>Sculptor Group</u> (of galaxies)

(40.157.035) - two cl. candidates in A0142-43.

NGC 147

(38.154.072) - integr. IR phot. of cl.

NGC 185

(38.154.072) - integr. IR phot. of cl.

NGC 205

(38.154.072) - integr. IR phot. of cl.

NGC 224 (M 31)

(39.157.067) - slitless sp.: 109 new cl. candidates.

(39.157.057) - cl. used as probes of I/S medium in gal.; obtain $E_{U-B}/E_{B-V} = 1.01 \pm 0.11$; confirm NW side is near side.

(38.154.053) - flattening and lum. profile of Mayall II.

(37.154.036; Engl. transl.: 38.154.045) - cl. show metal. gradient radially outward from center of gal.

(40.154.005; Engl. transl.: 41.154.025) - integr. p.e. UBV obs. of 53 cl.

(38.154.015) - p.e. UBV phot. of 21 cl.   (38.154.072) - integr. IR phot. of cl.

(37.154.037; Engl. transl.: 38.154.019) - ellipticities and maj. axis orientations of 5 cl.

(40.154.055; Engl. transl.: 42.154.007) - ellipticities and orientation of 30 cl.

(39.157.088) - statistically significant excess of br. images in lum. range for cl.

NGC 253

(41.157.006) - search for cl. candidates: of 25, 3 are bl ., 4 are redder than Gal. cl.; est. $N_{tot} = 80 \pm 42$ cl.

NGC 524

(39.157.119) - st. cts.; cl. system more populous and more extended than in NGC 1052.

NGC 598 (M 33)

(37.157.226) - 4 true cl., all more m.-p. than 47 Tuc (NGC 104).

NGC 1052

(39.157.119) - st. cts.; cl. system less populous and less extended than in NGC 524.

NGC 1705

(40.157.038) - super st. cl.: integr. $M_B = -15.4$, <Sp. Type> = B3 V.

NGC 2403

(37.154.001) - find 19 obj. br. than V = 20 which are cl. candidates.

NGC 2683

(40.157.132) - st. cts.; cl. system as populous as Galaxy's; from $N_{cl} = 100 \pm 31$ cl. obs. , est. $N_{tot} = 321 \pm 108$ cl.

NGC 3109

(40.157.035) - 6 cl. candidates: one positive.

NGC 3115

(41.157.190) - st. cts. reveal spatial distrib. of cl. population.

NGC 3379 (M 105)

(40.157.062) - $N_{tot} = 140 \pm 35$ cl.; peak lum. of cl. distrib. may be 'standard candle'.

NGC 3557

(39.157.059) - cl. population is sparse.

NGC 4486 (M 87)

(41.158.280) - CCD obs. of gal. core; distrib. of cl. br. than B = 24.4; no systematic radial trend in cl. LF.

(41.154.002) - [Fe/H]-values for 5 cl.   (41.158.195) - 124 cl., B < 23.6, from CCD obs.

(39.158.082) - LF for cl. to B = 25.4; est. $N_{tot} = 2 \times 10^4$ cl.

NGC 4594 ('The Sombrero')

(37.154.018) - st. cts. reveal 1200(±100) cl. to $V_{lim} \approx 23.1$ and 515(±100) cl. to U_{lim} about 23.

NGC 5128 (Centaurus A)

(37.158.105) - IR obs. of 12 cl.; 5 are more m.-r. than Sun, and quite lum.

(38.158.270) - est. $1200 \leq N_{cl} \leq 1900$ from UVR st. cts.

(38.158.271) - dist. of cl. is typical for ell. gal.   (37.158.041) - sp. confirmation of 20 cl.

7 .Dynamical theory

Over the last three years the dynamical theory of star clusters and associations has been dominated by research into the evolution of globular clusters, especially the role played by tidal binaries in the phase after collapse of the core. It is some measure of the progress made that these topics occupy only a small part of the volume of proceedings of IAU Symposium 113 (39.012.070), which was devoted to the dynamics of star clusters. And yet important progress has been made in other areas too. The modelling of open clusters is now almost as realistic as it can be, and at last a respectable amount of attention is being paid to the dynamics of very young clusters. The theory of the stability of stellar systems is another area which has produced interesting new results in the last three years, though it still remains to integrate this field properly with the theory of dynamical evolution.

We begin with general developments of theory, i.e. those with no specific application in view. In view of its fundamental importance, it is surprising how much discussion there has been about the basic timescale for the dynamical evolution of collisional systems (37.151.044, 39.151.098, 39.151.121, 41.151.043, 42.151.040). New formulations have been proposed for a kinetic equation to describe collisonal relaxation (39.151.114, 90.88554), and for the virial theorem (40.151.059; see also 38.151.058 and 40.151.069 for applications of the standard virial theorem). New studies of the role played by violent relaxation (38.151.002, 39.151.143, 40.151.001, 41.151.065) have been supplemented by work on entropy and its analogues (41.151.027, 90.88546). Another interesting paper describes a modification of the Holtsmark distribution at small separations (42.042.002). Finally, it is very useful to have a new general introduction to a wide range of problems in stellar dynamics (40.003.011).

As was mentioned above, developments in the theory of stability have been a feature of the review period. One publication, however, goes back much further; this is an English translation of Antonov's classic 1962 paper on the thermodynamic stability of isothermal clusters (39.151.133; see also 40.151.022 on this topic). Another important but newer translation is that of the very comprehensive study by Fridman and Polyachenko (38.003.008, 009). Some of the new research has sought to clarify the role of anisotropy in generating dynamical instabilities (39.151.111, 40.151.086, 41.151.002, 90.66683), but there have been interesting developments also in the study of isotropic systems (38.151.045, 39.151.070, 39.151.095, 40.151.074, 42.151.051,079), multi-component systems (39.151.019, 42.151.025), and those exhibiting rotation (90.101933).

One of the traditional tasks of theory is the construction of models for the interpretation of observations. In view of the long history of this topic, the development which it has undergone in the last three years is remarkable. Among the simplest models are those with a phase-space cutoff (37.154.041, 90.72556), as in King's models, of which a useful set of tables has been published (41.151.097). Recent work has added greatly to our knowledge of analytic anisotropic models (37.151.042, 39.151.096, 40.151.008, 41.151.112, 90.55741), and new models have been described for multi-component systems (39.151.115) and for slowly rotating ones (39.151.118, 42.151.034, 90.101865). Developments in the solution of Poisson's equation in three dimensions may be useful for this purpose (41.151.051). The possibility of contructing models with discontinuous distribution functions has also been raised (39.151.142, 40.151.039).

Let us turn now to the dynamical evolution of stellar systems, and especially to young systems such as newly formed open clusters, which have not yet reached dynamical equilibrium. Dynamical aspects of their formation have been discussed several times recently (38.151.048, 39.151.130, 41.131.082, 41.154.021). If the loss of gas pressure on star formation leads to a collapse of the protocluster, this collapse will be unstable (37.151.052, 41.151.069, 42.151.059). Models have been constructed for non-equilibrium clusters (40.151.046, 40.151.106, 41.151.079), and other studies on such systems have concentrated on mechanisms of relaxation (40.151.045,.107) and stellar escape (40.153.011, 41.153.024). Other studies relevant to the dynamical theory of young clusters which have also appeared in the review period are 39.118.027, 40.151.009 and 42.153.003. The first two deal with dynamically isolated subgroups of stars, and the third is a review, including associations.

Open clusters have continued to attract theoretical attention, and reviews of different aspects of their dynamics can be found in the proceedings of IAU Symposium No.113 (39.151.129, 39.153.045), along with a translation of one of Ambartsumian's classic papers on the subject (39.151.132). A considerable amount of work has been devoted to their interaction with their environment, with regard to tidal effects (40.151.044,070, 41.151.067, 42.151.084) and also capture of field stars (41.153.012). Other studies have dealt with the motion of open clusters, relative to each other (41.153.019) and within the Galaxy (41.153.028).

The computer simulation of open clusters has now reached a remarkable level of realism (90.55748), and somewhat more idealised simulations have clarified our understanding of the role of hard binaries (39.151.039,116, 40.151.094), relaxation (39.151.113), and the measurement of core parameters (40.151.-077, 42.151.052). In a sense, then, the simulation of open clusters is now a manageable problem, essentially because open clusters are so small. But attempts have been made to interpret results from small simulations in the context of globular clusters (41.151.007, 42.151.107,.108). As yet, however, the only simulations which can really handle such large systems are fluid ones (the validity of which is studied in

39.151.001) or else hybrids which deal with the innermost parts by means of direct N-body simulation (38.042.083, 38.151.034,.035, 39.151.109). The present state of the art has been reviewed in detail recently (39.151.108, 42.151.078), but there are many promising ideas for extending the size of simulated N-body systems to that of a small globular cluster (39.042.026,.088, 39.151.110, and the book *The Use of Supercomputers in Stellar Dynamics*, eds. P. Hut & S. McMillan, Springer-Verlag, 1986). One of the problems is the handling of small subsystems, and a convenient treatment of these is now available (40.042.013).

Most of the work on globular clusters during the review period has made use of simplified models for their dynamical evolution, such as the Fokker-Planck equation (39.151.102), and not of N-body methods. Introductory articles will be found in 38.151.013, 39.154.046 and 41.154.035, and a simple article on stellar orbits in globular clusters is given in 38.151.046. The early evolution is dominated by core collapse (37.151.098, 39.151.100), and it is very useful now to have some results on the collapse of systems started as lowered Maxwellian models (40.151.093), since these are popular models for the interpretation of globular clusters. To say that core collapse is now understood obscures the fact that this is true only for idealised models. For example we know little of the effect of rotation (38.151.050, 42.151.021). On the other hand good progress has been made in understanding the effects of mass-loss by stellar evolution (41.151.029), close encounters (39.151.103), anisotropy (39.151.112, 41.151.072,.119), and, most notably, a mass spectrum (37.151.057,.103, 38.151.091,.094, 39.151.038,.077,.104, 40.151.010, 41.151.108, 42.151.114). Consult these papers if you want to construct a multi-mass model and are tempted to assume equipartition.

During the last three years there has been a steady increase in the attention paid to what happens after core collapse, since some clusters should be old enough dynamically (37.154.033). Several methods have been used for these studies, including Monte Carlo simulation (39.151.124, 41.154.061), fluid models (37.151.009,.085, 38.151.095, 39.151.101), a hybrid model (39.151.119, 42.151.028) and even a simplified homological model (39.151.168). One of the surprises of these studies was the discovery of gravothermal oscillations after core collapse; the nature of these is still under investigation (37.151.074, 39.151.105,.106, 39.154.087, 90.101859).

In these studies of post-collapse evolution the role played by binary stars is crucial, and there has been an explosion of interest in their dynamics. We now have predictions for their rate of formation both by three-body interactions (39.151.107) and by tidal two-body encounters, using reasonably realistic stellar models (37.154.016, 39.151.127, 40.151.057,.073, 41.151.109, 42.117.227, 42.151.038,.101). Having formed, the binaries evolve by interactions with each other (39.151.120, 42.042.043, 42.151.113) or with single stars (38.154.025,.026, 39.117.310,.311, 42.151.103). For interactions with tidally formed binaries it is important to take into account their finite sizes (38.117.205, 40.151.088, 42.154.005). For this reason, and because of the estimated rate of collisions between single stars in globular clusters, attention has been turned to the description and numerical simulation of this phenomenon (39.151.041,.042, 41.154.040, 42.151.068, 42.154.004). The importance of all these processes for the dynamics of the cluster itself is implicit in most of the papers referred to above on post-collapse evolution, and is also specifically addressed in 37.154.004, 39.151.123,.091 and 41.151.039.

In the last three years relatively little work has been done on the once fashionable area of cluster dynamics in the presence of a black hole (37.154.003, 39.067.103, 39.151.125,.126, 90.115529), though small black holes may be a significant constituent of some clusters (38.154.027, 39.154.067).

Finally we turn to the theory of the system of Galactic globular clusters. A little new work has been done on their orbits (38.154.013, 39.154.066, 41.151.048), but much more on their stability against tidal disruption. Some of this concerns details of the effects of tides on an individual cluster (37.154.008, 39.151.122,.128), but other investigations have concentrated on the dependence of tidal effects on galactocentric distance (37.154.007, 38.151.025, 38.154.044), or their influence on cluster lifetimes (39.154.033) and internal evolution (42.151.080).

Very little work has been done on the dynamical theory of cluster systems in other galaxies, except for the interesting phenomenon of cluster exchange between members of a cluster of galaxies (38.160.046, 41.151.028, 42.151.003, 90.101943). This has been an area of intense activity recently on the observational side, and it is to be hoped that this will be matched by an appropriate effort on the theoretical side in the coming years. Another probable growth area may well be improvements in the modelling of tidal interactions and collisions and their influence on the internal evolution of individual globular clusters. But it is to be hoped that more will be done to provide more realistic models for globular clusters. We still use models developed in the 1960s, and much progress has been made in the understanding of the dynamical evolution of clusters since then. These developments should be incorporated in the models we use now.

COMMISSION 38. <u>EXCHANGE OF ASTRONOMERS</u> (ECHANGE DES ASTRONOMES)

(Committee of the Executive Committee)

President: Edith A. Müller
Vice-President: Sir Francis Graham Smith
Organizing Committee: A.A.Boyarchuk, A.Florsch, Y.Kozai, K.CH.Leung, G.Swarup,
C.R.Tolbert, H.H.Voigt, F.B.Wood (ex off.), Ye Shu-Hua.

The present report covers the period 1 February 1985 to 30 November 1987.
During this period a total of 23 travel grants were awarded as follows:

Name	Origin	Destination
A. Aiad	-Cairo University, Arab Rep. Egypt	-ESO, Garching, FRG
B. Barbuy	-Univ. Sao, Paulo, Brazil	-Obs. de Paris-Meudon, France
E. Brandi	-Univ. La Plata, Argentina	-Inst. d'Astrophys.Liège,Belgium
S. Caganoff	-Mt.Stromlo & Siding Springs Obs., Australia	-VLA of NRAO, New Mexico, USA
Cheng Fuhua *	-Univ. Science & Technology Heifei, China (Nanjing)	-NASA Goddard Space Flight Center Greenbelt, USA
Chun Young-Woo *	-Yonsei Univ. Obs. Seoul, Rep. Korea	-Dominion Astrophys. Obs., Victoria, Canada
H. Deasy	-Dunsink Observatory Dublin, Ireland	-Astronomical Center, Warsaw, Poland
J. C. Forte	-Astronom. Obs., La Plata, Argentina	-CTIO, La Serena/Cerro Tololo, Chile
S. Giridhar	-Indian Inst. Astrophys., Bangalore, India	-Univ. of Mexico, Mexico
I. Hubeny	-Univ. Vienna, Austria	-Obs. de Paris-Meudon, France
K. Iwasaki	-Kwasan & Hido Obs. Kyoto, Japan	-Bosscha Observatory, Lembang, Indonesia
A. Kus	-Radio Astronomy Obs. Torun, Poland	-CALTECH, Pasadena, USA
J.McCarthy	-CALTECH, Pasadena, USA	-Univ. of Munich, FRG
T. Ozkan	-Univ. Obs.Istanbul, Turkey	-Univ. of Munich, FRG
E. Pehlemann	-Göttingen Obs., FRG	-Sacramento Peak Obs., New Mexico, USA
N. Ramamani *	-Osmania Univ. Hyderabad, India	-Univ. of Edinburgh, U.K.
A. Ray	-TIFR, Bombay, India	-Univ. of Virginia, Charlottesville, USA
K. C. Sahu *	-Phys. Research Lab., Ahmedabad, India	-Kapteyn Lab. Groningen, NL, and Inst.d'Astrophys.,Paris,France
D. J. Saika	-Radio Astronomy Center, TIFR, Bangalore, India	-NRAL Jodrell Bank, U. K.
S. Sukumar	-Radio Astronomy Center TIFR, Bangalore, India	-Univ. of Illinois, Urbana, USA
P. Teague *	-Mt.Stromlo & Siding Springs Observatory, Australia	-Center for Astrophysics, Cambridge, Mass., USA
P.Venkatakrishnan	-Indian Inst. Astrophys. Bangalore, India	-Center for Astrophysics, Cambridge, Mass., USA
Yao Bao-an	-Shanghai Observatory China (Nanjing)	-ESO, Garching, FRG

The 5 travel grants attributed during the period 1 February - 30 November 1985 (i.e. before the XIXth IAU General Assembly in Delhi) are marked with an asterisk.

One additional colleague from China (Nanjing) was given a travel grant to visit the Raman Institute in Bangalore, India, for half a year. Unfortunately, he had to change his plans and cancel his project before he could undertake his travel to Bangalore. Consequently, he returned his travel grant he had received from the Commission.

Out of the 23 astronomers receiving a travel grant 33% came from India. In total the beneficiaries came from 14 different countries. The host institutes were located in 11 different countries, 33% being in the USA. We are most grateful to all host institutes for providing our grantees with the necessary research facilities and, in most cases, with a grant or fellowship for covering their living expenses during their prolonged visits. Quite a few more institutes would be willing to host a foreign astronomer with similar research interests as the host astronomers but they cannot pay for the visitor's living expenses. It is becoming more and more difficult for a host institute to provide a visitor with such a financial support. It is, therefore, recommended to prospective travel grant applicants wishing to spend several months or a year at a foreign astronomical institute to make every effort possible in searching for a grant or fellowship, either international or national (in the home or the host country), if the host institute has no means to offer a subsistence support.

Almost all those having received a travel grant have submitted, upon their return to the home institute, a report on their scientific activities, their plans for continued collaboration and, possibly, their publications produced during their visit abroad.

A total of 17 applications had to be rejected because they did not meet with the requirements of the travel grant scheme. Some of the reasons were that the applicant had no means or grant for covering the living expenses at the host institute, or the applicant wished to go on a short observing mission only, or wished to attend a meeting, etc.

Quite a number of inquiries were received but no formal application followed up when explanations were given concerning the Commission's guidelines.

Details of the guidelines for the travel grant scheme and the application procedure were published in the IAU Information Bulletin No. 55, February 1986.

The Commission President greatly appreciated the perfect and most valuable collaboration she had with the Commission's Vice-President, and the useful advice she received from various Members of the Organizing Committee. The excellent cooperation of the IAU General Secretary, Jean-Pierre Swings, and of Brigitte Manning at the IAU Secretariat in Paris is most gratefully acknowledged.

Edith A. Müller
President of the Commission

40. RADIO ASTRONOMY (RADIO ASTRONOMIE)

PRESIDENT: J E Baldwin

VICE-PRESIDENT: P G Mezger

ORGANISING COMMITTEE: A Barrett, A Baudry, R Booth, D Jauncey, V Kapahi, K Kellermann, L Matveyenko, G Nicholson, E Seaquist, G Setti, R Strom, S Wang, T Wilson.

The following commission members have contributed to this report:

M Birkinshaw, R J Cohen, J J Condon, T J Cornwell, J R Dickel, P A Feldman, R Genzel, M Goss, V Kapahi, Gopal Krishna, M Kundu, A G Lyne, C R Masson, A C Readhead, W Reich, J M Riley, A J Turtle, J M van der Hulst, T L Wilson.

INSTRUMENTATION

Aperture Synthesis Radio Telescopes - Vigorous progress has continued in this field over the last three years with proposals for new telescopes, first operations of those under construction and enhancements of existing telescopes.

In Japan the Nobeyama interferometer comprising five 10m antennas is now operational at 1.3 cm and it is planned to start observations near 3 mm soon.

Construction of the Australia Telescope is well advanced with the rail track and the first of the six 22 m antennas on a 6 km baseline completed whilst the other antennas are under construction. Tests on the first interferometer are planned for the end of 1987 with completion of the array in 1988. Initial operation is expected to cover four frequency bands between 1.4 and 10 GHz.

The three-antenna IRAM interferometer for mm wavelengths is under construction in southern France. The first of the 15 m carbon-fibre antennas is now completed and under test.

In the United States notable progress has been made on the Very Long Baseline Array (VLBA) with the completion of construction of the first of the ten 25-m antennas at a site close to the VLA in New Mexico. It is hoped that the array will be completed in 1992.

The proposal for a Giant Metre-Wavelength Radio Telescope (GMRT) in India has been approved and a site near Pune adopted. The array will comprise 34 45 m fully steerable paraboloids, 16 being concentrated in a 1 km central array and the remaining 18 being distributed along arms of a Y-shaped array with each arm extending to about 14 km. The design includes operation in five frequency bands between 38 and 610 MHz and completion is planned for 1992.

Significant improvments to existing arrays include:- the addition of a third 10.4 m antenna to the Berkley mm array at Hat Creek; the operation of the VLA at 327 MHz and trials on four of the antennas at 75 MHz; operation of the MERLIN array at 151 MHz and extension of the array by the inclusion of an existing 18 m antenna at Cambridge; conversion of the Cambridge 5 km telescope to cooled broad-band operation for microwave background work and operation of the low frequency synthesis telescope at 38 MHz.

<u>Single Dish antennas</u> - The principal developments in this field have been for mm
and submm antennas.

The IRAM 30 m antenna sited near Granada in Spain has been in full operation
during the last three years, with full efficiency at 2 mm wavelength.

The UK-Netherlands James Clerk Maxwell Telescope (JCMT), a 15 m antenna for
mm and submm wavelengths, sited on Mauna Kea, Hawaii, has been completed in 1987
and first operations at 1.3 mm have started. Full efficiency at 0.8 mm is
expected after further surface adjustment following current holographic tests.

The 10.4 m Cal-Tech antenna, also sited on Mauna Kea, has been commissioned
in 1987 and is in operation at mm and submm wavelengths.

Another 10.4 m antenna, based on the same design, has been completed at
Bangalore, India and has been in operation since 1985. A Swedish-European 15 m
antenna using the design for the IRAM dishes has been built in Chile and is now in
operation, although not yet at the shortest wavelengths.

<u>Imaging Techniques</u> - The last triennium has seen continued improvement in the
imaging performance of radio-telescopes, due largely to research in algorithms for
image processing.

The CLEAN algorithm for image deconvolution has been investigated critically
by Tan (1986, MNRAS, 220, 971-1001), particularly with respect to the well-known
instability which produces corrugations on extended structure. A now widely-used
variant of CLEAN (Steer et al., 1984, A & A, 137, 159-165) avoids the instability
and also provides much faster reconstruction of extended emission.

The Maximum Entropy method of image deconvolution is becoming more popular,
mainly because of its flexibility in handling different types of data, and its
speed advantages over CLEAN when processing very extended objects. The properties
of MEM solutions are better-understood than those of CLEAN solutions, mainly
because the former are defined by a condition rather than as the end result of a
sequence of very non-linear operations as in CLEAN. Narayan and Nityananda (1986,
Ann. Rev. Astron. & Astrophys., 24, 127-170) have reviewed MEM in image
reconstruction ably and comprehensively. Recent work has focused on algorithms
(e.g. Wilczek and Drapatz, 1985, A & A, 124, 9-12, Cornwell and Evans, 1985, A &
A, 143, 77-83, Reiter and Pfleiderer, 1986, A & A, 166, 381-392) rather than
philosophy. Cornwell (1984, NRAO mm array memo 34) has used MEM for the imaging
of very large objects by interferometric arrays.

Braun and Walterbos (1985, A & A, 143, 307-312) developed a simple method for
estimating the short-spacings, around about a dish diameter, which are normally
inaccessible to interferometric arrays. Conway (see Wilkinson et al, 1986, Proc.
IAU symp. on VLBI, Boston) has developed a technique of "Multi-frequency
Synthesis" in which the u, v plane is filled in by observing at a number of
frequencies and spectral effects are implicity corrected in the deconvolution
step. This will allow VLBI imaging of sources of more complexity than previously
possible.

Recent developments in methods for obtaining good phase information from
interferometers are mainly refinements of the now widely-accepted
self-calibration/closure phase techniques (see e.g. Pearson and Readhead, 1984,
Ann. Rev. Astron. & Astrophys., 22, 97-130). Linfield (1986, A.J., 92, 213-218)
has described a difficulty in applying these techniques to arrays with poorly
mixed spacings. Cornwell (1987, A & A 180, 269-274) has shown that the
speckle-masking technique, which is used in high-resolution imaging at optical

wavelengths, relies upon closure phase, and also that it can be adapted to
radio-interferometry to allow imaging of weaker objects than previously possible.
Cornwell and Napier (1986, Proc. NRAO workshop on "Radio Astronomy in Space") have
shown how, with the aid of a correlating focal plane array, self-calibration may
be applied to single-aperture observations to remove the effects of aberrations.

Finally, as a sign that imaging techniques developed in radio astronomy are
spreading, aperture synthesis imaging has been demonstrated at optical wavelengths
by Haniff et al., (1987, Nature, 328, 694-696). By measuring true amplitudes and
closure phases with a simple, rotatable linear mask in a conventional optical
telescope, they were able to image a simple binary at high-resolution (about 50
milliarcsecond) with low-light levels.

SOLAR SYSTEM RESEARCH

The Sun - Microwave Images of solar active regions and flares obtained with the
VLA (Kundu and Lang, 1985, Science, 228, 9) and WRST (Alissandrakis and Kundu,
1984, A & A 271, 39), were used to study the preflare build-up (Kundu et al.,
1985, Ap. J. Suppl. 57, 621) and magnetic field topology in the vicinity of
flaring regions. It seems that the appearance of new flux, which interacts with
pre-existing regions to form neutral or current sheets is essential to trigger the
onset of a flare (Kundu and Shevgaonkar, 1985, Ap. J., 291, 860; Kundu, 1986, Adv.
Sp. Res. 6, 93). Observational evidence for magnetic reconnection in microwave
bursts was obtained from VLA images of a flare (Kundu 1985, Proc. IAU Symp. No.
107, p. 185). Three-dimensional structures of active regions were obtained using
the VLA at 2, 6 and 20 cm wavelengths (Shevgaonkar and Kundu, 1984, Ap. J., 283,
413). The 2 cm radiation is mostly thermal bremsstrahlung from loop foot points,
while 6 and 20 cm radiation is dominated by low-harmonic gyroresonance radiation
from the upper part of the loop.

Simultaneous observations with the VLA at 2 and 6 cm of a flare showed
different locations of 2 cm (foot points) and 6 cm (loop top) sources, and time
differences in the peak times (Shevgaonkar and Kundu 1985, Ap. J. 292, 733).
Using a DC electric field flare model, this delay in the peak times was
interpreted and the strength of the electric field in the flaring region was
estimated to be 0.2-4 μ stat volt cm^{-1}. Detailed computations have been made of
the microwave structure of hot coronal loops and compared with VLA and WSRT
observations (Holman and Kundu 1985, Ap. J. 292, 291).

Interpreting solar active region emission at centimeter wavelengths as
gyroresonance radiation, Westerbork and VLA images have been used to estimate
coronal magnetic fields (Alissandrakis and Kundu, 1984, A & A, 139, 271; Gary and
Hurford, 1987, Ap. J., 317, 522). Using a new method of analysis and identifying
certain features on the radio maps, both the vertical and horizontal components of
the sunspot magnetic field have been mapped at specific locations in the low
corona.

VLA images of the quiet sun have been made at 20 and 6 cm with particular
emphasis on coronal bright points (Habbal et al., 1986, 306, 740; Fu et al., 1987,
Solar Phys. 108, 99) and filaments (Kundu et al., 1986 A & A, 167, 166; Gary,
1986, NASA Publ. No. 2442, p.121). Coronal bright points appear to fluctuate in
intensity over time scales of a few minutes, similar to what is observed in EUV.
Filament observations show good correspondence between 20 cm temperature
depressions and large coronal cavities as observed in He spectroheliograms.

Meter-decameter wavelength imaging observations have been made with the Clark
Lake and Nancay radioheliographs (Kundu 1986, Highlights in Astronomy, 7, 725,
Lantos and Alissandrakis, 1986, Highlights in Astronomy, 7, 761). In particular

the Clark Lake instrument has been used as a radio coronagraph for studies of
coronal streamers and coronal holes (Kundu et al., 1986, Solar Phys. 108, 113;
Gopalswampy et al., 1986, Solar Phys. 108, 333). Type III burst emitting
electrons propagate in dense coronal streamers, and coronal holes at lower
frequencies (50 and 38.5 MHz) are radially displaced relative to their higher
frequency locations and chromospheric positions as determined from He
spectroheliograms. The discrepancy that exists in electron density values as
determined from radio and UV data has not been resolved. Microbursts which are
low brightness temperature type III-like bursts have been discovered with the
Clark Lake radioheliograph (Kundu et al., 1986, Ap. J. 308, 436). The
observations imply that energy releases on the Sun continue to be impulsive with
nonthermal electron distributions, for small releases of energy, similar to the
hard X-ray microbursts. They seem to be produced by plasma emission process at
the fundamental plasma frequency (White et al., 1986, Solar Phys. 107, 135).
Raoult et al., (1985, Astrophys. J., 299, 1027) showed that in the flare preflash
phase type III and type V bursts occur, characteristic of events associated with
high energy electron acceleration mechanisms.

On the theoretical side, type IV bursts have been explained as due to
harmonic plasma radiation (Gary et al., 1985, A & A, 152, 42). Electron-cyclotron
maser stability has also been proposed as the source of type IV bursts at
decimeter and longer wavelengths, specifically to Zebra-stripes and continuum
bursts as well as type IV continuum bursts (Winglee and Dulk 1986, Ap. J. 307,
808, and 310, 432).

Planets, Satellites, Comets Asteroids - Largely because of the general interest in
Halley's Comet, studies of comets have dominated solar system radio astronomy
during the past triennium. Collected results appear in two symposium volumes:
"Exploration of Halley's Comet" - 20th Eslab symposium (ESA report SP250) and
"Cometary Radio Astronomy" - proceedings of an NRAO workshop, September 1986.
Continuum measurements have resulted in detection of Halley's comet (Altenhof et
al., Astron. Astrophys., 164, 227, 1986) and indications of possible effects of
plasma in the tail of Comet Austin 1982g during occultation of a radio source (de
Pater and Ip, Ap. J., 283, 895, 1984). Spectroscopy has provided much new
information about molecular production in these objects. Highlights include the
unambiguous detection of HCN in Halley's comet (Despois et al., Astron.
Astrophys., 160, L11, 1986; Schloerb et al., Ap. J. Lett., 310, L55, 1986;
Winnberg et al., Astron. Astrophys, 172, 335, 1987) and a number of detailed
studies of OH. The pre-Halley OH-observations of 16 comets are summarized by
Snyder (AJ, 91, 163, 1986). The first radio image of a comet was made in OH by de
Pater et al., (Ap. J. Lett., 304, L33, 1986). The emission is concentrated in
clumps surrounding a hole which maybe the first direct observation of OH maser
quenching in the inner cometary coma. Five comets have now been detected by radar
(summarized by Ostro, Encyclopedia of Science and Technology, 10, 611, 1987)
including Halley by the Arecibo group (Harmon et al., in preparation).

With the detection of Pluto at mm wavelengths by the 30-meter telescope at
Pico Veleta, thermal disk emission has now been recorded from all nine official
planets. Other work on the planets has included the first radio resolution of the
disk of Mercury using the VLA at a wavelength of 6 cm (Burns, et al., Nature, Oct.
1987) and studies of the surface temperature of Mars where a surface anomaly
appears to be present between about 10° and 40° south latitude (Rudy, Ph.D.
thesis, Caltech 1987). CO in the atmosphere of Venus is being studied by teams of
observers using both the Hat Creek and OVRO millimeter arrays.

Data on Saturn and its rings at wavelengths ranging from 2.7 mm (Dowling, et
al., Icarus, 70, 506, 1987) to 20 cm now exist at all inclination angles of the
rings from edge-on to the maximum tilt of 26°. A Grossman (Caltech) is analyzing

all the VLA data to derive dielectric and other properties of the ring particles
plus the filling factors in the rings. In an interesting experiment to obtain
both spatial and Doppler resolution, R. Goldstein, et al., have transmitted an 8
GHz radar signal from the Goldstone antenna and received it in the line mode with
the VLA. The results of this powerful technique look very promising.

The first observation of the zone-belt structure in Jupiter's atmosphere was
reported by de Pater and Dickel (Ap. J., 308, 459, 1986). de Pater (Icarus, 68,
344, 1986) analyzed these data at 1.3, 2, and 6 cm to conclude that there is a
significant depletion of ammonia in the Jovian belts at altitudes above a level
where the pressure is about 2 atmospheres. Below that, ammonia is overabundant
relative to the solar value in both zones and belts. Jupiter's radiation belts
continue to be monitored at 1.4 GHz by de Pater for correlation with other
solar-system phenomena. Radio astrometry of the Jovian satellites has indicated
an error of 475 km in the relative positions around the orbits of Callisto and
Ganymede (Muhleman et al., AJ, 92, 1428, 1986).

The Planetary Radio Astronomy experiment on the Voyager 2 spacecraft
performed with full reliability during the Uranus encounter in January 1986. By
analysis of the decametric radiation, Warwick el al., (Science, 233, 102, 1986)
have determined a rotation period of 17.24 hours for the dipolar magnetic axis of
that planet which is tilted 60° from the rotational axis.

Observations of asteroids with the VLA at 2-cm have now provided data on at
least one asteroid in each of the C, S, and U optical classifications (Webster et
al., Icarus, 60, 538, 1984 and 69, 29, 1987). They all appear to be covered with
a thin layer of impact-hardened material.

GALACTIC RESEARCH

Surveys - The second part of an absolutely calibrated 1.42 GHz survey of the
entire northern sky has been published (Reich and Reich, 86 AA Sup. 63, 205).
These observations with 35' angular resolution cover the area $24° \gtrsim \delta \gtrsim 19°$. Jonas et
al., (85 AA Sup. 62,105) have mapped the area $12^h \leq \alpha \leq 22^h$, $-63° \leq \delta \leq -24°$ of the
southern sky at 2.3 GHz with 20' angular resolution. An analysis of these data
has been made by Jonas (86 MN 219,1) reporting new loops, spurs and arcs with
angular dimensions between 2° and 6°.

Sensitive observations of the Galactic plane at 2.7 GHz have been made by
Reich et al., (84 AA Sup. 58, 197). Total intensity maps with 4.3' angular
resolution are presented for the region $357.4° \leq 1 \leq 76°$ and $-1.5° \leq \beta \leq 1.5°$. In
addition, a catalogue lists the data for 1212 small diameter sources. Maps
showing the linear polarized emission obtained from the same observations have
been published by Junkes et al., (86 AA Sup. 69,451). The Cygnus region has been
mapped at 4.8 GHz with 2.6' angular resolution by Wendker (84 AA Sup. 58, 291).
Sofue et al., (84 PASJ 36, 297) discussed a first section ($21° \leq 1 \leq 26°$) of a 10 GHz
survey of the Galactic plane with 2.7' angular resolution. The Galactic plane has
been surveyed with 0.5° angular resolution at submillimeter wavelength (0.7-2 mm)
between 0° and 30° longitude by Pajot et al., (86 AA 154, 55). Results of
decameter observations at four wavelengths covering the Galactic plane between
147° and 153° longitude have been discussed by Abramenkow (85 AZh 62, 1057; Sov
Ast 29, 616).

A source survey covering the northern sky between $-5° \leq \delta \leq 82°$ at 1.4 GHz was
carried out by Condon and Broderick (85 AJ 90, 2540; 1986 AJ 91, 1051). The maps
have 12.7'x11.1' angular resolution and contain also Galactic structures with an
extent of less than 1° in declination. About 3×10^3 sources per sr stronger than
0.15 Jy are contained in the maps. The RATAN-600 radio telescope has been used at

various wavelengths between 2 cm and 31 cm to catalogue and to investigate
Galactic radio sources in the centre and anticentre area (Pyantunina, 85 AZh 62,
218; Sov Ast 29, 125; Gosachinskii, 85 AZh 62, 226; Sov Ast 29, 128; Berlin et
al., 85 AZh 62, 229; Sov Ast 29, 130; Aliakberov et al., 85 AZh 62, 482; Sov Ast
29, 281). Fich (86 AJ 92, 787) has made a complete high resolution survey at 6 cm
of sources with an extent less than 2' and a flux density larger than 0.3 Jy at 21
cm wavelength in the area $93° \leq |\leq 163°$, $-4° \leq B \leq 4°$. The sources are either
HII-regions or of extragalactic origin. No compact supernova remnant could be
identified. High resolution observations of 44 compact radio sources located in
the Galactic plane have been made by Green (85 MN 216, 691) in a search for young
but distant supernova remnants. Most of the sources seem to be extragalactic.
Repeated 6 cm observations of discrete sources in the Galactic plane were done by
Geregory and Taylor (86 AJ 92, 371) to detect and to study variable sources. From
their list of 1274 sources, which is estimated to be complete down to 70 mJy, they
find 32 variable sources and 27 sources which are possible variable.

Radio Stars - The high sensitivity and resolving power of the VLA has been used
increasingly to study radio emission from a variety of stellar classes.

Multi-wavelength radio observations of red dwarf (dM_e) flare stars (cf. Kundu
and Shevgaonkar 85 ApJ 297, 644) can be interpreted as indicating that the
quiescent flux is non-thermal gyrosynchrotron radiation, but the narrow-band
structure of the slowly varying decimetric emission reported for YZ CMi (Lang and
Willson 86 ApJL 302, L 17) may also require a coherent radiation mechanism. Some
coherent emission process is favoured for radio flares of red dwarfs: a) 20 cm
flaring from AD Leo and L726-8A exhibited narrow-band differences (White et al. 86
ApJ 311, 814); b) radio bursts from AD Leo contained ms "spikes" which are up to
100% circularly polarized and have $T_B \geq 10^{16}$ K at 1.4 GHz (Lang and Willson 86 ApJ
305, 363); c) the first dynamic specra of stellar microwave flares from UV Cet =
L726-8B showed distinct variations with frequency (Bastian and Bookbinder 87
Nature 326, 678).

Important advances have been made in the study of a variety of radio-emitting
pre-main-sequence (PMS) stars. A luminosity-limited 6 cm survey of the Tau-Aur
dark clouds (Bieging et al. 84 ApJ 282, 699) indicated that, by the time a PMS
star is visible as a T Tau variable, detectable continuum emission (either
free-free from an ionized stellar wind or non-thermal radiation) is rarely found.
However, the PMS star DoAr21 in the ρ Oph star-formation cloud was found to
exhibit strong, extremely variable radio flaring (Feigelson and Montmerle 85 ApJL
289, L19). This added credence to the idea that at least some PMS stars exhibit
high degrees of atmospheric magnetic activity. More recent work by Kutner et al.
(86 AJ 92, 895) and by Cohen and Bieging (86 AJ 92, 1396) on the radio variability
of some PMS stars has strengthened this conclusion.

Understanding of the radio emission of symbiotic stars has improved following
the VLA survey by Seaquist et al. (84 ApJ 284, 202) and a binary model (Taylor and
Seaquist 84 ApJ 286, 263) in which the radio flux arises form the portion of the
stellar wind of the cool, mass-losing star that is photoionized by the hot
companion. Observations of the symbiotic star CH Cyg by Taylor et al. (86 Nature
319, 38) have revealed a strong radio outburst and the production of a
multi-component radio jet (cf. R. Aqr?) expanding at the rate of 1.1 arcsec/yr.
This has been interpreted as the result of accretion at, or near, the Eddington
limit from the M7III star onto a white dwarf companion.

In 1985 the first radio detection of a recurrent nova was made by Padin et
al. (85 Nature 315, 306) only 18 days after the optical outburst. Subsequently,
an extensive series of multifrequency VLA observations (Hjellming et al. 86 ApJL
305, L71) has revealed complex decay of the radio emission which is probably

non-thermal.

The radio emission of Wolf-Rayet stars has been carefully studied by Abbott et al. (86 ApJ 303, 239,). Most are thermal wind sources but some are non-thermal emitters. Non-thermal radio emission, which is much less common in W-R stars than in luminous OB stars, is most plausibly explained as synchrotron emission from electrons accelerated to relativistic energies by the strong shocks that permeate the chaotic stellar winds close to the star (White 85 ApJ 289, 698). In this model non thermally emitting OB stars (e.g. Abbott et al. 84 ApJ 280, 671; Persi et al. 85 AA 142, 263) would possess much larger ratios of rotational to shock velocities and/or stronger magnetic fields than W-R stars.

Results on radio flaring from Cyg X-3 are given by Molar et al. (84 Nature 310, 662), Johnston et al. (86 ApJ 309, 707) and Spencer et al. (86 ApJ 309, 694). A 5-year study of ScoX-1 (Geldzahler and Fomalont 86 ApJ 311, 805) has shown that all 3 radio components are approximately co-moving with the optical binary and are thus associated with it. Radio detections of galactic-bulge X-ray sources have been made by Grindlay and seaquist (86 ApJ 310, 172).

Valuable surveys of RSCVn and RSCVn-like stars have been made by Mutel and Lestrade (85 AJ 90, 493), Drake et al. (86 AJ 91, 1229) and Mutel et al. (87 AJ 93 1220). Useful results for individual stars were given by Innis et al. (85 ProcASA 6, 160). Simon et al. 85 ApJ 295, 153), Little-Maranin et al. (86 ApJ 303, 780) and Bunton et al. (86 ProcASA 6, 316).

Results of a survey of cool giants and supergiants were given by Drake and Linsky (86 AJ 91, 602). Variable radio emission from the 4 Dra system (M3 III + cataclysmic binary) was detected by Brown (87 ApJL 312, L51). The first radio detection of a conventional Be star was made by Taylor et al. (87 MN, 228, 811).

Dual polarisation VLVI observations enabled the milliarcsec radio structures of 7 stellar binaries to be determined (Mutel et al., 85 ApJ 289, 262). Stellar radio astrometry has been actively pursued by Johnston et al. (85 AJ 90, 1343 and 2390).

The theory of long-wavelength radiation from isothermal stellar winds has been reformulated as a useful curve of growth method by Lamera and Waters (84 AA 136, 37). Finally, Wendker has done the field a good service by bringing out a new edition of his catalogue of stars with radio continuum emission (87 AA Suppl. Ser 69, 87).

Pulsars - The discovery of the millisecond pulsar PSR 1937+21 and uncertainty over the evolution of young pulsars has led to several surveys with good sensitivity to pulsars with short period. Two new millisecond pulsars have been discovered. Segelstein et al., (86 IAUC 4162) used Arecibo in discovering PSR 1855+09, a binary pulsar with a period of 5.4 ms. Lyne et al., (87 Nature 328, 399) used the Jodrell Bank antenna to discover PSR 1821-24 with period 3.1 ms in the core of the globular cluster M28. This followed a two year trail of investigation (Hamilton et al., 85 AJ 90, 606; Mahoney & Erickson 85 Nature 317, 154; Erickson et al., 87 ApJ, 314, L45). Both discoveries give strong support for the accretion spin-up mechanism for millisecond pulsars (Srinvasan & Radhakrishnan 84 Proc. 2nd Asian Pac. Reg. Meeting on Astronomy, Bandung, Eds. Hidayat and Feast, 423).

440 pulsars are now known. 100 have been discovered in the past 3 years in major surveys (Dewey et al., 85 ApJ 294, L25; Stokes et al., 86 ApJ 311, 694; Clifton & Lyne 86 Nature 320, 43). Few of the new pulsars have periods less than 100 ms. Clifton & Lyne at 1400 MHz overcame galactic background noise and interstellar scattering and detected 40 new pulsars towards the inner galaxy in

only 200 square degrees. Three new pulsars have been discovered in binary
systems, making a total of seven: the millisecond pulsar PSR 1855+09, PSR 1831-00
(Dewey et al., 86 Nature 322, 712) and PSR 2303+46 (Stokes et al,m 85 ApJ 294$_L$,
L21). Van den Heuvel (84 JAA, 5, 209) has reviewed binary evolution scenarios.
Kulkarni (86 ApJ 306, L85) has identified the companions of PSR 0655+64 and PSR
0820+02 as white dwarfs. The low white dwarf temperature suggests great age for
PSR 0655+64 but the magnetic field has not decayed accordingly.

Lyne et al. (85 MNRAS 213, 613) have summarised current understanding of the
galactic distribution and evolution of normal pulsars. The small number of short
period pulsars have led Chevalier & Emmering (86 ApJ 304, 140) and Narayan 87 IAU
Symp 125, 67) to make cases for pulsar birth at long period.

Gwinn et al. (86 AJ 91, 338) have used VLBI to measure the annual parallaxes
of PSR 0950+08 and PSR 0823+26.

Davis et al. (85 Nature 315, 547) have shown that PSR 1937+21 is comparable
in stability to the best man-made atomic clocks. They demonstrate its value as a
possible gravitational wave detector. A value of braking index n = 2.83 ± 0.03
has been obtained for PSR 1509-58, only the second so far measured (Manchester et
al., 85 Nature 313, 374). One small and one massive glitch have occurred in the
relatively old pulsar PSR 0355+54 (Lyne 87 Nature 326, 569). In a monitoring
program in Hartebeesthoek, Flanagan observed a glitch in the Vela pulsar in July
1985 (87 IAU Symp 125, 64). Crab pulsar monitoring continues at Jodrell Bank.
Backer & Hellings (86 Ann Rev. AA. 24, 537) have summarised the use of pulsar
timing observations and General Relativity. Pines & Alpar (85 Nature 316, 27)
have reviewed superfluidity in NS.

Hankins and Rickett (86 ApJ 311, 684) have measured the frequency dependence
of the profiles of 12 pulsars, showing spectral differences between core and conal
emission. Rankin (86 ApJ 301, 901) shows how single pulse properties differ
between core and cone. Valuable single pulse polarisation studies have been made
by Stinebring et al., (84 ApJ Supp 55, 247 and 279). Remarkable subpulse drifting
through the whole of the period has been observed in PSR 0826-34 (Biggs et al., 85
MNRAS 215, 281.

Rickett et al. (84 AA 134, 390) recognised that the long term, intensity
variations of pulsars are due to refractive effects in the interstellar medium.
Refractive effects were observed by Hewish et al., (85 MNRAS 213, 167), Smith &
Wright (85 MNRAS 214, 97) and Cordes & Wolszczan (86 ApJ 307, L27). Theoretical
discussion was provided by Blandford & Narayan (85 MNRAS 213, 591), Romani et al.,
(86 MNRAS 220, 19) and Cordes et al. (86 ApJ 310, 737). Cordes (86 ApJ 311, 183)
estimated the speeds of 71 pulsars using ISS data. Lyne & Hamilton (87 MNRAS 224,
1073) have determined the rotation measures or 163 pulsars.

Many observational and theoretical aspects of pulsars were reviewed at IAU
Symposium 125 on the Origin and Evolution of Neutron Stars (87 Eds Helfand &
Huang) and at the Nato ASI on High Energy Phenomena Around Collapsed Stars (86 Ed
Pacini). Taylor & Stinebring (86 Ann Rev AA 24, 285) have reviewed recent
progress in the understanding of pulsars.

Supernovae (SN) and Supernova Remnants (SNR) - SN 1987A in the LMC has provided
the first opportunity for radio observations to be made soon after a SN outburst.
A weak but immediate burst of emission with a time scale of a few days was
detected at several Australian observatories (87 Nature 327, 38). Early VLBI
observations apparently resolved the SN which is consistent with the initial
expansion velocity measured optically. A review of radio SN in more distant
galaxies (86 ApJ 301, 790) includes detailed light curves for two Type II events.

Radio emission was detected only after some months as it rose rapidly to a peak around a thousand times more luminous than for SN 1987A to date. Reduced absorption in the circumstellar material as the region of non-thermal emission expands can explain this behaviour. The weakness and promptness of the radio burst for SN 1987A may be due to a low circumstellar density in the immediate vicinity of its blue progenitor. It could remain radio quiet for decades until either it interacts with the surrounding medium or a central pulsar is revealed.

More than 40 radio sources have been detected within 300 pc of the nucleus of M82 (85 ApJ 291, 693). Most are presumed to be radio supernovae or remnants but are more luminous than comparable objects in the Magellanic Clouds. They may be due to a recent burst of star formation deep within giant molecular clouds. VLBI observations of expanding remnants may be be used for the determination of extragalactic distances (85 SN as distance indic. pp 107, 123, 130).

Radio properties of SNRs are described in general reviews (84 Ann Rev Astron Astrophys 22, 75; 84 Ap Sp Sci 3, 35). Statistical studies and Σ-D relations are discussed in 85 ApJ Lett 295, L13; 86 AA 166, 257 and 86 Astrophys 24, 232.

Most work continues to be on individual SNRs in the Galaxy. A comprehensive review by Caswell will appear in the proceedings of IAU Colloquium 101 'Interaction of SNRs with the Interstellar Medium'. A few newly identified remnants have been reported but studies have concentrated on the structure of known remnants. Maps with high resolution and sensitivity obtained from the VLA and Westerbork telescopes combined with information from other wavelengths have permitted detailed investigation of interactions with circumstellar and interstellar material. At lower resolution the Molonglo (MOST) and Penticton (DRAO) telescopes have provided maps with a wide field of view.

All types of SNR morphology have received attention - shells, centrally concentrated Crab-like plerions, shell-plerion combinations and unusual morphologies where special circumstances may enhance a particular property. The classification of some non-thermal galactic objects as SNRs has been questioned (85 Nat 313, 115; 314, 720; MNRAS 225, 329).

The broad structure of many remnants has been interpreted as due to 'barrels' - regions with axial symmetry and only weak radiation from the end caps (Kesteven and Caswell, 87 AA in press) or as double loops (87 AA 171, 205). Distinctive features in individual remnants include jets (85 Nature 316, 44; Kesteven et al., IAU Symp. 125), markedly different expansion on opposite sides (87 ApJ 315, 580) and expansion into pre-existing interconnected spherical shells (86 MNRAS 221, 809; 86 AA 164, 193).

The last reference deals with IC 443 which is also the subject of other studies (85 ApJ 290, 596; 86 MNRAS 221, 473; 86 AJ 92, 1349; 87 AA 173, 337). Detailed maps of other well known objects have been used for measurements of expansion and interactions with the surrounding medium:-
the Cygnus Loop (84 MNRAS 211, 433; 85 AA 148, 52; 86 ApJ 306, 266); SN 1006 (86 AJ 92, 1138); Tycho (85 MNRAS 216, 949); Kepler (84 ApJ 287, 295); Cas A (86 MNRAS 219, 13; 87 Nature 327, 395); CTB 80 (84 AA 139, 43; 85 AA 145, 50) and the Crab Nubula (85 MNRAS 212, 359; 86 Sov Astron Lett 12, 112).

Observations of the emission and absorption of radio spectral lines associated with SNRs are valuable for estimating distances and studying interactions. Some results are contained in the references above; others are reported in 86 AA Supp 63, 345; 86 MNRAS 219, 427; 86 MNRAS 221, 809.

The increasing use of mm wavelengths is expected. There are recent reports of continuum observations of Cas A and the Crab Nebula (86 ApJ 298, 644; 86 AA

167, 145) and spectral line measurements (86 ApJ 309, 667; 87 AA 173, 337).
Another forthcoming development is the completion in 1988 of the Australia (radio
synthesis) Telescope which will permit improved mapping of southern objects in the
Galaxy and the Magellanic Clouds.

Galactic Center - An important new result has been the first direct determination
with VLBI of the distance to the galactic center molecular cloud Sgr B2 from the
proper motions of a cluster of H_2O maser spots (Reid et al., 87, IAU Symp. 115:
Star Forming Regions, ed. Peimbert). The derived distance is 7.1±1.5 kpc, where
the uncertainlty is dominated by the number of maser features measured. Sgr B2 is
almost certainly within a few hundred parsec of the galactic center.

     Investigations at high spatial resolution (VLA, 100 m telescope, 45 m
Nobeyama telescope) of the central ≈10^2 pc region of the Galaxy have given
evidence for large scale, poloidal magnetic field structures. Nonthermal
filaments in the radio "arc" (1^{II}≈0.1°) stretch 40 pc above and below the galactic
plane and show extended, highly polarized plumes at their ends (Tsuboi et al., 85,
Publ. Astr. Soc. Japan 37, 359; Seiradakis et al., 85, Nature 317, 697;
Yusef-Zadeh 86, Ph.D. Thesis University Columbia). Various explanations of the
phenomena have been proposed, typically based on gas which is accelerated along
the field lines and away from the center and the galactic plane. Faint threads of
30 pc length and ≈0.3 pc width near Sgr A have been interpreted as magnetic flux
tubes or, in a more speculative vein, as superconducting cosmic strings (Morris
and Yusef-Zadeh 86, BAAS 18, 1023). Thermal radio emission in the curved
filaments of the "arc" now appear to be the photoionized edges and surfaces of
dense, tidally disrupted molecular clouds which may fall toward the center
(Yusef-Zadeh et al., 87, AIP Conf. 155: "The Galactic Center", ed. Backer; Güsten
et al., 87b, AA, in press). A high quality, well sampled survey at ≈100"
resolution of the molecular gas in the galactic center region has become available
from work at the Bell Labs' 7 m telescope (Bally et al., 87a ApJ Sup. 65,; 87b
ApJ, in press), giving detailed information on distribution, density structure,
mass and kinematics of molecular clouds near the galactic center. These and other
observations show that the molecular clouds in the central few hundred parsec are
denser and warmer than clouds in the disk, possibly due to turbulent heating
(Güsten et al., 85, AA 142. 381).

     A circum-nuclear ring of neutral gas and dust (diameter a few pc) originally
discovered in the far-infrared (cf. Becklin et al., 82, ApJ 258, 134), has now
been studied in the rotational transitions of several molecular lines at 3 mm at
resolutions between 20" and 3" with the 30 m IRAM telescope, the 45 m Nobeyama
telescope and the Hat-Creek mm-interferometer (Serabyn et al., 86, AA 169, 85;
Güsten et al., 87a, ApJ 318, 124; Kaifu et al., 87, AIP Conf. 155: The Galactic
Center, ed. Backer). The ring is almost complete, is highly clumped, turbulent
and is in strongly perturbed rotation about the center. The excitation, density
and temperature of the atomic and molecular gas in the ring is unusually high (cf.
Genzel et al., 85, ApJ 297, 766; Harris et al., 85, ApJ Lett. 294, L93). Ionized
gas in the radio "mini-spiral" is associated with the ring's sharp inner edge.
The velocity field of other ionized gas streamers in the central 2 pc have been
studied with the VLA and mid-IR spectroscopy. These data indicate that gas is
streaming toward (Serabyn and Lacy 85, ApJ 293, 445) or away from the center (van
Gorkom et al., 87, ApJ, in press). The dynamics of neutral and ionized gas in the
central 5 pc give substantial, but not fully convincing evidence for a ≈ 10^6 $M_\odot$
central black hole (for a review see Genzel and Townes 87, Ann. Rev. Astr. Ap.
25). New information on the central mass distribution has also come from the
velocity field of OH-IR stars (Winnberg et al., 85, ApJ Lett. 291, L45). The best
candidate for a black hole in the center may be the compact radio source Sgr A*
(Lo et al., 85, Nature 315, 124). Its proper motion against background quasars
indicates that it is located in the galactic center and must be rather massive

(Backer and Sramek 87, AIP Conf. 155: *The Galactic Center*, ed. Backer).

HII Regions and Recombination Lines - High sensitivity radio recombination line
(RRL) surveys of galactic HII regions using single dish radio telescopes continue
to be important in determining the kinematics and properties of a large sample of
galactic HII regions. A northern hemisphere H85, H87 and H88 alpha RRL survey
using the 140 foot radio telescope of NRAO with an anglar resolution of 3 arc min
at 3 cm has been carried out by Lockman (87, in press). 462 HII regions in the
longitude range 350-0-60° were detected. A southern hemisphere survey in the H109
alpha line using the 64 m Parkes telescope with a 4 arc min beam has detected 316
HII regions in the longitude range 210 to 360° (Caswell and Haynes 87 AA 171,
26). Both the Parkes and the NRAO surveys have detected a number of nebulae with
low electron temperature; these objects have in addition smaller Doppler line
widths (<15 km/s).

     Low frequency RRLs have gained in significance in the last years.
Anantharamaiah (85 JAA 6, 177 and 203; 86 JAA 7, 131) and Anantharamaiah and
Bhattacharya (86 JAA 7, 141) have observed 53 directions in the galactic plane
including the galactic centre using the Ooty Radio Telescope. The observations of
even lower frequency lines (below 100 MHz) opens up a fascinating new field of RRL
research. The carbon line is seen in absorption in the direction of Cas A at
these low frequencies (16-30 MHz: Konovalenko 84 Sov Astron Lett 10, 353; 26-68
MHz: Anantharamaiah et al., 85 Nature 315, 647; 42-84 MHz: Ershov et al., 84, Sov
Astron Lett 10, 348; 35-240 MHz: Payne et al., 87, in press). The latter
observations show that these carbon lines go into emission at frequencies greater
than 200 MHz. Additional detections of the carbon lines in the direction of the
galactic centre and M16 have been reported by Anantharamaiah et al., (87, in
press).

     High resolution studies of HII regions using the VLA and the WSRT form the
major part of the Ph.D. thesis of P.R. Roelfsema (Groningen, 1987). Hydrogen,
carbon and helium RRLs at frequencies from 327 MHz to 14 GHz were observed (H272
to H76 alpha). Spatial variations of the apparent abundance of ionized helium in
several compact HII regions have been seen. Apparent helium abundances well in
excess of 20 percent have been observed in portions of W3-A. Narrow RRLs of both
hydrogen and carbon from the H^O regions were mapped in the direction of several
sources (DR21, K350 and W3).

     High frequency mm line observations of RRLs have provided information on the
He/H ratio in Orion A and M17 at the 53 alpha line at 43 GHz at Nobeyama (Peimbert
et al., 87, Publ Astron Soc Japan, in press). Welch and Marr (87 ApJ 317, L21)
have observed the compact HII region W3 (OH) in the H42 alpha line at 86 GHz using
the Hat Creek mm interferometer. A systematic change of radial velocity with
observing frequency was detected. The H42 alpha velocity is in close agreement
with the OH maser and NH_3 absorption velocities. Gordon (87, ApJ, in press) has
observed the H40 and He 40 alpha lines from seven HII regions using the 12 m
telescope of NRAO at 99 GHz.

     VLA observations of the H76 alpha line at 2 cm with an angular resolution of
0.4 arc sec have been made of G34.3+0.2 and G45.07+0.13 (Garay et al., 86 ApJ 309,
553). Large velocity gradients have been observed. Churchwell et al., (87 ApJ
321, 516) have mapped the central 90 arc sec region of the Orion nebula in the
continuum at 2 cm using the VLA with an angular resolution of 0.1 to 0.2 arc sec.
Twenty nine ultracompact sources (EM $> 10^9$ pc cm^{-6}) with diameters in the range <
30 to 230 AU were detected.

     RRL observations of Sgr A West using both the WSRT and the VLA at 6 and 2 cm
are summarized by van Gorkom et al., (84 Proc IAU Sym 106, p 371). Mezger and

Wink (86 AA 157, 252) have mapped Sgr A West using the 100 m telescope of the MPI
in the H66 alpha line with an angular resolution of 42 arc sec. H110 alpha (6 cm)
emission from the galactic centre ARC has been observed by Yusef-Zadeh et al., (87
The Galactic Center, ed Backer p 190).

Radio observations of extragalactic HII regions are important in determining
global, extinction-free parameters. A WSRT survey of M33 at 1.4 GHz (Viallefond
et al., 86 AA Suppl. 64, 237) has resulted in the detections of over 100 radio
sources associated with H alpha nebulosity. Detailed high resolution VLA
observations of the giant HII region NGC 5471 in the galaxy M101 were carried out
by Skillman (85 ApJ 290, 449) in the HI line and in the continuum at 20 and 6 cm.
Evidence for a supernova remnant imbedded in the HII region was found. Sramek and
Weedman (86 ApJ 302, 640) have observed a number of HII regions in M33 and M101
using the VLA at 20 and 6 cm. RRL observations of extragalactic objects are quite
difficult due to the low intensities and the large line widths. The gas rich
galaxy M82 has been mapped with the VLA at 6 cm in the H110 alpha line (Seaquist
et al., 85 ApJ 294, 546) and in the H166 alpha line by Roelfsema (87 Ph.D. thesis,
Groningen).

Interstellar Molecules - With the normal operation of the IRAM 30-m telescope
(Baars et al. 1987 A&A 175, 316) and the imminent commissioning of the 15-m
Maxwell, 10-m Onsala-ESO SEST and the Caltech 10-m High Dish, the future of
molecular line observing is very bright. The searches for complex molecular
species have been hampered by the large number of mm transitions (see Blake et al.
1987 ApJ 315, 621) but the detection of acetone at 3 mm has been reported by
Combes et al. (1987 A&A 180, L13). At shorter wavelengths the lines of HCl (Blake
et al. 1985 ApJ 295, 501) and H_2D^+ (Phillips et al. 1985 ApJ 294, L45) are
reported. The laboratory measurements of C_3H_2 by Thaddeus et al. (1985 ApJ 299,
L63) allowed the identification of many previously unidentified lines, from this
very abundant, widespread organic ring molecule. Further progress in
identifications was made by measurements of lab spectra for "non-terrestrial"
molecules such as C_2S (Saito et al. 1987 ApJ 317, L115) and C_3S (Yamamoto et al.
1987 317, L120). Identifications of C_5H (Cernicharo et al. 1986 A&A 164, L1) and
C_6H (Guélin et al. A&A 175, L5) as well as a heavy radical (Gomez-Gonzales et al.
1986 A&A 157, L17) were accomplished without the aid of previously known
frequencies. A related development is the possible connection of very small
grains with polymers of aromatic ring molecules such as $C_{25}H_{50}$ (Léger and Puget
1984 A&A 137, L5). Compelling evidence for the evaporation of complex molecules
from the surface of dust grains has been given by Walmsley et al. (1987 A&A 172,
311) and Plambeck and Wright (1987 ApJ 317, L101). Caution in understanding
interstellar chemistry is required, since fundamental problems related to the
abundance of CH^+ are still present (Crutcher and Federman 1987 ApJ 316, L71), even
after 36 years of study. Complete maps covering large areas in the J=1-0 line of
^{13}CO have been made for the Orion cloud (Bally et al. 1987 ApJ 312, L45) and the
Taurus dust cloud (see Kleiner and Dickman (1987 ApJ 312, 839). An analysis for
large CO surveys of our galaxy in the J=1-0, ground state line is found in Waller
et al. (1987 ApJ 314, 397). The investigation of hot molecular clouds has begun
with the detection of the J=7-6 line of CO (Schultz et al. 1985 ApJ 291, L61;
Harris et al. 1985 ApJ 294, L93). Fifteen sources were measured in the (7,7) line
of NH_3 (Mauersberger et al. 1986 A&A 162, 199).

Observations of CO near the galactic center have been used to study the mass
distribution, but no definite proof of the presence of a "Black Hole" has yet been
obtained (Crawford et al. 1985 Nature 315, 467; Serabyn et al. 1986 A&A 169, 85).

Using molecular lines, the search for potential protostars such as IRAS 1629A
(Walker et al. 1987 ApJ 309, L47; but see Mundy et al. 1986 ApJ 311, L75, and
Menten et al. 1987 A&A 177, 231) has been extended to angular scales of 16"-20".

Measurements of many stellar molecular envelopes in CO are in Knapp and Morris (1985 ApJ 292, 640). The use of large single telescopes (see e.g. Likkel et al. A&A 173, L11) and interferometers (see Heiligmann et al. 1986ApJ 308, 306) allows a spatial resolution of structures and provides for the first time the possibility to map out the distribution of different species. Bipolar outflows also give a connection between molecules and stellar evolution. Results are summarized in Lada (1985 Ann Rev. A&A 23, 267). A detailed investigation of the first discovered bipolar outflow source, L1551, shows that on a scale of 20" the molecular disk is not rotating (Batrla and Menten 1985 ApJ 294, L125) or collimating the flow (Walmsley and Menten 1987 A&A 179, 213). In the most energetic outflow, Orion, the line connecting the red and blue shifted lobes is 6"±2" north of the supposed source IRC2 (Wilson et al. 1986 A&A 167, L17).

In external galaxies, the J=2-1 line of CS found in dense galactic regions has been detected by Henkel and Bally (1985 A&A 150, L25). Extragalactic C_3H_2 was found by Seaquist and Bell (1986 ApJ 303, L67), as well as OH absorption from rotationally excited levels (Henkel et al. 1986 A&A 168, L13). Maps of CO have been extended to higher resolution by interferometers (see e.g. Scoville et al. 1986 ApJ 311, L49). Although much of the single-dish flux density is lost, and in extended sources many fields must be mapped, these data could be combined using methods described in A&A 143, 77.

Interstellar Masers – This field has seen tremendous growth in the past three years because of advances in receiver technology and interferometric techniques, the availability of the IRAS data base, and because of the increasing number of telescopes available at cm and mm wavelengths. The major developments in the study of circumstellar masers have been reviewed by Bowers 1985 (Mass Loss from Red Giants p189) and Cohen 1987 (IAU Symp No. 122. 229), and masers in star-forming regions are discussed extensively in the Haystack Symposium Proceedings "Masers, Molecules and Mass Outflows in Starforming Regions" 1986 (particularly reviews by Genzel p233 and Elitzur p299). The impact of QUASAT on study of galactic masers, and the interstellar scattering problem, are discussed by Booth 1984 (ESA-SP 213, 171).

The number of circumstellar maser sources nown now exceeds one thousand. IRAS-based surveys for OH and H_2O masers have been carried out by Engels et al. 1984 (A&A 140, L9), Lewis et al. 1985 (Nature 313, 200), Sivagnanam & Le Squeren 1986 (A&A 168, 374), Zuckerman & Lo 1987 (A&A 173, 263) and others. The detection rate is typically 30%. Surveys of a more traditional type have been carried out by Bowers & Hagen 1984 (ApJ 285, 637, Jewell et al. 1985 (ApJ 298, L55), Slootmaker et al. 1985 (A&A Suppl. 59, 465). Dickinson et al. 1986 (AJ 92, 627), Nyman et al. 1986 (A&A 160, 352), Bujarrabal et al. 1987 (A&A 175, 164) and others. Winnberg et al. 1985 (ApJ 291, L45) have detected a large concentration of 1612 MHz OH-IR sources in the direction of Sgr A.

Extensive monitoring programmes of the major maser lines have been undertaken. Monitoring of OH 1612 MHz OH-IR sources is especially valuable because of its applications to the galactic distance scale. Phase-lag measurements of 1612 MHz maser shell diameters have been made by Herman & Habing 1985 (A & A Supp, 59, 523), and combined with VLA and MERLIN maps to obtain source distances (Herman et al. 1985, A&A 143, 122; Diamond et al. 1985, MNRAS 212, 1). This technique when refined may ultimately give the most accurate measurement of the galactic centre distance. The detection of the first extragalactic OH-IR star in the LMC by Wood et al. 1986 (ApJ 306, L81) encourages even grander hopes.

Results of monitoring circumstellar OH mainline, H_2O and SiO masers have been published by Berulis et al. 1984 (Sov.Inf.Astr.Council. 56, 92), Le Squeren & Sivagnanam 1985 (A&A 152, 85), Gomez Balboa & Lepie 1986 (A&A 159, 166), Clark et

al. 1985 (ApJ 289, 756), Nyman & Olofsson 1986 (A&A 158, 67), Snyder et al. 1986
(AJ 82, 416) and others. The data give insight on the physical processes
occurring nearer to the star. The theory of SiO masers has been comprehensively
discussed by Langer & Watson 1984 (ApJ 284, 751), and Cooke & Elitzur 1985 (ApJ
295, 175) have presented a detailed model for circumstellar H_2O masers. Work on
pumping the OH mainlines has not yet recovered from the revelation by Andersen
1986 (A&A 154, 42) that the Λ-doublet states of OH have hitherto been incorrectly
assigned.

VLA, MERLIN and VLBI maps of circumstellar OH mainline and H_2O masers which
probe the main inner regions of circumstellar envelopes have been published by
Chapman & Cohen 1985 (MNRAS 212, 375), Diamond et al. 1985 (MNRAS 216, IP),
Johnston et al. 1985 (ApJ 290, 660) and Diamond et al. 1987 (A&A 174, 95). The
observations of S Per by Diamond et al. resolve H_2O maser spots only ~1 mas in
extent. But the most comprehensive data set is that by Chapman & Cohen 1986
(MNRAS 220, 513) for VX Sgr. Observations of OH, H_2o and SiO masers enable the
velocity field through this envelope to be studied in detail. Multi-frequency
studies of this type hold great promise.

Alcock & Ross have given new impetus to the study of radiative transfer in
circumstellar masers in a series of papers on saturation and beaming (ApJ 290,
433; 299, 763; 305, 837, 306, 649 and 310, 838). They point out major defects in
the standard shell model, and conclude that the outflow must be clumpy.
Observational evidence for this comes from study of SiO masers (Miller et al.
1984, ApJ 287, 892) and the detection of fine structure in the OH profiles of
OH-IR sources (Fix 1987, AJ 92, 433 and Cohen et al. 1987, MNRAS 225, 491).

The highlight of the study of interstellar, as opposed to circumstellar
masers has been the detection of powerful new methanol masers by Batrla et al.
1987 (Nature 326, 49). The $2_0-3_{-1}E$ maser line at 12 GHz is both widespread and
comparable in its photon rate with galactic OH and H_2O masers. It is usually
found in association with compact HII regions, but sometimes only absorption is
observed. The new maser seems very suitable for interferometric studies. Other
methanol masers have been reported by Wilson et al. 1985 (A&A 147, L19), Morimoto
et al. 1985 (ApJ 288, L11) and Menten et al. 1986 (A&A 157, 318). VLA
observations of W3OH by Menten et al. 1985 (ApJ 293, L83) show that the $9_2-10_1A^+$
methanol masers at 23 GHz are excited in the same cloud as the OH masers.

Surveys for interstellar OH and H_2O masers were made by Matthews et al. 1985
(A&A 149, 227), Wouterlout & Walmsley 1986 (A&A 168, 237), Caswell & Haynes 1987
(Aust.J.Phys.40, 215), Braz & Epchtein 1987 (A&A 176, 245), and others. Caswell &
Haynes also studies the galactic distribution of OH masers. Two more SiO masers
similar to those in Orion-KL where reported by Ukita et al. 1985 (IAU Symp.No.
115, p178).

Interferometric studies of OH and H_2O masers and their relationship to
compact HII regions have been carried out by Cohen et al. 1984, (MNRAS 210, 425),
Baart & Cohen 1985 (MNRAS 213, 641), Garay et al. 1985 (ApJ 289, 681), Forster &
Caswell 1985 (IAU Symp.No. 115, p.174) and others. Garay et al. also measured the
$H76\alpha$ and $H66\alpha$ recombination lines from compact HII regions associated with OH
masers, to try to establish the dynamics of the maser regions. Maps of excited OH
at 6 cm have been published by Palmer et al. 1984 (MNRAS 211, 41P) and Gardner et
al. 1987 (MNRAS 225, 469). The highlight of interferometer maping has been the
dtermination of the distance to Sgr B2 from VLBI measurements of H_2O proper
motions by Reid et al. 1985 (IAU Symp.No. 115, p.554).

Magnetic fields in star-forming regions have been measured from Zeeman
spliting of OH maser lines by Benson et al. 1984 (AJ 89, 1391), Baart et al. 1986
(MNRAS 219, 145) and others. The transfer of polarized radiation in cosmic masers

in the presence of Zeeman splitting has been treated theoretically by Deguchi &
Watson 1986 (ApJ 300, L15 and 302, 750). Rapid variations and flares of masers in
starforming regions have been reported by Lehkt & Sorochenko 1984 (Sov.Ast.Let.
10, 307), Haschick & Zisk 1984 (AJ 89, 1387), Mattila et al. 1985 (A&A 145, 192),
Rowland & Cohen 1986 (MNRAS 220, 233), Abraham & Vilas Boas 1986
(Rev.MexAstr.Astrof. 12, 228) and others. These studies have been most successful
when accompanied by simultaneous interferometer meaaurements to remove the
ambiguity of spectral blending. Tarter & Welch 1986 (ApJ 305, 467) have presented
a model for H_2Oflares which can account for the rapid timescales observed.
However there is still a problem in explaining the luminosities of the most
powerful sources. An alternative model by Kylafis & Norman 1986 (ApJ 300, L73)
invokes collisions with neutral and charged particles at different temperatures to
overcome this difficulty.

EXTRAGALACTIC RESEARCH

Extragalactic source surveys - Large-scale surveys have been reported over a wide
range of frequencies. The UTR-2 telescope survey of six frequencies in the 10-25
MHZ band has continued, detecting 313 sources with $52° < \delta < 60°$ (Braude et al. 85
Ap. Space Sci. 111, 1). The 151 MHz 6C survey of 1761 sources stronger than 120
mJy and north of $\delta = 80°$ (Baldwin et al. 85 MN 217, 717) was made with $4!2 \times$
$4!2cosec\delta$ resolution and is especially sensitive to old, extended radio
components. The angular sizes, optical identifications, and size evolution of 6C
sources with $S \approx 2$ Jy have been studied (Eales 85 MN 217, 149; 217, 167; 217,
179). The upgraded Bologna telescope produced the B3 408 MHz survey of 13 354
sources complete to 0.1 Jy in a 0.78-sr region $37°15' < \delta < 47°37'$ (Ficarra,
Grueff, and Tomassetti 85 AA Suppl. 59, 255). The NRAO 91-m telescope was used to
make confusion-limited ($\sigma \approx 30$ mJy) 1400 MHz maps covering $-5° < \delta < +82°$ with
$11!1 \times 12!7$ resolution (Condon and Broderick 85 AJ 90, 2540; 86 AJ 91, 1051).

Wall and Peacock (85 MN 216, 173) combined several high-frequency surveys to
generate an "all sky" catalogue of 233 sources stronger than 2 Jy at 2.7 GHz. At
3.9 GHz the RATAN-600 telescope surveyed 0.92 sr in the $0° < \delta < 9°$ band with
$4^s.6sec\delta \times 52'$ resolution and detected 3 255 sources stronger than 80 mJy
(Amirkhanyan et al. 85 Comm. Spec. Ap. Obs. USSR 47). The MIT-Green Bank 5 GHz
survey was made with the NRAO 91-m telescope to detect gravitational lens
candidates. A total of 5 974 sources stronger than $5\sigma = 53$ to 106 mJy were found
in the 1.87-sr region defined by $-00°30' < \delta < +19°30'<\delta<+19°30', |b|>10°$ (Bennett
et al., 86 ApJ Suppl. 61, 1). source counts were obtained by Bennett, Lawrence,
and Burke (85 ApJ 299, 373); VLA radio structures and optical identifications of
nearly 1000 sources were reported by Lawrence et al., (86 ApJ Suppl. 61, 105).

The narrow strip $32°54'<\delta<33°30'$ was resurveyed with the NRAO 91-m telescope
at 5 GHz during 1981 November to detect variations since the original 1970-71
survey (Altschuler 86 AA Suppl. 65, 267). The spectral-index distributions and
counts of sources selected at 4850 MHz were investigated by three surveys made
with the MPIfR 100-m telescope. A $9,2 \times 10^{-3}$-sr portion of the 1400 MHz GB survey
region (Maslowski et al., 84, AA 141, 376) and 6.6×10^{-3} sr of the 408 MHz 5C2
survey area were resurveyed. (Maslowski et al, 84, AA 139, 85). Benn et al., (84
MN 209, 683) mapped 4.7×10^{-3} sr of the 5C 12 field. These surveys confirm the
convergence of the 5 GHz source counts and the declining fraction of flat-spectrum
sources at low flux densities. Aperture-synthesis surveys have covered a number
of small areas with higher sensitivity and resolution. The 5C 20 survey centred
on Abell 2218 found discrete sources at 408 and 1407 MHz that might affect
measurements of the Sunyaev-Zeldovich effect (Birkinshaw 86 MN 222, 731). Four
Einstein deep X-ray survey fields were mapped at 608.5 and 1412 MHz with the WSRT,
(Katgert-Merkelijn et al., 85 AA Suppl. 61, 517). Sensitive ($\sigma\approx0.2$ mJy) MOST maps
of $25\approx1°$-diameter fields made with $43" \times 43"$ cosecδ resolution have been used to

determine the 843 MHz source count to S=1mJy (Subrahmanya and Mills 87 Proc. IAU Symp. 124, 569).

The Leiden-Berkeley Deep Survey (LBDS) is a major radio/optical project to study faint radio galaxies. The WSRT 3-km array mapped four high-latitude fields covered by deep multicolor KPNO 4-m telescope plates, detecting 306 sources stronger than 5σ($\geqslant$0.6 mJy) at 1412 MHz [with 12".5 x 12".5 cosecδ resolution] (Windhorst, van Heerde, and Katgert 84 AA Suppl. 58, 1). Optical identifications (Windhorst, Kron and Koo 84 AA Suppl. 58, 39), broadband photometry, and spectroscopy (Kron, Koo and Windhorst 85 AA 146, 38) reveal an increasing population of blue galaxies associated with sources fainter than 10 mJy. The Lynx.2 field (α=8^{h}41^{m}46^s, δ=+44°46'50") was mapped at higher sensitivity with both the WSRT 3-km array (Oort and Windhorst 85 AA 145, 405) and the VLA C-array (Windhorst et al., 85 ApJ 289, 494). High resolution VLA A-array snapshot maps of 133 LBDS sources show that their angular sizes decrease with flux density in the 1-10 mJy range (Oort et al., 87 AA 179, 41).

The deepest 1.49 GHz VLA map of a single primary-beam area (30' FWHM) at α=13^{h}00^{m}37^s, δ=+30°34' reached the confusion limit (σ_n=10 μJy, σ_c=11μJy) for 17".5 resolution, yielding direct source counts to 84 μJy and statistical P(D) counts down to 10 μJy (Mitchell and Condon 85 AJ 90, 1957). Significantly deeper surveys can be made only with higher resolution. The angular-size distribution of sub-mJy sources was estimated from a 5".8-resolution VLA B-array map with σ=21 μJy [at α=8^{h}52^{m}15^s, δ=+17°16']; the median angular size of sources fainter than 400 μJy atr 1.465 GHz is $\langle\theta\rangle$<3" (Coleman and Condon 85 AJ90, 1431), significantly lower than the value $\langle\theta\rangle\approx$10" found for S $\geqslant$ 10 mJy.

The source counts and angular-size distributions obtained by different groups observing independent areas of sky with both the WSRT and the VLA agree within the $\sqrt{N}$ statistical uncertainties in samples of N sources, suggesting that clustering in the small areas covered ($\approx10^{-4}$ sr) does not distort the results. The unexpectedly large number of compact sub-mJy sources is well established at 1.4 GHz, but their nature and space distribution remain to be determined (cf. Wall et al., 86 Highlights Astron. 7, 345; Windhorst 86 Highlights Astron. 7, 355).

Some 6 cm VLA D-array surveys approach the discrete-source confusion limit and may also be affected by small-scale (<1') fluctuations in the 3K cosmic background radiation. One deep field ($\geqslant$=11 μJy) and ten intermediate-sensitivity fields (σ=68 μJy) were mapped at 18" resolution, yielding source counts to 60 μJy (Fomalont et al,m 84 Science 225, 23). Partridge, Hilldrup, and Ratner (86 ApJ 308, 46) reached 100 μJy with VLA C- and D-array maps. Source counts in the 25-400 μJy range (Kellermann et al., Highlights Astron. 7, 367) were obtained from the deepest 6 cm VLA D-array survey ($\sigma_n\approx$4.5 μJy). the sub-mJy 6 cm counts agree within their $\sqrt{N}$ errors, which are large because N is still small. More deep 6 cm surveys are needed, with overlapping 20 cm maps to define the spectral-index distribution of the faintest sources.

Cosmological tests - Counts of radio sources at the mJy and sub-mJy level have been reported from surveys near 1.4 GHz using the VLA (Coleman et al 85 AJ 90, 1437; Mitchell et al 85 AJ 90, 1957) and the Westerbork array (Katgert-Merkelijn et al 85 AA Supl 61, 547; Oort and Windhorst 85 AA 145, 405; Windhorst et al 85 ApJ 289, 494) and near 5 GHz using the VLA (Kellermann et al 86 Highlights of Astronomy, Vol 7, p 367; Partridge et al 86 ApJ 308, 46). Subrahmanya and Mills (87 IAU Symp 124, p 569) have estimated the counts at 843 MHz for the entire range of flux density down to aproximately 1 mJy based on observations with a single instrument, the Molonglo Synthesis Telescope.

The deepest counts now reach source densities of approximately $10^6/$ steradian. There is reasonably good agreement on the observed upturn in the normalized differential counts below approximately 1 mJy, at both 1.4 and 5 GHz in different fields over the sky, suggesting the emergence of a new population of sources at these flux leves. the nature of this population is however not well understood and remains controversial (Windhorst 86 Highlights of Astronomy Vol 7, p355; Wall et al. 86 Highlights of Astronomy Vol 7, p 345). It is becoming clear from deep optical identifications and spectroscopy (Windhorst et al. 84 AA Supl. 58, 39; Kron et al. 84 AA 146, 38; Windhorst et al. 87 IAU Symp. 124, p 573; Downes et al. 86 MNRAS 218, 31; Mitchell and Condon 85 AJ 90, 1957; Weistrop et al. 87 AJ 93, 805), however, that at low flux levels one is seeing more and more of the moderately active galaxies of relatively lower radio and optical luminosity.

V/V_m tests on large samples of bright sources at 2.7 and 5 GHz have confirmed that strong evolution is required for high luminosity sources of both steep as well as flat spectrum (Wall and Peacock 85 MNRAS 216, 173; Marrides and Mutus 84 AA 131, 81). Free-form models of the epoch dependence of the radio luminosity function also support this conclusion (Peacock 85 MNRAS 217, 601; Zawislak-Raczka and Kumor-Obryk 86 MNRAS 222, 487). This is true also in the evolutionary model reported by Condon (84 ApJ 287 461) which uses a combination of density and luminosity evolution but treats all sources in the same way.

The need for evolution with epoch in the linear sizes of extended radio sources based on angular size - redshift (θ-Z) tests has in the past been difficult to establish because of the inability to distinguish such evolution from the effects of a possible inverse correlation between radio luminosuity and linear size. From studies of the θ-Z relations for radio galazies of similar luminosity, it now appears that strong evolution in linear size is indeed necessary (Kapahi 85 MNRAS 214, 19; Eales 85 MNRAS 217, 179; Gopal-Krishna et al. 86 IAU Symp 119, p 193; Oort et al. 87 AA 179, 41). It is also becoming clear that there is a weak but direct correlation between luminosity and size for radio galaxies (Machalski and Condon 85 AJ 90, 973; Kapahi 86 Highlights of Astronomy Vol. 7, p 371; Oort et al. 87 AA 179, 41).

In the case of quasars, although the observed θ-Z relation is similar to that for galaxies (Kapahi 87 IAU Symp 124, p251), interpretation is less simple because of the difficulty of defining reasonably complete and unbiased samples.

The effect of various selection effects on the interpretation of the angular size - flux density (θ-S) relation has been discussed by Allington-Smith (84 MNRAS 210, 611). Comparison of the observed θ-S relation at 408 MHz with predictions based on several strong-source samples and evolution models of the RLF shows that the data cannot be fitted without invoking size evolution (Kapahi et al. 87 JAA 8, 33).

Gopal-Krishna and Wiita (87 MNRAS 226, 531) have investigated the interaction of jets with hot galactic haloes and an intergalactic medium in order to explain the observed evolution in linear sizes.

High resolution VLA observations of mJy and sub-mJy sources from surveys at 1.4 GHz show that the median angular size at these flux levels drops to $\leq 3"$ arc (Coleman and Condon 85 AJ 90, 1431; Oort and Windhorst 85 AA 145, 405; Oort et al. 87 AA 179, 41) compared to the value of approximately 10" arc at higher flux levels. This appears to be related to the emergence of the new population of galaxies at the faintest levels which have a smaller radio size than the ellipticals.

Flux densities and spectra - Measurements of flux densities and spectra of active
radio sources have been reported over a wide range of frequencies. Braude et al
(85 Ap Sp Sci 111, 237) have extended the spectra of 114 sources down to 12 MHz.
Roger et al. (86 AA Sup 65, 485) have reported flux densities of 395 sources at
22 MHz. Tovmassian et al (84 AA Sup 58, 317) have observed 464 Markarian galaxies
at 2.7 GHz leading to 14 detections. Flux densities of sources in the 5C2 survey
region have been measured at 5 GHz and used to investigate the dependence of
spectral index on flux density for samples defined at 1.4 and 5 GHz (Maslowski et
al 85 AA 139, 85). Kulkarni and Mantovani (85 AA Sup 61, 1) have reported
measurements of flux density at 5 GHz for 184 sources selected from the B2.3
survey, which have been used to improve the spectral index-flux density
correlation for source samples defined at 408 MHz (Kapahi and Kulkarni 86 AA 165,
39). Lawrence et al (86 ApJ Sup 61, 105) have published spectral indices for a
large sample of 632 sources from the MIT-Green Bank Survey at 5 GHz.

At shorter wavelengths, Tabara (84 PAS Jap 36, 297) has measured at 10 GHz
flux densities of 135 weak flat-spectrum sources. A sample of 95 sources has
been observed at 22 and 37 GHz by Efanov et al (84 Izv Krymskoy Astrofiz 69, 78).
Valtaoja et al (85 Izv. Krymskoy Astrofiz. obs. 70, 144) have reported coordinated
observations of 20 sources in the 11-37 GHz range. A second list of 25
'gigahertz-peaked-spectrum' sources has been prepared, to facilitate searches for
compact doubles and high-redshft sources (Spoelstra et al 85 AA 152, 38).

Weak radio nuclei of powerful radio sources are shown to have distinctly more
curved radio spectra than strong "flat-spectrum" sources selected at 5 GHz and,
thus, they may not be merely scaled-down versions of the latter (Rudnick et al 86
AJ 91, 1011). Synchrotron spectra of the hotspots of powerful radio sources have
been shown to extend to near-infrared wavelenths in cases of 3C33 and 3C273 and
perhaps to X-ray band in Pictor A (Röser and Meisenheimer 87 ApJ 314, 70).

Relationship of the compact and extended radio emission of quasars to their
optical and X-ray emission and the possible role of relativistic beaming at all
these bands has been quantitatively examined (Browne and Murphy 86 MN 226, 601,
see also Kembhavi et al 86 MN 220, 51 and Feigelson et al 84 AJ 89, 1464). Robson
et al. (85 MN 213, 355) have reported non-detection of a dozen radio-quiet
quasars at 1 mm wavelength down to a flux density generally below 1 Jy.

Much effort has been devoted to combine the radio flux measurements of the
active galactic nuclei with their observations (often, quasi-simultaneous) at
higher frequencies reaching up to the X-ray band in several cases. Considerable
improvement has thus been attained in understanding the physical mechanisms and
the inter-relationship of the nuclear emission in the different wave bands, in
disentangling the thermal and non thermal components and in assessing the role of
opacity effects and of the postulated bulk relativistic motion of the radiating
plasma. Such studies comprise those involving special individual objects such as
NGC1275 (Longmore et al. 84 MN 209, 373; Gear et al 85 MN 217, 281), 3C345
(Bregman et al 86 ApJ 301, 708), 3C446 (Brown et al 86 MN 219, 671), Mrk421
(Makino et al 87 ApJ 313, 662) and 3C273 (Courvoisier et al 87 AA 176, 197), as
well as samples of blazars/flat-spectrum radio sources (e.g., Mufson et al 84 ApJ
285, 571; Worrall et al 84 ApJ 286, 711; Gear et al. 85 ApJ; 291, 511; Ledden and
O'Dell 85 ApJ 298, 630; Lépine et al (85 AA, 149, 351); Roellig et al 86 ApJ 304,
646). Based on similar multi-band observations, substantial differences of
spectral and other properties have been reported between the radio-selectred and
X-ray selected blazars (Stocke et al. 85 ApJ 298, 619; Ghisellini et al. 86 ApJ
310, 317, Marasch et al. 86 ApJ 310, 325). Landau et al. (86 ApJ 308, 78) have
used their near-simultaneous, multifrequency observations of 15 blazars over the
radio-optical range to conclude that the global spectra of all these sources can
be well fitted with smoothly varying functions like parabolae and that interesting
correlation is present between the fit parameters. This hints towards a common

regulating mechanism for active extraglactic sources.

<u>Structure of Radio Galaxies and Quasars</u> - Multifrequency mapping and
high-sensitivity, high-dynamic range observations with a variety of synthesis
telescopes - the VLA, MERLIN, Cambridge, Westerbork, Fleurs, Molonglo and Ooty -
have produced a wealth of new information on extragalactic radio sources on scales
from about 0.1 arcsec to many minutes of arc. Excellent reviews have appeared in
several conference proceedings e.g. the NRAO Workshop No. 9 on "Physics of Energy
Transport in Extragalactic Radio sources" (ed. A H Bridle & J A Eilek) (hereafter
PETER), IAU Symposium No. 119 on "Quasars" (ed. G Swarup & V K Kapahi), the
Canadian Institute for Theoretical Astrophysics conference on "Jets from Stars and
Galaxies" (ed. R N Henriksen) (Canadian J Phys 64, 351), the 1986 Erice Summer
School on "Astrophysical Jets and their Engines" (ed W Kundt).

Surveys of source structure - Statistical analysis of the structural properties of
complete samples of radio sources leads to useful constraints on physical models
of these objects. Fanti et al (87 AA Supp 69, 57 and references therein) have now
completed VLA observations of the B2 sample of about a hundred low luminosity
radio galaxies and Machalski & Condon (85 AJ 90, 5; 85 AJ 90, 973; 86 AJ 91,998)
have made VLA observations of a complete sample of intermediate-strength radio
sources selected from the GB/GB2 1400-MHz surveys. Thirty extragalactic 3C
sources in the Galactic plane have been mapped with the Cambridge 5-km telescope
(Pooley et al 87 MNRAS 224, 847). Pearson et al (85 AJ 90, 738) have mapped 36
compact sources from the 3C catalogue with the VLA with resolutions of 0.4
arcsec. These are mostly steep-spectrum objects, many of which look like small
angular-size doubles, though there is a variety of structures; however none of the
steep-spectrum objects are like the few flat-spectrum 3C sources which have
dominant cores and one-sided arcsecond jets. There have been a number of VLA
surveys of the structures of quasars. Gower & Hutchings (84 AJ 89, 1658) have
mapped low redshift quasars, Rogora et al (86 AA Supp 64, 557; 87 AA Supp 67, 267)
have looked at 74 B2 quasars and Owen & Puschell (84 AJ 89, 932) have mapped 26
quasars from the Jodrell Bank 966-MHz survey.

Jets - Bridle (84 AJ 89, 979; 86 Canadian J Phys 64, 353) and Bridle & Perley (84
ARAA 22, 319) summarize and discuss the overall properties of extragalactic jets.
The jets in lower luminosity sources (less than 10^{25} WHz^{-1} at 1.4 GHz) are
two-sided and smooth and dominated by perpendicular magnetic fields whilst those
in the more powerful double sources are one-sided and knotty with parallel
magnetic fields. The jets in the lower luminosity sources also spread more
rapidly.

   Multifrequency mapping has revealed much about the physics of these jets.
Perhaps the most exciting observations are those of the double-lobed source 3C 120
(Walker et al 87 ApJ 316, 546). A sequence of maps with increasing resolution
shows that the jet is continuously connected from within 1 pc of the central
engine to over 100 kpc away, well outside the associated galaxy, and gradually
bends through about 90°. The source exhibits superluminal motions on parsec
scales with features which move outwards, and the fact that this is continuous
with the large scale jet is the best evidence available to date that material <u>is</u>
moving outwards in these kpc-scale jets; the velocities on parsec scales are
presumably relativistic but there is little constraint on the velocity of the
large scale jet. Synchrotron radiation from jets probably results from particle
acceleration in shocks produced by the interaction between the jets and the
surrounding medium. VLA maps of the inner 700 pc of the Cen A jet (Clarke et al
86 ApJ 300, L41) which show it to consist of limb-brightened knots which alternate
from side to side strongly support this suggestion. Observations also indicate
that jets are light and not usually free; limits on the surrounding hot gas
indicate that they can in general be confined by thermal pressure (e.g. 3C 219,

Bridle et al 86 AJ 92, 534) though this is almost certainly not the case for the jet in Cygnus A (Dreher et al 84 PETER, 57) for which magnetically-assisted collimation is proposed.

Multifrequency maps of the jets in the narrow-angle tail source NGC 1265 in the Perseus cluster (O'Dea and Owen 86 ApJ 301, 841; 87 ApJ 316, 95) are consistent with a model in which the jets have been bent by the motion of the galaxy through the cluster gas. The magnetic field near the core is parallel to the jet as expected if a tangled magnetic field is sheared tangentially to the jet surface. In the outer parts of the jets, where they spread rapidly, the magnetic field is perpendicular to the jet axis - this can be attributed to expansion and deceleration. Similar behaviour is observed in the jets of 3C66B (Leahy et al 86 AA 156, 234). The jets in high-power sources also exhibit bends and wiggles. The bends, which are often quite sharp e.g. 1759 + 211 (Saikia et al 87 MNRAS 224, 53), 4C 29.50 (Lonsdale & Barthel 86 ApJ 303, 617), are attributed to collisions with dense clouds of inter- or intra-galactic gas. The wiggles and oscillations, such as those observed in 3C 273 (see below), 1759+211 and 0800+608 (Shone & Browne 86 MNRAS 222, 365) are generally thought to be due to hydrodynamical instabilities.

As already noted, the jets in powerful sources are one-sided; of the one hundred known quasar jets none is clearly two-sided (Owen 86 Quasars, 173). There is a correlation between jet and core power (Burns et al 84 ApJ 283, 515) and the strengths of both core and jet are the outstanding differences between radio galaxies and quasars. Given that some of these cores are also superluminal, there is, at first sight, strong evidence that the one-sidedness of the kpc-scale jets in powerful sources is due to relativistic beaming. Contrary evidence is provided by the fact that the projected linear size distributions for sources with and without jets are not significantly different and by the asymmetries in the outer lobes (e.g. Saikia 84 MNRAS 209, 525). Further, observations of the quasar 3C 273 with MERLIN at 151, 408 and 1666 MHz (Foley & Davis 85 MNRAS 216, 679; Flatters & Conway 85 Nature 314, 425; Davis et al 85 Nature 318, 343) indicate that it is intrinsically one-sided. The jet is continuous from the core to beyond the optical jet, and there is an extended lobe associated with this jet; there is however no counter lobe. Whilst the jet motion is likely to be relativistic throughout, Doppler beaming factors must be too small to hide the counter lobe. The jet one-sidedness in the inner regions of some low-luminosity sources is also intrinsic (Laing 84 PETER 119). There are faint counter jets in the powerful radio galaxies 3C 219 (Bridle et al 86) and Cygnus A (Dreher et al 84) which could be accounted for if the jets are symmetric but contain relativistically moving shocks. It is interesting however that the counter jet components are at the same distance from the core as a gaps in the jet so that the jet - counter jet asymmetry could be a result of episodic energy transport or flip-flop (Rudnick 84 PETER, 35).

Core-dominated sources and unified schemes - In the unified schemes the differences between core- and lobe-dominated sources is simply a function of the angle that they make with the line-of-sight - the core-dominated sources being Doppler boosted. High resolution, high dynamic range observations of core-dominated sources have revealed extended structure on arcsecond scales in the majority of them (e.g.Browne & Perley 86 MNRAS 222, 149; Antonucci & Ulvestad 85 ApJ 294, 158; Antonucci et al 86 AJ 92, 1; Antonucci 86 ApJ 304, 634; Saikia et al 86 Quasars, 219) and are in broad agreement with the predictions of the unified schemes as far as the observed strength and size of the radio emission is concerned. However the facts that the extended emission around core-dominated sources is sometimes very large (> 100 kpc) (de Bruyn & Schilizzi 86 Quasars, 203) and that some double sources with projected sizes of several hundred kiloparsecs have superluminal cores (Shone et al 85 Nature 314, 603; Zensus et al 87 Nature 325, 36) are difficult to accommodate in the schemes. They can remain tenable

only if there is gross misalignment between the inner and outer structure in a few
sources, such as observed in 3C 120 (Walker et al 87), Mark 501 (van Breugel &
Schilizzi 86 ApJ 301, 834) and 1928+738 (Johnston et al 87 ApJ 313, L85). Other
tests of the schemes come from comparison of optical, X-ray and radio properties
of quasars. Kembhavi et al (86 MNRAS 220, 51) and Browne & Murphy (87 MNRAS 226,
601) have looked at the correlations between radio luminosity and X-ray luminosity
and conclude that their results are consistent with a two-component origin for the
optical and X-ray emission, one of which is isotropic and one relativistically
beamed.

Lobes and Hotspots - Multifrequency observations of radio sources have provided a
wealth of data on spectral ageing, magnetic field configurations, flow directions
and matter densities in the lobes, interactions of sources with their environments
and on matter distributions in the intervening regions.

The radio spectra in different regions of double radio sources are, in
general, well fitted by synchrotron-loss curves (e.g. Alexander 87 MNRAS 225, 27)
from which relative ages for the regions can be deduced. The results are
consistent with models in which particle acceleration occurs in the hotspots which
advance into the intergalactic medium leaving the radiating material behind (e.g.
Myers & Spangler 85 ApJ 291, 52; Alexander & Leahy 87 MNRAS 225, 1). If the
hotspots are fed by light jets, backflow of the lobe material also occurs.
Supporting evidence for the existence of backflow comes from detailed mapping of
the lobes and bridges (Leahy & Williams 1984 MNRAS 210, 929) which show that they
are markedly distorted near the central galaxy. The rate of separation of hotspot
and lobe material is typically $\sim 10^3 - 10^4$ km s^{-1}; it correlates with luminosity
and may be as high as ~ 0.2 c in the most powerful sources. The bridges get
narrower relative to their length at higher powers, consistent with the greater
expansion speeds (Leahy et al 86 Quasars, 189).

The magnetic fields in these double sources follow the contours of total
intensity and are either parallel or perpendicular to ridge lines (Leahy et al 86
MNRAS 222, 753; Miller 85 MNRAS 215, 773; Spangler et al 84 AJ 89, 1478) and the
fractional polarization is high. The magnetic field geometry is that expected for
fluid distortions associated with backflow. The depolarization and rotation
measure observations do not provide definite evidence on the existence of cold
matter within the sources. There is however strong evidence for differential
rotation produced by the medium surrounding a source and in our Galaxy. Strom &
Conway (85 AA Supp 61,547) find from polarization maps made at 49 cm that there is
little or no polarized emission within a 50 kpc radius of the associated object,
suggesting that the thermal gas surrounding the object is providing the Faraday
depolarization. Leahy et al (86) confirm that the rotation measure and
depolarization correlate with luminosity and find that rotation measure variations
are usually larger on the brighter side of a source; it seems likely that this is
because the brighter component is interacting with denser material. Leahy (87
MNRAS 226, 433) has investigated the effect of our Galaxy on rotation measure and
concludes that the observed variations on scales of 5 - 200 arcsec may be due to
small faint HII regions along the line of sight.

Detailed observations of Cygnus A have produced a number of remarkable
results. Dreher et al (85 PETER, 57) present a multiconfiguration VLA map showing
the jet, counter jet, complex fine structure in the lobes with wispy structure
near the heads, thin arcs and apparent sideways shocks, and very hard edges to the
lobes suggesting ram-confinement. Spectral variations in the lobes at high
frequencies (Alexander et al 84 MNRAS 209, 851) and between 81.5 and 2700 MHz
(Spinks et al 86 Nature 319, 471) imply that the speed of the lobe material
relative to the hotspots in ~ 0.1 c. Dreher et al (87 ApJ 316, 611) have made a
detailed study of Faraday rotation in Cygnus A. The rotation, which must be in a
Faraday screen near the radio source, in the cluster gas or possibly in a denser

sheath round the lobes, varies over an enormous range between -4000 and +3000 rad m^{-2} and requires field reversals over scales of 20-40 kpc. The external magnetic field required is very high and its origin is a mystery.

A number of classical double sources including Cygnus A have double hotspots, one of which is markedly more compact than the other and lies closer to the source axis (Lonsdale & Barthel 86 AJ 92, 12). Williams & Gull (85 Nature 313, 34) have found that jet deflection in a single-jet fluid-dynamical model produces such structures; the compact subcomponent occurs where the jet meets the cocoon wall and the more diffuse subcomponents then result as "splatter spots".

Tailed and other low luminosity sources - Observations of Abell clusters (e.g. O'Dea & Owen 85 AJ 90, 927) have resulted in the discovery of many narrow-angle tailed (NAT) and wide-angle tailed (WAT) radio galaxies. There are now sufficiently large numbers of these objects to investigate their global properties. O'Dea & Owen (85, AJ 90, 954) find that the dominant influence on the radio structure of NATs is the motion of the associated galaxy through the cluster and O'Dea et al (87 ApJ 316, 113) conclude that neither gravity nor buoyancy predominates in determining the curvature of the tails. Burns (86 Can J Phys 64, 373) reviews the properties of WATs which he divides into three subclasses by linear size. He suggests that the smallest (e.g. that in the Coma cluster galaxy NGC 4874, Feretti & Giovanni 85 AA 147, L13), which are limited to the optical extent of the associated galaxies, are bent by dynamic pressure as a result of the motion of the radio galaxy through the cluster. The largest WATs are about 1 Mpc in size and often associated with cD galaxies; the jets exhibit abrupt bends (e.g. 1919+479, Burns et al 86 ApJ 307, 73) which probably result from an interaction between the jet and clouds in the cluster gas. Stocke et al (85 ApJ 299, 799) and Saikia et al (87 MNRAS 224, 911) propose a similar mechanism to explain the abrupt bends in the radio structure of some quasars. Moderate sized WATs, such as that associated with NGC 2329 (Feretti et al 85 AA 147, 321), may bend and decollimate as a result of the passage of the jets through sharp pressure gradients in galaxy halos. One of the most exciting discoveries is the WAT source, 3C 75, in Abell 400 (Owen et al 85 ApJ 294, L85) which has a pair of twin jets originating in the double nucleus of the associated galaxy; the jets bend in the same directions and on larger scales appear to merge.

Detailed optical and radio observations by van Breugel et al (85 ApJ 290, 496) of 3C 277.3 show that the brightest emission-line gas is found next to the brightest radio features and a shell-like Hα structure bounds most of the northern radio lobe. The presence of emission-line gas and depolarization are correlated. They suggest that the jets are propagating through a cloudy medium and that this material is excited and accelerated by these jets. A similar correlation between unpolarized emission and emission-line gas is observed in 4C 29.30 (van Breugel et al 86 ApJ 311, 58).

Leahy et al (86) have made a multifrequency study of the twin-jet source 3C 66B. It is the first source for which the correlation between depolarization and rotation measure implies that some of the rotation must be occurring within the lobes; the inferred density is $>10^3$ m^{-3}, very similar to that in the external medium detected in X-rays, indicating that matter may have entered the lobe by entrainment.

VLBI - The Proceedings of IAU Symp 119 "Quasars" (86 eds. Swarup and Kapahi) of IAU Symp 129, "The Impact of VLBI on Astrophysics and Geophysics" (1988, eds. Moran and Reid) and the volume on "Sumperluminal Radio Sources" (1987, eds. Zensus and Pearson, Cambridge University Press) contain many recent accounts of research relevant to this section.

Over the last four years much of the work in this section has focussed on the role of relativistic beaming in theories of powerful extragalactic radio sources and the superluminal phenomenon (see, eg. Begelman et al. 84 Rev Mod Phys 56, 255). There have been two major lines of attack on these problems, viz. VLBI surveys of large samples of powerful objects and detailed studies of a few objects at many epochs and frequencies.

VLBI surveys: A number of major VLBI surveys have been actively pursued over the last four years. These fall into two categories - surveys of compact objects, of which a high fraction are detectable with the Mark II VLBI system, and surveys of the central components of extended triple sources which require the sensitivity of the Mark III system.

The major programs in the first class are those of Pearson and Readhead (88 Ap. J. in press), Witzel and his collaborators (87 in Superluminal Radio Sources, p. 83), and the high frequency survey of Lawrence et al (85 Ap.J. 296, 458). These programs have the broad objectives of discovering the full range of source morphologies, providing statistically complete samples for testing theoretical models and searching for new superluminal sources.

A carefully selected group of triple radio sources suitable for statistical tests of theoretical models has been defined by Cawthorne et al. (86 MNRAS 219, 883). There are three major ongoing programs on the central components of triple sources, by Barthel et al (84 AA 140, 399), Hough and Readhead (87 "Superluminal Radio Sources", P. 114) and Zensus and Porcas (IAU Symp 119 p. 167).

Through these surveys a number of new superluminal sources have been found (Barthel et al. 86 Ap.J. (Letters), 296 L23; Eckart et al 85 Ap.J. Letters), 296 L23; Eckart et al 86 AA 168, 17; Hough and Readhead 87 Ap.J.(Letters) in press; Pearson et al. 86 Ap.J)Letters), 300, L25; Zensus et al. 87 Nature 325, 36). It is now clear that all classes of powerful extragalactic radio sources, except the compact doubles with steep high frequency spectra, display superluminal motion.

Thus far only three superluminal sources have been found in the surveys of the central components of triple sources, but the mean apparent velocity of this class of object is significantly lower than that of the compact flat spectrum sources. This result is consistent with simple beaming models for these objects, but there are a number of results which show that the simplest models are no longer tenable. In particular, they predict that for each compact flat spectrum source, identified on this model with a nuclear jet whose axis is close to the line of sight, there are about fifty objects whose jet axes are not close to the line of sight. There appear to be too few quasars with the correct properties to constitute this parent population. Phinney (85 in "Astrophysics of Active Galaxies and Quasi-Stellar Objects" ed. Miller, University Science Books) and Readhead et al. (88 IAU Symp 129) have summarised these arguments. In addition the overall sizes of superluminal sources when deprojected according to the simplest beaming models are uncomfortably large (Barthel 87 in "Superluminal Radio Sources".

There is one other important VLBI survey which has been made of the steep spectrum compact objects (Fanti et al. 85 AA 143, 292; Fanti and Fanti 87 "Superluminal Radio Sources" p. 174; Fanti et al. 88 IAU Symp 129). This has shown these objects comprise a class of randomly oriented, intrinsically small sources, and are not simply extended triple objects viewed end-on.

Detailed Observations of Individual Objects: The most extensively studied compact radio source is 3C345 (Biretta et al. 86 Ap.J 308, 93: Biretta and Cohen 87 "Super luminal Radio Sources" p. 40). In this object a new radio component was observed to increase in apparent speed from a subluminal speed to an apparent

speed of 10c between 1978 and 1985. In addition the new component itself expanded
superluminally. There is now some evidence that successive components in this
object are following the same path as earlier components, but this has not yet
been definitively demonstrated. If this result is confirmed it will provide very
important constraints on theories of superluminal sources.

Another object which has been intensively studied is 3C120 (Walker et al. 87
"Superluminal Radio Sources" p. 48). In this object a continuous jet can be
traced from the most compact structures observed in VLBI observations, on parsec
scales, out to the outer lobes.

It has now been establised that BL Lac is a superluminal source (Mutel and
Phillips 87 "Superluminal Radio Sources"), and that some compact steep spectrum
sources are superluminal (Wilkinson et al 86 IAU Symp 119 p. 165).

In 4C39.25 and 3C395 superluminal motion has been detected in a component
which is straddled by two apparently stationary components (Shaffer and Marscher
87 "Superluminal Raido Sources" p. 67; Simon et al. 88 IAU Symp 129). It is
clearly of great importance to monitor these objects and to determine whether the
most widely separated components are indeed stationary, since this could strongly
constrain the theories of superluminal motion.

In 3C 84 (Maars et al. 88 IAU Symp 129) the existence of a compact component
at a projected distance of 2 pc from the flat spectrum core has been confirmed,
and it has been found that this component is moving at an apparent speed of 0.4c
relative to the core. In addition, a jet-like feature close to the core has shown
significant position angle changes over a five year period, and it is clearly
important to monitor this motion to determine whether it repeats periodicaley, as
might be expected for example, if it were caused by precession.

<u>Extragalactic Molecules</u> - The release of the IRAS data spurred much work,
including searches for CO in highly luminous infrared galaxies, such as Arp220 and
NGC6240 (Sanders and Mirabel 85 ApJ 298 L31; Young et al., 84 Ap.J. 287 L65). The
most distant galaxy detected in CO is VIIZw31 at z=0.054 (Sage and Solomon 87
preprint). The report of the second IRAS Conference (C.J. Persson 8 Conf Pub 2466
NASA: Washington) includes papers by most groups active in extragalactic molecular
research and is a good summary of the state of the field in mid 1986.

High Infrared Luminosity Galaxies. Studies of these galaxies, whose
luminosities can equal those of quasars, have shown that the CO luminosity, which
traces molecular gas, is well correlated with the far infrared luminosity, though
the most luminous galaxies have higher values of L_{FIR}/L_{CO}, probably due to higher
dust temperatues caused by increased star formation (Young et al., 86 ApJ 304
443). Sanders et al., 86 (ApJ 305 L45) present evidence that the cause may be
galaxy collisions. The relation betwen active galaxies and starbursts is still
unclear. Although L_{FIR}/L_{CO} for highly luminous galaxies is similar to that in
starburst galaxies such as M82, interferometry (Scoville et al., 86 ApJ 311 L47)
shows more extreme conditions in Arp 220, with a CO emitting region coincident
with the compact nucleus and less than 1.5 kpc in extent containing 10^{10} solar
masses of molecular material.

Normal Galaxies. There have been many investigations of normal galaxies,
with the aims of correlating CO emission with other properties, and searching for
spiral structure. Samples of Virgo spirals (Young et al., 85 ApJ 288 487; Stark
et al., 86 ApJ 310 660) show a good correlation between L_{FIR} and L_{CO}, and little
variation of L_{CO} with spiral type, although CO is more centrally peaked in late
type spirals. Unlike HI, CO content does not vary with position in the cluster.
High resolution observations with interferometers and large single dishes are

beginning to detect molecular structure in many nearby galaxies. For example, the CO gas in the nucleus of IC342 was shown to lie almost entirely in a bar with a length of 1.5 kpc (Lo et al., 84 ApJ 282 L59) and molecular spirals arms were detected in the inner part of M51 (Lo et al., 87 ApJ 317 L63; Rydbeck et al., 85 AA 144 282) though most of the gas forms a smooth distribution between the arms. A similar result is found in M31, in which CO spectra show a narrow, spatially resolved feature, due to giant molecular clouds, superposed on a ubiquitous, weak, broad line which is presumably due to an ensemble of smaller clouds (Boulanger et al., 84 AA 140 L5; Blitz et al., 85 ApJ 296 481; Ryden and Stark 86 ApJ 305 823; Ichikawa et al., 85 PASJ 37 439). An individual GMC in M31 has been imaged by Vogel et al., 87 (preprint).

VLA maps of OH absorption in M82 (Weliachew et al., 84 AA 137 335) showed that the dense molecular gas in M82 was located predominantly in a ring of radius 250 pc, also seen later in CO (Nakai et al., 85 IAU Symp 115; Lo et al., 87 ApJ 312 574). Studies of isotope ratios (Richard and Blitz 85 ApJ 292 L57; Young and Sanders 86 ApJ 302 680), which show a range of values for 13CO/CO, indicate that molecular masses derived assuming a constant CO/H_2 ratio are uncertain by a factor of 2.

Extragalactic Masers – In the past three years OH megamasers have emerged as a new phenomenon. They are OH 18 cm masers associated with the nuclei of active galaxies, and are over a million times more powerful than galactic OH/HII masers such as W3OH. Sixteen are now known primarily infrared galaxies, with enormous FIR luminosities (10^{11}-10^{17} L_0) and distinctive FIR colours. Systematic surveys based on these properties are underway at several observatories (Baan et al. 1985, ApJ 298, L51; Bottinelli et al. 1985, A&A 151, L7; Norris et al. 1986, MNRAS, 221, 51P; Schmeltz et al. 86 AJ, 92, 1291; Staveley-Smith, et al. 1987, MNRAS 226, 689). Early work has been reviewed by Baan 1985 (Nature 315, 26), but more recent results are available only as preprints and IAU Telegrams (Circulars 4106, 4231, 4248, 4268, 4357, 4362). Models of megamasers are heavily weighted towards the prototype IC4553=Arp 220, for which there are detailed radio optical and infrared data (Norris et al. 1985, MNRAS 213, 821 and 216, 701; Scoville et al. 1986, ApJ 311, L47; Baan et al. 1987, ApJ 313, 102). Excited OH at 5 cm has been detected in absorption IC4553 by Henkel et al. 1986 (A&A 168, L13), who discuss a possible FIR pump for megamasers.

Intermediate between normal masers and megamasers are the OH masers in the starburst galaxies M82 and NGC253. These have been mapped using the VLA by Weliachew et al. 1984 (A&A 137, 335) and Turner 1985 (ApJ 299, 312). NGC253 is unique in showing OH maser emission in all four 18 cm lines from extended regions over a kiloparsec in size. M33 was searched unsuccessfully for OH masers by Fix & Mutel 1985 (AJ 90, 736) using the VLA. It is still hard to detect normal OH masers in extragalactic systems.

H_2O masers are rather easier to detect. New extragalactic H_2O masers have been reported by Claussen et al. 1984 (Nature 310, 298), Henkel et al. 1984 (A&A 141, L1), Haschick & Baan 1985 (Nature 314, 144), Henkel et al. 1986 (A&A 155, 193) and Whiteoak & Gardner 1986 (MNRAS 222, 513). The most luminous extragalactic H_2O masers are less than 1000 times as powerful as the galactic source W49. They present a challenge to the theoretician because of their compact sizes (Claussen & Lo 1986, ApJ 308, 592), which constrain pump schemes.

Extragalactic Hydrogen – Thousands of galaxies have now been observed with the large single dishes of Arecibo, Green Bank and Effelsberg. Such data is widely used to study the dependence of properties of galaxies on environment (Haynes et al. 84 Ann. Rev.AA 22, 445; Helou et al. 84 ApJ Suppl 55, 433; Hoffman et al. 85

ApJ 289, L15; 87 ApJ Suppl 63, 247; Giovanelli and Haynes 85 ApJ 292, 404; 86 ApJ
306, 466; Williams and Rood 87 ApJ Suppl 63, 265; and several publicaitons in
"Clusters and Groups of Galaxies" 84 Ap & Sp Sci Lib vol 111 and "The Virgo
Cluster of Galaxies" 85 ESO Workshop no 20). The evidence that galaxies lose part
of their HI disk through stripping in the cores of the denser clusters is quite
clear now (see also Warmels, 86 PhD Thesis, University of Groningen) and simple
orbit analyses (Dressler 86 ApJ 301, 35; Giraud 86 AA167,25) seem to support this.

Several large surveys have been used to extract information about the
clustering of galaxies on large scales (Giovanelli et al 86 ApJ 300, 77 and 86 AJ
92, 250; Bicay and Giovanelli 87 AJ 93, 1326) revealing a very filamentary
structure of the universe. The regions devoid of galaxies have been looked at but
no HI has yet been found (Krumm and Brosch 84 AJ 89, 1461, Hulsbosch 87 AA Suppl
69, 439).

In particular the VLA has contributed to studies of the faint HI emission in
Elliptical galaxies (Appleton et al. 85 MNRAS 217, 779; van Gorkom et al. 86 AJ
91, 791; Lake et al. 87 ApJ 314, 57) in addition to continued studies with the
WSRT (Knapp and Raimond 84 AA 138, 77; Sancisi et al. 87 ApJ 315, L39). A variety
of HI morphologies appear: rings, disks and asymmetric distributions. The general
impression is that HI is more abundant in low luminosity systems and that there is
more HI than can be accounted for by stellar mass loss. The latter suggests an
external origin for the HI such as capture of small, gas rich systems.

The ring distributions are very reminiscent of the HI found in SO galaxies,
where both very small and very extended HI rings have been found (van Driel 87 PhD
Thesis, Univ. of Groningen; Bajaja et al.84 AA 141, 309; Krumm et al. 85 AA
144,202; Knapp et al. 85 AA 142, 1; Gottesmann and Hawarden 86 MNRAS 219, 759;
Shostak 87 AA 175, 4). In addition several more SO galaxies with polar rings have
been observed (van Gorkom et al. 87 ApJ 314, 457). HI in these objects is
associated with the polar rings, which appear to be stable, rotating structures.

Studies of individual galaxies made with the VLA and WSRT include large,
nearby galaxies (M31: Brinks and Burton 84 AA 141, 195; Brinks and Bajaja 86 AA
169, 14; Walterbos and Schwering 87 180, 27; M33: Deul and van der Hulst 87 AA
Suppl 67, 509; NGC 6946: Tacconi and Young 86 ApJ 308, 600, NGC 55: Hummel et al.
86 AA 166, 97) and several peculiar or interacting systems (NGC 1023: Sancisi et
al. 84 MNRAS 210, 497; NGC 3718: Schwarz 85 AA 142, 273; NGC 4725/47 Wevers et al.
84 AA 140, 125). Puzzling in the case of interacting galaxies remains the
occurrence of large, outlying HI complexes like in Stephan's Quintet (Shostak et
al. 84 AA 139, 15) or the Leo group of galaxies (Schneider et al. 86 AJ 91,13).
These large "intergalactic" gas complexes are probably tidal debris, though a
primordial origin cannot be ruled out entirely.

A few barred galaxies have been studied (Gottesmann et al. 84 ApJ 286, 471;
Ball 86 ApJ 307, 453). There appears to be little promise for studying the
kinematics of the bar using the HI line, often because of lack of detectable HI or
lack of resolution in the bar region.

The question of dark matter in galaxies has received increased attention (see
"Dark Matter in the Universe" IAU Symp. 117). The use of HI to determine the
rotation curves of galaxies at large galactocentric radii is very important and a
first detailed study of NGC 3198 (van Albada et al. 85 ApJ 295, 305) has been
followed by studies of a large number of objects using data from the literature
(Athanassoula et al. 87 AA 179, 23; Kent 87 AJ 93, 816). In general more than
half the total mass of a galaxy appears to reside in a dark halo component. A
study of binary galaxies where the orbital motion gives another estimate of the
total mass confirms this picture (van Moorsel 87 AA 176, 13). A very massive, or
at least fast rotating galaxy has been found by Giovanelli et al. 86 ApJ 301, L7.

Detailed HI maps suggest the existence of a surface density threshold for
star formation (Skillman 86 "Star Formation in Galaxies" NASA Conf. Pub. 2466,
263), and such a threshold phenomenon may have inhibited recent star formation in
low surface brightness galaxies (van der Hulst et al. 87 AA 177, 63). A study of
the star forming regions in a prominent spiral arm of M83 (Allen et al. 86 Nature
319, 296) suggests that if the density wave picture holds molecular clouds must
dissociate on a kpc large scale.

Active and low redshift QSOs have also been studied in HI in the past few
years. HI emission in QSOs is faint but detectable (Bothun et al. 84 AJ 89, 1293;
Condon et al. 85 AJ 90, 1642; Hutchins et al. 87 AJ 93, 2) and suggests HI
properties similar to normal spiral galaxies, supporting the idea that QSOs live
in normal host galaxies. HI absorption in the direction of active nuclei (Dickey
86 ApJ 300, 190) suggests inflow of HI into the active nucleus in a number of
objects and the presence of rotating systems, probably disks (Haschick and Baan 85
ApJ 289, 574). Wolfe et al. reported HI absorption at a very high redshift of
z = 2.04, possibly originating in a distant galactic disk. Searches for
redshifted HI have not been very successful thus far (de Waard et al. 85 AA 145,
479).

The Microwave Background Radiation - *Spectrum*.

Substantial progress on measuring
the detailed spectrum of the microwave background radiation has been achieved
through careful radiometric measurements at centimetre wavelengths (De Amici et
al., 1985, Astrophys. J., 298, 710; Mandolesi et al., 1986, Astrophys. J., 310,
561; Witebsky et al., 1986, Astrophys. J., 310, 145; Johnson & Wilkinson 1986,
Astrophys. J., 313, L1; Smoot et al., 1987, Astrophys. J., 317, L45), through
bolometric work from balloons at millimetre wavelengths (Peterson et al., 1986,
Phys. Rev. Lett., 55, 332), and through improved measurements of optical
transitions of CN molecules (Meyer & Jura 1985, Astrophys. J., 297, 119; Crane et
al., 1986, Astrophys. J., 309, 822; Mandolesi et al., 1986, IAU Symposium 124,
59). These results are consistent with a temperature T_r = 2.74 K from 20 cm to 1
mm. The deviations from a black-body spectrum reported in 1981 (Woody & Richards,
Astrophys. J., 248, 18) have not been confirmed. To an accuracy of better than
five per cent, there are no spectral distortions in the background radiation over
this wavelength range. This fact severely constrains events causing large energy
releases in the early Universe and non-standard models of the origin of the
background radiation.

Polarization. Linear polarization in the microwave background radiation may be
caused by anisotropic expansion of the Universe near epochs of substantial
ionization. On scales >7° there have been no improvements to the limits of about
0.2 mK on linear polarization measured by Lubin et al., (1983; Astrophys. J., 273,
L51).

Dipole and Quadrupole Anisotropy. Re-analyses of data taken earlier by Halpern
(1986; IAU Symposium 124, 63) and Lukash & Novikov (1986; IAU Symposium 124, 73),
and the newer data of Lubin et al., (1985; Astrophys. J., 298, L1), are consisent
with a dipole anisotropy of amplitude 3.3 mK oriented towards α = 11^h, δ = -7°.
There is no significant evidence for any quadrupolar anisotropy > 0.2 mK, or any
higher-order terms in these data. These results are interpreted in terms of a 370
kms^{-1} motion of the local system of rest. Lubin et al.'s data show the expected
30 kms^{-1} annual modulation of the dipole anisotropy caused by the motion of the
Earth around the Sun.

Anisotropies on scales 1° - 60°. Sensitive measurements of the anisotropy of the
microwave background radiation on scales >2° have been made by Davies et al.,
(1987; Nature, 326, 462). These data indicate the possible detection of intrinsic
fluctuations of about 0.16 mK on intrinsic scales of about 4°. Results from the

Prognoz 9 satellite (Lukash & Novikov 1986; IAU Symposium 124, 73) indicate
correlated fluctuations less than 0.07 mK on scales above 20°. Mandolesi et al.
(1986; Nature, 319, 751) found limits of 2 mK for fluctuations on scales of
2°-10°. This homogeneity of the Universe on scales larger than the horizon at
recombination has been used as evidence for inflationary models of the Universe.

Anisotropies on scales 1' - 1°. Scales <20' are amenable to single-dish
observations of the background radiation, and a number of useful measurements of
limits to fluctuations are available on arcminute scales. The recent measurements
by Uson & Wilkinson (1984; Nature 312, 427) and Readhead et al., (1987; private
communication) limit fluctuations on angular scales 5' to 15' to be less than
about 0.1 mK. the absence of such fluctuations is becoming an embarassment to
many theories of the formation of galaxies and clusters of galaxies, unless the
Universe went through a late, quiescent, hot, phase in which early fluctuations
were erased, or unless the Universe contains large quantities of non-baryonic
matter.

Anisotropies on scales less than 1'. Observations with interferometers provide
the only limits to the fluctuation amplitudes on the smallest angular scales.
Fomalont et al., (1984; Astrophys. J., 277 L23), and Knoke et al., (1984,
Astrophys. J., 284, 479) achieved limits to the fluctuation amplitude of about
$1(\theta/arcmin)^{-1}$ mK on angular scales of 0.1 - 1 arcmin. Kellermann et al., (1986,
Highl. Astron., 7, 367) reported the tentative detection of fluctuations at a
level 0.13 mK on a scale of 1 arcmin and 0.54 mK on a scale of 0.3 arcmin. More
recent observations (Martin & Partridge 1987, Astrophys. J., in press; Fomalont &
Kellermann 1987, Astrophys. J., in press) have achieved higher sensitivities, and
limit the fluctuation amplitudes to be less than 0.12 mK on scales of 1 arcmin,
and less than 0.5 mK on scales of 0.3 arcmin. On these sub-arcminute scales,
fluctuations related to galaxy formation by shocks might appear.

The Sunyaev-Zel'dovich effect. Reliable detections of the Sunyaev-Zel'dovich
effect, the changes in the temperature of the background radiation caused by its
passage through the hot gas in clusters of galaxies, have now been reported, with
two independent groups working at 1.5 cm reporting consistent detections at the
level of about 0.3 mK towards several clusters of galaxies (Birkinshaw 1986, Green
Bank Workshop, 16, 261; Uson 1986, Green Bank Workshop, 16, 255). Preliminary
data on the angular sizes of the effects in a few clusters are also available.
Attempts to detect the effect at millimetre wavelengths (e.g. Chase et al., 1987,
Mon. Not. R. astr. Soc., 225, 171) have, as yet insufficient sensitivity.

A variety of other effects, such as cosmic strings and moving gravitational
lenses, are also expected to produce signals in the microwave background
radiation. These effects are generally too small to be observationally
interesting, although the absence of an anisotropy near the candidate lensed
double quasar 1146+111B,C has been used to argue against its interpretation as the
effect of a cosmic string (Stark et al., 1986, Nature, 322, 805; Lawrence et al.,
1986, Astr. J., 92, 1235.

In summary, the microwave background has an accurately thermal spectrum, is
not strongly polarized, and is largely isotropic. The 0.1 per cent dipole
anisotropy is interpreted in terms of motion of the local system of rest, and
shows annual variations caused by the orbital motion of the earth, but no
quadrupolar anisotropy has been found. On smaller angular scales the radiation
field remains remarkably isotropic, although there are tentative indications that
structures of cosmological origin may be appearing, and the Sunyaev-Zel'dovich
effect appears to have been detected.

41. HISTORY OF ASTRONOMY
(HISTOIRE DE L'ASTRONOMIE)

PRESIDENT: J.A. Eddy
VICE-PRESIDENT: J.D. North
ORGANIZING COMMITTEE: S. Debarbat, H. Eelsalu, O. Pedersen, Xi Ze-Zong.

Commission 41 has been involved in one colloquium and one symposium since the last report:

IAU Colloquium 91 on "The History of Oriental Astronomy" was held in New Delhi, November 13-16, 1985, preceding the XIXth General Assembly. Members of the scientific organizing committee were S.M.R. Ansari, E.S. Kennedy, D. King, R. Mercier, O. Pedersen, D. Pingree, G. Saliba, Xi Ze-Zong and K. Yabuuti. The colloquium was co-sponsored by the International Union for the History and Philosophy of Science, and by a number of organizations in India: the Council of Scientific and Industrial Research, New Delhi, the Department of Science and Technology, New Delhi, the Indian Institute of Astrophysics, Bangalore, the Indian National Science Academy, New Delhi, the Tata Institute of Fundamental Research, Bombay, and the University Grants Commission, New Delhi. The local organizing committee, chaired by G. Swarup, made possible a number of local excursions, including a conducted tour of the great stone open air observatory, built in the city by the enlightened Maharadjah Jai Singh in the 18th century. The colloquium brought 84 participants from 19 countries. 46 papers were presented of which 10 were invited, covering aspects of astronomy in the far east and middle east since the earliest civilizations. Papers from Colloquium 91 have now been published in book form: History of Oriental Astronomy, G. Swarup, A.K. Bag, and K.S. Shukla, editors, Cambridge University Press, Cambridge, England, 1987. Contributions are divided into three broad categories: ancient astronomy and its characteristics, ancient elements and planetary models, and medieval astronomy. Within these are papers on the characteristics and achievements of early astronomy in the eastern half of the world, including inter-regional development and mutual influences, ancient data relating to eclipses, supernovae and comets, medieval astronomical developments, instruments and early observatories, and the interplay between observational and theoretical astronomy. A short introductory paper by the revered historian E.S. Kennedy opens the book, as it set the stage for the colloquium in New Delhi: "We find (astronomy) originating a few centuries before the Christian era in two disparate cultures, Mesopotamia and the Hellenistic world. From the Mediterranean it passed to India, there to flourish. Thence the centroid of activity moved westward, residing in the lands of Islam during medieval times, more recently in Europe. Now astronomical research is carried out throughout the entire world."

IAU Symposium 133 on "Mapping the Sky: Past Heritage and Future Directions" was held in Paris, June 1-5, 1987, as a joint endeavor of Commissions 41 and 24 (photographic astrometry). The International Union for the History and Philosophy of Science was again a co-sponsor. The occasion for the symposium, and the choice of site, was the centenary of the first great international collaboration in astronomy, the "carte du ciel" which was initiated at the Paris Observatory in 1887. The symposium further commemorated the sesquicentenary of the first proof of stellar parallax, by Friedrich Wilhelm Bessel, in 1837: the first reliable measurement of distance to any object outside the solar system. The purpose of the symposium was to provide a forum for astronomers and historians of astronomy to survey past accomplishments, current progress and future plans in the area of celestial cartography. Papers were divided into 8 sessions, 3 of which covered historical matters, from contributions on the carte du ciel and early astrographic telescopes and measurement machines to the making, with naked eye, of early oriental star maps. A total of 46 papers were presented orally. These with papers based on 31 poster presentations are being prepared for publication by D. Reidel in 1988 as a volume in the I.A.U. Symposium series (S. Debarbat, J. Eddy, H. Eichhorn, and A. Upgren, editors). The scientific organizing committee included V. Abalakin, S. Debarbat, Ch. de Vegt, R. Duncombe, J. Eddy, H. Eichhorn, M. Hoskin, J. Hughes, C. Jaschek, P. Kulikovsky, C. Murray, A. Upgren and P. Wayman. The local organizing committee was led by S. Debarbat and the Director of the Paris Observatory, P. Charvin.

was to provide a forum for astronomers and historians of astronomy to survey past
accomplishments, current progress and future plans in the area of celestial cartography.
Papers were divided into 8 sessions, 3 of which covered historical matters, from contributions
on the carte du ciel and early astrographic telescopes and measurement machines to the
making, with naked eye, of early oriental star maps. A total of 46 papers were presented
orally. These with papers based on 31 poster presentations are being prepared for publication
by D. Reidel in 1988 as a volume in the I.A.U. Symposium series (S. Debarbat, J. Eddy,
H. Eichhorn, and A. Upgren, editors). The scientific organizing committee included
V. Abalakin, S. Debarbat, C. Devegt, R. Duncombe, J. Eddy, H. Eichhorn, M. Hoskin,
J. Hughes, C. Jaschek, P. Kulikovsky, C. Murray, A. Upgren and P. Wayman. The local
organizing committee was led by S. Debarbat and the Director of the Paris Observatory,
P. Charvin.

42. CLOSE BINARY STARS (ETOILES BINAIRES SERREES)

PRESIDENT: J. Smak
VICE-PRESIDENT: R.H. Koch
ORGANIZING COMMITTEE: K.D. Abhyankar, J. Andersen, A.H. Batten, E. Budding,
A.M. Cherepashchuk, D.M. Gibson, M. Kitamura, Y. Kondo, K.-C. Leung, J. Rahe,
M. Rodonó, G. Shaviv

1. Introduction

During the XIXth General Assembly of the IAU in Delhi the number of members
of Commission 42 increased to 260. This simply reflects the growing interest and
importance of our field. Growing is not only the number of astronomers involved in
research on CBS but also the number of papers resulting from that activity. As an
example one can quote the numbers of papers listed during the last few years in
Sections 117 (Close Binaries), 119 (Eclipsing Binaries), and 120 (Spectroscopic
Binaries) of the *Astronomy and Astrophysics Abstracts*: 705(1982), 775(1983),
836(1984), 1080(1985), and 911(1986); note that many additional references could
be added to these numbers from other sections. Naturally, such numbers alone do
not reflect the quality and even less so the position and significance of the CBS
field. Here one could perhaps mention an impressive record of successful research
proposals involving requests for the observing time on large, ground based telescopes and on space instruments. Indeed, in spite of a very strong competition
from other fields, programs involving CBS are usually placed very high on the
priority lists (cf. Sections 2D and 2E). Obviously, the close binary systems,
their evolution, and the physical processes which occur in them (accretion,
stellar winds, nuclear burning, etc) appear interesting and important not only to
those who are involved in their studies but also to astronomers from other fields.

No large symposium devoted exclusively to CBS has been organized during the
past triennium. A proposal for an IAU Symposium on *Circumstellar Matter in Close
Binary Systems*, to be held in Canada, in 1987, has been submitted in 1985 but did
not meet an approval by the IAU Executive Committee. The Commission's Organizing
Committee, while considering its possible re-submission, decided not to do so, but
rather to concentrate in the future on smaller, more specialized meetings. (In
fact, some OC members expressed an opinion that our field is already too large for
a successful, working meeting involving *all* types of CBS). The Commission co-sponsored IAU Colloquium No.93: *Cataclysmic Variables*, Bamberg, June 1986; IAU Colloquium No.97: *Wide Components in Double and Multiple Stars*, Brussels, June 1987;
and IAU Colloquium No.103: *The Symbiotic Phenomenon*, Toruń, August 1987. The Commission is also a sponsor of the IAU Coll. No.107: *Algols*, to take place in Victoria, in August 1988.

There have been very many meetings not sponsored by our Commission, but devoted - partly or completely - to the field of CBS and attended or co-organized by
members of our Commission. Their list, obviously incomplete, includes: IAU Symp.
125: *The Origin and Evolution of Neutron Stars*, Nanjing, May 1985; IAU Coll. 89:
Radiation Hydrodynamics in Stars and Compact Objects, Copenhagen, June 1985; IAU
Coll. 92: *Physics of Be Stars*, Boulder, August 1986; IAU/COSPAR Coll.: *Physics of
Compact Objects: Theory vs. Observations*, Sofia, July 1987; Japan/US Seminar: *Com-
pact Galactic and Extragalactic X-Ray Sources*, Tokyo, January 1985; ESA Workshop:
Recent Results on Cataclysmic Variables, Bamberg, April 1985; NATO Workshop on the
Evolution of Galactic X-ray Binaries, Rottach-Egern, June 1985; Symp.: *Critical
Observations vs. Physical Models for Close Binary Systems*, Beijing, November 1985;
ESA Workshop: *Physics of Accretion onto Compact Objects*, Tenerife, April 1986;

Workshop: *High Energy and Ultra High Energy Accreting Sources*, Vulcano, May 1986;
AAVSO Variable Star Symposium, Cambridge, August 1986.

Many good review articles have been published in various journals and in the
conference proceedings volumes (cf. references in sections below). New books were:
J.E. Pringle and R.A. Wade, eds.: *Interacting Binary Stars* (Cambridge U.Press,
1985) and J. Frank, A.R. King, and D.J. Raine: *Accretion power in astrophysics*
(Cambridge U.Press, 1985). Among the catalogues one should mention the appearance
in 1985 of Vol.1 (Andromeda-Crux) and Vol.2 (Cygnus-Orion) of the 4th edition of
the *General Catalogue of Variable Stars*, published by Dr. P.N. Kholopov and his
collaborators.

As described in the last Report, our Commission endorsed a proposal for an
IAU grant to support the publication by AAVSO of its observations of cataclysmic
variables. It is now satisfying to report the publication by J.A. Mattei *et al.* of
AAVSO Monographs No.1, SS Cyg and No.2, U Gem.

The *Bibliography and Program Notes on Close Binaries*, edited, published, and
distributed by Dr. Tibor Herczeg (University of Oklahoma and Remeis-Sternwarte),
continues to play its very important and useful role. During the time covered by
this Report Nos. 41-44 were published, the contributors to the *Bibliography* being:
K.D. Abhyankar, B. Cester, O. Demircan, D.S. Hall, M. Kitamura, H. Mauder, R.
Olowin, J. Papousek, M.B.K. Sarma, C.D. Scarfe, A. Schulberg, R.F. Sistero, F.
Van'tVeer, M. Vetesnik, and A. Yamasaki.

This Report, covering the interval from July 1984 to June 1987, is essentially similar to previous ones, with some sections having been merged and some
new sections being added. The emphasis is on major highlights and most crucial
developments and prospects in the field. With few exceptions, no full lists of
references are being given; for a complete bibliography the reader is referred to
the *Astronomy and Astrophysics Abstracts*. The President acknowledges with grati-
tude the cooperation of the co-authors, who prepared various sections, as indicated below.

Throughout the Report the following abbreviated references are used:

AA	= Acta Astron.		BullAbas	= Bull. Abastumani Astrophys. Obs.
AAp	= Astron. Astrophys.			
AApSup	= Astron. Astrophys. Suppl. Ser.		CommAp	= Comments on Astrophys.
			IAUColl	= IAU Colloquium, followed by a number
AASin	= Acta Astron. Sinica			
AApSin	= Acta Astrophys. Sinica		IAUC	= IAU Circular
AdvSpRes	= Adv. Space Research		IAUSymp	= IAU Symposium, followed by a number
AJ	= Astron. J.			
AN	= Astron. Nachr.		IBVS	= Inf. Bull. Variable Stars
AnnRevAAp	= Ann. Rev. Astron. Astrophys.		IzvKrym	= Izv. Krymskoi Astrofiz. Obs.
			JApA	= J. Astrophys. Astron.
AnnTokyo	= Ann. Tokyo Astron. Obs. Second Ser.		JRASC	= J. R. Astron. Soc. Canada
			MittAG	= Mitt. Astron. Ges.
ApJ	= Astrophys. J.		MN	= Mon. Not. R. Astron. Soc.
ApJSup	= Astrophys. J. Suppl. Ser.		Obs	= Observatory
ApL	= Astrophys. Lett.		PASJ	= Publ. Astron. Soc. Japan
ApSpSc	= Astrophys. Space Sci.		PASP	= Publ. Astron. Soc. Pacific
ATs	= Astron. Tsirk.		PisAZh	= Pis'ma v Astron. Zh.
AZh	= Astron. Zh. Akad. Nauk USSR		PRL	= Phys. Rev. Letters
BAAS	= Bull. American Astron. Soc.		PubDAO	= Publ. Dominion Astrophys. Obs.
BAC	= Bull. Astron. Inst. Czechoslovakia		QJRAS	= Quart. J. R. Astron. Soc.
			RevMex	= Revista Mexicana Astron. Astrofis.

2. Observational Data

A. PHOTOMETRIC OBSERVATIONS AND METHODS
   AND RESULTS OF LIGHT CURVE ANALYSIS (R.H. Koch)

This summary was compiled from materials held at Pennsylvania no later than June 30, 1987. It is styled so as to be similar to Sections 2 and 3A of the 1985 Report.

The recent history of programs leading to photoelectric light curves is compiled in Table 1. It is important to understand why photoelectric activity seems not to have continued its historical increase during the last triennium. An acceptance criterion somewhat more severe than hitherto has been imposed on the literature references counted in the first entry for 84-87: unless abstracts, administrative reports and announcements give a quantitative expression of the data that have been obtained, they have been excluded from the count. This achieves nearly a 6% diminution with respect to the last previous tally. Under-reporting must also be considered. Once more it is true that not all literature citations are among the Pennsylvania holdings or those that are otherwise accessible. However, this deficiency appears not to be significantly worse than previously. The 81-84 reporting interval is about 11% longer than usual and the present interval correspondingly shorter. When these factors are accounted for and coupled to the 6% diminution described above, it is seen that activity has actually increased but by a sensibly smaller percentage than familiarly for almost the ten previous years.

Table 1. Photoelectric Observing for the Past Four Triennia

	75-78	78-81	81-84	84-87
References for photoelectric data	346	564	713	565
Close binaries observed	209	342	455	339
Binaries not observed previous 3 years	184	240	313	187
Northern systems (δ > +23°)	88	120	197	149
Equatorial systems	50	77	135	123
Southern systems (δ < -23°)	35	76	123	67

Prorated as just described, the number of binaries observed has remained nearly constant at about 400 per triennium for 6 years. This has been possible because sustained productivity from the northern sky and a slight increase from the equatorial region have compensated for the considerable decrease of the contribution from the far southern sky. (It must be remembered that more systems than formerly are not observed by ground-based workers so simply counting by declination leads to a small bias). Perhaps many southern systems have not yet emerged from the pipeline. But it does seem clear that the growth rate sustained for the decade beginning in 1975 is over at least for now. A more detailed understanding of this effect will emerge later in this Report, but a final remark will forecast the result. Even though observers have chosen not to repeat light curve coverage for many familiar binaries, in the last 3 years they have actually observed more than 90 systems never observed before.

As in previous reports it is possible to summarize the light curve analyses that have appeared in the last triennium. As before also, some small number of references could not be checked from local holdings and these were omitted from the listing which appears in Table 2.

Table 2. Photometric Solutions

AB And *ApSpSc* 127,153, *AN* 307,17; AN And *AnnTokyo* 21,311; BL And *AA* 35,327; CN And *AA* 33,345, *PASP* 97,310; DS And *AN* 304,263; RY Aqr *AAp* 172,155; ST Aqr *AA* 35,327, *ApSpSc* 117,375; CX Aqr *MN* 223,607; DV Aqr *PASP* 97,62; OO Aql *ApSpSc* 114,23;

V1343 Aql *MN 210,279*;   SS Ari *AA 34,445*; SX Aur   *MN 224,649*; TT Aur *MN 211,39, MN
211,229, MN 224,649, AAp 162,62*; ZZ Aur *AJ 90,115, AASin 24,217*; AH Aur *AA 33,159*;
HS Aur *AJ 91,383*; IM Aur *AApSup 60,389*; LY Aur *ApJ 298,345*; ∈ Aur *BAAS 15,925,
ApSpSc 123,31*; XY Boo *ApSpSc 107,347*; TU Cam *AnnTokyo 21,229*; AO Cam *ApSpSc 113,
25, PASP 97,648*; AT Cam *AAp 141,266, AApSin 5,26, ChinAAp 9,139*; AZ Cam *AAp 141,
266, AApSin 5,26, ChinAAp 9,139*; S Cnc *AJ 90,504*; AC Cnc *AZh 63,123*; AH Cnc *AApSup
58,405*; RS CVn *ApSpSc 105,259*; BI CVn *BAAS 18,850*; GZ CMa *AJ 90,1324*; XZ CMi *AAp
96,415*; GL Car *AAp 161,275*; OY Car *AAp 130,81*; V348 Car *MN 213,75*; SX Cas   *BAC
36,153*; TV Cas *AA 34,47*; TX Cas   *MN 216,663*; XX Cas *IBVS 3001*; YZ Cas *AZh 63,690*;
AE Cas *AA 34,281*; AO Cas *ApSpSc 105,259*; DO Cas *AA 35,327, ApSpSc 119,381*; HT Cas
ApJ 305,740, AA 36,395, AAp 130,81; MN Cas *AnnTokyo 21,229*; V364 Cas *ApSpSc 106,
273*; V368 Cas *IBVS 2944*; V375 Cas *AApSin 6,185*; V523 Cas *AAp 170,43*; RR Cen *ApSpSc
127,153*; ST Cen *AnnTokyo 21,229*; SV Cen *ApSpSc 117,351*; SZ Cen *AnnTokyo 21,311*;
V757 Cen *AApSup 58,405, AA 34,217*; V758 Cen *ApSpSc 109,271*; V779 Cen *AZh 63,494*;
U Cep *PASP 96,162, ApSpSc 125,219*; RS Cep *AJ 89,562, AJ 91,1421*; VW Cep *ApSpSc
105,259, AApSup 58,261*; WX Cep *AnnTokyo 21,229*; WZ Cep *AA 36,105*; XX Cep *AAp 156,
38*; XZ Cep *MN 211,39*; AH Cep *MN 223,513*; BE Cep *PASP 98,662*; DH Cep *IBVS 2932*;
EG Cep *AA 34,433*; EI Cep *Izv.Engel.Kazan 47,19*; EK Cep *AJ 89,1256*; ER Cep *AApSup
58,405*; GW Cep *AA 34,217*; TX Cet *ApSpSc 117,375*; VY Cet *AAp 161,264, RevMex 10,
283*; YY Cet *MN 218,159*; Z Cha *AA 36,211, AAp 130,81*; RS Cha *AnnTokyo 21,311*;
YZ Cha *AJ 92,1420*; RW Com *BAAS 18,696*; RZ Com *PASP 96,646*; RT CrB *ApSpSc 112,133*;
α CrB *AJ 91,1428*; RV Crv *MN 223,595*; SS Cyg *PisAZh 12,219*; CG Cyg *AJ 90,761, PASP
99,410*; CI Cyg *IzvKrym 68,108, AA 33,403*; KR Cyg *ApSpSc 117,351*; MR Cyg *ApJ 316,
754*; V367 Cyg *ApJ 313,801*; V380 Cyg *AAp 141,39*; V388 Cyg *ApSpSc 117,351*; V453 Cyg
AnnTokyo 21,229; V478 Cyg *AnnTokyo 21,229*; V541 Cyg *ATs 1270*; V548 Cyg *AA 33,163*;
V909 Cyg *ApSpSc 123,305*; V1143 Cyg *AAp 141,1*; V1357 Cyg *AZh 60,727*; V1727 Cyg *MN
218,63*; DM Del *AApSup 67,87*; AA Dor *AA 34,381*; AR Dra *AnnTokyo 21,229*; BV Dra *AJ
92,666*; BW Dra *AJ 92,666*; RU Eri *PASJ 36,277*; WX Eri *AnnTokyo 21,229*; YY Eri *AA
36,79, AAp 159,142*; AS Eri *AAp 141,1*; U Gem *AA 34,93*; RY Gem *ApSpSc 105,259*;
IR Gem *ApJ 282,236*; TT Her *AA 35,327*; AK Her *PASP 97,1005*; V624 Her *AJ 89,1057*;
u Her *MN 211,943*; SY Hor *RevMex 10,283, AAp 161,264*; FG Hya *ApSpSc 127,153*; KW Hya
AAp 130,102 (not KM Hya), AAp 175,355; Chi² Hya *AnnTokyo 21,229*; Y Hyi *ApSpSc 119,
345*; SW Lac *PASP 96,634, PASP 96,646, ApSpSc 114,23, AA 36,79, AApSup 67,365*;
VY Lac *AA 34,207*; AR Lac *AA 34,291, AJ 91,1438*; CM Lac *Izv.Engel.Kazan 47,19*;
CO Lac *Izv.Engel.Kazan 47,19, AnnTokyo 21,229*; T Leo *ApJ 276,305*; UV Leo *AnnTokyo
21,229*; UZ Leo *ApSpSc 127,153*; XY Leo *AA 33,277*; AM Leo *PASP 96,646*; GX Lib *AJ
90,2581*; FT Lup *AA 36,113, MN 208,135, MN 220,883*; SW Lyn *AA 35,327*; TZ Lyr *AA
35,327*; FL Lyr *AJ 91,383*; TZ Men *AAp 175,60*; RU Mon *AZh 63,288*; RZ Oph *PASP 96,
737, AAp 168,72*; V451 Oph *ApSpSc 106,93, AAp 167,287*; V566 Oph *ApSpSc 114,23, AA
36,275*; V839 Oph *ApSpSc 114,23*; V2051 Oph *MN 222,871*; VV Ori *AnnTokyo 21,229, AJ
93,950, ApSpSc 127,79*; ER Ori *AAp 155,46*; EW Ori *AJ 91,383*; U Peg *AApSup 57,487,
PASP 96,646, ApSpSc 121,61*; BB Peg *AJ 90,515*; BO Peg *PASP 98,1325*; BX Peg *AJ 90,
515, AA 34,217*; IP Peg *MN 222,655, MN 224,1031, PisAZh 11,696, AAp 130,81*; AG Per
ApSpSc 129,187; DM Per *MN 222,167*; IQ Per *ApJ 295,569*; KR Per *AJ 90,1855*; LX Per
ApSpSc 112,273; β Per *ApSpSc 108,227, ApSpSc 123,305*; AI Phe *ApJ 282,748*; AU Phe
RevMex 10,283; SZ Psc *ApSpSc 105,23*; PV Pup *AAp 132,219*; U Sge *ApSpSc 125,219*;
V Sge *ApJ 306,618*; WY Sge *ApJ 282,763*; V760 Sco *AAp 151,329*; V906 Sco *AnnTokyo
21,229*; RT Scl *MN 223,581*; VZ Scl *MN 225,43*; AU Ser *AA 36,113*; CD Tau *ApSpSc 123,
305*; GR Tau *PASJ 36,175*; 33 Tau *AJ 90,1334*; AQ Tuc *MN 223,581*; CF Tuc *ApSpSc 133,
45*; W UMa *ApJ 316,389, AA 36,79*; XY UMa *ApSpSc 128,369*; BE UMa *ApJ 316,399*; DN UMa
ApSpSc 125,181; RU UMi *AA 35,327*; GP Vel *ApJ 280,259, AZh 63,690*; AG Vir *AA 36,
121, AApSup 61,313*; AH Vir *PASP 96,646, AA 34,217*; AX Vir *AJ 90,115, AApSin 5,124*;
FO Vir *AJ 91,1221*; HD1826 *AApSup 66,303*; HD27130 *AJ 93,1471*; HD47755 *AJ 91,590*;
HD149779 *AAp 167,53*; HD164270 *PASP 98,1170*; HD184035 *IBVS 2552*; HD199497 *ApSpSc
115,309*; HR3337 *JRASC 79,119*; HR7551 *AAp 139,123*; ADS 9019B *AJ 90,346*; A0620-00
ApJ 308,110; PG1012-029 *ApJ 276,233*; 1E1048.5+5421 *ApJ 314,641*; Sk188 *ApJ 292,511*;
AGK3-0 965 *RevMex 13,149*.

All the analyses enumerated in Table 2 have been achieved by one or more analytical/synthetic procedures. Over the past 20 years there has been a substantial flux in the availability and acceptability of these methods. A portion of this history may be seen in Table 3.

Table 3. Percentages of Light Curves Studied by Different Computational Procedures

	75-78	78-81	81-84	84-87
A – Budding	2 %	0 %	2 %	0 %
B – Eaton-Hall	1 %	0 %	1 %	2 %
C – Hill	2 %	2 %	1 %	6 %
D – Kitamura	3 %	4 %	0 %	2 %
E – Kopal (alpha functions)	1 %	0 %	0 %	0 %
F – Kopal (frequency domain)	7 %	17 %	15 %	12 %
G – Lavrov	2 %	1 %	0 %	2 %
H – Miscellaneous	5 %	14 %	20 %	19 %
I – Mochnacki-Binnendijk-Nagy	3 %	0 %	0 %	3 %
J – Ruciński	3 %	1 %	3 %	3 %
K – Russell-Merrill	23 %	11 %	11 %	6 %
L – Soderhjelm	4 %	0 %	0 %	0 %
M – Nelson-Davis-Etzel (sphere, EBOP)	1 %	2 %	3 %	8 %
N – Wilson-Devinney	13 %	14 %	22 %	27 %
O – Wood	30 %	34 %	21 %	8 %
P – Yamasaki	0 %	0 %	1 %	2 %

Three preliminary remarks may be made: (a) because of redundant applications the number of analyses is measurably greater than the number of binaries analyzed; (b) the *Miscellaneous* category percentage for 81-84 is diminished from its value in the preceding Report because 1% contributors appear explicitly above; and (c) a zero entry may not mean that the method was completely unused but only that usage did not amount to 1% of the total. Most of the obvious trends reflect the convenience and availability of, and confidence in the assorted codes. During the 78-84 interval something approaching production line use of the Wilson-Devinney and Wood codes was brought to bear on numerous archival light curves. It appears that students of light curves subsequently concluded that the flexibilities and modeling returns offered by Wilson-Devinney procedures more than compensate for the less-than-modern documentation associated with this program. There also remain serious concerns about the parameter resolution possible with this code for contact and over-contact pairs. The only other significant contenders for abundant use in the near future appear to be the frequency domain procedure and EBOP. The former has found favor with the Manchester staff and their present and former students and in Japan. For limited applications EBOP offers two impressive advantages: (a) it has been forcefully recommended after careful tests on feebly-interacting systems and (b) on diskette it can run efficiently at PC terminals. Finally, since 1985 60% of the catchall category of *Miscellaneous* procedures has been applied to binaries containing collapsed objects. Almost all of these modelings have concerned themselves with the structure and photometric activity of the disk embedding the collapsed member.

One may make some judgments on why the trends in Table 3 have run their courses. The student of light curve must have confidence in the integrity of the code that he is using and thus the numerous tests performed on archival light curves by the Trieste group have served this useful purpose. He should also have convinced himself that the mathematical order of the procedure and grid fineness are sufficient for the light curve under study and that an acceptable criterion prevents over-discussion. This statement should not mislead one into supposing Table 3 (which is not utterly exhaustive) to be concerned only with computational procedures (and it must be admitted that differences between some of the proce-

dures are slight). For light curve study physical realism is the only endpoint
concern and so the analyst of light curves makes decisions which are not intrinsically mathematical or calculational. By reference to the letter key in Table 3
consider the possible bases for the figures of the component stars: (a) spheres
with no specified structure (e.g. M); (b) similar ellipsoids of rotation, similarly oriented with no specified structure (e.g. K); (c) tri-axial ellipsoidal
polytropes (e.g. O); and (d) Roche geometrical, gravitationally-specified volumes
(e.g. N). Even if all the categories of Table 3 were expressed in this enumera-
tion, a complete understanding of the procedures of Table 3 would still not be
summarized so succinctly. The entries of that table further distinguish themselves
mutually by the choice of function for the stellar irradiance, the device for handling the radiative interactions between the stars, and the method(s) chosen for
handling non-theoretical effects. It is very common to make choices of procedure
on the basis of information (e.g. spectrographic or photoelectric mass ratios,
magnetic field strength) that cannot be recovered even indirectly from the
photometric measures.

Table 4 shows the trends of the kinds of binaries which have engaged the
interests of workers for more than a decade. In view of the asymmetries which are
so conspicuous for contact systems and for above-the-MS pairs, it isn't clear that
the slightly evolved identification with RS CVn-type display is justified, but
this category is a small one anyhow. For the rest it is clear what has happened:
systems which one used to call Algol- and β Lyr-type objects are not viewed with
the interest which they once commanded but attention is sustained for contact
pairs and binaries with collapsed stars.

Table 4. Percentages of Evolutionary Configurations
among Solutions of Binary Light Curves

	75-78	78-81	81-84	84-87
Numbers of binaries	173	211	215	183
Mostly non-contact, ZAMS to TAMS	33 %	22 %	16 %	28 %
Near MS contact	21 %	29 %	37 %	35 %
Slightly evolved, e.g. RS CVn syndrome	4 %	7 %	6 %	5 %
Substantially evolved, but still non-degenerate	37 %	34 %	24 %	20 %
Substantially evolved, with collapsed component	5 %	9 %	17 %	12 %

It is impossible to avoid calling explicit attention to the last citation in
Table 2. Any interpretation of light variability as due to orbiting or passing
clouds occulting a star or stars represents a possible application of light curve
procedures reminiscent of those used to study atmospheric eclipse phenomena.

B. SPECTROSCOPIC AND SPECTROPHOTOMETRIC OBSERVATIONS (A.H. Batten)

Spectroscopic and spectrophotometric studies during the past triennium are
listed in Table 5. The format is the same as has been used in recent reports. An
asterisk after a reference indicates that the paper contains an "orbital study".
The quality of such studies, however, varies widely. Some of the papers cited present evidence *against* the supposed duplicity of the stars concerned. Some of the
stars are included in the Table, as binaries, on the strength of weak or indirect
evidence; e.g. the symbiotic spectra of some objects seem best explained by the
existence of a companion, even if no certain velocity variations have been detected. The Table is about half as long again as that in the last volume of these
Transactions, impressive evidence - despite the foregoing *caveats* - of the level
of activity in the field. Investigations of the X-ray and γ-ray sources that are
based only on observations from those two regions of the electromagnetic spectrum
have been omitted from the Table. This somewhat arbitrary limitation of the term
"spectroscopy" is based on the assumption that such studies will be more adequately dealt with elsewhere in the report. Studies of novae and nova-like var-

iables that are concerned only with the progress of individual outbursts, rather than with the presumed duplicity of the star, have also been omitted. There are no references to the IAU Circulars or to abstracts which have been superseded by an easily indentifiable full paper. The Cepheid variables included are not there by error. A growing number of such stars is now recognized to belong to binary systems.

Table 5. Spectroscopic and Spectrophotometric Studies

Z And *IBVS* 2844; RX And *AAp* 140,345; DS And *BAAS* 18,682; EG And *ApJ* 295,620*, *AJ* 91,1400; ET And *AN* 305,79, 306,329*; KX And *BAC* 36,313; ∂ And *MN* 224,93; Lambda And *AA* 36,369, *ApJ* 281,286; o And *ApSpSc* 100,13*; QS Aql *BAAS* 19,709*; V794 Aql *PASP* 97,1189; V1182 Aql *MN* 225,961*; V1315 Aql *ApJ* 301,240*; V1343 Aql (SS433) *ApJ* 308,152, *AZh* 63,94; 5 Aql *ApJSup* 59,229*; R Aqr *PASP* 98,118*; BW Aqr *AApSup* 69, 397*; CX Aqr *MN* 223,607*; FF Aqr *BAAS* 18,983; μ Aqr, 32 Aqr *ApJSup* 59,229*; TT Ari *PASP* 97,847*, 98,507*, *ApJ* 290,707*; UX Ari *AJ* 92,1403; VY Ari *BAAS* 19,709; 29 Ari *PASP* 98,468*; 66 Ari *Obs* 104,69; SS Aur *AJ* 92,658*; TT Aur *AAp* 162,62*, *MN* 224, 649*; HS Aur *AJ* 91,383*; IM Aur *AJ* 92,441; α Aur *AJ* 90,1503*, *JApA* 7,45, *IBVS* 2937; ε Aur *PASP* 97,1163, 98,389 and 637, *AAp* 144,395, *PASJ* 39,135, *AASin* 26,144, *AApSin* 5,180; Dzeta Aur *BAAS* 17,552, *AAp* 170,70; V363 Aur (Lanning 10) *ApJ* 307, 760*; TY Boo *BAAS* 19,643*; Kappa^2 Boo *AJ* 91,1416*; 6 Boo *JApA* 6,77*; 44 i Boo *AApSin* 5,36; AN Cam *AApSup* 67,161*; α Cam *PisAZh* 12,305*; S Cnc *ApJ* 285,208*; AC Cnc *ApJ* 280,235*; Y CVn *BAC* 35,74; TX CVn *AJ* 91,1400*; R CMa *ApJ* 297,250*; EZ CMa *AJ* 91,925, *MN* 222,809, *JApA* 7,305; GZ CMa *AJ* 90,1324*; ZZ CMi *PASP* 96,894; Dzeta Cap *PASP* 96,226; DW Car *RevMex* 10,323*; EM Car *PASP* 98,788*; HH Car *RevMex* 11,99*; V348 Car *MN* 213,75*; RX Cas *MittAG* 62,275, *PisAZh* 12,212*; PV Cas *AJ* 93, 672*; V373 Cas *AAp* 171,123*; V523 Cas *AJ* 90,354*; V651 Cas *IBVS* 2868*; RR Cen *MN* 209,645*; V834 Cen *MN* 224,987, *ApJ* 285,214, *ApSpSc* 131,613; α Cen *AAp* 158,273, 165,126; Nu Cen *AA* 35,395*, *ApJSup* 64,487*; 4 h Cen *ApJSup* 64,487*; RS Cep *AJ* 93, 171; VW Cep *IBVS* 2711, *AJ* 93,672*; XZ Cep *ATs* 1275, *BullAbas* 58,45*, *AZh* 62, 938; AH Cep *MN* 223,513*; CQ Cep *IAUSymp* 88,117*, *AAp* 134.45*, *BullAbas* 58,25*, *ApJ* 265, 961*, *JApA* 7,171*; DH Cep *BAAS* 19,714; WW Cet *AJ* 90,2082*; YY Cet *MN* 218,159*; AY Cet· *ApJ* 295,153*; 5 Cet *PASP* 97,355*; Z Cha *MN* 225,551*; TV Col *ApJ* 280,729; RW Com *AJ* 90,109*; T CrB *AJ* 91,125*; α CrB *AJ* 91,1428*; Θ CrB *IBVS* 2801; σ^2 CrB *BAAS* 16,473, *AJ* 89,1740*; RV Crv *MN* 223,595*; W Cru *MittAG* 62,275; VY Cru *ApJSup* 56,295; AI Cru *MN* 226,879*; SS Cyg *BAAS* 18,945, *ApJ* 286,747*, 300,794*, 305,732*; CG Cyg *AJ* 90,761*; CH Cyg *ApSpSc* 102,123, 116,355, 131,733, *IBVS* 2610, 2866, 2921, 2935, *PASJ* 36,567, *AnnTokyo* 20,75, *AAp* 156,186, 159,117, *ESA SP-218*,k407; CI Cyg *AA* 35,65, *AAp* 140,91; V367 Cyg *ATs* 1284; V380 Cyg *AAp* 141,39*; V444 Cyg *ApJ* 313, 358; V448 Cyg *AZh* 63,702; V729 Cyg *AAp* 143,209; V1016 Cyg *AAp* 142,85; V1143 Cyg *AAp* 174,107*; V1357 Cyg (X-1) *BAAS* 16,407, *PisAZh* 10,756; V1727 Cyg *MN* 218,613*; 31 Cyg *ApJ* 281,751, *BAAS* 19,707, *AAp* 170,70; 32 Cyg *ibid.*, *AASin* 25,56; CM Del *AJ* 90,643*; UX Dra *BAC* 35,65; UZ Dra *AApSup* 65,97*; AB Dra *AJ* 90,2082*; AG Dra *AJ* 91, 1400*; BF Dra *AApSup* 59,357*; BP and BW Dra *PASP* 98,92*; α Dra *PASP* 99,130; Nu^2 Dra *ApJSup* 59,229*; Chi Dra *AJ* 93,1236*; RU Eri *PASJ* 36,277*; YY Eri *AAp* 159, 142*; EF Eri *MN* 212,609*; EI Eri *BAAS* 19,708; Dzeta Eri *ApJSup* 59,229*; RX Gem *PASP* 99,274; RY Gem *AJ* 93,440; IR Gem *ApJ* 282,236*; σ Gem *BAAS* 17,588, *IBVS* 2937; 65 Gem *JRASC* 80,91*; Z Her *BullAbas* 58,163; YY Her *AAp* 135,410; AH Her *MN* 219, 791*, *AAp* 172,187; DQ Her *ApJ* 281,194; HZ Her *ApJ* 292,670*; V533 Her *PASP* 99,57*; V795 Her *AJ* 91,940; Iota Her *AJ* 89,1876; 68 u Her *MN* 211,943*; 89 (V441) Her *PASP* 96,641*; 96 Her *IBVS* 2778*; 105 Her *ApJ* 302,764*; 108 Her *ApJSup* 59,229*; 112 Her *BullAbas* 58,265*; RW Hya *AJ* 91,1400*; TT Hya *BAAS* 18,976, 19,708; EX Hya *ApJ* 317, 765; EZ Hya *MN* 209,645*; 21 (KW) Hya *AAp* 130,102*; VW Hyi *MN* 212,645, 225,113; RT Lac *AJ* 90,499, 91,583*; AR Lac *IBVS* 2579, *AAp* 176,267; X Leo *AJ* 92,658*; XY Leo *ApJ* 285,683*, *AAp* 35,29, *ApJ* 317,333* (quadruple system); DH Leo *AJ* 89,683*; 10 Leo *Obs* 105,7*; 93 Leo *IBVS* 2937; ST LMi *MN* 226,209*; UZ Lib *ApJ* 285,202*; GX Lib *AJ* 90,2581*; ∂ Lib *ATs* 1420; SZ Lyn *AApSup* 57,249*; FL Lyr *AJ* 91,383*; β Lyr *BAC* 37,42, *AJ* 90,773; Dzeta^1 Lyr *ApJSup* 59,229*; TY Men *AA* 34,345; TZ Men *AAp* 175,60*; AU Mon *BAAS* 19,713; AX Mon *RevMex* 10,229; BX Mon *AAp* 153,35*;

V641 Mon *ApJ 288*,731;   V644 Mon *BAAS 19*,708; 2 Mon *ApJSup 59*,229*; 18 Mon *Obs 104,* 267*;   SY Mus *PASP 97*,268; GR Nor *Obs 104*,221; U Oph *BAAS 19*,709*; RS Oph *AAp 167,* 91,   *PASP 98*,875,   *AJ 91*,1400*;   RZ Oph *AAp 168*,72*; V380 Oph *AJ 90*,643*; V426 Oph *ApJ 301*,L29*; V451 Oph *AAp 167*,287*; V502 Oph *MN 209*,645*; V508 Oph *PASP 98*,577*; V986 Oph *JRASC 79*,236*; V2051 Oph *AAp 154*,197*; 70 Oph *PASP 96*,903*; CN Ori *ApSpSc 131*,501*;   EW Ori *AJ 91*,383*; Iota Ori *Obs 107*,5*; Psi Ori *PASP 97, 428*; 64 Ori *AJ 92*,1162*; AR Pav *MittAG 62*,275; U Peg *PASP 97*,1086*; EZ Peg *Obs 105*,81*, *PASP 97,* 72; II Peg *AAp 176*,267; IP Peg *MN 224*,1031*; Iota Peg *PASP 96, 537*; 1 Peg B *Obs 107*,1*; 75 Peg *PASP 97*,280*; X Per *ApJ 299*,653*; RW Per *PASP 99*,159; RY Per *AJ 92,* 1168; AW Per *BAAS 19*,710*; AX Per *BAAS 16*,506; DM Per *MN 222*,167*; GK Per *ApJ 300,* 788*; IQ Per *ApJ 295*,569*; β Per *RevMex 10*,257, *PASP 97*,51; Phi Per *ApSpSc 107,* 323;   AE Phe *AA 34*,345; Al Phe *ApJ 282*,748*;   UV Psc *BAAS 16*,473; VZ Psc *BAAS 17,* 584; AO Psc *PASP 98*,104*; Dzeta Psc B *PASP 96*,179*; Omega Psc *AApSup 61*,363*; RX Pup *MN 208*,161; PV Pup *AAp 132*,219*; U Sge *PASP 97, 138*, *AJ 92*,1168*; V Sge *MN 219*,809,   *ApJ 306*,618; WY Sge *ApJ 282*,763*; WZ Sge *ApJ 301*,252*; HM Sge *ApJ 280,* 695, *AAp 139*,296, *142*,85; ∂ Sge *Highlights Astr. 7*,207*; W Sgr *PASP 96*,630; RS Sgr *PASP 98*,1342*; V1223 Sgr *ApJ 289*,300*; V1647 Sgr *AAp 145*,206*; V3885 Sgr *AAp 151,* 157*, *MittAG 62*,281*; Ypsilon Sgr *AAp 166*,237; U Sco *MN 213*,443; V701 Sco *MN 226,* 889*; V760 Sco *AAp 151*,329*; V818 Sco (X-1) *AJ 90*,2077*; Nu Sco, π Sco and Rho Sco *ApJSup 64*,487*; RT Scl   *MN 223*,581*; VY Scl *PASP 96*,559*; VZ Scl *MN 225*,43; AL Scl *AAp 179*,141*; RS Sct *MN 209*,645*; RY Sct *Bull Abas 58*,101*; RZ Sct *ApJ 289*,748; W Ser *MittAG 62*,275, *63*,194; RT Ser *PASP 97*,151; EG Ser *PASP 98*,1312*; MR Ser *MN 226*,209*; 39 Ser *PASP 97*,355*; 41 Sex *ApJSup 59, 229*, *PASP 98*,238*; V471 Tau *BAAS 18*,978, *JRASC 79*,235; V711 Tau *BAAS 16*,473, *AJ 92*,1403, *AAp 160*,73, *176*,267, *Pub. Beijing Obs. 6*,211; σ¹ and 63 Tau *ApJSup 59, 229*; AQ Tuc *MN 223, 581*; SU UMa *ApJ 309*,721*; SW UMa *ApJ 308*,765*; UX UMa *AJ 89*,1555*, *PisAZh 11*,617; AN UMa *AZh 63,* 516*; AW UMa *AJ 90*,767*; BE UMa *PASP 97*,328, *ApJ 316*,399; CH UMa *AJ 91*,940*; DN UMa *PASP 98*,1312*; Xi UMa *PASP 99*,38; RR UMi *PASP 98*,650*; τ² Vel *AJ 91*,1386*; FO Vir *AJ 91*,1221*; α Vir *AJ 90*,92; VW Vul *AJ 90*,643*; ER Vul *BAAS 16*,473; PU Vul *AJ 91*,563, *ApSpSc 131*,487, *IBVS 2576*; 22 Vul *PASP 97*,725*, *AAp 166*,252; HR 152 *PASP 97*,740*; HR 1023 *PASP 96*,609*; HR 1105 *Obs 104*,224*; HR 1120 *PASP 97*,637*; HR 1878 *JRASC 78*,151*; HR 2214 *ApJSup 59*,229*; HR 2259 *Obs 106*,16*; HR 2692 *PASP 97*,355*;   HR 3337 *JRASC 79*,119*; HR 3523 *ApJSup 59*,229*; HR 3725 *PASP 97*,355*; HR 4550   *Obs 104*,192; HR 5053 *Obs 106*,35*; HR 5273 *AJ 93*,683*; HR 6384 *IBVS 2686*; HR 6469 *IBVS 2937*; HR 6902 *JApA 7, 195*; HR 7038 *PASP 97*,637*; HR 7041 *Obs 107,* 58*; HR 7551 *AAp 139*,123*; HR 7617 *PASP 97*,637*; HR 8708 *ApJSup 59*,229; HD 434 *JRASC 79*,49*; HD 1383 *PubDAO 16*,193*; HD 3950 *ApJSup 61*,419*; HD 7272 *MN 210*,745*; HD 7331 *AAp 178*,114*; HD 8358 *ApJ 297*,691*; HD 9974 *AJ 91*,1392*;   HD 11246 *PASP 98*,238*;   HD 14346 *Obs 105*,126*; HD 14985 *Obs 105*,201*; HD 16909 *AJ 90*,609*; HD 17198 *Obs 106*,197*; HD 20126 *MN 212*,663*; HD 23838 *AAp 175*,136*; HD 25099 *Obs 105*,29*; HD 25799 *BAAS 18*,985; HD 27935, 28291, 28394, 28634, 29608, 29896 and 30197 *AJ 90*,609*; HD 30869 *AJ 90*,609*, *Obs 106*,13*; HD 37737 *ApJSup 61*,419*; HD 37847 *IBVS 2669*; HD 44172 *ApJSup 61*,419*; HD 44780 *JRASC 80*,91*; HD 46407 *ApJ 268*,264*; HD 47129 *Obs 107*,68*; HD 47755 *AJ 91*,590; HD 52533 *ApJSup 61*,419*; HD 53299A *BAAS 18*,986*;   HD 54371 *PASP 97*,355*; HD 55510 *Obs 106*,108*; HD 56429 *AJ 90*,1324*;   HD 63099 *PASP 96*,549*; HD 64503 *IBVS 2242*; HD 65195 and 68874 *MN 212,* 663*; HD 72754 *IBVS 2949*; HD 77581 *ApJ 314*,634*, *317*,746; HD 83065 *Obs 105*,226*; HD 89249 *AAp 177*,105; HD 91948 *IBVS 2542*; HD 94546 *RevMex 11*,143*; HD 96342 *Obs 106*,154*; HD 102010 and HD 102465 *AApSup 57*,99*; HD 102928 *AAp 144*,403*; HD 105982 *JApA 6*,71*; HD 106225 *IBVS 2543*; HD 106760 *JApA 5*,181*; HD 110195 *JApA 6*,159*; HD 112486 *PASP 98*,238*; HD 120710 *AApSup 64*,487*; HD 123058 *MN 211*,793; HD 128220 *MN 226*,249*; HD 137432, 138690, 139160, 139365 *ApJSup 64*,487*; HD 140629 A and B *BAAS 18*,683*; HD 142096, 142165, 142315, 142883 *AApSup 64*,487*; HD 145206 *AAp 144,* 403*;   HD 145482 and 145519 *AApSup 64*,487*; HD 145677 *MN 210*,745*; HD 146227 *MN 212*,663*;   HD 149162 *ApJ 310, 354*; HD 149240 *MN 210*,745*, *ApJ 302*,764*; HD 153919 *BAAS 18*,946; HD 158393 *MN 226*,813*; HD 166478 *MN 210*,745*; HD 174853 *IBVS 2848*; HD 176435 *MN 210*,745*; HD 178428 *PASP 97*,355*; HD 182593 *Obs 106*,67*; HD 184728 *AAp 163*,326; HD 191567 *ApJSup 61*,419*; HD 192163 *AAp 149*,337; HD 193793 *ApJ 312,* 807*;   HD 194056 *Obs 107*,114*; HD 196795 *AAp 178*,114*; HD 197406 *ApJ 304*,188*;

HD 210737 *AApSup 62*, 355*; HD 214608 *AAp 178*,114*; HD 214850 *JRASC 79*,167*; HD 218393 *IBVS* 2519; HD 219018 *AAp 178*,114*; HD 219634 *AAp 151*,254*; HD 220057 *ApJSup 61*,419*; HD 224113 *AAp 179*,141*; HDE 245770 *AAp 177*,91*, *BullAbas 58*,282*; HDE 284414, 285766, 285947, 285970, 287116 and J331 *AJ 90*,609*; BD 61°1211 *IBVS* 2669, *AJ 92*,1403; BD 37°444 *PASP 98*,1321*; BD 26°730 *AJ 90*,609*; BD 23°635 *PASP 98*,457; BD 13°3683 *AAp 170*,55*; BD -3°2525 *ApJ 304*,721*; BD -16°6074 *AApSup 69*,397*; CD -31°10727 *ApJSup 64*,487*; CPD -48°1577 *ApJ 292*,601, *ApSpSc 99*,145*, *MN 204*,35P; CPD -58°271 *MN 225*,1005*; W 33436, 60232, 42576, 42574, 12965, 42404, 40480 *ApJ 281*,L41*; Gliese 268 *AJ 92*,1424*; Sk 188 *ApJ 292*,511*; G82-23 *ApSpSc 110*,162; M28 V7 *PASP 97*,962; R31 *PASP 96*,811*; R130 *ApJ 309*,714*; LSS 2018 *ApJ 294*,L107*; MS4 (WR29a) *AApSup 58*,117; MWC 560 *BAAS 16*,516; PHL 227 *AAp 149*,L4; PSR 2303+46 *ApJ 294*,L21*; 1502+09 *PASP 97*,41; 0623+71 *PASP 97*,990*; NS 105-67 *ApJ 310*,715*; A 0538 -66 *MN 210*,855, 212,565, *PASP 97*,418*; A 0620-00 *ApJ 308*,110*; 3A 0729+103 *MN 210*,663, *ApJ289*,300*; 1E 15487+1125 *PASP 97*,1096; 1E 1048.5+5421 *ApJ 314*,641; 1E 1145.1-6141 *PASP 99*,420*; E 1013-477 *ApSpSc 131*,613; E 1114+182 *ApJ 293*,303; E 2003 +225 *MN 221*,823*, 226,209, *PisAZh 12*,468; EXO 0748-676 *ApJ 306*,599*; H 0139 -68 *PASP 97*,423, *ApJ 286*,328; PG 0834+488 *AJ 91*,940*; 2S 0114+650 *ApJ 299*,839*; 4U 1223 -62 *ApJ 304*,241*, 287,856*; 4U 1258-61 *MN 221*,961; 4U 1538-52 *ApJ 314*,619*; 4U 1907 +09 *PASJ 38*,463; 4U 1957+11 *ApJ 312*,739*; V 0332+53 *AAp 162*,117.

A number of papers surveying specific characteristics of several members of a group of binaries have not been included in the Table, since the frequent repe- tition of the same reference would have unduly lengthened it. Studies of RS CVn stars (*AASin 27*,130 and 259; *AApSin 6*,277) and of the rotation of binary stars (*AApSin 6*,154) may be of interest to several readers. Students of cataclysmic variables may be interested in the spectrophotometric study of such objects in *ApJSup 63*,685 and in many more papers in *ApSpSc 130* and *131* than are cited in the Table.

C. POLARIMETRIC STUDIES (A.M. Cherepashchuk)

The most important results are: spectroscopic and polarimetric investiga- tions of the runaway WN7 star HD197406 - a possible binary with an X-ray quiet black hole (Drissen *et al. ApJ 304*,188); polarimetric investigations of ϵ Aur (Kemp *et al. ApJ 300*,L11); circular spectropolarimetry of VV Pup (Wickramasinghe *et al. MN 210*,37); calculations of polarization of intrinsic radiation of tidally distorted stars (Bochkarev and Karitskaya, *ApSpSc 109*,1); calculation of angular distribution and polarization of X-ray burster radiation (Lapidus and Sunyaev *MN 217*,291).

Other publications are: *RevMex 10*,267, *AAp 142*,333, *MN 212*,709, *MN 215*,83, *MN 218*,201, *ApSpSc 118*,291, *ApJ 301*,881, *RevMex 12*,332, *AAp 162*,99, *RevMex 12*,407, *MN 220*,663, *ApJ 306*,215, *MN 222*,225, *Mitt AG 67*,310, *ApSpSc 131*,657, *AAp 130*,197, *PASP 99*,62, *AdvSpRes 3*,265, *AZh 63*,71, *Bull Abas 58*,273, *Astrofizika 23*,503, *IzvKrym 75*,120, *AZh Letters 11*,623, *CommAp 12*,1, and papers in conference proceed- ings.

D. X-RAY OBSERVATIONS (Y. Kondo)

During the present reporting period two X-ray satellites were in operation. The European satellite EXOSAT, which was launched in May 1983, ended its produc- tive operation in April 1986. The Japanese satellite *Ginga* (galaxy), which was launched in February 1987, was operating successfully as of the closing date of this report. The American satellite *Einstein* (HEAO-2) ended its operation in June 1981. In view of its ongoing archival research program, we will also describe it briefly.

Einstein

The archival data include about 100 binaries with compact objects, about 100 cataclysmic variables, about 80 "ordinary" binaries, and about 60 RS CVn objects.

The X-ray data are mostly spectroscopic and are in the spectral range 0.3-3.0 keV. A complete listing of all the objects observed with the *Einstein* can be found in the booklet *A listing of All Targets Observed by the Einstein Observatory*, which is available from: Dr. Frederick D. Seward, Head, Einstein Guest Observer Program, Center for Astrophysics, 60 Garden Street, Cambridge, MA 02138, U.S.A. He can also provide information about the *Einstein* archival data research program.

EXOSAT

All the objects observed with EXOSAT, numbering 2000, are listed in *The EXOSAT Observing Log*. The catalogue contains hundreds of binary objects. Among the X-ray binaries listed are: SMC X-1, LMC X-1, 2, 3, 4, 6, Cen X-3, Cir X-1, Her X-1, AM Her, Cyg X-1, 2, 3, and Vela X-1. A number of cataclysmic variables, such as U Gem, SS Cyg and WZ Sge, have also been observed. Among the "ordinary" binaries observed are Algol and W UMa. Information concerning the use of the EXOSAT archival data may be obtained from: Dr. N. E. White, EXOSAT Project Scientist, ESTEC-ESA, Postbus 299, Noordwijk, The Netherlands.

Ginga

The primary objective of the *Ginga* is to observe variabilities in X-ray sources. As of the time of the preparation of this report, Algol and UX Ari have been observed. Compact X-ray sources observed include X 0540-693, 1E 2259+586, SMC X-1, GX 1+4, GX 3+1, NGC 6624, X 1916-056, GC Transient, GX 5-1, Cyg X-2, LMC X-1, LMC X-2, SS 433 and MX 1353-645, some of which are known binary systems. The project scientist for *Ginga* is Dr. F. Makino, Institute of Space and Astronautical Science, 4-6-1 Komaba, Meguro-ku, Tokyo 153, Japan.

E. ULTRAVIOLET OBSERVATIONS (B.J.M. Hassall)

Results from the satellite, the International Ultraviolet Explorer (IUE), continue to dominate the literature for the wavelength range 1200-3200Å. Close binaries comprise a large fraction of this work; of the 61966 IUE observations obtained by 1 June 1987, 1324 exposures are of symbiotic stars, 1167 of dwarf novae, 880 of classical novae and 960 of X-ray binaries alone. The maturity of IUE is illustrated by the publication of the book *Exploring the Universe with the IUE Satellite* which reviews the advances in UV astronomy over the last ten years. Chapters on symbiotic stars (Nussbaumer and Stencel), novae (Starrfield and Snijders), cataclysmic variables and X-ray binaries (Cordova and Howarth), interacting binaries (McCluskey and Sahade), Wolf-Rayet stars (Willis and Garmany) and supergiant eclipsing systems (Hack and Stickland), cover their respective topics in greater depth than the limited space here allows.

Further UV observations of close binaries are reported in the three proceedings of IUE symposia published during the triennium, namely: NASA CP2349 (1984), ESA SP218 (1984) and ESA SP263 (1986). Proceedings of conferences devoted specifically to interacting binaries also include contributions on UV observations: *Interacting Binaries*, Reidel, 1985, Eds. Eggleton and Pringle; *Recent Results on Cataclysmic Variables*, 1985, Bamberg, ESA SP236; *Cataclysmic Variables*, ApSpSc *130* and *131*. Of the other recent UV satellites, Copernicus, ANS and Voyager, the last is currently the most important with regard to observational papers. The paragraphs below address each of the major groups of close binaries, listing the more important UV references for each.

IUE observations of recent nova outbursts have confirmed the existence of the class of neon novae in which a TNR outburst on a ONeMg white dwarf leads to enrichment of Ne and Mg in the ejecta. The signature in the UV is provided by strong lines of [NeIV] 1602,2426 and [NeV] 3346. Thus V1500 Cyg is joined by V693 CrA 1981 (Williams *et al. MN 212*,753) with $Z \approx 0.4$, V1370 Aql 1982 (Snijders *et al. MN 211*,7p) and N Vul 1985 No.2 (Starrfield *et al. ApJ 303*,L5). The 1985 outburst

spectra of RS Oph revealed strong coronal lines of FeXI, FeXII and FeXIII, the
first to be observed in a recurrent nova in the UV (Snijders in Proc. Manchester
Conf.: *RS Oph and the Recurrent Nova Phenomenon*, Ed. Bode, p.51). They are presumed to originate in shocked gas where the nova shell meets the companion's
stellar wind. The ejecta abundance appears to be compatible with TNR rather than
an accretion event.

Long term IUE monitoring of several symbiotics has been undertaken, in some
cases giving direct support for binarity. Fernandez-Castro *et al.* (ESA *SP236*,225)
observed the prototype Z And in both outburst and quiescent phases, and attribute
quiescent variability to orbital motions with a period of 760 days. Periodic photometric variations and wavelength shifts in the UV emission lines in HBV 475 led
Mueller *et al.* and Nussbaumer *et al.* to a binary interpretation (*AAp 154*,313; *AAp
169*,154). A similar conclusion was reached by Viotti *et al.* (*ApJ 283*,226) for the
high velocity symbiotic AG Dra. Other observations of symbiotics are reported in
BAAS 17,886 (Voyager data of AG Peg); *MN 212*,939; *AAp 140*,317 and *AAp 161*,287.

During the principal eclipse of Algol-type systems (Plavec *et al.* NASA *CP
2349*,420) the existence of a high temperature region and the so-called SiIV/CIV
flux reversal has been demonstrated. Peters and Polidan (*ApJ 283*,745) identify the
source as a high temperature accretion region (HTAR) where the accretion stream
impacts directly on the stellar surface, although differences in line profiles
between ion species imply more than one region may be involved (Sahade and
Hernandez *RevMex 10*,257). Plavec and Dobias continue their series on this group
(e.g. *PASP 99*,159 and 274; *AJ 92*,171 and 440).

Rucinski (*MN 215*,615) finds that the chromospheric MgII line flux of W UMa
stars, while independent of orbital phase, may be stronger at lower effective temperatures. This relation is extended to RS CVn systems by Fernandez-Figueroa *et
al.* (*AAp 169*,237). Conversely, the transition region lines in W UMa's behave the
same as in single stars (Oranje *AAp 154*,185). Further rotational effects in RS
CVn's are investigated by Vilhu and Heise (*ApJ 311*,937) and the Armagh group and
co-workers (*AAp 180*,172 and refs. therein).

Two dwarf novae monitored during quiescence (VW Hyi, Verbunt *et al.* *MN
225*,113, WX Hyi, Hassall *et al.* *MN 216*,353) undergo a secular decrease in UV luminosity between outbursts. This is more easily accommodated within a mass transfer
burst model than a disc instability, unless the source of the UV luminosity is a
cooling white dwarf rather than the accretion disc. Extreme UV (Voyager) observations of a superoutburst of VW Hyi (Polidan and Holberg *MN 225*,131) show that the
fading between precursor and outburst proper is more marked than at optical or IUE
wavelengths. Otherwise, the precursor closely resembles an ordinary outburst in
the extreme UV. Phase resolved studies of line profiles indicating a wind are
reported for RW Tri and DQ Her (Cordova and Mason *ApJ 290*,671), for OY Car in
superoutburst (Hassall *et al.* *ApSpSc 130*,371) and for Z Cam (Szkody and Mateo *ApJ
301*,286). Observations of several other CV's are reported in *AJ 90*,9; *MN 210*,197;
MN 212,231; *MN 212*,645.

The UV flux of the "3-period" intermediate polar TV Col was observed to be
modulated on the orbital period of 5.5h, which Bonnet-Bidaut *et al.* (*AAp 143*,313)
associate with an X-ray illuminated bulge on the accretion disc. Mateo *et al.* (*ApJ
288*,292) also detect the 4-day period. Other magnetic CV's are reported in *ApJ
293*,321 and *ApJ 315*,L123 and references therein.

McClintock *et al.* (*ApJ 283*,794) argued from a reddening study that the distance to Cyg X-2 is greater than 1.1kpc and hence that the degenerate object is
indeed a neutron star in this LMXB. High resolution spectroscopy of the high mass
X-ray binaries Vela X-1 and SMC X-1 show substantial variations with orbital phase
of the P Cygni profiles (Hammerschlag-Hensberge *et al.* *ApJ 283*,249; Sadakane *et*

al. *ApJ 288*,284), in general agreement with the models by McCray *et al.* *(ApJ 282,* 245) for a strong stellar wind interacting with the X-ray source. Other early-type X-ray binary observations are reported in *ApSpSc 109*,175; *AAp 141*,279; *MN 225*,985.

Observations of atmospheric eclipses in supergiant systems are reported for ∈ Aur (Ferluga and Hack *AAp 144*,395) which has no completed its 1982-84 eclipse. A series of papers by the Hamburg group (*AAp 138*,333; *AAp 147*,103; *AAp 156*,172; *AAp 170*,70) investigate the wind in Dzeta Aur and its relatives 31 Cyg and 32 Cyg. The new member of the class 22 Vul, with the shortest binary period, has been much studied. The fact that the B star is permanently embedded in the dense wind of its G companion, leads to a much hotter (300,000K) electron temperature than in the longer period systems (Ake *et al.* *ApJ 298*,772; Reimers and Che-Bohnenstengel *AAp 166*,252).

Papers relating to binary aspects of Wolf-Rayet stars concentrate on the search for compact companions and what change, if any, binarity imposes on the wind structure (e.g. *ApJ 296*,222; *ApJ 313*,358; *AAp 146*,307).

3. Derived Physical Data

A. ABSOLUTE DIMENSIONS AND APSIDAL MOTION (D.M. Popper)

Observational Results on Absolute Dimensions

There appear to have been no reviews of absolute dimensions of close binaries within the three-year period. New determinations of masses and radii, based on observations leading to new spectroscopic or photometric orbits or both have been obtained for the following detached systems. Only results of good accuracy are listed here. HS Aur *AJ 91*,383; GZ CMa *AJ 90*,1324; PV Cas *AJ 93*,672; WX Cep *AJ 93*,672; AH Cep *MN 223*,513; V1143 Cyg *AAp 174*,107; V624 Her *AJ 89*,1057; FL Lyr *AJ 91*,383; TZ Men *AAp 175*,60; EW Ori *AJ 91*,383; V451 Oph *AAp 167*,287, *ApSpSc 106*,93; IQ Per *ApJ 295*,569; LX Per *ApSpSc 112*,273; AI Phe *ApJ 282*,748; V1647 Sgr *AAp 145*,206; V760 Sco *AAp 151*,329; AL Scl *AAp 179*,141; DM Vir *AAp 137*,281.

New determinations for semi-detached systems are: TT Aur *AAp 162*,62; S Cnc *ApJ 285*,208; R CMa *ApJ 297*,250; HH Car *RevMex 11*,99; YY Cet *MN 218*,159; RY Gem *AJ 92*,440; u Her *MN 211*,943; while W UMa systems with newly determined masses and radii are: XY Boo *ApSpSc 107*,347; BV Dra *AJ 92*,666; BW Dra *AJ 92*,666; and XY Leo *ApJ 290*,696. Standards of accuracy for systems in these two categories are lower than for detached systems.

Masses and luminosities have been obtained by combining spectrographic and interferometer observations for α Aur (*AJ 90*,1503) and Chi Dra (*AJ 93*,1236).

Techniques

Techniques for spectroscopic observations and for their analysis continue to develop. While most of the results listed here are based on photographic spectrograms, an increasing proportion is employing linear digital detectors of considerably greater sensitivity and lower noise (Reticons, CCD's). Radial velocity spectrometers are starting to be employed in this endeavor, although the relatively large rotational velocities in most eclipsing binaries are an obstacle for this kind of observation. For photographic observations, as well as those using the newer types of detectors, cross-correlation techniques are being used increasingly, with the scheme developed at Victoria (*PubDAO 16*,159) receiving considerable application for the analysis of photographic plates. The method has the potential of detecting and measuring velocities of components not amenable to more conventional techniques and thus of providing results for stars of classes not heretofore available. Examples are the evolved early B stars HR 7551 (B0.5Ib, *AAp*

132,123) and V380 Cyg (B1.5 II-III, *AAp 141*,39). The feature due to the secondary in the cross-correlation function in each of these systems is very poorly defined, and troublesome systematic effects are present. Further studies are required (e.g. high resolution, high S/N profiles) in order to test the reliability of this type of analysis. The same comment holds for any of the newer techniques, particularly in cases where the pertinent details of the spectrum are not displayed directly. An example of a particularly thorough test of cross-correlation results is that for the W UMa system XY Leo (*ApJ 285*,683).

Some Noteworthy Results

The early B systems V348 Car (*MN 213*,75) and V1182 Aql (*MN 225*,961), though not yet definitive enough for the list of detached systems given above, have the potential of replacing LY Aur as the most massive system with directly determined masses. The secondary of EK Cep, mass 1.12MΘ, is found to be in a state of pre-main sequence contraction (*ApJ 313*,L81), unique among stars with well established properties. A light curve of HD 27130, the eclipsing binary in the Hyades cluster, has been obtained and analyzed (*AJ 93*,1471) to accompany the earlier double-lined spectrographic orbits. It is a matter of judgment whether the results for this important system are better employed for refining the distance to the Hyades, as in the cited reference, or to serve as a direct test of interior models (*ApJ 307*, L61), the distance being assumed known.

During the period of this report, noteworthy progress was made for the first time in extending our knowledge of fundamental properties down the main sequence to stars less massive than the sun. In addition to HS Aur, FL Lyr, and EW Ori, listed under "observational results", systems with less definitive results are α CrB (*AJ 91*,428) and CG Cyg (*AJ 90*,761). The former is particularly noteworthy for the orbit of the secondary, which is approximately 5 mag fainter than the primary, its detection and measurement being an accomplishment with the McDonald Reticon.

One use made of absolute dimensions is, by comparison with interior models, to deduce the age and chemical composition of a "best" model. In some cases, the He abundances so deduced have been at variance with abundances (about 0.24 to 0.30 by mass fraction, Y) generally considered to prevail on the basis of spectrographic evidence of various kinds. Examples of unusually low He abundances (Y$\approx$0.18) are for IQ Per (*ApJ 295*,269) and DM Vir (*AAp 137*,281). In such cases, with excellent observational results, it may be prudent to call into question the models employed in the comparisons. Another example is the system AI Phe. The more massive component (1.2MΘ) is well evolved into the Hertzsprung gap. Comparisons with models (*ApJ 291*,270) yield Y$\approx$0.38, an exceptionally large value. But preliminary revised velocities (unpublished), increasing the velocity amplitudes by only a few km s^{-1}, bring Y to 0.34, and it remains to be seen what the outcome of a definitive study will be.

Another interesting and important case is Chi Dra, a binary studied by speckle interferometry (*AJ 93*,1236). Earlier work had led to the conclusion that the primary had a mass of only 0.88MΘ, unprecedentedly low for an F7 star. Small revision in the radial velocity of the secondary (*AJ 93*,1236) increases the mass to 1.03MΘ, an excellent example of the care with which radial velocities must be obtained in order to avoid improper results. Even this mass is low for the luminosity, leading to the conclusion that the star is well evolved, with a system age $\approx$ 8x10^{9}y, an almost unique case of an old disk star ([Fe/H] $\approx$ -0.3), appreciably older than the sun, with well determined fundamental properties.

Finally, we note two contributions by R. and R. Griffen, based on results from their radial velocity spectrometer. *JRASC 80*,91 contains a review of the current position with respect to masses of cool giants. The preponderance of masses well above 2.0M⊙ is strengthened. In *JApA 7*,195 they discuss a new program for radial velocities of systems with composite spectra, the results from which could increase our meager information on masses of evolved stars.

Apsidal Motion

New results of high quality on apsidal motion are for EK Cep (*AJ 90*,358) and V1143 Cyg (*AJ 90*,348, *AAp 174*,107). Both show good agreement with the predictions of generally accepted gravitational theory. The results for EK Cep (see also *ApJ 297*,405) may provide the best example to date for agreement between theory and observation for a case in which the general relativistic and Newtonian contributions are comparable. On the other hand, the agreement in the case of AS Cam (*BAAS 19*,578), for which the Newtonian contribution is dominant, is very poor, the observed motion being only about 40% of the predicted Newtonian and 30% of the predicted total motion. This disagreement is of a much different nature from that in the much-discussed case of DI Her, where the relativistic term should dominate, and both the Newtonian and the observed values are vanishingly small. Suggested effects not considered in the standard theory (spin-orbit coupling with the rotational and orbital axes at large angles to each other – Shakura *SovAstrLett 11*, 244; viscosity – Hosokawa *ApSpSc 115*,403) would appear inadequate to explain the discrepancies.

B. PROXIMITY EFFECTS AND LIMB DARKENING (M. Kitamura)

The reflection effect for gray and nongray atmospheres has been investigated in detail by several workers with application to close binary systems. Of these, Vaz and Nordlund (*AAp 147*,281) studied the effect for gray atmospheres with convection and for the particular case of Algol they showed that the theory is in good agreement with observation. A critical review on the treatment for the reflection effect of eclipsing binaries was also presented by Vaz (*ApSpSc 113*,349). Yamasaki (*PASJ 38*,449) carried out a detailed numerical calculation for the reflection effect on nongray, plane-parallel, LTE, radiative equilibrium atmospheres with application to a model close binary consisting of early-type components and obtained rigorous values of monochromatic albedo.

The gravity-darkening of highly distorted stars in close binary systems has been studied as series work by Kitamura and Nakamura (*AnnTokyo 21*,229, 311 and 387). They showed that for main-sequence components in detached systems the empirical values of the gravity-darkening exponent are almost unity for O9.5 to early A and 0.35±0.03 for late A down to G1, while for normal giants in detached systems the corresponding empirical values are found to agree with the ones expected from model atmospheres. Most conspicuous result would be for secondary components filling the Roche lobe in semi-detached systems in which the empirical values of the exponent are found to be significantly greater than unity. In order to explain such observational evidence of excessive gravity-darkening, an additional darkening by mass loss in semi-detached systems has been studied by Unno *et al.* (*IAUColl 108*).

Non-linear limb-darkening laws have been studied by Rubashevskij (*Astrometr.Astrofiz. 51*,23 and 28, *52*,18) with two parameter representation; he also presented new values of theoretical limb-darkening for application to classical eclipsing binaries (*ATs* 1275,1297,1299). Another study on non-linear limb-darkening laws for detached systems was made by Goncharskij *et al.* (*AZh 63*,725).

4. Structure and Models of Close Binaries

A. EARLY TYPE SYSTEMS (K.-C. Leung)

Early type contact binaries come in two different varieties: evolved contact and zero-age contact systems. So far, the number of the evolved systems discovered has far exceeded the number of zero-age systems, of which very few are known, among them V701 Sco and BH Cen and, possibly, TU Mus, PZ Pyx and AW Lac. It is of vital interest for our understanding of binary formation and for the modeling of the structure of contact systems that these systems be studied. Their mass ratios are very important for testing the current theories of the structure of zero-age early type contact systems. Some progress has been made in the past three years. Bell and Malcolm (*MN 226*,399) obtained good phase coverage radial velocity curves as well as new light curves of V701 Sco. Their results confirmed the earlier result (with only a few spectra) reported on this system. There is excellent agreement between the spectroscopic and photometric mass ratios, both being essentially unity. The same authors (*MN 227*,481) also secured spectroscopic and photometric observations of the suspected (from period and spectral relation) zero-age contact system RZ Pyx. Again, there is good agreement between the spectroscopic and photometric mass ratio, 0.82. Unfortunately, the long awaited spectroscopic result for BH Cen is still not available (a photometric mass ratio of 0.84 having been reported earlier). TU Mus, with a mass ratio of 0.72, may be a marginal zero-age system (Leung, *Beijing Coll.*). Spectroscopic work is urgently needed for BH Cen and AW Lac (a photometric mass ratio of unity having been reported in an earlier study). Evidently, the mass ratios for early-type zero-age systems are not limited to unity.

If we accept the standard view of close binary evolution, we expect the mass gainer in semi-detached binaries to evolve toward asynchronous rotation due to the rapid accretion of mass and momentum. In some cases, the gainers may reach critical rotation while the losers are in contact with their standard critical potential surfaces. These systems were named double contact binaries by R.E. Wilson. Investigations of this phase of close binary evolution are important to our understanding of rapid phase of mass transfer. Wilson *et al.* (*ApJ 289*,748) reported another system of this type, RZ Sct. It is recommended to spectroscopists that radial velocity studies of the Rossiter effect are very much needed for the study of asynchronous rotation and for the investigation of truly double contact systems.

At present, there are about 15 early-type contact systems with spectral types B or earlier reported in the literature. Some of them present severe challenges in interpretation and may require multi-dimensional approaches. For example, the research group at the Abastumani Astrophysical Observatory (Cherepashchuk, Babaev, Kumsiashvili, Karetnikov and Skul'skij) tackled RY Sct by means of spectrophotometry, optical, intermediate band-pass photometry, emission at Hα and radio observations. Other systems could well deserve similar approaches. The systematic spectroscopic and photometric observational program on early-type systems by Bell, Hilditch, Adamson and Malcolm at St.Andrews is particularly noteworthy. This effort will help to build a good data base for absolute dimensions of early-type systems, reliable temperatures for their components and more accurate locations in the conventional H-R diagram. Such work is very useful for the study of close binary evolution.

Among early-type binaries, some consist of one component with spectral type A and a companion of later spectral type. Their light curves are similar to that of β Lyr, i.e. with large differences between the depths of eclipses (primary and secondary). The polar temperature differences between components are huge, typically several thousands of degrees. Their photometric solutions indicate contact configurations. Many systems with modern Roche lobe light curve analyses indicating contact configurations have appeared in the literature. How can the huge jump (or drop) in temperature be maintained at the interface of such a contact binary? Usually, we would expect A stars to have radiative atmospheres and late-type stars to have convective atmospheres. There does not seem to be any debate about two early-type stars forming an early-type contact and two late-type stars forming a W UMa system. Can we form a mixed-type contact system with a large temperature difference? Are the published (contact) solutions incorrect?

B. ALGOLS (M. Plavec)

The term "Algols" is becoming a convenient and fairly accurate synonym for "detached binary systems with non-degenerate components". The once contemplated subgroup with "undersize subgiants" consisted partly of stars now included among RS CVn systems, and partly of inadequately observed semi-detached systems; the demise of the latter group, advocated by Hall, is further supported by a more recent statistics by Budding (*PASP 97*,594). The status of our knowledge on Algols, current trends in the field, and main problems have been very nicely reviewed by Budding (*ApSpSc 118*,241). He also formulated the importance of studying the Algols: Among the eclipsing binaries, this is the most populous type; they are relatively easy to observe, and their fundamental parameters can be established with fair accuracy; and they represent a crucial link between the virtually non-interacting detached binary systems and the generally much more complex inter-acting binaries of advanced age (usually containing compact components). I wish to add that some phenomena associated with "interaction", i.e. with the process of mass transfer between the components and mass loss from the system, are present in the "classical Algols", albeit on a diminished scale; but just this circumstance makes it useful to study some aspects of the processes on Algols. The position of the Algols in the grand evolutionary scheme of interacting close binaries is best seen if you read the comprehensive treatment by Iben and Tutukov (*ApJSup 58*,661; *ApJ 313*,727), or the overview by Iben (*QJRAS 26*,1). The possible post-Algol evo-lution was also discussed by de Loore (*ApSpSc 99*,199).

Budding compiled a catalog of 414 Algols and possible Algols (*Bull.d'Inform.* No.27, ed. C. Jaschek). Together with the earlier statistics by Giuricin *et al.* (*ApJSup 46*,1; *ApJSup 52*,35; *AApSup 45*,85), this is a rich source of statistical information, and good reference for further studies. However, the material is of necessity of very inhomogeneous quality and reliability. The creation of a much shorter but homogeneous list of well-determined parameters will eventually result from the efforts of many people; a good standard is set e.g. by Etzel and Olson (S Cnc: *AJ 80*,504) and by Etzel's work on TT Hya. Plavec and Dobias (*AJ 93*,171 and 140; *PASP 99*,159 and 274; Plavec, 10th ERC-IAU, Prague) find from the IUE spectra that the Algols usually classified as A1-A5 are quite often late B stars, the op-tical spectra being contaminated by shell lines and optical photometry being at times affected by an incomplete subtraction of the flux of the secondary compo-nent. An opposite phenomenon was found for the B5 primary in RY Per, whose spec-trum resembles a B9 star. This star is very rapidly rotating (Van Hamme and Wilson *AJ 92*,1168) and has a circumstellar shell. If the latter is the cause of the dis-crepancy, then RY Per is similar to the non-eclipsing interacting binary KX And, for which Štefl (*BAC 36*,313) demonstrated the presence of an optically thick shell in the UV. Thus, caution is needed in evaluating the statistics, and for compari-son with theory, it is often better to deal with individual cases.

A systematic effort at comparing the theoretical models with actually ob-served Algols is under way in Brussels. De Greve (*SpScRev 43*,139) presented a thorough review of the current status of theoretical modeling of the evolution of Algols, based on the very extensive work done by him and the entire, very active group headed by de Loore in Brussels. The most recent novelty introduced in the modeling is the inclusion of the convective overshooting of the core; De Greve finds that it makes the end masses larger and increases the probability of the oc-currence of case A of mass transfer. Case B remains the most usual case for Al-gols, though. De Greve finds once again that observed systems can be matched only if significant loss of mass and of angular momentum from the evolving systems is postulated.

These loses presumably occur mainly during the rapid phase of mass transfer, therefore the search for such systems continues. The accompanying presence of large and dense structures of circumstellar and circumbinary matter seriously com-plicates the analysis of such systems, but also calls attention to them. The problems associated with studying these complex systems (called W Serpentis stars by Plavec) have been discussed by Plavec (in Eggleton and Pringle: *Interacting Binaries*, p.155) and, from a different point of view, by Sahade and McCluskey (in Kondo: *Ultraviolet Astronomy with the IUE*), and by Sahade (in *New Insights in Astrophysics*, p.267). Foremost among them is β Lyrae, reviewed comprehensively by Sahade (*SpScRev 26*,349). Plavec (10th ERC-IAU, Prague) finds, from a combination of optical and IUE spectra, that the secondary star definitely has a flux distribution typical for an accretion disk, while the primary is also anomalous; in the optical, it can be reasonably well fitted by a model atmosphere at T_{eff} = 13,000 K, log g = 2.5, but with respect to any model hotter than 11,500 K, its far UV flux is too low. The above specification of parameters of the primary star in the optical region comes from the work by Balachandran *et al.* (*MN 219*,479), who found from spectral synthesis of the line profiles that the surface material in the primary star of β Lyr underwent the CNO-cycle processing and is enriched in He and N, and underabundant in H, C, and O. Indications of the same abundance anomalies have been found in Lambda Tau and β Per by Cugier and Hardorp (*BAAS 17*,553).

Although it is tempting to find and understand a system near the phase of rapid mass transfer, it shows that ordinary Algols display similar phenomena, on a milder scale, but easier to analyze. Thus the emission lines found in the Serpentids can also be seen in the Algols, provided they are observed at or near the total eclipse of the primary component. Systematic search by Plavec (10th ERC-IAU, Prague) netted 10 systems with emission lines. Absolute and relative intensities of the super-ionized emission lines differ; in particular, early-type systems (V356 Sgr, RY Per) tend to display NV much stronger than CIV, while in systems near A0 (TT Hya, RS Cep), the opposite is true. Interesting information about electron temperatures and chemical composition can be extracted from such data, following the pioneering work by Peters and Polidan (*ApJ 283*,745). The problem of the energy source and location of this super-ionized plasma is related to the general problem of the geometry and physics of the circumstellar structures in Algols. Do genuine accretion disks exist in Algols? Are they optically thin or thick?

The problem of the circumstellar structures is being attacked from several directions. Observationally, Kaitchuck *et al.* (*PASP 97*,1178) continued their search for semi-permanent and transient disks in Algols, using fast spectrophotometry of the Hα line. More oriented towards detecting structure emitting in the continua is the systematic work by Olson (in Eggleton and Pringle: *Interacting Binaries*, p.155). Another approach is to attempt to solve the light curves of more complicated systems by including the effects of a disk; this has been done by Wilson for several objects (e.g. for RW Per, subm. to *PASP*), by Pavlovski and Křiž for SX Cas (*BAC 36*,153), by Pustylnik and Einasto (*ApSpSc 105*,259) for a circumbinary envelope in A0 Cas and RY Gem. Another approach is to study theoretically the interaction between the mass transferring stream, the disk, and the accreting star. Among these studies, let me mention the two-dimensional hydrodynamical treatment that reveals the presence of spiral shocks, by Sawada *et al.* (*MN 219*, 75), and the three-dimensional study by Hadrava (*BAC 35*,335).

An important part of the problem is the interaction between the rotation of the accreting star and the impacting stream. Wilson and collaborators repeatedly point out the probable existence of "double-contact binaries", in which the accreting star spins so fast that it has filled its critical tidal lobe and cannot accept any more matter (Van Hamme and Wilson *AJ 92*,1168 and references therein).

Another useful approach to the problems of Algols is to realize that they are related in many ways to other types of binary stars. The connection to the RS CVn stars has been pointed out by Olson (*IAPPP 19*,6) who started a search for spots on the cooler components. Another close relation exists between Algols and Be stars, and was stressed by Harmanec, Plavec, etc. (in Slettebak and Snow: *Phy-*

sics of Be Stars). The third and perhaps most important connection is to cataclysmic variables, with which the Algols share the accretion processes, circumstellar structures, and super-ionized UV lines. This connection is systematically studied by Polidan (10th ERC-IAU, Prague). Polidan also uses his own observations with the Voyager spectrometer, which extend the spectral coverage down to 912Å.

C. W UMA SYSTEMS (K.-C. Leung)

For more than a decade since the development of the Roche model for light curve analysis we have been able to derive relatively reliable photometric solutions for close binary systems. These methods, developed by many authors, introduce a new parameter, the mass ratio, to our photometric solutions. This is an important parameter, since it enables us to calculate absolute dimensions for single-lined spectroscopic binaries. The potential for gaining a much larger set of fundamental astrophysical quantities (e.g. mass and radius) is enormous. However, in practice, reality has proven to be not as rosy as we had hoped. As more modern photometric solutions are published, we find that there are irreconcilable differences between the spectroscopic and newly obtained photometric mass ratios in a considerable number of close binary systems. There are doubts in the minds of many astronomers, especially among the spectroscopists, about the reliability of the photometric values.

There have been three independent studies, by Kałużny (*AA 35*,313), Maceroni *et al.* (*MN 217*,843) and Leung (*Beijing Coll.*), using very different criteria in compiling data from the literature, to compare the mass ratios from spectroscopy and photometry. All of these investigations conclude that there is good agreement between accurate spectroscopic results (especially from cross correlation methods) and modern photometric mass ratios (where well-determined) for both A and W type W UMa systems. No matter how well things fit together, however, there always seems to be an exception to the rule. Maceroni (*AAp 170*,43) found that for V523 Cas the mass ratio determined by the cross correlation method differed significantly from the photometric value. He attempted many numerical variations on the light analysis but still could not account for the discrepancy. We wonder if the photometrists and spectroscopists should re-examine this system.

One of the outstanding problems in late-type contact binaries is the relation/difference between A and W type systems. Kilmartin *et al.* (*ApJ 319*,334) found V677 Cen most unusual, classifiable either as an A or W type system. Further investigation of this system may be of great interest to the connection between these two types of W UMa binaries.

Asymmetry in light curves (O'Connell effect) has been a major problem for the photometrists. Its astrophysical meaning is still a great mystery. Milone *et al.* (*ApJ 319*,325) interpreted this effect in terms of starspots on the contact system RW Com. This was the first attempt to include spot parameters into the Wilson-Devinney computing codes.

The importance of studying eclipsing systems in star clusters has been re-emphasized by Baliunas and Guinan (*ApJ 294*,207) and Kałużny and Shara (*ApJ 314*, 585) in their papers on W UMa systems in NGC 188. Work on this cluster is of great importance to the evolution of close binaries since it contains so many systems.

The St.Andrews group (Hilditch, McFarland and King) continue to observe and publish spectroscopic and photometric results on many W UMa systems. They have also tried to determine the absolute temperatures of the components. There is no doubt that their continued effort will make great contributions to the accumulation of basic uniform astrophysical quantities for the late-type contact and near-contact binaries. These data will be important for modeling late-type systems and testing the reliability of spectroscopic and photometric mass ratios. Of course, there are many individuals and groups around the world who continue excellent work on these systems also. If this trend of activity continues, there is hope that we may have a better understanding of the nature of A and W types by the time of the next Commission report.

Unfortunately, we cannot describe the situation for long-period contact or near-contact binary stars as optimistically as we did for W UMa systems. There are many reasons for the general reluctance to observe long or very long period systems, but these stars are just as important as the short period systems as far as our study of the total picture of binary evolution is concerned. In order to study Case B or Case C mass transfer systems we must look to longer periods. There are systems which have light curves very similar to those of W UMa and β Lyr stars with spectral types of F and K, but with periods ranging from about one hundred to several hundred days. Li and Leung (*Highlights of Astron.* 7,217) reported that three of these systems, 5 Cet, PW Pup and HD104901B, are contact or near-contact systems. Clearly, these binaries must be the result of Case B mass transfer. The light curve of UU Cnc reported by Nha *et al.* (*Korean J.Astron.Sp.Sci.* 3,1) is very similar to that of 5 Cet (both systems have periods of about 96 days). It is believed that the components of these systems are supergiants or giants. We hope that observers will take new interest in longer period systems.

D. CATACLYSMIC VARIABLES

Review articles covering different aspects of CV's can be found in conference proceedings. Other publications include *Catalogues* by Duerbeck (*SpScRev 45*,1) and Ritter (*AApSup 57*,385). Below we shall discuss only some of the highlights; reader is also referred to Sections 2 and 6 of this Report.

CV's are being vigorously observed in the X-ray, UV, optical and IR parts of the spectrum. An attempt to detect the EUV (<912Å) flux from CV's turned unsuccessful (Polidan and Carone *ApSpSc 130*,235). On the other hand the study of their radio-emission becomes an important topic (Chanmugan *ApSpSc 130*,53). The targets of observing programs are both new objects which have previously been unobserved in a given spectral region and selected individual systems of particular importance. For example, VW Hyi was observed in a broad range of wavelengths, from X-ray (EXOSAT), through UV (Voyager and IUE), to the visual (*MN 225*,73 and ff.). Of particular value are observations of dwarf novae throughout their outburst cycles. Photometry of Z Cha (Cook *MN 216*,219) revealed that the disk radius increased during outburst by 40%; there was also a change in the shape of the white dwarf eclipse which can be interpreted as being either due to variable, non-uniform surface brightness distribution of the white dwarf (Smak *AA 36*,211) or due to variable optical thickness of the inner parts of the disk (Wood *MN* in press). Photometry during eclipses is being used for mapping of the surface brightness distribution on the accretion disk (e.g. Horne *MN 213*,129; Horne and Stiening *MN 216*,933; Warner and Donoghue *MN 224*,733; Włodarczyk *AA 36*,395). Mapping of the disk in the emission lines also becomes increasingly important (Marsh and Horne *ApSpSc 130*,85).

Irradiation of the secondary by the boundary layer and/or the white dwarf is becoming an important topic. Observational evidence is available for Z Cha (Wade and Horne *ApJ* in press) and SS Cyg (Hessman *et al. ApJ 286*,747, Robinson *ApSpSc 130*,113), where enhanced irradiation during outbursts apparently leads to an increased mass outflow (Hessman *ApJ 300*,794). Variable irradiation is important in many types of systems, including SU UMa (Osaki *AAp 144*,369), novae (Shara *et al. ApJ 311*,163; Kovetz *et al. ApJ* in press), soft X-ray transients (Hameury *et al. AAp 162*,71) and magnetic CV's with high and low states (King and Lasota *AAp 140*,L16).

Warner (*MN 227*,23) re-determined absolute magnitudes of CV's and discussed several important relations for dwarf novae, involving orbital periods and parameters of the outburst cycles. There have also been new statistical studies of dwarf nova light curves (Szkody and Mattei *PASP 96*,988; Gicger *AA* in press; Smak *AA 35*, 357; van der Woerd and van Paradijs *MN 224*,271).

Many authors continued to construct time-dependent accretion disk models applicable to dwarf novae (recent reviews: Meyer *IAUColl 89*, Bath *ApSpSc 130*,293) and to compare them with observations. The most sensitive test involves the rise to outburst. Pringle *et al* (*MN 221*,169) concluded that the disk instability models are unable to reproduce observations of VW Hyi during that phase but Meyer-Hofmeister (*AAp 175*,113) and Meyer-Hofmeister and Meyer (*AAp* in press) showed that when proper opacities are used at low temperatures the agreement becomes satisfactory.

Progress has been reported in the theory of the boundary layer and its X-ray emission (Patterson and Raymond *ApJ 292*, 535 and 550; King and Shaviv *Nature 308*; King *et al*. *Nature 313*,290; review: Shaviv *ApSpSc 130*,303) and in the theory of accretion in magnetic CV's (reviews: King in *ESA Workshop on Recent Results on CV's*; Hameury *et al*. *ApSpSc 131*,583).

The TNR theory of the nova outbursts has been significantly modified. Shara *et al*. (*ApJ 294*,271) identified Nova CK Vul 1670, the oldest of all historic novae and determined its absolute magnitude at $M_R=10.4$. Thus CK Vul is now more than 100 times fainter than post-novae of the 20th century, its present accretion rate being accordingly also very low ($\approx 10^{-12}$ MΘ/yr). Together with the less extreme case of Nova WY Sge 1783 (Shara *et al*. *ApJ 282*,763), this was used as an evidence by Shara *et al*. (*ApJ 311*,163), Prialnik and Shara (*ApJ 311*,172), and Livio and Shara (*ApJ* in press) to formulate the hibernation theory for novae: The high accretion rates (10^{-8} MΘ/yr) are maintained only briefly (100 yrs) before and after outburst, while during millenia between outbursts novae hibernate at much lower rates. The long duration of the hibernation phase provides an explanation for the CNO enrichment and explains the observed diversity among novae. Longer hibernation times lead to stronger enrichments and more violent outbursts of the fast nova type.

E. ∈ AURIGAE (R.E. Stencel)

The following is a brief update to reports published in *Highlights of Astronomy 7*,143. There have been several important post-eclipse observations which are affecting overall interpretation.

1. Voyager re-observation confirms presence of hot star in far ultraviolet. Altner and Polidan report a new far UV spectrum obtained recently with the UVS on one of the Voyager spacecraft now in the outer solar system. Although of low S/N, it matches a previous Voyager spectrum of Dzeta Oph, a reddened late O-type star. This supports previous analysis by Altner *et al*. (*AApSup 65*,199) of a hot continuum in IUE observations. Comparison observations of Canopus are planned. Alther also reports continuing IUE observations and failure to detect continuum changes on very short timescales.

2. Infrared monitoring fails to detect increasing temperatures as hotter part of disk moves into view with approaching quadrature. Backman reports, in a private communication, that his continued shortwave infrared monitoring of the system has failed to show any color temperature changes since the end of eclipse, as predicted quadrature approaches. He estimates that nearly 30% of the F-star illuminated disk is visible and that a temperature increase would have been measurable by mid 1987.

3. Continued optical photometry and polarimetry reveal that the F supergiant is clearly a non-radial pulsator with various periods. Hopkins, Kemp and collaborators report that their monitoring shows unambiguous power spectrum peaks corresponding to 61, 80, 96, 129, 233 and 485 days, which support the suggestion by

Ferro (*MN 216*,571) of irregular variations. The new work is reported in Kemp *et al.* *ApJ 300*,L11, and Krause *et al.* *BAAS 19*,752.

4. Some additional important papers are: Saito *et al.* *PASJ 39*,135 – low mass solutions for ε Aur, but is it consistent with non-radial pulsations reported by Kemp *et al.*? See also a series of papers which appeared in *ApSpSc 120–123*. Quantitative optical spectroscopy: Lambert and Sawyer *PASP 98*,389; Thompson *et al.* 1987 preprint.

F. BINARY PULSARS (V. Trimble)

For the first seven years after their discovery, all pulsars were known (from absence of variable Doppler shifts of their periods) to be single stars. This was not surprising since, on the one hand, the violence of a supernova explosion might well disrupt even a close system, and, on the other hand, the wind of a normal companion would have a sufficiently high plasma frequency to keep pulsed radio emission from reaching us. Then, in late 1974, came "the" binary pulsar, 1913+16 (*ApJ 195*, L51). After several years of splendid isolation, it was joined by others, and data have now been published for seven binaries (*Nature 322*,712 and 714) and two single millisecond pulsars (*Nature 300*,615; *IAUC* 4401), thought to be closely related to the binaries. Their status relative to pulsars in general has been elegantly reviewed by Taylor and Stinebring (*AnnRevAAp 24*,285).The companions are, in all cases, white dwarfs or second neutron stars, so that ambient gas does not prevent radio wave propagation.

Binary (and msec) pulsars comprise only about 2% of currently catalogued objects, but, because they are fainter than average by a factor of about 10 (that is the known ones all have rather small dispersion measures; Taylor, *IAUSymp 125*) their real incidence must be larger, probably about 10%, the actual distribution of pulsar periods probably being bimodal (*ApJ 311*,694; *CurrSci 55*,327). The faintness is associated with smaller-than-average period derivatives and calculated magnetic fields of $10^{8.5-11.5}$ G *vs.* 10^{12} G or more typical of single objects. The binary and msec pulsars tend also to have rather small velocities and distances from the galactic plane.

The binary and msec pulsars are important in two ways. First, they provide exceedingly accurate clocks, the precision of terrestrial time standards now setting the limit to how well they can be measured (*Nature 315*,547). The absence of jitter in the phase and period show that any sea of gravitational radiation in which these pulsars are immersed must have energy density over a wide range of wavelengths considerably less than that needed to close the universe (*ApJ 265*,L35; *ApJ 315*,149; Taylor, *IAUSymp 125*). "The" binary 1913+16 shows a relativistic rotation of the line of apsides of about 4°/yr (vs. 43"/century for Mercury) and a gradual shrinkage of its orbit attributable to energy lost by gravitational radiation, as well as other relativistic effects (*PRL 52*,1348; Taylor, *IAUSymp 125*). These provide confirmation that general relativity (including the quadrupole formula for gravitational radiation emission) is an accurate picture of the way gravitation really works. In addition, the relativistic effects depend sensitively on the component masses, telling us that 1913+16 contains neutron stars of 1.45 and 1.38M☉. Similar effects in 2303+46 currently place the total system mass at 3±1M☉; precision of the determination will improve as t^2 but probably never rival that of the shorter period system. These masses require the neutron star equation of state to be somewhat harder than might have been thought on other grounds (*PRL 57*,1120).

Second, properties of the binary and millisecond pulsars tell us a good deal about the evolution of pulsars and of binary systems in general. A number of discussions have been published (*JApA 5*,235; *ApJ 308*,680; *AAp 173*,279; van den Heuvel in *IAUSymp 125*; *SovAstrAJ 29*,645; *Nature 300*,720) including some that disagree with the majority (*AAp 177*,163, *ApSpSc 128*,363). There are several main points, each with a certain amount of observational support. Pulsars are born (not necessarily as very fast rotators, $JApA$ $1,25$) with fields in excess of 10^{12} G, which decay in 10^{6-7} yrs to below 10^{10} G but not less than $10^{8.5}$ G ($Nature$ $322,153$) at the same time that the rotation periods lengthen to several seconds; single pulsars become extinct at this point. But a significant subset can (a) form in close binaries without the supernova event disrupting the system, (b) form peacefully in close binaries through mass transfer driving a white dwarf above the Chandrasekhar limit (ApJ $305,235$), or (c) acquire companions by capture in dense environments like globular clusters. Mass transfer onto these neutron stars will not only produce the observed X-ray binaries but also spin the neutron star back up to rotation periods in the msec range. Thus, when the companion evolves to a white dwarf or neutron star and ceases to provide interfering gas, the rejuvenated neutron star will once again function as a short-period (but weak field) pulsar. The known binary pulsars can all be fitted into various phases of this picture for systems of widely varying initial masses. The identity of the companion as WD or NS can be predicted from the orbit parameters ($Nature$ $325,416$) and is confirmed by optical identification for three of the WDs (ApJ 306,L85; $Nature$ $324,127$) and absence thereof for the two NSs. Two of the three WDs are quite cool (old) providing confirmation that neutron star magnetic fields need not disappear completely even in several billion years. In addition, it is possible for the two stars to coalesce or be tidally disrupted, leaving isolated msec pulsars like the two seen so far (Romani, Kulkarni and Blandford, subm. to $Nature$).

5. Statistical Investigations (A.H. Batten)

A number of studies of binary incidence in specific groups of stars have been published (e.g. $ApJSup$ $59,229$; $61,419$; and $64,487$). One of the more important is the new study, by Stryker et $al.$, of binary frequency among population II stars. From discussion of new data and re-discussion of older data, and by application of more refined statistical tests for the variability of velocities, Stryker et $al.$ conclude that the binary frequency amongst these stars is around 30% - distinctly higher than earlier estimates.

Griffin (in $Interacting$ $Binaries$, p.1) considered the distribution of the orbital periods of binaries observed by himself. He finds that the median period of this group of binaries is appreciably longer than that of the well-observed systems listed in the $Seventh$ $Catalogue$ and the addition of the new systems changes noticeably the period distribution of spectroscopic binaries as a whole. This shows that observational selection can be very specific to the technique of observation that is used. It now appears less likely that there is a sharp division between spectroscopic and visual binaries.

Morbey and Griffin (ApJ $317,343$) have published an important paper in which they develop statistical tests of the reliability of orbital elements (especially the periods) derived from a limited number of observations. They apply these tests to the study of duplicity among solar-type stars published some years ago by Abt and Levy ($ApJSup$ $30,273$) and are led to question the reality of many of the orbital elements published there. By implication, the results cast doubt on deductions made from similar surveys by several authors. Abt (ApJ $317,353$) argues that the effect on our estimates of binary incidence will not be large, because if stricter criteria are adopted in assessing the variability of the velocity of a given star, their effect will to some extent be balanced by the need to make larger corrections for undetected binaries. Nevertheless, the new analysis re-opens the question of binary frequency in various groups of stars and underlines the need for an adequate number of observations of each star included in any survey.

Scarfe ($JRASC$ $80,257$) has examined again the distribution of mass-ratios in spectroscopic binaries, found by Trimble (Obs $98,163$) to be bimodal with peaks near mass-ratios of unity and 0.3 - although recently (in $Interacting$ $Binaries$, p.393) she suggested that we do not yet know the true distribution. Scarfe suggests that binaries showing two spectra are much more likely to have their orbits determined than those showing only one, and that this will create an artificial excess of binaries with nearly equally luminous components in our catalogues. This excess will also, for the most part, consist of binaries with mass-ratios near unity. Proper allowance for this selection effect, Scarfe argues, may appreciably change the observed distribution of mass-ratios.

A common theme in these investigations seems to be the re-opening of questions that we thought had been, at least provisionally, answered. A combination of new observing techniques and more refined statistical analyses has made this re-opening possible. The next decade or so may well see appreciable advances in our knowledge of binary-star statistics.

6. Origin and Evolution

A. ORIGIN OF BINARIES (R.H. Durisen)

As reviewed in Black and Matthews (*Protostars and Planets II*) and in Hollenback and Thronson (*Interstellar Processes*), rapid strides have been made in understanding the mechanisms and conditions for star formation. However, the processes which determine whether single or multiple stars form remain elusive, in part because they seem to occur during the enshrouded collapse and accretion phases. By the time Young Stellar Objects (YSO's) become visible, their binary frequency for semimajor axes $\gtrsim 1$ AU is roughly similar to that of Main Sequence stars to within statistical uncertainties (Hartmann, private communication). Detections among pre-Main Sequence stars now include several close spectroscopic binaries (e.g. Popper *ApJ 313*,L81, Marschall and Mathieu *BAAS 19*,707).

Dynamic collapse from uniformly rotating, uniform density protostellar clouds can certainly produce binary and multiple systems (e.g. Tohline *Fund.Cosm. Phys. 8*,1; Boss *ApJSup 62*,519). Progress is being made in formulating general criteria for fragmentation (Miyama *et al. ApJ 279*,621; Larson *MN 214*,379; Hachisu and Eriguchi *AAp 140*,259; Tohline *Icarus 61*,10; Hachisu *et al. ApJ* in press). However, Shu and his collaborators (*AnnRevAAp 25*) argue for inside-out collapse from centrally condensed molecular cloud cores produced quasistatically by ambipolar diffusion. With centrally condensed initial conditions, single stars form first and later become surrounded by massive disks as the accretion proceeds. Boss (*ApJ 319*,149) has confirmed that no direct fragmentation then occurs during collapse.

Production of very close binaries by classical fission of a rapidly rotating star seems unlikely for both observational and theoretical reasons. Rotation rates of YSO's are too low for them to reach fission instabilities during pre-Main Sequence contraction (Bouvier *et al. AAp 165*,10; Hartmann *et al. ApJ 309*,275). Numerical hydrodynamic simulations to date show that barlike fission instabilities do not produce binaries (Durisen *et al. ApJ 305*,281; Williams and Tohline *ApJ 315*, 594). A detached ring or disk forms instead via spiral arm ejection, while gravitational torques suppress binary formation. Nevertheless, Lebovitz (*ApJ 275*,316, *ApJ 284*,364, *Geo.Ap.F.Dyn. 38*,15) and Eriguchi and Hachisu (*AAp 142*,256) continue to elucidate other bifurcations and instability points that have yet to be tested, and it remains possible that some of these could produce binary fission.

As cited in the Shu *et al.* review, abundant evidence now exists for Solar System-sized disks around both protostars and YSO's. The spectra of some disks cannot be explained by simple viscous dissipation or reprocessed starlight (Adams *et al. ApJ 312*,788). These "active" disks are probably massive (a few tenths M⊙ or more) and may be undergoing some form of instability or wave transport. Perhaps stellar companions can form from massive circumstellar disks by mechanisms yet to be understood. Unexpected instabilities leading to fragmentation in rotating disks and tori continue to be discovered (e.g. Papaloizou and Pringle *MN 208*,721, *MN 213*,799; Goldreich *et al. MN 221*,339; Hawley *MN 225*,677). Comparisons of the binary statistics of classical versus naked T Tauri stars (Walters *PASP 99*,31) should be very useful in this context.

The above discussion suggests that, during the process of star formation, binaries might form by direct fragmentation during collapse if initial cloud conditions are uniform or by fragmentation of a massive circumstellar disk if the protostellar clouds are centrally condensed. The first mode of binary formation is relatively well established theoretically. Detailed mechanisms for the second mode have yet to be proposed. Observations of protostars and the clouds in which they are embedded should help to clarify whether initial cloud conditions favor one, the other, or a mix of the two modes. Once a wide ($\gtrsim 1$ AU) binary forms by either mode, gravitational torques (Larson *MN 206*,197; Boss *MN 209*,543) or magnetic torques (e.g. Tutukov *Astrofizika 20*,573; Iben and Tutukov *ApJ 284*,719; Moss *MN 218*,247) could cause the system to evolve to a more compact state.

Close binaries can also be formed by tidal capture in dense stellar systems. In star forming regions, the stellar densities are too low to form an appreciable number of close binaries by this method. The application to globular clusters is the best studied case (Krolik *ApJ 282*,452; van der Woerd and van den Heuvel *AAp 132*,361). Direct evidence of binarity for two globular cluster X-ray sources (see the review by Grindlay *IAUSymp 126*) confirms that capture processes have been active (Verbunt *ApJ 312*,L23; Bailyn and Grindlay *ApJ 316*,L25). Several groups have reinvestigated the physics of tidal capture (Lee and Ostriker *ApJ 310*,176; McMillan *et al. ApJ 318*,261). New cross sections and rates have been applied to dynamical models of globular cluster evolution (Hut and Inagaki *ApJ 298*,502; Statler *et al. ApJ 316*,626; Lee *ApJ 319*,772).

B. EVOLUTION OF CLOSE BINARIES (R.F. Webbink)

Overviews of the structure and evolution of binary stars have been published by Boyle (*Vistas in Astr. 27*,149), Webbink (*Interacting Binary Stars*, p.39), and Eggleton (*The Evolution of Galactic X-Ray Binaries*, p.87). Notable reviews of the evolution of W UMa binaries were published by Smith (*QJRAS 25*,405) and Mochnacki (*Interacting Binaries*, p.51), of X-ray binaries by van den Heuvel (*The Evolution of Galactic X-Ray Binaries*, p.107) and Sutantyo (*ApSpSc 118*,257), and of semi-detached binaries generally, but Algol-type systems in particular, by De Greve (*SpScRev 43*,139).

New approaches to a number of difficult physical problems connected with binary star evolution have been made recently. The hydrodynamical structure and stability of mass flow through the inner Lagrangian point was modeled by Edwards (*MN 212*,623; *226*,95), who found Bath-type instability in the outflow from a 1MΘ star, and by Gilliland (*ApJ 292*,522), who however found no such instability. Papaloizou and Savonije (*MN 213*,85) and Rocca (*AAp 175*,81) studied the excitation of g-mode pulsations by tides, and their role in synchronization, subjects reviewed by Savonije and Papaloizou (*Interacting Binaries*, p.83). We note also renewed interest in the tidal mass transfer in binaries with eccentric orbits, the subject of numerical simulations by Brown and Boyle (*AAp 141*,369) and by Boyle and Walker (*MN 222*,559), and of analytic work by Dolginov and Smel'chakova (*AZh 62*,301).

Evolutionary sequences of massive binaries, including the effects of mass loss in a stellar wind and of convective overshooting, were published by Doom (*AAp 138*,101) and Sybesma (*AAp 142*,171; *159*,108; *168*,147). Hellings (*ApSpSc 104*,83) used thermal equilibrium model sequences to study the evolution of a grid of massive close binaries through case B mass transfer. Nakamura and Nakamura (*ApSpSc 104*,367; *134*,161; *134*,219) evolve both components of a number of models of a 20.4M0 (total mass) binary, differing in mass ratio and orbital angular momentum, but all evolving in case A; all evolve rapidly into deep contact.

The evolution of intermediate-mass binaries leading to helium-star mass transfer remnants, and the subsequent evolution of those remnants, are the subjects of studies by Iben and Tutukov (*ApJSup 58*,661), van der Linden (*AAp 178*, 170), and Habets (*AAp 165*,95). Remnants of intermediate mass (0.75-2.1M⊙ in the calculations by Iben and Tutukov) fill their Roche lobes a second time (case BB mass transfer), whereas lower-mass remnants do not. Habets finds that a more massive remnant, 2.5M⊙, evolves to neon ignition, and presumably beyond to core collapse and neutron star formation; he suggests that low-eccentricity Be star X-ray binaries may originate in this way. Uomoto (*ApJ 310*,L35) proposes that Type Ib supernovae are produced by this type of core collapse of a helium star. Among slightly less massive progenitors, Iben (*ApJ 304*,201) finds that if the initial mass transfer is delayed beyond the core helium burning phase, but occurs before the second dredge-up begins, the CO white dwarf remnant may be greater in mass than that produced by single stars of the same initial mass. Iben *et al.* (*ApJ 304*, 217) show that helium star remnants of mass transfer occupy the same reaches in the Hertzsprung-Russell diagram as subdwarf O and B stars. Tutukov (*IAUColl 87*, p.483) reviews the binary origin of helium rich stars of this type, among others.

The role of magnetic stellar winds in the evolution of Algol-type binaries (and other low-mass binaries as well) has been explored by Iben and Tutukov (*ApJ 284*,719) and Krajcheva *et al.* (*Astrofizika 24*,287), and seems to offer a plausible explanation for the degree of angular momentum loss needed to understand the evolutionary status of observed systems. Pastetter (Thesis, Munich) has modeled the early phases of mass transfer from a thermally-pulsing asymptotic branch giant. In the event that the accretor is an object of planetary mass, and becomes engulfed by the envelope of the giant, Soker *et al* (*MN 210*,189) suggest it may grow to stellar mass, evolving ultimately to become the donor star of a short-period cataclysmic variable. Iben and Tutukov (*ApJ 311*,753) have derived a white dwarf mass distribution produced by binary evolution.

The secular evolution of cataclysmic binaries remains a topic of great interest. A number of studies (Verbunt *MN 209*,227; Fedorova and Yungelson *ApSpSc 103*,125; *107*,207; *Nauchn.Inf. 57*,64; Ritter *AAp 145*,227) have concentrated on constraining the form of the magnetic breaking law so as to reproduce the absence of orbital periods between 2 and 3 hours among cataclysmic binaries. According to van Paradijs (*MN 218*,31P), magnetic torques on the secondaries in these systems could be so great that tidal torques cannot maintain synchronous rotation, but Czerny and King (*MN 221*,55P) respond that the observed mass transfer rates cannot then be explained. The possibility that cataclysmic binaries evolve through cycles of activity, appearing variously in nova-like, dwarf nova, and perhaps even inactive states, was elaborated in the hibernation model of Shara *et al.* (*ApJ 311*,163). Prialnik and Shara (*ApJ 311*,172) showed that such cyclic accretion is capable of producing nova outbursts which sustained rapid accretion suppresses. Sion (*ApJ 297*,538) finds that the luminosities of white dwarfs in low-$\dot{M}$ systems are consistent with accretion rates of 10^{-11} to 10^{-9} M⊙/yr over nova outburst recurrence time scales, and MacDonald (*ApJ 305*,251) estimates that angular momentum loses in nova explosions can dominate secular evolution at orbital periods below 6 hours. Nelson *et al.* (*ApJ 299*,658) studied the influence of rotation and tidal distortion on the structure of short-period cataclysmics, confirming that the 81-min. orbital period limit is produced by gravitational-radiation-driven evolution. Ritter (*AAp 148*,207) showed that absorption of angular momentum by the accreting white dwarf can enhance mass transfer rates, and also explored the evolutionary implications of the properties of pre-cataclysmic binaries. He has comprehensively reviewed the secular evolution of cataclysmic binaries (*High Energy Astrophysics and Cosmology*, p.207; *The Evolution of Galactic X-Ray Binaries*, p.271).

The secular evolution of magnetic cataclysmic variables poses special problems. There is a growing consensus that asynchronous rotators (the DQ Her stars) and synchronous rotators (the AM Her stars) have comparable magnetic moments, and that the former tend to evolve toward the latter state (Chanmugan and Ray *ApJ 285,* 252; King *et al. MN 213,*181; Hameury *et al. ApJ 316,*275). Campbell (*MN 211,*69; *211,*83; *215,*509; *219,*589) continued his studies of the nature of magnetic synchronization; Kaburaki (*ApSpSc 119,*85) proposed a new electrodynamical synchronization mechanism. King and Lasota (*AAp 140,*L16) suggest that the high/low state dichotomy of these systems is produced by X-ray heating of the secondary.

Several papers explore the dependence of the minimum orbital period of compact binaries on the degree of hydrogen-deficiency of the donor star (Rappaport and Joss *ApJ 283,*232; Sienkiewicz *AA 34,*325; Nelson *et al. ApJ 304,*231). These results have been applied to low-mass X-ray binaries and systems like GP Com, but Tutukov *et al.* (*PisAZh 11,*123) and Savonije *et al.* (*AAp 155,*51) suggest that their donor stars may not be degenerate at all, but helium main sequence stars. Papers by Khokhlov and Ergma (*PisAZh 12,*366), Iben and Tutukov (*ApJ 313,*727), and Iben *et al.* (*ApJ 317,*717) explore the possibility that helium star + CO white dwarf binaries produce Type Ib supernovae.

Considerable interest has been generated in the possibility that double white dwarfs may be progenitors of Type Ia supernovae. MacDonald (*ApJ 283,*241) showed that the white dwarfs in most cataclysmic binaries probably decrease in mass in the long term, and so are unlikely progenitors. Published studies of accretion of carbon and oxygen onto CO white dwarfs (Nomoto and Iben *ApJ 297,*531; Saio and Nomoto *AAp 150,*L21; Khokhlov *PisAZh 11,*755; Kawai *et al. ApJ 315,*229) agree that, for accretion rates in excess of 2×10^{-6} M$\odot$/yr, up to Eddington-limited rates, lead to non-degenerate carbon ignition in the accreted envelope, and not to a carbon deflagration supernova. Merger models by Hachisu *et al.* (*ApJ 308,*161) however indicate rapid formation of a common envelope and overflow of the outer Lagrangian point. They suggest that CO white dwarf pairs exceeding 2.4M$\odot$ may yet reach carbon deflagration. Tornambe and Matteucci (*MN 223,*69) nevertheless find the SN I rate predicted by the double degenerate CO dwarf model falls short of the local rate by a factor of ten.

The very small period derivatives of known millisecond pulsars indicate anomalously weak magnetic fields, prompting the suggestion that they have been spun up in binary systems. Evolutionary models involving mass transfer from a low-mass giant successfully reproduce the properties of millisecond pulsars known to be in long-period binary systems (de Kool and van Paradijs *AAp 173,*279), but PSR 1937+214 is almost certainly now single. Ruderman and Shaham (*ApJ 289,*244) suggest that it formerly had a very low mass degenerate donor, which became dynamically unstable owing to angular momentum loss to an accretion disk. However, Taam and Wade (*ApJ 293,*504) find this instability very model-dependent, and Bonsema and van den Heuvel (*AAp 146,*L3) argue that this millisecond pulsar could only have been spun up by dynamical merger of a massive (>0.66M$\odot$) white dwarf companion.

Krolik (*ApJ 282,*452), Krolik *et al.* (*ApJ 282,*466), Verbunt (*ApJ 312,*L23), and Bailyn and Grindlay (*ApJ 316,*L25) have explored the tidal or collisional formation of binaries in globular star clusters, the latter two papers specifically addressing the 685-second X-ray binary in NGC 6624. In such a dense stellar environment, encounters with a third star can be an important mechanism driving evolution (Hut and Paczyński *ApJ 282,*675; Donnison *MN 210,*915). Bailyn and Grindlay (*ApJ 312,*748) and Bailyn (*ApJ 317,*737) also examine the evolutionary consequences of a bound third star.

Finally, we note the introduction of interacting triple star models to account for the peculiar binaries: ϵ Aur (Eggleton and Pringle *ApJ 288,*275), A0620-00 (Eggleton and Verbunt *MN 220,*13P), and SS 433 (Fabian *et al. ApJ 305,*333).

Józef Smak
President of the Commission

44. ASTRONOMY FROM SPACE
(L'ASTRONOMIE A PARTIR DE L'ESPACE)

PRESIDENT: Y. Kondo
VICE-PRESIDENT: K.A. Pounds
ORGANIZING COMMITTEE: A.A. Boyarchuk, G.W. Clark, G. Courtes, M. Grewing,
E.B. Jenkins, F.D. Macchetto, M. Oda, J. Rahe,
G.B. Sholomitzcy, Y. Tanaka, J. Truemper,
K.A. van der Hucht, A.J. Willis.

Introduction

Yoji Kondo

The tragic loss of Space Shuttle _Challenger_ and her 7 crew members in January 1986 has seriously affected the astronomical research from space in the U.S.A. and, to a lesser extent, in Western Europe. The incident has caused setbacks in a number of space projects, including the delay in the launching of the Hubble Space Telescope. Nevertheless, the field of space astronomy remained active during the present reporting period (1984 July through 1987 June).

The orbiting astronomical telescopes that were productive during the trienneial period include IUE, EXOSAT, Tenma, Ginga, and several experiments aboard Mir. Five interplanetary probes to Comet Halley and Voyagers 1 and 2 also returned important astronomical data. In addition, a number of rocket, balloon and airplane payloads yielded valuable astronomical results.

In 1984-1987, we witnessed two extraordinary astronomical events, i.e., the bright supernova 1987a that exploded in our sister galaxy, Large Magellanic Cloud, and the perihelion approach of Comet Halley in 1985-6. Astronomical observations from space played important roles in both of these events.

At the time of Comet Halley's perihelion passage, five interplanetary probes, the European Giotto, the Russian Vega, the Japanese Sakigake and Suisei, and the American ICE intercepted the comet for in-situ observations. The first space observation of Comet Halley was obtained with the geosynchronous ultraviolet satellite observatory International Ultraviolet Explorer (IUE), which monitored the comet for about a year. Many advances in our understanding of the comet have resulted from those space observations.

The brightest supernova since Johannes Kepler's naked-eye observations of a galactic supernova in 1604 was detected on 1987 February 24. Within a few hours of the notification of the discovery, the first far-ultraviolet spectrum of the supernova was obtained with IUE. Because of its advantageous geo-synchronous orbit, IUE was the only telescope available to observe the supernova during the early hours of the crucial first day after the discovery. Aided by the results of the sophisticated analysis of the pre-explosion photographic plate of the supernova field, the progenitor of the supernova was identified as a blue supergiant from the IUE spectrum.

X-ray emission from the supernova was detected in August by the Japanese satellite Ginga and also from the Soviet space station Mir, which carried the West German and British X-ray telescopes.

1. GAMMA RAY ASTROPHYSICS BEYOND THE SOLAR SYSTEM
C. E. FICHTEL

A. INTRODUCTION

There have been several interesting new results, interpretations, and
theoretical developments in gamma-ray astrophysics over the last three years,
although the pace has slowed as the analysis of the data from previous satellites
is being finished, and new spacecraft are not yet launched. Some of the new data
are related to the nature of the central galactic region, SS433, the detection of
a different type of gamma ray burst than detected previously, the large high
energy emission of Cygnus X-3, and gamma radiation from galaxies. Regarding the
future, the Soviet satellites GAMMA I and GRANAT will be launched in 1988, and
the U.S. NASA Gamma Ray Observatory (GRO) is now planned for 1990. There will
also be high altitude balloon flights over the next few years that will be of
interest both for their potential new scientific results and the demonstration of
new instrument techniques.

B. OUR GALAXY

From the earliest satelite measurements, it was clear that the gamma ray sky
was dominated by the galactic plane. Although these data did not have the
angular accuracy to separate the diffuse galactic radiation from that of point
sources, the theoretical analyses of the time (Bignami and Fichtel, 1974, and
Bignami et al., 1975) suggested that the majority of this galactic radiation was
diffuse, coming from cosmic ray interactions with the galactic matter (e. g.,
Stecker, 1970) and photons. Although there was some later speculation that point
sources might make a quite large contribution, recent studies with improved
matter distribution estimates (e. g., Strong et al., 1987) support the original
view, and Simpson and Mayer-Hasselvander (1987) even show that the number of
detectable point sources at a given sensitivity level is probablly notably less
than some of the larger numbers proposed. Strong et al. also conclude that, on
the average, there is a variation of the cosmic ray intensity with galactic
radius, as several earlier studies had suggested. This gradient is consistent
with the galactic cosmic-ray-matter coupling hypothesis of Bignami and Fichtel.

There have been several reports of lines being detected from the general
region of the galactic center in the last few years. Since the low energy
detectors that have flown have wide fields, typically from 15° to 35°, precise
location of the orgin of this radiation is not possible. The most extensive
recent results related to the half MeV line from this general direction are those
of Share et al. (1987), who reported a series of measurements using a telescope
on SMM over the period from 1981 to 1986 consistent with the several earlier
results, but of higher statistical weight in total. Other recent reports
(Leventhal et al., 1982; Paciesas et al., 1982; and MacCallum and Leventhal,
1985) give only upper limits just consistent with the positive results if the
emission is diffuse. With the exception of the Bell Sandia results (Leventhal,
1985, and earlier papers), all appear now to be consistent with a constant,
diffuse galactic emission. The 1.81 MeV line, presumed to be ^{26}Al originally
seen by HEAO-3 (Mahoney et al., 1982) has now also been seen by instruments flown
on SMM (Share et al., 1985) and on a balloon (von Ballmoos, Diehl, and
Schonfelder, 1986). If this line comes from the debris of supernovae as
suggested by Ramaty and Lingenfelter (1977), its distribution should be broad.
Hence, better positional accuracy is greatly to be desired.

The HEAO-3 gamma ray group (Lamb et al., 1983) reported gamma ray lines at
1.5 and 1.2 MeV during the period of 1979 and 1980 from SS443. More recent
attempts to detect these lines during the period from 1980 to 1985 by MacCallum,

et al. (1985) and Geldzahler et al. (1985) have led only to upper limits, in some cases well below the previously reported intensity.

A new type of burst phenomena seems to have emerged, which in photon energy seems to be about an order of magnitude larger than the soft x-ray bursts, but much lower than the now well known gamma-ray bursts. Thus far, three, and possibly four, have been observed. The most distinctive feature of these soft gamma bursts is that they are repetitive; they are also relatively brief and free of the complex temporal structure often seen with the hard gamma ray bursts. The most recently discovered series of bursts seen by instruments on PROGNOZ-9, the International Comet Explorer and the Solar Maximum Mission (Hurley et al., 1987; Laros et al., 1987; Kouveliotou et al., 1987) have a source consistent with an earlier January, 1979 event in the direction of the central part of the galaxy. The other two seen with gamma-ray burst type instrumentation consisted of at least twelve bursts. One was in the plane of the Galaxy (Golenski et al., 1983; Mazets et al., 1981) and the other in the direction of N49 in the Large Magellanic Cloud (Mozets and Golenski, 1979). It is also possible that an unusual transient series seen by other instruments (Babushkina et al., 1975) may belong to this class. The properties of these bursts have been summarized by Cline, Kouvelitou, and Norris (1987).

Several reports of very high energy gamma rays from Cygnus X-3 in the last several years (See, e.g., Weekes, 1987) have stimulated considerable theoretical study of the interesting question of how can a source that is approximately ten kiloparsecs from the Earth produce energetic particles of very high energy in sufficient quantity to lead to this gamma ray emmission. Cygnus X-3 which is a remarkable highly variable source is the only gamma ray source that has been seen by more than one observation team in the high energy region (>30 MeV), the very high energy range (>10^{11} eV), and the ultra high energy realm (>10^{14} eV). Three different groups (Danaher, 1981; Lamb, 1982; and Dowthwaite, 1983) have all reported evidence for activity in the TeV gamma ray energy region centered on the 0.6 to 0.7 phase of the 4.8 hour period seen in the x-ray resion. A reexamination of the SAS-2 high energy gamma ray data (Fichtel et al., 1987) confirmed the 4.8 Hour periodic emission, but indicated no evidence of the ten minute type bursts seen both at very high energy gamma ray energies (Dowthwaite et al., 1983; Weekes, 1983) and at infrared Wavelengths (Mason, Cordova, and White, 1986). A further study of the COS-B high energy gamma ray data (Hermsen et al., 1987) obtained at a later time than the SAS-2 data, which was collected just after a large radio flare, again gave only an upper limit to the high energy gamma radiation. One theory that seems to be able to explain the very high energy gamma ray emission reasonablly well is that of Chanmugam and Brecher (1985) who suggest that the high energy particles which produce the gamma rays are accelerated in an accretion disk, surrounding the neutron star, by the unipolar induction mechanism. Thus, in this model accretion is the ultimate source of energy.

In addition to Cygnus X-3, detection of very high energy gamma radiation by ground level detectors has now been reported for the Crab pulsar, the Vela pulsar, PSR 1953+29, PSR 1802-23, PSR 1937+21, Hercules X-1, 4U0115+63, and Vela X-1 in our galaxy (See Weekes, 1987, for a summary). In addition, ultra-high gamma radiation has been reported from three of these sources.

A recent analysis of the entire set of COS-B data from the Crab pulsar and Nebula (Clear et al., 1987) showed that the spectrum of the pulsar emission can be represented by a single power law of index 2.00 ± 0.10 over the entire energy range of the instrument, 50 to 3×10^3 MeV. Unpulsed emission was measured up to 5×10^2 MeV with a spectral index of 2.7 ± 0.3, which supports the steepening of this spectrum observed in the hard X-ray region.

C. EXTRAGALACTIC GAMMA RADIATION

Looking beyond our galaxy, no new sources beyond those known three years ago have been added to the observed list. The extragalactic gamma ray sources that are generally accepted as detected are the Sefert galaxies NGC 4151 and MCG 8-11-11, the quasar 3C 273, and Centaurus A (NGC 5128), although other marginal posibilities are candidates for further future study. There have been two new confirmations of the low energy continuum from Centaurus A (Gehrels et al. 1984 and Ballmoos et al., 1985), but no confirmation of the very high energy radiation. The new low energy data strengthen the already established dramatic drop in intensity between about 15 and 40 MeV, with there being only the severe upper limits established by SAS-2 and COS-B existing above about 40 MeV.

In addition to expanding our knowledge of active galaxies with the larger satellite instruments to fly in the future, several calculations (e.g. Fichtel and Trombka, 1981; Houston, Riley, and Wolfendale, 1983; and Ozel and Berkhuijsen, 1987) show that several of the closest normal galaxies should be visible in gamma rays. The large Magellenic Cloud and the Andromeda galaxy M 31 should be of particular interest because their level of intensity and proximity may permit at least a crude study of their structure in the gamma ray realm.

One of the most interesting considerations for nuclear gamma ray spectroscopy is that it should be possible to detect gamma ray lines from material synthesized in a supernova explosion under certain conditions. Among other considerations, including of course the correctness of the model, the ability to see the gamma ray lines depends on how soon the overlying material becomes transparent to the gamma rays being omitted at any particular level and on how far away the supernova occurs. For the recent supernova 1987A in the Large Magellenic Cloud, the line most likely to be seen is the one at 0.847 MeV from Co^{56} with the next most likely line being 0.511 MeV (See, e.g., Clayton, Colgate, and Fishman, 1968). Lines from nuclei decaying more quickly probably will not escape at detectable levels; even the lines just mentioned will probably not be at observable levels if they cannot be seen in the first three years from the time of the explosion in early 1987.

There have been no new measurements on the isotropic diffuse radiation. More detailed information on this subject of potentially very significant cosmological implications will probably have to await results from GRO.

D. FUTURE PROSPECTS FOR GAMMA RAY ASTRONOMY

The satellite opportunities for γ-ray astronomy in the near future are the Soviet GAMMA-I and GRANAT planned to be launched in 1988 and the NASA Gamma Ray Observatory, currently scheduled for launch in 1990. The information to be obtained from these satellites will be supplemented by that from instruments carried on high altitude balloons and ground based very high energy telescopes.

i. GAMMA-I and GRANET

The next gamma ray satellite expected to fly is GAMMA I. The major instrument on GAMMA-I is Galper, which is similar to SAS-2 and COS-B in the sense that its central element is a multilayer spark chamber system, triggered by a directional counter telescope, and surrounded on the upper end by an anticoincidence system. The upper spark chamber system is a twelve-level wide gap Vidicon system. The directionality of the electrons is determined by a time-of-flight system rather than a directional Cerenkov counter. The sensitive area is about 1400 cm^2 or about 2½ times that of SAS-2 or COS-B. The area solid angle factor is about the same, because the viewing angle is smaller. The gamma-ray arrival direction measurements are expected to be an improvement over those of SAS-2 and COS-B. GAMMA-I will also carry a 0.2 to 20 MeV NaI telescope using a modulated anticollimator. GRANAT is primarily an x-ray satellite, but will carry an instrument called SIGMA using a coded mask system to study the 0.04 to 2 MeV range. It will have an energy resolution of about 7% and an angular resolution between 0.1° and 0.2°.

ii. The Gamma Ray Observatory (GRO)

There are four instruments on GRO covering the energy range from 0.03 MeV to 3×10^4 MeV, all having a major increase in sensitivity over previous satellite experiments. The Gamma Ray Observatory will be a shuttle-launched, free-flyer satellite. The nominal circular orbit will be about 400 kilometers with an inclination of 28.5°. Celestial pointing to any point on the sky will be maintained to an accuracy of ±0.5°. Knowledge of the pointing direction will be determined to an accuracy of 2 arcminutes. Absolute time will be accurate to better than 0.1 milliseconds to allow precise comparisons of pulsars and other time varying sources with observations at other wavelengths from ground observations and other satellites.

The four instruments to fly on GRO are:

The Gamma-Ray Observatory Scintillation Spectrometer (OSSE)

This experiment utilizes four large actively-shielded and passively-collimated Sodium Iodide (NaI) Scintillation detectors, with a 5° x 11° FWHM field of view. The large area detectors provide excellent sensitivity for both Y-ray line and continuum emissions. An offset pointing system modulates the celestial source contributions to allow background subtraction. It also permits observations of off-axis sources such as transient phenomena and solar flares without impacting the planned Observatory viewing program. The energy range is from below 0.1 to over 30 MeV.

The Imaging Compton Telescope (COMPTEL)

This instrument employs the signature of a two-step absorption of the gamma-ray, i.e., a Compton collison in the first detector followed by total absorption in a second detector element. This method, in combination with effective charged particle shield detectors, results in a more efficient suppression of the inherent instrumental background. Spatial resolution in the two detectors together with the well defined geometry of the Compton interaction permits the reconstruction of the sky image over a wide field of view (~ 1 steradian) with a resolution of a few degrees. The energy range is 1 to 30 MeV.

The Energetic Gamma-Ray Experiment Telescope (EGRET)

The High Energy Gamma-Ray Telescope is designed to cover the energy range from about 20 MeV to over 10^4 MeV. The instrument uses a multi-level magnetic core spark chamber system to detect and record gamma-rays coverted by the electron-positron pair process. A total energy counter using NaI (T1) is placed beneath the instrument to provide good energy resolution over a wide dynamic range. The instrument is capped with a plastic scintillator anticoincidence dome to prevent readout on events not associated with gamma-rays. The combination of high energies and good spatial resolution in this instrument should provide the best source positions of any GRO instrument.

The Burst and Transient Source Experiment (BATSE)

The Burst and Transient Source Experiment for the GRO is designed to continuously monitor a large fraction of the sky for a wide range of types of transient gamma-ray events. The monitor consists of eight wide field detector modules. Four have the same viewing path as the other telescopes on GRO and four are on the bottom side of the instrument module viewing the opposite hemisphere. This arrangement provides maximum continuous exposure to the unobstructed sky. The capability provides for 0.1 msec time resolution, a burst location accuracy of about a degree and a sensitivity of 6×10^{-8} erg/cm^2 for a 10 sec burst.

iii. Other Possibilities

Among the other experiments capable of producing new results in the next several years are those to be flown on balloons. The Compton Telescope instruments appear to hold hope for a better understanding of the spectral region where marked changes in spectral shape are occuring. Several nuclear spectroscopy detectors will be flown over the next few years both to look at supernova 1987A and to study further the gamma ray lines that have already been reported.

At very high energies (above about 10^{11} eV), photons can be detected by instruments at sea level. These measurements are an important extension of the space measurements because the very low intensities at these energies make collection of the photons in space impractical, at least for the present. In the region above about 10"eV, the telescope records the Cerenkov light produced in the atmosphere from a series of interactions initiated by a single incident gamma-ray. A technique using two parallel large reflectors, each equipped with multiple detector channels to provide two images of the shower in Cerenkov light, appears to be one of the more promising approaches for the future (e.g., Weekes and Turver, 1977). In the ultra high energy region ($E \geq 10^{14}$ eV), careful analysis or extensive air showers to pick out point source enhancements due to gamma rays has yielded some indications of gamma rays even at these extreme energies as noted earlier here, and this work will be continued in the future.

References

Atteia, J., et al., 1987, submitted to Ap. J. (Lett.).
Ballmoos, P.V., et al., 1985, Proc. 19th Internat. Cosmic Ray Conf. 1, 273.
Bignami, G.F., and Fichtel, C.E., 1974, Ap. J. (Letters), 189, L65.
Bignami, G.F., Fichtel, C.E., Kniffen, D.A., and Thompson, D.J., 1975, Ap.
    J., 199, 54.
Chanmugan, G., and Brecher, K., 1985, Nature, 313, 767.
Clayton, D.D., Colgate, S.A., and Fishman, G.J., 1968, Ap. J.
Clear, J., et al., 1987, Astron. Astrophys., 174, 85-94.
Cline, T.L., Kouveliotou, C., Norris, J., 1987, 20th Inter. Cosmic Ray Conf.,
    OG-1.
Danaher, S., Fegan, D.T., Porter, N.A., Weekes, T.C., 1981, Nature, 289, 568.
Dowthwaite, J.E., et al., 1983, Astron. Astrophys., 126, 1.
Fichtel, C.E., Thompson, D.J., and Lamb, R.C., 1987, Astrophys. J., to be
    published in August 1 issue.
Fichtel, C.E., and Trombka, J.I., 1981, "Gamma Ray Astrophysics, New Insight
    into the Universe," NASA SP-453.
Gehrels, N., et al., 1984, Ap. J., 278, 112.
Geldzahler, B.J., et al., 1985, Proc. 19th Inter. Cosmic Ray Conf., 1, 187.
Golenetskii et al., 1983, Nature, 290, 379.
Hermsen, W., 1987, Astron. Astrophys., 175, 141.
Houston, B.P., Riley, P.A., and Wolfendale, A.W., 1983, 18th Inter. Cosmic Ray
    Conf. Proceedings, Vol. 1, 89.
Kouveliotou, C., et al., 1987 in preparation.
Lamb, R.C., Godfrey, C.P., Wheaton, W.A., Tumer, T., 1982, Nature 296, 543.
Lamb, R.C., et al., 1983, Nature 305, 37.
Laros, J., et al., 1987, submitted to Ap. J. (Lett.).
Leventhal, M., MacCallum, C.J., Huters, A.F., and Stang, P.D., 1982, Ap. J.
    Letters, 260, L1.
MacCallum, C.J., and Leventhal, M., 1985, Proc. 19th International Cosmic Ray
    Conference, 1, 213.
MacCallum, C.J., et al., 1985, Ap. J., 291, 486.
Mason, K.O., Cordova, F.A., and White, N.E., 1986, Ap. J., TBD.
Mazets et al., 1981, Nature, 290, 379.

Mazets and Golenetskii, S., 1979, A. F. Ioffe, preprint No. 632.

McHoney, W.A., Ling, J.C., Jacobson, A. S., and Lingenfelter, R.E., 1982 Astrophys. J., 262, 742.

Ozel, M.E., and Berkhuizsen, E.M., 1987, Astron, and Astrophys., 172, 378.

Paciesas, W.S., et al., 1982, Astrophys. J., 260, L7.

Ramaty, R., and Lingenfelter, R.E., 1977, Astrophys. J., (Letters), 213, L5.

Share, G.H., et al., 1985, Astrophys. J., 292, L61.

Share, G.H., et al., 1987, 20th Int. Cosmic Ray Conf., OG2.3-5.

Simpson, G., and Mayer-Hasselwander, H., 1987, 20th International Cosmic Ray Conf., OG-2 p.p 1-14.

Stecker, F.W., 1970, Ap. and Space Sci., 6, 377.

Strong, A.W., et al., 1987, "Gamma-Ray/Gas Correlations Over the Whole Galaxy."

Von Ballmoos, P., Diehl, R., and Schonfelder, V., 1986, Adv. Space Res.

Weekes, T.C., 1983, in Proc. of the Workshop on Very High Energy Cosmic Rays.

Weekes, T.C., 1987, Physics Reports, to be published.

Weekes, T.C., and Turver, K.E., 1977, ESLAB Symposium on Recent Advances in Gamma Ray Astronomy, Frascotti, Italy, 279.

2. X-RAY ASTRONOMY

K.A. Pounds

1.   Introduction

X-ray astronomy has continued to flourish in the three years covered by the present report (to June 1987) despite the continuing scarcity of new missions. The European EXOSAT has probably made the greatest impact during this period, carrying out over 2000 separate observations up to it's loss of attitude control in April 1986. A major reason for the success of EXOSAT was the unusual spacecraft orbit which provided uniquely long source exposures, uninterrupted by Earth occultation, of up to 70 hours duration. The continuous light curves of many galactic and extragalactic sources have proved particularly valuable in studying details of time variability over a broad frequency range and in this respect the EXOSAT data archive is unlikely to be superceded in this century. For several months after the demise of EXOSAT, astronomers had no operational X-ray satellite for the first time since the launch of Uhuru in 1970. This unfortunate state of affairs ended in February 1987 with the successful launch of the Japanese ASTRO-C satellite. Three months later, observations began with several X-ray instruments on board the Soviet MIR space station and at the time of writing both GINGA (the post-launch name of ASTRO-C) and the MIR instruments are working well.

Looking ahead, future X-ray missions are generally no closer to realisation than they were 3 years ago, a gloomy situation which, in major part, is a consequence of the Shuttle Challenger accident in January 1986. The German ROSAT X-ray telescope now appears likely to be the next major launch, probably early in 1990 on a Delta 2 vehicle. Present plans indicate a probable launch of the US Extreme Ultraviolet Explorer mission, also on a Delta 2 rocket, in $\sim$ 1990. The two 'world class' X-ray projects, NASA's AXAF and ESA's XMM 'cornerstone' mission, are now unlikely to fly before the latter half of the 1990's. A major new Soviet initiative, Spectrum-X, involving substantial participation by W. European groups, may now precede both.

In overview, although launch delays have inevitably slowed the pace of X-ray Astronomy, remarkable and important discoveries continue to be made (examples in the present report include, QPO's, rapid variability in Seyfert nuclei, and red-shifted X-ray line emission from SS 433) and a healthy development is the much broader international community now active in X-ray astronomy than was the case in earlier years. The following sections provide a brief progress report of X-ray astronomy for the period since the Commission's 1982 report.

2.   The X-ray Background and Cosmology

The nature of the intense and near-isotropic sky background X-ray flux discovered in the first rocket exposures of the early 1960's has remained an intriguing puzzle to the present day (see Boldt, 1987, for a recent review). Although it is widely believed that the X-ray background (XRB) is primarily due to the integrated emission of many unresolved sources at high redshift, present data require a substantial extrapolation in the observed source counts of different classes of active galaxies (AGN), the favoured candidates. In addition, the present spectral data of AGN and the XRB appear incompatible. A re-analysis of fluctuations in the XRB based on Einstein Observatory IPC data by Hamilton and Helfand (1987) has recently supported an earlier suggestion that the luminosity function of the constituent sources flattens just beyond the Einstein Deep Survey limit (at present the faintest sources directly resolvable), implying that a major fraction of the XRB must arise from a strongly evolving population of very faint (and presently unknown) sources, or be of truly diffuse nature (which requires an uncomfortably large total energy density of the baryonic component of the Universe). It seems unlikely that this fundamental question will be clearly answered until future, large and high resolution, X-ray telescopes are available to directly resolve the XRB. In the meantime, useful progress may be expected from smaller missions, such as GINGA, which may be capable of checking whether the X-ray spectra of distant quasars are more like that of the XRB than the nearer AGN and the deep exposures of ROSAT which will extend the direct limits of the Einstein Observatory by almost an order of magnitude (Trumper, 1984).

3.   Active Galactic Nuclei (AGN) and clusters of galaxies

Continued study of the Einstein Medium and Deep Survey data has yielded X-ray fluxes of many quasar-like objects and allowed quantitative assessment of the X-ray emission of different sub-classes of AGN, such as radio-loud and radio-quiet QSO's (Tananbaum et al, 1986; Worrall, 1987).

Perhaps the most significant recent progress in the AGN area has resulted from EXOSAT observations. First, a number of Seyfert 1 galaxies have been found to have a complex X-ray spectrum, with an up-turn to low energies near 1 keV (Arnaud et al, 1985; Pounds et al, 1986a). The discovery that the 'soft excess' varies on timescale as short as a few hours in some cases supports the view suggested in both the above references that the low energy X-ray component arises as thermal emission from the innermost regions ($T \sim 10^5$ K) of a accretion disc. The second EXOSAT discovery, that rapid variability (down to the typical detection limit $\sim 10^3$ s for faint AGN) is common in Seyfert 1 galaxies, has lent further support to the view that these objects are indeed powered by accretion onto a massive black hole. Light travel time arguments have been used to suggest masses of the central hole in the range 10^6 - 10^7 $M_\odot$ (Pounds et al, 1986a,b).

Improved X-ray spectra of many bright clusters of galaxies, obtained with EXOSAT, have shown strong correlations between the intercluster gas temperature, the X-ray luminosity and the velocity dispersion of the constituent galaxies (Edge et al, 1988) Such data are important to obtaining an understanding of the evolution of the cluster gas and spectral data on more distant clusters (perhaps with GINGA) are eagerly awaited.

Further studies of the cooling flow phenomenon in clusters have revealed one remarkable example, PKS 0745791, where the mass flow rate is of the order of 1000 $M_\odot$ per year (Fabian et al, 1985) and a second with a cooling flow of red shift 0.39 (Fabian et al, 1987).

4.   Normal galaxies

Analysis of Einstein data for complete samples of spiral and elliptical galaxies have shown (Fabbiano and Trinchieri, 1985; Trinchieri and Fabbiano, 1985) that strong correlations exist in both types, for example, between the X-ray and blue-band luminosities. The major component of the X-ray emission in the spirals is thought to arise in X-ray binary systems (as in our Galaxy), whilst in ellipticals the emission is mainly from hot gas at $kT \sim 1$ keV. In the latter case the amount of hot gas (typically $\sim 10^9$ $M_\odot$) is too large to associate with a galactic wind and probably must be gravitationally bound. This leads, in turn, to the implication of massive haloes of dark matter in elliptical galaxies (Forman et al, 1985; Nulsen et al, 1984).

The typical densities of X-ray-emitting gas in the elliptical galaxies are such that cooling flows are likey to be important here too, with the intriguing consequence that star formation is probably still occurring in elliptical galaxies, contrary to the conventional view. The EXOSAT observations of M87 provide a particularly clear case of a large 'dark halo', an extended spectral map of the galaxy showing a constant gas temperature (kT $\sim$ 2 keV) out to $\sim$ 500 kpc (Edge, Smith and Stewart, 1987), requiring a mass-to-light ratio $\sim$ 500.

5. Supernova remnants and the ISM

The wealth of X-ray images and spectral data from many galactic and Local Group supernova remnants provided by the Einstein Observatory has now been extended with resolved spectra out to $\sim$ 10 keV. The GSPC instruments on EXOSAT and TENMA have yielded the best recent data in this band, showing the expected strong line emission of helium-like Fe, Ca, Ar and S from young remnants such as Tycho, Kepler, Cas A, RCW 103 and W49 B (Smith, 1987; Tsunemi et al, 1986). Analysis of all available X-ray data has shown simple models of the X-ray emission to be invalid, with clear evidence for non-equilibrium ionisation (NEI) in remnants even as old as Puppis (Winkler et al, 1983). The general trend is that fitting with NEI models considerably reduces the large mass estimates of the heated gas (to $\sim$ 1 - 3 solar masses in most cases), while still leaving a significant over-abundance of metals. Undoubtedly, much remains to be learned about the nature and evolution of SNR from spatialy resolved X-ray spectroscopy offered by future missions such as AXAF and XMM.

6. Low mass X-ray binaries

X-ray binary systems containing a neutron star component, which include most of the brightest X-ray sources in the night sky, conventionally are sub-divided by mass of the optical counterpart. Progress over the past 3 years has been most rapid in the study of the low mass systems (LMXRB's), mainly due to the broad band spectroscopic data from TENMA and EXOSAT and the long, continuous observations and high time resolution data from EXOSAT. The detection of periodic 'dips' in the X-ray light curves of several LMXRB's has yielded the first orbital periods (as expected, typically of the order of a few hours), in addition to offering an important probe of the accretion disc structure in such systems. For example, in the 'dip' source 4U1755-33, White et al (1984) have used X-ray spectral data from EXOSAT to show the accreting material to be under-abundant in heavy elements by a factor > 100. Recent broad band X-ray spectra of several LMXRB's have been shown to contain two separate components, a black body component (arising from the heated neutron star surface?) and a softer component, which probably originates in the inner accretion disc (see review by White, 1986).

Probably the most intriguing new result in the study of the LMXRB's has been the discovery of quasi-periodic-oscillations (QPO's), with frequencies in the range $\sim$ 1 - 30 Hz. GX5-1 was the first source found to exhibit QPO's (van der Klis et al, 1985) and the correlation of the frequency and width of the QPO peak with source intensity led to an initial explanation in which the QPO was related to 'beating' between the rotation of material at the inner edge of the accretion disc and the neutron star surface. More recent observations have shown a much more complex picture and it now seems unlikely that one description applies to all cases (see Lewin, 1986, for a review).

Further details on the properties of X-ray bursts, generally associated with LMXRB's having neutron stars of low magnetic field, have come from both EXOSAT and TENMA. The simple thermonuclear flash model has been shown to be inadequate, with the discovery of multiple peaks in several bursts (e.g. van Paradijs et al, 1986) and also series of bursts too closely spaced to allow complete fuel replenishment by accretion (Lewin et al, 1987). Fujimoto et al (1987) have proposed that hydromagnetic instabilities in the accreting matter cause pockets of unburnt material to spiral deep into the neutron star's atmosphere, producing partial ignition. Other bursts have shown evidence of radial expansion of the entire X-ray photosphere, with a tendency for luminosity saturation. Separate saturation levels have been associated with Eddington-limited accretion in helium-and hydrogen-dominated atmospheres. The X-ray absorption feature identified in several burst spectra by TENMA (Waki et al, 1984) and EXOSAT (Turner and Breedon, 1984) has continued to inspire theoretical study, mostly based on an interpretation as a gravitationally red-shifted Fe-K feature. The existence of this spectral feature has recently been confirmed by GINGA observations, but the red-shift is apparently too large for standard neutron star models (Lewin, 1984) and the feature too strong for production in a conventional atmosphere (Foster et al, 1987).

7.   High mass X-ray binaries/black hole candidates

For over a decade Cygnus X-1 had the unique status of being the only X-ray binary for which there is compelling evidence for the compact object having a mass > 3 M$_\odot$ - thus probably being a black hole. New optical data of the remarkable Ariel-5 X-ray transient A0620-00 now strongly suggest that this system too contains a black hole (McClintock et Remillard, 1986). Interestingly, one of the previously favoured black hole candidates, Circinus X-1, has now been shown almost certainly to be a neutron star source, following the discovery of X-ray bursts in a long EXOSAT observation (Tennant et al, 1986).

Although SS 433 is almost certainly a high mass XRB, it is well known primarily for its relativistic jets. New insight on the jets has been provided by EXOSAT and TENMA observations which show that the Fe-K line moves in energy in a manner consistent with Doppler shifts from material in the inner jet regions (Watson et al, 1986). Analysis of these data has provided constraints on the dimensions of the emission region and on the energetics and acceleration mechanisms in the jets.

8.   Globular cluster sources

The first optical identification of an X-ray source in a globular cluster was made in the case of M15 (NGC 7078) by Charles et al (1986). This identification was based on coordinated EXOSAT observations and optical photometry and spectroscopic studies of the star AC 211. The remarkable discovery of an 11 minute (!) modulation in the X-ray flux of a second globular cluster source, 4U 1820-30 (by Stella et al, 1987) and a lack of period variations has yielded the shortest known orbital period of any stellar system. The difficulty of such a system evolving from a longer period binary suggests it must have been formed by neutron star capture - a possibility presumably most likely to occur in the dense core of a globular cluster.

9.   White dwarf binaries /Cataclysmic variables (CV's).

A considerable amount of EXOSAT observing time was devoted to the study of cataclysmic variables, both in outburst and in quiescence. As a result, it is now clearly established that CV's containing a magnetised white dwarf star typically have substantially higher X-ray luminosities than non-magnetic systems. Also, the strongly magnetised systems, or polars, have a strong soft X-ray flux which - in some cases at least - is much greater than the hard X-ray emission. X-ray light curves of intermediate polars (magnetic systems with a lower field) show the X-rays to arise from large polar caps (e.g. Watson and King, 1985).

Important new results for non-magnetic CV's include evidence for coronal emission in SS Cygni (King et al, 1985) and the detection of 14 sec. pulsations in the superoutburst of the SU UMa system VW Hyi (van der Woerd et al, 1987).

10. Stellar X-rays

The most active, and X-ray brightest, cool stars are the RS CVn binaries. Several of these have been well studied for one or more orbital periods. A lack of evidence for X-ray eclipses in Algol (White et al, 1986) and ER Vul (White et al, 1987b) indicate that the coronal scale heights are > 1 $R_\odot$. By contrast the existence of a deep minimum in the flux observed by the EXOSAT LE telescope, centred on the primary eclipse of AR Lac (White et al, 1987a), has allowed a detailed model of the system to be developed, where a plasma of 5 - 7 x 10^6 K is closely confined to the two small regions of the G star and a higher temperature component (15 - 30 x 10^6 K), associated with the K star extends to $\sim$ 1 $R_\odot$.

Broad band EXOSAT observations of RS Cvn's suggest that coronal X-ray emission typically arises from components at two different temperatures ($\sim$ a few x 10^6 K and $\sim$ a few x 10^7 K). While this simple model is very useful, EXOSAT transmission grating observations (Mewe et al, 1986; Schrijver and Mewe, 1985) show a more complex picture of plasma having a continuous temperature distribution, but with the differential emission measure distribution peaking at the above temperatures. This is incompatible with simple static loop coronal models and other models (e.g. quasi-static loops of varying cross-sectional area or dynamic loops with strong downward flow) must be considered.

A number of high resolution spectra of hot DA white dwarfs were also obtained by EXOSAT (e.g. Paerels et al, 1986), allowing improved temperature measurements of their objects and a detailed study of their atmospheres, in particular placing tight constraints on the abundances. The photosphere of HZ43 is found to consist solely of H, presumably the result of uninterrupted downward diffusion of heavy elements in the strong gravitational field. A photometric survey of 21 DA white dwarfs (Heise, 1987) has revealed that many of these objects have traces of He in their atmospheres. Further analysis of this sample should allow determination of the relative importance of radiative support, convective mixing and accretion from the interstellar medium on white dwarf evolution.

Finally, EXOSAT observations have enabled the first soft X-ray detections of the hottest, He-rich DO white dwarfs (Barstow, 1987), at temperatures exceeding 10^5 K. Some of these objects are optical pulsators and one, PG1159-035, was also found to be an X-ray pulsator (Barstow et al, 1986). This is the first observation of photospheric X-ray pulsations. The relative phases and amplitudes of the various X-ray and optical modes are a sensitive probe of the atmospheric structure of white dwarfs.

References

Arnaud, K.A. et al: 1985, Mon. Not. R. astr. Soc. 217, p. 105.
Barstow, M.A., Holberg, J.B., Grauer, A.D. and Winget, D.E.: 1986,
    Ap. J., 306, L25.
Barstow, M.A.: 1987, Proceedings of the Nato AS1 "Hot Thin Plasmas in
    Astrophysics, in press.
Boldt, E.: 1987, Phys. Rep. 146, p. 216.
Charles, P.A., Jones, D.C. and Naylor, T.: 1986, Nature, 323, p. 417.
Edge, A.C., Stewart, G.C. and Smith, A.: 1987, NRAO Workshop, 16, p. 105.
Edge, A.C., Stewart, G.C. and McHardy, I.M.: 1988, Proc. NATO Adv. St. Inst.,
    Cargese.
Fabbiano, G. and Trinchieri, G.: 1985, Astrophys. J., 296, p. 430.
Fabian, A.C. et al: 1985, Mon. Not. R. astr. Soc., 216, p. 923.
Fabian, A.C., Crawford, C.S., Johnson, R.M. and Thomas, P.A., Mon. Not. R. astr.
    Soc., in press.

Foster, A.J., Ross, R.R. and Fabian, A.C.: 1987, Mon. Not. R. astr. Soc. 208, p. 29.

Fujimoto, M.Y., Sztajno, M., Lewin, W.H.G. and van Paradijs, J.: Astrophys. J. 319, p. 902.

Hamilton, T.T. and Helfand, D.: 1987, Astrophys. J. 318, p. 93.

Heise, J.: 1987, Proceedings of the Nato AS1 "Hot Thin Plasmas in Astrophysics, in press.

Jones, C. and Forman, W., Jones, C. and Tucker, W.: 1985, Astrophys. J., 293, p. 102.

King, A.R., Watson, M.G. and Heise, J.: 1985, Nature, 313, p. 290.

Lewin, W.H.G.: 1984, in "X-ray Astronomy 84". Eds. M. Oda and R. Giacconi, p. 157. ISAS.

Lewin, W.H.G.: 1986, in "The Physics of Accretion onto Compact Objects". Eds. K. Mason, M.G. Watson and N.E. White, p. 377, Springer-Verlag.

Lewin, W.H.G. et al: 1987, Astrophysics J., 319, p. 893.

McClintock, J.E. and Remillard, R.A.: 1986, Astrophys. J., 308, p. 110.

Mewe, R., Schrijver, C.J., Lemen, J.R. and Bentley, R.D.: 1986, Adv. Res., 6, p. 133.

Nulsen, P.E., Stewart, G.C. and Fabian, A.C.: 1984, Mon. Not. R. astr., 208, P. 185.

Paerels, F.B.S., Bleeker, J.A.M., Brinkman, A.C., Gronenschild, E.H.B.M., and Heise, J.: 1986, Ap. J., 308, p. 190.

Pounds, K.A., Stanger, V.J., Turner, T.J. and King, A.R. and Czerny, B.: 1986a, Mon. Not. R. astr. Soc., 224, p. 443.

Pounds, K.A., Turner, T.J. and Warwick, R.S.: 1986b, Mon. Not. R. astr. Soc., 221, 7P.

Schrijver, C.J. and Mewe, R., 1985, in "Cool Stars, Stellar Systems and the Sun", ed. Zeilik and Gibson, Springer-Verlag, p. 300.

Smith, A.: 1987, Proc. IAU Colloquium 101.

Stella, L, Priedhorsky, W. and White, N.E.: 1987, Astrophys. J., 312, L17.

Sztajno, M. et al: 1985, Astrophys. J. 299, p. 487.

Tananbaum, H. et al: 1986, Astrophys. J. 305, p. 57.

Tennant, A.C., Fabian, A.C. and Shafer, R.A.: 1986, Mon. Not. R. astr. Soc., 221, 27P.

Trinchieri, G. and Fabbiano, G.: 1985, Astrophys. J., 296, p. 447.

Trumper, J.: 1984, in "X-ray and UV emission from AGN", Eds. W. Brinkmann and J. Trumper, p. 254. M.P.E., Garching.

Tsunemi, H., Yamashita, K., Masai, K., Hayakawa, S. and Koyama, K.: 1986, Astrophys. J., 306, p. 248.

Turner, M.J.L. and Breedon, L.M.: 1984, Mon. Not. R. astr. Soc., 208, p. 29.

Van der Klis, M. et al: 1985, Nature, 316, p. 225.

Van der Woerd et al: 1987, Astron. and Astrophys., 182, p. 219.

Van Paradijs, J. et al: 1986, Mon. Not. R. astr. Soc., 221, p. 617.

Waki, I. et al: 1984, Pub. astr. Soc. Japan, 36, p. 819.

Watson, M.G. and King, A.R.: 1985, Mon. Not. R. astr., 212, p. 917.

Watson, M.G., Stewart, G.C., Brinkmann, W. and King, A.R.: 1986, Mon. Not. R. astr. Soc., 222, p. 261.

White, N.E. et al: 1984, Astrophys. J., 283, L9.

White, N.E.: 1986, in "The Physics of Accretion onto Compact Objects", Eds. K. Mason, M.G. Watson and N.E. White, p. 377.

White, N.E., Culhane, J.L., Parmar, A.N., Kellet, B.J., Kahn, S. van der Oord, G.H.J. and Kuipers, J.: 1986, Ap. J., 301, p. 262.

White, N.E., Shafer, R. and Parmar, A.N., 1987a, Proceedings of the 5th Cambridge Workshop, "Cool Stars, Stellar Systems and the Sun", in press.

White, N.E., Culhane, J.L., Parmar, A.N. and Sweeney, M.A., 1987b, MNRAS.

Winckler, P.F., Canizares, C.R. and Bromley, B.C.: 1983, in Proc. IAU Symposium 101, "Supernova Remnants and their X-ray Emission", p. 245. Reidel.

Worrall, D.M.: 1987, Astrophys. J., 318, p. 188.

3A. Ultraviolet Astronomy (Non-IUE)

Edward B. Jenkins

I. INTRODUCTION

Results from the IUE satellite, summarized in the section which follows this
one, continue to dominate the literature for research topics which rely on
observations in the ultraviolet. This trend may be accentuated in the near future,
as we experience the natural attrition of papers based on results from previous
major missions which are no longer operating, such as TD-1, Copernicus, ANS and
BUSS. The Challenger accident on January 28, 1986 abruptly halted flights of new
orbital facilities which depend on the Space Shuttle and has created long and
somewhat indefinite postponements in the eventual manifesting of payloads ranging
in size from simple experiments in Getaway Special (GAS) and Spartan carriers, to
telescopes of intermediate size on Spacelab (such as those which were to fly on
the Astro mission in March 1986) to the Hubble Space Telescope. Suborbital
missions, i.e., sounding-rockets and balloons, will probably dominate the
extra-IUE uv astronomy scene until there is a re-establishment of a vigorous
launch schedule for expendable vehicles and/or the Space Shuttle.

II. OPERATING MISSIONS

The ultraviolet spectrometers (20.032.579) on Voyagers 1 and 2 continue to
deliver spectrophotometric scans of stars over the wavelength range 500 -1700Å at
a resolutions of 15-30Å. These observations provide important information on the
fluxes from hot sources over wavelength ranges not covered by IUE (38.131.253),
and a catalog of observations of some 300 objects has been created (41.002.032).
The UFT experiment on the high-apogee Astron satellite has an 80-cm diameter
telescope and a spectrometer with three channels which scan from 1100 to 3500Å
(37.035.022; 39.051.066; 42.035.005). On a flight of the Columbia Space Shuttle in
January 1986 (STS-61C), two spectrometers [one built by U. C. Berkeley
(38.035.079), the other by Johns Hopkins U. (Tennyson, et. al. COSPAR XXVI)] were
flown in GAS canisters for the UVX mission to monitor the diffuse cosmic
background emissions from 600 to 3200Å and to obtain more definitive measurements
of the background emission associated with the shuttle glow reported from a
previous uv imaging mission (38.142.007).

Payloads launched on sounding rockets include instruments which recorded
stellar spectra in the windowless ultraviolet (i.e. below 1150Å) at resolving
powers $\lambda/\Delta\lambda$ = 60,000 (U. Colorado) and 200,000 (Princeton U). The primary
objectives were to observe interstellar lines toward δ and π Sco, respectively.
A payload built at U. C. Berkeley to record EUV line emissions between 80 and 650Å
was flown on a rocket in 1986 (38.035.080; 42.035.141; 42.106.059).

III. SPECTRA OF STARS

1. Absolute Photometry and Stellar Spectral Features

Catalogs of fluxes over a bandpass 1200-1600Å for stars in Cygnus and
Sagittarius have been presented by Carruthers and Page (37.155.042; 37.155.079),
who used wide-field images recorded by an electrographic camera during the Apollo
16 mission. Savage, Massa and Meade (40.131.228) published a catalog of uv fluxes
for 1415 stars of spectral type B7 or earlier for the 5 passbands between 1500 and
3300Å of the ANS satellite. They analyzed this ANS color survey to study the
relationships at different wavelengths for uv extinctions by dust grains.

Absolute spectrophotometry of (mostly) normal, early-type stars has been
reported by Tanaka, et al. (37.142.056), Woods, Feldman and Bruner (39.113.041),
and Carruthers, Heckathorn and Opal (39.114.014). Faraggiana and Malagnini
(38.114.018) outlined some discrepancies between S2/68 and OAO-2 measurements of

fluxes at 2740Å for stars with spectral types earlier than F8. Hua, et al.
(42.114.152) discuss a factor of 3 discrepancy in the flux of an O8f star,
compared with earlier measurements. However Polidan, Carone and Campbell
(39.115.027) caution that most stars of spectral type B5 or earlier have fluxes at
λ < 1100Å which vary over periods of several days. Polidan and Holberg
(40.114.085) stated that at these wavelengths main sequence stars are variable
(also see Polidan and Stalio 41.114.142), but subluminous stars have stable
fluxes whose distributions are smooth extrapolations of spectral distributions
measured at longer wavelengths. This conclusion underscores Holberg's suggestion
(39.113.067) that subluminous stars are the most desirable photometric standards.

Rogerson has supplemented earlier uv spectral atlases recorded at high
resolution by Copernicus for τ Sco (20.114.542), ι Her (27.114.062), and β Ori
(32.002.005) with those for two new stars, γ Peg (39.002.059) and α CMa (Ap. J.
Suppl 63, 369). All of these papers contain line identifications except the one
for τ Sco; identifications of features for this star have been published by
Rogerson and Ewell (39.114.131). Rogerson is currently preparing an atlas for α
Lyr. Peters and Polidan (40.114.075) have combined Voyager fluxes and Copernicus
line scans to evaluate new effective temperatures, surface gravities and element
abundances for ι Her and τ Sco. Oegerlie and Polidan (38.112.083) have analyzed
Copenicus spectra of rapidly rotating B stars and concluded that not all Be stars
display shell lines in the ultraviolet, contrary to some earlier findings. They
state that uv shell lines appear only in "classical" shell stars and are formed in
low-velocity, outflowing, flattened disk structures. Boyarchuk, et al.
(37.035.022) studied spectra of several stars recorded by the UFT on Astron to
derive velocity shifts of C IV and Al III lines caused by mass loss. They also
examined spectra of Ap stars for features from heavy elements (Pb, Th, U).

Various classes of stars have been monitored by the UVS on Voyager: the Be
stars ζ Tau and α Eri (39.112.116; 39.112.117; 40.114.118), the symbiotic star AG
Peg (41.117.124), the β Cep variable BW Vul (37.122.200; 39.122.163), the
eclipsing binaries μ Sag and β Lyr (39.117.017), and several cataclysmic variables
[see Polidan and Carone (Astr. Space Sci. 130, p235) for a brief summary] —
primarily SS Cyg, U Gem, and VW Hyi (37.117.196; 37.117.261; 39.117.217;
39.117.224; 39.124.382; 41.117.086; Polidan and Holberg, MNRAS 225, p131). Models
of dwarf nova outbursts have benefited substantially from the contributions by
Voyager which supplemented observations of these phenomena at other wavelengths
(Carone, Polidan and Wade 42.117.293; Pringle, et al. MNRAS 225, p73).

The observations by Voyager permit the spectral energy distributions derived
by IUE to be expanded to shorter wavelengths, permitting more accurate
determinations of effective temperatures of very hot objects. The combined
observations are less prone to errors from absolute flux calibrations and
reddening corrections. Studies using such combined observations have improved our
understanding of subdwarf B stars (Wesemael, et al. 40.114.126), subdwarf O stars
(Drilling, Holberg and Shonberner 39.126.002 and 38.126.042), the DA white dwarfs
Sirius B (Holberg, Wesemael and Hubeny 37.126.082) and CD −38°10980, (Holberg, et
al. 39.126.092), and an x-ray source (H1504+65) thought to be a metal-rich, near
degenerate object with a temperature of order 160,000 K (Nousek, et al.
42.126.052).

2. Interstellar Absorption Lines

de Boer, et al. (41.131.092) analyzed Interstellar lines between 2000 and
3000Å for 22 stars observed by the BUSS echelle spectrograph and telescope which
flew on a balloon, and they derived abundances of Mg, Cr, Mn, Fe and Zn along the
lines of sight. Bruhweiler, et al. (38.131.246) studied Mg I and Mg II lines in
Copernicus and IUE spectra in 5 stars ranging from 2 to 40 pc from the Sun to
investigate the local interstellar medium. Both of these teams made use of the
expected abrupt increase in Mg I abundance caused by the onset of dielectronic

recombination at T > 5000 K to derive the amount of gas above this temperature.

Federman (42.131.288) has performed a general study of an unidentified spectral feature near 1088Å in Copernicus spectra of many stars. He concluded that this feature is caused by Cl I.

Eder (39.132.032) analyzed high resolution Copernicus spectra of λ and υ Sco to work out the geometry, ionization equilibrium and local densities of gas ionized by radiation from λ Sco. A comparison of the two lines of sight indicates that there are fluctuations in local densities. A more definitive analysis of this region was performed by Eder and York (42.131.140) who deconvolved separate velocity components in the spectra. Ionized material in a much lower density regime was analyzed in the spectrum of β CMa by Gry, York and Vidal-Madjar (40.131.099). Most of the path to this star contains H II material with n ~ 0.1 cm^{-3} or a diffuse coronal gas.

There have been several, more specific studies of neutral gas. Meyers, et al. (39.131.074) isolated a low velocity (~10 km/s) shock toward the ρ Oph cloud and concluded that the post-shock gas has greater element depletions and more molecules than the material ahead of the shock. Snow, McClintock and Voels (Ap. J., in press) analyzed H_2 lines in a high-resolution spectrum of δ Sco recorded by a spectrograph on a sounding rocket. They concluded that lines from different J levels all have identical radial velocities and thus are not created in an expanding circumstellar shell. Snow, Lamers and Joseph (PASP, in press) have used BUSS results to refine a previous analysis of abundances along the line of sight to ζ Per.

IV. PHOTOMETRY AND IMAGERY OF EXTENDED OBJECTS

Onaka, et al. (38.131.296) observed the Orion reflection nebulosity using a rocket-borne photometer which performed a raster scan of the region and recorded fluxes in 5 wavelength channels between 1300 and 2000Å. These observations are complementary to the pictures in 4 colors between 1400 and 2620Å of dust reflection in Orion obtained earlier by Bohlin, et al. (31.134.009) from an image tube on a rocket. Onaka, et al. derived values for the albedo and scattering asymmetry of the dust at different wavelengths, although the results are very dependent on the assumed distribution of the dust with respect to the stars.

Two wide-field cameras were flown on Spacelab-1 to obtain images of extended objects in the ultraviolet. Both experiments were badly compromised by background light associated with the Shuttle orbit. One, the FAUST experiment, obtained pictures of the Cygnus Loop and the galaxy cluster Abell 2634 from 1300 to 1800Å (Bixler, et al. 38.142.007). The other, an all-reflecting uv Schmidt camera, obtained photographs of the Large and Small Magellanic Clouds in 3 passbands between 1650 and 2530Å (Courtes, et al. 38.142.006; Viton, et al. 39.156.015). The uv images of the clouds highlighted hot stars in Shapley's wing of the SMC, along a bridge of matter to the LMC. Pierre, et al. (41.156.001) used the uv fluxes as a sensitive way to confirm the correctness of the luminosity and initial mass functions for stars in the SMC based on UBV photometry from the ground.

An image intensifier on a sounding rocket recorded images (1' resolution) at 1500 and 1900Å of associations in the LMC and enabled Smith, Cornett and Hill (Ap. J., in press) to model the sequence of star formation at different locations in the system's spiral structure. Carruthers and Page (37.142.086) show uv pictures of the LMC in their general summary of results from the S201 mission.

Bohlin et al. (39.154.029) derived the uv brightnesses of 144 stars stars in the M5 globular cluster, using a rocket-borne telescope which was a prototype for the UIT facility to fly on the Astro Spacelab mission (see also 33.154.044 and

34.157.151). In conjunction with stellar evolution models, the results for 50 horizontal-branch stars allowed them to infer an initial helium abundance and distance modulus for the cluster, along with a minimum mass for stars in the horizontal branch.

From a lack of discrete sources in a rocket uv image of the bulge of M31, Bohlin et al. (40.157.146) concluded that the upturn of flux in the uv is caused by unresolved, post-AGB stars, rather than a young stellar population. This same group has also obtained two-color uv images of M33, and detected several hundred sources (Landsman, et al. 39.157.191). Israel, de Boer and Bosma (42.157.054) measured fluxes (dominated by OB associations) in the 5 ANS bands (1550 -3300 Å) within selected 2.5X2.5 arc min fields covering M31, M33, M81, M101, NGC 2403 and a number of compact (Zwicky) galaxies. In a follow up of an earlier investigation using OAO-2 results (Donas and Deharveng 38.157.145), Donas, et al. (Astr. Ap. 180, 12) used images recorded by a balloon telescope to determine the 2200 Å fluxes from 149 spiral and irregular galaxies. They concluded that the star formation rates in these galaxies are well correlated with total gas contents, and they studied variations in this relationship as a function of morphological type. Holberg and Barber (39.160.072) used upper limits for the flux between 912 and 1150 Å from the Coma Cluster measured by the Voyager spectrometer to place lower limits for the lifetimes of neutrinos decaying into massless products.

V. DIFFUSE BACKGROUND

Recent observations of the uv diffuse background (excluding contributions from stars) show a correlation of measured fluxes with column densities of neutral hydrogen. At some wavelengths, fluorescence from molecular hydrogen may be important, while backscattering from dust is important over the entire spectrum above the Lyman limit. From the slope of this relationship shown in their UVX results, Hurwitz, Martin and Bowyer (42.131.350) concluded that the albedo times (1-g) for dust scattering from 1450 to 1850 Å is of order 0.07. Jakobsen, et al. (38.142.040) observed the background with a telescope on an Aries sounding rocket and found intercepts at N(H) = 0 of 550 and 900 phot cm^{-2} s^{-1} sr^{-1} Å^{-1} over bandpasses 1450-1780 and 1610-1950 Å, respectively. Although the wavelengths are somewhat different, it seems difficult to understand why these results contrast so sharply with Holberg's (42.142.041) upper limit toward the north galactic pole of 100-200 phot cm^{-2} s^{-1} sr^{-1} Å^{-1} over the range 500-1150 Å, obtained from a spectrophotometric scan by Voyager 2 over a very long integration time. (The latter measurement supercedes by a large margin Bixler, Bowyer and Grewing's (38.131.181) upper limit of 9700 phot cm^{-2} s^{-1} sr^{-1} Å^{-1} at 1060 Å.)

A sounding rocket instrument described by Labov, Bowyer and Martin (42.035.141) was flown by Labov and Bowyer (42.106.059) to study diffuse emission between 80 and 650 Å. In addition to detecting the interplanetary He I 584 Å emission, they registered some broad, unidentified emission features centered on 110 and 190 Å (plus a weaker, narrow feature at 630 Å). Martin and Bowyer (42.131.351) identified C IV λ1550, O III] λ1663, and O IV or Si IV λ1400 diffuse emission in data accumulated by the UVX experiment. Martin and Bowyer conclude the C IV emission originates from collisionally excited gas in the galactic halo, since it is too bright to be of local origin (without an unreasonably high gas pressure) and it is anti-correlated with H I column densities (i.e., the accompanying dust causes foreground absorption).

Finally, Opal and Weller (38.155.025) reported a flux of 1.4X10^5 phot cm^{-2} s^{-1} Å^{-1} over the band 912 -1050 Å registered by a photometer aboard the STP 72-1 satellite. This flux is primarily from O and B type stars in Gould's Belt, and its magnitude is somewhat higher than most previous estimates. This measurement is an important datum for calculating photoionization and photodissociation rates in the local interstellar medium.

3B. UV: International Ultraviolet Explorer

W. Wamsteker

The continued functioning of the International Ultraviolet Explorer (IUE), supported by the funding Agencies NASA, ESA and the SERC has been very important for Astronomy in quite unanticipated ways. After the serious launcher problems encountered over the reporting period a shortage of observing capabilities for space astronomy, especially in the UV, could be foreseen. The extended life of IUE has clearly softened the impact of this problem. The IUE presents an interesting first in space astronomy, by being the first pointed space telescope with an incomplete inertial reference system (only 2 gyroscopes of the original six remain operational) which retains its full three-axes stability. The continued availability of IUE cannot be expected to fill the gap caused by the delays in the launch of the next major UV facility: the Hubble Space Telescope, but the IUE has remained a continuing source of new and exiting data in the ultraviolet for many fields of Astronomy. Together with the Astron satellite (USSR) no space facilities were available to Astronomy after the Challenger accident. The impact of the Astron satellite has been significantly less then that of the IUE, in part this is due to its limited sensitivity, but mainly because of its restricted availability to the general Community.

As a facility for observational Astronomy IUE has been successful without precedent and has set standards for the use of such highly efficient space observatories which will be difficult to surpass by future projects, both ground-based and space-borne. The efficiency of the IUE Observatory is clearly understood if one realizes that the IUE has already supplied, in its 9 years of operation, more observing time then most ground-based telescopes will supply in 30 years of operation. Compared with the modern good quality sites the available observing time per year on the IUE is more then 3 times as large. Of this time more then 50% is used for actual photon collection even in the present non-optimized mode of operations. Also the IUE Data Archive is a continuing source of data for many Astrophysical studies. The importance of the easily accessible data is clear from the fact that during the reporting period the number of spectra retrieved from the archive became larger then the number of spectra taken by the satellite (> 80000 spectra de-archived vs 60000 spectra taken).

It is not surprising that essentially no field of astronomy has been unaffected by such a rapid data and information flow. The distribution of the observations over the reporting period was:

Solar system	1094 spectra	Variable stars	1362 spectra
O-type and related	2036 spectra	CV's	2164 spectra
B stars	3480 spectra	Nebulae	490 spectra
A-K stars	3526 spectra	Extragalactic	1091 spectra

Including calibrations this amounts to 15243 spectral images. The number of publications based on data of the IUE in the main refereed journals is by now well over 1200. The international nature of the IUE facility was reinforced by the merging of the various IUE conferences held under different sponsorship into one single conference cosponsored by NASA, ESA and SERC. The first such conference was held in London (Rolfe, 1986). Important reference atlases have been produced on the UV spectra of Supernovae (Benvenuti et al., 1982), Normal stars (Heck et al, 1984), O-type stars (Walborn et al., 1985) and Late Type stars (Wing et al., 1983) and extragalactic HII regions (Rosa et al., 1984). Also special mention can be made of the first reference book on UV astronomy "Exploring the Universe with the IUE satellite" (Kondo, 1987).

A quite important aspect of such long-lived relatively "simple" observatory-type satellite was high-lighted by the many UV observations made under the IHW with the IUE in support of the spacecraft encounters with Comet Halley. Of the many results I would like to mention the dramatic changes in the CO_2+/OH ratio during outburst (Feldman et al., 1986) and very extensive (1 year) monitoring of the gas production rate. Similarly, the explosion of the first Supernova to reach naked eye brightness since Kepler's SN in 1604, has reinforced the importance of space borne facilities which are capable of rapid response and flexible scheduling. Although instrumentation and cost considerations drive at present the trend away from flexibility, as is also the case in many other sciences, it is worth while to remember than in Astronomy the schedule of many "cosmic experiments" is not under human control. The first results on UV observations of SN 1987A showed quite a few unexpected results. Early results are described by Wamsteker et al. (1987), Panagia et al. (1987), Cassatella et al. (1987), de Boer et al. (1987) and Fransson et al. (1987).

For a detailed and extensive overview of the many important results obtained with the IUE the reader is referred directly to Rolfe (1986) and Kondo (1987). However, it is considered to be justified to illustrate here with some examples the importance of the "workhorse" of UV astronomy: IUE. Although such a summary is unavoidably strongly biased it is a good indication of the width of the impact of the first general user space facility.

In the solar system area the detection of molecular sulfur in the cometary evaporates and the extensive series of observations of the IO Torus - the only region where HII region-like conditions have been studied both by classical means and through in situ observations, clearly stand out. Before IUE only 4 comets were observed in the UV, at this time an extensive data base is available with data for 26 comets.

In the area of stellar studies it is worth noting that before IUE only a single UV observation was available for classical novae, while at present a wealth of data, also at high resolution, has been obtained for 13 novae allowing detailed abundance analysis. From these results it has become obvious that the ejecta of all novae show abundance anomalies. We can also mention the considerable variations in the CIV lines' in Be stars; the discovery of absorption lines at the velocity plateau in the winds of WR stars; the first insights and consistent understanding of the symbiotic star phenomenon; the detailed studies of chromospheric activity in late type stars through Doppler imaging and the discovery of extended coronal regions around dwarf novae.

For the ISM we would like to mention the results on interstellar Zn depletion showing that although Zn and S contribute only a small mass fraction to the ISM, they play a crucial role in the surface chemistry of interstellar grains.

In the extragalactic area the detection of the HeI line (584A) in a QSO at Z = 1.21 represents the first observation of this line in any astronomical object apart from the Sun. The accumulation of many spectra on the highly variable Seyfert I galaxies presents the prospect that we may begin to understand the processes driving Active Galactic Nuclei. The observations of CIV and Ly-α in absorption at redshifts lower then Z (emission) in QSO's do not indicate the presence of strong heavy element depletion, but may show some evidence of evolutionary effects in the intergalactic cloud number densities.

References

Benvenuti, P., Sanz, L., Wamsteker, W., Macchetto, F., Palumbo, G.C., Panagia, N., 1982, ESA SP-1046.
Cassatella, A. et al., 1987, Astron. Astrophys., 177, L29.
de Boer, K. et al., 1987, Astron. Astrophys., 177, L37
Feldman, P.D., A'Hearn, M.F.A., Festou, M.C., McFadden, L.A., Weaver, H.A., Woods, T.N., 1986, Nature, 324, 433.
Fransson, C. et al., 1987, Astron. Astrophys., 177, L33.
Heck, A., Egret, D., Jaschek, M., Jaschek, C. 1984, ESA SP-1052.
Kondo, Y., 1987, Astrophysics and Space Sciences Library vol. 129, D. Reidel Publ. Co.
Panagia, N. et al., 1987, Astron. Astrophys., 177, L25.
Rolfe, E.J., 1986, New insights in Astronomy, ESA SP-263.
Rosa, M., Joubert, M., Benvenuti, P., 1984, Astronom. Astrophys. Suppl. Ser., 57, 361.
Walborn, N.R., Nichols-Bohlin, J., Panek, R.J., 1985, NASA RP #1155.
Wamsteker, W. et al., 1987, Astron. Astrophys., 177, L21.
Wing, R.F., Carpenter, K.G., Wahlgren, G.M., 1983, Perkins Observatory special Publication No. 1.

4. Infrared Space Astronomy
Terry Herter

I. INTRODUCTION

The period from 1984 through 1987 saw no new infrared space missions. Investigations, however, continued with airborne, balloon, and rocket activities, and follow-up analysis of data from the Infrared Astronomical Satellite (IRAS). Two major ephemeral events occurred during this time, the Comet Halley apparition and SN 1987a, the supernova in the Large Magellanic Cloud. With improved analysis on the database and follow-up observations new results have been forthcoming from IRAS. This work may help astronomers understand everything from the origin and evolution of quasars to the composition of material in cometary nuclei.

No short review can do justice to all the remarkable science that has resulted from the efforts of the many investigators involved in near space and space missions. As such only a few highlights can be given below.

II. NEAR SPACE ACTIVITIES

The year 1986 marked the return of Halley's Comet. The Kuiper Airborne Observatory (KAO), a modified Lockheed C-141 cargo transport airplane operated by NASA-Ames which carries a 91-cm telescope to altitudes above 12.8 km, was deployed from Moffett Airfield in Mountain View, California, and from Christchurch, New Zealand, to provide 2-200μm observations of the comet. The result was the first detection of water in a comet (Mumma et al. 1986, Weaver et al. 1986). Water is expected to be the primary volatile constituent in cometary nuclei. Previous evidence, however, had been circumstantial based on abundances and velocities of H, O and OH seen in the coma and taken to result from photodissociation of H_2O that sublimed from the nucleus. The deduced abundance of water indicates its primary role. Observations of the dust continuum emission were made from 5 through 160μm (Campins et al. 1986b, Herter et al. 1986, Glaccum et al. 1986, Campins et al. 1986a). This is the first such comprehensive coverage. The data fit a greybody emissivity for wavelengths greater than 15μm indicating the existence of large grains. These measurements provide an excellent comparison for the Halley fly-by missions which make in situ measurements of the dust.

In light of the success of the Comet Halley mission, a Southern Hemisphere expedition to observe the recently discovered Comet Wilson was undertaken in April 1987. Comet Wilson had the virtue of being a nonperiodic comet that was detected well in advance of perihelion and as such represented an excellent candidate to perform a comparative study to Comet Halley. Again water was detected in Comet Wilson (Mumma, 1987). Whereas the water and dust production rates for Comet Halley were very variable (sometimes on the timescale of hours) Comet Wilson displayed no such variability.

During the planning stages of the comet mission another spectacular event occurred, SN 1987a, the Supernovae in the Large Magellanic Cloud. As a result observations of SN 1987a were made during the Comet Wilson expedition. A spectrum of the SN from 5.3 to 12.5μm at a resolving power of about 100 was obtained by Rank et al. (1987). They find strong evidence for an infrared excess (over photospheric emission alone) possibly indicating an early infrared echo. Hydrogen recombination lines are also seen as well as a broad feature at 5.3μm which may be due to a blend of iron lines. Larson et al. (1987) measured the 1.5-3.0μm spectrum of SN 1987a using a Michelson interferometer at a resolution of 2000 (subsequently degraded to a resolution of 300). They see Pα and several Brackett recombination lines. Pα shows an asymmetric double peaked structure separated by approximately 4000 km/sec. At 2.7 to 3μm a broad absorption feature, believed to be due to dust, is also seen. Over the next few years further Southern Hemisphere missions to observe the SN with the KAO are planned.

The KAO has been used to obtain solar limb intensity profiles at 30, 50, 100, and 200μm with arc second resolution during a solar occultation (Lindsey et al. 1986). The brightness of the limb agrees roughly with plane-parallel model predictions at all wavelengths. However, at 100 and 200μm the limbs are extended significantly beyond that expected from models. This is taken to be strong evidence for large departures from gravitational-hydrostatic equilibrium almost immediately above the chromospheric temperature minimum.

Jaffe et al. (1986) recently report the detection of the $^3P_2$-$^3P_1$ [CI] fine structure line at 370μm in the Orion molecular cloud. This line complements a longer wavelength line at 609μm due to the $^3P_1$-$^3P_0$ transition observed previously (Phillips et al. 1980). Although at present the emission region cannot be unique-ly identified, the observations are consistent with the observed lines emanating from the warm optically thin interface between the Orion HII region and the molecular cloud which lies behind it. Earlier results with the 609μm line alone have indicated that the C I abundance is substantial in many clouds and possibly could be comparable to the CO abundance (Phillips et al. 1980, Phillips and Huggins 1981). If these early results were proved valid, the implication is that a large neutral carbon abundance exists throughout the clouds, requiring a revision of current interstellar carbon chemistry models and/or of the usual accepted lifetimes of molecular clouds ($\gg 10^6$ years). With the new 370μm observations this now does not appear to be required.

A joint effort between the Japanese and the United States using sounding rockets has resulted in the measurement of the infrared background from 100μm to 1mm (Matsumoto et al. 1987). This resulted in the first detection of the deep minimum expected in this spectral region due to the falling off of the 2.7K cosmic microwave background (CMB) with increasing frequency and a subsequent rise in emission due to interstellar dust emission. Unlike IRAS these measurements are not influenced by interplanetary dust emission because of the elongation angle difference for the two sets of observations, 180 deg. for the rocket measurements versus 90 deg. for IRAS. An excess of an order of magnitude is seen in the 2.7K CMB emission over that expected from the Wien falloff of the Planck function. This excess appears to be a real effect and is explicable in terms of several different cosmological scenarios; distortion of the CMB due to Compton scattering at z > 8 from a hot ionized nonrelativistic gas; excess radiation due to dust (created and) heated by population III stars at z = 15-20; or enhancement of emission through the decay of exotic particles (Hayakawa et al. 1987).

The excess in the CMB discussed above should not be confused with the earlier reported excess near the peak of the 2.7K radiation by Woody and Richards (1981). Subsequent balloon work remeasuring the spectrum near the peak shows that these deviations are not present (Peterson, Richards, and Timusk 1985).

III.   The Infrared Astronomical Satellite (IRAS):

IRAS was a cryogenically cooled 57-cm telescope conceived as a joint venture between the United States, the Netherlands and the United Kingdom. It was launched in early 1983 and performed a sensitive survey of the sky at 12, 25, 60 and 100μm until its cryogen was depleted some ten months later. The spectacular success of IRAS and its results revolutionized many aspects of our views of the universe. Among many other discoveries the accomplishments of IRAS include our first view of the stellar distribution of the Milky Way; the discovery of excess emission due to dust around stars such as a Lyra; the discovery of highly lumi-nous, infrared active galaxies; and the discovery of galactic infrared cirrus emission. Many of the initial results and discoveries are contained in the March 1984 issue of The Astrophysical Journal (Letters), Vol. 278. The science resulting from this short duration mission did not stop, however, after these initial discoveries. Through a vigorous data archival program and the application of new data-processing techniques, IRAS continues to generate new surprises.

New data products which promise to enhance the IRAS database further include the Faint Source Survey (FSS) and Super Sky Flux Plates (SSFP). The FSS will co-add approximately three-fourths of the sky, focussing on filtering the data to enhance point and slightly extended sources. This should result in a survey approximately three times deeper than the original point source catalog and the addition of about 250,000 new sources. Image format plates and a catalog of observations will be available. The SSFP will provide higher (1 arc minute) spatial resolution and improved calibration plates. Destriping and removal of the zodiacal emission are also being performed on the SSFP.

Much of the IRAS "follow-up" work is summarized in three international conferences. The first, entitled "Light on Dark Matter," was held on 1985 June 10-14 in Noordwijk, The Netherlands (Israel, 1986). This was followed by "Star Formation in Galaxies" on 1986 June 16-19 in Pasadena, California, (Lonsdale Persson 1987) and "Comets to Cosmology" on 1987 July 6-10 at Queen Mary College, London. Two review papers summarize the IRAS view of the extragalactic sky (Soifer, Houck and Neugebauer 1987), and of the galaxy and the solar system (Beichman, 1987). It is not possible to cite the wealth of scientific results emanating from IRAS and only a few selected examples are given below. The reader is directed to the above references for a more complete and in-depth summary of IRAS-related work.

IRAS has resulted in the discovery of a new class of extragalactic sources -- ultraluminous infrared galaxies (Sanders et al. 1988). At the very highest luminosities, infrared loud galaxies dominate and an abundance of advanced merger systems is evident. In addition these galaxies show evidence for both star formation and nonthermal power sources and may represent the initial dust-enshrouded stages of quasars. As dust from the merger is shed, the active galactic nuclei (AGN) visually dominate the decaying starburst. The discovery of these objects gives astrophysicists new clues to understanding quasars and the AGN phenomena.

Jura et al. (1987) and Tytler (1987) have examined the infrared emission from Shapley-Ames elliptical galaxies using the IRAS database. More than half of the bright "normal" ellipticals emit detectable $100\mu m$ emission due to cool dust and therefore contain a significant amount of interstellar gas, often 10^7 or 10^8 $M_{\odot}$. The origin and evolution of this interstellar matter is not fully understood. Certainly interstellar matter is expected from stellar mass loss. In fact, estimates of the amount of ejected material should greatly exceed that observed if the galaxies have existed for a Hubble time. This situation is not relieved by outflows since interstellar material does not appear to be flowing out of these galaxies as winds. Sinks of material include the possibility that the gas lost resides in a very hot extended halo or goes into the formation of new low-mass stars.

IRAS 16293-2422 is an extremely cold source in the Rho Ophiuchi molecular cloud which follow-up observations reveal is associated with a high-velocity molecular flow (Walker et al. 1986). Line asymmetries associated with this source appear best explained by the presence of in-falling material in the inner regions of the cloud. The observed luminosity, density structure and velocity profiles agree well with predictions of collapsing protostar models for low-mass stars. This indicates that IRAS 16293-2422 may represent the first discovery of a true protostar, a young stellar object in the process of acquiring mass through accretion of an in-falling envelope.

Initial examination of the zodiacal emission demonstrated the presence of a band structure shown to be associated with material produced by multiple collisions between certain families of main belt asteroids (Low et al. 1984). The zodiacal dust bands are now thought to have substructure (a breaking up into finer bands) which is interpreted as preliminary evidence for a recent collision between two asteroids (Sykes 1987). In addition, recent work has found comet trails that appear as large trails of infrared radiation due to large dust parti-cles which remain in orbit for long periods of time (Sykes 1986; Sykes, Hunten and

Low 1986). These trails extend over 2 AU in some cases. The largest particles thus far seen are those associated with Temple-2. Particle sizes of at least 1 cm (on the basis of dynamical arguments) must be present. Although this source of particles cannot replenish the zodiacal dust, it does represent the refractory component of the nucleus, independent of the gas and ice components of the comet, and provides information on the nature of the dark material. Albedos of these particles are on the order of a few percent, providing additional confirmation of the (necessary) low albedo of cometary nuclei.

IV. FUTURE MISSIONS

The wealth of information gained from suborbital programs and the revolution in infrared astronomy brought about by IRAS clearly demonstrates the necessity of extending existing capabilities. Some of the major missions being carried forward now or planned for the future are discussed below.

A. The Infrared Space Observatory (ISO):

ISO is an astronomical satellite containing a cryogenically cooled 60-cm telescope with four focal plane instruments to be used for imaging, photometric, spectroscopic and polarimetric observations at wavelengths from 3 to $200\mu m$. The observatory, which is an approved and funded project of the European Space Agency (ESA), will be launched around 1993 and operate for 18 months. The instruments are being built by an international consortia of scientific institutes.

B. The Stratospheric Observatory for Infrared Astronomy (SOFIA):

SOFIA is a proposed 3-meter class telescope in a Boeing 747 airplane, anticipated as a joint development by NASA and the West German Science Ministry (BMFT). The concept is an extension of the KAO which operates a 91-cm telescope in a Lockheed C-141 jet transport. Focal plane instruments providing imaging, photometry, and spectroscopic capability over the range from $0.3\mu m$ to 1.5mm will be provided mainly by the investigators. A program of roughly 120 flights/year is planned and with an anticipated operational lifetime of 20 years SOFIA will complement future astronomical space missions.

C. The Cosmic Background Explorer (COBE):

COBE is a mission to make measurements of the spectrum and large-scale anisotropy of the cosmic microwave background (CMB) and to search for a diffuse cosmic infrared emission. This satellite is being developed by the National Aeronautics and Space Administration (NASA). It utilizes three instruments: a Far Infrared Absolute Spectrophotometer to measure the CMB spectrum from $100\mu m$ to 1 cm; Diffuse Microwave Radiometers to search for anisotropies in the CMB at frequencies of 31, 93 and 90 GHz on scales of seven degrees and larger; and a Diffuse Infrared Background experiment to search for a diffuse cosmic infrared background from 1 to $300\mu m$. COBE was originally designed for launch by the Space Shuttle, but it is now being modified for a Delta launch in early 1989.

D. Next Generation Hubble Space Telescope (HST) Instruments:

As part of an approved program of ongoing maintenance and refurbishment for HST, a second generation of focal plane instruments is being developed. Included in these instruments will be an infrared imager/spectrograph operating from roughly 0.8 to $2.5\mu m$. Selected in a competitive review with instruments at other wavelength ranges, two near-infrared instruments designs are being studied by groups headed by Rodger Thompson of the University of Arizona and by Don Hall of the University of Hawaii. Present planning is to have one of these instruments ready for installation into the HST during the second planned maintenance mission six years after launch.

E. The Space Infrared Telescope Facility (SIRTF):

SIRTF is envisioned to be a long-lived, meter-class cryogenically cooled space observatory for infrared astronomy. The goal of SIRTF is to achieve the highest sensitivities, limited only by the faint infrared background of the earth's (space) environment. The focal plane instruments consist of cameras, photometers and spectrographs covering the range from 2.5 to 200μm, and a 700μm photometer. Being developed by NASA, SIRTF is expected to be launched in the mid-1990s and have a lifetime of greater than ten years through cryogen refills.

References

Beichman, C. A., 1987, Ann. Rev. Astr. and Astrophy., **25**, 521.

Campins, H., Joy, M., Harvey, P. M., Lester, D. M., and Ellis, H. B., Jr. 1986a, 20th ESLAB Symp. (ESA, SP-250), Vol. II), p. 107.

Campins, H., Bregman, J. D., Witteborn, F. C., Wooden, D. H., Rank, D. M., All'amandola, L. J., Cohen, M., and Tielens, A. G. G. M. 1986b, 20th ESLAB Symp. (ESA, SP-250), Vol. II), p. 121.

Glaccum. W., Moseley, S. H., Campins, H., and Loewenstein, R. F. 1986, 20th ESLAB Symp. (ESA, SP-250), Vol. II), p. 111.

Hayakawa, S., Matsumoto, T., Matsuo, H., Murakami, H., Sato S., Lange, A. E., and Richards, P. L. 1987, submitted to Ap. J. (Letters).

Herter, T., Gull, G. E., and Campins, H. 1986, 20th ESLAB Symp. (ESA, SP-250), Vol. II), p. 117.

Israel, F. P. (ed.) 1986, Light on Dark Matter, (Reidel: Dordrecht, Holland).

Jaffe, D. T., Harris, A. I., Silber, M., Genzel, R. and Betz, A. L. 1986, Ap. J. (Letters), **290**, L59.

Jura, M., Kim, D. W., Knapp, G. R., and Guhathakurta, P. 1987, Ap. J. (Letters), **312**, L11.

Larson, H. P., Drapatz, S., Mumma, M. J., and Weaver, H. A. 1987, Proc. of ESO Supernova Conf., in press.

Lindsey, C., Becklin, E. E., Orrall, F. Q., Werner, M. W., Jeffries, J. T., and Gatley, I. 1986, Ap. J., **308**, 448.

Low, F. J. et al. 1984, Ap. J. (Letters), **278**, L19.

Lonsdale Persson, C. J. 1987, Star Formation in Galaxies, (NASA, Conf. Publ.2466).

Matsumoto, T., Hayakawa, S., Matsuo, H., Murakami, H., Sato S., Lange, A. E., and Richards, P. L. 1987, in preparation.

Mumma, M. J., Weaver, H. A., Larson, H. P., Davis, D. S. and Williams, M. 1986, Science, **219**, 1523.

Mumma, M. J. 1987, private communication.

Peterson, J. P., Richards, P. L. and Timusk, T. 1985, Phys. Rev. Letters, **55**, 332.

Phillips, T. G. and Huggins, P. J. 1981, Ap. J., **251**, 533.

Phillips, T. G., Huggins, P. J., Kuiper, T. B. H., and Miller, R. E. 1980, Ap. J. (Letters), **238**, L103.

Rank, D. M., Bregman, J., Witteborn, F. C., Cohen, M., Lynch, D., and Russell, R. 1987, in preparation.

Sanders, D. B., Soifer, B. T., Elias, J. H., Madore, B. F., Matthews, K., Neugebauer, G. and Scoville, N. Z. 1988, Ap. J., in press.

Soifer, B. T., Houck, J. R., and Neugebauer, G. 1987, Ann. Rev. Astr. and Astrophys., **25**, 187.

Sykes, M. V. 1986, "IRAS Observations of Asteroid Dust Bands and Cometary Dust Trails," Ph.D. Thesis, Univ. of Arizona.

Sykes, M. V. 1987, B.A.A.S., in press.

Sykes, M. V., Hunten, D. M., and Low, F. J. 1986, Advances in Space Research, **6**, 67.

Tytler, D. 1987, in preparation.

Walker, C. K., Lada, C. J., Young, E. T., Maloney, P. R. and Wilking, B. A. 1986, Ap. J. (Letters), **309**, L47.

Weaver, H. A., Mumma, M. J., Larson, H. P., and Davis, D. S. 1986, in 20th ESLAB Symp. (ESA, SP-250, Vol. I), p. 329.

Woody, D. P. and Richards, P. L. 1981, Ap. J., **248**, 18.

5. Radioastronomy from Space

A. Boischot

Radioastronomy can use space technology either to go close to the source of the radioemission to increase the sensitivity of the observations and have different viewing geometry, or to avoid the effect of the Earth's environment, i.e. to extend the useful frequency range below the ionospheric cut-off of a few megahertz and at submillimeter wavelengths. A third field which will certainly develop in the future is the use of space to increase the spatial resolution of radio-interferometry for imaging radiosources.

Exploration of the solar system

The exploration of the solar system, where most of the results of space radioastronomy have been obtained recently includes outer planets, comets and the Sun.

The continuous success of the NASA Voyager mission brought the spacecraft V2 to an encounter with Uranus on January 24, 1986. Radioemission below 800 KHz has been detected from that planet only three days before the time of closest approach, and has been received by the Planetary Radio Astronomy (PRA) instrument for about three weeks (Warwick et al, 1986 ; Gurnett et al, 1986). Uranus is then the forth planet, after the Earth, Jupiter and Saturn, known to be strong source of radioemissions. Its study is particularly interesting because of the special geometry of the planet : rotation axis in the ecliptic plane and magnetic dipole axis highly inclined on the rotation axis.

Between the encounter and June 1987 a dozen of papers have been published, or are in press to date in a special issue of the Journal of Geophysical Research to appear at the end of 1987. The subjects are the auroral emissions, emissions from electrostatic discharges and several related topics like the study of grain impacts, low frequency whistlers and plasma waves, etc ...

Other papers have been published between June 1984 and June 1987 about results of the PRA experiment on Jovian and Saturnian emissions. Review papers can be found in Gehrels and Matthews, 1984, and Rucker and Bauer, 1984. Voyager 2 is now on its way to Neptune. Their encounter will take place in August 1989.

The study of the radioemissions of the outer planets is complemented by that of the Earth, specially the Auroral Kilometric radiation (AKR). Several Earth oribting satellites have been used with great success, particularly the Dynamic Explorer (NASA) and Viking (Sueden).

Other results have been obtained by the NASA "ICE" spacecraft (ex ISEE 3) on Comet Giacobini-Zinner by the on board radioastronomy experiment. They refer principally to the plasma environment of the comet (Meyer-Vernet et al, 1986).

Finaly solar radioemissions, which have been the first detected from space at subionospheric frequencies in the 70's have been studied from different spacecraft. Type III solar bursts have been used to study the interplanetary medium (Dulk et al, 1985 and references herein).

Space radioastronomy projects

The second part of this report concerns the plans for future space experiments which are presently under study by the different Space Agencies.

In the field of the Solar System two missions, Galileo and Ulysses, equiped with radioastronomy instruments, were planned to be launched in the mid 80's. They have been postponed to around 1990 due to the Chalenger tragedy and will arrive at Jupiter around 1995. Galileo will remain in a Jovian orbit for several years while Ulysses will continue its Journey in an out-of-ecliptic trajectory from which it will study the solar radiobursts. Other projects of interest for radioastronomers are CRAFT, Cassini (Saturn orbiter and Titan probe), and Cluster (Study of Earth's magnetosphere). They are planned for the mid 90's.

Going out of the solar system, the main radioastronomy project is to use space for VLBI, together with a ground based network of radiotelescopes. "Quasat" is discussed jointly by NASA and ESA (Quasat, 1984) and "Radioastron" is a soviet project which includes a radiotelescope on a highly excentric orbit. The space VLBI technique has been tested using a TDRSS-NASA satellite and several ground based radiotelescopes : interferometric fringes have been successfully obtained (Levy et al, 1986). Another "première" in this field is the obtention of interferometric fringes from the source of Auroral kilometric radiation by the Iowa group using the two Earth orbiting satellites ISEE1 and 2 (Baumback et al, 1986).

The last topic we must include in this report is the submillimeter radioastronomy, which must be essentially performed from space. The progress of the heterodyne technology will certainly extend astronomical observations in this domain during the next few decades. Submillimeter high performant satellites are presently studied, and will be launched only in the next century. But, preliminary experiments are done from aircraft or planned in the near future with balloon born instruments.

References

Baumbach, M.M., Gurnett, D.A., Calvert, W., Shawhan, S.D. : 1986, Geophys. Res. Lett. 13, p. 1105.

Dulk, G.A., Steinberg, J.L., Lecacheux, A., Hoang, S., MacDowall, R.J. : 1985, Astron. Astrophys. 150, L 28.

Gehrels, T., Matthews, M.S., : 1984, "Saturn", The University of Arizona press, Tucson, U.S.A.

Gurnett, D.R., Kurth, F.L., Scarf, W.S., Poynter, R.L. : 1986, Science, 233, p. 106.

Quasat "a VLBI Observatory in Space" : 1984, ESA Report SP 213, Paris, France.

Levy, G.S. et al. : 1986, "VLBI observations made with an orbiting radiotelescope", Science, 234, p. 117.

Meyer-Vernet, N., Couturier, P., Hoang, S., Perche, C., Steinberg, J.L., Fainberg, J., Meetre, C. : 1986, Science, 232, p. 370.

Rucker, H.O., Bauer, S.J. : 1984, "Planetary Radio Emissions", Proceedings of an International Workshop held at Graz, Austria.

Warwick, J.W., Evans, D.R., Romig, J.H., Sawer, C.B., Desch, M.D., Kaiser, M.L., Alexander, J.K., Carr, T.D., Staelin, D.H., Gulkis, S., Poynter, R.L., Aubier, M., Boischot, A., Leblanc, Y., Lecacheux, A., Pedersen, B.M., Zarka, P. : 1986, Science, 233, p. 102.

6. SPACE OBSERVATION OF SOLAR SYSTEM OBJECTS

Jurgen Rahe

The program of exploration of solar system objects from space is now in a more
mature phase where third generation missions to inner as well as outer solar
system bodies are developed. Emphasis in the inner solar system is on Mars and
Venus, in the outer-solar system on Jupiter (Galileo Mission). The exploration
of the primitive small bodies (comets, asteriods) has lagged in the past,
partly due to technological considerations. Following the spectacular, but
still reconnaissance-level flybys of Comets Giacobini-Zinner and Halley in 1985
and 1986, respectively, ISAS considers a cometary flyby and coma sample return
mission, and NASA plans to initiate an even more comprehensive exploration by
conducting a close flyby of a main-belt asteroid, followed by a multi-year
rendezvous with a short-period comet (CRAF) in the early 1990s.

During its past apparition 1985/1986, Comet Halley was the focus of an
unparalleled global scientific effort of exploration from the ground, from
Earth orbit, from Venus orbit, from interplanetary space, and from within the
comet itself.

Four space agencies - the European Space Agency (ESA), the Intercosmos of the
USSR Academy of Sciences, the Japanese Institute of Space and Astronautical
Science (ISAS), and the National Aeronautics and Space Administration (NASA) -
sent six spacecraft to Halley's Comet. ESA launched Giotto, Intercosmos
launched Vega-1 and Vega-2, ISAS launched Sakigake and Suisei, and NASA used
its ICE spacecraft. The different missions complemented each other in their
flyby distances, ranging from about 600 (Giotto) to about 30 million (ICE)
kilometer, and comet heliocentric distances ranging for the time of the
encounters from 0.8 to 0.9 A.U. The scientific experiments on the various
spacecraft provided the full complement of experiments that can be flown on a
flyby mission. In addition, there was a large overlap between the experiments
on the different spacecraft allowing a comparison of data after the encounters.
The missions also extended the total time of in-situ measurements in the
cometary environment.

Realizing that many aspects of mission planning, spacecraft and experiment
design, and data evaluation are essentially the same for all missions, and that
the overall scientific return could be increased through cooperation, the four
agencies formed in the fall of 1981 the Inter-Agency Consultative Group (IACG)
for Space Science. The IACG undertook the task to coordinate matters related
to the space missions to Halley's Comet, similar to the International Halley
Watch (IHW) which coordinates the ground-based Halley observations. IHW and
IACG are described in more detail e.g., in ESA Bulletin Nos. 38 and 39, 1985.

First results of these missions have been published in Nature, Vol. 321, 1986,
and in the proceedings of the 20th ESLAB Symposium "Exploration of Halley's
Comet" (ESA (SP-250).

Following a series of highly complicated orbital manuevers about the Earth and
the Moon, the International Sun-Earth Explorer 3 (ISEE-3) which had been
launched in 1978, was renamed International Cometary Explorer (ICE) and went
through the tail of Comet Giacobini-Zinner on September 11, 1985 at a distance
from the nucleus of 7800 km. The results of this flyby are reported in the
April 18, 1986, issue of Science.

The comet nucleus was observed from the Vega and Giotto Missions. It appeared as peanut-shaped object both larger (about 16x8x8 km) and darker (albedo lower smaller than 4%) than previously thought, making it one of the darkest objects in the solar system. Gas and dust emanate in form of jets from only a few regions of the nucleus, while the rest is covered with a dark crust. Much of the cometary dust is made of organic material. As expected, water vapor was the dominant parent molecule in the coma. At the time of Giotto flyby, the total mass production rate was 6.9x10 mol/sec of which 80% were water molecules. The water vapor production amounts to 15 tons/sec.

The interaction between the solar wind plasma and the cometary ionosphere is characterized by two distinct boundaries, the bow shock and the contact surface. Outside the bow shock, is the undisturbed supersonic solar wind. Inside the contact surface, close to the nucleus, there are only cometary ions. In between these two boundaries is a mixture of cometary and subsonic solar wind ions. Some cometary neutral particles can travel large distances from the nucleus before they are ionized and picked up by the solar wind. Such "pick-up ions" were observed by the Vega and Giotto spacecraft out to distances of 10 million km, by ICE even at distances of 30 million km.

Since 1983, substantial contributions to Venus research have been made. High resolution radar images obtained by Venera 15 and 16 revealed a variety of volcanic, impact, and tectonics features. Vega 1 and 2 deployed a pair of landers and instrumented atmospheric balloons, providing information on the surface composition and atmospheric conditions. The Pioneer Venus Orbiter continues to make excellent measurements, indicating, e.g., that the amount of SO above the cloud tops has declined by more than a factor of ten since these measurements started in 1978.

There is growing evidence that water has played an important role in the climate and geological evolution of Mars. A new international effort to explore Mars is underway which includes plans for Mars orbiters, balloons, landers, rovers, and Mars sample returns.

The two Voyager spacecraft have successfully explored Jupiter and Saturn and their satellites. Voyager 1 is leaving the solar system on a trajectory inclined above the ecliptic plane. Voyager 2 reached Uranus on January 24, 1986, and is now proceeding to Neptune where it will arrive in August 1989. All 11 Voyager 2 instruments operated perfectly at the Uranus encounter and sent to Earth volumes of data and thousands of images of uncalculable scientific value. In addition to the five known moons, 10 others were found with diameters averaging about 50 km. Oberon and Titania showed crater-packed surfaces with wide cracks, Miranda exhibited virtually all geological forms found elsewhere in the solar system. Uranus has a surprisingly strong magnetic field with a 55 degree tilt of the magnetic pole axis from the axis of rotation.

It is expected that the next report, covering the period through June 1990, will be dominated by results from the two Soviet Phobos missions which are scheduled to land on Mars in early 1989, from the Voyager 2 encounter with Neptune on August 25, 1989, and by measurements of Solar System objects obtained from the Hubble Space Telescope.

PLANETARY MISSIONS LAUNCHED SINCE 1983

NAME	COUNTRY	TARGET/MISSION	LAUNCH DATE	ARRIVAL DATE		HALLEY FLYBY(km)
VENERA 15	USSR	VENUS-ORBITER	6/2/83	10/10/83		
VENERA 16	USSR	VENUS-ORBITER	6/7/83	10/14/83		
VEGA 1	USSR	VENUS COMET HALLEY	12/15/84	6/11/85 3/6/86	V H	 8.9x10
VEGA 2	USSR	VENUS COMET HALLEY	12/21/84	6/15/85 3/9/86	V H	 8.0x10
SAKIGAKE	JAPAN	COMET HALLEY	1/9/85	3/11/86		7x10
GIOTTO	ESA	COMET HALLEY	7/2/85	3/14/86		605
SUISEI	JAPAN	COMET HALLEY	8/19/85	3/8/86		1.5x10
ICE	USA	GIACOBINI-Z COMET HALLEY	8/12/78	10/11/85 3/25/86		 2.8x10

7. Solar Research From Space 1985 - 1987

S. D. Jordan

I. INTRODUCTION

The three year period 1985 - 1987 was a time of both continuing advances in our understanding of the Sun, as well as a period of planning at least three future space missions for solar physics during the next decade. The greatest advances in the study of solar physics from space observations during 1985 - 1987 probably came in the general area of solar flares, thanks to data from the Solar Maximum Mission (SMM) and HINOTORI satellites. In addition, significant new insights into small-scale photospheric turbulence (granulation) and into mass balance in the chromosphere-corona transition region were obtained from experiments deployed on Spacelab II. There is some concern over the difficulties the solar physics community has experienced in obtaining a suitable High Resolution Solar Observatory (HRSO). Nevertheless, significant progress was made toward realizing a Solar and Heliospheric Observatory (SOHO), to be launched in 1995 for studies of acceleration mechanisms for the solar wind and studies of the solar interior via observations of solar pulsations (helioseismology). Particularly encouraging during this period was the announcement that a SOLAR-A high-energy flare mission would definitely be developed for operation during the period 1991-1993 of the next solar maximum. In general, the world-wide scientific community seems to be developing a growing interest in solar physics, and in obtaining solar observations from space.

II. RECENT PROGRESS

1. Solar Flares

Great progress toward a more complete understanding of solar flares was made during and following the last solar maximum, thanks in large measure to two orbiting spacecraft, the Solar Maximum Mission (SMM), and the HINOTORI. This progress continued during the period of this review, both from continuing analysis of data taken during the early 1980's and by the extended operation of the SMM, following its repair by a Shuttle mission in 1984. Areas of significant advance included: evidence for beams of suprathermal electrons, prompt acceleration of protons, improved element abundance determinations, and the determination of a 154-day periodicity in flare activity.

A major problem in solar flare physics is the development of a satisfactory model for the flare process for a given flare or for a class of flares. One aspect of this problem is the discrimination between two different models which are sometimes proposed to explain the same flare phenomena. These are the thick-target model and the thermal model. The former involves the acceleration of a beam of suprathermal electrons in the corona, while the latter involves bremsstrahlung-producing electrons with a thermal velocity distribution. While the latter model can be considerably less taxing on the energy requirements for the electrons, evidence for beams of suprathermal electrons would demonstrate that certain features of the thick-target model are probably realized in at at least some flares. These features are either the existence of an "external" driver such as a shock wave to produce the acceleration or a high degree of current filamentation within the acceleration region. Such evidence for suprathermal beams has been found. During some flares, the SMM Hard X-Ray Imaging Spectrometer and the HINOTORI Solar X-Ray Telescope obtained images which are consistent with a "footpoint" hard X-Ray structure at energies around 20 KeV, indicative of the non-thermal impact signature of the thick-target model. The thick-target model is described in Lin and Hudson (1976) and Hoyng et al. (1976). The thermal model is described in Brown et al. (1979) and Smith and Harmony (1982). Recent observations bearing on this problem are described in Dennis at al. (1986).

Evidence for the prompt acceleration of protons comes from gamma-ray lines
in the 4- to 8- MeV range, observed with the Gamma-Ray Spectrometer (GRS) on SMM.
These lines, produced by nuclear deexcitations, were observed to be emitted
simultaneously with the hard X-Ray bremsstrahlung continuum to within the 1-s
timing accuracy of the data. Since the excited nuclei responsible for the emission
of these lines are produced by high-energy proton impact, it follows that
impulsive proton acceleration is a significant feature of these gamma-ray
producing flares (Murphy and Ramaty, 1985). Gamma-ray spectra obtained with the
GRS have also proved useful in relative abundance determination, particularly in
the chromosphere (Murphy et al. 1985). These analyses suggest lower abundances of
C and O, relative to Mg, Si, and Fe, in the chromosphere compared to the
photosphere. This in turn implies a variation in chemical composition with space
and time in the solar atmosphere, in general, and may even have implications for
our knowledge of cosmic abundances. Soft X-ray spectra are also useful in this
context (Doschek et al. 1985).

Finally, evidence has been found in several energy regimes for a 154-day
periodicity in solar flare activity. Recent Fourier analyses of gamma-ray, hard
X-Ray, radio microwave, and H- alpha event-occurrence rates clearly show a
variability in the rates with a period of approximately 5 months (Rieger et al.
1984, Dennis, B. R., 1985, Bogert and Bai, 1985, Ichimoto et al. 1985).

2. Photospheric Turbulence and Magnetic Fields

One of the most spectacular recent observations in solar physics was the
series of granulation pictures taken at sub-arcsecond spatial resolution and time
intervals ranging from 10 to 60 seconds with the Solar Optical Universal
Polarimeter (SOUP) instrument flown on Spacelab II during August 1985. This time-
resolution permitted the creation of a movie which reveals clearly the difference
between the large-scale flows in the lower solar photosphere and the five-minute
oscillation (Title et al. 1986). The granulation was observed at higher spatial
resolution than ever before achieved for a time longer than that required to
produce a single snapshot. The phenomenon of "exploding granules," heretofore
known to characterize only a, possibly small, fraction of the granules, was shown
to be ubiquitous. Stated simply, the turbulent convective elements are seen to
rise in the photosphere until contact with the overlying zone of predominantly
radiative transport acts like an almost impenetrable wall, which forces the
granule to impart its forward momentum to the atmosphere as a whole, while the
material of the granule itself moves out in all directions perpendicular to the
upward motion.

A second achievement of the operation of the SOUP on Spacelab II is the
remarkable correlation of the photospheric flow field with the small-scale
photospheric magnetic field, where the latter was observed simultaneously from the
Big Bear Observatory (Simon et al. 1987). In this experiment, the white-light
granules act as "corks," or tracers of the flow field. It is clear from comparing
the flow field so mapped with the SOUP to the magnetic field mapped at Big Bear
that the magnetic features move with the flow field, and thus congregate at the
same locations where the flows converge, i.e., at the edges of the supergranulation
pattern on the Sun. This confirms in great detail and with exceptional clarity
a picture of solar surface convection and magnetic field migration which provide
boundary conditions for modeling solar convection and dynamo activity.

The data set from the SOUP was acquired at the end of the Spacelab II mission,
after a number of initial problems were successfully solved. The excellence
of the brief data set, of about one hour's duration, demonstrated convincingly
that sub-arcsecond solar observations, in this case with about a quarter of an
arcsecond angular resolution, will reveal both new phenomena and also a convincing
confirmation (or disconfirmation) of already studied phenomena to a remarkable
degree.

3. Outer Atmospheric Energetics

The Spacelab II mission also featured a solar High Resolution Telescope and Spectrograph (HRTS) for obtaining high resolution ultraviolet spectra of the solar chromosphere and transition regions. The HRTS consisted of a 30 cm Gregorian telescope, a slit spectrograph covering the 1190 - 1680 A region with 0.05 A spectral resolution, a broadband (90 A FWHM) spectroheliograph, and an H-alpha filter system (Brueckner et al. 1986). This experiment obtained sufficient spectra and images in the strong resonance doublet of C IV (1548 A and 1550 A) to permit statistical studies to be performed on the possible contribution of upward moving features in the solar transition region to coronal heating. It also revealed strong downflows in the transition region, thus contributing to a better understanding of the mass-balance problem in the outer solar atmosphere.

The upward moving features observed in the C IV resonance lines were considered as a possible mechanism for coronal heating. These features were of 1-8 arcseconds in angular extent, and moved with velocities up to 400 km/sec in the upward direction (as well as, in many cases downward, but never with such high velocities). From the frequency of occurrence of these upward moving "bullets," and local values of temperature and density in the solar transition region, the investigators estimated the contribution of these events to radiative loss, enthalpy flow, and kinetic energy flow in the solar atmosphere. The results were perhaps disappointing, but enlightening, with respect to coronal heating. From Withbroe and Noyes (1977), the energy loss of the quiet corona is estimated at 3×10^{5} ergs cm^{-2} sec^{-1} for the quiet corona, and up to 10^{7} ergs cm^{-2} sec^{-1} for coronal active regions. These events, on the average, provide an estimated 2.5×10^{4} ergs cm^{-2} sec^{-1}, clearly insufficient to heat the corona.

The HRTS results have considerable significance for understanding the transition-region mass balance, however. It has long been known that the upward mass flow in spicules alone is more than an order of magnitude greater than the mass loss measured at 1 a. u. in the solar wind, if there is no downflow within the spicule itself. The upward and downward motions observed in the transition region, like upward and downward motions measured earlier in chromospheric Mg II from satellites, show that mass balance in the solar atmosphere is a complex phenomenon, with upflows almost balanced by downflows, as required by mass conservation.

III. FUTURE PLANS

1. High Resolution Solar Observatory

The world-wide solar physics community has for over a decade accorded a very high priority to the realization of a sub-arcsecond spatial resolution on the Sun. This spatial resolution, with correspondingly high spectral and, in some cases of strong lines, temporal resolution will provide the capability of addressing many problems in the fundamental physics of the Sun's surface magnetic field and convection, and of the heating and mass balance in the overlying atmosphere up to the corona. A spatial resolution approaching 0.1 arcseconds is needed for these studies, because this is the approximate scale (0.1 arcseconds = 73 km on the Sun) of the hydrodynamic scale height and of a continuum photon mean-free - path in the solar photosphere, i.e., the scale on which collective energy transport occurs in much of the solar atmosphere. Realization of this scale from the ground is virtually impossible for more than a fraction of a second, because of atmospheric effects. Thus dynamical phenomena on this scale, in particular, must be observed from space, even in the visible.

The High Resolution Solar Observatory (HRSO) is the current version of a program planned during the 1970's by a group of American and European solar physicists in cooperation with the National Aeronautics and Space Administration (NASA) to provide this capability. The HRSO is a somewhat descoped version of an earlier, more ambitious program called the Solar Optical Telescope (SOT).

The HRSO retains the very high spatial resolution capability of the SOT, approaching 0.1 arcseconds, but does not offer an ultraviolet capability below about 2200 A. (The SOT went down to about 1175 A.) As such, it is more limited than its predecessor for studies of the solar atmosphere above the temperature minimum, but it still provides an outstanding and needed capability for studying the fine structure and dynamics of solar photospheric convection and the Sun's surface magnetic field, in which a number of fundamental solar-stellar astrophysical processes occur.

2. Solar and Heliospheric Observatory

The Solar and Heliospheric Observatory (SOHO) has been planned as a major joint mission of the European Space Agency (ESA) and NASA. The mission will be part of a major Solar Terrestrial Science Programme (STSP) consisting of at least four free-flying satellites, of which three will focus on magnetospheric and outer-solar-wind problems and one, the SOHO, will concentrate on the physics of the inner solar wind and on solar pulsations. The experiments will fly on a satellite to be launched in 1995 to achieve a halo orbit around the first Lagrangian point in the Earth-Sun system. This way the satellite will be exposed to the unobstructed solar windstream well in front of the terrestrial magnetosphere. Measurements of comparatively small oscillatory motions on the solar surface will also be facilitated by this orbit, due to the small relative velocity between the satellite and the Sun.

The SOHO will address the following three fundamental and interrelated questions in solar and heliospheric physics:
   - How is the solar corona heated?
   - Where and how are the solar-wind streams accelerated?
   - What is the structure of the solar interior?
The first question, on heating the solar corona, has been a fundamental question in solar physics since the mid-century. To date, no completely satisfactory answer has been offered. About the only thing now known reliably is that acoustic waves generated by convective turbulence in the lower photosphere do not heat the corona. On the other hand, certain magnetohydrodynamic waves cannot be ruled out, particular in the low-density coronal hole regions. Various current-dissipation mechanisms for extracting energy from ambient magnetic fields are attractive for coronal active regions, and an attempt has been made to identify the magnitude of the heating from the observed photospheric velocity power spectrum without determining the exact mechanism for the dissipation. All of these theories, including certain features of the latter, require a better determination of densities and associated energy fluxes in the corona than are currently available (Ionson, 1985). High-resolution extreme-ultraviolet spectroscopy on SOHO is expected to provide the needed data.

The second question concerns the need for a mechanism for accelerating the solar wind to the high velocities observed at 1 a. u. by satellites in Earth orbit. Acceleration by momentum transfer from Alfven waves and other mechanisms have been proposed, but no definitive solution has been found as yet (Axford, 1985). Since solar wind conditions must be measured out to many solar radii to cover a significant fraction of the acceleration region, a coronograph capable of providing information on the solar wind velocity at these distances is needed on the SOHO spacecraft.

The third question, on the structure of the solar interior, depends upon applying the methodology developed during the previous decade for inferring the structure of the solar interior from careful monitoring of velocity and luminosity oscillations of the solar surface (Gough, 1985). A Solar Oscillations Instrument on SOHO will obtain the needed data. One of the most promising possibilities here is the determination of the velocity structure of the solar convection zone, where the convective motions, combined with the Sun's differential rotation, are thought to be responsible for generating the Sun's magnetic field.

3. SOLAR-A

A major space mission, to be called SOLAR-A has been planned for the next solar maximum in the early 1990's by the Institute of Space and Astronautical Science (ISAS) in Japan. The emphasis of this mission will be on the high-energy phenomena occurring in association with flares, particularly during the flares' early phases. Both X-ray and gamma-ray instruments will be carried on a spacecraft built and launched by ISAS. The proposed payload for the SOLAR-A consists of: a hard X-ray imaging instrument, a hard X-ray continuum spectrometer, a soft X-ray imaging telescope, a soft X-ray continuum spectrometer, a Bragg crystal spectrometer, a gamma-ray and neutron spectrometer, and a solar intensity monitor. All of the instruments but one will be built in Japan. The single exception is the soft X-ray imaging telescope, to be built in the United States and provided by NASA under terms of a joint agreement between the two governments. The SOLAR-A mission will continue the systematic study of high-energy solar processes begun by the SMM and HINOTORI missions of the early 1980's.

The core of the Solar-A payload is to be the hard X-ray imaging instrument and the soft X-ray imaging telescope. The hard X-ray instrument will yield new information above 30 keV (full range is to be 7 - 70 keV) on nonthermal electron acceleration in the flare, a process known to be one of the most important energetically among several energy-release mechanisms. The spatial resolution of 5 arcseconds will permit the flare kernel to be identified against the background. The soft X-ray telescope will complement the hard X-ray data by revealing the magnetic structures within which the flash phases of flares are known to occur, both before flaring and also thereafter, when these loop structures tend to fill up with hot, dense plasma. Imaging with an angular resolution of 2-3 arcseconds is the goal of this instrument. Obtaining simultaneous images of flares with comparable high temporal and spatial resolution in both hard X-rays and soft X-rays should reveal the morphology of the flaring plasma in sufficient detail to provide a deep understanding of the physical processes occurring during the flares' early phase.

Continuous measurement of the Sun's irradiance during the scheduled mission period of 1991-1993 will extend the data base already provided by SMM. The SOLAR-A solar intensity monitor will also introduce a new feature into solar irradiance measurements, a modest degree of spatial resolution across the solar disk. The data will be used to study active-region evolution and low-order global oscillations, as well as continuing the study of secular changes in the solar constant.

4. Other Programs

A large number of smaller programs and at least one other of major consequence are currently in various stages of planning or development. These include sounding rocket and balloon-borne solar experiments, possible solar payloads to be launched into orbit on Scout rockets by NASA, and Russian plans for a Mars mission named after the moon to be studied, PHOBOS. Of these, the PHOBOS mission is the most ambitious, for the spacecraft, during its journey to Mars, would obtain data on the solar wind and the interplanetary medium between Earth and Mars.

Proposals for high-energy solar experiments to be flown on small satellites launched by Scout rockets are likely to be submitted. Operation in the period of the next solar maximum is planned. One possibility being studied at this writing would involve flying a spacecraft developed by Argentina, launched by NASA on a Scout rocket, and carrying a solar high-energy experiment provided by American investigators.

It was evident from the scientists who attended the XIX General Assembly for the IAU in New Delhi during November 1985 that both India and China are developing significant groups of scientists pursuing solar physics. The discipline is rapidly achieving world-wide scope, and it is to be expected that major groups everywhere will be interested in pursuing solar physics from space in the near future.

References

Axford, I., 1985, Solar Phys., 100, 575.
Bogert, R. S., Bai, T., 1985, Astrophys. J. Lett., 299, L 51.
Brueckner, G. E., Bartoe, J. -D. F., Cook, J. W., Dere, K. P., Socker, D.
    G., 1986, Adv. Space Res., 6, 263.
Brown, J. C., Melrose, D. B., Spicer, D. S., 1979, Astrophys. J., 228,
    592.
Dennis, B. R., 1985, Solar Phys., 100 465.
Dennis, B., Chupp, E., Crannell, C. J., Doschek, G., Hudson, H. Hurford,
    G., Kane, S., Lin, R., Prince, T., Ramaty, R., Shane, G.,
    Tandberg-Hanssen, E., "Max-91, An Advanced Payload for the Exploration of
    High Energy Processes on the Active Sun, NASA Report, June 1986.
Doschek, G. A., Feldman U., Sheely, J. F., 1985, M. N. R. A. S., 217,
    317.
Gough, D., 1985, Solar Phys., 100, 65.
Hoyng, P., Brown, J. C., van Beek, H. F., 1976, Solar Phys., 48, 197.
Ichimoto, K., Kubata, J., Suzuki, M., Tohmura, I., Kurokawa, H., 1985,
    Nature, 316, 422.
Ionson, J. A., 1985, Solar Phys., 100, 289.

Lin, R. P., Hudson, J. S., 1976, Solar Phys., 50, 153.
Murphy, R. G., Ramaty, R., 1985, Adv. Space Research, 4, 127.
Murphy, R. G., Ramaty, R., Forrest, D. J ., Kozlovsky, B., 1985, "Proc.
    19th Intern. Cosmic Ray Conf.," La Jolla, 4, 249, 253.
Rieger, E., Share, G. H., Forrest, D. J., Kanbach, G., Repping, C., Chupp,
    E. L., 1984, Nature, 312, 623.
Simon, G. W., Title, A. M., Topka, K. P., Tarbell, T. D., Shine, R. A.,
    Ferguson, S., Ziring, J., 1987, in preparation.
Smith, D. F., Harmony, D. W., 1982, Astrophys. J., 252, 800.
Title, A. M., Tarbell, T., Simon, G., 1986, Adv. Space Research, 6, 253.
Withbroe, G. L., Noyes, R. W., 1977, Ann. Rev. Astron. Astrophys., 15 363.

45. STELLAR CLASSIFICATION (CLASSIFICATION STELLAIRE)

PRESIDENT: R.F. Garrison
VICE-PRESIDENT: M. Golay
ORGANIZING COMMITTEE: J. Clariá, A. Heck, N. Houk, T. Lloyd-Evans,
D.J. MacConnell, E.H. Olsen, A. Slettebak, V. Straižys.

I. Introduction

Stellar Classification is an important activity for astronomers, since it provides "systems" for comparison with new types of stars. A good classification scheme can be used cannily to segregate "peculiar" objects and to gain insight into the processes which generate "normal" objects. Eventually, when there are enough objects in a given "peculiar" class, the definition of "normal" can be extended to include them. Through the process of classification, prototypes can be isolated for detailed study, providing a short-cut to the lengthy process of studying all stars in detail. Thus, classification of stars and maintenance of the reference frames are important on-going processes in astronomy.

In recognition of increased activity in the subtopics of automatic classification and calibration, they have been assigned separate headings and reporters.

II. Classification Using Slit Spectra(C.J. Corbally)

A) O- AND B-TYPE STARS

Wolf-Rayet stars in the Galaxy were classified in several studies: Moffat and Seggewiss (38.114.029); Lundstrom and Stenholm (38.114.047); Downes (38.114.081), who included a star with strong Fe II emission; Lortet *et al.* (38.132.031), who found a WC star while observing O-stars; Torres *et al.* (41.114.003), who revised WC types using red region spectra; Aller and Keyes (41.134.010), who found W-R stars during a survey of the central stars of planetary nebula; Gomez and Niemela (MNRAS **224**, 641), who were using both blue and red region spectra to classify supergiant O-stars (as also in RevMex AA **14**, 293).

From normal O-stars with the same effective temperature but different spectral types, Underhill (38.114.138) has argued for their having similar photospheres but different mantles. Ruban (39.114.039) compared classifications from line information with those from the continuum distribution for O9-A0 stars.

The star exciting the RCW 34 nebula was classified from line ratios as O9.5 I by Vittone *et al.* (AA **179**, 157). Be stars were classified by Goraya and Padalia (37.112.114), Corbet and Mason (37.114.024), Wolf and Stahl (40.114.023), and Finkenzeller (40.121.023). The extensive survey of 1874 UV excess objects by Green *et al.* (41.002.074) contains mainly subdwarf sdB stars. A helium rich SdO star, CD -24° 9052, was classified by Kilkenny *et al.* (41.126.011), and Kilkenny (42.126.061) also studied WD 1225-079, a DZA$_4$ star.

B) A- AND F-STARS

The peculiar stars have dominated studies in these spectral classes. Hauck (41.002.012) has produced a 3rd catalogue of Am stars. Zverko (38.114.063) has reclassified HR 830 and 21 CVn as silicon stars. Gray (AJ, in press) has found that λ Bootis stars form a morphologically distinct set of stars that subdivide naturally into normal hydrogen-line or peculiar hydrogen-line types. High radial velocity stars, generally weak-lined, were studied by Stock *et al.* (38.111.021). White dwarfs, classified by Greenstein (41.126.060) at high signal-to-noise, proved to be mainly of DA type. The star exciting the HH57 object was classified by Cohen et al. (41.121.035) as F8 III, implying 5 magnitudes of extinction.

Abt (38.114.071) has made spectroscopic tests on photometric stellar classifications for 169 abnormal A5-G0 stars. Similar spectroscopic tests were made for the weak-line stars by Abt (42.114.079) and by Corbally (AJ **94**,161) who both used Houk's extension of the MK System. Corbally and Boyle (AA, in press), while testing classifications predicted from Vilnius photometry, found unacceptable disagreement in the A- to F-star region, though satisfactory agreement was shown among late-type stars.

C) LATE TYPE-STARS

Keenan and Yorka (40.114.147) have continued to revise and extend their list of classifications for stars later than G0, with an eventual goal of more than 800 stars. Yorka is observing M-stars in the yellow-red region to improve classifications, especially in the range M2 to M5. Other M-star classifications were made by Dahn et al. (41.115.023), Ianna and Bessell (42.111.002), and Sabbadin et al. (AA Suppl **67**, 541). The differentiation of J type carbon stars from N-type stars is discussed by Lloyd Evans (41.065.121), and carbon stars were also classified by Azzopardi et al. (40.114.073), Hartwick and Cowley (40.155.046), and Lloyd Evans (40.114.040).

Cool stars with excesses of heavy elements were treated by several speakers at the Strasbourg Observatory Colloquium (39.012.101). The classification of giant barium stars was discusssed by Yorka and Keenan (39.114.141). Keenan et al. (PASP **99**, 629) have commented on the recognition and classification of strong-CN giants.

Rose, using quantitative classifications, has found red horizontal branch stars (39.114.082) and strong-lined G-dwarfs (39.114.083) in the galactic disk. Friel (AJ **93**, 1388) has also applied quantitative methods to low resolution spectra of yellow giants in a galactic structure study. MacConnell (RevMexAA **14**,367) is using 4Å/pixel spectra of the Ca II triplet in the near infrared to classify possible cool supergiants along the southern galactic plane. The pre-main-sequence star HDE 283572 was classified as G5 IV by Walter et al. (ApJ **314**, 297), and the variable shell star HD 50845 was found to be K0 by Sahade and Ringuelet (39.112.125). The puzzling solar-type twin system, ζ^1 and ζ^2 Ret, was studied by Da Silva and Foy (AA **177**, 204).

Solar analog stars were investigated by Neckel (42.114.011). Keenan is constructing a small, Garrison Type, slit spectrograph to allow monitoring of the spectral type of the sun.

Silva et al. (40.114.140) have classified three X-ray-selected stars. An optical candidate for the Geminga γ-ray source received a rough classification from Halpern et al. (40.143.007), and Bertre (AA **180**, 160) found that two type-II OH/IR sources have a Mira variable and a K-star optical counterpart, respectively.

D) BINARIES AND MULTIPLES

The comprehensive studies include the 1000 MK types of visual multiples by Abt (40.118.017), the companions to Am stars by Abt and Levy (40.120.019), the companions to Be and B stars by Abt and Cardona (38.112.059), the trapezium systems by Abt (41.118.027), the cool components of symbiotic stars by Kenyon and Fernandez-Castro (AJ **93**, 938), and the search for symbiotic stars among PN objects by Stenholm and Acker (AA Suppl **68**, 51). Composite spectra of rather subtle appearance were presented by Corbally (ApJ Suppl **63**, 365), who listed clues to help their discovery. Decompositions of composite spectra were made by Noordanus et al. (39.120.003), Fekel and Scarfe (42.118.040), and Stahl and Leitherer (AA **177**, 105). Torres et al. (140.117.087) propose dM03 and dM4e for the components of Gl 425, a BY Dra type star.

Eclipsing binaries were classified by Yamasaki et al. (37.117.145), Krzeminski (IAU circ 4014), Nakamura *et al.* (38.117.080), Milone *et al.* (39.119.009), Etzel and Olsen (39.119.016), Kartasheva and Snezhko for a Wolf Rayet system (40.117.103), St. Cyr and St. Cyr (41.119.053), Milone (41.119.071), and Davidge (42.119.069). Spectroscopic binary classifications by Harmer are cited in Griffin (39.120.002). The symbiotic star AG Dra was confirmed to have a G7 V component by Iijima *et al.* (AA **178**, 203), but Ipatov *et al.* (37.117.219) cannot reject the single star hypothesis for the symbiotic CH Cyg. Close binary systems were classified by Howell and Bopp (39.117.087), Lu (41.117.016), Mochnacki *et al.* (41.117.240), Martin *et al.* (MNRAS **224**, 1031), and Mukai and Charles (MNRAS **226**, 209).

E) VARIABLE STARS

Celis (37.122.093) has related classifications of late-M Mira variables with photometry. Margon and Anderson (40.122.151) claim M28 V7 is a Mira variable not a cataclysmic variable, and Gosset *et al.* (39.123.002) have classified the red variable CPD -59° 2857.

Lloyd Evans has classified RV Tauri variables (38.122.070, 40.112.067), giving new Preston spectral types for some of the stars. Timoshenko (39.122.096) has used quantitative classification techniques for irregular variables. HR 7671, a possible UU Her star, was studied by Fernie (41.122.021). Tsvetkov (41.122.164) has investigated luminosities of two δ Scuti stars. The spectral type of AH Her's secondary was determined by Bruch (AA **172**, 187), while UY Phe was confirmed as an RR Lyrae variable, not a dwarf nova, by Warner and Barrett (42.122.018) using spectra from Lloyd Evans.

F) CLUSTERS AND ASSOCIATIONS

Open cluster stars have received classifications from Loden for M7 (38.153.037), from Christian for King 8 and Be 19 (38.153.047), from Abt for Praesepe's bright stars (41.153.022), from Corbally and Garrison for Praesepe's faint stars (42.153.017), and from Sowell (ApJ Suppl **64**, 241). Specific investigations have been made into Be stars in 12 open clusters by Slettebak (40.115.034), into cocoon stars in M17 by Chini and Krugel (39.112.071), into OB stars in Puppis by Stetson and FitzGerald (39.114.119), into ZAMS stars in the Cha I association by Wesselius *et al.* (40.121.060), into blue stragglers by Abt (40.153.009), and into CP2 stars by Maitzen *et al.* (42.153.010). The new edition of Mermilliod's catalogue (Bull.Inf.CDS **31**, 175), giving UBV data and spectral types for stars in open clusters, has drawn together current information in this field.

Garrison and Albert (41.114.015) have found that good spectra of UV bright stars in globular clusters cannot be confused with normal Population I OB stars. An UV-bright star in M22 has been classified as SdO by Glaspey (39.114.040).

Stars near dark clouds were classified by Vrba and Rydgren (38.131.112 and 40.131.018) to determine total-to-selective extinction. Low-resolution spectra were similarly used by Whittet *et al.* (MNRAS **224**, 497). Two stars towards the Taurus dark cloud were investigated by Straižys *et al.* (39.113.052), and some towards the Bok globule B361 were classified by Hasegawa and Seki (42.131.312). The ionizing O-stars in IC 2944 were reconsidered by Walborn (AJ **93**, 868).

G) STARS IN GALAXIES

The increasing availability of digital detectors has fostered extra galactic stellar classification. Regrettably, some workers omit observing standard stars, thus making their results quite approximate.

Wolf-Rayet stars in the LMC were classified by Cowley *et al.* (38.111.026), Azzopardi and Breysacher (40.156.007), Moffat *et al.* (ApJ **312**, 612), and Morgan and Good (MNRAS **224**, 435). Other OB stars in the LMC were classified by Conti *et al.* (42.156.014) and Fitzpatrick (ApJ **312**, 596); in the SMC by Walborn and Blades (41.114.102), Garmany and Walborn (PASP **99**, 240), and Garmany *et al.* (AJ **93**, 1070); and in M31 and M33 by Massey *et al.* [no standards] (40.157.096) and Bianchi *et al.* (42.114.117). Melnick (AA **153**, 235) and Walborn (41.114.143) have found a spectacular clustering of very early O type stars surrounding R136 in 30 Doradus. The early O-type stars in 30 Doradus continue to be investigated by Walborn and Blades (ApJ Lett, in press) for their bearing on recent massive star formation. A massive double-lined O type binary in the LMC was found by Niemela and Morrell (42.120.030).

Supergiant candidates in the Clouds were investigated by Baird and Flower (41.111.017, AJ **93**, 851) and Sowell (AJ **91**, 79). Crampton (39.142.022) examined X-ray candidates in the LMC. By using MK catalogues, Morel (Bull Inf CDS **31**, 167) identified some HD "O, P, Pec" objects as stars in the LMC.

Late-type stars in the Clouds were classified by Elias *et al.* (39.115.004), and Reid and Mould (40.156.044). Lloyd Evans (39.156.004) has compared large amplitude red variable stars in the Clouds with those in the Galaxy, and Cepheids were studied by Wallerstein (38.122.121) and Welch [no standards] (ApJ **317**, 672). Aaronson et al (39.157.123) discovered the first S-type star in NGC 6822, while several M- and C-type stars were confirmed there.

A young open cluster in the LMC was studied by Niemela *et al.* (42.153.050), and stars in the LMC association LH39 were classified by Schild (AA **173**, 405).

H) GENERAL

MK dwarf-giant standards cooler than the sun are being re-examined by Morgan, Abt, and Garrison (Std Star Newsletter **9**, 15) to determine the "strong points" of the MK system between F8 and M2. These will be used to define (a) parallel sequences of strong-line and weak-line standards, and (b) two sets of faint secondary standards, at about 10th and 15th magnitudes, to be located near the equator (Std.Star Newsletter **6**, 5).

Kuiper's spectral classifications of over 3200 stars, mainly of large proper motion, are listed by Bidelman (40.114.119), who points out that, while these are not MK classes, they are valuable.

Jaschek has reviewed the classification of all groups in "The Classification of Stars", which book's strength lies in the many references included. Kaler, starting with the spectral morphology of each classification group in his book "Stars and their Spectra", has broadly introduced the different kinds of stars encountered in the Universe.

III. Objective-Prism and Slitless Spectral Classification(D.J. MacConnell)

A) WORK IN THE GALAXY
1) General and galactic latitude-independent.

Vol. 4 of the Michigan Spectral Catalogue appeared (Houk 1987) containing MK classifications for 33,124 HD stars between -26° and -12° Houk and Sowell (41.002.034) reported on some results from the first three volumes and part of the fourth, and Houk (42.114.136) and MacConnell (42.114.135) discussed the problems of recognition and classification of weak-lined stars on plates of moderate dispersion. Bidelman (39.114.027) published the second in a series listing 244 newly recognized peculiar/interesting stars from blue spectrum plates taken for the re-classification of the northern HD stars. Plate-taking continues on this project and Bidelman continues to search them for interesting objects; the plate collection remains in Cleveland.

Rajamohan (38.114.100) demonstrated the usefulness of very low dispersion spectra for studies of star clusters, and Kharadze, *et al.* (40.114.061) compared the classifications in the Abastumani catalogue with those of Morgan and colleagues and found a high degree of correlation. Savage *et al.* (39.002.109,.110,.111) discussed the properties of the UKST prisms and gave illustrations of various types of objects distinguishable on spectrum plates.

Kilkenny and Kelly (39.114.104,42.114,083,.084) present several hundred stars earlier than F0 found at very low dispersion on UKST plates in two high latitude and one low latitude fields. Savage, *et al.*(39.002.109,.110, .111) discussed the properties of the UKST prisms and gave illustrations of various types of objects distinguishable on the plates.

Microdensitometer scans and image-processing software were used by Fuenmayor and Bulka (41.114.136) to recognize M and C stars. Stephenson (39.114.081) presented a list of 105 new cool C stars to V=13.5 and indicated (41.114.122) that three stars in his S-star catalogue are in fact M dwarfs with exceptionally strong CaH bands. Robertson and Jordan (39.114.157 and BAAS **19**, 703) classified nearly 600 late K and M stars in the 24 equatorial Selected Areas in the range 10<V<16. Voroshilov *et al.* published B,V mags and spectral classes for 6000 stars (40.002.010).

2) In the Galactic plane.

All spectral types: Reed and FitzGerald (38.155.035) classified >3000 stars in a low-absorption field in Pup. Fehrenbach, *et al.* (AASuppl **68**, 515) listed types and radial velocities for 258 stars near α Per from plates taken with the OHP Schmidt.

Early-type stars: Balazs and Paparo (41.155.095) gave temperature types for nearly 1000 stars <F7 to mag 12.5 near the open cluster NGC 7686. Gieseking (40.153.002) estimated types for 76 stars in the M7 field on plates from the ESO/GPO astrograph, and Geyer and Nelles (40.153.019) gave types for 153 stars in 7 open clusters. Radoslavova (41.114.127) used plates at 166Å/mm taken with the Abastumani meniscus telescope to find 22 Am and Ap stars in Vul OB4. Wiramihardja, *et al.*(41.114.081, 115.002) surveyed the CMa complex using plates from the Kiso Schmidt; they report 1800 OBA stars and 128 emission stars.

Late-type stars: Blanco (41.155.014,.087, AJ **93**, 321) has surveyed cool giants in Baade's Window and other clear bulge regions near l=0°, b=−2.4° to b=18° using a grism and IV-N plates at the prime focus of the CTIO 4-m. Ichikawa and Sasaki (41.155.080) classified about 1500-late M giants near l=116 to study the warping of the old stellar disk. MacConnell (41.002.033) continues classifying southern stars in the IRAS Point-Source Catalogue and continues a search for K/M supergiant candidates (41.113.015, RevMex AA **14**, 367) on 3400Å/mm I-N plates. Stephenson has begun a similar survey at 1700Å/mm along the northern plane over the interval 0° <l<240°. Azzopardi, *et al.*(39.114.073,.107) report the discovery with a grens of 15 C stars toward the galactic center. Digitized spectra from Kiso Schmidt plates were used by Maehara (40.114.053) to classify 56 known and 3 new C stars in Cas on Yamashita's T,A system. Mechara and Soyano (AnnTokyo **21**, #3,4) reported finding 98 new C stars in Cas and 21 in the anticenter direction on Kiso 4°, I-N plates. Zlakomanova (42.114.147) found 7 new faint C stars toward l=178°, b=0° on near-ir plates. Alksnis *et al.* (Sun and Red Stars No. 25, p42, 1987; Sci Inf USSR Riga No.65) have found 55 new carbon stars in Cygnus.

H-α emission stars: Kun (42.113.006) reports 155 new stars of V>13 near the IC 1396 HII region on Konkoly plates. Tsvetkov and Semkov (39.114.086) list six new stars in the region of the Khavtassi 193 dark cloud. Ogura and Maehara (41.131.223) discuss several on-going surveys in the Bosscha-Kiso collaboration, and Wiramihardja, *et al.*(42.121.069) found 157 new objects in the Ori B and Belt regions. Parsamyan and Khodzhaev (40.114.064) identified 20 certain and 18 suspected stars in the Taurus region.

3) Out of the plane.

Markarian, et al. (41.002.014,42.002.075) list blue galactic stars found in a search for emission-line QSO's, and Pesch and Sanduleak (41.002.013, ApJ Suppl **163**, 809; IBVS No. 2989) in a similar survey gave blue stars of various types, suspected subdwarfs of intermediate type, and faint C and late M stars. Sion, et al.(40.126.052) present two peculiar, subluminous stars near B=18 found on grens plates. Morton, et al. (39.114.015, 41.155.083) studied spectra of 753 objects in a small field in Aqr to B∼19; 3 new white dwarfs were found. Philip (39.114.161) reported finding many faint A stars at the SGP and at b=−60° which may be 10 to 20 kpc off the plane.

Beers, et al. (40.114.042) searched for extremely metal-poor stars to B∼16 using short-band-pass filters around the Ca II doublet. Corbally and Garrison isolated the natural group of F8-G5 dwarfs on IIIa-J plates taken with the thin prisms on the Burrell and Curtis Schmidt telescopes and are publishing lists for 2 sq. deg. areas toward the NGP (192 stars) and the SGP (214 stars) to mag 16. McNeil (39.002.105) presents a list of >2200 G5-M stars to V∼13.5 at the SGP for a study of their spatial distribution. Stephenson (41.114.085, 42.111.007) lists more than 3800 dwarf K and M stars and (41.002.011) 206 new H-α-emission stars over a large area of sky. His survey was done in the green-red region at 750Å/mm (D-lines). Stock's catalogue (40.002.083) of more than 10,000 stars giving types, positions, and indicative radial velocities appeared on microfiche. Bidelman (IBVS **2993**) gives types for 60 red, named variables which are in the IRAS Point Source Catalogue and classified a few thousand IRAS sources in 71 5x5 sq. deg. northern areas; these types as well as those by MacConnell mentioned above are in the database at the IRAS processing and analysis center at Cal Tech.

B) WORK IN OTHER GALAXIES
1. Magellanic clouds

Morgan and Good (40.114.039) presented 6 new W-R stars in the LMC found on IIIaJ plates taken with the UKST, while Kontizas, et al. (41.156.011,.037,.040, 42.156.007,.010, AASuppl **67**, 25; 68,357; 69,213) gave classifications of stars in SMC and LMC clusters over the range 14<B<18.5 from film copies of UKST plates. Westerlund, et al. (42.156.004) used the grism technique on IIIaJ plates at 2200Å/mm to find more than 450 C-type stars in two 3/4° areas of the SMC, and A. Cowley and Hartwick (BAAS **18**, 997) searched for C and CH stars to B∼19 in a 400 sq. deg. area around both Magellanic Clouds. Azzopardi, et al. (AASuppl **69**, 421) discussed the classification of 195 luminous SMC stars using the H-γ equivalent width/luminosity relation found from galactic standards.

2. Other galaxies

Bohannan, et al. (39.157.078) used grism plates from the KPNO 4-m to find several W-Rs in M33, and Lequeux, et al. (AASuppl **67**, 169) searched the local-group galaxies M33 and IC1613 for emission-line objects on CFHT grens plates, finding W-Rs among them. Westerlund, et al.(AA **178**, 41) identified 47 C, 30 M, and one S star in the Fornax dwarf galaxy using the red grism at the ESO 3.6-m. Azzopardi, et al.(41.157.213) found new C stars in several spheroidal galaxies using an objective grating at the CFHT.

IV. Automatic Spectral Classification (M.J. Kurtz)

A) AUTOMATED MK CLASSIFICATION

No automated MK classification has yet been achieved. Kurtz (39.036.063) has suggested an algorithm consisting of the iterative application of multiple weighted linear discriminant functions, which he claims can be used to automate the MK Process and with spectra of the proper dispersion to automate the MK Classification. One feature of the algorithm is that it requires the entire (large) learning set to be considered as equal standards. This he calls the machine equivalent of experience. One advantage of this technique is that one may obtain additional local classification dimensions through normal statistical classification techniques, and one may obtain easy measures for the degree of peculiarity of any spectral feature in a classified object. Rybski (39.034.035) has suggested that any algorithm requiring the use of multiple standards is incapable of classifying spectra according to the MK-78 prescription, as set forth in the MAT atlas.

Zekl (30.036.064) discussed his program for quantitative spectral classification. His is by far the most fully realized program for classification at MK dispersion. Unfortunately work on it has ceased. LaSala, in *Astronomy from Large Data Bases*, reports on his preliminary work in establishing an automated MK classification. He is using the APM machine and its spectra extraction software, plates loaned from the Michigan survey by N. Houk, and the algorithm described by Kurtz.

B) CONTINUING WORK

Einasto, Malyuto, and Karchenko (41.155.064) report automatic classifications for 3000 spectra in their continuing galactic structure program and have developed new algorithms (Bull Abast Ap Obs). Fehrenbach and Burnage (37.036.190) report the routine use of Simien's classification program as a step in their radial velocity measurements. Ratnatanga (41.036.063) is continuing his study of K giant halo stars using low resolution objective prism spectra.

C) NEW LOW-RESOLUTION WORK

Schucker (41.036.232, 42.036.211) has developed a system to classify low-resolution objective-prism spectra. Especially noteworthy is his use of a fuzzy-rule-based, continuum-finding algorithm. Adorf (42.036.206) has developed a cross-correlation classifier for low resolution spectra, primarily as a tool to search for QSOs. Ichikawa (42.036.203) and Timoshenko (41.034.104) have developed new quantitative classification software at low dispersion. Ruban (38.114.133) discusses the limits of low resolution classification.

D) OTHER OBJECTIVE-PRISM WORK

Much work involving objective-prism spectra involves the use of techniques which are very similar to the techniques used in automated spectral classification, indeed a rough classification is often a step in the reductions. Cooke *et al.* (41.036.017) gives a general description of the use of objective prism data by the COSMOS group. Objective prism redshift surveys are discussed by Beard *et al.* (41.160.019) (Edinburgh), Seitter (42.161.352) and Schucker and Horstmann (42.036.123) (Munster), and by Borra *et al.* (preprint Laval University). QSO surveys are discussed by Hewett *et al.* (39.036.068) and by Clowes (42.036.120), who gives a general review.

E) CLASSIFICATION USING THEORETICAL SPECTRA

Comparing model-atmosphere calculations with data can often be very similar to spectral classification. Cairney, Laird and their collaborators (preprint) have obtained metallicities for late type stars by global comparison of echelle spectra with a grid of model atmospheres. This procedure is essentially identical to the unweighted euclidean-distance metric used by Kurtz in his thesis (32.031.649) to classify spectra by comparison with standard stars. McMahan (42.126.043) has actually used portions of Kurtz' software to classify white-dwarf spectra to a two-dimensional grid of his model atmospheres. Malagini and Morossi (37.036.006) also discuss fitting observations to theoretical spectra.

F) MISCELLANEOUS METHODS

Ramella *et al.* (38.036.032) have discussed new algorithms for automated line identification. LaSala and Kurtz (40.036.083) have discussed their Fourier technique for rectification of spectra.

G) MATHEMATICS AND THE THEORY OF CLASSIFICATION

Kurtz (37.021.004) has discussed the epistemology of the classification process, and has given an overview of some of the major mathematical classification techniques, along with an extensive bibliography. Murtagh and Heck in *Multivariate Data Analysis* have considerably expanded on Kurtz' introductory survey paper and the earlier monograph of Bijaoui (27.021.032) with a detailed survey of current methods and a very extensive bibliography of astronomical applications. Kurtz in *Astronomy from Large Databases* discusses the current state of classification in large datasets.

H) IN A CLASS BY ITSELF

Heck *et al.* (41.114.026) have independently established a classification scheme for low-dispersion IUE spectra based on a multivariate statistical analysis of measured features in the spectra themselves. They have shown that the classification scheme developed corresponds very closely with that obtained by using the classical morphological approach, thus confirming both methodologies. This work has the potential to lead to the first real working automated classification for stellar spectra, as opposed to the programs which have actually been predictions of spectral class on the basis of similar measurements.

V. Classification from Extra-Atmospheric Spectra (A. Heck)

All the work in this field during the period covered by this report has been based on spectra collected by the International Ultraviolet Explorer (IUE).

A synthesis of the IUE stellar spectral classification work has been published by Heck in the IUE memorial book (*Exploring the Universe with the IUE Satellite*, eds. Kondo *et al.*, Reidel, 1987, pp. 121-137). A statistical classification of IUE low-dispersion spectra for normal stars has been carried out by Heck *et al.* (40.114.071 + 41.114.026) confirming the classification system introduced in the IUE Low-Dispersion Spectra Reference Atlas (Heck *et al.*, 38.002.015 + 38.114.002). Walborn and Panek (38.114.134 + 39.114.088) and Walborn and Nichols-Bohlin (42.114.081 + PASP **99**, 1987, pp. 40-53) have studied IUE short-wavelength high-resolution spectra of O-type stars (main sequence, ON and OC stars), and OB supergiant stars respectively. They illustrate standard sequences and introduce classification criteria. Rountree *et al.* (40.114.073 + IAU Comm 45 Meeting, New Delhi, Nov 1985) announced a programme of spectral classification of B stars, using high-resolution IUE spectra. Nandy (IAU Comm 45 Meeting, New Delhi, Nov 1985) investigated the measurements of major stellar features in the UV over a range of luminosity and spectral types for a large number of B stars in our Galaxy and the Magellanic Clouds.

Heck *et al.* (38.002.015 + 39.002.101 + 42.002.091) have continued their classification work of peculiar IUE low-dispersion stellar spectra slowed down by the lack of correspondence between peculiarities in the visible and UV ranges. Jaschek *et al.* (40.114.107) have investigated the peculiarities in UV IUE range of λ Bootis and HB stars. Cacciari (40.114.026) studied the UV fluxes of Population II stars from IUE low resolution spectra.

Parsons and Ake (AAS Bull. **19**, 1987, 708) have used IUE low-resolution spectra of binaries to derive estimates of luminosities and spectral classes of the components. Gurzadian has used ORION-2 spectra (3800-2000 Å) for a catalogue (41.002.027,.047).

VI. Classification Using Multicolor Photometry (E.H. Olsen)

A) WIDE-BAND SYSTEMS
1) The UBVRI System

a) Relationships: The calibration of UBV photometry, utilizing theoretical and observed spectra, has been discussed by Buser *et al.* (40.113.037,038). The influence of metallicity and interstellar reddening on the position of stars in the two-color diagram was studied by Cameron (39.115.010). Saxner & Hammarback (40.113.047) have published an empirical T_e calibration of B-V for F- and early G-type dwarfs. A period-color relation for dwarf novae has been found by Echevarria & Jones (37.117.024). Gliese & Jahreiss discuss the use of nearby stars in calibrating UBV photometry (40.115.012).

b) Field Stars and General Surveys: UBV catalogues of faint field stars were published by Oja (38.113.011; 39.113.007; 40.113.003; 42.113.014; AA Suppl. **68**, 211).

The "Hyades supercluster" stars were studied by Eggen (39.153.008,009); common proper motion pairs by Caldwell *et al.* (38.113.001); proper motion stars by Carney & Latham (AJ **93**, 116) and by Eggen (AJ **93**, 379); FK4 stars by Moreno & Carrasco (42.113.013); stars in HII regions by Lahulla (40.113.012); population II stars by Norris *et al.* (40.155.017) and by Sandage & Kowal (41.113.053); and nearby stars by Torra *et al.* (40.113.018 and AA Suppl. **67**, 157).

The UBVRI system has been used to investigate helium stars and hydrogen-deficient stars by Drilling *et al.* (37.113.046), and early-type stars for galactic structure by Forbes (37.155.051).

A list of possible solar analogs was given by Neckel (42.113.020). Other late-type dwarfs have been studied by Hartwick *et al.* (38.114.121, 42.113.031), by Eggen (42.113.032), by Stauffer & Hartmann (42.116.002), and by Robertson & Hamilton (AJ **93**, 959). Photometry of late-type giants has been carried out by Celis (41.113.016) and of supergiants by Arellano (40.122.035). Fekel *et al.* (41.116.009) studied chromospherically active stars.

A search for white dwarfs in the Praesepe cluster has been continued by Anthony-Twarog (37.153.011). Shaw & Kaler (40.134.010) studied the nuclei of planetary nebulae Kilkenny *et al.* (40.113.065) looked at early-type shell and pre-main-sequence stars. A UBV catalogue of cataclysmic variables was compiled by Bruch (37.117.183).

Cepheids in the Magellanic Clouds have been studied by Wayman *et al.* (37.122.084), and Freedman *et al.* (40.122.111). In our galaxy Cepheids have been studied by Barnes & Moffett (39.122.045,046, 40.122.051), by Gieren (40.122.030, 42.122.014,045), by Madore (40.122.098), by Petersen & Diethelm, (41.122.016) and by Coulson *et al.* (41.122.103).

T Tauri stars were the subject of work by Vrba *et al.* (39.121.010, 040). RR Lyrae stars have been studied by Cacciari & Clementini (42.122.173), and RV Tauri stars by Goldsmith *et al.* (MNRAS **227**, 143).

c) Open clusters and associations: Many investigations of stars in open clusters have been made (del Rio, (37.153.020); Janes & Smith, (37.153.022; 38.153.001,049,050); Fenkart & Schroder, (39.153.004); Clariá, (39.153.005,021); Richtler, (39.153.007); Stauffer et al., (39.153.015); Turner & Pedreros, (40.153.006); Upgren et al., (40.153.015); Tokhtas'ev, (41.002.039); Forbes, (41.153.021); Sagar et al., (40.153.026; 41.153.025; 42.153.008); Clariá & Lapasset, (41.153.027); Reimann & Pfau, AN **308**, 111). The detection of a possible new cluster was made by Turner (37.153.071), while a search for clusters around five Cepheids was unsuccessful (van den Bergh et al., 39.152.004).

Associations have been studied by Gasparyan (39.113.030) by Nurmanova (39.152.002) by Heske & Wendker (40.152.008) and by Whittet et al. (MNRAS **224**, 497).

d) Globular clusters: Broad-band photometry in globular clusters has been carried out by A. Wehlau & Hogg, (40.154.027).

e) Magellanic Clouds and other galaxies: Stars in the Magellanic Clouds were studied by Ardeberg et al. (40.156.004), by Bernazzam et al. (40.156.010) and by Grieve & Madore (42.156.027,028). Studies in the direction of the Clouds were carried out by Robin et al. (AA Suppl. **68**, 63).

2) The RIJHKLMNQ system.

The calibrations of the system were discussed and improved by Rieke et al. (38.113.057,058; 39.113.027,028), Tapia et al. (Rev. Mexicana **13**, 115), and Glass (39.113.022). A classification system for late-type giants has been developed by Tignanelli & Feinstein (40.113.015). The temperature calibrations for cool stars were reviewed by Bessell et al. (40.113.041). Color-metallicity relations for population II red giants were derived by Martinez Roger (AA **171**, 77). A discussion of intrinsic colors of hot stars was given by Moreno & Chavarria (41.131.206).

The system was used to detect and investigate winds from and shells around early-type stars (Abbott et al., 37.112.077; v.d. Hucht et al., 37.112.108; Stahl et al., 37.122.012; Chini & Krugel, 39.112.071), and to study the properties of Ap stars (Groote & Kaufmann, 37.113.005; Kroll et al., AA Suppl. **67**, 195), Be stars (Goraya, 39.113.051), helium stars and hydrogen-deficient stars (Drilling et al., 37.113.046), and other early-type stars (Kilkenny et al., 40.113.065; The et al., 42.002.040).

The system was used to study late-type giants in the nuclear bulge of our galaxy (Frogel et al., 37.155.105; 38.113.027; Jones et al., 37.155.105; 38.122.015; Whitelock et al., 42.133.006), and the population structure of low-mass dwarfs (Reid & Gilmore, 37.115.004; Hartwick et al., 38.114.121). Other studies included carbon stars (Gao et al., 39.133.053), late-type giants (Noguchi & Akiba, 42.113.043), and late-type dwarfs (Stauffer & Hartmann, 42.116.002).

The system was used to detect and investigate winds from and shells around RV Tauri stars (Lloyd Evans, 40.112.067; Goldsmith et al., MNRAS **227**, 143), and to study the properties of Cepheids (Welch et al., 37.122.131; 39.122.133; ApJ **317**, 672; Fernley et al., MNRAS **225**, 451).

The system and similar systems were used extensively to identify, classify and study infrared sources (Ghosh, 37.133.005; Persi et al., 37.133.007; Danks et al., 37.155.023; Gao, 38.112.094; 40.112.081,105; Herman, 38.122.081; Gehrz et al., 39.112.065; Band et al., 39.112.092; Whitelock, 39.133.002; Vrba et al., 40.121.010; Hrivnak et al., 40.133.002; Elias et al., 41.112.059; Kwok et al., 41.112.145; Churchwell & Koornneef, 41.121.005; Th et al., 41.121.011; Melnick et al., 41.131.246; Jones & Hyland, 42.112.101; Kawara, 42.131.062; Le Bertre & Epchtein, AA **171**, 116; Braz & Epchtein, AA **176**, 245).

Open clusters and associations were studied by Tapia *et al.* (37.153.065; 38.153.008), Whittet *et al.* (MNRAS **224**, 497) and Wilking *et al.* (42.153.002), while Caputo *et al.* (40.154.036) studied the CNO abundances in both globular and open clusters.

A review of the properties of red giants in the Magellanic Clouds was given by Aaronson (37.156.039). Other stars in the Magellanic Clouds were also studied: (Frogel, 38.156.017; Welch & Madore, 37.156.044; Feast & Whitelock, 38.156.001; Wood *et al.*, 39.156.014; Laney & Stobie, 42.156.009). A comparison between cluster supergiants in the galaxy and in the Magellanic Clouds was made by McGregor & Hyland (37.156.081).

3) The RGU System.

The calibration of RGU colors in terms of MK classes has been investigated by Labhardt & Buser (40.113.038). The investigations of galactic structure by classification of faint stars in galactic fields have continued (Fenkart *et al.*, 38.113.004,005,012; 39.113.004; 40.155.024; 41.155.011; AA Suppl. **67**, 245; 68,397; 69,33,281; Alfaro & Garcia-Pelayo, 38.113.019; Topaktas, 38.113.051; Becker *et al.*, 39.002.010; 39.155.073; Spaenhauer *et al.*, 39.113.012).

4) The Washington system.

By adding the DDO 51 filter to the Washington filter system, Geisler (38.113.026) has improved its luminosity classification. A comparison between the classification properties of the Washington system and the medium-band DDO system was made by Smith (41.113.030). The abundance indices have been recalibrated (Canterna *et al.*, 42.153.016; Geisler, 42.113.027) and applied to LMC cluster giants (Geisler, AJ **93**, 1081).

The system has been used to investigate open cluster giants (Geisler & Smith, 38.153.045; Clariá & Lapasset, 39.153.021; 41.153.017,027; Canterna et al., 42.153.016) and six metal-rich globular clusters (Geisler, 42.154.024).

5) Far-infrared systems.

For stars of all types, the relations between Johnson's BVRIN colors and IRAS colors were studied by Waters *et al.* (AA **172**, 225), and the relations between MK classifications and IRAS colors by Cohen *et al.* (AJ **93**, 1199).

In the range 53-200 microns, shells around carbon stars have been studied by Goebel & Moseley (39.112.068), who propose solid MgS to be present. Circumstellar disks around exciting stars of Herbig-Haro objects were studied at six wavelengths between 40 and 160 microns (Cohen *et al.*, 40.121.012). Mass-loss rates have been estimated for evolved stars by observations of thermal emission at 400 microns (Sopka *et al.*, 40.112.004; Werner, 40.112.038).

IRAS observations have been utilized to study a wide variety of stellar objects and their environments: Supergiants later than type F0 (Stickland, 40.112.069; 41.112.124), stars with disks of possibly protoplanetary material (Aumann, 40.112.107; Gillett, 41.112.134; Sadakane & Nishida, 42.118.004), symbiotic objects (Whitelock, 41.117.329; Kenyon *et al.*, 42.117.257) and binary systems containing compact objects (Schaefer, 41.117.297).

IRAS has also been used to study hydrogen-deficient stars (Walker, 40.112.072; 42.114.171); Wolf-Rayet stars (v.d. Hucht *et al.*, 40.112.099, 40.112.137, 41.113.045); O, B and A stars (Waters *et al.*, 41.113.064; 41.116.011); Be stars (Waters *et al.*, 41.112.136; AA **176**, 93); and planetary nebulae (Pottasch, 41.134.065; 42.134.054; Iyengar, AA Suppl. **68**, 103),

IRAS observations of variable and other late-type stars include RV Tauri stars (Jura, 42.122.163), T Tauri stars (Beichman *et al.*, 42.133.004), other cool stars (Rowan-Robinson *et al.*, 41.112.140; Odenwald, 41.113.063; 42.112.042; Perrin & Karoji, AA **172**, 235; Herman *et al.*, 42.112.049; Glass, 42.133.005), and carbon stars (Willems, 41.113.065),

B) MEDIUM-BAND SYSTEMS
1) The Strömgren uvbyβ system.

Strömgren has reviewed the properties of population-II stars of types F and G. He summarizes the most recent calibrations and surveys in the context of galactic structure and evolution (39.155.020; 40.113.026).

Cousins has published secondary standards in the E regions (41.113.061; 42.113.019). Bell & Oke (42.114.039) discuss scans and colors of four F subdwarfs defined as spectrophotometric standards. General problems of observation and reduction were discussed by Manfroid (40.113.032,035) and Kilkenny & Menzies (42.113.015). The intrinsic colors of early-type supergiants were determined by Kilkenny & Whittet (40.113.010).

Several authors have re-discussed the uvbyβ calibrations (Moon & Dworetsky, 38.113.024; 40.114.098; Balona & Shobbrook, 38.115.011,016; Shulov, 40.113.067; 42.115.004; Alexander, 41.113.036; McNamara & Powell, 41.114.008; Olsen, Observatory 105, 99; Lester, MNRAS 227, 135).

An improved theoretical calibration based on unpublished models by Kurucz and the secondary spectrophotometric standards was given by Lester et al. (42.113.002). Saxner & Hammerback have published an empirical T_e calibration of b-y and gb for F- and early G-type dwarfs (40.113.047), which has been extended to population II by Magain (AA 181, 323). Luminosity calibrations have been published for white dwarfs (Greenstein, 37.126.042) and for F-type supergiants (Antonello, 40.115.020). Tables and diagrams giving luminosities, radii, and MK types over a large area of the HR diagram were published by Moon (40.115.031; 41.114.067).

Gray and Garrison (Ap.J.Suppl. 1988) have published detailed comparisons between the MK system and the uvbyβ system and have discussed the effects of rotation on the two systems. The photometric effects of rotation in A-type stars were computed from model atmospheres by Collins & Smith (39.116.016), and also studied by Schmidt & Forbes (37.153.023). Empirical calibrations for late-type stars, also in terms of MK classes, were presented by Olsen (38.113.014), Nelles et al. (40.113.033) and Ardeberg & Lindgren (40.113.036). Abt (42.114.079) finds that 97% of stars photometrically predicted to be weak-lined F- and G-type dwarfs, are indeed so.

For horizontal-branch stars masses and other properties have been determined (Philip et al., 37.115.023; 40.115.027). The zero point of the PLC relation for classical Cepheids was discussed by Schmidt (38.122.199) and Balona & Shobbrook (39.122.070). The distance to the LMC was determined by observations of non-supergiant B stars (Shobbrook & Visvanathan, MNRAS 225, 947).

Studies and classifications have been published for the galactic poles (Hilditch et al., 40.155.022; Philip, 41.155.021), for four fields at b = −60° degrees (Andersen & Jensen, 39.113.005), for SA 132 (Knude, 41.113.003), and for stars on the Hipparcos observing program (Manfroid et al., AA Suppl. 69, 505),

The system was also used to study early-type stars (Wade & Smith, 39.113.001; Kilkenny et al., 40.113.065; Eggen, 42.153.035; vander Linden & Sterken, AA Suppl. 69, 157), β Cephei stars (Shobbrook, 39.122.106), B-type supergiants in the Magellanic Clouds (Shobbrook, 41.156.008), white dwarfs (Howell, 41.113.039; Fontaine et al., 39.126.089), hot subdwarfs (Wade, 38.117.083), and faint or high-latitude blue stars (Kilkenny et al., 38.113.048; 41.114.006; Tobin, 39.113.037; 40.113.034),

RR Lyrae stars (Alania, Ap Sp Sci. 132, 313), horizontal-branch stars (Philip, 42.113.047,048), A-type stars (Eggen, 38.114.020), 38.114.020), Ap stars (Schneider, 41.113.037), Am stars (Dworetsky & Moon, 41.114.129), and Cepheids (Eggen,

40.122.004,005,006; 40.122.004,005,006; Kim, Ap Sp Sci **133**, 1) have been studied. Late-type dwarfs have been observed by Eggen, (42.113.032),

Visual binaries have been classified by Duncan (37.113.039), while Lindroos has continued his investigation of young binary systems with O-or B-type primaries and pre-main-sequence secondaries (39.118.013; 41.118.013).

Blue stragglers in open clusters have been studied by Twarog & Tyson, 40.153.007). Other observations in open clusters include: (Lynga and Wramdemark, 37.153.008; Schmidt & Forbes, 37.153.023; Jakobsen, 37.153.064; Schmidt, 38.122.199; 38.153.-001,049; Shobbrook, 38.153.034, 39.153.002; 41.153.038; MNRAS **225**,999; Delgado *et al.*, 38.153.036; 39.153.006; Richtler, 39.153.007; Schneider, 40.153.001; AA Suppl. **67**, 545; Eggen, 39.153.009,039; 40.111.024; Reimann & Pfau, AN **308**, 111) and in associations include: (de Zeeuw & Brand, 40.152.013; Perry & Landolt, 42.152.005).

A very detailed study of M67, including a rediscussion of the calibrations for F-and early G-type stars, was made by Nissen *et al.* (AJ **93**, 634). CCD photometry in M67 and the globular cluster NGC 6397 was published by Anthony-Twarog (AJ **93**, 647,1454).

2) The Geneva System.

A discussion of the intrinsic colors of A- and F-type supergiants were given by Meynet & Hauck (40.113.-013). Metallicism among A and F giants was studied by Hauck (41.113.007), who also gives a list of λ-Bootis-type candidates (41.114.005) and discusses Be and shell stars (AA **177**, 193) and population II stars (42.113.046). The discovery of a new class of variable stars with mid-B-type classifications was announced by Waelkens & Rufener (40.122.074). They identified them with the so-called 53 Persei variables, which show line-profile variations. Accurate physical parameters were derived for pulsating stars, by combining the photometry with precise radial velocity curves (Meylan *et al.*, 41.122.010,014,027,028; Grenon & Waelkens, 41.122.012). The main-sequence gap around F0 and its possible relation to onset of convection has been discussed by Jasniewicz (38.113.035). Cramer continued his study of B-type stars (38.113.036). Bp stars were studied by Lanz (39.114.031) and Ap stars by North *et al.* (38.116.027; 40.116.002; 42.116.026; AA Suppl. **69**, 371) and Waelkens (40.113.001). The relation between mean surface magnetic field and D(V1 - G) for Ap stars was questioned by Oetken (40.116.061).

3) The Vilnius System.

Belyaeva has published a theoretical calibration of this system (42.113.012). An automated two-dimensional classification method based on photographic Vilnius photometry was developed by Smriglio *et al.* (42.113.018). B-type stars were studied by Straižys *et al.* (39.113.052). Bright stars in open clusters containing red giants were measured by Dzervitis & Paupers (41.113.019,020,023; 42.113.051) and by Kazlauskas (Vilnius No.75,p18). Stars in the Cygnus standard region were reobserved by Zdanavicius & Cerniene (41.113.024) and in SA 92,108, and 112 by Cernie (Vilnius No 75, p31). Large-proper-motion stars at the SGP and NGP were observed by Bartasiute (41.113.025) (Vilnius No 74, p15). The determination of metallicity and O/C ratio for late-type giants was discussed by Straižys & Sleivyte (41.113.054), while solar analogs were studied by Glushneva *et al.* (42.113.005). Ap stars were investigated by Nikolov & Iliev (42.113.017). Metal weak giants were studied by Straižys *et al.* (Vilnius No 75, p3); CH, barium stars by Sleivyte (41.114.063); carbon stars by Sleivyte (Vilnius No 74, p24, No 75, p36 and No 77); Ap stars by Zitkevicius (Vilnius No 76, p8); and subdwarfs by Straižys *et al.* (Vilnius No 77).

4) The DDO system.

The red giants in several open clusters have been investigated (Janes & Smith, 37.153.022; 38.153.050; Clariá et al., 39.153.005,021; 41.153.017,027) as also yellow giants and supergiants (Schmidt, 38.153.001,049). A large sample of barium stars was studied by Lu & Upgren (39.114.143). Population II stars were investigated by Norris et al. (40.155.017), who also studied red giants of the old disk (AJ **93**, 616). Observations of red giants in two globular clusters were discussed by Smith & Hesser (42.154.023). Data on very strong-lined K giants were given by Johnson et al. (ApJ Suppl. **63**, 983).

5) The Walraven system.

Pel discussed the fundamental parameters of classical Cepheids (39.122.036; 41.122.107). The system was used to investigate associations (v. Genderen, 38.152.007; de Geus et al., 41.152.002), open clusters (Th et al., 40.153.028; Steemers & v. Genderen, 41.113.001; van Leeuwen et al., 42.153.009), globular clusters (Nelles & Seggewiss, 38.154.024), stars in the Magellanic Clouds (v. Genderen et al., 40.156.006; 41.156.006), fields around planetary nebulae (Gathier, 39.113.036; 41.134.006), supergiants (Steemers & v. Genderen, 41.113.001), F- and G-type stars at the SGP (Trefzger et al., 37.155.026; 39.113.060; 39.155.019), Cepheids (Diethelm, 41.122.062), solar-type stars (Greve & v. Genderen, 41.113.017), G- and K-type dwarfs in the Pleiades (v. Leeuwen et al., AA suppl. **67**, 483) and OB-type stars (v. Genderen et al., 40.113.002,046; 40.122.099; 41.113.009; v. Paradijs et al., 41.113.002; The et al., 42.002.040).

6) The Arizona 13-Color System.

Mitchell & Schuster have investigated the solar colors on this system. They present photometry of 63 solar-like stars, and an improved absolute calibration of relative colors (40.113.016). A calibration of the system in terms of T_e and M_v, and applicable to B-type stars, was derived by Conconi & Mantegazza (40.113.043). A discussion of intrinsic colors of hot stars was given by Moreno & Chavarria (41.131.206). Stars associated with HII regions were studied by Chavarria (AA **171**, 216).

7) The Thuan-Gunn uvgr system.

This four-color system has effective wavelengths similar to the wide-band UBVR system, but half-widths between 400Å and 900Å, and thus essentially non-overlapping bands. Thuan & Gunn (18.113.025) published the list of standard stars that define the system, but so far it has had its main application in CCD photometry of galaxies. However, Kent (39.113.013) has published a revised set of standard stars, and he discusses uvgr photometry of more than 400 field and cluster stars covering a wide range of stellar properties. A mean main-sequence and reddening curves are derived. Bell & Oke (42.114.039) discuss four F subdwarfs defined as spectrophotometric standards.

C) NARROW-BAND SYSTEMS

Narrow-band H-β photometry was used together with medium-band uvby photometry (see section b1 of this report). H-α photometry for 150 dM and dK stars was reported by Layden & Herbst (42.113.054). Mendoza (41.153.043) has used the $\alpha\,\lambda$ system to measure intensities of H-α and OI in Hyades and Coma cluster stars, as well as Ap stars (42.113.009).

The Lockwood five-color narrow-band system and its application to M-type giants was discussed by Mennessier (40.113.024; 40.114.090). Maitzen et al. searched for Ap stars in open clusters by measuring the l5200 depression (38.153.015; 40.153.016; 41.153.047;

42.153.010,013; AA **178**, 313). This depression was also measured in field Ap stars (Schneider, 41.113.037).

CaH and TiO indices have been determined for late-type dwarfs (Hartwick et al., 38.114.121). A CO index (2.3 μ) was measured for 200 bright M giants in an unsuccessful attempt to discover extremely low carbon abundances (McWilliam & Lambert, 38.114.127). The same index was utilized in a comparison between late-type supergiants belonging to clusters in the galaxy and in the Magellanic Clouds (McGregor & Hyland, 37.156.081). Frogel, Cohen et al. continued to measure the infrared bands of CO and H2O in globular cluster giants (39.154.023; 41.154.065), M giants in the galactic nuclear bulge (41.155.049,124) and red giants in Local Group galaxies (39.115.004; 39.157.073).

Model atmospheres have been used to calibrate the Wing system (Steiman-Cameron & Johnson, 41.064.022). Frisk & Bell (40.114.088) determined Te for G and K subgiants by models and photometry in three filters at ll5900,7800 and 10600Å. Narrow-band 1-5 micron magnitudes have been measured for dwarf stars (B8-A3), relative to Vega, by Leggett et al. (42.113.029).

D) GENERAL

Buser (39.113.059) and Straižys (40.113.030) have reviewed photometric methods for determining stellar metallicity and stressed their importance in the context of galactic structure and evolution. The photometric properties of peculiar red giants were reviewed by Wing (39.113.062). Rufener has examined the experimental conditions which must be satisfied by photometric systems, if the observational parameters are to be correlated unequivocally with physical quantities of a star (40.113.028).

The significance of the closed loops described by classical Cepheids in color-color diagrams, was emphasized by Onnembo et al. (40.122.077). A new optimal four-filter medium-band system for F-and G-type dwarfs was suggested by Park & Lee. A combination of DDO and Strömgren filters is close to the optimal system (41.113.059). Photometric systems for classification of stars of population II were reviewed by Ardeberg & Lindgren (42.113.049).

· VII. Astrophysical Calibration of Classifications (V. Straižys)

The UBV system has been calibrated in terms of temperatures, gravities, and magnitudes by Buser and Kurucz (40.113.037) and Straižys (Bull Vilnius Obs No 76,**17**,1987), in terms of temperatures and metallicities by Cameron (39.115.010) and in temperatures by Saxner and Hammarback (40.113.047). The M_V, B-V diagram has been calibrated in age by van den Berg and Bell (40.65.053, 40.154.014), the zero-age main sequence has been revised by Shulov (37.113.028, 40.113.067) and Kopylov (39.113.025). The solar UBV values have been discussed by Hayes (40.115.011), Makarova and Kharitonov (41.071.022), and Neckel (42.113.020, 42.114.011, 42.115.010).

A new calibration of infrared magnitudes in absolute fluxes has been given by Campins et al. (39.113.027), Rieke et al. (39.113.028), and Beichman et al. (IRAS Cats&Atlases, Explanatory Suppl, JPL, 1985). Infrared color indices have been used for T_e determination by McGregor and Hyland (37.156.081), Aaronson and Mould (39.154.012, 39.157.089), Frogel (39.154.023), Steiman-Cameron and Johnson (41.064.022), and Wing et al. (40.114.092).

The Washington system has been calibrated in terms of luminosities and abundances for late-type giants by Geisler (38.113.026, 38.154.082, 42.113.027, 42.154.024), Geisler and Smith (38.153.045), and Canterna et al. (42.153.016).

The uvbyβ system has been calibrated in effective temperatures, luminosities, gravities, and metallicities for B-F stars by Moon and Dworetsky (40.114.097), for B-type stars by Balona and Shobbrook (38.115.011, 38.115.016), Lester et al. (42.113.002), and

Shulov (42.115.004), for A-type stars by Schmidt and Forbes (37.153.023), Anthony-Twarog (37.153.035), and Olsen (Obs105,**99**, 1985), for F-type stars by Laird (39.114.032), Saxner and Hammarback (40.113.047), Shulov (40.113.067), Antonello (40.115.020), McNamara and Powell (41.114.008), Lester *et al.* (42.113.002), for G and K type stars by Olsen (38.113.014), Nelles *et al.* (40.113.033), Laird (39.114.032), Ardeberg and Lindgren (40.113.035), Moon (40.115.031), Eggen (42.113.032), and for metallic-line stars by Dworetsky and Moon (41.114.129).

In the Geneva system the $B_2 - V_1$ color index has been calibrated in temperatures by Hauck (40.113.029). Instrinsic color indices of supergiants have been determined by Meynet and Hauck (40.113.013). Effective temperatures and bolometric corrections of Bp stars have been studied by Lanz (38.114.050, 39.114.031).

The DDO system has been calibrated in element abundances by Norris et al.(40.115.017), Smith and Hesser (42.154.023), and Rego *et al.*(41.113.044).

The Walraven system has been calibrated in temperatures, gravities, and metallicities by Nelles *et al.* (38.113.021), Greve and van Genderen (41.113.017), and Gathier *et al.* (Astron Astrophys **157**,171,1986).

The MK spectral types have been calibrated in temperatures and/or absolute magnitudes by Grenier *et al.* (39.115.009), Keenan and Pitts (39.115.022), Keenan (40.114.066), Gliese and Jahreiss (40.115.012), Couteau (40.115.015), Rakos (40.115.016), Mikami (41.115.003), and de Jager and Neiuwenhuijzen (Astron Astrophys **177**, 217,1987). The Hγ absorptions have been related to absolute magnitudes by Millward and Walker (39.115.003, 39.115.006, 40.115.013) and Hill *et al.* (42.114.160). The luminosity dependence of the Mg II K-line emission widths has been studied by Parthasarathy (40.115.014).

VIII. Catalogues and Atlases (D. Egret)

A) SPECTROSCOPIC CATALOGUES

Lists of new machine-readable catalogues and atlases are regularly published in the "CDS Information Bulletin", Strasbourg, and by the NSSDC, NASA, Greenbelt. We give here a short list of catalogues made available on tape during the period, together with their reference number in the lists of the data centres:

Henry Draper Catalogue and Extension I (HD, HDE)(3099): revised version with a number of errors corrected. McCook and Sion (3100): A Catalogue of Spectroscopically Identified White Dwarfs, 2nd edition (37.002.084). Page (3110): Catalogue, Spectrum and Magnitude Data Bank of Be, Bp and Bpe Stars (Mt Tamborine Obs., 1984). Rousseau *et al.* (3111): Studies of the Large Magellanic Cloud Stellar Content III. Spectral Types and V Magnitudes of 1822 Members (21.159.002). Bartaya (3112): Catalogue of Spectral and Luminosity Classes of 10396 Stars in Kapteyn Areas NN 2-43 (BAAO 51, 1979). Sanduleak (3113): A Deep Objective-Prism Survey for Large Magellanic Cloud Members (1969; ADC Version 1987).

The 6th edition of "MK Spectral Classifications- General Catalogue" was published by Buscombe in 1984: the 7th edition is in preparation. Neither contain references and are thus less useful.

Stock published a Catalogue of Radial Velocity and Position from Objective-Prism Plates (1984, RevMex AA **9**, 77) (available on tape: 3101). Osborn and MacConnell have submitted to publication a list of metal-poor stars from the above catalogue.

Work with the Kiso 105-cm Schmidt telescope is reported by Maeara and collaborators: this includes catalogues of Cool Carbon Stars (40.114.053, and Annals Tokyo Astron Obs 2nd Ser **21**, 293 and 423, 1987), 598 Ultraviolet-Excess Objects (38.113.055), and a Catalogue of M-Type Stars (37.002.075).

The following catalogues are available on tape from the Soviet Data Center of the Astronomical Council, Academy of Sciences: Bartaya, Karadze (Catalogue of Spectral and Luminosity Classes for 5900 stars in Kapteyn areas 44-67). Chargeishvili (Catalogue of Spectral and Luminosity Classes for 5500 stars in the direction of the anticenter).

B) PHOTOMETRIC CATALOGUES

A large number of photometric catalogues has been published in machine-readable form, and an exhaustive list appears in the report by Warren to Commission 25. We will mention especially the compilations made in Lausanne by Mermilliod (UBV data), North (Vilnius photometry), Hauck and Mermilliod (uvby) and Lanz (UBVRI).

Hauck and collaborators have discussed the photometry of nearby stars (37.002.027), Be stars (38.113.053) and bright stars (CDS Inf Bull **31**, 131). Hauck has published a Third catalogue of Am stars with known spectral types (41.002.012), and Hauck *et al.* presented a review of photometric data files (42.002.029).

A new edition of the photometric catalogue in the Geneva system is announced by Rufener: the catalogue now contains 190,000 measurements for 28,200 stars, and is certainly the largest set of homogeneous photometric observations presently available. A new absolute calibration of the passbands is available.

A catalogue giving uvbyβ photometry for 650 stars of spectral types B0-A0 in magnitude range V=6.5 to 10 is in final preparation at Stockholm Observatory by Loden.

C) ATLASES

The following atlases have been published or are in preparation:
Walborn *et al.*: IUE Atlas of O-type Spectra from 1200 to 1900Å. (NASA, 1985; on tape: 3115). Oliversen *et al.*: An Atlas of IUE Spectra of Planetary Nebulae and Related Objects (announced in AAS Bull **19**, 1987). Corbally: An atlas of 12 rather subtle composite spectra (ApJ Suppl **63**, 365). The second volume of the IUE Reference Atlas devoted to peculiar stars is in preparation by Heck *et al.*. Ferland *et al.* have published the Spectrophotometry of Nova Cygni 1975 (41.124.101): the tape version is available under reference number 3109.

D) MISCELLANEOUS

A Critical Catalogue of Stellar Abundance Analyses including more than 400 analyses for 700 stars was presented by Koeppen at the Paris Symposium in June 1987.

A list of IUE meeting bibliographies for peculiar stars has been established by Heck (38.002.055).

COMMISSION 46 : TEACHING OF ASTRONOMY
(ENSEIGNEMENT DE L'ASTRONOMIE)

PRESIDENT : C.Iwaniszewska
VICE-PRESIDENT : A.Sandqvist
ORGANIZING COMMITTEE:
    L.Gouguenheim - ICSU-CTS representative
    L.Houziaux - Past-president of Commission, publisher
                 of Newsletter
    J.Kleczek - Secretary of ISYA
    J.R.Percy - Editor of Newsletter
    M.Gerbaldi,E.V.Kononovich,R.R.Robbins - Astr.Educat.Materials
    D.G.Wentzel,S.Ferraz-Mello,B.Hidajat,Y.K.Miao - Visiting
                 Lecturers Programme

1.Introduction

    The President is happy to report that a considerable increase
of interest in astronomy teaching problems seems to have taken place in many countries during the period 1985-1987. Some countries
organized local teachers courses and meetings, other - national
conferences devoted to establishing modern school programmes and
training teachers methods including astronomical notions. An international workshop for physics teachers showing possibilities of
practical introducing astronomy when teaching physics / GIREP 1986 /
brought into contact astronomers and physicists. And, last but not
least, teaching problems have been also introduced during the IAU
Colloquium No.98, "The Contribution of Amateur Astronomers to Astronomy", in 1987, while the first IAU meeting dedicated to teaching
problems - the IAU Colloquium No.105 "The Teaching of Astronomy" -
is planned for 1988 in Williamstown /USA/. Let us hope that in the
near future all astronomers will be well aware of their own responsibility in helping astronomical education all over the world.

    The Commission 46 various forms of activity have been run by
the Organizing Committee members; here are some details.

2.Membership

    Much attention has been devoted to the preparation of new mem-
bership rules. These rules are not the same as in other IAU Commissions since Commission 46 is a subcommittee of the IAU Executive
Committee. A set of rules accepted by the Executive Committee will
be presented during the Commission meetings at the IAU General
Assembly in Baltimore in 1988.

3. International Schools for Young Astronomers - ISYA

    The ISYA project has been continued with the financial help of
IAU, UNESCO, ICSU and local authorities. The XIVth ISYA took place
in August 1986 in Beijing /China/; 52 students from China, Japan,
Hong Kong, Thailand, Iran and UK attended. The XVth School has been
held in Espinho /Portugal/ in September 1986; 30 participants came
from Angola, Czechoslovakia, Greece, Guinea, Spain, Portugal and
Turkey. The main advantages of these Schools lie in the possibility
of discussing by the participants their own scientific problems

with the lecturers - eminent astronomers from their own country and from abroad, and in the high level of seminars where students get acquainted with scientific works of many abroad institutions. The 1988 ISYA will be held in Cuba.

4. Visiting Lecturers Programme - VLP

The VLP for University San Marco in Lima /Peru/ is nearly finished, the last lectures will be held in the fall of 1987. The results of the whole three-year programme may be summarized as follows: three master degrees in astronomical subjects are being prepared, astrophysics has been included into the curriculum of the Faculty of Physics, there is a possibility of creating a department of astrophysics in future. It has not been possible to arrange for a VLP in Nsukka /Nigeria/. A VLP is planned for Asuncion /Paraguay/ in 1988.

5. Newsletter

The Commission Newsletter is issued twice a year, approximatively in January and June; the numbers 18 to 22, a total of 115 pages, have been published and distributed until end of 1987. Beginning with No.19, 1986, J.R.Percy of Toronto / Canada/ has acted as editor, while the printing and distributing has been arranged by L.Houziaux in Liège /Belgium/. The articles printed contained either general informations on astronomy education in a given country or details on proposed teaching aids, school programmes, meetings and conferences, the contributors have been nearly from every part of the world.

6. National Reports

As it has been customary, National Reports prepared for 1983 - 1985 by the National Representatives to Commission 46 have been printed in Newsletter No.18, p.1 and 2. They are prepared every three years and they present a worldwide overview of what the cur - rent situation is in astronomy teaching at all educational levels, from primary schools to universities and general public programmes. A report for the period 1985-1987 based on National Reports will be published in the Newsletter in 1988.

7. Astronomy Education Material

Every three years an updated list of most important teaching publications /books, lecture notes, films, etc/ is published in three language groups. The 1985 publication comprised materials in Slavic languages prepared by E.V.Kononovich, in non-English languages /with an extensive list of Chinese publications/ - by M.Gerbaldi, in English - by R.R.Robbins. Similar lists will be prepared by the same authors for the period 1985-1987.

8. Travelling Telescope Project

In 1985 two Commission 46 members, D.McNally and R.M.West, put forward the idea of a "travelling telescope". It had to be a small instrument with an auxilliary equipment, easily transportable to a given country in conjunction with other teaching projects /VLP,ISYA/, in order to give opportunities of gaining direct experience in mo- dern observing techniques. J.R.Percy got in 1987 the necessary funding for this project from the Canadian Commission for UNESCO. The purchasing of the telescope and of the additional instruments, the preparation of documents and organization of maintenance of this telescope will be the task of J.R.Percy and D.Brückner of Toronto /Canada/. It is to be hoped that through the project of Travelling Telescope developing countries can gain more observing experience and will start having their own small observatories.

References
Proceedings of the GIREP Conference 1986"Cosmos - an Educational Challenge", Copenhagen, 18-23 August,1986,(ESA SP-253, Nov.1986)

C.Iwaniszewska

President of the Commission

COMMISSION 47: COSMOLOGY (COSMOLOGIE)

PRESIDENT: G. Setti   VICE-PRESIDENT: K.Sato

ORGANIZING COMMITTEE: J. Audouze, G. de Vaucouleurs, J.E. Gunn, S. Hayakawa,
                       L. Zhi Fang, M.S. Longair, I.D. Novikov, G.A. Tammann,
                       V. Trimble

Introduction

(G. Setti)

The number of pages allocated to the commission report has been very limited
and certainly not sufficient to cover in any exhaustive manner the wide range of
topics relevant to cosmology and to provide also extensive bibliographies. Because
of the vast amount of material to be covered, the report is based on a number of
contributions from different colleagues who have been asked to highlight the main
trends in the triennium (mid 1984 - mid 1987), together with a list of references
sufficiently comprehensive to serve as a guideline for further reading. Unfortunately, two of the expected contributions did not reach me in time for inclusion in
the report, and consequently topics such as the large scale structure and streaming motions, the clusters of galaxies and the counts of extragalactic radio
sources are not included. However, it is my understanding that a large portion, if
not all, of these topics will be covered in the reports of Commissions 28 and 40,
and if true, this will at least avoid unnecessary overlaps. It should also be
mentioned here that several proceedings of very recent IAU conferences provide
excellent, updated and exhaustive reviews of the research work relevant to
cosmology. These are:

IAU Symp. 117 on "Dark Matter in the Universe", J. Kormendy and G.R. Knapp (eds.),
    Reidel, Dordrecht, 1987.
IAU Symp. 124 on "Observational Cosmology", A. Hewitt, G. Burbidge and L. Zhi Fang
    (eds.), Reidel, Dordrecht, 1987
IAU Symp. 130 on "The Structure of the Universe", J. Audouze and A. Szalay (eds.),
    Reidel, Dordrecht, to be published.

The Cosmological Parameters

(V. Trimble)

The traditional parameters of general relativistc cosmology number about
five. H_0 (Hubble's constant) measures the current expansion rate. Its value is
probably between 30 and 120 km/s/Mpc, implying a characteristic time scale (its
reciprocal, the Hubble time) of 8-30 billion years. The deceleration parameter,
q_0, probably falls somewhere between -1 and $+3$, a value of $\frac{1}{2}$ marking the line
between continued expansion and eventual recontraction. The density parameter, Ω_0,
is the ratio of total mass-energy density, ρ_0, to $\rho_c = 3H_0^2/8\pi G$, and is probably
between 0.1 and 1. The cosmological constant, Λ, enters the equations like a
vacuum energy density (positive or negative) and, if expressed in units of
H_0^2/c^2 is almost certainly in the range $+10$ to -10. Finally, the curvature
constant, k, takes on values of $+1$, 0, or -1 for positively curved, flat, or
negatively curved space.

These quantities are not completely independent of each other, but neither does any one suffice to determine the others, unless $\Lambda = 0$ or $k = 0$ is assumed ab initio. For instance, $-q_O = -\Omega/2 + \Lambda c^2_O/3H^2_O$, and $k/3H^2_OR^2 = \Omega_O + \Lambda c^2/3H^2_O - 1$.

A factor of two uncertainty in H_O has persisted for some 30 years, almost all of it coming from the difficulty of measuring accurate distances to objects far enough away for their velocities to reflect primarily uniform expansion. The very promising Tully-Fisher method continues to have difficulties with establishing a zero point that does not depend on galaxy morphology[1] and with proper removal of Malmquist bias[2]. Various considerations of supernovae have yielded small[3], medium[4] and large[5] H_O's, not uncorrelated with the values found by the same authors by other methods. Type I supernovae are not, in any case, the perfect standard candles once hoped for[6]. A couple of relatively new methods, using globular cluster populations[7] and novae[8] yield intermediate values, but the discovery of very large scale streaming motions[9] leaves one in some doubt about whether these are really penetrating deep enough to see pure Hubble flow.

Ages of globular clusters and radioactive nuclides set lower limits to the age of the universe ($2/3 \, H^{-1}_O$ if $q = \frac{1}{2}$). Globular clusters at about 16×10^9 yrs[10] are traditionally the most severe constraint, but this number can be reduced a couple of billion years if $[CNO/Fe] \sim +1$ and to as low as $6-8 \times 10^9$ yrs if there is significant mass loss on the main sequence[11]. Age limits set by the radioactive nuclides have at least as wide a distribution, from 10^{10} yrs[12] through intermediate values[13] to 18×10^9 yrs[14] or more[15]. Calculations of the cooling time for the faintest white dwarfs[16] should be taken to mean that the age of the Milky Way disc could be as small as $8-10 \times 10^9$ yrs, not that it has to be.

The traditional method of probing q_O, deviations from linearity in a Hubble diagram, continues to be plagued by uncertain corrections for galactic evolution and has nearly been abandoned in the past triennium. The surface brightness test[17] has the same problem and picks out a narrow range of values, centered unfortunately right around the critical $\frac{1}{2}$. Direct detection of dz/dt of some object[18] remains merely promising more than a decade after the discovery of narrow radio absorption lines made it cease to seem impossible. The use of galaxy counts to measure co-moving volume as a function of redshift[19,20] leads to very narrow error bars around $\frac{1}{2}$, modulo certain assumptions about the evolution of the galaxy luminosity distribution, but strictly this technique measures k, and says that space is nearly flat, which is not quite the same thing. One slightly non-standard approach leads to a firm value $q_O = 1.6$ [21].

The present writer has recently reviewed determinations of Ω_O [22] and will say here only that there are, on the one hand, ways around the nucleosynthetic limit on baryon density[23] and, on the other hand, some observational arguments against $\Omega_O = 1$ in any form[24] as well as the many theoretical arguments for it.

We have, at present, no direct observational handle on Λ, even very crudely. Where non-zero values have been suggested[25] it has been for the sake of reconciling otherwise inconsistent limits on H_O, ages, and q_O or k. The inflationary scenario, while it requires Λ to have been very large in the past and much smaller now, does not in fact predict zero or any other definite present value[26]. Attempting to calculate Λ from the vacuum energy of the electromagnetic field implied by the Lamb shift and the Casimir effect leads to numbers much larger than permitted by the dynamics of the universe. Gravitation or some other field must contribute a nearly equal and opposite density. Discussions of Λ are bedeviled by units; the limits are about ± 10 in H^2_O/c^2 or $\pm 10^{-56}$ in cm^{-2} or $\pm 10^{-119}$ in Planck (dimensionless) units.

Finally, the geometric parameter, k, is, in principle, measurable, for instance via the distance-dependence of apparent angular diameters of standard-sized objects. But, just as evolutionary effects keep us from having good enough standard candles to determine q_0 directly, evolutionary changes in sizes of both radiosources and clusters of galaxies[27] dominate the cosmological effects in the angular diameter test. Measurement of comoving volume vs. redshift may possibly work better[28] and a first attempt [0] has found the observations consistent with flat space. A useful review of the relationships among the cosmological parameters and the functional shapes of R(t) implied by various possible combinations can be found in ref. 28.

References

1 - Giraud, E.: 1986, Astrophys. J. 309, pp. 512.
    Bothun, G.D., and Mould, J.R.: 1987, Astrophys. J. 313, pp. 629.
2 - Bottinelli, L., et al.: 1986, Astron. Astrophys. 166, pp. 393.
    Giraud, E.: 1987, Astron. Astrophys. 174, pp. 23.
3 - Arnett, W.D., Branch, D. and Wheeler, J.C.: 1985, Nature 314, pp. 337.
4 - Bartel, N., et al.: 1985, Nature 318, pp. 25.
5 - de Vaucouleurs, G.: 1985, Astrophys. J. 289, pp. 5.
    de Vaucouleurs, G. and Corwin, H.G.: 1985, Astrophys. J. 297, pp. 23.
6 - Fu, A.J. and Arnett, W.D.: 1986, Astrophys. J. 307, pp. 726.
    Frogel, J.A., et al.: 1987, pp. Astrophys. J. 315, pp. L129.
7 - van den Bergh, S., Pritchet, C. and Grillmair, C.: 1985, A.J. 90, pp. 595.
    van den Bergh, S.: 1985, Astrophys. J. 297, pp. 361.
8 - van den Bergh, S.: 1987, P.A.S.P., in press.
9 - Collins, C.A., Joseph, R.D. and Robertson, N.A.: 1986, Nature 320, pp. 506.
    Dressler, A., et al.: 1987, Astrophys. J. 313, pp. L37.
10 - Caputo, F.: 1987, Astron. Astrophys. 172, pp. 67.
11 - Willson, L.A., Bowen, G.H. and Struck-Marcell, C.: 1987, Comm. Astrophys. 12, pp. 17.
12 - Slish, V.I.: 1985, Sov. Astron. Lett. 11, pp. 126.
13 - Fowler, W.A.: 1987, Q. Jl. Roy. Astr. Soc. 28, pp. 87.
    Winters, R.R., Macklin, R.L. and Hershberger, R.L.: 1987, Astron. Astrophys. 171, pp. 9.
14 - Beer, H. and Macklin, R.L.: 1985, Phys. Rev. C32, pp. 738.
15 - Thielemann, F.-K. and Truran, J.W.: 1986, in "Nucleosynthesis and its Implications on Nuclear and Particle Physics", J. Audouze and N. Mathieu (eds.), Reidel, Dordrecht, NATO ASI Series C, Vol. 163, pp. 373.
16 - Winget, D.E., et al.: 1987, Astrophys. J. 315, pp. L77.
17 - Phyllips, S.: 1985, Astrophys. Lett. 24, pp. 225.
18 - Teuber, J.: 1986, Astrophys. Lett. 25, pp. 139.
19 - Loh, E.D. and Spillar, E.J.: 1986, Astrophys. J. 307, pp. L1.
20 - Loh, E.D.: 1986, Phys. Rev. Lett. 57, pp. 2865.
21 - Narlikar, J.V. and Seshadri, T.R.: 1985, Astrophys. J. 288, pp. 43.
22 - Trimble, V.: 1987, Ann. Rev. Astron. Astrophys. 25, in press.
23 - Ramadurai, S. and Rees, M.J.: 1985, Mon. Not. R. astr. Soc. 215, 53p.
    Applegate, J.H., Hogan, C.J. and Scherrer, R.J.: 1987, Phys. Rev. D 35, pp. 1151.
    Kasper, U.: 1986, Astron. Nachr. 307, pp. 271.
24 - Koo, D.C. and Szalay, A.S.: 1984, Astrophys. J. 282, pp. 390.
    Saslaw, W.C., Antia, H.M. and Chitre, S.M.: 1987, Astrophys. J. 315, pp. L1.
25 - Peebles, P.J.E.: 1984, Astrophys. J. 284, pp. 439.
    Klapdor, H.V. and Grotz, K.: 1986, Astrophys. J. 301, pp. L39.
26 - Barrow, J.D. and Tipler, F.J.: 1985, Mon. Not. R. astr. Soc. 216, pp. 395.
27 - Allington-Smith, J.R.: 1984, Mon. Not. R. astr. Soc. 210, pp. 611.
    Johansen, K.T., Florentin-Nielsen, R. and Teuber, J.: 1985, Astron. Astrophys. 152, pp. L21.
28 - Felten, J.E.: 1986, Rev. Mod. Phys. 58, pp. 689.

The Very Early Universe

(K. Sato)

In recent years, the research on the very early universe has shown quite remarkable developments. As is well known, this development was brought about by the introduction of the Grand Unified Theories (GUTs) into cosmology. These theories have not only enabled us to trace the evolution of the Universe back to the very early stage at temperatures of 10^{16} GeV or higher, but also introduced various new aspects into cosmology, such as baryogenesis, phase transitions and topological defects (monopoles, etc.). In particular, inflation, which grew out of the study of GUT phase transition, is the most important and fascinating outcome.

INFLATION

At the start, the inflationary universe model attracted people as a model which explains some global features of the present Universe, such as homogeneity and isotropy. Now inflation becomes a much more fascinating idea which may explain everything: the origin of matter as well as the origin of detailed structures, such as galaxies and their distribution. The inflationary universe model has now almost become the standard model for the early stage of the Universe[1].

Along with such developments, however, some problems were pointed out. One of them is that it turned out that we must move the inflationary epoch back to the Planck time and make the interaction of the inflation driving field extremely weak in order to account for the observed inhomogeneity of the Universe. Hence we are obliged to cut off the inflation model from GUT. This drastic revision of the model has given rise to a kind of chaos in the field, since it implies for us to lose the fundamental ground we stand on. At the same time it forces us to study the birth of the Universe itself[2], which is possible only in the framework of quantum gravity. Cosmology of the very early Universe is now waiting for the next big development in particle physics. Superstring theories which try to unify gravity and other non-gravitational gauge interactions may be the one. At the present stage, however, we can say nothing definite.

The second problem is the inflation in an anisotropic and inhomogenous universe. The important consequence of an inflationary universe model is that it explains why our present Universe is so homogeneous and isotropic (the horizon problem). Paradoxically, inflation has been usually analyzed in the context of the homogeneous and isotropic Robertson-Walker model. If the inflation solves the horizon problem under general conditions, we can expect that all the inhomogeneous and anisotropic universes with cosmological constants (the vacuum energy density) evolve towards the de Sitter universe. This conjecture is called "cosmological no hair theorem". Concerning the anisotropic and homogeneous universe, Wald had shown that all the Bianchi types except IX evolve towards the de Sitter solution, and generalization and more detailed investigations have been done by many people (see ref. 3 and papers cited therein). For an inhomogeneous universe, Jensen and Stein-Schabes[4] showed that any inhomogeneous universe will tend towards the de Sitter universe if i) the dominant energy condition and ii) the strong energy condition are satisfied, and iii) the scalar spatial curvature is never positive. Unfortunately, generality of their justification is almost lost by the third condition, because usual inhomogeneities always contain positive curvature regions.

In spite of many efforts to justify the no hair theorem, however, there exists a simple counter example against the no hair theorem, i.e. the existence of the Schwarzschild-de Sitter solution[5]. This is a black hole or a wormhole solution in the de Sitter universe. It is obvious that the no hair theorem does not hold in the strict meaning, because the universe cannot evolve to a homogeneous de Sitter universe in the classical level, if holes exist from the start or once holes have formed.

Recently, Linde[6] considered the evolution of the density fluctuations in the chaotic inflation model and discussed that the high density domains evolve to causally disconnected universes and proposed an eternally existing self-reproducing universe, which is essentially an extension of the multi-production of the universe in the original inflation model (K. Sato et al., see paper cited in ref. 5). It must be, however, mentioned that the "weak no hair theorem" may hold despite the fact that the large amplitude and large scale fluctuations evolve to disconnected universes, because exact solutions which evolve to homogeneous de Sitter universes were found[7] and it was shown that the class of the solutions does not measure zero. This shows the importance of a careful investigation in order to clarify the conditions for the no hair theorem.

BARYOGENESIS

As is well recognized, the interaction of the scalar field which drive inflation must be extremely weak in order to account for the observed large scale structure. This weakly interacting nature leads in general to such low reheating temperatures that the conventional baryogenesis scenario based on GUT becomes difficult[8]. As a possible solution to this problem, recently a new non-GUT mechanism of baryogenesis has been proposed which utilizes the baryon non-conserving process by the anomaly of the electro-weak interactions[9]. Such theories are interesting in that they are pointing out a possible importance of the physics of the GeV to TeV region in cosmology.

QUARK-HADRON PHASE TRANSITION

Recent numerical simulations of lattice QCD strongly indicate the first-order nature of the quark-hadron phase transition[10]. If this is the case, it leads to interesting cosmological consequences. In particular so-called strange quark nuggets will be produced at the cosmological quark-hadron phase transition. As pointed out by Witten[11], such nuggets have planetary mass and could be candidates for dark matter, if they are stable at zero temperature. The interesting point is that it is naturally explained why the ratio $\Omega_{Dark\ Matter}/\Omega_{Baryon}$ is o(1), neither o(10^{-20}) nor o(10^{20}), if dark matter is the quark nuggets[12]. Unfortunately, its zero temperature stability is not established. Furthermore, there are some strong arguments suggesting that they will evaporate thermally at or just after the formation even if they are stable at zero temperature[13]. However, since the production of these nuggets may affect strongly the primordial nucleosynthesis[14] (hence the estimate of the present photon-baryon ratio is changed), study of the quark-hadron phase transition is cosmologically very important.

WEAKLY INTERACTING MASSIVE PARTICLES (WIMPs) AND COSMIC STRINGS

Though not so drastic, some important interplay with particle physics has been observed on the rather recent stage of the Univese. From astrophysical considerations it is shown that dark matter may consist of weakly interacting particles (WIMPs), which are the remnants of the early fire ball stage. Particle theory at present provides lots of candidates for WIMPs[15]. An interesting point is that these candidates are strongly constrained from cosmology and astrophysics. In fact, though not decisive, the only remaining candidates are axions and SUSY ions. They are both intimately connected with the concepts which have played important roles in the recent development of particle pyhsics: gauge anomaly and supersymmetry.

Recent developments in cosmic string theory, in particular, the theory of galaxy formation in terms of cosmic strings[16] should not be overlooked. It seems that except for this type of galaxy formation theory, there is no clear theory at present which gives a consistent scenario of galaxy formation. In this sense, it deserves further study, though there exists as yet no natural model which provokes inflation and at the same time produces cosmic strings of cosmological significance.

References

1 - See for example: 1986, Inner Space/Outer Space, Chicago Univ. Press.
2 - Linde, A.D.: 1984, Rep. Prog. Phys. 47, pp. 925.
3 - Turner, M.S. and Widrow, L.M.: 1986, Phys. Rev. Lett. 57, pp. 2237.
4 - Jensen, L.G. and Stein-Schabes, J.A.: 1987, Phys. Rev. D35, pp. 1046.
5 - Sato, K.: 1987, Proc. of the IAU Symposium 130, "The structure of the
    Universe", to be published.
6 - Linde, A.D.: 1986, Phys. Lett. 175B, pp. 395.
7 - Starobinski, A.: 1983, JETP Lett. 30, pp. 66.
  - Stein-Schabes, J.A.: 1987, Phys. Rev. D35, pp. 2345.
  - Barrow, J.D. and Grón, O.: 1986, Phys. Lett. 182B, pp. 25.
8 - Steinhardt, P.J. and Turner, M.S.: 1984, Phys. Rev. D29, pp. 2162.
9   Kuzmin, V.A., Rubakov, V.A. and Shaposhnikov, M.E.: 1985, Phys. Lett. 155B,
    pp. 36.
10 - See for example:
    Fukugita, M. and Ukawa, A.: 1986, Phys. Rev. Lett. 57, pp. 503.
    Guputa, R., et al.: 1986, Phys. Rev. Lett. 57, pp. 2621.
    Karsch, F., et al.: 1987, Phys. Lett. 188B, pp. 353.
11 - Witten, E.: 1984, Phys Rev. D30, pp. 272.
12 - Carr, B.J. and Turner, M.S.: 1987, Mod. Phys. Lett. A2, pp. 1.
13 - Alcock, C. and Farhi, E.: 1985, Phys. Rev. D32, pp. 1273.
    Madsen, J., et al.: 1986, Phys. Rev. D34, pp. 2947.
14 - Applegate, J.H., Hogan, C.J. and Scherrer, R.J.: 1987, Phys. Rev. D35, pp.
    1151.
15 - Turner, M.S.: 1987, Proc. of the IAU Symp. 130, "The structure of the
    Universe", to be published.
16 - See for example: Turok, N.: 1987, Proc. of the IAU Symp. 130, "The structure
    of the Universe", to be published.

Primordial Nucleosynthesis

(J. Audouze)

Primordial nucleosynthesis which is responsible for the formation of the lightest elements (D, ^{3}He, ^{4}HE and ^{7}Li) might be as important as the overall expansion of the Universe and the cosmic background radiation to prove the occurrence of a dense and hot phase for the Unvierse about 15 billion years ago. As recalled in many reviews (e.g. refs. 1, 2) the standard Big Bang nucleosynthesis leads to two important conclusions regarding (i) a limitation of the baryonic density such that the corresponding cosmological parameter $\Omega_B \lesssim 0.1$; (ii) a limitation of the number of neutrino flavours to 3-4 consistent with the results concerning the widths of the Z_O and $W^{\pm}$ particles[3].

The most recent progresses concerning this important problem deal with (i) some recent abundance determinations of the light elements; (ii) the discussion of the validity of the standard Big Bang model; (iii) the chemical evolution of the D and ^{3}He abundances; (iv) the elaboration of models taking into account either the decay of non baryonic particles or the inhomogeneities resulting from the quark-hadron phase transition.

RECENT ABUNDANCE DETERMINATIONS OF THE LIGHTEST ELEMENTS

An excellent review of the D abundances can be found in ref. 4. There is a tentative determination[5] of the D/H ratio in (z ~ 3) absorption line QSOs. Concerning ^{3}He a recent reconsideration of the interstellar ^{3}He$^+$/H ratio from radio lines has reduced somewhat but not eliminated the large abundance range reported in previous analyses[6,7].

The primordial ^{4}He abundance (Y_p) has been thoroughly discussed in a recent conference[8]. There seems to be a slight tendency towards lower values of Y_p (e.g. refs. 9 and 10). Finally, regarding ^{7}Li the discovery of ^{7}Li/H ~ 10^{-10} in Pop II stars[11] is confirmed by two different groups [12,13].

THE VALIDITY OF THE STANDARD BIG BANG NUCLEOSYNTHESIS

The so-called Chicago-Bartol group is still strongly arguing about the striking validity of the simple (canonical) Big Bang model. This group has also studied the implications on their models of new nuclear reaction rates which could affect the ^{7}Li abundance[14] (see also ref. 15) and found no reason to abandon their views regarding the success of such models[16]. These views are challenged in part by the Paris group since there seems to be a growing discrepancy between the baryonic density deduced from low Y_p values on the one hand and low $\left(\frac{D+^3He}{H}\right)_p$ values on the other hand (e.g. ref. 2).

SPECIFIC MODELS OF GALACTIC EVOLUTION

In order to overcome this difficulty the Paris group[17] have considered models implying for instance varying rates of star formation where D can be destroyed thoroughly during the galactic history and where a low Y_p value would correspond to a high $\left(\frac{D+^3He}{H}\right)_p$ value. There is an observational test which could discriminate between the Chicago-Bartol and the Paris views depending on the non-variability or the variability of the D/H ratio observed in different regions of our Galaxy.[18]

PRIMORDIAL NUCLEOSYNTHESIS AND PARTICLE PHYSICS

Since standard Big Bang models put very strong constraints on the baryonic density of the Universe, many attempts have been made to alleviate such an important constraint. Among them one can quote (i) the partial photo-disintegration of ^{4}He and ^{7}Li induced by photons coming from the decay of massive non-baryonic particles such as massive neutrinos and gravitinos[19], photinos[20], and WIMPS of any kind[21]; (ii) the consideration of an anisotropic universe[22] although the ^{7}Li abundance puts severe constraints on this specific model[23,24]. (iii) the possible effect of the quark-hadron phase transition on the primordial nucleosynthesis, a most exciting proposal made first by Applegate et al.[25], followed by Alcock et al.[23]. This phase transition might induce the formation of neutron and proton rich zones, the existence of which could affect the outcome of the primordial nucleosynthesis. In that frame it has been argued[25] that this model could allow the possibility of having $\Omega_B = 1$ consistent with the results of the primordial nucleosynthesis, while other investigations show that the primordial abundance of ^{7}Li rules out this most exciting idea.[23,26]

References

1 - Boesgeard, A.M. and Steigman, G.: 1985, Ann. Rev. Astron. Astrophys. 23, pp. 319.
2 - Audouze, J.: 1987, in "Dark Matter in the Universe", J. Kormendy and G.R. Knapp (eds.), Reidel, Dordrecht, pp. 499.
    Audouze, J.: 1987, in IAU Symp. 124 on "Observational Cosmology", A. Hewitt, G. Burbidge and L. Zhi Fang (eds.), Reidel, Dordrecht, pp. 89.
3 - Cline, D.B., Schramm, D.N. and Steigman, G.: 1987, Comm. Nucl. Part. Phys., in press.
4 - Vidal-Madjar, A.: 1987, in "Space Astronomy and Solar System Exploration", W. Burke (ed.), ESA-SP 268.
5 - Webb, J.K., et al.: 1987, Astron. Astrophys., in press.
6 - Bania, T.M., Rood, R.J. and Wilson, T.L.: 1987, Astrophys. J., in press.
7 - Rood, R.T., Bania, T.M. and Wilson, T.L.: 1984, Astrophys. J. 280, pp. 629.
8 - Shields, G.A.: 1986, P.A.S.P. (special issue) 98, pp. 956.

9 - Beckman, J. and Pagel, B.E.J.: 1987, Proc. of the IAU Symp. 134. G. Cayrel
    and M. Spite (eds.), to be published.

10 - Gallagher, J.S.(III), Schramm, D.N. and Steigman, G.: 1987, Astrophys. J., in
    press.

11 - Spite, F. and Spite, M.: 1982, Astron. Astrophys. 115, pp. 357.

12 - Hobbs, L.M. and Duncan, D.K.: 1987, Astrophys. J., in press.

13 - Rebolo, R., Beckman, J. and Molaro, P.: 1987, Astron. Astrophys. 172, pp.
    L17.

14 - Kawano, L., Schramm, D.N. and Steignman, G.: 1987, Astrophys. J., in press.

15 - Kajino, T., Toki, H. and Austin, S.M.: 1987, MSUCL 574 preprint

16 - Steigman, G., et al.: 1987, Phys. Lett., in press.

17 - Vangioni-Flam, E. and Audouze, J.: 1987, Astron. Astrophys., in press.

18 - Delbourgo-Salvador, P., Salati, P., Reeves, H. and Audouze, J.: 1987, in
    preparation.

19 - Audouze, J., Lindley, D. and Silk, J.: 1985, Astrophys. J. 293, pp. L53.

20 - Salati, P., Delbourgo-Salvador, P. and Audouze, J.: 1987, Astron. Astrophys.
    173, pp. 1.

21 - Scherrer, R.J. and Turner, M.S.: 1987, Astrophys. J., in press.

22 - Matzner, R.A.: 1986, P.A.S.P. 98, pp. 1049.

23 - Alcock, C.R., Fuller, G.M. and Mathews, G.J.: 1987, UCRL 95896 preprint.

24 - Reeves, H.: 1987, Proc. of the Varenna Summer School, J. Audouze and F.
    Melchiorri (eds.), to be published.

25 - Applegate, J.H., Hogan, C. and Scherrer, R.J.: 1987, Phys. Rev. D, in press.

26 - Delbourgo-Salvador, P., Audouze, J. and Vidal-Madjar, A.: 1987, Astron.
    Astrophys 174, pp. 365.

Background Radiation in the Universe

(G. De Zotti)

MICROWAVE BACKGROUND

a) _Spectrum_. A collaboration between US and Italian groups performed accurate
observations at five wavelengths[1]. The experiment was particularly conceived to
achieve the highest possible relative accuracy, allowing an effective search
for spectral distortions. The Berkeley and the Milano groups further improved
the spectral coverage[2].

Johnson and Wilkinson[3] avoided the main problems of ground-based experiments
(primarily the atmospheric emission) by flying a special radiometer operating
at λ = 1.2 cm on a balloon. Thus they arrived at the most precise measurement
reported to date: T_O = 2.783 ± 0.025 K.

Accurate determinations of T_O at 2.64 mm and estimates at 1.32 mm were also
obtained through high-resolution observations of the CN absorption lines[4].

The good agreement between all results listed above, involving very different
systematic effects, is encouraging. The brightness temperature in the Rayleigh-
Jeans region is now known to better than 1%; Bose-Einstein distortions with a
chemical potential larger than a few times 10^{-3} are ruled out. The ensuing
constraints on processes of cosmological interest have been recently reviewed[5].

The information on the Wien tail of the spectrum has also been growing fast in
the last few years. The balloon-borne photometer flown by Peterson and co-
workers[6] has made measurements at five wavelengths, ranging from 3.5 to 1 mm.
The new results do not confirm the strong excess around the peak[7]. Most
recently, a rocket-borne radiometer, designed to measure the background
radiation in six passbands between ≈ 1 mm and ≈ 100 μm, was launched by a

collaboration between the Nagoya and Berkeley groups. According to preliminary reports, the brightness temperature at 1 mm is consistent with that measured at lower frequencies. At .68 mm and .46 mm, however, a strong excess is observed, that might be interpreted in terms of a Comptonization distortion by a non-relativistic plasma. It remains to be seen whether the data may also be consistent with the distortion produced by a mildly relativistic gas that could produce the X-ray background[8].

b) Isotropy. The two independent maps at 3 mm [9] and at 12 mm [10] have been combined to produce a combined map which is better connected than either and has a sky coverage increased to $\approx$ 90%[11]. The small discrepancy in the dipole amplitudes ($\approx 2\sigma$) does not affect the direction significantly: the two experiments agree within 1.6°. The results are in excellent agreement with those from the RELIKT experiment[12]. No signals were found for higher harmonics. The tightest upper limits were set by the RELIKT experiment; for the quadrupole they find (95% confidence) $\Delta T/T < 3 \ 10^{-5}$.

Significant fluctuations with an observed standard deviation of $3.7 \ 10^{-5}$ on scales of $\approx$ 8°-10° have recently been reported[13]. While recognizing that structure in the radio continuum emission from our galaxy may contribute appreciably, the authors argue that a substantial part of the signal is probably intrinsic. Similar fluctuations were previously observed at $\lambda = 1$ mm [14]. Again the interpretation depends on an uncertain correction for the galactic contribution.

A graininess which could not be attributed to known instrumental effects was detected, on scales $\leq$ 1', in recent VLA maps[15]. It is still unclear, however, which fraction of the detected signal is due to unresolved discrete sources.

High sensitivity observations by several groups have led to remarkably tight upper limits. The most stringent yet published is $\Delta T/T \leq 2.5 \ 10^{-5}$, on a scale of 4.5'![16]

The variety of models on the origin of structure in the Universe that can be found in the recent literature reflects the lack of sufficient data to discriminate between them, and translates into a variety of predictions for $\Delta T/T$ [17]. Particularly helpful in defining the best observational strategy for testing models are the discussions of the statistical properties of radiation patterns generated by density fluctuations[18].

c) Sunyaev-Zeldovich effect. The small dips in the directions of three rich clusters of galaxies were the first small scale anisotropies for which detection at a high level of significance has been claimed[19]. Many years of experimental work, however, have revealed several astrophysical and instrumental effects that may distort the results[20]. The exploitation of the astrophysical information provided by these data is probably still premature.

d) Polarization. Until very recently both observational and theoretical studies dealt with polarization on large angular scales, introduced by anisotropic expansion[21]. On the other hand, the predicted polarization associated to small scale anisotropies induced by adiabatic perturbations is rather high, $\approx$ 10% [22]. Polarimetry may then be decisive in determining the origin of anisotropies in the presence of confusion from faint sources. Limits on polarization on scales from 18" to 180" are $\approx$ 3 times lower than those on temperature fluctuations on the same scales[23].

INFRARED BACKGROUND
     The absolutely calibrated Nagoya-Berkeley rocket experiment (cf. Sect. 1) will help in establishing the IRAS zero point and, hence, to check the reality of

the isotropic background component suggested by the analysis of 100 μm IRAS data[24]. The new data at 280 μm, 145 μm and 110 μm are consistent with interstellar plus interplanetary dust emissions; on the other hand, since these observations refer to a relatively low galactic latitude (b ≈ 33° ± 2°), their interpretation requires a detailed modelling of the galactic emission.

The diffuse near-IR (1 to 5 μm) radiation intensity has recently been measured with a rocket experiment[25]. A significant isotropic component was found, whose intensity exceeds by a substantial amount that predicted even by extreme galactic evolution models[26]. If it is of truly extragalactic origin, a substantial activity at early epochs would be called for[27].

THE X-RAY BACKGROUND (XRB)

Its origin is still a puzzle. Contrary to earlier expectations, the most recent estimates seem to converge in indicating a quite modest contribution from QSOs[28], consistent with the growing evidence that these objects have, on the average, X-ray spectra substantially steeper than the XRB[29]. A recent reanalysis of the HEAO 1 A-2 database[30] has led to conclude that, barring the case of unexpectedly strong cosmological evolution, low luminosity Active Galactic Nuclei are unlikely to make up the bulk of the XRB. Thus AGNs, which constitute the dominant population of extragalactic sources in the Einstein Deep Survey[31], might not be the dominant constituents of the background.

Strong constraints on the properties of sources that account for the latter come from: a) the high precision measurements of its spectrum[32]: b) the analysis of surface brightness fluctuations on data from the Einstein IPC[33] which indicates that the surface density of point sources must be ≥ 5000 deg^{-2}, far in excess of the estimated surface density of QSOs.

Two classes of sources that could meet the above requirements have been proposed: precursor Active Galactic Nuclei[34] and actively star-forming galaxies[35,30].

References

1 - Smoot, G.F., et al.: 1985, Astrophys. J. 291, pp. L23.
2 - Sironi, G.F., et al.: 1987, Proc. 13th Texas Symp., in press.
    Smoot, G., et al.: 1987, Astrophys. J. 317, pp. L45.
3 - Johnson, D.G. and Wilkinson, D.T.: 1987, Astrophys. J. 313, pp. L1.
4 - Meyer, D.M. and Jura, M.: 1985, Astrophys. J. 297, pp. 119.
    Crane, P., et al.: 1986, Astrophys. J. 309, pp. 822.
5 - De Zotti, G.: 1986, Progress in Particle and Nuclear Physics 17, pp. 117.
6 - Peterson, J.B., Richards, P.L. and Timusk, T.: 1985, Phys. Rev. Lett. 55, pp. 332.
7 - Woody, D.P. and Richards, P.L.: 1981, Astrophys. J. 248, pp. 18.
8 - Guilbert, P.W. and Fabian, A.C.: 1986, Mon. Not. R. astr. Soc. 220, pp. 439.
9 - Lubin, P., et al.: 1985, Astrophys. J. 298, pp. L1.
10 - Fixsen, D.J., Cheng, E.S. and Wilkinson, D.T.: 1983, Phys. Rev. Lett. 50, pp. 620.
11 - Lubin, P. and Villela, T.: 1986, in "Galaxy Distances and Deviations from Universal Expansion", B.F. Madore and R.B. Tully (eds.), Reidel, Dordrecht, pp. 169.
12 - Lukash, V.N. and Novikov, I.D.: 1987, in IAU Symp. 124 on "Observational Cosmology", A. Hewitt, G. Burbidge and L. Zhi Fang (eds.), Reidel, Dordrecht, pp. 73.
13 - Davies, R.D., et al.: 1987, Nature 326, pp. 462.
14 - Melchiorri, F., et al.: 1981, Astrophys. J. 250, pp. L1.
15 - Martin, H.M. and Partridge, R.B.: 1987, Astrophy. J. Lett., submitted.
    Kellermann, K.I., et al.: 1986, in Highlights of Astronomy, Vol. 7, pp. 367.

16 - Uson, J.M. and Wilkinson, D.T.: 1984, Nature 312, pp. 427.

17 - Kaiser, N. and Silk, J.: 1986, Nature 324, pp. 529.
Efstathiou, G. and Bond, J.R.: 1986, Phil. Trans. R. Soc. Lond. A 320, pp. 585.
Fabbri, R., Lucchin, F. and Matarrese, S.: 1987, Astrophys. J. 315, pp. 1.
Lukash, V.N.: 1987, in IAU Symp. 117 on "Dark matter in the Universe", J. Kormendy and G.R. Knapp (eds.), Reidel, Dordrecht, pp. 379.
Peebles, P.J.E.: 1987, Astrophys. J. 315, pp. L73.

18 - Vittorio, N. and Juszkiewicz, R.: 1987, Astrophys. J. 314, pp. L29.
Bond, J.R. and Efstathiou, G.: 1987, Mon. Not. R. astr. Soc. 226, pp. 655.

19 - Birkinshaw, M. and Moffet, A.T.: 1986, in Highlights of Astronomy, Vol. 7, p. pp. 321, and ref. therein.

20 - Partridge, R.B., et al.: 1987, Astrophys. J. 317, pp. 112, and ref. therein.

21 - Tolman, B.W.: 1975, Astrophys. J. 290, pp. 1, and ref. therein.

22 - Kaiser, N.: 1983, Mon. Not. R. astr. Soc. 202, pp. 1169.
Milaneschi, E. and Valdarnini, R.: 1986, Astr. Astrophys. 162, pp. 5.

23 - Partridge, R.B., Nowakowski, J. and Martin, H.M.: 1987, preprint.

24 - Rowan-Robinson, M.: 1986, Mon. Not. R. astr. Soc. 219, pp. 737.

25 - Matsumoto, T., Akiba, M., and Murakami, H.: 1987, in IAU Symp. 124 on "Observational Cosmology", A. Hewitt, G. Burbidge and L. Zhi Fang (eds.), Reidel, Dordrecht, pp. 69.

26 - Yoshii, Y. and Takahara, F.: 1987, Astrophys. J., submitted.

27 - Bond, J.R., Carr, B.J. and Hogan, C.J.: 1986, Astrophys. J. 306, pp. 428.

28 - Setti, G.: 1987, in IAU Symp. 124 on "Observational Cosmology", A. Hewitt, G. Burbidge and L. Zhi Fang (eds.) Reidel, Dordrecht, pp. 579.
Giacconi, R. and Zamorani, G.: 1987, Astrophys. J. 313, pp. 20.

29 - Marshall, F.E.: 1986, Bull. AAS 18, pp. 914.
Elvis, M., et al.: 1986, Astrophys. J. 310, pp. 291.

30 - Persic, M., et al.: 1987, Astrophys. J. submitted.

31 - Murray, S.S.: 1987, Invited paper presented at AAS Meeting No. 169.

32 - Gruber, D.E., et al.: 1984, in "X-Ray and UV Emission from Active Galactic Nuclei", W. Brinkmann and J. Trümper (eds.), MPE Report 184, pp. 129.

33 - Hamilton, T. and Helfand, D.: 1987, Astrophys. J. 318, pp. 93.

34 - Boldt, E. and Leiter, D.: 1986, in "Structure and Evolution of Active Galactic Nuclei", G. Giuricin et al. (eds.), Reidel, Dordrecht, Astrophys. Spa. Sci. Library 121, pp. 383.

35 - Bookbinder, J., et al.: 1980, Astrophys. J. 237, pp. 647.

Formation and Evolution of Galaxies

(A.A. Klypin, V.N. Lukash, I.D. Novikov)

GENERAL TRENDS

At the beginning of this review period a number of arguments were put forward against the neutrino model which became popular in 1980-1983[1]: too high a rate of the structure evolution at the non-linear stage and the same difficulty in the galaxy formation. As a consequence, many other schemes of the structure origin have been elaborated: models with "cold" particles, with unstable missing mass, etc. In these models the missing mass is in the form of weakly interacting particles (axion, photino, gravitino, etc.), or of usual particles (e.g., neutrino) but with properties that are out of the ordinary (e.g. instability). However, the standard neutrino model cannot yet be regarded as rejected[2], the more so in view of the recent data on the large-scale peculiar velocities[3].

The "cold"-particle hypothesis has been actively developed. In its simplest version this hypothesis contradicts many observational data and demands biasing, a process of galaxy formation where the distribution of visible matter does not reproduce the distribution of missing mass. Attempts to modify the old neutrino model have brought about a large family of models with unstable missing mass. Some of these models rapidly develop and look quite vital. A tendency towards composite (hybrid) schemes of the formation of galaxies is becoming rather obvious. These are for instance, neutrino models with a "cold" component[4] or with cosmological strings and explosions, a "cold" model with Λ-term[5], models with unstable particles and Λ-term[6], etc.

The problems we have mentioned here were treated in several review papers[7] and in recent IAU simposia (see the Introduction by G. Setti).

COLD DARK MATTER MODELS (CDM) AND BIASED GALAXY FORMATION

Davies et al.[8] found that N-body simulations of CDM with $\Omega = 1$ adequately represent the observational picture of superclusters and voids if galaxies were located at the high (2.5 standard deviations) mass density peaks. Different schemes for biased galaxy formation were suggested by other authors[9]. The three-point correlation function was estimated by Melott and Fry[10], while the N-point correlation function was discussed by Jensen and Szalay[11].

Statistical properties of high mass density maxima were studied by Peacock and Heavens[12], and Bardeen et al.[13]. This approach is doubtlessly one of the most promising on the way to disclosing the mystery of the galactic creation.

UNSTABLE DARK MATTER MODELS (UDM)
First variants of UDM with decaying neutrinos were proposed by several authors[14]. It was found[15] that the structural parameters would contradict the observations if galaxies were formed before the particle decayed. UDM with formation of the non-linear structures at the epoch of decays ($Z_d \simeq 3 - 10$) and later were discussed by Doroshkevich et al.[16]. Virgocentric infall velocities in UDM were estimated by Hoffman[17]. CDM and UDM meet difficulties in the explanation of bulk velocities on scales ~ 100 Mpc.

HIERARCHICAL EXPLOSIONS (EM) AND STRING (SM) MODELS
The physical aspects of EM were discussed by Carr and Ikeuchi[18], and Ikeuchi and Ostriker[19], while numerical simulations for galaxy distribution were presented by Saarinen et al.[20]

The development of SM has been much pursued[21] and applications to formation of structures were made by many authors[22]. It was found[23] that the results of non-linear simulation of SM with cold particles does not agree with observations.

NEW APPROACHES

The problem of the difference of the correlation functions for galaxies and for rich clusters is still puzzling. New approaches to the problem have been suggested[24].

New methods for studying and modelling the large-scale galaxy distribution were proposed, such as the sponge-like structure and the Euler characteristics[25]. A new approach to the percolation method[26] provides sensitive tests and enables the application of the method to catalogues with non-cubical boundaries. A new method for large-scale simulations based on Burger's equation was suggested[27].

References

1 - Hut, P. and White, S.D.M.: 1984, Nature 310, pp. 637.
    White, S.D.M., Davis, M. and Frenk, C.S.: 1984, Mon. Not. R. astr. Soc. 209, pp. 271.

2 - Melott, A.L.: 1985, Astrophys. J. 289, pp. 2.

3 - Vittorio, N., Luszkievich, R. and Davis, H.: 1986, Nature 323, pp. 132.

4 - Achilli, S., Occhionero, S. and Scaramella, R.: 1985, Astrophys. J. 299, pp. 577.

5 - Kofman, L.A. and Starobinsky, A.: 1985, Pis'ma Astron. Zh. 11, pp. 643.

6 - Doroshkevich, A.G., Klypin, A.A. and Khlopov, M.U.: 1985, Pis'ma Astr. Zh. 11, pp. 483.

7 - Blumenthal, G.R., et al.: 1984, Nature 311, pp. 517.
    Dekel, A.: 1986, Comm. Astrophys. 11, pp. 235.
    Peebles, P.J.E.: 1987, Nature 321, pp. 27.

8 - Davis, M., et al.: 1985, Astrophys. J. 292, pp. 371.

9 - Couchman, H.M.P. and Rees, M.J.: 1986, Mon. Not. R. astr. Soc. 221, pp. 53.
    Dekel, A. and Rees, M.J.: 1987, Nature 326, pp. 455.
    Hoffman, Y. and Shaham, J.: 1985, Astrophys. J. 297, pp. 16.
    Rees, M.J.: 1985, Mon. Not. R. astr. Soc. 213, pp. 75P.
    Schaeffer, L. and Silk, J.: 1985, Astrophys. J. 292, pp. 319.
    Silk, J.: 1985, Astrophys. J. 297, pp. 1.

10 - Melott, A.L. and Fry, J.N.: 1986, Astrophys. J. 305, pp. 1.

11 - Jensen, L.G. and Szalay, A.: 1986, Astrophys. J. 305, pp. L5.

12 - Peacock, J.A. and Heavens, A.F.: 1985, Mon. Not. R. ast. Soc. 217, pp. 805.

13 - Baardeen, J.M., et al.: 1986, Astrophys. J. 304, pp. 15.

14 - Doroshkevich, A.G., and Khlopov, M.U.: 1984, Mon. Not. R. astr. Soc 211, 277
     Turner, M.S., Steigman, G., and Krauss, L.: 1984, Phys. Rev. Lett. 52, 2090

15 - Gelmini, G., Schramm, D. and Valle, J.: 1984, Phys. Lett. 146B, pp. 311.
     Olive, K., Seckel, D. and Vishniac, E.: 1985, Astrophys. J. 292, pp. 1.
     Suto, Y., Kodama, H. and Sato, K.: 1985, Mon. Not. R. astr. Soc. 218, pp. 637.
     Suto, Y., Kodama, H. and Sato, K.: 1985, Phys. Lett. B 157B, pp. 259.
     Flores, R.A., et al.: 1986, Nature 323, pp. 781.

16 - Doroshkevich, A.G., Klypin, A.A., and Kotok E.V.: 1986, Astron. Zh. 63, 417

17 - Hoffman, Y.: 1986, Astrophys. J. 305, pp. L1.

18 - Carr, B.J. and Ikeuchi, S.: 1985, Mon. Not. R. astr. Soc 213, pp. 497.

19 - Ikeuchi, S. and Ostraiker, J.P.: 1986, Astrophys. J. 301, pp. 522.

20 - Saarinen, S., Dekel, A. and Carr, B.J.: 1987, Nature 325, pp. 598.

21 - Vilenkin, A.: 1985, Phys. Reports 121, pp. 263.

22 - Sato, H.: 1986, Mod. Phys. Lett. A1, pp. 9.
     Stebbins, A.: 1986, Astrophys. J. 303, pp. L21.
     Stebbins, A., Brandenberger, R., Veeraraghavan, S., Silk, J. and Turok, N.: 1987, Astrophys. J., preprint.
     Turok, N.: 1985, Phys. Rev. Lett. 55, pp. 1801.
     Turok, N., and Brandenberger, R.: 1986, Phys. Rev. D33, pp. 2175.

23 - Melott, A.L. and Sherrer, R.: 1987, Nature, preprint.

24 - Dekel, A.: 1984, Astrophys. J. 284, pp. 445.
     Kaiser, N.: 1984, Astrophys. J. 284, pp. L9.
     Politzer, D. and Wise, M.: 1984, Astrophys. J. 285, pp. L1.

25 - Gott, J.R., Melott, A.L. and Dickinson, M.: 1986, Astrophys. J. 306, pp. 341.
     Hamilton, A.J.S., Gott, J.R. and Weinberg, D.: 1986, Astrophys. J. 309, pp. 1.

26 - Klypin, A.A.: 1987, Astron. Zh. 64, pp. 15.

27 - Gurbatov, S.M., Saichev, A.I. and Shandarin, S.F.: 1985, Doklady Acad. Nauk 285, pp. 323.

Quasars: Their Evolution, Absorption Lines and the Intergalactic Gas

(G. Setti)

QUASAR EVOLUTION

For many years now it has been known that the number vs. magnitude counts of quasars is the prima facie evidence of a cosmological evolution of this class of objects. Down to an apparent magnitude B ~ 19 the number of quasars increases by almost a factor 8 per magnitude interval, compared to a factor 4 obtained in an Euclidean universe filled with a uniform distribution of sources, and correspondingly less for the classical Friedmann models of Gen. Relativity due to the redshift effects. This strong evolution has been recently questioned[1] because of several biases which may artificially steepen the slope of the counts in the optical surveys. Deep surveys of quasars selected via multi-colour techniques down to B ~ 23 have confirmed the long standing inference that the number count relationship must flatten beyond B ~ 20 [2,3]. As a consequence, pure density evolution models, where only the number density of quasars increases with the redshift z, leaving unchanged the shape of the local luminosity function, are ruled out, since too many quasars are predicted at faint magnitudes compared to the dramatic flattening of the counts (see, e.g., ref. 4).

These findings have been fully confirmed by a recently published survey of UVX selected quasars[5] which, most importantly, is large enough (170 objects with B < 20.9, z < 2.2 and M_B < -23) to permit a somewhat detailed description of the evolution of the luminosity function. It is found that the luminosity function, best parameterized by two power laws (the steeper one applicable to the most luminous objects), is globally shifted toward higher luminosities according to a power law of the form $(1 + z)^{3.6}$, almost a hundredfold increase at z = 2.2, while the derived local (z = 0) luminosity function is consistent with the luminosity function of Sy 1 nuclei[6]. The flattening of the counts is a direct consequence of the break in the luminosity function.

These results very much strengthen earlier conclusions based on studies of different combinations of optical samples[2,4,7,8] and provide some conclusive evidence on the long debated question of the precise nature of the evolution. Luminosity dependent density evolution models[9] and luminosity (or luminosity dependent) evolution models in which the luminosity increase is an exponential in the look-back time[10] do not appear to be consistent with present data. The question of the precise form of the luminosity function and its evolution is important also because it may shed light on the physics of quasars. Thus, a pure luminosity evolution poses a severe energetic problem, since it indicates that the order of 10^{9-10} $M_\odot$ must be radiated away during the life-time of a typical object[9,11], while a break in the luminosity function may indicate the presence of competing emission mechanisms[7].

The extension of the studies to samples complete at higher redshifts is probably important to pin down the precise form of the evolution. While most authors agree on the decline of the evolution beyond a redshift z ~ 2.5, the cut-off does not appear to be as sharp as it was thought to be[12,13]. The quasar distribution at high redshifts remains very uncertain. Attempts to interpret the decline in the number of quasars in terms of an increased amount of dust absorption at higher z in the host galaxies or in the surrounding intergalactic medium[14] seems to be contradicted by the observations that do not indicate the presence of a gradual increase of the reddening of quasar spectra at high z [15]. The redshift record is now set at z = 4.43 [16], close to the limit of the multi-colour technique used to discover this quasar[17]. This raises the important question of how to discover quasars at even higher redshifts where the optical detection techniques so far used seem to fail, while at the same time the detection of even one object with a redshift significantly larger than 5 would have profound implications on the long debated question of the epoch at which structures in the Universe are believed to have formed.

Since most quasars, if not all, are X-ray loud[18], it is clear that X-ray observations provide a powerful tool to investigate the cosmological properties of these objects, independent from the sometimes uncertain selection effects which may bias the optical surveys. The Einstein Medium Sensitivity Survey has provided a complete sample of X-ray selected quasars and Sy 1 nuclei[19]. The slope of the X-ray source counts is steep, clearly showing the presence of a strong cosmological evolution within the framework of the Friedmann models, but not as steep as that observed in the optically selected samples since it corresponds to an increase in the number of objects of only about a factor 5 per magnitude interval. Although a number of effects may satisfactorily account for the different slopes, it appears that the X-ray source counts derived from the optical counts (via the observed X-ray to optical flux ratios) overpredict the observed one by a factor 1.5-2.5 [20,21]. Some ways out have been proposed, but this discrepancy has not been satisfactorily solved as yet[21,22].

Although the cosmological evolution of radio loud quasars (about 10% of total) is well established, the evolution rate does not appear to be as strong as the one derived from the optically selected samples[23]. Within the statistics now available there is no significant difference between the evolution rates of quasars with flat and steep radio spectra[23,24]. Similarly, the radio luminosity function of flat spectrum radio quasars appears to fall at relatively large redshifts ($z \geq 2.5$) by a substantial factor, while this remains uncertain for the steep spectrum subclass[25,26]. The precise forms of the evolution of the radio luminosity functions are not yet constrained by the observations, but a pure luminosity evolution may be difficult to understand in view of the fact that the radio source lifetimes are believed to be much shorter than the Hubble time.

Following earlier work, it has been argued that the luminosity functions and surface densities of quasars can be noticeably affected by gravitational lensing due to compact objects (such as stars, Jupiters and black holes) either in galaxies or randomly distributed[27,28]. The brightening of quasars by minilensing can strongly influence the source counts, if their intrinsic luminosity function is steep[29]. Until now, however, there is no statistical and/or astrophysical evidence that this may indeed be the case.

Finally, it should be mentioned that the quasar counts (and also the extragalactic radio source counts) apparently could be interpreted within the framework of the chronometric cosmology without any evolution[30], but the validity of this result has been recently questioned[31].

CLUSTERING OF QUASARS

The clustering properties of quasars remain a somewhat controversial subject. While clustering on scales $\leq 10h^{-1}$ Mpc ($h = H_O/100$) has tentatively been detected in a UVX selected sample[32], clustering on these small scales has not been found in a much larger sample selected from IIIa-J objective prism plates[33] and in a comparable size sample of objects selected from blue grens plates[34]. There is no indication of general clustering at scales $> 10\ h^{-1}$ Mpc, although one has found[34] a group of seven quasars confined within a volume of size ~ 50 h^{-1} Mpc and with a dispersion in redshift consistent with velocity dispersions expected in an expanding supercluster.

QUASAR ABSORPTION LINES

The absorption lines found in the quasar spectra provide a unique tool to investigate the properties of the intergalactic gas at large distances, the physical parameters, composition and evolution of the intergalactic clouds and of the interstellar gas in galaxies at large redshifts, the large-scale distribution

of matter over a wide range in redshift, and other properties relevant to cosmology and cosmogony. An excellent review of this subject can be found in ref. 35.

Several absorption systems are of interest here since they are generally
interpreted as due to intervening absorbers:

a) The Lyα-only systems (sharp absorption lines blueward of the Lyα emission
line) are usually interpreted as due to intervening intergalactic clouds with
typical column densities $N(HI) < 10^{16}$ cm^{-2}. These clouds are commonly thought to
be tenuous condensation of highly ionized hydrogen with mass of the order of a few
times 10^7 M$_\odot$ and size ~ 10 kpc in pressure equilibrium with a hotter intergalactic
medium or, maybe, gravitationally bound by cold dark matter[36].

b) The heavy element systems present, in addition to Lyα, narrow lines of heavier
elements consistent with a composition not too different from solar and are
typically found at intermediate column densities, $10^{16} < N(HI) < 10^{20}$ cm^{-2}. These
systems are almost always found also in the Lyman limit systems, so called because
of the Lyman discontinuity becoming detectable when the optical depth at λ < 912 Å
is $\tau \geq 1$, that is for column densities $N(HI) \geq 1.5 \times 10^{17}$ cm^{-2}. It has been assumed
that the heavy element systems are formed in the outer regions of intervening
galaxies. However, it has been recently argued[37] that the similarity of the column
density distribution function to that of Lyα-only systems strongly supports the
hypothesis of one absorber population, intergalactic clouds with (perhaps) chemical composition ~ 0.1 the solar value. The fact that by definition the heavy elements absorption lines are not found in the Lyα systems implies a small degree of
ionization, hence total column densities $\leq 10^{17}$ cm^{-2} and, therefore, clouds of
small size ($\leq 10^{18}$ cm). It should be stressed here that while this suggestion may
explain a number of properties of the absorption system samples, it still remains
difficult to understand how the enrichment in heavy elements has occurred and the
lower limit to the absorber size ($\geq$ 3 kpc) derived from Lyα systems in the two
components of a double quasar due to gravitational lensing[38].

   ˙ It should be noted that not all the heavy element systems necessarily
originate in intergalactic clouds. For instance, the predicted number of Lyman
limit systems due to absorption in intervening spiral galaxies is a factor ~ 6
smaller than the ones observed for column densities $\leq 3 \times 10^{18}$ which comprise ~ 85%
of the heavy element systems, while for larger column densities the predictions
agree with observations[37]. It should also be noted that the sky coverage of these
absorbers is ~ 50% [37], and thus any background radiation capable of ionizing
hydrogen can be largely affected. The origin of the intergalactic clouds is
unknown.

c) The damped Lyα systems (very strong Lyα absorption with associated heavy element lines of low ionization) are found at large column densities $N(HI) > 10^{20}$
cm^{-2} and are believed to be associated with intervening HI disk galaxies[39]. The
density of these systems implies that ~ 20% of the sky is covered by gaseous disks
with large column densities in the redshift interval z (2-3), a factor five in
excess of the one predicted[40].

   Finally, it is generally believed that the Lyα systems show a clear
indication of cosmological evolution in the sense that their co-moving density
appears to increase with z [41], an effect which would entail an increase of the
number density and/or of the average size of the Lyα absorbing clouds. Since the
same does not appear to be true for the heavy element systems, this has been
considered as a further evidence against a common origin of the two types of
absorption systems. However, a number of anomalies elucidated in the analysis of
published samples indicate the possible presence of (poorly understood) selection
effects which may render any statement concerning the evolution rather
uncertain[42].

Clarification of these matters is important also because the Lyα absorption lines can provide a very powerful tool to probe the large scale structure of the Universe at large redshifts.[43]

INTERGALACTIC GAS

Spectrophotometric observations of a sample of high redshift (z ~ 3) quasars, combined with the statistical properties of the Lyα absorption systems to correct for the apparent depression of the continuum shortward of Lyα, has led to a downward revision of more than an order of magnitude on the Gunn-Peterson upper limit to the amount of neutral hydrogen present in the diffuse intergalactic gas.[44] The new limit corresponds to $n(HI) < 9 \times 10^{-14}$ h cm^{-3} (at z = 0). Under the hypothesis that the intergalactic gas is maintained fully ionized by the UV radiation from quasars (at z ~ 2.6) and that it is in pressure equilibrium with the Lyα clouds one finds $n(H_{tot}) < 4.6 \times 10^{-7}$ h cm^{-3} and a total contribution to the density of the Universe of $\Omega_0(H) < 0.05$ h.

References

1 - Wampler, E.J. and Ponz, D.: 1985, Astrophys. J. 298, pp. 448.

2 - Koo, D.C.: 1986, in "Structure and Evolution of Active Galactic Nuclei", G. Giuricin et al. (eds.), Reidel, Dordrecht, Astrophys. Spa. Sci. Library 121, pp. 317.

3 - Koo, D.C., Kron, R.G. and Cudworth, L.M.: 1986, P.A.S.P. 98, pp. 285. Marano, B., Zamorani, G. and Zitelli, V.: 1986, in "Structure and Evolution of Active Galactic Nuclei", G. Giuricin et al. (eds.), Reidel, Dordrecht, Astrophys. Spa. Sci. Library 121, pp. 339.

4 - Setti, G.: 1984, in "X-Ray and UV Emission from Active Galactic Nuclei, W. Brinkmann and J. Trümper (eds.), MPE Report 184, pp. 243.

5 - Boyle, B.J., et al.: 1987, Mon. Not. R. astr. Soc. 227, pp. 717.

6 - Weedman, D.W.: 1986, in "Structure and Evolution of Active Galactic Nuclei", G. Giuricin et al. (eds.), Reidel, Dordrecht, Astrophys. Spa. Sci. Library 121, pp. 215.

7 - Marshall, H.L.: 1985, Astrophys. J. 299, pp. 109.

8 - Koo, D.C., and Kron, R.G.: 1987, in IAU Symp. 124 on "Observational Cosmology", A. Hewitt, G. Burbidge and L. Zhi Fang (eds.), Reidel, Dordrecht, pp. 383.

9 - Schmidt, M.: 1987, in IAU Symp. 124 on "Observational Cosmology", A. Hewitt, G. Burbidge and L. Zhi Fang (eds.), Reidel, Dordrecht, pp. 619.

10 - Cavaliere, A., Giallongo, E. and Vagnetti, F.: 1985, Astrophys. J. 296, pp. 402.

11 - Setti, G. and Zamorani, G.: 1984, in COSPAR/IAU Symp. on "High Energy Astrophysics and Cosmology", G.F. Bignami and R.A. Sunyaev (eds.), Advances Spa. Res. 3, pp. 175.

12 - Schmidt, M., Schneider, D.P. and Gunn, J.E.: 1987, Astrophys. J. 316, pp. L1.

13 - Véron, P.: 1986, Astron. Astrophys. 170, pp. 37.

14 - Ostriker, J.P. and Heisler, J.: 1984, Astrophys. J. 278, pp. 1.

15 - Wrigth, E.L.: 1986, Astrophys. J. 311, pp. 156. Malkan, M.: 1986, in IAU Symp. 119 on "Quasars", G. Swarup and V.K. Kapahi (eds.), Reidel, Dordrecht, pp. 453.

16 - Hewett, P.C.: 1987, private communication.

17 - Hewett, P.C.: 1987, in IAU Symp. 124 on "Observational Cosmology", A. Hewitt, G. Burbidge and L. Zhi Fang (eds.), Reidel, Dordrecht, pp. 664.

18 - Avny, Y. and Tananbaum, H.: 1986, Astrophy. J. 305, pp. 83.

19 - Gioia, I.M., et al.: 1984, Astrophys. J. 283, pp. 495.

20 - Schmidt, M. and Green, R.F.: 1986, Astrophys. J. 305, pp. 68.

21 - Setti, G.: 1987, in IAU Symp. 124 on "Observational Cosmology", A. Hewitt, G. Burbidge and L. Zhi Fang (eds.), Reidel, Dordrecht, pp. 579.

22 - Franceschini, A., Gioia, I.M. and Maccacaro, T.: 1986, Astrophys. J. 301, pp. 124

23 - Wall, J.V., and Peacock, J.A.: 1985, Mon. Not. R. astr. Soc. 216, pp. 173.

24 - Savage, A., et al.: 1987, in IAU Symp. 124 on "Observational Cosmology", A. Hewitt, G. Burbidge and L. Zhi Fang (eds.), Reidel, Dordrecht, pp. 673.

25 - Peacock, J.A. and Dunlop, J.S.: 1986, in IAU Symp. 124 on "Observational Cosmology", A. Hewitt, G. Burbidge and L. Zhi Fang (eds.), Reidel, Dordrecht, pp. 455.

26 - Downes, A.J.B., et al.: 1986, Mon. Not. R. astr. Soc. 218, pp. 31.

27 - Vietri, M.: 1985, Astrophys. J. 293, pp. 343.

28 - Ostriker, J.P. and Vietri, M.: 1986, Astrophys. J. 300, pp. 68.

29 - Schneider, P.: 1986, Astr. Astrophys., submitted (MPA 218, preprint).

30 - Segal, I.E. and Nicoll, J.F.: 1986, Astrophys. J. 300, pp. 224.

31 - Wright, E.L.: 1987, Astrophys. J. 313, pp. 551.

32 - Shanks, T., et al.: 1987, Mon. Not. R. astr. Soc. 227, pp. 739.

33 - Clowes, R.G., Iovino, A. and Shaver, P.: 1987, Mon. Not. R. astr. Soc. 227, pp. 921.

34 - Crampton, D., Cowley, A.P. and Hastwick, F.D.A.: 1987, Astrophys. J. 314, pp. 129.

35 - Sargent, W.L.W.: 1987, in IAU Symp. 124 on "Observational Cosmology", A. Hewitt, G. Burbidge and L. Zhi Fang (eds.), Reidel, Dordrecht, pp. 777.

36 - Rees, M.J.: 1986, Mon. Not. R. astr. Soc. 218, pp. 25p.

37 - Tytler, D.: 1987, Astrophys. J. 321, pp. 49.

38 - Foltz, C.B., et al.: 1984, Astrophys. J. 281, pp. L1.

39 - Wolfe, A.M., et al.: 1986, Astrophys. J. Suppl. 61, pp. 249.

40 - Smith, H.E., Cohen, R.D. and Bradley, S.E.: 1986, Astrophys. J. 310, pp. 583.

41 - Murdoch, H.S., et al.: 1986, Astrophys. J. 309, pp. 19.

42 - Tytler, D.: 1987, Astrophys. J. 321, pp. 69.

43 - Carswell, R.F. and Rees, M.J.: 1987, Mon. Not. R. astr. Soc. 224, pp. 13p.

44 - Steidel, C.C. and Sargent, W.L.W.: 1987, Astrophys. J. 318, pp. L11.

COMMISSION N°48 : HIGH ENERGY ASTROPHYSICS (ASTROPHYSIQUE DES HAUTES ENERGIES)

President   : CESARSKY C.J.
Vice-president   : SUNYAEV R.A.

Organizing Committee : CLARK G.W., GIACCONI R., QU WIN-YUE, SALPETER E.E.,
                     : SCHEUER P.A., SCHRAMM D.N., TRIMBLE V.L., TRUEMPER J.,
                     : WOLFENDALE A.W., WOLTJER L.

I.   X-RAY ASTRONOMY

The european X-ray observatory (EXOSAT), which was launched in 1983 and which finished operations in April 1986, has brought a rich harvest of results in the period 1984-1987, surveyed here. The EXOSAT payload consisted of three sets of instruments : two low energy imaging telescopes (LE:E<2 KeV), a medium-energy experiment (ME:E=1-50KeV) and a gas scintillation proportional counter (GSPC:E=2-20KeV). Over most of the energy range covered, EXOSAT was not more sensitive than its predecessor, the american EINSTEIN satellite. But the EINSTEIN satellite is far from having exhausted the treasures of the X-ray sky. And EXOSAT, thanks to its elliptical 90-hour orbit, had the extra advantage of being able to make long, continuous observations of interesting objects, lasting up to 72 hours. Thus, EXOSAT was very well suited for variability studies, and many of its most important findings are in this area. EXOSAT observations sample a wide range of astrophysical sources : X-ray binaries, cataclysmic variables and active stars ; supernova remnants and the interstellar medium ; active galactic nuclei, and clusters of galaxies. Among the highlights, let us mention :

-    the detection of quasi-periodic oscillations (QPO) in seven well known X-ray sources, starting with the galactic bulge source GX 5-1, and including Sco X-1, the brightest X-ray source in the sky, Cyg X2 and the Rapid Burster. The majority of the QPO sources are very luminous ($>10^{38}$erg/sec), and the oscillations are strong (5% rms variation in flux), persistent ($> 10^{5}$cycles), and take place at high frequency (5-50 Hz). Other characteristics of the QPO, however, such as their relation to the source intensity or the spectral shape, vary widely from source to source. QPO sources are believed to be binary systems containing a neutron star, and the models proposed generally involve interactions between the accretion disk and the neutron star magnetosphere.

-    the mapping out of the X-ray galactic ridge in the 2-6 KeV band. The "ridge" is a disk of radius 10 to 12 Kpsec and a height of a few hundred parsecs ; its total luminosity is 10^{38}erg/sec.

- the detailed observations of active galactic nuclei (AGN), revealing, in some cases, the presence of a soft X-ray component, in the .05-1 KeV range. Long term monitoring of AGN allowed to study their variability on timescales from minutes to years. For NGC 4151, the flux in the 2-10 KeV band can vary by a factor 2 over periods of six months, while the soft component remains constant. A possible interpretation is that the soft component originates in a hot intercloud medium in the narrow-line region, while the hard X-ray flux is emitted by regions surrounding the nucleus. EXOSAT observations indicate that rapid variability, on time scales of the order of an hour, is frequent in AGNs.

The results obtained by EXOSAT have found a useful complement in those obtained by the japanese satellite TENMA, which was launched in 1983. The main instrument on board of TENMA consists of two sets of four Gas Scintillation Proportional Counters, which attain maximum efficiency in the range 1.5-35 KeV, and yield a spectral resolution ~ 9% at 6 KeV. The most interesting results obtained with the TENMA satellite concern the iron line spectroscopy.

Iron lines in emission have been detected in a wide variety of sources, including AGN, clusters of galaxies, SNR and X-ray binaries. TENMA has also taken spectra of the galactic ridge ; the spectra have a characteristic thermal shape, and in most of them the helium-like iron line is present at 6.7KeV, at about the expected intensity for a normal Fe abundance. The temperatures derived from the spectra are in the 5-10 KeV range. It is difficult to explain the galactic ridge as a superposition of small sources ; it seems more likely that the existence of a new, perhaps transient, very hot phase of the interstellar medium has been revealed.

In many sources, the iron K-edge absorption feature at ~ 7.2 KeV has also been measured by TENMA. This is important because, by comparison with the soft X ray photoelectric absorption, it allows to measure the iron abundance. TENMA observations of galactic massive binaries and of a few AGN indicate that their Fe abundance is normal.

II. COSMIC GAMMA RAYS

The analysis of the data of the Cos B satellite has been pursued by various groups ; the activity has been centered on the detailed comparison of the gamma ray data with observations of atomic and molecular hydrogen, taking advantage of the new CO survey of the Columbia group. In September 1985, the Cos B data base has been released ; it is now available for the whole scientific community.

New results have been obtained in gamma ray spectroscopy. Line emission due to the decay of the radioactive isotope ^{26}Al, at 1.809 MeV, has been observed by the gamma ray spectrometers on board of the HEAO 3 and SMM satellites, and confirmed by several balloon experiments. A diffuse ^{26}Al line is expected from the interstellar medium ; novae, Wolf-Rayet stars, and, to a lesser extent, supernovae should contribute to it.

Recent observations by SMM of the electron-positron pair annihilation radiation around 511 KeV suggest the existence of an extended steady source of this radiation, super imposed to the central, highly variable point source which had been discovered earlier. This emission implies the presence of a background of positrons, whose origin is still open to debate. They may be products of the decay of radioactive nuclei, such as ^{44}Ti, or perhaps ^{56}Ni from supernovae, or ^{26}Al from novae and Wolf-Rayet stars ; or they may be accelerated and ejected by pulsars.

Much excitement has been generated in the high energy community these last years over the detection of very high energy (TeV range) and ultra high energy (PeV range) gamma ray emission from X-ray binaries such as Cyg X-3 and Her X-1. At such energies, the observations are made from ground, using the atmospheric Cerenkov technique for TeV gamma rays, and arrays of particle detectors for PeV gamma rays. The fluxes measured for X-ray binaries are always close to the limit of detection of the instruments used, and these observations await for confirmation from more powerful installations.

Two groups have reported detection of muons from Cygnus X-3 with proton decay experiments, but other experiments, obtained upper limits lying below the detections claimed.

III. COSMIC RAYS

Many aspects of the mechanism of particle acceleration by diffusive shock waves have been examined in detail : effect of oblique waves ; of stellar wind terminal shocks ; dynamic effects and damping of cosmic-ray generated waves, etc.

A complete, but approximate, time-independent solution to the problem of cosmic-ray dominated shocks has been developed ; for interstellar conditions, the spectrum predicted is close to a power law of index (-2), and the efficiency of acceleration is ~ 25%.

IV. PROSPECTS FOR THE NEAR FUTURE

In the very near future, much attention will be devoted to observations of high energy radiation from supernova 1987A, and their interpretation. Predictions have been made by many groups for the flux and flux variations expected in the framework of various models : comptonisation of the cobalt decay line radiation, of the radiation from a central pulsar ; radiation associated to newly accelerated particles, etc. The japanese-british large area proportional counter, on board of the japanese Ginga satellite, and the european-soviet payload Roentgen on board of the Qvant module attached to a MIR station, are operational now, and can measure this radiation respectively in the range 0.5-20 KeV and 2-1500 KeV.

In the next year and a half, many hard-X and γ ray experiments (most often american) will be flown by balloons over the southern hemisphere, to measure the continuum and, hopefully, the gamma-ray line emission from SN 1987A. The SMM satellite may also contribute to this search.

In 1988, two satellites will be launched by the Soviet Union, which are devoted to high energy astrophysics : the soviet-french-polish experiment GAMMA 1, for observations of gamma rays of energy >50 MeV, and GRANAT, which will carry soviet and danish X-ray experiments, soviet and french gamma ray-burst experiments, and the french gamma-ray camera SIGMA (30 KeV-2 MeV).

Important advances are also expected for ground gamma ray astronomy in the TeV range, since new and sophisticated detector systems will be put in operation all over the world.

It is only towards the end of the decade that other missions, delayed by the consequences of the Challenger accident, will be launched, such as the german ROSAT soft X-ray satellite, the american Gamma Ray Observatory, the european ULYSSES which carries the most promising cosmic-ray experiments to be flown in the near future...

More details about these and other topics of high energy astrophysics can be found in the proceedings of the following meetings :

X-Ray Astronomy in the EXOSAT era.
    (edited by A. Peacock)
    Space Science Reviews, N°1-4 (1985)

The physics of accretion onto compact objects.
    (edited by K.O. Mason, M.G. Watson and N.E. White)
    Springer-Verlag, Lecture Notes in Physics, n°266 (1986)

High Energy Phenomena around Collapsed Stars.
    (edited by F. Pacini)
    NATO ASI Series, Reidel (1987)

    Proceedings of the 19th International Cosmic Ray Conference,
    La Jolla, August 1985

    Proceedings of the 20th International Cosmic Ray Conference,
    Moscow, August 1987.

Gamma-Ray Astronomy
    (edited by K. Hurley and G. Vedrenne)
    Advances in Space Research, Vol.6 n°4, Pergamon Press (1986).

C.J. CESARSKY

PRESIDENT: S. Grzedzielski   VICE-PRESIDENT: L.F. Burlaga

Introduction
S. Grzedzielski

The area of interest to the Commission includes:
1. Solar wind composition and dynamics;
2. Interaction of solar wind with extended interplanetary sources of plasma and gases of non-solar origin;
3. Structure and dynamics of the three-dimensional heliosphere;
4. Interaction of heliosphere with the local interstellar medium.
     The following reports summarize recent developments in the aforementioned fields.

Dynamic Phenomena in the Solar Wind
L.F. Burlaga

In this review a few topics have been selected which are being actively investigated and which are particularly important for understanding the heliosphere. The reader is also referred to the following reviews: Burlaga (1984, 1986, 1987a), Gosling (1986), Hundhausen (1985), Hundhausen et al. (1984), Kahler (1985), Klein (1987), Mihalov (1987), Pizzo (1985, 1986), Richter et al. (1985), Schwenn (1986) and Smith (1985).

LARGE-SCALE MAGNETIC FIELD AND PLASMA

Slavin et al. (1984) and Thomas et al. (1986) reported that at 10 AU the azimuthal component of the magnetic field is 25% lower than predicted by Parker. Klein et al. (1987), who considered the effect of temporal variations of the bulk speed, found from an analysis of Voyager data that the difference is only (10±7)% for the interval from 1977 to 1981. They found that the difference between the magnitude of the magnetic field observed at 10 AU and that predicted by Parker was less than 1% . Good agreement between observations and the predictions of Parker between 1 AU and 9.5 AU was also found by Burlaga et al. (1984).

Suess et al. (1985) and Nerney and Suess (1985) calculated that a 25% flux deficit might be produced in an axially symmetric solar wind as a result of the fact that the magnetic field is higher near the ecliptic than at higher latitudes. However, Pizzo and Goldstein (1987) point out that the solar wind conditions assumed by Suess are unrealistic. They show that a deficit of 10% can be produced by certain flow configurations which might be observed during the declining phase of solar activity, but they suggest that the deficit should be smaller when solar activity is increasing or high, which was the situation during which the radial variations discussed above were determined.

Thomas et al. (1987) report a significant meridional component of the magnetic field at 1 AU, corresponding to a flaring angle of 1.3°. Pizzo and Goldstein (1987) note that this implies a meridional flow speed of 10 km/s, for which there is no observational or theoretical basis.

The radial variation of the plasma parameters shows no surprises. The density falls of as R^{-2}, and $T_p \propto R^{-\alpha}$, where $\alpha = 0.5 \div 0.7$ (Gazis (1984)). There

is no appreciable difference in the median values and distributions of the
density, speed and temperature for two parts of the solar cycle at 10 AU
(Barnes and Gazis, 1984). However, the distribution of dynamic pressure shows
a larger tail when solar activity is high (Barnes and Gazis (1984)).

MAGNETIC CLOUDS, CMEs AND BIDIRECTIONAL ANISOTROPIES

Goldstein (1983) suggested that magnetic clouds are force-free
configurations, but he did not present a specific solution. Marubashi (1987)
computed magnetic field profile that might be observed by a spacecraft passing
through a magnetic cloud with a force-free configuration, based on the
arbitrary assumption that the pitch angle of the magnetic field increases as
the square of the distance from the axis of the magnetic cloud. He found
qualitative agreement between the model and the observations for two magnetic
clouds. Woltjer's (1958) configuration of lowest energy force-free field with
α = const gives magnetic field profiles which resemble those observed in
magnetic clouds (Burlaga, 1987b).

A dynamical model of magnetic clouds was presented by Ivanov and
Harshiladze (1984). Suess (1987) presented a model in which the magnetic field
in a magnetic cloud is driven by an axial current as in the pinch effect.

Further evidence that magnetic clouds may be associated with disappearing
filaments and coronal mass ejections was presented by Wilson and Hildner
(1984, 1986) and Burlaga et al. (1987b). It is not likely that every coronal
mass ejection is associated with a magnetic cloud. Magnetic clouds are
associated with geomagnetic storms (Zhang and Burlaga, 1987; Wilson, 1987).
Magnetic clouds which are associated with compound streams can produce
relatively large geomagnetic storms (Burlaga et al. ,1987; Zhang and Burlaga,
1987), because the magnetic field is amplified in the interaction between a
magnetic cloud and a stream or shock. A magnetic cloud has only small effect
on galactic cosmic rays, unless it is preceded by a shock and a turbulent
interaction region (Zhang and Burlaga, 1987).

Bidirectional anisotropies of energetic particles have been interpreted
as evidence of closed magnetic loop structures. Gosling et al. (1986, 1987a)
identified "bidirectional solar wind heat flux events" in ISEE-3 data, which
they assume are signatures of coronal mass ejections. Gosling et al. (1987b)
found that for such events there is a westward deflection of the flow just
ahead of the event followed by an eastward deflection in the event itself.
Some events are magnetic clouds, but many others do not show the rotation of
the magnetic field which is characteristic of a magnetic cloud even though
they typically have low temperature, low magnetic field variance, and high
magnetic field intensity, which are characteristic of magnetic clouds. Marsden
et al. (1987) identified bidirectional anisotropies in low energy protons.
Some of their events are related to magnetic clouds. There does not appear to
be a one to one relation between bidirectional solar wind heat flux events and
bidirectional anisotropies in the protons.

SHOCKS AND FLOWS IN THE INNER HELIOSPHERE

MHD models of transient flows and shocks associated with coronal
transients and flare ejecta in the corona were developed by Dryer and Smart
(1984) and Dryer et al. (1984). Three-dimensional flows within 1 AU were
modeled by Dryer et al. (1986). A kinematic model of interplanetary
disturbances associated with shocks produced by flares was further discussed
by Akasofu et al. (1985a, 1985b), Olmsted (1985, 1986) and Hakamada (1987).
Smart and Shea (1985) discussed a method of "timing" a solar flare, based on
the assumption that a shock moves at constant speed out to some distance and
thereafter moves at a speed proportional to $R^{-1/2}$. An improved method for
calculating shock normals was developed by Vinas and Scudder (1986).

Interplanetary shocks observed at 1 AU in association with Type II radio bursts (Cane, 1984) are relatively fast and are almost always associated with a solar flare (Cane and Stone, 1984) and a long duration solar X-ray event (Cane, 1985). Shocks related to solar filament eruptions outside active regions are not associated with Type II bursts, move relatively slowly, and are accompanied by only weak soft X-ray bursts (Cane et al., 1986). The shocks are spherical over a wide range of longitudes (Cane et al., 1987). The three-dimensional geometry of shocks associated with coronal mass ejections was discussed by Schwenn (1986).

A slow forward shock wave was observed at 0.3 AU by Richter et al. (1985). Whang (1987) has shown that a slow shock can transform into a fast shock as it moves away from the sun.

SHOCKS AND INTERACTION REGIONS IN THE OUTER HELIOSPHERE

Near the sun, isolated streams are the dominant dynamical feature (see, e.g., Richter and Luttrell, 1986), whereas far from the sun large pressure waves unaccompanied by streams are dominant (Burlaga, 1984).

At intermediate distances (1 to ~10 AU) fast streams overtake slower streams to form "compound streams", and their separate interaction regions coalesce to form "merged interaction regions" (Burlaga et al., 1985a). This process of "entrainment" was modeled by Burlaga et al. (1985b) and Whang and Burlaga (1985a,b). Burlaga et al. (1986a) showed one case in which five streams and interaction regions at 1 AU coalesced to form one compound stream and two merged interaction regions at 6.5 AU, when an unusually fast stream entrained the flows ahead. Whang and Burlaga (1986) showed that the two merged interaction regions themselves coalesced between 6.5 AU and 9.5 AU.

Whereas a shock wave at 1 AU is usually driven by a stream, a shock wave far from the sun may be detached from the stream which originally produced it or the stream may damp out, leaving the shock to propagate alone. Two forward shocks or two reverse shocks may coalesce to form a single forward shock or reverse shock, respectively; and a forward shock may interact with a reverse shock, in which case both shocks emerge from the collision weaker than they were initially (Whang and Burlaga, 1985a,b). Shocks may form as far as 7 AU from the sun (Gazis et al., 1985). Shocks and transient streams are observed out to at least 29 AU (Kayser, 1985; Kayser et al., 1984). However, shocks associated with flares are less frequent, slower, and weaker far from the sun than those close to the sun (Mihalov, 1985). The topological configurations of corotating shocks in the outer heliosphere for the case of one or two corotating streams at 1 AU were computed by Burlaga and Klein (1986).

To determine the structure of the solar wind on a scale of 30 AU one must examine data over an interval of the order of 120 days. During such an interval one typically sees at least several interaction regions and the variations of the magnetic field and plasma parameters appear as "large-scale fluctuations" (Burlaga, 1984). Burlaga et al. (1984) showed that the characteristic period of the large-scale fluctuations in the magnetic field intensity apparently increased with distance from the sun between 1 AU and 9.5 AU, which they attributed to entrainment. This was confirmed by Burlaga and Mish (1987) using simultaneous data from ISEE-3 and Voyager. Burlaga and Goldstein (1984) found that the radial evolution of the spectra of the magnetic field strength may be different for systems of corotating flows and systems of transient flows. Evidence for period-doubling in the solar wind was found by Burlaga and Lazarus (1987), who showed that Voyager 2 observed recurrent interaction regions with a period of 25 days at ~15 AU when IMP-8 observed recurrent interaction regions with a period of 12.5 days relative to a fixed frame at 1 AU. This behavior is contrary to predictions for strictly periodic flows (Smith et al., 1985), but such behavior is observed in driven damped nonlinear oscillators.

REFERENCES

(JGR = Journal of Geophysical Research)

Akasofu, S.I. et al.: 1985a, JGR 90, p 8193; Akasofu, S.-I. et al.: 1985b, JGR 90, p 4439; Barnes, A., Gazis, P.R.: 1984, Uranus and Neptune, ed. J.T. Bergstralh, NASA-CP 2330, NASA, Washington DC, p 527; Burlaga, L.F.: 1984, Space Sci. Rev. 39, p 255; 1986, The Sun and the Heliosphere in Three Dimensions, ed. R.G. Marsden, D. Reidel, Dordrecht, p 191; 1987a, Solar Wind VI, submitted; 1987b, JGR, submitted; Burlaga, L.F., Goldstein, M.L.: 1984, JGR 89, p 6813; Burlaga, L.F., Klein, L.W.: 1986, JGR 91, p 8975; Burlaga, L.F., Lazarus, A.J.: 1987, JGR; Burlaga, L.F., Mish, W.H.: 1987, JGR 92, p 1261; Burlaga, L.F. et al.: 1984, JGR 89, p 10659; Burlaga, L.F. et al.: 1985a, JGR 90, p 12127; Burlaga, L.F. et al.: 1985b, JGR 90, p 7377; Burlaga, L.F. et al.: 1986a, JGR 91, p 13331; Burlaga, L.F. et al.: 1986b, JGR 91, p 2917; Burlaga, L.F. et al.: 1987b, JGR 92, p 5725; Cane, H. V.: 1984, Astron. Astrophys. 140, p 205; 1985, JGR 90, p 191; Cane, H. V., Stone, R.G.: 1984, Astrophys. J. 282, p 339; Cane, H. V. et al.: 1986, JGR 91, p 13321; Cane, H. V. et al.: 1987, JGR, in press; Dryer, M.S., Smart, D.F.: 1984, Adv. Space Res. 4, p 291; Dryer, M.S. et al.: 1984, Astrophys. Space Sci. 105, p 187; Dryer, M.S. et al.: 1986, The Sun and the Heliosphere in Three Dimensions, ed. R.G. Marsden, D. Reidel, Dordrecht, p 135; Gazis, P.R.: 1984, JGR 89, p 775; 1987, JGR 92, p 2231; Gazis, P.R. et al.:1985, JGR 90, p 9454; Goldstein, H.: 1983, Solar Wind V, ed. M. Neugebauer, NASA Conf. Proc. SP 2280, p 731; Goldstein, M.L. et al.: 1984, JGR 89, p 3747; Gosling, J.T.: 1986, Magnetospheric Phenomena in Astrophysics, eds. R.I. Epstein and W.C. Feldman, American Institute of Physics, N.Y., N.Y., p 124; Gosling, J.T., McComas, D.J.: 1987, Geophys. Res. Lett. 14, p 355; Gosling, J.T. et al.: 1986, JGR 91, p 352; Gosling, J.T. et al.: 1987a, JGR 92, p 8519; Gosling, J.T. et al.: 1987b, JGR, in press; Hakamada, K.: 1987, JGR 92, p 4339; Hundhausen, A.J.: 1985, Collisionless shocks in the Heliosphere, eds. R.G. Stone and B.T. Tsurutani, AGU Geophysical Monograph 34, p 37; Hundhausen, A.J. et al.: 1984, eds. D.M. Butler and K. Papadopoulous, NASA Ref. Publ. 1120, pp 6.1-6.32; Ivanov, K.G., Harshiladze, A.F.: 1984, Solar Phys. 92, p 351; Kahler, S.: 1987, U.S. National Report to International Union of Geodesy and Geophys. 1983-1986, American Geophysical Union, Washington D.C., p 663; Kayser, S.E.: 1985, JGR 90, p 3967; Kayser, S. E. et al.: 1984, Astrophys. J. 285, p 339; Klein, L.: 1987, Solar Wind VI; Klein, L.W. et al.: 1987, JGR, in press; Marsden, R.G. et al.: 1987, JGR; Marubashi, K.: 1987, Adv. Space Sci.; Mihalov, J.D. 1985, JGR 90, 210; 1987, U.S. National Report to International Union of Geodesy and Geophys. 1983-1986, American Geophysical Union, Washington D.C., p 697; Nerney, S.F., Suess, S.T.: 1985, Astrophys. J. 296, p 259; Olmsted, C., Akasofu, S.-I.: 1985, Planet. Space Sci. 33, p 831; 1986, JGR 91, p 13689; Pizzo, V.J.: 1985, Collisionless Shocks in the Heliosphere: Reviews of Current Research, eds. B.T. Tsurutani and R.G. Stone, Geophysical Monograph 35, American Geophysical Union, Washington D.C., p 51; 1986, Adv. Space Res. 6(1), p 353; Pizzo, V.J., Goldstein, B.E.: 1987, JGR 92, p 7241; Richter, A.K., Luttrell, A.H.: 1986, JGR 91, p 5873; Richter, A.K. et al.: 1985, Collisionless Shocks in the Heliosphere: Reviews of Current Research, eds. B.T. Tsurutani and R.G. Stone, Geophysical Monograph 35, American Geophysical Union, Washington D.C., p 33; Richter, A.K. et al.: 1985, JGR 90, p 7581; Schwenn, R.: 1986, Space Sci. Rev. 44, p 139; Slavin, J.A. et al.: 1984, Geophys. Res. Lett. 11, p 279; Smart, D.F., Shea, M.A.: 1985, JGR 90, p 183; Smith, E.J., 1985, Collisionless Shocks in the Heliosphere: Reviews of Current Research, eds. B.T. Tsurutani and R.G. Stone, Geophysical Monograph 35, American Geophysical Union, Washington D.C., p 69; Smith, Z.K. et al.: 1985, JGR 90, p 217; Suess, S.T.: 1987, JGR, in press; Suess, S.T. et al.:

1985, JGR 90, p 4378; Thomas, B.T. et al.: 1986, JGR 91, p 6760; Vinas, A.F., Scudder, J.D.: 1986, JGR 91, p 39; Whang, Y.C.: 1984, JGR 89, p 7367; 1987, JGR, in press; Whang, Y.C., Burlaga, L.F.: 1985a, JGR 90, p 10765; 1985b, JGR 90, p 221; 1986, JGR 91, p 13341; Wilson, R.M., Hildner, E.: 1984, Solar Phys. 91, p 168; 1986, JGR 91, p 5867; Wilson, R.M.: 1987, Planet. Space Sci. 35, p 329; Woltjer, L.: 1958, Proceedings of the National Academy of Sciences 44 No 5, p 489; Zhang, G., Burlaga, L.F.: 1987, JGR, submitted

Minor Ions in the Solar Wind
Peter Bochsler

It has been recently recognized that the presence of ions heavier than hydrogen determines to a large extent the dynamics of the expanding solar corona. In the following I shall give a brief account of the most recent results related to minor ions (i.e. ^{3}He and ions heavier than helium).

ELEMENTAL ABUNDANCES

Table 1 gives elemental abundances as obtained by in situ measurements.

Table 1
Abundances relative to oxygen

	Solar Wind		Solar Energetic Particles		Solar System	
H	1900±400	[1]	---		1400	[11]
He	75±20	[2]	72±3	[8]	108	[11]
^{3}He/^{4}He	$(4.9\pm0.5)\cdot10^{-4}$	[3]	---			
C	0.43±0.02	[4]	0.435±0.040	[9]	0.60	[11]
N	0.15±0.06	[4]	0.124±0.010	[9]	0.12	[11]
O	≡1		≡1		≡1	
Ne	0.17±0.2	[2,5]	0.142±0.014	[9]	0.14	[12]
^{20}Ne/^{22}Ne	13.7±0.3	[5]	$9.2^{+1.9}_{-2.2}$	[10]		
Si	0.22±0.07	[6]	0.161±0.009	[9]	0.050	[10]
Ar	$(4.0\pm1.0)\cdot10^{-3}$	[5]	$(3.3\pm0.6)\cdot10^{-3}$	[9]	0.0048	[12]
Fe	0.19±0.07	[7]	0.154±0.015	[9]	0.045	[10]

[1] - Bame et al., 1975   [7] - Schmid et al., 1987
[2] - Bochsler et al., 1986   [8] - Cook et al., 1984
[3] - Coplan et al., 1984   [9] - Brenemann, Stone, 1985
[4] - Gloeckler et al., 1986   [10] - Mewaldt et al., 1984
[5] - Geiss et al., 1972   [11] - Anders, Ebihara, 1982
[6] - Bochsler, 1987   [12] - Meyer, 1985

These data remain incomplete since they do not include information on isotopic compositions except for helium and neon. Isotopic compositions of several additional elements, mostly noble gases, are available from the analysis of lunar soils. Recently, Wieler and co-workers (1986) have shown that lunar soil contains a surface implanted component with a ^{20}Ne/^{22}Ne ratio of 11.3±0.3 which they ascribe to Solar Energetic Particles (SEP). This result confirms the difference of the solar wind isotopic ^{20}Ne/^{22}Ne ratio (=13.7±0.3 - Geiss et al., 1972) from SEP and it supports evidence for a secular decrease of the flux ratio of SEP to solar wind.

The ISEE 3/ICI (K.W. Ogilvie, P.I.) results have established strong correlations of the fluxes of the heavier elements with helium fluxes over time scales of several years. Undoubtedly there exist strong variations of these fluxes and their respective ratios as is well known for the case of helium abundances (Neugebauer, 1981). It is difficult to find significant changes of elemental ratios among minor species and to associate these changes with specific features of the solar corona or the solar surface. The clearest evidence of changes of elemental ratios emerges again from ratios involving helium which has an unfavorably low Coulomb drag factor and which requires a large ionization energy to feed it into the corona.

The composition of solar wind particles in the terrestrial magnetosheath has been measured with the CHEM instrument on AMPTE/CCE (G. Gloeckler, P.I.). It has been possible to unambiguously distinguish C^{6+} from the $^4He^{++}$ and N^{5+} from Si^{10+}, etc. and thus to add C and N to the list of measured elements. A remarkable general feature is the close agreement of the solar wind elemental composition with the composition of SEP. There appears to be a fundamental mechanism ordering abundances of elements with respect to their first ionization potentials. The process is not understood at present, but it seems that separation of ions from neutrals in the chromosphere or transition region occurs by diffusion (Geiss and Bochsler, 1985), by gravitational settling out of magnetic structures (Vauclair and Meyer, 1985), or by acceleration of magnetic loops filled with neutrals and ions (Bochsler, 1987).

CHARGE STATE DISTRIBUTIONS
The resolution of ions not only according to mass per charge but also according to mass opened a new dimension in solar wind studies. Although only measurements in the terrestrial magnetosheath which is occasionally compressed by the solar wind are available, the charge state distributions of several elements have been measured (Ipavich et al., 1987). Surprisingly, in coronal hole associated flows, freezing-in temperatures as low as $1 \cdot 10^6$ K have been found for carbon and oxygen, whereas iron shows significantly higher freezing-in temperatures as predicted by a theoretical study (Bürgi, 1987). The solution of the mass and momentum conservation equation on the basis of electron densities observed in the inner corona has yielded a consistent picture of acceleration and the freezing-in process of charge states in the inner corona (Bürgi and Geiss, 1986).

VELOCITY DISTRIBUTIONS OF MINOR IONS
The rule $T_{kin}(i) \sim m(i)$ (equal velocity spread) has been confirmed in normal solar wind regimes for ions as heavy as iron (Bochsler et al., 1985). This result, generally interpreted as consequence of wave heating action (Isenberg and Hollweg, 1983), should be revisited in regions of high collisionality in view of the work by Livi and Marsch (1987), who - without invoking wave action - find strongly skewed velocity distributions for electrons and protons in their simulation of an expanding coronal plasma. Another somewhat surprising result from ISEE 3/ICI is the fact that Fe (Schmid et al., 1987) and Si (Bochsler, 1987) tend to lag behind $^4He^{++}$ in high speed streams, whereas for normal solar wind a good agreement among the speeds of all minor ions is found (Schmid et al., 1997; Ogilvie et al., 1982).

OTHER SOURCES OF PLASMA IN THE INNER HELIOSPHERE
By means of the SULEICA instrument on AMPTE/IRM it was possible to identify $^4He^+$ arising from interstellar pick-up ions (Möbius et al., 1985). Comets as sources of weakly ionized atoms and molecules have been investigated and the contribution of extended planetary magnetotails to the interplanetary plasma has been studied (Macek and Grzędzielski, 1986).

REFERENCES
Anders, E., Ebihara, M.: 1982, Geochim. Cosmochim. Acta 46, pp 2363-2380; Bame, S.J. et al.: 1975, Solar Physics 43, pp 463-473; Bochsler, P.: 1987, Abstracts International Union of Geodesy and Geophysics (IUGG), Vancouver, p 658; 1987, Physica Scripta, in press; Bochsler, P. et al.: 1985, J. Geophys.

Res. 90, pp 10779- 10789; Bochsler, P. et al.: 1986, Solar Physics 103, pp 177-201; Brenemann, H.H., Stone, E.C.: 1985, Astrophys. J. Lett 299, pp L57-L61; Burgi, A.: 1987, Abstracts *International Union of Geodesy and Geophysics (IUGG)*, Vancouver, p 657; Burgi, A., Geiss, J.: 1986, Solar Physics 103, pp 347-383; Cook, W.R. et al.: 1984, Astrophys. J. 279, pp 827-838; Coplan, M.A. et al.: 1984, Solar Physics 93, pp 415-443; Geiss, J., Bochsler, P.: 1985, in: *Rapports isotopiques dans le système solaire*, Cepadues-éditions, pp 213-228; Geiss, J. et al.: 1972, Section 14 in: *Apollo 16 Preliminary Science Report*, NASA SP-315; Gloeckler, G. et al.: 1986, Geophys. Res. Lett. 13, p 793; Ipavich, F.M.: 1987, Abstracts *International Union of Geodesy and Geophysics (IUGG)*, Vancouver, p 657; Isenberg, P.A., Hollweg, J.V.: 1983, J. Geophys. Res. 88, pp 3923-3935; Livi, S., Marsch, E.: 1987, J. Geophys. Res. 92, pp 7255-7261; Macek, W., Grzedzielski, S.: 1986, ESA SP-251, pp 155-161; Mewaldt, R.A. et al.: 1984, Astrophys. J. 280, pp 892-901; Meyer, J.P.: 1985, Astrophys. J. Suppl. 57, pp 151-171; Mobius, E. et al.: 1985, Nature 318, pp 426-429; Neugebauer, M.: 1981, Fundamentals of cosmic physics 7, pp 131-199; Ogilvie, K.W. et al.: 1982, J. Geophys. Res. 87, pp 7363-7369; Schmid, J. et al.: 1987, submitted to Astrophys. J.; 1987, J. Geophys. Res., in press; Vauclair, S., Meyer, J.P.: 1985, in: *Proceedings of 19th International Cosmic Ray Conference 4*, La Jolla, p 233; Wieler, R. et al.: 1986, Geochim. Cosmochim. Acta 50, pp 1997-2017

Solar Wind Interaction with Venus, Mars and Comets
M.K. Wallis

To model and quantify the comet-like interactions with the atmosphere of Venus the extent and time-variability of the suprathermal exospheric coronas need to be established. An extensive suprathermal O corona hypothesized to arise *via* dissociative recombination of the dominant O_2^+ ion was confirmed by the Orbiter's UV spectrometer. Uncertainties remain, firstly because the O-corona's extension beyond the observed limit of 1500 km altitude depends on the poorly-known partition between dissociation channels, secondly because the O_2^+ ionosphere varies substantially with solar UV inputs, and thirdly because the ionosphere and therefore O-corona decrease strongly from dayside to nightside (Kliore *et al.*, Adv. Space Res. 5(11), 1985).

Copious data available from Pioneer Venus Orbiter has allowed detailed study of atmospheric modification at Venus (Luhmann, Space Sci. Rev. 44 p 241, 1987). Confirming suggestions from the early Venus missions (Wallis 1972), the bow-shock is displaced sunwards from the position given by MHD modelling: this has been put on a firm statistical basis (Alexander *et al.*, GRL 13, p 917, 1986) and a solar cycle dependence demonstrated. Whether there is some weakening of the shock due to atmospheric ions created upstream of it − as strongly evident in Halley's comet − is unclear. The sunwards displacement and increased divergence (flaring) of the shock limbs has been demonstrated by gasdynamic modelling (Krymskii & Breus, Kosm. Issled. 24, p 778, 1986) with the atmosphere treated as sources of mass within the flow.

On one side of Venus, O^+ ions tend to be injected into atmosphere, but on the opposite side ejected further out. The precipitated flux from 30-80° zenith angle is calculated at 1-2% of the solar wind (Wallis, Geophys. Res. Lett. 9, p 427, 1982). Solar wind protons also probably penetrate into the ionosphere in similar fluxes, "diffusing" through to the ionopause under fluctuating fields (Gombosi *et al.*, JGR 85, p 7747, 1980). Other evidence for permeability of the 'magnetosheath' of enhanced $\mathbb{B}$-field (adjacent to the ionopause) are the small-scale "flux ropes" of twisted $\mathbb{B}$ found in the ionosphere. Consequent on the large O^+ gyroradii, asymmetry in the flow as registered by the bow shock distances on the flanks has now been demonstrated statistically (Alexander *et al.*, GRL 13, p 917, 1986). These authors also find dependence on the inclination of the $\mathbb{B}$-field: "pick-up" of the O^+ ions appears less efficient when $\mathbb{B}$ is roughly parallel to the flow. Phillips *et al.* (1986) have developed O^+ ion trajectory calculations in the specific magnetized flow model, demonstrating both the ionosphere precipitation and that "mass addition" is likely to be a poor approximation (Wallis, Adv. Space Res. **6**(1), p 195, 1986) for extensive atmospheric sources. Luhmann *et al.*, (JGR **92**, p 2544, 1987) have demonstrated that pick-up does operate with parallel mean $\mathbb{B}$, *via* low frequency fluctuating fields propagating from the ionopause or adjacent mantle (ionosheath).

While a distinct ionopause is confirmed, dividing ionosphere plasma from the external flow with a current layer a few gyroradii thick, it turns out to be highly variable. The combined external plasma plus magnetic pressure is balanced by the internal ionospheric pressure, so the sub-solar position varies significantly with solar wind dynamic pressure. But Phillips *et al.* (JGR **89**, p 10676, 1984) found that it varies even more strongly at zenith angles 60-90°, between 300 and 1200 km altitude. Such a range presumable reflects fluctuations and instability of the ionopause and surrounding "mantle" loaded with atmospheric ions, corresponding to the clouds of thermalized O^+ intruding from the ionosphere (Brace *et al.*, Planet. Space Sci. **30**, p 29, 1982). This mantle or ionosheath of planetary ions that extends into the tail thus appears to be a mixture of ionospheric clouds and implanted O^+ ions. The Venera spacecraft have given a lot of detail on the ionosphere and ion sources (Breus *et al.*, Kosm. Issled. 24, 1986) and particularly on the strongly variable polar ionosphere (Savich *et al.*, Kosm. Issled. **24**, 1986). The longstanding problem of what maintains the nightside ionosphere now seems related to this ionopause variability. When it is high — generally around solar UV maximum, O^+ flows from the dayside ionosphere driven by pressure gradients (Theis *et al.*, JGR **89**, p 1477, 1984) and perhaps by effective viscous forces near the ionopause (Perez-de-Tejada, JGR **91**, p 6765, 1986). When the ionopause is low, irregular precipitation of tail electrons are dominant. The magnetic "flux ropes" of 10 nT intensities and widths up to ~10 km were an anomalous discovery. They arise from penetration of the dayside ionopause, probably due to MHD instability, but some fluid turbulence is thought necessary to twist and concentrate the field (Luhmann & Elphic, JGR **90**, p 12047, 1985). To explain their evolution, diffusion-convection modelling of material transport down in the ionosphere is being developed (Phillips *et al.*, JGR **89**, p 10676, 1984; Shinagawa & Cravens, JGR **92** (July), 1987). The wake of Venus contains much more magnetic flux than in these 'ropes'; its double-lobed structure arises from the draped interplanetary field, strengthened from being hung-up on slower planetary plasma analogously to the cometary case. Saunders & Russell (JGR **91**, p 5589, 1985) showed that part of the field closes across the tail, while Russell (Adv. Space Res. **6**(1) p 291, 1986) demonstrated that Maxwell stresses suffice to accelerate the planetary ion. The tail is flattened, corresponding to asymmetry in ion pick-up if not to anisotropy in magnetosonic speeds (Vaisberg & Zeleny, Icarus **58**, p 412, 1984). The pick-up ions modelled by Phillips *et al.* (1986) are too energetic for PVO detection, while those detected spasmodically (Mihalov & Barnes, GRL **8**, p 1277, 1981) suggest an erratic tail structure of rays or clouds.

In case of Mars, Vaisberg & Smirnov (Adv. Space Res. **6**(1), p 301, 1986) reaffirm their strong influence in the ionotail with comet-like properties. Compared with Venus, the Martian shock and ionosheath lie further out, but not necessarily due to a planetary magnetic field (Vaisberg *et al.*, 1972). Breus (Adv. Space Res. **6**(1), p 167, 1986) argues that suprathermal oxygen provides an extensive enough exosphere under the low gravity to "mass load" the plasma flow and displace the bow shock out by 200-400 km. Whether or not this oxygen-corona exists, the solar wind is strong enough at times to press down to the ionosphere and presumably induces magnetic fields to resist its penetration analogous to Venus under enhanced solar wind (Wallis & Ip, Nature **298**, p 229, 1982). In modelling the exterior plasma flow mass sources have recently been introduced (Breus 1986, Breus *et al.*, Pl. Space Sci., 1987). Large gyroradii of picked-up O^+ ions make this modelling unsatisfactory. The magnetic field is treated as weak, which may not be appropriate for studying unsteady situations with enhanced interplanetary field (Luhmann *et al.*, JGR **91**, p 3001, 1986). A possible magnetic barrier has been studied on an approximate analytic scheme (Krymskii & Breus, Kosm. Issled. **24**, p 778, 1986).

The wealth of *in situ* data from the cometary probes yields a picture to a first approximation surprisingly close to theoretical predictions. Cometary ions "picked-up" far out in the coma by solar wind fields are accelerated to high energies and strongly affect the supersonic flow. The bow shock is weak (~Mach-2), positioned at a few times 10^5 km in front of Halley, and the flow deflects around the comet rather than being absorbed. The energetic ions are lost *via* charge exchange, allowing cometary ion structures ("envelopes") to form inside 10^5 km (Galeev, ESA-SP 250, 1, p 3, 1986). The interplanetary magnetic field adopts the draped topography and is enhanced around the tail core but only to the 50 nT level comparable to dynamic pressures. Cometary pick-up ions deviate strongly from the mean flow in such large-scale draped fields, which might explain structure and deviations seen at Giacobini-Zinner (Wallis, Adv. Space Res. 6(1), p 239, 1986; Luhmann, GRL **14(8)**, 1987). Several field reversals registered on GIOTTO's inbound path can be matched to ones outbound (Raeder *et al.*, ESA-SP 250, 1, p 173), suggesting that solar wind field reversals are convected "frozen-in" the plasma flow with negligible magnetic diffusion ("reconnection").

Identification of the bow shock at both comets has been controversial. At Giacobini-Zinner, gyroradii of pick-up ions were comparable to fluid scales, so a gradual "bow-wave" transition was plausible. But even at Halley with the 10 times larger fluid scale there was no classical bow shock (Riedler *et al.*, Nature **321**, p 288). SUISEI detected the clearest example (Mukai *et al.*, Nature **321**, p 299), with $20°$ deflection and 50% speed reduction, but that may be confused with a discontinuity in solar wind. Of VEGA's shock passages, one is identified by changed level of fluctuations, the two others by flow speed changes but over many gyroradii thickness (Galeev *et al.*, GRL **13**, p 841, 1986). GIOTTO's ion, electron and magnetometer experiments (having better resolution) agreed on the inbound shock crossing, though it is clear that the weak jumps in speed, density or field by 10-20% are readily confused with large and widespread fluctuations associated with the overall interaction. There has been no demonstration that Rankine-Hugoniot relations are satisfied – and they appear not to be at GIOTTO's inbound shock (Coates *et al.*, ESA-SP 1, p 263). No heating of the protons and electrons is discernible; deflection of the flow and phase scattering (pitch angle) of the ions are apparent, but take place too at other locations. The energetic cometary ions constitute a source of free energy and unstable beams in $v_\perp$ and $v_\parallel$ space, and so doubtless generate the observed fluctuations, whether *via* ion-cyclotron, lower hybrid, or Alfvèn wave instabilities (Galeev, ESA-SP 250 1, p 3). While they must lose some energy, they do not thermalize, but primarily suffer scattering in pitch angle, converting "ring" to "shell" distributions (Neugebauer *et al.*, ESA-SP 250 1, p 10; Johnstone *et al.*, Astron. Astrophys., October 1987). The energetic ions are rather lost, spiralling away in the outer regions of draped magnetic field, or *via* charge exchange with neutrals in the inner region. That many gain high energies, up to several 100 keV were detected, was a surprise – explained as the operation of Fermi acceleration (Sagdeev *et al.*, GRL **13**, p 85, 1986; Gribov *et al.*, ESA-SP 250, p 271) *via* the large amplitude fluctuations.

Interstellar Neutrals in the Heliosphere and Their Interaction with
the Solar Wind
H. J. Fahr

LISM PENETRATION INTO THE HELIOSPHERE

The neutral component of the local interstellar medium (LISM) can deeply
penetrate into the heliosphere. This fact is both observationally well
confirmed by interplanetary EUV glow data and theoretically well established
as shown e.g. in the more recent reviews on this field by Fahr (1983), Bertaux
(1984), Fahr (1986a/b), Ajello et al. (1987).

The puzzling problem that turns out from all up-to-now helium and
hydrogen glow interpretations concerns the fact that the derived thermodynamic
parameters for the LISM helium and hydrogen do not seem to be rooting back
into one common thermodynamic status of the unperturbed LISM. Especially the
magnitude of the bulk flow velocity for hydrogen ($\sim$20 km/s) appears to be
appreciably smaller than that of helium ($\sim$27 km/s) as raised by Bertaux
(1984). Furthermore the hydrogen temperature ($\sim$8000 K) is pointed out to be
substantially lower than the helium temperature ($\sim$15000 K) when evaluated with
the help of conventional interplanetary glow models (Dalaudier et al., 1984,
Chassefière et al., 1986). These results call for plausible physical
explanations. Possibly the most promising idea to solve the problem of
seemingly incompatible LISM parameters is to look for changes that the neutral
LISM flows may experience in the plasma interface region ahead of the
heliosphere. This idea has first been quantitatively evaluated in the works by
Ripken and Fahr (1983) and Fahr and Ripken (1984). There it was shown that the
LISM hydrogen due to its strong interaction with the perturbed LISM plasma in
the interface region via resonant charge exchange processes is subject to
substantial change of its momenta of the velocity distribution function, like
density, bulk velocity and temperature. In a simplified approach to this
problem also Wallis (1984) confirmed the prediction of appreciable density
depletion in the LISM hydrogen flow. A comprehensive study of the interface
effect has recently been given by Fahr (1986b). Of special interest, also for
pick-up ion problems and the question of the origin of the anomalous component
of the cosmic rays (Fisk et al., 1974, Jokipii, 1985, Cummings et al., 1984),
is the fact that the "interface effect" operates in a very gas-specific way.
LISM oxygen suffers the highest losses in the interface due to the lack of any
production processes there for O-atoms. The LISM hydrogen will be depleted by
a much smaller amount (about 50%) and helium nearly penetrates the interface
unmodified due to the very low charge exchange interaction rate. LISM helium
is possibly only subject to some, however weak interaction with the interface
plasma due to elastic collisions with the interface ions (Chassefière and
Bertaux, 1987).

MODELLING OF LISM NEUTRALS IN THE INTERPLANETARY SPACE

As recently reviewed by Fahr (1986a) there existed some flaws in the
theoretical modelling of the distributions of interplanetary neutrals and the
related EUV resonance glow intensities. For instance it was felt that part of
the discrepancy of the derived LISM helium and hydrogen temperatures could be
dissolved by treating the dynamics of interstellar He atoms more carefully,
taking into account a kind of drag force connected with elastic collisions of
neutrals and solar wind ion species (Fahr et al., 1985). The amount of
temperature reduction that was derived on the basis of this process, however,
was questioned by Gruntman (1986) and Chassefière and Bertaux (1986) with the
argument that the cross section used by the above authors probably is too
large. Since the elastic interaction potential at large impact parameters
still is a matter under debate, Nass and Fahr (1986) have shown for
illustrative purposes how much the density structure in the helium cone and
the derived LISM helium temperature is changed by the elastic drag effect for
a large range of cross section values.

A further complication has been pointed out in the work by Lallement *et al.* (1985), who showed that a satisfactory fit of interplanetary Ly-α glow intensities as observed with the satellites Prognoz 5 and 6 requires to take into account the solar wind latitudinal flux asymmetries. This point was already stressed in earlier papers by Witt *et al.* (1979, 1981) in connection with Ly-α data obtained with Mariner 10. It seems therefore that the LISM hydrogen loss rate needs a solar latitudinal modulation, disregarding its physical nature.

Furthermore it was criticized by Fahr *et al.* (1987) that up to now only stationary models have been used for the resonance glow interpretations. As the authors show in their recent paper the strongly time-dependent solar EUV-emissions can only be adequately taken into account by a time-dependent modelling. Especially the upwind-to-downwind helium density and He-584Å resonance intensity ratios are strongly influenced by solar cycle variation of the helium photoionization rate. This may easily have led earlier He-glow interpreters to derive too high LISM helium temperatures by about 3000 K (Dalaudier *et al.*, 1984, Ajello, 1978, Weller and Meier, 1981).

A subject of deeper investigations presently is the relevance of solar wind electron impact ionization collisions for neutrals in the inner solar system. In contrast to earlier work in this field (see Holzer, 1977) it is now felt that due to the pronounced core-halo structure in the solar wind electron velocity distribution functions, electron impact ionization is important for He inside the orbit of the earth (Ruciński and Fahr, 1987). Since this ionization process has a complicated dependence on the solar distance, its inclusion into modelling of interplanetary densities of LISM neutrals requires enhanced computational efforts which, however, can be shown as a necessary prerequisite for a reliable derivation of the LISM helium temperature.

The region inside the orbit of the earth not only leads to intensive ionization processes, by which secondary (pick-up) ions are produced, but also gives rise to deionization processes. These latter ones are connected with neutralization of solar wind ion species at zodiacal dust surfaces. The production rates of neutrals and ions connected with this interaction process are strongly piling up towards small solar distances. This context is reviewed as a whole in Fahr and Ripken (1985).

PICK-UP IONS IN THE SOLAR WIND
An interesting feature has been detected in the He^+-pick-up ion fluxes in the solar wind when observed from the orbit of the earth (Möbius 1986). The fluxes strongly increase in magnitude while the ion-detecting earthbound satellite AMPTE is moving from upwind towards downwind regions. This feature can be understood as action of the enhanced neutral helium densities in the downwind helium cone structure. There exists now a new independent method to determine the LISM He temperature from the upwind-to-downwind He^+-flux ratio. As raised by Fahr and Ruciński (1987) it is, however, indispensable to correctly take into account all relevant helium ionization processes inside the orbit of the earth, especially also electron impact ionization processes, since otherwise a theoretical fit of this He^+-flux ratio would lead to much too high LISM helium temperatures.

ASSIMILATION OF PICK-UP IONS TO THE SOLAR WIND BULK
It has been recognized since quite some time that the initial distribution function developing immediately after the pick-up of freshly generated secondary ions in the solar wind rest frame is unstable with respect to linear plasma wave growth. A conclusive solution for all the aspects of this problem has not yet been reached (Winske *et al.*, 1985, Price and Wu,

1987, Winske and Gary, 1986) and much more work in this field is still in progress (Lee et al., 1987, Gaffey et al., 1987, Fahr and Ziemkiewicz-Dabrowska, 1987).

The pick-up process in its first step leads to the formation of a population of secondary ions that initially move with a velocity $V_{0\parallel} = V_{sw} \cdot \cos\beta$ parallel and gyrate with a velocity $V_{0\perp} = V_{sw} \cdot \sin\beta$ perpendicular to the local magnetic field, with V_{sw} and β being the solar wind bulk flow velocity and its inclination with respect to the local magnetic field. Due to some minor thermal spread in the initial velocities of the freshly picked up ions, this scenario leads to a primary toroidal velocity distribution function. This function when convected outwards with the solar wind bulk is subject to a series of interaction processes: 1) pitch-angle scattering at intrinsic and selfgenerated Alfvén waves, 2) energy diffusion due to nonlinear coupling to selfgenerated plasma waves, 3) adiabatic cooling due to the tendency to conserve the magnetic moments, and 4) relaxation processes due to Coulomb collisions between the different ion populations. To all of these processes typical time periods τ_1 through τ_4 can be ascribed. It is thought at present that the following hierarchy amongst these periods may be valid: $\tau_4 > > \tau_2 > \tau_3 > \tau_1$. This view is only unquestionable in what concerns the fact that, at least at regions not too far from the sun (<2 AU), pitch angle scattering with intrinsic Alfvénic turbulences has the smallest time period. According to this recognition the initially toroidal distribution is predicted to quickly be transformed into a shell distribution which then as such drives hydromagnetic waves (Lee et al., 1987, Isenberg, 1987). Due to the relatively weak effectiveness of processes 2), 3) and 4) competing with pitch angle scattering, the pick-up ion population is not likely to become assimilated to the solar wind ion bulk, rather the solar wind expanding to larger distances may require to be considered as a tri-fluid plasma consisting of electrons, primary thermal ions and secondary suprathermal ions. Therefore the effect of pick-up ion heating of the distant solar wind, predicted in papers by Holzer and Leer (1973), Fahr (1973), Ripken and Fahr(1980), Petelski et al. (1980), Isenberg et al.,(1985), Grzędzielski and Ratkiewicz, (1975) may now need some revision. The attempt to model the distant solar wind as a tri-fluid plasma expansion with no thermal coupling between primary and secondary solar wind ion populations has recently been undertaken by Isenberg (1986). As it turns out from his theoretical approach, no solar wind temperature minimum can be expected in the outer solar system, unless unreasonably high neutral LISM hydrogen densities are considered. It appears, however, not to be assured that no energy dissipation from suprathermal ions to the solar wind bulk ions is occurring. Possibly this may also be a question of what regions in the solar system – ecliptic, low or high latitude, polar, upwind or downwind regions – are considered.

REFERENCES
Ajello, J.M.: 1978, Astrophys. J. 222, p 1068; Ajello, J.M. et al.: 1987, Astrophys. J. 317, p 964; Bertaux, J.L.: 1984, IAU Colloquium No. 81 eds. Y. Kondo, F.C. Bruhweiler and B.D. Savage, p 3.; Chassetiere, E. et al.: 1986, Astron. Astrophys. 160, p 229; Chassefiere, E., Bertaux, J.L.: 1986, Astron. Astrophys. 174, p 239; Cummings, A.C. et al.: 1984, Astrophys. J. Lett. 287, p L99; Dalaudier, F. et al.: 1984, Astron. Astrophys. 134, p 171; Fahr, H.J.: 1973, Solar Phys. 30, p 193; 1983, Solar Wind V, NASA CP-2280, p 541; 1986a, in: The Sun and the Heliosphere in Three Dimensions, ed. by R.G. Marsden, Reidel Publ. Comp., pp 420-434; 1986b, Advances in Space Res. 6(2), p 13; Fahr, H.J., Ripken, H.W.: 1984, Astron. Astrophys. 139, p 551; 1985, IAU Colloquium No. 85, eds. by R. Giese and P. Lamy, pp 205-233; Fahr, H.J. et

al.: 1985, Astron. Astrophys. 142, p 476; Fahr, H.J. et al.: 1987, Ann. Geophysicae: Space Phys., 5(4). p 316; Fahr, H.J., Ruciński, D.: 1987, submitted to Ann. Geophysicae; Fahr, H.J., Ziemkiewicz-Dabrowska, J.: 1987, submitted to Astron. Astrophys.; Fish, L.A.: 1986, in: *The Sun and the Heliosphere in Three Dimensions*, ed. by R.G. Marsden, Reidel Publ. Comp., pp 401-413; Gaffey, J.D. et al.: 1987, submitted to J. Geophys. Res.; Gruntman, M.A.: 1986, Planet. Sp. Sci. 34, p 387; Grzedzielski, S., Ratkiewicz, R.: 1975, Acta Astron. 25, p 177; Holzer, T.E., Leer, E.: 1973, Astrophys. Space Science 24, p 335; Holzer, T.E.: 1977, Rev. Geophys. Space Physics 15, p 467; Isenberg, P.A.: 1986, J. Geophys. Res. 91, p 9965; 1987, J. Geophys. Res. 92, p 1067; Isenberg, P.A. et al.: 1985, J. Geophys. Res. 90, p 12040; Jokipii, J.R.: 1986, J. Geophys. Res. 91, p 2929; Lallement, R. et al.: 1985, J. Geophys. Res. 90, p 1413; Lee, M.A. et al.: 1987, submitted to J. Geophys. Res.; Mobius, E.: 1986, Advances in Space Res. 6(2), p 26; Nass, H.U., Fahr, H.J.: 1986, Advances in Space Res. 6(1), p 365; Petelski, E.F. et al.: 1980, Astron. Astrophys. 87, p 20; Price, C.P., Wu, C.S.: 1987, Geophys. Res. Lett. 14(8), p 856; Ripken, H.W., Fahr, H.J.: 1980, Mitteilungen Astron. Gesselschaft 50, p 316; 1983, Astron. Astrophys. 122, p 181; Ruciński, D., Fahr, H.J.: 1987, submitted to Astron. Astrophys.; Wallis, M.K.: 1984, Astron. Astrophys. 130, p 200; Weller, C.S., Meier, R.R.: 1981, Astrophys. J. 246, p 386; Winske, D.C. et al.: 1985, J. Geophys. Res. 90, p 2713; Winske, D.C., Gary, S.P.: 1986, J. Geophys. Res. 91, p 6825; Witt, N. et al.: 1979, Astron. Astrophys. 73, p 272; 1981, Astron. Astrophys. 95, p 80; Wu, C.S. et al.: 1986, Geophys. Res. Lett. 13, p 865

COMMISSION 50: IDENTIFICATION AND PROTECTION OF EXISTING AND POTENTIAL
OBSERVATORY SITES

PRESIDENT: S. van den Bergh   VICE-PRESIDENT: D.L. Crawford

ORGANIZING COMMITTEE: C. Blanco, V. Blanco, G. Coyne, S-Y Jiang, Y. Kozai,
P. Murdin, V. Pankonin, P.V. Shcheglov, C.A. Torres, J. Tremco, and M. Walker

1. New Delhi Meeting

At the XIXth General Assembly our Commission held two meetings during the
course of which the following papers were presented:

V.M. Blanco:	Climatology and Site Selection
E. Brosterhus:	Optical Site Evaluation in Saudi Arabia
J. Davis:	An Interferometric Seeing Monitor – Measuring r_o Directly
P. Murdin, F. Sanchez:	Site Selection and Protection – Canary Islands
J.P. Osorio:	Radio Frequency Interference Problems
G. Swarup:	Radio Noise Surveys for India's Giant Meter-Wavelength Radio Telescope
A. Bhatnagar:	Site Testing for an Infrared Telescope at Mt. Laddakh
G.W. Lockwood:	1955-1985: The Effects of Volcanic Eruptions
C. Blanco:	Volcanic Activity and Astronomical Observations
B. Hidajat:	Change of Coefficients of Extinction at Lembang Caused by Volcanic Eruption
L. Huang:	The Extinction Coefficients at the Xinglong Station of the Beijing Observatory

2. Conferences

In honour of the retirement of our past president Dr. A.A. Hoag the Lowell
Observatory organized a symposium May 22-23 1986 on "Identification, Optimization,
and Protection of Optical Observing Sites" (Eds. R.L. Millis et al. – Lowell
Observatory). This conference presented authoritative reviews on the physics of
seeing, modern methods of site testing, characteristics of promising observatory
sites, telescope and dome design factors relating to seeing and combating site
degradation.

Commission 50 will sponsor a colloquium, to be held in Washington D.C., on
light pollution, radio interference and the effects of man-made space debris on
astronomy immediately following the XXth General Assembly of the IAU in Baltimore.

3. Activities of the Commission

During the period 1985-1987 activities of the Commission centred on dangers
posed to all branches of observational astronomy by light pollution, radio
interference and "space junk." A proposal to orbit a ring of satellites to
celebrate the centenary of the Eiffel Tower was withdrawn following intense
pressure by the French and international astronomical communities.
Representations were also made to the US Department of Transportation regarding
the environmental impact of the proposed launch of cremated human remains into
Earth orbit by the Celestis Corporation of Florida. The proposed launch of huge
satellites to convert sunlight into electricity for cities and industries on Earth
by the USSR is also a source of grave concern.

During the report period the Commission provided information to assist a
search for observing sites in Western China and the foothills of the Himalayas,
and for a study of astronomical seeing in Saudi Arabia. The Commission also acted
as a channel for expressions of concern regarding the effects of drilling for
natural gas near the site of the Vienna Observatory. Commission 50 also
maintained close contact with the Commission Internationale de l'Eclairage and
continued its educational efforts directed towards the reduction of light
pollution in cities near major observatories.

Under the active leadership of D.L. Crawford progress continues on the
campaigns to control the adverse effects of light pollution on astronomical
observatories. In Arizona, there are now more than 40 city or county ordinances
to control outdoor lighting and minimize the urban sky glow. Several such
ordinances also exist in California, near the major observatories, and on the
island of Hawaii, where the Mauna Kea Observatory is located. In all these areas,
there is increasing use of low pressure sodium lighting sources. LPS is the
preferred light source to minimize adverse effects of urban sky glow, and it is
the most cost-effective light source to operate as well.

4. Radio Interference

The principal international concern of radio astronomers during this
reporting period is related to transmissions from USSR GLONASS satellites
interfering with observations of the OH spectral line near 1612 MHz. Reports of
serious interference have been received from several observatories worldwide. At
the latest count (June 1987), nine satellites in this system are transmitting at
frequencies in the range 1603.125 - 1614.375 MHz, but the system is still
evolving. Periodic monitoring of the system status continues. Written inquiries
have been made to Soviet officials to get more information on the system, and to
try and open a dialogue to mitigate some of the problems. To date these inquiries
have not been successful.

PRESIDENT: F. D. Drake

VICE-PRESIDENT: George Marx

SECRETARY: M. D. Papagiannis

ORGANIZING COMMITTEE: R. D. Brown, P. Connes, G. D. Gatewood, J. Jugaku,
                       P. Feldman, J. M. Greenberg, N. S. Kardashev,
                       P. Morrison, and V. S. Troitsky

INTRODUCTION

The past three years have seen not only a growth in the activities of our commission, but an extension of its activities into important areas which have heretofore motivated too little activity. Of particular interest have been the many activities directed towards elucidating the question of the abundance of extrasolar planetary systems. There have been a number of observations showing the presence of disks of dust around nearby stars, disks which fit the idea that stars are often formed with an accompanying disk of dust which may in many or perhaps all cases produce a planetary system. Infra-red evidence for dust disks exists for something like twenty stars. The disk of Beta Pictoris has even been clearly imaged at optical wavelengths, showing without a doubt that such disk structures exist. One very impressive detection of an apparent brown dwarf object has also been made; should this be confirmed by other observations, it would be clear evidence for the existence of planet-like bodies in the systems of other stars.

The development and pursuit of radio searches for extraterrestrial intelligent radio signals has proceeded at an impressive rate. Of special interest has been the development of systems which can monitor more than 8 million radio frequency channels at once. One of these has been placed into operation in a continuous search with a high sensitivity radio telescope. Even more powerful systems, using specially designed VLSI chips, are under construction. These will bring to our work the most sophisticated equipment permitted by modern technology.

Over the past few years, programs in bioastronomy have received increased support both from within the scientific community and from governmental and private funding agencies. As a result, the programs of all kinds now underway or under development are more stable, better organized, and far more powerful than the programs of the past. It is realistically expected that the future searches for extraterrestrial radio signals, now being organized, will be millions of times more powerful than past searches.

ACTIVITIES OF COMMISSION 51

Through the good offices of our secretary, M. Papagiannis, a newsletter concerning Commission 51 activities, "Bioastronomy News" was distributed as warranted to the members of the Commission.

The major activity of the commission since its last report was IAU Colloquium 99, which was held at Balatonfured, Lake Balaton, Hungary from June 21 through June 27, 1987. The title of the Colloquium was "Bioastronomy: The Next Steps". This Colloquium was very ably organized by our Vice-President, George Marx, and a number of his colleagues from Hungarian scientific

institutions.   The Chairperson of the Scientific Organizing Committee was Jill
Tarter.   Approximately 120 scientists participated in the Colloquium, including
many who were new to the activities of our Commission; we applaud this spread of
interest in our subject.

The subjects discussed at the colloquium covered the usual broad range which is
characteristic of our meetings.   They included new results on prebiotic
chemistry on earth-like planets; the possible nature of and evidence for the
existence of organic molecules in comets, particularly Halley's Comet; ideas
about the origins of chirality in biological systems; and some proposed exotic
theories of the origin of life. Also discussed were specialized optical systems
for the detection of planets or spectral evidence for life; the cognitive
systems of non-human large brained creatures, such as dolphins; and some
speculations on the social structures of other civilizations.   There was a
special meeting to discuss protocols for the handling of the report of an actual
detection of another civilization; a number of very helpful suggestions were
made to those actually conducting searches.   Some hypotheses regarding the
galactic distribution of life were presented, as well as some ideas as to the
location of possible biological habitats within the solar system, such as in the
hypothetical ocean of Europa.

A very interesting group of papers was presented describing the efforts of
a number of groups to measure very accurate stellar radial velocities. The goal
of these projects is to detect small periodic deviations in the radial velocity
caused by the gravitational effects of companions, particularly those of
planetary mass.   Several groups reported success in measuring radial velocities
to an accuracy of about 10 meters per second, an accuracy adequate to detect the
presence of planets of Jupiter mass with many stars.   Some tantalizing early
results were given in which radial velocity variations consistent with the
presence of planet appear in the data, but not yet with very high statistical
significance.

Papers were given describing the observations of circumstellar disks at
infra-red wavelengths, and the observation of the Beta Pictoris disk at optical
wavelengths by two groups. The optical photographs of the Beta Pictoris disk are
of much better quality than the earlier ones, and show the disk to be an
extremely thing structure.

There were a number of papers describing ongoing attempts to detect
extrasolar radio signals. There were a group of related papers dealing with the
data analysis aspects of such searches; it was clear that considerable effort
must be devoted to optimizing data analysis procedures when millions of channels
are being monitored, and when a large variety of signal types need to be
searched for in the data.

The living and working accommodations at this Colloquium were excellent.
In addition, the organizers provided a host of excellent social occasions which,
as always, contributed to the transfer of ideas and results, and built a spirit
of international cooperation.

SOME SCIENTIFIC HIGHLIGHTS OF RECENT YEARS

Considering the many topics which are dealt with by Commission 51, it is
not possible to write a comprehensive summary of all the research relevant to
the work of the Commission. However, certain areas are of particular relevance,
and we give here some of the highlights of work in these areas.

The Detection of Circumstellar Disks

D. E. Backman and F. C. Gillett of the Kitt Peak National Observatory have examined the observations of IRAS for evidence of excess infra-red radiation from stars, excess radiation which is indicative of the presence of a circumstellar disk. 136 stars have been studied from a group of A, F, G, and K stars lying within 20 pc. of the earth. Of this sample, 24 have very significant infra-red excesses. The largest excesses are exhibited by Alpha Lyrae, Alpha Pisces Austrinus, Beta Pictoris, and Epsilon Eridani. Three of these, Beta Pictoris, Alpha Pisces Austrinus, and Alpha Lyrae have disks which are actually resolved in the IRAS data. Typical disk temperatures are 80K. Estimates of total dust mass range from 0,2 to 4 earth masses. The disks seem to have central voids.

Bradford Smith of the University of Arizona and Richard Terrile of the Jet Propulsion Laboratory have obtained improved optical images of the Beta Pictoris disk. These images show that the disk has a diameter of about 2500 A. U. and a width which may be as small as 50 A. U.; the disk is very thin, and is evidently being seen almost edge-on. They have searched for similar disks in more than 28 other nearby stars and have found none. As a result of their experience with such searches, they have designed improved instrumentation which should make it possible to detect much fainter examples of such disks. Observations of much higher sensitivity will be very important for they may reveal whether the disks are truly absent in other stars, or whether the disk material is being consumed by the process of planetary formation.

Searches for Other Planetary Systems

In a remarkable development, at least three groups have been attempting to detect the existence of planets through very accurate measurements of stellar radial velocities. The group with the most extensive data is probably B. Campbell, U. of Victoria, G. Walker, U. of British Columbia, and S. Yang of the U. of British Columbia. They have made frequent observations of 16 solar type stars over a period of six years using the Canada-France-Hawaii telescope. The errors in their measurements are 13 meters/second. Seven of their stars exhibit long term velocity variations in the range 25 to 65 m/sec., with implied companion orbital periods of more than about ten years. These perturbations can not be caused by brown dwarf companions, since conventional astrometry would have detected them. Companions of a few Jupiter masses are implied. They suggest that these companions are the tip of the planetary mass spectrum. The stars in their sample showing the most significant variations are Epsilon Eridani and Gamma Cephei.

Similar accuracies are being obtained by R. McMillan of the Lunar and Planetary Laboratory of the University of Arizona. The observations have not continued long enough to give good evidence of companions. Similar observations are being made by G. Marcy and V. Lindsay of San Francisco State University. Again, accuracies of the order of ten meters/second are being obtained, but the observations have not continued long enough to give definitive results.

Several groups, including in particular a group led by G. Gatewood at the Allegheny Observatory, are attempting to detect planets through the detection of small perturbations in proper motion. All of these groups are using modern photometric devices to obtain very accurate differential measurements of stellar positions. The accuracies obtained in these instruments are of the order of 0.001 arcsecond, which is sufficient to detect the presence of planets of the mass of Jupiter accompanying nearby solar type and less massive stars.

Late in 1987, B. Zuckerman, University of California, Los Angeles, and E. E. Becklin, University of Hawaii, reported the remarkable discovery of excess infra-red radiation from the location of the white dwarf star Giclas 29-38. The infra-red color temperature of the excess radiation is 1200K, with an estimated uncertainty of 200K. If the radiation is emitted by a single spherical black body of 1200K, then its radius equals 0.15 solar radius. These parameters are very similar to the theoretical values of the substellar objects called brown dwarfs. After testing a number of alternative explanations, Zuckerman and Becklin have concluded that the object is almost certainly a brown dwarf in orbit around G29-38. This is a very notable discovery. In previous searches of white dwarf stars for evidence of cool companions, the authors did not find evidence for such objects (Zuckerman, B. and Becklin, E. E., 1987, Ap. J.), thus indicating that they are rare. A previous report of the detection of a similar object in orbit around the M dwarf Van Biesbroeck 8 was not confirmed by subsequent searches for the object.

Considering the large amount of activity directed towards the detection of extrasolar planets, and the prospects for more powerful instruments, such as the Hubble Space Telescope and the Keck 10-meter telescope, it is possible to be optimistic that the question of the abundance of planetary systems will be answered in the near future.

Studies of Biologically Relevant Molecules

The "Astrochemistry" group at the University of Massachusetts continued studies of comparative molecular abundances in different types of interstellar clouds; the investigation of reaction pathways; mapping of physical, chemical, and dynamical properties of cloud regions which may be sites of solar-type star formation; and searches for new molecular constituents of dense interstellar clouds. Phosphorus nitride, the first interstellar molecule to contain phosphorus, a biologically important element, was discovered (Ziurys, 1987). Other detections have been made of $HCNH+$, $SO+$, C_3H, HC_2CHO. A study of C_3H_2, the first interstellar hydrocarbon ring, has been carried out.

An important question for bioastronomy is whether the molecules of the interstellar clouds survive the stellar and planetary formation processes. An additional question is whether there is some mechanism which delivers the molecules to the surfaces of planets after they are cool enough that the molecules will not be destroyed. A possible storage place and means of delivery is provided by comets. Thus there has been considerable interest in determining the content of large molecules in comets. An important goal has been to find formaldehyde in comets. Formaldehyde is one of the prime intermediaries in the production of the basic molecules of terrestrial biology. Its presence would suggest that the other intermediaries are present, and that planets will be commonly seeded with the materials which produce an earth-like biochemistry.

Radio emission from OH in Comet Halley has been mapped with high resolution by de Pater, Palmer, and Snyder (1986). This work showed that the Very Large Array was an excellent instrument for detecting molecules in resolved comets. The work on OH was followed by similar observations in the 6-cm formaldehyde line. This search yielded a weak formaldehyde emission line. Model fitting to the observational results suggests that the formaldehyde abundance in the cometary gas is about 5%. This supports the idea that the molecular precursors of biologically relevant molecules exist in interstellar and cometary material in substantial abundances. This suggests strongly that comets could have provided much prebiological material to planetary surfaces.

Radio Searches for Extraterrestrial Intelligent Radio Emissions

Very encouraging major advances have been made in the instrumentation for the search for extraterrestrial intelligent radio signals. As has long been recognized, there have been two obstacles to the pursuit of extensive radio searches: 1) lack of substantial observing time on large radio telescopes; and 2) the inability to observe very large numbers of frequency channels simultaneously. Two programs are addressing these problems very successfully.

The first program is that of P. Horowitz of Harvard University, who has acquired the dedicated use of the 84-foot radio telescope at the Oak Ridge Station. Horowitz constructed a signal processor, based on conventional computer integrated circuit technology, which could act as a radio spectrum analyzer for more than 100,000 channels simultaneously. This signal processor was used for several years in a project named "Project Sentinel" to search the entire sky visible to the radio telescope for signals at the 21-cm wavelength of neutral hydrogen. The channel bandwidth used was of the order of 0.01 Hz, so that the total bandwidth covered in this search was very small. Success in the search would have required the extraterrestrial transmitter to transmit at a frequency which was corrected for Doppler effects both at the transmitters location and for the motion of the solar system. A small number of candidate signals were detected in this search at frequencies whose use on earth is not permitted by international convention. Subsequent efforts to recover these signals from the same celestial location and on the same frequency did not succeed. With only this evidence, the assumption must be made that these signals were of terrestrial origin and were the product of improperly tuned transmitters or were harmonic radiations from transmitters.

This original system has now been replaced with a new system which can analyze over 8 million channels simultaneously. This system is being used in a new search of the entire sky called "Project Meta". This search will be carried out at the 21-cm frequency again, harmonics of the 21-cm line frequency, and also at frequencies associated with the OH line and possibly others. The increased frequency coverage provided by the new spectrum analyzer allows signals to be detected even if the frequency has not been corrected for the Doppler effects imposed by the transmitter's motion and the motion of the solar system.

A more ambitious program is the SETI (Search for Extraterrestrial Intelligence) program of NASA. This program is based at the NASA Ames Research Center and the Jet Propulsion Laboratory. The Program Manager is B. Oliver. The key instrument in this program is a Multi-Channel Spectrum Analyzer, or "MCSA", which will be able to analyze at least 8 million frequency channels simultaneously. This analyzer will use customized VLSI technology to achieve high speed, therefore large overall frequency coverage, at low cost. It will have the ability to observer several bandwidths simultaneously, and will have minimum bandwidths of the order of one Hertz, allowing it to achieve very high sensitivities for narrow-band signals, and yet allowing it to cover a large fraction of the low-noise radio window in reasonable observing times.

A very important feature of this program will be the sophisticated computer analysis system which will operate in real-time with the MCSA. At present, it is planned that the output of all frequency channels will be read out at one second intervals. The computer analysis system will then search this ensemble for a large variety of signals, including narrow band continuous wave signals, broadband signals, and narrow band signals which drift in frequency, as could be the result of Doppler effects or intentional signal protocols. The system will also search for pulsed signals, where the pulses may come at uniform intervals or irregular intervals, and pulses which drift in frequency. To accomplish all of this with so many data points and at reasonable expense has been the greatest challenge to this project. However, many of the demanding computer algorithms required to accomplish all of this have been developed.

Upon completion of the instrumentation, it is planned to use this equipment on dedicated radio telescopes, and the world's largest radio telescopes to the extent observing time is available on them. Two distinct searches will be made. One, the "All Sky Survey", will survey the entire sky for signals over a wide oberall band of frequencies covering about 10 Ghz of the radio spectrum. Smaller antennas will be used in this search, as well as limited integration time, leading to a good but not ultimate sensitivity for signals. The other "Targetted Search" will use large antennas and long integration times to search for signals from about 800 nearby stars and other promising objects. The sensitivity will be several orders of magnitude higher than with the All Sky Survey. It is expected that these searches will be commenced several years from now.

A search for signals from 78 stars is being made with a 30-meter telescope at the Instituto Argentino de Radioastronomia by F. Colomb, M. Martin, and G. Lemarchand. So far 34 stars have been examined in the frequency range 1415.4 to 1425.4 Mhz. A spectrometer providing 74 channels of 2.5 Khz width each was employed. With the observing parameters employed, the minimum detectable flux was $5 \ 10^{-23}$ watts/m^2. The search will be completed for the entire sample of 78 stars, and then will be repeated at the OH line. In the search a number of signals were detected; however, in every case observational tests revealed that the sources of the signals was terrestrial interference.

I. Mirabel of the Instituto de Astronomia Y Fisica del Espacio, of the University of Buenos Aires, has conducted a search at the 4829 Mhz frequency of formaldehyde using the 140-foot radio telescope of the NRAO. A spectrum analyzer providing 1024 channels of 76.2 Hz each was used. The galactic center was observed for four hours. Thirty-three stars were also observed for about two hours each. For one hour the system was tuned to the formaldehyde frequency tuned to rest in the reference system of the star, and the other hour had the system tuned to the rest frequency relative to the sun. No evidence for signals was found. The typical noise after one hour of integration was 1.27 Jy.

An ongoing seach has been carried out for many years at the Ohio State University under the direction of R. Dixon. Recent activities there have been directed towards upgrading the instrumentation. A new computer is being installed which will allow the observatory to search a very wide frequency range near the H and OH lines continuously. An automatic "frequency zoom" is being implemented which will be triggered by a signal detection, and will lead to the automatic detailed recording of the signal. Telescope tracking is being increased to one to two hours, and will be automatically activated upon detection of a signal. In addition to these developments, a phased array is being developed to produce many observing beams simultaneously. The goal is to be able to observe the entire sky with large collecting area at any given time.

At the University of California, Berkeley, S. Bowyer, D. Wertheimer, and V. Lindsay have developed an instrument which searches for extraterrestrial radio transmissions by utilizing the cosmic radio power collected in the course of other radio astronomy observations. The instrument searches approximately 65,000 contiguous radio channels, each about one Hertz in bandwidth, for signals. The observing project, called "Serendip II" has operated about 2000 hours at the NRAO 300-foot telescope. When a narrow peak in the radio spectrum is detected, the instrument automatically records power, telescope direction, and the time and frequency of the event for further study. So far about 37 signals have been detected by the system, and these will be reobserved to identify, if possible, their source. In addition, a real-time scheme to reject terrestrial radio interference has been completed.

At the Nancay Observatory of the Paris Observatory, a search is being conducted by F. Biraud and J. Tarter for narrow band signals in the bands at 21-cm and 18-cm wavelength. The resolution is 50 Hz, and a large number of nearby solar-type stars are being searched for evidence of intelligent radio transmissions.

At the Algonquin Radio Observatory, a search has been made at 10.6 Ghz., (2.8-cm wavelength) by J. Vallee and M. Simard-Normandin of the Herzberg Institute of Astrophysics. In this special search, a special arrangement of the telescope is used such that only signals with a very high percentage of linear polarization are accepted by the system. The hypothesis is that intelligent signals will be very highly polarized, in contrast to most natural cosmic emissions. A single channel of about 200 Mhz bandwidth is used. The region of the galactic center has been observed with this system. No highly linearly polarized source of flux density greater than about 30 mJy was found.

PLANS FOR THE FUTURE

There will be an international symposium on bioastronomy in Toronto in the summer of 1988. Commission 51 expects to hold its next Symposium or Colloquium in the summer of 1990 in a location yet to be decided. We look forward to the important results to be obtained from projects now in progress.

Frank D. Drake
President of the Commission

REFERENCES

We give here a list of some papers published by members of our commission over the last few years. Of necessity, this list is not complete.

Searches for Extraterrestrial Intelligent Emissions

Biraud, J. 1988, Proc. of IAU Colloq. 99, ed. G. Marx, D. Reidel, in press.
Bowyer, W., Wertheimer, D., and Lindsay, V. 1988, Proc. of IAU Colloq. 99, ed. G. Marx, D. Reidel, in press.
Colomb, F., Martin, M., and Lemarchand, G. 1988, to be published.
Dixon, R. 1986, in Problems in the Search For Life in the Universe, ed. V. A. Ambartsumian, et al, Moscow Science Press.
Dixon, R. 1986, in Second Symposium on Chemical Evolution and the Origin and Evolution of Life, NASA Conference Publication 2425.
Dixon, R. 1986, in The Search for Extraterrestrial Intelligence, Proceedings of the NRAO Workshop, National Radio Astronomy Observatory.
Heidman, J. 1986, Comp. Rendus, 303, Serie II, 47.
Heidman, J. 1988, Proc. of IAU Colloq. 99, ed. G. Marx, D. Reidel, in press.
Shostak, G., and Tarter, J. 1985, Acta Astronautica, 12, 369.
Sullivan, W. 1984, Icarus 60, 675.
Tarter, J. 1985, Proc. of NRAO Workshop on SETI, NRAO.
Tarter, J. 1985, Proc. of 1985 Meeting of the IAA. To be published in Acta Astronautica.
Vallee, J. P. 1985, in The Search for Extraterrestrial Life: Recent Developments, ed. M. Papagiannis, D. Reidel Publ. Co., 321.
Vallee, J. P. 1985, J. Roy. Astron. Soc. Canada, 79, 9.

Searches for Extrasolar Planets, Dust Disks, and Brown Dwarfs

Aumann, H., Gillett, F., Beichman, C., de Jong, T., Houck, J., Low, F.,
      Neugebauer, G., Walker, R., and Wesselius, P. 1984, Ap. J. Lett, 278, L23.
Aumann, H. 1985, P. Astro. Soc. Pacific, 97, 885.
Backman, D., Gillett, F., and Low, F. 1986, Adv. in Space Res., 6 (7), 43.
Backman, D., and Gillett, F. 1988, in Proc. of IAU Colloq. 99, ed. G. Marx,
      D . Reidel, in press.
Benest, D. 1974, Astron. Astrophys. 32, 39.
Benest, D. 1975, Astron. Astrophys. 45, 353.
Benest, D. 1976, Astron. Astrophys. 53, 231.
Bracewell, R. 1984, in The Moon and Planets, 30, 75.
Campbell, B. , Walker, G., and Yang, S. 1988, in Proc. of IAU Colloq. 99,
      ed. G. Marx, D. Reidel, in press.
Smith, B. and Terrile, R. 1988, Proc. of IAU Colloq. 99, ed. G. Marx, D. Reidel,
      in press.
Weintraub, D., Masson, C., and Zuckerman, B. 1987, Ap. J. 320, 336.
Williams, I. 1987, in Physics of the Planets, ed. S. Runcorn, J. Wiley & Sons,
      401.
Zuckerman, B. and Becklin, E. 1987, Ap. J. Lett., 319, L99.
Zuckerman, B. and Becklin, E. 1987, Nature, in press.

Observations of Interstellar and Cometary Molecules

de Pater, I., Palmer, P., and Snyder, L. 1986, Ap. J. Lett, 304, L33.
Friberg, P., Irvine, W., Madden, S., and Hjalmarson, A. 1986, in Astrochemistry
      (IAU Symp. 120), ed. M. Vardya and S. Tarafdar, D. Reidel, 201.
Irvine, W. 1986, in Astrochemistry (IAU Symp. 120), ed. M. Vardya and S.
      Tarafdar, D. Reidel, 245.
Irvine, W., Avery, L., Friberg., Matthews, H., and Ziurys, L. 1987, in
      Interstellar Matter, Gordon and Breach, in press.
Irvine, W. Goldsmith, P., and Hjalmarson, A. 1987, in Interstellar Processes,
      ed. D. Hollenbach and H. Thronson, D. Reidel, in press.
Matthews, H., and Irvine, W. 1985, Ap. J. 290, 609.
Turner, B., and Ziurys, L. 1987, in Galactic and Extragalactic Astronomy, ed.
      K. Kellerman and G. Verschurr, Springer-Verlag, in press.
Wallis, M., Rabilizirov, R., and Wickramasinghe, N. 1986, ESA Sp-250, 251.
Wallis, M., Rabilizirov R., and Wickramasinghe, N., and Al-Mufti, S., 1987, in
      ESA Special Publication SP-278.
Ziurys, L. 1987, Ap. J. Lett, in press.

Hypotheses About Extraterrestrial Life

Balazs, B. 1986, Acta Astronautica 13, 123.
Fracassini, M., Pasinetti, L., Rosazza, S., and A. Pasinetti 1988, in Proc. of
      IAU Colloq. 99, ed. G. Marx, D. Reidel, in press.
Hajduk, A. 1988, Proc. of IAU Colloq. 99, ed. G. Marx, D. Reidel.
Hoyle, F. and Wickramasinghe, N. 1986, Nature, 322, 509.
Papagiannis, M. 1988, two papers in Proc. of IAU Colloq. 99, ed. G. Marx, D.
      Reidel, in press.
Schwartzman, D., and Rickard, L. 1988, in Proc. of IAU Colloq. 99, ed. G. Marx,
      D. Reidel, in press.
Soderblom, D. 1986, Icarus, 67, 184.
Sturrock, P. 1987, J. Sci. Exp., 1, in press.
Tough, A. 1988, Proc. of IAU Colloq. 99, ed. G. Marx, D. Reidel, in press.
von Hoerner, S. 1985, in Proc. of NRAO Workshop on SETI, ed. Kellerman and
      Seielstad, NRAO.

Instrumentation for SETI

Albrecht, R. and Balazs, B. 1988, Proc. of IAU Colloq. 99, ed. G. Marx, D.
    Reidel, in press.
Cohen, H. 1988, in Proc. of IAU Colloq. 99, ed. G. Marx, D. Reidel, in press.
Drake, F. 1988, in Proc. of IAU Colloq. 99, ed. G. Marx, D. Reidel, in press.
Rather, J. 1988, in Proc. of IAU Colloq. 99, ed. G. Marx, D. Reidel, in press.
von Hoerner, S. 1986, in Proc. of 37th IAF Congress, Pergamon.

WORKING GROUP FOR PLANETARY SYSTEM NOMENCLATURE
(Committee of the Executive Committee)

PRESIDENT:	H. Masursky
MEMBERS:	K. Aksnes, G.E. Hunt, M. Ya. Marov, P.M. Millman, D. Morrison, T.C. Owen, V.V. Shevchenko, B.A. Smith, V.G. Tejfel
CONSULTANTS:	J.M. Boyce, G.E. Burba, A.M. Komkov, J.D. Rosendhal

Since the General Assembly at New Delhi in November 1985, the Working Group held two meetings within six weeks of each other; most members of the Working Group and several members of Task Groups were able to attend at least one of these meetings. The thirteenth meeting of the Working Group was held at Toulouse, France on June 30 to July 2, 1986; the fourteenth meeting was held at Moscow, USSR, on August 10, 1986; the fifteenth meeting took place from August 13 to 15 in Soviet Armenia.

As a result of these meetings, and correspondence following the meetings, the following resolutions were passed:

1. Small Uranian satellites will be named for Shakespearean characters.
2. Seven craters on the farside of the Moon will be named for deceased American astronauts on the Challenger shuttle.
3. Features on the five large Uranian satellites will be named for Shakespearean, or bright- and dark-spirit personages and places.
4. A temporary nomenclature will be used for ground-based observation of events on Io.
5. The existing nomenclature will be retained for the rings of Uranus; new rings discovered by Voyager will be given Greek letters, beginning the sequence with lambda.
6. Future editions of the Gazetteer of Planetary Nomenclature will list the planets or satellites in order of distance from the sun or planet, according to IAU convention.
7. A new feature term, Unda (plural, Undae), was approved for dune-like features on Mars.

No decision was reached concerning additional names of Saturnian ring and ring-gap features. The name Colombo will be applied to a Saturnian ring feature of the meeting in Armenia.

The following names have approval from the Executive Committee (since September 1987):

MOON

Name	Lat	Long	Prev. Design.	Diameter (km)
Crater (Challenger crew members)				
Jarvis	34.9°S	148.9°W	Borman Z	38
McAuliffe	33.0°S	148.9°W	Borman Y	19
McNair	35.7°S	147.3°W	Borman A	29

Onizuka	36.2°S	148.9°W	Anon	29
Resnik	38.8°S	150.1°W	Borman X	20
Scobee	31.1°S	148.9°W	Barringer L	40
Smith	31.6°S	150.2°W	Barringer M	34

MARS

Name	Lat	Long	Attribute
Crater			
Amet	23.7°N	57.4°W	Small town in India
Canso	21.5°N	60.6°W	Town in Nova Scotia, Canada
Inuvik	78.6°N	21.9°W	Town in N.W. Territories, Canada
Jeki	23.9°N	52.3°W	Town in Ethiopia
Musmar	24.8°N	55.7°W	Small town in Sudan
Perrotin	2.9°S	77.9°W	Henri A. Perrotin (1845-1904) French astronomer
Rong	26.5°N	55.4°W	Small town in Norway
Sevel	78.2°N	39.5°W	Town in Denmark
Worcester	26.8°N	50.3°W	Town in New York, USA
Catena			
Labeatis Catenae	18.5-20.6°N	91.5- 94.6°W	Classical albedo feature Labeatis Lacus at 30N, 75W
Cavus			
Cavi Frigores	78.3-81.5°S	61.0- 67.5°W	Classical albedo feature Polus Frigoris at 84S, 30W
Hyperborei Cavi	79.2-80.0°N	50.1- 55.0°W	Classical albedo feature Hyperboreus Lacus at 76N, 60W
Dorsum			
Sinai Dorsa	9.7-15.0°S	77.0- 82.0°W	Classical albedo feature Sinai at 20S, 70W
Fossa			
Ecus Fossae	.5°S- 3.5°N	76.0- 79.0°W	Classical albedo feature Echus Lacus at 1N, 90W
Nilokeras	23.0-26.0°N	56.0- 59.5°W	Classical albedo feature Nilokeras
Tithoniae Fossae	6.0- 7.6°S	76.5- 83.5°W	Classical albedo feature Tithonius Lacus at 4S, 85W
Tractus Fossae	22.5-28.5°N	99.5-103.0°W	Classical albedo feature Tractus Albus at 30N, 80W
Labes			
Candor Labes	4.5- 5.3°S	74.8- 76.7°W	Classical albedo feature Candor at 5N, 75W
Ius Labes	7.2- 7.9°S	78.0- 78.9°W	Classical albedo feature Ius Lacus at 6S, 88W

Labyrinthus
Hyperboreus
Labyrinthus	77.5-82.1°N	57.0- 62.5°W	Classical albedo feature Hyperboreus Lacus at 76N, 60W

Planum
Planum Angustum	73.3-81.5°S	78.6- 92.7°W	Classical albedo feature Rima Angusta at 77S, 58W

Rupes
Rupes Tenuis	80.8-83.0°N	57.9- 72.5°W	Classical albedo feature Rima Tenuis at 88N, 60W

Unda
Abalos Undae	75.0-81.0°N	76.0- 98.0°W	Classical albedo feature Abalos at 72N, 70W
Hyperboreae Undae	77.3-84.5°N	20.0- 57.5°W	Classical albedo feature Hyperboreus Lacus at 76N, 60W

Vallis
Lobo Vallis	26.5-28.0°N	60.8- 61.9°W	Modern river in Ivory Coast
Rhabon Valles	20.7-22.5°N	89.0- 93.5°W	Classical river, Dacia (Romania)
Termes Vallis	10.9-11.6°S	156.9-157.2°W	Classical river, Lusitania
Varus Valles	8.2- 9.7°S	155.4-156.4°W	Classical river, Cisalpine Gaul

----

JUPITER

Name	Lat	Long	Attribute

GANYMEDE
Crater
Anubis	83.0°S	120.0°W	Egyptian jackal-headed god who opened the underworld to dead
Bau	24.5°N	53.0°W	Sumerian fertility goddess
Hapi	31.0°S	212.5°W	Egyptian god of the Nile
Khônsu	38.0°S	189.0°W	Egyptian Moon god
Kingu	35.5°S	227.5°W	Assyro-Babylonian leader of Tiamat's forces
Min	29.0°N	303.0°W	Egyptian fertility god
Neith	29.0°N	9.0°W	Egyptian warrior goddess
Ninkasi	56.5°N	56.0°W	Sumerian goddess of brewing
Ptah	67.0°S	214.5°W	Sovereign god of Memphis (Egypt)
Sati	31.0°N	14.5°W	Egyptian wife of Khnum
Seker	41.0°S	351.0°W	God of dead at Memphis (Egypt)
Shu	43.0°N	348.0°W	Egyptian god of air
Ta-urt	26.5°N	306.0°W	Egyptian goddess of maternity

Facula
Sais Facula	38.0°N	14.0°W	Egyptian capital; mid 7th C., BC

Sulcus

Bubastis Sulci	65-90°S	125-302°W	Egyptian town where Bast was worshipped

CALLISTO

Crater

Holdr	44.0°N	19.5°W	Son of Karl and Snor in Rigdismal
Ilma	31.5°S	168.0°W	Finno-Ugric divinity of air
Lempo	26.0°S	319.5°W	Finno-Ugric evil spirit
Njord	13.0°N	134.0°W	Third Aesir (Nordic god)
Vidarr	11.5°N	194.0°W	One of Aesir; Odin's son

URANIAN SATELLITE FEATURES

Name	Lat	Long

MIRANDA

Crater (Human char., Tempest)

Alonso	47°S	345°E
Ferdinand	36°S	208°E
Francisco	70°S	246°E
Gonzalo	13°S	75°E
Prospero	35°S	323°E
Stephano	36°S	239°E
Trinculo	67°S	168°E

Corona (Shakespearean place)

Arden	10-60°S	30-120°E
Elsinore	10-42°S	215-305°E
Inverness	38-90°S	0-360°E

Regio (Shakespearean place)

Dunsinane	20-75°S	345- 65°E
Mantua	10-90°S	75-300°E
Silicia	10-50°S	295-340°E

Rupes (Shakespearean place)

Argier	40-50°S	310-340°E
Verona	10-40°S	340-350°E

ARIEL

Crater (Light spirit, indiv.)

Abans	16°S	251°E
Agape	47°S	336°E
Ataksak	53°S	225°E
Befana	17°S	32°E
Berylune	23°S	328°E
Deive	23°S	23°E

Djadek	12°S	251°E
Domovoy	72°S	339°S
Finvara	16°S	19°E
Gwyn	78°S	23°E
Huon	39°S	33°E
Laica	22°S	44°E
Mab	39°S	353°E
Melusine	53°S	9°E
Onagh	22°S	244°E
Rima	18°S	260°E
Yangoor	68°S	280°E

Chasma (Light spirit, class)

Brownie	05-21°S	325-357°E
Kachina	24-40°S	210-280°E
Kewpie	15-42°S	307-335°E
Korrigan	25-46°S	328-353°E
Kra	32-36°S	355- 2°E
Pixie	18-25°S	350- 20°E
Sylph	45-50°S	328- 15°E

Vallis (Light spirit, class)

Leprechaun	05-15°S	350- 25°E
Sprite	12-17°S	332-355°E

FEATURE NAMES
LARGE URANIAN SATELLITES

Name	Lat	Long

UMBRIEL

Crater (Dark spirit, individual)

Alberich	31°S	43°E
Fin	36°S	41°E
Gob	9°S	6°E
Kanaloa	11°S	351°E
Malingee	22°S	15°E
Minepa	41°S	13°E
Peri	9°S	6°E
Setibos	31°S	350°E
Skynd	1°N	335°E
Vuver	2°S	311°E
Wokolo	33°S	7°E
Wunda	6°S	274°E
Zlyden	24°S	330°E

TITANIA
Crater (Shakespearean hero)

Adriana	20°S	4°E
Bona	55°S	351°E
Calphurnia	42°S	292°E
Elinor	44°S	334°E
Gertrude	15°S	288°E
Imogen	25°S	318°E
Iras	19°S	339°E
Jessica	55°S	286°E
Katherine	51°S	333°E
Lucetta	9°S	277°E
Marina	15°S	316°E
Mopsa	11°S	302°E
Phrynia	24°S	309°E
Ursula	13°S	44°E
Valeria	34°S	4°E

Chasma

Belmont	4-25°S	25-35°E
Messima	8-28°S	325- 5°E

Rupes

Rousillon	7-25°S	17-38°E

OBERON
Crater (Shakespearean hero)

Antony	28°S	65°E
Caesar	27°S	61°E
Coriolanus	11°S	345°E
Falstaff	22°S	19°E
Hamlet	46°S	46°E
Lear	5°S	31°E
Macbeth	59°S	112°E
Othello	65°S	44°E
Romeo	28°S	88°E

Chasma (spirit place)

Nommur Chasma	10-20°S	300-340°E

SMALL URANIAN SATELLITES

Name	Temporary Designation	Distance from Uranus (km)
Puck	1985 U1	86,000
Portia	1986 U1	66,085
Juliet	1986 U2	64,352
Cressida	1986 U3	61,777
Rosalinda	1986 U4	69,942
Belinda	1986 U5	75,258
Desdemona	1986 U6	62,676
Cordelia	1986 U7	49,771
Ophelia	1986 U8	53,796
Bianca	1986 U9	59,173
